INSIDE THE STARS

IAU Colloquium 137

A SERIES OF BOOKS ON RECENT DEVELOPMENTS IN ASTRONOMY AND ASTROPHYSICS

Printed by BookCrafters, Inc.

First published 1993

Library of Congress Catalog Card Number: 93-70129
ISBN 0-937707-59-7

D. Harold McNamara, Managing Editor of Conference Series
408 ESC Brigham Young University
Provo, UT 84602
801-378-2298

A SERIES OF BOOKS ON RECENT DEVELOPMENTS IN ASTRONOMY AND ASTROPHYSICS

Vol. 1-Progress and Opportunities in Southern Hemisphere Optical Astronomy: The CTIO 25th Anniversary Symposium
ed. V. M. Blanco and M. M. Phillips ISBN 0-937707-18-X

Vol. 2-Proceedings of a Workshop on Optical Surveys for Quasars
ed. P. S. Osmer, A. C. Porter, R. F. Green, and C. B. Foltz ISBN 0-937707-19-8

Vol. 3-Fiber Optics in Astronomy
ed. S. C. Barden ISBN 0-937707-20-1

Vol. 4-The Extragalactic Distance Scale: Proceedings of the ASP 100th Anniversary Symposium
ed. S. van den Bergh and C. J. Pritchet ISBN 0-937707-21-X

Vol. 5-The Minnesota Lectures on Clusters of Galaxies and Large-Scale Structure
ed. J. M. Dickey ISBN 0-937707-22-8

Vol. 6-Synthesis Imaging in Radio Astronomy: A Collection of Lectures from the Third NRAO Synthesis Imaging Summer School
ed. R. A. Perley, F. R. Schwab, and A. H. Bridle ISBN 0-937707-23-6

Vol. 7-Properties of Hot Luminous Stars: Boulder-Munich Workshop
ed. C. D. Garmany ISBN 0-937707-24-4

Vol. 8-CCDs in Astronomy
ed. G. H. Jacoby ISBN 0-937707-25-2

Vol. 9-Cool Stars, Stellar Systems, and the Sun. Sixth Cambridge Workshop
ed. G. Wallerstein ISBN 0-937707-27-9

Vol. 10-The Evolution of the Universe of Galaxies. The Edwin Hubble Centennial Symposium
ed. R. G. Kron ISBN 0-937707-28-7

Vol. 11-Confrontation Between Stellar Pulsation and Evolution
ed. C. Cacciari and G. Clementini ISBN 0-937707-30-9

Vol. 12-The Evolution of the Interstellar Medium
ed. L. Blitz ISBN 0-937707-31-7

Vol. 13-The Formation and Evolution of Star Clusters
ed. K. Janes ISBN 0-937707-32-5

Vol. 14-Astrophysics with Infrared Arrays
ed. R. Elston ISBN 0-937707-33-3

Vol. 15-Large-Scale Structures and Peculiar Motions in the Universe
ed. D. W. Latham and L. A. N. da Costa ISBN 0-937707-34-1

Vol. 16-Atoms, Ions and Molecules: New Results in Spectral Line Astrophysics
ed. A. D. Haschick and P. T. P. Ho ISBN 0-937707-35-X

Vol. 17-Light Pollution, Radio Interference, and Space Debris
ed. D. L. Crawford ISBN 0-937707-36-8

Vol. 18-The Interpretation of Modern Synthesis Observations of Spiral Galaxies
ed. M. Duric and P. C. Crane ISBN 0-937707-37-6

Vol. 19-Radio Interferometry: Theory, Techniques, and Applications, IAU Colloquium 131
ed. T. J. Cornwell and R. A. Perley ISBN 0-937707-38-4

Vol. 20-Frontiers of Stellar Evolution, celebrating the 50th Anniversary of McDonald Observatory
ed. D. L. Lambert ISBN 0-937707-39-2

Vol. 21-The Space Distribution of Quasars
ed. D. Crampton ISBN 0-937707-40-6

Vol. 22-Nonisotropic and Variable Outflows from Stars
ed. L. Drissen, C. Leitherer, and A. Nota ISBN 0-937707-41-4

Vol. 23-Astronomical CCD Observing and Reduction Techniques
ed. S. B. Howell ISBN 0-937707-42-4

Vol. 24-Cosmology and Large-Scale Structure in the Universe
ed. R. R. de Carvalho ISBN 0-937707-43-0

Vol. 25-Astronomical Data Analysis Software and Systems I
ed. D. M. Worrall, C. Biemesderfer, and J. Barnes ISBN 0-937707-44-9

Vol. 26-Cool Stars, Stellar Systems, and the Sun, Seventh Cambridge Workshop
ed. M. S. Giampapa and J. A. Bookbinder ISBN 0-937707-45-7

Vol. 27-The Solar Cycle
ed. K. L. Harvey ISBN 0-937707-46-5

Vol. 28-Automated Telescopes for Photometry and Imaging
ed. S. J. Adelman, R. J. Dukes, Jr., and C. J. Adelman ISBN 0-937707-47-3

Vol. 29-Workshop on Cataclysmic Variable Stars
ed. N. Vogt ISBN 0-937707-48-1

Vol. 30-Variable Stars and Galaxies, in honor of M. S. Feast on his retirement
ed. B. Warner ISBN 0-937707-49-X

Vol. 31-Relationships Between Active Galactic Nuclei and Starburst Galaxies
ed. A. V. Filippenko ISBN 0-937707-50-3

Vol. 32-Complementary Approaches to Double and Multiple Star Research, IAU Colloquim 135
ed. H. A. McAlister and W. I. Hartkopf ISBN 0-937707-51-1

Vol. 33-Research Amateur Astronomy
ed. S. J. Edberg ISBN 0-937707-52-X

Vol. 34-Robotic Telescopes in the 1990s
ed. A. V. Filippenko ISBN 0-937707-53-8

Vol. 35-Massive Stars: Their Lives in the Interstellar Medium
ed. J. P. Cassinelli and E. B. Churchwell ISBN 0-937707-54-6

Vol. 36-Planets and Pulsars
ed. J. A. Phillips, S. E. Thorsett, and S. R. Kulkarni ISBN 0-937707-55-4

Vol. 37-Fiber Optics in Astronomy II
ed. P. M. Gray ISBN 0-937707-56-2

Vol. 38-New Frontiers in Binary Star Research
ed. J. C. Leung and I. S. Nha ISBN 0-937707-57-0

Vol. 39-The Minnesota Lectures on the Structure and Dynamics of the Milky Way
ed. Roberta M. Humphreys ISBN 0-937707-58-9

Vol. 40-Inside the Stars, IAU Colloquim 137
ed. Werner W. Weiss and Annie Baglin ISBN 0-937707-59-7

Inquiries concerning these volumes should be directed to the:
Astronomical Society of the Pacific
CONFERENCE SERIES
390 Ashton Avenue
San Francisco, CA 94112-1722
415-337-1100

ASTRONOMICAL SOCIETY OF THE PACIFIC
CONFERENCE SERIES

Volume 40

INSIDE THE STARS

IAU Colloquium 137

Edited by
Werner W. Weiss and Annie Baglin

TABLE OF CONTENTS

Posters

IV Input physics and basic stellar data
Chair: D. Schönberner

Posters

VI Asteroseismology: theory
Chair: H. Shibahashi

IX Prejudiced conclusion

X Round Table

XI Supplement

XII PRISMA

XIII INDEX

PREFACE

The purpose of this Colloquium was the confrontation of realistic models of simple stars with high quality observations in order to obtain quantitative information about fundamental physics. Consequently, this meeting was focussed on *stars on the main sequence and close to it.* Their structure is the simplest possible among stars and these stars are barely influenced by the complexity of many yet unknown parameters. They populate the solar neighbourhood and can be observed with the highest accuracy.

Stars provide the basis for most of the calibrations of the parameters describing our Universe, like age, distances and masses. The precise knowledge of stellar parameters is therefore fundamental for understanding our Universe.

The next few years will provide us with new powerful computational facilities and observational data which will be either completely new or much more precise than ever before. We refer here only to few of the issues:

- Pulsation frequency spectra of eigenmodes of a significant number of stars will be determined by space experiments and ground based networks.
- Stellar distances will be very accurately determined by Hipparcos.
- Rotational velocities and magnetic fields will be measured by Doppler- and Zeeman imaging techniques.
- New detectors with increasing size and sensitivity will dramatically improve the signal-to-noise ratio of photometric and spectrometric observations.
- Several new and exciting observing facilities are currently under consideration, in particular space missions.
- Dramatic advances in computer technology will permit increasingly complex modelling, for example in hydrodynamics, radiation transfer and statistical physics.

A further motivation for organizing this conference provided a decision of the European Space Agency (ESA) to elect PRISMA (described in this volume) as a candidate for the next medium-size mission. It was felt by the Science Team that a concentration of theoretical and observational efforts is crucial to achieve an optimally designed space project.

The Colloquium was opened in the large ceremonial hall of the University of Vienna with a review subtitled *Outside the Stars*. The Scientific Organizing Committee intentionally invited an outstanding scientist working in the field of extragalactic research and cosmology to present the opening speach. Malcolm Longair was asked to highlight topics of current importance in the astrophysics of galaxies, active galaxies and cosmology in which an understanding of the origin and evolution of stars is crucial and to demonstrate how reliable astrophysical conclusions critically depend on how secure our understanding is of the structure and evolution of the stars.

After the stage was set in this manner, two reviews on the controversial sun and on the impact of solar space experiments followed. These two presentations linked our conference to IAU colloquium *Inside the Sun*, which was focussed on a very particular star. At that conference in 1989 it was demonstrated that it is necessary to study other stars with different physical parameters (temperature, luminosity, chemical composition, rotation rate, magnetic field, etc.), and to confront realistic stellar models with high quality observations. Otherwise, a significant improvement of our understanding of normal stars, which, after all, are the most frequent and easily observable ingredients of the universe, will not be possible.

The conference continued with seven sections: Atmospheres, a window to the interior - Input physics for stellar structure - Basic stellar data - The changing interior - Asteroseismology, Theory - Activity, a break of spherical symmetry - Asteroseismology; Observation. A total of 30 invited speakers covered the subject of the conference with three to five reviews per section. More than 200 abstracts for short communications were submitted by closely 280 participants to the Scientific Organizing Committee.

Editing the proceedings was a major task not only due to the size of the volume. Although e-mailing and text processing is generally used nowadays, communication with various authors turned out to be quite cumbersome due to a high failure rate in the mailing system. A considerable fraction of submitted manuscripts did not follow the style rules for the conference proceedings. The help of Rainer Kuschnig in transforming the input from the participants into a useable output for the publisher is greatly acknowledged.

Since communication within the scientific community was a prime goal for the conference, it was decided to include the manuscripts from all posters to this proceedings. With about 150 expected manuscripts, refereeing would have delayed the publication of the proceedings and was therefor abandoned. The responsibility for the contents of the contributions lies fully with the authors.

It would not have been possible to organize this conference without sponsors and many dedicated helpers who contributed numerous working hours and much effort, most of it in their free time. In addition to the members of the Scientific and Local Organizing committees we have to express our thanks also to students of the Institute for Astronomy, M. Gelbmann, R. Fensl, F. Hiesberger, Mag. W. Ostermann and G. Reiss. Last, but not least, we are grateful for all the support we have received from Werner Däppen for scientific matter, who was visiting professor in Vienna while this conference was being prepared.

Werner W. Weiss
Institute for Astronomy
University of Vienna

Annie Baglin
DASGAL
Observatoire de Paris-Meudon

Vienna, September 1992

PATRONAGE:

Dr. Erhard Busek, Vice Chancellor and Minister of Science and Research

Dr. Helmut Zilk, Mayor and Governor of the City of Vienna

Mag. Siegfried Ludwig, Governor of Lower Austria

Prof. Dr. Alfred Ebenbauer, Rector of the University of Vienna

European Astronomical Society

SPONSORS:

Bundesministerium für Wissenschaft und Forschung

International Astronomical Union

NOVA Aerospace

Österreichische Forschungsgemeinschaft

Österreichische Gesellschaft für Weltraumfragen (ASA)

SCIENCE program of the European Community

Universität Wien

Scientific Organizing Committee:

Th. Appourchaux (ESA-ESTEC), A. Baglin (France, chair), A. Boyarchuk (Russia), W. Dziembowski (Poland), S. Frandsen (Denmark), D.Gough (U.K.), H. Hudson (U.S.A.), S. Kawaler (U.S.A.), Ph. Lemaire (France), M. Rodono (Italy), H. Shibahashi (Japan), W. W. Weiss (Austria) and representatives of the supporting IAU commissions: M. Breger (Austria), E. Jenkins (U.S.A.), D. Lambert (U.S.A.), A. Maeder (Switzerland), I. McLean (U.K.).

Local Organizing Committee:

M. G. Firneis, R. Kuschnig, G. Polnitzky, A. Schnell, W. W. Weiss (chair), *Institute for Astronomy, Vienna*; COMMUNICA, *Vienna*

LIST OF PARTICIPANTS

Alecian Georges	DAEC , Obs. de Meudon	France
Andersen Johannes	Copenhagen University	Denmark
Antillon Mario	Universita Degli Studi di Trieste	Italy
Antonello Elio	Osservatorio Astronomico di Brera	Italy
Appourchaux T.	ESTEC	Netherlands
Aquilano Robert.O.	Inst. de Fisica Rosario	Argentinia
Arevalo M.J.	Inst. de Astrofisica de Canarias	Spain
Arnould Marcel L.	Univ. Libre de Bruxelles	Belgium
Audard Nathalie	Departement Cassisi,Nice	France
Babel Jacques	Inst. d'Astronomie	Switzerland
Baglin A.	DASGAL, Obs. de Meudon	France
Baker Norman H.	Columbia Univ.	USA
Balachandran S.	Univ. of North Carolina	USA
Ballester J.L.	Universitat Illes Balears	Spain
Barone Fabrizio	Universita' di Napoli	Italy
Barylak Michael	ESA IUE Obs., Villafranca	Spain
Baturin Vladimir A.	Sternberg State Astronomical	Russia
Becker Stephen A.	Los Alamos National Laboratory	USA
Belvedere Gaetano	Citta` Universitaria, Catania	Italy
Berthomieu G.	Observatoire de Nice	France
Bertiau F.	Sterrenkundig Inst. Waversebaan	Belgium
Blöcker Thomas	Inst. f. Theoretische Physik, Kiel	Germany
Bocchia Romie	Obs. de l'Universite de Bordeaux	France
Böhm Torsten	Observatoire de Paris	France
Bouchet Patrice	ESO/La Silla	Germany
Boyarchuk A.	Astronomical Inst. R.A.S, Moskow	Russia
Breger Michael	Inst. für Astronomie, Wien	Austria
Brown Timothy M.	High Altitude Observatory	USA
Budaj Jan	Slowak Academy of Sciences	CSFR
Castelli Fiorella	Osservatorio Astronomico, Trieste	Italy
Catala Claude	Observatoire de Paris-Meudon	France
Catalano Santo	Catania Astrophysical Observatory	Italy
Chabrier Gilles	Ecole Normale Superieure de Lyon	France
Charbonneau Paul	High Altitude Observatory	USA
Charbonnel Corinne	Université Paul Sabatier, Toulouse	France
Chechetkin Valerij	Keldysh Inst.tute of Applied Math.	Russia
Christensen-Dalsgaard J.	Inst. of Astronomy, Aarhus	Denmark
Claret dos Santos Antonio	Inst. de Astrofisica de Andalusia	Spain
Clausen Jens Viggo	Copenhagen University	Denmark
Clement Leon	Dept. Matematica Aplicada	Spain
Collados Manuel	Inst. de Astrofisica de Canarias	Spain
Cuypers Jan	Koninklijke Sterrenwacht van Belgie	Belgium
Däppen Werner	Inst. für Astronomie, Wien	Austria
De Greve J.P.	Vrije Universiteit Brussel	Belgium

Deliyannis Constantine	University of Hawaii	USA
Demarque Pierre	Yale University, New Haven	USA
Dimitrijevic Milan S.	Astronomical Observatory, Beograd	Yugoslavia
Doazan V.	Observatoire de Paris	France
Donati J.-F.	Observatoire de Meudon	France
Dorfi Ernst	Inst. für Astronomie, Wien	Austria
Durisen Richard H.	Indiana University	USA
Dziembowski Wojciech	Astronomical Center, Warsawa	Poland
D'Antona Francesca	Osservatorio Astr. di Roma	Italy
Elkin Vladimir	Special Astrophysical Observatory	Russia
Ergma Ene	Tartu University	Estonia
Ferluga Steno	Universita Degli Studi di Trieste	Italy
Feuchtinger Michael	Inst. für Astronomie, Wien	Austria
Figueiredo J.	Astrophys. Inst. VUB, Brussels	Belgium
Firneis Maria G.	Inst. für Astronomie, Wien	Austria
Frandsen Soeren	University of Aarhus	Denmark
Frantsman Ju.L.	Radioastrophysical Observatory, Riga	Latvia
Friel Eileen D.	Maria Mitchell Observatory	USA
Fröhlich Claus	Phys. Meteor. Observatorium Davos	Switzerland
Gabriel Maurice	Inst. d'Astrophysique, Liége	Belgium
Gaige Yves	Université Paul Sabatier, Toulouse	France
Gal Janos	Jate University, Szeged	Hungaria
Gameiro Jorge F.S.	Universidade do Porto	Portugal
Gautschy Alfred	MPI für Astrophysik, Garching	Germany
Gelbmann Michael	Inst. für Astronomie, Wien	Austria
Gerbaldi Michele	Inst. d'Astrophysique, Paris	France
Gerth Ewald	Zentralinst. für Astrophysik, Postdam	Germany
Gimenez Alvaro	VILSPA, Villafranca	Spain
Glagolevskij Y.V.	S.A.O., Nizhni Arkhys	Russia
Godlowski Wlodzimierz	Uniwersytet Jagiellonski, Krakow	Poland
Gokhale M.H.	Indian Inst. of Astrophysics, Banglore	India
Gomez-Stephanovich Ana	Observatoire de Meudon	France
Gough Douglas	Inst. of Astronomy, Cambridge	U.K.
Goupil Marie Jo	Observatoire de Paris-Meudon	France
Grandpierre Attila	Konkoly Observatory	Hungaria
Grenon Michel	Observatoire de Geneve	Swizerland
Grevesse Nicolas	Université de Liege	Belgium
Guerrero Gianantonio	Osservatorio di Brera	Italy
Gustafsson Bengt	Astron. Observatory Uppsala	Sweden
Handler Gerald	Inst. für Astronomie, Wien	Austria
Harmanec Petr	Astronomical Inst. , Ondrejov	CSFR
Hegedüs Tibor	Baja Observatory	Hungaria
Helt Bodil E.	University Observatory, Copenhagen	Denmark
Hiesberger Felicitas	Inst. für Astronomie, Wien	Austria
Höfner Susanne	Inst. für Astronomie, Wien	Austria
Horner Scott	NCAR/HAO	USA

Houdek Günter	EDV-Zentrum, TU-Wien	Austria
Hoyng P.	Laboratory for Space Research	Netherlands
Hric L.	Astronomical Inst. Slovak	CSFR
Hudson Hugh S.	Univ. of California, La Jolla	USA
Huisong Tan	Yunnan Observatory	China P.R.
Isaak G.R.	University of Birmingham	U.K.
Israelian Garegin	Byurakan Astrophysical Observatory	Armenia
Jankovics Istvan	ELTE, Szombathely	Hungaria
Janku Vladimir	Astronomical Inst. Slovak	CSFR
Jarzebowski Tadeusz	Pedagogical University, Slupsk	Poland
Jones Andrew	Inst. of Astronomy	Denmark
Katsova Maria	Sternberg Inst., Moscow	Russia
Kellett Barry John	Rutherford Appleton Laboratory	U.K.
Khoklova Vera L.	Inst. of Astronomy, Moscow	Russia
Klacka Josef	Astronomical Inst. Slovak	CSFR
Klvana Miroslav	Czechoslovak Academy of Sciences	CSFR
Könighofer Gerhard	Inst. für Astronomie, Wien	Austria
Kopal Zdenek	University of Manchester	U.K.
Kosovichev Alexander	University of Cambridge	U.K.
Koubsky Pavel	Astronomical Inst. Slovak, Ondrejov	CSFR
Krause Fritz	Zentralinst. für Astrophysik, Potsdam	Germany
Kreidl T.J.	Lowell Observatory, Flagstaff	USA
Krivodubskij Valerij N.	Kiev University	Ukraine
Kuschnig Rainer	Inst. für Astronomie, Wien	Austria
Kusnetzov O.	Keldysh Inst. of Appl. Math., Moscow	Russia
Langer Norbert	Universitätssternwarte, Göttingen	Germany
Lazaro C.	Inst. de Astrofisica de Canarias	Spain
Lebreton Yveline	DASGAL, Meudon	France
Lee Umin	University of Rochester	USA
Leifsen Torben	Inst. of Theoret. Astrophysics, Oslo	Norway
Lemaire P.	Inst. d'Astrophysique Spatiale	France
Leon Rosa Clement	Dept. Astronomia, Valencia	Spain
Longair Malcolm	Univ. of Cambridge	U.K.
Lopes Ylidio Pereira	Commissariat a l'Energies Atomique	France
Lopez Ramon J. Garcia	Inst. de Astrofisica de Canarias	Spain
Lopez de Coca P.	Inst. de Astrofisica de Andalucia	Spain
Ludmany Andreas	Heliophysical Observatory, Debrecen	Hungary
Lyubimkov Leonid S.	Crimean Astrophysical Observatory	Ukraine
MacGregor Keith B.	High Altitude Observatory	USA
Maceroni Carla	Osservatorio Astronomico di Roma	Italy
Magazzu Antonio	Osservatorio Astrofisico, Catania	Italy
Maggio Antonio	Osservatorio Astronomico di Palermo	Italy
Malyuto V.	Tartu Astrophysical Observatory	Estonia
Mancuso Santi	Universita' di Napoli	Italy
Mantegazza L.	Osservatorio Astronomico di Brera	Italy
Massacrier Gerard	Observatoire de Meudon	France

Matthews Jaymie	Univ. de Montreal	Canada
Mayr Bernhard Johannes	Inst. für Astronomie, Wien	Austria
Megessier Claude	Observatory de Paris-Meudon	France
Michalec Adam	Jagiellonian University	Poland
Michaud Georges	Univ. de Montreal	Canada
Mikulasek Zdenek	Copernicus Observatory, Brno	CSFR
Mitskevitch A.	Osservatorio Astrofisico di Arcetri	Italy
Mktrtichian David E.	Astronomical Observatory of Odessa	Ukrainia
Monier Richard	ESA IUE Ground Station	Spain
Monteiro Mario Joao	University of London	U.K.
Morrison Douglas R.O.	CERN, Geneve	Switzerland
Mowlavi Nami	Université Libre de Bruxelles	Belgium
Müller Heinrich F.	Inst. für Astronomie, Wien	Austria
Muthsam Herbert	Mathematisches Inst., Wien	Austria
Myrrha Maria Leticia M.	Federal University of Minas Gerais	Brazil
Natta Antonella	Osservatorio Astrofisico Arcetri	Italy
Neff James E.	Penn. State University	USA
Nelson Matthew J.	University of Wisconsin	USA
Neuforge Corinne	Univ. de Liége	Belgium
Nikoghossian Arthur	Byurakan Astrophysical Observatory	Armenia
Nissen Poul Erik	Univ. Aarhus	Denmark
Noels Arlette	Univ. Liége	Belgium
Novotny E.	Inst. of Astronomy, Cambridge	U.K.
Noyes Robert W.	Center for Astrophysics, Cambridge	USA
Nugis Tiit	Tartu Astrophysical Observatory	Estonia
Ocana Gerardo	Queen Mary College, London	U.K.
Odell Andy	N.A.Z. University	USA
Oedegaard Knut J.R.	University of Oslo	Norway
Oprescu Gabriela	Romania Academy	Roumania
Osaki Yoji	University of Tokyo	Japan
Palla Francesco	Osservatorio Astrofisico Arcetri	Italy
Pamyatnykh Alosha	Inst. of Astronomy, Moscow	Russia
Paunzen Ernst	Inst. für Astronomie, Wien	Austria
Pavlovski Kresimir	Hvar Observatory	Croatia
Petrovay Kristof	Eötvös University	Hungaria
Pfleiderer J.	Inst. für Astronomie, Innsbruck	Austria
Pijpers Frank P.	Astronomy Unit	U.K.
Pinsonneault M.H.	Yale University	USA
Podsiadlowski Philipp	Inst. of Astronomy, Cambridge	U.K.
Polosukhina-Chuvaeva N.	Crimean Observatory	Ukraine
Pottasch Edward	Kapteyn Sterrenwacht	Netherlands
Praderie Francoise	Observatoire de Paris-Meudon	France
Proffitt Ch.	Space Telescope Science Inst.	USA
Provost Janine	Observatoire de Nice	France
Pustylnik Izold	Tartu Astrophysical Observatory	Estonia
Qingyao Liu	Copenhagen Unicersity Observatory	Denmark

Rebolo Rafael	Inst. de Astrofisica de Canarias	Spain
Reglero V.	Dept. Astronomia, Valencia	Spain
Richer Harvey B.	University of Cambridge	U.K.
Ripodas Pilar	Inst. de Astrofisica de Canarias	Spain
Roca Cortes T.	Inst. de Astrofisica de Canarias	Spain
Rodono M.	Inst. d'Astronomia, Catania	Italy
Rodriguez Eloy	Inst. de Astrofisica de Andalusia	Spain
Rogl Jadwiga	EDV-Zentrum, TU- Wien	Austria
Rolland A.	Inst. de Astrofisica de Andalucia	Spain
Romanyuk Iosif	S.A.O., Nizhni Arkhys	Russia
Roxburgh Ian W.	Queen Mary and Westfield College	U.K.
Runacres Mark	Astrofysisch Inst. V.U.B.	Belgium
Ruusalepp Mare	Tartu Astrophysical Observatory	Estonia
Ruzic Z.	Inst. Astron., Zagreb	Kroatia
Ruzdjah V.	Inst. Astron., Zagreb	Kroatia
Rybak Jan	Astronomical Inst. Slovak	CSFR
Saniga Metod	Astronomical Inst. Slovak	CSFR
Sarna Marek J.	University of Sussex	U.K.
Schaerer Daniel	Observatoire de Geneve	Switzerland
Schatzman E.	Observatoire de Meudon	France
Schneider Hartmut	Universitäts-Sternwarte, Göttingen	Germany
Schnell Anneliese	Inst. für Astronomie, Wien	Austria
Schober Hans Josef	Inst. für Astronomie, Wien	Austria
Schönberner D.	Universität Kiel	Germany
Schrijver C.J.	Sterrekundig Inst.	Netherlands
Schou J.	HAO NCAR	USA
Seaton Michael J.	Univ. College London	U.K.
Sekii T.	Inst. of Astronomy, Cambridge	U.K.
Selam Selim	Ankara University	Turkey
Semel M.	Observatoire de Meudon	France
Severino Guiseppe	Osserv. Astronomico di Capodimonte	Italy
Sharp Christopher Martin	Service PTN	France
Shibahashi Hiromoto	University of Tokyo	Japan
Simmons John C.	Queen Mary and Westfield College	U.K.
Simon Klaus P.	Universitäts-Sternwarte, München	Germany
Singh Mahendra	Uttar Pradesh State Observatory	India
Smeyers P.	Katholieke Universiteit Leuven	Belgium
Soderblom David	Space Telescope Science Inst.	USA
Steffen Matthias	Sternwarte der Universität, Kiel	Germany
Stehle C.	Obs. Paris-Meudon	France
Stepien Kazimierz	Warsaw University Observatory	Poland
Stolzmann Werner	Sternwarte der Universität, Kiel	Germany
Strassmeier Klaus	Inst. für Astronomie, Wien	Austria
Suran Doru Marian	Romanian Academy	Roumania
Szatmary Karoly	Jate University	Hungaria
Takeuti Mine	Tohoku Univ.	Japan

Tanzer Christian	Inst. für Astronomie, Wien	Austria
Tautvaisiene Grazina	Inst. of Astronomy, Vilnius	Lithuania
The P.S.	University of Amsterdam	Netherlands
Thomas Richard N.	Radiophysics Inc., Boulder	USA
Thompson Michael John	Queen Mary and Westfield College	U.K.
Torelli Maria	Osservatorio Astronomico, Roma	Italy
Toutain Thierry	Phys. Meteor. Observatorium Davos	Switzerland
Trullols Enric	Universitat de Barcelona	Spain
Ulrich Roger	University of California	USA
Uyaniker Bulent	Middle East Technical University	Turkey
Van den Bliek Nicole	Sterrewacht Leiden	Netherlands
Vanbeveren D.	Vrije Universiteit Brussel	Belgium
VandenBerg Don A.	University of Victoria	Canada
Vaz Luiz Paolo Ribeiro	Federal University of Minas Gerais	Brazil
Ventura Rita	Osservatorio Astrofisico, Catania	Italy
Vera Rafael A.	Universidad de Ciencias	Chile
Vetesnik Miroslav	Dept. of Astrophysics, Brno	CSFR
Vidal Silvestre Inma	Inst. de Astrofisica de Canarias	Spain
Voloshina Irina	Sternberg Astronomical Inst.	Russia
Vorontsov Sergei	Inst. of Astronomy, Cambridge	U.K.
Vrsnak Bojan	Hvar Observatory	Croatia
Walton Sarah Jane	Queen Mary Westfield College	U.K.
Wehlau William	Univ. of Western Ontario	Canada
Weiss Werner W.	Inst. für Astronomie, Wien	Austria
Wenzel K.-P.	ESA/ESTEC, Space Science Dept.	Netherlands
Winget Don	Univ. Texas	USA
Wlodarczyk Krzysztof	Mt.Suhora Astronomical Observatory	Poland
Wuchterl G.	Inst. f. Theor. Astrophysik, Heidelberg	Germany
Xuefu Liu	Beijing Normal University	China P.R.
Zahn Jean-Paul	Obs. du Pic-du-Midi	France
Zboril Milan	Astronomical Inst. Slovak	CSFR
Zharkova Valentina V.	Kiev University	Ukraine
Ziznovsky Jozef	Astronomical Inst. Slovak	CSFR
Zola S.	Jagiellonian University	Poland
Zverko Juraj	Astronomical Inst. Slovak	CSFR

Inside the Stars, IAU Colloquium 137
ASP Conference Series, Vol. 40, 1993
Werner W. Weiss and Annie Baglin (eds.)

OUTSIDE THE STARS

MALCOLM S. LONGAIR
Cavendish Laboratory, Madingley Road, CAMBRIDGE CB3 0HE, England.

INTRODUCTION

It is a great pleasure and privilege to give the opening lecture at this IAU Colloquium "Inside the Stars". It is particularly appropriate that it is held in Austria, the country of Ludwig Boltzmann whose name will appear explicitly or implicitly in every lecture.

My task is to describe the astrophysical and cosmological setting within which our discussions will take place. I emphasise that I am an **outsider** at this colloquium in all possible senses. My own research interests are in the areas of high energy astrophysics, extragalactic research and astrophysical cosmology. In lecturing to my students, however, I emphasise that the subject of the present colloquium is at the very heart of virtually all astrophysics and these studies are quite essential in order to make sense of galaxies and extragalactic systems. If we did not have this confidence in our ability to understand the stars, at least in principle, we would worry about the reliability of the enormous astrophysical edifice which has been built up to explain the large scale features of our Universe. I also emphasise to my students that the study of the stars is among the most exact of the astrophysical sciences — in my enthusiastic moments, I claim that, in the very best of these studies, astrophysics approaches the precision of laboratory experiment. I hope to find many examples this week to reinforce this belief.

From my perspective, what I need is a **User's Guide to Stars and Stellar Evolution**, in other words, a reliable set of rules about the origin and evolution of stars in order to diagnose the physical properties of the systems I am trying to understand. I will illustrate the types of information we need by discussing three case studies in the areas of (i) high energy astrophysics, (ii) classical cosmology and (iii) astrophysical cosmology and the origin of galaxies. Necessarily, these studies will be far from complete, but I hope they will illustrate some of the issues which come up in these disciplines. In the course of the discussion, it will become apparent that I will touch upon essentially all branches of contemporary astrophysics. I will take very different approaches to the three case studies.

HIGH ENERGY ASTROPHYSICS

One of the reasons I was delighted to accept the invitation of the organisers to deliver this lecture was that, at this very moment, I am writing Chapter 14 of my text-book *High Energy Astrophysics: Volume 2.* Here are its title and contents:

14 Aspects of stellar evolution relevant to high energy astrophysics

- *14.1 Introduction*
- *14.2 Nucleosynthesis in stars*
  - *14.2.1 Observations of solar neutrinos*
  - *14.2.2 Solar and stellar oscillations*
  - *14.2.3 Nucleosynthesis of the heavy elements*
- *14.3 Stellar evolution in globular clusters - the age of the Galaxy*
- *14.4 Post main-sequence evolution and mass-loss*
- *14.5 Binary stars and stellar evolution*

These topics will be familiar to everyone in this room but let me run through them from my position as an outsider. This may give some impression to the insiders of our perception of their disciplines.

Solar neutrinos

The first thing we ask is "Do we really understand our own Sun?" All of us would worry very deeply if the combination of nuclear physics and stellar structure could not get our own Sun right. The subject is dominated by two complementary studies, the observations of solar neutrinos and helioseismology. Let me deal with my impression of the solar neutrinos first. What most of us are familiar with are the various assessments by Bahcall, admirably summarised in his monograph "Neutrino Astrophysics" (Bahcall 1989). There is much excitement in this area because of the recent results from the Kamiokande II and SAGE experiments and the imminent publication of the results of the GALLEX experiment. Let me summarise what I believe to be the situation from the outsider's point of view.

The longest-running experiment is the famous Davis experiment in which the high energy neutrinos from the decay of ^{8}B are detected by the nuclear transmutations which they induce in ^{37}Cl. According to Bahcall, the discrepancy between the observed and predicted fluxes of the high energy neutrinos is as follows:

$$\begin{aligned} \text{Observed flux of solar neutrinos} &= 2.1 \pm 0.9 \text{ SNU} \\ \text{Predicted flux of solar neutrinos} &= 7.9 \pm 2.6 \text{ SNU} \end{aligned} \qquad (1)$$

Confirmation that this flux of high energy neutrinos indeed originates in the Sun is provided by the beautiful Kamiokande II experiment for which a deficit

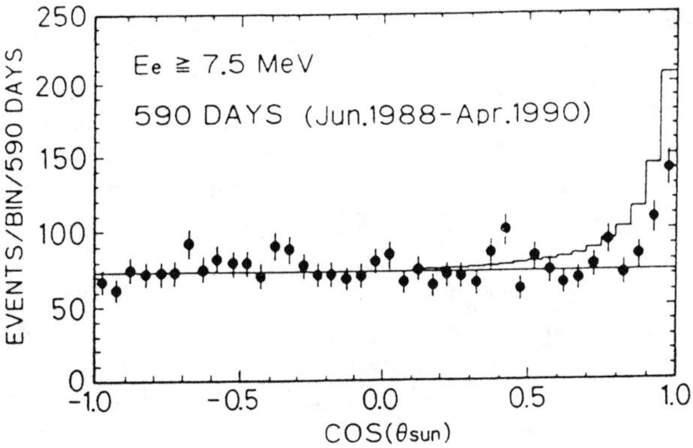

FIGURE 1 The distribution in $\cos\theta_{\mathrm{Sun}}$ for the 590-day sample for $E_e \geq 7.5$ MeV. θ_{Sun} is the angle between the momentum vector of an electron observed at a given time and the direction of the Sun. The isotropic background which is roughly 0.1 events day^{-1} bin^{-1} is due to spallation products induced by cosmic ray muons, γ-rays from outside the detector and radioactivity in the detector water. The angular resolution of the detector system has been taken into account in calculating the expected distribution of arrival directions of the neutrinos from the Sun (Hirata et al 1990).

of solar neutrinos is again reported relative to Bahcall's Standard Solar Model. The deficit is quoted to be

$$\frac{\text{Measured flux of neutrinos}}{\text{Predicted flux of neutrinos}} = 0.46 \pm 0.13\ (\text{stat}) \pm 0.08\ (\text{syst}) \qquad (2)$$

The beauty of the Kamiokande experiment is the fact that it provides information about the arrival directions of the neutrinos. The excess of neutrinos is found only in the direction of the Sun (Figure 1: from Hirata et al 1990). The discrepancies illustrated by the results (1) and (2) are the famous **Solar Neutrino Problem.**

Less well-known than Bahcall's analysis are those of other solar modellers and, generally, they have predicted somewhat smaller neutrino fluxes than Bahcall's Standard Solar Model. For example, the models of Turck-Chièze et al (1988) predict a solar neutrino flux of 5.8 ± 1.3 SNU which would reduce considerably the discrepancy between the Kamiokande observations and the theoretical predictions. According to some workers, there may not be a solar neutrino problem at all (see, for example, Morrison 1992). These are technical issues which need to be addressed by the experts. It is encouraging that this key problem is now being addressed by a number of independent solar modellers.

The most eagerly awaited results are those of the SAGE and GALLEX experiments which seek to detect the much more numerous low energy neutrinos associated with the weak p-p reaction which forms deuterium

$$\mathrm{p} + \mathrm{p} \rightarrow {}^{2}\mathrm{D} + \mathrm{e}^{+} + \bar{\nu}_{\mathrm{e}}$$

The importance of this reaction is that it is directly related to the primary energy source in the Sun, unlike the ^{8}B neutrinos. The SAGE consortium has already reported an preliminary low flux as compared with the predictions of the standard solar model (Abazov et al 1992). The expected rate of arrival of low energy antineutrinos is estimated to be 132 $^{+20}_{-17}$ SNU where the errors are 3σ errors. The preliminary result from 5 runs during the period January to July 1990 was a neutrino flux of 20 $^{+15}_{-20}$ (stat) $\pm$32 (syst) SNU. This results in a upper limit to the flux of low-energy antineutrinos from the Sun of less than 79 SNU at the 90% confidence level. The GALLEX team will release their first results in June 1992.

My interest in these results is whether or not we can use the standard solar model as a baseline for stellar evolution. If the GALLEX experiment finds no deficit of solar neutrinos, most of us would regard that as the end of the story and go on to other problems. If the flux is low, we will have to incorporate some new features into the physics of the neutrino. If this is necessary, the best bet at the moment seems to be the MSW effect which is discussed in some detail in Bahcall's book and in the paper by Bahcall and Bethe (1990).

Solar and stellar seismology

Solar and stellar seismology have revolutionised the study of the Sun and stars and are, of course, one of the most important motivations for the present colloquium. Until the discovery of solar oscillations, the solar neutrinos provided the only way of studying the interior of the Sun. Otherwise, the stellar astrophysicist was restricted to studying the surface properties of the stars and inferring indirectly from these their internal structures. I have always been impressed by how well the observations and the theory of stellar evolution can explain the evolution of stars from one part to another of the H-R diagram and many astrophysical studies depend upon the reliability of these results.

As everyone at this meeting is aware, the resonant frequencies of the Sun provide detailed **dynamical** information about its internal structure. What I find quite incredible is the phenomenally good agreement between the theory of the internal structure of the Sun and the structure inferred from inverting the solar oscillation data. Each time I discuss this problem with Douglas Gough, he tells me that the oscillation data have confirmed the predictions of stellar structure with exquisite accuracy. The most recent data of which I have knowledge from the Birmingham group indicate that the models of stellar structure remain very good indeed right into the central nuclear burning regions of the Sun (Elsworth et al 1990). This is the reason why we take the point of view that, if there is a solar neutrino problem, it is probably in the nuclear physics rather than in the astrophysics of the solar interior. This makes many of us very happy because we feel that we can at least trust the astrophysics.

I eagerly await the most recent developments in this story, not only for our own Sun but also for the nearby stars. These data already provide glimpses of what might be achieved by a dedicated space mission such as PRISMA. Among topics of the greatest interest will be the impact of the new opacity data provided by the OPAL and OP programmes to be discussed at this meeting by Michael Seaton. The reason for stressing the importance of these studies is that outsiders such as myself take the results of these studies as the datum for extrapolation to more distant systems and to much more exotic environments.

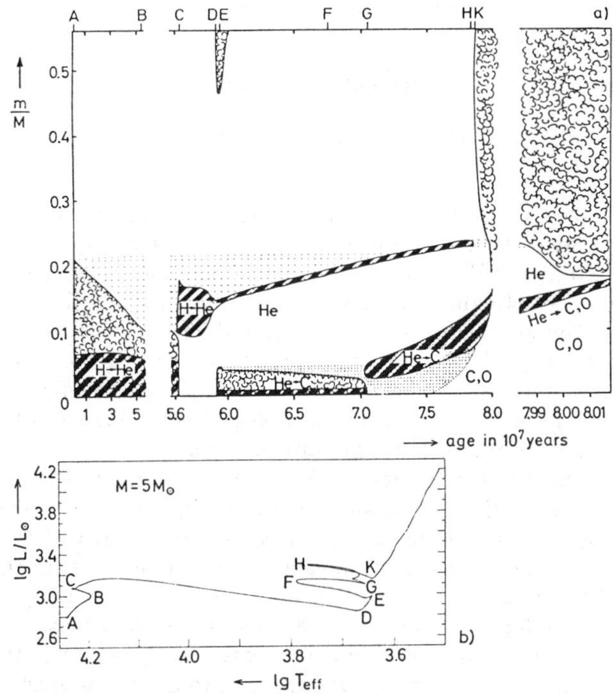

FIGURE 2 The evolution of the internal structure of a star of 5 $M_\odot$ of extreme population I illustrating the synthesis of carbon and oxygen in the core of the star. The abscissa shows the age of the model star after the ignition of hydrogen in units of 10^7 years. The ordinate shows the radial coordinate in terms of the mass m within a given radius relative to M the total mass of the star. The cloudy regions indicate convective zones. The corresponding positions of the star on the H-R diagram at each stage in its evolution are shown in the bottom diagram. (from R. Kippenhahn and A. Weigert 1990)

Nucleosynthesis of the heavy elements

The next issues we outsiders need to address for a vast number of different reasons are the sites of synthesis of the heavy elements. Thus, all of us give lectures showing the "onion-skin" model of stellar evolution showing heavier and heavier elements being created in the most massive stars until the synthesis proceeds through to iron in the cores of the most masive stars. The famous picture of Kippenhahn et al (1965) shows us how the heavier elements are built up in massive stars (Figure 2). What we need are the best modern versions of these diagrams with the corresponding yields of the heavy elements when the stars die. These data are crucial for many aspects of contemporary cosmology. Examples include

(1) The chemical evolution of galaxies;

(2) The chemical composition of dust grains formed in stellar explosions and in the interstellar gas;

(3) The abundances of the elements in the cosmic rays;

(4) Clues about the epochs when the metals were first formed during the evolution of galaxies.

We also need to be able to answer somewhat more subtle questions. A good example concerns the origin of the light elements, lithium, beryllium and boron. The standard Big Bang picture can account remarkably successfully for the abundances of light elements such as deuterium, helium-3 and helium-4 as well as for the abundance of lithium found in old cool stars, provided the mean baryon density of the Universe Ω_b is about one tenth of the critical value. The determination of the primaeval abundances of these elements in old stars is obviously of the greatest interest for cosmology because traces of beryllium and boron are created as well in the standard picture.

There has been great interest recently in studying the cosmic abundance of beryllium. It is synthesised in small amounts in the early stages of the standard Big Bang but its abundance can increase significantly if there were inhomogeneities in the distribution of matter during the epochs of nucleosynthesis. Any information, however indirect, on fluctuations in the early Universe is of crucial importance for cosmology. There is therefore the greatest interest in understanding whether all the beryllium is created in stars or whether there is an excess which might be attributed, for example, to an inhomogeneous Big Bang. Figure 3 shows a summary by Pagel (1991) of some recent observations of the abundance of beryllium in which it can be seen that there may be a small excees. The origin of this excess is of considerable interest and importance.

One of the long-standing problems in cosmic ray physics is exactly why the overall abundances of the elements in the cosmic rays are not so different from the cosmic abundances of the elements. This is strongly tied up with the acceleration mechanisms for the cosmic rays. Precise element and isotope abundances can help disentangle the most important sites for the injection of particles into the acceleration regions.

Stellar evolution in globular clusters - the age of the Galaxy

In the last subsection, we have already touched upon the evolution of stars off the main sequence. What we need are evolutionary tracks for stars of different masses so that we can build models of galactic evolution under different assumptions about their star formation rates. The type of picture we need is shown in Figure 4 which displays the evolutionary tracks of stars of different masses from the main sequence onto the giant branch. This classic picture assumes that there is no mass loss from the stars during evolution from the main sequence onto the giant branch.

There are at least three reasons why these tracks are important for outsiders.

(1) The H-R diagrams of open and galactic clusters provide tests of the theory of evolution onto the giant branch. An aspect of special importance is the use of these diagrams in detemining the ages of globular clusters. In Figure 5, I show the beautiful analysis of Hesser et al (1989) of the H-R

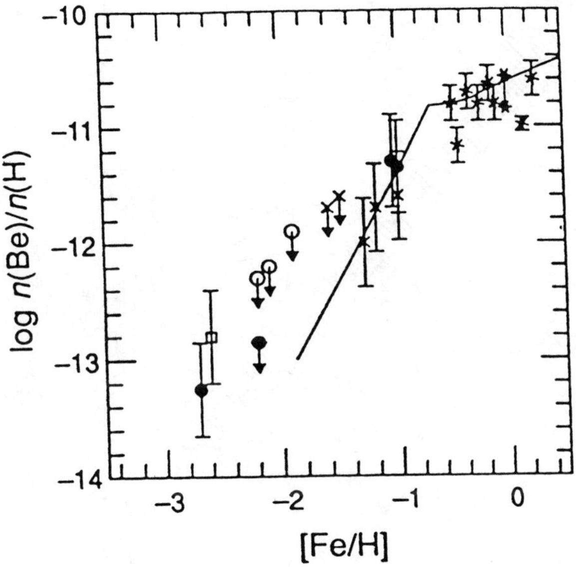

FIGURE 3 The beryllium to hydrogen ratio as a function of the iron to hydrogen ratio for a sample of stars studied by Ryan et al, Gilmore et al and others. The two left-hand points are for HD 140283, a nearly primordial subdwarf. The solid line represents a standard model of beryllium evolution through normal galactic nucleosynthesis. Is the slight excess of beryllium cosmological or not? (from B. Pagel 1991)

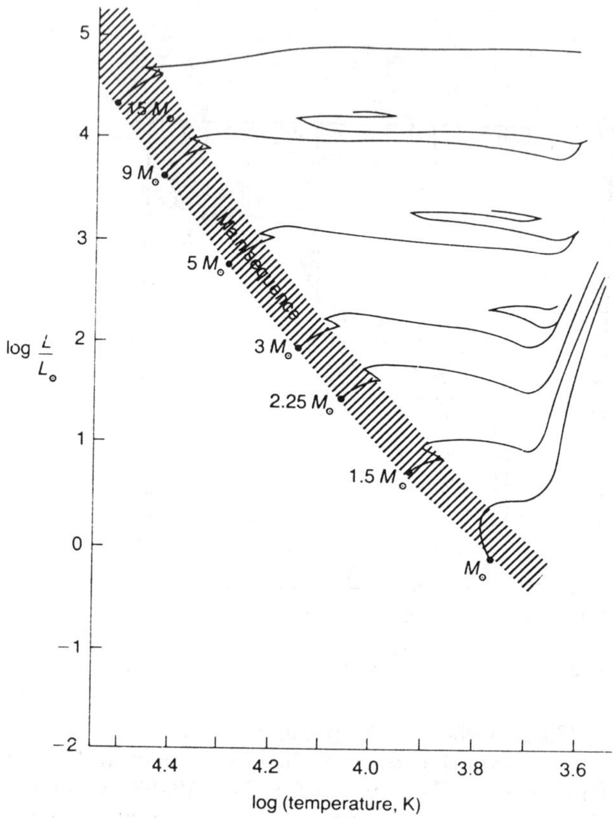

FIGURE 4 Evolutionary tracks for stars of different initial masses on the H-R diagram. These models assume that there is no mass loss from the stars. Most of the lifetime of the star is spent on the main sequence and all the subsequent phases of evolution on the giant branch take place over much shorter timescales. (After D. Michalas and J. Binney *Galactic Astronomy: Structure and Kinematics* 1981)

diagram of the globular cluster 47 Tucanae. The isochrones shown on the diagram have been derived by VandenBerg. It is apparent that, in this well-studied cluster, the theoretical isochrones can provide an excellent fit to the data.

(2) It follows that the study of the H-R diagram provide an age for the cluster and, according to VandenBerg's analysis, the cluster must be between 12 and 14×10^9 years old. This figure is of great cosmological significance since it sets a lower limit to the age of the Galaxy and to the age of the Universe. This is a crucial piece of cosmological data and, in the simplest world models, implies that Hubble's constant should be closer to 50 than to 100 km s^{-1} Mpc^{-1}. It is crucial for cosmology to know exactly how accurately this age can be determined. By exactly the same token, the study of such diagrams for open clusters provide ages for much younger systems. The general point of importance for extragalactic astrophysics is that the study of the H-R diagrams and the synthesised colours from populations of stars provide one of the few ways of determining ages for the events observed in galaxies. I will come back to this point in Section 3 on classical cosmology.

(3) I emphasise again how important these studies are for understanding the circulation of the heavy elements from the sites of their synthesis inside stars to the interstellar medium and to the next generation of stars. Much of the enrichment is thought to be due to supernova explosions of massive stars but it is far from clear how much is produced per explosion. Indeed, we need to know much more about the evolution of the most massive stars which burn their fuel all the way through to iron and beyond.

Post main-sequence evolution and mass loss

The above discussion leads naturally to the role of mass loss in stellar evolution. When stars begin their evolution up the giant branch there is observational evidence, supported by theoretical arguments, that mass-loss plays a key role in the evolution of stars from the giant branch to the horizontal branch, at the tip of the red giant branch and finally in the final demise of the star as a white dwarf, neutron star or black hole. This much is well established and leads to the circulation of matter between the stars and the interstellar gas.

However, mass-loss is also likely to be important for massive stars on the main sequence. Figure 6 shows the modifications to the evolutionary tracks of massive stars if strong mass loss from their surfaces plays an important role. In a simple interpretation of what is happening, it can be seen that the effect of mass loss is to remove the surface layers exposing the hotter interior and thus moving the stars to the left across the H-R diagram. This behaviour, which has been described in detail by Maeder (1981), can account for a number of important features for other aspects of astronomy. One example is the explanation of the observation that there seem to be only hot blue stars on the H-R diagram and not the corresponding luminous red stars predicted by the tracks without mass-loss. It may be that it is the process of mass loss which leads to only massive blue stars being observed in galaxies. This idea has been used to support the use of the most luminous blue stars in galaxies as distance indicators.

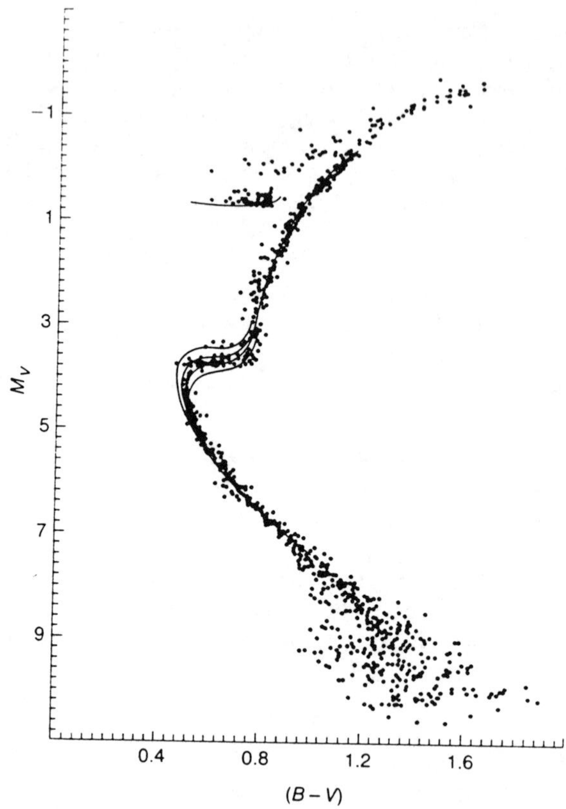

FIGURE 5 The H-R diagram for the globular cluster 47 Tucanae. The scatter in the points increases towards faint magnitudes because of the increase in observational error associated with the photometry of faint stars. The solid lines show best fits to the data using theoretical models for the evolution of stars from the main sequence onto the giant branch due to VandenBerg. For this cluster, the best-fit isochrones have ages between about 12 and 14×10^9 years and the cluster is metal rich relative to other globular clusters, the metal abundance corresponding to about 20% of the solar value. (From J.E. Hesser, W.E. Harris, D.A. VandenBerg, J.W.B. Allwright, P. Stetson 1989)

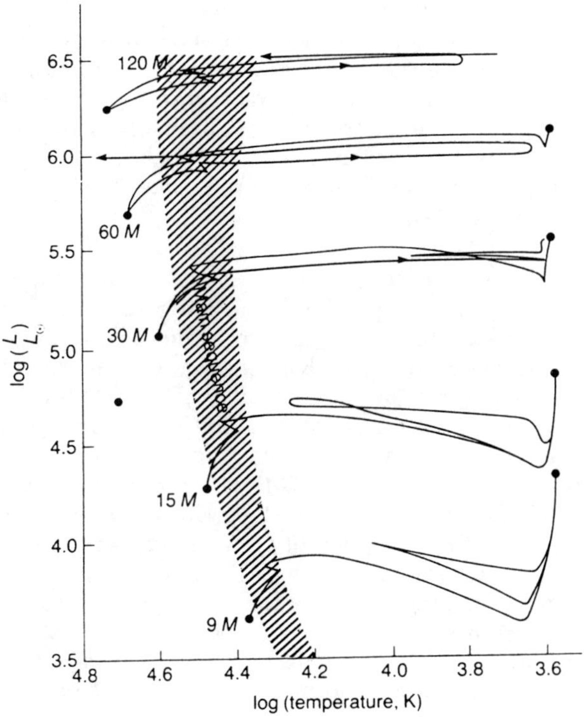

FIGURE 6 Examples of theoretical evolutionary tracks of massive stars once account is taken of the effects of mass loss during their evolution on the main sequence and during the red giant phase. The solid dot at the end of each track corresponds to the onset of carbon burning. Those tracks which end at the right of the diagram include no mass loss. The lines which show major excursions to the left across the H-R diagram include the effects of strong mass loss. (After A. Maeder 1981)

A second example of the importance of these studies for high energy astrophysics is that the Wolf-Rayet stars, which are among the most luminous stars known, have chemical over-abundances of the isotope of neon ^{22}Ne, similar to the isotope anomalies found in the cosmic rays. It might be that these stars are important progenitors for these heavy elements in the cosmic rays.

The general point of importance is that mass loss from stars is a key issue for many of the most important processes in high energy astrophysics and astrophysical cosmology. Processed gas clouds are observed very close to the massive black holes in active galactic nuclei and we know that the accretion of mass onto supermassive black holes is a very attractive energy source for these objects. Likewise, the enrichment of the interstellar gas from which new generations of stars form is a key process in the early evolution of galaxies and potentially the observation of these processes can help define the sequence of events which lead to the galaxies as we know them now.

Binary stars and stellar evolution

The key role of binary stars in providing the basic calibration of the mass-luminosity relation for stars is very well-known and to the outsider pictures such as Figure 22.3 of the text-book by Kippenhahn and Weigert (1990) give us a lot of confidence that the basic physics of main sequence stars is in good shape. Obviously, there will be an enormous potential for improving the accuracy with which this comparison can be made with the magnificent data-base which will become available from the Hipparcos satellite of the European Space Agency.

Binary star evolution is, of course, of the greatest interest in its own right from the point of view of the formation of systems such as the low-mass X-ray binaries and the more exotic systems in which there is evidence for black holes. I consider this topic to be an integral part of the theme of this colloquium because the white dwarfs, neutron stars and black holes are the ultimate products of what is going on inside the stars. It would be a great advance if we had secure estimates of what types of dead star form from which types of star and why.

Star formation

Finally, I have to say something about the process of star formation and, in particular, the internal structure of pre main-sequence stars. In this case, the internal structure is very different from that of a main-sequence or post main-sequence star. According to the standard picture of Shu et al (1987), the energy source in protostars and pre main-sequence stars is accretion rather than nuclear energy generation. The removal of the gravitational binding energy of the accreted matter is effected by the reradiation of heated dust at far infrared wavelengths at which the protostellar cloud is transparent. The signature of this process is to be seen everywhere in infrared maps of our Galaxy and of regions of star formation. Whilst no one would claim that the relation of these observations to the detailed processes of star formation are fully understood, infrared and sub-millimetre observations of these regions have revolutionised our understanding of the necessary ingredients of an acceptable theory. Besides the intrinsic importance of understanding the processes of star formation, these studies are crucial for galactic evolution.

To give only one example, we know that the IRAS galaxies which are the most powerful far infrared emitters are associated with interactions or collisions between galaxies. The implication is that the strong gravitational interaction between the galaxies results in a high rate of star formation which in turn results in the extreme far-infrared luminosities observed in these galaxies. But this concept is not only important for galaxies now. Among the favoured scenarios for the formation of galaxies is the hierarchical clustering picture which is a necessary part of the cold dark matter scenario of galaxy formation. Presumably the strong IRAS galaxies are examples of the types of process which lead to the formation of more massive galaxies out of the coalescence smaller galaxies.

Summary

My intention has been to show in these examples how the understanding of the internal structures of the stars is directly related to processes which are of importance in a wide range to different astrophysical disciplines. Although in some cases, the relation may appear quite distant, there is no question but that the foundations of many of these disciplines depend upon the reliability of our understanding of the internal processes which go on in stars. I will be intrigued to learn of how many of my preconceived notions survive this meeting intact.

CLASSICAL COSMOLOGY

In the second case study, I look at the old problem of determining the deceleration of the Universe from observations of distant galaxies. Let me treat first the traditional problem of the Hubble diagram and then look at the problem of determining timescales at large redshifts.

The infrared K-z relation

I will tell this story from a personal point of view. The traditional way of tackling this problem is through the redshift-magnitude relation for the brightest cluster galaxies (see, for example, Gunn 1977). The problem with this approach is that the sample of clusters extends only to redshifts of about 0.5 at which the differences between the expectations of different world models is still small. In the early 1980s, Simon Lilly and I (1984) began studying the infrared properties of radio galaxies in the J, H and K infrared wavebands (that is, the 1.2, 1.65 and 2.2 μm infrared windows). It had been known for some time that the radio galaxies have more or less the same absolute magnitudes and are among the most luminous giant elliptical galaxies known. By the early 1980s, complete samples of radio galaxies were available with redshifts which extended up to about 2. Figure 7 shows what we found when we plotted the K (2.2 μm) apparent magnitudes against redshift. I still find this an astounding diagram.

However one wishes to interpret the diagram, the key points are that the dispersion in apparent magnitude about the mean line through the points does not change with redshift and that the diagram extends to redshifts well beyond 1. As more and more points have been added to this diagram, the miracle continues (Figure 8). For whatever reason, there is clearly some systematic behaviour going on in these galaxies. In the most naive interpretation one might say that it is only the most massive galaxies which can become radio galaxies.

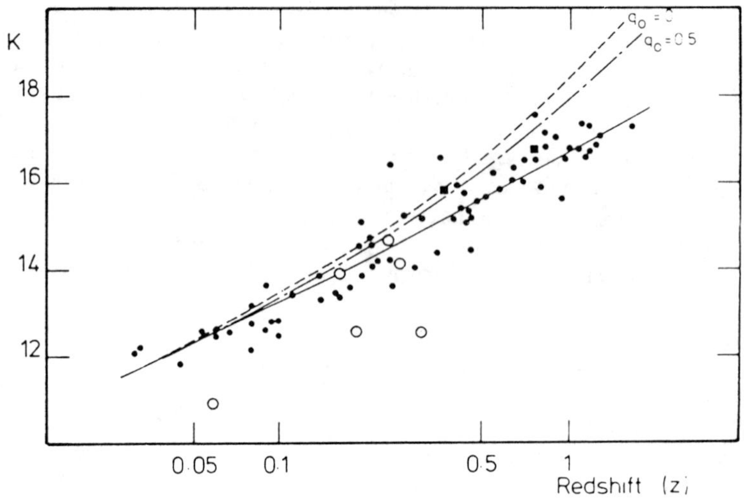

FIGURE 7 The infrared K-z relation for a complete sample of 3CR radio galaxies. The sample consists of those galaxies in which the infrared flux is the starlight of galaxies and is not contaminated by non-thermal optical or infrared emission. (S.J. Lilly and M.S. Longair 1984)

The second key point is the importance of the infrared waveband for carrying out this analysis. If this diagram were redrawn using optical instead of infrared magnitudes, there would be a much greater dispersion in apparent magnitude at redshifts greater than about 0.5 because these galaxies display a wide range of optical-to-infrared colours, reflecting the different amounts of star formation going on in these galaxies. However, if one uses infrared magnitudes, most of the light is due to giant stars which are derived from the old stellar population rather than the more transient populations which can dominate the optical spectrum. Thus, the great advantage of using the 2.2 μm waveband is that the absolute magnitudes should reflect the gross properties of the underlying stellar population averaged over cosmological time-scales.

Also shown on Figure 7 are the expectations of uniform world models with deceleration parameters $q_0 = 0$ and 0.5, corresponding to the empty and critical Universes respectively. It can be seen that both lines lie significantly above the mean line through the points in the sense that the galaxies at redshifts about 1 seem to be more luminous than is expected in these models. If one attempts to find a best fitting Friedmann model, the answer is $q_0 \approx 3.5$ corresponding to 7 times the critical value of the deceleration parameter.

Of course, what we have not taken into account is the effect of stellar evolution upon the luminosities of the galaxies and this is where the significance of the considerations of this colloquium comes in. At a redshift of 1, the galaxies are less than half the age they are now and so the whole of the stellar population was much younger. We need to be able to make corrections for the evolution of the stellar populations of the galaxies as a function of cosmic epoch. In fact, by restricting our attention to the infrared luminosity of the galaxy, this correction becomes much simpler than one might imagine. If we assume that the only

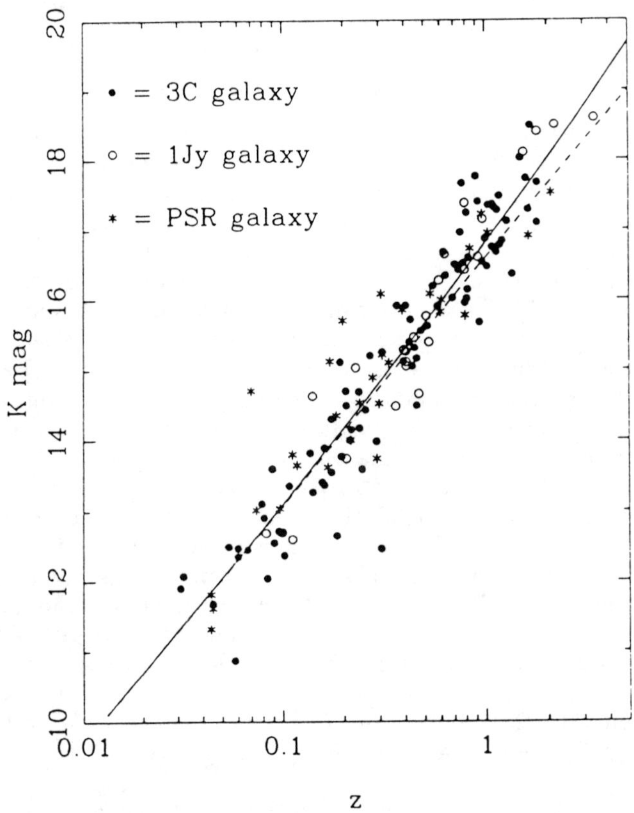

FIGURE 8 The K-z relation for radio galaxies. This compilation includes radio galaxies from the 3CR sample (solid circles), from the 1 Jy samples (open circles) and the 0.1 Jy samples (asterisks) (From J. Dunlop and J.A. Peacock 1990)

correction we have to make is for the ageing of the stellar population, we can work it out in a simple way. Since the bulk of the luminosity is simply the integrated light of stars on the giant branch, the evolution of the luminosity in the infrared waveband is determined by the rate at which stars evolve from the main sequence onto the giant branch. In a simple calculation, we can write that the luminosity of galaxy is proportional to the number of red giants $N_{\rm rg}$ which can be written

$$N_{\rm rg} = \left[\frac{dN(M)}{dM}\right]_{MS} \left(\frac{dM_{\rm to}}{dt}\right) \tau_{RG}$$

where $dN(M)/dM$ is the main-sequence mass function, $M_{\rm to}$ is the mass at the main-sequence termination point and τ_{RG} is the lifetime of a star on the giant branch. Putting in reasonable values for the main sequence mass function and the mass-luminosity relation for stars, we find that, at a redshift of 1, the galaxies are expected to be about 1 magnitude brighter than they are at the present epoch. It can be seen from Figure 7 that this is exactly the difference we find betweeen the observed locus of points and the expectations of models with $q_0 \sim 0 - 0.5$. We interpret these results as meaning that we have observed the effects of stellar evolution upon the luminosities of galaxies at redshifts of 1 and greater.

If one is then feeling very ambitious, one can try to solve simultaneously for the effects of stellar evolution and for the value of the deceleration parameter. Doing this, we found that there is a 90% probability of the value of q_0 lying in the range 0.1 to 0.9. In other words, so far as we can tell q_0 has the value 0.5 ± 0.5. Now, this may not look all that wonderful but at least we believe that we have done as reasonable a job as one can using straightforward methods and that the result is not implausible. The relevant comparison is with the density parameter Ω which, according to the Friedmann world models should satisfy the relation $q_0 = \Omega/2$. Most workers would agree that the value of Ω derived from observation is at least 0.1 and so it is likely that the equality $q_0 = \Omega/2$ is correct within about a factor of 10. Optimists would argue that if they really are so similar, they must be the same. Most of us would, however, like to improve the accuracy with which this equality is known to be correct but that is a very big job indeed.

The central role which studies of stars and stellar evolution play in this type of work can now be appreciated. I have included only the simplest possible correction to the luminosities of the galaxies for the effects of stellar evolution. Obviously, one would like to do much better than this by including the best models available for the evolution of stars. My own view is that this type of analysis is indicative of the studies which will unquestionably be important in the future. It is still very difficult to make high quality observations of these galaxies but, with the next generation of large telescopes, it should begin to be possible to undertake proper spectroscopic studies of the populations of stars in these galaxies and so much better models of the evolution of the stellar populations will be needed. I have no illusions about the fact that there will certainly be many other effects to take into account but we can be certain already that stellar evolution is one of them.

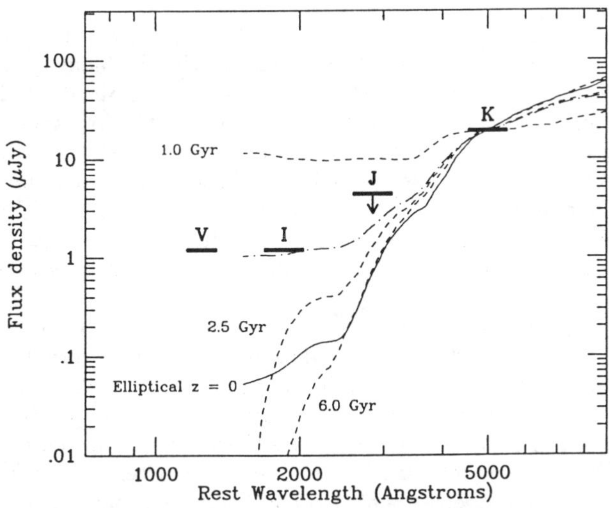

FIGURE 9 The rest-frame broad-band spectral energy distribution of the radio galaxy 0902+34 from 100 to 500 nm. The curves represent different stages of evolution Bruzual's C model and a zero redshift elliptical galaxy. The flat spectrum for V to I indicates substantial star formation. The steep rise into the observed infrared is interpreted as representing a massive older population. (From S.J. Lilly 1988)

The ages of stellar populations at large redshifts

There is a similar way in which observations of this type can be used to set important limits to cosmological parameters and which depend in a central way upon the understanding of the stellar evolution in galactic populations as a whole. Again, I use this example as an illustration of the types of issue which cosmologists can now address and which are revelant to the theme of this colloquium.

The example concerns the work of Simon Lilly who has continued the studies described above and found some radio galaxies with very large redshifts among the 1 Jy sample of radio sources. An example of interest is the radio galaxy 0902+34 which has redshift 3.395 (Lilly 1988). It is important that this galaxy lies precisely on the K-z relation shown in Figure 8. From broad band photometry, Lilly was able to define a broad band spectrum for this galaxy which is shown in Figure 9. Also shown on the diagram are the results of modelling the evolution of a stellar population which is formed with the standard initial mass function and then simply allowed to age. The aim of the calculation is to estimate the minimum permissible age for the population responsible for the infrared emission. It can be seen from these models that, if the galaxy were any younger than about 2×10^9 years, the optical emission would be relatively too bright as compared with the infrared emission.

The argument then proceeds as follows. The observation of the strong infrared emission is interpreted as the radiation of a popultion which is at least 2×10^9 years old. If we were to adopt cosmological parameters $\Omega = 1$ and $H_0 = 100$ km s^{-1} Mpc^{-1}, then the Universe would be only 7.3×10^8 years old at a redshift of 3.4! In other words, this simple observation can lead to significant constraints on cosmological parameters. Plainly, the problem is relieved if we adopt a value of $H_0 = 50$ km s^{-1} Mpc^{-1} and a smaller value of Ω. From the astrophysical point of view, the important point is that the argument hangs upon the question of the reliability of the lower limit to the age of the stars emitting in the K waveband from studies of stellar evolution. In principle, it is not so different from VandenBerg's analysis of the properties of the H-R diagrams for globular clusters although now we are dealing with very faint galaxies without the possibility at the moment of obtaining much more than photometric spectra for these objects.

Not surprisingly, this work has been the subject of some controversy. Perhaps the most fascinating development has been the discovery that the radio structures of some of these distant radio galaxies are aligned with their optical continuum and line emission (e.g. Chambers, Miley and van Breugal 1990). Since the radio source lifetimes are much less than 10^9 years, this suggested that the aligned components must have lifetimes much shorter than 10^9 years. This has led to a number of models in which the aligned component is created by the passage of the radio jet through the interstellar gas and thus has nothing to do with the old stellar population (Rees 1989, Chambers and Charlot 1989). The key quesion is the degree to which these phenomena impact the idea that there is an old stellar population in the radio galaxies as a whole. The recent detailed analysis of Rigler et al (1992) strongly suggests that the aligned component has a flat spectrum and is independent of the old stellar population which represents the bulk of the stars in the galaxy and which shows no alignment with the radio structure.

I personally find the argument of Rigler et al (1992) quite compelling that there indeed exist old stellar populations in these very distant galaxies. If this is the case, the observation has important consequences for the origin of galaxies. Evidently, in a few cases, the galaxies have developed to a considerable state of maturity by a redshift of 3. Adopting the most straightforward interpretation of the data, if $H_0 = 50$ km s^{-1} Mpc^{-1} and $\Omega \approx 0$, the stars in the galaxy would have to be formed at redshifts much greater than 4, at least at a redshift of 7 in the empty world model.

THE ORIGIN OF GALAXIES

The big problem in understanding the origin of galaxies is to reconcile the quite remarkable smoothness of the Cosmic Microwave Background Radiation with the gross irregularity in the distribution of galaxies, most vividly portrayed in the three-dimensional pictures of the local distribution of galaxies provided by the Harvard-Center for Astrophysics survey of almost 30,000 galaxies. As is well known, the problem arises because, whereas the Jeans' instability in a static medium grows exponentially with time, in an expanding medium, the rate of growth of the perturbation is only *algebraic*. The result first derived by Lifshitz can be written

$$\frac{\Delta\rho}{\rho} \quad \propto \quad R \quad = \quad \frac{1}{(1+z)}$$

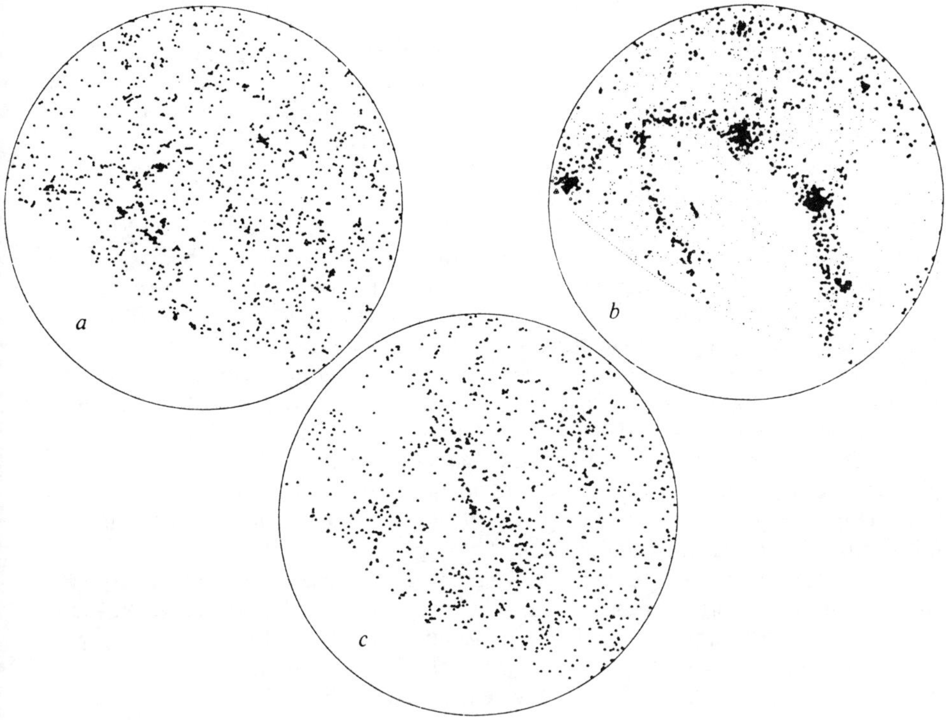

FIGURE 10 Equal area projections of the galaxy distribution on the northern sky and in artificial catalogues made from n-body simulations. (*a*) The standard cold dark matter model including biased galaxy formation; (*b*) the neutrino-dominated model in which galaxy formation began at a redshift of 2.5; (*c*) the galaxies observed in the Harvard-Center for Astrophysics northern sky survey. The outer circle represents galactic latitude +40° and the empty regions lie at declinations below 0°. (From M. Davis, G. Efstathiou, C.S. Frenk and S.D.M. White 1992)

provided $\Omega z \gg 1$. In this expression $\Delta\rho$ is the density enhancement relative to the mean background density ρ. R is the scale factor of the Universe which describes how the distance between two points expanding uniformly with the Universe changes with time, normalised to $R = 1$ at the present epoch, and z is redshift. In the limit $\Omega z \gg 1, R \propto t^{2/3}$ and the above relation describes the algebraic time development of the density perturbations. The problem is that the matter and radiation decoupled when the scale factor was about 1/1000 of its present value and so, since we observed lots of collapsed structure now, there must have been significant density perturbations in the matter distribution at the time when the photons of the Cosmic Microwave Background Radiation were last scattered. There should therefore be fluctuations in the intensity of the Cosmic Background Radiation in different directions on the sky. This is the line of reasoning which leads to the difficulty of reconciling the smoothness of the background radiation with the irregularity of the distribution of galaxies.

The most popular solution is to build dark matter into the model universe so that it is dominant gravitationally. Then, the fluctuations are assumed to be present in the dark matter at the moment of decoupling but these are not reflected in the distribution of the baryonic matter which is closely coupled to the background radiation. Although the nature of the dark matter is unknown, this is the current orthodoxy and the question is whether or not astrophysical arguments can cast more light on its nature.

The most popular picture is the *Cold Dark Matter* model in which the dark matter fluctuations on a wide range of scales collapse and begin to form galaxies by a process of hierarchical clustering. Large scale structures are all rather recent features of the Universe. A less popular version is the *Hot Dark Matter* model in which the neutrino has a finite rest mass. In this case, all fine scale structure is obliterated by the free streaming of the neutrinos and the large scale structures form first. Galaxies form within the potential wells created by the large scale structure. Neither picture gives exactly the right answer without some additional astrophysical considerations (Figure 10). The cold dark matter picture can account for the correlation function for galaxies but has difficulty producing enough power on the very largest physical scales. The hot dark matter picture produces too much large scale structure.

Recently, there has been great excitement in the press and in the astronomical community because at last fluctuations in the Cosmic Microwave Background have been discovered by the COBE satellite at an intensity level about one part in 100,000 on angular scales of 7° and greater. These angular scales correspond to physical scales in the Universe very much greater than the large holes in the distribution of galaxies. What has encouraged many theorists is that this is roughly the level of intensity fluctuations expected in the most popular variant of the Cold Dark Matter picture, if the initial spectrum of fluctuations is extrapolated in the most natural way to these large physical scales.

From the point of view of the present discussion, the interesting question is when the galaxies first formed. Unfortunately, this is not a well posed question because it is not clear what we mean by the expression "When did the galaxies form?". Do we mean when the system was first bound, or when the first stars formed? Do we mean when the bulk of the stars formed or do we mean the time by which it had more or less attained its present mass? This list simply emphasises that one must be careful when one tries to quantify what one means by the epoch of galaxy formation.

The area in which these considerations bear upon the subject matter of this colloquium concerns the question of when the elements were created. There is a variety of evidence to suggest that the process of galaxy formation and metal formation must have taken place over a considerable redshift range. None of these pieces of evidence is unambiguous but taken together they are suggestive.

(1) Galaxies and quasars at large redshifts seem to have remarkably normal chemical abundances;

(2) There exist radio galaxies at redshifts of 3 and greater which seem to be younger counterparts of radio galaxies at the present epoch;

(3) The cosmological evolution of the quasar population shows a maximum or plateau at redshifts betweeen about 2 and 3:

(4) The abundances of the elements in G-dwarfs suggest there may have been an initial burst of element enhancement;

(5) The intergalactic gas needs more than simply the ultraviolet emission of quasars to ionise it fully at redshifts 2 to 4 and the emission of young galaxies in these redshift ranges could be responsible for the ionisation of the gas.

(6) Young galaxies may be able to produce a significant fraction of the metals which we observe today.

This list is not complete and certainly not uncontroversial but it makes the point that there is evidence for metal formation over a wide range of redshift. If we were able to pin down the history of chemical enrichment as a function of cosmic epoch, this would help constrain the global pictures of galaxy formation.

A very elegant argument was presented recently by Lilly and Cowie (1987) which indicates how the rate of element formation can be related to the surface density of young galaxies. In its simplest form, the argument goes as follows. When a density ρ_m of metals is created, an energy $0.007\rho_\mathrm{m}c^2$ is released, corresponding to the binding energy liberated when helium is synthesised from hydrogen. Now models of galaxies with continuous star formation have very flat spectra. Figure 11 shows the predicted spectra of a "young galaxy" in the sense that the star-burst lasts 12×10^9 years and the spectra at different ages during that period are illustrated.

It can be seen from Figure 11 that, to a good approximation, the intensity spectrum can be taken to be flat, F_ν = constant, for all wavelengths longer than the Lyman limit. A simple calculation then shows that there is a simple relation between the integrated intensity of young galaxies in a particular redshift interval and the contribution which these galaxies make to the density of metals now. In simplified form, the relation is

$$I_\nu = 7.5 \times 10^{-25} \left(\frac{\rho_\mathrm{m}}{10^{-31}\ \mathrm{kg\ m^{-3}}} \right) \quad \mathrm{W\ m^{-2}\ Hz^{-1}\ sr^{-1}}$$

The meaning of this formula is that, if one measures a certain intensity I_ν from a given redshift interval from these young galaxies, then the density of metals ρ_m observed now was synthesised in that redshift interval.

In fact, Cowie (1989) claims to have discovered a class of such flat spectrum galaxies in very deep multicolour surveys. The integrated flux density of these objects suggests that they can account for a significant fraction of the metals observed now, possibly most of it. The redshift at which most of the metals were formed is not certain but it appears that a significant fraction occurred at redshifts of the order 1.

The important point from the perspective of this conference is the way in which processes occurring in the interiors of stars have a key role to play in providing further evidence on the origin and evolution of galaxies. This type of argument needs to be refined using better models of star forming galaxies and the yields of metals from stars of different types.

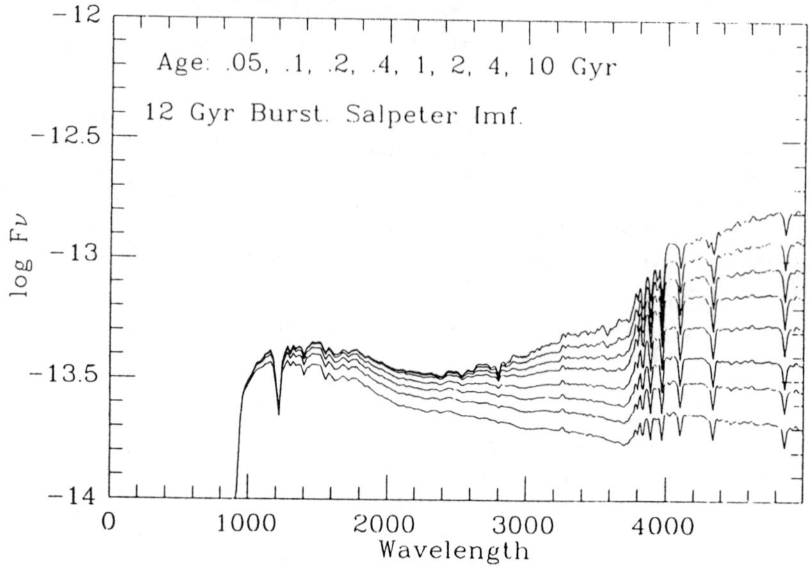

FIGURE 11 Synthetic spectra for a region with constant star formation rate at the ages indicated. A Salpeter mass function is assumed with cut-offs at 75 and 0.08 $M_{\odot}$. The spectra were generated by Gustavo Bruzual from a recent version of his evolutionary synthesis programmes. (From S.D.M. White *The Epoch of Galaxy Formation* 1989)

CONCLUSIONS

It is my firm conviction that the whole of the edifice of modern astronomy and cosmology is only as strong as the foundations upon which it is built. Central to the whole of astronomy is the understanding of the stars and the processes which occur in their interiors. As I have indicated, these studies have significance far beyond interest in the stars themselves.

What no one could have predicted is the spectacular way in which the subject has been rejuvenated by solar and stellar seismology. It is always rash to make predictions about the birth of new eras in astronomy but I would be surprised if, in ten years time or sooner, stellar seismology had not joined photometry and spectroscopy as one of the standard tools of the astronomer for the study of the stars. In my view, in the case of projects such as PRISMA, it is not a question of "If it will fly" but rather "When it will fly". The scientific significance seems to me unassailable.

REFERENCES

Abasov, A.I., Anosov, O.L., Bowles, T.J., Cherry, M.L., Cleveland, B.T., Davis, Jr., R., Elliott, S.R., Faizov, E.L., Gavrin, V.N., Kalikhov, A.V., Knodel, T.V., Knyshenko, I.I., Kornoukhov, V.N., Kouzes, R.T., Lande, K., Mezentseva, S.A., Mirmov, I.N., O'Brien, H.A., Ostrinsky, A.V.,

Pshukov, A.M., Revzin, N.E., Shikhin, A.A., Timofeyev, P.V., Veretenkin, E.P., Vermul, V.M., Wark, D.L., Wilkerson, J.F. and Zatsepin, G.T. (1991). Proc. 22nd International Cosmic Ray Conference, Dublin. **3**, 724.

Bahcall, J.N. (1989). *Neutrino Astrophysics*, Cambridge: Cambridge University Press.

Bahcall, J.N. and Bethe, H.A. (1990). *Phys. Rev. Letters*, **65**, 2233.

Chambers, K.C. and Charlot, S (1990). Astrophys. J., 348, L1.

Chambers, K.C., Miley, G.K. and van Breugel, W.J.M. (1990). Astrophys. J., **363**, 21.

Davis, M., Efstathiou, G., Frenk, C.S. and White, S.D.M. (1992). *Nature*, **356**, 489.

Dunlop, J. and Peacock, J.A. (1990). *Mon. Not. R. Astr. Soc.*, **247**, 19.

Cowie, L.L. (1989). In *The Epoch of Galaxy Formation*, eds. C.S. Frenk, R.S. Ellis, T. Shanks, A.F. Heavens and J.A. Peacock, 31. Dordrecht: Kluwer Academic Publishers.

Elsworth, Y, Howe, R., Isaak, G.R., McLeod, C.P. and New, R. (1990). *Nature*, **347**, 536.

Gunn, J.E. (1978). In Observational Cosmology, 8th Advanced Course, Swiss Society of Astronomy and Astrophysics, Saas-Fee, Geneva Observatory Publications.

Hesser, J.E., Harris, W.E., VandenBerg, D.A., Allwright, J.W.B., Schott, P. and Stetson, P. (1989). *Publ. Astron. Soc. Pacific*, **99**, 739.

Hirata, K.S., Inoue, K., Katija, T.,Kifune, T., Kihara, M., Nakahata, M., Nakamura, K., Ohara, S., Sato, N., Suzuki, Y., Totsuka, Y., Yaginumi, Y., Mori, M., Oyama, Y., Suzuki, A., Takahashi, K., Yamada, M., Koshiba, M., Suda, T., Tajima, T., Miyano, K., Miyata, H., Takei, H., Fukuda, Y., Kodera, E., Nagashima, Y., Takita, M., Kaneyuki, K., Tanimori, T., Beier, E.W., Feldscher, L.R., Frank, E.D., Frati, W., Kim, S.B., Mann, A.K., Newcomer, F.M., Van Berg, R. and Zhang, W. (1990). *Phys. Rev. Letts*, **65**, 1297.

Kippenhahn, R, Thomas, H.-C. and Weigert, A. (1965). *Z. Astrophys.*, **61**, 241.

Kippenhahn, R. and Weigert, A. (1990). *Stellar Structure and Evolution*. Heidelberg: Springer-Verlag.

Lilly, S.J. (1988). Astrophys. J., **333**, 161.

Lilly, S.J. and Cowie, L.L. (1990). In *Infrared Astronomy with Arrays*, eds. C.G. Wynn-Williams and E.E. Becklin, 473. Honolulu: University of Hawaii, Institute of Astronomy Publications.

Lilly, S.J. and Longair, M.S. (1984). Mon. Not. R. astr. Soc., **211**, 833.

Maeder. A. (1981). *Astron Astrophys.*, **102**, 405.

Michalas, D. and Binney, J. (1981). *Galactic Astronomy: Structure and Kinematics*. New York: W.H. Freeman and Co.

Morrison, D.R.O. (1992). *Particle World*, **3**, 30.

Pagel, B. (1991). *Nature*, **354**, 267.

Rees, M.J. (1989). Mon. Not. R. astr. Soc., **239**, 1P.

Rigler, M.A., Lilly, S.J., Stockton, A., Hammer, F. and Le Fèvre, O. (1992). *Astrophys. J.*, **385**, 61.

Shu, F.H., Adams, F.C. and Lizano, S. (1987). *Ann. Rev. Astron. Astrophys.*, **25**, 23.

Turck-Chièze, S., Cahen, S., Cassé, M. and Doom, C. (1988). *Astrophys. J.*, **335**, 415.

White, S.D.M. (1989) *The Epoch of Galaxy Formation*, (eds. C.S. Frenk, R.S. Ellis, T. Shanks, A.F. Heavens and J.A. Peacock), 15. Dordrecht: Kluwer Academic Publishers.

Inside the Stars, IAU Colloquium 137
ASP Conference Series, Vol. 40, 1993
Werner W. Weiss and Annie Baglin (eds.)

THE CONTROVERSIAL SUN

ROGER K. ULRICH
Department of Astronomy, University of California at Los Angeles, Los Angeles, CA 90024

ABSTRACT The sun serves as an important test case for a variety of problems related to stellar structure and evolution as well as fundamental physics. The sun also influences the terrestrial environment through its varied outputs. These two aspects of the solar interior combine to generate a surprising level of controversy for such an inherently simple star. I review three topics each of which is the subject of some degree of controversy: 1) the solar neutrino problem, 2) the status of modeling and observational efforts to understand the solar cycle of activity and 3) observational efforts to detect and identify solar $g-$modes.

Keywords: Solar Neutrinos, Solar Activity Cycle, $g-$mode oscillations

INTRODUCTION

The title of this review which was assigned by the scientific organizing committee is perhaps surprising. As a middle-age main sequence star of modest mass, the sun is an unlikely candidate for controversy and would be considered one of the dullest astrophysical objects were it not the center of our solar system. There are two basic reasons that the sun is able to generate controversy: it is the primary source of energy for our planet earth and its proximity allows a level of detailed examination unique among the stars. These two factors motivate much of solar physics — we know the sun is not perfectly stable and want to estimate how it might change and we can use the relatively simple state of the solar interior to test theories under conditions not found in terrestrial laboratories. In this review I discuss three topics which are currently under active discussion in the community — the solar neutrino problem, observations and theory of the solar cycle of activity and the search for solar $g-$mode oscillations. Aspects of each of these topics remain unsettled and controversial.

SOLAR NEUTRINO OBSERVATIONS

The "solar neutrino problem"

The sun is a site to test physical theories under conditions not readily attainable in terrestrial laboratories. The best known test of our understanding of the supposedly simple physics of the solar interior has been through the observations of the neutrino capture rate on ^{37}Cl carried out by Davis (1978, 1987, Davis *et al.* 1989) in the Homestake gold mine. These observations have resulted in "the solar neutrino problem" in which the observed rate of capture of neutrinos of 2.1 ± 0.3 SNU (Rowley, Cleveland & Davis 1985) has been roughly a factor of $3\frac{1}{2}$ smaller than the predicted rate of 7.4 ± 2.8 SNU (Bahcall &

Ulrich 1988, Bahcall & Pinsoneault 1992, the ± 2.8 here is a 3σ error). The last reference above contains a detailed comparison between the theoretical results of Sienkiewicz *et al.* (1990), Sackmann *et al.* (1990) and Turck-Chièze *et al.* (1988) with the result that when each calculation is corrected to the same input physics, the predicted counting rates disagree by less than 3%. Reported results normally show a larger range than this because different investigators choose different best values for the uncertain parameters. The uncertainty in the theoretical counting rate is defined as an equivalent 3σ error largely because some of the quantities such as the opacity do not have an experimentally measured error.

For many years the ^{37}Cl experiment was the only direct probe of the deep solar interior and it provides just one measured quantity. With a single quantity a variety of mechanisms could be proposed to resolve the discrepancy between theory and observation. In part because of this long-standing discrepancy, a number of additional experiments to measure solar neutrinos are either being planned, under development or are in progress. Most significantly in the last category is the Kamiokande II Collaboration from which a positive detection is now available (Hirata *et al.* 1989, 1990a,b, 1991). This experiment uses a large purified water tank with photomultiplier tubes on the walls to detect the Cherenkov radiation from electrons recoiling from a neutrino-electron interaction. The Kamiokande II result is summarized by the equation: $r_{e-\nu} = [0.46 \pm 0.05(\text{stat}) \pm 0.06(\text{syst})]$ where the ratio $r_{e-\nu}$ is given by:

$$r_{\nu-e^-} = <\sigma\phi>_{\nu-e^-} / <\phi\sigma>_{\text{standard}} \ .$$

Theoretical and experimental errors

It can be misleading to discuss both neutrino experiments simultaneously due to the way that the Kamiokande II scattering experiment have reported their results as a ratio $r_{\nu-e^-}$ of the observed rate to the predicted rate from a standard solar model. Theoretical uncertainties from the standard model are not included in the ratio; however, when the standard model is changed so must be the quoted value of $r_{\nu-e^-}$. This adds uncertainty to the quoted $r_{\nu-e^-}$ and reduces the significance of the difference between the observed value of $r_{\nu-e^-}$ and unity. In order to achieve a uniformity of presentation and illustrate the important point that the relative values of r for different experiments change in a correlated way, I express the results of the ^{37}Cl experiment in the terms of $r_{^{37}\text{Cl}} = <\sigma\phi>_{^{37}\text{Cl}} / <\sigma\phi>_{\text{standard}}$ which for the current best numbers given above is $r_{^{37}\text{Cl}} = [0.28 \pm 0.04]$. As with the $\nu - e^-$ experiment this experimental range does not include the theoretical errors.

The above two experiments at present have the best established results. Two experiments using ^{71}Ga as the target nucleus are in the early stages of operation. Predictions for this target nucleus include a large counting rate from the $p - p$ neutrinos which are not sensitive to the details of the solar model. The early results for one of these, the SAGE experiment, give a capture rate of $20^{+15}_{-20}(\text{stat}) \pm 32(\text{syst})$ SNU as compared to a predicted counting rate of 128^{+19}_{-16}SNU. As a ratio to the standard model this result can be expressed: $r_{^{71}\text{Ga}} = <\sigma\phi>_{^{71}\text{Ga}} / <\sigma\phi>_{\text{standard}} = [0.16^{+0.12}_{-0.16}(\text{stat}) \pm 0.25(\text{syst})]$.

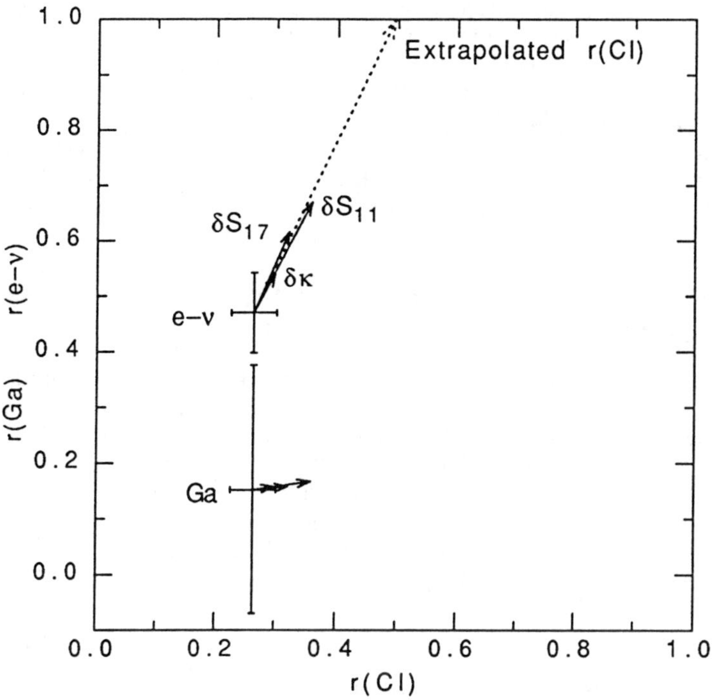

Fig. 1. This figure shows the ratios of observed counting rates to theoretical counting rates as computed from a standard solar model. Three different experiments are shown in this figure with the well established case of the ^{37}Cl serving as the reference. The error bars shown are just for the observations and do not include errors in the standard model. Theoretical errors are shown by the three solid arrows which represent 3σ changes in the r's resulting from the three most important theoretical uncertainties. This figure illustrates that errors in the standard model produce correlated changes in the r's. The dashed arrow indicates the direction and location of the extrapolated point where a change in the standard model would produce agreement with the $\nu - e^-$ scattering experiment. Note that the desired value of unity for the ^{37}Cl experiment is well away from the extrapolated point.

Neutrino propagation

Bahcall & Bethe (1990) have argued that the Kamiokande II result combined with the Homestake result probably requires a modification of the standard electroweak model with zero neutrino masses. The kernel of the argument is that the e-scattering Kamiokande II experiment has a threshold of about 7.5 MeV and is more dependent on the ^{8}B neutrinos than is the ^{37}Cl experiment. This means that in general for a change in some solar model parameter such as the $p - p$ or ^{7}Be$(p,\gamma)^8$B cross-section, the changes in the two r values will be related by $|\delta r_{\nu-e^-}| > |\delta r_{^{37}\mathrm{Cl}}|$. Since $r_{^{37}\mathrm{Cl}} < r_{\nu-e^-}$ a model based on a modified solar structure cannot yield unity for both r values simultaneously. Figure 1 shows this quantitatively where the r values for all three experiments are plotted with their experimental errors. The three arrows illustrate the changes in the r values which are brought about by taking $S_{1,1}$, $S_{1,7}$ and opacity variations to have their equivalent 3σ errors in a direction which goes to bring all the r's toward unity. The extrapolated point where $r_{\nu-e^-} = 1$ leaves $r_{^{37}\mathrm{Cl}} = 0.5$ which is significantly different from unity according to the errors of this experiment. Within the quoted detection rates and their errors this result is quite robust.

The strong correlation between changes in the different r's comes from the fact that the shape of the solar neutrino energy spectrum is not modified substantially by a change in the solar model. Only the relative number of each type of neutrino is modified and not the detailed energy dependence of each type. This is especially important for the ^{8}B neutrinos which dominate the counting rates for both the $\nu - e^-$ and ^{37}Cl experiments. Only by modifying the energy dependent shape of the ^{8}B neutrino flux can the r's be moved off the straight lines on Figure 1. The MSW effect (Mikheyev & Smirnov 1986; and Wolfenstein 1979) accomplishes this modification because according to this model the electron neutrinos ν_e emitted at the solar center are converted into a second flavor in an energy dependent manner. On the basis of this model the flux of neutrinos at the detector $(\phi_{\nu_e})_{\mathrm{detector}}$ is related to the flux at the source reaction $(\phi_{\nu_e})_{\mathrm{source}}$ by:

$$(\phi_{\nu_e})_{\mathrm{detector}} = (\phi_{\nu_e})_{\mathrm{source}} \exp(-\frac{C_{\mathrm{jump}}}{E})$$

where the value of C_{jump} is given by:

$$C_{\mathrm{jump}} = \pi \left| \frac{1}{n_e} \frac{dn_e}{dr} \right|^{-1} \Delta m^2 \sin^2\Theta_V \ ,$$

n_e is the electron density in the sun, Δm^2 is the square of the mass difference between the neutrino flavors and Θ_V is the neutrino mixing angle in vacuum. Bahcall and Bethe fit the ^{8}B result by choosing:

$$C_{\mathrm{jump}} = 10.5 \pm 5.5 \text{ MeV}$$

Since the value of the electron scale height is not model dependent, the value of C_{jump} then leads to:

$$\Delta m^2 = 1.0 \times 10^{-8} \sin^{-2}\Theta_V \ \ \mathrm{eV}^2 \ .$$

The modification of the high energy spectrum implied by this model is detectable by the Kamiokande experiment but at present the observational errors are too large to yield a constraint.

THE SOLAR CYCLE

$\alpha - \omega$ dynamos and internal rotation

The relative quiescence of the sun allows us to take the stability of the sun as assured yet we do not satisfactorily understand the sun's magnetic cycle in which any variability must certainly be rooted. The solar cycle of activity is a complex process. Throughout each cycle sunspots and magnetic fields appear in a well-known pattern which is reasonably regular from one cycle to the next. This pattern has been the subject of many dynamo model studies which range from the semi-empirical, kinematic ones like the $\alpha - \omega$ model of Babcock (1961) and Leighton (1964, 1969) or more fundamentally based models like the calculations of Gilman & Miller (1986) and Glatzmaier (1984, 1985a,b). The primary distinction between kinematic and dynamic dynamo models is the treatment of the velocity field — in the kinematic models it is assumed as input whereas for the dynamical models, the velocity field is part of the modelling process. Reviews of dynamo theory with discussion of such issues have been given by Parker (1979), Krause and Rädler (1980), Gilman (1986) and Levy (1992).

Helioseismology now permits the measurement of the sun's internal rotation law and this new data provides either a critical check on the validity of the assumptions in the former case or a test of the derived results in the latter case. In fact virtually all kinematic $\alpha - \omega$ dynamo studies carried out prior to the availability of the helioseismology data successfully reproduced the solar cycle by assuming that the rotation rate in the solar convection zone increases inward (Gilman 1986). While there have been a number of studies of the sun's rotation law (Duvall *et al.* 1984, Duvall, Harvey & Pomerantz 1986, Brown *et al.* 1989 and Rhodes *et al.* 1990), the errors in the deductions from these earlier results have been large enough to permit the dynamo models to remain viable. Improved helioseismology data have permitted Thompson (1991), Schou (1991) and Korzennik (1991) to deduce internal rotation laws with smaller errors. Thompson (1991) and Schou (1991) both used the data of Libbrecht (1989) from Big Bear Solar Observatory (BBSO) while Korzennik (1991) used combined data from the Mt. Wilson 60-foot tower project described by Rhodes *et al.* (1990) as well as the BBSO data. Thompson (1990) and Schou (1990) obtain very similar results from the same data base from observations made in 1988 and 1986 while the Korzennik (1991) analysis uses data from 1988 with higher spatial resolution. The Korzennik (1991) analysis loses its ability to probe the latitude dependence of the rotation law due to its use of averages over all observed modes at each value of spherical harmonic degree ℓ. Thompson (1990) only gives results for 1988 while Schou (1990) gives results for both 1988 and 1986. The results from these two authors for 1988 are very similar. Figure 2 shows the rotation curves for 1988 from Schou (1990) in the upper panel along with a comparison between the equatorial results of Schou (1990) and Korzennik (1991) in the lower panel. In addition to the analysis of the combined data, Korzennik (1991) also compared the BBSO and Mt. Wilson data in the interior region where both give valid results. These show very similar depth dependence although the curve based on Mt. Wilson alone lies about 8 nHz below the combined curve shown in figure 2. This difference is larger than the errors. The indicated surface rotation points indicated by the arrows are from the doppler measurements made at the 150-foot tower. These are an extension of the results previously published by Ulrich *et al.* (1988).

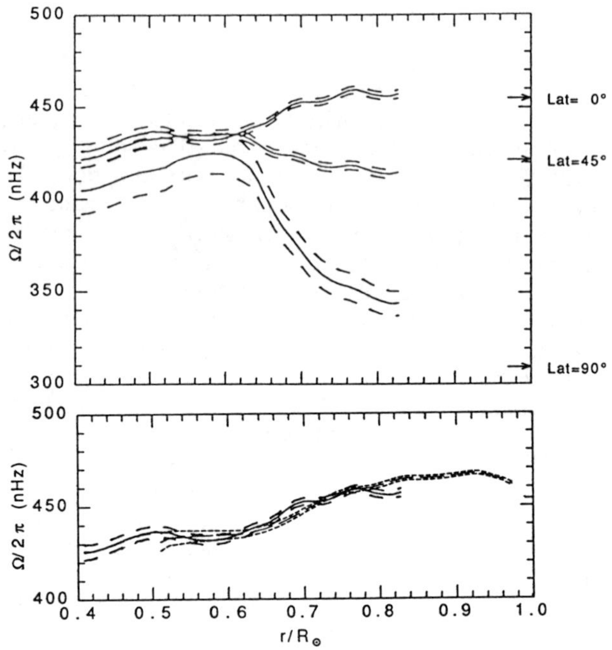

Fig. 2. A comparison between the deductions by Schou (1991) and by Korzennik (1991) of the 1988 rotation curve. The upper panel shows the latitude dependent results deduced by Schou (1991) which are based on observations by Libbrecht (1989). The lower panel repeats the equatorial results by Schou (1991) and also shows Korzennik's (1991) results which are based on combined observations from Mt. Wilson (Rhodes *et al.* 1990) and from BBSO (Libbrecht 1989). In both panels the most probable result is shown as a solid line and the errors of the deduction are indicated by the adjacent dashed lines. For Schou's results the dashes are long while for Korzennik's results, the dashes are shorter.

Based on the above deduced internal rotation pattern, it is clear that models based on the assumption of an inwardly increasing rotation rate must now be rejected. These depend on an increasing inward rotation rate and although there is a small zone with $d\Omega/dr < 0$ just below the surface of Korzennik's curve, this region is far smaller than previously assumed. Boyer and Levy (1992) have used a flat rotation law and shown that the oscillatory magnetic field loiters in mid-latitudes. Such behavior represents a serious deficiency of dynamo theory. The failure of the kinematic $\alpha-\omega$ dynamos as a result of the new helioseismology data is a manifestation of the lack of constraints from the full system of MHD equations typical of the kinematic approach. More fundamentally based dynamic models as discussed by DeLuca & Gilman (1991) generally produced dynamo behavior incompatible with the solar dynamo. Alternative ideas have been discussed by Gilman, Morrow & DeLuca (1989). Durney, DeYoung & Passot (1990) and DeLuca & Gilman (1991) have argued that the solar dynamo operates at the base of the convection zone. Recently a kinematic model based on the observed constraints of the solar velocity field which uses meridional circulation in place of the ω gradient has been developed by Wang, Sheeley & Nash (1991, WSN hereinafter). Since this model does not rely on a gradient in ω, many of the degrees of freedom associated with the $\alpha - \omega$ models are avoided. This model includes several other changes to the Leighton (1964, 1969) description of the kinematic dynamo for which reference is best made to the WSN paper.

Meridional circulation

The WSN model produces stable oscillations only for surface meridional flow at a rate of roughly 10 m/s and a return subsurface flow of about 1 m/s. While some questions about the nature of solar velocity observations remain, the analysis by Ulrich *et al.* (1988) which extends the earlier results of Duvall (1979) and Howard & LaBonte (1982a) indicates there is little doubt about the reality of an effect which very strongly resembles meridional circulation. The Ulrich *et al.* (1988) analysis derives the meridional circulation velocity by comparing the north-south trend of limb shift to the east-west limb shift. Points are paired in the east-west direction to cancel solar rotation and when the observed magnetic field in λ5250 exceeds 20 gauss for either point, both points were not used.

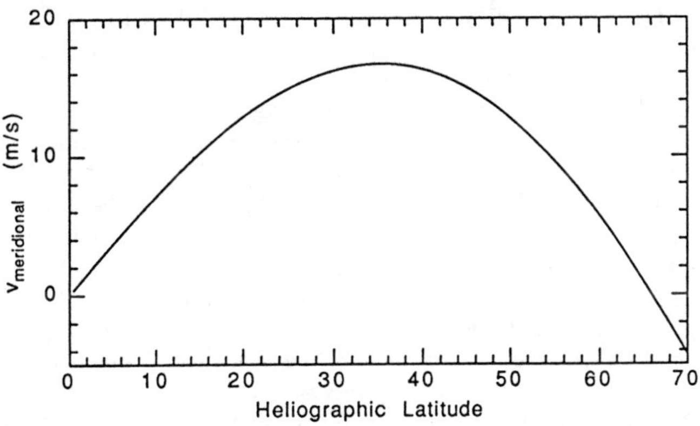

Fig. 3. The meridional circulation law determined from the Mt. Wilson 150-foot tower synoptic velocity observations. This law provides the baseline velocity pattern against which deviations are studied below. The velocity shown here is parallel to the local solar surface. Deviations from this law are calculated including projection effects and the apparent tilt of the solar axis of rotation.

In Ulrich *et al.* (1988) we discussed the meridional circulation for the entire 21-years available at that time. The data prior to 1982 had been influenced by a less well controlled entrance slit assembly and exhibited a substantial time dependence. Following the upgrade of the entrance slit assembly in 1982 we noted that much of the variation ceased but due to the declining phase of the solar cycle this could have been a result of less activity on the solar surface. With additional data we are now able to state that the meridional circulation pattern is relatively stable and has not shown continuing large variations. Figure 3 shows the average meridional circulation law which we have now fixed so that we can study deviations in a separate format presented below. This circulation law is appropriate to $\lambda 5250$ and may not apply to other lines. Deviations from this law are significant but are at the 20% level instead of the factor of 3 level found in the pre-1982 data. This circulation law is generally consistent with the WSN assumptions although it does differ in detail. For the purpose of the kinematic calculations of WSN this circulation law can be regarded as consistent with their model although it would be desirable to examine the consequences of including our current rotation law. Note however that we do not have reliable observations of the circulation rate near the poles due to foreshortening and the difficulty of removing the effects of scattered light.

Although doubts have been expressed by Gilman (1992) that the meridional circulation is a result of material motion, only by invoking a limb shift function which depends on latitude could this circulation velocity be absent from the solar surface. There are only two ways that high latitudes differ from lower latitudes: their higher level of magnetic activity and their lower centripetal acceleration. Since magnetically affected points are not used and the meridional circulation is present during 1986 when there was little activity, the centripetal acceleration difference is the only remaining distinction and this effect is so small that is unclear how it could affect the line formation process. I conclude that material motion in the form of meridional circulation is the most likely explanation for the limb shift difference.

Comparison of the WSN dynamo to observations

The dynamo model of WSN assumes symmetry between the northern and southern hemisphere and provides the time dependent magnetic field pattern for only one hemisphere. In their publication they show the time dependence over a number of full magnetic cycles after the dynamo has stabilized and compare the results with the data from the Wilcox Solar Observatory after artificially extending that data set to one full 22-year cycle. The Mt. Wilson synoptic program now has 24 years of fully reduced data which can be compared to the WSN model without extension. Since the model does not include any longitudinal information and has been symmetrized, we have averaged the magnetic field as a function of latitude over each solar rotation and then multiplied the southern hemisphere by -1 and added it to the northern hemisphere. The publication by WSN includes figures which show the time and latitude dependence of the magnetic field in an overly compressed format. The authors of that paper have kindly provided a modified display of their model in a format which permits a more satisfactory comparison to the observations.

The WSN model results are shown in the top panel of figure 4. Nearly matching observed results are shown in the middle panel of this figure while a more natural display of the observed magnetic field is given in the bottom panel. The observed fields over this extended period of time include some systematic effects at the 20- 40% level as a result of changes in the instrument calibration procedure. Prior to 1982 the fields should be multiplied by 1.4. Between 1982 and April, 1987 a variable factor between 1.2 and 1.4 should be applied. Overall due to saturation in the $\lambda 5250$ line the observed field in these plots needs to be multiplied by a scale factor change of 3.5 (Ulrich 1992). Due to the fluxtube character of the solar field and the low spatial resolution of the observations, the absolute magnitude of the field is not easy to compare to a theoretical model. Within the above limitations of the observations, the agreement between the

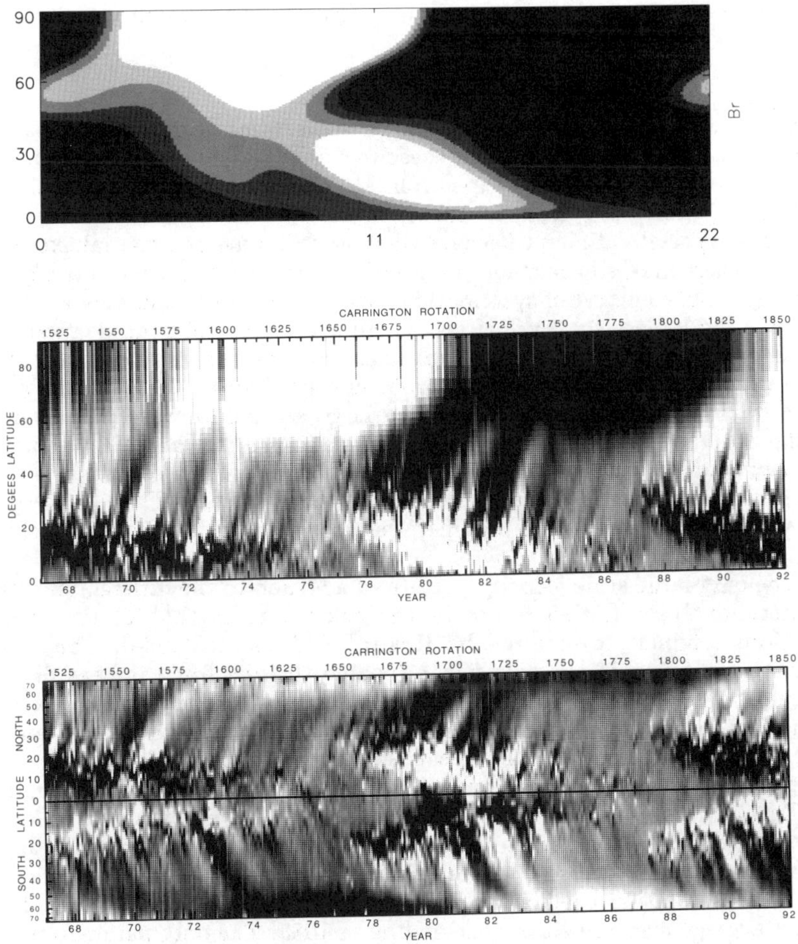

Fig. 4. Comparison of the Wang, Sheeley & Nash (1991) dynamo model to 24 years of magnetic observations from Mt. Wilson. The top panel gives the WSN model for B_r from their figure 1. The time interval shown is 22 years. The center panel gives the observations for 24 years after folding the north and south hemispheres together in an antisymmetric manner. Each resolution element in the center panel plot represents a bin of latitude width originally 1/34th of the projected solar diameter and temporal duration of one equatorial solar rotation period. The pixels have been stretched in latitude to match the WSN format. The bottom panel shows the unfolded average magnetic field without the latitude stretching. This panel indicates that some of the distinct features of the middle panel actually are found only in one hemisphere while others are more nearly antisymmetric.

model and the observations is quite striking and indicates that this approach to the solar dynamo is promising. The primary deficiency in the model is the lack of the strong surges of the magnetic field toward the poles as were seen in 1970-72, 1981-83 and which appear to be starting again this year. We have looked for a corresponding change in the meridional circulation velocity and do not find any clear signal.

Large scale velocity details

Associated with the solar cycle should be some large scale velocity patterns. Some dynamo theories indicate that these velocities should have the form of giant cells in the shape of a banana shaped roll (see for example Gilman & Miller 1986 and Gilman, Morrow & DeLuca 1989). In fact the search for such large scale patterns is especially difficult because their long lifetime and spatial extent over the solar image makes it easy for systematic instrumental effects to masquerade as solar signals. Removal of systematic instrumental effects can very easily alter or remove the largest scale velocities. There is also the effect of magnetic fields on both the properties of the spectral lines and the dynamics of the material. Bogart (1987) has discussed this last problem while Ulrich (1991) has shown that a very good correlation exists between a large scale downward flow of matter and the absolute value of the magnetic field. Scherrer *et al.* (1986) concluded that they could not find giant cell velocities because of the confusion from the magnetic effects. High spatial resolution observations by Title *et al.* (1987) show that the magnetic field modifies the properties of the velocity field in a fundamental way.

The only large scale velocity pattern in addition to the differential rotation and meridional circulation found in the solar data is that of the torsional oscillations originally discovered by Howard & LaBonte (1980). This velocity pattern is not found in the dynamical dynamo theories but has been discussed in a qualitative way by Snodgrass & Wilson (1987). An important case of the difficulty of correcting systematic effects is that of the study of the torsional oscillations where there was a disagreement between LaBonte & Howard (1982b) who found two features in each hemisphere and Snodgrass (1985) who found only one. The LaBonte & Howard (1982b) results may have contained an artifact from the fitting function used to correct for systematic effects. In order to avoid such artifacts, we now use a procedure in which all differential rotation and meridional circulation coefficients are held fixed based on their values derived from an average over the solar cycle ending in 1985. The only parameters which are now adjusted for each observation are the overall rotation rate and the scattered light corrections.

Beginning in 1986 the Mt. Wilson synoptic program has been pursuing a program of regular observations throughout each day instead of just once per day as was the case previously. We refer to this effort as the "fast-gram" program because we have modified the usual setup to use a larger entrance aperture and a faster scanning speed. The purpose of the larger number of observations is to reduce the noise in the average velocity which results from the supergranulation and 5-minute oscillations. Ideally the solar velocity should be sampled continuously so that these more rapidly varying velocities would cancel. In our case we are only sampling the velocity field at intervals varying between 30 minutes and several days so that the velocity from both oscillations and supergranulation is essentially random. The net effect of these rapidly varying velocity fields then decreases as $N^{-1/2}$ where N is the number of redundant observations. Due to this nature of the cancellation, we obtain a reduction by factor of three in the effect of these velocity fields with 10 redundant observations but would have to make 100 redundant observations to achieve another similar reduction. Each observation is set up individually and reduced individually rather than as an automated image sequence. Consequently, the practical maximum has proven to be just over 20 observations in a single day. Even an automated observing sequence would only improve this number slightly and would introduce the penalty of poorer calibration. With 15 to 20 observations we have in fact realized the anticipated reduction in the noise effect of the rapidly varying velocity fields and are now able to study the torsional oscillations and meridional circulation in detail over the time period of 1986 to present. There has been a modest improvement in the quality of the magnetic observations as a result of the additional data but because there is no analogue to the oscillations for the magnetic field, the gain in noise is not crucial.

The results of the fast-gram program are shown in figure 5. The velocities have been resolved into zonal and sectoral components where the zonal velocities are parallel to the solar equator and represent deviations from the steady differential rotation rate and the sectoral velocities are parallel to lines of constant heliographic longitude and are a combination of any residual downdraft effect from magnetic regions, giant cells if present and variations in the meridional circulation rate. This resolution was made using the method described by Ulrich *et al.* (1988). One feature of the algorithm not mentioned in that earlier paper is the fact that points for which the absolute value of magnetic field exceeds 20 gauss in λ5250 are omitted from the calculation. This test should prevent the bulk of the magnetically induced motions from influencing the results however, there may be some effect from weaker fields which are included. In order to obtain a result which is sensitive primarily to the largest scale motions, the circumference of the sun is divided into bins 60° wide in longitude and the average zonal and sectoral velocity within the bin is computed from equations (11) and (12) of Ulrich *et al.* (1988). The resulting averages are then plotted as a function of the time of the central meridian passage of the bin center. Each point includes all observations for which the bin was visible during that particular solar rotation and is thus an average over roughly 10 days.

The top panel of figure 5 shows the zonal velocity with the grey-level plot being set up so that full black denotes 7.5 m/s faster than average rotation and full white denotes 7.5 m/s slower than average rotation. The torsional oscillation pattern is very clear in the dark diagonal band approaching the equator. The peak of this rotation enhancement moves in a straight line versus time on this plot and is surprisingly smooth. The previous torsional oscillation results which did not have the benefit of the fast-gram data suggested a much more chaotic pattern. Successive Carrington rotation numbers are indicated along the top of each panel and assist in the search for periodic effects such as might be the signature of giant cells. The only feature which readily appears is the series of black/white patches near the north pole between 1986 and 1987.5 and again near the south pole between 1990 and 1991.5. These represent a polar crossing flow. The sectoral velocities described below also show evidence of this flow and furthermore the phasing of the approach velocity is consistent with the phasing of the zonal velocity. The highest latitudes in 1991 begin to show an increase in the rotation rate that may be the beginnings of the next torsional oscillation

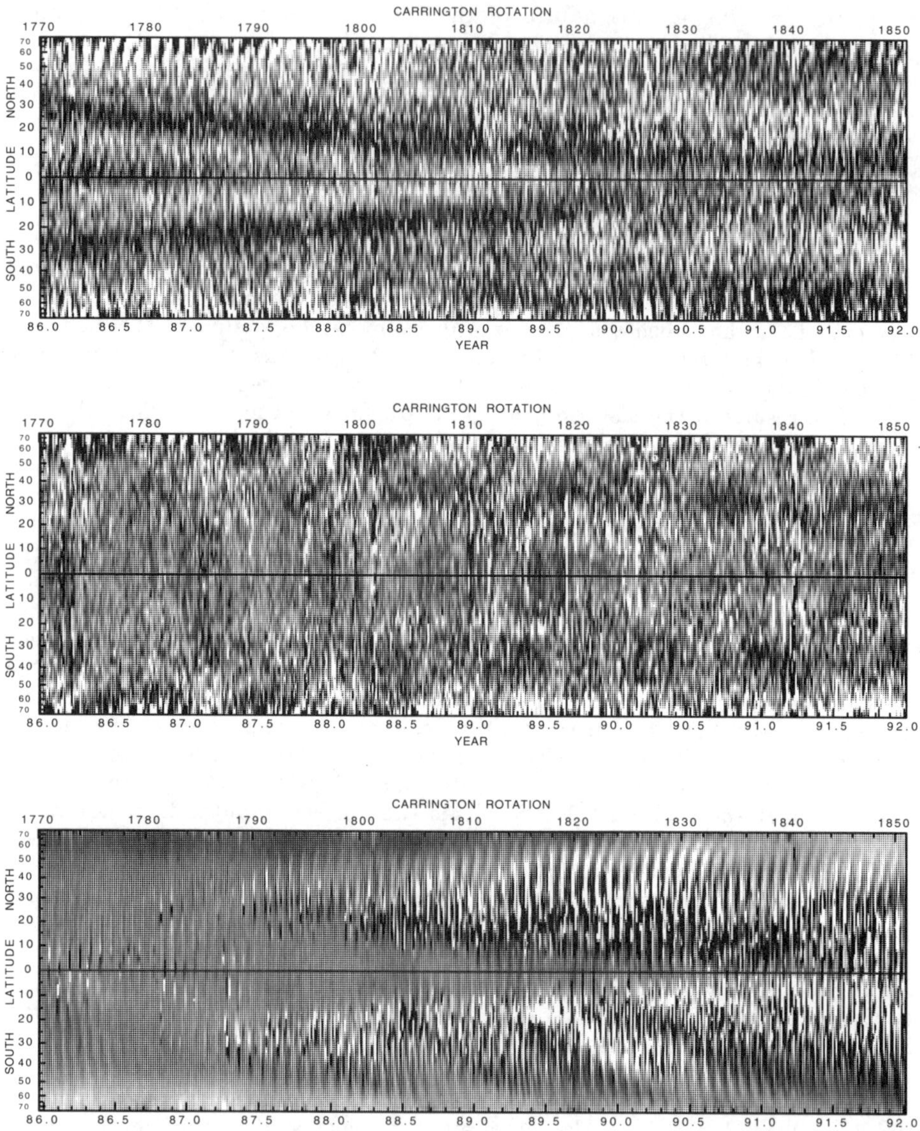

Fig. 5. This figure shows the results of the "fast-gram" program at Mt. Wilson. Longitudinal information is retained through the use of bins 60° wide. The top panel gives the zonal velocity which is the change in rotation rate. Black → 7.5 m/s faster than average rotation, white → 7.5 m/s slower than average rotation. The middle panel gives the sectoral velocity which corresponds to changes in the average line-of-sight velocity. Black → 7.5 m/s approach velocity, white → 7.5 m/s recession velocity. The bottom panel gives the magnetic field averaged over the same longitude bins.

band. At the same time the remnants of this cycle's rotation enhancement band have not dissipated near the solar equator. Thus there appear to be two torsional oscillation bands of enhanced rotation present at the same time.

The middle panel of figure 5 shows the sectoral velocity. The encoding of the velocity has receding velocity of 7.5 m/s as white and approach velocity of 7.5 m/s as black. The regular occurrence of noisiness in the January to March time of year is related to the typical weather conditions which reduce the number observations that we can make. During the period of low activity in 1986, there is little evidence of any velocity. Based on the appearance of this figure, giant cells must have a surface velocity less than about 2 m/s. As solar activity begins in 1988, a large scale pattern appears in this figure with a structure which is not expected from leakage of the directly induced downdraft velocity from the magnetic fields which are less than our limit of 20 gauss. Because we do not alter the global meridional circulation solution, the approach velocities at high latitude represent an actual change in the spectral line position rather than a shift in the global fit. Interpreted literally, the motion at this phase represents a net flow of material towards the regions of greatest activity with excess poleward motion at low latitude and a reduction in poleward motion at the highest latitude. Note however that maximum velocities represented by these deviations are smaller than the meridional circulation velocity shown in figure 3.

Finally the bottom panel of figure 5 shows the magnetic field averaged over the same bins as the upper panels. This figure is very similar to the bottom panel of figure 4 except that some longitude information is now retained. This longitude information is actually of interest because the tilt associated with the generation of the polar field is quite evident. In fact the gradual rotation of the position angle of the bipolar regions clearly plays the anticipated role in the reversal of the solar dipole field. Also of considerable interest is the fact that the very quiet region below the zone of solar activity follows the position of the maximum in rotation rate enhancement seen in the top panel. This maximum seems to be the boundary between the active and quiet parts of the solar surface.

After differential rotation, the torsional oscillations are the most prominent large scale velocity feature on the solar surface. Dynamic dynamo models do not show such a pattern of motion and this may be a clue as to deficiencies in these models. The regularity of the torsional oscillation pattern shown in figure 5 indicates that some underlying organization is present in the solar interior. Previous torsional oscillation maps were too noisy to permit the recognition of this regularity.

THE SEARCH FOR g-MODES

General considerations

The field of helioseismology has matured to the point where its controversies are at a very detailed level as long as only the five-minute band data is used. It is in the search for the g-modes that much controversy as well as promise remains. The value of detection and identification of a large number of g-modes could be enormous. Because the modes should have great purity due to the concentration of their kinetic energy in the quiet deep interior, the frequencies may be measurable with very great precision. This in turn will provide unprecedented quality information about the solar interior structure and dynamics. In addition, the gravity wave frequency is sensitive to the density and pressure gradients and will be strongly influenced by the presence of any irregularities in composition that may be left from the time of solar system formation.

In a general way the need for observations of $g-$modes to analyze the solar interior is well known. Figure 2 illustrates the problem. This figure shows the rotation rate only for those parts of the solar interior where the inversion is reliable. The plot stops at $r = 0.4R_\odot$ because the errors become too large due to the restriction of the available data to the $p-$modes alone. The inversions for the sun's internal sound speed also show a growth in the error for the same part of the interior although present results are able to distinguish between models and the sun at an interesting level (Gough 1984, 1986, Christensen-Dalsgaard *et al.* 1985, Brodsky & Vorontzov 1988, Vorontsov 1988, Thompson 1991). Hill, Gao and Rosenwald (1988) discussed an inversion method of studying composition gradients through their effect on the mean molecular weight.

Mode identification

The identification of those parts of the observed signal which correspond to globally coherent solar oscillations has proven to be a very challenging part of studying low frequency solar oscillations. This task is called mode identification. It is complicated by the absence of any simple tools for distinguishing between solar oscillation modes and other non-coherent sources of signal such as terrestrial atmospheric effects, instrumental drifts, solar supergranulation and solar active regions. Depending on the details of the observational dataset, each of these could produce a time varying signal which is comparable to what is observed. Theoretically the solar oscillation modes should have such a large number of frequencies that their spectrum should not be superficially different from a noise spectrum. This is in marked contrast to the situation with the $p-$mode oscillations where the solar signal stands out clearly against the noise background.

Three means of distinguishing between a solar mode signal and a noise signal have been used: a) mode frequencies should have long term stability, b) mode frequencies should obey asymptotic relationships, and c) the highest amplitude peaks in the power spectrum may be of solar origin. Of these criteria, only a) is reliable due to the high inertia of the $g-$modes. Identifications based on criteria b) and c) have been published by Scherrer and Delache (1983), Frölich and Delache (1984), Isaak *et al.* (1984), Pallé and Roca Cortés (1988) and van der Raay (1988). Application of criterion a) has led to a serious questioning of the identifications. Henning and Scherrer (1988) compared successive years of applying the procedure of Scherrer and Delache (1983) and found that few of the mode frequencies agreed. Garcia, Pallé and Roca Cortés (1988) carried out a cross-correlation between the power spectra from the first and second half of the 1984 to 1987 and found no correlation. A good discussion of $g-$mode observations and analysis has been given by Pallé (1991).

Although there has been some general agreement between the asymptotic parameters derived from all these observations, the frequencies found by the different groups do not agree with each other to within the errors of measurement. Fossat *et al.* (1988) have provided a very useful discussion of the statistical testing for $g-$modes. Hill *et al.* (1991) give a very extensive discussion of these searches for $g-$modes as well as the studies by the Santa Catalina Laboratory for Experimental Relativity by Astrometry (SCLERA) and studies of the 160-minute oscillation. I will not review the SCLERA results here because many of the essential publications of the data from this effort are not generally available and because the power spectrum sample published by Bos and Hill (1983) is very difficult to reconcile with other observations of global solar velocities. First, the internal rotation law provided by Hill, Bos and Goode (1982) from the identified multiplets is in strong disagreement with the $p-$mode results shown above. Second, the amplitudes of the signal seen in the SCLERA system are substantially larger than found in any other global solar oscillations. The coherence analysis by Bos and Hill (1983) indicated that 98% of the power they detect is globally coherent so that essentially every peak in their power spectrum is to be treated as a candidate for identification. The power spectral density (PSD) given by Bos and Hill (1983) for these peaks is equivalent to a velocity PSD of 8×10^5 $(m/s)^2$ Hz^{-1} at a frequency of $465\mu Hz$. The integrated sunlight velocity systems give a PSD of 10^2 $(m/s)^2$ Hz^{-1} at this same frequency (Jiménez *et al.* 1988). This difference is attributed by the SCLERA project to the dependence of the eigenmodes on height in the solar atmosphere (see the Hill *et al.* 1992 review for references and further discussion). If the vertical mode structure in the solar atmosphere involved such a large amplification factor, there should be similar differences between spectral lines formed at different depths and in the infrared continuum. Comparisons between the Na and K spectral lines by Isaak *et al.* (1988) and infrared brightening observations by Deming *et al.* (1988) and Kopp *et al.* (1992) do not show any large effects.

Solar noise reduction

The GOLF experiment on SOHO (Gabriel *et al.* 1991) will be a space-based search for low velocity amplitude solar oscillations using a Na resonance cell. One important component of this experiment is a study of solar noise sources with the hope of at least partial compensation for the incoherent or non $g-$mode portion of the solar signal. This is a delicate project due to the risk of removing the $g-$mode signal along with the solar noise. To avoid this risk the compensation for incoherent signals needs be derived from information other than the observed velocity. Following the model proposed by Harvey (1985) for these noise sources, we identify two processes which are major contributors: active regions and supergranulation. In the case of the active regions the magnetic fields and associated line profile changes can provide the needed additional information. Early studies of the global velocity signal by Claverie *et al.* (1982) detected a variation with a 13 day period. These were quickly interpreted by Andersen & Maltby (1983) and Edmonds & Gough (1983) as being the result of the passage of active regions across the solar disk. Both interpretations recognized the fact that the effect depends on the spectral line used for the observations. Also the only effect included in the model was the darkening associated with the sunspot groups and the velocity was derived from the positions and areas of the sunspot assuming the intensity of the line radiation is changed only inside the umbra and penumbra. Both studies were able to account for the observed effects. In particular Andersen & Maltby (1983) model the velocity signal with errors generally less than half the size of the induced changes. The task for modelling the effect for the Na D lines is complicated by the fact that even small magnetic fields can alter the line profile. Studies to develop a reliable model of this effect for the GOLF experiment on SOHO are currently in progress.

The supergranulation is more difficult because its signal is a pure velocity with a disorderly spatial structure. In this case no correlated non-velocity signal is available from which an equivalent velocity correction can be calculated. Unfortunately, the supergranulation is probably the largest contributor of an incoherent signal. The only strategy which might help with the supergranulation is to take advantage of the center-to-limb amplitude dependence and isolate that part of the global signal which comes from the limb. Such a plan might be possible through the analysis of the imaged data from the Solar Oscillation Imager on SOHO.

CONCLUSIONS

It is clear there will continued intensive study of the solar interior. Our understanding of past controversies is improving but as with any field with new techniques of observation, new controversies are generated more quickly than the old ones are put to rest. New solar neutrino observations will clarify the nature of the solar neutrino problem, improved internal differential rotation studies from helioseismology will guide the development of dynamo models and we look forward to a more definitive observational search for the $g-$modes.

ACKNOWLEDGEMENTS

I would like to thank John Boyden, Larry Webster, Tom Shieber, Carl Henney, Pam Gilman and Steve Padilla for their help in obtaining and analyzing the Mt. Wilson data. This work has been supported by NASA grant NAGW-472, NSF grant AST90-15108 and ONR:N00014 91-J114.

REFERENCES

To conserve space the following abbreviations are used in this reference list:

IAU No. 123 → J. Christensen-Dalsgaard & S. Frandsen 1988, *Advances in Helio- and Asteroseismology*, (D. Reidel, Boston)

Tenerife → V. Domingo & E.J. Rolfe 1988, *Seismology of the Sun & Sun-like Stars*, (ESA Publications Division, SP-286, Noordwijk)

COSPAR 28 → B.H. Foing 1991, *Helioseismology from Space, COSPAR 28th Plenary Meeting, Adv. Space Res., vol. 11, no. 4*, (Pergamon Press, Oxford)

CSW 7 → M. Giampappa 1992, *Proceedings of the 7th Cool Star Conference*, (Ast. Soc. Pacific, San Franscisco)

SBITP → D. Gough & J. Toomre 1991, *Challenges to Theories of the Structure of Moderate-Mass Stars, Lect. Notes in Phys., 388*, (Springer-Verlag, Berlin)

Andersen, B.N. & Maltby, P. 1983, *Nature,* **302**, 808.

Babcock, H.W. 1961, *Astrophys. J.,* **133**, 572.

Bahcall, J.N. & Bethe, H.A. 1990, *Phys. Rev. Letters,* **65**, 2233.

Bahcall, J.N. & Pinsoneault, M.H. 1992, *Rev. Mod. Phys.,* , in press.

Bahcall, J.N. & Ulrich, R.K. 1988, *Rev. Mod. Phys.,* **60**, 297.

Bogart, R.S. 1987, *Solar Phys.,* **110**, 23.

Bos, R.S. & Hill, H.A. 1983, *Solar Phys.,* **82**, 89.

Boyer, D.W. & Levy, E.H. 1992, *Astrophys. J.,* , in press.

Brodsky, M.A. & Vorontzov, S.V. 1988, in: IAU No. 123, p. 137.

Brown, T.M., Christensen-Dalsgaard, J., Dziembowski, W.A., Goode, P., Gough, D.O. & Morrow, C.A. 1989, *Astrophys. J.*, **343**, 526.

Christensen-Dalsgaard, J., Duvall, T.L., Gough, D.O., Harvey, J.W., & Rhodes, E.J.,Jr. 1985, *Nature,* **315**, 378.

Claverie, A., Isaak, G.R., McLeod, C.P., van der Raay, H.B., Pallé, P.L. & Roca Cortés, T. 1982, *Nature,* **299**, 704.

Davis, R., Jr. 1978, in: *Proceedings of Informal Conference on Status and Future of Solar Neutrino Research*, ed. G. Friedlander, (Brookhaven National Laboratory, Upton, N.Y.), Report No. 50879, Vol. 1, p. 1.

Davis, R., Jr. 1987, in: *Proceedings of Seventh Workshop on Grand Unification, ICOBAN'86, Toyama, Japan*, ed. J. Arafune, (World Scientific, Singapore), p. 518.

Davis, R., Jr., *et al.* 1989, in: *Proceedings of the 21st International Cosmic Ray Conference*, ed. R.J. Protheroe, (Univ. Adelaide Press, Adelaide), p. 143.

DeLuca, E.E. & Gilman, P.A. 1991, in: *The Atmosphere and Interior of the Sun*, ed. A.N. Cox, W.C. Livingstone & M.S. Matthews, (Univ. of Arizona Press, Tucson), p. 275.

Deming, D., Glenar, D.A., Käufl, H.U. & Espenak, F. 1988, in: IAU No. 123, p. 425.

Durney, B.R., DeYoung, D.S. & Passot, T.P. 1990, *Astrophys. J.*, **362**, 709.

Duvall, T.L. 1979, *Solar Phys.*, **63**, 3.

Duvall, T.L., Dziembowski, W.A., Goode, P.R., Gough, D.O., Harvey, J.W. & Leibacher, J.W. 1984, *Nature,* **310**, 22.

Duvall, T.L., Harvey, J.W. & Pomerantz, M.A. 1986, *Nature,* **321**, 500.

Edmonds, M.G. & Gough, D.O. 1983, *Nature,* **302**, 810.

Fossat, E., Grec, G., Gavrjusev, V. & Gavrjuseva, E. 1988, in: Tenerife, p. 393.

Frölich, C. & Delache, P. 1984, *Mem. Soc. Astr. Ital.*, **55**, 99.

Gabriel *et al.* 1991, in: COSPAR 28, 103

Garcia, C., Pallé, P.L. & Roca Cortés, T. 1988, in: Tenerife, p. 353.

Gilman, P.A. 1986, in: *Physics of the Sun, Vol. 1*, ed. P.A. Sturrock, T.E. Holzer, D.M. Mihalas & R.K. Ulrich, (Reidel, Dordrecht), p. 95

Gilman, P.A. 1992, Paper presented at the Twelfth NSO/Sacramento Peak Summer Workshop - "The Solar Cycle", to appear in P.A.S.P.

Gilman, P.A. & Miller, J. 1986, *Ap. J. Suppl.*, **61**, 585.

Gilman, P.A., Morrow, C.A. & DeLuca, E.E. 1989, *Astrophys. J.*, **338**, 528.

Glatzmaier, G.A. 1984, *J. Comput. Phys.*, **55**, 461.

Glatzmaier, G.A. 1985a, *Astrophys. J.*, **291**, 300.

Glatzmaier, G.A. 1985b, *Geophys. Ap. Fluid Dyn.*, **31**, 137.

Gough, D.O. 1984, *Phil. Trans. Roy. Soc. Lond. A,* **313**, 27.

Gough, D.O. 1986, in: *Seismology of the Sun and the distant stars*, ed. D.O. Gough, (Reidel, Dordrecht), p. 125.

Harvey, J.W. 1985, in: *Future Missions in Solar, Heliospheric and Space Plasma Physics*, ed. E. Rolfe & B. Battrick, (ESA Publications Division, Noordwijk), SP-233, p. 199.

Henning, H.M. & Scherrer, P.H. 1988, in: Tenerife, p. 419.

Hill, H.A., Bos, R.J. & Goode, P.R. 1982, *Phys. Rev. Letters,* **49**, 1794.

Hill, H.A., Fröhlich, Gabriel, M. and Kotov, V.A. 1991, in: *The Atmosphere and Interior of the Sun*, ed. A. Cox, W. Livingstone & M.S. Matthews, (Univ. of Arizona Press, Tucson), p. 562

Hill, H.A., Gao, Q. & Rosenwald, R.D. 1988, in: Tenerife, p. 403.

Hirata *et al.* 1989, *Phys. Rev. Letters,* **63**, 16.

Hirata *et al.* 1990a, *Phys. Rev. Letters,* **65**, 1297.

Hirata *et al.* 1990b, *Phys. Rev. Letters,* **65**, 1301.

Hirata *et al.* 1991, *Phys. Rev. D,* **44**, 2241.

Howard, R.H. & LaBonte, B.J. 1980, *Astrophys. J.,* **239**, 133.

Isaak, G.R., van der Raay, H.B., Pallé, Roca Cortés, T. & Delache, P. 1984, *Mem. Soc. Astr. Ital.,* **54**, 91.

Isaak, G.R., McLeod, C.P., van der Raay, H.B., Pallé, P.L. & Roca Cortés, T. 1988, in: IAU No. 123, p. 53.

Jiménez, A., Pallé, P.L., Pérez Hernandez, F., Régulo, C. & Roca Cortés, T. 1988, *Astr. Ap.,* , **192**, L7.

Kopp, G., Lindsey, C., Roellig, T.L., Werner, M.W., Becklin, E.E., Orrall, F.Q. & Jefferies, J.T. 1992, *Astrophys. J.,* **388**, 203.

Korzennik, S.G. 1991, Ph.D. Thesis, UCLA.

Krause, F. & Rädler, K.-H. 1980,*Mean-Field Magnetohydrodynamics and Dynamo Theory*, Pergamon Press, Oxford.

LaBonte, B. J. and Howard, R. 1982a, *Solar Phys.,* **80**, 361.

LaBonte, B. J. and Howard, R. 1982b, *Solar Phys.,* **80**, 373.

Leighton, R.B. 1964, *Astrophys. J.,* **140**, 1547.

Leighton, R.B. 1969, *Astrophys. J.,* **156**, 1.

Levy, E.H. 1992, in: CSW 7, in press.

Libbrecht, K.G. 1989, *Astrophys. J.,* **270**, 288.

Mikheyev, S.P. & Smirnov, A.Yu. 1986, *Nuovo Cimento C,* **9**, 17.

Pallé, P.L. & Roca Cortés, T. 1988, in: IAU No. 123, p. 79.

Pallé, P.L. 1991, in: COSPAR 28, p. 29.

Parker, E.N. 1979,*Cosmical Magnetic Fields*, Oxford Univ. Press, Oxford.

Rhodes, E.J., Jr., Cacciani, A. & Korzennik, S.G. 1990, in: *Progress of Seismology of the Sun and Stars, Proc. Oji Int. Seminar, Lecture Notes in Physics, 367*, ed. Y. Osaki & H. Shibahashi, (Springer-Verlag, Berlin), p. 163.

Rowley, J.K., Cleveland, B.T. & Davis, R., Jr. 1985, in: *Solar Neutrinos and Neutrino Astronomy, Conf. Proc. No. 126*, ed. M.L. Cherry, W.A. Fowler & K. Lande, (AIP, New York), p. 1.

Sackmann, I.-J., Boothroyd, A.I. & Fowler, W.A. 1990, *Astrophys. J.,* **360**, 727.

Scherrer, P.H., Bogart, R.S. & Hoeksema, J.T. 1986, *Bull. A.A.S.,* **18**, 702.

Delache, P. & Scherrer, P.H. 1983, *Nature,* **306**, 651.

Schou, J. 1991, in: SBITP, p. 81.

Sienkiewicz, J. Bahcall, J.N. & Paczyński, B., *Astrophys. J.,* **349**, 641.

Snodgrass, H.B. 1985, *Astrophys. J.,* **291**, 339.

Snodgrass, H.B. & Wilson, P.R. 1987, *Nature,* **328**, 697.

Thompson, M.J. 1991, in: SBITP, p. 61.

Title, A.M. 1992, in: CSW 7, in press.

Title, A.M., Tarbell, T.D. & Topka, K.P. 1987, *Astrophys. J.,* **317**, 892.

Turck-Chièze, S., Cahen, S., Cassé, M. & Doom, C. 1988, *Astrophys. J.,* **335**, 415.

Ulrich, R.K., Boyden, J.E., Webster, L., Snodgrass, H.B., Padilla, S.P., Gilman, P. & Shieber, T. 1988, *Solar Phys.,* **117**, 291.

Ulrich, R.K. 1991, in: COSPAR 28, 217

Ulrich, R.K. 1992, in: CSW 7, in press.

van der Raay, H.B. 1988, in: Tenerife, p. 339.

Vorontsov, S.V. 1988, in: Tenerife, p. 475.

Wang, Y.-M., Sheeley, N.R., Jr. & Nash, A.G. 1991, *Astrophys. J.,* **338**, 431.

Wolfenstein, L. 1978, *Phys. Rev. D,* **17**, 2369.

Wolfenstein, L. 1978, *Phys. Rev. D,* **20**, 2634.

Inside the Stars, IAU Colloquium 137
ASP Conference Series, Vol. 40, 1993
Werner W. Weiss and Annie Baglin (eds.)

ASTEROSEISMOLOGY: THE IMPACT OF SOLAR SPACE OBSERVATIONS

H.S. HUDSON
Institute for Astronomy, University of Hawaii, Honolulu,
Hawaii, USA 96822

ABSTRACT Observations from space relevant to solar global properties (oscillations, magnetic activity, *etc.*) are helpful both scientifically and technically in preparing for stellar observations. This paper summarizes the results from the main previous experiments (ACRIM, SOUP, and IPHIR), and also give an initial technical report from the SXT instrument on board *Yohkoh*, launched in August 1991. The solar observations to date demonstrate the existence of several mechanisms for low–level variability: spots, faculae, the photospheric network, granulation, and p–mode oscillations. The observations of oscillations have been particularly helpful in setting limits on solar interior rotation. In addition to the solar processes, stars of other types may have different mechanisms of variability. These may include the analogues of coronal holes or solar flares, modes of oscillation not detected in the Sun, collisions with small bodies, duplicity, and probably mechanisms not invented yet but related in interesting ways to stellar convection and magnetism.

INTRODUCTION

Observations from space relevant in any direct way to our knowledge of the solar interior have been few in number, even given the three–decade span of space research and the relative simplicity of some of the basic observations. This paper summarizes the main experiments of the past, emphasizing optical observations bearing on irradiance variability and helioseismology. The data have shown the existence of low–level irradiance variability on all time scales sampled, especially including the p–mode frequency range of a few mHz. These "Sun as a star" observations encourage us very strongly to obtain comparable stellar observations. The fundamental limit of photon arrival statistics does not pose a problem for large numbers of stars of all spectral types, and there appears to be no real limitation imposed by the instrumentation.

The success of these observations illustrates a general point. Traditionally we use observations from space (*e.g.*, in X–ray astronomy) to avoid atmospheric extinction and thus to open a new wavelength ranges for astrophysics. We can also use space to bypass atmospheric seeing effects, for example to improve angular resolution and therefore sensitivity. The experiments described below have made orders–of–magnitude improvements in a different direction, namely in photometric precision, as a result of their deployment in space. These advantages largely remain to be exploited in the stellar domain.

The main contributions have come from the three experiments (ACRIM, SOUP, and IPHIR) described in the next section. The *Yohkoh* spacecraft, launched in August, 1991, is currently returning relevant data, and we give below an assessment of the data quality. Finally, the first sophisticated observations intentionally emphasizing solar global properties will come with the launch of SOHO, and all of the results described here should probably be regarded as precursors to these data.

ACRIM, SOUP, AND IPHIR

These three experiments (see Table I for interpretation of the acronyms) were all launched within the decade of the 1980's. Briefly, ACRIM provided the first credible broad–band observations of solar luminosity variability; SOUP provided the first solar optical imaging observations with good cadence and image stability; and IPHIR provided definitive observations of the solar low–degree p–modes. We note that the total irradiance measurements from *Nimbus–7* preceded those of ACRIM by more than a year and continue to the present, and that NASA has agreed to re–fly ACRIM or otherwise maintain a continuous monitoring of the total solar irradiance. The *Stratoscope* images should also be mentioned here, since these images, with excellent resolution, were obtained from a solar telescope in space (*e.g.* Danielson, 1961). Each experiment made remarkable contributions. In some cases the advances have been duplicated or extended with ground–based data, but the message of their success is clear enough to encourage us towards comparable stellar observations.

TABLE I Solar Space Experiments

Experiment	Full Name	Spacecraft	Time Range
ACRIM	Active Cavity Radiometer Irradiance Monitor	*Solar Maximum Mission*	1980–89
SOUP	Solar Optical Universal Polarimeter	*Spacelab*	1986
IPHIR	InterPlanetary Helioseismology by IRradiance measurements	*Phobos*	1988–89
SXT	Solar X–ray Telescope	*Yohkoh*	1991–??

The above experiments have revealed several types of variability of the solar luminosity. A "frequency domain" view gives a summary of the mechanisms that have been identified to date, as listed in Table II (for more complete reviews, see Hudson, 1988, or Fröhlich *et al.*, 1991). several broad–band noise components modeled as power–law spectral distributions; g–modes; p–modes, and a high–frequency white noise component. The observational identification of the mechanisms for solar variability began with the ACRIM experiment, launched in 1980 and continuing until 1989. From ACRIM we quickly learned that sunspots

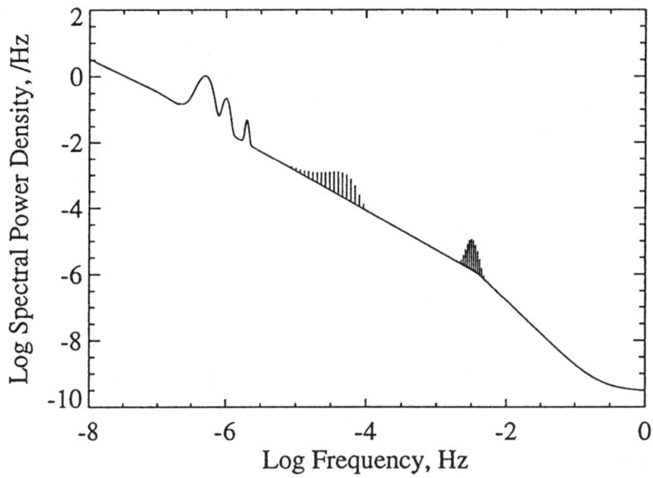

FIGURE I Sketch of the overall power spectrum of a star, modeled on the solar observations. The features represent line and continuum contributions as listed in Table II.

decrease the solar luminous output, essentially in the measure expected from their projected areas (Willson *et al.*, 1981). Subsequently the p–mode oscillations were found to make tiny variations in the solar luminosity (Woodard and Hudson, 1983). These were the easy effects to identify. The effects associated with faculae are much more complicated and still do not have a complete resolution. The irradiance perturbations come from facular elements at the smallest size scales, and yet they are closely associated with the relatively long time scale of the solar–cycle modulation, or even longer time scales (Foukal and Lean, 1990). The longer time scales have direct significance for stellar observations, since the Wilson program has revealed the existence of photometric cycles on solar–type stars with time scales comparable to the solar time scale of 11 years (*e.g.* Baliunas, 1985).

At frequencies above the 3 mHz vicinity of the p–modes, the solar variability spectrum is not known well. The power spectra of ACRIM data in Figure II (Fisher *et al.*, 1992) show that the spectrum falls off rapidly above the p–mode band. These spectra come from the incoherent sum of many individual orbits over the period from July 15 to October 16, 1989, during a period of relatively high solar activity. The right panel of Figure II shows the result of subtracting a best–fit constant term (white noise) from the raw spectrum of the left panel. The ACRIM white noise presumably is the instrumental noise floor, not a separate component of solar variability.

The definitive helioseismic observations in integrated sunlight were then carried out by the IPHIR photometers on board *Phobos* (Toutain and Fröhlich, 1992). These simple measurements are an excellent precursor to asteroseismology. The IPHIR data analysis overcame long–term detector degradation and demonstrated that this would not preclude obtaining the high frequency resolu-

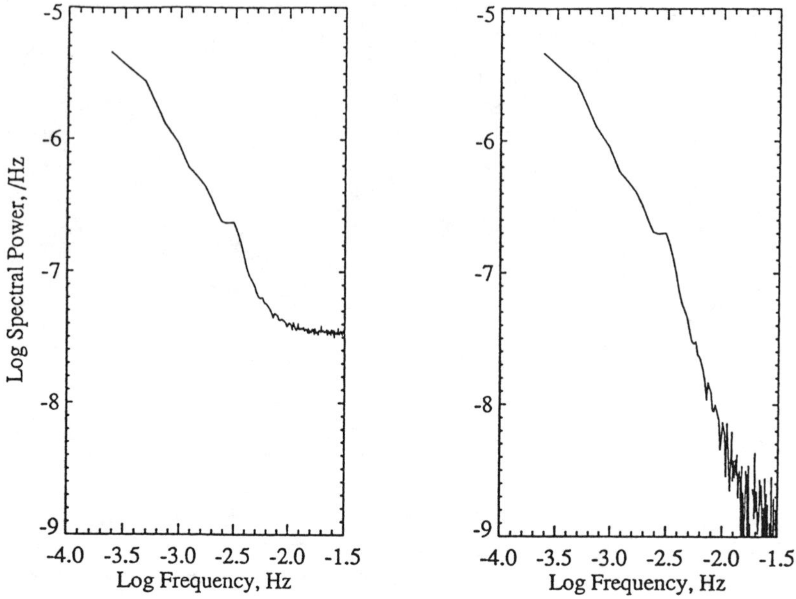

FIGURE II Low–resolution power spectra from the ACRIM experiment on board the *Solar Maximum Mission* — left, a sum of raw spectra; right, the sum with a fit to the white noise at high frequencies subtracted.

tion enabled by long, continuous time series.

TABLE II Variability of Solar Luminosity

Mechanism	Time scale	Amplitude
Oscillations	5 min	few parts per million
Granulation	tens of min	tens of parts per million
Sunspots	few days	< 0.2% peak–to–peak
Faculae	tens of days	< 0.1% peak–to–peak
Solar cycle	11 years	~0.1% peak–to–peak
Trend	tens of years	~0.1% peak–to–peak

The SOUP observations from *Spacelab*, while restricted to a narrow field of view and the use of film as a detector, demonstrated excellent stability. There were significant results in "solar astrometry," feature tracking, in spite of the limited amounts of data, and these results are instrumental in the newly developing view of the nature of solar convection (Title *et al.*, 1989).

YOHKOH AND SXT

Yohkoh ("Sunshine" in Japanese) is at present the main spacecraft with an experiment relevant to the general topics of stellar activity and global properties (Ogawara *et al.*, 1992). This is SXT, a soft X–ray telescope that will be discussed briefly here (see, *e.g.*, Tsuneta *et al.*, 1991, for fuller details) . The general properties of SXT appear in Table III. The X–ray imager itself consists of a grazing–incidence telescope with a set of broad–band filters for spectral analysis in the vicinity of 1 keV photon energy. This telescope forms images on a 1024 × 1024 pixel CCD with pixel subtense 2.46 arcsec. In addition to the X–ray imager, SXT contains a concentric 5–cm lens that forms optical images on the same CCD, principally for alignment of the X–ray images relative to solar optical features.

TABLE III The SXT aspect camera on board *Yohkoh*

Orbital inclination	31°
Orbital period	~97 min
Focal Length	1538.4 mm
Aperture	50 mm
Exposure times	0.1 – 1 sec
Pixel size	2.46 arcsec (1024 × 1024 CCD)
Wavelength	Two broad-band filters
Sampling	A few 512 × 512 images per day

The data from this aspect camera are the relevant item here, since they are providing the first long–term optical observations of the Sun from space with a photometrically stable imaging detector. The typical observational sequence has been to obtain one 2 × 2 image, *i.e.* an image with 4.92 arcsec pixels, approximately 10–15 times per day. Additional optical images for special purposes are also taken from time to time, for example at a high cadence in searching for white–light flare emission during the occurrence of a flare, or by special programming.

The data in Figure III show some results from a special set of observations carried out on Feb. 25, 1992, to test the photometric stability of the *Yohkoh* CCD camera. The observations consisted of a continuous sequence of 256 × 256 images of the whole Sun (9.84 arcsec pixels) taken through a 140 Å filter with a central wavelength of 4580 Å. The *Yohkoh* telemetry system permits one such image to be taken every 32 sec, but introduces gaps as can be seen in the time series. Nevertheless, these data can be analyzed to illustrate the CCD stability. In the initial data reduction, we subtracted the dark–current background but did not attempt to "flat–field" the images. The sum of the data from the central 64 × 64 pixels, as a fraction of the total signal, is shown in Figure III. The rms amplitude of this time series is 2.55×10^{-4}, approximately consistent with the amplitude expected from the ACRIM broad–band data. At this frequency, the ACRIM fluctuations are modulated by the level of solar activity. Much of this variation is solar in origin. There is also a presently–unknown contribution from

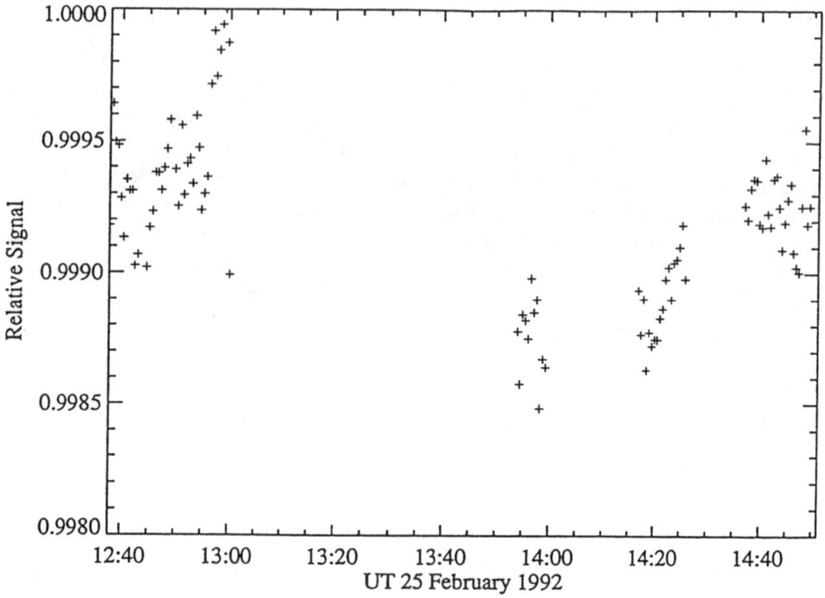

FIGURE III Time series of data from whole–Sun images obtained with the CCD detector of the SXT instrument on board *Yohkoh*; the minimum sample spacing is about 32 sec and the time series contains 97 images.

flat–field imperfections modulated by the spacecraft pointing jitter (a total range of approximately 10 arcsec in the present data set). We thus conclude that on this time scale (about two hours total), and by inference on shorter time scales, the *Yohkoh* CCD exhibits adequate stability for asteroseismology investigations.

The *Yohkoh* data, in the test time series shown in Fig. II, achieve a high–frequency noise level consistent with the statistics expected for the total number of photons detected ($\sim 10^{11}$) during the two–hour duration.

CONCLUSIONS: THE IMPACT OF SOLAR SPACE EXPERIMENTS

This paper discusses the current solar observations as a guide towards stellar observations related to global properties, and in particular to asteroseismology. We summarize first the main scientific conclusions that have been drawn from the existing data:

- There are several mechanisms of low–level variability of the solar luminosity, mostly broad–band in character.
- The solar p–modes are easily visible as small perturbations of the solar luminosity (*i.e.*, using integrated sunlight as a source).
- There is a solar–cycle variation of p–mode frequencies.

- The frequency splittings of the low–degree p–modes establish that the solar core is not rotating extremely rapidly.

The lessons that should be drawn from these successful solar observations might be summarized as follows:

- Relatively simple instrumentation can produce significant results.
- Observations from space are superior to ground–based observations because they avoid fluctuations in brightness and image shape from seeing effects, and because they allow long, interrupted time series.
- Optical observations are the most important for stellar work, since the photon–statistics limit can be approached. The solar observations suggest that this limit is achievable also with CCD detectors.

These results are a promising beginning and there is no reason to suppose, on the basis of the present solar observations, that a photometric approach to asteroseismology will not be feasible. On the other other hand, there are mechanisms of stellar variability not presently detected on the Sun, and these are potential sources of "noise." Some of these may be oscillatory, and others random in nature. Before the analysis of the ACRIM data we had hoped to detect solar flares in the total solar irradiance values, for example, but this search has not yet been successful (*e.g.* Hudson and Willson, 1985). Additional possible mechanisms for stellar variability include duplicity, occultation by dark external bodies, collisions with small bodies, non–solar modes of oscillation, the stellar analogs of coronal holes, and probably many others.

ACKNOWLEDGMENTS

This work has been supported at the Institute for Astronomy, University of Hawaii, by NASA under contract NAS 8-37334 to the Lockheed Palo Alto Research Laboratory, and at UCSD by NASA under grant NAG. I would like to thank the *Yohkoh* team members for permission to obtain the test data from the SXT aspect camera reported here.

REFERENCES

Baliunas, S.L., and Vaughan, A.H., 1985, *Ann. Revs. Astron. Astrophs.* **23**, 374.

Danielson, J., 1961, *Ap. J.* **134**, 275.

Fisher, B.M., Gruber, D.E., Hudson, H.S., and Willson, R.C., 1992, paper in preparation.

Foukal, P., and Lean, J., 1990, *Science* **247**, 556.

Fröhlich, C., Foukal, P.V., Hickey, J.R., Hudson, H.S., and Willson, R.C., 1991, in C.P. Sonett *et al.* (eds.), *The Sun in Time* (Arizona), p. 11.

Hudson, H.S., 1988, *Ann. Revs. Astron. Astrophys.*, **26**, 473.

Hudson, H. S., and Willson, R. C., 1983, **86**, 123.

Hudson, H.S., Acton, L.W., Hirayama, T., and Uchida, Y., 1992, *Publ. Astr. Soc. Japan*, to be published.

Ogawara, Y., and 14 co–authors, 1992, *Publ. Astr. Soc. Japan*, to be published.

Title, A., Tarbell, T.D., Topka, K.P., Ferguson, S.H., Shine, R.A., and the SOUP team, 1989, *Ap. J.***336**, 475.

Toutain, T., and Fröhlich, C., 1992, *Astron. Astrophys.* **257**, 287.

Tsuneta, S., and 12 co–authors, 1991, *Solar Phys.*, **136**, 37.

Willson, R. C., Gulkis, S., Janssen, M., Hudson, H. S., and Chapman, G. A., 1981, *Science*, **211**, 700.

Woodard, M., and Hudson, H. S., 1983, *Nature* **305**, 589.

Inside the Stars, IAU Colloquium 137
ASP Conference Series, Vol. 40, 1993
Werner W. Weiss and Annie Baglin (eds.)

MICROSCOPIC PHYSICS IN THE SOLAR STANDARD MODEL

ILIDIO LOPES[1] AND SYLVAINE TURCK-CHIEZE
Service d'Astrophysique, DAPNIA, CE Saclay,
91191 Gif-sur-Yvette Cedex 01, France

ABSTRACT The microphysics of the solar standard model is revised. We discuss the contribution of composition, nuclear reaction rates, screening effect and opacity (OPAL project) to the theoretical predictions on neutrino flux and acoustic modes.

INTRODUCTION

The multiplication of the observed solar neutrino fluxes coupled with their sensitivity to the temperature ($\Phi(pp) \propto T_C^{-1.2}$, $\Phi(^7Be) \propto T_C^8$ and $\Phi(^8B) \propto T_C^{18}$) and the high accuracy in the measurement of the five-minute oscillations (leading to an accurate sound speed, $c_s^2 \propto T/\mu$) impose a revision of the nuclear and atomic physics strongly dependent of the temperature.

MICROPHYSICS IMPROVEMENTS

Reaction Rates and Screening Factor

We follow the reaction rates compiling by Caughlan and Fowler 1988, except for (p,p), $(^3He,^3He)$ and $^7Be(p,\gamma)$ (Turck-Chièze and Lopes 1992). The two calculations recently published to redetermine the $p+p$ reaction by Gould and Guessoum 1992 and Carlson et al 1991 present a difference of about 5% on the astrophysical factor $S(0)$. This difference has small consequences on the solar structure and on the gallium prediction, but leads to a 10 % difference on the ^8B neutrino flux. In the central region of the Sun, the atoms are, except for iron, completely stripped and the shielding effect of the free electrons reduces the coulomb potential and increases the thermonuclear reaction rates within the star by a factor $f = exp\Lambda$, where Λ is the ratio of the Coulomb interaction energy to the kinetic energy. If this hypothesis of weak screening (WS) ($\Lambda = E_D/(kT) << 1$, Salpeter and Van Horn 1969), seems justified in the Sun for the main *pp* reaction, Λ may reach 0.2, 0.4 in a proton interaction with ^7Be or ^{14}N, so the hypothesis of *weak screening* is questionable. Graboske et al 1973, considers an intermediate regime(IS) defined for $\Lambda > 0.1$. Their general treatment in terms of chemical potentials was considered in our previous calculation (Turck-Chièze et al 1988) but the limitation of such approach comes from the negligence of nuclear and dynamical effects. As this effect concerns high Z ions, its main consequence is a slight modification of the branching ratio of the *pp* chains and the CNO cycle (*see table 1: model c and d*). It mainly affects the neutrinos connected to high Z reactions, and pratically not the present structure of the Sun and consequently the acoustic frequencies. We consider in a recent work that this treatment underestimates the effect of screening and suggest a better adapted treatment which is compatible with the mean value of WS and IS (Dzitko et al 1992).

---

[1] Also, Centro de Astrofísica, Universidade do Porto, Rua do Campo Alegre, 823, 4100 Porto - PORTUGAL

Solar Model	model *a*	model *b*	model *c*	model *d*	model e
$T_c(10^6 K)$	15.47	15.33	15.49	15.50	15.43
$\rho_c(g/cm^3)$	141.80	142.80	147.30	147.20	146.89
$Y_{initial}$	0.2736	0.2664	0.2762	0.2760	0.2714
r_{BCZ}	0.729	0.733	0.721	0.721	0.725
$T_{BCZ}(10^6 K)$	2.00	2.10	2.16	2.16	2.11
$\Phi(pp)/10^{10}(cm^{-2}s^{-1})$	6.05	6.08	6.02	5.99	6.01
$\Phi(pep)/10^8(cm^{-2}s^{-1})$	1.29	1.30	1.38	1.38	1.39
$\Phi(^7Be)/10^9(cm^{-2}s^{-1})$	4.16	4.08	4.42	4.62	4.42
$\Phi(^8B)/10^6(cm^{-2}s^{-1})$	4.04	4.28	4.51	5.15	4.71
$\Phi(^{13}N)/10^8(cm^{-2}s^{-1})$	3.20	3.40	3.49	3.99	4.00
$\Phi(^{15}O)/10^8(cm^{-2}s^{-1})$	2.62	2.82	2.91	3.45	3.39
chlorine detector(SNU)	5.88	5.22	6.51	7.30	6.78
water *detector*$^*/10^6$)	4.04	3.51	4.59	5.15	4.71
gallium detector(SNU)	120	116	123	126	124

TABLE I Different predictions of the present Sun : Solar models computed with pp reaction rate proposed by Gould and Guessoum and intermediate screening with LAOL tables for high (*model a*) and low iron (*model b*) composition. The other models are computed with pp reaction rate calculated by Carlson et al with OPAL tables and the low iron composition. *Model c* and *d* use respectively IS or WS for reactions involving nuclei with $Z > 3$. *Model e* (comparable to d) introduces the new CNO composition.($1SNU = 10^{-36}$ captures per second per target atom, $^{(*)}$ *in* $cm^{-2}s^{-1}$)

Initial composition and opacity

The radiative energy transport of the low mass stars needs precise estimates of the photon interactions (Thomson scattering, free-free processes, bound-free and bound-bound processes) with the low and heavy elements. For a precise calculation, one needs a proper determination of the initial composition (the heavy elements contribute to 90% at the basis of the convective zone). Solar models have been recomputed with the Los Alamos (LAOL) Opacity library for the high and low iron ccmposition adjusting the ratio Z/H accordingly to the observed values (*see table 1: model a and b*). One notices a difference of 13% on the chlorine and water neutrino flux detection. In the acoustic mode frequency predictions, we have a degradation of the order of $2\mu Hz$ for modes $l \leq 40$ and $\nu \approx 4500\mu Hz$ which are trapped in the intermediate region (0.4 - 0.7 R⊙). The new CNO composition recently proposed by Grevesse et al 1991 reduces the ratio Z/X by 9 % and produced an effect of the same order on neutrino detection (*model d and e*), (Turck-Chièze and Lopes 1992).

The improvements realised by the Livermore group (Rogers and Iglesias 1992) using OPAL code concern the equation of state and also the treatment of the mixture with coupled equations for the full solar composition. In the case of the Sun, their results differ from the LAOL estimate by less than 5% in the center and 20 % in the intermediate region. The first consequence is a better determination of the intermediate region, illustrated in *figure 1* by the determination of the $p - modes$ prediction. The second consequence is an increase on the central temperature of the same order but with an opposite sign than the effect of improved composition.

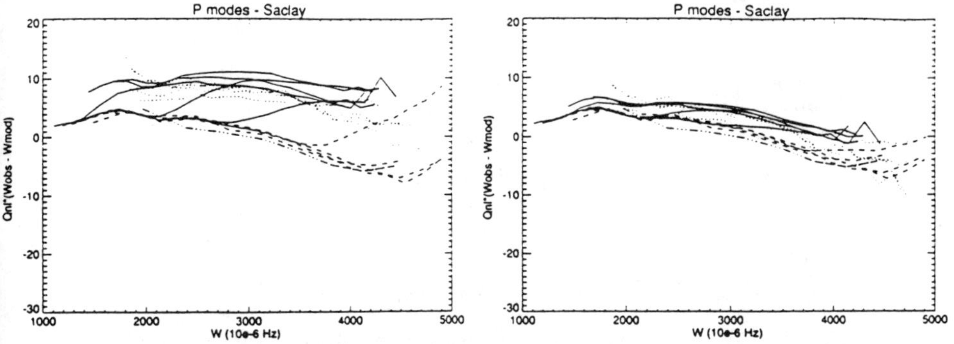

FIGURE I Scaled Frequency differences between observed frequencies of solar oscillations from the compilation by Libbrecht et al and theoretical frequencies of Solar Standard Model for selected values of l. Points correponding to a given value of l have been connected, according to the following convention: $l = 0 - 3$ (*doted line*), $l = 4, 5, 10, 20, 30$ (*continuous line*), $l = 40, 50, 70, 100$ (*dashed line*), $l = 150$ (*3 doted line*). The figure 1a belongs to the *model b* and the figure 1b belongs to the *model d.*

DISCUSSION

Improvements in the microscopic physics help us to better understand solar modelling and the determination of acoustic mode frequencies is a guide to qualify such model. The next step will be the revision or confirmation of the nuclear reaction rates and the treatment of the outerpart (molecular contribution to the opacity), before the introduction of the dynamical effects.

ACKNOWLEDGMENTS

We are particularly grateful to Jørgen Christensen-Dalsgaard who supplied us the pulsation code and to Werner Däppen who has introduced us to the helioseismological field. This work was partly financed by grant BD/749/90-RM from JNICT, Portugal.

REFERENCES

J. Carlson, D. O. Riska, R. Schiavilla, and R. B. Wiringa, Phys. Rev. C, 1991 , 44, 619.
G. R. Caughlan and W. A. Fowler, 1988, Atomic Data and Nucl. Data Tab., 40, 283.
H. Dzitko, P. Delbourgo-Salvador, C. Lagrange, S. Turck-Chièze, 1992, International Symposium, June $22 - 25^{th}$, I.A.P, Paris.
H. C. Graboske, H. E. DeWitt, A.S. Grossman and M. S. Cooper, 1973, *ApJ*, 181 , 457.
N. Grevesse, D. L. Lambert, A. J. Sauval, E. F. van Dishoeck, C. B. Farmer, and R. H. Norton, Astron. Astrophys.,1991, 242, 488.
R. J. Gould and N. Guessoum, 1992, *ApJ*, 359 , L67.
F. J. Rogers and C. A. Iglesias, 1992, *ApJS*, 79, 507.
K.G. Libbrecht, M. F. Woodard and J. M. Kaufman, 1990, *ApJS*, 74, 1129.
E. E Salpeter and H. M. Van Horn, 1969, *ApJ*, 155, 183
S. Turck-Chièze, S. Cahen, M. Cassé and C. Doom, 1988, *ApJ*, 335 , 415.
S. Turck-Chièze and I. Lopes, 1992, *ApJ*, to be published

Inside the Stars, IAU Colloquium 137
ASP Conference Series, Vol. 40, 1993
Werner W. Weiss and Annie Baglin (eds.)

CESAM SOLAR MODELS

P. MOREL[1] , G. BERTHOMIEU[1], J. PROVOST[1]
[1] Cassini , URA CNRS 1362, Observatoire de la Côte d'Azur, Nice

Y.LEBRETON[2]
[2] DASGAL, URA CNRS 335, Observatoire de Paris, Paris

ABSTRACT Solar models have been computed using the code CESAM constructed at Nice (Morel, 1992), and the new opacities of Livermore for different mixtures. Their global characteristics, predicted capture rates of neutrinos for the chlorine and gallium experiments and their seismological properties are given.

I - GLOBAL PROPERTIES AND NEUTRINO FLUXES

A set of solar calibrated models has been computed with the CESAM code, insuring the solar radius and luminosity ($L_\odot = 3.846\ 10^{33}$ ergs s^{-1}) at the age $t_\odot = 4.7510^9$ years.

TABLE I

Models	Y	α	$T_c 10^{-6}$	ρ_c	^{37}Cl	^{71}Ga
reference:ST	0.275	1.90	15.55	151.9	7.43	127.7
ST1	0.276	1.89	15.51	149.6	7.19	126.5
ST2	0.277	1.90	15.53	150.1	7.35	127.5
ST3	0.276	1.89	15.51	149.6	6.52	124.9
Bahcall *et al.*	0.271		15.6	148.	7.9	132.
Lebreton *et al.*	0.278	2.16	15.54	148.	7.6	126.9.
Turck Chièze *et al.*	0.276	1.55	15.51	147.7	5.8	125.
Sackman *et al.*	0.278	2.07	15.43	146.6	7.68	125.
EFF-LAOL	0.283	2.57	15.62	154.2	7.83	129.4
CEFFH-LAOL	0.274	2.60	15.56	151.9	7.31	126.3
OPAL-RA	0.284	2.56	15.68	152.5	8.43	132.2
OPAL-AG	0.285	2.60	15.75	151.2	8.89	134.2
OPAL-AGFe (1)	0.277	2.60	15.58	152.2	7.59	127.9
OPAL-AGFe (2)	0.277	1.85	15.58	152.3	7.65	128.0
OPAL-AG-KU	0.285	1.91	15.73	151.2	8.90	134.3

ST: Model with equation of state with the Debye Huckel corrections CEFF and low iron OPAL opacities extended with Kurucz opacities at the surface (KU) and Z= 0.0194; ST1: Model ST calibrated at a lower age $t_\odot = 4.610^9$ years; ST2: Model ST with $t_\odot = 4.610^9$ years and $L_\odot = 3.8610^{33}$ ergs s^{-1}; ST3: Model ST1 with the lower value of the cross section for the reaction ^{7}Be(p,γ)^{8}B used by Turck-Chièze *et al.* EFF: Eggleton (*et al.*, 1973) equation of state; LAOL: Los Alamos opacities; OPAL: Livermore opacities with different mixtures: Ross-Aller (RA), Anders and Grevesse mixture with photospheric iron content (AG) and low iron close to the meteoritic abundance (AGFe).

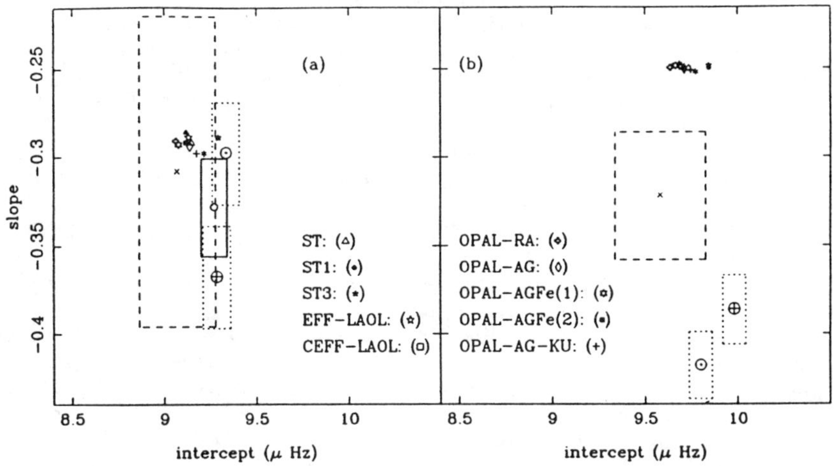

FIGURE I Values of the differences $\nu_{00} - \nu_{02}$ and $\Delta\nu_0 - \Delta\nu_2$ (intercept and slope for $l = 0$, (a)) and $3/5(\nu_{01} - \nu_{03})$ and $3/5(\Delta\nu_1 - \Delta\nu_3)$ (3/5 times the intercept and slope for $l = 1$, (b)) for different solar models and for the observations of Toutain *et al* (full line box), Anguera Gubau *et al.* (two dotted line boxes) corresponding to the solar maximum and minimum and Elsworth *et al.* (dashed line box).

The comparison of our models with other published models and their sensitivity to the physical inputs (opacities and equation of state) are shown in Table I where the initial helium abundance, mixing length, central temperature and density, capture rates for chlorine and gallium experiments are given. The new OPAL opacity tables are extended to low temperature either by L'AOL tables at $T < 70000^0K$ (except at T < 6000 ^{0}K for OPAL-AGFe(2)) or by Kurucz opacity tables.

Our models are in agreement with the previous solar model computations. They have a higher central density due to the choice of a larger value for the solar age. The neutrino capture rates for chlorine experiment vary by 20% with the changes in the opacities while they vary by 5% for the gallium experiment.

II - SEISMOLOGICAL P-MODES PROPERTIES

Low degree modes give informations on the properties of the core of the Sun. According to the method proposed by Scherrer *et al.* (1983) the frequencies are fitted to the following polynomial expression:

$$\nu_{n,l} = \nu_{0l} + \Delta\nu_l(n + \tfrac{l}{2} - n_0) + a_l(n + \tfrac{l}{2} - n_0)^2$$

where $n_0 = 21$. The difference $\Delta\nu_{l,l+2} = \nu_{n,l} - \nu_{n-1,l+2}$ is essentially sensitive to the sound velocity in the core of the sun and it is close to a linear function of n characterized by its slope and the intercept ($\sim$ value for $n = n_0$).

The values of the slope and the intercept for $\Delta\nu_{0,2}$ and for $\Delta\nu_{1,3}$ of our different models and the observations including an estimation of the observational error bars are plotted in Figure I. The modes with radial nodes between 17 to 26 have been used. The datas of Anguera Gubau *et al.* are splitted into solar maximum and solar minimum observations. Two different values of the slope and intercept are derived with error boxes which do not overlap. The results of IPHIR are in between which is coherent with the dates at which the observations have been made.

The figure shows that the two characteristic parameters derived from the low degree p-modes are close to the observation for $l = 0$ and systematically larger for the slope derived from $l = 1$ in agreement with Christensen-Dalsgaard (1992) calculations, but they are not much sensitive to the different opacities and equation of state, contrarily to the neutrinos fluxes.

Concerning the modes of intermediate degrees, the new opacities decrease but do not suppress the discrepancy which appears between low ($l < 20$) and high ($l > 50$) degree modes in the difference between observed and theoretical frequencies. The frequencies with degree ranging from 20 to 50 correspond to the modes having their turning point just below and just above the convection zone and the persistency of this observed discrepancy may be the signature of the presence of some penetration taking place at the lower boundary of the convection zone.

In conclusion the global properties of our models are in agreement with the previous results. The causes of the differences are pointed out in a more complete discussion in Berthomieu *et al.* (1992). With the new opacities of Livermore with low photospheric iron content we find for the chlorine experiment $\Phi_{^{37}Cl} = 7.43 \pm 2.7$ SNU far from the measure of Davis and for the gallium experiment $\Phi_{^{71}Ga} = 127.7 \pm 20$ SNU.

REFERENCES

Anders E., Grevesse N. 1989, *Geochim. Cosmochim. Acta* **53**, 197

Anguera Gubau M., Palle P., Perez Hernandez F., Regulo C. Roca Cortes T. 1992, *Astron. Astrophys.* **255** 363-372

Bahcall J., Ulrich R.K. 1988, *Rev. Mod. Physics* **60**, 296

Berthomieu G., Provost J., Morel P., Lebreton Y. 1992, submitted to *Astron. Astrophys.*

Christensen-Dalsgaard J. 1992, *Challenges to Theories of the Structure of Mode rate--Mass Stars*, D.O. Gough, J. Toomre (Eds), Springer Verlag, p 11-36

Eggleton P.P., Faulkner J., Flannery B.P. 1973, *Astron. Astrophys.* **23** 325-330

Elsworth Y., Howe R., Isaak G.R., McLeod C.P., New R. 1990, *Nature* **345**, 322-324

Huebner, W.F., Merts, A.L., Magee, N.H.Jr, Argo, M.F. 1977, Astrophysical Opacity Library, Los Alamos Scientific Laboratory Report LA-6760-M

Iglesias C. A., Rogers F. J. 1991, *Astrophys. J.* **371**, 408-417

Lebreton Y., Däppen W. 1998, *Seismology of the Sun & Sun-like Stars*' ed. E.J. Rolfe, ESA SP-286, pp 661-664

Libbrecht K.G., Woodard M.F., Kaufman J.M. 1990, *Astrophys. J. Supp.* **74**, 1129

Morel P., 1992, these proceedings

Ross J.E., Aller L.H. 1976, *Science* **191**, 1223

Sackmann I.J., Boothroyd A.I., Fowler W.A. 1990 *Astrophys. J.* **360**, 727-736

Scherrer P.H., Wilcox J., Christensen-Dalsgaard J. Gough D.O. 1983 *Solar Phys.* **82**, 75-87

Toutain Th., Fröhlich C. 1992, *Astron. Astrophys.* to appear

Turck-Chièze S., Cahen S., Cassé M., Doom C. 1988, *Astrophys. J.* **335**, 415

Inside the Stars, IAU Colloquium 137
ASP Conference Series, Vol. 40, 1993
Werner W. Weiss and Annie Baglin (eds.)

RESTORATION OF THE ATMOSPHERE IN SOLAR MODELS

P. MOREL[1], C. VAN'T VEER[2]
1 Cassini, URA CNRS 1362, Observatoire de la Côte d'Azur, Nice
[2] DASGAL, URA CNRS 335, Observatoire de Paris, Paris

G. BERTHOMIEU[1], R. ÇAYREL[2], F. CASTELLI[3], M.J. GOUPIL[2], Y. LEBRETON[2], J. PROVOST[1]
[3] Observatorio Astronomico, Trieste

ABSTRACT From models of the Sun atmosphere obtained with ATLAS 9 (Kurucz 1992) it appears that the diffusion limit of the transfer equation becomes valid only at depths larger than $\tau_R \sim 10$. Even with physics as consistent as possible one restores the atmosphere with an accuracy not better than $\sim 40\%$, but with small consequences on solar calibrated models. Despite the eigenmodes have turning points located in the atmosphere, the study of the oscillations reveals differences less than 2μHz when the atmosphere is restored with various assumptions.

I – OPTICAL DEPTH REQUIRED BY THE DIFFUSION LIMIT

An accurate description of the atmosphere is needed for precise stellar interior calculations. Since it is out of possibilities to include full atmosphere calculations into a stellar model, one separates the interior where the diffusion approximation is valid from the atmosphere which is restored from a $T(\tau)$ law, the limit between the two parts beeing located at a large Rosseland optical depth τ_b.

The so-called diffusion approximation only valid at great optical depths writes (Mihalas, 1978):

$$\lim_{\tau_\nu \to \infty} H_\nu = -\frac{1}{3}\frac{\partial B_\nu}{\partial \tau_\nu}$$

If, in some parts of the spectrum such as in spectral lines, this limit is reached at depths that correspond to a few tenths in the Rosseland optical depth τ_R, in other parts where the medium is less opaque, it is far to be fullfilled even if $\tau_R > 1$. Hence, for the calculation of the radiative flux one needs an estimate τ_b of the optical depth above which the value of the integrated radiative flux H reaches its optically thick limit.

Along the solar evolutionary track from the ZAMS to now: the gravity is almost constant and the effective temperature T_{eff} varies from $\sim 5700°K$ to $\sim 5780°K$. With the code ATLAS 9 (Kurucz, 1992) we have calculated atmosphere models of a star of $1M_\odot$ with $T_{eff} = 5650°K$, $5700°K$, $5750°K$, $5800°K$ ($\log g = 4.47$, $\alpha = l/Hp = 1.7$, $X = 0.70$, $Y = 0.28$, $Z = 0.02$) and compared the radiative flux derived from the diffusion approximation to its exact value. From the Fig.I the relative difference of these two fluxes remains greater than 1% above $\tau_R = 10$. As a consequence one should set $\tau_b \geq 10$, provided the $T(\tau)$ law includes the convective transport. However from table I one sees that the global parameters of the model are not significantly affected by changes of τ_b, except for the radiative Eddington's law.

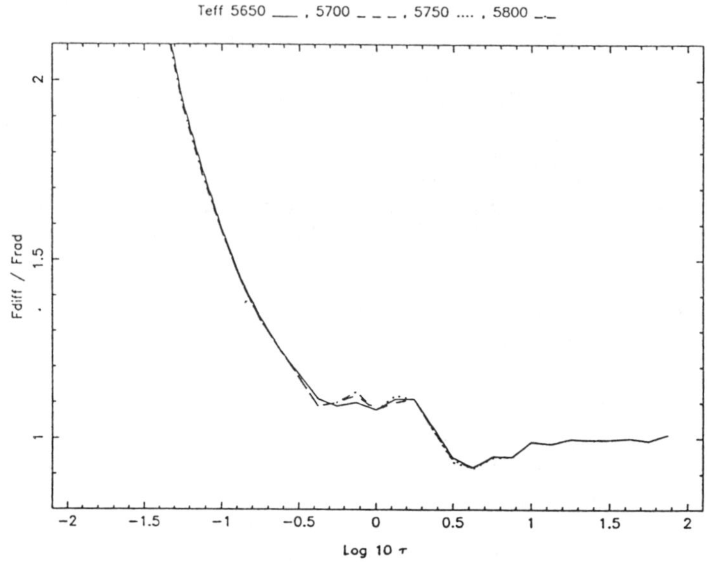

FIGURE I Ratio between the radiative flux and its optically thick value derived by the diffusion equation. The curves are drawn for various T_{eff}. Only for $\tau_R \geq 10$ the relative difference is less than 1%.

II – RESTORATION OF AN ATMOSPHERE WITH A $T(\tau)$ LAW

In stellar evolution calculations the derivation of the temperature needs the knowledge of the radiative and convective fluxes. In the optically thick medium the radiative flux can be derived from the diffusion equation but, for $\tau_R \leq 10$, much more sophisticated radiative transfer computations are needed. These heavy calculations can be avoided by the knowledge *a priori* of the temperature as a function of some parameters of the model. For the Sun one writes the temperature as a function of the effective temperature and of the Rosseland optical depth $T(\tau, T_{eff})$. The $T(\tau)$ laws are derived either from theory (Eddington, Hopf) or from models (HSRA). The temperature dependence with respect to the optical depth is smooth enough so that, within ±0.5%, for any T_{eff}, in the range $5700°K \leq T_{eff} \leq 5800°K$ the temperature can be restored by a "tangent" Hopf's law:

$$T = T_{eff} \left[\frac{3}{4}\left(\tau + \bar{q}\left(\tau\right)\right)\right]^{1/4}, \quad \text{with the function} \quad \bar{q}(\tau) = \frac{4}{3}\left(\frac{T}{\overline{T}_{eff}}\right)^4 - \tau$$

obtained from a $1M_\odot$ model of atmosphere with $\overline{T}_{eff}(= 5750°K$ hereafter).

Throughout the atmosphere the luminosity is equal to its value at the bottom: $L = L(\tau_b) = L_*$. The star radius R_* is defined as the radius at τ_* where $T = T_{eff} = (L_*/4\pi\sigma R_*^2)^{1/4}$ which reduces to $\tau_* = 2/3$ with the Eddington law. The set of equation to be solved can be found in Morel *et al.* (1990).

The degree of accuracy of the restored atmosphere can be assessed by comparing the values of physical quantities such as pressure, temperature and sound speed as provided by the full atmosphere model calculation and the restored atmosphere. Even if the restoration is made with the maximum of consistency, as we attempted, with respect to the geometrical depth the pressure (respt. the sound speed) is hardly restored with an accuracy better than 40% (respt. 5%).

III - RESULTS AND DISCUSSION

A set of calibrated solar models with different $T(\tau)$ laws and different values of $\tau_\star$ and τ_b and their p modes oscillations frequencies have been computed, using the Nice CESAM code (these proceedings). Their global properties are given in Table I.

The model SS is obtained with a boundary condition at $\tau_* = 2/3$: $P = 2g/3\kappa$ (κ is the opacity) derived from the equation of hydrostatic equilibrium, which avoids the complication of the atmosphere modelisation.

From comparison of models 1, 3 and 6, 7 it is seen that the use of $\tau_\star = 2/3$ for another law than Eddington's one does not introduce significant modification on the model. The simplified assumption SS gives a model close to the other ones not accurate enough for helioseismoly but sufficient for some problems of stellar evolution. The effect

TABLE I

models	$T(\tau)$	τ_*	τ_b	α	Y	$R_{zc}/R_\odot$
1	5750	2/3	10	1.803	0.2744	0.721
2	5750	2/3	2	1.794	0.2744	0.721
3	5750	0.548	10	1.794	0.2744	0.721
4	Edding	2/3	10	2.585	0.2744	0.721
5	Edding	2/3	2	1.678	0.2744	0.722
6	HSRA	2/3	2	1.899	0.2743	0.721
7	HSRA	0.420	2	1.869	0.2743	0.721
8	SS	2/3	2/3	1.798	0.2742	0.721

of different treatments of the atmosphere on the sound speed is localized in the external layers. The differences of the p modes frequencies for degrees l= 0 to 400 between models with same realistic $T(\tau)$ law (taking into account the convection) and different values of $\tau_\star$ and τ_b are less than 0.2μHz. The differences of the p modes frequencies between models with three different $T(\tau)$ laws (models 2 5 6) vary from zero for low frequency modes to 2 μHz for frequencies around 4 mHz.

In conclusion we have shown that the transfer equation reaches its asymptotic diffusion approximation at optical depths larger than ten. At such depths the energy is transported by convection and the $T(\tau)$ law needs to be derived from a model of atmosphere including convection and using consistent physics. However we find that the global properties and the oscillations frequencies of the solar calibrated models are not significantly sensitive to the optical thickness of the atmosphere provided it is larger than 2. The effect on the helioseismology of different $T(\tau)$ laws is rather small but not negligeable. More detailed analysis of the sensitivity of the solar model to the description of the atmosphere will be given in Morel *et al.* (1992).

REFERENCES

Ando H., Osaki Y. 1975, *P.A.S.J.*, **27**, 581

Kurucz R.L. 1992, private communication

Mihalas D. 1978, Stellar Atmosphere, 2d Ed. Freeman and Cie

Morel P., Provost J., Berthomieu G. 1990, *Solar Physics* **128**, 7

Morel P., Van't Veer C., Berthomieu G., Cayrel R., Castelli F., Goupil M.J., Lebreton Y., Provost J. 1992, in preparation

Inside the Stars, IAU Colloquium 137
ASP Conference Series, Vol. 40, 1993
Werner W. Weiss and Annie Baglin (eds.)

SEISMOLOGICAL CONSTRAINTS ON CONVECTIVE PENETRATION IN THE SUN

G. BERTHOMIEU[1], P. MOREL[1], J. PROVOST[1]
[1] Cassini , URA CNRS 1362, Observatoire de la Côte d'Azur, Nice

J.-P. ZAHN[2,3]
[2]Observatoire Midi-Pyrénées, Toulouse
[3] Columbia University, New York

ABSTRACT Penetrative convection is expected below stellar convection zones, where it should achieve a nearly adiabatic stratification. A theoretical prediction of the penetration depth has been recently made by Zahn (1991) which includes an arbitrary parameter ζ depending on the properties of the convective motions. We use the helioseismological constraints to calibrate the value of this parameter.

I. SOLAR MODELS WITH CONVECTIVE PENETRATION.

A theoretical prediction for the depth of the convective penetration has been made recently by Zahn (1991). It is largely inspired by laboratory experiments and by recent numerical simulations of convection in a stratified medium. These show that convective penetration is due mainly to plumes which originate in the superadiabatic region and penetrate into the stable domain, which they render nearly adiabatic over some distance. The linearization of the variation of the radiative flux, which is due mainly to the increase with depth of the radiative conductivity χ ($16\sigma T^3/3\rho\kappa$ with the usual notations), leads to the following estimate for the extent of penetration L_p : $L_p/H_P = \zeta/\chi_P$, where H_P is the pressure scale-height and $\chi_P = (\partial \log \chi/\partial \log P)_{ad}$.

The parameter ζ is related to the third moment of the convective velocity; it should be close to unity, but presumably a little less, since the plumes are probably less sharply

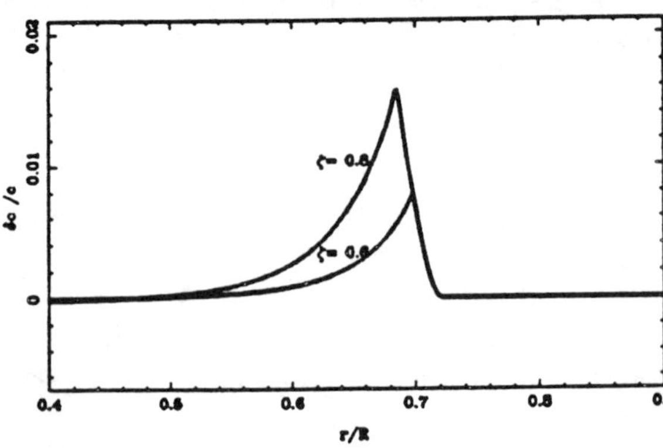

FIGURE I Variation of the difference of sound velocity between models with convective penetration and the reference solar model below the convection zone for different values of ζ.

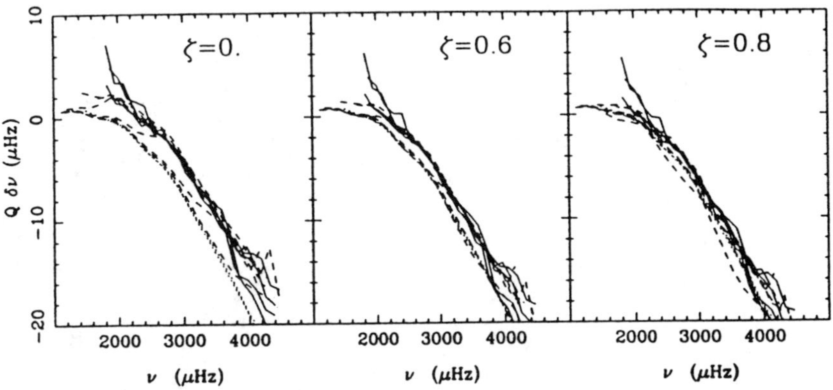

FIGURE II Relative difference between the p-mode frequencies of three models with convective penetration $\zeta = 0$ (reference model), $\zeta = 0.6, \zeta = 0.8$ and the observed periods given by Libbrecht *et al.* as a function of the frequency for degrees $l = 0, 1, 2, 3, 4, 5$ (full line), $l = 10, 20, 30, 40, 50$ (dashed line), $l = 70, 100$ (dotted line).

focused in the penetration layer than above. Therefore, the extent of penetration is about half a pressure scale-height in the Sun, where $\chi_p \approx 1.8$.

A series of solar models have been built including convective penetration at the base of the convection zone, with different values of the parameter ζ. The penetration region is nearly adiabatic, and its base is very sharply defined, since the temperature gradient there changes abruptly from the adiabatic to the (lesser) radiative slope within a thermal boundary layer whose thickness if of order of 1 km (see fig. 2 in the review by Zahn, this meeting). Results for the sound velocity are given in figure I .

II. SEISMOLOGICAL SIGNATURE OF THE CONVECTIVE PENETRATION.

A) Calibration of ζ with low and intermediate degree p-modes.

The parameter ζ can be calibrated by comparing the acoustic frequencies of models with penetration with those observed on the Sun by Libbrecht *et al.* (1990). The relative differences between the frequencies of the standard model and the Sun show a significant gap between the low degree ($l < 20$) and high degree ($l > 50$) modes. It was generally attributed to an underestimate of the opacity below the convection zone, which would be filled with better determinations.

The new opacities computed by Iglesias and Rogers for the Anders Grevesse mixture are indeed larger than the Los Alamos ones in that region, and they partly fill the gap. But on closer examination, they do not suppress it, as can be seen in the first plot of figure II. The two other plots of figure II show the effect of convective penetration: it brings the low degree and high degree curves even closer together. The best agreement is achieved for a value of about $\zeta \sim 0.6$. However it is necessary to keep in mind that this value depends on the physics which is included in the models, and particularily on the opacities at the base of the convection zone. It may also depend on other physical processes which have been ignored so far, such as thermal diffusion below the convection

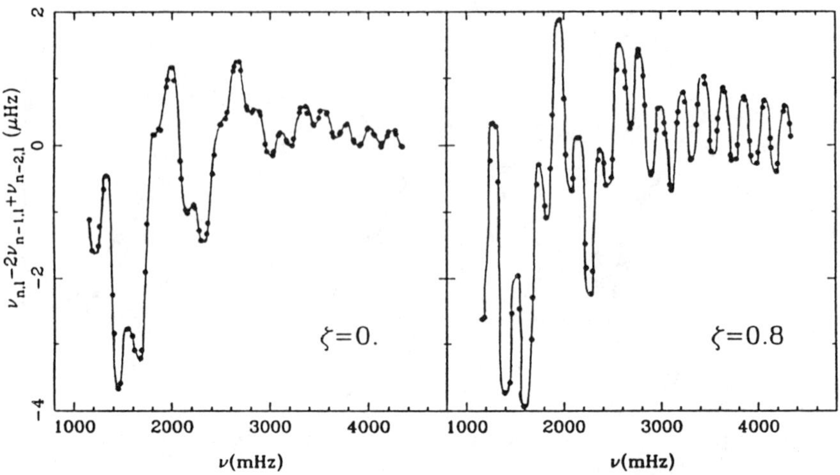

FIGURE III Variation of the differences of frequencies $\nu_{n,l} - 2\nu_{n+1,l} + \nu_{n+2,l}$ for p-modes of degree $l = 0, 1, 2, 3$ as a function of the frequency corresponding to two models: the reference model (a) , and a model with $\zeta = 0.8$ (b).

zone.

B) Second order differences of low degree p-modes frequencies.

To distinguish between several competing effects, on can turn to the second order difference $\Delta = \nu_{n,l} - 2\nu_{n+1,l} + \nu_{n+2,l}$, as pointed out by Gough (1990). When plotted against the frequency, this quantity has a sinusoidal behavior with two different "periods" of the order of 230 μHz and 700 μHz. Theses oscillations have been interpreted by Gough: the larger "period" is due to the rapid variation of γ in the ionisation zones, and the lower "period" is an indication for a discontinuity in the derivative of the sound velocity at the base of the convection zone.

Such a discontinuity is expected when there is convective penetration, as we have seen in figure I. Therefore it should be possible to calibrate the penetration parameter ζ by fitting the second order quantity Δ to the observations, as soon as sufficiently precise determinations of the low degree modes frequencies will become available.

REFERENCES

Anders E., Grevesse N.: 1989, *Geochim. Cosmochim. Acta* **53**, 197

Gough D.O.: 1991, in *Progress of seismology of the sun and stars*, Eds Osaki Y. Shibahashi H., Springer Verlag, p 283-318

Iglesias C. A., Rogers F. J.: 1991 *Astrophys. J.* **371**, 408-417

Libbrecht K.G., Woodard M.F., Kaufman J.M.: 1990, *Astrophys. J. Supp.* **74**, 1129

Zahn J.P.: 1991, *Astron. Astrophys.* **252** 179-188

Inside the Stars, IAU Colloquium 137
ASP Conference Series, Vol. 40, 1993
Werner W. Weiss and Annie Baglin (eds.)

TEST OF A NEW THEORY FOR STELLAR CONVECTION USING HELIOSEISMOLOGY

L. PATERNÒ
Istituto di Astronomia dell'Università di Catania, Italy

R. VENTURA
Osservatorio Astrofisico di Catania, Italy

V.M. CANUTO
NASA - Goddard Institute for Space Studies, New York, USA

I. MAZZITELLI
CNR - Istituto di Astrofisica Spaziale, Frascati, Italy

ABSTRACT Two evolutionary models of the Sun have been tested using helioseismological data. The two models use the same input microphysics (nuclear reaction rates, opacity, equation of state) and the same numerical evolutionary code, but differ in the treatment of turbulent convection. The first model employs the standard mixing - length theory of convection, while the second one employs a new turbulent convection model which overcomes some basic inconsistencies of the standard theory of convection.

The test rests on the calculation of p-mode eigenfrequencies and on the comparison with the helioseismological data.

The comparison shows an overall improvement of the eigenfrequencies calculated with the new model with respect to those calculated with the standard model, although it appears that both models still suffer from inaccuracies especially in the treatment of the surface layers.

INTRODUCTION

The problem of stellar convection has been treated, for decades, by means of the mixing-length theory (Böhm - Vitense 1958; henceforth cited as MLT) which, in spite of the practical advantage of simplicity and the apparent ability to fit observation by adjusting a free parameter, makes use of drastic assumptions, physically unsuitable to be applied to the almost inviscid stellar interior.

A new approach to the problem of stellar turbulent convection has been recently proposed by Canuto and Mazzitelli (1991; henceforth cited as CM). The CM model overcomes some basic limitations of the standard mixing-length theory and can be employed in stellar codes without major modifications.

In order to test to what extent the CM model has improved the description of turbulent convection vis à vis the standard MLT model, two models of the Sun have been adopted. The first of them is based on the standard mixing-length theory, the second one on the new CM model in its more recent version (Canuto and Mazzitelli 1992).

A comparison between the helioseismological data (Libbrecht et al. 1990) and the p-mode eigenfrequencies calculated for the two models has been carried out.

MODELS

The convective turbulence in the interior of the stars is compressible and is characterized by a wide spectrum of eddies of all sizes.

The mixing-length theory treats the energy spectrum of turbulence as if it consisted of only one large eddy and provides an expression for the mixing length Λ which is assumed to be proportional to the local scale height through a free parameter adjusted to reproduce the observed position of a given star in the H-R diagram.

The CM treatment employs a new turbulent convection model which accounts for the full spectrum of turbulent eddies and derives a new expression for the turbulent flux as well as for the turbulent pressure, usually ignored in the MLT approach. While retaining the assumption of incompressibility , the CM model also provides an expression for Λ, given as a function of local and non-local variables, substantially independent of free parameters.

The two MLT and CM approaches have been then applied to construct two evolutionary models of the Sun adopting a numerical code described in detail elsewhere (Mazzitelli 1989). The two models have the same $L_\odot$ and $R_\odot$, use the same microphysics (equation of state, opacity, nuclear reaction rates) and the same numerical code. The difference lies only in the treatment of convective transport. Thus differences in the predicted p-mode eigenfrequencies should be ascribed to the different treatment of convection since it reflects on differences in physics variables of helioseismological interest.

In view of a helioseismological analysis, radial and non-radial p-mode adiabatic eigenfrequencies have been computed for the two models.

RESULTS AND DISCUSSION

A first comparison has been based on the analysis of modes with $0 \leq \ell \leq 11$, following the procedure described in Dziembowski et. al. (1988). For $\ell < 20$ deviations in frequencies $\delta\nu$ of theoretical eigenfrequencies either from those of another model or from the observed ones, can be expressed as:

$$\delta\nu(\nu, \ell) = [f(\nu, \ell)]_{0,r_e} + [a\nu + b]_{r_e,r_i} + [g(\nu)]_{r_i,R_\odot} \qquad (1)$$

where the different functional forms describe the dependence of $\delta\nu$ on ν and ℓ in different regions located at different depths inside the Sun. In the present analysis $r_e = 0.735R_\odot$ and $r_i = 0.95R_\odot$.

In so doing the source of inaccuracies in the models can be identified by looking at the behavior of $\delta\nu$ as frequency function for the modes considered.

The dependence of the frequency differencies between CM and MLT models and observations on the observed frequencies, for modes with $0 \leq \ell \leq 11$, is dominant in both the models, thus indicating inaccuracies in the treatment of surface layers, especially above r_i=0.95 $R_\odot$ where $\delta\nu$ is a non-linear function of ν. However, the ν-dependence appears less pronounced in the CM model than in the MLT model. This result indicates that the CM model represents an improved approach to the problem of stellar convection with respect to the classical one. Moreover, since all the curves are not parallel, a small ℓ-dependence cannot be ruled out, thus indicating that also the physics of the interior needs revisions.

When the MLT and CM eigenfrequencies are compared with the observed values for some high degree modes with ℓ ranging from 40 to 1300, the CM eigenfrequencies appear, in general, closer to the observations than the MLT eigenfrequencies for all the ℓ values examined. The deviations of the theory from observations for $\ell > 400$ are in any case positive. This indicates that the theoretical sound speed in the convection zone is overestimated by both the CM and the MLT, although in the CM model the deviations are sensibly smaller. The largest deviations occur in the $1000 < \ell < 1300$ range, which concerns modes trapped in a very thin surface layer (0.006 $R_\odot$).

CONCLUSION

The present analysis has shown that both the CM and the MLT models still suffer from inaccuracies especially in the treatment of the surface layers, although some problems concerning the interior cannot be ruled out.

As a final result we infer that, compared with the MLT model, the new CM model predicts eigenfrequencies closer to the observed values. In any case some improvements to the model are required.

REFERENCES

Böhm - Vitense, E. 1958, *Z.Astrophys.*, **46**, 108.

Canuto, V.M., Mazzitelli, I. 1991, *ApJ*, **370**, 295 (cited as CM).

Canuto, V.M., Mazzitelli, I. 1992, *ApJ*, **389**, 724

Dziembowski, W.A., Paterno', L., Ventura, R. 1988, *A&A*, **200**, 213.

Libbrecht, K.G., Woodard, M.F., Kaufman, J.M. 1990, *ApJS*, **74**, 1129.

Mazzitelli, I. 1989, *ApJ*, **340**, 249.

Inside the Stars, IAU Colloquium 137
ASP Conference Series, Vol. 40, 1993
Werner W. Weiss and Annie Baglin (eds.)

STEADY AND FLUCTUATING PARTS OF SUN'S INTERNAL MAGNETIC FIELD

M. H. GOKHALE and K. M. HIREMATH
Indian Institute of Astrophysics, Bangalore 560 034, India

ABSTRACT We have modeled the 'steady' part of sun's internal poloidal magnetic field in the form of a central dipole and a central hexapole with strengths $(0.6\pm0.1)B_oR_o^3$ and $(0.16\pm0.05)B_oR_o^5$ embedded in an asymptotically uniform field B_o. A small deviation from isorotation seems to indicate: (i) a slow build-up of toroidal field near the base of the convection zone and (ii) presence of torsional MHD perturbations in the outer radiative core, with latitudinal structure and time scales which may be similar to those of solar activity.

1. INTRODUCTION

In this paper we determine a likely configuration of an axi-symmetric poloidal magnetic field that can remain in a 'steady' state with the helio-seismologically determined internal rotation of the sun.

According to Cowling's theorem, the resistive and the inductive terms of this steady part of the real field must vanish separately. Thus the steady part of the internal field must be in isorotation (Ferraro 1937) with the steady part of the rotation of the plasma. This requires existence of functional relation between the 'observed' rotation rate $\Omega(\mathbf{r})$ at a point $\mathbf{r}$ and the flux function $\phi(\mathbf{r})$ of the 'steady' poloidal field.

2. METHOD

The relation between the rotation velocity $\Omega(r,\vartheta)$ and the magnetic flux due to the internal and external sources is assumed to be linear, so that :

$$\Omega(r,\vartheta)=\Omega_o+ \Omega_1x^2\sin^2\vartheta + \Omega_2[(2\mu_1x^{-1} + 4\mu_3x^{-3} +\dots)\sin^2\vartheta + (-5\mu_3x^{-3}+\dots)\sin^4\vartheta+\dots] \quad (1)$$

where x is the distance from the center in the unit of the solar radius R_0, ϑ is the co-latitude, $\Omega_o, \Omega_1, \Omega_2$ are constants and μ_ℓ, ℓ=1,3,5,...are the strengths of the multipoles, in the standard expansion of the magnetic potential, in units of $B_o R_0^{\ell+2}$ where B_o is the asymptotically uniform field at large distances. Isorotation requires $\Omega_1 = \Omega_2$.

3. THE DATA USED

We use the rotation data from the internal rotation $\Omega(\mathbf{r})$, given by Christensen-Dalsgaard and Schou (1988). We determine the coefficients $\Omega_o, \Omega_1, \mu_1, \mu_3$, etc., by obtaining least square fits for successive combinations of terms in equation (1) using the data points in convective envelope (CE), outer radiative core (ORC) and combination of these data ('ORC+CE').

4. RESULTS AND CONCLUSION

The least squre fit to the 'observed' $\Omega(r,\vartheta)$ in 'CE' yields:

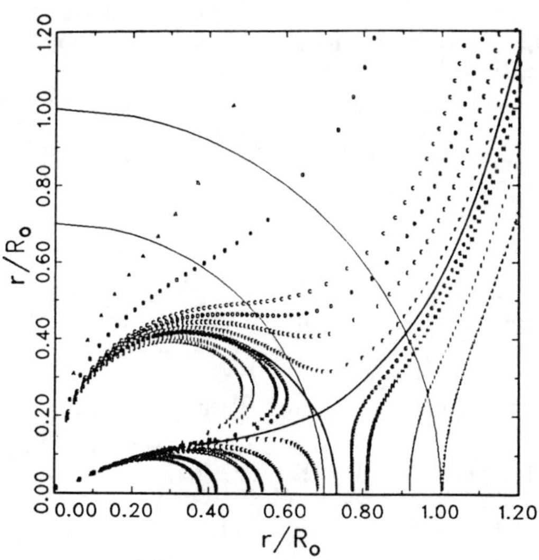

Fig.1. Structure of the 'steady ' part of the poloidal field given by μ_1=0.6 and μ_3=0.16 in one quadrant of a meridian plane. The field lines A,...,J correspond to various flux values from 0.5 to 2.1 units. The continuous lines represent branches of the separatrix : ϕ = 1.84 units.

Ω_o=326±12 nHz/unit flux, $\Omega_1=\Omega_2$= 68±11 nHz/unit flux, μ_1=0.6±0.1 and μ_3=0.16±0.05 . The strength B_o cannot be determined from this fit itself, but can be independently estimated to be in the range of 10^{-4} - 1G.

The resulting field structure is shown in Fig. 1. Field directions will be relative to that of B_o which is presently unknown. Outer part of the structure will be deformed by several processes. The field structure also contains a separatrix. Revolution of its branch running pole-wards near the center, and along the inner boundary of 'CE' in low latitudes, gives the critical surface S_* , which rotates with angular velocity Ω_* =430 nHz. Inside S_* the rotation is 'rigid', with $\Omega(r,\vartheta) \approx \Omega_*$. However the fit in 'ORC+CE' indicates that near S_* it can be approximated by equation (1) with Ω_2=75 nHz/unit flux. The 'non-isorotation' due to the difference ~ 7 nHz/unit flux between Ω_1 and Ω_2 gives a time scale ~ 10^5 - 10^9 year for 'winding' of the poloidal field into a toroidal field B_T ~ 10^6G near S_* . The fit in 'ORC+CE' also yields a small μ_5, implying presence of time dependent torsional MHD perturbations, in ORC, with dominant term ℓ = 5. For these perturbations the non-isorotation near S_* yields 'periods' in the range ~ 1-100 year depending upon the ratio of perturbations in the toroidal and poloidal fields. It is known that in solar magnetic cycle the dominant term is ℓ = 5 (Gokhale, et. al 1992 ; Stenflo, J.O. 1988).

On long time scales, the field may also provide outward transport of angular momentum and magnetic flux.

REFERENCES

Christensen-Dalsgaard, J., & Schou, J. 1988, in Seismology of the Sun and Sun-like Stars, ed. V. Domingo and E.J. Rolfe ESA : SP-286, p.149.

Ferraro, V.C.A. 1937, M.N.R.A.S., 97, 457.

Gokhale, M. H., Javaraiah, J., Kutty, K. N., & Varghese, B. A. 1992, Solar Phys, 138, 35.

Stenflo, J. O. 1988, Astrophys. & Space Sci., 144, 321.

Inside the Stars, IAU Colloquium 137
ASP Conference Series, Vol. 40, 1993
Werner W. Weiss and Annie Baglin (eds.)

HELIOSEISMIC CONSTRAINTS ON THE STRUCTURE OF THE PRESENT SUN

V.A.BATURIN AND S.V.AJUKOV
Sternberg Astronomical Institute, Moscow 119899, Russia

INTRODUCTION

Given the chemical composition (essentially the hydrogen abundance), mass and the physics of the interior, the structure of a star is uniquely determined. The usual method for the determination of the hydrogen-abundance profile of the present Sun is *via* evolutionary calculations. But the existing uncertainties in the physics and evolutionary history suggest considering a more extended set of possible solar models. Here, we examine models with solar values of radius, luminosity and mass, but with arbitrary profiles of the hydrogen abundance in the core. Of course, the profiles $X(m/M)$ must satisfy some (rather weak) restrictions, so that the solar values (determined by fitting) can be attained.

We assume the hypothesis that within fixed physics (opacity, thermodynamics, nuclear reactions) there is for all solar models a rigorous relation between the surface helium abundance and the entropy of the adiabatic part of the convective zone. We verify this hypothesis with models using somewhat simplified physics (Christensen–Dalsgaard 1988), and with six evolutionary models computed by Christensen–Dalsgaard (1992). The relation mainly depends on the opacity in the radiative zone between the lower boundary of convection and the region of nuclear reactions. We discuss the possibility of an agreement between these relations and the helioseismic determination of the surface helium abundance.

SOLAR MODELS WITH FIXED ENVELOPE

Let us consider solar models with fixed chemical composition in the convective envelope (hydrogen and helium abundance X_0, Y_0, respectively). To obtain the solar radius R and luminosity L we vary the hydrogen profile $X(m/M)$ in the core as well as the entropy s of the convective envelope. We use the specific entropy s in the adiabatic part of the convective zone as the parameter of the solar envelope instead of the more conventional mixing-length parameter of convection α. The reason is that the value of α depends on the structure of the outermost layers (in particularly on opacity). The specific entropy s is in addition related to the depth of the convection zone depth, but this quantity is also affected by opacity. To compute the specific entropy we use the expression for the free energy of Mihalas et al. (1988), but with omitted electron degeneracy:

$$s = -\left(\frac{\partial F}{\partial T}\right)_{V,N} = \left(k\sum_s n_s\left(\frac{5}{2}+\ln\frac{g_s}{n_s\Lambda_s^3}\right)+\frac{4}{3}aT^3\right)\Big/\rho$$

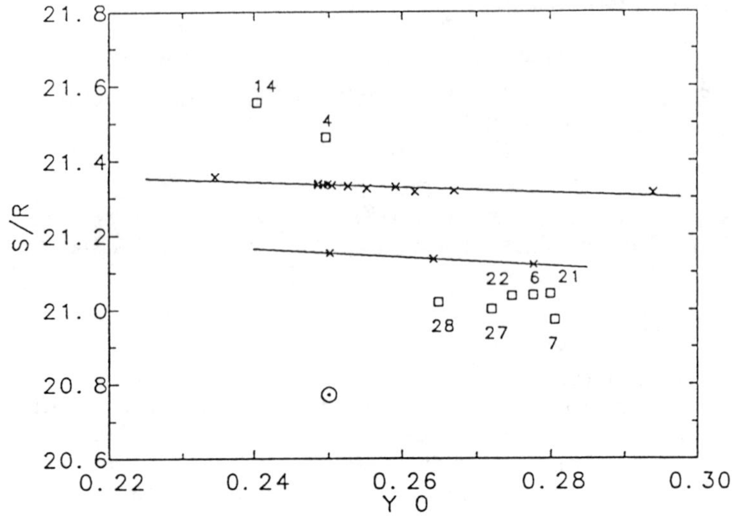

FIGURE I Specific entropy s/R (R: gas constant) and surface helium abundance. Crosses connected by lines represent the $Y_0 - s$ relation of our models. Labelled squares denote the models from Christensen–Dalsgaard 1992. ⊙ represents the helioseismic determination of the envelope parameters (Vorontsov et al. 1991).

where the sum is taken over all particles, g_s are the partition functions for atoms and ions ($g_s = 2$ for electrons). Λ_s denotes the thermal wavelength, equal to $h/\sqrt{2\pi kTm_s}$.

We found as a result that the best fit of s turned out to be the same for these models, and all our attempts to construct a solar model with another value of entropy, but constant Y_0 in the convective envelope, failed. In other words, although there are many models with different $X(m/M)$ profiles in the interior, and the same Y_0 in the envelope, they have all the same entropy s.

RELATION BETWEEN HELIUM AND ENTROPY IN ENVELOPE

We have also computed solar models with various values of the helium abundance Y_0 of the envelope. Fitting to solar R and L is again possible, but these models have another entropy s in the convective zone (and consequently another depth of the convective zone d_b). However, for a given helium abundance of the convective zone Y_0 only one s is allowed, i.e. there is a relation between Y_0 and the entropy s for a set of models with various $X(m/M)$ profiles in the interior (plotted on Figure I by crosses).

To investigate the sensitivity of this $Y_0 - s$ relation to the internal physics we have computed models with the opacity altered locally a) in the energy-generating core and b) in the radiative zone (denoting the region between the energy-generating core and the bottom of the convective zone). If the opacity is changed in the core (where luminosity is not constant), the $Y_0 - s$ relation is preserved. However, changing the opacity in the radiative zone shifts the $Y_0 - s$ curve as a whole. Thus the opacity in the radiative zone is one of the factors determining the position of the curve on $Y_0 - s$ plane.

Our calculations were performed with simplified model physics (according to Christensen–Dalsgaard, 1988). The helium abundance and entropy of our evolutionary model do not agree with those obtained in recent evolutionary calculations (Kim et al. 1991, Sackmann et al. 1990, Courtaud et al. 1990, Christensen–Dalsgaard 1991). To reveal the $Y_0 - s$ relation for other models we have calculated specific entropy for 8 models computed by J. Christensen–Dalsgaard (1992). We find a similar relation for three models with different age (22, 6, 21) and for three models with corrected opacity in the core (7, 27, 28). Model 4 differs from model 6 due to another opacity and model 14 from model 4 because of the inclusion of Coulomb interactions. But all these models have too-high an entropy compared with the value of the helioseismic determination by Vorontsov et al. (1991) (marked by ⊙ on figure). Preliminary investigations indicate that to reach a low entropy and the required helium content it is necessary to increase opacity in the radiative zone (by a factor of 1.5 or 2) compared with our opacity (based on the simple formulae from Christensen–Dalsgaard 1988). Such an increase would entail a change of opacity and (or) the profile of the hydrogen abundance in the core.

ACKNOWLEDGMENTS

We thank J.Christensen–Dalsgaard for stimulating discussions and for making his models available to us. We also thank W. Däppen for his help with this poster.

REFERENCES

Christensen–Dalsgaard, J. 1988, *Computational Procedures for GONG Solar Model Project.*

Christensen–Dalsgaard, J. 1991, in *Challenges to Theories of the Structure of Moderate-mass Stars*, pp.11-36, ed. D.O.Gough and J.Toomre (Lecture Notes in Physics, **388**, Springer, Heidelberg, 1991).

Christensen–Dalsgaard, J. 1992, *Geophys. Astrophys. Fluid Dynamics*, in press.

Courtaud, D., Damamme, G., Genot, E., Vuillemin, M., and Turck-Chièze, S. 1990, *Solar Phys.*, **128**, 49-60.

Kim, Y.-C., Demarque, P., and Guenther D.B. 1991, *ApJ*, **377**, 407-412.

Mihalas, D., Däppen, W., and Hummer, D.G. 1988, *ApJ*, **331**, 815-825.

Sackmann, I.-J., Boothroyd, A.I., and Fowler W. 1990, *ApJ*, **360**, 727-736.

Vorontsov, S.V., Baturin, V.A., and Pamyatnykh, A.A. 1991, *Nature*, **349**, 49-51.

Inside the Stars, IAU Colloquium 137
ASP Conference Series, Vol. 40, 1993
Werner W. Weiss and Annie Baglin (eds.)

TWO-DIMENSIONAL HELIOSEISMIC INVERSIONS

J. SCHOU
Institut for Fysik og Astronomi, Aarhus Universitet, Denmark, and
High Altitude Observatory, NCAR *, Boulder, Colorado, USA

J. CHRISTENSEN-DALSGAARD
Institut for Fysik og Astronomi, Aarhus Universitet, Denmark

M. J. THOMPSON
Astronomy Unit, Queen Mary and Westfield College, London, England

INTRODUCTION

We present results of investigations of different inversion methods for inferring from helioseismic observations the rotation rate in the solar interior as a function of radius and latitude, using a mode set similar to that which is expected from the GONG network.

The rotation of a star perturbs the frequency ω_{nlm} of a normal mode by

$$\Delta\omega_{nlm} = \omega_{nlm} - \omega_{nl0} = \int_0^R \int_0^\pi K_{nlm}(r,\theta)\Omega(r,\theta) r dr d\theta \,, \qquad (1)$$

where n, l and m are the mode 'quantum' numbers, θ the colatitude, Ω the rotation rate and K_{nlm} a known function. Until now the observations have not been in terms of individual $\Delta\omega_{nlm}$'s; rather, coefficients $a_i(n,l)$ of a parametrization of the m-dependence of $\Delta\omega_{nlm}$ have been given. By representing the latitudinal dependence of the rotation rate in terms of an expansion in $\cos\theta$, an inversion of the a_i can be carried out as a series of 1-dimensional inversions. Such inversions have been dubbed "1.5-dimensional". Our present implementation of a 1.5D procedure follows Schou *et al.* (1992), and uses a_1, a_3 and a_5.

By contrast a full 2-dimensional inversion uses all the individual $\Delta\omega_{nlm}$. Our implementation minimizes a discretized version of

$$\sum_i \left(\frac{\int K_i \bar{\Omega} r d\theta dr - \Delta\omega_i}{\sigma_i}\right)^2 + \mu \int \left[f(r,\theta)\left(\frac{\partial^2\bar{\Omega}}{\partial r^2}\right)^2 + g(r,\theta)\left(\frac{\partial^2\bar{\Omega}}{\partial\theta^2}\right)^2 \right] d\theta dr \qquad (2)$$

(Schou 1991; cf. Sekii 1990, 1991). Here $_i$ stands for the combination $_{nlm}$, μ is a trade-off parameter, σ_i are the errors on the observed splittings, f and g are functions of r and θ that can be specified (here $f = 1$ and $g = r^{-4}$), and $\bar{\Omega}$ is the rotation rate inferred from the inversion.

---

* NCAR is sponsored by the National Science Foundation.

RESULTS

Figure 1 shows the inferred rotation rate $\bar{\Omega}(r_0, \theta_0)$ obtained with the two methods from splittings computed for an assumed "true" rotation rate. We have used a mode set comprising all modes with $l = 0 - 200$ and $\nu = 1 - 4$mHz, which is roughly what is expected from the GONG network. The assumed errors were estimated from the mode linewidths, and should represent the errors in real data. The trade-off parameters have been adjusted to obtain the same error for both methods. Superficially both methods reproduce the original rotation rate with reasonable precision. Error estimates show that with a realistic dataset one should be able to determine the rotation rate with considerable precision: for example, the standard error in $\bar{\Omega}$ is approximately 1 nHz at $r_0 = 0.5R$ on the equator.

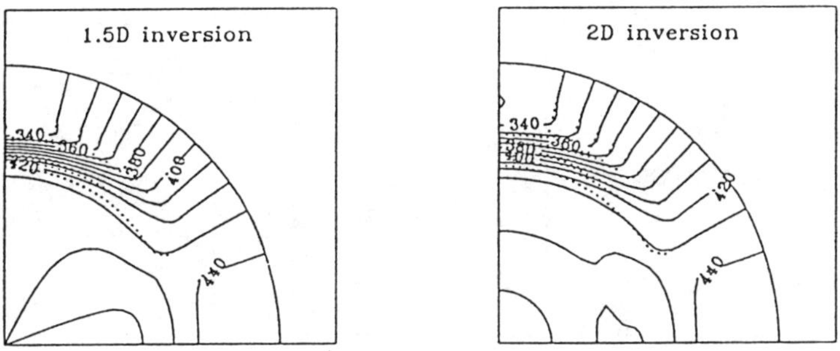

Fig. 1. The inferred rotation rate from 1.5D and 2D inversions, in a single quadrant with the North pole towards the top. Contours are labelled by the rotation rate, in nHz. Dotted contours show the "true" rotation rate.

For both methods *averaging kernels* $\mathcal{K}(r_0, \theta_0, r, \theta)$ exist which relate the inferred and true rotation rates (Schou *et al.* 1992):

$$\bar{\Omega}(r_0, \theta_0) = \int_0^R \int_0^\pi \mathcal{K}(r_0, \theta_0, r, \theta)\Omega(r, \theta) r dr d\theta \,. \tag{3}$$

As illustrated in Figures 2 and 3 these provide a more sensitive impression of the properties of the inversions. It is evident that the 2D inversion has better latitudinal resolution: the reason is the 2D method uses information that is effectively thrown away when expanding in a-coefficients. Note that the improvement in resolution is achieved with the same errors in the data, and similar errors in the inferred $\bar{\Omega}$, as for the 1.5D case.

REFERENCES

Schou, J., 1991, In *Lecture Notes in Physics* vol. **388**, 93 – 96, eds D. O. Gough & J. Toomre, Springer, Berlin.

Schou, J., Christensen-Dalsgaard, J., & Thompson, M. J., 1992, *Ap. J.*, **385**, L59 – L62.

Sekii, T., 1990. In *Lecture Notes in Physics*, vol. **367**, 337 – 340, eds Osaki, Y. & Shibahashi, H., Springer, Berlin.
Sekii, T., 1991. *Publ. Astron. Soc. Japan*, **43**, 381 – 411.

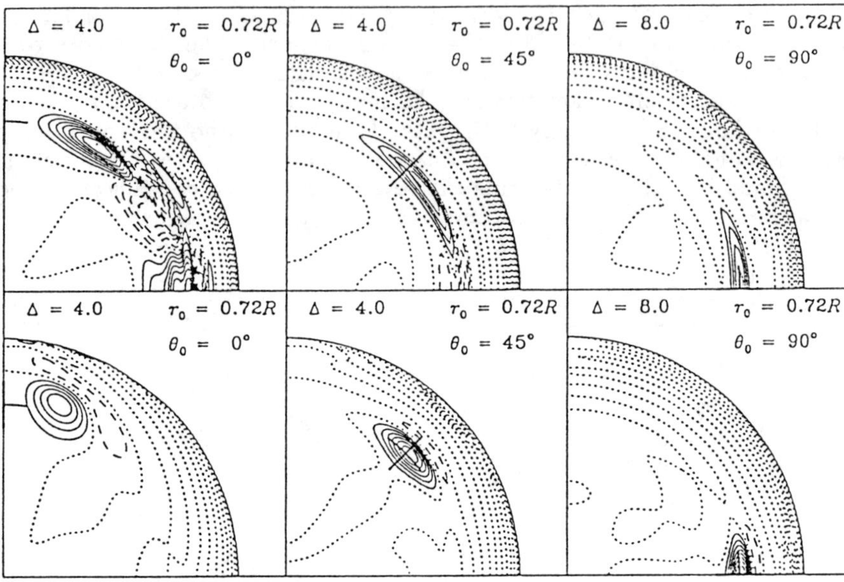

Fig. 2. Averaging kernels $\mathcal{K}(r_0, \theta_0, r, \theta)$ for the 1.5D (top) and 2D (bottom) inversions. Solid, dotted and dashed lines show contours for positive, zero and negative values, respectively. Heavy crosses show the location of the target position (r_0, θ_0); Δ is the contour spacing.

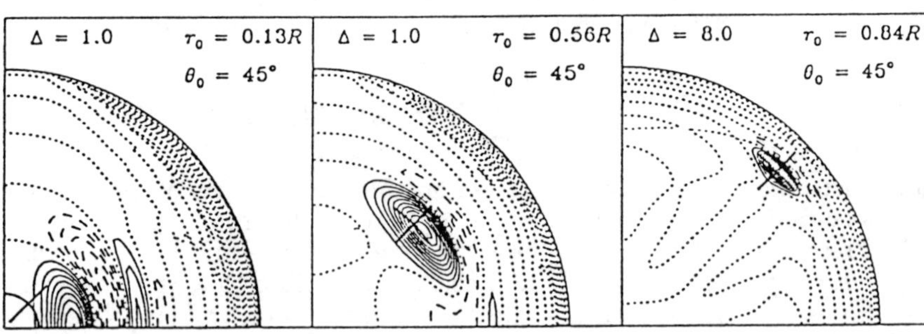

Fig. 3. Averaging kernels $\mathcal{K}(r_0, \theta_0, r, \theta)$ for the 2D inversion, at $\theta_0 = 45°$, for three different target radii. See the caption to Figure 2 for explanations.

Inside the Stars, IAU Colloquium 137
ASP Conference Series, Vol. 40, 1993
Werner W. Weiss and Annie Baglin (eds.)

ACTIVITY CYCLE IN THE SOLAR CORE

ATTILA GRANDPIERRE
Konkoly Observatory, 1525 Budapest, P.O. Box 67, Hungary

ABSTRACT I analyze here the recent solar seismology data relevant to the core regarding the energetics. The result is that the magnetic torsional oscillations are hardly able to support the observed rate of change of the solar core rotation. Interpreting the solar neutrino data as directly coming from the core and showing the activity cycle of the core, the energetics of the core can be balanced. I show here a simple model for the global and local activity-related changes of the solar core. The parameters of this model are consistent with the values derived by the Convective Flare Theory. This model is able to interpret the anticorrelation of the sunspot number with the neutrino flux, the related changes in the rotation of the core and the first which interprets the observed activity-related element abundance anomalies with a model for the solar core.

ROTATIONAL ENERGY CHANGES OF THE CORE

Goode and Dziembowski(1991) inverting the solar seismology data suggest that the rotation of the solar core anticorrelates with the solar cycle. By their data the rotation rate of the solar core at 0.4 solar radius is $\omega_1 = 485$ nHz in 1986, near to solar minimum, and $\omega_2 = 430$ nHz in 1988, close to the solar maximum. I estimate here roughly the related changes in the rotational energy, ΔE_ω, of the core if the core rotates rigidly.

$$\Delta E_\omega = \frac{1}{2} \times \frac{2}{5} m R_c^2 \times (\omega_1^2 - \omega_2^2) = 4.3 \times 10^{40} erg.$$

Here m is the mass of the rotating core $R \leq 0.4 R_c$, $R_c = 0.4 R_\odot$. If we want to argue, that the rotational energy changes are supplied by the magnetic fields, we would need a value of B = 100 000 G for the average magnetic field strength in the whole core! This magnetic field as a whole should have to be produced again in every cycle always with different polarities. This would give in an 11 year cycle a magnetic luminosity, assumed to be equal with the rotational luminosity, to be $1.2 \times 10^{32} ergs/sec$, 3% of the total solar luminosity! It is clear, that much more plausible to think, that the rotational changes of the core does not involve rotational energy changes.

A PULSATING CORE MODEL

Let us assume, that the rotation of the core is changing because of isobaric expansion and contraction of the core. In that case the temperature of the core should have to change with the core radius. If the core really changes with the cycle, than the activity cycle has to have on origin from the core. This temperature change of the core has to be present in the thermonuclear reaction rates. I propose that exactly this temperature change is the cause of the observed anti-correlation of the sunspot number with the solar neutrino flux (shown by Bieber et al.,1990). Let us assume, that the core of the Sun consists from two parts with different character. One is the main part of the core, which is as a whole, expands and contracts, and small parts around its surface, which produce from time to time local thermonuclear runaways (Grandpierre, 1990, 1991). In this case the luminosity of the Sun is produced by these two parts together, by the luminosity of the quiet core, L_c, and the luminosity of the flaring part of the core, L_f:

$$L_\odot = L_c + L_f .$$

The high energy neutrinos, detected by the Homestake neutrino detector, are produced with the 18th power of the temperature. For these neutrinos the observed neutrino flux Φ_ν^{obs} is about 1/4 of the theoretical value Φ_ν^{theor}.

$$\frac{\Phi_\nu^{obs}}{\Phi_\nu^{theor}} \approx \frac{1}{4} \approx \left(\frac{T_{q.c}}{T_{c.st}}\right)^{18} .$$

where $T_{q.c}$ and $T_{c.st}$ are the average temperatures of the quiet part of the 'flaring core' and of the theoretical, standard core. The flaring part of the core does not produce neutrinos, because on these high temperatures the proton - proton cycle cannot be completed. One can estimate the luminosity of the quiet part of the flaring core assuming that the luminosity goes with the fourth power of the temperature,

$$L \propto T^4 \qquad L_{q.c} \approx \left(\frac{1}{4}\right)^{\frac{4}{18}} \times L_\odot \approx 0.735 L_\odot .$$

The flare luminosity of the core L_f is the solar luminosity $L_\odot$ minus the luminosity of the non-flaring part of the core,

$$L_f = L_\odot - L_c = L_\odot \times \left(1 - \left(\frac{1}{4}\right)^{\frac{4}{18}}\right) \approx 0.265 L_\odot = m \times \epsilon_f = \frac{4\pi}{3} r^3 \rho_c \times \epsilon_f ,$$

where ϵ_f is the specific energy production coefficient of the flaring volume and its value is $5 \times 10^{17} erg/sec$, m is its mass and ρ_c is the average density of the core, $60 g/cm^3$. From the ratio $\Phi_\nu^{obs}/\Phi_\nu^{theor}$ we got L_f, with m and ϵ_f we can got the linear size r of the flaring volume, and with the density its mass as

$$r(\text{from the neutrino flux}) \approx 2 \times 10^4 \; cm, \quad and \quad m = 1.8 \times 10^{15} g.$$

And the change in the temperature is only $\Delta T = 7.4\%$.

These flares in the solar core are triggered by some kind of movement of electric charges in a local magnetic field (Grandpierre, 1990), e.g. by planetary tides. Then the runaway shoots up the material of the flaring volume when the local actual Rayleigh number surmounts the critical value of the Rayleigh number as the linear size of the flaring volume grows (Grandpierre, 1984). We can estimate the size of the flaring volume from this theory, too, as

$$d_{crit} = 2\pi \left(\frac{\kappa \nu}{g \alpha \beta} \right)^{\frac{1}{4}},$$

$$d_{crit} \approx 2\pi \left(\frac{2.64 \times 10^{11} \times 1.36 \times 10^{-3}}{2.4 \times 10^3 \times 10^{-7} \times 4 \times 10^{-4}} \right)^{\frac{1}{4}},$$

$$r(\text{from the Convective Flare Theory}) \approx 2.45 \times 10^4 \; cm.$$

We can see that the two values for the size of the flaring volume are in good agreement. In taking for $\Phi_\nu^{obs}/\Phi_\nu^{theor} = 1/3$ only 3% changes is necessary in the temperature of the core. At the same time, it is clear, that the flare luminosity of the core is large enough to initiate the observed changes in the rotation of the core.

This phenomenon of the flaring core is the one which is able to interpret the flare related chemical anomalies of the Sun as the enhancement of the He^3/He^4 ratio with a factor of 10^4. The observed Fe-enhancement also calls for a process arising above $10^9 K$ and high density.

REFERENCES

Bieber, J. W., Seckel, D., Stanev, T. and Steigman, G., 1990, *Nature*, **348**, 407.

Goode, Ph. R. and Dziembowski, W.A., 1991, *Nature*, **349**, 223.

Grandpierre, A.,1984, in *Theoretical Problems in Stellar Stability and Oscillations*, eds. Noels, A. and Gabriel, M., Liege, 48.

Grandpierre, A., 1990, *Sol. Phys.*, **128**, 3.

Grandpierre, A., 1991, *Mem. Soc. Astr. Ital.*, **62**, 401.

Inside the Stars, IAU Colloquium 137
ASP Conference Series, Vol. 40, 1993
Werner W. Weiss and Annie Baglin (eds.)

GENERATION OF STRONG TOROIDAL MAGNETIC FIELD NEAR THE BOTTOM OF THE SOLAR CONVECTIVE ZONE

V.N.KRIVODUBSKIJ
Kiev University Astronomical Observatory
Observatorna St. 3, Kiev-53 254053, Ukraine

Abstract. The generation mechanism of the toroidal magnetic field by the angular velocity radial gradient acting on the relict poloidal magnetic field on the boundary between the convective and radiative zones is proposed. The magnetic induction magnitude of the toroidal field reaches about 2×10^6 G, the limiting effect of the magnetic buoyancy being taking into account. This value conforms to the estimation of toroidal field obtained from helioseismological data.

Studying the helioseismological data, recently Dziembowski and Goode (1989) have shown that there is strong axisymmetric quadrupole toroidal field of magnetic induction $(2 \pm 1) \times 10^6$ G centred near the base of the convective zone. The question about the origin of this magnetic field arises. In connection with fast upwelling to the surface of intensive magnetic field due to the magnetic buoyancy, it is necessary to show also how this field may reserve inside the Sun during long time. Here it is made the attempt to solve this problem with help proposed by us the mechanism of generation of the strong toroidal magnetic field in deep layers of the Sun.

According this mechanism, the radial gradient of angular velocity acts on the relict poloidal magnetic field $\underline{B}_p$ and excites the toroidal magnetic field $\underline{B}_t = B_\varphi \underline{i}_\varphi$ the magnitude of which is limited by ohmic dissipation and magnetic buoyancy (Dudorov et al. 1989; Krivodubskij 1990)

$$\partial \underline{B}_t / \partial t = r \sin\theta \, (\underline{B}_p \underline{i}_r \nabla \Omega) \underline{i} + \nu_m \Delta \underline{B}_t + \mathrm{rot}(\underline{V}_b \times \underline{B}_t). \quad (1)$$

Here $\nabla\Omega = \Delta\Omega/\Delta r$ is the radial gradient of angular velocity, $\Delta\Omega$ is the increment in the angular velocity over step Δr along solar radius $R_\odot$, ν_m is the magnetic viscosity, $\underline{V}_b \simeq U_T(\lambda/a)^2(B^2_\varphi/8\pi P_e)\underline{i}_r$ is the velocity of upwelling of a toroidal magnetic flux tube (MFT) in the region of the radiative energe transfer, $\underline{i}_\varphi$ and $\underline{i}_r$ are the azimuthal and radial unit vectors in the spherical coordinat system, θ is the complement of the latitude, r is the distance from the centre of the Sun, λ is the temperature scale height, a is the transverse radius of the MFT U_T is the mean velocity of upward transport of thermal energy, P_e is the electron pressure.

On the present stage of solar evolution the magnetic buoyancy plays the main role in the constraining of the magnetic induction of the toroidal field. Then from the condition of stationarity, $\partial B_\varphi/\partial t = 0$, neglecting ohmic dissipation, from Eq.(1) we obtain the following expression for maximum value of the resulting stationary toroidal field (Krivodubskij 1990, 1992):

$$\max|B_\varphi| = B^o_\varphi \simeq \left(\frac{\Gamma}{\gamma}\right)^{1/3} = \left[r\,\frac{\Delta\Omega}{\Delta r}\,B_p\,\frac{8\pi P_e L}{U_T}\left(\frac{a}{\lambda}\right)^2\right]^{1/3} \qquad (2)$$

which is excited in the radiative zone at the distance r from the centre of the Sun. Here $\Gamma = rB_p\Delta\Omega/\Delta r$ is the generation parameter, $\gamma = U_T/8\pi P_e L^p(\lambda/a)^2$ is the parameter of the rate of loss of induction due to magnetic buoyancy, L is the space scale of the magnetic field. The characteristic time of reaching of stationary magnetic induction B^o_φ is $t^o = B^o_\varphi/\Gamma = (1/\gamma\Gamma^2)^{1/3}$.

Dziembowski, Goode and Libbrecht (1989) from the oscillation data of Libbrecht (1989) found that there is the transition from differential rotation to riget rotation at the convective zone boundary with the radiative interior. Our analysis of the dependence $\Omega(r,\theta)$ showed that the maximum jump of angular velocity, $\Delta\Omega \simeq 1\times10^{-7}$ rad/s, occurs in the equatorial plane along solar radius on short piece length of radius, $\Delta r \simeq 0{,}74\ R_\odot - 0{,}67\ R_\odot = 0{,}07\ R_\odot \simeq 5\times10^9$ cm. Id est, there is the sharp radial gradient of angular velocity, $\Delta\Omega/\Delta r \simeq 2\times10^{-17}$ rad/s cm, in the sufficient narrow layer Δr. This evidences that at transition from convective zone to radiative interior there is the favourable conditions for effective generation of toroidal magnetic field in accordance with the mechanism above - mentioned.

When for physical parameters U_T, λ, P_e at the upper part of radiative zone we take values from internal structure model of the Sun (Allen 1973); for space parameters of toroidal field B^o_φ we take $L = 2a = \Delta r$; and for quadrupole mode of the observed poloidal component B_p we take value ~ 1 G in accordance with Stenflo (1988), then we obtain the following estimates for the parameter generation, Γ, and the parameter of loss of magnetic induction, γ, near the radiative zone top ($r \simeq 0{,}7 R_\odot$) at narrow layer $\Delta r \simeq 5\text{x}10^9$ cm: $\Gamma \simeq 1\text{x}10^{-6}$ G/s, $\gamma \simeq 2\text{x}10^{-25} G^{-2} s^{-1}$. For relatevely short time space $t^o \simeq 70{,}000$ years magnetic induction of excited toroidal field reaches maximum stationary value $B^o_\varphi \simeq 2\text{x}10^6$ G (at the diameter of MFT $d_t \leqslant 5\text{x}10^9$ cm which are oriented along the parallels of solar latitude with opposite sign on either side of equatorial plane). This value agrees with the helioseismological finding of Dziembowski and Goode (1989). The presence of such intensive toroidal field near convective zone bottom is essential for maintenance of dynamo-process in the region of turbulent convection.

REFERENCES

Allen, C.W. 1973, Astrophysical Quantities, Athlone Press, London.

Dudorov, A.E., Krivodubskij, V.N., Ruzmaikina, T.V. and Ruzmaikin, A.A. 1989, Astron. Zh., 66, 809 (Sov. Astron., 33, 420).

Duvall Jr., T.L. et al. 1984, Nature, 310, 22.

Dziembowski, W.A., Goode, P.R. 1989, Ap.J., 347, 540.

Dziembowski, W.A., Goode, P.R. 1990, Inside the Sun, Dordrecht, p.341.

Dziembowski, W.A., Goode, P.R., Libbrecht, K.G. 1989, Ap. J. (Letters), 337, L53.

Krivodubskij, V.N. 1990, Vestn. Kiev. Univ. Astronomia, 32, 3 (Russian).

Krivodubskij, V.N. 1992, Visn. Kyijv. Univ. Fiz.-mat. nauky, 3, 84 (Ukrainian).

Libbrecht, K.G. 1989, Ap. J. 336, 1092.

Stenflo, J.O. 1988. Ap. Space Sci., 144, 321.

Inside the Stars, IAU Colloquium 137
ASP Conference Series, Vol. 40, 1993
Werner W. Weiss and Annie Baglin (eds.)

DETECTION POSSIBILITY OF THE GIANT ROLLS IN THE SUN

TÜNDE BARANYI and ANDRÁS LUDMÁNY
Heliophysical Observatory of the Hungarian Academy of Sciences, H-4010 Debrecen, P.O.Box 30., Hungary

ABSTRACT The expected giant convection pattern has been studied by using its probable twisting influence on the emerging magnetic fields. The studies carried out on extended computations and material have confirmed the previously found evidence for a pattern with longitudinal wavenumber l=11 in the year 1977.

INTRODUCTION

The present work is the extended version of the study (Baranyi and Ludmány, 1992) which will be referred to as Paper I henceforth. As is described therein the work is based on the assumptions that (1) the internal magnetic field is basically toroidal and (2) the diverging and converging parts of the giant convection cell system can twist the azimuthally directed emerging magnetic fluxes on account of the Coriolis-force. The orientation data of the emerging active regions were used to look for the mentioned distributions in the year 1977 using Debrecen photoheliograms (Dezső et al., 1987). Clear spatial separation of positive (equatorward) and negative (poleward) tilts was searched.

SEARCH FOR DISTRIBUTION PATTERNS

The possible longitudinal patterns are considered to be columnar giant convection cell systems expected by the theory (Glatzmaier, 1984; Gilman and Miller, 1986). Many possibilities have been examined by varying the following parameters: the longitudinal wavenumber (l), difference of the internal and outer (Carrington) rotation rates ($\Delta\omega$) and curvature of the rolls. The search means a procedure for finding high coincidence of positive (negative) tilts with diverging (converging) fields respectively in the possible configurations.

The coincidence is described by a suitable parameter. Each sunspot group is characterized by a W weight. Consider various

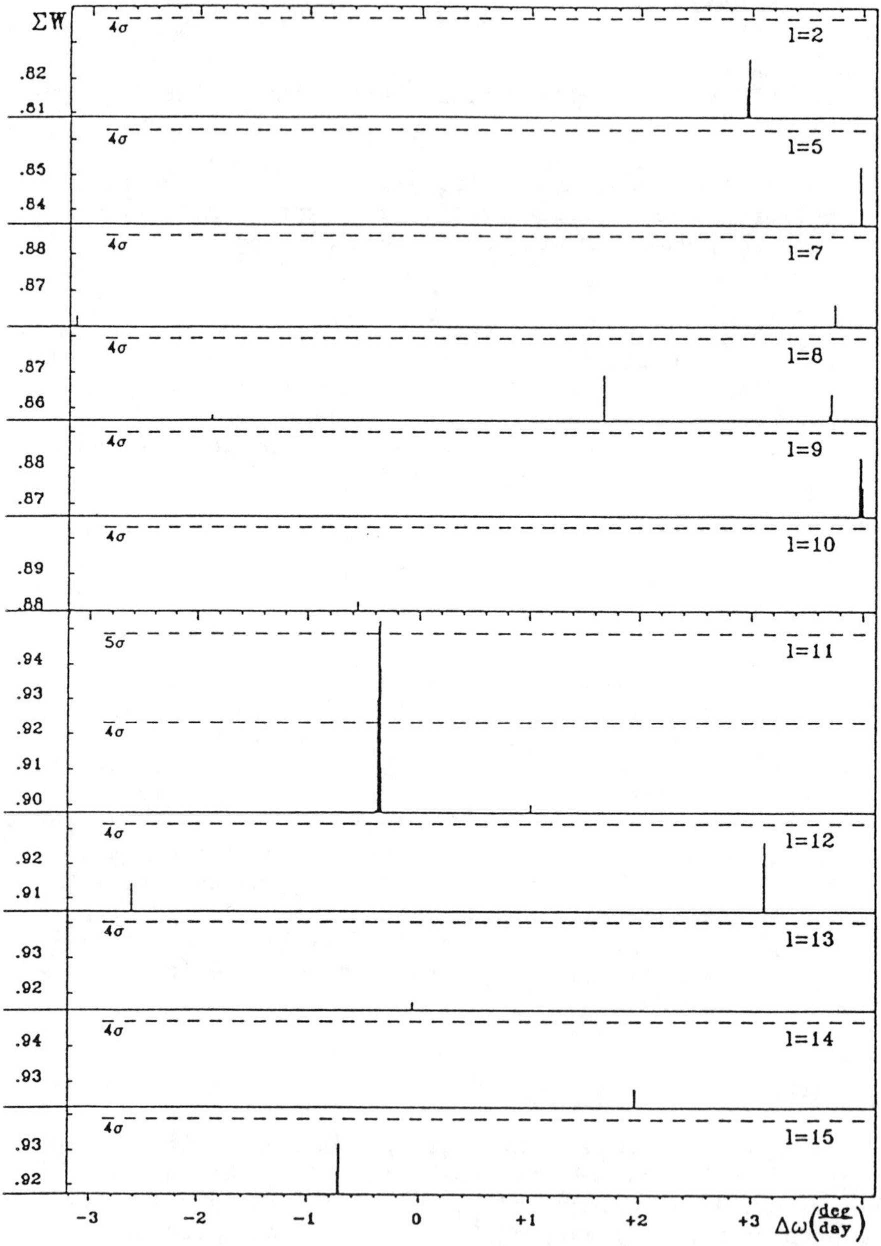

Fig. 1. Normalized sums of orientation weights (ΣW) characterizing the coincidence of the sunspot group tilts with various supposed internal sector structures as functions of the longitudinal wavenumber (l) and the difference of the internal-outer rotation rates (Δω). Only the values above the 3σ level are displayed. The band of high values at l=11 and Δω=-0.37 deg/day implies a probable internal structure.

hypothetical sector structures with alternating positive-negative sectors. If the sign of the W corresponds to the sign of the sector in which the given sunspot group emerges then the W is added to a sum of weights (ΣW), if not, it is omitted. Three kinds of definition were used and averaged for the W.

Definition 1. (used also in Paper I, allowing 76 groups):

$$W = \Sigma\ \alpha_i (A_i)^{0.5} \qquad i=1,2,3$$

α_i: tilt of the sunspot group dipole to the azimuth (degrees), A_i: area of the spots of the group (millionths of the solar hemisphere) on the i-th day of its existence. Tolerance strips of 2.5 degrees are allowed beyond either sector borders.

Definition 2. The W is the same as in definition 1, but only for the first bipolar day. This allows 88 sunspot groups.

Definition 3. corresponds to the definition 2, but the strip of tolerance is equal to one sixth of the sector width.

Two sector parameters (difference of internal-outer rotation rates ($\Delta\omega$), and sector curvature) have been computed for the same values as in the Paper I, but the wavenumbers (l) were varied from 2 (instead of 9) through 15. The highest values of ΣW have been chosen for all l and $\Delta\omega$ values, they have been normalized to the highest achievable ΣW values and averaged for the three definitions.

The Fig. 1 shows only the ΣW values exceeding the 3σ level (no such cases for l=3,4,6). The band of high maxima at l=11 and $\Delta\omega$=-0.37 degrees/day makes probable an internal sector structure characterized by these parameters and slight curvature in the given year and confirms the results and conclusions of Paper I which is based on more restricted data.

ACKNOWLEDGEMENTS

Thanks are due to J. Christensen-Dalsgaard, J. Schou and T. M. Brown for helpful discussions and to the "Foundation for the Hungarian Science" for financial support to Baranyi T.

REFERENCES

Baranyi, T. and Ludmány, A. 1992, Solar Phys., in press.

Dezső, L., Kovács, Á. and Gerlei, O. 1987, Debrecen Photoheliographic Results 1977, Publ. Debrecen Obs., Heliographic Series No.1.

Gilman, P. A. and Miller, J. 1986, Astrophys.J.Suppl. 61, 585.

Glatzmaier, G. 1984, J.Comp.Phys. 55, 461.

Inside the Stars, IAU Colloquium 137
ASP Conference Series, Vol. 40, 1993
Werner W. Weiss and Annie Baglin (eds.)

ASYMMETRY AND PERIODICITIES IN SOLAR ACTIVITY

M. CARBONELL, R. OLIVER AND J.L. BALLESTER
Departament de Fisica. Universitat de les Illes Balears. E-07071 Palma de Mallorca. Spain

ABSTRACT The historical record of daily sunspot areas (1874 - 1989) has been analysed, looking for the short (155 days) and intermediate (323 and 540 days) term periodicities. Also the North - South asymmetry during those years has been studied.

INTRODUCTION

During recent years the presence of short and intermediate term periodicities in several features of solar activity has been reported.

A 155-day periodicity has been detected in the solar flare occurrence rate, the sunspot area variations,the Zurich sunspot number and the aurorae record.

A 323-day periodicity has been reported in the Zurich sunspot number, in the solar diameter measurements, in the sunspot blocking function, in the 10.7 cm radio flux, and the CaII K plage index.

The 540-day periodicity has been found in the 10.7 cm radio flux, in solar flares, in sunspot areas during solar cycle 21 and the observations of the solar horizontal diameter.

On the other hand, the existence of a North-South asymmetry in solar activity is generally accepted, and it has been detected in most of its manifestations (solar flares, sunspots, magnetic flux, etc.) although a detailed study of its behaviour is still missing.

Sunspots are key features of solar activity, being a measure of the amount of magnetic flux emerging through the photosphere, and they have been one of the main indices of solar activity for a long time. For this reason, we have chosen the historical record (1878-1989) of daily sunspot areas to assess the presence of those periodicities and to study the behaviour of its North - South asymmetry.

DATA AND TECHNIQUES

The Greenwich Photoheliographic Results (1874-1982) and the Unites States Air Force (1983-1989) have provided us with the total daily area of visible sunspots expressed in millionths of the Sun's visible hemisphere. To look for the periodicities we have adopted the Lomb-Scargle Periodogram (Lomb, 1976; Scargle 1982), modified by Horne & Baliunas (1986), which is appropiate for the analysis of unevenly-spaced data. Scargle (1982) and Horne & Baliunas (1986) have provided with a convenient estimate (the false alarm probability, FAP) for the probability that a given peak be a true signal. Moreover, we have set our FAP at a height that implies a very restrictive condition for the peak's significance, i.e. the probability that a peak that exceeds the FAP level due to chance is smaller than 0.01%. The periodogram has been computed at 400 equally spaced frequencies over the frequency interval 57-115 nHz (100-200 days) and 19-39 nHz (300-600 days). This number of frequencies is sufficient to obtain the resolution of all the peaks present in the portion of the periodogram under study.

To analyse the behaviour of the asymmetry, we have constructed an asymmetry time series and studied it by means of the following methods: 1) Power spectrum and statistical analysis. 2) Montecarlo simulations. 3) Chaotic analysis. 4) Multifractal analysis.

RESULTS

The picture we have obtained from our study of daily sunspot areas is that a short periodicity between 150-160 days seems to be present, in a significant way, from cycle 16 to 21. The period changes from one cycle to the next when whole individual cycles are considered. However, during cycles 12, 14, 16, 17, 18, 19, 20, 21, a significant peak at 155.6 days is found in some time intervals, this being a feature suggesting intermittency. Also, a non coherence in the phase is found during solar cycles 16-18 which could be due to different times in the trigger of the periodicity suggesting, again, intermittency. If this short periodicity is a true feature of solar activity, it would imply that the emergence of magnetic flux presents a periodicity, perhaps intermittent. Thus, the periodic emergence of magnetic flux could trigger, via magnetic reconnection with the old flux, flares, giving place to the periodicity found in them.

Concerning the periodicity at 323 days, the results indicate that it is only confirmed for solar cycle 21 while its presence throughout the historical record of solar activity is very doubtful.

The situation is different for the periodicity around 540 days. Its presence is well established for cycle 19 and appears clearly in cycle 18. Also, during cycles 12, 14 and 17 there are indications of its presence.

The study of the North-South asymmetry in sunspot areas has produced the following results : 1) The asymmetry is a real feature in the North-South distribution of sunspot areas. 2) The asymmetry time series can be modelled by means of a multicomponent model made by a linear trend, a deterministic cycle with a period around 12.5 years and a dominant purely random component which accounts for most of the variance of the time series. 3) Statistical rank correlation tests confirm the existence of an anticorrelation between the asymmetry and the solar cycle. 4) Montecarlo simulations indicate that the real asymmetry cannot be obtained randomly from a binomial or uniform distribution of probability between hemispheres. 5) Chaotic analysis reveals that the asymmetry time series has a stochastic behaviour. 6) The multifractal properties of the asymmetry time series indicate that it presents temporal intermittency.

A more extense information about this research can be found in Carbonell & Ballester, (1990); Carbonell & Ballester, (1991); Oliver et al., (1992); Carbonell & Ballester, (1992) and Carbonell et al., (1992).

ACKNOWLEDGMENTS

We would like to acknowledge financial support from DGICYT under grant PS89-0148.

REFERENCES

Carbonell, M., Ballester, J.L., 1990, A&A 238, 377.

Carbonell, M., Ballester, J.L., 1991, A&A 249, 295.

Carbonell, M., Ballester, J.L., 1992, A&A 255, 350.

Carbonell, M., Oliver, R., Ballester, J.L., 1992 (Submitted).

Horne, J.H., Baliunas, S.L., 1986, ApJ 302, 757.

Lomb, N., 1976, Ap&SS 39, 447.

Oliver, R., Carbonell, M., Ballester, J.L., 1992, Sol. Phys. 137, 141.

Scargle, J.D., 1982, ApJ 263, 835.

Inside the Stars, IAU Colloquium 137
ASP Conference Series, Vol. 40, 1993
Werner W. Weiss and Annie Baglin (eds.)

23rd CYCLE OF SOLAR ACTIVITY IN THE LIGHT OF 34 YEARS OF CRACOW OBSERVATIONS OF SOLAR RADIO EMISSION

ADAM MICHALEC
Astronomical Observatory of the Jagiellonian University
ul. Orla 171, 30-244 Cracow, Poland

Systematic daily observations of solar radio emission were started at the Fort Skała Observatory on 1st October 1957. The observations were made at the frequency of 810 MHz first with a 5m radio telescope - and since 1964 with a 7m one; at present they comprise already almost 70 thousand hours of observations (Michalec 1991). During that period there were two longer interruptions in the observations: from 23 September 1963 to 2 February 1964 - due to transferring the radio telescope's antenna from the top of the Fort to a new site, and from 6 November 1973 to 11 March 1974 - connected with the reconstruction of the receiver. Other, fortunately short, interruptions in the course of observations had no influence on the homogeneity of the series.

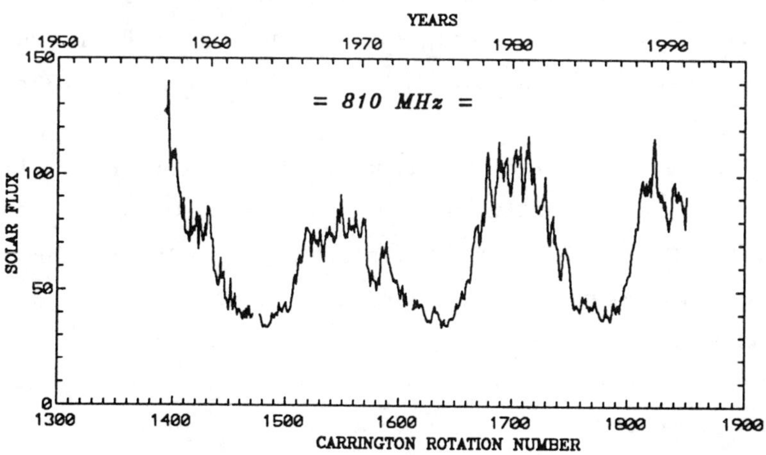

Fig.1 represents the mean values of solar radio emission in

the individual synodical rotations covering the period of 34 years of Cracow observations. In this diagrams the 11-year cycle of solar activity is distinctly pronounced.

Now, let us draw our attention to the secondary maxima appearing in the rotations: 1430 (year ~ 1960.6), 1589 (year ~ 1972.5), and 1746 (year ~ 1984.2) at the intervals of 159 - 157 rotations. If we accept the mean value 158, as the period of recurring secondary maxima, then the next expected secondary maximum will occur in the rotation no. 1904, i.e. in the year 1996.0.

The occurrence of secondary maxima was pointed out by Mc Intosh (Wilson 1988), on the basis of the Canadian observations of solar radio emission - on the 10.7 cm wave-length (2.8 GHz), started 10 years earlier than the Cracow observations.

These secondary maxima - if to accept the idea - constitute a reflection of the violent process of remagnetization of the old cycle's solar spots and of the appearance at high heliographic latitudes of the first spots belonging already to the new cycle of activity (Vitinsky at al. 1986)

The secondary maxima appear about 1.5 year after the phase of maximum activity and 3.5 years before the minimum (Michalec 1989). If this hypothesis were true, then on its base it could be forecasted that the minimum of the 22/23 cycle of solar activity will fall on about 1998.5 and the maximum of 23rd cycle on the year 2002.5.

The determination of fundamental moments in the 11-year cycles of solar activity, such as the minimum or maximum (Withbroe 1989) is extremely important, if only for purposes of planning long-term cosmic flights, manned or unmanned (Sky & Telescope 1990).

ACKNOWLEDGEMENTS

I am deeply grateful to all the observers who by performing the prolonged and painstaking observations of the Sun contributed to establishing an observational series long enough to make those far-reaching conclusions plausible.

REFERENCES

Michalec, A. 1989, *Solar Activity,* **7**, 8 *(Materials from Astronomical Sessions in Dąbrowa Górnicza - in Polish).*

Michalec, A. 1991, *Postępy Astronomii,* **39**, 90 *(in Polish).*

.... 1990, *Sky & Telescope,* **79**, 584.

Vitinsky Yu.I., Kuklin G.V., and Obridko, V.N. 1986, *Solnetschnye Dannye* **3**, 53 *(in Russian).*

Wilson, P.R. 1988, *Solar Physics,* **117**, 205.

Withbroe, G.L. 1989, *Journal of Spacecraft and Rockets,* **26**, 394.

Inside the Stars, IAU Colloquium 137
ASP Conference Series, Vol. 40, 1993
Werner W. Weiss and Annie Baglin (eds.)

PRELIMINARY RESULTS FROM OBSERVATIONS WITH THE FOURIER TACHOMETER

J. SCHOU
Institut for Fysik og Astronomi, Århus Universitet, Denmark, and High Altitude Observatory, National Center for Atmospheric Research *, Boulder, Colorado, USA

T.M. BROWN
High Altitude Observatory

K.T. BACHMANN
High Altitude Observatory

INTRODUCTION

Recently several interesting results concerning the time dependence of solar p-mode frequencies have appeared. In particular it has been shown that the mode frequencies change with solar activity (Libbrecht and Woodard 1991).

We present preliminary results from an analysis of two $\approx$ 3 month observation runs with the Fourier Tachometer (which was operated by HAO and NSO) from 1987 and 1989. The analysis was done with two different methods in order to test for systematic errors. We show that the mode frequencies change in a manner similar to that reported by Libbrecht and Woodard. We also present results for the frequency splittings caused by the solar rotation.

RESULTS

In both analysis methods the apodized velocity images are first projected onto a set of masks to isolate individual (m, l)'s. These time-series are then Fourier transformed. In the analysis method used for the results presented here a simultaneous maximum likelihood fit is done to all m's in a frequency range around each target (l, n). The fit is at each m a background plus a Lorenzian with a linewidth independent of m, a mode amplitude and a frequency of

$$\nu_{nlm} = \nu_{nl} + L \sum_{i=1}^{6} a_i P_i^l(\frac{m}{L}),$$

* The National Center for Atmospheric Research is sponsored by the National Science Foundation.

where the P's are polynomials similar to Legendre polynomials but designed to be orthogonal with respect to summation. The leakage from modes with different n, l or m is taken into account together with the correlations between m's introduced by this leakage. The other method does a simpler least squares fit not taking all the leakages etc. into account. The two methods generally give similar results. The differences, although systematic, are not large enough to be seen on most plots, and the interpretations given here do not depend on the method used.

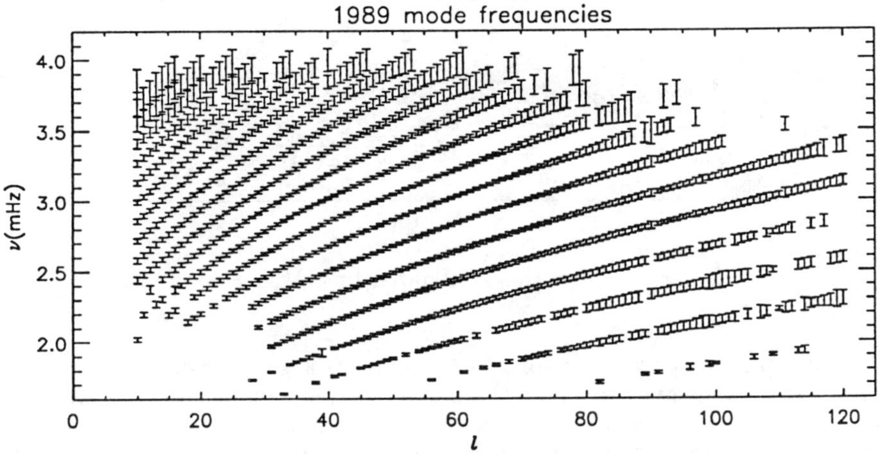

Fig. 1. The modes successfully fit for 1989. The error bars are 250 times the estimated standard errors.

DISCUSSION

The frequency differences in figure 2 are qualitatively similar to those of Libbrecht and Woodard, confirming their results. In particular it appears that most of the frequency changes scale like the inverse mode mass and hence are caused by perturbations in the structure close to the solar surface.

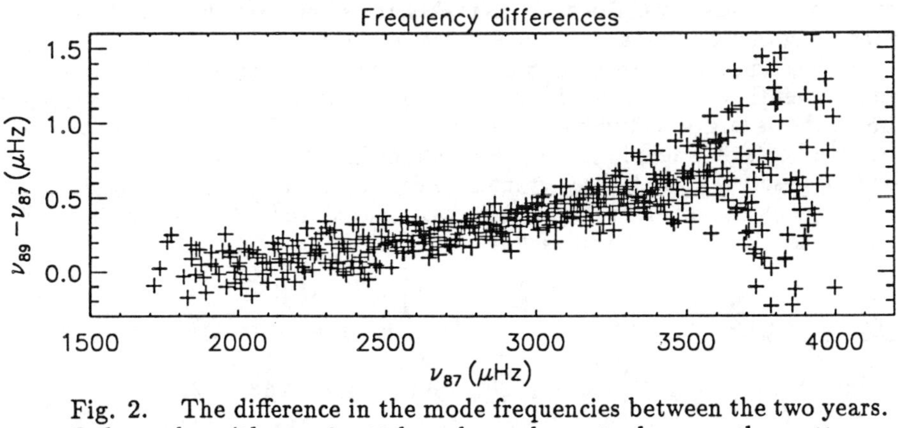

Fig. 2. The difference in the mode frequencies between the two years. Only modes with $20 \leq l \leq 60$ have been shown to decrease the scatter.

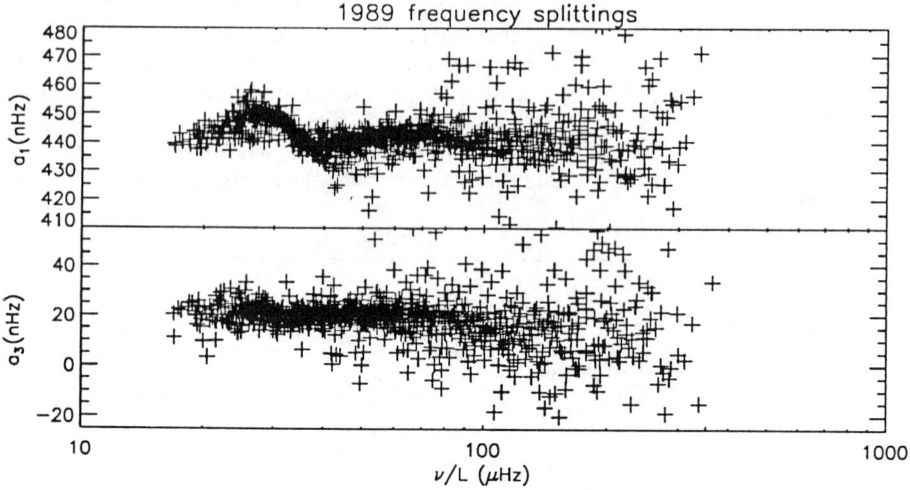

Fig. 3. a_1 and a_3 from the 1989 dataset as a function of ν/L.

The plot of a_1 as a function of ν/L (which is a measure of the turning point depth, and thus the region the mode samples) contain some interesting features below $\nu/L < 40\mu Hz$. We have not yet been able to identify the cause of these features. In particular we have no evidence that they are artifacts of the analysis methods used. A reason to be concerned, however, is that the region of the bump $25\mu Hz \le \nu/L \le 40\mu Hz$ corresponds with the range in which the frequency spacing between two neighbouring l's is equal to $(1\text{day})^{-1} = 11.57\mu Hz$ within a mode linewidth. Since the daily observation cycle causes sidelobes to appear around a given mode with this spacing and the individual (l,m)'s are not perfectly isolated, this could potentially cause problems. For the 1989 data we have tried such tests as dividing the time-series in three parts and analyzing them individually, decreasing the duty cycle of the observations (thereby increasing the 1/day sidelobes), changing the initial estimates for the mode parameters, change the assumed radius of the Sun, and distorting the images in various ways. None of this makes any significant changes. It is also interesting that the feature is very different in 1987.

The standard interpretation of the a's is that they are caused by the solar rotation. If they are inverted to obtain the internal solar rotation it turns out that it shows large oscillations as a function af radius. If this result for the rotation holds, it would indeed be very interesting. It is, however, such an extreme result that we feel reluctant to accept it. A comparison of results from several years and different instruments would seem useful before drawing wide-ranging conclusions from this result.

REFERENCES

Libbrecht, K. G. and Woodard, M. F., 1991, Science, **253**, 152.

Inside the Stars, IAU Colloquium 137
ASP Conference Series, Vol. 40, 1993
Werner W. Weiss and Annie Baglin (eds.)

SEISMIC EVIDENCE OF MODULATION OF THE STRUCTURE AND ANGULAR VELOCITY OF THE SUN ASSOCIATED WITH THE SOLAR CYCLE

D. O. GOUGH[1], A. G. KOSOVICHEV[2] AND T. SEKII
Institute of Astronomy, Madingley Road, Cambridge CB3 0HA, England

K. G. LIBBRECHT, M. F. WOODARD
Big Bear Solar Observatory, California Institute of Technology, Pasadena, CA 91125, USA

INTRODUCTION

Solar p-mode multiplet oscillation frequencies and expansion coefficients for degeneracy splitting have been measured in the summers of 1986 and 1988–1990. Temporal variations not unlike the differences between 1988 and 1986 reported by Libbrecht & Woodard (1990) continue into 1990. These differences are possibly associated with the solar cycle. We report here on our first findings about the variation of the internal angular velocity and asphericity deduced from these frequency differences.

INVERSION RESULTS

Figure ?? illustrates the angular velocity of the sun in the years 1986, 1988, 1989 and 1990 inferred from inverting the odd coefficients in the degeneracy-splitting expansion

$$\nu_{n,l,m} - \nu_{n,l,0} = l \sum_{k=1}^{5} a_k P_k(m/l)$$

of data acquired at Big Bear Observatory and analysed by the procedure described by Libbrecht (1989). The angular velocity is plotted as a function of radius for three values of colatitude θ.

Because the differences are small, we have inverted the data both as optimally localized averages and by a regularized least-squares procedure.

A contour plot for 1989 is shown in Figure ??*a*. Notice that Ω is approximately constant on cylinders in the very equatorial regions. The averaging kernels are insufficiently localized at high latitudes for us to be certain that that behaviour does not persist further than suggested by the figure.

Although the differences from year to year are comparable with the differences between these methods of inversion, those methods yield comparable

---

[1]and Department of Applied Mathematics and Theoretical Physics, University of Cambridge

[2]and Crimean Astrophysical Observatory

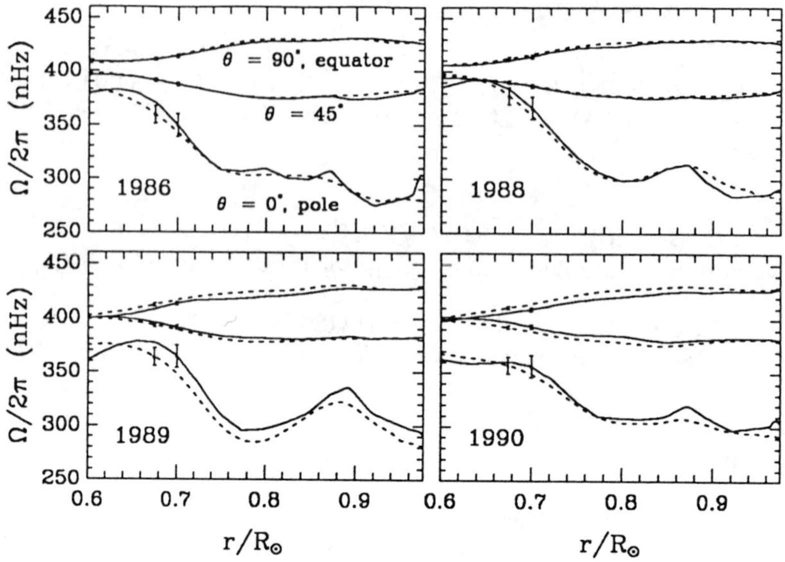

FIGURE I The solar angular velocity $\Omega/2\pi$ against r for $\theta = 0, \pi/4$ and $\pi/2$. The continuous lines are optimally localized averages, the dashed lines are regularized fits by least-squares.

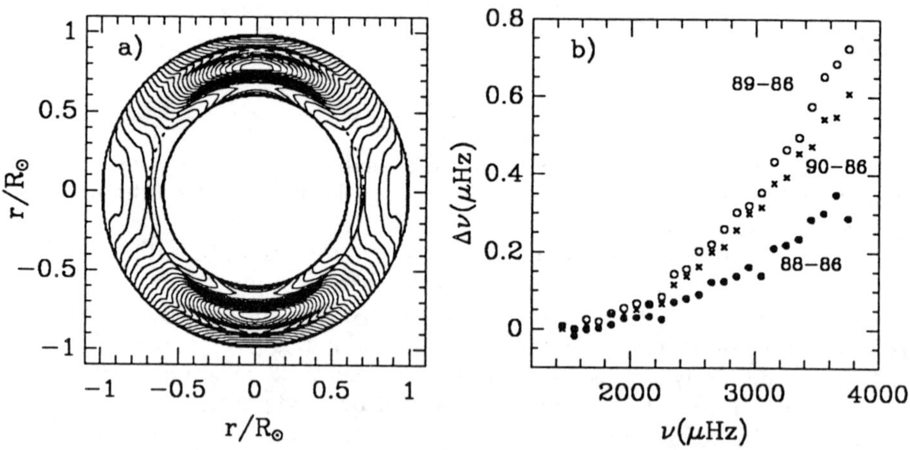

FIGURE II a) Contours of constant $\Omega/2\pi$ in 1989. Adjacent contours are separated by 5 nHz. The dashed circle shows the bottom of the convection zone. b) Multiplet cyclic frequency deviations from the corresponding values in 1986.

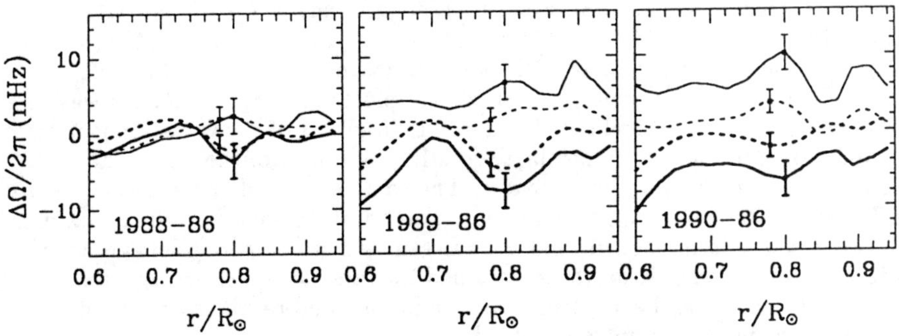

FIGURE III Deviations of $\Omega/2\pi$ from its value in 1986 for $\theta = \pi/4$ (thin lines) and $\theta = \pi/2$ (thick lines). Continuous lines are optimal averages and dashed lines are least-squares fits.

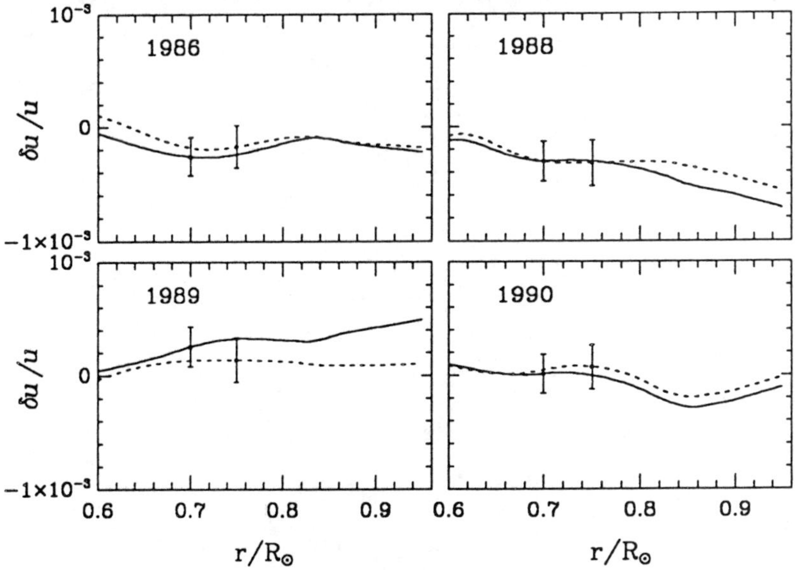

FIGURE IV Relative difference $\delta u/u$ between the value of u in the sun and that of a standard reference model for $\theta = \pi/4$ (dashed lines) and $\theta = \pi/2$ (solid lines)

temporal trends, as shown in Figure III, suggesting, in agreement with the statistical analysis of 1986 and 1988 data by Gough & Stark (1992), that the sun's angular velocity does change with the solar cycle.

The multiplet frequency variations are plotted in Figure II*b*. As was previously reported by Libbrecht & Woodard (1990), most of the variation is inversely proportional to the modal inertia, suggesting that most of the temporal variation in the sun is in the very outer layers. The same is true of the even splitting coefficients. After removing that dominant trend, inversions for δu, the difference between $u = p/\rho$ in the sun and u in a standard reference model, are shown in Figure IV. Although it has been assumed that the acoustic propagation speed c is given by $c^2 = \gamma u$, the apparent variation in u could result from a modification to c by, for example, a magnetic field.

REFERENCES

Gough, D.O. & Stark, P.B. 1992, *Astrophys. J.* (in preparation)

Libbrecht, K.G. 1989, *Astrophys. J.*, **336**, 413

Libbrecht, K.G. & Woodard, M.F. 1990, in *Progress of seismology of the sun and stars* (eds. Y.Osaki & H.Shibahashi) Heidelberg: Springer, p.145

Inside the Stars, IAU Colloquium 137
ASP Conference Series, Vol. 40, 1993
Werner W. Weiss and Annie Baglin (eds.)

SOLAR 5-MIN OSCILLATIONS AT 2.23 μm

TORBEN LEIFSEN[1]
Institute of Theoretical Astrophysics, University of Oslo, P.O. Box 1029, Blindern, N-0315 Oslo, Norway

INTRODUCTION

Large amplitude infrared 2.23 μm solar intensity oscillations were detected in photometer observations obtained at Oslo Solar Observatory in 1987 and 1988. Five wavelength regions ranging from 0.67 μm to 2.23 μm and 7 circular entrance apertures ranging from 0.5 to 4.3 arc min were observed simultaneously at all wavelengths with the same detector. The 2.23 μm region showed remarkably higher amplitudes than the other 4 wavelength-regions. The observed power was concentrated to the 2.5 - 3.5 mHz region suggesting that we observe the well known 5 min oscillations (Leifsen and Maltby, 1990).

OBSERVATIONS AND DATA REDUCTION

Two main questions arise from the photometer observations; one relates to the high power observed at 2.23 μm and the other concerns the nature of the power spectrum. We have choosen to address these questions by using the McMath telescope with its Fourier Transform Spectrometer at National Solar Observatory at Kitt Peak. Two observing modes were used: A 50 arc sec circular area at the solar disk center was observed using the McMath East Auxilliary Telescope. Spatially integrated light was observed using the McMath main heliostat and two flat mirrors. Active guiding was used in the imaged mode. Two timeseries were obtained in May 1991; one in the imaged mode at the solar disk center over a 4 day long period and one with integrated light lasting 6 days. These observations consist of timeseries of spectra separated in time by 1 minute and with a spectral resolution of 0.014 Å in the spectral range 2.0 - 2.5 μm. Each spectrum contains 825 Kilobytes of data. Twelve hours of observations result in 580 Megabytes of data. In order to handle these long timeseries each spectrum was divided in shorter spectral intervals (typically 630 Å for a timeseries of four days duration) and stored on several optical disks. Each series of intensity measurements in time at a given wavelength element was then treated as a separate timeseries. Each timeseries was normalized by dividing by a 20 min running mean, thus removing all long period variations, and cosine bells were added at the ends and before and after interruptions in the timeseries. Then power spectrum analysis was applied to each timeseries. Line identifications were taken from *An atlas of*

[1]Visiting Astronomer, National Solar Observatory, National Optical Astronomy Observatories, operated by the Association of Universities for Research in Astronomy, Inc. (AURA) under cooperative agreement with the National Science Foundation.

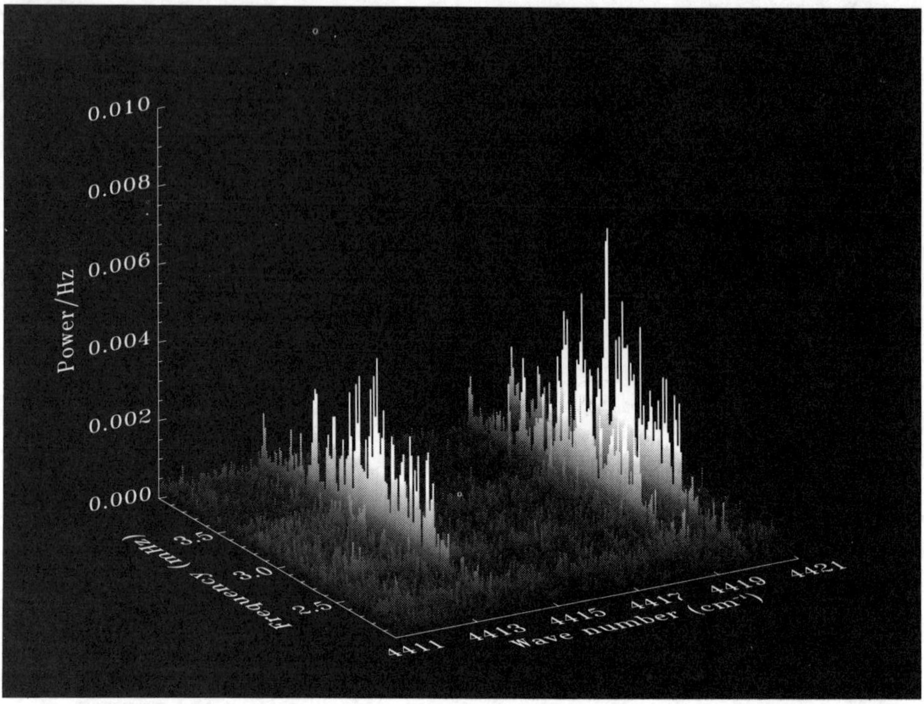

FIGURE I Power as a function of wavelength (x-axis) and oscillation frequency (y-axis) in a four day long time series at the solar disk center with a 50 arc sec entrance aperture. The ridges are solar intensity oscillations in solar Si (4410.7 cm^{-1}), CaI (4413.1 cm^{-1}), CaI (4413.6 cm^{-1}), CaI (4418.4 cm^{-1}), CaI (4418.7 cm^{-1}) and Fe (4419.7 cm^{-1}) lines. Note that terrestrial CH_4 lines in this spectral region show no intensity oscillations.

the Solar Spectrum from 1850 to 9000 cm^{-1} (Livingston and Wallace, 1991) and the Hall (1973) atlas for solar lines, and from the HITRAN molecular database (Rothmann *et al.*, 1992) for terrestrial lines. In order to study the variation of the power spectra as a function of wavelength, each powerspectrum was plotted in a 3-dimensional shaded-surface image (Figure 1). The x-z plane represents power spectra at given wavelength elements. In this way we can study the variation of the power spectra by comparing the results obtained for solar atomic and molecular lines, the continuum and for lines formed in the Earths atmosphere.

RESULTS

A study of the four days of disk center observations revealed the following: A large number of spectral lines in the observed wavelength region show large intensity oscillations. A majority of these lines show a frequency distribution in the power spectrum that resembles the well known solar p-mode oscillations. A few lines show a distinctly different frequency distribution, with some power in the 2-4 mHz region but also with the power increasing strongly towards lower frequencies. These lines were indentified as water vapor lines originating in the terrestrial atmosphere. All the other lines showing high power were identified as solar lines. There are several hundred solar spectral lines in the observed wavelength region, including the 2-0 and 3-1 bands of CO and several atomic lines. Figure 1 shows a power spectrum of an extract of the solar spectrum containing several atomic lines. All terrestrial molecular lines (expect water vapor), i.e. mainly CH_4, show no intensity oscillations above noise. This excludes the possibility of bands of terrestrial lines as the source of the observed high amplitude oscillations. The wavelength region shown in Figure 1 includes several CH_4 lines. None of these show intensity oscillations. Even though water-vapor lines in the region show intensity oscillations, they are far to few and have a distinctly different power distribution to account for the observed intensity oscillations. A comparison of timeseries integrated over wavenumber both including and excluding the water-vapor lines show that the water vapor lines have little or no influence on the observed power spectra. The main reason for this is that relatively few water vapor lines contribute to the power and they tend to be concentrated to the edges of the 2.23 μm window. Both molecular and atomic solar lines also show velocity oscillations with the familiar p-mode power distribution. We are currently analysing the velocity data and the phase between the velocity and line intensity oscillations. There is no doubt that the observed intensity and velocity oscillations in the solar lines are of solar origin.

CONCLUSION

The new observations, obtained with different equipment at an other site show intensity oscillations of solar origin in the 2.23 μm region. This conclusion is strengthened by the fact that terrestrial spectral lines (except water vapor) show no sign of oscillations above noise. The possible influence of water vapor absorption in the Earth's atmosphere is of minor importance due to the fact that the wavelength region in question contains only a few lines. The disk center observations show both intensity and velocity oscillations in solar lines from the CO 2-0 and 3-1 bands as well as in atomic lines in the wavelength region. We are currently working on velocity - intensity phase studies.

REFERENCES

Hall, D. N. B.: 1973, *An Atlas of Infrared Spectra of the Solar Photosphere and of Sunspot Umbrae.* (Tucson: Kitt Peak National Observatory)

Leifsen, T. and Maltby, P: 1990, *Solar Phys.* **125**, 241.

Livingston, W., and Wallace, L.: 1991, *An Atlas of the Solar Spectrum in the Infrared from 1850 to 9000 cm^{-1} (1.1 to 5.4 μm)*, National Solar Observatory Technical Report no. 91-001.

Rothmann, L. S., Gamache, R. R., Tipping, R. H., Rinsland, C. P., Smith, M. A. H., Chris Benner, C., Malathy Devi, V., Flaud, J.-M., Camy-Peyret, C., Perrin, A., Goldman, A., Massie, S. T., Brown, L. R., and Toth, R. A.: 1992, *J. Quant. Spect. and Rad. Transf.*, to appear in special edition.

Inside the Stars, IAU Colloquium 137
ASP Conference Series, Vol. 40, 1993
Werner W. Weiss and Annie Baglin (eds.)

BRIEF REVIEW OF SOLAR MODELS AND SOLAR NEUTRINO EXPERIMENTS

DOUGLAS R.O. MORRISON
CERN, Geneva, Switzerland

ABSTRACT Solar Evolutionary Models are briefly reviewed and while the models are robust, there are uncertainties in the input data which justify rather larger errors. The 1992 experimental results from GALLEX, SAGE II and Kamiokande are shown to be consistent with calculated fluxes of solar neutrinos whereas the Chlorine results continue to be significantly low though this experiment has a problem with the high variability with time of its results in contradiction to Kamiokande. It is concluded that the evidence for a solar neutrino problem is not compelling and New Physics are not demanded. Further experiments are essential to search for neutrino masses and to study the Sun.

CONVENTIONAL WISDOM

The Conventional Wisdom for 20 years has been that the experimentally measured flux of neutrinos from the Sun was very significantly less than that predicted by calculations of the Standard Solar Model, SSM. Further this neutrino flux varied with time according to the inverse of the sunspot activity (Bahcall, 1989).

For the Chlorine experiment in the Homestake mine (Lande, 1990), the neutrino flux was found to be 2.2±0.3 SNU whereas the SSM value calculated by Bahcall and Ulrich (1989) was 7.9 SNU. This result was confirmed in 1989, by the Kamiokande experiment (Hirata et al. 1989) using a different technique, who obtained the ratio;

$$Data/SSMvalue = 0.46 \pm 0.05 \pm 0.06 = 0.46 \pm 0.08 \quad (1)$$

In 1990, the Soviet-American Gallium Experiment, SAGE, reported (Gavrin et al. 1990) results using a gallium target for which the predicted flux of 132 SNU (Bahcall and Ulrich, 1989) is much higher as this includes the fundamental proton-proton reaction. They found a best fit of 20 SNU and an upper limit of 70 SNU with 90% confidence - significantly less than predicted.

Thus in 1990 there was clearly strong evidence that there was a major problem. This belief is examined here - earlier and fuller examinations with drawings have been published (Morrison, 1992). This brief review was completed in July 1992.

However things are not so simple and clear. There is not one SSM calculation by Bahcall et al., but there have been many. Another important calculation available in 1990 was by the French-Belgian Collaboration (Turck-Chieze et al. 1988) who found for the Chlorine experiment a much lower value of 5.8 SNU instead 7.9 SNU and what is more important gave a much larger error of 22% instead of 11%. With this lower calculated value, the Kamiokande result is now

$$Data/SSM = 0.70 \pm 0.12 \tag{2}$$

Recalculating (2) but now including also the theoretical error as well as the experimental error, gives;

$$Theory - Expt. = (1.00 \pm 0.22) - (0.70 \pm 0.12) = 0.30 \pm 0.25 \tag{3}$$

Thus a significant result has become a non-significant 1.2 standard deviation effect. This shows that the errors on the model calculation are important and 10% effects must be seriously considered.

SOLAR EVOLUTIONARY MODEL

In 1957 Schwartzschild et al. (1957) calculated a model of the Sun where its evolution was followed from its birth from a protostar 4.5 Gyr ago to the present. Several hundred shells are taken and are followed for a series of intervals of time. The input is only the present luminosity, mass, and radius, plus the age and the abundance of the elements. The Sun is considered to have three main regions;

(1) the Core where the nuclear energy is produced,

(2) the Intermediate or Radiation Zone and

(3) an outer Convection Zone. The energy created in the Core is transferred to the Convection Zone essentially by radiation.

The convection in the outer zone is described by a single Mixing Length Parameter, α (it would be better to have more parameters but there is not enough pieces of input information).

Since radiation is dominant, it is essential to know the opacity. This requires a detailed knowledge of all the isotopes, their state of ionisation at each depth, and their energy levels. After a plea (Stone, 1982) to resolve a stellar problem in 1982, the opacity has been recalculated. The new calculation, called OPAL (Inglesias and Rogers, 1991), has resulted in better fits to stellar data and has also changed the neutrino fluxes.

The Solar Evolutionary Model, called the standard model, SSM, has a number of problems.

(1) the fractions of lithium and beryllium are 0.005 and 0.5 resp. of the predicted value.

(2) screening effects are not well known and affect the choice of the Equation of State, and calculations of opacity and of nuclear reactions. The conditions of the Sun's plasma do not exist on Earth. These effects could give errors of about 15% on the neutrino flux for gallium and about 20% for the high energy neutrinos (Turck-Chieze, 1992)

(3) it is assumed that there is no rotation though initially the Sun was probably a T Tauri star, rotating strongly as mass was attracted to it from the rest of the protostar - this angular momentum and some mass would have been lost by a stronger solar wind than now; again this not in the SSM.

(4) diffusion exists at some level; recently Bahcall and Pinsonneault (1992) included 2 forms of diffusion (gravitational settling and thermal diffusion) which raised the neutrino flux estimates, but did not include turbulent diffusion which is expected to be important and which would lower the neutrino flux.

(5) some nuclear reaction rates are poorly known such as for the basic proton-proton reaction, ^{3}He-^{3}He and most importantly for high energy ^{8}B neutrinos, the ^{7}Be(p,γ)^{8}B reaction for which there is only one reliable measurement (see Morrison ref. b). A new measurement is being made at Riken by a Japanese-American collaboration. Thus there still exists considerable uncertainty even though the model is generally robust.

Early 1992, four new calculations were performed by Lopes and Turck-Chieze (1992), the Nice group of G. Berthomieu et al. (1992), the Yale group of Guenther et al. (1991) and by Bahcall and Pinsonneault (1992). Assuming the following standard conditions (best at this time); OPAL opacities, the astrophysical S(0) value for ^{7}Be(p,γ)^{8}B of 20.2 keV (see Morrison ref. 6b), age = 4.5 Gyr, low iron abundance in agreement with the meteorite value (see Morrison ref. 6b) and no partial diffusion, then the spread of neutrino rates are found to be:

(1) for chlorine targets, 5.8 to 6.4 SNU (Bahcall and Pinsonneault (1992) find 8.0 SNU which becomes 6.4 SNU for the standard conditions and 7.2 with the addition of their partial diffusion calculation).

(2) for gallium targets, the spread is 119 to 128 SNU (ref 12 gives 132 SNU which falls to 128 SNU under the standard conditions).

Overall the Stellar Evolutionary Models are surprisingly good and stable and the Yale group (Guenther et al. 1991) are even able by using 1800 shells, to fit the structure of the low ℓ, p-waves found in helioseismological measurements! However there are still some uncertainties and errors are probably larger than some models suggest, especially for the ^{8}B neutrinos.

EXPERIMENTS

Introduction

There are two experimental techniques used;

(1) a water Cerenkov detector where the neutrino elastically scatters on an electron which gives a ring of Cerenkov light detected by arrays of photomultipliers on the walls.

(2) chemical extraction experiments where a few atoms of a target of about 10^{30} atoms are converted by the neutrinos to a radioactive element which is extracted and the decays of these few atoms are counted.

The Kamiokande group are using a water Cerenkov detector; the threshold is high so that only the ^8B neutrinos are measured.

At the Homestake mine, the ^{37}Cl in a chlorine target can produce ^{37}Ar. The neutrinos are mainly from the higher energy ^8B decays in the Sun.

Gallium is used as a target where the ^{71}Ga isotope produces ^{71}Ge atoms. This has a low threshold so that the basic proton-proton reaction is a main contributor to the neutrino flux. Thus results from such experiments are of major importance whereas the ^8B neutrinos are from a minor branch of the chains of reactions in the Sun.

There are two gallium detectors - the Soviet-American Gallium Collaboration, SAGE, taking data in the Baksan Lab, and the GALLEX Collaboration in the Gran Sasso Lab.

The KAMIOKANDE Experiment

The Kamiokande experiment is run by a strong well-funded group and their experiment and results are fully described in refereed journals. Their first result called Kamiokande II, was based on 1040 days of operation where some 100 events were obtained above background, ie one event per 10 days. They observe a peak in the direction of the Sun confirming that the neutrinos are of solar origin. They found no variation with time, not with the sunspot number, nor with the season nor with the day/night cycle.

After improving their detector, the Kamiokande III run has begun and first results (Totsuka, 1992) from 220 days gave a ratio

$$data/SSM = 0.59 \pm 0.11 \pm 0.06 \tag{4}$$

which is slightly higher than previously but not significantly. The Cold Fusion cells of Steve Jones et al. at the centre of the Kamiokande detector have not seriously interfered with the neutrino work.

SAGE I

The first results for five runs of the SAGE detector with an average about 27 tons of gallium, between January and July 1990 were presented in August 1990. They (Gavrin et al. 1990) analysed their data using the method for low statistics of Cleveland et al. (1983) to obtain a best fit of 20 SNU which was substantially below a standard solar model calculation (Bahcall and Ulrich, 1989) of 132 SNU.

However as pointed out in Morrison (1992), the Cleveland method assumes that if the run data give a negative amount of ^{37}Ar atoms, then this is unphysical and the value is put to 0.0. Three of the five runs were given values of 0.0 SNU. Looking at the counts as a function of time, as shown in Morrison (ref 6a,b), it can be seen that the counting rates are so low that it is better to assume that the best fit to the five runs is zero.

There can be two main explanations of this null result - either this is New Physics or there is some unexpected problems with the extraction of a few atoms of ^{71}Ge from some 10^{30} atoms of ^{71}Ga. The SAGE Collaboration plan to calibrate their detector and extraction process by exposing it to a mega-curie source of neutrinos - probably late in 1993.

GALLEX

The GALLEX Collaboration spent considerable time preparing their experiment and announced that they would not publish until they had performed all calibrations and had sufficient statistics. In June 1992 they presented (Anselmann et al. 1992) the results of 14 runs with 30 tons of gallium, where they found

$$83 \pm 19 \pm 8 \text{ SNU} \tag{5}$$

which is 1.3 to 2 standard deviations from the four recent SSM calculations and indicated there was no significant disagreement with the models.

They observed peaks in the electron energy spectrum at 10.3 and 1.3 keV which is what would be expected from decays from the K and L shells. They had a total of 65 counts attributable to ^{71}Ge which corresponded to about one count per 5 days. They also plan to calibrate their detector with a strong neutrino source.

SAGE II

In June 1992, the SAGE Collaboration presented (Gavrin, 1992) the preliminary results of five runs in 1991 - for the four later runs, the mass of gallium was doubled to 57 tons. They obtained rates of 27, 300, 48, 75 and 93 SNU. This would appear to be consistent with the expectations from SSM calculations, but the authors say that these results are preliminary, so did not combine the five runs nor draw any conclusions from this experiment which is called SAGE II.

The CHLORINE Experiment

The Chlorine experiment has been running since 1967 but the first three years data were excluded as it was found that by requiring the pulse of the ^{37}Ar count to have a fast rise-time, the background was suppressed and more reliable results were obtained (if there is a decay in the counter, the ions produced are clustered close together and hence reach the wire in a short interval of time giving a fast rise-time to the pulse whereas if a background cosmic ray particle traverses the counter, the ions produced are dispersed and give a slow pulse rise-time).

Since 1970, the Homestake mine experiment has given rates which have stayed close to the latest value (Lande, 1992) of 2.25 $\pm$ 0.3 SNU which is significantly less than the recent SSM values of 5.8 to 6.4 (or 8.0). This series of results have been the basis of the Solar Neutrino Problem.

However there is a major worry about these results - this is their extreme variability. Thus after a prolonged shut-down to replace pumps, the results were for some time generally close to the Kamiokande II values (see Morrison ref 6b) This variability has been interpreted in terms of the solar sunspot cycle of 11 years though the best fit is obtained (Filippone and Vogel, 1990) with a 4.5 year cycle. Perhaps the fullest analysis using Poisson statistics, has been made by Filippone and Vogel who found that if one assumed a constant flux the probability of the fit was 3.9% whereas if one assumed there was a variation with the inverse of the sunspot number, the fit had a 8.3% probability. On the other hand another analysis (Davis, 1987) gave a 5 standard deviation effect. Another worry is the abnormal number of runs with zero or one count which is inconsistent with the claimed average of 5.6 counts per run (from 1970 to 1984 there were (Bahcall, 1989) 339 counts of ^{37}Ar in 61 runs giving one count per 15 days).

However in 1992, the data have been re-analysed (Fiorini, 1992) with wider selection criteria and almost all values have been changed and it is claimed that there are 8.1 counts per run. For 93 runs there are now only 6 zeros, but this number still seems high.

It is intended to start a new experiment with iodine which should give a substantially higher counting rate(about 5 times) - one awaits their results with interest.

NEUTRINO MASS

There is no reason to exclude neutrinos from having masses theoretically. Neutrino masses would be welcome, in particular they could be an important component of Dark Matter in the Universe, for example a value of 7 eV as suggested by the COBE data (Smoot et al. 1992). The errors of the theory and of the experiments are at present, both sufficiently large so that neutrino masses of up to about 10 eV are possible.

DETAILED CONCLUSIONS

a) There are many Standard Solar Model calculations who broadly agree for the same input, however there are important uncertainties so that errors of 15 to 20% at least must be considered in comparing with data.

b) The Kamiokande rate is consistent with most SSM calculations. No time variation is observed.

c) The SAGE I experiment found essentially no evidence for solar neutrinos or had extraction problems

d) the GALLEX experiment is consistent with all SSM calculations within 1.3 to 2 standard deviations

e) The SAGE II preliminary results appear to be consistent with the SSM

f) The rate found by the Chlorine experiment is significantly lower than SSM estimates, however the data show a disturbing variability with time.

g) Extraction experiments with gallium or chlorine, are difficult.

GENERAL CONCLUSIONS

- The History of Science is full of cases where a first pioneering experiment opened the way but the result was controversial. Later experiments built on the pioneering one and also corrected the problem.
- In the past there may have been a Solar neutrino Problem. In the future there may a Solar Neutrino Problem. But today the evidence for a Problem is not compelling - not for a rate problem nor for a time variation.

- There is still considerable space for Neutrino Masses.
- It is essential to KEEP SEARCHING - for neutrino masses and to study the Sun which is the only star near us

ACKNOWLEDGMENTS

It is a pleasure to acknowledge helpful discussions with Drs. J.N. Bahcall, G. Berthomieu, T. Bowles, V.N. Gavrin, W. Hampel, K. Lande, A. Maeder, M.H. Pinsonneault, A. Suzuki, Y. Totsuka, S. Turck-Chieze, D. Vignaud, and J.F. Wilkerson, though the responsibility for the content is the author's.

REFERENCES

Anselmann P. et al. Phys. Lett. B283(1992)376 and 390.

Bahcall J. N., 'Neutrino Astrophysics', CUP, Cambridge (1989).

Bahcall J. N. and Pinsonneault M. H., a) Princeton preprint IASSNS-AST 92/10 (1992) and to be publ. in Rev. Mod. Phys.; b) to be publ. Astrophys. J. (1992).

Bahcall J. N. and Ulrich R. K., Rev. Mod. Phys. 60(1989)297-372.

Berthomieu G. et al., subm. to Astrom. and Astrophys. (1992).

Cleveland B. T., Nucl. Instrum. Meth. 214(19783)451.

Davis R. Jr., Proc. Informal Conf. on Status and Future of Solar Neutrino Research, Ed. E. Friedlander, (1987), BNL report No. 50879 Vol. 1 p.1.

Filippone B. W. and Vogel P., Phys. Lett. B246(1990)546.

Fiorini E., Workshop on Neutrino Telescopes, Ed. M. Baldo-Ceolin, Palazzo Loredan, Venice, (1992), to be publ. and K. Lande, priv. comm., (1992).

Gavrin V. N. et al, Phys. Rev. Lett. 67(1990)3332.

Gavrin V. N., Particle Astrophysics Colloq. Blois, June 1992.

Guenther D. B. et al. Astrophys. 46(1991)377-393.

Hirata K. S. et al. Phys. Rev. Lett. 63(1989)16, 65(1990)1297, 65(1990)1301 and Phys. Rev. D44(1991)2241.

Inglesias C. A. and Rogers F. J., Astrophys. J. 371(1991)L73; 371(1991)408 and Rogers F. J. and Inglesias C. A., to appear in Astrophys. J. Suppl.(1992).

Lande, 25th Intl. Conf. on High Energy Physics, Singapore, 1990, Eds. K.K. Phua and Y. Yamaguchi.

Lande K., Particle Astrophsics Colloq. Blois 1992.

Lopes I. and Turck-Chieze S., IAU Colloquium 137 'Inside the Stars' Vienna, April 1992, and sub. to Astrophys. J. (1992).

Morrison D. R. O., a) CERN-PPE/92-109; b) Particle World, 3(1992)30-39; c) Joint Intl. Lepton-Photon Symp. and Europhysics Conf. on High Energy Physics, Geneva, 1991, Eds. S. Hagerty, K. Potter, and E. Querigh, World

Scientific, Singapore (1992) Vol. I p599. d) Int. Conf. on High Energy Physics, Singapore, 1990, Eds K.K. Phua and Y. Yamaguchi.

Schwartzschild M. et al., Astrophys. J. 125(1957)233.

Smoot G. F. et al., subm. to Astrophys. J. (1992).

Stone N. , Astrophys. J. 260(1982)L87.

Totsuka Y., Particle Astrophysics Colloq., Blois, June 1992.

Turck-Chieze S., talk at at Moriond Workshop on Massive Neutrinos and Tests of Fundamental Symmetries, Jan. 1992.

Inside the Stars, IAU Colloquium 137
ASP Conference Series, Vol. 40, 1993
Werner W. Weiss and Annie Baglin (eds.)

PROGRESS IN HIGH S/N SPECTROSCOPY AND ITS IMPORTANCE FOR STELLAR MODELLING

P. E. NISSEN
Institute of Physics and Astronomy, University of Århus, DK-8000 Århus C, Denmark

ABSTRACT Recent advances in high S/N spectroscopy are reviewed with particular emphasis on new data for the abundances of chemical elements of importance for the modelling of stellar interiors.

It seems well established that young, nearby B-type stars have abundances of CNO elements that are about a factor of two lower than in the Sun except for a small fraction of nitrogen-rich stars. The existence of such stars among main sequence stars remains to be explained.

Among normal A-type stars (excluding A_m and A_p stars) large deviations from solar abundance ratios occur with interesting anticorrelations of the abundances of C and Si. This suggests that diffusion processes and/or gas-dust separation of the chemical elements in the protostellar cloud play an important rôle.

For F and G main sequence stars the atmospheric abundance ratios of elements heavier than boron probably represent the original interior composition. For metal-poor stars the abundance ratios are non-solar, and overabundances of oxygen and other α-elements have important effects on stellar opacities and energy generation.

The observed abundances of Li and Be in F and G stars show a complicated behaviour as a function of T_{eff}, g, [Fe/H], and age partly due to depletion of these elements by (p, α) reactions. This provides interesting information on stellar structure, e.g. the depth of the convection zone and rotational mixing. At the same time it is very important that the depletion of ^{6}Li, ^{7}Li and ^{9}Be can be modelled so that reliable primordial abundances can be derived.

INTRODUCTION

The advent of new detectors and efficient spectrographs has led to remarkable improvements in observations of stellar spectra at high S/N and high resolution. With 4 meter class telescopes it is now possible to obtain spectra of 12th magnitude stars with a resolution of $R \simeq 50.000$ and $S/N \simeq 200$ for exposure times of a few hours, and brighter stars ($V < 9.0$) may be observed with $R \geq 100.000$ and $S/N \simeq 500$. This has led to important new knowledge about stellar properties like:

i) rotation and macroturbulence,

ii) magnetic fields, star spots and activity,

iii) pulsations and oscillations,

iv) convection, granulation and velocity fields,

v) effective temperatures, gravities and chemical composition.

The first three topics are discussed by other speakers at this colloquium and will not be dealt with in the present review. Studies of convection, granulation and stellar velocity fields will be briefly summarized in the next section, but the main part of the review will deal with new abundance data for various classes of normal main sequence stars and the impact of these data on stellar modelling. As a general theme it is discussed if the observed surface abundances represent the interior composition of the star at zero age.

STELLAR CONVECTION AND GRANULATION

Spectral signatures of stellar convection and granulation can be observed as asymmetries in photospheric absorption lines, usually described by the line bisectors. Recent work dealing with observations and interpretation of bisectors has been reviewed by Gray (1990) and Dravins (1990a). In general, solar-type stars have $\subset$ shaped bisectors, that can be explained by numerical, three-dimensional hydrodynamical models of stellar surface convection (Dravins & Nordlund 1990). It is remarkable that these models are free from classical *ad hoc* parameters like mixing length, micro- and macro-turbulence.

Based on a survey of bisectors in late A to late G type stars of all luminosity classes Gray & Nagel (1989) have shown that a granulation boundary exists in the HR diagram. The boundary runs from F0V to G1Ib. On the cool side of the boundary the spectral line bisectors have $\subset$ shapes corresponding to a granulation pattern as known from the Sun with characteristic velocities of a few km s^{-1}. On the hot side of the boundary the bisectors are inverted having $\supset$ shapes. Similar bisectors have been observed in early-type stars like Sirius (A1 V; Dravins 1987) and τ Sco (B0 V; Smith & Karp 1978, 1979). As discussed by Dravins (1990b) these bisectors suggest that the stars are covered by small granules corresponding to upwelling currents in "geysers" with velocities of perhaps 10-20 km s^{-1} balanced by a slower downdraft in more extended areas. The source of these strong velocity fields remains an open question, but it is interesting that non-LTE analyses of stellar equivalent widths in B and A type stars also suggest that strong velocity fields exist in their photospheres. Thus, Kilian (1992) and Gies & Lambert (1992) derive microturbulence velocities of 5-10 km s^{-1} in B0-B3 V stars, and Lemke (1989) finds velocities of 2-3 km s^{-1} in A0-A2 V stars.

In addition to new information about stellar convection obtained from the study of spectral line bisectors, such data also provides important checks on inhomogeneous models of stellar atmospheres based on hydrodynamical codes (e.g. Nordlund & Dravins 1990). The construction of such models for a large range of the basic atmospheric parameters, T_{eff}, g, and [Fe/H], would be important in order to improve the accuracy of stellar abundance determinations. Thus, the puzzling differences between oxygen abundances of metal-poor stars derived from the infrared OI triplet at 7774 Å and the forbidden [OI] line at

6300 Å may well be due to the neglect of inhomogeneities in the models used to analyze the observed equivalent widths (Kiselman 1991; Tomkin et al. 1992).

THE CHEMICAL COMPOSITION OF MAIN SEQUENCE STARS

In this section we discuss recent work on the abundances of some key elements in various classes of normal main sequence stars. These works are based on high resolution spectra observed with high S/N and analyzed by the aid of non-LTE model atmosphere computations. At the same time the works are characterized by including a relative large number of stars so that statistical significant results are obtained.

CNO and Si abundances in early B-type stars

Two important studies of CNO abundances in early B main sequence stars have recently appeared. The first one by Gies and Lambert (1992) includes 34 'northern' stars mostly of spectral types B0-B3 and luminosity classes V and IV. The stars were observed with the McDonald 2.1 m telescope with a resolution of $R \simeq 20.000$ and a $S/N \simeq 300$ in four bands of about 100 Å in the spectral region 4200-5200 Å. Equivalent widths of about 70 absorption lines - some as faint as 5 mÅ - were measured for each star. Effective temperatures of the stars were determined from the Strömgren Balmer discontinuity index $[c_1]$ and surface gravities from the profile of the H_γ line. LTE abundances were first derived using the line blanketed model atmospheres of Kurucz (1979) and then compared with non-LTE abundances based on the equivalent tables of Eber & Butler (1988) (CII), Becker & Butler (1988a) (NII) and Becker & Butler (1988b) (OII). The two sets of abundances agree well with maximum differences of about 0.20 dex.

The second study of abundances of CNO elements in early B-type stars has been carried out by Kilian (1992), and is based on spectra of 20 'southern' B0-B3 main sequence stars obtained with the ESO 3.6m telescope and its CASPEC spectrograph with a resolution of 20.000 and a $S/N \simeq 150$. The whole spectral region 4000-5000 Å was observed and equivalent widths of up to 250 lines per star are published by Kilian & Nissen (1989). Atomic data including oscillator strengths for the lines are given by Kilian, Montenbruck & Nissen (1991). Effective temperatures of the stars have been determined from a non-LTE analysis of Si II, Si III and Si IV lines and gravities are determined from the profiles of H_γ lines (Kilian et al. 1991). The derived abundances of the CNO elements are based on the non-LTE computations of Eber, Becker & Butler referred to above. The errors of the abundances are estimated to be about ±0.10 dex.

The B stars analyzed by Gies & Lambert and Kilian lie within 1 kpc from the Sun and include several stars in the Orion and Sco-Cen associations. In order to avoid severe blending of spectral lines the sample has been restricted to stars with projected rotational velocities $V\sin i < 80\ \mathrm{km\,s^{-1}}$. Five stars are in common between the two studies and their derived CNO abundances agree within a rms deviation of about 0.15 dex.

Fig. 1 shows the frequency of stars as a function of $[X/H] \equiv \log(N_X/N_H)_{star} - \log(N_X/N_H)_{Sun}$ for X = C, N and O. As seen the results of the two works described above agree very well. The widths of the histograms of C and O abun-

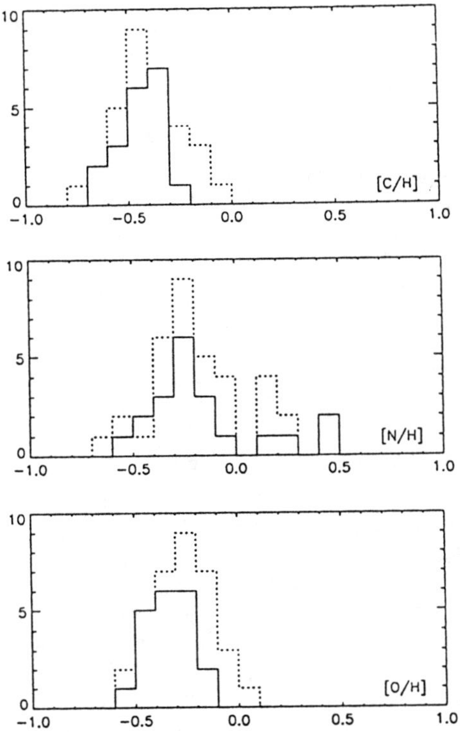

FIGURE I Number of B0-B3 main sequence stars versus the logarithmic abundances of C, N and O relative to the solar abundances with data taken from Gies & Lambert (1992) (dotted line) and Kilian (1992) (solid line)

dances are consistent with the errors of the abundance determinations, whereas the histograms of the N abundances are much broader with a tail of positive values of [N/H], which cannot be explained by errors in the abundance analyses.

Excluding for a moment the nitrogen-rich stars, it is seen that the mean values of the abundances of C, N and O are about a factor of two lower than the solar abundances, i.e. $[X/H] \simeq -0.3$. This surprising result can hardly be due to systematic errors in the model atmosphere analysis of the observed equivalent widths or in the atomic data – the calculated oscillator strengths of the lines are estimated to be accurate to about 10%. It means that nearly all nearby B main sequence stars have metallicities of $Z \simeq 0.01$ instead of the solar value, $Z_{\odot} = 0.02$. It remains to be investigated if this low metallicity is also representative for the stellar interiors or if the surface layers have been depleted in CNO by e.g. selective mass loss. The fact that the local HII regions are also deficient in oxygen by a factor of two relative to the Sun (Peimbert

1987) suggests, however, that the low abundance of CNO elements is a general phenomenon of gas and young stars in the solar neighborhood of the Galaxy. Information on other heavy elements is scarce. Note, however, that Kilian (1992) also derives a mean value of the Si abundance 0.3 dex below the solar value.

In stellar modelling it is normally assumed that the metallicity of nearby B stars is higher than or equal to the solar value, $Z_\odot = 0.02$, because one expects the metallicity of the interstellar gas to increase as a function of time due to chemical evolution in the Galaxy. Recent, accurate data on abundances and ages of F dwarfs (Edvardsson et al. 1992) shows, however, that there is no smooth age-metallicity relation for the galactic disk. At a given time and galactocentric distance the variation in the abundance of heavy elements amounts to almost a factor of two. Such variations may be related to infall of relatively metal-poor gas that triggers star formation before mixing in the galactic disk. In any case, model builders should be open for the possibility that the interior composition of young, nearby stars is characterized by a metallicity of $Z \simeq 0.01$ with the interesting consequences this may have for stellar opacities, energy generation, mass loss etc. As an example we refer to the recent work by Moskalik & Dziembowski (1992) on the origin of β Cephei pulsation. Using the new opacities of Iglesias & Rogers (1991a,b) they suggest that the driving of the pulsations is caused by the κ-mechanism acting in a zone with temperatures near $2 \cdot 10^5$ K where there is a bump in opacity. However, as shown in their Fig. 1 this mechanism works only if $Z > 0.02$. In this connection we note that several β Cephei stars are included in the abundance studies of Gies & Lambert and Kilian and have C, N, O, and Si abundances, corresponding to $Z \simeq 0.01$.

The existence of nitrogen-rich stars is another puzzling problem. Lyubimkov (1991) has presented evidence that the nitrogen abundance of B main sequence stars is correlated with stellar age and has suggested that mixing exists between the outer layers and the interior leading to a gradual transfer of CNO-cycle products from the stellar interior to the atmosphere. Gies & Lambert, however, find no significant correlation between stellar age and nitrogen abundance, and as discussed by Lambert (1992) other parameters than mass, age and initial composition seem to control the mixing of CNO-cycled material into the atmosphere. Rotation is a good candidate. Although the nitrogen-rich stars observed have low projected rotational velocities they may be fast rotators seen close to pole on. In any case, it is a surprise that mixing can occur in main sequence stars and a detailed explanation remains to be presented.

C, Si and Fe abundances in early A-type stars

A thorough study of the chemical composition of early A-type stars has been conducted by the stellar atmospheric group in Kiel (Holweger et al. 1986a,b). The spectra analyzed have primarily been obtained with the ESO Coudé Echelle Spectrometer at a resolution of 50.000. The sample is restricted to 17 B9.5V-A2V stars classified as normal in the Bright Star Catalogue (Hoffleit & Jaschek 1982). Detailed non-LTE model atmosphere analyses of the equivalent width data have been presented by Gigas (1986, 1988) and Lemke (1989, 1990), and have led to the discovery of surprisingly large variations from star to star in the abundance of iron peak elements accompanied by parallel variations of α-elements (Si and Ca). Thus, Sirius has twice the solar abundance of these elements, whereas Vega is underabundant with about a factor of four with respect to the Sun. As recently discussed by Holweger (1992) the carbon abundance is anticorrelated with the Si and Fe abundance in the sense that [C/Si] is high when [Si/H] or [Fe/H] is low and vice versa. Holweger points out that C differs from these elements by having a low condensation temperature, $T_c = 80$ K, while the other elements have $T_c > 1200$ K. Carbon is therefore less depleted in the interstellar gas than Si and Fe. Hence, the large variations in the Si and Fe abundances as well as the anticorrelation of [C/Si] with [Si/H] could be due to a variable gas/dust ratio in the interstellar matter from which the stars were formed. This would probably mean that the initial interior composition of these A-type stars is rather different from the solar composition with interesting consequences for the stellar models. On the other hand it may also be that the peculiar composition of the A-type stars has been created during later accretion of gas and dust in variable ratios. Then the atmospheric composition is different from the initial composition of the star. The same would be the case if the peculiar composition is due to diffusion processes.

The large abundance variations (almost a factor of 10 in metallicity) among normal A-type stars is a very surprising result. In young F or B stars the variation in metallicity is less than a factor of 2. Also there is no sign of an anticorrelation between the abundance of C and Si in B and F stars. Why are the early A-type stars behaving in a special way?

Oxygen, α-element and iron abundances in F and G stars

For F and G main sequence stars the observed abundances of elements heavier than boron probably represent the initial stellar composition. There are two observational arguments for this statement. First, the solar photospheric composition agrees very well with the composition of meteorites (Anders & Grevesse 1989). In particular, we note that the recent revision of the iron abundance in the solar photosphere (Biémont et al. 1991; Holweger et al. 1991) has led to a nearly perfect agreement with the meteoritic iron abundance. Second, the abundance of a given element in open cluster stars seems constant in the effective temperature range $5500 < T_{eff} < 7000$ K. Thus, Cayrel et al. (1985) and Boesgaard & Friel (1991) find [Fe/H] to be constant within ± 0.05 dex for the Hyades. A similar small upper limit on the intrinsic abundance dispersion of C is found by Friel & Boesgaard (1991) for several young open clusters, whereas Garcia Lopez et al. (1992) find the oxygen abundance in the Hyades stars to be constant within ± 0.10 dex for $5800 < T_{eff} < 7400$ K. If atomic diffusion or hydrodynamical processes played a rôle in modifying the surface composition for elements heavier than boron one would expect to observe some T_{eff} dependence of cluster abundances of these elements like in the case of Li and Be.

Assuming then that F and G stars reflect the composition of pre-stellar clouds they can be used for mapping the chemical composition of the Galaxy as a function of age and galactocentric distance. Furthermore, trends and scatter of various abundance ratios as a function of metallicity provide important information about nucleosynthesis in stars and supernovae. At the same time it is essential to know the number ratios of the most abundant elements for stars with different metallicities in order to construct reliable models to be used for e.g. age determinations. In the following some interesting new results concerning O

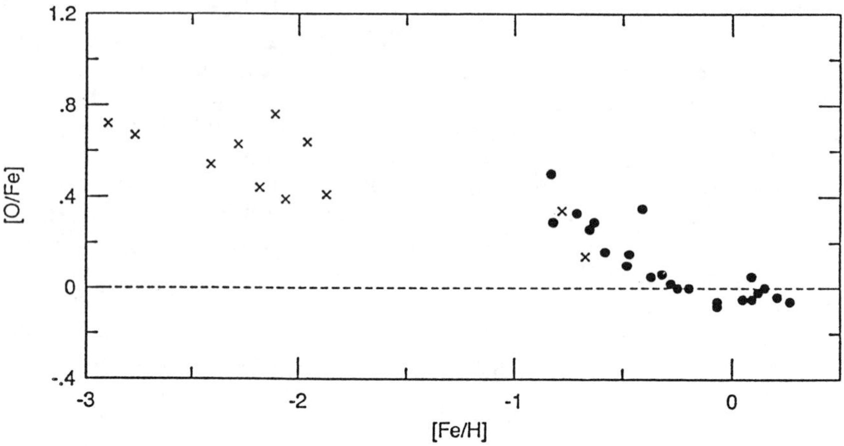

FIGURE II The oxygen-to-iron ratio as a function of metallicity in F and G dwarf stars. •, refers to oxygen abundance determinations based on equivalent widths of the faint, forbidden oxygen line at 6300 Å by Nissen & Edvardsson (1992). ×, refers to oxygen abundances derived from OH lines in the near ultraviolet by Gilmore et al. (1992)

and other α-element abundances as a function of Fe/H are discussed.

Existing data on the O-Fe relation in the galactic disk and halo are in severe disagreement. From analyses of the OI triplet at 7774 Å in F and G dwarfs a steep relation, $[O/Fe] \simeq -0.5$ [Fe/H], is found both for disk stars (Andersen et al. 1988) and halo stars (Abia & Rebolo 1989), whereas the forbidden lines at 6300 and 6363 Å in K giants lead to a shallower relation for disk stars (Barbuy & Erdelyi-Mendes 1989) and a constant value, $[O/Fe] \simeq 0.35$, for halo stars (Barbuy 1988, Sneden et al. 1991). The discrepancy amounts to about 1.0 dex in [O/Fe] for the most metal-deficient halo stars. Recent observations of the [OI] line in metal-poor dwarf stars (Spite & Spite 1991, Spiesman & Wallerstein 1991) show that the O abundances derived from the triplet lines are 0.3 to 0.5 dex too high. Apparently, this cannot be explained in terms of non-LTE effects (Kiselman 1991, Tomkin et al. 1992). Instead, the discrepancy may be due to the neglect of thermal inhomogeneities in the model atmospheres, because the population of the high excitational energy levels corresponding to the triplet lines is very sensitive to temperature. In any case, it seems that the O abundances derived from the triplet are not reliable. In Fig. 2 we therefore show some new results based on accurate observations of the faint [OI] line in F and G dwarfs by Nissen & Edvardsson (1992) as well as oxygen abundances derived from high resolution observations of OH lines in the near ultraviolet region, 3138 –3155 Å by Gilmore et al. (1992). As seen the O/Fe ratio is close to the solar ratio for the metallicity range $-0.3 < [Fe/H] < +0.3$. Below $[Fe/H] = -0.3$ the O/Fe ratio increases steeply and reaches $[O/Fe] \simeq 0.4$ at $[Fe/H] \simeq -0.8$. The halo stars in Fig. 2 have a mean ratio of $[O/Fe] \simeq 0.5$ i.e. three times the solar O/Fe ratio. It is, however, unclear from the data whether the O/Fe ratio in halo stars

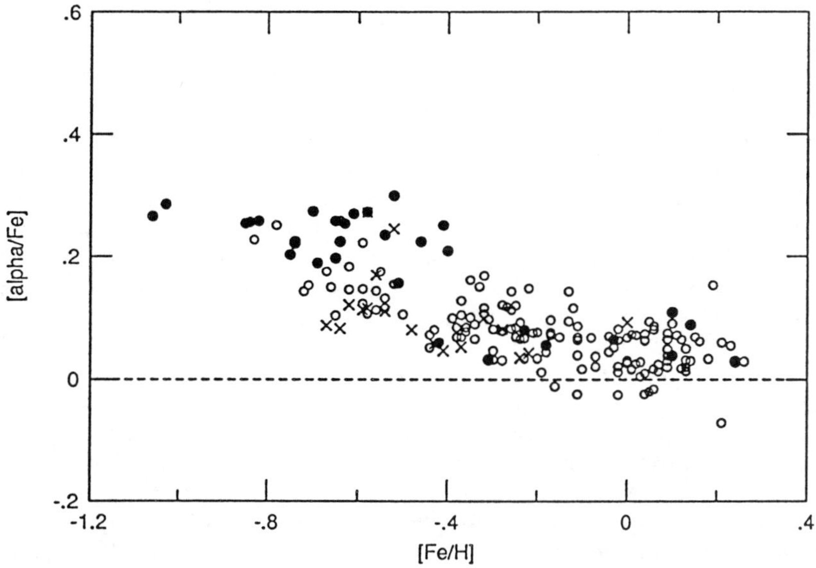

FIGURE III The abundance ratio between α-elements and iron as a function of metallicity for F and G stars with different mean galactocentric distances: (•) $R_m \leq 7$ kpc; (o) $7 < R_m < 9$ kpc; (×) $R_m \geq 9$ kpc. $[\alpha/\mathrm{Fe}]$ is defined as the mean value of [Mg/Fe], [Si/Fe], [Ca/Fe] and [Ti/Fe]

is constant or depends on metallicity. Furthermore, there may be a significant cosmic dispersion in O/Fe at a given [Fe/H]. Such a dispersion is indeed found for other α-elements, as discussed in the next section.

A survey of the chemical composition of 189 F and G dwarf stars in the galactic disk has recently been completed by Edvardsson et al. (1992). The survey is based on high resolution observations in selected spectral bands with the McDonald 2.7 m and the ESO 1.4 m CAT telescopes. The stars were selected without kinematical bias from the big catalogues of $uvby-\beta$ photometry (Olsen 1983) and cover the metallicity range $-1.0 < [\mathrm{Fe/H}] < +0.3$ uniformly. Accurate radial velocities, proper motions and distances are available for the stars so that galactic orbits and mean galactocentric distances R_m can be calculated. As shown by Grenon (1987), R_m is a good estimate of the galactocentric distance at which the star was formed.

The equivalent widths measured were analyzed by the aid of new, fully line-blanketed model atmospheres. T_{eff} and g were determined from the $uvby-\beta$ photometry. The excitation and ionization balance of iron lines suggest that these two parameters are accurate to ± 100 K and ± 0.15 dex, respectively. The derived abundances assume LTE population of the atomic levels, but non-LTE calculations were carried out to ensure that this simplifying assumption introduces negligible errors.

The analysis of the data has led to the determination of abundances of O, Na, Mg, Al, Si, Ca, Ti, Fe, Ni, Y, Zr, Ba and Nd relative to the solar abundances.

Interesting information on the trends and scatter of various abundance ratios and correlations with stellar age and kinematical parameters are obtained. Here we only comment on the results for Mg, Si, Ca and Ti, which all behave in a similar way as oxygen. The mean abundance of these four elements is plotted in Fig. 3 as a function of [Fe/H] with different symbols for stars born in the inner parts of the Galaxy, in the solar 'cylinder' and in the outer parts. For $[Fe/H] > -0.4$ the three groups are well mixed with an average value $[\alpha/Fe] = 0.05$. Apparently, this small offset from the solar ratio cannot be explained as due to errors in the analysis. For $[Fe/H] < -0.4$ the three groups separate. At a given [Fe/H] stars born in the inner part of the Galaxy tend to have higher values of $[\alpha/Fe]$ than stars born in the outer part. This behaviour can be explained by assuming that the rate of chemical evolution in the galactic disk is a decreasing function of galactocentric distance (see discussion in Edvardsson et al. 1992). As a direct consequence for stellar modelling we conclude that it is not sufficient to characterize the interior composition by only two independent parameters, say X and Z. One needs at least one additional parameter specifying the α/Fe ratio, where α refers to elements synthesized by the α-process in massive, short-lived stars exploding as Type II supernovae (O, Ne, Mg, S, Si etc.).

The effect of variations in the O/Fe ratio on models for metal-poor stars has been investigated in detail by VandenBerg (1992). At $[Fe/H] \simeq -2.0$ an increase of [O/Fe] from 0.0 to 0.75 leads to a decrease of globular cluster ages derived from turnoff luminosities by 2.2 Gyr primarily due to changes in the H-burning nucleosynthesis in the CNO cycle. The opacity effect arising from enhanced oxygen are considerably less important than its nucleosynthesis effects at this low metallicity. At higher metallicities, say $[Fe/H]= -0.5$, one would expect that the opacity effect plays a larger rôle especially if the enhancement of the other α-elements (Ne, Mg, Si, S, etc.) is taken into account. In particular, it would be interesting to see if the high ages (15 Gyr) of thick disk stars with $[Fe/H]\simeq -0.5$ determined by Edvardsson et al. (1992) will be reduced when isochrones corresponding to enhanced α-element abundances are applied.

Li, Be and B abundances in F and G stars

The abundances of light elements (Li, Be and B) in late type stars depend on nuclear processes in at least three different environments: (*i*) nucleosynthesis in the Big Bang phase of the Universe, (*ii*) interaction between cosmic rays and interstellar gas, (*iii*) (p, α) reactions in stellar interiors. Hence, by studying the abundances of these light elements as a function of stellar parameters like T_{eff}, g, [Fe/H], age, mass, and rotation we have the possibility to obtain information about the physical conditions in the early Universe, in the Galaxy, and in the stars. It is, however, difficult to disentangle the various processes leading to the observed surface abundances of Li, Be and B in a given star.

Observations of Li in Pop. I and II stars have recently been reviewed by Spite (1991), Molaro(1991), Balachandran (1991) and Soderblom (1991). Most notable are the constancy of Li abundance among halo stars with $T_{eff} > 5500$ K, the dip in Li abundance in open cluster stars at $T_{eff} \simeq 6600$ K, and the complicated dependence of Li abundances on age and [Fe/H] for $T_{eff} < 6000$ K. From theoretical studies it is well known that the temperature at the bottom of the convection zone is not high enough to explain the observed degree of Li depletion. Additional mixing caused by overshooting, gravity waves, mass loss, rotation or diffusion is needed. For a recent discussion of these mechanisms the reader is referred to Schatzman (1991) and Michaud & Richer (1991).

A recent accurate survey of lithium in 81 main-sequence stars with $5800 < T_{eff} < 6700$ by Lambert, Heath & Edvardsson (1991) has augmented our knowledge to Li abundances in disk stars considerably. A novel result of this survey is that, for the large majority of old disk dwarfs too low in mass to belong to the Li-gap, Li is only mildly depleted. In a small fraction (about 10 %) of the stars Li is, however, strongly depleted by a factor of 10 or more. Apparently, stars of intermediate Li abundance do not exist. This suggests that Li depletion in stars is influenced by more than the standard stellar variables of mass, age and metallicity. It is of great interest to find out which additional parameter (e.g. rotation, mass loss or binarity) plays a decisive rôle for the degree of Li depletion.

Beryllium abundances in disk stars have been reviewed by Boesgaard (1991). For field stars a dichotomy seems to exist. Most stars have the same Be abundance as the Sun (log(Be/H)= −10.85), but some F dwarfs in the Li-gab are markedly deficient in Be. The Be-deficient stars all have severe Li deficiencies, but many Li-deficient stars have solar Be abundances. This makes sense because Be destruction by (p, α) reactions requires a higher temperature than Li destruction. Again, other parameters than mass, age and metallicity seem to be decisive for the occurrence of Be depletion.

Recently, beryllium and boron abundances have been determined for very metal-poor stars. Gilmore, Edvardsson & Nissen (1991) succeeded to detect the Be II resonance doublet at 3131 Å in HD140283 ([Fe/H]$\simeq -2.7$) using the Anglo-Australian Telescope and the UCL Echelle Spectrograph with a resolution of $R \simeq 60.000$. Later Ryan et al. (1992) and Gilmore et al. (1992) have determined Be abundances in other metal-poor stars. Furthermore, Duncan, Lambert & Lemke (1992) have derived Boron abundances in three metal-poor stars from observations of the BI resonance lines at 2497 Å observed with the Goddard High Resolution Spectrograph of the Hubble Space Telescope. The data on Be and B is shown in Fig. IV. As seen both elements have abundances proportional to the abundance of iron. The scatter in the relations can be explained by errors in the abundance determinations, and the linear dependence of Be and B on Fe (and O) as well as the constant B/Be ratio of about 10 are compatible with cosmic ray spallation production of Be and B, although current models for this production have to be modified (Gilmore et al. 1992, Duncan et al. 1992). Hence, there is no evidence for primordial Big Bang production of Be and B. Neither, there is evidence for depletion of Be and B in these metal-poor stars, which have $5500 < T_{eff} < 6200$ K. *Lithium* depletion in metal-poor halo stars is, however, a controversial problem. As shown by Deliyannis et al. (1990) classical, non-rotating models predict a negligible degree of ^{7}Li depletion in metal-poor stars with $T_{eff} > 5500$ K. Models of halo stars with rotation predict, however, a depletion factor between 5 and 10 depending on the initial angular momentum of the star (Pinsonneault et al. 1992). If correct, the primordial Big Bang lithium abundance may be as high Li/H $\simeq 10^{-9}$ instead of the canonical value of $1.5\,10^{-10}$. Observations of the ^{6}Li/^{7}Li ratio in metal-poor stars would be helpful in solving this crucial problem. Pilachowski et al. (1989)

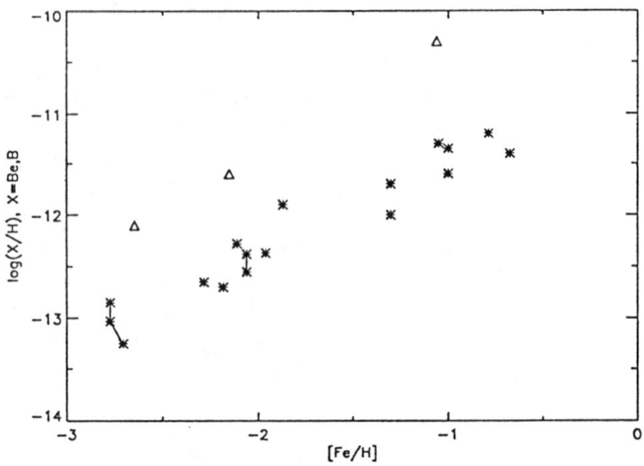

FIGURE IV Beryllium and boron abundances in metal-poor stars as a function of metallicity. $*$, Be abundances from Rebolo et al. (1988), Ryan et al. (1992) and Gilmore et al. (1992). $\triangle$, B abundances from Duncan et al. (1992). Points referring to different observations of the same star are connected with straight lines

have searched for ^{6}Li in HD84937 ($T_{eff} \simeq 6200$ K, [Fe/H] $\simeq -2.4$) by observing the profile and wavelength of the Li doublet at 6707 Å and have obtained an upper limit $^6\mathrm{Li}/^7\mathrm{Li} < 0.10$. Recently, Smith, Lambert & Nissen (1992) have reported a detection of ^{6}Li in the same star at a level of $^6\mathrm{Li}/^7\mathrm{Li} \simeq 0.05$ based on very high S/N observations of the doublet with the McDonald 2.7 m telescope. If confirmed, this would support the standard models, because the fragile ^{6}Li isotope is completely destroyed in models with rotational mixing (Deliyannis 1990).

CONCLUSIONS

As emphasized in this review and as discussed in more detail by other authors in these proceedings high S/N spectroscopy provides information on many aspects of stellar modelling. Much new information relating to stellar rotation, convection, magnetic fields and chemical composition has been obtained during the last few years. In particular, we have discussed some recent work on abundances of key elements in main sequence stars. Several novel results have been obtained. CNO abundances in young B stars turn out to be a factor of two lower than in the Sun with the exception of the (unexplained) occurrence of a small fraction of nitrogen-rich stars. Among the A-type stars large (up to a factor of 10) variations of iron-peak and α-element abundances are found with interesting anticorrelations of C and Si. For the F and G stars the trends and scatter in various abundance ratios as a function of metallicity have been studied in detail with important consequences for nucleosynthesis and opacity in stellar interiors. In addition, observations of ^{6,7}Li, Be and B abundances are providing new insight in Big Bang nucleosynthesis, cosmic rays in the early Galaxy and stellar structure. We conclude that the field of high S/N spectroscopy is likely to continue to contribute significantly to our knowledge about stellar structure and to other branches of astrophysics as well.

REFERENCES

Abia, C., Rebolo, R. 1989, *ApJ*, **347** 186.

Anders, E., Grevesse, N. 1989, *Geochim. Cosmochim. Acta*, **53**, 197.

Andersen, J., Edvardsson, B., Gustafsson, B., Nissen, P.E. 1988, in *Proc. IAU Symp. No. 132*, eds. G. Cayrel de Strobel and M. Spite (Dordrecht: Kluwer), p. 485.

Balachandran, S. 1991, *Mem.S.A.It.*, **62**, 33.

Barbuy, B. 1988, *A&A*, **191**, 121.

Barbuy, B., Erdelyi–Mendes, M. 1989, *A&A*, **214**, 239.

Becker, S.R., Butler, K. 1988a, *A&AS*, **76**, 331.

Becker, S.R., Butler, K. 1988b, *A&AS*, **74**, 211.

Biémont, E., Baudoux, M., Kurucz, R., Ansbacher, W., Pinnington, E.H. 1991, *A&A*, **249**, 539.

Boesgaard, A.M. 1991 in *The Formation and Evolution of Star Clusters*, Ed. K. Janes, *PASPC*, **13** (in press).

Boesgaard, A.M., Friel, E.D. 1991, *ApJ*, **351**, 467.

Cayrel, R., Cayrel de Strobel, G., Campbell, B. 1985, *A&A*, **146**, 249.

Deliyannis, C.P. 1990, *Ph.D. dissertation*, Yale University.

Deliyannis, C.P., Demarque, P., Kawaler, S.D. 1990. *ApJS*, **73**, 21.

Dravins, D. 1987, *A&A*, **172**, 211.

Dravins, D. 1990a, *Mem.S.A.It*, **61**, 513.

Dravins, D. 1990b, *A&A*, **228**, 218.

Dravins, D., Nordlund, Å. 1990, *A&A*, **228**, 203.

Duncan, D.K., Lambert, D.L., Lemke, M. 1992, *ApJ* (submitted).

Eber, F., Butler, K. 1988, *A&A*, **202**, 153.

Edvardsson, B., Andersen, J., Gustafsson, B., Lambert, D.L., Nissen, P.E., Tomkin, J. 1992 (in preparation).

Friel, E.D., Boesgaard, A.M. 1991, *ApJ*, **351**, 480.

Garcia Lopez, R.J., Rebolo, R., Herrero, A., Beckman, J.E. 1992 (these proceedings).

Gies, D.R., Lambert, D.L. 1992, *ApJ*, **387**, 673.

Gigas, D. 1986, *A&A*, **165**, 170.

Gigas, D. 1988, *A&A*, **192**, 264.

Gilmore, G., Edvardsson, B., Nissen, P.E. 1991, *ApJ*, **378**, 17.

Gilmore, G., Gustafsson, B., Edvardsson, B., Nissen, P.E. 1992, *Nature*, **357**, 379.

Gray, D.F. 1990, *Mem.S.A.It.*, **61**, 503.

Gray, D.F., Nagel, T. 1989, *ApJ*, **341**, 421.

Grenon, M. 1987, *JA&A*, **8**, 123.

Hoffleit, D., Jaschek, C. 1982, *The Bright Star Catalogue*, Yale University Observatory, New Haven.

Holweger, H. 1992, in *Atmospheres of Early-Type Stars*, Eds. U. Heber & C.S. Jeffery, Lecture Notes in Physics (Heidelberg: Springer) (in press).

Holweger, H., Bard, A., Kock, A., Kock, M. 1991, *A&A*, **249**, 545.

Holweger, H., Gigas, D., Steffen, M. 1986a, *A&A*, **155**, 58.

Holweger, H., Steffen, M., Gigas, D. 1986b, *A&A*, **163**, 333.

Iglesias, C.A., Rogers, F.J. 1991a, *ApJ*, **371**, 408.

Iglesias, C.A., Rogers, F.J. 1991b, *ApJ*, **371**, L73.

Kilian, J. 1992, *A&A*, (submitted).

Kilian, J., Becker, S.R., Gehren, T., Nissen, P.E. 1991, *A&A*, **244**, 419.

Kilian, J., Montenbruck, O., Nissen, P.E. 1991, *A&AS*, **88**, 101.

Kilian, J., Nissen, P.E. 1989, *A&AS*, **80**, 255.

Kiselman, D. 1991, *A&A*, **245**, L9.

Kurucz, R.L. 1979. *ApJS*, **40**, 1.

Lambert, D.L. 1992 in *High Resolution Spectroscopy with the VLT*, ed. M.-H. Ulrich, ESO Workshop, February 11–13, 1992 (ESO: Garching) (in press).

Lambert, D.L., Heath, J.E., Edvardsson, B. 1991, *MNRAS*, **253**, 610.

Lemke, M. 1989, *A&A*, **225**, 125.

Lemke, M. 1990, *A&A*, **240**, 331.

Lyubimkov, L.S. 1991 in *Evolution of Stars: The Photospheric Abundance Connection*, eds. G. Michaud & A. Tutukov, IAU Symp. 145 (Dordrecht: Kluwer), p. 125.

Michaud, G., Richer, J. 1991, *Mem.S.A.It.*, **62**, 151.

Molaro, P. 1991, *Mem.S.A.It.*, **62**, 17.

Moskalik, P., Dziembowski, W.A. 1992, *A&A*, **265**, 65.

Nissen, P.E., Edvardsson, B. 1992, *A&A*, (in press).

Nordlund, Å., Dravins, D. 1990, *A&A*, **228**, 155.

Olsen, E.H. 1983, *A&AS*, **54**, 55.

Peimbert, M. 1987, in *Star Forming Regions*, eds. M. Peimbert & J. Jugaku (Dordrecht, Reidel), p.111.

Pilachowski, C., Hobbs, L.M., De Young, D.S. 1989, *ApJ*, **345**, L39.

Pinsonneault, M.H., Deliyannis, C.P., Demarque, P. 1992, *ApJS*, **78**, 179.

Rebolo, R., Molaro, P., Abia, C., Beckman, J.E. 1988, *A&A*, **193**, 193.

Ryan, S.G., Norris, J.E., Bessell, M.S., Deliyannis, C.P. 1992., *ApJ*, **388**, 184.

Schatzman, E. 1991, *Mem.S.A.It.*, **62**, 111.

Smith, M.A., Karp, A.H. 1978, *ApJ*, **219**, 522.

Smith, M.A., Karp, A.H. 1979, *ApJ*, **230**, 156..

Smith, V.V., Lambert, D.L., Nissen, P.E. 1992 (in preparation).

Sneden, C., Kraft, R.P., Prosser, C.F., Langer, G.E. 1991, *AJ*, **102**, 2001.

Soderblom, D.R. 1991, *Mem.S.A.It.*, **62**, 43.

Spiesman, W.J., Wallerstein, G. 1991, *AJ*, **102**, 1790.

Spite, M., Spite, F. 1991, *A&A*, **252**, 689.

Spite, F. 1991, *Mem.S.A.It.*, **62**, 11.

Tomkin, J., Lemke, M., Lambert, D.L., Sneden, C. 1992 (preprint).

VandenBerg, D.A. 1992. *ApJ*, **391**, 685.

Inside the Stars, IAU Colloquium 137
ASP Conference Series, Vol. 40, 1993
Werner W. Weiss and Annie Baglin (eds.)

IS SUPERFICIAL KNOWLEDGE ABOUT STELLAR SURFACES SUFFICIENT?

BENGT GUSTAFSSON
Astronomiska Observatoriet, Box 515,
S-751 20 Uppsala, Sweden

ABSTRACT The significance of a detailed knowledge of the atmosphere of a star for a proper understanding of its interior is illustrated and discussed. It is concluded that significant progress in the understanding of stellar atmospheres is needed in order to match methods, discoveries and needs in the study of stellar internal structure and evolution.

INTRODUCTION

In his classical textbook, *Stucture and Evolution of the Stars*, Martin Schwarzschild (1958) discussed the question of the outer boundary conditions (at r=R). The author found that, although the zero condition (T=0, P=0) was adequate for stars with radiative envelopes (of spectral types earlier than, say, A5), for later spectral types it was not satisfactory due to convection. In that case he defined the "convective zero solution", the straight extrapolation of the simple adiabatic relation all the way out to the surface. He then argued that the exact solution only differs appreciably at temperatures lower than fifty thousand degrees and, "*since errors restricted to such very low temperatures will have virtually no consequences for our results for the stellar interior*, we may use the simple adiabatic relation throughout the convection zone to the very edge." If this were true my task here, having been asked to review "the role of model atmospheres for understanding stellar structure" would have been simply reduced to that of the interested listener. However, things are, 34 years later, not so easy any longer.

In 1958 only few models of stellar atmospheres that considered the effects of the wavelength dependence of the opacity sources (so-called "non-gray models") had been calculated. Now, very extensive grids of such models, with consideration of a number of continuos opacity sources and of the blanketing of millions of spectral lines, are available for late-type stars. In spite of this sophistication, these models are still quite primitive in the sense that they are static and stratified in plane-parallel layers. Convective energy transport is generally modelled with the mixing-length theory (MLT), and at least in most cases LTE is assumed to prevail. The question has been much debated to which extent this "speculative-classical theory" (Thomas, 1989) is useful at the interpretation of observational data like stellar spectra. We shall see that its use may be even more questionable when boundary conditions for contemporary studies of stellar interiors are needed.

Models of stellar atmospheres are needed for a number of different purposes when one wants to study the interior of stars. First of all, the understanding of diagnostics for effective temperature, gravity, and atmospheric chemical composition is not only vital when putting the stars onto the log L-T_{eff} diagram for comparison with stellar models but also when tracing the results of evolutionary processes in the dredge up, burning or settling of various chemical elements. That is illustrated in Poul Erik Nissen´s contribution to the present proceedings and I shall only add a few comments on this important aspect of application of model atmospheres. Instead, I shall dwell on more physical connections between the outer and the inner regions of stars, such as the determination of radius (or estimation of mixing length) and the significance of atmospheric opacities, and the relation between the atmospheric processes and stellar oscillations, magnetic fields, angular momentum and internal mixing processes. I shall essentially confine my discussion to solar-type stars, in spite of the temptation to glance at stars in other parts of the HR diagram to find more drastic examples to illustrate my leading theses. Obviously, my discussion will be very far from comphrehensive; any reader who would wish a more systematic presentation of stellar atmospheres and related problems for solar-type stars is advised to consult the excellent monograph by Cram and Kuhi (1989). Also, many of the contributions at this conference will contain additional details and further examples along the lines of my presentation - such as the reviews by Ulrich, Zahn, Michaud and Proffitt, Christensen-Dalsgaard, Krause, Schriver and MacGregor, and several others.

THE DETERMINATION OF FUNDAMENTAL PARAMETERS

The art of deriving stellar fundamental parameters from stellar spectra, using model atmospheres, has been repeatedly summarized in recent years (see, e.g., Gustafsson and Jørgensen (1985), Spite and Spite (1985), Gehren (1988), Lambert (1989), and Wheeler, Sneden and Truran (1989)). Also, a great number of impressive results have been obtained, as was also illustrated in the Nissen´s presentation. Here, only a few aspects will be discussed, of particular significance for solar-type stars.

It was since long recognized that the ultraviolet flux of solar-type stars was not properly reproduced by classical model atmospheres. Even if all known spectral lines of atoms and molecules were taken into account the theoretical fluxes were too strong, in the solar case by typically 10%, and in cooler stars by much more (cf. Gustafsson et al. 1975, Gustafsson and Bell 1979). Following Holweger (1970) and others we ascribed this failure to the presence of a "veil" of many very weak metal lines in the late-type stellar spectra. Recently, Bob Kurucz has completed his calculation of a very detailed line list with more than 2×10^7 metal lines, giving identifications, wavelengths, excitation energies, gf-values, damping constants and Landé-factors. He has also shown that when this line list is used a good agreement is obtained between observed and calculated fluxes of the Sun in the near ultraviolet (Kurucz 1991). We have found similar results, using Kurucz´s list but otherwise a completely different computer code. It could be remarked that, when scrutinizing synthetic solar spectra calculated with Kurucz´s data for the metal lines in the 3300 Å region we have found very great discrepancies (typically a factor of 10) between calculated and observed metal line strengths. It seems, however, that statistically the veil of weak metal lines is very vell described. We have also found a resonably good agreement between calculated and observed uv fluxes for F stars of different metallicities (cf. Edvardsson et al. 1992).

When introducing the new ultraviolet opacity data we also found a gratifying agreement between surface gravities for F stars, normally deduced from photometric criteria calibrated basically on parallaxes, and those one would derive from ionization equilibria. This is illustrated in Fig. 1, where the iron abundances deduced from Fe I and Fe II lines, respectively, are plotted relatively to each other for our sample of solar-type disk stars with different metal abundances. The agreement is much better than one could derive from our previous set of models and not dependent on the stellar temperature or metallicity. Likewise, the abundances deduced from lines of different excitation agree very well. These findings and a number of other results suggest that the effects of overionisation due to the non-local radiative field, or of temperature inhomogeneities, are not very severe for abundance analysis of solar-type disk stars, as long as a strictly differential approach is taken relative to the Sun and weak spectral lines are used. One may still suspect, however, that such problems may plague, e.g., detailed analyses of Pop. II dwarfs.

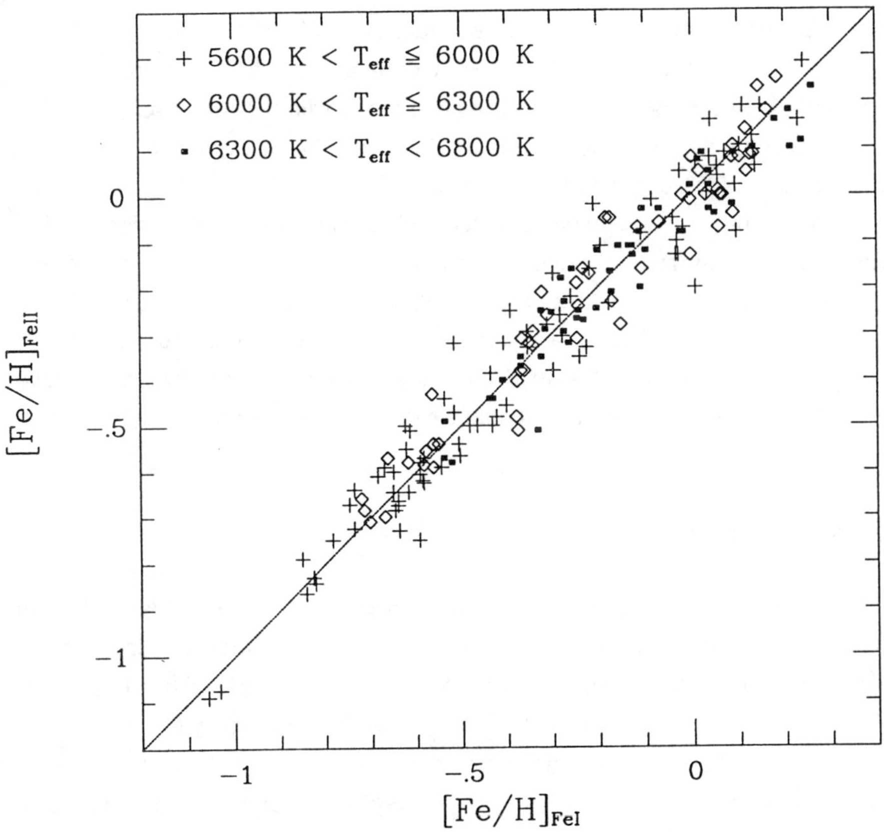

Fig. 1. Iron abundances derived for almost 200 solar-type disk dwarfs from Fe I and Fe II lines, respectively. The symbol for each star is depending on the stellar effective temperature. The line is a one-to one line. Obviously, there is no significant difference, depending on stellar temperature or metallicity, in the abundances from neutral or ionized Fe lines. (From Edvardsson et al. 1992.)

An interesting result of our extensive abundance survey for disk dwarfs (Edvardsson et al. 1992) is the location of the Sun in the abundance space. We thus find that the solar abundances of, e.g., Mg and Si, seem low by typically 25%, while the Ca abundance seems normal and O is high, with respect to Fe. Is this effect really real? We do not think it can be due to errors in the estimated colours of the Sun (i.e. in the zero-point in the temperature scale), nor in the gf values, which were estimated from the solar equivalvent widhts, measured in the same way as those of the stars. Nor should it be the result of errors in the model atmospheres, if the solar atmosphere is not systematically different from those of stars with similar effective temperatures, gravities and iron abundances. We argue that the reason probably is that the Sun is somewhat poor by typically -0.1 dex in many chemical elements, relative to iron, as compared with the mean abundances of similar stars in the solar neigbourhood. This may be of significance when one relies on our star as a calibration standard.

Presumably these examples, as well as Nissen´s contribution to these proceedings, show that rather interesting results may be obtained from a careful use of classical models to calibrate spectroscopic observations of stellar photospheres. Yet, one should note that the classical and static models were designed mainly to be used for predicting photospheric radiation fluxes, while they will fail totally in describing most dynamic phenomena.

MIXING LENGTH AND OPACITIES

For some time model atmospheres have been used as outer boundary conditions for models of stars with extensive convection zones. In fact, this is connected with estimating the mixing length parameter $\alpha = l/H$, with l being the mixing length and H the scale hight. If l is fixed by using the Sun with its well known radius, age and mass one can then use the temperature-pressure scales of a suitable grid of model atmospheres as boundary conditions, assuming α to remain the same.

One example of this is the extensive grid calculated by VandenBerg (1983) where the models by Bell et al. (1976 and later) were used for establishing the "photospheric pressures", i.e. the gas pressures at the depth τ_{eff} where $T = T_{eff}$. (In principle, one should sooner use a boundary point deeper into the atmosphere where the optical depth at the most transparent wavlength is >>1, but the choice of τ_{eff} was indeed found to be satisfactory, at least for T_{eff} around 5000 K, by Pedersen, VandenBerg and Irwin (1990.) However, VandenBerg and Laskarides (1987) preferred to revert back to the more primitive boundary conditions for the following reason: If model atmospheres were used, the properties of red giants were well discribed by the models over a wide range in metallicity with the value of the mixing-length parameter $\alpha = 1.5$. However, for the Sun this grid of models with α = 1.5 gave a predicted T_{eff} for a solar model which was significantly too hot, by more than 100 K. Instead of fixing this by lowering the solar value of α VandenBerg and Laskarides argued that this failure might be due to the missing ultraviolet opacity in the model atmospheres. This conjecture has been checked with the new model atmospheres (Edvardsson et al. 1992) as boundary conditions. Indeed, there is a very substantial improvement for the Sun. As is seen in Fig. 2 the solar model now falls almost exactly in the right position in the theoretical HR diagram.

In spite of this success one should give a strong warning concerning these plane-parallel model atmospheres, with mixing-length convection. The atmospheres of true stars in this parameter range, like that of the Sun, are known to be strongly structured by non-local convection and radiation cooling in intricate interplay, leading to considerable inhomogeneities in temperature and velocity fields (granulation). The detailed numerical 3D simulations by Nordlund and Dravins (1990) clearly showed the inadequacy of standard models, even for determining the mean temperature structure of these photospheres. What the traditional mixing-length theory succeeds in describing is the entropy jump accross the shallow super-adiabatic layer near the stellar surface. Rather different local convection theories with free parameters may be calibrated fit the global solar parameters (cf. Gough and Weiss, 1976) but there is no guarantee that this parameter will be the same for different stars. Nordlund and Dravins demonstrated that the standard MLT *with a reasonably constant value of α* (close to 1.5) can be used at least within a parameter range characteristic of solar-type stars (see, however, Steffen´s contribution to the present proceedings). For red giant stars, it still remains to be investigated whether this is true.

No doubt, the considerable improvements that have taken place in recent years concerning atomic opacities affect the stellar interior models considerably. This is further commented on by M. Seaton in the present proceedings, and I shall therefore limit my comments to a few words about molecular opacities (for a more general review, see Jorgensen 1992). The effects of molecules are mainly limited to the cool end of the HR diagram, and

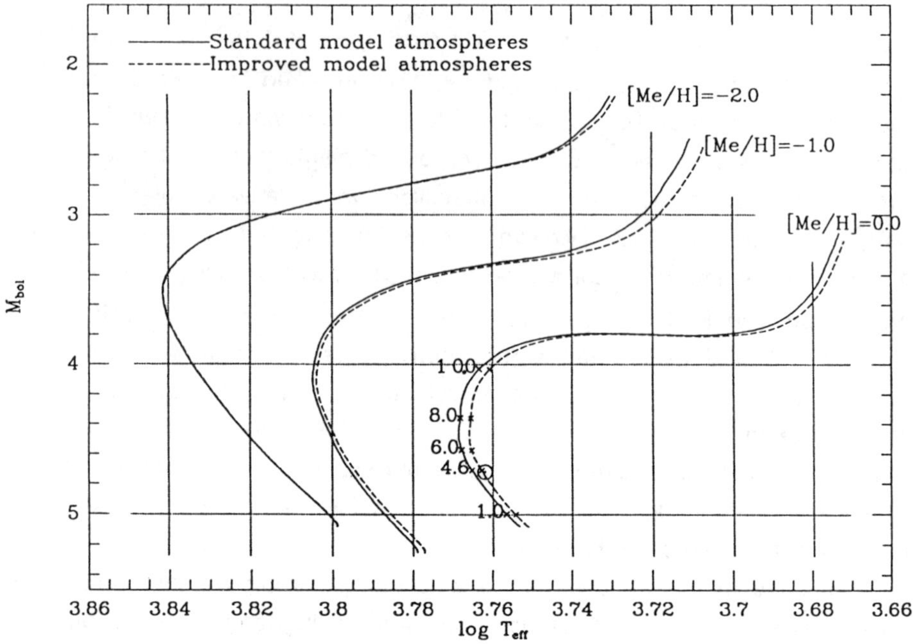

Fig. 2 Evolutionary tracks for α = 1.5, calculated by D. A. VandenBerg (1992) using model atmospheres from the grid of Bell et al. (1975, full lines) and a grid with improved uv blanketing (cf. Edvardsson et al. 1992). Time in Gyears and the solar position is shown at the [Me/H]=0.0 tracks. The tracks are still preliminary; the opacities can be further improved. Note the different effects of the new boundary conditions for different metallicities.

in particular to the red giants, whose diameters are very considerably affected by the molecular opacities. (We found an extreme effect of this kind some years ago when introducing polyatomic molecular opacities into model atmospheres for carbon stars (Eriksson et al. 1984). The introduction of HCN for T_{eff} around 3000 K expanded the model atmosphere such that the radius of the star could grow by a factor of two or more, as compared with a model without any polyatomics.) For interior models of ABG stars Sackman and Boothroyd (1991), using the new LAOL opacity tables which included opacities from CO, CN and H_2O, also found very significant effects on the radii. In order to restore agreement with observed giant branches it was found necessary to increase α on the AGB to more than 2.0. For further studies of these effects it is obviously important that reasonably realistic model atmospheres, including all relevant molecular opacities and sphericity effects, are used for setting the boundary conditions. As soon as possible, dynamic models with more realistic representation of convection similar to contemporary simulations for the solar type stars should be attempted.

For models of stars of about solar temperatures the effects of the molecular opacities are presumably so limited that the errors in them should not lead to very significant model errors, in spite of the fact that also for diatomic molecules the present situation as regards absorption coefficients is far from satisfactory (cf. Jørgensen 1992).

ATMOSPHERES AND DYNAMIC PROCESSES IN THE INTERIOR

a. Stellar oscillations

Although the effect of the outer layers on the interior of static stellar models is often negligible the corresponding is not necessarily true when dynamic processes are taken into account. For solar-type stars the question of the driving source of the solar "five-minute" oscillations has been debated intensively. Although these oscillations probably do not affect the grand structure of the interior layers, they are very significant diagnostics of this structure . It is therefore important to understand their physical origin, not the least when judging the possibilities of observing corresponding phenomena for other stars (see Christensen-Dalsgaard and Frandsen 1983).

The power spectrum of the solar oscillations is practically independent of the degree (ℓ) of the modes for $\ell<100$ (Christensen-Dalsgaard and Gough 1982, Libbrecht et al. 1986). This means that the oscillation amplitudes are not dependent on the horizontal wavenumbers, suggesting that the amplitudes are determined in layers where the acoustic waves propagate vertically, i.e. close to the surface. Several recent studies have stressed the probably very significant role that the upper boundary of the convection zone plays for exciting and damping the oscillations. In a series of theoretical papers Balmforth (1992a-c) has investigated the physical character of solar oscillations - their thermodynamics, frequencies and amplitudes. Although the question whether the solar pulsation modes are self-excited or stochastically excited by turbulent convection is not totally solved, Balmforth´s study gives most support to the latter mechanism. He also finds a good agreement between calculated and observed damping rates, except for waves of high frequencies which propagate out into the atmosphere before being reflected back into the interior. Balmforth suggests that that mismatch is because the mixing-length theory does not take horizontal inhomogeneities into consideration. High-order oscillation modes, with wavelengths comparable to the scales of these inhomogeneities, should experience incoherent scattering by these inhomogeneities in the superficial layers.

Recently extensive numerical simulations have been carried out for the Solar atmosphere by Nordlund and Stein (1991a and b) of convection and waves with a method that allows compression and shocks. They find that the switch from convective to radiative energy transport at the solar surface causes non-adiabatic fluctuations in the gas pressure that stochastically excite the p-modes. In fact, both the excitation and the linear damping mechanisms of these waves are concentrated near the surface.

The detailed structure of the surface layers, which is only partially known mainly because of the unsatisfactory theory of convection, also effects the frequencies. Balmforth (1992b) has investigated the small but significant errors on the 0.5 percent level between calculated and observed frequencies for the 5 min oscillations. He suggests that the discrepancy is due to the modelling of the upper solar boundary layers with the conventional mixing-length theory. Indeed, the recent simulations by Stein and Nordlund (1991b) give such effects of the right order of magnitude due to convective overshoot and turbulent pressure support. Also at the calculation of amplitudes the details of the upper layers play a decisive role (cf. Christensen-Dalsgaard and Frandsen, 1983, Balmforth 1992c).

We thus find that the shortcomings of the understanding of the upper stellar layers have direct consequences for the interpretation of solar, and stellar, oscillations. When once measured, the stellar oscillation power spectra should contain new information on the stellar surface gravity and effective temperature. It is, however, not clear to me how useful this information may be, in view of the uncertainties discussed above. In their recent review Gough and Toomre (1991) conclude: "The greatest helioseismological uncertainties lie in the superficial layers of the Sun." On the other hand this means that observations of oscillations, when combined with reasonably sound theoretical models, will give further valuable information as regards the structure of the surface layers.

b. Magnetic fields, mass loss, and momentum transport

The relation between convection, rotation, magnetic field, mass loss and mixing in stars is complex and interesting. One could ask to which extent the structure of the outer layers of the star play an active role in this interplay.

In outlining this we shall start with the magnetic fields and convection, and then proceed to the angular momentum transfer and mixing processes.

Magnetic fields may connect the structure and evolution of the interior of a star to its boundary in intricate ways. Recently, Charbonneau and MacGregor (1992a) have simulated the angular momentum transport in stellar magnetic and, initially, nonrotating radiative zones, around a spinning core. They find a rich variety of behaviour: time scales and evolution are critically dependent on the way the magnetic field is anchored in the core and its behaviour at the surface (i.e. the bottom of the convective envelope).

The convection simulations of Stein and Nordlund (1989) and Nordlund and Stein (1991a) led to the interesting dicovery that the topology of solar convection at great atmospheric depths is qualitatively very different from that closer to the surface. The subsurface structure is characterized by thin filaments of cool gas, downdrafts decending from the surface in a background of slowly ascending, very nearly isentropic gas. Moreover, the cold filaments seem to merge into larger and more widely separated filaments at successively larger depths. This structure of downdrafts, containing magnetic flux-tubes, may be of great significance for the way the magnetic fields are generated by the dynamo in the lower boundary layer of the convection zone. Still, one may speculate that this structure could be critically dependent on the conditions in the atmosphere where the radiative cooling of the hot rising gas takes place. We note that Cattaneo et al. (1991) find similar structures in their 3D simulations of compressible convection, but that it is not clear yet how sensitive these structures are to the boundary conditions.

Interesting simulations of dynamos in 3D convection models of this type have recently been produced by Nordlund et al. (1992). In these simulations, spontaneous dynamo action occurs at the lower boundary of the convection zone, with most of the magnetic field being generated as coherent flux tubes in the vincinity of strong downdrafts. The mechanism is closely related to the "stretch-twist-fold" dynamo, suggested by Vainshtein and Zeldovich (1972).

Although the applicability of the results discussed above is very uncertain they point at the possibility that the stellar boundary region , i. e. the atmosphere and convective envelope, and its structure, may be of great significance in affecting the evolution of magnetic fields and angular momentum in the interior.

The wellknown observations that the mean rotation speed of main sequence (MS) stars drops drastically at spectral types around F 0 is generally ascribed to the fact that stars with masses below about 1.5 solar masses have relatively deep convective envelopes. These convective zones are thought to promote dynamo action and magnetic braking. The latter is a consequence of the torque applied to the star by its magnetically coupled wind, as was first suggested by Schatzman (1962). This magnetic braking in the pre-MS or early MS life of the stars is probably a key significance in determining the stellar rotation pattern.

The theory of angular momentum loss from the star due to magnetic braking was summarized by Mestel (1984) and further developed by Kawaler (1988). The angular momentum loss rate depends on the mass loss rate $\dot{M}$, the surface magnetic field strength and the magentic field configuration in the upper atmosphere. There is still no convincing theory for winds from solar-type stellar atmospheres (including the pre-MS stage) that would allow us to successfully predict $\dot{M}$ from more basic quantities, such as fundamental parameters of the photosphere and a measure of the mechanical energy flux through it. It is not even known whether such a theory is physically correct at all, or whether $\dot{M}$ should also be added to the list of fundamental atmospheric parameters (cf. the discussion referred to by Cram 1989 and references cited therein, Thomas 1989, Cannon and Thomas 1977). However, in the calculations of the transfer of angular momentum in the Sun Pinsonneault et al. (1989) find a rather small effect of "moderate" changes in $\dot{M}$ on the resulting distribution of angular momentum. These authors show, on the other hand, that the magnetic field strength and other features of the wind model have a dominant influence on the surface rotation for a mature main-sequence star, like the Sun, as well as for rotationally induced mixing. Clearly, considerable progress has to be made in the modelling of stellar magnetic fields and winds before the magnetic braking can be predicted with certainty.

The solar model evolved by Pinsonneault et al. included angular momentum redistribution by rotationally induced hydrodynamical instabilities of various kinds and at various scales. Some free parameters were set such that a good agreement with solar data was obtained. A rapidly rotating central core persisted. Zahn (1992) has recently developed a more detailed analytical treatment of the meridional circulation and turbulence in a rotating model, assuming turbulence to be anisotropic with a greater horizontal than vertical transport. He predicts a rapidly spinning solar core, concludes that the slowly rotating old Pop. II dwarfs all should have more or less depleted Li abundances and argues for the "wind-driven circulation" as the relevant concept - the wind brakes and that causes the circulation.

As an alternative to the transfer of angular momentum through the stars by hydrodynamical means mediation by internal magnetic fields has been proposed (e.g., by Mestel and Weiss 1987). Recently Charbonneau and MacGregor (1992a and b) have studied this further on the basis of numerical simulations and obtained significantly lower core rotation velocities due to the magnetic copuling between the core and the layers outside it. They demonstrate the significane of the internal magnetic field configuration for the rotational behaviour of solar-type stars, but also find that the present solar surface rotation rate is a very poor indicator of the strength and geometry of the internal fields. The surface rotation in the early ages, or the present internal differential rotation, are much more dependent on these fields.

c. Diffusion and settling

Another example of processes where the atmosphere and its structure affects the interior are the diffusion processes. While, for early-type stars, helium settles gravitationally, the heavier elements either diffuse inwards or outwards, depending on whether the outward radiative acceleration is greater or smaller than the acceleration of gravity (cf., e.g., Michaud 1991, and references cited therein). This leads to interesting consequences for the internal structure - e.g. that the He II convection zone vanishes when diffusion proceeds and He gets strongly depleted. The atmosphere plays a significant role, setting the boundary conditions for this process. Significant atmospheric properties that may determine what happens are mass loss and the evolution of surface magnetic fields.

Even if similar phenomena may occur at greater depths in cooler stars, the importance of the atmosphere for these processes is probably much less, since it is situated on the top of the deep hydrogen convection zone. We note, however, that Michaud (1991) seems to find significant settling effects on surface abundances for main sequence stars with T_{eff} as low as 6000 K.

CONCLUSIONS

I hope that this presentation has demonstrated that the interrelation between stellar interiors and atmospheres is far more intricate than the atmosphere just being a single-valued boundary condition. It was early appreciated that static model atmospheres are not very helpful for the analysis of criteria showing the presence of stellar chromospheres, coronae and winds. Likewise, if the range of new methods for exploring stellar structure and evolution are to be exploited, the description of the atmosphere as a static and inactive plane-parallel zone on top of the star is insufficient. One way to circumvent this problem is to develop parametrized "theories" where the parameters are determined from observations. Examples of simple such approaches are the mixing-length theory or the modelling of magnetic braking discussed above. For many tasks these recipes may be acceptable, although one will have to wait for a detailed physical theory before this can be proven. The free parameters, however, may not have a sound physical interpretation, and there is no guarantee that they stay constant for a given star or equal for different but similar stars. Basically worse is that it may not lead to a correct understanding of the underlying physics, not even to the formulation of new relevant hypotheses. The physical connections between the atmosphere and the interior are dynamical and complex phenomena and need systematic studies, aiming at detailed semi-empirical modelling of the dynamic behaviour of a number of stars. Reasonably realistic numerical simulations, compared with high-resolution (spectral and if possible spatial) monitoring will continue to play the key role here. Fortunately, the methodological development in these areas is presently very promising. Hopefully, in some years we shall know much more, also about how far from complex realities our ideas are.

ACKNOWLEDGEMENT

Thanks are due to D. A. VandenBerg for calculating the evolutionary tracks shown in Fig. 2.

REFERENCES

Balmforth, N. J. 1992a, b, c, *M.N.R.A.S.* **255**, 603, 632, 639.
Bell, R. A., Eriksson, K., Gustafsson, B., Nordlund, Å. 1976, *A.&Ap. Suppl.* **23**, 37.
Cannon, C. J., Thomas, R. N. 1977, *Ap.J.* **211**, 910.
Cattaneo, F., Brummell, N. H., Toomre, J., Malagoli, A., Hurlburt, N. E. 1991, *Ap.J.* **370**, 282.
Charbonneau, R., MacGregor, K. B. 1992a, *Ap.J.* **387**, 639.
Charbonneau, R., MacGregor, K. B. 1992b, *Ap.J. Letters*, submitted.
Christensen-Dalsgaard, J., Gough, D. O. 1982, *M.N.R.A.S.* **198**, 141.
Christensen-Dalsgaard, J., Frandsen, S. 1983, *Solar. Phys.* **82**, 165.
Cram, L. E. 1989, in Cram and Kuhi 1989, p. 181.
Cram, L. E., Kuhi, L. V. 1989 (eds.), *FKG stars and T Tauri Stars, Monograph Series on Nonthermal Phenomena in Stellar Atmospheres, NASA SP*-**502**.

Edvardsson, B., Andersen, J., Gustafsson, B., Lambert, D. L., Nissen, P. E., Tomkin, J. 1992, *A.&Ap.*,to be submitted.

Eriksson, K., Gustafsson, B., Jørgensen, U. G., Nordlund, Å. 1986, *A.&Ap.***132**, 37.

Gehren, T. 1988, in *Rev. Mod. Astron.* **1**, 52, G. Klare (ed.) (Springer: Heidelberg).

Gough, D. O., Weiss, N. O. 1976, *M.N.R.A.S.* **176**, 589.

Gough, D., Toomre, J. 1991, *Ann. Rev. A. Ap.* **29**, 627.

Gustafsson, B., Bell, R. A., Eriksson, K., Nordlund, Å. 1975, *A.&Ap.* **42**, 407.

Gustafsson, B., Jørgensen, U. G. 1985, in *Proc. IAU Symp. 111: Calibration of Fundamental Stellar Quantities*, D. S. Hayes, L. E. Pasinetti, A. G. D. Philip (eds.) (Reidel: Dordrecht) p. 303.

Holweger, H. 1970, *A.&Ap.* **4**, 11.

Jørgensen, U.G. 1992, *Rev. Mex. A. Ap.*, in press.

Kawaler, S. D. 1988, *Ap.J.* **333**, 236.

Kurucz, R. 1991, in *Stellar Atmospheres: Beyond Classical Models*, L. Crivellari, I. Hubeny, D. G. Hummer (eds.), *NATO ASI-Series C* **341**.

Lambert, D. L. 1989, in *Cosmic Abundances of Matter*, C. Jake Waddington (ed.), *AIP Conf. Proc.* **183**, 168.

Libbrecht, K.G., Popp, B.D., Kaufman, J.M., Penn, M.J. 1986, *Nature* **323**, 235.

Mestel, L. 1984, in *3rd Cambridge Workshop on Cool Stars, Stellar Systems and the Sun*,S. L. Baliunas, L. Hartmann (eds.) (Springer: New York), p. 49.

Mestel, L., Weiss, N. O. 1987, *M.N.R.A.S.* **226**, 123.

Michaud, G. 1991, in *Proc. IAU Symp. 145, Evolution of Stars: The Photospheric Abundance Connection*, G. Michaud, A. Tutukov (eds.) (Kluwer Acad.: Dordrecht), p. 111.

Nordlund, Å., Dravins, D. 1990, *A.&Ap.* **228**, 155.

Nordlund, Å., Stein, R. F. 1991a, in *Stellar Atmospheres: Beyond Classical Models*, L. Crivellari, I. Hubeny, D. G. Hummer (eds.), *NATO ASI-Series C* **341**, 263.

Nordlund, Å., Stein, R. F. 1991b, *Proc. of the Mini Workshop on the Solar 5 Minutes Oscillations*, Univ. of Oslo.

Nordlund, Å., Brandenburg, A., Jennings, R. J., Rieutord, M., Ruokolainen, J., Stein, R. F., Tuominen, I. 1992, *Ap.J.* **392**, 647.

Pedersen, B. B., VandenBerg, D. A., Irwin, A. W. 1990, *Ap.J.* **352**, 279.

Pinsonneault, M. H., Kawaler, S. D., Sofia, S., Demarque, P. 1989, *Ap.J.* **338**, 424.

Sackmann, I.-J., Boothroyd, A. I. 1991, *Ap.J.* **366**, 529.

Schatzman, E. 1962, *Ann. d'Astrophys.* **25**, 18.

Schwarzschild, M. 1958, *Structure and Evolution of the Stars*,(Princeton Univ. Press).

Spite, M., Spite, F. 1985, *Ann. Rev. A. Ap.* **23**, 225.

Stein, R. F., Nordlund, Å. 1989, *Ap.J. Letters* **342**, L95.

Stein, R. F., Nordlund, Å. 1991, in *Challenges to Theories of the Structure of Moderate Mass Stars*, J. Toomre and D. O. Gough (eds.), *Lecture Notes in Physics* **388**, 195 (Springer: Heidelberg).

Thomas, R. N. 1989, in Cram and Kuhi (1989), p. ix.

Vainshtein, S. I., Zeldovich, Ya., B. 1972, *Sov. Phys. Usp.* **15**, 159.

VandenBerg, D. A. 1983, *Ap.J. Suppl.* **51**, 29.

VandenBerg, D. A. 1992, private communication.

VandenBerg, D. A., Laskarides, P. G. 1987, *Ap.J. Suppl.* **64**, 103.

Wheeler, J. C., Sneden, C., Truran Jr., J. W. 1989, *Ann. Rev. A. Ap.* **27**, 279.

Zahn, J.-P. 1992, *A.&Ap.*, submitted.

Inside the Stars, IAU Colloquium 137
ASP Conference Series, Vol. 40, 1993
Werner W. Weiss and Annie Baglin (eds.)

TEMPERATURE, ABUNDANCE AND MAGNETIC MAPPING OF STELLAR ATMOSPHERES

J.-F. DONATI
University of Western Ontario, London Ontario, N6A 3K7 Canada

ABSTRACT In this paper, I review recent progress in the domain of indirect imaging of stellar atmospheres. Since all stars other than the Sun are too far away for us to disclose any feature on their surface directly, one has to use an indirect tool to image their atmosphere. This method (Doppler Imaging) takes advantage of a star's rotation to identify and characterise inhomogeneities on its surface. This is done by studying *both* how inhomogeneities (carried in and out of the observer's view by rotation) modulate photometric and spectroscopic data with *time*, and how surface features distort the *shape* of line profiles at a given rotational phase. In the last decade, Doppler Imaging has first been applied to abundance mapping of magnetic Ap stars, then extended to temperature and magnetic mapping of cool active stars. Allowing a solar-like way of studying the surface of stars, this technique provides a wealth of information about stellar atmospheres and, in particular, about the physical processes which dictate their dynamics and energy balance.

INTRODUCTION

In ancient Greece and China, astronomers first observed that the surface of our Sun was non uniformly bright, and covered with dark spots. Since then, we have learned a lot about the inhomogeneous and stratified aspect of the solar atmosphere; for instance, kG magnetic fields have been found in spots, granules are observed in the quiet photosphere, plages are seen above spots at the chromospheric level. In particular, we know now that, despite their small relative size, these features are extremely important and witness the interaction between dynamical and radiative processes in the solar atmosphere, as well as their structuration by the magnetic field. Therefore, in the same way as the interpretation of some physical phenomenon at a macroscopic scale may require a microscopic description, the understanding of the overall physics of the Sun's atmosphere calls for unravelling the processes at work in small size inhomogeneities.

Most information we have already and we expect to get soon about fine solar atmospheric structures comes from angularly resolved observations, allowed by the proximity of the Sun relative to the Earth. Among stars other than the Sun, the M2 supergiant star Betelgeuse (α Ori) is the only one for which a recent interferometry study (Buscher et al. 1990) concludes that a large scale bright feature is probably present on its visible hemisphere. However, due to the resolution limit of the method, no fine detail of the stellar surface is revealed, though this star has the largest angular diameter after the Sun. For all other stars, we cannot yet investigate details on their surfaces with direct imaging methods and will likely not be able to do so in a close future. However, for some type of stars and in particular for magnetic Ap and cool active stars, we already have ample evidence, through photometric and spectroscopic observations, that their surface is highly non uniform.

This inhomogeneous aspect of stellar surfaces was first suggested by the fact that certain indexes (like the equivalent width of given spectral lines, the disc-integrated longitudinal component of the magnetic field or UBVRI luminosities of the star) are showing periodic modulation. Since different types of indexes in a given star are varying with the same frequency, the observed modulation was attributed to the star's rotation, carrying features successively in and out of the observer's view (Stibbs 1950; Kron 1952). For rapid rotators, more information about the non uniform surface distribution at a given rotational phase may be found in the often asymmetric *shape* of rotationally broadened spectral lines; each profile can indeed be seen as a one-dimensional map of the star's visible disc, resolved by Doppler effect in a direction perpendicular to both rotation axis and line of sight, but blurred in the orthogonal direction (Pyper 1969; Fekel 1983).

Both time modulation and line profile distortions of stellar spectra have been used since early attempts to map surfaces of magnetic Ap stars (Pyper 1969) or solar-type stars (Dorren et al. 1981; Vogt & Penrod 1983). In cool active stars, light curves and profile distortions are attributed, as in the Sun, to temperature (i.e. brightness) spots unevenly distributed over the stellar surface and probably caused by the emergence of magnetic fields out of the photosphere. On magnetic Ap stars, by contrast, it is now widely accepted that chemical abundance rather than temperature varies from one point of the atmosphere to another, in relation with the surface magnetic field. In both cases, mapping techniques aim at recovering an image of the atmosphere from the observations. Computing light curves or line profiles produced by a rotating star with a *known* atmospheric model, geometry and distribution, is a relatively straightforward task. The problem at hand is the inverse of this.

The very first method was developed for magnetic Ap stars (Deutsch 1970) and consisted in a direct analytical inversion. From the observed equivalent width and mean Doppler shift periodic variations of a disc-integrated line, one could, in principle, determine the coefficients of a spherical harmonics expansion of the distribution of *local* equivalent widths at the surface of the star. However, for two main reasons, this technique turned out to be ineffective. Since the expansion series was truncated after a few terms to keep the analytical problem tractable, the type of surface feature that could be imaged was severely restricted. Moreover, only the *integral* characteristics of the line profiles were used, while the very rich information contained in their *shape* was neglected. Although several improvements were brought to this first method to take better account of the profile shape of lines (e.g. Pyper 1969; Falk & Wehlau 1974, Mégessier 1975), it was finally abandoned and replaced by a new generation of more powerful tools.

Instead of a direct inversion, several groups began trying to image stars by manual trial and error iterative procedures (Khokhlova 1975; Vogt & Penrod 1983). At the same time, pioneering work by Goncharskij et al. (1982)

in attempting to define stellar imaging more rigorously led to the understanding that this problem is fundamentally ill-posed, i.e. has multiple solutions. An additional criterion must be invoked to select one of them, for instance the "simplest". Recently, robust and efficient algorithms have been developed to make the whole imaging process (Doppler Imaging) fully automatic. Extensive simulations were also carried to determine when test distributions are correctly reconstructed, i.e. what sort of data requirements are needed to recover successfully an input image (Vogt et al. 1987; Rice et al. 1989; Piskunov et al. 1990; Brown et al. 1991).

In this paper, I first outline the principles of Doppler Imaging and point out the individual peculiarities of temperature, abundance and magnetic mapping. I then describe the important results that have been obtained to date in this domain, and their implications for our understanding of stellar atmospheres. To conclude, I mention which orientations stellar imaging could take soon, as well as the possible astrophysical fallouts.

IMAGING METHOD

Principles

As mentioned in the introduction, the profile *shape* of spectral lines is the richest source of information about non uniform surface distributions on a rapidly rotating star. Due to Doppler effect induced by *rotation*, the visible disc of the star is sliced into strips of constant Doppler shift. Each inhomogeneity on the disc will then affect in some way the spectral contribution of the corresponding isovelocity strip, and thus distort the disc-integrated profile of the line at this given Doppler shift. Hence, each profile can be seen as a 1D distribution of the visible hemisphere of the star. From a time series of similar 1D distributions, one can follow how widely and how fast these distortions move across the line profile during a rotational cycle, and infer some accurate information about the location and shape of the parent inhomogeneities, as well as about their physical properties. Using profile sets in all Stokes parameters (intensity I, linear polarisations Q and U, circular polarisation V) and for various spectral lines should, in principle, allow for the simultaneous reconstruction of abundance/brightness and magnetic distributions.

Formally, the relation between the image $\mathbf{I}$ to be reconstructed and the observed data set $\mathbf{D}$ may be written $\mathcal{R}(\mathbf{I}) = \mathbf{D}$ where $\mathcal{R}$ notes a function which includes all line formation calculations required when computing profiles from a known atmospheric distribution $\mathbf{I}$, model structure and geometry. Since the problem is ill-posed (Goncharskij et al. 1982), a regularisation function S is used to select one of the solutions compatible with the data, for instance the "simplest" or "smoothest" one. For that purpose, the Entropy function (Skilling & Bryan 1984) and the Tikhonov function (Tikhonov 1963) have been invoked most of the time, both yielding very similar results (Piskunov 1991). The inversion hence resumes to maximise a Lagrange function $Q = S - \lambda C$ (λ is a conventional Lagrange multiplier), i.e. to maximise S under the constraint of a fit to the data at an accuracy of $C = C_{aim}$ where C is a statistic (for instance a chi-square) and C_{aim} the desired accuracy of the fit. An iterative algorithm (often based on a conjugate gradient method) is used to converge towards the solution.

To compute the function $\mathcal{R}$ and its gradient at each iteration, LTE line formation routines with atmospheric models are used. Since local profiles need to be obtained at each grid cell on the stellar surface, at each rotational phase and at each iteration of the code, direct integration of the transfer equation as part of the imaging process requires an enormous amount of computing time, that only super computers can handle. To overcome this problem, tabulated LTE profiles corresponding to different abundances, temperatures and limb angles are first built prior to the imaging process. Then, during the inversion, the code interpolates in this table to obtain the profiles needed. Though sophisticated already, intrinsic profiles calculations could be wrong by a significant amount, due for instance to the unrealistic LTE atmospheric models used. However, in these rapidly rotating stars, the rotational profile is usually much wider than the intrinsic one, whose main effect is thus line broadening. The synthetic *shape* of local profiles has therefore little influence on the imaging process.

Of course, Doppler Imaging relies on several assumptions. It assumes first that the observed profile variations are entirely due to rotational modulation. If the observed profile variations are correctly reproduced by the imaging process and sufficient phase coverage is obtained, it usually implies that this assumption is verified *a posteriori* (e.g. Hatzes 1990). This can also be checked independently by comparing profiles recorded at a given phase, but from different rotational cycles. In addition, Doppler Imaging often assumes that the star is spherical and rotates as a rigid body, but could easily cope with ellipsoidal stars and differential rotation if needed. However, these effects are likely to be quite small on the inferred maps.

In case of a slow rotator ($v \sin i < 10$ km/s where v notes the star's equatorial velocity and i the inclination angle of its rotational axis to the line of sight) or poor spectral resolution [instrumental profile with HWHM (km/s) $> v \sin i$], spectral profiles can no longer be seen as 1D distribution of the star, since they mainly reflect a local profile averaged over the disc and smeared by the instrumental broadening. Only disc-integrated quantities, such as BV luminosity, line equivalent width, average longitudinal magnetic field, can then be measured throughout the rotational cycle. If such variations can be accounted for with simple spotted (e.g. Dorren et al. 1981) or off-centred dipole (Borra & Landstreet 1980) structures, this sort of data modelling has to assume *a priori* a given kind of structure to be imaged (e.g. one with two cool circular spots of variable size and location). Since such models can be totally wrong [e.g. starspots could be bright rather than dark (Vogt 1981; Donati et al. 1992b)] and, at the same time, perfectly account for the observed data, the derived results highly depend on the assumed structure rather than just on the observations. For this reason, such modelling processes are not considered as *true* imaging and will not be discussed herein. However, some of the results obtained will be mentioned.

Before working with real data, it is important to check with adequate simulations that imaging behaves well in the particular case one plans to use it, and to test for instance that the data to be inverted contain enough *spatial* information to characterise the parent structure. A basic simulation consists in two steps. A synthetic image is first produced, from which sets of profiles are computed and artificially noised. In a second step, a map is recovered from these profiles and compared to the original one. Simulations with different kinds of synthetic maps have been presented to date in the literature. Most of them describe reconstructions of sharp-edged features such as spots, deviating from a uniform background and usually covering a small fraction of the total stellar surface. Some others simulate reconstructions of structures smoothly varying from one surface point to another and covering most of the stellar surface, such as dipoles. In the following sections, I review the major results obtained from simulations with these two different kinds of images, referred to as "spotted" and "dipolar-like" distributions.

Temperature and abundance mapping

On a cool star, a dark spot causes a *deficit of photons* to the isovelocity strip in which it is located, therefore decreasing its intensity-weighted area. Since the intensity-weighted area of an isovelocity strip is proportional, at first order, to the *depth* of a disc-integrated profile at the corresponding Doppler shift, it implies that a dark spot shows up as a hump in a disc-integrated profile. For the strong lines commonly used in cool stars imaging, line equivalent width variations with temperature have usually smaller effects than continuum variations and can be seen as second order perturbations. On magnetic Ap stars by contrast, the temperature is roughly constant over the photosphere and local abundance inhomogeneities generate local increase or decrease in line equivalent widths, producing respectively notches or humps in disc-integrated profiles.

Several papers (Vogt et al. 1987; Rice et al. 1989; Hatzes et al. 1989; Piskunov et al. 1990; Piskunov & Wehlau 1990; Piskunov 1991; Brown et al. 1991, Jankov & Foing 1992) present simulated imaging of synthetic brightness and abundance maps. For both brightness and abundance imaging, spotted distributions are found to be reasonably reconstructed; in particular, both shape and location of spots are well recovered, while the brightness or abundance deviation from the uniform background is sometimes slightly underestimated. If brightness imaging of dipolar-type structures behaves quite well (Brown et al. 1991), abundance distributions of similar type are usually less accurately recovered, with poles being dragged slightly but significantly to the subobserver latitude.

The influence of low S/N, low resolution or badly phase-sampled data have been discussed extensively in the papers quoted above, the main consequence being a progressive lessening and smearing of details in the reconstructed map. More problematic are errors resulting from a wrong guess of some atmospheric parameters like $v \sin i$, i or the limb darkening constant. These may cause spurious features in the derived images (Vogt et al. 1987). Such artifacts are easy to detect though and are actually used for a fine tuning of atmospheric parameters.

The strongest limitation in Doppler Imaging is that the profile corresponding to a uniform photosphere, from which distortions are measured and interpreted, is basically *unknown*. Indeed, this "undisturbed photospheric profile" is scarcely observed, since such stars often show at least one spot at each rotational phase. We thus have to take the classical round-shape rotational profile as granted. If this round-shape profile correctly accounts for lines of unspotted rapid rotators (such as Am stars), it could be affected in spotted stars by, say, the presence of a chromosphere. Since such perturbations tend to distort the profiles at all different phases the same way, latitude independent features (i.e.

rings) encircling the star and polar artifacts are expected. For this reason, it is safer to confirm each feature reconstructed by Doppler Imaging, whenever independent results, such as magnetic maps (see below), are available.

In a general case, temperature/abundance and magnetic mappings should be done simultaneously. Indeed, magnetic fields can broaden significantly and even split Stokes I profiles of spectral lines. For fields of several kG, such distortions are comparable in size to those produced by temperature/abundance inhomogeneities. However, in the case of surface fields weaker than about 1 kG, unpolarised profiles are only slightly affected and temperature/abundance mapping can be dissociated from magnetic imaging (Rice & Wehlau 1990).

Magnetic mapping

To study stellar magnetic fields, we usually analyse Zeeman distortions of spectral lines. Since Zeeman effect is, by far, the likeliest cause of circular (Stokes V) or linear (Stokes Q and U) polarisation *in line profiles*, observing in polarised light is considered as the optimal tool for investigating stellar magnetic structures. In this kind of study, circular polarisation signatures are more often measured than linear polarisation profiles, which are usually about an order of magnitude smaller and thus only accessible for stars with fields of at least several kG. In case of very strong fields, Stokes I profiles may exhibit Zeeman splitting if rotational broadening is small (i.e. $v \sin i < 20$ km/s). For most stars rotating rapidly enough to be Doppler imaged, Zeeman signatures in unpolarised profiles mainly reduce to line broadening, which can be identified reliably only for fields stronger than several kG.

Magnetic mapping from polarisation profiles of spectral lines (Zeeman-Doppler Imaging or ZDI) was first proposed by Semel (1989) and studied more extensively in following papers (Donati et al. 1989; Donati & Semel 1990; Donati 1990; Brown et al 1991; Semel et al. 1992). It has several advantages compared to conventional Doppler Imaging. As magnetic field is the likeliest source of polarisation *in line profiles*, the "undisturbed photospheric profile" involved in the imaging process is now perfectly known; for a null field structure, Stokes Q, U and V profiles are indeed flat *by definition*. No spurious features, and in particular, no *polar artifacts* can be generated in magnetic images, which are thereby much more reliable than temperature or abundance maps. Such images can be used to check independently the reality of brightness/abundance features reconstructed in conventional Doppler images.

However, ZDI has certain limitations as well. Brown et al. (1991) showed that only spotted magnetic structures (Fig. I) can be reconstructed from sets of circular polarisation Stokes V profiles. If the information coming from sets of linear polarisation profiles is not incorporated, dipolar-like structures are not correctly recovered. In particular, the poles of an originally dipole field distribution are both reconstructed at the subobserver latitude; hence, distributions "simpler" than the original one are compatible with the data. This means essentially that sets of Stokes V profiles do not contain *enough spatial information* about magnetic dipolar structures, making them impossible to be recovered unambiguously when sets of Stokes Q and U profiles are not used.

Another drawback is the fact that fields in very dark or under-abundant regions may easily be dimmed or even totally obscured since the polarisation

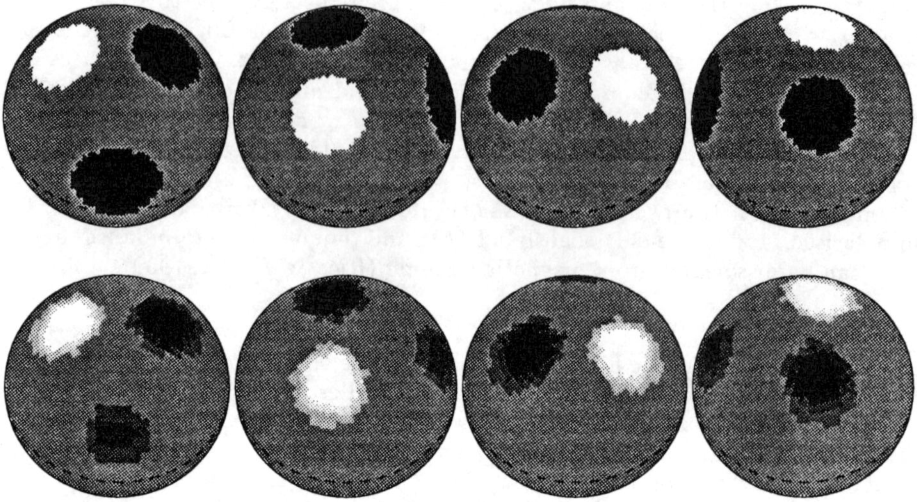

FIGURE I Maximum entropy reconstruction (lower row) of a synthetic spotted magnetic distribution (upper row) from Stokes V data. From left to right, the star is shown at rotational phases 0.00, 0.25, 0.50 and 0.75 (as in all following figures). The location, field intensity, and shape of spots are reasonably well recovered (Brown et al. 1991).

rate generated by such features is very small, and partially hidden by photon noise. Using a brightness or abundance image derived from corresponding Stokes I data sets may partly compensate for this problem, but cannot, of course, regenerate the missing information drowned by noise.

RESULTS

Magnetic Ap stars

In magnetic Ap stars, elements are spread inhomogeneously over the stellar surface, in a distribution which often reflects the magnetic structure. According to the diffusion theory (Michaud 1970), the strong fields observed on such stars inhibited large scale mixing processes like convection, meridional circulation and turbulence in the outer layers of the star, leading to a very stable photosphere. Microscopic diffusion then becomes the dominant transport process at the stellar surface, with neutral elements migrating radially while ions follow the field lines. However, this so called "simple diffusion model" can only explain the average abundance anomalies observed for instance on 53 Cam (Landstreet 1988), but fail to account for the abundance *variations* over the surface of this star (Babel & Michaud 1991a). The effects of magnetic fields on abundance distributions (Vauclair at al. 1979; Michaud et al. 1981; Mégessier 1984; Babel & Michaud 1991b) is not well understood yet. In particular, inhomogeneous mass loss seems required to match observations better. What have we learned to date from surface imaging of magnetic Ap stars?

Fields of magnetic Ap stars have been extensively studied since Babcock (1958) detected them. Till recently, works on fields of magnetic Ap stars mainly consisted in measuring the disc-averaged longitudinal and surface field throughout the rotational cycle of a star and fitting them with centred or off-centred dipole models (e.g. Borra & Landstreet 1980; Mathys 1991) or more complex non axisymmetric models (Stift & Goosens 1991). Very few studies measured polarisation *profiles* of spectral lines and used their *shape* to derive the parameters of an assumed dipolar magnetic distribution (Borra 1980; Piskunov 1985; Glagolevskij et al. 1985; Donati et al. 1990b). Though such fits with dipole models are not true imaging methods (since they assume *a priori* a given large scale topology of the field), they still remain the best way to characterise magnetic structures of Ap stars till magnetic maps of dipolar-like distributions can be recovered. This should be possible, in principle, when linear polarisation profiles of spectral lines can be measured (Brown et al. 1991).

In stars with strong magnetic fields, Stokes I profiles show clear Zeeman distortions. Since magnetic fields cannot be mapped accurately yet, abundance mapping is also subject to caution; hence, only rough information about the abundance distribution of elements may be inferred (e.g. Landstreet 1988). However, even such limited spatial information proved to be very useful to constrain better diffusion models, and to show that mass loss seems required to explain the geographic distribution of elements on the surface of 53 Cam (Babel & Michaud 1991a; Babel 1992).

In the case of weak field Ap stars (surface field < 1 kG), Stokes I profiles are predominantly distorted by abundance inhomogeneities so that Doppler Imaging becomes valid. Within the last three years, considerable effort has been realised in trying to recover accurate abundance maps of magnetic Ap stars, and images have been derived for 6 different objects, mainly from elements like Fe, Cr and Si (Hatzes et al. 1989; Rice & Wehlau 1990, 1991; Hatzes 1990; Donati 1990; Hatzes 1991a, 1991b). There is some general agreement on some typical features found in these maps. All images show evidence of rings and spots distributed in a basically axisymmetric distribution. The observed rings are often irregular (e.g. Rice & Wehlau 1991). For a given star, distributions in different elements have the same axis of symmetry (Hatzes 1991a). Note that, for stars on which a field was detected, the symmetry axis of the abundance distribution sometimes differ from the magnetic axis derived from dipole modelling (e.g. Rice & Wehlau 1991). However, the derived abundance and magnetic poles, whenever non coincident, always agree in *longitude*. The remaining discrepancy is likely attributable to the large *latitude* uncertainties in the process of dipole modelling of disc-integrated magnetic data (such as Hβ polarimetric measurements).

Inhomogeneous element distributions in such stars are therefore very likely induced by the magnetic structure, as first proposed by Stibbs (1950). In such a context, the Si-depleted spots and Si-enhanced irregular rings observed on θ Aur (Rice & Wehlau 1991; Hatzes 1991a), γ^2 Ari (Hatzes et al. 1989) and BP Boo (Hatzes 1990) could indicate respectively the magnetic poles and equator. By contrast, on ϵ UMa (Rice & Wehlau 1990; Hatzes 1991b), θ Aur (Rice & Wehlau 1990; Hatzes et al. 1991a), ω Her and 45 Her (Hatzes 1991b), Cr seems depleted at the magnetic equator *and* at one pole at least, while intermediate regions are over-abundant. Fe is sometimes depleted at the magnetic poles and enhanced at the magnetic equator as in α^2 CVn (Khokhlova & Pavlova 1984), or vice versa as in ϵ UMa (Donati 1990; Rice & Wehlau 1991). Little has been done to date about rare-earth abundance distributions in magnetic Ap stars. Early results on α^2 CVn (Goncharskij et al. 1983) suggest that these elements concentrate near one or both magnetic poles.

Solar-type stars

Dynamo models describe tentatively the MHD processes observed at the surface of the Sun and other cool active stars. A weak radial/poloidal magnetic field is first generated by rotation in the convection zone of a cool star. Differential rotation then winds field lines in a mainly toroidal structure. Cyclonic motions in the convective zone act to bring back the initial radial/poloidal magnetic structure through reconnection of field lines (Parker 1970). As this process goes on, many activity demonstrations are observed, such as photospheric sunspots witnessing the emergence of flux tubes out of the convective zone by buoyancy, or strong emission lines produced by a high-temperature chromosphere and transition region. If observations have shown us already that such phenomena also occur on many other active stars, only stellar imaging can bring accurate information about their *local* physical properties.

However, most cool dwarfs, subgiants or giants of spectral type later than early G rotate very slowly ($v \sin i < 10$ km/s) due to some kind of magnetic breaking or/and angular momentum redistribution process starting at about this evolutionary stage. As a result, indirect surface imaging as described above does not apply for such slow rotators since the spatial information contained in the profile *shape* of spectral lines is greatly reduced. In this case, rotational modulation (e.g. light curves) yields most of what we can learn about temperature distributions of such stars (e.g. Vogt 1981; Rodonò et al. 1986). Although the simple spot modelling often used for such data is not true imaging, it can still be very informative when analysing *long term* variability, like *changes* in light curves shape (e.g. Strassmeier & Bopp 1992); such a method indeed evidenced that differential rotation in cool active stars *decreases* when rotation rate increases (Hall 1991). In addition to rotational modulation, one can also use the very small variability in profile shape by looking at changes in line *bisectors* (Toner & Gray 1988). If this can bring further information and support on the location and physical properties of surface inhomogeneities, the somewhat arbitrary spot model is still required.

Even if less numerous, rapidly rotating cool active stars can yield very important clues about activity phenomena, observed to strengthen with increasing rotation rates (Strassmeier at al. 1990; Saar 1990). Most rapid rotators in this spectral class are associated in close binary systems, whose *orbital* angular momentum sustains, through tidal coupling (Zahn 1977), a high rotation rate for both components. For 6 objects, most of them being RS CVn systems, temperature images have been reconstructed to date (Vogt 1988; Piskunov et al. 1990; Strassmeier et al. 1991; Vogt & Hatzes 1991; Donati et al. 1992b; Strassmeier et al. 1992). Recently, ZDI has been applied to RS CVn binaries (Donati et al. 1990a), leading to the detection of circularly polarised Zeeman signatures in 4 systems (Donati et al. 1992a), and to the first magnetic mapping of a solar-type star other than the Sun (Donati et al. 1992b).

Although discrepancies still exist between images reconstructed by different

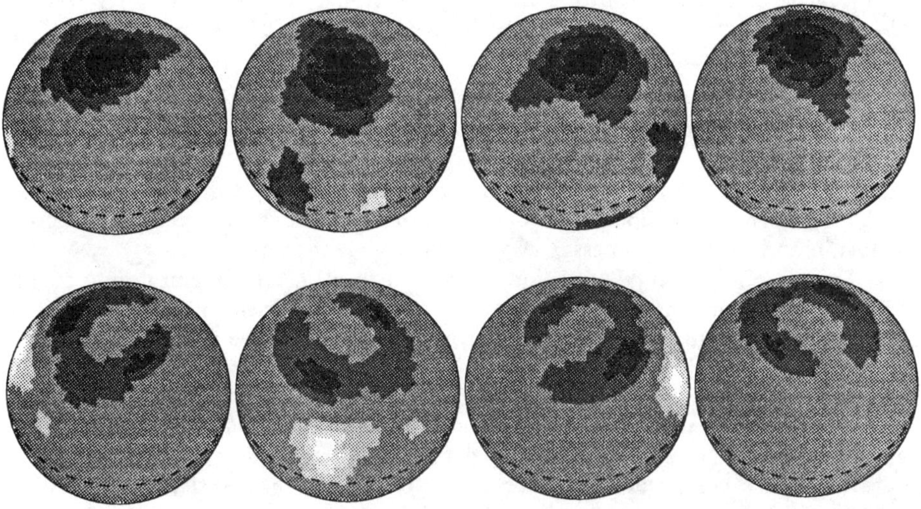

FIGURE II Temperature (upper row) and magnetic (lower row) photospheric image of the K1 subgiant of HR 1099 obtained with Zeeman-Doppler Imaging (Donati et al. 1992b). Magnetic regions are found to be associated with toroidal field (indicated by darker and brighter grey levels with steps of 200 G) and to coincide with cool spots in the temperature image.

groups from the same data set (Strassmeier et al. 1991), some results seem relatively well established now. Large starspots, about 1000 K cooler than the photosphere, often develop at high latitudes (and sometimes even on the pole) in such objects, while sunspots are confined to equatorial belts. Since Doppler Imaging can produce polar artifacts, the concept of polar spot has been very controversial over the last few years. However, new temperature and magnetic images obtained for the K1 subgiant of the RS CVn system HR 1099 by Donati et al. (1992b) suggest that magnetic fields of several hundred G are associated with polar spots, thereby strongly supporting the reality of such features. Apparently, polar spots are long-lived and bear some resemblance with solar coronal holes, as noted by Vogt (1988); being associated with fields at least 100 times stronger than solar, these features could become prominent at the photospheric level in these stars, while they are only visible at the coronal level in the Sun. Both bright and dark equatorial features are also detected on these stars (e.g. Piskunov et al. 1990), though often smaller and more transient than polar ones.

The magnetic image of HR 1099 (Donati et al. 1992b) shows that starspots are not active solar complexes made of many small spots of opposite polarities [as proposed earlier by Vogt (1988)], but large unipolar magnetic regions (see Fig. II). Moreover, the field *orientation* in these regions varies with time; Stokes V spectra of HR 1099 suggest that the magnetic structure of the K1 subgiant evolved from a radial/poloidal structure at epoch 1989.6 to a mainly toroidal configuration at epoch 1990.9, probably due to the winding of field lines around the star by differential rotation (Donati et al. 1992b). From a time series of images obtained at nearby epochs, Vogt & Hatzes (1991) find that differential rotation in the rapidly rotating K0 subgiant of UX Ari is significantly smaller than solar [in agreement with the result of Hall (1991)], but with a pole rotating *faster* than the equator.

FUTURE ORIENTATIONS IN STELLAR IMAGING

If stellar imaging produced many valuable results already, it is still in its infancy and many other directions shall be explored soon and reveal new clues about diffusion and dynamo processes at work in stellar atmospheres.

Several observational strategies may lead to a better understanding of diffusion mechanisms in magnetic Ap stars. Firstly, for each observed star, it is important to obtain abundance maps from many different elements in both iron-peak and rare-earth groups, having very different sensitivities to diffusion processes. At the same time, true mappings of the magnetic structure of such stars (from a simultaneous inversion of all four Stokes profiles) could tell us how local abundance inhomogeneities relate to variations in both field strength and orientation. Such comprehensive information could help in identifying, modelling and constraining the main process (possibly mass loss) leading to surface inhomogeneities in the atmosphere of such stars (Babel & Michaud 1991a; Babel & Michaud 1991c; Babel 1992). Finally, abundance maps at different wavelengths (from UV to IR) could be derived to obtain further observational evidence and accurate measurements of the stratification of elements in the photosphere of magnetic Ap stars (Lanz 1987; Babel & Lanz 1992). In particular, this should also allow quantitative tests of diffusion theories and models.

Indirect imaging can also bring valuable information about dynamo processes in cool active stars. For instance, one key-project of the coming years could consist in monitoring a few stars throughout their activity cycle, i.e. getting several atmospheric images per year for a total period of about a decade. From such data, one could first study how spots appear, migrate and disappear on the stars' surface. In particular, this will allow measurements of differential rotation which could be used to clarify the role of this parameter in dynamo models. Simultaneous magnetic images will also show how flux tubes emerge from the photosphere, and how the global magnetic structure is progressively distorted during an activity cycle. Particularly interesting should be the changes in field orientation when the star evolves from activity minimum to activity maximum, and then back to activity minimum (Donati et al. 1992b). Finally, indirect imaging could be extended to the upper atmosphere of such stars, as initiated by Gondoin (1986) and Neff et al. (1989), but in a much more sophisticated way. Simulations suggest that chromospheric and transition region maps could be reconstructed (see for instance Fig. III) using UV PRISMA data ($S/N = 50$ and $\lambda/\Delta\lambda = 30000$). Images from several lines could yield temperature and density distributions with height in active regions on various stars. In particular, this could bring new clues in our understanding of chromospheric and coronal heating in cool active stars.

As instrumentation improves, other classes of stars could be studied with surface imaging. In particular, investigations of T Tauri, BY Dra and Ae/Be Herbig stars could bring an unprecedented highlight of atmospheric phenomena in early stages of stellar evolution.

Fig. IIIa

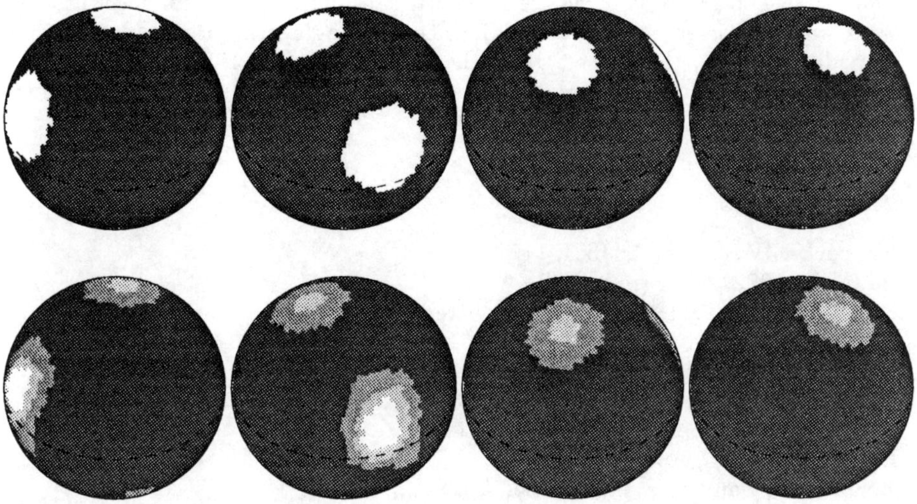

Fig. IIIb

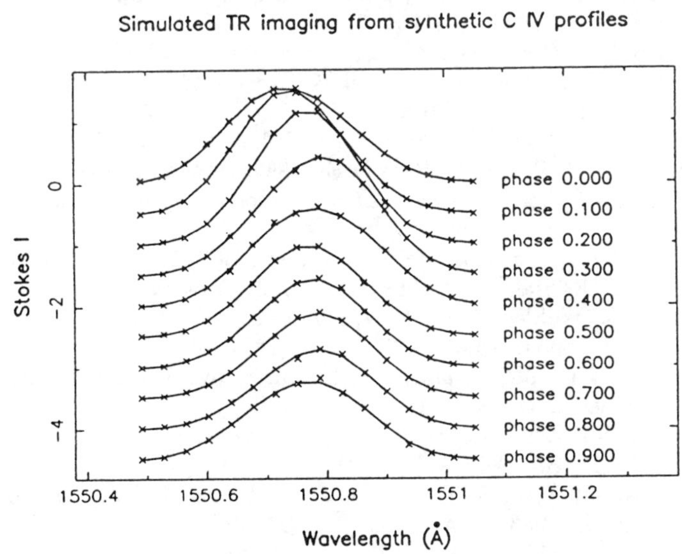

FIGURE III Simulated maximum entropy reconstruction (lower row of **a**) of a synthetic transition-region spotted structure (upper row of **a**) from synthetic C IV line profiles. Grey levels indicate line strength with steps of 200% while medium grey indicates the quiet transition region. In **b**, the synthetic C IV profiles (crosses) used for the maximum entropy fit (full line) correspond to UV data as PRISMA could collect them.

ACKNOWLEDGMENTS

I would like to thank F. Praderie for a critical reading of the manuscript. This work was supported by the Natural Science and Engineering Research Council of Canada. I warmly thank the SOC of the IAU Colloquium 138 "Inside the Stars" and the Centre National de la Recherche Scientifique for providing additional travel funds.

REFERENCES

Babcock H.W., 1958, ApJS 3, 141
Babel J., 1992, A&A (in press)
Babel J., Lanz T., 1992, A&A (submitted)
Babel J., Michaud G., 1991a, ApJ 366, 560
Babel J., Michaud G., 1991b, A&A 241, 493
Babel J., Michaud G., 1991c, A&A 248, 155
Bopp B.W., Fekel F.C., 1976, AJ 81, 771
Borra E.F., 1980, ApJ 235, 915
Borra E.F., Landstreet J.D., 1980, ApJS 42, 421
Brown S.F., Donati J.-F., Rees D.E., Semel M., 1991, A&A 250, 463
Buscher D.F., Haniff C.A., Baldwin J.E., Warner P.J., 1990, MNRAS 245, 7P
Deutsch A.J., 1970, ApJ 159, 985
Donati J.-F., 1990, PhD thesis, Université Paris VII
Donati J.-F., Brown S.F., Semel M., Rees D.E., et al., 1992b, A&A (in press)
Donati J.-F., Semel M., 1990, Solar Phys. 128, 227
Donati J.-F., Semel M., Praderie F., 1989, A&A 225, 467
Donati J.-F., Semel M., Rees D.E., 1992a, A&A (in press)
Donati J.-F., Semel M., Rees D.E., Taylor K., Robinson R.D., 1990a, A&A 232, L1
Donati J.-F., Semel M., del Toro Iniesta J.C., 1990b, A&A 233, L17
Dorren J.D., Siah M.J., Guinan E.F., McCook G.P., 1981, AJ 86, 572
Falk A.E., Wehlau W.H., 1974, ApJ 192, 409
Fekel F.C., 1983, ApJ 268, 274
Glagolevskij Y.V., Piskunov N.E., Khokhlova V.L., 1985, Pis'ma AZh 11, 371 (SvA Lett. 11, 154)
Goncharskij A.V., Stepanov V.V., Khokhlova V.L., Yagola A.G., 1982, AZh 59, 1146 (SvA 26, 690)
Goncharskij A.V., Ryabchikova T.A., Stepanov V.V., Khokhlova V.L., Yagola A.G., 1983, AZh 60, 83 (SvA 27, 49)
Gondoin Ph., 1986, A&A 160, 73
Hall D.S., 1991, in: The Sun and Cool Stars: Activity, Magnetism, Dynamos. IAU Coll. 130, 353. Tuominen I., Moss D., Rüdiger G. (eds.). Springer, Berlin Heidelberg New York
Hatzes A.P., 1990, MNRAS 245, 56
Hatzes A.P., 1991a, MNRAS 248, 487
Hatzes A.P., 1991b, MNRAS 253, 89
Hatzes A.P., Penrod G.D., Vogt S.S., 1989, ApJ 341, 456
Jankov S., Foing B.H., 1992, A&A 256, 533

Khokhlova,V.L., 1975, AZh 52, 950 (SvA 19, 576)
Khokhlova V.L., Pavlova V.M., 1984, Pis'ma AZh 10, 377 (SvA Lett. 10, 158)
Kron G.E., 1952, ApJ 115, 301
Landstreet J.D., 1988, ApJ 326, 967
Lanz T., 1987, PhD thesis, Observatoire de Genève
Mathys G., 1991, A&AS 89, 121
Mégessier C., 1975, A&A 39, 263
Mégessier C., 1984, A&A 138, 267
Michaud G., 1970, ApJ 160, 641
Michaud G., Mégessier C., Charland Y., 1981, A&A 103, 244
Neff J.E., Walter F.M., Rodonò M., Linsky J.L., 1989, A&A 215, 79
Parker E.N., 1970, Ann. Rev. Astron. Astrophys. 8, 1970
Piskunov N.E., 1985, Pis'ma AZh 11, 44 (SvA Lett. 11, 18)
Piskunov N.E., 1991, in: The Sun and Cool Stars: Activity, Magnetism, Dynamos. IAU Coll. 130, 309. Tuominen I., Moss D., Rüdiger G. (eds.). Springer, Berlin Heidelberg New York
Piskunov N.E., Tuominen I., Vilhu O., 1990, A&A 230, 363
Piskunov N.E., Wehlau W.H., 1990, A&A 233, 497
Pyper D.M., 1969, ApJS 18, 347
Rice J.B., Wehlau W.H., 1990, A&A 233, 503
Rice J.B., Wehlau W.H., 1991, A&A 246, 195
Rice J.B., Wehlau W.H., Khokhlova V.L., 1989, A&A 208, 179
Rodonò M., Cutispoto G., Pazzani V., et al., 1986, A&A 165, 135
Saar S.H., 1990, in: Solar Photosphere: Structure, Convection and Magnetic Fields. Proc. IAU Symp. 138, 427. Stenflo J.O. (ed.). Kluwer, Dordrecht
Semel M., 1989, A&A 225, 456
Semel M., Donati J.-F., Rees D.E., 1992, A&A (submitted)
Skilling J., Bryan R.K., 1984, MNRAS 211, 111
Stibbs D.W.N., 1950, MNRAS 110, 395
Stift M.J., Goosens M., 1991, A&A 251, 139
Strassmeier K.G., Bopp B.W., 1992, A&A (in press)
Strassmeier K.G., Fekel F.C., Bopp B.W., Dempsey R.C., Henry G.W., 1990, ApJS 72, 191
Strassmeier K.G., Rice J.B., Wehlau W.H., et al., 1991, A&A 247, 130
Strassmeier K.G., Rice J.B., Wehlau W.H., Hill G., Matthews J.M., 1992, A&A (submitted)
Tikhonov A.N., 1963, Sov. Math. Dokl. 4, 1624
Toner C.G., Gray D.F., 1988, ApJ, 334, 1008
Vauclair S., Hardorp J., Peterson D.M., 1979, ApJ 227, 526
Vogt S.S., 1981, ApJ 250, 327
Vogt S.S., 1988, in: The Impact of Very High S/N Spectroscopy on Stellar Physics. IAU. Symp. 132, 253. Cayrel de Strobel G., Spite M. (eds.). Kluwer, Dordrecht
Vogt S.S., Hatzes A.P., 1991, in: The Sun and Cool Stars: Activity, Magnetism, Dynamos. IAU Coll. 130, 297. Tuominen I., Moss D., Rüdiger G. (eds.). Springer, Berlin Heidelberg New York
Vogt S.S., Penrod G.D., 1983, PASP 95, 565
Vogt S.S., Penrod G.D., Hatzes A.P., 1987, ApJ 321, 496
Zahn J.-P., 1977, A&A 57, 383

Inside the Stars, IAU Colloquium 137
ASP Conference Series, Vol. 40, 1993
Werner W. Weiss and Annie Baglin (eds.)

PARAMETERS T_{eff} AND log g FOR LATE B AND EARLY A TYPE STARS

M. GERBALDI and J. ZOREC
Institut d'Astrophysique de Paris, 98bis, Bd. Arago, 75014 Paris, France

ABSTRACT: We compare the log g parameters of early A and late B type stars determined from model atmospheres with those issued from models of stellar evolution. There is a systematic deviation between both determinations for log g < 3.7 .

PARAMETERS T_{eff} AND log g OBTAINED FROM STELLAR BOLOMETRIC FLUX AND ATMOSPHERE MODELS

As there is no measured stellar angular diameters of stars from B5 to early A type studied in this paper, we used an indirect procedure to determine T_{eff} based on: a) an iterative, Blackwell and Shallis' (1977) like method described in Zorec and Mercado (1987); b) visible and near-IR ground based flux observations (mainly 13-colour photometry of Johnson and Mitchel (1975)), and far-UV fluxes obtained by TD1 or ANS space observations; c) classical LTE line-blanketed model atmospheres for normal abundances (Kurucz 1979).

The main error on the T_{eff} determination is due to uncertainties of the adopted ISM-absorption colour excess E(B-V). We have compared the E(B-V), derived from the uvby photometry and using the Moon's (1985) code, to that adopted in this paper, determined mainly from the ISM λ2000 absorption bump and from the method of neighbouring stars. We saw that the uncertainty on E(B-V) is $|\delta E(B\text{-}V)| \simeq 0.02$ mag, which implies errors $\delta T_{eff}/T_{eff}$ ranging unsymmetrically from 5 % to 0.7 % according to the sign of δE(B-V).

To give a test of our T_{eff} determinations, we compared the parameters obtained by this method for *Vega* (HD 172167) to those generally admitted for this star. We found T_{eff} = 9467 K, log g = 4.10, while Kurucz gives T_{eff} = 9400 K and log g = 3.95. Vega is considered by some authors as a mild λ Boo type. We give then the values of T_{eff} obtained for some other stars belonging to the λ Boo group: HD 31295, T_{eff} = 9300 K; HD 1104111, T_{eff} = 9260 K; HD 125162, T_{eff} = 9215 K; HD 192640, T_{eff} = 8780 K. Comparing these values with those given by Venn and Lambert (1990), we see that ours are about 500 K higher. We remind that the temperature determined by this method, does not sensitively depend on the log g value chosen to perform the computation. It may however depend on the chemical abundances. Underabundant models lead for T_{eff} < 10000 K to higher values of θ, and so, lower values of T_{eff} of 200 to 300 K when log (metal abundance) ~ -1.0.

To derive the log g parameter, two approaches were used: 1) The first one depends on the Strömgren's Hβ Balmer line photometric index β taken from the Hauck and Mermilliod's (1990) compilation and its calibration in terms of log g done by Lester et al. (1986). Let us note that the stars used in this study have $T_{eff} \geq 10000$ K for which the β index is still a gravity sensitive parameter;

2) The second determination comes from the Moon and Dvoretsky's (1985) grid and its extension computed by Castelli (1991). The adopted log g parameter is the mean of both determinations. We have checked that both determinations are almost the same but they present a slight systematic difference. This difference may be probably due to calibration effects in the grids produced by Lester et al. (1984) and those by Moon and Dworetsky (1985), which were already discussed by Castelli (1991).

PARAMETERS log g OBTAINED FROM STELLAR EVOLUTION COMPUTATIONS

The log g parameter can be derived from the stellar evolution models using the (observed) bolometric luminosity $L/L_\odot$ and T_{eff}. The most important errors affecting $L/L_\odot$ are due to the distance d and E(B-V) determinations. For d we used the mean value obtained from ground based trigonometrical parallaxes and the distances reported in the literature derived by several independent methods. Our sample is made of stars having distances smaller than 200 pc and more than 60 % of them have $d \leq 100$ pc. The mean dispersion of compilated distances for each star is $\sigma(d) \simeq 25$ %. We could finally use the relation $\log g = \log g_\odot + \log (M/M_\odot) - 2 \log (R/R_\odot)$ where the masses $M/M_\odot$ were interpolated in the Maeder and Meynet's (1988) evolutionary tracks using our $(L/L_\odot, T_{eff})$ parameters. For each star the $R/R_\odot$ was calculated combining the angular diameter θ determined at the T_{eff} iteration with the adopted stellar distance d .

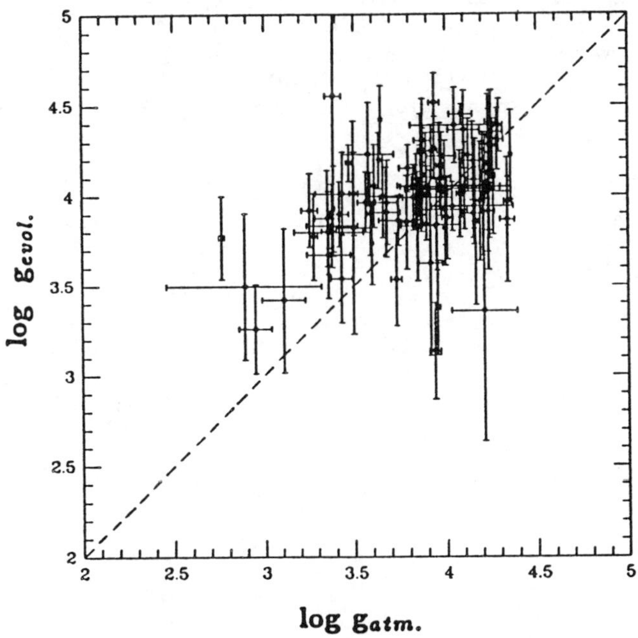

Figure 1: Comparison of log g parameters determined from stellar evolution calculations ($\log g_{evol.}$) with those obtained from stellar atmospheric parameters ($\log g_{atm.}$).

RESULTS AND DISCUSSION

In Figure 1 we compare both determinations of log g for the sample of stars studied in this paper. We see that there is a systematic deviation when log g $<$ 3.7 which cannot be attributed only to distance uncertainties and to log g determinations using stellar atmospheric parameters. The vertical error bars are due both to distance δd and absorption δE(B-V) uncertainties. The horizontal ones, represent the difference to the mean value of log g obtained by the two determinations issued only from atmospheric parameters.

To account for the systematic deviation δ(log g) $\simeq$ 0.5 seen in Figure 1 of somewhat evolved stars, we should admit a mass overestimation of a factor 3. Perhaps, difficulties of the stellar structure calculation at the Helium ionization region, could possible produced underluminous models, which should be partially responsible for this result (Maeder, pivate communication).

REFERENCES

Blackwell, D.E., Shallis, M.J.: 1977, *M.N.R.A.S.*, **180**, 177
Castelli, F.: 1991, *Astr. Ap.*, **251**, 106
Hauck, B., Mermilliod, J.C.: 1990, *Astr. Ap. Suppl.*, **86**, 107
Johnson, H.L., Mitchel, R.I.: 1975, *Rev. Mexicana Astr. Astrof.*, **1**, 299
Kurucz, R.L.: 1979, *Ap. J. Suppl.*, **40**,1
Lester, J.B., Gray, R.O., Kurucz, R.L.: 1986, *Ap. J. Suppl.*, **61**, 509
Maeder, A., Meynet, G.: 1988, *Astr. Ap. Suppl.*, **76**, 411
Moon, T.T.: 1985, Comm. London Observatory n^o **78**
Moon, T.T., Dworetsky, M.M.: 1985, *M.N.R.A.S.*, **217**, 305
Venn, K.A, Lambert, D.L.: 1990, *Ap. J.*, **363**, 234
Zorec, J., Mercado-Ibañez, R.: 1987, *C.R. Acad. Sci. Paris*, **305**, Serie II, 1279

Inside the Stars, IAU Colloquium 137
ASP Conference Series, Vol. 40, 1993
Werner W. Weiss and Annie Baglin (eds.)

EFFECTIVE TEMPERATURES AND ANGULAR DIAMETERS OF A-G MAIN SEQUENCE STARS

I.N. GLUSHNEVA, I.B. VOLOSHINA
Sternberg Astronomical Institute, Moscow 119899, Russia

L.N. KNYAZEVA
Fessenkov Astrophysical Institute, Alma-Ata 480068, Kazakhstan

ABSTRACT For 9 stars of A5-G2 spectral type interval IV and V luminosity classes effective temperatures, bolometric corrections, radii, angular diameters and luminosities were obtained by means of the infrared flux method in the modification by Blackwell et al. (1980).

The infrared flux method (Blackwell and Shallis, 1977) is used often for determination of the stellar effective temperatures and angular diameters, when reliable spectrophotometric measurements in the ultraviolet and visible ranges and infrared photometry are at our disposal.
9 stars were chosen of A5-G2 spectral type interval IV and V luminosity classes which are the members of the catalogue of 238 secondary spectrophotometric standards (Glushneva et al. 1992). In this catalogue energy distribution data are the mean of monochromatic fluxes in the range 3200-7600 A obtained independently at the Sternberg Astronomical Institute and Fessenkov Astrophysical Institute, Alma-Ata. Besides this catalogue contains also near infrared (6300-10800 A) energy distribution data for 99 stars obtained at the Sternberg Institute Crimean Station. So the possibility is to have homogenuous spectrophotometric data for the range wide enough from 3200 to 10800 A and it is especially important for A5 - G2 stars where the fraction of the total flux in this range is about 70-75 %.
Total fluxes of the investigated stars were obtained by means of the integration of monochromatic fluxes in the UV, visual and IR ranges. In the ultraviolet for l < 3200 A spectrophotometric data of IUE were used (Heck et al.,1984; Wu et al.,1991). For BS 21, BS 2943, BS 5986 and for BS 8162 (in the

range 1720-3200 A),where IUE low-dispertion data are absent, OAO-2 measurements were taken (Code and Meade, 1979).In the range 3200-10800 A spectrophotometric data obtained at the Sternberg Institute Crimean Station and Alma-Ata Astrophysical Institute from the catalogue of 238 secondary spectrophotometric standards (Glushneva et al., 1992) were used. For BS 483, BS 4540, BS 5986 and BS 6561 in the near-infrared range 7650-11084 A 13- color photometry by Johnson and Mitchell (1975) was taken using calibration given in this paper. In more longwavelength region up to 4 mm the fluxes were calculated using trapezium formula and absolute fluxes obtained from J_n, K_n, L_n photometry by Selby et al. (1988) and calibration of J, K, L bands by Saxner and Hammarback (1985) for the stars BS 483, BS 1543 and BS 2943. This calibration is based on the model of a Lyr by Dreiling and Bell (1980). In the visible and infrared up to 1 mm the calibration of a Lyr by Hayes (1985) was used. For BS 6561 where infrared photometry data are absent the flux for l >11084 A was obtained using model atmospheres data by Kurucz (1979) for T_{eff}=7500 K, lg g=4.00.In this case monochromatic flux was taken for l 11084 A from Johnson and Mitchell (1975). The effective temperatures were obtaned using infrared flux method in the modification by Blackwell et al. (1980) for the calibration at J, K, L bands by Koornneef (1983) and for the calibration at I band by Johnson (1966). For BS 483 T_L was excluded from the averaging because of too large differences with all the other T_{eff} determinations at I, J and K bands. Table 2 contains the main physical parameters of the investigated stars: total fluxes (W/m^2), bolometric correction B.C., angular diameters (milliarcsec), radii, luminosities and log g. log $\mathfrak{M}/\mathfrak{M}_\odot$ were taken according to Straizys and Kuriliene (1981).

TABLE I. EFFECTIVE TEMPERATURES

BS	T_I	T_J	T_K	T_L	T_{mean}
21	6990	6900	6960	-	6950
269	8010	-	-	-	8010
483	6130	6020	5925	5730	6025
1543	6430	6520	6450	-	6467
2943	6550	6600	6585	6600	6584
4540	6110	6035	6095	-	6080
5986	6520	-	-	-	6520
6561	7580	-	-	-	7580
8162	7820	7770	7725	7730	7774

TABLE II. Physical parameters of stars.

BS	F_{tot}	B.C.	θ''	$R/R_\odot$	lg $L/L_\odot$	lg $\mathfrak{M}/\mathfrak{M}_\odot$	lg g
21	$3.114\ 10^{-9}$	-0.01	2.00	2.92	1.270	0.18	3.68
269	$6.963\ 10^{-10}$	+0.01	0.71	1.93	1.152	0.26	4.12
483	$3.014\ 10^{-10}$	-0.16	0.83	1.08	0.156	0.00	4.36
1543	$1.392\ 10^{-9}$	-0.06	1.55	1.19	0.364	0.08	4.36
2943	$1.868\ 10^{-8}$	-0.07	5.46	1.97	0.832	0.10	3.94
4540	$9.571\ 10^{-10}$	-0.07	1.45	1.47	0.436	0.04	4.14
5986	$6.684\ 10^{-10}$	-0.08	1.05	2.18	0.900	0.11	3.86
6561	$9.536\ 10^{-10}$	0.00	0.93	3.27	1.516	0.20	3.60
8162	$2.702\ 10^{-9}$	-0.03	1.49	2.32	1.260	0.22	3.92

REFERENCES

Blackwell,D.E., Shallis,M.J.1977, M.N.R.A.S.,180, 177
Blackwell,D.E., Petford,A.D., Shallis,M.J. 1980, Astr.Ap., 82, 249.
Code, A.D. and Meade,M.R. 1979, Ap. J. Supplements, 39, 195.
Dreiling,L.A., Bell,R.A. 1980, Ap.J., 241, 736.
Glushneva,I.N., Kharitonov,A.V., Knyazeva,L.N., Shenavrin,V.I.1992, Astr.Ap.Suppl., 92, 1.
Hayes,D.S. 1985, Calibration of Fundamental Stellar Quantities, in IAU Symposium 111, ed. D.S.Hayes, L.E.Pasinetti and A.G.Davis Philip, p.247.
Heck,A., Egret,D., Jaschek,M., Jashek,C. 1984, IUE Low-dispersion Spectra Reference atlas. Part 1: Normal Stars. ESA SP-1052.
Johnson,H.L., Mitchell,R.I.,Iriarte,B., Wisniewski, W.Z. 1966, Comm.Lun.Planet.Lab., 4, 99.
Johnson,H.L. 1966, Ann.Rev.Astr. Ap., 4, 193.
Johnson,H.L., Mitchell,R.I.1975, Rev.Mex.Astr.Astrof. 1, 299.
Koornneef,J. 1983, Astr. Ap., 128, 84.
Kurucz,R.L.1979, Ap.J.Supplements, 40, 1.
Selby,M.J., Hepburn,I., Blackwell,D.E. et al. 1988, Astr.Ap.Suppl,.74, 127.
Saxner,M., Hammarback,G. 1985, Astr. Ap., 151, 372.
Straizys,V., Kuriliene, G. 1981, Astrophys. and Space Sci., 80, 353.
Wu,C.C., Crenshaw,D.M., Blackwell,J.H., et al. 1991, The IAU Ultraviolet Spectral Atlas, Addendum 1, International Ultraviolet Explorer (IUE) NASA Newsletter

Inside the Stars, IAU Colloquium 137
ASP Conference Series, Vol. 40, 1993
Werner W. Weiss and Annie Baglin (eds.)

ENERGY DISTRIBUTIONS AND FUNDAMENTAL PARAMETERS OF 6 A STARS

R. MONIER
ESA IUE OBSERVATORY, Apartado 50727, 28 MADRID, SPAIN

ABSTRACT The energy distributions (EDs) of 6 peculiar (3 Aps and 3 λ Bootis) stars are presented from 1200Å up to about 12000Å. For those stars having infrared photometry, the integrated flux from 1200Å to 22000Å has been calculated and the Infrared Flux Method (IRFM, Blackwell and Shallis, 1977) applied to derive effective temperatures and angular diameters. For all stars, the effective temperature T_{eff}, the surface gravity *logg* and metallicity [M/H] were also derived by matching the EDs to a grid of model atmospheres. A chisquare techniques is used to sort out the best fit to the observed EDs.

OUTLINE OF THE METHOD

The stars studied here are 3 early Ap stars: 78Vir (A2p), 52HerA (A2p) and CS Vir (A0p) and 3 λ Boo stars: λ Boo, π^1 Ori and 29 Cyg (all A0). The Ap stars are known to have overabundances of heavy elements (metals) whereas the λ Bootis have underabundances of these elements. The comparison of the EDs of these stars to model atmospheres computed for peculiar metallicities (Kurucz, 1979) is expected to yield information on their temperature, gravity and chemical composition.
To construct the EDs, low resolution IUE spectra (1200Å to 3200Å) were merged with published optical spectrophotometry (3300Å to 8000Å) (Pyper and Adelman,1983, Pyper and Adelman,1985, Breger,1975) and ,for the 3 Aps, with infrared photometry in the J and K bands from Groote and Kaufmann (1983) (no infrared photometry is available for the λ Bootis). The optical spectrophotometry has been transformed into absolute fluxes using Hayes and Latham's (1975) calibration of Vega and the infrared magnitudes were converted into absolute fluxes using Hayes' (1979) infrared calibration of Vega. The IFRM (Blackwell and Shallis,1977) was then applied to the three Ap stars. This method allows a direct determination of the effective temperature from the integrated flux, F, which defines it. The later was measured by fitting cubic splines throughout the entire ED and integrating under the splines. The infrared fluxes were then used together with predicted fluxes to derive the angular diameter using the relation:

$$f_\lambda = 0.25\theta^2 H_\lambda \ (1)$$

f_λ is the observed infrared flux and H_λ is the model flux at the infrared wavelength computed for a first guess effective temperature. Once θ is derived, the effective temperature is found from its definition:

$$F = 0.25\theta^2 \sigma T_{eff}^4 \quad (3)$$

For all stars,a grid of models was then computed varying the IRFM effective temperature (or a guessed temperature derived from Stromgren's indices for the λ Bootis stars) inside its allowed range $T_{eff} \pm \delta T_{eff}$ (δT_{eff} is typically 200K to 300K) , the surface gravity in the range 4.25 $\pm$ 0.25, and for different abundance sets ($\odot$, 3 $\odot$ and 10 $\odot$ for the Ap stars and 0.10 $\odot$ and 0.033 $\odot$ for the λ Bootis). The observed and model fluxes were ratioed to their values at 5000Å to allow for comparison. The best fit to the observed energy distribution was then searched by minimizing the chisquare χ^2 of the model flux ,H_λ, versus the observed flux, f_λ. The chisquare was calculated as:

$$\chi^2 = \sum_i (w_i/\sigma_i)^2$$

where

$$w_i = \frac{f_{\lambda_i}}{f_{5000}} - \frac{H_{\lambda i}}{H_{5000}}$$

from 2000Å up to the longest wavelength of the optical spectrophotometry (the far ultraviolet range is discarded because of the poor knowledge of conitnuous opacity in that range).

To obtain an even number of data over all wavelengths, the observed and model fluxes were interpolated to a common and even wavelength scale. The ultraviolet, optical and infrared fluxes were weighted by their respective uncertainties σ_i : $\pm$ 5% in the ultraviolet , $\pm$ 1.5% in the optical and $\pm$ 3.5% in the infrared . Also, the hydrogen lines were removed from the model in the chisquare calculation since the optical spectrophotometry carefully avoids these strong features. *Note that the best fit is not expected to be unique since a slight adjustment in each parameter around the optimal solution usually allows to reproduce the energy distribution within the uncertainties on the data.*

RESULTS

The found parameters are collected in table 1 and the energy distribution of one star, CS Vir, is compared to its best fit in figure 1.

Table 1: Found parameters

Star name	Integrated flux	θ	$ModelT_{eff}$	$IRFMT_{eff}$	$logg$	[M/H]
78 Vir	2.68 E-7	0.343	9250K	9200K	4.50	10$\odot$
52HerA	2.76 E-7	0.390	8900K	8550K	4.50	3 $\odot$
CS Vir	1.29 E-7	0.210	9700K	9600K	4.25	10$\odot$
λ Boo	-	-	8600K	none	4.20	0.033 $\odot$?
π^1 Ori	-	-	8600K	none	4.20	0.033 $\odot$?
29 Cyg	-	-	7800K	none	4.00	0.033 $\odot$?

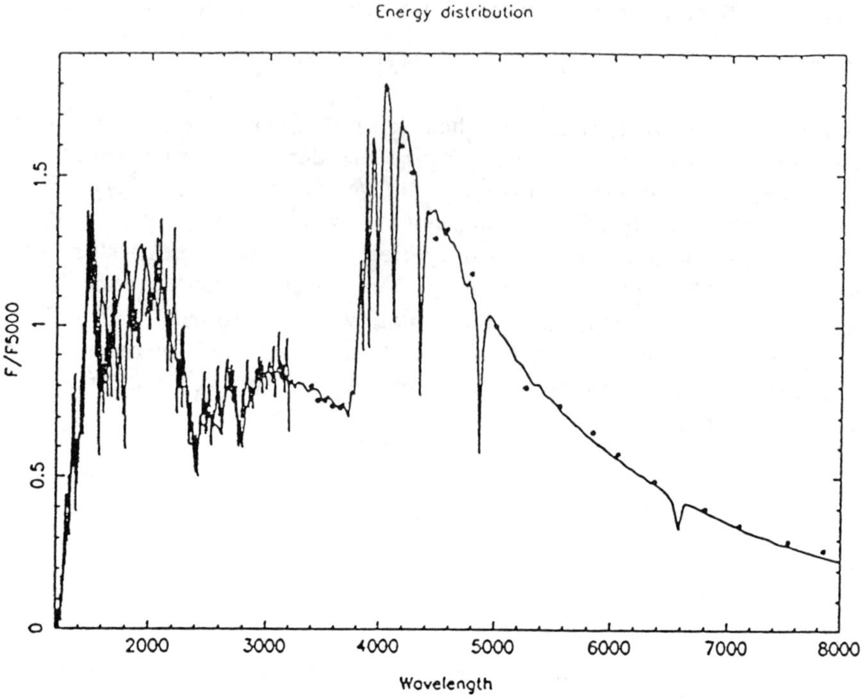

Figure 1: Comparison of the energy distribution of CS Vir (thin line and dots) with its best fit (thick line)

<u>REFERENCES</u>

Blackwell D.E. and Shallis M.J., 1977, *Monthly Notices of the Royal Astronomical Society*,174,489

Breger M., 1975,*ApJ Sup*,32,7

Groote and Kaufmann, 1983,*A and A Sup*,53,91

Hayes D.S., 1979,*in Problems of calibration of Multicolor photometric systems,Dudley Observatory Report*,14,29

Hayes D.S. and Latham M.R.,1975,*ApJ*,197,593

Kurucz R.L., 1979,*ApJ Sup*,40,1

Pyper D.M. and Adelman S.J.,1983,*A and A Sup*,51,365

Pyper D.M. and Adelman S.J.,1985,*A and A Sup*,59,369

Venn K.A. and Lambert D.A., 1990,*ApJ*,363,234

Inside the Stars, IAU Colloquium 137
ASP Conference Series, Vol. 40, 1993
Werner W. Weiss and Annie Baglin (eds.)

THE INFRARED EXCESS OF EARLY-TYPE STARS: MODELLING THE IRAS OBSERVATIONS

M. RUNACRES
Astrophysical Institute, Vrije Universiteit Brussel, Pleinlaan 2, B-1050 Brussel, Belgium

R. BLOMME
Koninklijke Sterrenwacht van België, Ringlaan 3, B-1180 Brussel, Belgium

ABSTRACT The IRAS satellite has observed some early-type stars in the infrared part of the spectrum. Most of this infrared radiation is due to free-free scattering in the dense stellar wind of these stars. We modelled this phenemenon by solving the equations of hydrodynamics as well as the radiative transfer equation (for a H and He mixture). This yields continuum fluxes at visible and infrared wavelengths which have been compared with the observations.

INTRODUCTION

Mass loss is an important and often unknown parameter of massive star evolution. Therefore, it is crucial to gain thorough understanding of the mass loss phenomenon. One aspect we study is the continuum radiation of these early-type stars. This continuum radiation differs significantly from what traditional atmosphere models (Kurucz, 1979) predict. Additional radiation is caused by free-free scattering in the material surrounding the star. The cross section for free-free scattering is proportional to the square of the wavelength. Hence, at larger wavelengths (infrared and radio) we see a larger radiating body.

We have modelled this radiation by superimposing a stellar wind model on the "classical" atmosphere model. For this stellar wind we solved the equations of hydrodynamics and of radiative transfer. The calculated continuum fluxes were compared with observations.

THE MODEL

Following the radiatively driven stellar wind theory (Castor et al., 1975; Abbott, 1982), we have solved the time independent equations of mass and momentum conservation for a spherically symmetric wind at constant temperature. This yields the dependence of velocity and density on radius. We have not solved the complete set of NLTE statistical equilibrium equations. Instead, we used LTE equations, taking into account however the geometrical dilution. Finally, we solved the equation of radiative transfer with electron scattering and free-free scattering by H and He as contributors to the opacity and emissivity. A classical Kurucz atmosphere provided the radiation input at the bottom of the wind.

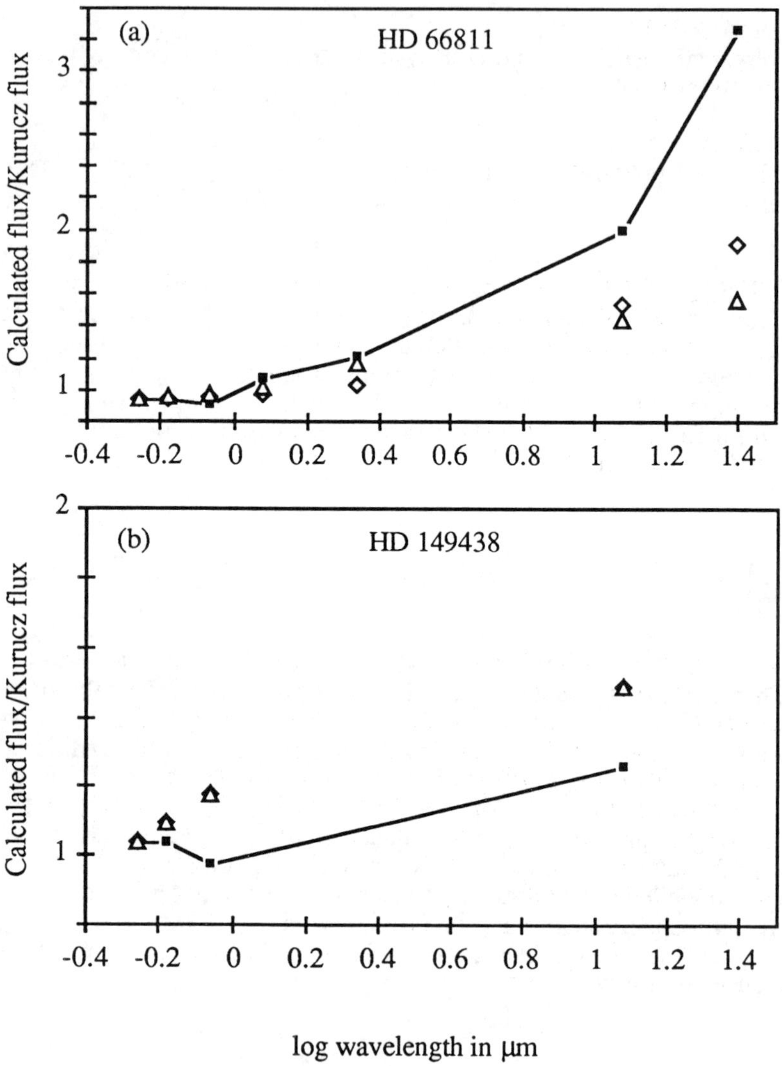

Fig. 1. Observed fluxes (▪) and model fluxes relative to the Kurucz flux at the same wavelength. Normal mass loss rates are indicated by a Δ, those three times higher by a ◇.

RESULTS

We have applied our model to the O4If star ζ Pup (HD 66811). As a distance gauge, we assumed the distance of the star to be such that the theoretical flux equals the observed one for the V filter (0.555 μm). With the mass loss rate as predicted by the radiatively driven stellar wind theory (Kudritzki et al, 1989), we cannot fit the observations. Using a mass loss rate which is about 2.5 times higher we can model the radio fluxes fairly well.

Besides ζ Pup we also studied the following stars: HD 24912, HD 38771, HD 149438, HD 207198 and HD 218376. For these stars there are no radio observations. We have modelled their infrared fluxes, using both the mass loss rate predicted by the radiatively driven stellar wind theory and a mass loss rate which is 3 times higher, corresponding to what we needed to fit the radio observations for ζ Pup.

For some stars (HD 66811 - see fig.1a, HD 38771) observed fluxes are higher than the model fluxes, especially at larger wavelengths. There appears to be a problem with the distance gauge for some of the stars studied (e.g. HD 149438 - see fig.1b). Agreement between theory and observations would be a lot better if we had used the I filter (0.87 μm) instead of the V filter (0.555 μm).

CONCLUSIONS AND FUTURE WORK

The fact that calculated infrared fluxes are lower than the observed ones in some cases may point to additional mechanisms in the wind, e.g. shocks as proposed by Owocki et al. (1988). This is an important challenge to both the atmospheric and evolutionary theory : is there an inherent wind instability or are the shocks caused by non-radial pulsations at the base of the wind? The latter case would be relevant to the study of the stellar interior. However, detailed error analysis is required to judge the significance of the differences between observed and calculated fluxes.

In the near future, we shall extend the sample of stars and include the influence of the variation of temperature with radius. Furthermore, the contributions of bound-free transitions to the opacity and emissivity will be investigated. Finally, our code for calculating H and He in NLTE will be included in this model.

REFERENCES

Abbott, D.C.: 1982, ApJ **259**, 282
Castor, J.I., Abbott, D.C., Klein, R.I: 1975, ApJ **195**, 157
Kudritzki, R.P., Pauldrach, A., Puls, J., Abbott, D.C.: 1989, A&A **219**, 205
Kurucz, R.L.: 1979, ApJS **40**, 1
Owocki, S.P., Castor, J.I., Rybicki, G.B.: 1988, ApJ **335**, 914

Inside the Stars, IAU Colloquium 137
ASP Conference Series, Vol. 40, 1993
Werner W. Weiss and Annie Baglin (eds.)

INFLUENCE OF ROTATION ON THE STELLAR PARAMETERS

V. WENSKE AND D. SCHÖNBERNER
Institut für Theor. Physik und Sternwarte der Universität Kiel
Olshausenstr. 40, W-2300 Kiel, Germany

For several years it has become quite common to derive stellar parameters like effective temperature, $T_{\rm eff}$, and surface acceleration, g, by means of properly calibrated photometric indices, and to use these values for the derivation of important properties of stellar aggregates (viz. ages, star formation history, distances, etc.). Photometric observations, however, fail to give informations about one important property of a star: *its rotational rate!*

The main effect of rotation is to increase the size of the star, mainly in the equatorial region, leading to lower surface temperatures and accelerations. Since in non–spherical stars $T_{\rm eff}$ and g depend on the latitude, the observed values, which are, of course, averages over the visible hemisphere, depend on the angle of inclination, i. Collins & Sonneborn (1977) utilized the rigidly rotating stellar models of Sackmann and Anand (1970) to compute emergent fluxes and photometric indices for various stellar masses, inclination angles i and rotational parameters $w = \frac{\omega}{\omega_{\rm crit}}$, $\omega_{\rm crit}$ being the break–up rotational rate. These indices, viz. c_0 and β, represent then averages over the visible part of the rotally distorted stellar models. Closer inspections indicated also that emergent fluxes and line profiles of rotationally distorted stars can be matched by the predictations of standard model atmospheres to a very high accuray, even for w close to unity (Wenske 1992, Diplom Thesis).

We observed nearly all B–type stars in the two open clusters NGC 2264 and α Per (Mel 20). The spectrogrammes, covering $\approx$ 130 Å, centered at H_γ, were obtained with the Coudé spectrograph (with CCD) of the 2.2 m telescope at the Calar Alto observatory. The resolution is $\approx$ 0.25 Å, and S/N better than 50 in most cases. These observations were supplemented by $uvby\beta$ photometry from the literature, and by IUE low–resolution spectrogrammes if available.

First we determined the effective temperatures from the photometry according to the prescription of Moon & Dworetsky (1985). The use of the IUE spectrogrammes via the total flux yielded the same results: $\overline{\Delta T_{\rm eff}} = (5 \pm \frac{150}{\sqrt{16}})$ K!

Once the temperature is given, surface acceleration and projected rotational velocity can be determined by matching theoretical line profiles based on Kurucz's fully line–blanketed model atmospheres to the observations.

NGC 2264

Figure 1 shows analysed objects of the young cluster NGC 2264, together with the ZAMS ($X = 0.70, Z = 0.03$) from Hejlesen (1980). Since that particular cluster is very young, (age $< 5 \cdot 10^6$ yrs) one expects that its stars would define an observational ZAMS, hopefully close to the theoretical one. This is not the case. Some cluster members deviate considerably from their expected position on the theoretical ZAMS. However, as the figure demonstrates, this occurs only for the fast rotators with $v \sin i \gtrsim 200\ \frac{\mathrm{km}}{\mathrm{sec}}$. Thus we tried to estimate the rotational shift ΔT_{eff}, Δg (from Δc_0, $\Delta\beta$, taken from Collins & Sonneborn) for each individual object in order to find the corresponding position of the non-rotating configuration. Firstly we assumed random orientation of the rotation axes, which corresponds to a mean inclination angle $\bar{i} = 52°$. This come close to $i = 60°$, one of the angles used by Collins & Sonneborn (1977). With i fixed to 60°, we derived $w = \frac{\omega}{\omega_{\mathrm{crit}}} = \frac{v_{\mathrm{eq}}}{\omega_{\mathrm{crit}}} \cdot R_{\mathrm{eq}}(w)$ by iteration since the equator radius R_{eq} depends explicitly on the rotation parameter w itself (see Collins et al. 1991). ω_{crit}, taken from the theoretical models, depends on mass and radius. With w known, the expected displacements in surface acceleration and temperature could be estimated and corrected for. The result is given in Figure 2. All objects are now close to the theoretical ZAMS.

Fig.1

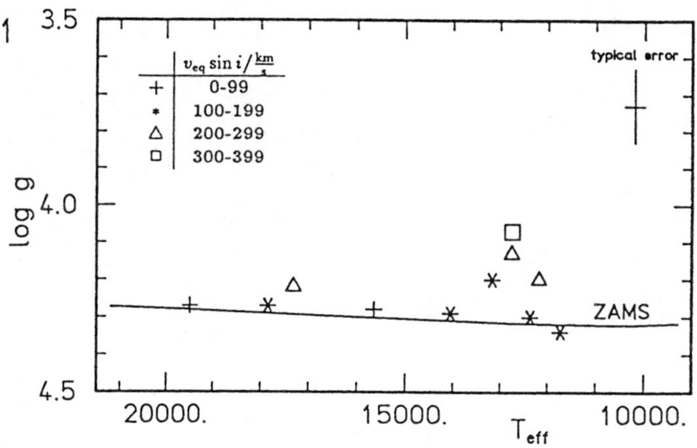

Fig.2

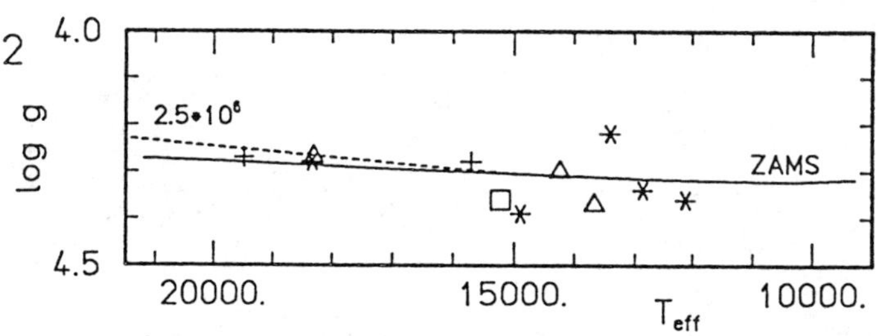

<u>α Per</u>

Figure 3 gives the T_{eff}, g - plane for the α Per stars together with the appropiate isochrones from Maeder & Meynet (1991) which include a moderate amount af overshooting. Figure 3 seems indeed to support the existence of two distinct populations, $\approx 6 \cdot 10^7$ and $2 \cdot 10^8$ yrs old, just as proposed by Eggen & Iben (1988) and confirmed spectroscopically by Wolff (1990).

However, Figure 4 shows a different picture: There we have corrected for rotational displacements in the same manner as shown above for the NGC 2264 cluster, and one can now assigne one single cluster age ($5.5 \cdot 10^7$ yrs).

In conclusion, we found by means of accurate spectroscopic determination of stellar parameters and application of simple rotating stellar models:

- Fast rotation ($w > 0.5$) mimics evolution and widens the main sequence!
- Observed stellar parameters must be corrected for rotational effects before interpreting them!
- The rigidly rotating stellar models of Collins & Sonneborn are sufficient for interpreting positions of rotating stars in the H–R diagramme.

Fig.3

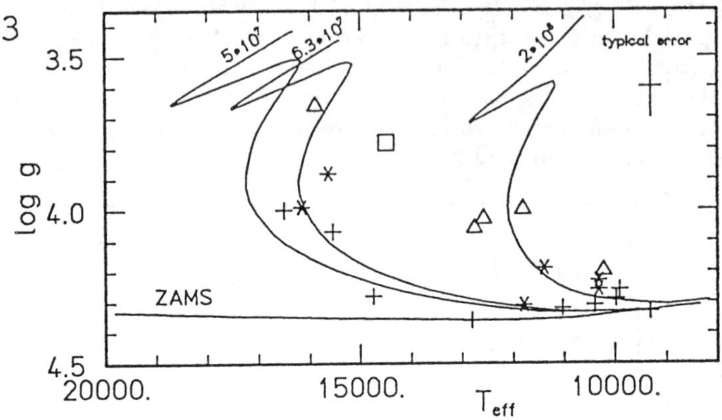

Fig.4

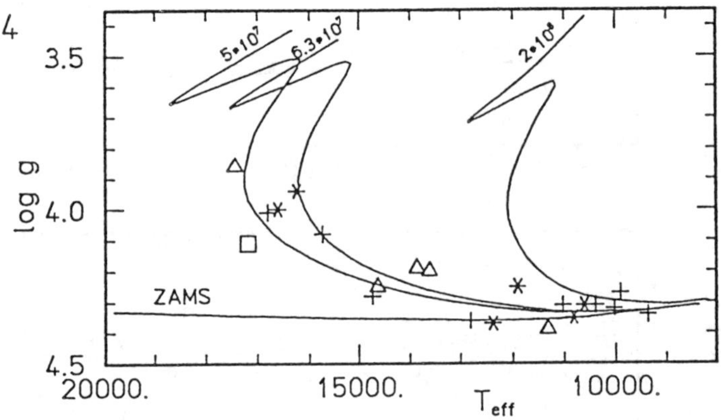

References

Collins, G.W., Sonneborn, G.: 1977, *Astrophys. J. Suppl.* **34**, 41
Collins, G.W., Truax, R.J., Cranmer, S.R.: 1991, *Astrophys. J. Suppl.* **77**, 541
Eggen, O.J., Iben, I.Jr.: 1988, *Astron. J.* **96**, 635
Hejlesen, P.M.: 1980, *Astron. Astrophys. Suppl.* **39**, 347
Maeder, A., Meynet, G.: 1991, *Astron. Astrophys. Suppl.* **89**, 451
Moon, T.T., Dworetsky, M.M.: 1985, *Monthly Not. Roy. Astron. Soc.* **217**, 305
Sackmann, I.-J., Anand, S.P.S.: 1970, *Astrophys. J.* **162**, 105
Wenske, V.: 1992, *Diplom Thesis,* Christian Albrechts Universität, Kiel
Wolff, S.C.: 1990, *Astron. J.* **100**, 1994

Inside the Stars, IAU Colloquium 137
ASP Conference Series, Vol. 40, 1993
Werner W. Weiss and Annie Baglin (eds.)

OBSERVATIONAL PATTERNS OF LITHIUM DEPLETION IN PRE-MAIN SEQUENCE STARS*

EDUARDO L. MARTIN, RAFAEL REBOLO and RAMON J. GARCIA LOPEZ
Instituto de Astrofísica de Canarias, E-38200, La Laguna, Spain

ANTONIO MAGAZZU
Osservatorio Astrofisico di Catania, I-95125 Catania, Italy

YAKOV V. PAVLENKO
The Principal Observatory of Ukraine, Kiev, Ukraine

ABSTRACT We present results based on the analysis of lithium abundances in a sample of ~ 50 pre-main sequence stars covering a wide range of masses (from 2 to 0.3 $M_{\odot}$) and luminosities (corresponding to ages of 1-100 Myr). Stars with masses estimated to be $\gtrsim 1\ M_{\odot}$ show lithium abundances close to cosmic with little scatter (± 0.3 dex). Stars with masses less than Solar present a wide range of lithium abundances, with a clear trend to lower abundances for lower luminosities (greater age). The observed Li abundances constrain theoretical predictions of lithium depletion in rotating pre-main sequence stars.

INTRODUCTION

Observations of the light element lithium are considered as a tool to gain insight on conditions inside young stars and the efficiency of matter transport between the surface and the interior. We are conducting a long-term observational program with the aim of constructing a statistically significant database for the study of lithium in the pre-main sequence (PMS). So far we have accumulated data on ~ 80 PMS stars, covering a wide range of masses (1.2-0.2 $M/M_{\odot}$), ages (1-100 Myr), rotational periods (1-12 d) and locations (field, star-forming regions, open clusters).

In the present paper we report results based on the analysis of the LiI resonance doublet at 670.8 nm in a sample of ~ 50 stars, including T Tauri stars, and K and M-type stars in two young open clusters: αPer (age 50-80 Myr) and Pleiades (age 70-100 Myr). Using the most recent model atmospheres available for cool high-gravity stars and new estimates of atmosphere parameters for each star we calculate the lithium abundance in

*Based on observations made with the 4.2 m William Herschel Telescope and the 2.5 m Isaac Newton Telescope, operated on the island of La Palma by the Royal Greenwich Observatory in the Spanish Observatorio del Roque de los Muchachos of the Instituto de Astrofísica de Canarias.

Recently, Magazzù *et al.* 1992 have shown that the lithium abundances in most classical T Tauri stars (CTTs) is similar to the abundance of this element in the interstellar medium (log N(Li)= 3.2, in the customary scale of log N(H)=12). In this paper we concentrate in weak T Tauri stars (WTTs) and post T Tauri stars (PTTs) for two reasons. One is that we can derive more reliable lithium abundances for WTTs and PTTs than for CTTs because we do not have to make veiling corrections, and we can be more confident on the stellar parameters and model atmospheres that are adopted (Basri *et al.* 1991). On the other hand, we expect that the masses and ages of WTTs and PTTs derived from theoretical tracks and isochrones are more reliable than for CTTS, and the comparison of observed Li abundances with theoretical predictions (cf. Proffitt & Michaud 1989, and Pinsonneault *et al.* 1990; hereafter PM and PKD respectively).

RESULTS

Our main results are conveniently summarized by dividing the stars in 3 bins of masses as shown in Figure 1.

1.- For $M/M_\odot \geq 1$ we do not find evidence for lithium abundances significantly different from the cosmic Pop.I value. The spread in the observed abundances is about 0.6 dex, and the estimated uncertainty in each individual abundance is about 0.3 dex (1σ). There is no correlation between Li abundance and luminosity. The mean Li abundance is close to the cosmic value of log N(Li)$\approx$3.2 . These results are consistent with the models of PM and the non-rotating models of PKD, but some of the rotating models (Case B3 for example) of PKD predict depletions of about one order of magnitude for masses equal to the Sun and ages about 30 Myr. Our data do not support those models.

2.- In the range of masses 0.9-0.6 $M/M_\odot$ we find very young WTTs with undepleted lithium, and PTTs (20-30 Myr) with lithium depletions of up to one order of magnitude (see Martín *et al.* 1992 for more details) . For Pleiades stars the Li depletion increases with decreasing mass, being larger than three orders of magnitude for stars with $M \sim 0.7$-$0.6\ M_\odot$. The observed depletion is consistent with the models of PM, but not so much with the rotating models of PKD which tend to predict higher Li depletions.

3.- The lithium abundances derived for the lowest masses considered by us are plotted at the bottom of Fig. 1 as a function of stellar luminosity. We find abundances close to the cosmic value for $\log(L/L_\odot) \gtrsim -0.4$. The turning point where lithium abundances start declining towards lower luminosities is very important for comparison with the interior models. This point appears to occur at lower luminosities for the very low mass stars than for the higher mass stars, which is qualitatively consistent with the models. For the lowest mass considered by PKD (0.4 $M_\odot$) the lithium depletion starts at $\log(L/L_\odot)=-1.1$, in the model without rotation, and for the rotating models the situation is similar. We find four M2,3 WTTs without detectable lithium lines (upper limits on the Li abundances in Fig. 1) and luminosities higher than $\log(L/L_\odot)$=-1, implying ages younger than 10 Myr. Hence, for these 4 stars the upper limits on the observed Li abundances are not consistent with theoretical expectations based on their apparent masses and ages. We note that the 4 objects are NTTs in two binary systems. It is a subject of further study if their binary nature may affect their lithium abundances and/or their estimated masses and ages. For the less luminous stars in our sample (those in the αPer cluster) we have detected the lithium line in two stars and place an upper limit on one star. We infer Li abundances about a factor 100 below the initial value. Such depletion is smaller by about one order of magnitude than predicted by the non-rotating models of PKD for 0.4 $M_\odot$ and 50 Myr.

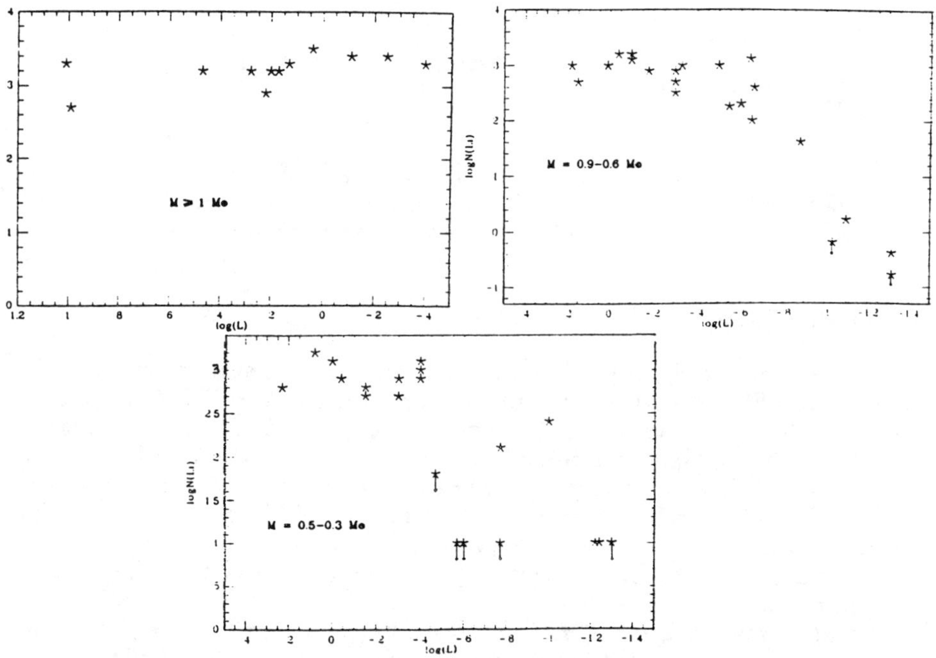

Fig. 1. Lithium abundances, defined as logN(Li)= 12 + $\log(N_{Li}/N_H)$, for PMS stars in three ranges of masses versus $\log(L/L_\odot)$.

REFERENCES

Basri, G., Martín, E.L., Bertout, C. 1991, *Astr. Ap.*, 252, 625.
Magazzù, A., Rebolo, R., Pavlenko, Ya.V. 1992, *Ap. J.*, in press
Martín, E.L., Magazzù, A., Rebolo, R. 1992, *Astr. Ap.*, 257, 186.
Pinsonneault, M.H, Kawaler, P., Demarque, P. 1990, *Ap. J. Suppl.*, 74, 501.
Proffitt, C.R., Michaud, G. 1989, *Ap. J.*, 346, 976.

Inside the Stars, IAU Colloquium 137
ASP Conference Series, Vol. 40, 1993
Werner W. Weiss and Annie Baglin (eds.)

OBSERVED AND COMPUTED FEATURES AROUND Li I 6708 IN NORMAL A AND EARLY F STARS

M. GERBALDI
Institut d'Astrophysique - 98bis, Bd. Arago - 75014 Paris - France

R. FARAGGIANA
Dipartimento di Astronomia - v. Tiepolo 11 - 34131 Trieste - Italy

F. CASTELLI
CNR/GNA c/o Osservatorio Astronomico - v. Tiepolo 11 - 34131 Trieste - Italy

ABSTRACT For a set of normal A and early F type stars, we discuss the possible contributors which can affect the Li feature and their effect on the determination of the Li abundance.

INTRODUCTION

A previous study of the Li abundance in some Ap stars shows that the behaviour of the spectrum around the Li I position at 6707.8 A may vary considerably from one star to another (Faraggiana et al., 1986). So far, we raised the problem of a possible confusion between the true Li doublet and very near lines produced by other elements (Gerbaldi and Faraggiana, 1991). So, before to deal with the determination of Li abundance in such a peculiar context, we have observed a sample of non-peculiar stars.

In the present paper we are giving the results obtained for these stars and in particular we discuss the possible contributors which can affect the Li I feature and their effect on the determination of log N_{Li}. Model atmosphere, atomic data and the code for the computation of synthetic spectra are those of Kurucz, implemented at Trieste and Paris-Meudon by one of us (F. C.).

OBSERVATIONS AND BASIC DATA OF THE SELECTED STARS

The observations were made with various instrumentations of similar capabilities : the Coudé Echelle Spectrometer (CES) of European Southern Observatory (Chile) and with the ISIS and AURELIE spectrographs at the Observatoire de Haute Provence (France). The list of the observed stars and their parameters are given in Table I.

We have derived T_{eff} and log g (columns 2 and 3, Table I) from the grid c_o versus $(b\text{-}y)_o$ of the new theoretical colours computed by Kurucz (1991, Data on magnetic tapes). The observed colours (Hauck and Mermilliod,1990) have been dereddened by using an extension of Moon (1985) code UVBYBETA in order to include stars of spectral type later than A3.

The vsini value has been estimated for each star, when possible, by comparing the lines of the observed spectrum with the computed ones broadened for various values of vsini. (Column 4 of Table I)

For each spectrogram, we measured the EW of the feature centered at 6707.8 A or, when no line was detectable, the upper limit of the possible feature at the Li I wavelength.

The EWs, in mA, are collected in Table I (column 5), a mean value is given when several spectrograms have been measured for a star.

COMPUTED Li FEATURE

Since the abundance determination is based on the fit between observed and computed fluxes, it is important to know by which factor lines of other elements could contaminate the spectrum at the location of the Li I feature. We have checked that the Kurucz line list does not overestimate the absorption features in this spectral domain by comparing the observed and computed spectrum for the Sun, Procyon and Vega. None of them contains detectable spurious lines.

The Fe I 6707.44 is not included in the Kurucz line list. We have added it; the energy levels have been taken from Sugar and Corliss (1985) and the log gf=-2.30 from Müller et al. (1975) has been adopted as it gives an upper limit. In measuring the EW of the computed Li I feature we have to take into consideration the contamination by lines in an interval depending on the vsini of the star. The chosen interval is λ6707.0-6708.6 representative of a vsini of 25 km/s.

An array of synthetic spectra with Li abundances ranging from log N[Li]=-8 to log N[Li]=-12 has been computed for each T_{eff}.

TABLE I The programme stars

HD	T_{eff}	log g	vsini	EW
739	6700	4.20	5	24
23878	8700	3.75	20	4
40136	7200	3.95	18	32
60179	9900*	4.00*	15	2
61421	6780	3.75	6	3:
82328	6550	3.80	12	107
128167	6970	4.15	10	2
156208	9700	3.00	25*	4
172167	9500	3.80	20	2
186689	8220	4.20	35	8:
203096	7250	2.35	30	5:
205939	7800	3.40	20	19

* : values taken from the literature

TABLE II Computed EW [λ6707.0-6708.6]

T_{eff}	EW	EW
6250	5.9	1.4
6500	4.9	1.4
7000	3.7	1.7
7500	4.3	2.9
8000	4.3	3.5
8500	3.8	3.5
9000	3.2	3.0
9500	2.5	2.4
10000	2.0	1.9

column 2 : EW with Fe I line
column 3 : EW without Fe I line

DISCUSSION

We notice that, whatever the value of T_{eff} is, when the Li abundance decreases towards low values, the limit of the total EW of the λ6708 feature is not zero.

The contribution of the Fe I 6707.44 line to the asymptotic value has been estimated by computing the EW of the Li I feature with and without the Fe I contribution and for a Li/H abundance of -12. These EW are given Table II. We notice that the Fe I intensity decreases with T_{eff} and is negligible for T_{eff} larger than 8500 K but for A-type stars the contribution of the others contributors cannot be forgotten.

Only 4 stars out of the 12 observed have a measured value of the 6708 feature higher than 10 mA. For these stars, in the hypothesis that the Kurucz' line list, with the addition of FeI line, is good and our choice of T_{eff} is correct, we obtain the following Li abundances (log N[Li]) :

HD 739 : -9.4; HD 40136 : -9.0; HD 82328 : -8.6; HD 205939 : -8.9.

For the remaining 8 stars the value of the λ6708 feature is less than 5 mA, except for HD 186689 for which we have only a noisy spectrum. The determination of Li abundance is practically meaningless in view of the above discussion.

For stars with T_{eff} higher than 8500 K, Li abundances lower than the cosmic value log N(Li)=-9 are very difficult to determine; they may be possible only for very low rotators and they require spectra at very high resolution and taken with a high S/N.

Acknowledgments

One of us (R.F.) gratefully acknowledges the financial support offered by the Centre National de la Recherche Scientifique for a stay at the Institut d'Astrophysique de Paris where most of this research has been accomplished and is grateful to the Director of this Institute, A. Omont, for the warm hospitality received there. We profoundly thank Claude van 't Veer for introducing us to the use of the programs implemented at the Centre de Calcul de Meudon. This research has made use of the Simbad database operated at the CDS, Strasbourg, France.

REFERENCES

Faraggiana, R.,Gerbaldi, M., Castelli, F., Floquet, M. : 1986, A.A. 158,200
Gerbaldi, M., Faraggiana, R. : 1991, Mem. Soc. Astron. Ital. 62, 97
Hauck, B., Mermilliod, J.C : 1990, A.A.Supp. 86, 107
Kurucz, R.: 1991, Precision Photometry : Astrophysics of the Galaxy A.G. Davis Philip, A.R. Upgren, and K.A. Janes, editors L. Davis Press, Schenectadty
Moon, T.T.: 1985, Comm. University of London Observatory No 78
Muller,E.,A., Peytremann, E., de la Reza, R.: 1975, Sol. Phys. 41, 53
Sugar, J., Corliss, Ch.: 1985, J. of Phys. and Chem. Ref. Data 14, Suppl.2

Inside the Stars, IAU Colloquium 137
ASP Conference Series, Vol. 40, 1993
Werner W. Weiss and Annie Baglin (eds.)

A SEARCH FOR LITHIUM IN BROWN DWARF CANDIDATES

RAFAEL REBOLO and EDUARDO L. MARTIN
Instituto de Astrofísica de Canarias, E-38200, La Laguna, Spain

ANTONIO MAGAZZU
Osservatorio Astrofisico di Catania, I-95125 Catania, Italy

ABSTRACT

We are conducting a search for lithium in very low mass objects with the aim of discriminating between stellar and substellar objects. Lithium is expected to be preserved in brown dwarfs with $M/M_\odot \le 0.06$, while it is known to be efficiently destroyed in low mass stars. In this paper we present high resolution observations in the region of the λ 6708 Li I resonance line of 5 very low mass dwarfs. In none of them lithium is detected, implying a Li destruction in their atmospheres of about four orders of magnitude. Our results suggest that these objects are probably very low mass stars rather than brown dwarfs.

INTRODUCTION

Very low mass stars (0.2-0.08 $M/M_\odot$) are thought to be fully convective and, since they reach the internal temperatures for efficient lithium burning before settling on the Main Sequence, we expect Li to rapidly disappear from their atmospheres. However, the maximum central temperature of these stars decreases the lower the mass. Below a certain mass no lithium burning should occur. Using a polytrope as a rough approximation to describe a brown dwarf (BD) interior (see Stevenson 1991), it can be shown that the limiting mass for lithium burning is about 0.07 $M/M_\odot$. A more refined estimate taking into account the cross section of Li (p,α) reactions (Caughlan & Fowler 1988) and the interior densities and temperatures of BDs (D'Antona & Mazzitelli 1985), leads us to the conclusion that this limiting mass must be close to 0.06 $M/M_\odot$ (Magazzù, Martín & Rebolo 1992). Such value is only slightly below the minimum mass for stable hydrogen burning ($M/M_\odot \sim 0.08$). Rebolo, Martín & Magazzú 1992 have claimed that a strong λ 6708 Li I resonance doublet should be present in the spectrum of brown dwarfs with a cosmic Li abundance (the Li abundance of young stars and the interstellar medium). We proposed that the detection of this feature is an observational test to discriminate between very low mass (VLM) stars and brown dwarfs.

Considering these expectations, we have conducted a search for Li in several VLM objects, namely GL 65A, GL65B, GL 406 and Gl 473AB. They were selected from the compilation by Burrows, Hubbard & Lunine 1989, who give mass estimates of 0.11 $M/M_{\odot}$ for GL 65 A and B, of 0.08 $M/M_{\odot}$ for GL 406, and of 0.06 for each component of GL 473. We note that this last object has an astrometric mass determination by Heintz 1989, giving 0.06 and 0.05 $M/M_{\odot}$ for each component. The small error bar (± 0.01 $M/M_{\odot}$) assigned to this determination make GL 473AB very strong candidates for being substellar objects. The effective temperature of all these objects are in the range 2600-2800 K, and therefore we expected to detect the presence of lithium in their atmospheres.

OBSERVATIONS

The spectra were obtained with the 2.5 m Isaac Newton Telescope and the 4.2 m William Herschel Telescope at La Palma. The reciprocal dispersion was between 0.22 and 0.37 Å/pixel. In Figure 1 we plot the spectra of GL 473 AB and GL 406 in the Li region. We also plot for comparison the spectrum of UX Tau C, the coolest component of the pre-main sequence triple system UX Tau (Magazzú, Martín & Rebolo 1991), where a strong Li feature is clearly seen in contrast with the other two objects. The similarity of the molecular bands in the three spectra indicates that they also have similar effective temperatures.

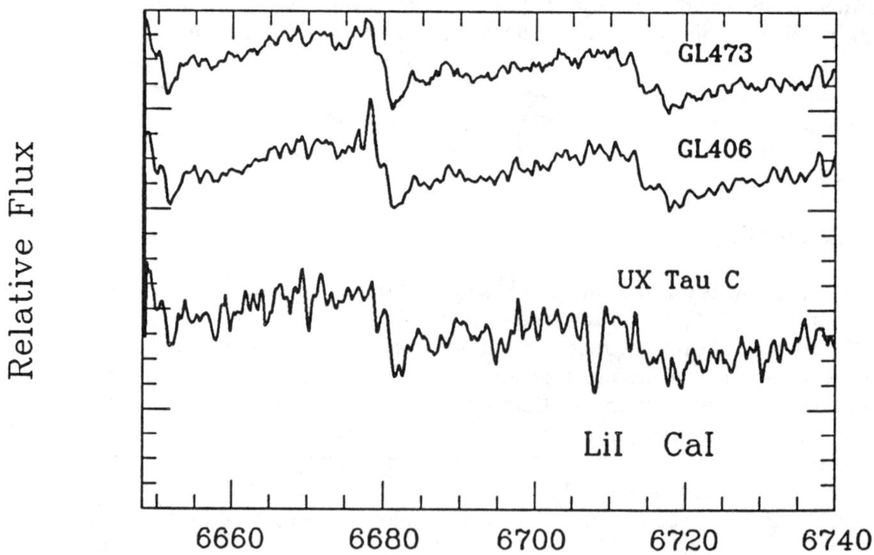

Fig. 1. Observed spectra of the very cool dwarfs GL 473AB and GL 406. The bottom spectra corresponds to the coolest component of the pre-main sequence triple system UX Tau. Note that the LiI line is detected only in UX Tau C.

ANALYSIS

Using recently produced model atmospheres for brown dwarfs (Allard 1990) and a spectral synthesis technique, we were able to reproduce most of the features in a region 20 Åaround the Li doublet. We adopted models of solar metallicity, high gravity (log g=5) and effective temperatures 2750 and 2500K, adequate to cover the range of our program objects. Several Li abundances were considered in the computations. The comparison between synthetic and observed spectra allowed us to establish an upper limit of log N(Li)=-1.5 (in the scale log N(H)=12) on the atmospheric lithium abundance in these objects. From the work by Tsuji (1973) it is not expected that any significant fraction of Li is bounded in molecular form (LiH, LiO, LiCl, etc.). It is unlikely that molecular effects prevent the formation of the Li doublet.

DISCUSSION AND CONCLUSIONS

The present abundance of Lithium in our objects is probably a factor 10^4 lower than their initial abundance. While such a large depletion in the case of GL 65 A and B, and GL 406 is consistent with their estimated masses, it is particulary striking that GL 473 AB, one of the best brown dwarf candidates, has not preserved a detectable amount of lithium in its atmosphere. Either each of the components has larger masses than measured using astrometric techniques (i.e. $M \geq 0.06\ M_{\odot}$), or else the predictions of maximum internal temperature and central density (the critical parameters concerning lithium destruction) for such low masses may have to be revised. It is important to remark that while the non-detection of the lithium resonance line in very cool high gravity objects cannot give a definitive answer on its substellar nature, the detection of the lithium feature can be certainly used as a criterion for positive identification of real brown dwarfs.

ACNOWLEDGEMENTS

This work is based on observations made with the Isaac Newton Telescope and the William Herschel Telescope, operated on the island of La Palma by the Royal Greenwich Observatory in the Spanish Observatorio del Roque de los Muchachos of the Instituto de Astrofísica de Canarias.

REFERENCES

Allard, F. 1990, Ph.D. Thesis, Univ. Heidelberg
Burrows, A., Hubbard, W.B., Lunine J.I. 1989, *Ap. J.*, 345, 939.
Caughlan, G., Fowler, W.A. 1988, Atom. data and Nuc. data tables, 40, 283.
D'Antona, F., Mazzitelli, I. 1985, *Ap. J.*, 296, 502.
Heintz, B.D. 1989, *Astr. Ap.*, 217, 145.
Magazzù, A., Martín, E.L., Rebolo, R. 1991, *Astr. Ap.*, 249, 149.
Magazzù, A., Martín, E.L., Rebolo, R. 1992, *Ap. J.*, submitted
Rebolo, R., Martín, E.L., Magazzù, A. 1992, *Ap. J.*, 389, L83.
Stevenson, D.J. 1991, *Ann. Rev. Astron. As.*, 29, 163.
Tsuji, T. 1973, *Astr. Ap.*, 23, 411.

Inside the Stars, IAU Colloquium 137
ASP Conference Series, Vol. 40, 1993
Werner W. Weiss and Annie Baglin (eds.)

BERYLLIUM IN LITHIUM-DEFICIENT F STARS : CONSTRAINTS ON STELLAR EVOLUTION

CONSTANTINE P. DELIYANNIS
Institute for Astronomy, University of Hawaii, 2680 Woodlawn Drive, Honolulu, HI 96822, USA

M. H. PINSONNEAULT, Center for Solar and Space Research, Yale University, P.O. Box 6666, New Haven, CT, 06511, USA

ABSTRACT Standard stellar evolution predicts that F stars should retain their initial surface lithium (Li) abundance because their convection zones are too shallow to destroy it at their base. Yet, observations reveal a severe Li depletion (the "Boesgaard Gap"), perhaps by as much as about two orders of magnitude, in a narrow T_{eff} range. Several physical mechanisms, not usually included in stellar evolution calculations, have been proposed to account for this Li deficiency. These include diffusion, mass loss, meridional circulation, and rotationally-induced mixing driven by angular momentum loss. Identifying which of these (if any) might really be at work is not only of vital interest to stellar evolution, but may also have serious implications elsewhere (e.g. cosmology, Deliyannis *et al.* 1991). We bring attention to beryllium (Be) observations in F stars, which are crucial for discriminating between scenarios. Particularly important is the star 110 Her, which is depleted in Be by about a factor of 5 - 10, but still has a *detectable* Li abundance (depleted by a factor of 100 - 200). Depleting surface Be without having depleted nearly all of the surface Li requires specific circumstances ; we discuss how *this depletion property severely constrains or eliminates most of the proposed mechanisms.* One mechanism, rotationally-induced mixing, predicts relative depletions for these elements that agree well with what is observed.

1. INTRODUCTION

Boesgaard and Tripicco (1986) discovered that, in a narrow T_{eff} region of a few hundred degrees, F stars deplete severely their surface Li abundance, and do so primarily during the main sequence (Boesgaard *et al.* 1988). This startling discovery blatantly contradicted standard stellar evolution theory. Already at the ZAMS, the model surface convection zone (SCZ) occupies only a small fraction of the Li preservation region ; hence, surface Li is not affected. To explain the Li gap, proposed additional physical mechanisms have proliferated, and it is not yet clear which (if any) are correct. Using Li and Be together to decipher the signature of the responsible mechanism(s) is powerful because Be survives to about twice the ZAMS depth that Li does. Thus, the effect of each mechanism on surface Be will in general be different.

We assume that solar metallicity stars typically formed with the Li and Be abundances measured in meteorites : 3.31 ± 0.04 and 1.42 ± 0.04 (Anders and Grevesse 1989, where abundance = $12 + \log [N_{element} / N_H]$). In the Hyades, mid-F stars are depleted in Be by a factor of 0 – 2 relative to late-F stars, and by a factor of 1.5 – 4 relative to Be_{met}. Field G dwarfs are depleted by 0 – 4 and in the same stars Li depletion ranges from 5 - 200 (Fig. 1). *F stars show a Be gap,* and Be depletion factors range from 2 – >100. *Especially important is the star 110 Her, which is (definitively) depleted in both Li and Be,* **but still preserves a detectable amount of Li.** Its depletion factors of 100 - 200 and 5 - 10 provide serious constraints for stellar evolution theory.

2. CONSTRAINTS ON STELLAR EVOLUTION

Simple Microscopic Diffusion. Diffusion mechanisms not related to rotation ("simple diffusion") include gravitational settling and thermal diffusion (downward), and radiative acceleration (upward) ; their efficiency increases with model radius, but can be inhibited (completely) by sufficient mixing. At a given age, the depth of the SCZ decreases with increasing T_{eff}. For Li, settling and thermal diffusion dominate for internal $T > 2 \times 10^5$ K where Li is fully ionized, creating the cool side of the gap (Michaud 1986). For smaller T, Li retains an electron so radiative acceleration dominates, creating the hot side (the availability of more Li to be pushed into the SCZ also helps). However, for $T_{eff} > 7000$ K, this creates an overabundance of Li by at least an order of magnitude, in contradiction with observation. A small rate of (finely tuned) mass loss has been proposed to eliminate the Li overabundance. *Below 6400 K, Be and Li are predicted to diffuse downward at comparable rates (Michaud and Charbonneau 1991), in contradiction with the observations of field stars (Fig. 1).* Particularly striking is 110 Her, which is at least 10 times more depleted in Li than in Be. At higher T_{eff}, an overabundance is also predicted for Be, *which is not observed.* Since Be retains an electron at higher T than Li, its radiative acceleration begins at lower T_{eff}. It is thus difficult for mass loss to be invoked to eliminate simultaneously both the Li and Be overabundances.

Main Sequence Mass Loss. To produce the morphology of the Li gap (Schramm *et al.* 1990), and with the correct timing, mass loss rates must vary across the Li gap in an exceedingly finely tuned manner. To deplete Be, mass loss must first deplete **all** the observable Li. *This is in stark contrast to 110 Her.* Furthermore, the high Li abundances of the stars in Fig. 1 contradict mass loss if ANY of these stars are in fact depleted in Be by a factor of 2-3.

Meridional Circulation plus Radiative Acceleration. In this model (Charbonneau and Michaud 1988), circulation increases with model radius and rotation rate, creating the cool side of the Li gap. For hotter stars, radiative acceleration must be invoked to overcome circulation and create the hot side of the gap ; the objections raised above apply here as well.

Rotationally-Induced Mixing. The Yale models spin-down through angular momentum (J) loss, and include a comprehensive treatment of instabilities that result in the transport of J and mixing (see Pinsonneault in this volume). Li destruction is related to J(total) lost. Initial J (J_o) in F stars is observed to increase with mass ; its loss results in the cool side of the gap. Hotter stars are not observed to spin down ; this results in the hot side of the Li gap. A spread in Li abundances is the natural result of a spread in J_o. Mixing time scales are roughly comparable to the evolutionary time scale, so mixing can resemble a dilution process. Thus, Be and Li can both be depleted at the same time, without catastrophic depletion of Li. *The Yale models match the depletion of Li and Be in 110 Her* (e.g. for an age of ~ 5 Gyr and a normal J_o, or younger ages and a higher J_o). The models also agree well with stars that have (Li,Be) depletion of (5-50,1.5-3), and can severely deplete Be at advanced ages.

REFERENCES

Anders, E., and Grevesse, N. 1989, Geochim. et Cosmo. Acta, 53, 197.
Boesgaard, A. M., and Tripicco, M. J. 1986, ApJL, 302, L49.
Boesgaard, A. M., Budge, K. G., and Ramsay, M. E. 1988, ApJ, 327, 389.
Charbonneau, P., and Michaud, G. 1988, ApJ, 334, 746.
Deliyannis, C. P., Demarque, P., and Pinsonneault, M. H. 1991, "Primordial Lithium, Stellar Evolution, Cosmology, and Dark Matter", BAAS, 22, 1214.
Michaud, G. 1986, ApJ, 302, 650.
Michaud, G., and Charbonneau, P. 1991, Space Sci. Rev., 57, 1.
Schramm, D. N., Steigman, G., and Dearborn, D. S. P. 1990, ApJL, 359, L55.

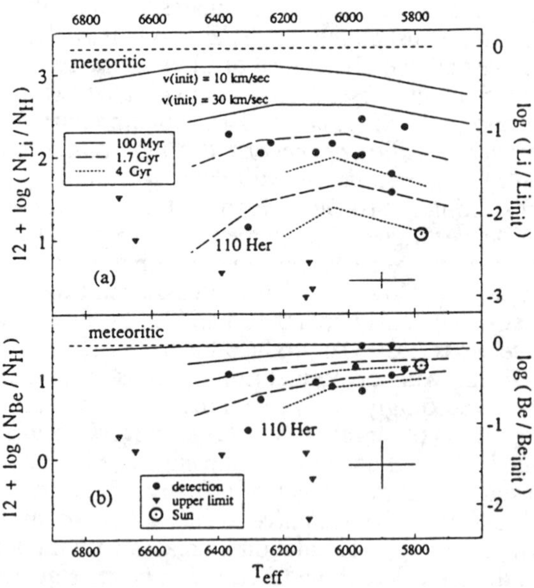

Figure 1. Li (a) and Be (b) abundances (left scale) in field F and early G stars (Boesgaard and Lavery 1986, ApJ, 309, 762, and references therein) plotted against isochrones from models with rotationally-induced mixing (right scale), with a likely range in v(initial) at each age. It is crucial that new observations determine whether other stars have the properties of 110 Her.

Inside the Stars, IAU Colloquium 137
ASP Conference Series, Vol. 40, 1993
Werner W. Weiss and Annie Baglin (eds.)

[Fe/H] AND LITHIUM ABUNDANCE FOR DETACHED ECLIPSING BINARIES

J.V. CLAUSEN, Copenhagen University Observatory, Denmark

1. Background

We are now seeing a new generation of stellar models based on the recent opacity tables by Rogers and Iglesias (1992). Various changes in other important physical ingredients like nuclear reaction rates, convection parameters, mass loss rates, helium/metal ratios, relative abundances of heavy elements etc. are also included. Besides the Sun the most obvious objects for confrontation of the new grids with observations are binary components, pulsating stars, and well-studied open and globular clusters. Among these candidates the detached, double-lined eclipsing binaries play a unique role as our only source of absolute dimensions which 1) can be considered valid for single stars and 2) are accurate enough for serious tests of the input physics of modern stellar models.

At least during central hydrogen burning an accuracy of 1-2% is needed, and today masses and radii of that precision are available for about 90 eclipsing binary components (Andersen, 1991). The great majority of these stars are within or near the rather well, but still not fully covered main sequence band. More data are under way (Andersen et al., 1992).

At one fundamental point there is, however, a clear lack of data for these binaries. Quite obviously detailed confrontation must be based on stars of known [Fe/H] matching Z of the models, but accurate results from detailed spectroscopic analyses have been published for only 4 systems: V818 Tau in the Hyades, AI Phe, UX Men and TZ For. As for masses and radii the demand on [Fe/H] is rather strick, an accuracy of 0.10-0.15 dex is needed.

2. The [Fe/H] project and its present status

On this background a spectroscopic program on [Fe/H] determination for additional binaries is presently being carried out. High resolution (R = 60000), high S/N (100-200) spectra have been observed for the about 25 double-lined eclipsing binaries shown in Fig. 1. The ESO Coudé Echelle Spectrograph (CES) was used, and spectra from several 50Å wide regions have been obtained.

For F and G type systems preliminary LTE analyses of weak (≈ 50mÅ) FeI lines show that an accuracy of 0.10-0.15 dex can be reached. Only six candidates have been studied so far; the individual [Fe/H] values range between -0.20 and 0.10.

LTE analyses of TiII, CrII, FeI and FeII lines indicate some difficulties for the B and A type systems. In general abundances better than about 0.2 dex can not be expected unless more precise determinations of microturbulence velocities become possible. NLTE effects are noticed in the FeI results but have not yet been examined.

A detailed confrontation with models will be done when abundances for all systems are available.

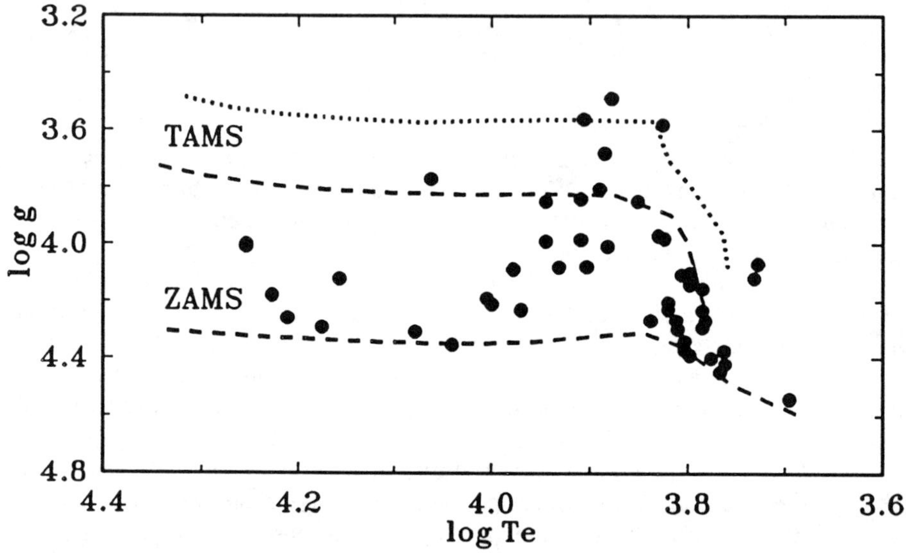

Fig. 1. log T_e - log g diagram for binary components observed in the [Fe/H] project. Main sequence band defined by standard (*dashes*) and overshooting (*dots*) models (Claret & Giménez, 1992).

3. Lithium

As part of the project Li-abundances are being determined for the components of 13 main sequence binaries. Most of the objects are F-type stars with masses between 1.1 and 1.6 $M_\odot$; two G stars (0.85 and 1.05 $M_\odot$) and one Am star (2.0 $M_\odot$) are included. Temperatures range between 5000 and 8000 K and rotational velocities from 7 to 40 km/s. Compared to existing Li studies based on cluster members or single field stars this sample has the clear advantage that masses and surface gravities are very accurately known.

Results from a *visual inspection* of the LiI resonance line at 6707.8 Å are shown in Fig. 2. The stars have been sorted in two groups; those with a clearly visible and often strong Li line (*open circles*), and those with either no Li line detected or eventually just a very weak feature (*filled circles*). The limit between the two groups is estimated to be located around log N(Li) $\approx$ 2.0.

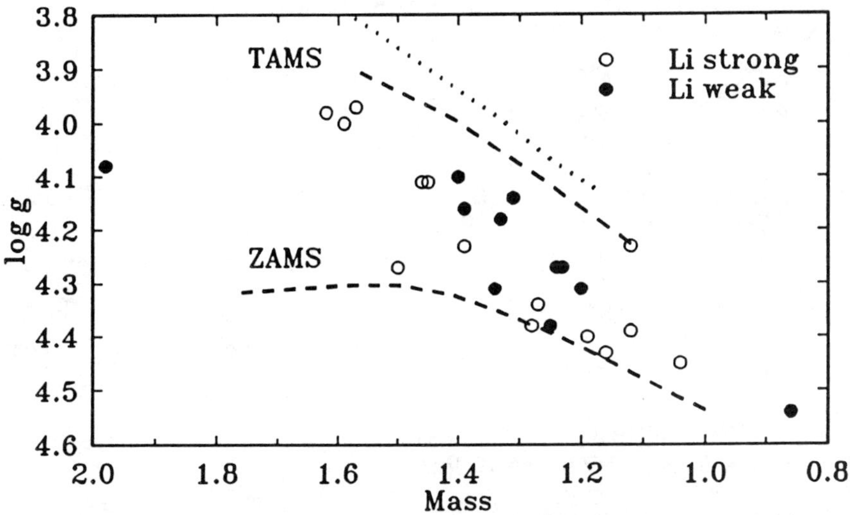

Fig. 2. Preliminary Li results for components of 13 main sequence binaries.

Sorting the stars according to mass some interesting trends are noticed already from the visual inspection of the spectra:

1. Independent of age a strong Li line is observed in all stars with masses between about 1.40 and 1.65 $M_\odot$. Due to evolution three of the stars are now located right in the center of the Li-gap at 6700 K observed for the Hyades and other not too young open clusters (Boesgaard, 1991). Ages for these binary components with no Li depletion are well above that of the Hyades.

2. For all stars from about 1.20 to 1.40 $M_\odot$ and older than about 2.5 10^9 yr the Li line is not detected or very weak. Among the younger candidates with ages below about 1.5 10^9 yr a tendency of slow rotating stars ($v\sin i \approx$ 10 km/s) in wide orbits to be Li-strong, and faster rotating stars ($v\sin i \approx$ 40 km/s) in narrower orbits to be depleted is noticed.

3. In all stars from about 1.00 to 1.20 $M_\odot$ a strong Li line is observed.

A definitive discussion of the data within the framework of current theories for depletion, dilution or destruction of Li has to await a proper determination of the Li abundances and [Fe/H] values. This will be done in the near future.

References

Andersen, J., 1991, *Astr. Ap. Rev.* **3**, 91.

Andersen, J., Clausen, J.V., García, M., Giménez, A., Helt, B.E., Johansen, K.T., Nordström, B., Stefanik, R.P., Vaz, L.P.R., 1992, in *Inside the Stars*, eds. W. Weiss and A. Baglin.

Boesgaard, A.M., 1991, *Ap. J.* **370**, L95, 1991

Claret, A., Giménez, A., 1992, private communication

Rodgers, F.J., Iglesias, C.A., 1992, *Ap. J. Suppl.*, in press

Inside the Stars, IAU Colloquium 137
ASP Conference Series, Vol. 40, 1993
Werner W. Weiss and Annie Baglin (eds.)

BERYLLIUM AND LITHIUM ABUNDANCES IN STARS WITH A RANGE OF METALLICITIES

RAMON J. GARCIA LOPEZ, EDUARDO L. MARTIN, RAFAEL REBOLO, AND JOHN E. BECKMAN
Instituto de Astrofísica de Canarias, E-38200, La Laguna, Spain

CONAL D. MCKEITH
Department of Pure and Applied Physics, The Queen's University of Belfast

JOHN K. WEBB
University of New South Wales

YAKOV V. PAVLENKO
The Principal Observatory of Ukraine

ABSTRACT We present new beryllium abundances for three unevolved stars with [Fe/H] in the range −0.4 to −1.6. The Be II doublet is detected in the three stars. A spectral synthesis analysis has been conducted to derive the abundances. We briefly discuss their Be and Li abundances, together with a wide sample of stars from the literature, in the context of mixing processes below the convection zone.

INTRODUCTION

Measurements of beryllium abundances in F and G stars with different metallicities provide information on mixing processes below their convective envelopes. Like lithium, beryllium is a fragile element which is destroyed in stellar interiors at relatively low temperatures (Be burns at $\sim 3.5 \times 10^6$ K while Li burns at 2.5×10^6 K). These temperatures are typical of the layers close to the base of the convective zone in F and G stars. The measurement of lithium in such stars has shown the important effect of metallicity on the efficiency of sub surface particle transport processes. The stars with [Fe/H] < -0.5 are able to preserve a large amount of lithium in their convective zones. On the contrary, stars with solar metallicity strongly deplete this element during their evolution, showing both an abundance decrease with decreasing mass and the "Li gap" found in middle F-type stars. There are a large number of lithium measuarements available in the literature (relatively easy to achieve using the resonance doublet at λ 6708 Å), but only very few of beryllium. This poor situation for beryllium determinations is mainly due to the fact that the only observable Be feature (the Be II resonance doublet at λ 3131 Å) lies in a very dense spectral region close to the atmospheric transmission cut-off. This paper presents new beryllium abundances for three stars in the metallicity range −0.4 to −1.6. In order to use these Be abundances to constrain the role of mixing processes below the convection zone, we have incorporated them in a larger sample of stars from the literature with known Be and Li abundances.

OBSERVATIONS AND ANALYSIS

The observations of two of the stars (HD132475 and HD38510) were carried out using the UCL echelle spectrograph attached to the AAT, providing a dispersion of 0.024 Å/pixel at λ 3130 Å. The third star (HD 165908) was observed with the IACUB (Instituto de Astrofísica de Canarias-University of Belfast) echelle spectrograph, at the 2.5 m Nordic Optical Telescope on La Palma. The dispersion obtained in this case was 0.017 Å/pixel. These observations belong to a large programme designed to measure light element abundances to test nucleosynthesis and mixing processes. We carried out a standard data reduction using the IRAF package.

To derive the beryllium abundance for each star we performed a spectral synthesis in order to fit the profile of the weaker line of the Be II doublet, which is clearly resolved in our spectra. Details on the stellar parameters and analysis procedure are given in Rebolo et al. (1992). Our error bar estimate for a beryllium determination is 0.3 dex, which includes errors in the synthesis fit as well as uncertainties in the adopted stellar parameters.

DISCUSSION AND CONCLUSIONS

Our results can be summarized as follows:

We have detected the Be II doublet in the three stars observed. The abundances inferred are in the range log N(Be)= 0.6 – 0.8 (in the scale log N(H)= 12).

In Figure Ia we plot the beryllium abundance versus effective temperature for the compiled Be data, while in Figure Ib we show lithium abundances versus the same parameter. In both figures we have divided the stars into three groups as a function of their metallicity: (i) [Fe/H] > −0.5, (ii) −0.5 > [Fe/H] > −1.6, and (iii) [Fe/H] < −1.6. In group (i) there is no evidence of Be depletion, except in the case of the hottest star (6900 K). Their Li abundances show the well known decline towards lower temperatures. The absence of Be depletion sets constraints on the mechamisms which cause the Li depletion in the lower Main Sequence (pre-Main Sequence depletion, mixing due to rotation and angular momentum loss, stellar winds, etc.). On the other hand, the hottest star in the sample shows both Be and Li depletion. This object is located in the range of F-type stars where the Li gap is present. Several mechanisms have been proposed to explain this phenomenon, and some of these predict also a Be depletion for the stars located in the Li gap (see García López and Spruit 1991 and references therein).

The stars belonging to group (ii) show fairly constant abundances for Li and Be, and are in an effective temperature range of 5500–5900 K. Their Be abundances are slightly below those of group (i), and considerably above those of group (iii). This shows the strong dependence of Be on metallicity, mainly for stars with [Fe/H] < −1.6 (Rebolo et al. 1992). The Li abundances in groups (ii) and (iii) reflect the presence of the Li plateau (Spite and Spite 1982) for metal-poor stars. The increasing Be abundance with metallicity and the constancy of the Li abundances can be interpreted in terms of pure galactic evolution and primordial nucleosynthesis, with no significant influence of mixing processes in the stellar interiors.

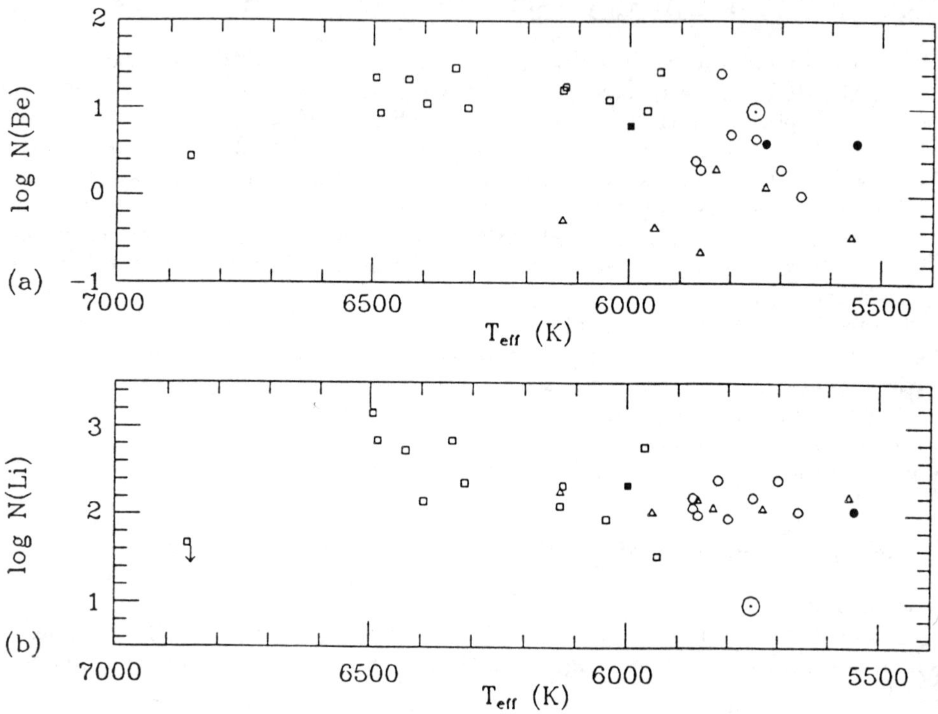

FIGURE I (a) Be abundances against effective temperature for unevolved stars divided in three metallicity bins: squares [Fe/H] > −0.5, circles −0.5 > [Fe/H] > −1.6, triangles [Fe/H] < −1.6. Our stars are filled symbols. (b) Li abundances against effective temperature for the same stars.

REFERENCES

García López, R. J., and Spruit, H. C. 1991, *Ap. J.*, 377, 268.

Rebolo, R., García López, R. J., Martín, E. L., Beckman, J. E., McKeith, C. D., Webb, J. K., and Pavlenko, Y. V. 1992, *"On the Origin of the Elements"*, Eds. N. Prantzos, M. Cassé, and E. Vangioni-Flam, in press.

Spite, F., and Spite, M. 1982, *Astr. Ap.*, 115, 357.

Inside the Stars, IAU Colloquium 137
ASP Conference Series, Vol. 40, 1993
Werner W. Weiss and Annie Baglin (eds.)

ON THE NITROGEN ABUNDANCE - AGE RELATION FOR MAIN SEQUENCE EARLY B-STARS

LEONID S. LYUBIMKOV
Crimean Astrophysical Observatory, Crimea, p.Nauchny, 334413, Ukraine

The results of the He, N and C abundance determinations for MS early B-stars, summarized in the review [1], lead to the inference that there are evolutionary variations in atmospheric contents of these elements. The conclusion was made that CNO-cycle products emerge on a surface of B-stars during their MS phase yet and such appearance can be considered as a probable manifestation of some mixing process between outer layers and stellar interiors. In particular the N enrichment with the age was found, which reachs about 1.0 dex for most massive B-stars (see [1,2]). However Gies and Lambert [3] (further GL) investigated recently the C, N and O abundances in a sample of early B-stars and did not confirm such great N enrichment. Since the problem of early mixing is very important for the understanding of first evolutionary stages of massive stars, the GL data need further and thorough analysis.

It is necessary to take into account some essential circumstances. Firstly, the comparison of GL nitrogen abundances with earlier results [2] shows a systematic difference (Fig.1). The discrepancy is small enough for most of 16 common stars (-0.17 dex on the average, dashed line in Fig.1), but for spectroscopic binaries $\pi^4 Ori$ and $\pi^5 Ori$ it is about -1.0 dex. It is possible that there is a marked influence of secondary components on spectra of these binaries. Moreover in the case of $\pi^4 Ori$ and $\pi^5 Ori$ there are great differences in used equivalent widths between GL and [2], and this can testify to probable spectral variability of both stars. Therefore these two stars must be excluded from further analysis.

Secondly, investigating the correlation between the nitrogen abundance $log\ \epsilon(N)$ and the stellar age t we should take into account that errors in surface gravity g have significant influence on t determination. The comparison of GL $log\ g$ values with results of Lyubimkov [2], Kane et al. [4] and Remie and Lamers [5] shows small systematic differences, which are correspondingly equal to 0.26 dex (14 common stars), 0.12 dex (11 stars) and 0.21 dex (7 stars) on the average. Consequently the values

$$[\Delta\, log\, g]_I = log\, g_{\rm GL} - log\, g_{[2]} - 0.26, \quad (1)$$
$$[\Delta\, log\, g]_{II} = log\, g_{\rm GL} - log\, g_{[4]} - 0.12, \quad (2)$$
$$[\Delta\, log\, g]_{III} = log\, g_{\rm GL} - log\, g_{[5]} - 0.21, \quad (3)$$

can be considered as residual discrepancies. These values are compared in Fig.2. We can see that discrepancies, which are greater than ±0.1, evidently cannot be accidental and should be regarded as real errors in GL results. Corresponding corrections will be introduced into relation between the N abundance and the age.

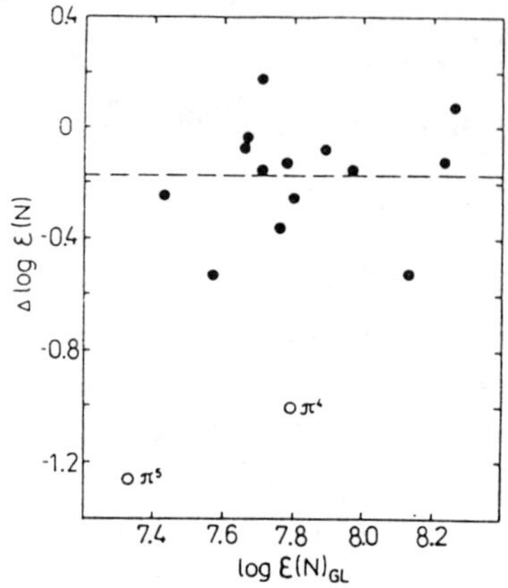

Fig.1. Difference in nitrogen abundances between GL data and our results [2]. Dashed line - the mean difference for all stars except $\pi^4 Ori$ and $\pi^5 Ori$ (open circles).

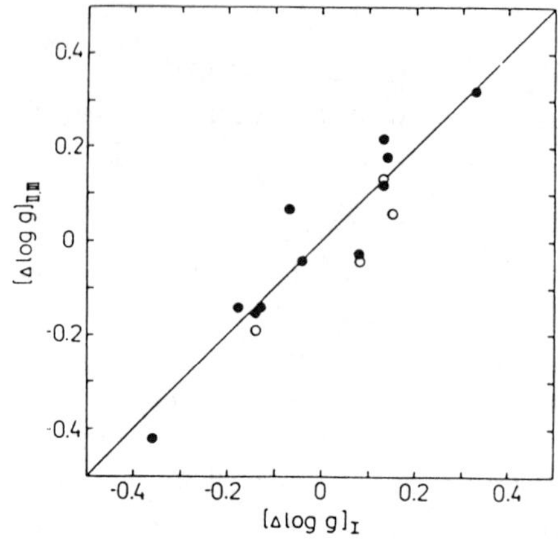

Fig.2. Residual discrepancies in *log g* calculated from formula (1) in comparison with the va lues determined from (2) and (3) (filled and open circles correspondingly).

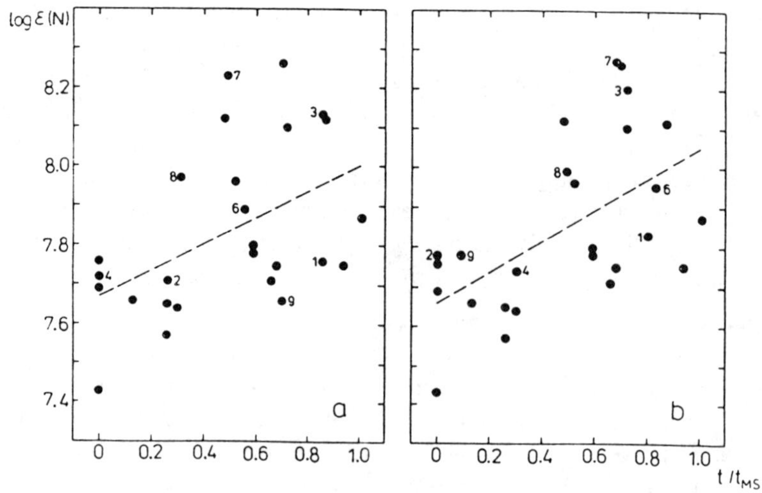

Fig.3. Nitrogen abundance in stars with masses 8.3 - 13.9 $M_\odot$ as a function of age in the cases of incorrected (a) and corrected (b) $log\ g_{GL}$ values for some stars. Dashed lines are drawn by the least squares method.

Thirdly, unlike previous works we will use the relative age t/t_{MS} instead of t value, where t_{MS} is the lifetime on main sequence. Such substitution allows to reduce a scatter, which is due to a difference in masses (both the mass M and the t and t_{MS} values are determined from evolutionary tracks).

In Fig.3a we have plotted the N abundance as a function of t/t_{MS} for early B-stars with masses M from 8.3 $M_\odot$ to 13.9 $M_\odot$ (26 stars, the mean mass $\overline{M}$ = 10.3 $M_\odot$). These stars are located in MS region as followed from evolutionary tracks [6] used in GL paper (all data in Fig.3a were taken from this paper, too). The dashed line drawn by the least squares method shows that $log\ \epsilon(N)$ value increases by 0.33 dex, when t/t_{MS} changes from 0 to 1.

For 8 stars the corrections to $log\ g_{GL}$ are known (numbered dots in Fig.3). Corresponding corrections in t/t_{MS} and $log\ \epsilon(N)$ values were found on a basis of calculations [6] and [7] (corrections in the N abundance are small). A new relation is shown in Fig.3b. The nitrogen enrichment on the MS phase is 0.39 dex now. It is interesting how the correlation coefficient q changes: it equals 0.36 for dependence on absolute age t, 0.48 for dependence on relative age t/t_{MS} (Fig.3a), and 0.56 for relation with corrected $log\ g$ values for some stars (Fig.3b). Consequently the correlation is strengthened.

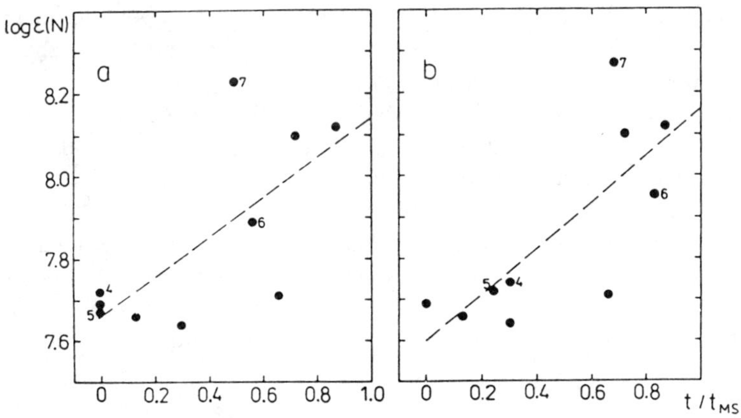

Fig.4. Same as Fig.3, but for stars with masses 11.1 - 14.8 $M_\odot$.

It is shown in [2] that the N enrichment intensifies with M increase. In Fig.4 the more massive stars from the GL list are considered. Their masses M vary from 11.1 $M_\odot$ to 14.8 $M_\odot$ ($\overline{M}$ = 12.5 $M_\odot$). This group consists of 10 stars, most of which were present already in Fig.3. In Fig.4 two cases are shown again: for incorrected (a) and corrected (b) $log\ g_{GL}$ values. The correction is included for 4 stars (numbered dots). The correlation coefficient is high enough in both cases: $q = 0.70$ (a) and $q = 0.76$ (b). The nitrogen enrichment on the MS phase is 0.48 dex and 0.56 dex correspondingly; this is found from dashed lines in Fig.4 drawn by least squares method.

So we see that there is the nitrogen enrichment during the MS phase, which is about 0.4 dex for stars with mean mass $\overline{M}$ = 10.3 $M_\odot$ and almost 0.6 dex for stars with $\overline{M} = 12.5\ M_\odot$ (in more accurate case b). Both the correlation with age and the dependence on mass are confirmed. However a total N enrichment is less than in our earlier paper [2]; for example, in the case of the star with $M = 13\ M_\odot$ the former value was about 1.0 dex and the new value is 0.6 dex. The explanation is that in the N abundance determination by GL unlike [2] the differences in *log g* between investigated stars were taken into account. Another probable cause suggested by GL is different approach to the microturbulence. Meanwhile our analysis shows that this explanation is not suitable, because there is no correlation between microturbulent velocities (non-LTE values in GL paper) and age $t/t_{\rm MS}$.

Besides nitrogen abundance it is interesting to consider the He and C abundances obtained by GL. Our analysis of 24 B-stars with known He/H values led to the conclusion that the He/H value increases during the MS phase from 0.08 to 0.13, i.e. by about 0.05 (for 7 stars the corrections in *log g* were taken into account). This result is in good accordance with previous one obtained for B-stars with masses M = 8 - 15 $M_\odot$ (see [1,8]). Consequently GL data confirm our He abundance - age relation for MS early B-stars.

As concerns carbon abundance in these stars, in GL paper the mean value $log\ \epsilon(C) = 8.20$ was found, which is by 0.4 dex lower than solar abundance. In the review [1] general carbon underabundance of about 0.3 dex was noted. We can see a good agreement here, however according to GL data the observed C deficiency is unexpectedly independent of age t/t_{MS}. As pointed out in [1], there is a simple method for testing the $log\ \epsilon(C)$ values: the total abundance C + N must be approximately conserved during the MS phase and the following stage of yellow supergiants. For 26 B-stars from the GL list the mean value is $log\ \epsilon(C+N) = 8.36 \pm 0.03$, whereas for supergiants from the list [9] it equals 8.62 (the latest solar value is 8.70). Obviously the condition is not fulfilled. Further investigations are needed to solve the problem of carbon underabundance in early B-stars.

REFERENCES

1. L.S.Lyubimkov. In *Evolution of Stars: The Photospheric Abundance Connection*, ed. G.Michaud and A.Tutukov, IAU Simp. N 145, Kluwer, Dordrecht, 1991, p.125.
2. L.S.Lyubimkov. Astrofizika, **20**, 475, 1984.
3. D.R.Gies, D.L.Lambert. Ap.J. (in press).
4. L.Kane, C.D.McKeith, P.L.Dufton. Astron. Astrophys., **84**, 115, 1980.
5. H.Remie, H.J.Lamers. Astron. Astrophys., **105**, 85, 1982.
6. A.Maeder, G.Meynet. Astron. Astrophys. Suppl., **76**, 411, 1988.
7. S.R.Becker, K.Butler. Astron. Astrophys. Suppl., **76**, 331, 1988.
8. L.S.Lyubimkov. Astrofizika, **29**, 479, 1988.
9. R.E.Luck, D.L.Lambert. Ap.J., **298**, 782, 1985.

Inside the Stars, IAU Colloquium 137
ASP Conference Series, Vol. 40, 1993
Werner W. Weiss and Annie Baglin (eds.)

OXYGEN ABUNDANCES IN LI-DEPLETED F-TYPE STARS

RAMON J. GARCIA LOPEZ, RAFAEL REBOLO, ARTEMIO HERRERO AND JOHN E. BECKMAN
Instituto de Astrofísica de Canarias, E-38200, La Laguna, Tenerife, Spain

ABSTRACT We have obtained the oxygen abundance in a sample of 44 F-type dwarfs in the Ursa Major group and the Hyades open cluster, using an NLTE analysis of the O I infrared triplet at $\lambda\lambda$ 7771-7775 Å. These clusters show the "lithium gap" in the F spectral range, and it has been suggested that if this phenomenon were due to microscopic diffusion, their Li-depleted stars could also show oxygen and/or nitrogen depletion. Our results here show a substantial measure of uniformity in the oxygen abundances. The data might indicate a very small dip (< 0.1 dex) in the O abundance for stars located in the Li gap. Microscopic diffusion seems to be the only known mechanism able to produce an oxygen dip. If this mechanism were responsible for the Li gap, it would have to account for a depletion of up to two orders of magnitude in Li and, at the same time, less than 0.1 dex in oxygen. We note that turbulent mixing could modify a microscopic diffusion pattern in order to yield the observed abundances of the two elements.

INTRODUCTION

The strong dip in the lithium abundance found in F-type stars with ages greater than $\sim 10^8$ yrs, belonging to different open clusters and the field, in the effective temperature range 6400 to 6900 K, is a phenomenon of considerable value for studying the structure and evolution of F stars. It was first discovered in the Hyades (Boesgaard and Tripicco 1986) and it is clearly present in the Ursa Major group (Boesgaard, Budge and Burck 1988). To explain the gap, several transport mechanisms have been proposed. These mechanisms can be divided into two different groups: those which invoke the nuclear burning of the Li atoms and those in which this is not necessary (see García López and Spruit 1991). Microscopic diffusion (Michaud 1986) is contained in this latter group. Michaud (1988) suggested that if the observed Li depletion in the F stars were due to microscopic diffusion, we could expect to find a similar phenomenon for oxygen and/or nitrogen, which would not be predicted for the other mechanisms proposed. With this test in mind, we set about measuring the oxygen abundance in the F stars of the Hyades (age $\sim 6 - 8 \times 10^8$ yrs.) and of Ursa Major (age $\sim 3 \times 10^8$ yrs.).

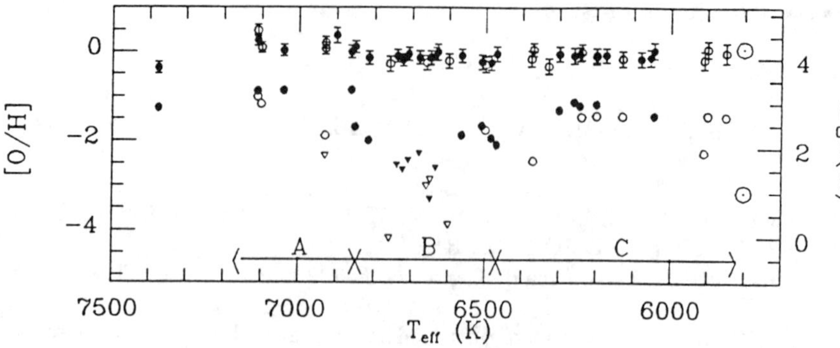

FIGURE I Oxygen and lithium abundances for the stars in our sample v. effective temperature. Filled circles: Hyades; open circles: Ursa Major group.

DERIVATION OF THE ABUNDANCES

The observations were performed during a series of observing runs on the 2.5 m Isaac Newton Telescope of the Observatorio del Roque de los Muchachos (La Palma) and the 2.2 m telescope of the Observatorio de Calar Alto (Almería). We selected the permitted triplet of oxygen I comprising the three lines at $\lambda\lambda$ 7771.944, 7774.167 and 7775.388 Å. The oxygen triplet was analyzed in NLTE, as the sensitivity of these lines to non-LTE effects is well known. For details of the analysis see García López et al. (1992).

RESULTS AND DISCUSSION

The results of our NLTE analysis of the infrared triplet are presented in Figure I, where we plot [O/H] against T_{eff}. We also show the lithium abundances for the same stars. A notable feature of Fig. I is that, in contrast to lithium, oxygen shows no major abundance "gap" in the effective temperature range considered. To examine the dependence of [O/H] on T_{eff} we can divide the Figure into three different zones (neglecting the hottest star in the sample), viz: zone A, with T_{eff} between 7150 and 6830 K, zone B between 6830 and 6450 K, and zone C between 6450 and 5800 K. The mean abundances for the three zones are:

A	B	C
$+0.18 \pm 0.16$	-0.15 ± 0.09	-0.08 ± 0.10
9 stars	17 stars	17 stars

where we have combined the data from the Hyades and the Ursa Major group. The quoted errors are the standard deviations with respect to the mean. We see that the mean abundance in group B is some 2σ lower than that of group A, and $\sim 1\sigma$ lower than for group C.

We must also draw attention here to the notable absence of oxygen in the hottest star of the sample VB38 (T_{eff} = 7375 K). This is related to the object being a chemically peculiar star, of type Am-Fm. VB38 figures as a well known

Am star in the *General Catalog of Ap and Am Stars* recently published by Renson, Gerbaldi and Catalano (1991).

For the stars in temperature range B, the data show a slightly lower abundance, 0.07 dex, than the stars in the cooler range C. If there were a real underabundance of oxygen in stars of zone B compared with both cooler and hotter objects, this phenomenon would very probably be due to microscopic diffusion of the oxygen atoms below the base of the convection zone, which is the only known mechanism able to produce this kind of depletion in the abundance of a chemical element which is not destroyed by nuclear burning in the stellar interior. Our present oxygen measurements place a clear observational restriction on the working of the microscopic diffusion hypothesis to explain the Li gap: it must be capable of reducing the Li abundance by up to 2 orders of magnitude, while leaving the oxygen unaffected, or almost unaffected, with reductions less than 0.1 dex. These restrictions apply quantitatively, only, of course, to the two clusters observed which have ages less than 10^9 years. Detailed computations for oxygen diffusion to compare directly with the observations are still required.

We note than the presence of turbulence will affect the simplest schemes of microscopic diffusion in which hydrostatic equilibrium is upset by an imbalance between gravitational and radiative accelerations. We can qualitatively suggest a possible scenario to explain the Li and O observations, in which this interaction would play a role. Given a degree of turbulence within the star, associated with rotation, or with rotational braking, or with the presence of internal gravity waves, the effect of the combination of microscopic diffusion and turbulence would be, in the $T_{\rm eff}$ range of the Li gap, to pull the Li atoms down to the level where they are destroyed by nuclear burning. On the other hand, the turbulent mixing would attenuate the effect of the microscopic diffusion on the O atoms. Since these are not destroyed, mixing with the interior will not notably affect the surface abundance. If the slight dip in the O abundance we appear to detect in range B were real, this could be due to the residual effect of the microscopic diffusion, as would the slightly raised abundance in range A. With observational effects of such small amplitude, however, more quantitative and testable conclusions must await more observations.

REFERENCES

Boesgaard, A. M., and Tripicco, M. J. 1986, *Ap. J.*, **302**, L49.

Boesgaard, A. M., Budge, K. G., and Burck, E. E. 1988, *Ap. J.*, **325**, 749.

García López, R. J., and Spruit, H. C. 1991, *Ap. J.*, **377**, 268.

García López, R. J., Rebolo, R., Herrero, A., and Beckman, J. E. 1992, *Ap. J.*, submitted.

Michaud, G. 1986, *Ap. J.*, **302**, 650.

Michaud, G. 1988, in *"Atmospheric Diagnostic of Stellar Evolution: Chemical Peculiarities, Mass Loss, and Explosion"*, IAU Coll. No. 108, Ed. K. Nomoto (Springer-Verlag), p. 3.

Renson, P, Gerbaldi, M., and Catalano, F. A. 1991, *Astr. Ap. Supp.*, **89**, 429.

Inside the Stars, IAU Colloquium 137
ASP Conference Series, Vol. 40, 1993
Werner W. Weiss and Annie Baglin (eds.)

HIGH S/N Na I D LINE PROFILES IN LATE-TYPE MS STARS AND THEIR IMPORTANCE FOR STELLAR MODELLING [1]

E. Covino[1], M.T. Gomez[1], G. Severino[1], and M. Franchini[2]

[1] *Osservatorio Astronomico di Capodimonte, Via Moiariello 16, Napoli*
[2] *Osservatorio Astronomico di Trieste Via Tiepolo 11, Trieste*

1 Introduction

Progress in observational techniques (i.e. high S/N spectroscopy) and modelling (e.g. RE models including convection, nonLTE codes, accurate atomic data) make significant to carry to extremes the numerical fit of a portion of the stellar spectrum and try to interpret the residual differences between calculations and observations in terms of stellar characteristics *new* respect to the employed model.

We have started to apply this synthesis method to a set of high resolution, high S/N spectra of late-type dwarfs in the region of the Na I D lines, obtained at ESO with the 1.4m CAT on May 1989 (Franchini et al. 1992). The dwarf stars are modelled in radiative equilibrium (RE), and the Na I D line synthesis accounts for nonLTE and blend effects.

We aim to interpret the residual differences between observations and calculations in terms of non-radiative heating in the stellar chromosphere. There we present our first results for the star 70 Oph A and, as a comparison, for the Sun as a star.

2 Method

As a standard atmosphere we assume a 1-dimensional radiative equilibrium (RE) model. Linsky (1985) lists a number of approximations inherent in RE models: (1) description of line blanketing (including molecules and nonLTE effects), (2) theory of convection (including overshoot), (3) treatment of atmospheric (magnetic) inhomogeneity, and (4) account for the CO cooling in non-magnetic regions. However the only alternative to 1-d RE model is 1-d semiempirical model but the latter already includes a temperature - rise in the chromosphere that is what we want specifically to study by the use of the synthesis technique. The RE models are built with the code MARCS which has been developed by Bengt

[1] Based on observations collected at the European Southern Observatory, La Silla, Chile

Gustafsson and Åke Nordlund. The models account for: (i) convection (Nordlund 1974), (ii) line blanketing by means of opacity distribution functions for metals and (iii) molecular opacities (see also Gustafsson et al. 1975).

Our main line formation parameters are the sodium abundance A_{Na} and the instrumental + macroturbulence + rotation broadening profile. We fit the stellar sodium abundance to the outer wings of the Na I D lines. This procedure yields $A^{\star}_{Na} = 2.35 \ 10^{-6}$, which is substantally equal to the solar value. Furthermore, we assume that instrumental + macro (granulation) + rotation broadening is accounted for by a normalized gaussian profile. So, we have only the halfwidth at half maximum of the gaussian profile, $\Delta\lambda$, as a free parameter. We estimate $\Delta\lambda$ "empirically", from the line widths of the weakest lines in the spectrum, and from the fit to the observed profiles of Fe I 5905 line. Both these procedures yield $\Delta\lambda_{\star} \sim 75$ mÅand $\Delta\lambda_{\odot} \sim 40$ mÅ.

Neutral sodium energy levels do not follow the Saha-Boltzmann law in the high photosphere and low chromosphere (e.g. Caccin et al. 1980). We solve the Na I statistical equilibrium equations coupled to the radiation transfer equation with the nonLTE code MULTI (Carlsson 1986). The sodium atomic model is similar to the one used by Caccin et al. but the Van der Waals damping constants are fixed after Andretta et al. (1991).

The departure from LTE of the line bound levels and continuum are the input for a successive program which accomplishes the spectral line synthesis accounting for the blend of the Na I D_1 and D_2 resonance lines.

3 Results and conclusion

For the comparison with the observations, we convolve the synthetic Na I D line profiles with a gaussian broadening normalized function whose HWHM is set to 40 mÅ for the Sun and 75 mÅ for 70 Oph A, as discussed above. Hence, we compute the ratio between observed and computed NaI D line profiles for both 70 Oph and the Sun; the FeI 5905 is also included in the calculation, as a matter of comparison . We find that while the calculations fit rather well the observations of both the iron line and the Na I D line wings (ratio $\sim$ 1), the Na I D line cores are brighter than those predicted on the basis of the RE models both for 70 Oph A and the Sun (ratio $\sim$ 1.5). In fact, the value of the ratio in the Na I D cores is roughly the same for 70 Oph A and the Sun. This result is indeed misleading and it is an artifact due to the different broadening functions pertaining to the two dwarf stars. When we deconvolve the observations with the broadening gaussian profiles discussed above , and then compute the ratio to the unconvolved RE synthetic line profiles, we obtain the results shown in Fig. 1, in which the maximum value of the ratio for 70 Oph A is roughly twice the solar one.

We think that differences between observed and computed Na I D line profiles, such as those reported in Fig. 1, contain evidence for non-radiatively heating in stellar chromospheres.

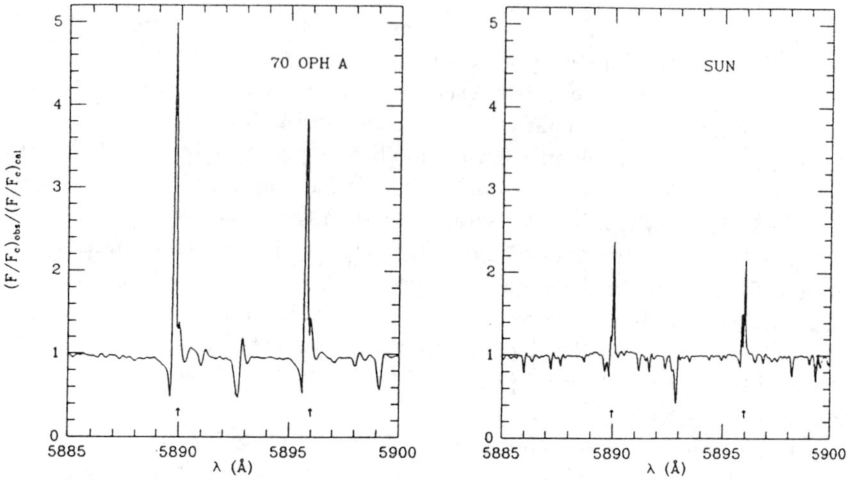

Figure 1. The observed to computed spectrum ratios for 70 Oph (left panel) and the Sun (right panel). The Na I D line cores are marked with arrows whilst the other negative peaks in the ratio correspond to absorption lines not included in the synthetic spectrum.

We plan to continue the work sketched in this poster expanding over the present stage in three respects: (i) processing the whole data set of late-type main sequence stars; (ii) improving the procedure, e.g. by the use of a radial-tangential model for macroturbulence broadening; (iii) calibrating different observed/computed Na I D line ratios in terms of non-radiative heated chromospheres.

References

Andretta, V., Gomez, M. T., and Severino, G.: 1991, Solar Phys., **131**, 1 3pt

Caccin B., Gomez M.T. and Roberti G.: 1980, Astron. Astrophys. **92**, 63 3pt

Carlsson M.: 1986, *A Computer Program for Solving Multi-Level Non-LTE Radiative Transfer Problems in Moving or Static Atmospheres*, Report No. 33, Uppsala Astronomical Observatory 3pt

Franchini M., Covino E., Stalio R., Terranegra L., Chavarria-K., C.: 1992, Astron. Astrophys., in press 3pt

Gomez, M.T., Rutten, R.J., Severino, G.: 1991, Astron. Astrophys., **244**, 501 3pt

Gustafsson B., Bell R.A., Eriksson K., Nordlund Å.: 1975, Astron. Astrophys., **42**, 407 3pt

Linsky J.L.: 1985, Solar Phys. **100**, 333 3pt

Nordlund Å.: 1974, Astron. Astrophys, **32**, 407 3pt

Inside the Stars, IAU Colloquium 137
ASP Conference Series, Vol. 40, 1993
Werner W. Weiss and Annie Baglin (eds.)

ABSOLUTE MEASUREMENTS OF STARSPOT AREA AND TEMPERATURE

JAMES E. NEFF AND DOUGLAS O'NEAL
Astronomy & Astrophysics, The Pennsylvania State University,
525 Davey Laboratory, University Park, PA 16801

STEVEN H. SAAR
Harvard/Smithsonian Center for Astrophysics, Mail Stop 58,
60 Garden Street, Cambridge, MA 02138

INTRODUCTION

Photometric and spectroscopic variability of late-type stars frequently has been interpreted as evidence of magnetic activity. The standard picture of stellar activity – inherited from solar observations – includes cool, dark "spots" in the photosphere and hot, dense regions in the chromosphere and coronae. The immediate cause of each of these phenomena is a closed topology of the local magnetic field. Because stars appear as mere points of light, these localized phenomena have not been directly resolvable on stars other than the Sun. Most observed effects are produced by an *asymmetric* distribution of starspots. If the distribution is symmetric, it would escape detection by most current techniques of light-curve and line-profile modeling. Even more troubling, the stellar properties measured with these techniques describe only a *difference* between contrasting hemispheres, not an absolute measure.

In order to devise an absolute measure, we searched the solar spectrum for spectroscopic features that would be produced only in starspots. The spectrum of sunspots in the region of the molecular bands of TiO near 7100 and near 8860 Å is radically different from that of the non-spotted photosphere. The presence of these features in the globally-averaged spectrum of a warmer star would be incontrovertible evidence for a cool (i.e., starspot) component of the photosphere.

We are using spectra of the TiO bands to *independently* measure the area and the temperature of starspots. The absolute strength of all the absorption bands is primarily a function of the total area of spots, while the relative absorption strength in different bands provides an independent constraint on the spot temperature.

OBSERVATIONS AND TARGET SELECTION

Our target stars consist of rapidly-rotating, late-type stars, primarily of BY Dra and RS CVn-type. Our method requires that we also obtain high-quality spectra of several categories of "inactive" stars. Spectra of inactive G and K dwarfs and subgiants represent unspotted photosphere at a variety of effective temperatures.

Spectra of M giants and M dwarfs represent spectra of starspots with various effective temperatures in stars with various gravities.

Our data were obtained during 9 observing runs at the National Solar Observatory's McMath telescope. They consist of 569 medium-resolution ($\Delta\lambda/\lambda \sim 20,000$), high-S/N (200–300) spectra with a range of approximately 200 Å centered on the TiO bands clustered near 7100 Å and on the band at 8860 Å. We observed 9 dwarf and 7 giant target stars, 13 inactive G and K stars, and 30 M dwarf and giant comparison stars.

ANALYSIS & MEASUREMENT TECHNIQUES

The two wavelength regions we selected contain TiO absorption bands with different temperature sensitivities: the 7100 Å bands first become visible for an immaculate star at spectral type about K5, while the 8860 Å band is not evident until M1 (Huenemoerder 1988; Ramsey and Nations 1980). Modeling the two bands simultaneously thus permits an estimate of both the size (from the absolute band strengths) and the temperature (from the ratio of the band strengths) of cool spots on the stellar surface. Our approach (patterned after Huenemoerder and Ramsey 1987 and Vogt 1981) is to model the normalized spectrum from a spotted star (F_{total}) as the weighted sum of F_{spot} and F_{phot}, the spectra of suitable standard stars with $T_{eff}=T_{spot}$ and $T_{eff}=T_{phot}$, respectively. The model is given by $F_{total} = [AR_\lambda F_{spot} + (1-A)F_{phot}]/[AR_\lambda + (1-A)]$, where A is the total fractional projected area of spots on the observed hemisphere (weighted by limb-darkening) and R_λ is the surface flux ratio between the spots and the photosphere. We assume that $R_\lambda = B_\lambda(T_{spot})/B_\lambda(T_{phot})$, the ratio of the respective Planck functions. We use simultaneous least-squares fits in both wavelength regions, with various standards to simulate F_{spot} and F_{phot}, until the best fit is achieved.

EXAMPLE: II PEG IN OCTOBER 1989

To illustrate our method, we have analyzed a series of spectra of the single-lined spectroscopic binary system II Pegasi (HD 224085) (see Saar and Neff 1990 for results for the dwarf star Gl 171.2A). We observed II Peg in both wavelength bands on each of five nights between 9 and 14 October 1989.

Because these spectra are being used to debug and develop our new fitting algorithm, the results are still uncertain. In particular, our current constraint on the net effective temperature is still quite primitive, and we have not yet codified the uncertainties in the derived properties.

Our preliminary results for II Peg yield $T_{quiet} = 5050 \pm 100$ K and $T_{spot} \sim 3250$ K. The fractional area coverage of such spots is 55 to 65% of the visible hemisphere. Both bands vary visibly in strength from night to night. The rotational period of this star is about 6.6 days, so we likely are witnessing rotational modulation of the spectrum from a star that is asymmetrically spotted. Huenemoerder, Ramsey, and Buzasi (1989) also saw variability in the 8860 Å band strength, but without observations at 7100 Å, they were unable to uniquely model the starspot properties.

WORK PLANNED AND IN PROGRESS

Thanks to the telescope allocation policy of the solar-stellar program at the U.S. National Solar Observatory, we were able to accumulate—over 45 nights spread over 2 years—a sufficient number of observations of comparison stars to piece together an appropriate "grid". This is the foremost requirement for our procedure to work. With the comparison grid in hand, we plan to continue monitoring our target stars to measure variability over rotational and cycle timescales.

There are still some difficulties in the analysis procedure that we must overcome: (1) The biggest problem is normalizing the 7100 Å continuum for the coolest M-type comparison stars. There are other TiO bands shortward of our ≈200 Å bandpass that could affect our normalization. (2) Our comparison grid is empirical, non-continuous, and non-evenly sampled. Our fitting procedure must account for this. (3) There is no evidence of spots on any of our comparison stars (though surely they have some small fraction of spottedness), but we must be wary of any anomalies in their spectra.

In March 1992 we began a new phase of our observational program using the Penn State fiber-optic echelle spectrograph at Kitt Peak. This instrument allows us to observe the entire spectrum simultaneously with the same sensitivity as our existing observations. We will have thousands of Angstroms of spectrum to constrain the synthesis, many more bands of differing temperature sensitivity, and a well-constrained continuum. The echelle data will produce superior results and will permit us to re-analyze our McMath data, solving any problems with normalization. These spectra also will simultaneously provide a measure of the chromospheric activity level.

REFERENCES

Huenemoerder, D. P. 1988, in *Cool Stars, Stellar Systems, and the Sun*, eds. J. Linsky and R. Stencel (Dordrecht: Reidel), p. 512.

Huenemoerder, D. P., Ramsey, L. W. 1987, *ApJ*, **319**, 392.

Huenemoerder, D. P., Ramsey, L. W., Buzasi, D.L. 1989, *AJ*, **98**, 2264.

Ramsey, L. W., Nations, H. L. 1980, *ApJ*, **239**, L121.

Saar, S.H., Neff, J.E. 1990, in *Cool Stars, Stellar Systems, and the Sun*, ed. G. Wallerstein, (ASP Conf. Series), p. 171.

Vogt, S.S. 1981, *ApJ*, **247**, 975.

Inside the Stars, IAU Colloquium 137
ASP Conference Series, Vol. 40, 1993
Werner W. Weiss and Annie Baglin (eds.)

A MULTILINE METHOD TO DETERMINE STELLAR MAGNETIC FIELDS

PILAR RÍPODAS, MANUEL COLLADOS,
JORGE SÁNCHEZ ALMEIDA, RAMÓN J. GARCÍA LÓPEZ
Instituto de Astrofísica de Canarias, E38200 La Laguna, Tenerife, Spain

INTRODUCTION

In the past, much effort has been devoted to determine stellar magnetic fields by means of the analysis of two spectral lines with different magnetic sensitivities (see, e.g., Robinson, 1980; Saar, 1988; Basri and Marcy, 1988). However, these methods are largely influenced by the uncertainties caused by, for instance, the presence of blends, uncertainties of atomic parameters, etc. Gray (1984) and, more recently, Basri *et al.* (1992) used a multiline analysis to measure stellar magnetic fields, reducing the importance of those uncertainties, thus increasing, in a statistical sense, the confidence of the results.

Here we present a new method based on the analysis of the equivalent widths of many Fe I spectral lines. The central idea is to obtain information on the magnetic field from the magnetic strengthening of the equivalent width of spectral lines. This technique has been applied to a weak-emission-line T Tauri star (TAP35) and a main sequence star (ξ Boo A), using data published by Basri *et al.*, (1992).

MAGNETIC FIELD DETECTION

The equivalent width (W^o) of a spectral line changes by an amount ΔW, when effective temperature (T_{eff}), microturbulence (ξ), (Fe/H) abundance (m) and/or magnetic field strength (B) are modified by ΔT_{eff}, $\Delta\xi$, Δm, and/or ΔB. In first order approximation, ΔW is given by

$$W = W^o + \frac{\partial W^o}{\partial T_{eff}}\Delta T_{eff} + \frac{\partial W^o}{\partial \xi}\Delta\xi + \frac{\partial W^o}{\partial m}\Delta m + \frac{\partial W^o}{\partial B}\Delta B.$$

The values of the derivatives can be calculated for a given model atmosphere at the same time the transfer equation is integrated, using the code developed by Ruiz Cobo and del Toro Iniesta (1992).

Thus, the method operates iteratively accordingly to the following scheme: starting with an initial guess model atmosphere one can estimate the amount by which these magnitudes must be changed, by a least-squares minimization, obtaining thus a second model atmosphere. Starting again with this new model atmosphere, this procedure is iterated until a convergence is reached. It is important to note that this technique provides all the free parameters simultaneously and so, they are self-consistent.

SIMULATIONS

We have carried out several numerical tests of the method with different model atmospheres, with $T_{eff} \approx 5500K$, $\log m \approx -4.5$, $\xi \approx 1.3$ km·s^{-1}, and $B \approx$ 1000 Gauss. Noise was added to the simulated data ($S/N \sim 100$). We have synthesized 16 spectral lines, which cover a large range of excitation potentials, equivalent widths and magnetic sensitivities, to ensure that they behave in a different manner to variations of the different free parameters of the model atmospheres. The results show that convergenge is reached after few iterations, with uncertainties of 100 K in T_{eff}, 0.10 in $\log m$, 0.30 km·s^{-1} in ξ and 60 Gauss in B. The standard deviation of the difference of the assumed true equivalent widths and the recovered ones is $\sim$ 1.5 mA.

OBSERVATIONAL RESULTS

TAP35

In a recent paper, Basri *et al.* (1992) have studied TAP35, a T Tauri star with weak-emission-lines, and give arguments which may suggest the presence of surface magnetic fields. We have used the equivalent widths measured by them on this star to compare their results with ours. As in their case, we have used 61 Uma, a low active G8V star to calibrate the oscillator strength of the lines observed. Using the above method, but neglecting the presence of magnetic field, we have obtained $T_{eff} = 5500$ K, $\xi = 0.5$ km $\cdot$ s^{-1}, and $\log(m) = -4.07$ (Basri *et al.* obtain $T_{eff} = 5250$ K and $\log(m) = -4.53$). This model atmosphere was used to re-calculate the oscillator strengths, so that the measured equivalent widths matched exactly those synthesized.

Our final model estimates $T_{eff} = 5350 \pm 200$ K, $\log m = -4.45 \pm 0.03$, $\xi = 2.3 \pm 0.3$ km ·s^{-1}, and $B = 1100 \pm 300$ Gauss for TAP35. Fig. 1 shows the relationship between the observed equivalent widths and those obtained with the best fit atmosphere. The standard deviation of the difference of equivalent widths is 11 mA. Basri *et al.* obtain a magnetic field value of 1000 ± 500 Gauss for TAP35, which is in agreement with our result.

ξ Boo A

The same analysis, with the same lines (taking again data from the paper by Basri *et al.*), and with the gf-values derived from 61 Uma, has been done on ξ Boo A (spectral type G8V). The final results show that $T_{eff} = 5400 \pm 70$ K, $\log m = -4.17 \pm 0.01$, $\xi = 0.15 \pm 0.15$ km ·s^{-1}, and $B = 330 \pm 90$ Gauss.

Fig. 2 shows the relationship between the simulated equivalent widths and those obtained with the best fit atmosphere. The standard deviation of the difference of equivalent widths is 3.2 mA.

Basri *et al.* can just give an upper limit to the mean magnetic field strength on this star ($B \leq 500$ Gauss).We detect a mean magnetic field of $B = 330 \pm 90$ Gauss, which is in agreement with Basri *et al.* again, and is a more accurate value.

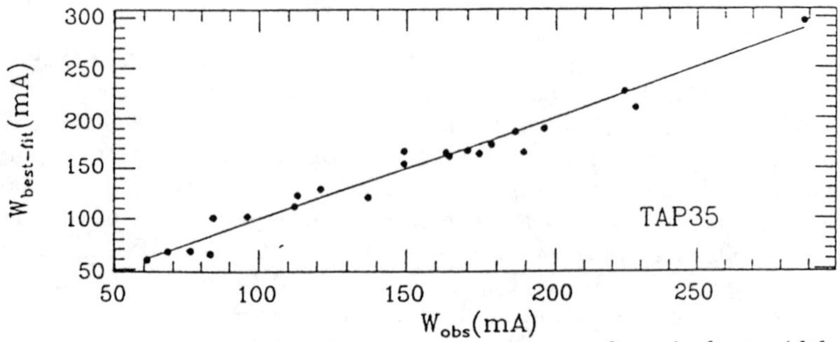

FIGURE I Relationship between the measured equivalent widths and those obtained with the best fit atmosphere for TAP35. The standard deviation of the difference of equivalent widths is 11 mA.

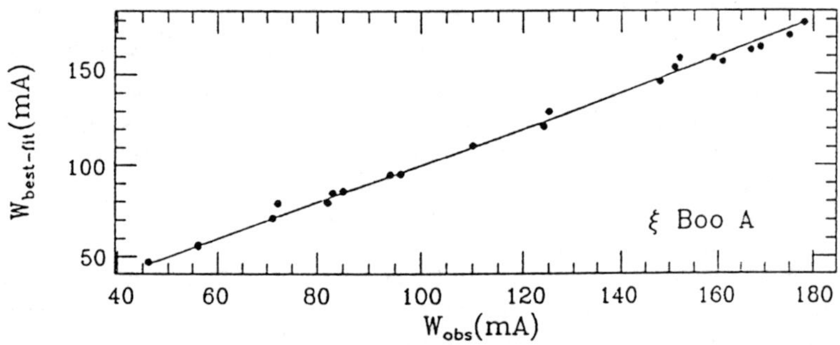

FIGURE II Relationship between the measured equivalent widths and those obtained with the best fit atmosphere for ξ Boo A. The standard deviation of the difference of equivalent widths is 3.2 mA.

ACKNOWLEDGMENTS

The authors wish to thank B. Ruiz Cobo and J.C. del Toro Iniesta for the implementation of their code to calculate fluxes and response functions, and for very fruitful comments.

REFERENCES

Basri, G., and Marcy, G.W.: 1988, *Astrophys. J.* **330**, 274.

Basri, G., Marcy, G.W., and Valenti, J.A.: 1992, *Astrophys. J.*, 390, 622.

Gray, D.F.: 1984, *Astrophys. J.* **277**, 640.

Robinson, R.D.: 1980, *Astrophys. J.* **239**, 961.

Ruiz Cobo, B., and del Toro Iniesta, J.C.: 1992, *Astrophys. J.* (preprint).

Saar, S.H.: 1988, *Astrophys. J.* **324**, 441.

Inside the Stars, IAU Colloquium 137
ASP Conference Series, Vol. 40, 1993
Werner W. Weiss and Annie Baglin (eds.)

ATMOSPHERIC STRUCTURE OF THE PULSATING Ap [CP2] STAR HR 3831 FROM RAPID MULTICOLOUR PHOTOMETRY

JAYMIE M. MATTHEWS[1]
Département de physique, Université de Montréal, C.P. 6128, succ. A, Montréal, Québec H3C 3J7 Canada

WILLIAM H. WEHLAU[1]
Department of Astronomy, University of Western Ontario, London, Ontario N6A 3K7 Canada

JOHN RICE[1]
Department of Physics, Brandon University, Brandon, Manitoba R7A 6A9 Canada

GORDON A. H. WALKER
Department of Geophysics & Astronomy, University of British Columbia, Vancouver, British Columbia V6T 1W5 Canada

OPENING A NEW WINDOW ON Ap ATMOSPHERES

Rapidly oscillating Ap (roAp) stars are cool magnetic CP2 stars which vary with periods of a few minutes and amplitudes less than 0.01 mag and 1 km/s in light and velocity. Analysis of their p-mode eigenfrequency patterns and splittings gives information about evolutionary state, rotation rate, magnetic field geometry and internal field strength (see Kurtz 1990; Matthews 1991). We present here an example of how roAp pulsations can be used to obtain an estimate of the temperature structure of an Ap atmosphere.

The pulsation amplitudes of roAp stars decline more rapidly with increasing wavelength than other known pulsators. Matthews *et al.* (1990) explained this by the wavelength dependence of limb darkening and its weighting effect on the integrated amplitude of an $(\ell, m) = (1,0)$ mode. This dipole mode – which appears to dominate roAp stars; in particular, HR 3831 – has two special properties:

- It does not change the projected area of the stellar disk, so all light variations can be attributed to flux (temperature) variations; and
- No matter what the inclination of the mode (except $i = 90°$), limb darkening *increases* the net amplitude measured on the visible disk.

[1] Visiting Astronomer, Cerro Tololo Inter-American Observatory, National Optical Astronomy Observatories, operated by the Association of Universities for Research in Astronomy, Inc., under contract with the National Science Foundation

FROM PULSATION AMPLITUDES TO ATMOSPHERIC STRUCTURE

Stronger limb darkening at shorter wavelengths enhances the amplitudes. By measuring amplitudes at various wavelengths, we can estimate the limb darkening coefficients and hence, the temperature structure of the atmosphere.

Using the CTIO 1.0- and 1.5-m telescopes simultaneously in November 1991, we obtained high-speed photometry of the roAp star HR 3831 in the optical (Strömgren *vby*, Cousins *RI*) and infrared (*JHK*). Figure 1 shows the Fourier amplitude spectra in the 8 bandpasses for the night of 25 Nov; the data cover ~18 pulsation cycles near phase 0.4 of the 2.85-d magnetic period of this star (Kurtz *et al.* 1990); i.e., just before magnetic and pulsation amplitude maximum. The prominent peak visible at shorter wavelengths occurs near 122 d^{-1} ($P = 11.8$ min): the dominant pulsation mode of HR 3831.

Models of the brightness oscillations of an $(\ell, m) = (1,0)$ mode for a given inclination i, mean temperature T_{eff} and temperature variation ΔT_{puls} were produced for all eight bandpasses. These were then weighted by the limb-darkening relation:

$$\frac{I_\nu(\theta)}{I_\nu(0)} = 1 - \beta_\nu(1 - cos\theta).$$

where β_ν is a free parameter. The resulting amplitude ratios were compared to the observed ratios and the coefficients β_ν estimated by least-squares fitting.

The resulting values of β_ν were used to solve the expression:

$$\frac{I_\nu(\theta)}{I_\nu(0)} = \int_0^\infty \frac{S_\nu(\tau_\nu)}{I_\nu(0)} e^{-\tau_\nu sec\theta} sec\theta \; d\tau_\nu$$

where the source function was modelled by the expansion

$$S_\nu(\tau_\nu) = a_0 + a_1\tau_\nu + a_2\tau_\nu^2 + a_3\tau_\nu^3 + a_4\tau_\nu^4 + a_5\tau_\nu^5$$

and $I_\nu(0)$ was sampled over the expected range of luminosity for an F0 V star. The coefficients a_j were obtained by *LU* inversion of the 6×6 matrix.

Finally, the source function was equated with the Planck function

$$S_\nu(\tau_\nu) = \frac{2h\nu^3}{c^2} \frac{1}{e^{h\nu/kt} - 1}$$

to produce the temperature structure as a function of optical depth.

AN EMPIRICAL $T - \tau$ RELATION FOR HR 3831

Model fits were performed for a range of mean effective temperatures [$7500\ K \leq T_{eff} \leq 8500\ K$] and inclinations of the pulsation pole [$30° \leq i \leq 80°$]. The observed amplitude ratios (and upper limits) of HR 3831 require a **steeper $T - \tau$ relation than for a solar atmosphere**. However, a monotonic $T - \tau$ curve cannot account for the data. The only solution is obtained by invoking a **temperature inversion near $\log\tau_{5000Å} \simeq -0.7$**, as shown in Figure 2. If real, this inversion suggests a strong source of opacity near that optical depth. If an artifact of the fit, it may mean that limb darkening is not the sole mechanism weighting the observed pulsation amplitudes. Flux redistribution by spots is a possibility, especially if the spots are linked to the magnetic (and pulsational) geometry. Even so, significant redistribution out to 2.2μ is unlikely, so the steepness of the $T - \tau$ curve is a robust result.

Figure 1. Fourier amplitude spectra of the pulsations of HR 3831.

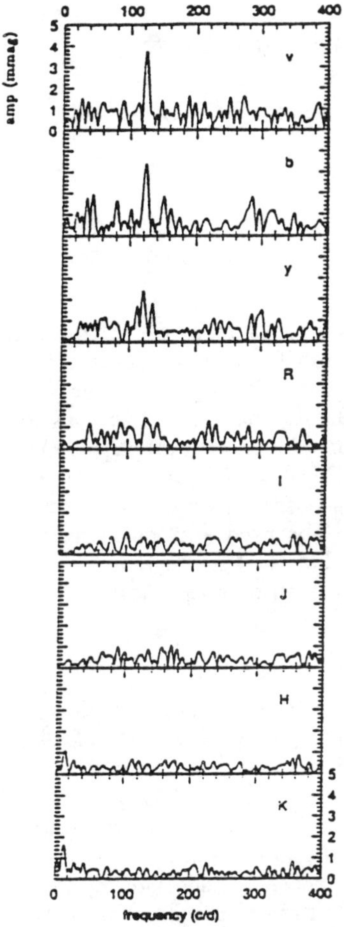

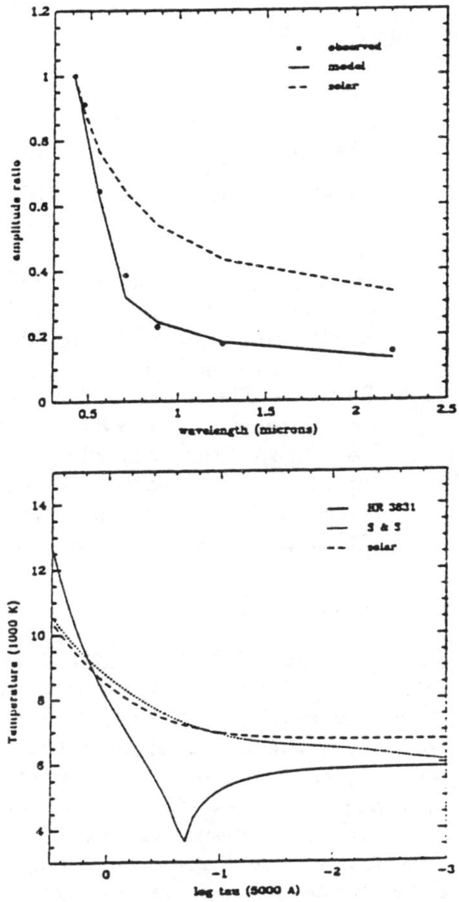

Figure 2. Is there a temperature inversion in the atmosphere of HR 3831? Above is the model fit to the observed amplitude ratios compared to that expected for a grey atmosphere. Below is the resulting $T - \tau$ relation, compared to the grey curve and a model by Shibahashi and Saio (1985) to account for roAp pulsation frequencies observed above the acoustic cutoff of a standard model.

REFERENCES

Kurtz, D.W. 1990, *Ann. Rev. Astron. Astrophys* **28**, 607.
Kurtz, D.W., Shibahashi, H. and Goode, P.R. 1990, *M.N.R.A.S.* **247**, 558.
Matthews, J.M. 1991, *P.A.S.P.* **103**, 5.
Matthews, J.M., Wehlau, W.H. and Walker, G.A.H. 1990, *Ap. J.* **365**, L81.
Shibahashi, H. and Saio, H. 1985, *P.A.S. Japan* **37**, 245.

Inside the Stars, IAU Colloquium 137
ASP Conference Series, Vol. 40, 1993
Werner W. Weiss and Annie Baglin (eds.)

THE ABUNDANCES OF VEGA FROM THE ULTRAVIOLET SPECTRUM

FIORELLA CASTELLI
CNR-GNA c/o Osservatorio Astronomico
Via G.B. Tiepolo 11 - 34131 Trieste - Italy

ABSTRACT We derived preliminary abundances from the ultraviolet spectrum of Vega by comparing high-resolution Copernicus spectra with spectra computed with the ATLAS9 code and with updated line lists. The T_{eff}=9400 K, log g=3.90 model, which fits the visual flux and the Balmer profiles is able to fit the ultraviolet flux either with [M/H]=−1 and microturbulent velocity ξ= 2 km s^{-1} or with [M/H]=−0.5 and ξ=0 km s^{-1}. The line spectrum is fitted by ξ=2 ± 0.5 km s^{-1} and by an iron abundance log (N(Fe)/N_{tot})=-5.2 ±0.1 dex. All the observed elements indicate underabundances ranging from about -0.35 dex (C) up to about -1 dex (Ni, Zn). The origin of the underabundances is an open question.

INTRODUCTION

Vega (α Lyrae = HR 7001 = HD 172167) is a Pop. I star of spectral type A0 V with a projected rotational velocity v*sini*=23 km s^{-1} (Gray,1980). Abundances derived from the visual range are lower than the solar ones from about 0 to -0.9 dex (Venn and Lambert,1990; Adelman and Gulliver, 1990). Vega is not unique, but is part of a group of metal-mild-underabundant slow-rotating A−type stars of luminosity class III−V for which abundance analyses based on high-resolution observations have been done only recently (see Cowley,1991). The observed underabundances could be due to physical mechanisms acting either inside the stars, as the differential diffusion (Michaud,1980), or acting outside the stars in the interstellar medium, as the separation of interstellar grains and gas (Venn and Lambert,1990), or acting both inside and outside the stars (Charbonneau, 1991). A quite attractive explanation could be the selective accretion by only the gas forming toghether with dust a circumstellar disk (Waters et al.,1992). Infrared observations have shown the existence of such a disk around Vega.

We have been studying the ultraviolet spectrum of Vega in order to add information to the abundance pattern of this star which nearly always is chosen as comparison standard in studies of A-type stars.

ANALYSIS AND RESULTS

Preliminary abundances were derived by comparing in the 130-135 nm and 200-318 nm regions LTE computed spectra with Copernicus U1 and V1 spectra with resolutions of 0.005 nm and 0.01 nm respectively. The U1 spectrum was observed by R.L. Kurucz in September 1974 and the V1 spectrum is published and exhaustively described by Rogerson (1989). Synthetic spectra were computed by using the SYNTHE code of Kurucz. The LTE model was computed by means of the ATLAS9 code with the new opacities (Kurucz,1991). The parameters T_{eff}=9400 K and log g=3.90 are obtained with the same methods and the same visual observed fluxes and observed Balmer profiles as decribed by Kurucz (1992). We derived the metallicity [M/H] and the microturbulent velocity ξ of the model from the comparison of the observed and computed ultraviolet fluxes. Observed fluxes are from IUE (Bohlin et al., 1990) and can be fitted either by a model with [M/H]=-1 and ξ=2 km s^{-1} or by a model with [M/H]=-0.5 and ξ=0 km s^{-1}.

TABLE I Preliminary abundances of Vega from the ultraviolet spectrum. $\log\epsilon=\log(N_{elem}/N_{tot})$

Elem	$\log\epsilon_{UV}$	$\log\epsilon_{Vis}$ VL	$\log\epsilon_{Vis}$ AG	$\log\epsilon_{Sun1}$ AG1	$\log\epsilon_{Sun2}$ G	$\log(\epsilon_{UV})-\log(\epsilon_{Sun1})$
C I	-3.88	-3.62	-3.81	-3.48	-3.44	-0.4
C II	-3.78			-3.48	-3.44	-0.3
N I	-4.49	-4.04		-3.99	-4.04	-0.5
O I	-3.71	-3.30		-3.11	-3.11	-0.6
Mg I	-4.96		-5.11	-4.46	-4.46	-0.5
Mg II	-4.96	-5.33	-5.15	-4.46	-4.46	-0.5
Si I	≤-5.09			-4.49	-4.49	≤-0.6
Si II	-5.09			-4.49	-4.49	-0.6
Si III	-4.49			-4.49	-4.49	0
S I	-4.83/-5.83	-4.83		-4.83	-4.77	0/-1
Ca II	-6.18	-6.20		-5.68	-5.70	-0.5
Ti II	-7.80	-7.49	-7.47	-7.05		-0.75
V II	-8.54			-8.04		-0.5
Cr II	-7.12		-6.80	-6.37		-0.5
Mn II	>-7.15		-7.24	-6.65		>-0.5
Fe I	-5.2	-5.06	-5.05	-4.37	-4.52	-0.83
Fe II	-5.2	-5.19	-5.16	-4.37	-4.52	-0.83
Co II	-8.12?			-7.12		-1 ?
Ni I	-6.79		-6.42	-5.79		-1
Ni II	-6.79		-6.33	-5.79		-1
Cu II	-8.83?			-7.83		-1 ?
Zn II	-8.44			-7.44		-1

VL=Venn and Lambert(1990); AG=Adelman and Gulliver(1990); AG1=Anders and Grevesse(1989); G=Grevesse(1991).

The computed line spectra are, however, independent from ξ and [M/H] adopted for computing the model, because both models (1) T_{eff}=9400 K, log g=3.90, ξ=0 km s^{-1}, [M/H]=−0.5 and (2) T_{eff}=9400 K, log g=3.90, ξ=2 km s^{-1}, [M/H]=−1 yield the same synthetic spectrum, provided that the same ξ and the same abundances are used for computing the spectrum. The adopted line lists are basically those of Kurucz (1989).

The iron abundance log (N(Fe)/N_{tot})=−5.2±0.1 dex was obtained from saturated strong Fe II lines, which weakly depend on ξ. After having fixed the iron abundance we derived ξ=2±0.5 km s^{-1} from the other Fe II lines depending on ξ. Preliminary abundances results are summarized in Table I. They are compared with the abundances from the visual spectrum (Venn and Lambert,1990; Adelman and Gulliver,1990), and with the solar photospheric abundances (Anders and Grevesse,1989; Grevesse, 1991). All the elements are underabundant ranging from −0.35 dex for carbon up to −0.83 dex for iron and to −1 dex for Ni and Zn. With the new solar iron determination of −4.52 dex the iron underabundance is reduced to -0.68 dex. Abundances of V, Mn, Co, and Cu may be affected by systematic errors, owing to the lack of hyperfine splitting in the calculations. Some lines, as, for instance, the blended features consisting mainly of N I (12) at 131.8998, 131.9005 nm, Mg I (1) at 285.2127 nm, S I (8) at 132.35156, 132.3523 nm, Fe II (1) at 259.837 and 259.9939 nm, and the strongest lines of Ti II Vis. mult. 5 at 307 nm require solar or nearly solar abundances in order to be correctly reproduced.

REFERENCES

Adelman,S.J., and Gulliver,A.F. 1990, *Ap.J.*, 348, 712.

Anders,E., and Grevesse, N. 1989, *Geochim. Cosmochim. Acta* 53, 197.

Bohlin,R.C.,Harris,A.W.,Holm,A.V.,Gry,C. 1990,*Ap.J. Suppl.*,73, 413.

Charbonneau,P. 1991, *Ap. J. (Letters)*, 372, L33.

Cowley,C.R. 1991, *IAU Symp.* No 145, p.183.

Gray,D.F. 1980, *Pub. A. S. P*, 92, 154.

Grevesse,N. 1991, quoted as private communication by Iglesias C.A.,Rogers, F.J.,and Wilson,B.G. in *Spin-Orbit Interaction Effects on the Rosseland Mean Opacity*, 1992,*Ap.J.*, preprint.

Kurucz,R.L. 1989, line lists on magnetic tapes

Kurucz,R.L. 1991, *New Opacity Calculations* in *Stellar Atmospheres beyond classical models*, ed. Crivellari et al., NATO ASI Series, p. 441.

Kurucz,R.L. 1992, *New atmospheres for modeling binaries and disks*, to be publised in *Improvements in Light Curve Modeling of Eclipsing Binary Stars*, Berlin,Springer-Verlag.

Michaud,G. 1980, *Astron. J.*, 85, 589.

Rogerson Jr.,J.B. 1989, *Ap. J. Suppl.*, 71, 1011.

Venn,K.A.,and Lambert,D.L. 1990, *Ap.J.*,363,234.

Waters,L.B.F.M.,Trams,N.R., Waelkens,C. 1992, *Astr. Ap.*, preprint.

Inside the Stars, IAU Colloquium 137
ASP Conference Series, Vol. 40, 1993
Werner W. Weiss and Annie Baglin (eds.)

LINE FORMATION IN STOCHASTIC WINDS OF T TAURI STARS

A. S. MITSKEVICH
Osservatorio di Arcetri, Largo Fermi 5, 50125 Firenze, Italy

A. NATTA
Osservatorio di Arcetri, Largo Fermi 5, 50125 Firenze, Italy

V. P. GRININ
Crimean Astrophysical Observatory, Nauchny, Crimea, 334413, Ukraina

ABSTRACT

We discuss in this poster the problem of the formation of Hα lines in the stochastic winds of T Tauri stars.

INTRODUCTION

The Hα profiles in T Tauri stars show a remarkable variety of line shapes. Nevertheless, it is possible to identify some typical features, which characterize them:
i) Hα has a double-peak shape, with a broad, symmetric emission and a narrow, slightly blue-shifted absorption component.
ii) The red emission peak occurs very near zero velocity, and it is in general two to five times stronger than the blue peak.
iii) Often the transition from the absorption to the red peak is very sharp.

Models which assume that the wind is continuous and spherically symmetric cannot reproduce these features. Recent attempts to reproduce the Hα profiles in conical winds (Calvet, Hartmann and Hewett 1991), as well as in accreting columns of gas (Calvet and Hartmann 1992) do account for some aspects of the line profiles only.

A new approach was suggested by Grinin and Mitskevich 1991, who assumed that T Tauri stellar winds are not continuous, but rather consist of clumps of gas (*stochastic winds*). We present here a new discussion of the formation of Hα profiles in stochastic winds.

WIND MODEL

We assume in our models that the wind is made of clumps moving in an empty space. The clump velocity increases to a maximum value of ~250 km s^{-1} near the stellar surface and then decreases to a terminal velocity v_∞. The wind is on average stationary, i.e., the mass loss rate at each radius is constant. The clump properties (velocity, line source function, optical depth, etc.) are only function of the distance from the star. Each clump keeps its individuality while moving outward. At each distance r from the star, the clumpy structure of the wind is characterized by the typical clump size l and the filling factor q.

The radiation transfer is solved following the approach developed by Lindsey and Jefferies 1990 for inhomogeneous, static stellar atmospheres, applied here to moving media.

Two classes of models have been computed, and the results are shown in Figure 1. In the first (*Case A* models) the filling factor is constant with r; in the second (*Case B* models), the clumps expand at sound speed and merge in the wind periphery. In this case, the filling factor is a function of r.

In all cases, we describe the line source function and opacity by simple parametric laws, which, nevertheless, agree well with the results of non-LTE calculations.

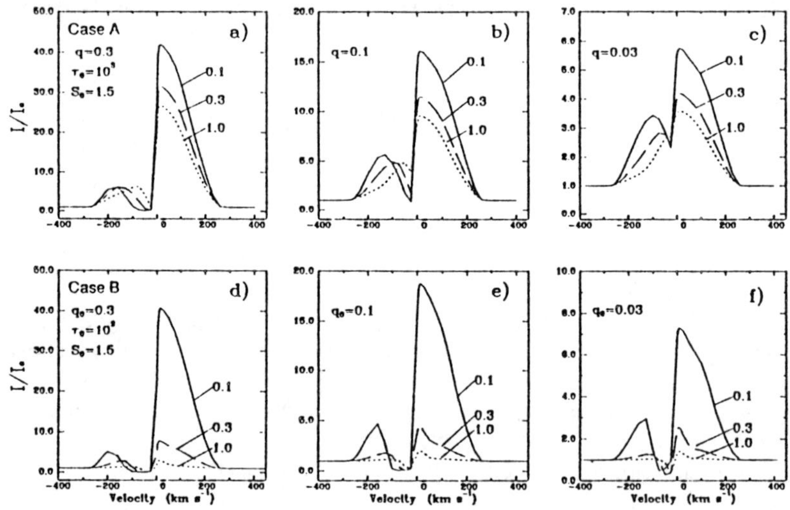

Fig. 1. Computed profiles for lines of high optical depth ($\tau_0 = 10^3$). Panels a, b and c are for *Case A* models, Panels d, e and f for *Case B* models. Each curve is labelled with the value of l_0 in units of the stellar radius.

RESULTS

The results indicate that most of the features characterizing the observed Hα profiles are indeed well accounted for by the models of stochastic, decelerating winds.

More specifically:

i) For a large range of the parameters q_0 and l_0, the computed lines have double peaked profiles, with an intensity of the blue peak two to five times smaller than that of the red peak.

ii) The position of the maximum of the red peak is very near zero velocity. The absorption peak is slightly blue-shifted; the width of the absorption depends on the model parameters, but is always much smaller than the width of the emission.

iii) The transition between the absorption and the red emission peak is very sharp. This is a typical feature of decelerating winds and is very difficult to obtain in any other type of models. The position of the transition depends mainly on the terminal wind velocity v_∞. Finite values of v_∞ may account for the fact that, in some stars, the transition from absorption to emission appears somewhat blue-shifted.

Few results have been obtained for lines of intermediate optical depth (τ_0=10) and are shown in Figure 2. In this case, the line profiles are symmetric, without any absorption feature. These results seem to provide an interesting solution to the puzzling properties of the Ca II infrared triplet lines at ~8600 Å. In many T Tauri stars, these three lines have broad and very symmetric, nearly gaussian, emission profiles, with no absorption (Grinin and Mitskevich 1991; Hamann and Persson 1992). At the same time, the fact that the three lines have comparable intensity requires them to be optically thick. Unless high turbulent velocity gas exists near the star, clumpy winds seem the only way to produce pure emission profiles in optically thick lines (cfr. Figure 2b).

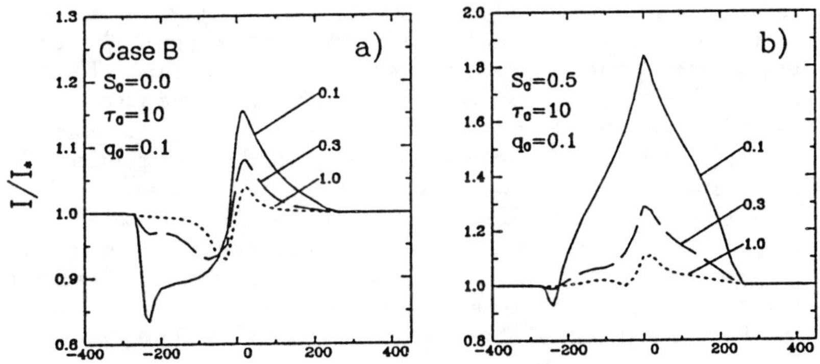

Fig. 2. Computed profiles for lines of intermediate optical depth. All curves refer to *Case B* models. The curves are labelled by the values of l_0 in units of the stellar radius.

From the results presented here, we think that stochastic winds can explain the main features of the various different line profiles observed in T Tauri stars. A detailed discussion of the method and the results will be published elsewhere.

REFERENCES

Calvet, N., & Hartmann, L., 1992, Ap.J. 386, 239.
Calvet, N., Hartmann, L., & Hewett R., 1992, Ap.J. 386, 229.
Grinin, V.P., & Mitskevich, A.S., 1991, Astroph.Sp.Sc. 185,107.
Hamann, F., & Persson, S.E., 1992, Ap.J., in press.
Lindsey, C., & Jefferies, J.T., 1990, Ap.J. 349, 286.

Inside the Stars, IAU Colloquium 137
ASP Conference Series, Vol. 40, 1993
Werner W. Weiss and Annie Baglin (eds.)

THE EQUATION OF STATE

WERNER DÄPPEN
Department of Physics and Astronomy, University of Southern California, Los Angeles, CA 90089-1342, U.S.A. and
Institut für Astronomie, Türkenschanzstr. 17, 1180 Vienna, Austria

ABSTRACT There are two basic approaches to the equation of state for stellar envelopes and interiors. The traditional method chooses the so-called "chemical picture", in which the notion of atoms is maintained despite the plasma environment. A mixture of atoms, molecules, ions, electrons and nuclei is considered, and the occurring ionization and dissociation reactions (thus the name chemical picture) are treated according to the entropy-maximum (or free-energy-minimum) principle. The alternative method is based on the so-called "physical picture", where only fundamental particles (electrons, nuclei) explicitly enter. Through the means of activity expansions, the problems of plasma physics and statistical mechanics are treated simultaneously and on the same footing. For helio- and asteroseismology, an accurate and precise equation of state is essential. Progress towards a better equation of state can be made in several ways: purely theoretical efforts, checks with experiments, including astrophysical data, and comparisons between different theoretical formalisms. Comparisons are useful to assess the domain of temperature and density where the theoretical complications matter, and to determine the diagnostic potential of astrophysical observables for equation of state issues.

INTRODUCTION

The three basic material properties required in stellar models are the equation of state, opacity, and the nuclear-energy generation rate. Opacity is discussed by Seaton (*these proceedings*). A recent review on the equation of state and opacity is Däppen, Keady & Rogers (1991). For a useful introduction to the equation of state see the book by Eliezer, Ghatak & Hora (1986). The term equation of state has a narrower and a larger meaning. In the restricted sense, the (thermal) equation of state relates pressure to density and temperature. (The caloric equation of state expresses internal energy as a function of density and temperature). Thermal and caloric equations of state are derived from a single thermodynamic potential, such as *e.g.* the free energy. In a slightly broader sense, the term equation of state often means the associated thermodynamic potential. The thermodynamic quantities must be consistent with each other, that is, their appropriate Maxwell relations have to be satisfied. Such *formal* consistency is always achieved if the equation of state and the thermodynamic quantities stem from the *same* thermodynamic potential. In trivial models (*e.g.* in a plasma assumed to be fully ionized everywhere) it is possible to write down a consistent equation of state and thermodynamic quantities independently. However, in more realistic cases, modelling a thermodynamic potential is the only practical way to obtain a consistent equation of state and thermodynamic quantities.

Improvements in the equation of state beyond the model of a mixture of ideal gases are difficult. This has both conceptual and technical reasons. As a fundamental conceptual reason I mention the fact that in a plasma environment already the idea of isolated atoms (and compound ions) has to be abandoned. A technical reason is the difficulty encountered when specific non-ideal effect are modelled. The three principal non-ideal effects are related to: (i) the internal partition functions of bound systems, (ii) pressure ionization, and (iii) collective interactions of the charged particles. The internal partition functions contain the difficult problem of excited states, where and how they are to be cut off. They are an important element in determining the ionization balances. Pressure ionization has to be provided by non-ideal interaction terms, because ideal gases would unphysically recombine in the central regions of stars.

The plasma of the interiors of "normal" stars is only weakly non-ideal. One would therefore think that finding a good equation of state is not too difficult. Indeed, simple models of the equation of state have been highly successful in many calculations of stellar structure and evolution. However, the inadequacy of simple equations of state has been clearly demonstrated by helioseismology. The observed solar oscillation modes are standing acoustic waves; hence the quantity most obviously probed is local sound speed. Since the oscillations are largely adiabatic (except very near the surface), the frequencies are determined predominantly by the local *adiabatic* sound speed, which is a thermodynamic quantity. In addition, the frequencies depend on the density distribution in the Sun. A good knowledge of the equation of state is therefore very important for any interpretation of solar observational data. However, the remarkable accuracy and wealth of these data provides, in addition, powerful constraints on the equation of state (for a review see Christensen–Dalsgaard & Däppen, 1992).

The smaller amount of expected *stellar* data will make a seismological determination of the equation of state more difficult. An accurate and precise equation of state is therefore a must. Obviously, all asteroseismic inferences will be subject to the uncertainty in the equation of state. However, not all applications require the same degree of accuracy. For instance, the calibration of solar evolutionary models to present-age radius and luminosity often reduces the influence on uncertainties in the equation of state. So, for instance, any uncertainty in the pressure in the solar centre is easily compensated with a slight change in the helium abundance. Many aspects of such a recalibrated model remain virtually unchanged and thus independent of the equation of state. However, if the same calibration is to yield the helium abundance, the result is affected by the whole uncertainty in the equation of state. As another example, the alternative helioseismic determination of the helium abundance in the solar convection zone (for recent progress see Kosovichev *et al.*, 1992) does not profit from favourable cancellations. Since the key idea of this helium abundance determination is based on the signature of the second ionization zone of helium on the adiabatic exponent Γ_1, and thus on sound speed, an absolutely accurate equation of state is essential. Helioseismology has, in principle, the potential to localize thermodynamical properties everywhere in the Sun. Therefore, Gough (1984a) suggested to use the power of helioseismology to determine at the same time the equation of state and the helium abundance.

In the following, I will discuss the statistical-mechanical foundations of the equation of state needed for the interiors of "normal" stars. I will then discuss the requirements of helio- and asteroseismology and present two representative formalisms of current interest. I will use detailed comparisons between these two formalisms to assess the present uncertainty in the equation of state and to localize those regions in temperature and density where progress is most urgent. Comparisons will also be helpful to test the influence of controversial features in the equation of state on observational quantities.

STATISTICAL MECHANICAL FOUNDATION OF THE EQUATION OF STATE

The typical task of quantum statistical mechanics consists in the calculation of the (total) partition function (see, *e.g.*, Huang, 1963, Reichl, 1980)

$$Z(T,V,\mathbf{N}) = \mathrm{Tr}(\mathrm{e}^{-H/kT}) \, . \tag{1}$$

Here, H is the Hamiltonian operator of the quantum system with N_i particles of species i, $i = 1,\ldots,m$ [*i.e.*, $\mathbf{N} = (N_1, N_2, \ldots, N_m)$] confined to a box of volume V, and k is the Boltzmann constant. (Note that this is the canonical point of view; the alternative, equivalent grand-canonical approach is another possibility, which is discussed in a subsequent section). The partition function directly leads to the thermodynamical potential of the free energy

$$F(T,V,\mathbf{N}) = -kT \ln Z(T,V,\mathbf{N}) \, , \tag{2}$$

from which all thermodynamic quantities can be derived. Given the large number of interacting particles, it is evident that various approximations are necessary before Eq. (1) can be evaluated. One begins with treating the motion of the heavy particles (nuclei, atoms, ions) according to classical mechanics, which is certainly appropriate as long one does not approach the conditions of the interior of neutron stars. Thus, only electrons have to be described by quantum mechanics.

Such a separation of effects leads to a factorization in Eq. (1), which is translated into a sum in Eq. (2). The free energy thus becomes modular, which is a very useful property of models based on the canonical partition function. The classical contribution in Eq. (1) is obtained from an integration over the Hamiltonian coordinates p_i and q_i ($i = 1,\ldots,3N$). Neglecting velocity-dependent interactions between the particles, the classical integrations over p_i and q_i factorize. The result is a product of the Maxwell-Boltzmann distribution for the momenta of heavy particles with the so-called configuration integral

$$Q(T,V,\mathbf{N}) = \int dp_1 \cdots dp_{3N} \, dq_1 \cdots dq_{3N} \, \mathrm{e}^{-V^{int}/kT} \, , \tag{3}$$

where $V^{int} = V^{int}(q_1,\ldots,q_{3N})$ is the interaction potential of the heavy particles. Equation (3) is at the base of the (negative) Coulomb-pressure correction for the charged particles and of excluded-volume effects of neutral particles (such as hard-sphere and Van-der-Waals type corrections). Excluded volume effects become very important in relatively cold plasmas of low-mass stars or planets (see Saumon & Chabrier, 1991,1992; Chabrier & Saumon, *these proceedings*). In the Coulomb case, charge neutrality requires the electrons to be included, though for consistency with the assumptions of Eq. (3) they have to be treated classically (but quantum correction factors can be included). An approximate evaluation of the Coulomb pressure correction is made in the Debye-Hückel theory, where it is assumed that each charge is surrounded by a cloud of opposite charge, which screens the Coulomb interaction outside a sphere with the so-called Debye radius. In many astrophysical equations of state the specific form of the Debye-Hückel Coulomb free energy used is that adopted by Graboske, Harwood & Rogers (1969). There, two additional effects are taken into account. First, with increasing partial degeneracy of the electrons, their influence in Eq. (3) is gradually reduced (by a factor involving Fermi integrals), and second, the configurational integral Q appears with a multiplicative correction factor τ that serves as a rough approximation for the non-vanishing size of the charged particles (see also Chabrier, 1990; Ebeling *et al.*, 1991). The Debye-Hückel approximation is basically valid as long as there are still relatively many particles in the so-called Debye sphere (with the Debye length as its radius).

Once the heavy particles are separated out, electrons remain, and they should be, of course, treated according to quantum mechanics. Here, there is a bifurcation into two distinct classes of approach, the "chemical picture" and the "physical picture" (Krasnikov, 1977). While in the more conventional chemical picture bound configurations (atoms, ions and molecules) are introduced and treated as new and independent species, only *fundamental* particles (electrons and nuclei) appear in the physical picture. In the chemical picture, reactions between the various species occur, and thus the thermodynamical equilibrium must be sought among the stoichiometrically allowed set of concentration variables by means of a maximum entropy (or minimum free-energy) principle. In contrast, the physical picture has the aesthetic advantage that there is no need for a minimax principle; the question of bound states is dealt with implicitly through the Hamiltonian describing the interaction between the fundamental particles. The different physical assumptions of the two pictures have been the source of considerable difficulties in recent comparisons and interpretations (see, *e.g.*, Rouse, 1983; Ebeling *et al.*, 1985; Däppen, Anderson & Mihalas, 1987).

There is an intuitive simplicity in the chemical picture: we usually take the existence of atoms in plasmas for granted, at least at densities that are not too high. However, this simplicity has to be paid for by additional minimization procedures in the multidimensional space of abundances of each species, restricted by the appropriate stoichiometrical relations and by mass and charge conservation. The physical idea behind this minimization is simple: the "internal" degrees of freedom, such as ionization degrees, are not adjustable by the experimenter; one can only control "external" parameters, like temperature, density, and mass fractions of each chemical element. The thermodynamic equilibrium is then determined as the one configuration, compared with those having different internal parameters, that minimizes the free energy, or equivalently, maximizes entropy. Once this minimum is found, the model free energy delivers all thermodynamic quantities in a straightforward way by differentiation.

It should be clear that the advantage of the chemical picture lies in the possibility to model complicated plasmas, and to obtain numerically smooth and consistent thermodynamical quantities (see below). Nevertheless, the heuristic method of the separation of the atomic-physics problem from that of statistical mechanics is not satisfactory, and attempts have been made to avoid the concept of a perturbed atom in a plasma altogether. Thus in the physical picture only fundamental particles (electrons and nuclei) enter. This has the advantage that the constraints appearing in the chemical picture (*i.e.*, mass and charge conservation) are automatically satisfied. In the absence of constraints, it becomes possible to use the powerful apparatus of the grand-canonical partition function. In the chemical picture, however, this road is barred, because the aforementioned constraints are expressed in numbers of particles and not in terms of the chemical potentials, the independent variables of the grand-canonical partition function. In the physical picture, one can build a theory of partially ionized plasmas similar to well-know cluster expansions for real gases (Rogers, 1981; for an introduction into cluster expansions see Huang, 1963).

STELLAR DEMANDS ON AN EQUATION OF STATE

It is important to realize that a stellar equation of state has to be *formally* precise and consistent, even before the question of the accuracy of the physical description is asked. It has to satisfy four conditions: *i*) a large domain of applicability (in ρ, T), *ii*) a high precision of its numerical realization, *iii*) consistency between the thermodynamic quantities, and *iv*) the possibility to take into account relatively complex mixtures with at least several of the more abundant chemical elements.

More specifically, the first condition demands that the formalism can be used from the stellar surface (the photosphere), where T is typically a few 10^3 K and ρ some 10^{-7} g/cm^3, to the center of a star where T is, again typically, about 10^7 K and ρ some 10^2 g/cm^3. The second condition demands that a given formalism can be cast in an algorithm that converges without ambiguity and with sufficient precision, so that all required thermodynamic derivatives (such as adiabatic gradients) can be computed. Note, that for this only *formal* precision is required: reality of the physical description is a different issue. The third condition, consistency, states that all thermodynamic quantities stem from a single thermodynamic potential. This condition is often violated in two- or more-zone formalisms, which contains a different physical theory in different parts of a star. An example is the *ad hoc* imposition of full ionization in the central region, in order to mimic a pressure-ionization device, in combination with a conventional Saha equation in the envelope of the star. Such a formalism leads to a discontinuous thermodynamic potential and a violation of thermodynamic identities. Thermodynamic identities are, however, often used in calculations of stellar structure and oscillations. As an example of the use of thermodynamical identities, I mention the transformation of the density fluctuation to the pressure fluctuation in linear adiabatic pulsation calculations. The connection between density and pressure changes is given by the adiabatic gradient Γ_1 , and it is therefore imperative that this gradient is consistent with the equation of state and other thermodynamic variables used in the model. This example illustrates the necessity of formal consistency. Finally, the third and last condition, *i.e.* the possibility to describe rather realistic chemical compositions, is less important for the equation of state itself. However, for opacity, heavy elements are important, and a good equation of state plays an important role in any opacity calculation.

It is in view of these specific requirements that astrophysicists dare to develop their own formalisms, which are often built around time-honoured intuitive ideas, and therefore lack a rigorous foundation. One of these "home-grown" astrophysical equations of state has been developed as part of the international "Opacity Project" (OP, see Seaton, 1987,1992, and *these proceedings*) by Mihalas, Hummer and Däppen (Hummer & Mihalas, 1988; Mihalas, Däppen & Hummer, 1988; Däppen *et al.*, 1988; hereinafter MHD). The MHD equation of state is written in the *chemical picture*. Plasma interactions are treated with modifications of atomic states, *i.e.* the quantum mechanical problem is solved before statistical mechanics is applied.

Of course, many more-or-less rigorous formalisms for strongly-coupled plasmas have been developed, and extensive numerical simulations have been made (see, *e.g.*, the book by Ebeling *et al.*, 1991, and references therein). However, since their motivation is different, they mostly aim at understanding qualitative phenomena, and they are normally not suited for the purposes of stellar models. That is, their results are not sufficiently precise, nor do they allow a description of realistic astrophysical mixtures. They would therefore not satisfy the four conditions mentioned above. A first attempt to develop a *stellar* equation of state with a better statistical mechanical foundation was made as part of an opacity project at Livermore (OPAL) (Rogers, 1986; Iglesias, Rogers & Wilson, 1987; Iglesias & Rogers, 1991; hereinafter Livermore equation of state).

TWO EXAMPLES OF EQUATIONS OF STATE: CHEMICAL AND PHYSICAL PICTURE

The MHD equation of state (chemical picture)

In the chemical picture, perturbed atoms must be introduced on a more-or-less *ad-hoc* basis to avoid the familiar divergence of internal partition functions (see *e.g.* Ebeling, Kraeft & Kremp, 1976). In other words, the approximation of unperturbed atoms precludes the application of standard statistical mechanics, *i.e.* the attribution of a Boltzmann-factor to each atomic state. The conventional remedy of the chemical picture against this is a modification of the atomic states, *e.g.* by cutting off the highly excited states in function of density and temperature of the plasma. Such cut-offs, however, have in general dire consequences due to the discrete nature of the atomic spectrum, *i.e.* jumps in the number of excited states (and thus in the partition functions and in the free energy) despite smoothly varying external parameters (temperature and density).

The MHD equation of state avoids these discontinuities (in the free energy) by introducing "soft" cut-offs in the form of occupational probabilities. These occupation probabilities have the same function as the "hard" cut-offs mentioned above. The occupational probabilities of a state simulate a result from quantum mechanics, denoting the fraction of atoms where the state can exist. Only then, these "available" states are populated according to statistical mechanics. It is clear that such an approach is largely intuitive. However, its advantage is that complicated plasmas can be modelled, with *detailed* internal partition functions of a large number of atomic, ionic, and molecular species. Also, full *thermodynamic* consistency is assured by analytical expressions of the free energy and its first- and second-order derivatives. This not only allows an efficient Newton-Raphson minimization, but, in addition, the ensuing thermodynamic quantities are of analytical precision and can therefore be differentiated once more, this time numerically. Reliable third-order thermodynamic quantities are thus calculated.

In the MHD occupation probabilities, perturbations by charged and neutral particles are taken into account. Correlations between the two effects are neglected (for lack of knowing how to describe them); thus the occupation probabilities due to charged and neutral perturbers are simply multiplied. The resulting weighted internal partition functions Z_s^{internal} of species s are (with is labelling the state i of species s)

$$Z_s^{\text{internal}} = \sum_i w_{is} g_{is} \exp\left[-\frac{E_{is} - E_{1s}}{kT}\right] \tag{4}$$

The coefficients w_{is} take into account charged and neutral surrounding particles. In physical terms, w_{is} gives the fraction of all particles of species s that can exist in state i with an electron bound to the atom or ion, and $1 - w_{is}$ gives the fraction of those that are so heavily perturbed by nearby neighbours that the state is effectively destroyed. Perturbations by neutral particles are based on an excluded volume treatment and perturbations by charges are calculated from a fit to a quantum-mechanical Stark-ionization theory. Hummer and Mihalas's (1988) choice has been

$$\ln w_{is} = -\left(\frac{4\pi}{3V}\right)\left\{\sum_\nu N_\nu (r_{is} + r_{1\nu})^3 + 16\left[\frac{(Z_s+1)e^2}{\chi_{is} k_{is}^{1/2}}\right]^3 \sum_{\alpha \neq e} N_\alpha Z_\alpha^{3/2}\right\} . \tag{5}$$

Here, the index ν runs over neutral particles, the index α runs over charged ions (except electrons), r_{is} is the radius assigned to a particle in state i of species s, χ_{is} is the (positive) binding energy of such a particle, k_{is} is a quantum-mechanical correction, and Z_s is the net charge of a particle of species s. Note that $\ln w_{is} \propto -n^6$ for large principal quantum numbers n (of state i), and hence provides a smooth (density-dependent) cutoff for Z_s^{internal}. As far as the energy levels are concerned, in the MHD equation of state no shifts due to plasma effects are assumed. This choice is based on experimental and theoretical arguments (see Hummer & Mihalas, 1988). Finally, the MHD equation of state also includes a Debye-Hückel term for the Coulomb-pressure correction, partially degenerate electrons, and radiation pressure.

The Livermore equation of state (physical picture)

There is an impressive body of literature on the physical picture. Important sources of information with many references are the books by Ebeling, Kraeft & Kremp (1976), Kraeft *et al.* (1986), Ebeling *et al.* (1991). However, the majority of work on the physical picture was not dedicated to the problem of obtaining a high precision equation of state for stellar interiors. Such an attempt was made for the first time by a group at Livermore as part of an opacity project (Rogers, 1986; Iglesias, Rogers & Wilson, 1987; Iglesias & Rogers, 1991,1992).

It is clear from the preceding subsection that the advantage of the chemical picture lies in the possibility to model complicated plasmas, and to obtain numerically smooth and consistent thermodynamical quantities. Nevertheless, the heuristic method of the separation of the atomic-physics problem from that of statistical mechanics is not satisfactory, and attempts have been made to avoid the concept of a perturbed atom in a plasma altogether. This has suggested an alternative description, the physical picture. In such an approach one expects that no assumptions about energy-level shifts or the convergence of internal partition functions have to be made. On the contrary, properties of energy levels and the partition functions should come out from the formalism.

To explain the advantages of this approach for partially ionized plasmas, it is instructive to discuss the activity expansion for gaseous hydrogen. The interactions in this case are all short ranged and pressure is determined from a self-consistent solution of the equations (Rogers, 1981)

$$\frac{p}{kT} = z + z^2 b_2 + z^3 b_3 + \ldots \tag{6}$$

$$\rho = \frac{z}{kT}\left(\frac{\partial p}{\partial z}\right) \tag{7}$$

where $z = \lambda^{-3}\exp(\mu/kT)$ is the activity, $\lambda \equiv h/\sqrt{2\pi m_e kT}$ is the thermal (de Broglie) wavelength of electrons, μ is the chemical potential and T is the temperature. The b_n are cluster coefficients such that b_2 includes all two particle states, b_3 includes all three particle states, *etc.*. More specifically, the second cluster coefficient for hydrogen includes the formation of H_2 molecules as well as the scattering states of the H atoms. The states of the H–H system are of different type according to the configurations of the electrons. Molecules are always in the singlet state, with total electron spin $S = 0$, and the lowest symmetrical orbital wave function, the ${}^1\Sigma_g$ bonding orbital. In this configuration, the interaction between the H atoms is described by the attractive ${}^1\Sigma_g$ bonding potential. The scattering states of the H atoms are either singlet states, with sufficient kinetic energy of the H atoms so that they are in the continuum of the ${}^1\Sigma_g$ bonding potential. Or, they belong to the triplet states (total electron spin $S = 1$), for which even the lowest orbital (the ${}^3\Sigma_u$ antisymmetric orbital) leads to a repulsive ${}^3\Sigma_u$ antibonding potential, which, of course, has only scattering states of the H–H system. Or, they belong to excited electronic states (singlet and triplet), which are never bound, either.

As a consequence, the second cluster coefficient for hydrogen includes the formation of H_2 molecules, the scattering states in the ${}^1\Sigma_g$ potential, in the ${}^3\Sigma_u$ potential, and in the potentials of all excited electronic states. The third cluster coefficient includes H_3 bound states, $H-H_2$ and $H-H-H$ scattering states. Eq. (6) demonstrates that the equation of state for reacting gases can be obtained without an explicit knowledge of the occupation numbers of the internal states of the composite particles.

For low-density gases the bound-state contributions to the b_n can be important at low temperature while the scattering contributions are too small to matter. Strict application of Eq. (6) would contain a large amount of unimportant information which is very hard to calculate. Consequently it is necessary to reorganize Eq. (6) such that the bound-state terms from each b_n are treated as being of the same order as the ideal gas term, *i.e.*, of order z. Terms of order z in the physical picture are roughly equivalent to what in the chemical picture is called the Saha equation.

In the case of partially ionized plasmas very similar steps are required, except that now even Eq. (6) must involve at least two species (nuclei and electrons) to assure electrical neutrality. In addition, due to the long range of the Coulomb potential each of the b_n is composed of a number of divergent terms, some of which are fictitious and some of which are real.

An example of real divergence is afforded by the classical ring diagrams occurring in each b_n (Mayer, 1950). They are individually divergent, but the many-body correlations introduced by summing over the b_n yield the Debye-Hückel correction [Eq. (6)]. This type of divergence occurs even for an electron gas in a neutralizing background for which there are no bound states. Although the original equations involved only even powers in the activity as a result of many-body Coulomb correlations, the Debye-Hückel term appears in the power of $z^{3/2}$ in the activity (also in density).

An important example of a fictitious divergence is that associated with the atomic partition function. This divergence is fictitious in the sense that the bound-state part of b_2 is divergent but the scattering state part, which is omitted in the Saha approach, has a compensating divergence. Consequently the total b_2 does not contain a divergence of this type (Ebeling, Kraeft & Kremp, 1976; Rogers, 1977). A major advantage of the physical picture is that it incorporates this compensation at the outset. A further advantage is that no assumptions about energy-level shifts have to be made (see the previous subsection); it follows from the formalism that there are none.

As a result, the Boltzmann sum appearing in the atomic (ionic) free energy is replaced with the so-called Planck-Larkin partition function (PLPF), given by (Ebeling, Kraeft & Kremp, 1976; Kraeft *et al.*, 1986)

$$\mathrm{PLPF} = \sum_{nl}(2l+1)\left[\exp(-\frac{E_{nl}}{kT}) - 1 + \frac{E_{nl}}{kT}\right] . \qquad (8)$$

The PLPF is convergent without additional cut-off criteria as are required in the chemical picture. I stress, however, that despite its name the PLPF is not a partition function, but merely an auxiliary term in a virial coefficient (see, for example, Däppen *et al.*, 1987).

EQUATION OF STATE COMPARISONS

So far there are no laboratory experiments that could distinguish between equations of state in the chemical and physical picture. Attempts to use constraints from a high-precision optical emission spectrum (*e.g.* Wiese, Paquette & Kelleher, 1973) have failed, because line-broadening effects were overshadowing the subtle details of statistical mechanics. The current equations of state (chemical and physical picture) all succeed in modelling the experiment (Däppen, Anderson and Mihalas, 1987; Seaton, 1990; Iglesias & Rogers, 1992). As another

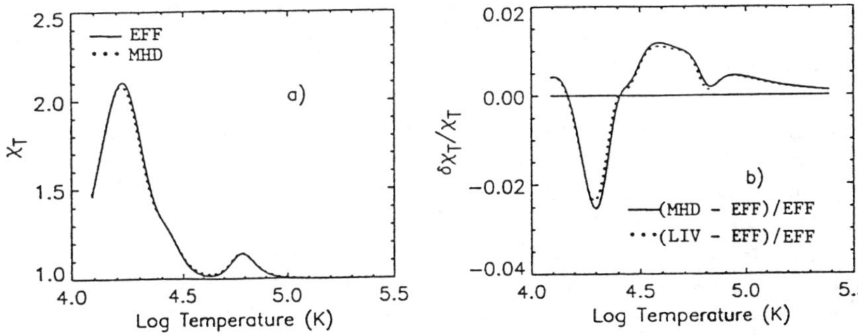

FIGURE I Comparison of $\chi_T = (\partial \ln p/\partial \ln T)_\rho$ for $\rho = 10^{-5.5}$ g cm^{-3}. Absolute quantities (a) and relative differences (with respect to EFF) (b) are shown. See text for more details.

type of observational diagnosis, solar oscillations promise to test the equation of state in the near future (see Christensen–Dalsgaard, Däppen & Lebreton, 1988; Christensen–Dalsgaard, 1991; Christensen–Dalsgaard & Däppen, 1992). Another way to learn about the merits of individual equations of states is with the help of theoretical comparisons. In addition, such comparisons also allow solar physicists to determine how uncertainties in the equation of state propagate into theoretically predicted oscillation frequencies. In this way, a "map" of the $T - \rho$ plane can be drawn, showing localized "interesting" regions, where non-ideal effects of one or another kind are important.

In the following, I discuss what has so far emerged from such comparisons. An early comparisons showed a striking agreement between the MHD and Livermore equation of state for conditions as found in the hydrogen-helium ionization zones of the Sun (Däppen, Lebreton & Rogers, 1990; Däppen, 1990). For convenience, a representative result from this early comparison is shown in Figure 1, which compares the MHD and Livermore results with that of the simple Eggleton, Faulkner & Flannery (1973; hereinafter EFF) formalism (which is essentially a consistent ground-state-only Saha equation of state including a – here irrelevant – arbitrary pressure-ionization device). The absolute curves of part *a* of Figure 1 are merely able to show the difference between MHD (or Livermore) and the simple EFF results. To see the difference between the MHD and Livermore results, one needs the magnified part *b*, which shows the *relative* differences between MHD and EFF, and between Livermore and EFF values, respectively. This relative plot now not only allows one to see the difference between MHD and Livermore results, but also their striking similarity.

Later, it turned out that this agreement was nearly accidental. Its *physical reason* was found by varying the parameters of the MHD equation of state. It followed that on the chosen isochore, all thermodynamical quantities are mainly dominated by the Coulomb pressure correction (Däppen, 1990; Christensen–Dalsgaard, 1991; Christensen–Dalsgaard & Däppen, 1992). The Coulomb correction overshadows the effect of the excited states (which are of course treated differently in the MHD and Livermore approach). However, the Coulomb term acts principally indirectly, at least in the language of the chemical picture, because it is not mainly the free-energy of the Debye-Hückel term itself, but rather the Coulomb-induced shift in the ionization equilibrium, which is responsible for the deviation from the unperturbed EFF result.

Of course, solar physicists were happy that two completely different formalisms delivered the same equation of state, but, by the same token, a first attempt to use the Sun as an equation-of-state test was also thwarted. This discovery suggested to upgrade the simple EFF equation of state with the help of the Coulomb interaction term. The resulting equation of state (called CEFF) has become a useful tool for solar physics (Christensen–Dalsgaard, 1991; Christensen–Dalsgaard & Däppen, 1992); at the same time, however, it became also clear that a helioseismic test of the important issue of chemical versus physical picture would be more difficult than first thought.

For reasons not yet fully understood it seems that in the chemical picture, the signature of internal partition functions, such as those employed in the MHD equation of state, is much less visible in the thermodynamic quantities than a naive estimation of the shift in the ionization equilibrium would predict. It is likely that there are accidental cancellations in the derivatives of the free energy. Notice that these cancellation would have nothing to do with those appearing in the physical picture, which lead to the Planck-Larkin partition function.

The cancellations of partition-function effects in the chemical picture seem to be greatest for the ionization zone of hydrogen and somewhat less for those of helium. A more recent comparison of MHD and Livermore values (Däppen, 1992) has examined selected cases of higher densities (where sizeable discrepancies appear) and a first case of a mixture involving a representative solar heavy element (oxygen). It appears that for the heavier elements, the internal partition functions finally lead to the intuitively expected consequences for the thermodynamic quantities. For convenience, Figure 2 shows the result of this comparison with oxygen for the quantity Γ_1. Density was chosen as $\rho = 0.005$ g cm^{-3}, a value suggested by a helioseismic study of the solar helium abundance (Kosovichev *et al.*, 1992). Here, not only do the large MHD partition functions cause shifts in the ionization balance but these shifts also significantly propagate into the thermodynamic quantities. The effect is large enough that it appears, despite the small relative number of the heavy elements in the mixture, to be within reach of helioseismology (Christensen-Dalsgaard & Däppen, 1992). Incidentally, it is precisely this effect that was mainly responsible for the discrepancies found between the different sets of MHD tables (MHD1 – MHD5) discussed in Kosovichev *et al.* (1992). Those different MHD tables were computed for various purposes (some of them with limited computing resources), and they differed by their choice of heavy-element composition (see also Däppen *et al.*, *these proceedings*).

To examine the MHD ionization fractions, a single case was examined ($T = 2.10 \times 10^5 K, \rho = 5.00 \times 10^{-3}$g cm^{-3}), once with the full MHD equation of state, once with a "stripped-down" version of MHD, which does not contain any excited states (but is otherwise identical). The resulting ionization fractions of O^{3+}, O^{4+}, O^{5+} were, respectively, 0.314, 0.248, 0.364 for the stripped-down MHD (without excited states), and 0.304, 0.476, 0.182 for the full MHD. (The result for the stripped-down very closely reflects the ground-state weights of the ions). Not unexpectedly in view of the Planck-Larkin partition function, the Livermore equation of state predicts ionization fractions close to those of the

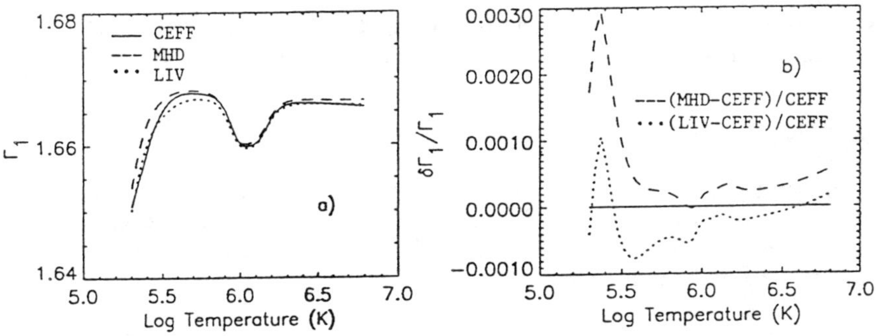

FIGURE II Γ_1 for $\rho = 5.00 \times 10^{-3}$g cm^{-3} and a representative solar mixture of H, He, and O. Parts (a) and (b) as in Fig.1, but here with CEFF instead of EFF. See text for more details.

stripped-down MHD equation of state (Rogers, *private communication*).

This comparison for the first time establishes a clear case of disagreement between the MHD and Livermore results. Clearly, the origin of the discrepancy in the ionization degrees is due to the treatment of the excited states. Of course, only some 2 percent of the matter in the Sun consist of elements heavier that H and He, and therefore the signature of the MHD-Livermore discrepancy (Figure 2) is small (of the order of 10^{-3}). Nevertheless, as has been demonstrated by Christensen-Dalsgaard & Däppen (1992), even the resulting tiny sound-speed differences are within reach of a helioseismic diagnosis.

CONCLUSION

Even weakly-coupled plasmas can pose tough problems if high accuracy is demanded. Solar oscillations are an example of a case where the present observational material is much better than the theoretical models. Stellar oscillations also promise interesting constraints. Their observations will not be as detailed as those of the Sun: however, a sample of more than one star will access a variety of different physical conditions and chemical compositions. A great many of astrophysical inferences will only be as good as the equation of state. The theoretical models used for the inferences will suffer from inconsistencies and bad numerical realizations of any formalism of the equation of state. These formal considerations are even above the question of an absolutely accurate representation of reality. At the moment, there is not much available to check the absolute accuracy of an equation of state. Solar and stellar seismology will undoubtedly play an important role. So far, theoretical comparisons between different formalisms are the only means to do this, but of course agreement does not mean correctness. Nevertheless, comparisons certainly give valuable indications, not to the least about the correct realization of the models. In addition, the comparisons are very important to localize the interesting regions in the T-ρ plane, where different formalisms lead to distinct thermodynamic quantities, which has observable consequences. For the Sun, such comparisons have already resulted in a better (and yet simple) equation of state (*i.e.* CEFF). There is hope that for stars, too, new data from ground-based and space projects will put constraints on the equation of state and shed light on controversial issues.

ACKNOWLEDGMENTS

I am grateful to Vladimir Baturin, Jørgen Christensen–Dalsgaard, Andreas Förster, Douglas Gough, and Forrest Rogers for stimulating discussions and helpful comments. Forrest Rogers is also thanked for providing the Livermore results displayed in the figures.

REFERENCES

Chabrier, G.: 1990, *J. de Physique (France)* **51**, 1607

Christensen-Dalsgaard, J.: 1991, In *Lecture Notes in Physics*, Vol. **388**: Challenges to Theories of the Structure of Moderate-mass Stars, eds Gough, D.O. & Toomre, J., Springer, Heidelberg, p. 11 – 36

Christensen-Dalsgaard, J. & Däppen, W.: 1992, *Astron. Astrophys. Review* submitted

Christensen-Dalsgaard, J., Däppen, W. & Lebreton, Y.: 1988, *Nature* **336**, 634 – 638

Däppen, W.: 1990, in *Lecture Notes in Physics*, Vol. **367**: *Progress of Seismology of the Sun and Stars*, eds. Osaki, Y. & Shibahashi, H., Springer, Berlin, p. 33 – 40

Däppen, W.: 1992, in *Astrophysical Opacities*, eds. C. Mendoza & C. Zeippen (*Revista Mexicana de Astronomía y Astrofísica*) 141 – 149

Däppen, W., Anderson, L.S. & Mihalas, D.: 1987, *Astrophys. J.* **319**, 195 – 206

Däppen, W., Keady, J. & Rogers, F.: 1991, in *Solar Interior and Atmosphere*, eds Cox, A.N., Livingston, W.C. & Matthews, M., Space Science Series, University of Arizona Press, Tucson, p. 112 – 139

Däppen, W., Lebreton, Y. & Rogers, F.: 1990, *Solar Physics* **128**, 35 – 47

Däppen, W., Mihalas, D., Hummer, D.G. & Mihalas, B.W.: 1988, *Astrophys. J.* **332**, 261 – 270

Ebeling, W., Kraeft, W.D. & Kremp, D.: 1976, *Theory of Bound States and Ionization Equilibrium in Plasmas and Solids*, Akademie Verlag, Berlin, DDR

Ebeling, W., Kraeft, W.D., Kremp, D. & Röpke, G.: 1985, *Astrophys. J.* **290** 24 – 27

Ebeling, W., Förster, A., Fortov, V.E., Gryaznov, V.K. & Polishchuk, A.Ya.: 1991, *Thermodynamic Properties of Hot Dense Plasmas*, Teubner, Stuttgart, Germany

Eggleton, P.P., Faulkner, J. & Flannery, B.P.: 1973, *Astron. Astrophys.* **23**, 325 – 330

Eliezer, S., Ghatak, A. & Hora, H.: 1986, *An introduction to equations of state: theory and applications*, Cambridge University Press

Gough, D.O.: 1984a, *Mem. Soc. Astron. Ital.* **55**, 13 – 35

Graboske, H.C., Harwood, D.J. & Rogers, F.J.: 1969, *Phys. Rev.* **A186**, 210

Huang, K.: 1963, *Statistical Mechanics*, John Wiley, New York, Chapt. 14

Hummer, D.G. & Mihalas, D.: 1988, *Astrophys. J.* **331**, 794 – 814

Iglesias, C.A., & Rogers, F.J.: 1991, *Astrophys. J.* **371**, 408 - 417

Iglesias, C.A. & Rogers, F.J.: 1992, in *Astrophysical Opacities*, eds. C. Mendoza & C. Zeippen (*Revista Mexicana de Astronomía y Astrofísica*) 161 – 170

Iglesias, C.A., Rogers, F.J. & Wilson, B.G.: 1987, *Astrophys. J.* **322**, L45

Kosovichev, A.G., Christensen-Dalsgaard, J., Däppen, W., Dziembowski, W.A., Gough, D.O., & Thompson, M.J.: 1992, *Mon. Not. R. astr. Soc.*, in the press

Kraeft W.D., Kremp D., Ebeling W. & Röpke G.: 1986, *Quantum Statistics of Charged Particle Systems*, Plenum, New York

Krasnikov Yu.G.: 1977, *Zh. Eksper. teoret. Fiz.* **73**, 516 (English translation: *Soviet Phys. - JETP* **46**, 270 – 274; author's name misspelt as "Karsnikov")

Mayer, J.E.: 1950, *J. Chem. Phys.* **18** 1426 – 1436

Mihalas, D., Däppen W. & Hummer, D.G.: 1988, *Astrophys. J.* **331**, 815 – 825

Reichl, L.E.: 1980, *A Modern Course in Statistical Physics*, University of Texas Press, Austin

Rogers, F.J.: 1977, *Phys. Lett.* **61A**, 358 – 360

Rogers, F.J.: 1981, *Phys. Rev.* **A24**, 1531 – 1543

Rogers, F.J.: 1986, *Astrophys. J.* **310**, 723 – 728

Rouse, C.A.: 1983, *Astrophys. J.* **272** 377 – 379

Saumon, D. & Chabrier, G.: 1991, *Phys. Rev* **A44**, 5122

Saumon, D. & Chabrier, G.: 1992, *Phys. Rev* **A**, (in press)

Seaton, M.: 1987, *J. Phys. B: Atom. Molec. Phys.* **20**, 6363 – 6378

Seaton, M.: 1990, *J. Phys. B: Atom. Molec. Phys.* **23**, 3255 – 3296

Seaton, M.J.: 1992, in *Astrophysical Opacities*, eds. C. Mendoza & C. Zeippen (*Revista Mexicana de Astronomía y Astrofísica*) 180

Wiese, W.L., Kelleher, D.E. & Paquette, D.R.: 1972, *Phys. Rev.* **A6** 1132 – 1153

Inside the Stars, IAU Colloquium 137
ASP Conference Series, Vol. 40, 1993
Werner W. Weiss and Annie Baglin (eds.)

RADIATIVE OPACITIES

M.J. SEATON
Department of Physics and Astronomy, University College London, Gower St., London WC1E 6BT, UK.

ABSTRACT A review is given of opacity calculations made during the past decade.

SOME DEFINITIONS

The net outward radiative flux inside a star is

$$F = -\frac{1}{\kappa_R} \frac{4\pi}{3} \frac{dB}{dT} \frac{dT}{dr} \tag{1}$$

where r is radius, T is temperature,

$$B = \frac{2\pi^4}{15c^2h^3}(kT)^4 \tag{2}$$

is the integrated Planck function and κ_R is the Rosseland mean opacity, defined as a harmonic mean weighted by the temperature derivative of the monochromatic Planck function. The working formula is

$$\frac{1}{\kappa_R} = \int \frac{f(u)}{\kappa(u)}\, du \tag{3}$$

where $u = h\nu/kT$, $\kappa(u)$ is the monochromatic opacity and

$$f(u) = (15/4\pi^4)u^4 \exp(-u)/[1 - \exp(-u)]^2. \tag{4}$$

This function is normalised to

$$\int f(u)\, du = 1, \tag{5}$$

is proportional to u^2 for u small, has a maximum value at $u = 3.830...$, and decreases exponentially for u large.

The monochromatic opacity is obtained on summing over all processes leading to extinction of radiation,

$$\kappa(u) = \sum_i N_i \sigma_i(u) \times E_i \tag{6}$$

where N_i is the number-density of particles of type i, $\sigma_i(u)$ is a cross section and E_i a correction-factor for stimulated emission: $E_i = [1-\exp(-u)]$ for absorption processes, $E_i = 1$ for scattering.

Use of equations such as (6) gives opacities per-unit-length. Those are the quantities which enter directly into the equation of radiative transfer and they are also the most convenient for general presentation of the theory. Numerical results, however, are usually given as opacities per-unit-mass (in units of cm^2 g^{-1}). The relation is κ(per-unit-length) $= \rho \times \kappa$(per-unit-mass), where ρ is the mass-density. In attempting to calculate opacities we face two main problems.

- **Determine the populations,** N_i. That is the problem of the equation of state (EOS).
- **Determine the cross sections,** σ_i. That is a problem of atomic physics.

One must consider very large numbers of atomic processes: bound-bound (spectrum lines); bound-free (photoionisation); free-free; and scattering (mainly by free electrons). For the line contributions one requires, in addition to line-frequencies and f-values, information about the line profiles determined by radiation damping, thermal motions and pressure-broadening.

We may distinguish between two regions "inside the stars", envelopes and interiors. So far as opacity work is concerned, I define envelopes to be those regions in which there are no major plasma perturbations of the systems i giving dominant contributions to opacities; and interiors those in which such perturbations are important. For the calculation of envelope opacities one requires an acceptable EOS and large amounts of accurate atomic data. For interiors one has the additional, and difficult, problem of calculating the cross-sections $\sigma_i(u)$ allowing for plasma perturbations.

BACKGROUND TO RECENT WORK.

I will not attempt to review the earlier literature: see Cox(1965) and Carson (1973) . For more than two decades extensive use has been made of opacities calculated at the Los Alamos National Laboratory in the USA (see Cox and Stewart, 1965 and 1970; Cox and Tabor, 1976; Hübner *et al.* 1977; Hübner, 1985; and Weiss *et al.* 1990). I shall refer to that work as LAOL (Los Alamos Opacity Library). Substantial use has also been made of the St. Andrews opacities (Carson *et al.*, 1968). The reasons for some discrepancies between the LAOL and the St. Andrews work have been explained (Carson *et al.*, 1984).

In his now-famous paper of 1982, Simon made a plea for the re-examination of the heavy-element contributions to opacities. He showed that an increase in the contributions from elements heavier than helium, by factors of two or three in envelope regions, would resolve a number of outstanding problems in the theory of pulsating stars. Although Magee *et al.* (1984) argued that such a large increase would be incompatible with atomic physics, two groups responded to Simon's plea. The first was at the Lawrence Livermore National Laboratory (F.J. Rogers, C.A. Iglesias and B.G. Wilson) and I will refer to it as OPAL, the name of their computer code. The second is a project with which I have been associated and which has involved quite a large team of workers from a number

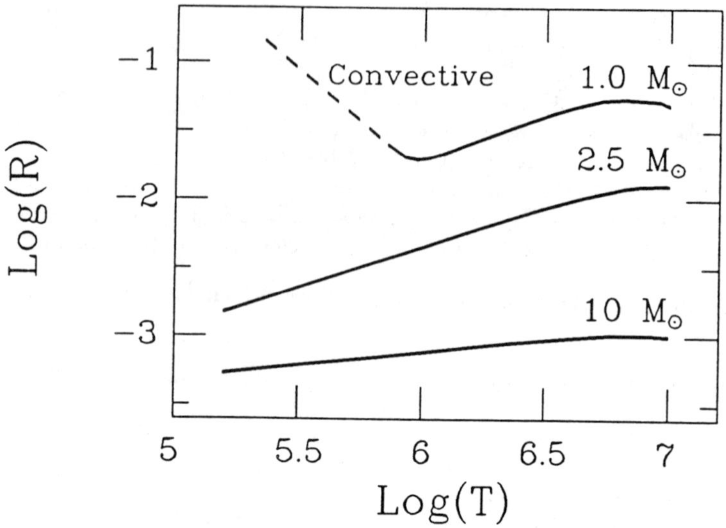

FIGURE I Values of $\log(R)$ against $\log(T)$ for three initial main sequence models, from Schwarzschild (1958). $R = \rho/T_6^3$ with ρ in g cm^{-3} and $T_6 = 10^{-6} \times T$ with T in K

of countries (France, Germany, the UK, the USA and Venezuela). We refer to it as *The Opacity Project* and I will refer to it here as OP. The remainder of the present review will be mainly concerned with the presentation of OPAL and OP results, and comparisons with those from LAOL.

The Rosseland mean, κ_R, depends on temperature T, density ρ and chemical composition. For any stellar model higher temperatures correspond to higher densities. The OPAL team use the variable $R = \rho/T_6^3$ with ρ in g cm^{-3} and $T_6 = 10^{-6} \times T$ with T in K. For a given stellar model, R is approximately constant — at least, to within a factor of 10 or so: Figure I shows $\log(R)$ against $\log(T)$ for 3 early stellar models (Schwarzschild, 1958). Thus a plot of κ_R against T for fixed R gives an impression of the variation of κ_R with depth inside a star. I will make use of such plots in the present review.

THE OPAL AND OP APPROACHES

The OPAL approach.

The EOS work, of both OPAL and OP, is discussed further by W. Däppen at this meeting and therefore need not be discussed in detail here. The OPAL work is based on a sophisticated approach to the plasma EOS problem developed over many years (see Rogers 1986, 1991 and references therein). The method is firmly the "physical picture" in which one starts by considering an assembly of electrons and nuclei. As the plasma conditions permit, the electrons can become bound to the nuclei to form states of atomic ions. In general such states are perturbed by the plasma environment and their wave-functions must be calculated with inclusion of plasma potentials. The simplest example of such a potential is the static screened Coulomb potential $\exp(-r/r_D)/r$ replacing the Coulomb potential $1/r$, r_D being the Debye radius. The use of such potentials has, however, been criticised by Hummer and Mihalas (1988).

For lower densities, and states which are not highly excited, the plasma perturbations become small and can be neglected. For unperturbed atomic states Rogers (1981) and Rogers *et al.* (1988) use parametric potentials of the type

$$V(r) = -\frac{2}{r}[z + \sum_i C_i \exp(-\alpha_i r)] \tag{7}$$

where z is the residual charge on the ion and C_i the number of electrons in shell i. With such potentials they use a spin-averaged Dirac equation to calculate atomic radial functions. For states with principal quantum numbers $n \leq 4$ they take full account of LS angular-momentum coupling and adjust the parameters α_i in (7) so as to obtain good agreement with experimental ionisation energies. For $n \geq 5$ they use a simple central-field model. In later work (Iglesias *et al*, 1992) they consider the use of intermediate coupling.

The OPAL atomic calculations are made on-line, that is to say the atomic data are not stored but are calculated as they are required. There is doubtless some cost in computer time but the approach does give the advantage of great flexibility and of allowing for the possibility to include plasma potentials as and when required. The OPAL EOS calculations are made on solving coupled equations for full mixtures at each density and temperature point. Their results for opacities are tabulated as functions of $\log(R)$ and T_6.

The OP approach

For a plasma, define a mean inter-particle separation r_0 by $(4\pi/3)r_0^3 N = 1$ and let r_i be the mean radius for absorber i (for hydrogenic systems with principal quantum number n, $r_n = (5n^2 + 1)/(4z)$ atomic units). For envelopes it is assumed that $r_i \gg r_0$ for all absorbers contributing significantly to the opacity. Hummer and Mihalas (1988) tentatively adopy $\rho = 10^{-2}$ g cm^{-2} for the lower boundary of envelopes. That corresponds to $r_0 \simeq 5.5$ atomic units.

The OP work is concerned with the determination of envelope opacities using results from extensive *ab initio* calculations of atomic data. The main atomic calculations are made R-matrix methods (see Berrington *et al.*, 1987) and are described in a series of papers "Atomic Data for Opacity Calculations" in the *Journal of Physics B* (to date, 17 papers published or submitted). For systems containing up to 12 electrons the OP atomic data can be considered to be "near definitive": with the exception only of a fairly small number of cases for which highly optimised calculations have been made, they are more accurate than other data available, and much more extensive. For systems with more than12 electrons, particularly those involving configurations $3s^x3p^y3d^z$, supplementary calculations are made using the code SUPERSTRUCTURE (Eissner and Nussbaumer, 1969; Nussbaumer and Storey, 1988).

The OP EOS work is described in four papers "The Equation of State for Stellar Envelopes" (Hummer and Mihalas, 1988; Mihalas *et al.*, 1988; Däppen *et al.*, 1988; and Mihalas *et al.*, 1990). The approach is firmly the "chemical picture" in which atomic particles are assumed to exist. In order to avoid divergences of partition functions (essentially, the exclusion of states with r_i not $\gg r_0$) it is necessary to introduce *occupation probabilities*, W_i. Equations for the determination of ionisation equilibria are obtained from minimisation of the free energy. It is found that, to a very good approximation, the equilibria depend only on temperature T and electron density N_e. Opacities are calculated separately for each chemical element as functions of T and N_e and are archived. Results for any mixture can then be calculated rapidly. The OP opacity tables give ρ and κ_R as functions of T and N_e and interpolation routines provide values of κ_R for any required value of T and ρ within the range of the tables.

Comparisons of the two approaches

Envelopes and interiors

The main thrust of the OP work is in the calculation of opacities for envelopes, say $\rho \leq 10^{-2}$ g cm^{-2}. For deeper interiors the OPAL results are to be preferred.

Equations of State

Comparisons between OPAL and OP EOS results have been made by Däppen, Lebreton and Rogers (1990). For envelope regions they find good agreement for basic thermodynamic quantities. Further comparisons of ionisation equilibria have also been made by members of the two teams and, again for envelopes, the agreement is quite good. Larger differences can be expected — and exist — if the OP EOS calculations are extended to interiors.

Atomic data

The OP R-matrix data are obtained with allowance for configuration interaction (CI), which is neglected in the OPAL work. For unperturbed atomic systems the OP atomic data should therefore be the more accurate.

When CI is neglected one can obtain bound states above the ionisation limit. Inclusion of CI leads to coupling between those states and continuum states and hence to auto-ionisation features. Many of the OP photo-ionisation cross sections contain very complicated auto-ionisation structures. An example is shown in Figure II. These features are included by OPAL as ordinary lines (transition to the "bound states" above the ionisation limits) but the auto-ionisation broadening is neglected. The figure also shows results from central-field calculations which are probably in close agreement with results used by OPAL.

Line profiles

The treatment of line profiles should be basically similar. A distinction is made between hydrogenic systems (nl states degenerate with respect to l) for which Stark effects are important (a detailed account is given by Seaton, 1990) and non-hydrogenic systems for which "isolated-line" approximations are used. For the isolated lines, both teams use methods related to those introduced by Griem (1968) which employ an extension to line-broadening theory of the $\overline{g}$-approximation of atomic collision theory. This gives expressions for line-widths in terms of oscillator strengths and some empirical factors. OPAL uses factors

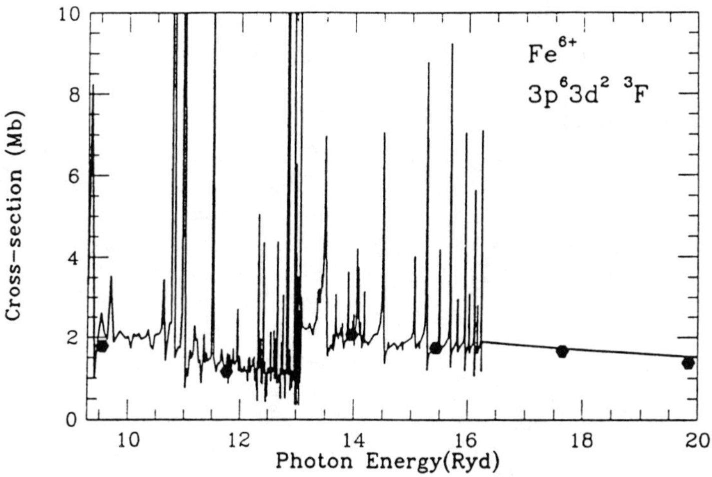

FIGURE II Cross section for photo-ionisation of the Fe^{6+} ground state, from Saraph *et al.* (1992). The solid hexagons are from central-field calculations by Reilman and Manson (1979).

determined by Dimitrijevič and Konijevič (1980) from fitting to experimental data while OP uses data fitted to results from R-matrix calculations. Table 12 of Seaton (1988) shows that the two approaches give close agreement. It may be desirable to consider further the use of the sharp distinction between the hydrogenic and non-hydrogenic cases.

For the widths of the auto-ionisation features, neither approach may be be entirely satisfactory. For those features, OP allows for auto-ionisation broadening but neglects other broadening mechanisms (Doppler and pressure), whereas OPAL includes those other mechanisms but neglects auto-ionisation.

Mesh points

There are two OP codes, one developed by Yu Yan and D. Mihalas in Illinois and the other by myself in London. The Illinois code uses a constant interval in $\log(u)$ and 10^5 mesh points, sufficient to resolve all Doppler profiles. I use a constant interval in u, $0 \leq u \leq 20$. If one reduces the number of frequency-mesh points some narrow lines may be missed altogether but, with trapezoidal-rule integrations, one will obtain over-estimates of contributions due to lines with centre-frequencies close to mesh points. These effects tend to cancel. I find that use of 10^4 frequency points, compared with 10^5, gives errors no larger than 1 % or so. These errors are not important in themselves but can produce a slight lack of smoothness which may be just discernable in some of the plots which I shall show later.

OP uses intervals of 0.25 in $\log(N_e)$ and 0.025 in $\log(T)$. OPAL uses 5×10^3 frequency points and intervals in $\log(\rho)$ and $\log(T)$ somewhat larger than those used by OP.

RESULTS FOR OPACITIES

OPAL results.

Results for a Cepheid model

The first published OPAL results (Iglesias *et al.*, 1987) were for $\rho = 10^{-5}$ g cm^{-2} and $kT = 20$ and 60 eV ($T = 2.32$ and 6.97×10^5 K) and showed that an improved treament of the atomic physics could lead to significant opacity enhancements. That these could be at least as large as those postulated by Simon (1982) was first shown by Iglesias *et al.* (1990), who gave results for a model of a Cepheid envelope previously considered by Rozsnayi (1989). The top part of Figure III shows $\log(R)$ against $\log(T)$ for the model (the highest density is $\rho = 0.0136$) and the lower part compares OPAL opacities with those from LAOL as used by Simon. The largest ratio, $\kappa_R(\mathsf{OPAL})/\kappa_R(\mathsf{LAOL})$, is 3.34.

In discussing Rosseland opacities one must consider the nature of the harmonic mean defined by (3) (failure to do so was the main source of error in the paper by Magee *et al.*, 1984). Any extended region in which $\kappa(u)$ is small will give a large contribution to the integrand in (3) and hence leads to a small value of κ_R (the radiation escapes in regions of high transparency). Iglesias *et al.* identify two effects in the LAOL work which lead to localised under-estimates of $\kappa(u)$. The first is neglect of $\Delta n = 0$ transitions, a consequence of the use of hydrogenic f-values. The second is the almost-complete neglect of term-splitting. Without such splitting transitions between two configurations are represented by a single line. When one allows for LS quantum numbers of initial and final states, and for all possible internal couplings of angular momenta, that single line is replaced by many lines with a significant spread in frequency. The most striking effects are for transitions involving configurations $3s^x3p^y3d^z$ in iron ions, which can give rise to very large numbers of lines.

The solar radiative interior

In constructing stellar models the net outward energy flux is generally a known quantity, determined by the rate of energy production in the deep interior. When conduction and convection can be neglected, the total flux is equal to the radiative flux and the temperature gradient, dT/dr, is then given by equation (1) and is proportional to the Rosseland mean opacity. Throughout much of the solar interior convection is important and in those regions knowledge of the Rosseland mean is not of such great significance. It is, however, again of importance in the solar radiative interior, the region below the convection zone. Accurate opacity determinations for that region are required in connection with studies of the solar neutrino problem and in considering sound-speeds determined by helio-seismology.

Iglesias and Rogers (1991) give a very careful and detailed discussion of opacities in the solar interior. They present OPAL results for $T \geq 10^6$ K, 3 values of $\log(R)$ and a number of different chemical mixtures. The differences between LAOL and OPAL for the solar interior are much smaller than those for the Cepheid envelope — for $T \geq 2 \times 10^6$ K the differences are no larger than 11 %. The revisions suggested by the OPAL work are, nevertheless, of considerable importance for problems in solar physics.

At the solar centre the opacity is mainly determined by free-free transitions

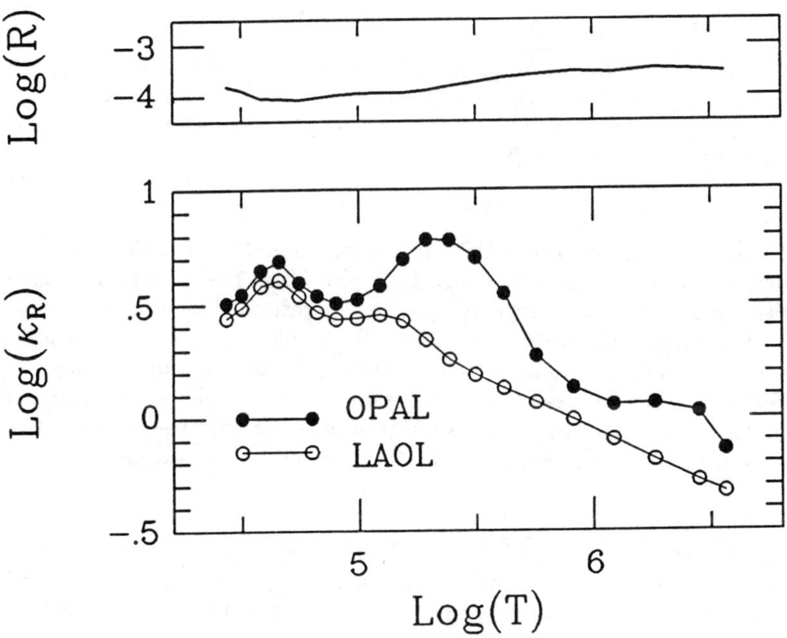

FIGURE III Opacities for a Cepheid model. OPAL results from Iglesias *et al.* (1990), LAOL results as used by Simon (1982). The top figure shows log(R) for the model.

and by electron scattering. In OPAL, free-free is calculated using parametric potentials screened for plasma correlations. The theory of Boercker (1987) is used for electron scattering. The OPAL/LAOL differences in the centre region are no more than 1 to 3 %.

The biggest OPAL/LAOL differences occur in the region just below the convection zone, with the OPAL results being the larger. Iglesias *et al.* conclude that these differences are mainly due to different treatments of the EOS problems. Different assumed chemical compositions can also lead to opacity differences of up to 5 % or so.

Comprehensive OPAL *opacity tables*

In a paper of major importance, Rogers and Iglesias (1992) give comprehensive tables of OPAL opacities. I have already given an outline description of the methods which they use. I will defer discussion of their results in order to compare them with those from OP.

OP opacities

First results for opacities from the OP work were presented at a Workshop on Astrophysical Opacities in Caracas, July 1991 (Yu Yan, 1992). Unfortunately those results were not entirely correct. I take full responsibility for the error, in one line of my code. Corrected results will be published in a post-script to the Workshop proceedings, added some months after the end of the meeting (Seaton, 1992). Further OP calculations have now been completed. Results can be provided, on request, for any required mixture together with the interpolation routine which give κ_R for all values of ρ and T covered by the tables.

Comparison and discussion of results

Mixtures are defined by X, Y and Z, fractions by weight of hydrogen, helium and all heavier elements. In the results presented here I will use the relative abundances of the heavier elements as given by Anders and Grevesse (1989).

Hydrogen and helium

Figure IV gives results for $X = 0.7, Y = 0.3, Z = 0.0$ and $\log(R) = -4, -3$ and -2. The OPAL and OP results are seen to be in close agreement. For OP I use a dashed line in regions with $\log(\rho) > -2$.

At the lower temperatures there are important contributions from H^- bound-free and free. From Figure IV it is seen that there are 3 maxima sometimes known, from left to right, as X-, Y- and Z-bumps. The first, at $\log(T) \simeq 4.1$, is due to absorption by H^0, and to a lesser extent He^0, in excited states. At these low temperatures the threshold for absorption by ground states comes at too high a frequency to make a significant contribution to the Rosseland mean. The next bump, at $\log(T) \simeq 4.7$, is due to absorption by the H^0 ground state with some contributions from the He^0 ground state and He^+ excited states. The third bump (more like a shoulder for He and He alone), at $\log(T) \simeq 5.2$, is due to absorption by the He^+ ground state.

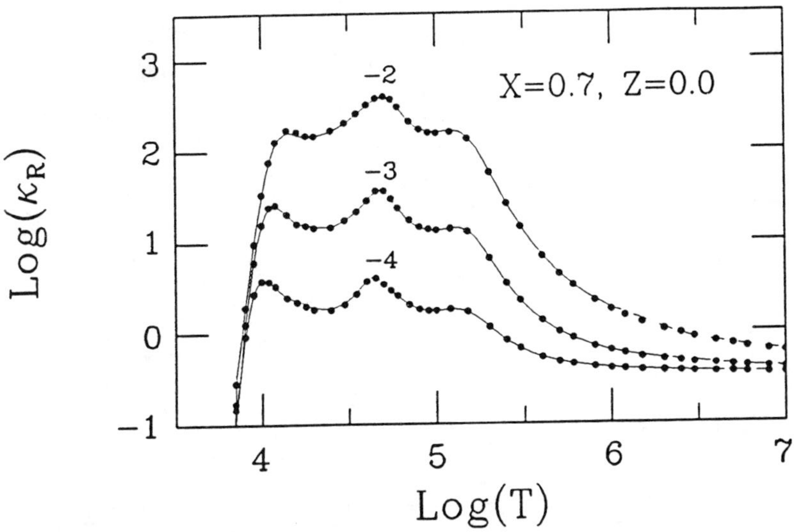

FIGURE IV Opacities for a mixture of hydrogen and helium. Full lines, OP; filled circles, OPAL. The curves are labelled with values of log(R)

The RR Lyrae case

Figure V shows, for the case of log(R) = −3, the effect of adding 0.1% of heavy elements. There are two curves, one for $Z = 0.000$ and the other for $Z = 0.001$. There is, again, quite close agreement between the two calculations. Even for this low value of Z there is a substantial opacity enhancement for $5.0 \geq \log(T) \geq 6.5$, much larger than the enhancement given by LAOL. These results present some problems in studies of RR Lyrae stars.

Solar abundances

Figure VI shows results for $X = 0.7$, $Z = 0.02$ and log(R) = −4, −3 and −2. Comparisons with LAOL for a similar composition have already been given in Figure III, and the effects producing an enhancement in metal contributions have been discussed in section 4.1.1. The agreement between OPAL and OP is fairly good although there are some differences with a tendency for the OP results to be the smaller. The reason for these differences should be studied further.

Figure VI shows that there is another bump in the opacities, which I will refer to it as the A-bump, at $\log(T) \simeq 6.3$.

Contributions from the different elements

Figure VII shows some OP results all made for the same mixture, $X = 0.7$, $Z = 0.02$, and log(R) = −3. In order to show the contributions of different groups of elements, their opacities have been successively "switched on". There are 4 curves: (a) H and He; (b) as (a) plus C, N, O and Ne; (c) as (b) plus Na, Mg, Al, Si, S, Ar and Ca; (d) as (b) plus Fe. The A-bump is mainly due to absorption

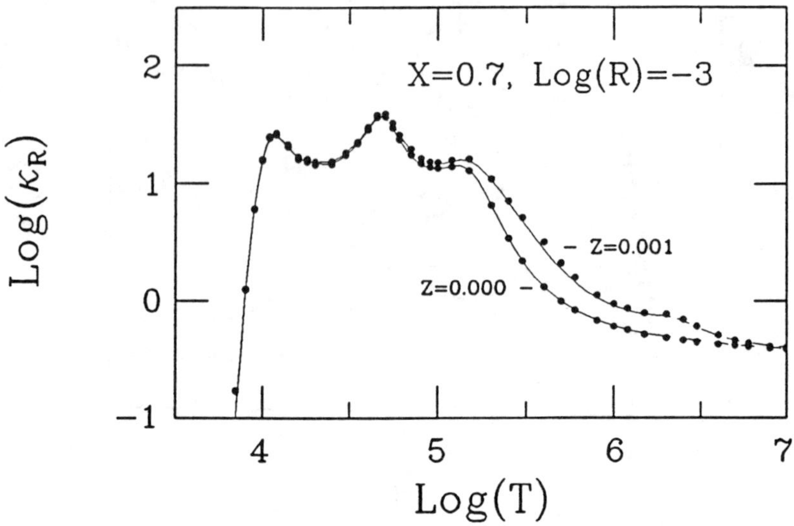

FIGURE V Opacities for $X = 0.7$, $\log(R) = -3$ and $Z = 0.000$ and 0.001. Curves as in Figure IV.

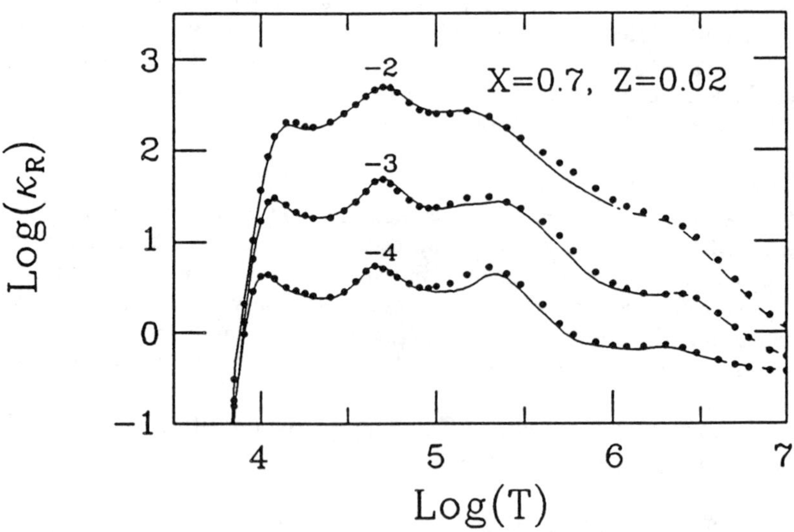

FIGURE VI Opacities for a solar mix, $X = 0.7$, $Z = 0.02$. Curves as in Figure IV.

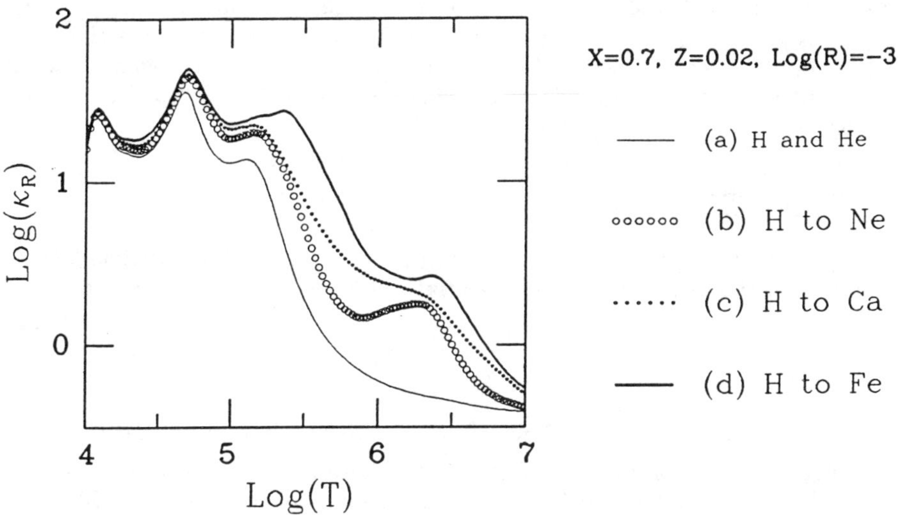

FIGURE VII Contributions to opacities from different groups of elements. OP calculations

from ground states in hydrogen-like and helium-like C, N and O. The elements from Na to Ca make important contributions in the vicinity of $\log(T) = 5.8$. The contribution of iron is important for all $\log(T) \geq 5$ but particularly in the vicinity of the Z-bump at $\log(T) \simeq 5.5$.

Inclusion of fine-structure

It has been seen that allowance for term-splitting (LS quantum numbers) produces major changes in results for opacities. It is therefore natural to ask whether allowance for fine-structure (introduction of J quantum numbers) will lead to further enhancements. This question has been considered recently by both the OPAL team (Iglesias *et al.*, 1992) and the OP team, with findings in good agreement. At high densities the line-widths are comparable to the LS-coupling line-separations, the lines form a quasi-continuum and allowance for fine structure hardly makes any change. At low densities the lines are much narrower and inclusion of fine structure does lead to opacity enhancements. Thus at $\log(R) = -5$, inclusion of fine-structure leads to an enhancement by nearly 50 % at the peak of the Z-bump; while at $\log(R) = -2$ inclusion of fine-structure produces negligible change.

SUMMARY

Work which has been done during the past decade, and which is now nearing completion, has led to major revisions in envelope opacities, by factors as large as 3 or 4. There are also some revisions in results for deeper layers, which are important but not so pronounced. Use of the new opacities will require a re-examination of a wide range of problems in stellar astrophysics.

It is very fortunate that two groups, OPAL and OP, took up the challenge. Their results do not agree exactly but there is generally close agreement, leaving no doubt about the magnitude of the changes in opacities compared with the earlier work.

ACKNOWLEDGMENTS

I thank Forrest Rogers and Carlos Iglesias for many helpful discussions, and all members of the OP team for their cheerful collaboration.

REFERENCES

Anders, E. and Grevesse, N., 1989. *Geochim. Cosmochim. Acta*, **53**, 197.

Berrington, K.A., Burke, P.G., Butler, K., Seaton, M.J., Storey, P.J., Taylor, K.T. and Yu Yan, 1987. *J. Phys. B: Atom. Molec. Phys.*, **20**, 6379.

Boercker, D.B., 1987. *ApJ*, **316**, L96.

Carson, T.R., 1973. *ARA&A*, **14**, 95.

Carson, T.R., Hübner, W.F., Magee, N.H. and Merts, A.L., 1984. *ApJ*, **283**, 466.

Carson, T.R., Mayers, D.F. and Stibbs, D.W.N., 1968. *MNRAS*, **140**, 483.

Cox, A.N., 1965. In *"Stars and stellar systems"* (ed. L.H. Aller and D.B. McLaughlin), Vol. 8, Univ. Chicago Press.

Cox, A.N. and Stewart, J.N., 1965. *ApJS*, **11**, 22.

Cox, A.N. and Stewart, J.N., 1970. *ApJS*, **19**, 243 and 261.

Cox, A.N. and Tabor, J.E., 1976. *ApJS*, **31**, 271.

Däppen, W., Mihalas, D., Hummer, D.G. and Mihalas, B.W., 1988. *ApJ*, **332**, 261.

Däppen, W., Lebreton, Y. and Rogers, F.J., 1990. *Solar Physics*, **128**, 35.

Dimitrijevič, M.S. and Konjevič, N., 1980. *JQSRT*, **24**, 451.

Eissner, W. and Nussbaumer, H., 1969. *J. Phys. B.: Atom Molec. Phys.*, **2**, 1028.

Griem, H.R., 1968. *Phys. Rev.*, **165**, 258.

Hübner, W.F., 1985. In *"Physics of the Sun"* (ed. P. Sturrock, T. Holzer, D. Mihalas and R. Ulrich), p. 33. Reidel.

Hübner, W.F., Merts, A.L., Magee, N.H. and Argo, M.F., 1977. *Los Alamos Laboratory Report* LA-6760-M.

Hummer, D.G. and Mihalas, D., 1988. *ApJ*, **331**, 794.

Iglesias, C.A. and Rogers, F.J., 1991. *ApJ*, **371**, 408.

Iglesias, C.A., Rogers, F.J. and Wilson, B.G., 1987. *ApJ*, **322**, L45.

Iglesias, C.A., Rogers, F.J. and Wilson, B.G., 1990. *ApJ*, **360**, 221.

Iglesias, C.A., Rogers, F.J. and Wilson, B.G., 1992. *ApJ*, submitted.

Magee, N.H., Merts, A.L. and Hübner, W.F., 1984. *ApJ*, **283**, 264.

Mihalas, D., Däppen, W. and Hummer, D.G., 1988. *ApJ*, **331**, 815.

Mihalas, D., Hummer, D.G., Mihalas, B.W. and Däppen, W., 1990. *ApJ*, **350**, 300.

Nussbaumer, H. and Storey, P.J., 1988. *A&A*, **193**, 327.

Reilman, R.F. and Manson, S.T., 1979. *ApJS*, **40**, 815.

Rogers, F.J., 1981. *Phys.Rev.A*, **23**, 1008.

Rogers, F.J., 1986. *ApJ*, **310**, 723.

Rogers, F.J., 1991. *Nuovo Cimento*, in press.

Rogers, F.J. and Iglesias, C.A., 1992. *ApJS*, in press.

Rogers, F.J., Wilson, B.G. and Iglesias, C.A., 1988. *Phys.Rev.A*, **38**, 5007.

Roznyai, B.F., 1989. *ApJ*, **341**, 414.

Saraph, H.E., Storey, P.J. and Taylor, K.T., 1992. *J. Phys. B: Atom. Molec. Phys.*, submitted.

Schwarzschild, M., 1958. *"Structure and evolution of the stars"*, Princeton.

Seaton, M.J., 1992. *Workshop on astrophysical opacities* (ed. A.E. Lynas-Gray, C. Mendoza and C.J. Zeippen), *Rev. Mex. Astron. Astrof.* (special issue, in press).

Seaton, M.J., 1988. *J. Phys. B.: Atom. Molec. Phys.*, **21**, 3033.

Seaton, M.J., 1990. *J. Phys. B.: Atom. Molec. Phys.*, **23**, 3255.

Simon, N., 1982. *ApJ*, **260**, L87.

Weiss, A., Keady, J.J. and Magee, N.H., 1990. *Atomic data and Nuclear data Tables*, Tables **45**, 209.

Yu Yàn, 1991. *Workshop on astrophysical opacities* (ed. A.E. Lynas-Gray, C. Mendoza and C.J. Zeippen), *Rev. Mex. Astron. Astrof.* (special issue, in press).

Inside the Stars, IAU Colloquium 137
ASP Conference Series, Vol. 40, 1993
Werner W. Weiss and Annie Baglin (eds.)

TURBULENCE IN STELLAR INTERIORS

JEAN-PAUL ZAHN[1,2]

[1]Observatoire Midi-Pyrénées, 14 ave. E. Belin, 31400 Toulouse, France
[2]Astronomy Department, Columbia University, New York, NY 10027, U.S.A.

ABSTRACT This review focusses on the most recent work which has been achieved concerning turbulence in stellar interiors. Among all possible causes for such turbulence, the most powerful is certainly the convective instability in unstably stratified regions, but little was known until now beyond the fact that thermal convection is capable of establishing an almost adiabatic stratification. The latest numerical simulations (Stein & Nordlund 1989; Cataneo *et al.* 1991) show that turbulent convection is highly intermittent, with strong downdrafts carrying most of the energy flux.

These plumes penetrate in the stable region below, which they render nearly adiabatic over some distance (Zahn 1991); the penetration depth should be measured soon through acoustic sounding of the Sun (Berthomieu *et al.* 1992). Such plumes are likely to exist also in a convective core, and to penetrate into the radiative region above (Roxburgh 1989).

Another type of turbulence apparently occurs in the radiation zones, due to differential rotation. It is much weaker, and strongly anisotropic, with more vigorous motions in the horizontal directions than in the vertical. Its role is negligible in the heat transport, but it acts to suppress the latitudinal variation of the angular velocity. This anisotropic turbulence also interfers with the meridian circulation, and it partly inhibits the advection of chemical elements (Chaboyer & Zahn 1992). The strongest indication so far for such turbulence is the thinness of the solar tachocline (Spiegel & Zahn 1992): this boundary layer, in which the rotation changes from differential in the convection zone to uniform below, has been detected through helioseismology, but it is still unresolved (Goode *et al.* 1991).

Assuming that such anisotropic turbulence enforces throughout the radiation zone a rotation state in which the angular velocity depends mainly on depth, it is possible to derive an explicit expression for the meridian flow, and thus to describe the evolution in time of the interior rotation (Zahn 1992). The main result is that both the meridian circulation and the turbulence are determined by the rate at which the star loses angular momentum. When there is no such loss, the meridian flow is very weak, and it can even vanish in slow rotators.

Keywords: Convection, convective penetration; meridian circulation; turbulence.

TURBULENT CONVECTION

Until recently, turbulence in stellar interiors has been synonymous with thermal convection, which indeed is the strongest of all dynamical instabilities, once the star has settled in hydrostatic equilibrium. Its contribution to the heat transport enforces a nearly adiabatic temperature profile, which is implemented in our models of stellar structure. We do this in general by resorting to the mixing-length treatment, for lack of a better prescription. And some of us still try to improve on that approach, with more or less sophistication.

But another, more promising road has been opened thanks to numerical simulation. The computers have progressed to a point which allows three-dimensional calculations, with a resolution of about 100^3. This is still insufficent to encompass a whole convection zone, and to correctly represent the kinetic energy cascade, but it permits to explore the organization of the motions on the largest scales. These resemble very much the granular pattern on the surface of the Sun, and their spectroscopic signature also agrees well with the observations (Nordlund & Dravins 1990).

Another striking result of these computer experiments is the presence of strong, long-lived downdrafts (Chan & Sofia 1986; Stein & Norlund 1989; Cattaneo *et al.* 1991). These behave like the thermal plumes which are well known in atmospheric sciences: they entrain some of the surrounding matter, and their life-time is much longer than the time the flow takes to travel through them. Such downdrafts were detected already in earlier two-dimensional calculations (Hulburt *et al.* 1984). They originate below the upper boundary layer, which is turbulent in the classical sense (the life-time of the eddies is of the order of their turn-over time). The plumes cross the whole computational domain, which is still limited by the possibilities of the present computers, but one can anticipate that some of them would traverse the whole convection zone. The picture which emerges today is sketched in fig. 1.

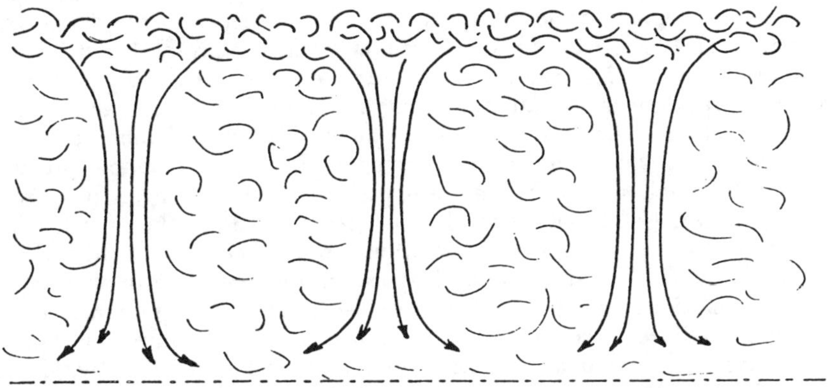

Fig. 1. A plausible picture of the convective envelope in solar-like stars, with downwards directed plumes traversing the whole convection zone. The dash-dotted line delineates the nearly adiabatic domain, including the region of convective penetration (see below).

Those plumes carry kinetic energy downwards and thermal energy (enthalpy) upwards, and surprisingly the two seem to cancel each other (Cattaneo *et al.* 1991). Therefore, it is the much quieter intersticial medium which is responsible for conveying the net energy flux to the surface. It remains to examine the full implications of this highly intermittent picture of the turbulent convection zone, both on the transport of heat and on that of angular momentum (cf. Rieutord & Zahn 1991; Simon & Weiss 1991).

Let us mention that such plumes are also observed in laboratory expermients, where the stratification is very weak: there they are directed either upwards or downwards, with a prefernce for the direction which allows for penetrative convection (cf. Whitehead & Chen 1970).

CONVECTIVE PENETRATION

Penetrative convection, or overshooting as it is sometimes called by astrophysicists, has been the subject of a long lasting debate (see Renzini 1987). However, there is overwhelming evidence for such penetration, in geophysical fluids, in the laboratory, and now also in the numerical simulations: the convective motions do not stop abruptly at the edge of the unstable region, but they travel beyond over some distance, and continue to enforce a nearly adiabatic stratification. The question is thus not whether penetration does occur in stellar interiors, but how important it is.

Stellar models that include this effect most often postulate a penetration depth of a fraction of the pressure scale height, or else of the radius of the convective core. And this fraction is calibrated by comparing the theoretical predictions with the observations. But everyone agrees that there is a need for a better prescription, on a sound physical basis.

It is appropriate to seek an answer in numerical simulations, like those which have been mentioned above, but with boundary conditions that allow for penetration. Those which have been performend in two dimensions show indeed that the conspicuous downdrafts penetrate rather deep into the stable region (Hurlburt *et al.* 1986). But the computer power which is presently available does not permit yet to explore the relevant parameter range. The goal is even more remote in three dimensions, although some attempts have been made already at low resolution (Brandenburg *et al.* 1991; Cattaneo, private communication).

In the meanwhile, one must make up with some reasonable prediction, based on minimal and plausible assumptions. One has been proposed by Roxburgh (1978, 1989), but its validity has been questioned because it rests on the assumption that the convective motions are confined in a nearly adiabatic domain (Baker & Kuhfuß 1987). This assumption therefore implies that the penetrative motions are efficient enough to establish such a temperature profile in the stable adjacent region. It can be shown that this requirement is met in a convective core, provided that rising, plume-like motions occur there, as it is suggested indeed by laboratory experiments (Zahn 1991).

Roxburgh's criterion states that the following integral vanishes at the edge of the penetration region:

$$\int (L_{nuc} - L_{rad})\, d\left(\frac{1}{T}\right) = 0, \tag{1}$$

where T is the temperature, L_{nuc} the total energy flux produced by the nuclear reactions, and L_{rad} the total flux carried by radiation in the nearly adiabatic temperature stratification. This predicts a penetration depth which is a fraction of about 20% of the core radius, and which increases with its mass (Roxburgh 1992; see also Zahn 1991). The effect of neglecting the viscous dissipation is now being investigated (see the poster by Roxburgh & Simmons, this meeting).

Unfortunately, this integral constraint cannot be employed for a convective envelope, because the departures from adiabaticity are too large at the surface boundary. One then needs to postulate the form of the convective motions, and the most natural assumption is that they are organized in plumes, as described above. This was first done by Schmitt *et al.* (1984), with application to the base of the solar convection zone. They found empirically that the penetration depth scales as $f^{1/2}W_i^{3/2}$, with W_i being the velocity at the top of the subadiabatic region and f a measure of the area occupied by the plumes.

The same scaling can be derived theoretically, as was shown by Zahn (1991). Furthermore, with a plausible estimate of the initial velocity W_i, one finds that the penetration depth is of the order of the scale height of the radiative conductivity: $|dr/d\ln\chi|$ ($\chi = 16\sigma T^3/3\rho\kappa$ with the usual notations). Thus the penetration depth L_p is given by

$$\frac{L_p}{H_P} = \frac{\zeta}{\chi_P}, \qquad (2)$$

where H_P is the pressure scale-height and $\chi_P = (\partial \log\chi/\partial \log P)_{ad}$. The unknown parameter ζ is related to the third moment of the convective velocity: it should be of order unity, but is presumably a little less. Once it has been calibrated with the Sun, it can be used in other stellar models with a deep convective envelope, much as one does with the familiar mixing-length parameter α.

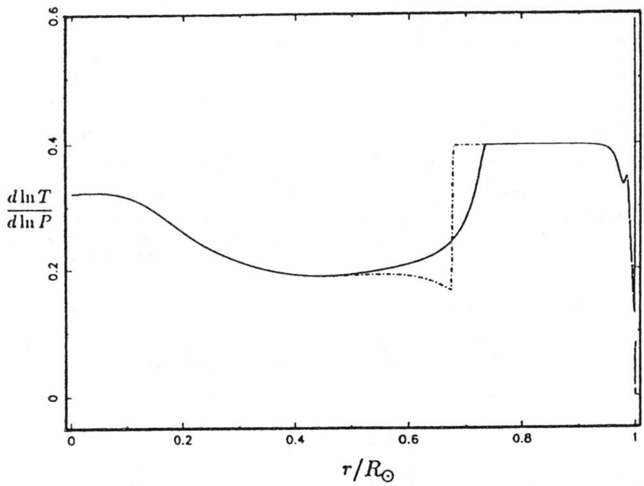

Fig. 2. The logarithmic temperature gradient in a standard solar model (continuous line), compared with a model allowing for a convective penetration of 30,000 km (dash-dotted line). Note the discontinuity at the edge of the penetration region (Berthomieu *et al.*, this meeting).

This calibration has been undertaken by Berthomieu *et al.* (1992). A series of solar models have been computed with the code CESAM built at the Nice Observatoire (see Morel *et al.* 1990); they include an increasing amount of penetration at the base of the convection zone, the parameter ζ varying from 0 to 1. These models were evolved from a ZAMS homogeneous model to the present age of the Sun, where they were fitted to the observed radius and luminosity. One of such models is depicted in fig. 2. The comparison of the eigenfrequencies between these models and the Sun (Libbrecht *et al.* 1990) is used to determine the value of the penetration parameter ζ. Work is still in progress, in order to include all relevant physical processes; presently, the best estimate is around 0.70, which corresponds to a penetration depth of about 20,000 km below the solar convection zone.

THE SOLAR TACHOCLINE

Thanks to helioseismology, it is now possible to probe the distribution of the angular velocity as a function of radius and latitude. The surface rotation, with its equatorial acceleration, carries on to the bottom of the convection zone with little depth dependence. In the radiation zone below, the equatorial and polar rotation rates approach a common value, within a layer which is too thin to be resolved as yet, i.e. which is less than about 1/10 of the solar radius (Brown *et al.* 1989; Goode *et al.* 1991).

The theory of this transition layer has been completed recently by Spiegel and Zahn (1992). Considering the observed rotation law in the convection zone as given, they predict the angular velocity in the radiative interior below by expressing the hydrostatic balance and the conservation of energy. The governing equations are simplified by approximations usually applied to thin layers with rapid rotation. A similar approach is taken by the oceanographers, when they study ocean circulation driven by the wind stress at the sea surface: for that reason the solar transition layer has been named *tacholine.*

The velocity boundary conditions applied to the top of the radiation enforce a horizontal pressure gradient, leading in turn to a horizontal temperature gradient in the pole-equator direction. This strong link between the temperature and differential rotation is well known in geophysical fluid dynamics as the thermal wind. In the Sun, because of radiation, the horizontal temperature distribution diffuses downwards, and it is accompanied by the differential rotation, which, in the inviscid case, can be modified only by advecting angular momentum.

But the meridian motions required for this are hindered by the stable stratification, and thus the transition layer spreads inward at a much slower rate. Instead of growing in time as in the normal diffusive way like $t^{1/2}$, and the thickness h of the layer increases approximately as

$$h \approx R_c \, (t/t_{ES})^{1/4} \, , \tag{3}$$

where the local Eddington-Sweet time is given by $t_{ES} = (\Omega/N)^{-2} \, R_c^2/K$, in terms of the mean rotation rate Ω, the buoyancy frequency N, and K the thermal diffusivity. A similar temporal evolution is found in the stratified spin-down problem (e.g. Sakurai 1970; Spiegel 1972). For the Sun, $h \approx 2\,10^5$ km after $4.6\,10^9$ yrs, according to this crude formula.

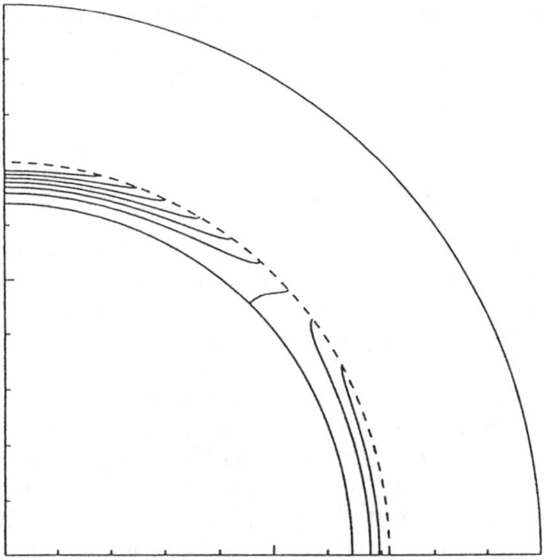

Fig. 3. The solar tachocline, whose thickness has been set arbitrarily to 50,000 mk (the actual value depends on the horizontal viscosity ν_h). The continuous lines show the contours of the angular velocity. The interior rotation is that of the convection zone at the latitude 42° (Spiegel & Zahn 1992; courtesy *Astr. Ap.*).

The actual calculation, allowing for the variation of the rotation rate with time, yields an even larger result: $h \approx 3\,10^5$ km. Hence, we conclude that the differential rotation of the convection zone would have extended well into the Sun, if radiative diffusion were alone responsible for its spread. The apparent discontinuity represented by the observed tachocline must therefore be ascribed to another physical process, capable of preventing the expansion of the layer.

In deriving the estimate above, the flow has been treated implicitely as laminar, while it is a well established fact that, at sufficiently high Reynolds numbers, virtually any flow becomes turbulent. Thus there is little doubt that the horizontal shear of the differential rotation will be unstable (Zahn 1975). An enhanced isotropic viscosity would spread the tachycline even more quickly into the interior than before, but one verifies that with sufficiently anisotropic turbulence one can construct solutions that resemble the observed tachycline (fig. 3).

It thus proves the presence, below the solar convection zone, of a predominantly horizontal tubulence, such as encountered in the geophysical context (Rhines 1979). In this regime, the horizontal advection of angular momentum is balanced by the turbulent stresses, and a steady state is achieved in a layer whose thickness can be well within the bound set by observations. That thickness is determined by the turbulent Prandtl number, which is here the ratio between the horizontal eddy viscosity ν_h and the thermal diffusivity K:

$$h = R_c\,(\Omega/N)^{1/2}\,(K/\nu_h)^{1/4}. \tag{4}$$

The theory also predicts that the interior uniform rotation is intermediate between the polar and equatorial rotation rates of the convection zone, in agreement with the observations. The vertical differential rotation changes sign at a latitude of about 42° (see fig. 3).

ROTATION-INDUCED TURBULENCE

In the solar tachocline, the differential rotation which is responsible for the turbulent motions is enforced by the convection zone above. But it can also arise from the advection of angular momentum through a large scale circulation. The origin of such meridian circulation has been known since Eddington (1925) and Vogt (1925): due to the centrifugal force, the radiation flux is no longer conservative, and the star cannot achieve thermal equilibrium. A large scale advection is thus induced to preserve the conservation of energy.

This meridian flow depends sensitively on the rotation law, which itself is continually modified by the flow. The differential rotation is liable to various instabilities, as we have mentioned above, and the turbulent motions that then occur do also interfer with the transport of angular momentum. The problem to solve is thus rather intricate, and until recently, it had not been treated in a self-consistent way, mainly because of our poor knowledge of the turbulent transport.

However, significant progress can be achieved by assuming that the turbulence is highly anisotropic, as observed in the solar tachocline: then it instaures a rotation state in which the angular velocity depends little on latitude, and may be considered as varying only with depth (Zahn 1992). With such a shellular rotation law, it is possible to derive an explicit expression for the meridian velocity, $u(r,\theta) = U(r)P_2(cos\,\theta)$, with $U(r)$ involving up to the third derivative of the angular velocity $\Omega(r)$ (r and θ are the polar coordinates).

The evolution of the rotation profile is governed by the conservation of angular momentum, which is then expressed by

$$\frac{\partial}{\partial t}\left[\rho r^2\,\Omega\right] = \frac{1}{5r^2}\frac{\partial}{\partial r}\left[\rho r^4 \Omega\, U\right] + \frac{1}{r^2}\frac{\partial}{\partial r}\left[\rho\nu_t r^4\frac{\partial\Omega}{\partial r}\right], \qquad (5)$$

with the usual notations, and ν_t being the vertical component of the turbulent viscosity. This viscosity is due essentially to the shear in the vertical direction, and ν_t is proportional to $(\partial\Omega/\partial r)^2$ (Zahn 1975). A similar equation has been used by Endal & Sofia (1976), and later by Pinsonneault *et al.* (1989) to model the transport of angular momentum in stellar interiors, but the advective term was lacking.

The p.d.e. above is of first order in time, and of fourth-order in space; it must be solved with suitable initial and boundary conditions, of which the most important expresses the loss of the angular momentum J at the surface of the star:

$$-\frac{1}{5}\,\rho r^4\Omega\, U - \rho\nu_t r^4\frac{\partial\Omega}{\partial r} = -\frac{3}{8\pi}\frac{\partial J}{\partial t}. \qquad (6)$$

When the star is not subject to a wind, the rotation rate settles rapidly into a state with little meridian circulation, the latter just compensating the adverse flux of angular momentum due to the turbulent motions. Moreover, if the rotation speed lies below some threshold, the shear of the differential rotation is too weak to generate such turbulence, and there is no circulation at all.

The situation is drastically different when the star emits a wind. Then a bulk circulation sets in to carry the angular momentum from the interior to the surface, and its speed is governed by the rate of angular momentum loss. The rotation profile adjusts slavishly to produce the required meridian flow, and so does also the turbulence. The core of the star rotates substantially faster than the surface, and one may expect that this is the case, in particular, for the Sun (fig. 4).

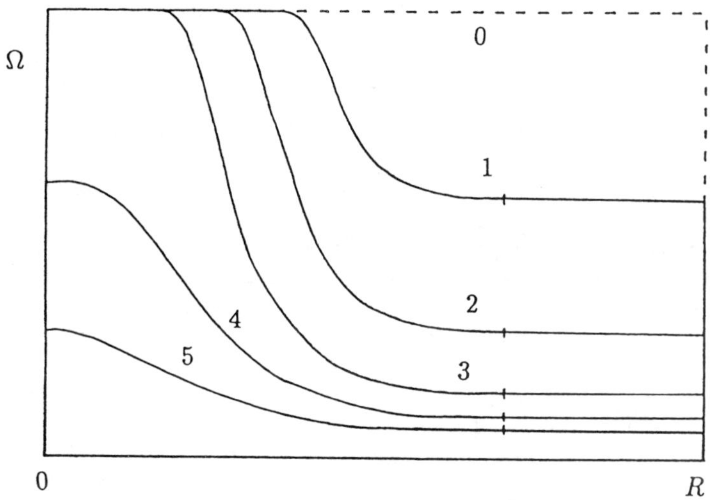

Fig. 4. Evolution of the rotation profile in a solar-like star which loses angular momentum (Zahn 1992). Work is in progress to refine this sketch.

This notion, that it is the spindown which drives the circulation, is well known for viscous fluids as the so- called Ekman pumping (Bondi & Littleton 1948; Greenspan & Howard 1963). Its relevance for stars has been recognized by Howard *et al.* (1967), with radiative diffusion taking the role of the viscous coupling, and with the whole convection zone acting as the turbulent Ekman layer (Bretherton & Spiegel 1968). But for some reason the astronomical community seems to have ignored these pionneering works.

The transport of angular momentum is not much affected by the turbulent motions, whose role is chiefly to enforce the shellular rotation state. In contrast, the advection of chemical species is partly inhibited by the horizontal component D_h of the turbulent diffusivity, which acts to smooth out the horizontal inhomogeneities that are built up by the circulation. When $D_h \gtrsim rU$, the vertical transport of matter behaves like a diffusion, with an effective diffusivity given by:

$$D_{eff} = \frac{1}{30} \frac{(rU)^2}{D_h}. \qquad (7)$$

This result was established by Chaboyer and Zahn (1992), who showed that the process is closely related to Taylor's shear dispersion (Taylor 1953). Their predictions were confirmed by Charbonneau (1992), who actually simulated, with a two-dimensional code, the interference of anisotropic turbulence with a meridian flow.

This turbulent erosion provides a straightforward explanation for the strong link between the transport of lithium and that of angular momentum, and moreover for the disparity which is observed between the efficiencies of these transports in the case of the Sun (Law *et al.* 1984). With some simplifying assumptions, it is possible to show that the surface abundance of lithium varies approximately as

$$-d\,log\,c = -k\,log\,J, \tag{8}$$

where k is a constant that depends only on the location of the base of the convection zone and that of the lithium destruction layer, but not on the spindown law itself (Zahn 1992).

Work is in progress to explore all consequences of the master equation (5) above, and in particular to predict the internal rotation of the Sun.

ACKNOWLEDGEMENTS

Support for this work was provided by the U.S. Air Force under grant AFOSR 89-0012 to Columbia University, and by the Centre National de la Recherche Scientifique (GDR 131).

REFERENCES

Baker, N.H., Kuhfuß, R. 1987, *Astr. Ap.*, **185**, 117

Berthomieu, G., Morel, P., Provost, J., Zahn, J.-P. 1992, *Astr. Ap.*, (preprint)

Bondi, H., Lyttleton, R.A. 1948, *Proc. Cambridge Phil. Soc.*, **44**, 345

Brandenburg, A., Jennings, R., Nordlund, Å., Rieutord, M., Ruokilainene, J., Stein, R., Tuominen, I. 1992, *Ap. J.*, (preprint)

Bretherton, F.P., Spiegel, E.A. 1968, *Ap. J.*, **153**, 277

Brown, T.M., Christensen-Dalsgaard, J., Dziembowski, W.A., Goode, P., Gough, D.O., Morrow, C.A., 1989, *Ap. J.*, **343**, 526

Cattaneo, F., Brummel, N.H., Toomre, J., Malagoli, A., Hurlburt, N.E. 1991, *Ap. J.*, **370**, 282

Chaboyer, B., Zahn, J.-P. 1992, *Astr. Ap.*, **253**, 173

Chan, K.L., Sofia, S. 1986, *Ap. J.*, **307**, 222

Charbonneau, P. 1992, *Astr. Ap.*, (preprint)

Eddington, A.S. 1925, *Observatory*, **48**, 78

Endal, A.S., Sofia, S. 1976, *Ap. J.*, **210**, 184

Goode, P.R., Dziembowski, W.A., Korzennick, S.G., Rhodes, E.J. 1991, *Ap. J.*, **367**, 649

Greenspan, H., Howard, L.N. 1963, *J. Fluid Mech.*, **17**, 385

Howard, L.N., Moore, D.W., Spiegel, E.A. 1967, *Nature*, **214**, 1297

Hurlburt, N.E., Toomre, J., Massaguer, J.M. 1984, *Ap. J.*, **282**, 557

Hurlburt, N.E., Toomre, J., Massaguer, J.M. 1986, *Ap. J.*, **311**, 563

Law, W.Y., Knobloch, E., Spruit, H.C. 1984, *Observational Tests of Stellar Evolution Theory* (ed. A. Maeder, A. Renzini; Reidel), 523

Libbrecht, K.G., Woodard, M.F., Kaufman, J.M. 1990, *Ap. J. Suppl.*, **74**, 1129

Morel, P., Provost, J., Berthomieu G. 1990, *Solar Phys.* **128**, 7

Nordlund, Å., Dravins, D. 1990, *Astr. Ap.*, **228**, 155

Rhines, P.B. 1979, *Ann. Rev. Fluid Mech.*, **11**, 401
Renzini, A. 1987, *Astr. Ap.*, **188**, 49
Rieutord, M., Zahn, J.-P. 1991, *The Sun and Cool Stars: Activity, Magnetism, Dynamos* (ed. I. Tuominen, D. Moss, G. Rüdiger; Springer), 193
Roxburgh, I.W. 1978, *Astr. Ap.*, **65**, 281
Roxburgh, I.W. 1989, *Astr. Ap.*, **211**, 361
Roxburgh, I.W. 1992, *Astr. Ap.*, (preprint)
Sakurai, T. 1970, *Stellar Rotation* (ed. A. Slettebak; Reidel), 329
Schmitt, J.H.M.M., Rosner, R., Bohn, H.U. 1984, *Ap. J.*, **282**, 316
Simon, G.W., Weiss, N.O. 1991, *M.N.R.A.S.*, **252**, 1p
Spiegel, E.A. 1972, *Physics of the Solar System* (ed. S.I. Rasool; NASA, Washington)
Spiegel, E.A., Zahn, J.-P. 1992, *Astr. Ap.*, (preprint)
Spruit, H.C., Nordlund, Å., Title, A.M. 1990, *Ann. Rev. Astr. Ap.*, **28**, 263
Stein, R., Nordlund, Å. 1989, *Ap. J.*, **242**, L95
Taylor, G.I. 1953, *Proc. Roy. Astron. Soc. London A2*, **19**, 186
Vogt, H. 1925, *Astron. Nachr.*, **223**, 229
Whitehead, J.A., Chen, M.M. 1970, *J. Fluid Mech.*, **40**, 549
Zahn, J.-P. 1975, *Mém. Soc. Roy. Sci. Liège*, 6e série, **8**, 31
Zahn, J.-P. 1991, *Astr. Ap.*, **252**, 179
Zahn, J.-P. 1992, *Astr. Ap.*, (preprint)

Inside the Stars, IAU Colloquium 137
ASP Conference Series, Vol. 40, 1993
Werner W. Weiss and Annie Baglin (eds.)

PARTICLE TRANSPORT PROCESSES

Georges Michaud
Département de Physique
Université de Montréal

C. R. Proffitt
Space Telescope Science Institute

ABSTRACT The effect of gravitational settling and radiation driven diffusion on the evolution of stars near the main sequence is reviewed. New simplified formulae for calculating diffusion are proposed that improve on previous such formulae. The reliability of available diffusion coefficients is discussed and areas where further work is needed are identified. Newly available opacity calculations are used to estimate the effects of radiative acceleration on Fe.

The size of the modifications to the evolution are shown to be modest: a reduction of order 10% on the evolutionary age of globular clusters. There are indications from the Li abundance in the high T_{eff} halo stars that some turbulence is present below the convection zone. Models calibrated using solar properties reproduce the T_{eff} at which the Li gap is observed in F stars as well as its depth without any arbitrary parameter. Similarly the presence of AmFm stars is explained over the T_{eff} range where they are observed.

I. ASTROPHYSICAL CONTEXT

While it is a fundamental physical process, atomic diffusion was put aside in stellar evolution following Eddington's (1926) argumentation that the absence of extreme heavy element abundances ruled out the equilibrium configurations to which diffusion would lead in stellar interiors. He suggested that meridional circulation was a competing process and probably ruled out the effects of atomic diffusion.

It is only in the 1960s that the role of atomic diffusion, driven by gravity (Aller and Chapman 1960) and differential radiation pressure (Michaud 1970), was reintroduced in main sequence stellar envelopes taking into consideration that the diffusion time scale is much shorter in external than in internal regions (Michaud *et al.* 1976). Since diffusion time scales are shorter than evolutionary time scales only in external stellar regions, anomalies should appear most strongly only in those stars where all competing hydrodynamical processes are reduced in external regions such as Ap stars. Equilibrium configurations for individual chemical species are never approached in stellar interiors.

It has since been shown that meridional circulation could compete with diffusion as Eddington suggested. Indeed, while the AmFm and the HgMn phenomena are observed up to equatorial rotation velocities of $V_e \sim 100$ km s^{-1}, a simple parameter free model taking diffusion and meridional circulation into account leads to an expected cut off velocity of $V_e \sim 100$ km s^{-1} for the AmFm and HgMn phenomena (Michaud 1982, Charbonneau and Michaud 1988, 1991).

II- DIFFUSION EQUATIONS

1. Basic Formulae

The most rigorous derivations of the equations for atomic diffusion start from the Boltzmann equation for the velocity distribution functions for each species of particle. A variety of techniques have been proposed to obtain approximate solutions to the Boltzmann equation (see, for instance, Chapman and Cowling 1970 and Burgers 1969). Usually the deviations from the ideal Maxwell-Boltzmann velocity distribution appropriate for statistical equilibrium in a uniform gas are small, but these deviations are essential for calculating higher order effects such as *thermal diffusion*.

There are many ways of expressing the resulting flow equations. It is often useful to treat two or three ions at a time and have a single diffusion velocity to solve as in Richer, Michaud and Proffitt (1992). For the general case of a multi-component plasma, the expressions given by Burgers (1969) are convenient and we will use them for our discussion here. For each ionic species there is a force equation (eq. 18.1 of Burgers 1969),

$$\frac{dP_i}{dr} + m_i n_i g + n_i Z_i (eE) = \sum_j K_{ij}(v_j - v_i) + \sum_j K_{ij} z_{ij} \frac{m_j r_i - m_i r_j}{m_i + m_j}. \quad (1)$$

A similar equation could be written for electrons or it may be deduced from the hydrostatic equation. For each species including the electrons there is a heat equation (eq. 18.2 of Burgers). To these is added the mass conservation equation (eq. 2.20 of Burgers) requiring that mass does not move with respect to the centre of mass. A similar equation may be written for the charge since only extremely small charge separations are allowed to develop. For s different ionic species, this gives 2s+3 equations. The unknown quantities to be solved for in this system of equations are the velocities v and heat flux vectors r, for both the ions and electrons, and the value for the electric field.

The quantities K_{ij} in the above equations are the resistance coefficients, and represent the effects of collisions between particles of types i and j. They can be written as,

$$K_{ij} = \frac{16}{3} n_i n_j \left(\frac{m_i m_j}{m_i + m_j} \right) \Omega_{ij}^{(11)}. \quad (2)$$

The heat flux terms involve additional collision integrals:

$$z_{ij} = 1 - \frac{2\Omega_{ij}^{(12)}}{5\Omega_{ij}^{(11)}}, \quad z'_{ij} = \frac{5}{2} - \frac{10\Omega_{ij}^{(12)} - 2\Omega_{ij}^{(13)}}{5\Omega_{ij}^{(11)}} \quad (3)$$

and a similar one for z''_{ij} (see eq. 11.7c of Burgers). The collision integrals Ω_{ij} are discussed in the following section.

The K_{ij}'s are simply related to Chapman and Cowling's (1970) *first approximation* for diffusion coefficients. It has been shown (Burgers 1969, and Roussel-Dupré 1981) that, when Chapman and Cowling's *second approximation* to the diffusion coefficient and thermal diffusion terms are included, the method of Chapman is equivalent to that of Burgers.

Radiative Accelerations

In the presence of a radiation flux, g should be replaced by $g-g_R$ in equation (1). The competition between radiative acceleration, g_R, and gravity determines whether diffusion leads to over- or underabundances. Accurate radiative accelerations have only been calculated for a few elements. General formulae are available (Michaud *et al.* 1976; Alecian and Artru 1990) but are accurate to at most a factor of 3 (see Michaud 1987 for a discussion). Calculations are needed using more accurate data banks. The required atomic data is becoming available from the Opacity project and from the large compilations of Kurucz (1991). Detailed calculations are under way for C, N, O.

For Fe, Alecian, Michaud and Tully (1992) have calculated radiative accelerations in detail in main sequence models with T_{eff} = 6700, 8000 and 10000 K and log g = 4.0 and 4.3 and a number of Fe abundances. The atomic data available from the Opacity Project for Fe IX to Fe XVII allows calculating the accelerations over an important fraction of the envelope. In stars of T_{eff} = 6700 K, radiative acceleration and gravity are nearly equal below the convection zone and only small Fe anomalies are expected. In cooler stars, underabundances of Fe are expected to materialize since gravity is always larger than g_R. They are expected to be small in so far as the progressive desaturation of the lines will lead to an increase of g_R. In stars of 8000 and 10000 K, g_R(Fe) is significantly larger than gravity over part of the envelope. If the envelope is stable enough, Fe overabundances will then appear over part of the envelope. Detailed evolutionary calculations are needed to determine where these overabundances appear. The Fe abundance is expected to modify the structure of the model since the recent results of Rogers and Iglesias (1992) show that over part of the envelope, iron is the main source of opacity.

2. The Collision Integrals

These integrals can be calculated by using the binary collision approximation, or the Fokker-Planck approximation, in which it is assumed that most collisions are sufficiently weak that the effects of two particle collisions can be summed, even if many particles are interacting simultaneously. When diffusion velocities are small compared to thermal velocities both approximations give the same result for sufficiently dilute plasmas.

In general, for interactions between two particles, collision integrals of the form

$$\Omega_{ij}^{(\alpha\beta)} = \int_0^\infty \int_0^\infty e^{-g^2} g^{2\beta+3} (1 - \cos^\alpha X_{ij})\; b\, db\, dg \qquad (4)$$

occur, with

$$X_{ij} = \pi - 2\int_{r_{min}}^\infty b\, dr\, \{r^2 [1 - \frac{b^2}{r^2} - \frac{V_{ij}(r)}{g^2 kT}]^{1/2}\}^{-1}. \qquad (5)$$

where g^2 is the kinetic energy in the centre of mass of the colliding particles in units of kT. For the Coulomb potential the above integrals become infinite due to the long range of the Coulomb force. In a real plasma the charged particles tend to polarize the surrounding medium, and the long range Coulomb interaction is cut off by this screening. This suggests using the screened Debye-Huckel potential,

$$V_{ij} = \frac{Z_i Z_j}{r} e^2 \exp(-r/\lambda_d), \qquad (6)$$

where the Debye screening length is (see DeWitt 1961):

$$\lambda_d = \left(\frac{kT}{4\pi e^2 \sum_\alpha n_\alpha Z_\alpha^2 \Theta_\alpha}\right)^{1/2}. \tag{7}$$

The sum in the denominator of the Debye length is over all charged particles in the plasma, including electrons. As the electrons become degenerate, they contribute less to polarization, and $\Theta_e < 1$ (see Fontaine *et al.* 1977).

The above integrals cannot be solved analytically when the Debye-Huckel potential is used. At sufficiently low density the collision integral is dominated by collisions with impact parameters less than the screening length, so the potential can be approximated as a pure Coulomb potential with the integration truncated at the Debye screening length. To find an analytic solution one must also replace g^2 in the integrals for X_{ij} with an assumed *average* value $<g^2>$ (typically a value of 2 or 3 is chosen). Analytically, it is then found that:

$$\Omega_{ij}^{(11)} = \pi \left(\frac{Z_i Z_j e^2}{2kT}\right)^2 \left[\frac{(m_i + m_j)kT}{2\pi m_i m_j}\right]^{1/2} \ln \Lambda_{ij}, \tag{8}$$

where

$$\ln \Lambda_{ij} \simeq \ln \left(\text{constant}\ \frac{2kT\lambda_d}{Z_i Z_j e^2}\right). \tag{9}$$

This quantity, $\ln \Lambda_{ij}$, is often referred to as the *Coulomb logarithm*. The constant inside the logarithm depends on the approximations made and is similar to the value chosen for $<g^2>$ in (5). Because the constant appears under the logarithm, the argument is usually made that, when $\ln\Lambda_{ij}$ is sufficiently large, uncertainties in the constant of a factor of a few are unimportant. When similar approximations are made for the other collision integrals it is found that, $z_{ij} = 0.6$, $z'_{ij} = 1.3$, and $z''_{ij} = 2$.

The above approximations are unnecessary and can lead to serious errors in the collision integrals for the conditions appropriate for stellar interiors (Paquette et al. 1986). Without making any of these approximations, the collision integral $\Omega_{ij}^{(11)}$ can be written in the form above, but with a function C_{ij} in place of $\ln\Lambda_{ij}$, that still depends only on the single parameter $2kT\lambda/Z_iZ_je^2$. This function can be evaluated numerically for the whole range needed. It is found that in the low density limit $\ln \Lambda_{ij}$ is an excellent approximation if the constant in (9) is chosen to be about 2/5.

In the interior of main sequence stars, the bulk of the ions are fully ionized hydrogen and helium, and heavier ions can be neglected when calculating the Debye screening length. This gives

$$\ln \Lambda_{ij} = -19.26 - \ln Z_i Z_j - \frac{1}{2}\ln\rho - \frac{1}{2}\ln\left[1 + \left(\frac{X+1}{2}\right)\Theta_e\right] + \frac{3}{2}\ln T, \tag{10}$$

when ρ is in gm cm^{-3} and T in K. Specializing to collisions between H and He and taking $\Theta_e = 1$ gives,

$$\ln \Lambda_{xy} = -19.95 - \frac{1}{2} \ln \rho - \frac{1}{2} \ln \left(\frac{X+3}{2}\right) + \frac{3}{2} \ln T. \qquad (11)$$

For conditions characteristic of the radiative interior of the Sun $\ln \Lambda_{xy} \simeq 2$, while in the centres of globular cluster turn-off stars it can be as small as 1.5. As can be seen from figure 1, using $\ln\Lambda_{xy}$ as an approximation for C_{xy} gives answers within about 10% when $\ln \Lambda_{xy} > 1$, but begins to diverge rather rapidly for $\ln \Lambda_{xy} < 1$.

For heavy elements diffusing in stellar interiors, Λ_{ij} is significantly smaller because of the $-\ln Z_iZ_j$ term in equation (10). For an element with $Z_i=8$, $\ln\Lambda_{ij} \sim 0$ for collisions with He nuclei in the solar interior. The collisional effects for heavy elements will therefore be substantially smaller than might be naively expected from the $(Z_iZ_j)^2$ factor in equation (8). While $\ln\Lambda_{ij}$ is clearly meaningless as an estimate of the integral C_{ij} in this case, the numerical values of Paquette et al. can be used. Figure 2 shows values for C_{ij} calculated using Paquette et al.'s fits for a model of the sun and for a model of a 0.75 $M_\odot$ star near core H exhaustion. Note that the values for collisions involving fully ionized O are much smaller than for collisions between H and He. If a simple formula is desired instead, the fitting formula

$$C_{ij} \simeq \frac{1}{1.2} \ln[e^{(1.2\ln \Lambda_{ij})} + 1] \qquad (12)$$

agrees with the numerically determined value for C_{ij} to within 15% for $\ln\Lambda_{ij} > -4$.

The ratios of collision integrals (z_{ij}, z'_{ij}, and z''_{ij}) also differ substantially from the values they assume in the low density limit (see figure 3). Near $\ln\Lambda_{ij}=2$, z_{ij} is 30% smaller than in the low density limit. As a result, for small $\ln\Lambda_{ij}$, thermal diffusion effects are smaller than given by estimates using the ratios appropriate for low densities.

For a given definition of the collision integrals there are well defined values for the resistance/diffusion coefficients. The arbitrary constant that is often found inside the Coulomb logarithm is a result of approximations that are only necessary if a simple analytic form for the integral is desired. It is wrong to take the range of formulae for the Coulomb logarithm that have appeared in the physics literature as giving the intrinsic theoretical uncertainty for diffusion in stellar interiors. (Especially as many of these formulae were derived by assuming that factors of 2 or so under the logarithm are unimportant). Numerical solutions to the collision integrals for a screened Coulomb potential have been calculated by Paquette et al. which avoid this uncertainty, and give the proper value in the low density limit where a simple analytic formula is meaningful.

This is not to say that there are no uncertainties in the diffusion velocities calculated with these collision integrals. The real uncertainties come from the physical assumptions involved in the initial definitions of the collision integrals and diffusion equations, not from the mathematical simplifications needed to derive simple analytic expressions. For instance, Roussel-Dupré (1982) estimates that including higher order moments of the velocity distribution function can change diffusion coefficients for H-He mixtures by about 10%. See Michaud (1991) for a discussion of other potential uncertainties.

The assumption that most of the collisions are weak (i.e. the Fokker-Planck approximation) is equivalent to the requirement that the kinetic energy of the ions is large compared to their interaction energy. This is true when $\Lambda_{ij} >> 1$, but becomes increasingly less valid as Λ_{ij} decreases. The collision integrals discussed above should therefore be reliable for H-He plasmas in the interior of main-sequence stars, while the results for heavy ions in the same H-He background are extrapolated

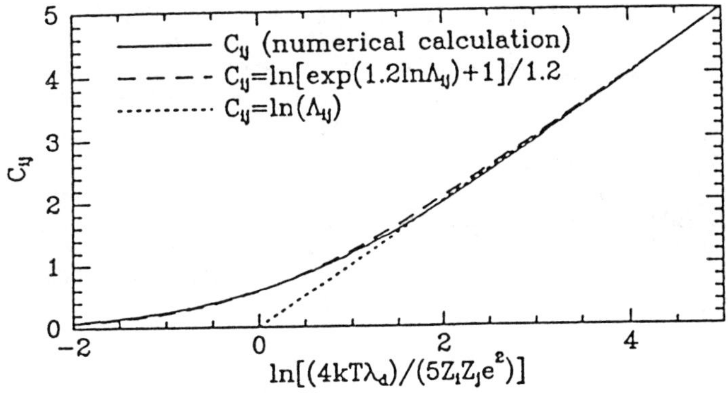

Fig. 1: Comparison of different approximations for C_{ij}.

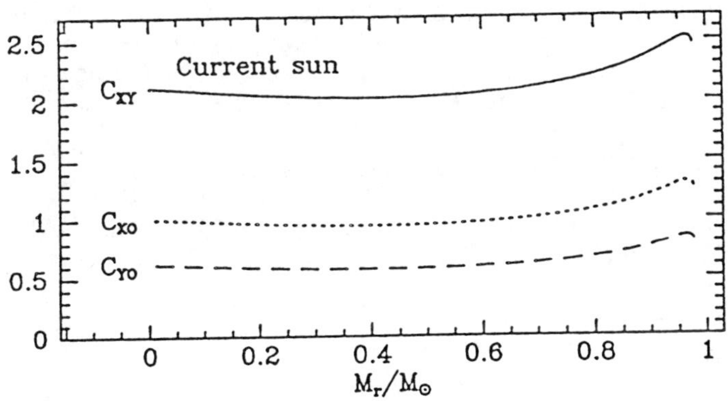

Fig. 2a: C_{ij} for H, He, and O in a model of the current sun.

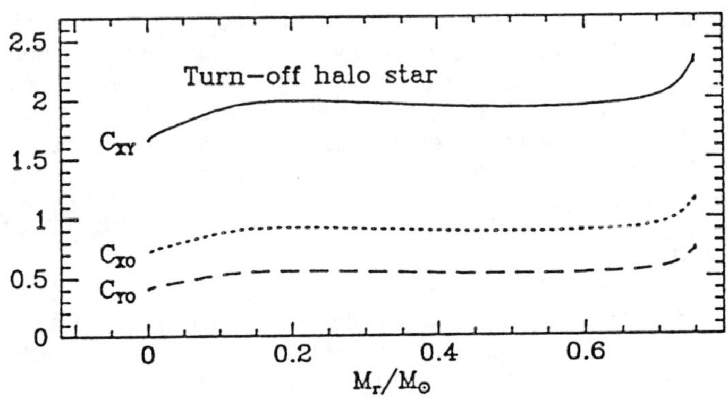

Fig. 2b: As in fig. 2a, but for a model of a 0.75 $M_\odot$ star.

beyond the regime where they are rigorously valid. Paquette et al. have shown that their results are in good agreement with calculations done in the limit of a very strongly coupled plasma. This suggests that the extrapolation may be valid for the intermediate coupling region. While this approach provides the best currently available estimates, it is very difficult to give quantitative estimates of accuracy.

3. Simplified Formula for Diffusion Velocities.

Given a value for the average velocity V_i with respect to the centre of mass of a given ionic species at each point in the stellar interior, the rate of change of the mass fraction abundance X_i of that species is given by

$$\frac{dX_i}{dt} = -\frac{1}{\rho r^2}\frac{d}{dr}(\rho r^2 V_i X_i). \qquad (13)$$

It is often useful to have available simple formulae for V_i for use in this equation. The formulae given below are for a mixture of fully ionized H and ^4He with other elements present only in trace amounts. Such formulae must be compared with the solutions of the full system of equations (1) to check their range of validity. Numerical results will be compared to a variety of simpler formulations.

3.1 A simplified formula for fully ionized H-He mixtures.

If the heat flux terms in equation (1) are ignored (i.e. $z_{ij} = 0$) and if it is assumed that C_{ij} can be approximated by $\ln\Lambda_{ij}$, then the diffusion velocity for H in a H-He mixture becomes,

$$V_H = -\frac{BT^{5/2}}{\rho \ln\Lambda_{ij}}\left[\frac{5}{4}(1-X)\frac{d\ln P}{dr} + \frac{3+X}{(1+X)(3+5X)}\frac{d\ln X}{dr}\right] \qquad (14)$$

where, to simplify writing this expression, the constant

$$B = \frac{15}{16}\left(\frac{2m_0}{5\pi}\right)^{1/2}\frac{k^{5/2}}{e^4} \qquad (15)$$

has been defined. This formula is identical to that implied by equation (41) of Bahcall and Loeb (1990) (B&L). However, for a trace H abundance, equation (14) underestimates the diffusion velocity obtained with a complete solution by about 30%. Examination of equation (18.2) of Burgers (1969) shows that the *heat flux terms*, the r_i's, are non-zero even in the absence of a temperature gradient. (The equivalent corrections are called *the second approximation* in Chapman and Cowling). An excellent approximation to the results of equation (1) can be obtained if we multiply formula (14) for V_H by the factor $1/(0.7+0.3X)$ (see figure 4 where $V_H - V_{He} = V_H/(1-X)$ is plotted).

When a non-zero temperature gradient is included, the rate of settling increases due to thermal diffusion. By comparing velocities calculated from the full system of equations with and without a temperature gradient, it is found that the fraction of the total diffusion velocity for H and He due to thermal diffusion,

$$R_{th} = \frac{v(\nabla\neq 0) - v(\nabla=0)}{v(\nabla=0)} \simeq 0.9\nabla \qquad (16)$$

in the interiors of main sequence stars. Typically, this implies that thermal diffusion increases the total diffusion velocity by about 30%. The value of R_{th} varies only slightly with the H-He ratio and $\ln\Lambda_{xy}$, and the formula

$$V_H = -\frac{B\,T^{5/2}}{\rho\ln\Lambda_{ij}(0.7+0.3X)}\left[\left(\frac{5}{4}+1.125\nabla\right)(1-X)\frac{d\ln P}{dr} + \frac{(3+X)}{(1+X)(3+5X)}\frac{d\ln X}{dr}\right] \qquad (17)$$

reproduces the total diffusion velocity as calculated for the full system to within 10% for $\ln\Lambda_{xy} > 2$. For smaller values of $\ln\Lambda_{xy}$ the accuracy begins to degrade. By using the approximation given by equation (12) in place of $\ln\Lambda_{xy}$ the usefulness of the above formula can be extended, and answers within 20% can be obtained for $\ln\Lambda_{xy}$ near 1. Figure 4 shows V_H-V_{He} calculated with B&L's assumptions together with our formula for $\ln\Lambda_{xy}$ (equation 11). Most of the difference between B&L's result and equation (17) is due to the extra heat-flux terms discussed above. If the formula for $\ln\Lambda_L$ given by Chaboyer *et al.*(1992) is used, smaller diffusion velocities are found. However there are numerous reasons for preferring our value for $\ln\Lambda_{ij}$.

3.2 Simple formula for a trace element in a H-He background

Ignoring the heat flux terms in equation (1), the velocity of a trace element i in a fully ionized H-He mixture can be written as,

$$V_i = -\frac{2BT^{5/2}}{5^{1/2}\rho Z_i^2}\left[\frac{\frac{d}{dr}\left\{\ln\left[\frac{X_i}{5X+3}\left(\frac{1+X}{5X+3}\right)^{Z_i}\right]\right\}+\left[1+Z_i-A_i\left(\frac{5X+3}{4}\right)\right]\frac{d\ln P}{dr}}{X(A_{ix}^{1/2}C_{ix}-A_{iy}^{1/2}C_{iy})+A_{iy}^{1/2}C_{iy}}\right] + XV_H\frac{(A_{ix}^{1/2}C_{ix}-A_{iy}^{1/2}C_{iy})}{X(A_{ix}^{1/2}C_{ix}-A_{iy}^{1/2}C_{iy})+A_{iy}^{1/2}C_{iy}}. \qquad (18)$$

Here X_i is the mass fraction of the trace element, and X, without a subscript, is the mass fraction of hydrogen while A_{ix} is the reduced mass (in atomic number units) of the trace element with hydrogen. If it is assumed that $C_{ix}=C_{iy}=\ln\Lambda_{xy}$ and $A_i >> A_y$, then equation (53) of B&L can be recovered (apart from an extra erroneous factor of X that B&L included in the composition gradient term). However, **these approximations should not be made, as they lead to qualitatively different results for the settling of heavy elements.** As can be seen from the discussion in section 2, the factor C_{ij} depends strongly on the product Z_iZ_j, especially when $\ln\Lambda_{ij}$ is small.

The first term in equation (18) can be identified as the settling velocity of the trace element in a fixed background, while the second term represents the net drag on the trace element from collisions with moving H and He ions. The approximation that B&L made results in a substantial overestimate of the strength of the collisions of the heavy elements with H and He. For typical parameters the settling term is underestimated by a factor of 2 to 3, and the drag term is overestimated by 3 to 5x. The drag term then dominates the total diffusion velocity. However, if formula (18) is used with either the numerically calculated values for C_{ij} or the approximation given by equation (12), then the drag term makes only a minor contribution to the total diffusion velocity.

While the results from formula (18) agree quite well with the solution of the

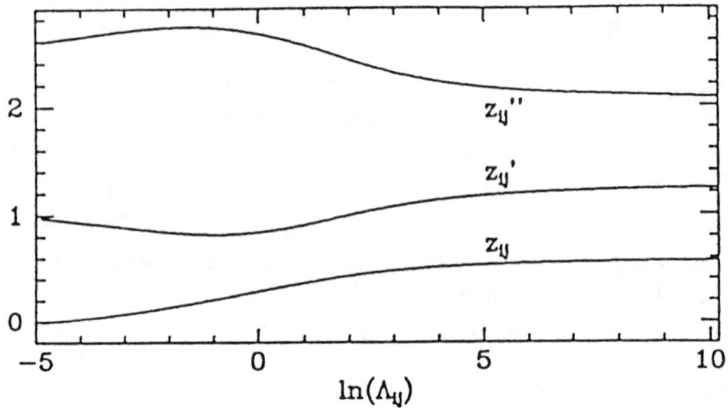

Fig. 3: The heat flux coefficients as a function of $\ln \Lambda_{ij}$.

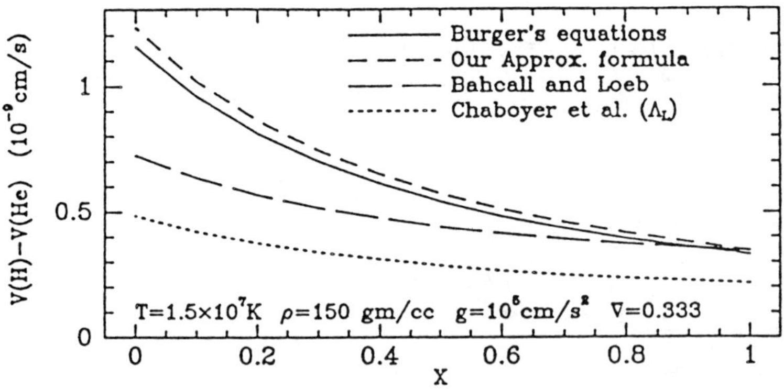

Fig. 4: Relative diffusion velocities of H and He as a function of X.

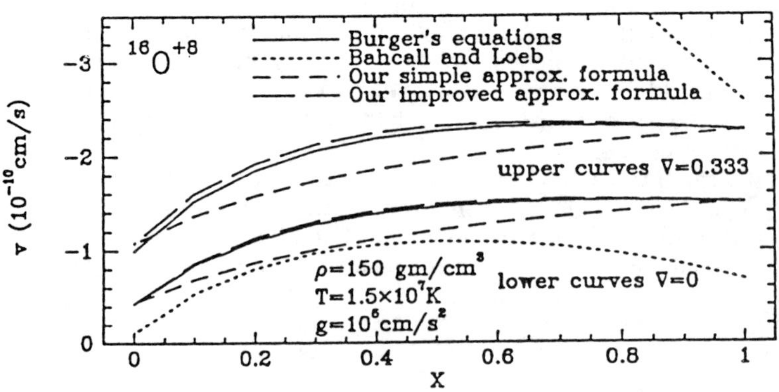

Fig. 5: Diffusion velocity for $^{16}O^{+8}$ in a H-He mixture.

full system of equations (1) for pure H or pure He, it significantly underestimates the velocity for intermediate values. Comparison to the full system of Burgers' equations reveals that the discrepancy comes from ignoring the heat flux terms in deriving equation (18). If equation (1) is solved while artificially setting z_{ij} to 0, then the results agree well with the values calculated by formula (18). This extra drag can be reasonably approximated by adding $\sim$-.23XV_X to the velocity given above.

For conditions characteristic for the interiors of main sequence stars, we find that the thermal diffusion velocity can be approximated (within 20%) by adding the empirically determined quantity,

$$V_{therm} = \frac{0.54B(4.75X + 2.25)T^{5/2}}{\rho(\ln \Lambda_{xy} + 5)} \nabla \frac{d \ln P}{dr} \qquad (19)$$

to the total diffusion velocity for heavy ions.

Figure 5 shows the V for ^{16}O as a function of X computed in a variety of ways. The *improved* formula curve includes the correction term -.23XV_X discussed above. Even in the absence of a temperature gradient, the approximations of B&L underestimate diffusion velocities by a factor of up to three, especially in pure H and pure He. The formula given by B&L for the thermal diffusion velocity of heavy elements is proportional to 1/X.

III- EVOLUTIONARY RESULTS IN POPULATION I STARS.

1. The Sun

A number of authors have estimated gravitational settling in the Sun (Aller and Chapman 1960, Noerdlinger 1977, Wambsganss 1988, Cox *et al.* 1989, Proffitt & Michaud 1991b, Bahcall & Pinsonneault 1992). Disagreements have been analyzed by Proffitt and Michaud (1991b) and it appears that, if diffusion is not inhibited by mixing below the convection zone, the current surface mass fraction helium abundance of the Sun, $Y_\odot$, should be about 0.03 (10%) lower than the initial solar helium abundance. Models with diffusion have slightly higher central densities (0.5%) and temperatures (2%). A slightly larger mixing length is also needed. Predicted neutrino fluxes are slightly increased.

The assumption that there is no mixing below the base of the surface convection zone is probably incorrect, as the current solar Li abundance is 100-200 times smaller than the initial value and it appears difficult to account for all of the depletion by pre-main sequence burning. While the cause and nature of main sequence mixing is controversial, observational data can constrain its magnitude. Figure 2 of Proffitt and Michaud (1991b) shows how two possible distributions of turbulent mixing that approximately reproduce the observed Li depletion affect the surface settling of He. It does not seem possible to reduce the surface settling by more than a factor of two without over-depleting Li, but the shape of the interior He profile and the steepness of the composition gradient can be quite sensitive to the details.

Observations of solar p-modes provide a means for directly measuring the He abundance of the surface convection zone and the sound speed as a function of radius (see the discussion in Dziembowski *et al.* 1992). These measurements are potentially sensitive enough to distinguish between the different profiles shown in Figure 2 of Proffitt and Michaud (1991b) and suggest a surface He abundance consistent with the diffusion predictions noted above (see also Guzik & Cox 1992). Once current uncertainties in other physics (opacities, abundances, and convection theory) are resolved, the oscillations should eventually provide a direct test of turbulent mixing in G dwarfs.

2. The Li Gap

The observation by Boesgaard ant Tripicco (1986) of a well defined gap in the Li abundance curve as a function T_{eff} in the Hyades (see their Fig. 2; see also Fig. 2.1 of Michaud and Charbonneau 1991) is suggestive of a competition between gravity and g_R(Li) as the convection zone becomes deeper with decreasing T_{eff} (Michaud 1986).

The T_{eff} interval covered by the Li gap stars in the Hyades is within a few hundred degrees of the region where the mass in the convection zone increases rapidly. The agreement between the observed and calculated T_{eff} of the deepest part of the gap depends on the value of α used for the models as well as on the opacity tables. Helium settling occurs at the same time as Li settling since the two diffusion velocities are nearly equal (Michaud 1986). To make precise comparisons between the observed Li gap in clusters and abundances to be expected from standard stellar evolution, it is necessary to take into account the settling of He (Richer *et al.* 1992).

In stars of $1.0 < M < 1.5\ M_\odot$, the evolutionary time scale varies from 10^{10} to 10^9 yr while the diffusion time scale within the core is about 10^{11} yr. While, in the Sun, gravitational settling has increased the He abundance in the He burning region by 5 % of the original value (Cox, Guzik and Kidman 1989, Noerdlinger 1977, Proffitt and Michaud 1991), the increase is negligible in $1.5\ M_\odot$ stars during the main sequence. The effects of evolution come from abundance changes in the centre, while diffusion causes abundance changes mainly in the envelope for stars of $M > 1.4\ M_\odot$. This allowed Richer *et al.* to calculate the effects of diffusion only in the envelope and use the evolutionary calculations without diffusion for the centre of such stars.

Using various opacity tables and detailed calculations of the radiative acceleration of Li, Richer and Michaud (1992) also calculated the Li gap as a function of cluster age. They calibrated the value of α using the Sun (see § III-1).

It is then found that: a) the position of the Li gap is nearly exactly the same for all opacity tables; b) the position of the Li gap is at T_{eff} = 6740 K, in agreement with observations; c) the depth of the Li gap varies with opacity tables. It is by a factor of 100 at the age of the Hyades using the OPAL tables but is only by a factor of 10 using the Los Alamos tables; d) the width of the gap is too narrow (about 150 instead of 250 K). However the very abrupt blue side of the gap is explained; e) as clusters become older the position of the gap shifts to slightly lower T_{eff}; f) for stars with T_{eff} > 7000 K, there is a need for an alternate hydrodynamical process competing with diffusion. This is the AmFm domain where there is a need for such a process to reduce the anomalies that uninhibited diffusion leads to.

3. A and B type stars

The evolutionary time scale for B and A stars vary from 10^8 to 10^9 yr while the diffusion time scale in the centre is around 10^{11} yr so that the effects of diffusion in the centre are small enough to be neglected. The effects are much larger in the external regions. Since the convection zone is of order $10^{-10}\ M_\odot$ and the helium settling time scale, Θ, varies approximately as (Michaud 1977) $2.3\ 10^{11}\ \Delta M^{0.545}$ yr, where ΔM is the mass in the convection zone in $M_\odot$, the time scale for the disappearance of the He convection zone is of order of 10^6 yr, much shorter than the evolutionary time scale. The diffusion time scales become even shorter once the He convection zone has disappeared. Since the largest observed abundance anomalies are observed in those stars, Michaud (1970) suggested that radiation driven diffusion was responsible for abundance anomalies observed in the ApBp stars.

The simplest of these stars to model are the AmFm stars. They are cool enough to have a H convection zone and they have no large organized magnetic field so that chemical separation occurs only below the convection zone. Immediately below the hydrogen convection zone, Ca and Sc are in a rare gas configuration so that the radiative acceleration on them is smaller than gravity and they sink as is observed (Watson 1971; Smith 1973; Cayrel Burkhart and Van't Veer 1991). This property is naturally explained by the parameter free evolutionary calculations, since these lead to the disappearance of the He convection zone because of gravitational settling. This also explains the absence of δ Scuti pulsators among the classical Am stars (Baglin 1972).

One may now also test the parameter free evolutionary calculations by comparing the T_{eff} where the AmFm phenomenon occurs with theoretical calculations. Then the effect of He settling on the T_{eff} and on the size of the convection zone (Vauclair, Vauclair and Pamjatnikh 1974), as well as evolutionary effects have to be taken into account (Richer *et al.* 1992). The red edge of AmFm phenomenon should be determined theoretically within at most 100 K. Calculations are in progress (Richer and Michaud 1992a).

At the same time as the Ca and Sc underabundances are explained, overabundances of most elements heavier than Mg are expected and observed (see Michaud *et al.* 1976). The model correctly predicts which elements are overabundant and which are underabundant, with few exceptions. The expected anomalies are however larger than observed unless a competing hydrodynamical process is present. Turbulence cannot be the only process, since it wipes out the Sc and Ca anomalies before reducing the overabundances (Vauclair Vauclair and Michaud 1978). Mass loss, of the order of 10^{-15} $M_\odot$ yr^{-1}, appears to be compatible with both (Michaud *et al.* 1983) though it remains to be seen if it can be compatible with the Li observations in clusters (Richer *et al.* 1992).

IV- DIFFUSION IN POP. II DWARFS

Pop. II stars are older than the Sun, and many have thinner surface convection zones, so that the effects of diffusion are larger than in Pop. I stars (Noerdlinger and Arigo 1980, Stringfellow *et al.* 1983). The central settling of He causes core H exhaustion to occur earlier and at a lower luminosity, while surface settling causes an increase in the stellar radius and a decrease in the T_{eff} relative to non-diffusion models. The total shift in the turnoff T_{eff} between diffusion and non-diffusion models is typically about 200 K. Stringfellow *et al.* concluded that this resulted in a 14-25% decrease in the ages of globular clusters. Proffitt and VandenBerg (1991) using Paquette et al.'s diffusion rates concluded that H-He diffusion results in a 10% decrease in the ages of globular clusters as measured by the luminosity of the turn-off (see also Chaboyer *et al.* 1992). The ^{7}Li is expected to settle at about the same speed as ^{4}He, and can be used to constrain surface settling and mixing. Spite and Spite (1982) discovered that metal poor halo stars with [Fe/H]$<$-1.3 and Teff $>$ 5500 K all have essentially the same Li abundance, with very little star to star scatter, and no obvious trend with T_{eff} or metallicity (Deliyannis *et al.* suggest an approximately 10% intrinsic scatter). However, models that include gravitational settling without any extra-mixing predict a noticeable decrease in the surface Li abundance towards the hot end of the plateau, due to gravitational settling of He and Li from the thin convection zones of these stars (see Proffitt and Michaud 1991a). To some extent this decrease towards high T_{eff} can be cancelled by increased pre-main-sequence depletion at cooler T_{eff} but the models still seem to predict an unobserved Pop. II analogue of the Pop. I F-star Li gap (but see G186-26 with Li down by at least 0.9 dex, Hobbs *et al.* 1991). Either (1) uncertainties, perhaps due to reddening, in the T_{eff} of the stars at the hot end of the halo sequence erroneously extend it, (2) the Paquette *et al.* diffusion calculations significantly overestimate atomic diffusion, (3) these models of Proffitt and Michaud significantly underestimate the thickness of the surface convection zone for the warmest halo dwarfs, or (4) some kind of mixing below the surface convection zone moderates the amount of gravitational settling.

We have already discussed why we believe that the Paquette et al. calculations are currently the best available.

The thickness of the surface convection zones is, however, uncertain for both theoretical and observational reasons. The expected mass of the surface convection zones in very metal poor dwarfs changes from about 10^{-2} $M_\odot$ near 6000 K to $< 10^{-4}$ $M_\odot$ at 6400 K, but the exact values are very sensitive to uncertain details of convection theory, opacities, assumed mixing length, and colour-T_{eff} conversions, including potential reddening corrections.

Calculating detailed interaction between gravitational settling and turbulence below the convection zone requires a model of turbulence. Pinsonneault, *et al.* (1992) have proposed a model for turbulent mixing in halo stars but they have not yet simultaneously considered the interaction between this turbulence and gravitational settling. Some constraints can be studied by simply enforcing differing minimum masses for the fully mixed surface region. We have constructed a number of models of 0.75 $M_\odot$ halo stars, in which we have set various minimum masses for the fully mixed surface region (i.e. we assumed a very large turbulent diffusion coefficient in the outer n% of the mass but allowed diffusion to be uninhibited below this). Eliminating a noticeable Li dip at the hot end of the Spite plateau requires keeping the outer 1-2% of the stellar mass well mixed, while forcing the outer 5% of the mass to always be well mixed will destroy 90% of the Li via nuclear burning. However, regardless of how the depth of the mixed surface zone is chosen, if the Paquette et al. diffusion rates are used, the combination of gravitational settling and nuclear burning ensures that the surface Li abundance at 15 Gyr will be AT MOST 70% of the ZAMS value. We do not believe that any distribution of turbulent mixing or overshooting would allow a higher Li abundance at this age. The reduction in surface settling caused by this assumed extra-mixing does not completely eliminate the larger radii and shift towards smaller T_{eff} that the He settling causes. Even when the outer 5% of the mass is thoroughly mixed, the difference in turn-off T_{eff} between diffusion and non-diffusion models is only reduced by half.

REFERENCES

Alecian, G., and Artru, M. -C., 1990, A&A, 234, 323.
Alecian, G., Michaud, G., and Tully, J. 1992, preprint.
Aller, L. H., and Chapman, S. 1960, ApJ, 132, 461.
Baglin, A. 1972, A&A, 19, 45.
Bahcall, J. N., and Loeb, A. 1990, ApJ, 360, 267.
Bahcall J. N., & Pinsonneault, M. H. 1991, ApJ, submitted
Boesgaard, A. M., and Tripicco, M. J. 1986, ApJL, 302, L49.
Burgers, J. M. 1969, Flow Equations for Composite Gases, (New York: Academic Press).
Cayrel, R., Burkhart, C., and Van't Veer, C. 1991, in IAU Symposium 145, **Evolution of Stars: the Photospheric Abundance Connection**, Golden Sands,

Bulgaria, 25-31 august, ed. G. Michaud and A. Tutukov (Kluwer), pp. 99-110.
Chaboyer, B., Deliyannis, C. P., Demarque, P., Pinsonneault, M. H., & Sarajedini, A. 1992, ApJ, 388, 372
Chapman, S., and Cowling, T. G. 1970, The Mathematical Theory of non-uniform Gases (3d ed.; Cambridge: Cambridge University Press).
Charbonneau, P., and Michaud, G. 1988, ApJ, 327, 809.
Charbonneau, P., and Michaud, G. 1991, ApJ, 370, 693.
Cox, A. N., Guzik, J. A., and Kidman, R. B. 1989, ApJ, 342, 1187.
Deliyannis, C. P., Pinsonneault, M. H. & Duncan, D. K. 1992, ApJ, submitted
Deliyannis, C. P., Pinsonneault, M. H. & Demarque, P. 1992, ApJS, 78,
DeWitt, H. E. 1961, J. Nucl. Energy C, Plasma Physics, 2, 27.
Dziembowski, W. A., Pamyatnykh, A. A., and Sienkeiwicz, R. 1992, preprint
Eddington, A. S. 1926, The Internal Constitution of the Stars (New York: Dover [1959] reprint), § 199.
Fontaine et al 1977.
Guzik, J. A., & Cox, A. N. 1992, ApJL, 386, 729.
Hobbs, L. M., Welty, D. E. & Thornburn, J. A. 1991, ApJ 373, L47
Kurucz, R. L. 1991, in Stellar Atmospheres: Beyond Classical Models, ed. by L. Crivellari, I. Hubeny and D. G. Hummer, NATO ASI Series (Dordrecht, Kluwer), p. 440.
Michaud, G. 1970, ApJ, 160, 641.
Michaud, G. 1977, Nature, 266, 433.
Michaud, G. 1982, ApJ, 258, 349.
Michaud, G. 1986, ApJ, 302, 650.
Michaud, G. 1987, Physica Scripta, 36, 112.
Michaud, G. 1991, Annales de Physique Fr., 16, 481.
Michaud, G., and Charbonneau, P. 1991, Space Science Reviews, 57, 1.
Michaud, G., Charland, Y., Vauclair, S., and Vauclair, G. 1976, ApJ, 210, 447.
Michaud, G., Tarasick, D., Charland, Y., and Pelletier, C. 1983, ApJ, 269, 239.
Noerdlinger, P. D. 1977, A&A, 57, 407.
Noerdlinger, P. D. & Arigo, R. J. 1980, ApJL, 237, L15
Paquette, C., Pelletier, C., Fontaine, G., and Michaud, G. 1986, ApJS, 61, 177.
Proffitt, C. R. & Michaud, G. 1991a, ApJ, 371, 584
Proffitt, C. R. & Michaud, G. 1991b, ApJ, 380, 238
Proffitt, C. R. & VandenBerg, D. A. 1991, ApJS, 77, 473
Richer, J., and Michaud, G. 1992, in preparation.
Richer, J., Michaud, G., and Proffitt, C. 1992, ApJS, in press.
Rogers, F. J., and Iglesias, C. A. 1992, ApJS, 79, 507.
Roussel-Dupré, R. 1981, ApJ, 243, 329.
Roussel-Dupré, R. 1982, ApJ, 252, 393.
Smith, M. A. 1973, ApJS, 25, 277.
Spite, F. & Spite, M. 1982, A&A, 115, 357
Stringfellow, G. S., Bodenheimer, P., Noerdlinger, P. D. & Arigo, R. J. 1983, ApJ, 264, 228
Vauclair, G., Vauclair, S., and Michaud, G. 1978, ApJ, 223, 920.
Vauclair, G., Vauclair, S., and Pamjatnikh, A. 1974, A&A, 31, 63.
Wambsganss, J. 1988, A&A, 205, 125.
Watson, W. D. 1971, A&A, 13, 263.

Inside the Stars, IAU Colloquium 137
ASP Conference Series, Vol. 40, 1993
Werner W. Weiss and Annie Baglin (eds.)

LINE BROADENING DATA: STARK BROADENING OF Ca II Sc III AND Ti IV LINES

MILAN S. DIMITRIJEVIĆ
Astronomical Observatory, Volgina 7, 11050 Beograd, Yugoslavia

SYLVIE SAHAL-BRÉCHOT
Observatoire de Paris-Meudon, 92195 Meudon CEDEX, France

ABSTRACT Electron- and proton-impact line widths and shifts for important Ca II, Sc III and Ti IV lines, have been calculated using the semiclassical-perturbation formalism. The obtained results were used to investigate the behaviour of Stark broadening parameters within the K I isoelectronic sequence.

INTRODUCTION

Stark broadening of spectral lines has been taking a new interest in astrophysics (Seaton, 1987), owing to the recent development of researches on the physics of stellar interiors: in subphotospheric layers, the modellisation of energy transport needs the knowledge of radiative opacities and thus, certain atomic processes must be known with accuracy.In order to provide a method for quick interpolation of new data along an isoelectronic sequence it is of interest to investigate if a sufficiently regular behaviour of Stark broadening parameters along such a sequence exists. Consequently, one of aims of this paper is to provide Stark broadening data for a number of transitions within several members of an isoelectronic sequence.

The present paper concerns Ca II, Sc III and Ti IV lines from the kalium isoelectronic sequence. Beyond the interest for the stellar atmospheres investigation and the modellisation of stellar interiors, the knowledge of Ca II, Sc III and Ti broadening parameters is important for a number of problems in astrophysics and plasma physics. Particularly is important Ca II which is among the most abundant elements in stellar plasma after hydrogen and helium.

RESULTS AND DISCUSSION

In order to provide reliable data for the mentioned lines broadened by collisions with all important charged perturbers in stellar plasmas, we have calculated electron-, proton-, and ionized helium-impact line widths and shifts for 28 Ca II (Dimitrijević et al, 1992ab), 10 Sc III and 10 Ti IV

TABLE I This table shows electron- and proton-impact broadening parameters for Ca II, Sc III and Ti IV lines, for perturber density of 10^{17} cm^{-3} as a function of temperature. Transitions and averaged wavelengths for the multiplet (in Å) are also given. By deviding c with electron-impact WIDTH we obtain an estimate for the maximum perturber density for which the line may be treated as isolated and tabulated data may be used. The asterisk identifies cases for which the collision volume multiplied by the perturber density (the condition for validity of the impact approximation) lies between 0.1 and 0.5.

PERTURBER DENSITY = 0.1D+18(cm-3)

		PERTURBERS ARE: ELECTRONS		PROTONS	
TRANSITION	T(K)	WIDTH(A)	SHIFT(A)	WIDTH(A)	SHIFT(A)
CA II 3D-4P	5000.	1.34	0.595E-01	0.397E-01	-0.642E-02
8581.1 A	10000.	1.05	0.308E-01	0.666E-01	-0.129E-01
C= 0.86E+21	20000.	0.856	0.228E-01	0.927E-01	-0.209E-01
	30000.	0.776	0.913E-02	0.101	-0.255E-01
	50000.	0.707	0.736E-02	0.112	-0.309E-01
	100000.	0.643	0.792E-02	0.125	-0.374E-01
CA II 3D-5P	5000.	0.303	-0.363E-01	0.238E-01	-0.737E-02
2132.3 A	10000.	0.241	-0.220E-01	0.308E-01	-0.115E-01
C= 0.17E+20	20000.	0.207	-0.224E-01	0.359E-01	-0.152E-01
	30000.	0.196	-0.182E-01	0.391E-01	-0.170E-01
	50000.	0.189	-0.146E-01	0.410E-01	-0.198E-01
	100000.	0.183	-0.132E-01	0.431E-01	-0.225E-01
CA II 3D-6P	5000.	0.413	-0.129		
1644.1 A	10000.	0.358	-0.988E-01		
C= 0.48E+19	20000.	0.333	-0.783E-01	*0.730E-01	-0.374E-01
	30000.	0.328	-0.634E-01	*0.762E-01	-0.415E-01
	50000.	0.332	-0.552E-01	*0.806E-01	-0.477E-01
	100000.	0.334	-0.436E-01	*0.885E-01	-0.558E-01
SCIII 3D-4P	20000.	0.278E-01	0.836E-03	0.105E-02	0.153E-03
1605.1 A	50000.	0.183E-01	0.894E-03	0.184E-02	0.331E-03
C= 0.95E+20	100000.	0.140E-01	0.785E-03	0.227E-02	0.477E-03
	200000.	0.113E-01	0.869E-03	0.257E-02	0.613E-03

TIIV 3D-4P	20000.	0.596E-02	0.758E-04	0.787E-04	0.128E-04
777.8 A	50000.	0.384E-02	0.913E-04	0.198E-03	0.326E-04
C= 0.29E+20	100000.	0.281E-02	0.912E-04	0.295E-03	0.563E-04
	200000.	0.215E-02	0.109E-03	0.373E-03	0.816E-04
TIIV 3D-5P	20000.	0.473E-02	0.553E-04	0.236E-03	0.127E-04
433.7 A	50000.	0.335E-02	0.891E-04	0.394E-03	0.298E-04
C= 0.35E+19	100000.	0.270E-02	0.891E-04	0.467E-03	0.457E-04
	200000.	0.227E-02	0.984E-04	0.525E-03	0.627E-04

multiplets (Dimitrijević and Sahal-Bréchot, 1992c), using the semiclassical-perturbation formalism (Sahal-Bréchot, 1969ab). This is a part of an effort to provide reliable Stark broadening data for stellar plasma research (see the review on up to now performed calculations for He I, Na I, K I, F I, Be II, Mg II, Ca II, Sr II, Ba II, Si II, Ar II, Ga II, Ga III and several lines of other light elements, in Dimitrijević and Sahal-Bréchot, 1991).

As an example, in Table I are presented results for some important Ca II, Sc III and Ti IV lines. The obtained results were used also to investigate the behaviour of Stark broadening parameters within the isoelectronic sequence in order to examine the use of such behaviour for the interpolation of new data of interest for the stellar plasma investigations. Our analysis shows that a regular behaviour exist but the mutual relation of the corresponding Stark broadening parameters depends on temperature. Additional experimental and theoretical work for the investigated case is needed as well as the extension to the other members of K isoelectronic sequence.

REFERENCES

Dimitrijević,M.S., and Sahal-Bréchot,S., 1991, *J.Phys.*IV **1**, C1-111.

Dimitrijević,M.S., and Sahal-Bréchot,S., 1992a, *JQSRT*, in press.

Dimitrijević,M.S., and Sahal-Bréchot,S., 1992b, *Bull. Astron. Belgrade* **145**, 83.

Dimitrijević,M.S., and Sahal-Bréchot,S., 1992c, *Astr. Ap. Suppl.*, in press.

Sahal-Bréchot,S., 1969a, *Astr. Ap.*, **1**, 91.

Sahal-Bréchot,S., 1969b, Astr. Ap., **2**, 322.

Seaton,M.J., 1987, *J.Phys.B*, **20**, 6363.

Inside the Stars, IAU Colloquium 137
ASP Conference Series, Vol. 40, 1993
Werner W. Weiss and Annie Baglin (eds.)

THE IMPORTANCE OF MOLECULAR OPACITIES IN STELLAR ATMOSPHERES

CHRISTOPER M. SHARP
Service P.T.N., Centre d'Etudes de Bruyères-le-Châtel, B.P. 12, F-91680 Bruyères-le-Châtel, France

INTRODUCTION

Opacities are of vital importance in astrophysics necessary to model stellar atmospheres and interiors, and below about 6000 K diatomic molecules make an increasing contribution to that opacity. Moreover, at around 3000 K polyatomic molecules, such as H_2O for an oxygen rich solar composition, start playing an important rôle, and additionally at high pressures collision induced $H_2 - H_2$ and $H_2 - He$ absorption is important for late-type main sequence stars.

In order to investigate the importance of molecular opacities, in Sharp (1992) we computed and compared opacities for a solar and a three times enhanced CNO abundance, and some further comparisons were made in Sharp et al. (1992). Some of these results are briefly summarised here.

RESULTS OF CALCULATIONS

In table 1, reproduced from Sharp (1992), Planck and Rosseland mean opacities are given for 104 temperature-density points for temperatures between $kT = 0.2$ and $1.0eV$, and densities between 10^{-10} and $10^{-3} gm\ cm^{-3}$ for a Cameron (1973) solar composition with a turbulent broadening of $1Km\ s^{-1}$ in all cases. At each temperature-density point, atomic and molecular equilibria were first obtained by iterating the Saha equations of atomic ionization and molecular dissociation, then the continuous absorption on a coarse grid of a few thousand points was calculated due to scattering, free-free, bound-free and collision induced absorption. To this was added a fine grid of several hundred-thousand points on which molecular bands were calculated on an individual line-by-line bases. Diatomic molecules were calculated using the Morse oscillator approximation (Doktorov et al. 1976, Sharp, 1984 and 1988), with the Hönl-London factors for the line strengths obtained from Kovács (1969), and CO_2 and H_2O were calculated using the harmonic oscillator approximation, (Auman 1966). Atomic lines were, however, neglected. Although molecules made a negligible contribution at the higher temperatures calculated, these points were included for continuity and to permit comparisons with other calculations.

In the equilibria mentioned above, calculations were performed for the 22 most abundant elements in the mixture with their associated ions, 38 diatomic molecules and 12 triatomic molecules, of which H_2O was by far the most important of the triatomics at the lowest temperatures. A subdset of these molecules was included in the opacity calculations, with the most important contribu-

Table 1. Planck (upper) and Rosseland (lower) mean opacities in cm^2/gm for a solar mixture.

kT (eV)	0.2	0.25	0.3	0.4	0.5	0.6	0.8	1.0
T (K)	2320	2900	3480	4640	5800	6960	9280	11600
ρ (gm/cc)								
1(-10)	1.725(-1)	8.009(-2)	4.668(-2)	3.468(-2)	1.796(-2)	2.436(-1)	3.678(1)	2.802(1)
	4.535(-4)	5.424(-5)	2.654(-4)	1.126(-3)	1.180(-2)	1.323(-1)	3.665(0)	2.756(0)
1(-9)	3.681(-1)	8.263(-2)	5.258(-2)	6.987(-2)	5.208(-2)	3.028(-1)	8.685(1)	2.180(2)
	1.428(-3)	3.311(-4)	7.743(-4)	3.537(-3)	2.633(-2)	1.951(-1)	8.378(0)	1.813(1)
1(-8)	4.753(-1)	1.075(-1)	5.822(-2)	9.806(-2)	1.684(-1)	5.176(-1)	1.184(2)	8.975(2)
	2.345(-3)	4.369(-3)	2.936(-3)	1.795(-2)	7.759(-2)	4.138(-1)	1.289(1)	7.447(1)
3(-8)	4.913(-1)	1.571(-1)	6.713(-2)	1.188(-1)	2.920(-1)	7.594(-1)	1.273(2)	1.286(3)
	2.720(-3)	1.042(-2)	6.257(-3)	4.016(-2)	1.370(-1)	6.299(-1)	1.551(1)	1.093(2)
1(-7)	5.001(-1)	2.399(-1)	9.090(-2)	1.722(-1)	5.289(-1)	1.250(0)	1.357(2)	1.630(3)
	3.150(-3)	1.945(-2)	1.626(-2)	9.426(-2)	2.665(-1)	1.033(0)	1.953(1)	1.456(2)
3(-7)	5.045(-1)	3.033(-1)	1.402(-1)	2.932(-1)	9.110(-1)	2.084(0)	1.444(2)	1.858(3)
	3.632(-3)	2.822(-2)	4.021(-2)	2.003(-1)	5.128(-1)	1.685(0)	2.507(1)	1.788(2)
1(-6)	5.079(-1)	3.478(-1)	2.436(-1)	5.938(-1)	1.680(0)	3.808(0)	1.584(2)	2.047(3)
	4.286(-3)	3.745(-2)	9.168(-2)	4.430(-1)	1.104(0)	3.009(0)	3.462(1)	2.225(2)
3(-6)	5.112(-1)	3.777(-1)	3.826(-1)	1.188(0)	3.118(0)	6.908(0)	1.804(2)	2.203(3)
	5.208(-3)	4.666(-2)	1.588(-2)	8.963(-1)	2.327(0)	5.400(0)	4.913(1)	2.800(2)
1(-5)	5.165(-1)	4.148(-1)	5.911(-1)	2.560(0)	6.659(0)	1.402(1)	2.254(2)	2.415(3)
	7.150(-3)	6.144(-2)	2.590(-1)	1.897(0)	5.412(0)	1.095(1)	7.745(1)	3.838(2)
3(-5)	5.275(-1)	4.920(-1)	1.003(0)	5.838(0)	1.533(1)	2.958(1)	3.112(2)	2.743(3)
	1.380(-2)	1.095(-1)	4.906(-1)	4.394(0)	1.282(1)	2.314(1)	1.299(2)	5.618(2)
1(-4)	5.835(-1)	8.217(-1)	2.063(0)	1.500(1)	3.942(1)	6.692(1)	4.928(2)	3.391(3)
	4.472(-2)	3.221(-1)	1.146(0)	1.165(1)	3.283(1)	5.246(1)	2.393(2)	9.188(2)
3(-4)	8.326(-1)	1.869(0)	5.483(0)	4.319(1)	1.030(2)	1.576(2)	8.724(2)	4.666(3)
	1.391(-1)	9.382(-1)	3.592(0)	3.538(1)	8.421(1)	1.220(2)	4.668(2)	1.620(3)
1(-3)	1.135(0)	2.920(0)	8.782(0)	5.865(1)	2.139(2)	6.874(2)	2.105(3)	8.350(3)
	1.982(-1)	1.358(0)	5.471(0)	4.668(1)	1.676(2)	4.269(2)	1.192(3)	3.631(3)

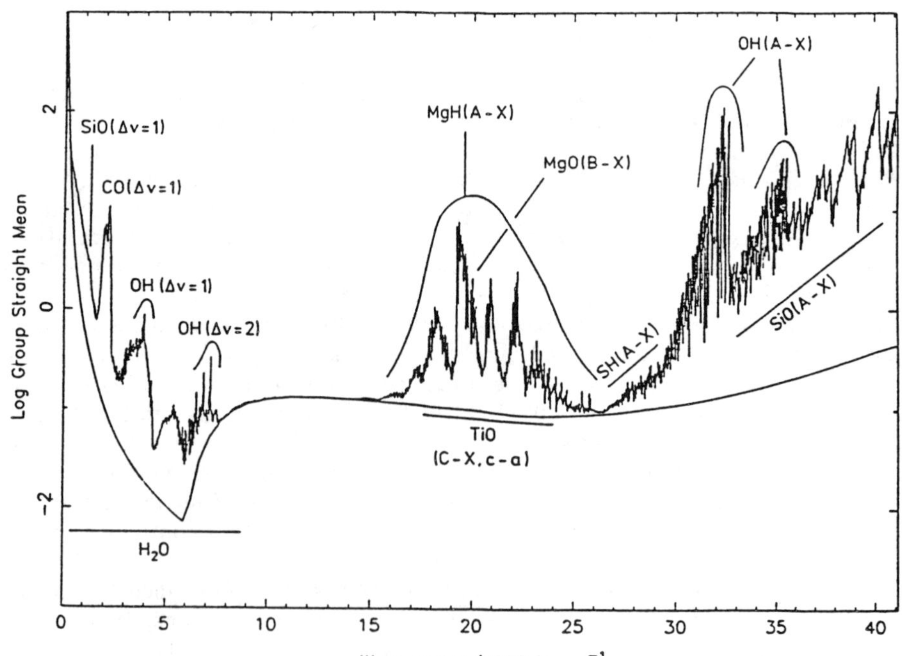

Figure 1. Plot of the group straight mean absorption coefficient at $kT = 0.3eV$, $\rho = 10^{-6} gm\ cm^{-3}$ and a turbulent velocity of $1Km\ s^{-1}$ for a solar composition. The absorption due to molecular bands plus continuum is shown superimposed on the continuum alone.

tors at a representative temperature-density point being indicated in figure 1 from Sharp (1992). In this figure for $kT = 0.3eV$ and a density of $10^{-6}gm^{-3}$, the group straight mean absorption coefficient over 3400 intervals in cm^2gm^{-1}, where the mean from the fine grid in each interval is represented, is plotted against the photon energy in units of 1000 cm^{-1}, with the lower curve being the continuum alone, (the minimum is at the H^-_{bf} absorption threshold). The vibration-rotation bands in the infrared, and the electronic systems in the visible and ultraviolet regions are identified.

At this temperature the maximum in the Rosseland weighting function is not far above in frequency from the minimum of the H^- opacity, so the contribution molecular bands make in this region is particularly important. At lower temperatures H_2O completely dominates this region and additionally the maximum of the weighting function is closer, resulting in H_2O increasing the Rosseland mean opacity by about a factor of 100 at the lowest temperatures tabulated. As discussed in Sharp (1992) and Sharp et al. (1992), a number of diatomic molecules and their ions, at least for CO^+, appear to be still important at $kT = 0.6eV$, however a number of important atomic processes were neglected.

CONCLUSIONS

Some extensive molecular opacity calculations discussed in Sharp (1992) are briefly summarised here. Scattering and some important atomic and molecular continuous opacity sources were included to obtain a background continuum, then molecular bands were calculated on a detailed line-by-line basis. As expected, it was found that molecular bands can make a very substantial contribution to the Rosseland mean opacity, and the graphs, one of which is included here, are particularly instructive in showing the importance of molecular bands.

However, we have had to make a number of important approximations in neglecting several atomic processes, calculating line strengths and assuming Gaussian profiles, together with using some relatively old data. Some of these problems are currently being addressed, and it is hoped that these improvements will be incorporated into later work.

REFERENCES

Auman Jr., J. 1966, *ApJS*, 14, 171.

Cameron, A.G.W. 1973, Space Science Rev., 15, 121.

Doktorov, E.V., Malkin, I.A., & Man'ko, V.I. 1976, J.Q.S.R.T., 16, 1021.

Kovács, I.: 1969, Rotational Structure in the Spectra of Diatomic Molecules, (Adam Hilger Ltd).

Sharp, C.M. 1984, *A&AS*, 55, 33.

Sharp, C.M. 1988, *A&AS*, 72, 355.

Sharp, C.M. 1992, *A&AS*, in press.

Sharp, C.M., Lebreton, Y., & Baglin, A. 1992, Comptes Rendus de la Academie des Sciences, in press.

Inside the Stars, IAU Colloquium 137
ASP Conference Series, Vol. 40, 1993
Werner W. Weiss and Annie Baglin (eds.)

CALCULATION OF IONIZATION EQUILIBRIUM FOR DENSE PLASMAS

MASSACRIER GERARD
DARC – UPR176 , Observatoire de Paris–Meudon, 92195 Meudon Cedex, France

ABSTRACT We propose a new approach to calculate the populations of the different ionic states and ionization stages that may exist in a plasma in thermodynamic equilibrium. A self-consistent scheme is solved for each ionization stage, which couples the local density of free electrons and the atomic structure of this stage.

INTRODUCTION

The calculation of equilibrium populations of the various species in a plasma is a necessary step to study opacities, radiation transfer or thermodynamic quantities. When the density is so high that the mean ionic volume is of the order of a few Bohr radii, Saha equation breaks down and models must be refined to take into account density effects like pressure ionization. Current models include free energy minimization (Graboske *et al.* 1969, Däppen 1980, Däppen *et al.* 1987) and average–atom approaches (Rozsnyai 1975, Feng *et al.* 1981, Davis and Blaha 1982). The first one leaves the bound states essentially unperturbed. Though very useful for a quick analysis of the equilibrium of such complicated systems, the average–atom approach leads to a fictitious ion with a non-integer number of bound electrons and to a one-electron energy spectra not relevant to the actual one. To proceed up to the individual populations of the different ionization stages is further hazardous. The model presented in this paper treats separately each degree of ionization. It has the advantage to lead at the same time to the equilibrium populations and to spectroscopic energies.

A MODEL PARTICULARIZING THE IONIZATION STAGES

The equilibrium of a pure plasma of a given element with nuclear charge Z is completely defined by two thermodynamic parameters, e.g. its temperature T and its total electron number density N_{elec}. We use instead of N_{elec} its associated intensive parameter μ, the electronic chemical potential.

The free electron number density N_{free} is a bad thermodynamic parameter for dense plasmas. It is indeed impossible to assign to all the electrons in these plasmas a definite bound or free character, because of those electrons delocalized on a few ionic sites (the hopping electrons, Dharma-Wardana and Perrot 1992). However, looking at a given nucleus, one might find bound ionic structures with N electrons strongly localized around this nucleus. These N–electronic states are similar to those of the isolated ion. The Coulombic interaction between them is very strong and must be solved through a Hamiltonian. On the contrary, their interaction with the other free (delocalized) electrons is weaker and might be treated through an averaged spherically symetric potential (energy) V^N_{free}. We hence obtain for each ionization stage with N bound electrons the energies $E_{N\alpha}$ and the antisymmetrical N–electronic wavefunctions $\Psi_{N\alpha}$ by solving (in a.u. here and after):

$$H_N\Psi_{N\alpha} = \left[\sum_{i=1}^{N}\left(-\frac{\nabla_i^2}{2} - \frac{Z}{r_i} + V^N_{\text{free}}(r_i)\right) + \sum_{i<j}^{N}\frac{1}{r_{ij}}\right]\Psi_{N\alpha} = E_{N\alpha}\Psi_{N\alpha}\,. \quad (1)$$

From the states $\Psi_{N\alpha}$, we get the mean bound electron density n^N_{bound} for this ionization stage by:

$$n^N_{\text{bound}}(\mathbf{r}) = \sum_{\alpha}\frac{e^{-E_{N\alpha}/T}}{Z_N}\sum_{i=1}^{N}\int|\Psi_{N\alpha}(\mathbf{r}_1\ldots\mathbf{r}_{i-1},\mathbf{r},\mathbf{r}_{i+1}\ldots\mathbf{r}_N)|^2\mathrm{d}\mathbf{R}_i \quad (2)$$

where $\mathrm{d}\mathbf{R}_i = \mathrm{d}\mathbf{r}_1\ldots\mathrm{d}\mathbf{r}_{i-1}\mathrm{d}\mathbf{r}_{i+1}\ldots\mathrm{d}\mathbf{r}_N$ and $Z_N = \sum_\alpha \mathrm{e}^{-E_{N\alpha}/T}$ is the partition function. Boltzmann statistics has to be used because we consider a system with a given (small) integer number N of particules. The mean potential due to these bound electrons V^N_{bound} is obtained from Poisson equation: $\Delta V^N_{\text{bound}} = -4\pi n^N_{\text{bound}}$.

V^N_{free} is in the same way connected to the mean free electron density n^N_{free} around the ion by the Poisson equation: $\Delta V^N_{\text{free}} = -4\pi n^N_{\text{free}}$. A free electron is in this region in the potential due to the nucleus, to the bound electrons and to the other free electrons V^N_{free}. We should normally use different bound potentials, one for each bound state $\Psi_{N\alpha}$. We instead use their thermal average V^N_{bound} due to computational constraint. We calculate the density of free electrons through:

$$\left(-\frac{\nabla^2}{2} - \frac{Z}{r} + V^N_{\text{bound}} + V^N_{\text{free}}\right)\psi_{Ni} = \varepsilon_{Ni}\psi_{Ni} \quad (3)$$

$$n^N_{\text{free}}(\mathbf{r}) = \int_0^\infty d\varepsilon\,\frac{1}{1+e^{-\beta(\varepsilon-\mu)}}\sum_i\delta(\varepsilon-\varepsilon_{Ni})|\psi_{Ni}(\mathbf{r})|^2 \quad (4)$$

We use here Fermi-Dirac statistics because the free electron gas is an open system of assumed independent particules. An improvement would be to add exchange and correlation potentials (Dharma-Wardana and Perrot 1982).

This set of equations is solved self-consistently for each ionization stage. From the total potentials $V^N_{\text{tot}} = -Z/r + V^N_{\text{bound}} + V^N_{\text{free}}$ we get the radii of neutrality R_N by $\mathrm{d}V^N_{\text{tot}}/\mathrm{d}r(R_N) = 0$. This null gradient condition allows to match the different ionization stages potentials and spheres in the whole plasma volume. The chemical potential μ defines the reference energy from which $V^N_{\text{tot}}(R_N)$ is positionned (see (3) and (4)). The probability p_N of a given ionization stage is given by:

$$p_N = \mathrm{e}^{\mu N/T}Z_N/\Xi \quad \text{where} \quad \Xi = \sum_N \mathrm{e}^{\mu N/T}Z_N \quad (5)$$

from which all quantities may be deduced: mean volume, ionic density, mean ionization charge, "free" electronic density...

PRACTICAL PROBLEMS AND PHYSICAL DISCUSSION

As written this set of equations is numerically unsolvable because of the great number of levels involved in the bound structure. It can however be shown that the majority of the N-electron wavefunctions $\Psi_{N\alpha}$ solution of H_N are quite unperturbed, so that their energies need not be computed at each step but are simply shifted by the amount $\langle \Psi_{N\alpha} | V^N_{\text{free}} | \Psi_{N\alpha} \rangle$. Moreover the $\Psi_{N\alpha}$ are built using a LS-coupling multiconfigurational scheme from one-electron bound wavefunctions $\varphi_{Ni}(\mathbf{r}) = P^N_{nl}(r) Y^m_l(\vartheta, \varphi)/r$ in such a way that configuration interaction is for most states negligible (Massacrier and Dubau 1990, 1991). As a consequence (2) is summed in a straighter way as a sum of one-electron wave-functions. Concerning Z_N, precise energies are needed in the lower part of the spectrum. This contribution is directly summed retaining the term splitting. The highest states contribute mainly through their huge number. They are totally summed with energies estimated from the one-electron spectrum.

A word must be said on the external boundary conditions imposed on the bound wavefunctions. Tightly bound states are not sensitive to them. The situation becomes much more confuse for the higher states which tend to extend on neighbouring ions. Our simplified point of view consists in solving the Schrödinger equation for φ_{Ni} with two boundary conditions: $P^N_{nl}(R_N) = 0$ and $r\mathrm{d}P^N_{nl}/\mathrm{d}r(R_N) = P^N_{nl}(R_N)$. We interpret the resulting two energies as the lower and upper bounds of an energy band (Rozsnyai 1972). It gives an idea of the sharpness of bound states and becomes noticeable only for those states close to the ionization limit. For bands sitting astride the limit, we decrease the degeneracy of this state by the amount of this band energy still in the bound part of the energy spectrum. Though crude this approach has the further advantage to avoid step–like behaviours of the partition functions, an important problem lenghtly studied (Hummer and Mihalas 1988).

CONCLUSION

Our model totally takes into account the strong interaction of bound electrons in each ionization stage and gives in a coherent way the populations and the energies of bound states. Its flexicibity allows to easily adapt it to treat mixtures or examine given levels. Improvements might be sought to the calculation of the free electron density and effects due to neighbouring ions included in V^N_{free}. Detailed results will be presented in a further paper.

REFERENCES

Däppen, W. 1980, *Astr. Ap.*, **91**, 212

Däppen, W. Anderson, L. and Mihalas, D. 1987, *Ap. J.*, **319**, 195

Davis, J. and Blaha, M. 1982, *J. Quant. Spectros. Radiat. Transfer*,, **27**, 307

Dharma-wardana, M. W. C. and Perrot, F. 1982, *Phys. Rev. A*, **26**, 2096

Dharma-wardana, M. W. C. and Perrot, F. 1992, *Phys. Rev. A*, **45**, 5883

Feng, I. J. Zakowicz, W. and Pratt, R. H. 1981, *Phys. Rev. A*, **23**, 883,

Graboske, H. C. Jr., Harwood, D. J. and Rogers, F. J. 1969, *Phys. Rev.*, **186**, 210

Hummer, D. G. and Mihalas, D. 1988, *Ap. J.*, **331**, 794

Massacrier, G. and Dubau, J. 1991, in *Proc. 4th Int. Workshop on Radiative Properties of Hot Dense Matter*, Sarasota, 1990, Ed. W Goldstein, p. 130

Massacrier, G. and Dubau, J. 1990, *J. Phys. B*, **23**, 2459S

Rozsnyai, B. F. 1972, *Phys. Rev.*, **145**, 1137

Rozsnyai, B. F. 1975, *J. Quant. Spectros. Radiat. Transfer*, **15**, 695

Inside the Stars, IAU Colloquium 137
ASP Conference Series, Vol. 40, 1993
Werner W. Weiss and Annie Baglin (eds.)

COULOMB INTERACTIONS IN THE EQUATION OF STATE

W. STOLZMANN AND T. BLÖCKER
Institut für Theoretische Physik und Sternwarte der Universität Kiel, Olshausenstr. 40, W-2300 Kiel, Germany

INTRODUCTION

Various models are used in order to develop an accurate equation of state (EOS) for stellar matter (see e.g. Saumon and Chabrier 1991). The complexity of physical effects, for instance degeneracy, relativity, Coulomb- and non-Coulombic interactions requires detailed knowledge from the many-particle physics.

For many applications it is desirable to have theories for these effects over a wide range of densities and temperatures. One of the main sources of uncertainties in the EOS models is the exchange-correlation contribution at any degeneracy and relativity. Previous investigations determine thermodynamic quantities of a partially (e.g. Graboske et al. 1969, Kraeft et al. 1986, Däppen et al. 1988) or fully ionized (e.g. Hubbard and DeWitt 1985, Ichimaru et al. 1987) plasma, resp. Interpolation procedures (Magni and Mazzitelli 1979, Fontaine et al. 1977) combine both cases in the EOS but show non-continuous behaviour.

We present an EOS for fully ionized stellar matter which includes non-ideal effects due to exchange-correlation interactions of charged particles at any degeneracy and which is applicable to arbitrary mixtures.

EQUATION OF STATE

For the ion-ion and the ion-electron terms we adopt the Padè formulae of Ebeling (1990). The Padè approximations consider the effect of weak (hydrogen-like) and strong (heavy ions) charge asymmetry. The electron-electron interaction is given by Ebeling et al. (1981) which we have modified at high densities.

Concerning the high density region in which relativistic effects are dominant the standard EOS of Salpeter (1961) is an excellent description at $T = 0$ K. Finite temperature corrections of the high density EOS are given approximately for the exchange pressure by Shaviv and Kovetz (1972). We calculate the electronic exchange pressure in a consistent manner, valid for any degeneracy and relativity. The consideration of these relativistic corrections for the Coulombic part of the EOS closes the gap between the low-to-moderate density (non-relativistic) and the high density (relativistic) region for any temperatures.

RESULTS

The figures show the relative pressure (ratio of total pressure to ideal pressure P^{id}) vs. density along representative isotherms. P^{coul} is the Coulomb contribu-

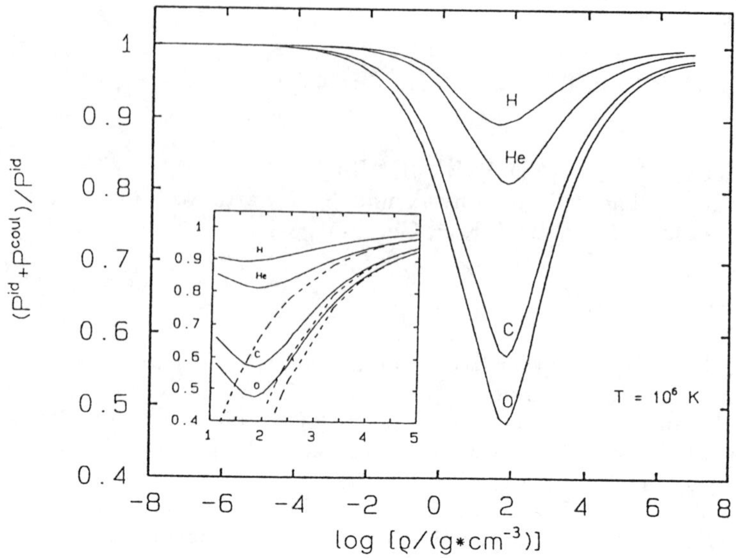

FIGURE I Relative pressure versus density for hydrogen, helium, carbon and oxygen at $T = 10^6\,$K. The density range consider both the non-degenerate limit ($\log \rho < -2$) and high-degenerate (relativistic) limit ($\log \rho > 5$). The inset shows the $T = 0\,$K - EOS of Salpeter (1961) [dashed lines].

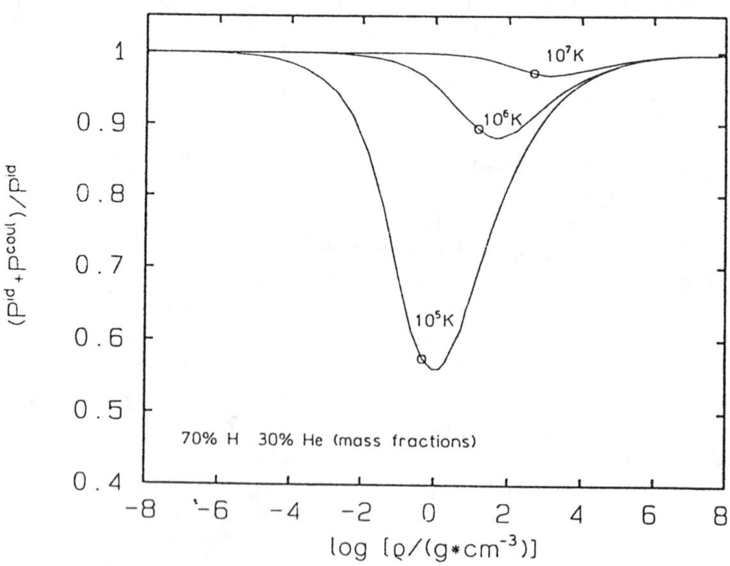

FIGURE II Relative pressure versus density for a H/He mixture of 70/30 by mass ($\approx 90/10$ by particle numbers) for three isotherms of $T = 10^5$, 10^6 and $10^7\,$K. The circles illustrate typical ρ, T-values of a $0.2 M_\odot$ main sequence star according to D'Antona and Mazzitelli (1982).

tion to the pressure, radiation pressure is not included.

The influence of the plasma composition is demonstrated in Fig. I. The strength of the Coulomb interaction increases strongly for heavy ions (e.g. carbon or oxygen). Three isotherms for a typical envelope composition of H/He = 70/30 by mass are presented in Fig. II. With decreasing temperature (at fixed density) the relative pressure decreases, too. Therefore, especially low-mass main sequence stars will be affected in their interiors by Coulomb interactions.

We have introduced a new formulation of the Coulomb interaction in the EOS which is valid in the whole temperature-density regime for fully ionized matter and applicable to arbitrary mixtures. Main features are:

- The major Coulomb contributions for high densities are the ion-ion interaction and the exchange term of the electrons.
- For low densities and moderate to high temperatures all considered contributions are of the same order of magnitude.
- Relativistic corrections influence the exchange term of the electrons significantly which becomes positiv for high densities ($\log \rho > 6$).
- The Coulomb interaction increases strongly for heavy elements.
- The new formulation provides a consistent connection between former low density/high temperature and high density/low temperature theories.

A detailed description of our investigations is given in Stolzmann and Blöcker (1992).

REFERENCES

Däppen, W., Hummer, D. G., Mihalas, D., Weibel-Mihalas, B.: 1988, *ApJ* **332**, 261

D'Antona, F., Mazzitelli, I.: 1982, *A&A* **113**, 303

Ebeling, W.: 1990, *Contr. Plasma Phys.* **30**, 553

Ebeling, W., Richert, W., Kraeft, W. D., Stolzmann, W.: 1981, *phys. stat. sol. (b)*, **104**, 193

Fontaine, G., Graboske, H. C., van Horn, H. M.: 1977, *ApJS* **35**, 293

Graboske, H. C., Harwood Jr., D.J., Rogers, F. J.: 1969, *Phys.Rev.A* **186**, 210

Hubbard, W. B., DeWitt, H. E.: 1985, *ApJ* **290**, 388

Ichimaru, S., Iyetomi, H., Tanaka, S.: 1987, *Phys. Rep.* **149**, 91

Kraeft, W. D., Kremp, D., Ebeling, W., Röpke, G.: 1986, *Quantum statistics of charged particle systems*, Plenum, New York

Magni, G., Mazzitelli, T.: 1979, *A&A* **72**, 134

Salpeter, E. E.: 1961, *ApJ* **134**, 669

Saumon, D., Chabrier, G.: 1991 *Phys.Rev.A* **44**, 5122

Shaviv, G., Kovetz, A.: 1972, *A&A* **16**, 72

Stolzmann, W., Blöcker, T.: 1992, *A&A* , submitted

Inside the Stars, IAU Colloquium 137
ASP Conference Series, Vol. 40, 1993
Werner W. Weiss and Annie Baglin (eds.)

MODELLING TIME-DEPENDENT DIFFUSION IN AM STAR ENVELOPES

G. ALECIAN
DAEC, URA173-CNRS,Observatoire de Paris, Section de Meudon,
F-92195 MEUDON CEDEX, FRANCE

ABSTRACT We study the diffusion process occurring just below the superficial convection zone of Am stars, improving the methods used sofar. We are now able to compute, in a more realistic way, the evolution of the concentrations during the stay of the stars on the main sequence for a large number of elements. This allows to better constrain the different properties (mass loss, depth of the superficial convection zone, transition between convection and radiative zone) entering the modelling of Am stars in the framework of a diffusion-dominant description.

NUMERICAL COMPUTATION OF THE TIME-DEPENDENT DIFFUSION

The diffusion process, in addition to the global mass flow (due to the stellar mass-loss), stratifies elements inside the stars. We have solved the continuity equation for the concentrations below the superficial convection zone using an improved version of the code of Alecian (1986, 1990).

The radiative accelerations are determined through the detailed computation of the number of photons absorbed by the ions of interest. This needs to often deal with a great amount of atomic data (here the data of Kurucz and Peytremann, 1975). To compute these accelerations we have used the method proposed by Alecian & Artru (1990).

STELLAR ENVELOPE MODELS

We have computed the envelope models for two effective temperatures (7400K and 8000K) with log g around 4.0, which correspond to the Am stars' characteristics. We have also decided to impose a low concentration of helium (Y=0.08) since He is supposed to be deficient in these stars (deficiency due to gravitational settling).

There are two important parameters concerning the diffusion process: the rate of mass loss and, the thickness of the superficial convection zone. The rate of mass loss may be considered as a free parameter since there is not any observational data about it for Am stars. On the contrary, the position of the superficial convection zone is not a free parameter. However, on the one hand there is some uncertainty in its determination and, on the other hand, the bottom of the convection zone may be followed by a thin mixing zone (overshooting, internal waves,...).

We have taken into account this uncertainty and we have quantified it through an exponential law with respect to the depth.

TABLE I The different cases (models) which have been considered

N°	T_{eff}	log g	Extension of the superficial CZ	mass-loss (In solar mass per year)
1	8000 K	3.94	no	1. 10^{-14}
2			no	3. 10^{-14}
3			yes	1. 10^{-14}
4	7400 K	4.05	no	1.10^{-14}

DISCUSSION

We have considered three elements: Al, Ca, Sc which present abnormal concentrations in Am stars. The time-dependent diffusion of these 3 elements has been computed (according to the different cases shown in Table I).

Our main conclusions are the following:

(i) - Elements go through successive phases of under/overabundances during the life of the star on the main sequence. We stress on the "long" computation for Ca (Fig. 1) which shows that one second small phase of overabundance is followed by a third phase of underabundance after less than one billion years of diffusion.

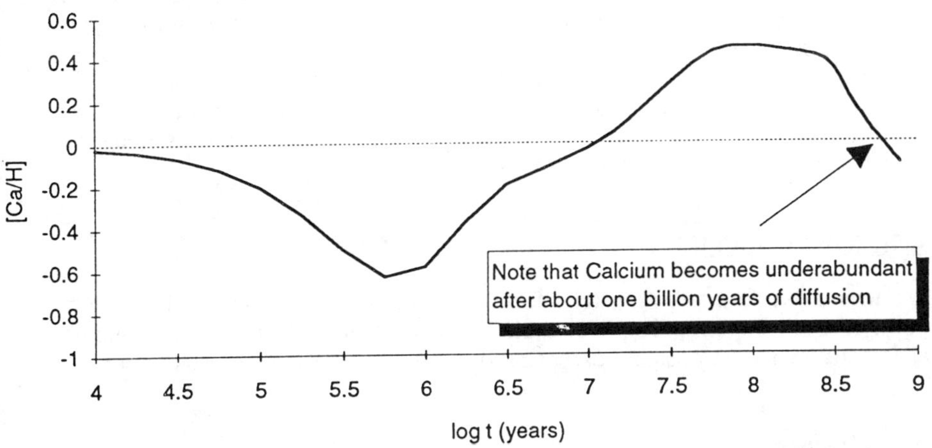

Fig. 1. Calcium aboundance during 10^9 years of diffusion (model 4).

(ii) - Some elements are more sensitive than others on the mass loss rate, depth of the convection zone or effective temperature (Fig. 2). However, in most cases the mass loss rate seems to be the most crucial one. Our computations suggest that the critical value for this rate may be around 10^{-14} solar mass per year for Am stars.

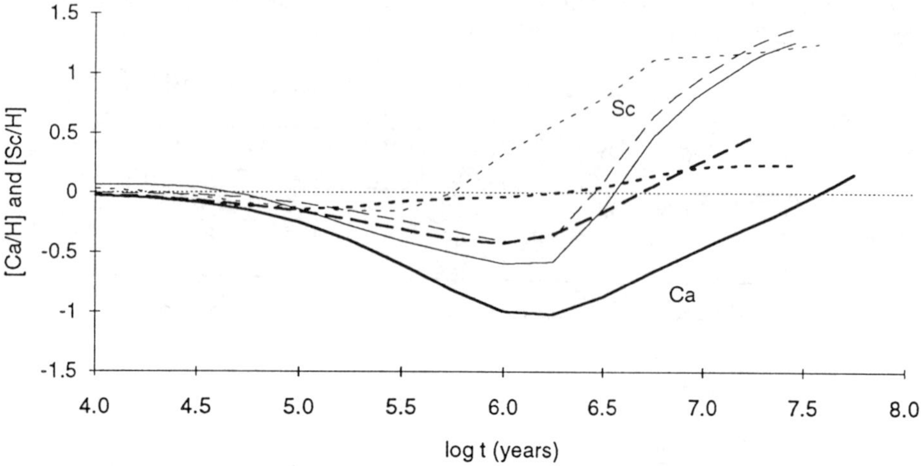

Fig. 2. Calcium and Scandium aboundances , T_{eff}=8000K. The thin lines are for Sc and the heavy ones are for Ca. The continuous lines correspond to model 1, the dashed ones to model 2 and the long-dashed ones to model 3.

(iii) - The results we have obtained are not satisfactory in the case of aluminium which is always found to evolve toward underabundance in our computations. Observations show a rather slight systematic overabundance (Burkhart and Coupry, 1991). However, since radiative acceleration on Al almost balances gravity just below the convection zone, this element is very sensitive to errors on radiative accelerations. The methods we have employed for these computations are precise and efficient. But the important improvements on atomic data such as the one provided by the OPACITY project for instance, will have strong consequences on calculations of time-dependent diffusion

ACKNOWLEDGEMENTS:

We would like to thank A. Baglin for having kindly communicated the code computing the envelope models. We would like also to thank E. Schatzman who has estimated for us the possible extension of the bottom of the convection zone.

REFERENCES

ALECIAN, G.: 1986, *Astron. Astrophys.*, 168, 204
ALECIAN, G., ARTRU, M.C.: 1990, *Astron. Astrophys.*, 234, 323
ALECIAN, G.: 1990, IAU Symposium N°145 on "Evolution of Stars: the photospheric abundance connection"
BURKHART, C., COUPRY, M.F.: 1991, *Astron. Astrophys.*, 249, 205
KURUCZ, R.L., PEYTREMANN, E.: 1975 , A table of semi Empirical gf Values, Smithonian Astrophys. Obs. Spec. Rept. N° 362

Inside the Stars, IAU Colloquium 137
ASP Conference Series, Vol. 40, 1993
Werner W. Weiss and Annie Baglin (eds.)

A DIFFUSION MASS-LOSS MODEL FOR THE AP STAR 53 CAM

J. BABEL
Institut d'Astronomie de l'Université de Lausanne, CH-1290 Chavannes-des-Bois, Switzerland

ABSTRACT The mechanism and properties of mass loss are poorly known for Ap stars. Present upper limits on the mass loss rate are of 10^{-10} $M_\odot yr^{-1}$, a value which does not permit any element separation. Abundance maps could be a very powerful tool to constrain the mass loss rate and the wind geometry of Ap stars, as surface abundances are sensitive to rates as small as 10^{-15} $M_\odot yr^{-1}$. We here propose a diffusion-mass loss model for 53 Cam and compute abundance distributions in the photosphere of 53 Cam. The mass loss geometry is determined from the Ca II K line profile and its time variation. We obtain that the diffusion-mass loss model explains many spectral features of 53 Cam, both in the UV and visible domains.

DIFFUSION AND MASS LOSS

In Babel & Michaud (1991a,b,c), it was shown that the parameter free diffusion model could explain the "mean" abundances observed on 53 Cam (Landstreet 1988), but not the abundance maps. We assume the presence of a weak stationary wind, which is inhomogeneous on the stellar surface. For 53 Cam, we obtain that most chemical elements rapidly reach (in about 10^6 yrs) an equilibrium solution with $V_{tot} = 0$. In contrast, some elements tend towards a stationary solution with constant positive outwards flux.

Equilibrium and stationary abundance distributions were calculated in detail for the atmosphere of 53 Cam, for Cr, Sr, Ca, Mn and Ti (see Fig. 1) and for mass loss rates lower than 4 10^{-15} $M_\odot yr^{-1}$ (see Babel 1992). We obtained that: 1) Over the mass loss range considered, chemical abundances of the considered trace elements in the convection zone raise with the mass loss rate. 2) Abundance distributions are strongly stratified in the atmosphere.

COMPARISON WITH OBSERVATION

The very large abundance stratification obtained and its effect on line profiles led us both to reanalyse the data on 53 Cam obtained by Landstreet (1988) (Babel 1992) and to perform abundance analysis on *IUE* high-resolution spectra (Babel & Lanz 1992)

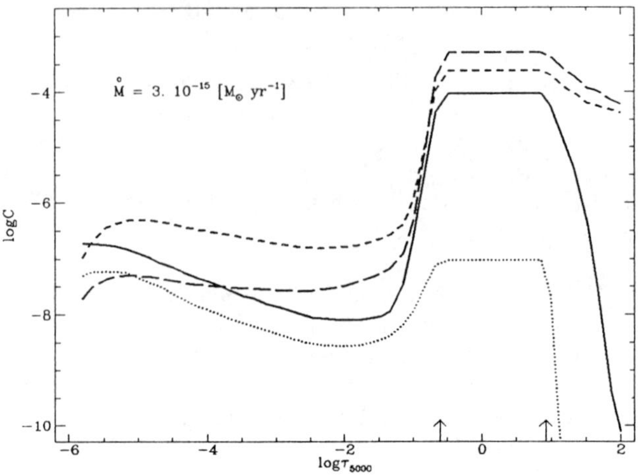

FIGURE I Equilibrium abundance distribution in 53 Cam, for a mass loss rate of $\dot{M} = 3\ 10^{-15}$ $M_\odot yr^{-1}$ (Ca:solid line, Sr:dots, Mn:long dashes, Cr:short dashes, Ti:dot-long dashes). The arrows indicate the borders of the H I convection zone.

Visible: The Ca II K line, was used to determine the wind geometry of 53 Cam, as the photospheric calcium abundance distribution varies strongly with the mass loss rate and as the K line is very sensitive to abundance stratification.

We obtain that the peculiar spectral profile of the K line and its time variation are very well reproduced by the diffusion-mass loss model, if a wind of $\dot{M} = 3\ 10^{-15}$ $M_\odot yr^{-1}$ is present in the two caps around the magnetic poles. The line profiles of other elements like Cr and Sr are also very well reproduced by this model. Elements like Ti and Mn are yet not well accounted (see Babel 1992).

UV: We investigated the UV domain (2500-3100 Å) with *IUE* spectra of 53 Cam (Babel & Lanz 1992) and obtained abundances of Cr, Fe, Mn, Ti and Ca for two rotation phases. The most striking result of our UV analysis is the considerably lower Cr and Fe abundances deduced (about 0.8 dex) from the UV domain than obtained by Landstreet (1988) from the visible. We obtained simultaneously a very good agreement between the predictions of the diffusion-mass loss model for chromium and both visible and UV Cr II line intensities. These results clearly show, in accordance with the diffusion model, that chemical elements have strongly stratified abundance distributions in the photosphere of 53 Cam.

CONCLUSION

We have developed a simple diffusion-mass loss model for 53 Cam. We assumed that H II convection was not affected by the magnetic field and introduced a

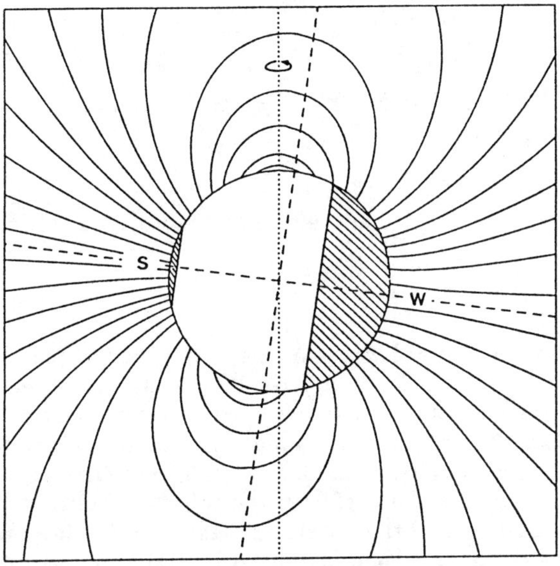

FIGURE II Mass outflow distribution on the surface of 53 Cam deduced from the profile and variation of the Ca II K line and from the diffusion-mass loss model ($\dot{M}$ =3 10^{-15} $M_{\odot}yr^{-1}$: hatched areas; $\dot{M} \simeq 0.0$, elsewhere). The magnetic geometry is from Landstreet (1988). Labels *S* and *W* indicate respectively the strong pole (−18000 G) and the weak pole (4100 G). The size of the mass loss caps are quite well constrained by the large phase coverage of the observations of Landstreet (1988).

small inhomogeneous wind as a free parameter. The diffusion-mass loss model explains simultaneously visible abundance maps, UV abundances of Cr and the profile of the Ca II K line. It gives the first evidence of a magnetically confined wind on the surface of a cold Ap star. It fails yet, in explaining the map of elements like Ti, which are sensitive to the transition region with the wind (see Babel 1992).

Our results finally show that the conjunction of detailed abundance maps and diffusion calculations could permit to reveal hydrodynamics phenomena which can not be detected at present by direct observation.

REFERENCES

Babel, J.:1992, *Astr. Ap.*, in press

Babel, J., Lanz, T.:1992, *Astr. Ap.*, in press

Babel, J., Michaud, G.:1991a, *Ap. J.*, 366, 560

Babel, J., Michaud, G.:1991b, *Astr. Ap.*, 241, 493

Babel, J., Michaud, G.:1991c, *Astr. Ap.*, 248, 155

Landstreet, J. D.:1988, *Ap. J.*, 326, 967

Inside the Stars, IAU Colloquium 137
ASP Conference Series, Vol. 40, 1993
Werner W. Weiss and Annie Baglin (eds.)

THE GENERATION OF INTERNAL WAVES

JOSEFINA MONTALBÁN
Observatoire de Meudon, 92195 Meudon-Cedex, France

INTRODUCTION

The generation of internal waves in the radiatively stable stellar region by the turbulent motion at the boundary of the overlaying convective zone is similar to the same case in the deep ocean or in the earth atmosphere (Townsend, 1965), and can be described in a simple way as following: When an turbulent fluid element arrives at the boundary of the convective region with a non-zero momentum, it beats and it deforms the interface between both regions. This disturbance of the equilibrium state excites a train of internal waves propagating below the convective zone in the horizontal and vertical directions for the frequencies lower than the characteristic one for the stable stratification (Brunt-Väisälä frequency).

MATHEMATICAL MODEL (FOURIER TRANSFORM)

We use the model proposed by Townsend (1965) to described the internal waves in the earth atmosphere. He considered that the disturbance at the boundary can be written by a gaussian function of space and time and describes the effect of this perturbation through a Fourier Transform:

$$\xi(z,t) = \iint_{-\infty}^{\infty} u(z \mid l, \omega) \exp\left(\frac{1}{2} \cdot (\alpha^2 l^2 + \tau^2 \omega^2)\right) \exp(\imath\,(lx - \omega t))\, dl\, d\omega$$

where α and τ are the characteristic size and period for an turbulent element of fluid; u is the vertical perturbation propagating in z-direction. We take u as the vertical velocity of the internal waves given by Press (1981) with the following assumptions:

- Plane-parallel geometry : $r_o - r \equiv z$, and Brunt-Väisälä frequency (N) constant below convective zone.
- ω^2 much smaller than N^2.
- The radiative damping term can be written :

$$\frac{1}{2} \int_{r_o}^{r} \frac{\sigma N^3}{\omega^4}\, k_H^3\, dr \simeq \frac{1}{2}\, F(z) \cdot \left(\frac{\omega_o}{\omega}\right)^4$$

where σ is the thermometric diffusivity, k_H the wave number, and ω_o the circular frequency attached to the largest wave length of the turbulent motion; involving that we consider for the different l-components the same damping, that one corresponding about to the pression scale hight.

COMPUTATIONAL METHOD

Fourier Transform given out by the application of Townsend' model has ben resolved using the method of steepest descent. This supposes first, that function behaves like a gaussian near its maximum, second, that the oscillation frequency is low enough compared to the gaussian decrease. In our case, this method is valid for values obeying the ω condition

$$\omega^4 \ll 2 \cdot 6^3 \omega_o^4 F$$

SUPERIMPOSITION OF GAUSSIAN PULSES

The motion below an turbulent region is the result of the superimposition of the effects of many fluids elements beating the boundary. As the arrival of the turbulent elements at the interface between the convective and stable regions is a random function of time and horizontal coordinates, the mean-square vertical velocity at any depth will arise from the product of the number of disturbances per unit of time and area and the mean-square vertical velocity due to a single pulse. Then

$$< u^2 >= \frac{3}{4}\frac{\pi}{\sqrt{6}} \left(\frac{\rho_o}{\rho}\right) \left(\frac{r_o}{r}\right)^3 \frac{u_{H_o}^2}{N^2} \cdot (2 \cdot F(z))^{\frac{1}{3}} \exp\left(-3 \cdot \left(\frac{F(z)}{4}\right)^{\frac{1}{3}} (\omega_o^4 \tau^4)^{\frac{1}{3}}\right)$$

where the index o refers to the boundary of the convective zone, ρ is the density and u_{H_o} is the imposed horizontal velocity at the boundary.

DIFFUSION COEFFICIENT

If we introduce the above vertical velocity in the expression of the diffusion coefficient given by Press for monochromatic internal waves, and assume that the turbulent flow in the convective zone can be described by the Kolmogorov spectrum, the diffusion coefficient that arise out is $F^{-\frac{3}{4}}$ times smaller than the one obtained by Schatzman (1991) using the Press' formalism and the same turbulence spectrum.

The validity of the results presented here are limited by the approximations employed in order to overcame the mathematical difficulties of the analytical method.

REFERENCES

Press, W.H.: 1981, *Astrophys. J.*, **245**, 286.

Schatzman, E.: 1991, in *Angular Momentum Evolution of Young Stars*, S. Catalano and J.R. Stauffer Eds., Kluwer Academic Plublisher, p.223.

Townsend, A.A.: 1966, *J. Fluid Mech.*, **24**, 307.

Inside the Stars, IAU Colloquium 137
ASP Conference Series, Vol. 40, 1993
Werner W. Weiss and Annie Baglin (eds.)

MIXING BY INTERNAL WAVES

JOSEFINA MONTALBÁN
Observatoire de Meudon, 92195 Meudon-Cedex, France

EVRY SCHATZMAN
Observatoire de Meudon, 92195 Meudon-Cedex, France

INTRODUCTION

That mixing take place in the radiative zone of many stars, is an event that cannot be forgotten when we try to explain observational results as the lithium abundance in the atmosphere of different stars, its dependence on spectral type, age or rotation velocity... During the last years many processes have been proposed as being responsible of this mixing: overshooting, turbulence induced by rotational instabilities, internal waves , etc... We will consider, following the results obtained by Press (1981), the role of this last mechanism in the transport of lithium to the burning level, not as generators of turbulence (in Press, 1981, and García-López and Spruit, 1991, it is shown that turbulence induced by internal waves decays very quickly inside the radiative zone), but as generators of a diffusive process due to non linear dissipative effects.

DIFFUSION COEFFICIENT

The motions in the convective zone can generate a velocity field which, at the boundary of the convective zone, generates gravity waves which propagate inside the star. Press has estimated the flux of mechanical energy carried in the radiative zone by gravity waves excited at the boundary of the convective zone and has also given an estimate of their radiative damping. For us, the most significative result of Press is the event that due to the radiative damping, the gravity waves generates a second order irreversible motion of fluid elements that can be interpreted as a macroscopic diffusive process which transports chemical elements.

The diffusion coefficient can be expressed as the product of a length by a velocity. The length is the distance reached by a fluid element after an overturn ω^{-1}; the velocity v is the product of the distance by the circular frequency, $v = l \cdot \omega$. For a gravity wave with a vertical wave number k_v and a vertical velocity u_v, Press gives as an order of magnitude

$$l = \frac{k_v u_v^2}{\omega^2} \frac{D_{th} k_v^2}{\omega}$$

where D_{th} is the thermal diffusivity, the first factor is the displacement in an overturn and the second is the ratio of the time of an overturn to the time of heat diffusion over a wave length k_v. It can be considered as a measure of the irreversible process, or as fraction of the entropy memory which is lost in the irreversible process. Schatzman (1991), following the works by Press (1981) and García-López and Spruit (1991) determined a diffusion coefficient due to internal waves. In order to obtain it he considered:

- The expression given by Press for the vertical velocity as a function of depth.
- That the horizontal velocity component is generated by the turbulent flow, and the relationship between the vertical and horizontal components of the motion is that corresponding to a sharp change of the Brunt-Väisälä frequency. That last hypothesis is justified by the new model of convective penetration (Zahn, 1991).
- The effects of the number of cells of size k which contribute to the horizontal displacements generating the wave (ω, k_H).
- That is possible to describe the effects of the turbulence spectrum by the Kolmogorov spectrum.
- That in the damping term is possible to separate the radius dependence (F(r)) from the k dependence.

The result is that the diffusion coefficient decreases very quickly from the bottom of the convective zone, like $(r_c - r)^{-4.6}$. Therefore, the value of the lithium depletion which will arise from the diffusion equation will be very senstive to the variation of the depth of the convective zone. And it will be important to take into account the events which can change this depth: convective penetration, age of the star...

COMPUTATION OF DEPLETION

The concentration of lithium in the radiative region obeys a diffusion equation where the nuclear reactions are taken into account:

$$\rho \frac{\partial c}{\partial t} = \frac{\partial}{\partial z}\left(D(z)\,\rho\,\frac{\partial c}{\partial z}\right) - K(\rho,T)\rho c$$

where z is the geometrical depth, $\rho(z)$ and $T(z)$ are the density and temperature given by the envelope model as a function of z, $K(\rho,T)$ is the nuclear destruction rate taken from Rolfs and Kavanagh (1986), and $D(z)$ is the above diffusion coefficient.

The method used to resolve the diffusion equation is that one used in Baglin et al. (1985), that is, the solution can be approximated by

$$c(z,t) = c(z)\,e^{\lambda t}$$

and with the boundary conditions, diffusion equation leads to an eigenvalue problem. Its solution, λ, gives us the depletion rate at a given age, for an given model and a given diffusion coefficient.

MODELS USED IN THE COMPUTING

The models employed in the computing of the lithium abundance in the Sun have been computed using the new code by P. Morel (1992) which introduces a recent model of convective penetration (Zahn, 1991). Opacities are from the Livermore opacity library (1991). The chemical composition is $X = 0.706, Z = 0.02$. And the value of the classical mixing length parameter is $\alpha = 1.9$.

We have determined the lithium depletion rate in the Sun at six different ages, from 0.0 Gyr. until 4.75 Gyr; and for several values of ζ, the parameter which describes the convective penetration.

RESULTS

For every ζ we obtained that lithium depletion rate is a function of time. That is, the Lithium has been burning faster in the early Sun than in the present Sun, in a factor that is almost constant for all the ζ taken. Then, the lithium abundance at time t, may be calculated by

$$[Li](t) = [Li](t=0) \exp\left(-\int_0^t \lambda(t)\, dt\right)$$

The fitting of the lithium abundance to the observed values will permit us to determinate the parameter ζ. The result ($\zeta = 0.3$) is in good agreement with the value obtained by G. Berthomieu et al. (1992) for the description of solar oscillations.

REFERENCES

Baglin, A., Morel, P.J., Schatzman E.: 1985, *Astron. Astrophys.* **149**, 309.

Berthomieu, G., Morel, P., Provost, J., Zahn, J-P.: 1992, (To be published in these proccedings).

García-López, R. and Spruit, H., 1991, *Astrophys. J.*, **337**, 268.

Press W.H.: 1981, *Astrophys. J.*, **245**, 286.

Rolfs, C. and Kavanagh, R.W.: 1986, *Nucl. Phys.*, vol. **A455**, p. 179.

Schatzman E.: 1991, in *Angular Momentum Evolution of Young Stars*, S. Catalano and J.R. Stauffer Eds., Kluwer Academic Publishers, p. 223.

Zahn, J-P.: 1992, *Astron. Astrophys.*, **252**, 179.

Inside the Stars, IAU Colloquium 137
ASP Conference Series, Vol. 40, 1993
Werner W. Weiss and Annie Baglin (eds.)

NEW CONSTRAINTS ON THE ROTATION-INDUCED MIXING IN STARS, FROM LITHIUM OBSERVATIONS IN MAIN SEQUENCE F-TYPE STARS AND SUBGIANTS

CORINNE CHARBONNEL, SYLVIE VAUCLAIR
Observatoire Midi-Pyrénées, 14,av.E.Belin, 31400 Toulouse, France

ABSTRACT Observations of lithium in F type main sequence stars and subgiants lead to specific constraints on the rotation-induced mixing. The fact that lithium is smoothly depleted in subgiants down to factors as large as 1000 at the end of the sequence can be explained if lithium suffered extra depletion inside the star while it was on the main sequence. We propose a simple consistent model, based on the discussion of rotation-induced mixing developed by Chaboyer and Zahn (1992), which could account for these observations, as well as the observations of lithium on the main sequence.

INTRODUCTION

Since the discovery of the lithium dip in the Hyades by Boesgaard and Tripicco (1986), different explanations have been proposed to account for this characteristic feature. The most severe constraints which may help chosing between the various theoretical models may be found in the study of stars with effective temperatures larger than those of the dip.

- In galactic clusters, the lithium abundance is normal (Log N(Li)=3.1) for main-sequence stars of the hot side of the dip, except for some peculiar A stars.
- The observed lithium abundance smoothly decreases along the subgiant sequence while the lithium abundance predicted by standard models would be constant in most of the sequence, and would fall at the end of the sequence when the convection zone reaches the region inside the star where lithium has been destroyed by nuclear reactions.
- Giants which belong to galactic clusters show a lithium depletion larger than the one expected from standard evolutionary models. This observation suggests that some extra lithium-depletion occurs inside the stars when they are on the main sequence, even if this extra-depletion does not appear at the surface.

A consistent treatment of hydrodynamical processes inside stars should explain not only the lithium dip, but also the lithium extra-depletion in main sequence stars on the hot side of the dip needed to account for the observations of giants and the smooth decrease in the subgiants.

MIXING INDUCED BY MERIDIONAL CIRCULATION IN ROTATING STARS

The thermal imbalance in a rotating star generates meridional flows known as "Eddington-Sweet meridional circulation". Zahn (1983) evaluated the turbulence induced by the circulation in terms of a turbulent diffusion coefficient, of the order of $|rU(r)|$, where r is the radius inside the star and U(r) the vertical velocity of the circulation at the radius r. In a uniformly rotating star, the turbulent diffusion coefficient may be written (Charbonnel et al. 1992a, hereafter CVZ) :

$$D_T = \gamma \left| \frac{L}{M^3} \frac{r^6 \Omega^2}{G^2(\nabla_{ad} - \nabla_{rad})} \left(1 - \frac{\Omega^2}{2\pi G \rho}\right) \right| \tag{1}$$

where Ω is the angular velocity in the star, G the gravitational constant, L, M and ρ the luminosity, the mass and the density at radius r, ∇_{ad} and ∇_{rad} the adiabatic and radiative gradients, and γ a factor of order one. In a radiatively stable star with a near solid body rotation, the meridional circulation splits up in two loops, which develop above and below the boundary layer for which $\Omega^2 = 2\pi G\rho$. The boundary between the two meridional loops lies inside the convection zone for stars cooler than the temperature of the lithium dip, and in the radiative zone for hotter stars (Vauclair 1988, Charbonnel et al. 1992a and b). This result suggests that the split up into two separate loops of meridional circulation is related to the fact that lithium is destroyed at the surface of stars below this effective temperature and not above. However, the screening effect of the "quiet zone" which lies between the two meridional loops is not sufficient to stop the lithium decrease at the surface of F stars (Charbonneau and Michaud 1990, Charbonnel and Vauclair 1992). Another effect must be introduced to account for the observations.

Expression (1) was obtained without taking into account the particle transport due to advection nor the mixing due to horizontal turbulence. Chaboyer and Zahn (1992) computed the combined effects of advection and mixing. With the assumption that the horizontal diffusion coefficient varies like rU(r), they obtained an effective diffusion coefficient which can be written :

$$D_{eff} = \gamma |rU(r)| \tag{2}$$

and is similar to equation (1), although the physics is somewhat different. The factor γ is unknown and depends on the efficiency of horizontal diffusion.

The cool side of the lithium dip is explained by meridional transport and nuclear destruction only if the efficiency of the meridional circulation in the deepest loop is not decreased (CVZ). We thus suppose that the horizontal turbulence in this loop is small enough to leave the advection unperturbated ($\gamma \simeq 1$). This may not be the case in the outer meridional loop. There we adjusted γ so that the lithium depletion at the surface is negligible as observed on the blue side of the dip. This result is obtained for $\gamma \leq 10^{-3}$ (Charbonnel and Vauclair 1992). While such a configuration prevents any observable lithium depletion at the surface of F stars at the age of the Hyades, it leads to more lithium destruction inside the star : lithium is now destroyed not only in the nuclear destruction region below 2.5×10^6K, but also up to the boundary beetween the two meridional

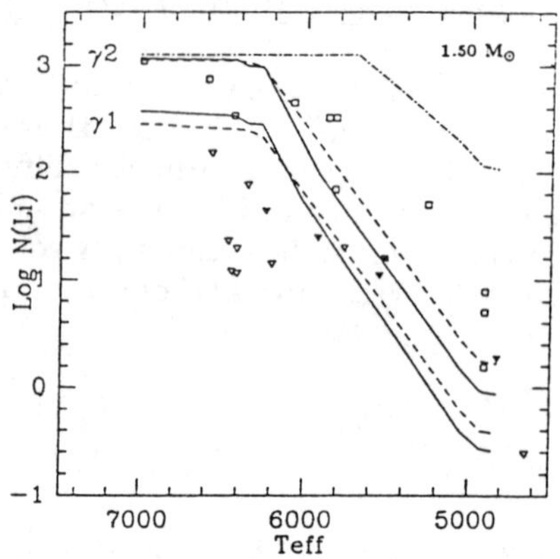

FIGURE I Observations of lithium abundance in field subgiants (open symbols) and in M67 subgiants (black symbols). The solid and dashed lines correspond respectively to a main sequence rotation velocity of 100 and 150km.s^{-1}. The dashed-dotted line represents the lithium abundance variation on the subgiant branch in case of no turbulence (our standard model).

loops. This prepares the star to the lithium abundance variations observed on the subgiant branch and at the bottom of the giant branch.

Using the Geneva stellar evolutionary code in which we have introduced the numerical method described in CVZ to solve the diffusion equation, we have computed the lithium abundance evolution on the main sequence and on the subgiant branch for a 1.5M$_\odot$ star. In figure 1 we compare the predicted lithium abundance variations along the subgiant branch for two values of γ in the upper loop ($\gamma_1 = 10^{-3}$, $\gamma_2 = 10^{-4}$) and two values of the rotation velocity of the star with the observations of lithium abundance in field subgiants and in M67 subgiants. These theoretical results account much better for the observations than the computations done in standard models.

REFERENCES

Boesgaard, A,M., and Tripicco, M.J., 1986, *Ap.J. Letters*, **302**, L49

Chaboyer, B., and Zahn, J.P., 1992, *Astr.Ap.*, **253**, 173

Charbonneau, P., and Michaud, G., 1990, *Ap.J.*, **352**, 681

Charbonnel, C., and Vauclair, S., 1992, *Astr.Ap.*, in press

Charbonnel, C., Vauclair, S., and Zahn, J.P., 1992a, *Astr.Ap.*, **255**, 191

Charbonnel, C., Vauclair, S., Maeder, A., Meynet, G., and Schaller, G., 1992b, *Astr.Ap.*, in preparation

Vauclair, S., 1988, *Ap.J.*, **335**, 971

Zahn, J.P., 1983, *Astrophysical Processes in Upper Main Sequence Stars*, ed. B.Hauck and A.Maeder, Publ. Geneva Observatory, 253

Inside the Stars, IAU Colloquium 137
ASP Conference Series, Vol. 40, 1993
Werner W. Weiss and Annie Baglin (eds.)

ANGULAR-MOMEMTUM TRANSFER BY NONRADIAL OSCILLATIONS IN MASSIVE MAIN-SEQUENCE STARS

UMIN LEE,
Department of Physics and Astronomy, University of Rochester, Rochester, NY 14627, U.S.A.
HIDEYUKI SAIO,
Astronomical Institute, Faculty of Science, Tohoku University, Sendai, Miyagi 980, Japan.

Angular mementum distribution is one of the most important factors for stellar structutre and evolution. Among other mechanisms, angular momentum is transfered by non-axisymmetric oscillations (nonradial oscillations). In this mechanism the angular momentum is carried mainly by the Reynolds stress, which is proportional to the product between radial and azimuthal components of oscillation velocity; i.e., $\upsilon_r'\upsilon_\phi'$ (ϕ direction is the direction of rotation velocity). In the linear oscillation analysis, the phase difference between υ_r' and υ_ϕ' is $\pi/2+\delta$ with $|\delta| << 1$. A finite value of δ, which arises from excitation or damping of the oscillation, makes the time average of $\upsilon_r'\upsilon_\phi'$ finite. Positive angular momentum is transfered from the driving zone to the damping zone by a prograde mode (Osaki 1986).

Lee & Saio (1986, 1987) and Lee (1988) have shown that low-frequency oscillations are excited in rotating massive main-sequence stars, and proposed that the oscillations thus excited are responsible for the light and line-profile variations observed in OB (Be) stars. In the present paper, we discuss the efficiency of angular momentum transfer by such low-frequency oscillations. The oscillation is generated by oscillatory convection (inertial oscillation) with negative energy in the convective core resonantly coupled with a g mode in the envelope (Lee & Saio 1990). They are prograde, although the phase velocity is very low, with respect to the core rotation. The oscillation energy flows from the core to the envelope, and hence positive angular momentum is carried from the core to the envelope, if the angular velocity of rotation of the core is equal to or faster than that of the envelope.

Let us begin with the angular momentum conservation equation given by

$$\frac{\partial(\rho\varpi\upsilon_\phi)}{\partial t}+\nabla\cdot(\rho\varpi\upsilon_\phi\boldsymbol{\upsilon})=-\frac{\partial p}{\partial\phi}-\rho\frac{\partial\Phi}{\partial\phi}, \tag{1}$$

where $\varpi=r\sin\theta$ and Φ is the gravitational potential. When we decompose each physical quantity of a rotating star into the axisymmetric and non-axisymmetric parts, the velocity and the density, for example, are written as

$$(\upsilon_r,\upsilon_\theta,\upsilon_\phi)=(0,0,\Omega r\sin\theta)+(\upsilon_r',\upsilon_\theta',\upsilon_\phi') \quad \text{and} \quad \rho=\bar{\rho}+\rho', \tag{2}$$

where the bar $(\bar{\ })$ denotes the (zonally averaged) axisymmetric part and the prime $(')$ the non-axisymmetric part ($\overline{f'}=0$). Then, taking the zonal average of equation (1) and integrating over the solid angle, we obtain

$$\frac{\partial j(r,t)}{\partial t}=-\frac{1}{r^2}\frac{\partial(r^2\Psi)}{\partial r}-\left\langle\rho'\frac{\partial\Phi'}{\partial\phi}\right\rangle-\frac{\partial\left\langle\varpi\overline{\rho' v'_\phi}\right\rangle}{\partial t}, \tag{3}$$

where the angular bracket means the integration of the enclosed quantity over the whole solid angle, and

$$j(r,t)\equiv\left\langle\omega\overline{\rho}\overline{v}_\phi\right\rangle,\quad \text{and}\quad \Psi\equiv\left\langle\varpi\left(\overline{\rho}\overline{v'_\phi v'_r}+\overline{v}_\phi\overline{\rho' v'_r}\right)\right\rangle. \tag{4}$$

Usually, the second and third terms in the right hand side of equation (3) have only small contributions if the tidal effects play no important role. Neglecting these terms, we have

$$\dot{\jmath}(r,t)\equiv\int_0^r 4\pi r^2\frac{\partial j(r,t)}{\partial t}dr\cong-4\pi r^2\Psi, \tag{5}$$

which represents the change in the equilibrium angular momentum within the sphere of the radius r. We assume that the non-axisymmetric parts of the physical quantities are caused by non-radial oscillations with $m\neq 0$. Then, equation (3) describe the angular momentum transfer by non-axisymmetric waves. We also assume that the functional form of the amplitude of non-axisymmetric oscillations can be represented by eigenfunctions for linear wave equations. Then, knowing that the Eulerian velocity perturbation generated by a non-radial oscillation is related to the displacement vector $\xi(r,\theta,\phi,t)$ as

$$\boldsymbol{v}'=i(\sigma+m\Omega)\boldsymbol{\xi}-r\sin\theta(\boldsymbol{\xi}\cdot\nabla\Omega)\mathbf{e}_\phi, \tag{6}$$

we can evaluate the right-hand-side of equation (3) by using the eigenfunctions of the non-radial oscillation. As in a previous study (e.g., Lee & Saio 1986), we employ the series expansion in terms of the spherical harmonic functions to represent the non-radial oscillation of rotating stars. For example, the radial component of the displacement vector is given as

$$\frac{\xi_r}{r}=\sum_{\ell\geq|m|}^{\infty}S_\ell(r)Y_\ell^m(\theta,\phi)e^{i\sigma t}, \tag{7}$$

where σ is the oscillation frequency in an inertial frame, $\ell=|m|+2j-2$ for even modes and $\ell=|m|+2j-1$ for odd modes and $j=1,2,3,\cdots$.

The equilibrium model we employ is the $10M_\odot$ zero-age main-sequence model, for which $\log(L/L_\odot)=3.73$, $\log T_{\rm eff}=4.41$ and $R=3.71R_\odot$. The model has a convective core, in which $\nabla-\nabla_{\rm ad}=10^{-3}$ is assumed. The outer boundary of the convective core is located at $x\equiv r/R=0.235$. We assume the angular frequency of rotation depends only on the coordinate r, and that

$$\overline{\Omega}=\overline{\Omega}_s\left\{1+\frac{b-1}{1+\exp[a(x-x_c)]}\right\}, \tag{8}$$

where $\overline{\Omega}_s$ is the angular velocity of rotation at the surface, $a=20$, $x_c=0.235$, and a parameter $b(\geq 1)$ is to specify the differential rotation. (The frequencies $\overline{\sigma}$ and $\overline{\Omega}$ are dimensionless ones normalized by $\sqrt{GM/R^3}=2.78\times10^{-4}\,\mathrm{s}^{-1}$.) We show an example of angular momentum transfer by the B_1 mode of $m=-2$ (see Lee & Saio 1986), which is an even prograde wave excited by a resonance coupling between an oscillatory convective mode in the core and a gravity mode in the envelope. The oscillatory convective mode has negative energy of oscillation, while the gravity mode has positive energy of oscillation (Lee & Saio 1990). For $\overline{\Omega}_s = 0.162$ and $b = 1.12$, we calculate the B_1 mode by taking accout of the non-adiabatic effects. Its frequency is $(\overline{\sigma}_R, \overline{\sigma}_I) = (0.383, -2.97\times10^{-4})$. The eigenfunctions of $|rS_2(r)|$ and $|rS_4(r)|$ are shown by the solid and the dotted lines in the upper panel, and the qunatity $\dot{\jmath}(r,t)$ is given as a function of r/R in the lower panel, where $\dot{\jmath}(r,t)$ is normalized by its maximum value. As shown by the figure, $\dot{\jmath}(r,t)$ change its sign at the outer boundary of the core and is negative throughout the envelope, which means there is a net angular momentum flux from the core to the envelope. The rapid increase in $\dot{\jmath}(r,t)$ near the surface means that a large amount of angular momentum deposition to the stellar rotation occurs because of large non-adiabatic effects (radiative dissipation) there. Since the angular momentum density is low, the rotation velocity near the surface would be changed on a short time scale. The rotation law can be modified until the phase velocity becomes equal to the rotation velocity.

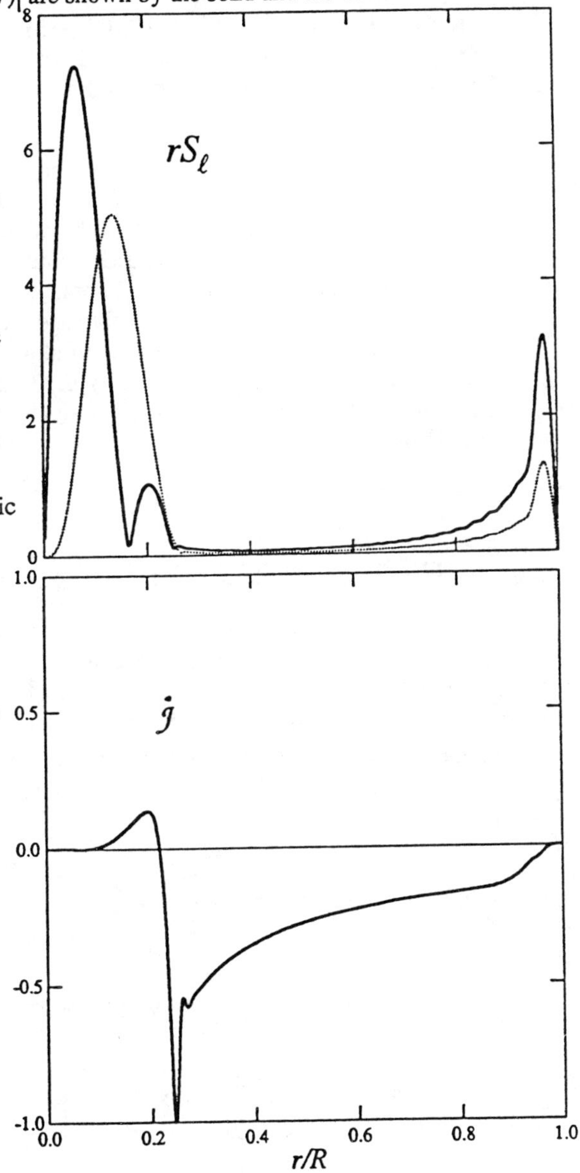

Reference

Lee,U. 1988, MNRAS, **232**, 711.

Lee,U. & Saio,H. 1986, *MNRAS*, **221**, 711.

Lee,U. & Saio,H. 1987, *MNRAS*, **225**, 643.

Lee,U. & Saio,H. 1990, *ApJ*, **360**, 590.

Osaki,Y. 1986, *PASP*, **98**, 32.

Inside the Stars, IAU Colloquium 137
ASP Conference Series, Vol. 40, 1993
Werner W. Weiss and Annie Baglin (eds.)

INTEGRAL CONSTRAINTS ON CONVECTIVE OVERSHOOTING TWO-DIMENSIONAL NUMERICAL STUDIES

IAN W. ROXBURGH AND JOHN SIMMONS
Astronomy Unit, Queen Mary and Westfield College, University of London, Mile End Rd., London E1 4NS, UK.

ABSTRACT The integral constraint on convective overshooting (Roxburgh 1976,1978,1989) is re-derived for for statistically stationary two-dimensional convection between stress-free horizontal boundaries maintained at constant temperatures, the vertical boundaries being periodic. We report on numerical simulations of such two-dimensional compressible convection in a fluid where the central regions are convectively unstable and the surrounding layers are stable. The results support the conclusion by Roxburgh (1989) that the contribution from the radiative flux is well approximated by the flux carried by the mean temperature gradient, but demonstrate that viscous dissipation is important, reducing the extent of overshooting.

THE INTEGRAL CONSTRAINT

We consider a compressible fluid confined between two horizontal stress free boundaries, $u_z = 0$ at $z = 0$, $z = h$; the acceleration due to gravity is a constant $-g$ in the vertical z direction, the top boundary is maintained at temperature T_1 and the bottom boundary at T_2 and the flow is two dimensional and is taken as periodic in the horizontal x direction, $U(x,z,t) = U(x+a,z,t)$ for all variables U. Since $g_i = -\partial\Psi/\partial x^i$ is independent of time the contribution to the entropy and total energy equations governing the fluid can be written in the form:

$$\frac{\partial}{\partial t}(\rho S) + \frac{\partial}{\partial x^i}(\rho u_i S) = -\frac{1}{T}\frac{\partial F_i}{\partial x^i} + \frac{\Phi}{T} \tag{1}$$

$$\frac{\partial}{\partial t}\left[c_v \rho T + \frac{1}{2}\rho u^2 + \rho\Psi\right] + \frac{\partial}{\partial x^i}\left[\rho u_i\left(c_p T + \frac{1}{2}u^2 + \Psi\right) - u_j \eta_{ij} + F_i\right] = 0 \tag{2}$$

$$\Phi = \eta_{ij}\frac{\partial u_i}{\partial x^j} = \frac{1}{2}\eta\left(\frac{\partial u_i}{\partial x^j} + \frac{\partial u_j}{\partial x^i} - \frac{2}{3}\delta_{ij}\frac{\partial u_k}{\partial x^k}\right)^2 \tag{3}$$

with η_{ij}, $F_i = -K(\rho,T)\,\partial T/\partial x^i$, $P = \mathcal{R}\rho T/\mu$, $E = 3\mathcal{R}T/2\mu$, $S = (\mathcal{R}/\mu)\,\ell n\,(T^{3/2}/\rho)$, the viscous stress tensor, radiative flux, pressure, internal energy and entropy per unit mass.

Let <U> denote the time average of $U(x^i,t)$, then for statistically stationary convection $\langle \partial U/\partial t \rangle = 0$. Integrating the entropy and total energy equations over the volume of the fluid, using Gauss's theorem to convert integrals over the volume V to integrals over the surface Σ, and using the boundary conditions that $u_z = 0$, $\eta_{xz} = 0$ on $z = 0$, h, and the periodic condition $U(x,z,t) = U(x+a,z,t)$, gives the two identities

$$\int_\Sigma \langle F_i \rangle d\Sigma_i = 0, \qquad \int_V \left\langle \frac{1}{T}\left(\frac{\partial F_i}{\partial x^i} - \Phi \right) \right\rangle dV = 0 \tag{4}$$

The contribution to the surface integral from the vertical boundaries ($x = 0$, a) cancel by the periodic conditions, so the time averaged flux entering the bottom boundary equals that leaving the top boundary and is constant ($= a\Gamma_0$). With T constant on $z = 0$, h, and periodic conditions on $x = 0$, a, it follows that $\int \langle F_i/T \rangle d\Sigma = a\Gamma_0(1/T_1 - 1/T_2)$. Integrating the second of equation (4) by parts and using Gauss's theorem yields

$$\int_V \left\langle \frac{1}{T}\frac{\partial F_i}{\partial x^i} \right\rangle dV = \int_V \left\langle \left(F_i - \Gamma_{0i}\right) \frac{1}{T^2}\frac{\partial T}{\partial x^i} \right\rangle dV = \int_V \left\langle \frac{\Phi}{T} \right\rangle dV \tag{5}$$

Defining the horizontal average $\overline{U(z,t)} = (1/a)\int U\, dx$ and $U_0(z) = \langle \overline{U(z,t)} \rangle$ then integrating over the horizontal direction gives the integral constraint in the form

$$\int_0^h \overline{\left\langle \left(F_i - \Gamma_{0i}\right) \frac{1}{T^2}\frac{\partial T}{\partial x^i} \right\rangle} dz = \int_0^h \overline{\left\langle \frac{\Phi}{T} \right\rangle} dz > 0 \tag{6}$$

Let $V(z) = \overline{\langle \Phi/T \rangle}$ and $I_1(z) = \overline{\langle (F_i - \Gamma_i)\,(1/T)^2\, \partial T/\partial x^i \rangle}$. In Roxburgh (1978), $I_1(z)$ was replaced by $I_0(z) = (F_0 - \Gamma_0)\,(1/T_0)^2\, dT_0/dz$, calculated using the mean values of T and the flux given by the mean field. This was criticised by Baker and Kuhfuß (1987) on the grounds that it neglected terms quadratic in the fluctuating quantities, but Roxburgh (1989) argued that these contributions are expected to be small if the mean stratification is almost adiabatic. In the numerical simulations reported here the difference between I_0 and I_1 is indeed small, but the viscous dissipation $V(z)$ is not small and cannot be neglected.

NUMERICAL RESULTS

Numerical simulations were made for a layer in which the variation of the conduction coefficient with temperature was $K(T) = K_o\,(\alpha T^5 + \beta/T^3)$, with (α,β) chosen so that the central regions of the layer are convectively unstable and the surrounding regions stable; the temperature at the bottom of the layer was twice that at the top and the viscosity was constant. The equations were integrated for 50 times the turn over time of the convection using a 64x64 array machine (DAP). A running time average was calculated at each mesh point and a horizontal average calculated at the end of a computing run. The integrands $I_0(z)$, $I_1(z)$ and $V(z)$ that enter the integral constraint were then determined.

Figure 1 shows the flow pattern for a case where the Rayleigh number at the centre of the layer $R_m = 120 R_c$, where R_c is the value at the onset of convection, and the Prandtl number $\sigma = \eta c_p/K = 1$. Figure 2 shows the values of the integrands $I_0(z)$ and $I_1(z)$ and the viscous dissipation term V(z), Figure 3 gives results for $\sigma = 0.1$

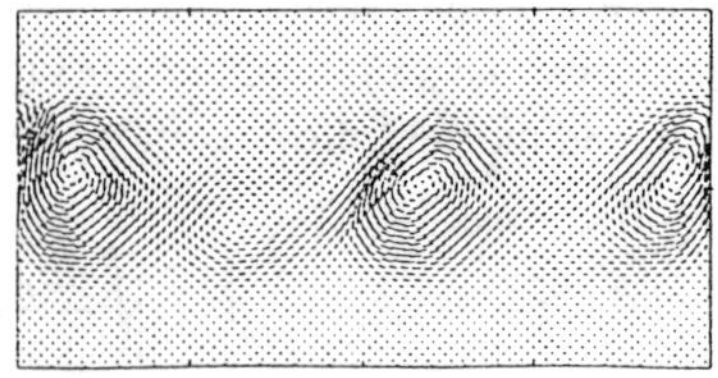

Figure 1. Flow pattern for case where $\sigma = 1$ and $R_m = 120 R_c$

As can be seen from these figures $I_1 \approx I_0$, and the contribution from terms quadratic in the fluctuating quantities is therefore small. The Integral Constraint is well approximated by

$$\int_0^h \left(F_0 - \Gamma_0 \right) \frac{1}{T_0^2} \frac{dT_0}{dz} dz = \int_0^h \overline{\left\langle \frac{\Phi}{T} \right\rangle} dz$$

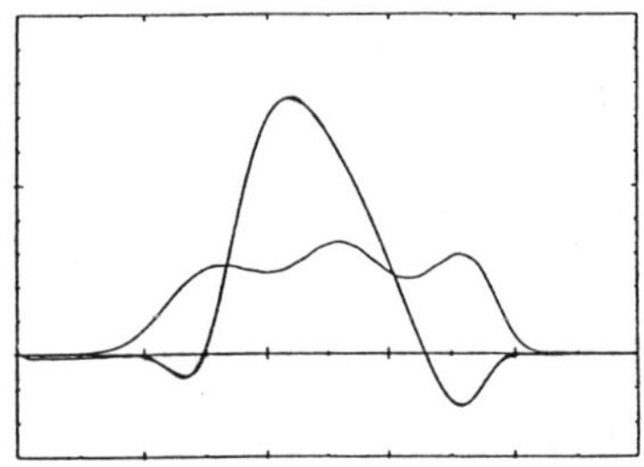

Figure 2. Variation of the Integrands I_0, I_1 and V with z for $\sigma = 1$. I_0 and I_1 are almost equal.

For the cases $\sigma = 0.1, 1$, studied here viscous dissipation cannot be neglected, although perhaps there is some indication that it may be less important for small σ ($\sigma \approx 10^{-6}$ in stars), but further calculations are needed on this point. Since $V > 0$, neglecting viscous dissipation can provide an upper limit on the extent of convective overshooting.

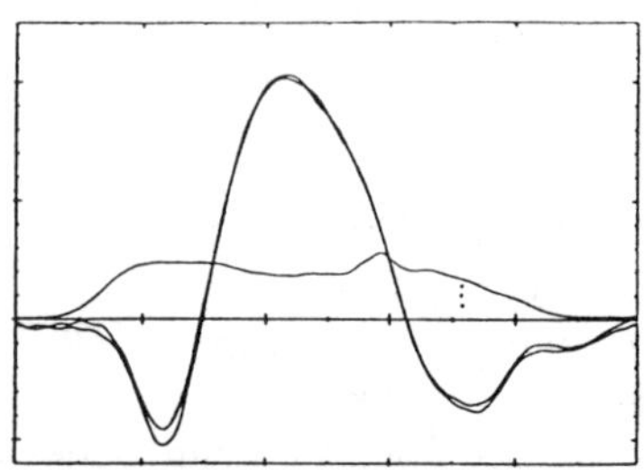

Figure 3. Variation of the Integrands I_0, I_1 and V with z for $\sigma = 0.1$. I_0 and I_1 are almost equal.

ACKNOWLEDGEMENTS

The authors gratefully acknowledge support from SERC; IWR for grant GR/H 09454, and JS for an SERC studentship.

REFERENCES

Baker, N H and Kuhfuß, R., 1987. 'On the Roxburgh Criterion for Convective Overshooting', *Astronomy and Astrophysics*, **185**, pp 117-123.

Roxburgh, I W, 1976, 'The Internal Structure of the Sun and Solar Type Stars', in *Basic Mechanisms of Solar Activity*, ed V Bumba and J Kleczeck, D Reidel, pp 451-456.

Roxburgh, I W, 1978, 'Convection and Stellar Structure', *Astronomy and Astrophysics*, **65**, pp 281-285

Roxburgh, I W, 1989, 'Integral Constraints on Convective Overshooting', *Astronomy and Astrophysics*, **211**, pp 361-364

Inside the Stars, IAU Colloquium 137
ASP Conference Series, Vol. 40, 1993
Werner W. Weiss and Annie Baglin (eds.)

LOCAL AND NON-LOCAL CONTRIBUTIONS TO HORIZONTAL MOTIONS IN STELLAR CONVECTIVE ZONES

K. PETROVAY
Eötvös University, Department of Astronomy, Budapest, Ludovika tér 2, H-1083 Hungary

ABSTRACT A method to determine the r.m.s. horizontal velocities in stellar convective zones (assuming that non-locality effects are not too strong) is given and applied to the solar case. The results may serve as useful input parameters for both convection theory and dynamo theory.

METHOD

Progress in stellar convection theory in the past decade has made it possible to break away from phenomenological models containing a large number of free parameters. The recent, more quantitative models (e.g. Unno and Kondo 1989, Canuto and Mazzitelli 1991) rely on the achievements of turbulence theory in their attempt to reduce the number of uncertain parameters. The velocity anisotropy of convection

$$s(z) := \frac{\overline{v_x^2}(z)}{\overline{v_z^2}(z)} \tag{1}$$

(x and y are horizontal coordinates, $\mathbf{v}$ is the velocity) is one such parameter to be specified. Here we present the results of a determination of the anisotropy as a function of depth in the solar convective zone on the basis of turbulence theory. (The effect of rotation will be neglected.)

Let us take the continuity equation (for simplicity in anelastic approximation and plane parallel geometry):

$$v_z \frac{d\rho}{dz} + \rho\frac{\partial v_z}{\partial z} + \rho\frac{\partial v_x}{\partial x} + \rho\frac{\partial v_y}{\partial y} = 0 \tag{2}$$

with z the depth and ρ the density.

Multiplying by ρv_z and averaging, we have

$$\frac{1}{2}\frac{d}{dz}(\rho^2 V_z^2) + 2\rho^2 \overline{v_z \frac{\partial v_x}{\partial x}} = \frac{1}{2}\frac{d}{dz}(\rho^2 V_z^2) - 2\rho^2 \overline{v_x \frac{\partial v_z}{\partial x}} = 0 \tag{3}$$

where the notation

$$V_i = (\overline{v_i^2})^{1/2} \tag{4}$$

was introduced for the r.m.s. velocities.

Now let us split v_x (and v_y) into two (local + non-local) components:

$$v_x = v_{x,l} + v_{x,n} \tag{5}$$

so that

$$\overline{v_{x,l}\frac{\partial v_z}{\partial x}} = 0 \qquad \text{while} \qquad \overline{v_{x,n}\frac{\partial v_z}{\partial x}} \neq 0. \tag{6}$$

Now let us *assume* that $V_{x,l}^2 \gg V_{x,n}^2$, i.e. that the non-locality effect is weak. In this case, supposing that $\rho(z)$, $V_z^2(z)$ and $l(z)$ are known from a convective zone model, the local contributions $s_l \equiv V_{x,l}^2/V_z^2$ and $X \equiv l^2/l_h^2$ (l_h is the horizontal correlation length) can be computed from a homogeneous anisotropic turbulence model (Petrovay 1992). The non-local contribution $s_n \equiv V_{x,n}^2/V_z^2$ can be computed from equation (3):

$$\frac{1}{2\rho^2}\frac{d}{dz}(\rho^2 V_z^2) \sim 2V_{x,n}\frac{V_z}{l}X^{1/2} \tag{7}$$

$$\Rightarrow \qquad V_{x,n} = \frac{l}{2\rho}X^{-1/2}\frac{d}{dz}(\rho V_z) \tag{8}$$

and $s = s_l + s_n$ due to the orthogonality of the decomposition ($\overline{v_{x,l}v_{x,n}} = 0$).

If $\rho(z)$, $V_z^2(z)$ and $l(z)$ are not *a priori* known, the above relations can simply be used together with the equations of the given convection theory to solve for all the unknown quantities.

Finally, using the above decomposition of the anisotropy the relations for the morphology of convection derived in Petrovay (1990) can more appropriately be written as

$$f_d = \frac{1}{2}\left[1 + \mathrm{sgn} F_{kin}\left(\frac{A^2}{A^2+4}\right)^{1/2}\right] \tag{9}$$

with

$$A^2 = \frac{F_{kin}^2}{\rho^2 V_z^6}\left(\frac{2}{2s+1}\right)^2, \tag{10}$$

and

$$f_{d,cr-lim} = \mathrm{sgn}\,\frac{d(\rho V_z)}{dz}\min\left[(1+20s_n)^{-1};0.5\right] + 0.5\left[1 - \mathrm{sgn}\,\frac{d(\rho V_z)}{dz}\right]. \tag{11}$$

Here, f_d is the downflow filling factor (in a horizontal cross-section), $f_{d,cr-lim}$ is the widest limit for the critical filling factor (topology reversal takes place in the regime $0.5 > f_d > f_{d,cr-lim}$ or $0.5 < f_d < f_{d,cr-lim}$), and F_{kin} is the kinetical energy flux.

APPLICATION TO THE SOLAR CASE

Our basic model will be the model D in Unno and Kondo (1989). ($l = H_P$ is supposed here). The functions $\rho(z)$, $V_z^2(z)$ are taken from this model. This is somewhat inconsistent as the model is based on the assumption that $s \equiv 0.5$,

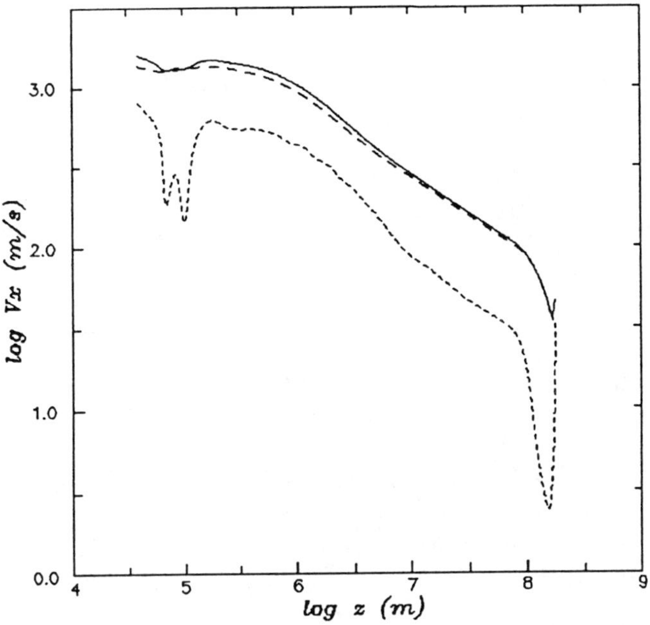

FIGURE I R.m.s. horizontal velocity as a function of depth in the solar convective zone. Solid: total r.m.s. velocity; long dashes: local contribution; short dashes: non-local contribution.

the results however show that this assumption is *a posteriori* confirmed with the exception of the neighbourhood of the boundaries of the solar convective zone. The two contributions to the horizontal velocities are shown on Figure I. Finally, it must be noted again that the present approach breaks down near the borders of the convective zone where strongly non-local effects come into play.

ACKNOWLEDGMENTS

This work was funded by the OTKA (grant no. 2135) as well as by the Hungarian Science Foundation (grant no. 23/91/III) and Hungarian High Education Foundation (grant no. 69/91) of the Hungarian Credit Bank.

REFERENCES

Canuto, V. M. and Mazzitelli, I. 1991, *Ap. J.* **370**, 295.

Petrovay, K. 1990, *Ap. J.* **362**, 722.

Petrovay, K. 1992, *Geophys. Astrophys. Fluid Dyn.*, in press

Unno, W. and Kondo, M. 1989, *P.A.S. Japan* **41**, 197.

Inside the Stars, IAU Colloquium 137
ASP Conference Series, Vol. 40, 1993
Werner W. Weiss and Annie Baglin (eds.)

NUMERICAL SIMULATIONS OF 3D COMPRESSIBLE CONVECTION

HERBERT J. MUTHSAM
JOHANN ZÖCHLING
Institute of Mathematics
University of Vienna
Boltzmanngasse 5
A–1090 Vienna/Austria
e-mail: A8131DAA@AWIUNI11.BITNET

ABSTRACT Numerical models of a convection zone embedded between two stable layers have been calculated; one is shortly presented and in particular a few items concerning the lower overshoot region are discussed.

1. INTRODUCTION

As one of the large, still in many a respect unsettled questions of stellar physics, the study of compressible convection has, in recent years, attracted the attention of a growing number of research groups using the approach of numerical modelling on the basis of the (more or less full) equations of hydrodynamics, with appropriate extension included. See e.g. Cattaneo et al. (1991), Chan and Sofia (1989), Hossain and Mullan (1991), Fox et al. (1991) for recent work.

We contribute here to such efforts and discuss a few items which have shown themselves when studying a 3D model of a convection zone which is bounded above and below by stable zones (thus essentially eliminating the effect of more or less artificial boundary conditions which arise when the convection zone is bounded by the margin of the computational domain). For 2D work see Hurlburt et al. (1986).

2. EQUATIONS

We have developed a numerical code ASCIC 90 (A Stellar Convection Investigation Code for the 90's) which allows to integrate in time the pertinent equations. In its present form it is based on central differencing on a staggered mesh and a leapfrog method in time. We are presently extending it to an ENO-scheme of high order, see Lafon and Osher (1991).

The equations are the compressible Navier–Stokes equations inside a box with a heat conduction term (= radiative transfer in the diffusion approximation) added. The boundary conditions are as in Hurlburt et al. (1986).

In order to get insight to the basic physics, we have adopted simple microphysics; we take as starting configuration polytropic layers situated atop of each other with parameters matched thus that consistency (same flux everywhere, etc.) is guaranteed. The coefficients of radiative conductivity and viscosity are held fixed at their initial values at each depth. The equation of state is that of a perfect gas.

Although our code would allow for such, subgrid modelling is not used in the model described in this note.

3. SHORT DISCUSSION OF A MODEL

We discuss model 003F in our designation. It is calculated on a 72*50*50 grid (72 depth points) and has an aspect ratio of 1:2:2. It is characterized by the following data:

depth	β	ρ/ρ_{top}	p/p_{top}	T/T_{top}
0.0	0.3	1.	1.	1.
0.2	0.8	3.11	4.34	1.39
0.3	1.25	4.28	7.51	1.75
0.55	1.25	7.12	20.5	2.90
0.65	0.3	9.1	27.3	3.11
1.0	0.3	19.8	69.0	3.47

The total depth ranges from 0 to 1. The values of ρ etc. are given in units of their values at the top; they are naturally only valid for the hydrostatic starting configuration. β is the temperature gradient in units of the adiabatic one. Thus, the main part of the convection zone is between depths 0.3–0.55. Its Rayleigh Number is 10^6.

We show in fig. 1 various fluxes. Of special importance are the radiative flux (FRAD) and convective flux FCONV. As is seen, the lower stable region has not yet quite converged to the nominal flux $(= -1)$. Of special interest is that at the bottom of the convection zone more than the nominal flux is transported by radiation whereas convection transports there heat downwards. We see this effect also in many other models.

We also hint at fig. 2. Here, correlation coefficients of pairs of quantities are evaluated at a horizontal plane and plotted as a function of depth. XC DG T (X-Correlation of Density of Gas with Temperature) shows a rather dull behaviour, being close to -1 over most of the depth-range. Thus, as expected, denser material is, on the average, cooler.

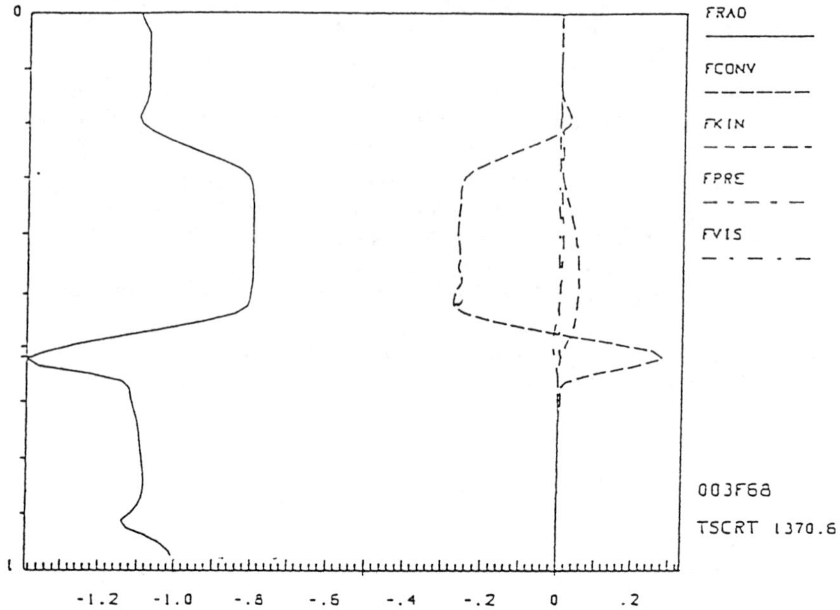

Fig. 1. Radiative and convective flux plus some fluxes of lesser importance plotted in dependency of depth (ordinate). The nomial flux is −1.

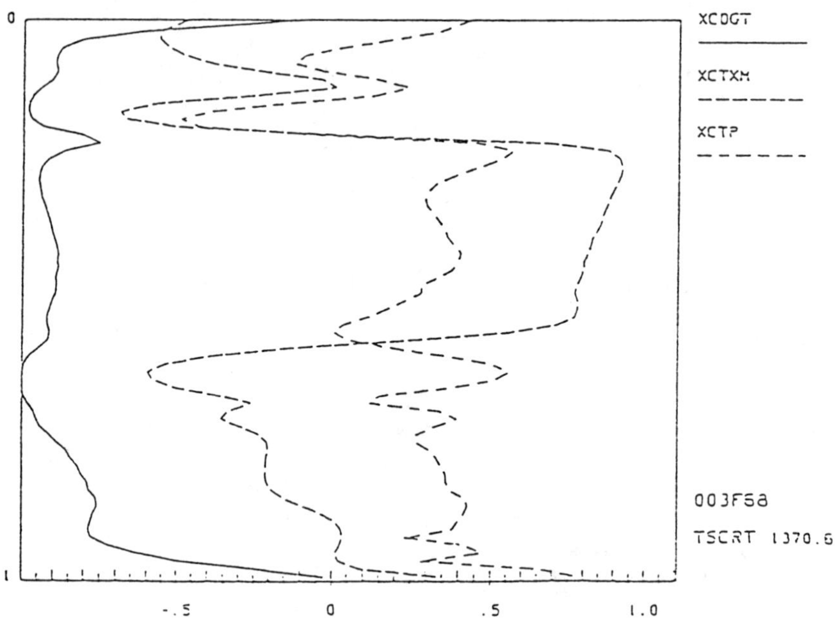

Fig. 2. Correlations of physical quantities over horizontal planes in dependency of depth (ordinate). For explanation see text.

XC T XM (X-Correlation of Temperature fluctuations about horizontal average with X-Momentum [vertical]) is much more interesting; in particular, it being negative at the lower boundary of the convection zone again tells that there heat is transported downwards by convection. This quantity and also XC T P (P=pressure) show considerable depth dependence. Effects that can be discerned by these and other data will be reported elsewhere, Muthsam, (1992).

ACKNOWLEDGEMENT

Calculations were performed at IBM 3090-400 VF, IBM 9021-720 computers and at an IBM RS 6000-550 workstation in the framework of EASI, European Academic Supercomputer Initiative, at the Computing Center, University of Vienna. Early stages of the work were granted by the Fonds zur Förderung der wiss. Forschung, project 4122.

REFERENCES

Cattaneo, F. et al. 1991, *Ap. J.*, 370, 282.
Chan, K.L. and Sofia, S. 1989, *Ap. J.*, 336, 1022.
Fox, P.A., Theobald, M.L. and Sofia, S. 1991, *Ap. J.*, 383, 860.
Hossain, M. and Mullan, D.J., 1991, *Ap. J.*, 380, 631.
Hurlburt, N.E. et al. 1986, *Ap. J.*, 311, 563.
Lafon, F. and Osher, S. 1991, *Journal of Computational Physics*, 96, 110.
Muthsam, H. 1992, in preparation.

Inside the Stars, IAU Colloquium 137
ASP Conference Series, Vol. 40, 1993
Werner W. Weiss and Annie Baglin (eds.)

THE DEPTH OF THE SOLAR CONVECTION ZONE INFERRED FROM HYDRODYNAMICAL MODELS OF THE SURFACE LAYERS

M. STEFFEN
Institut für Theoretische Physik und Sternwarte der Universität, Olshausenstr. 40, W-2300 Kiel 1, Germany

HYDRODYNAMICAL MODELS OF SOLAR SURFACE CONVECTION

We have obtained detailed 2-dimensional models of the surface layers of solar-type stars from extensive numerical simulations solving the time-dependent, non-linear equations of hydrodynamics for a stratified compressible fluid. The calculations take into account a realistic equation of state (including ionization of H and He as well as formation of H_2-molecules) and use an elaborate scheme to describe multi-dimensional, non-local, frequency-dependent radiative transfer. The hydrodynamical models include the photosphere as well as part of the convective subphotospheric layers, with an open lower boundary located about one pressure scale height ($\approx$ 320 km) below $\tau_{\rm Ross} = 1$, allowing a free flow of gas out of and into the model. A fixed specific entropy, s^*, is (asymptotically) assigned to the gas entering the simulation volume from below, which uniquely determines the effective temperature of the hydrodynamical model. For details about the physical assumptions, numerical method and characteristics of the resulting convective flow at the solar surface see Steffen (1991); Steffen and Freytag (1991).

The results presented in the next section are taken from one of our most recent hydrodynamical model calculations of convection at the solar surface, based on 40 (vertical) by 140 (horizontal) grid points, assuming 2-dimensional Cartesian geometry with periodic lateral boundaries enforcing a repetition of the flow pattern every 5250 km (comfortably larger than the observed maximum size of solar granules, $\approx$ 3000 km). The sequence spans a time interval of 8400 sec, enough to yield statistically stable results.

The purpose of this paper is to demonstrate that under certain assumptions concerning the topology of solar-type convection zones, our hydrodynamical models of stellar *surface convection* provide direct information about the entropy of the deep, adiabatic layers of the convection zone. In combination with standard stellar envelope models based on mixing length theory (MLT), this immediately determines the location of the *base of the convection zone*.

FROM THE SURFACE TO THE BASE OF THE CONVECTION ZONE

Fig. 1a shows the *mean* entropy as a function of depth obtained from the hydrodynamical solar surface model mentioned above by averaging over horizontal planes and over time. Clearly, this model does not extend deep enough to include those layers where the mean stratification of the convection zone is expected to

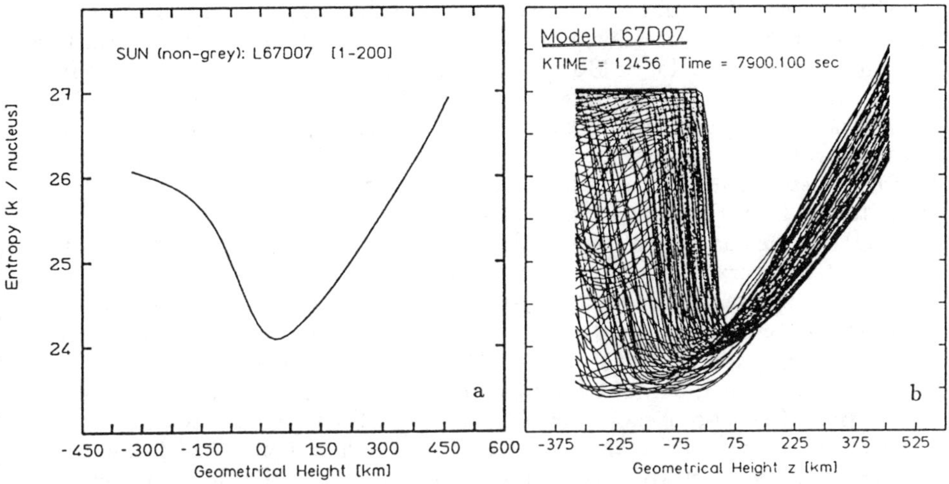

FIGURE I Depth dedendence of entropy in the solar surface layers as obtained from hydrodynamical calculations. The mean entropy (horizontal and temporal average) is shown at left (a), the spatially resolved entropy profiles after a simulated time of 7900 sec are displayed at right (b). Geometrical height zero corresponds to $\tau_{\rm Ross}=1$.

become adiabatic, more than 2000 km (5 to 6 pressure scale heights) below $\tau_{\rm Ross}=1$. While the mean entropy stratification of the hydrodynamical model does not permit a determination of the entropy corresponding to the adiabat of the deep convection zone, the *spatially resolved* entropy profiles, displayed in Fig. 1b for an arbitrary instant of the same sequence, contain additional information. We note a remarkable entropy plateau in the subsurface layers, indicating that (in contrast to the narrow downdrafts) the gas in the central regions of broad ascending flows is still thermally isolated from its surroundings in the deepest parts of the simulation volume due to the large optical depths until it reaches the layers around $\tau_{\rm Ross}=1$ where radiative losses become important.

The height of the entropy plateau is essentially independent of time and corresponds to s^*, the entropy specified with the lower boundary condition to adjust the effective temperature of the hydrodynamical model (see above). We found that the value of s^* required by the simulations to transport the solar surface flux is largely independent of the treatment of radiative transfer (grey / frequency-dependent), the horizontal and vertical extent of the computational domain, and of numerical details of the lower boundary condition. From a variety of different simulations we find for the solar case:
$s^* = 1.730 \pm 0.015 \cdot 10^9$ [erg/g/K] or 27.08 ± 0.25 [Boltzmann units per nucleus], where all results obtained so far lie within the indicated error bars.

Now we suggest that s^* may be identified directly with the entropy of the deep, adiabatic convective layers. This idea is based on the qualitative picture of solar-type convection zones proposed by Stein and Nordlund (1989) (see also Spruit et al., 1990; Chan et al., 1991), which is fundamentally different from MLT assumptions. According to this scenario the downdrafts continue all the way from the surface to the bottom of the convection zone, merging to fewer and stronger currents in successively deeper layers. The flow closes only near the base of the convective envelope. Most of the gas elements beginning to ascend from there overturn into neighbouring downflows before reaching the surface. Only a small fraction of gas continues to the surface, reaching the layers corresponding to the location of the lower boundary of our hydrodynamical models without entropy losses, following an adiabat almost up to the visible surface. Hence, s^* obtained from the simulations indicates the entropy of warm, ascending gas throughout the convection zone. This, in turn, is very nearly equal to the *mean* (horizontally averaged) entropy near the base of the convection zone because (i) the downflows are markedly entropy-deficient only near the surface and become continuously diluted by overturning entropy-neutral gas as they reach greater depths, and (ii) the fractional area of the downdrafts decreases with depth.

We have tested this idea by comparison with the results from a 3-dimensional simulation of solar convection by Nordlund and Stein (1991), extending much deeper than our models. Near the bottom of their model (2500 km below $\tau_{\rm Ross}=1$) the *mean* stratification becomes essentially adiabatic, approaching a limiting entropy which indeed closely agrees with the value of s^* derived from our convection models, indicating the validity of the above assumptions.

For comparsion with standard MLT we have calculated solar *envelope* models (not subject to central boundary conditions) including the entire solar convection zone for different values of the mixing length parameter $\alpha = l/H_p$, using the well-known Kippenhahn code in the version described by Schönberner (1983). For the outer layers ($T < 11600$ K) opacities were taken from Cox and Stewart (1970), while in the deeper layers ($T > 16000$ K) we used Los Alamos opacities from Huebner et al. (1977), with a smooth transition between the two regimes. We adopted a chemical composition as similar as possible to that used in the hydrodynamical model: X=0.7035, Y=0.2795, Z=0.017 (by mass). The choice of α uniquely determines the resulting entropy of the deep convective layers, s_0. The smaller α, the larger s_0. The mixing length model approaches the same adiabat as indicated by the hydrodynamical simulations ($s_0 = s^*$) if we choose $\alpha = 1.26 \pm 0.05$. The corresponding depth of the convection zone is $D = 199000 \pm 8000$ km ($2.05 \cdot 10^6 < T_0 < 2.30 \cdot 10^6$ K). This result is in perfect agreement with helioseismology, which according to Christensen-Dalsgaard et al. (1991) constrains the depth of the solar convection zone to $D = 200000 \pm 2000$ km ($2.17 \cdot 10^6 < T_0 < 2.23 \cdot 10^6$ K).

CONCLUDING REMARKS

The results presented above should not be interpreted as a unique calibration of α. We have only demonstrated that a solar mixing length model with $\alpha \approx 1.3$ gives the "right" entropy jump from the surface to the deep convection zone, which is consistent with both the results from hydrodynamical surface convection models and helioseismology. This does not imply that the superadibatic surface layers are also represented correctly by such a mixing length model. A detaild comparison with our hydrodynamical models shows that for no possible choice of α is standard MLT appropriate to quantitatively describe the thermal structure and dynamics of the surface layers of solar-type stars.

Finally, we would like to point out that one has to be very careful when intercomparing α-values of different authors. Unfortunately, the way α enters the formulation of MLT may be different in different codes, so that the numerical value of α required to desrcibe the same physical model may actually differ considerably from one code to another (up to a factor of 2). In addition, the numerical value of α is quite sensitive to the T-τ relation adopted for the atmosphere. Had we used a more realistic atmosphere instead of a grey T-τ relation, we probably would have needed $\alpha \approx 1.5$ to fit the same adiabat.

The resulting depth of the convection zone is sensitive to the choice of opacities. Using Cox and Stewart (1970) opacities through the whole depth range leads to D=188000 $\pm$ 8000 km. Moreover, it is clear that our method does not account for possible overshooting at the base of the convection zone.

ACKNOWLEDGMENTS

Mixing length models of the solar envelope for different vaules of α were calculated by T. Blöcker, using the Kiel Version of the Kippenhahn Code. This work was supported by the Deutsche Forschungsgemeinschaft (Ho 596/28-1).

REFERENCES

Chan, K.L., Nordlund, Å., Steffen. M., Stein, R.F.: 1991, in *Solar Interior and Atmosphere*, eds. A.N. Cox, W.C. Livingston, M.S. Matthews (University of Arizona Press, Tucson), p. 262-270

Christensen-Dalsgaard, J., Gough, D.O., Thompson, M.J.: 1991, *ApJ* **378**, 413

Cox, A.N., Stewart, J.N.: 1970, *ApJS* **19**, 243

Huebner, W.F., Merts, A.L., Magee, N.H.Jr., Arog, M.F.: 1977, Los Alamos Scientific Laboratory Report No. LA-670-M

Nordlund, Å., Stein, R.F.: 1991, in *Stellar Atmospheres: Beyond Classical Models*, eds. L. Crivellari, I.Hubeny, D.G. Hummer, NATO ASI-Series C, Vol. **341**, p. 263

Schönberner, D.: 1983, *ApJ* **272**, 708

Spruit, H.C., Nordlund, Å., Title, A.M.: 1990, *Ann. Rev. Astron. Astrophys.* **28**, 263-301

Steffen, M.: 1991, in *Stellar Atmospheres: Beyond Classical Models*, eds. L. Crivellari, I. Hubeny, D.G. Hummer, NATO ASI-Series C, Vol. **341**, p. 247

Steffen, M., Freytag, B.: 1991, in *Reviews in Modern Astronomy*, ed. G. Klare, Springer, Heidelberg, Vol. **4**, p. 43

Stein, R.F., Nordlund, Å.: 1989, *ApJ* **342**, L95-L98

Inside the Stars, IAU Colloquium 137
ASP Conference Series, Vol. 40, 1993
Werner W. Weiss and Annie Baglin (eds.)

ON THE INFLUENCE OF THE TREATMENT OF HEAVY ELEMENTS IN THE EQUATION OF STATE ON THE RESULTING VALUES OF THE ADIABATIC EXPONENT Γ_1

W. DÄPPEN[1,2], D.O. GOUGH[3,4], A.G. KOSOVICHEV[3,5], AND E.J. RHODES, JR[1,6]

[1] Department of Physics and Astronomy, University of Southern California, Los Angeles, USA

[2] Institut für Astronomie, Universität Wien, Austria

[3] Institute of Astronomy, University of Cambridge, UK

[4] Department of Applied Mathematics and Theoretical Physics, University of Cambridge, UK

[5] Crimean Astrophysical Observatory, Crimea

[6] Jet Propulsion Laboratory, California Institute of Technology, Pasadena, USA

INTRODUCTION

Helioseismology and asteroseismology put high demands on the accuracy and consistent numerical realization of the equation of state (for a review see Däppen, *these proceedings*). This is explicitly illustrated by the helioseismic determination of the helium abundance of the solar convection zone in a recent investigation by Kosovichev *et al.* (1992). In that work it was observed that details of the treatment of the heavy elements matter more than is intuitively expected. Naively, one would expect an uncertainty of less than 10^{-4} in the key thermodynamic quantity, the adiabatic gradient Γ_1. This is because in material of solar composition the heavy-element abundance is less than about 1.5×10^{-3} by number, and under solar conditions the dominant nontrivial contributions to the seismically relevant thermodynamic quantities predicted by modern equations of state agree to a few per cent, even for the much more abundant hydrogen and helium. However, Kosovichev *et al.* (1992) found that uncertainties in the treatment of the heavy elements translate into discrepancies in Γ_1 of the order of 10^{-3}, which is enough to disturb the helioseismic helium-abundance determination significantly. We briefly present the reason below. A forthcoming paper will show more detailed results, though some further information can already be found in papers by Kosovichev *et al.* (1992) and Christensen–Dalsgaard & Däppen (1992).

Kosovichev *et al.* (1992) considered solar models based on different equation-of-state tables, computed with the MHD partition-function formalism (Hummer & Mihalas, 1988; Mihalas, Däppen & Hummer, 1988; Däppen *et al.*, 1988). Such different sets of MHD tables were available from previous studies, which had, because of limited computing resources, a simplified treatment of the heavy elements. The simplification was accomplished either by reducing the number of heavy elements (but with a fixed overall mass fraction of 0.02) or by limiting

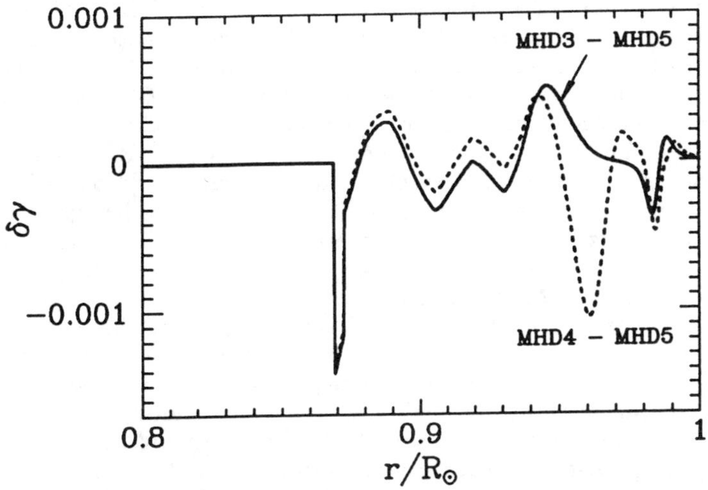

FIGURE I Differences of Γ_1 (denoted $\delta\gamma$) in a solar model

the number of atoms and ions of the heavy elements that were treated with full internal partition functions (the rest of them having ground states only). At first it would seem that the principal influence of the heavy elements is due to the type of mixture (*i.e.* the number of heavy elements and their distribution). However, it turned out that the resulting discrepancies in Γ_1 were really the result of the internal partition functions.

THE EQUATION-OF-STATE TABLES

Our nomenclature is the same as that of Kosovichev *et al.* (1992). We consider three sets of MHD tables, which all have the same mass fractions of the chemical elements H,He,C,N,O,Fe, but are different in their treatment of the internal partition functions of the heavy elements. While all three tables have full MHD partition functions for H, He and He^+, they are distinguished in the following way:

* MHD3: no detailed partition functions for C,N,O,Fe and all of their ions, (*i.e.* only ground-state contributions),
* MHD4: full MHD treatments for C,N,O and all of their ions, and for Fe and Fe^+; ground-state contributions for all other Fe ions,
* MHD5: full MHD treatment for all elements and their ions.

RESULTS AND DISCUSSION

Figure 1 shows, along a single thermodynamic path (provided by a solar model), the difference between Γ_1 computed with MHD3 and MHD5 (solid line) and the analogous difference between MHD4 and MHD5 (dashed line). Intuitively, one might expect some sort of a continuous transition from MHD3 to MHD5 *via* MHD4. However, this appears to be far from the case, because the difference between MHD4 and MHD5 exhibits a sharp spike between r/R = 0.95 and 0.97 that is not evident in the difference between MHD3 and MHD5. The reason for this unexpected behaviour was found by systematically switching on and off all internal partition functions, thus moving smoothly between MHD3 and MHD5. In an ensemble of several heavy elements (and their ions), each ionization zone produces a small well known "dip" in Γ_1 (relative to the ideal-gas value of 5/3). Since each ionization zone has its characteristic location in the Sun, Γ_1 is composed of a string of such dips. In the transition from ground-state-only (MHD3) to fully fledged MHD partition functions, each zone of ionization is shifted downwards (see, *e.g.*, Däppen, *these proceedings*), but the shifted string of all the Γ_1 dips still looks roughly the same. However, in the intermediate case (MHD4), the lower half of the dips (more precisely those belonging to Fe^{++} to Fe^{26+}) is not shifted downwards. Therefore, at the transition (in our case between the ionization zones of Fe^{+} and Fe^{++}) the upper and the lower halves of the string are pushed into each other, causing an artificial local dip from blending the inconsistently treated partition functions. This explains the spike at $r/R = 0.96$ in the dashed line of Figure 1.

CONCLUSION

In principle, it is clear that a full MHD treatment for only a part of the species can produce a certain unrealistic roughness in the thermodynamic quantities. The surprise of this study is the *size* of the effect. It is due to the strong shift of the ionization equilibrium of heavy elements caused by the MHD treatment. One should therefore definitely refrain from computing the equation of state without full partition functions for all species. Furthermore, since the treatment of bound states is controversial (see Däppen, *these proceedings*), the results shown in Figure 1 herald a diagnostic potential of helioseismology to probe this important issue of physics.

REFERENCES

Christensen-Dalsgaard, J. & Däppen, W.: 1992, *Astron. Astrophys. Review* submitted

Däppen, W.: 1992, in *Astrophysical Opacities*, eds. C. Mendoza & C. Zeippen (*Revista Mexicana de Astronomía y Astrofísica*) 141 – 149

Däppen, W., Mihalas, D., Hummer, D.G. & Mihalas, B.W.: 1988, *Astrophys. J.* **332**, 261 – 270

Hummer, D.G. & Mihalas, D.: 1988, *Astrophys. J.* **331**, 794 – 814

Kosovichev, A.G., Christensen-Dalsgaard, J., Däppen, W., Dziembowski, W.A., Gough, D.O., & Thompson, M.J.: 1992, *Mon. Not. R. astr. Soc.*, in the press

Mihalas, D., Däppen W. & Hummer, D.G.: 1988, *Astrophys. J.* **331**, 815 – 825

Inside the Stars, IAU Colloquium 137
ASP Conference Series, Vol. 40, 1993
Werner W. Weiss and Annie Baglin (eds.)

AN IMPROVED EQUATION OF STATE FOR HYDROGEN

D. SAUMON
Lunar and Planetary Laboratory, University of Arizona, Tucson AZ 85721

G. CHABRIER
Ecole Normale Supérieure, 69364 Lyon Cedex 07, France

An improved theory of fluid hydrogen at high density, based on a detailed treatment of inter-particle correlations and a self-consistent treatment of pressure ionization, has become available recently (Chabrier 1990, Saumon and Chabrier 1991,1992). We present a preliminary comparison between this new EOS (hereatfer SC) and equations of state frequently used in astrophysical contexts, namely: Fontaine, Graboske and Van Horn 1977 (FGVH), Däppen et al. 1988 (MHD) and Magni and Mazzitelli 1979 (MM).

The SC theory *predicts* a first-order phase transition in the region of pressure-ionization (the so-called Plasma Phase Transition, or PPT), between an essentially neutral mixture of atoms and molecules ($x_{e^-} < 10^{-2}$), and a partially ionized plasma ($x_{e^-} \approx 50\ \%$), with a critical point located at $P_c = 0.614\,\mathrm{Mbar}$, $T_c = 15300\,\mathrm{K}$ and $\rho_c = 0.35\,\mathrm{g/cm^3}$.

We have selected 6 isotherms common to all four EOS, except for the three hottest isotherms which required T-interpolation in the MM table. These isotherms are $\log T = 3.70, 4.10, 4.50, 5.30, 6.10$, and 6.90. The quantities compared are $\log P$ and $\log U$ for all six isotherms. Exceptions are the MHD EOS which is not plotted for the two hottest isotherms (the table available to us is limited to $\log T \leq 6$) and the MM EOS which does not provide $\log U$.

All quantities compared here are taken directly from the original EOS tables. In all cases, the zero of energy is chosen as the ground state of the H_2 molecule and the contribution of the photon gas is included. Note that the two lowest isotherms cross the PPT of the SC EOS which is represented by a gap near $\log \rho = -0.25$ in the corresponding curves.

As can be seen in the figures, the differences can be so large that it is not necessary to plot ratios or differences between the various EOS! Recall that hydrogen is ideal to better than 1% for $\log \rho < -2$ (based on the SC EOS). Surprisingly, significant differences are seen even for $\log \rho < -2$. Each quantity is briefly discussed below.

$\log P$:

MM has a rather high pressure at low-T in the ideal gas regime ($-4 < \log \rho < -2$). This feature is hard to explain. All four equations of state agree very well in the fully ionized regime where the plasma is weakly to moderately coupled and electrons are partially degenerate and begin to dominate

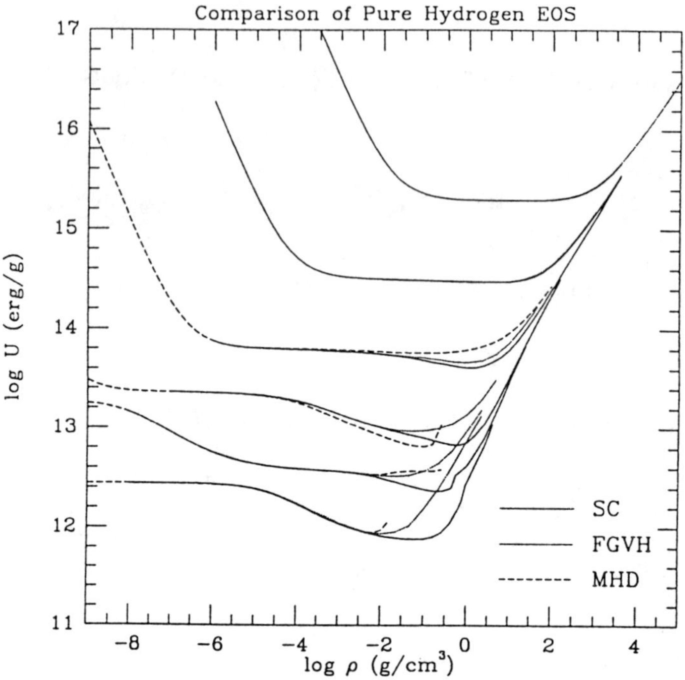

log T:
6.90
6.10
5.30
4.50
4.10
3.70

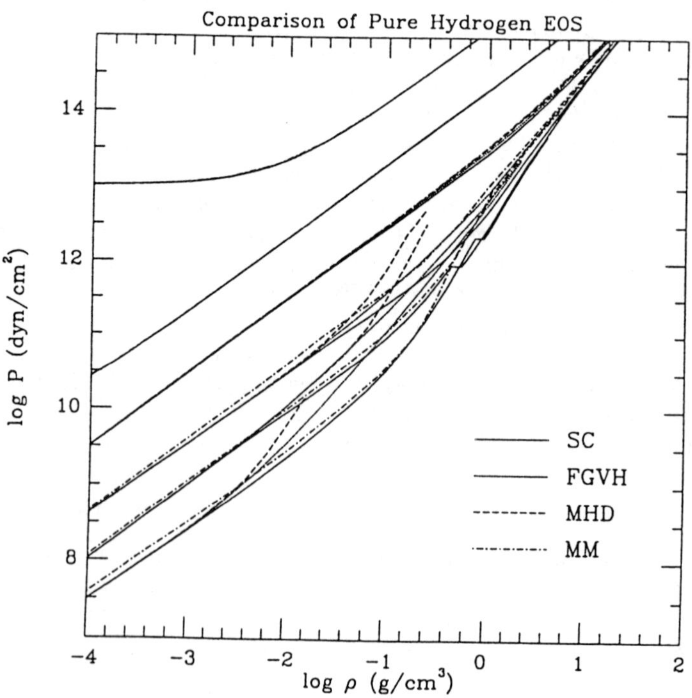

log T:
6.90
6.10
5.30
4.50
4.10
3.70

the total pressure. At low-T and at intermediate densities ($\log T \leq 4.50$, $-2 < \log \rho < -0.5$), MHD and FGVH are up to an order of magnitude above SC. A comparison with the results of shock tube experiments show that the SC EOS is quite good in that regime (Saumon and Chabrier 1991, 1992). FGVH and MHD overestimate the pressure because they use hard sphere interactions between neutral particles which are much too repulsive. Also, MHD uses a rather high value for the hard sphere diameter of H_2 (and a small one for H, see Fig. 6 of Saumon and Chabrier 1991), which enhances the effect.

$\log U$:

Most of the effects discussed above are also visible in the internal energy. The differences between SC and MHD for $\log T = 4.50$ and 5.30 are due to a different treatment of the bound states of H. The MHD theory for the treatment of bound states includes two contributions: a contribution from surrounding *neutral* particles (excluded volume), and a contribution from surrounding *charged* particles (Stark ionization). The SC model improves upon MHD for the first contribution (it includes non-linear terms in the hard sphere interactions for the ground state of the particles) but it does *not* include the second contribution, i.e. the dissolution of the upper levels due to Stark ionization by the microfield of the plasma (see Saumon and Chabrier 1992 for details). The effect of this missing physical effect in SC becomes noticeable when thermal excitation and ionization are significant.

The previous remarks still apply for the second derivatives (not shown here). Substantial disagreement again appears at low to intermediate temperatures in the $-3 < \log \rho < 1$ domain, where ∇_{ad}, χ_ρ and χ_T are very sensitive to the interaction models and the treatment of the internal partition functions.

The present hydrogen EOS has been combined with a more simple helium EOS, using the additive volume rule. In a first application of this H/He EOS, new static interior models and evolutionary models of Jupiter, Saturn and brown dwarfs were computed. They are presented in Chabrier et al. 1992 and Saumon et al. 1992, where the effect of the PPT on the mechanical and thermal structures of these objects is discussed in details.

REFERENCES

Chabrier G., *J. de Physique*, **51**, 1607 (1990)

Chabrier G., Saumon D., Hubbard W. B., and Lunine J. I., *ApJ*, **391**, 817, (1992)

Däppen, W., Mihalas, D., Hummer, D. G., and Mihalas, B. W. *ApJ*, **332**, 261 (1988)

Fontaine G., Graboske and Van Horn H. M., *ApJS*, **35**, 293 (1977)

Magni M. and Mazzitelli I., *A&A*, **72**, 134 (1979)

Saumon D., Chabrier G., *Phys. Rev. A*, **44**, 5122, (1991); *ibid.*, in press (1992)

Saumon D., Hubbard W. B., Chabrier G., and VanHorn H. M., *ApJ*, **391**, 827 (1992)

Inside the Stars, IAU Colloquium 137
ASP Conference Series, Vol. 40, 1993
Werner W. Weiss and Annie Baglin (eds.)

THE NON-EXPLOSIVE MODES OF HYDROGEN BURNING : A BRIEF REVIEW

MARCEL ARNOULD AND NAMI MOWLAVI
Institut d'Astronomie et d'Astrophysique
Université Libre de Bruxelles, C.P. 165 ; Av. F.D. Roosevelt, 50
1050 Bruxelles, BELGIUM

ABSTRACT The present status of our theoretical and experimental knowledge of the rates of hydrogen-burning reactions in hydrostatic stellar conditions is very briefly reviewed. Special emphasis is put on reactions of the proton-proton chain, CNO cycles, NeNa cycle and MgAl chain whose rates are still affected by uncertainties that can have an impact on predictions regarding astrophysical issues of interest. This concerns in particular the solar neutrino problem, or the nucleosynthesis accompanying the CNO, NeNa and MgAl chains of reactions, as it is revealed in particular by the first dredge-up episode. A brief discussion of the possibility of reducing the present rate uncertainties is presented.

1.INTRODUCTION

Throughout their lives, stars transform nuclear and gravitational energy into photons and neutrinos. Such a transformation takes place through a succession of "controlled" thermonuclear burning stages interspersed with phases of gravitational contraction. The latter stages are responsible for a temperature increase, while the former ones produce nuclear energy and lead to composition changes.

The first major nuclear burning phase able to lead to the "complete" (hydrostatic plus thermal) equilibrium of a star is the combustion of hydrogen in its central regions, settling the object on the main sequence of the Hertzsprung-Russell diagram. After its exhaustion in the stellar core, hydrogen continues to burn in peripheral layers at later evolutionary stages.

Since the pioneering works of von Weizsäcker (1937, 1938), Bethe and Critchfield (1938) and Bethe (1939), it has become clear that those central and peripheral burning episodes take place through the "cold" p-p chains or CNO cycles, the dominant reaction sequence depending upon the stellar mass and initial composition (e.g. Rolfs and Rodney 1988, hereafter RR88). Other hydrogen burning modes, the "cold" NeNa and MgAl reaction chains have been identified more recently. They play a minor role in the stellar energy budget. In contrast, they are of importance in the synthesis of the Na to Al isotopes, especially in massive stars (e.g. RR88; Prantzos et al. 1986).

Much theoretical and laboratory effort has been devoted recently to an improvement of our knowledge of many of the nuclear reactions involved in the cold H-burning modes. In that field, experiments face enormous problems and represent a real challenge. Many of these difficulties relate to the fact that the energies of astrophysical interest for the charged particle induced reactions involved in the H-burning reactions mentioned above are much lower than the Coulomb barrier energies. As a consequence, the corresponding cross sections can dive into the nanobarn to picobarn abysses. In general, it has not been possible yet to measure directly such small cross sections. Theoreticians are thus requested to supply reliable extrapolations from the lowest energies attained experimentally to those of astrophysical relevance.

Progress in our knowledge of the rates of H-burning reactions is indispensable in order to put various astrophysical predictions on safer grounds, like the solar neutrino flux, or the surface composition of low- or intermediate-mass Red Giant or Asymptotic Giant Branch stars having possibly experienced various dredge-up episodes able to pollute the outer convective zone with the products of central or shell H-burning (e.g. Sneden 1991; Iben 1991 for reviews). If freed from nuclear physics uncertainties to the largest possible extent, those calculated surface abundances could provide key constraints on the stellar structure from the main sequence to the Giant phase.

The aim of this contribution is to present a (non exhaustive) review of some recent experimental and theoretical efforts devoted to some important reactions involved in the cold modes of H burning. As such, it provides some update of more thorough presentations that can be found in e.g. RR88.

2. THE p-p CHAINS

The p-p chains, which involve the reactions displayed in Fig. 1, are expected to take place in main sequence $M \lesssim 1.3\,M_\odot$ stars of typical Pop I composition, as well as in more massive stars if their CNO content is low enough. Their net result is the transformation of H into ^{4}He.

Much experimental and theoretical efforts have been devoted to the reactions of the p-p chains, that flurry of activity being of course directly connected with the famed "solar neutrino problem" (e.g. Bahcall 1989; Bahcall and Pinsonneault 1992, hereafter BP92; Ulrich, this conference). The status of our knowledge concerning the pp reactions has already been reviewed in great detail recently (e.g. RR88; Bahcall 1989; BP92), so that we just limit ourselves here to some brief comments.

As no direct measurement of the $^1H(p,e^+\nu)^2H$ and $^1H(pe^-,\nu)^2H$ rates can be expected in any foreseeable future, their knowledge relies entirely on theoretical evaluations. Recent works on that subject have been exhaustively reviewed by BP92 (note that Kurucz (1992) has claimed that the $^1H(pe^-,\nu)^2H$ rate has been incorrectly evaluated in the past. This claim has to be disregarded, as shown in an unpublished note by Bahcall and Salpeter).

A recent reanalysis of the $^3He(^3He,2p)^4He$ rate taking into account laboratory electron screening corrections (see below) leads BP92 to adopt an astrophysical S-factor that is 3% lower than reported by Bahcall (1989).

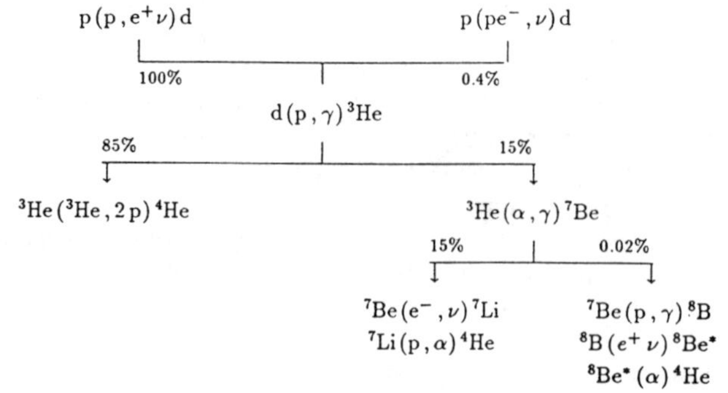

Figure 1. Reactions involved in the pp chains. The indicated percentages give the fractions of the $4p \rightarrow \alpha + 2e^+ + 2\nu$ transformations due to the various possible pp branchings in solar conditions (Bahcall 1989). The very weak (0.00002%) $^3He(p, e^+\nu)\,^4He$ termination is not indicated

Based on theoretical arguments, it has been suggested that a low-energy resonance might exist in that reaction (e.g. Fetisov and Kopysov 1975), leading to an enhancement of its rate and a corresponding lowering of the relative importance of the $^3He + {}^4He$ channel, and thus of the solar ^8B neutrino flux. The latest experimental data on that reaction (Krauss et al. 1987), as well as previous laboratory works, are consistent with a non-resonant reaction mechanism, at least down to $E_{cm} = 24.5$ keV. They cannot exclude the presence of a narrow ($\Gamma_R << 24$ keV) resonance at lower energies. Recent microscopic calculations (Vasilevskii and Rybkin 1989) do not show any sign of such a resonance. Even so, its existence cannot be firmly excluded, and it is of some interest to test its impact on solar models and on the predicted ^8B ν-flux. To this end, evolutionary computations have been performed with the addition to the $^3He(^3He, 2p)\,^4He$ non-resonant rate of the contribution of a resonance able to reduce by a factor of 2 the standard ^8B ν-flux while fulfilling the above-mentioned experimental constraints. Those calculations indicate that the required reduction is achieved with a resonance increasing the standard rate by a factor of ≈ 5. Concomitantly, a ≈ 55 % reduction in the ^7Be ν-flux is derived, while the total ν-flux and the other observable solar properties (including the ^3He surface abundance) can be recovered.

As far as the remaining ^3He burning channels are concerned, the $^3He(\alpha, \gamma)\,^7Be$ rate is estimated to be known to within a 3σ error of 6% (Bahcall 1989), while the recently studied very weak $^3He(p, e^+\nu)\,^4He$ branching is considered still to be uncertain by a factor of 6 (BP92). This affects only the corresponding (hep) neutrino flux.

Even if it just represents a weak branching in the pp chains, the ^7Be proton capture rate is of key importance in solar neutrino calculations.

Unfortunately, there are still significant uncertainties in its low-energy cross section. This is due in particular to the extreme difficulty of performing direct measurements on that reaction (e.g. RR88). In such conditions, indirect methods complemented with model predictions have to be used, this approach having uncertainties of its own. From a critical analysis of those problems, Barker and Spear (1986) propose an astrophysical S-factor for the reaction that is $\approx$ 30% lower than the latest model predictions (Johnson et al. 1992) adopted by BP92.

The ^{7}Be electron-capture rate, to which the solar ^{8}B ν-flux is inversely proportional, has been reexamined in detail by Johnson et al. (1992). Their proposed slight revision has an unsignificant impact on the ν-flux calculations.

We end this section with some comments on the role of electron screening on nuclear reaction rates at very low energies. Since the pioneering studies of Schatzman (1948) and Salpeter (1954) on stellar electron screening, it is well known that charged-particle induced reactions taking place in a stellar plasma are more rapid than if they were just involving bare nuclei. It has been a common practice to evaluate the stellar rates by multiplying the experimentally derived rates by some correction factor f_* that is just slightly in excess of unity (by some percent to some tens of percent) in conditions relevant to main sequence stars.

Painstaking laboratory efforts have made some cracks appear in this whole procedure. Experiments conducted on light species down to energies below $\approx$ 10 keV indeed reveal an unexpected rise of the astrophysical S-factor with decreasing energy (Engstler et al. 1988; Engstler et al. 1992). This behavior has been assigned to the effect of an electron screening in laboratory conditions, leading to an energy dependent laboratory correction factor f_{exp}. Model predictions do not fully account for the experimentally determined f_{exp}, even if they explain the S-factor rise in a qualitative way (Assenbaum et al. 1987; Bracci et al. 1990, 1991; Azrak 1992; Wauters 1992).

At this point, the situation thus appears to be more intricate than previously imagined, a multistep process being required in order to go from laboratory data to stellar rates : (1) the laboratory electron screening effect (f_{exp}) has to be extracted from the experimental cross section data in order to get the S-factors for bare nuclei, and (2) the latter factors have to be corrected for the stellar electron screening (f_*) in order to evaluate the stellar rates. This procedure obviously increases the uncertainties that can affect those rates.

3. THE CNO CYCLES

The reactions involved in the CNO cycles are represented in Fig. 2. The nuclear physics problems raised by their rate determinations have been discussed in several places (e.g. RR88), so that we limit ourselves here to some brief comments on the nuclear physics and astrophysics aspects of those reaction chains.

The net result of the CNO cycles is the transformation of H into ^{4}He using the C, N and O isotopes as catalysts. Cycle I (or CN cycle) is the

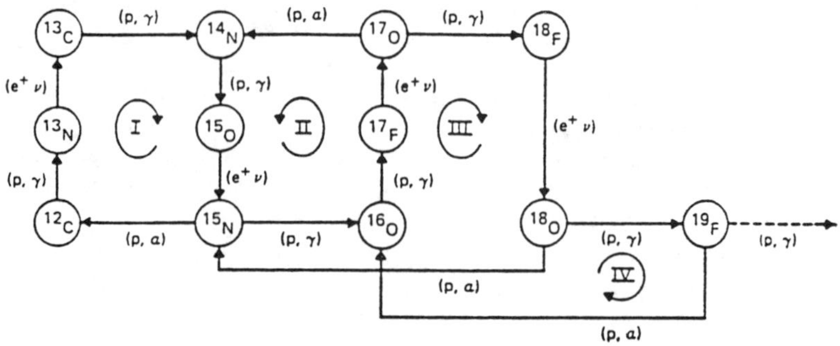

Figure 2. Illustration of the four CNO cycles involved in the transformation of H into ^{4}He. The weak $^{19}F(p,\gamma)^{20}Ne$ leak out of the cycles is also indicated

largest energy producer (e.g. RR88), while all cycles I to IV are important as far as nucleosynthesis is concerned. Figure 3 presents the evolution of the abundances of the C, N, O and F isotopes as the burning proceeds at a given temperature and density, as well as the final abundances of those isotopes at hydrogen exhaustion versus temperature (assumed to remain constant during the whole burning). It clearly appears that the initial catalysts are mainly transformed into ^{14}N. This well known effect results from the relative slowness of the $^{14}N(p,\gamma)^{15}O$ reaction. It now remains to be seen if those predictions concerning the nucleosynthesis and energy production can be significantly affected by remaining reaction rate uncertainties.

The status of our knowledge concerning the rates of the reactions involved in cycle I of Fig. 2 has been reviewed by RR88, while Bahcall (1989) provides an estimate of the remaining uncertainties.

The reactions of cycles II-IV of Fig. 2 have been discussed in detail by RR88. Since their review, much experimental work has been devoted to the $^{17}O(p,\alpha)^{14}N$ and $^{17}O(p,\gamma)^{18}F$ reactions in order to reduce the very large uncertainties that were still affecting their rates, as reported by e.g. Caughlan and Fowler (1988; hereafter CF88). Those experiments and their results are discussed in detail by Bogaert et al. (1989), Landré et al. (1989) and Berheide et al. (1992). At temperatures of astrophysical interest, it appears that the $^{17}O(p,\alpha)^{14}N$ and $^{17}O(p,\gamma)^{18}F$ rates could be substantially different from the widely used CF88 values. However, uncertainties still remain. They affect in particular the $^{17}O(p,\alpha)^{14}N$ / $^{17}O(p,\gamma)^{18}F$ branching ratio that determines the leakage from cycle II to cycle III. They also have a profound influence on the detailed CNO yield predictions. The standard results shown in Fig. 3 are obtained with values for the ^{17}O p-capture rates that are the geometrical means of the lower and upper limits provided by Landré et al. (1990). The impact of their estimated nuclear uncertainties on the oxygen isotopic composition is displayed in Fig. 4. Also indicated are the results obtained with the upper

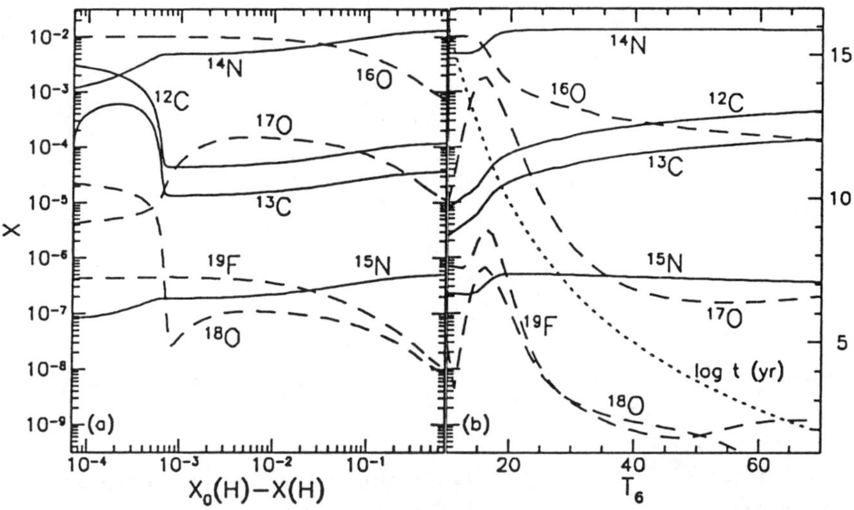

Figure 3. (a) Time evolution of the mass fractions of the stable nuclides involved in the CNO cycles versus the amount of hydrogen burned at constant temperature T = 25x10^6 K and density ρ = 100 g/cm^3. X_0(H) = 0.7 is the initial H mass fraction ; (b) same as (a), but versus T_6 (temperature in 10^6 K) at H exhaustion (X(H) = 10^{-9}). The short-dashed line indicates the logarithm of the H-burning time (in yr) to be read on the right-hand side ordinate

limits of the ^{17}O p-capture rates derived by Berheide et al. (1992), those upper limits being inferior to the lower bounds proposed by Landré et al. (1990). The impact of the revised ^{17}O p-capture rates has also been studied in the framework of more realistic stellar models by Landré et al. (1990).

The ^{18}O burning has been studied recently by Champagne and Pitt (1986) and Berheide et al. (1992). It proceeds predominantly through $^{18}O(p,\alpha)^{15}N$, $^{18}O(p,\gamma)^{19}F$ being roughly 200 times slower at relevant temperatures.

Recently, the ^{19}F proton capture rates have been studied experimentally and theoretically (Kious 1990 ; Bulski 1992, and references therein). Even if uncertainties remain in the astrophysical rates, it can be concluded that $^{19}F(p,\alpha)^{16}O$ is much faster than $^{19}F(p,\gamma)^{20}Ne$, preventing any important leakage out of the CNO cycles to ^{20}Ne. In addition, the (p,α) channel is so rapid that the ^{19}F amount that can emerge from cycle IV is too low for accounting for the solar system ^{19}F, in spite of its relatively very low abundance (Goriely et al. 1989 ; Kious 1990). In the same way, the

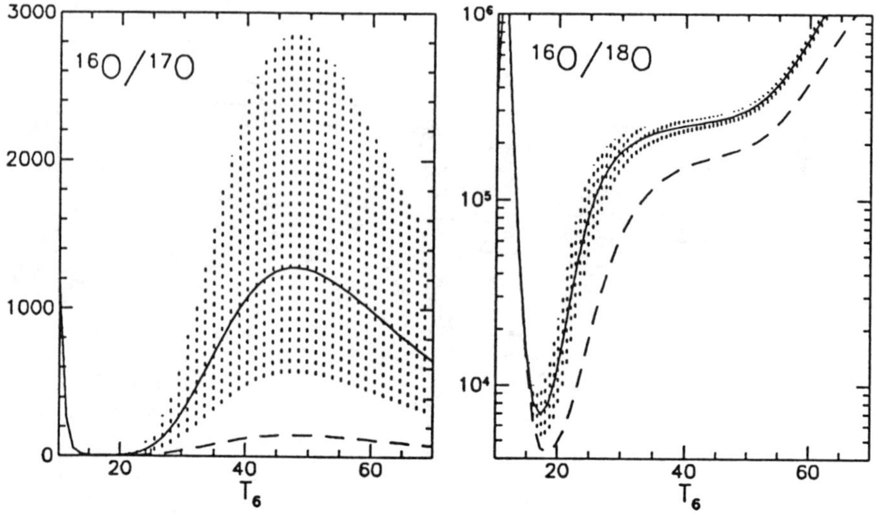

Figure 4. $^{16}O/^{17}O$ and $^{16}O/^{18}O$ abundance ratios versus (constant) temperature (in 10^6 K) at H exhaustion for various values of the $^{17}O(p,\alpha)^{14}N$ and $^{17}O(p,\gamma)^{18}F$ rates. The solid and dashed lines are obtained with the intermediate rates used in Fig. 3. The shaded areas indicate the impact of the remaining nuclear uncertainties claimed by Landré et al. (1990). The dashed lines result from the use of the upper limit put by Berheide et al. (1992) on the ^{17}O p-capture rates

CNO cycles cannot account for the ^{19}F recently observed in Asymptotic Giant Branch stars (Jorissen et al. 1992).

4. THE NeNa CYCLE

The nuclear information of relevance for that cycle, represented in Fig. 5, is reviewed by RR88. Let us just stress that still large uncertainties remain in the rates of many of the involved proton captures (see also the analysis of the status of the $^{20}Ne(p,\gamma)^{21}Na$ reaction by Prantzos et al. 1991). Those uncertainties cannot affect the conclusion that the $^{23}Na(p,\alpha)^{20}Ne$ reaction rate is large enough compared to the competing $^{23}Na(p,\gamma)^{24}Mg$ channel to guarantee cycling.

Figure 6 presents the NeNa cycle yield predictions derived with the CF88 rates. When their expressions contain an uncertainty factor F, "intermediate" rate values (F=0.1) are selected. Upper (F=1) and lower (F=0) limits to the rates involved in the production and destruction of ^{23}Na

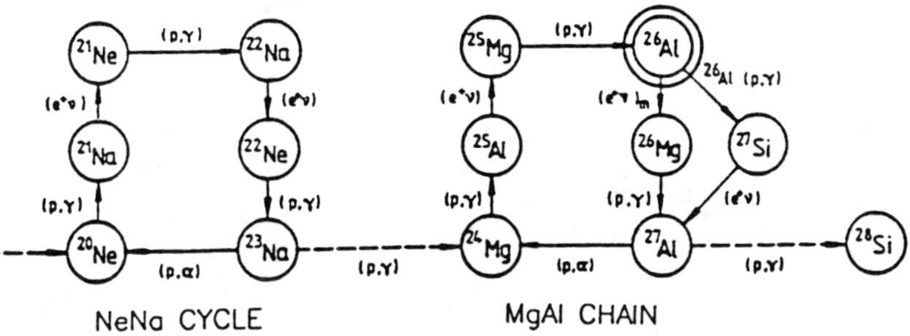

Figure 5. Reactions involved in the NeNa cycle and in the MgAl chain. The leakages into and out of those reaction sequences are also schematized. The ground ($^{26}Al^{g}$; $t_{1/2} = 7.05 \times 10^5$ yr) and isomeric ($^{26}Al^{m}$; $t_{1/2} = 6.3$ s) states of ^{26}Al are considered as two separate species, as they cannot reach thermal equilibrium in realistic conditions of operation of the MgAl chain (Nørgaard 1979 ; Ward and Fowler 1980).

are also used to evaluate the impact of the nuclear uncertainties on the ^{22}Ne and ^{23}Na yields. It is seen that only modest abundance spreads result.

The nucleosynthesis associated with the NeNa cycle has been computed in the framework of detailed stellar models, and its impact on the surface composition of Wolf-Rayet or of Red Giant stars has been analyzed (e.g. Prantzos et al. 1986, 1991). Let us note in particular that the Na excess observed in certain yellow supergiants (e.g. Boyarchuk et al. 1989) and seemingly correlated with the $^{12}C/^{13}C$ ratio has been interpreted in the framework of the NeNa cycle. However, that interpretation has a chance to account successfully for the observations only if some special assumptions are made on the initial abundance of ^{22}Ne (e.g. Denisenkov 1990), or if the $^{20}Ne(p,\gamma)^{21}Na$ rate is artificially enhanced by a factor of at least ≈ 10 over its currently adopted CF88 value (Prantzos et al. 1991). Both proposed solutions appear quite unlikely. It remains that the observed Na excess presents a new and interesting challenge for theorists of stellar evolution and nucleosynthesis.

5. THE MgAl CHAIN

Much nuclear physics effort has been devoted to the reactions of the MgAl chain displayed in Fig. 5, and in particular to those involved in the production or destruction of ^{26}Al and ^{27}Al. This directly relates to the recognition that $^{26}Al^{g}$ is of crucial importance in γ-ray astronomy and in cosmochemistry ($^{26}Al^{m}$ is so short-lived that it is of no practical interest in

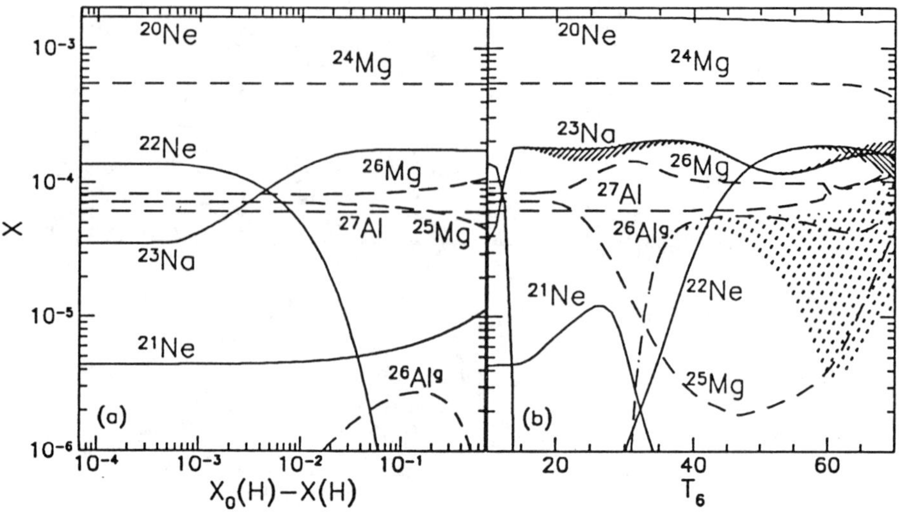

Figure 6. Same as Fig. 3, but for the nuclides involved in the NeNa cycle (solid lines) and MgAl chain (dashed lines). Data for $^{26}Al^{g}$ are also presented. The hatched zones indicate the ^{22}Ne, ^{23}Na or $^{26}Al^{g}$ abundance uncertainties of nuclear origin (see text and Fig. 7)

those respects). On the one hand, the 1.8 MeV γ-ray emission observed in the galactic disk is attributed to the decay of $\approx 3\,M_\odot$ of $^{26}Al^{g}$ that have been present in the interstellar medium over the last $\approx 10^6$yr (e.g. Schönfelder and Varendorff 1991 ; Clayton and Leising 1987 for reviews). On the other hand, there is now ample observational evidence that $^{26}Al^{g}$ has decayed in situ in various meteoritic inclusions, as well as in identified single grains of likely stellar origin (e.g. Wasserburg 1985 ; Zinner et al. 1991 ; Amari et al. 1992).

Recently, the rate of the ^{25}Mg proton captures leading to both $^{26}Al^{g}$ and $^{26}Al^{m}$ has been studied experimentally by Iliadis et al. (1990), and is now put on quite safe grounds. The derived results are significantly different (factor $\gtrsim 5$) from the previously adopted CF88 ones for temperatures T $\lesssim 10^8$ K. The branching ratio for forming $^{26}Al^{g}$ through ^{25}Mg p-capture has received much experimental attention as well (Endt and Rolfs 1987). The $^{26}Al^{g}(p,\gamma)^{27}Si$ reaction has also been studied recently (Vogelaar 1989). The deduced rate appears to be lower than the $^{26}Al^{g}$ β-decay rate for T $\lesssim$ $3 - 4\mathrm{x}10^8$ K. In other words, $^{26}Al^{g}$ is predominantly destroyed by β-decay in those conditions, and by $^{26}Al^{g}(p,\gamma)^{27}Si$ if the MgAl chain can indeed take place at higher temperatures, which may not be the case in realistic astrophysical situations. If indeed the $^{26}Al^{g}$ p-capture channel is significant, the uncertainties in its rate influence the reliability of the $^{26}Al^{g}$ yield predictions. In addition $^{26}Al^{g}$ is connected electromagnetically to $^{26}Al^{m}$.

The γ-ray transitions of relevance in astrophysical conditions have been discussed recently (Vogelaar 1989). The $^{26}Al^{m}$ isomer can also transform by $^{26}Al^{m}(p,\gamma)^{27}Si$. Due to the very short $^{26}Al^{m}$ β-decay half-life, that reaction has not been amenable yet to any experimental study, the evaluation of its rate thus relying on theoretical evaluations only (e.g. Arnould et al. 1980). Recent experiments have also been carried out on the ^{26}Mg and ^{27}Al proton capture rates. The new data obtained for $^{26}Mg(p,\gamma)^{27}Al$ (Iliadis et al. 1990) leave uncertainties that are less than a factor of ≈ 2 for T $\lesssim$ 3x10^7 K, but that can reach factors of $\approx$ 10 to 100 in the 4x10^7 $\lesssim T \lesssim$ 6x10^7 K range. As already stressed before, these temperatures may be a bit high for the operation of the cold MgAl chain in realistic astrophysical situations, so that the corresponding large uncertainties may have only little astrophysical consequences in practice.

From their experimental data concerning $^{27}Al(p,\gamma)^{28}Si$, Iliadis et al. (1990) suggest to use the CF88 rate. In contrast, the CF88 rate for $^{27}Al(p,\alpha)^{24}Mg$ is subject to important changes, following the laboratory work of Champagne et al. (1988) and Timmermann et al. (1988). From these recent data, it appears in particular that $^{27}Al(p,\alpha)^{24}Mg$ may be slower than $^{27}Al(p,\gamma)^{28}Si$, so that a MgAl cycle cannot set up, in contrast to what was thought previously. Further work is required in order to reduce the uncertainties still affecting the $^{27}Al(p,\alpha)^{24}Mg$ and $^{27}Al(p,\gamma)^{28}Si$ rates, and thus the precise extent of the leakage out of the MgAl region.

Figure 6 presents the abundances of the Mg and Al isotopes involved in the MgAl chain. These results are obtained in the framework of a schematic astrophysical model assuming constant temperature and density during hydrogen burning. On the other hand, the CF88 rates are used, except for the above-mentioned reactions for which more recent data are available. In this case, intermediate rates (geometrical means) are adopted when only upper and lower bounds can be provided experimentally. It appears in particular that a substantial amount of $^{26}Al^{g}$ can be produced at the expense of ^{25}Mg (Fig. 6(a)). However, not much $^{26}Al^{g}$ remains at hydrogen depletion for T $\lesssim$ 3x10^7 K (Fig. 6(b)). This is essentially due to the fact that the synthesized $^{26}Al^{g}$ has time to β-decay before the end of H burning.

The impact of the uncertainties remaining in the rates of the reactions involved in the production and destruction of $^{26}Al^{g}$ and ^{27}Al on the predicted abundances of those nuclides can be evaluated from Figs. 6 and 7 (see also Iliadis et al. 1990).

The MgAl chain has also been studied in the framework of detailed stellar models. In particular, the $^{26}Al^{g}$ yields from the core H-burning phase of Wolf-Rayet stars (e.g. Prantzos et al. 1986 ; Meynet and Arnould 1992), or from the shell H burning in low- and intermediate- mass thermally pulsing Asymptotic Giant Branch stars (Forestini et al. 1992) have been computed. It is concluded that those sites can be significant contributors to the present day interstellar medium ^{26}Al, even if they cannot account for the whole 3 $M_{\odot}$ of ^{26}Al inferred from observation (see also Prantzos 1991). They can also provide a satisfactory explanation for a large variety of $^{26}Al/^{27}Al$ ratios that are inferred at the time of formation of inclusions or grains found in meteorites (Meynet and Arnould 1992).

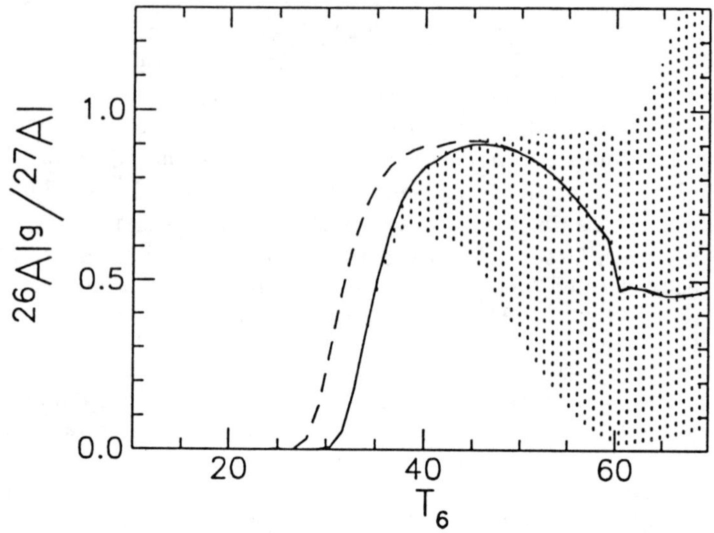

Figure 7. $^{26}Al^g/^{27}Al$ abundance ratio versus (constant) temperature (in 10^6 K) close to H exhaustion. The solid and dashed lines correspond to final H mass fractions $X(H) = 10^{-9}$ and 10^{-2}, respectively. They are obtained with the adoption of intermediate reaction rates, when appropriate (see text). The minimization or maximization of the abundance ratio is obtained by taking lower or upper bounds (as appropriate) on the rates of $^{26}Mg(p,\gamma)^{27}Al$ (Iliadis et al. 1990), $^{26}Al^g(p,\gamma)^{27}Si$ (Vogelaar 1989), $^{27}Al(p,\gamma)^{28}Si$ (CF88), and $^{27}Al(p,\alpha)^{24}Mg$. The lower bound on the rate of the latter reaction is deduced from the contribution of the $E_R \geq 505$ keV resonances evaluated by Timmermann et al. (1988). The maximum contributions of the resonances at lower energies reported by Champagne et al. (1988) are added in order to obtain the upper bound on that reaction rate

6. CONCLUSIONS AND PROSPECTS

Over the years, heroic laboratory efforts complemented with lengthy calculations have helped greatly improving our knowledge of the rates of a large variety of nuclear reactions of importance in stellar evolution and nucleosynthesis. A substantial fraction of those activities have focussed on the various cold modes of hydrogen burning briefly reviewed here.

In spite of that, nuclear uncertainties still remain, and can blur (in combination, of course, with uncertainties of astrophysical origin) our understanding of various questions of great astrophysical interest, like the solar neutrino problem, or the surface C, N, and O elemental and isotopic composition of stars having experienced dredge-up episodes. The puzzle raised by the Na excess observed in yellow supergiants, or the origin of the ^{26}Al amount inferred to exist in the present day interstellar medium or in meteoritic inclusions or grains at the time of their solidification are other exemples of fascinating questions.

In order to reduce the astrophysically important remaining uncertainties of nuclear origin, further theoretical efforts are required, along with the development of improved and new experimental techniques. Because the cross sections drop by many orders of magnitude when the energies decrease from the Coulomb barrier to the energy range of astrophysical interest, the experimental investigations require the availability of high ion-beam currents, of stable and pure targets, as well as of efficient detectors (e.g. RR88).

New accelerators able to deliver higher currents than previously have just been built or are currently tested. On the other hand, windowless gas targets or specially prepared solid targets can provide the required stability and purity. The implantation technique has also been shown rencently to be able to produce isotopically enriched and contaminant-free targets capable of withstanding high beam powers. As far as detectors are concerned, their choice and design require a careful study for each specific experiment. Much progress has been made recently, in particular through the development of new techniques in the detection of neutrons, charged particles, and γ-rays. Of course, the sensitivity of the best designed detectors is still limited by the cosmic-ray background. Its reduction by nearly three orders of magnitude can be achieved by carrying out experiments deep underground. Such an exciting possiblity is at the center of the LUNA project to be conducted at the Gran Sasso underground laboratory (Arpesella et al. 1991).

Methods for studying astrophysically important reactions through experiments carried on other reactions from which properties of astrophysical interest can be deduced have also proved in the past to be very helpful. They certainly deserve further developments. Exemples of such indirect techniques include the measurement of inverse reactions, or the Coulomb dissociation approach.

All those dedicated nuclear physics efforts will undoubtedly bring much progress in our knowledge of the stellar nuclear "alchemy". In particular, they will help reducing possible ambiguities in the interpretation of solar neutrino experiments (Bahcall 1989).

Aknowledgements. This work has been supported in part by the SCIENCE Program SC1-0065. M.A. is Chercheur Qualifié F.N.R.S. (Belgium), and N.M. is Boursier I.R.S.I.A.

REFERENCES

Amari, S., Hoppe, P., Zinner, E. and Lewis, R. 1992, *Lunar Planetary Science XXIII* , to appear

Arnould, M., Nørgaard, H., Thielemann, F.K. and Hillebrandt, W. 1980, *Astrophys. J.*, **237**, 931

Arpesella, C. et al. 1991, *LNGS Report 91/18*
Assenbaum, H.J., Langanke, K. and Rolfs, C. 1987, *Z. Phys.*, **A327**, 461
Azrak, Z. 1992, *Ph.D. Thesis* , Université Libre de Bruxelles (unpublished)
Bahcall, J.N 1989, *Neutrino Astrophysics* (Cambridge University Press, Cambridge)
Bahcall, J.N. and Pinsonneault, M.H. 1992, *Astrophysics Preprint Series IASSNS-AST 92/10* (to be published in *Rev. Mod. Phys.*) (BP 92)
Barker, F.C. and Spear, R.H. 1986, *Astrophys. J.*, **307**, 847
Berheide, M., Rolfs, C., Schröder, U. and Trautvetter, H.P. 1991, *preprint (Bochum)*
Bethe, H.A. 1939, *Phys. Rev.*, **55**, 103 and 434
Bethe, H.A. and Critchfield, C.L. 1938, *Phys. Rev.*, **54**, 248 and 862
Bogaert, G.L., Landré, V., Aguer, P., Barhoumi, S., Kious, M., Lefebvre, A., Thibaud, J.-P. and Bertault, D. 1989, *Phys. Rev.*, **C39**, 1972
Boyarchuk, A.A., Gubeny, I., Kubat, I., Lyubimkov, L.S. and Sakhibullin, N.A. 1989, *Astrophysics*, **28-29**, 197
Bracci, L., Fiorentini, G., Melerzhik, V.S., Mezzorani, G. and Quarati, P. 1990, *Nucl. Phys.*, **A513**, 316
Bracci, L., Fiorentini, G., Melerzhik, V.S., Mezzorani, G. and Pasini, P. 1991, *Phys. Lett.*, **A152**, 456
Bulski, G. 1992, *Ph.D. Thesis,* Universität Stuttgart (unpublished)
Caughlan, G.R. and Fowler, W.A. 1988, *Atomic Data Nucl. Data Tables*, **40**, 283 (CF88)
Champagne, A.E. and Pitt, M.L. 1986, *Nucl. Phys.*, **A457**, 367
Champagne, A.E., Cella, C.H., Kouzes, R.T., Lowry, M.M., Magnus, P.V., Smith, M.S. and Mao, Z.Q. 1988, *Nucl. Phys.*, **A487**, 433
Clayton, D.D. and Leising, M.D. 1987, *Phys. Rep.*, **144**, 1
Denisenkov, P.A. 1990, *Astrophysics*, **31**, 588
Endt, P.M. and Rolfs, C. 1987, *Nucl. Phys.*, **A467**, 261
Engstler, S. Krauss, A., Neldner, K., Rolfs, C., Schröder, U. and Langanke, K. 1988, *Phys. Lett.*, **B202**, 179
Engstler, S., Raimann, G., Angulo, G., Greife, U., Rolfs, C., Schröder, U., Somorjai, E., Kirch, B. and Langanke, K. 1992, *Phys. Lett.*, **B279**, 20, and *Z. Phys.*,**A342**, 471
Fetisov, V.N. and Kopysov, Y.S. 1975, *Nucl. Phys.*, **A239**, 511
Forestini, M., Paulus, G. and Arnould, M. 1991, *Astron. Astrophys.*, **252**, 597
Forestini, M., Goriely, S., Jorissen, A. and Arnould, M. 1992, *Astron. Astrophys.*, **261**, 157
Goriely, S., Jorissen, A. and Arnould, M. 1989, in *Proceedings of the 5th Workshop on Nuclear Astrophysics*, eds. W. Hillebrandt and E. Müller, **MPA/P1** (Max-Planck-Institut für Astrophysik, Garching), p. 60
Iben, I.Jr. 1991, in *Evolution of Stars : The Photospheric Abundance Connection*, IAU Symposium 145 (eds. G. Michaud and A. Tutukov (Kluwer Academic Pub., Dordrecht), p. 257
Iliadis, Ch., Schange, Th., Rolfs, C., Schröder, U., Somorjai, E., Trautvetter, H.P., Wolke, K., Endt, P.M., Kikstra, S.W., Champagne, A.E., Arnould, M. and Paulus, G. 1990, *Nucl. Phys.*, **A512**, 509

Johnson, C.W., Kolbe, E., Koonin, S.E. and Langanke, K. 1992, *preprint*
Jorissen, A., Smith, V.V. and Lambert, D.L. 1992, *Astron. Astrophys.*, **261**, 164
Kious, M. 1990, *Ph. D. Thesis*, C.S.N.S.M. Orsay (unpublished)
Krauss, A., Becker, H.W., Trautvetter, H.P. and Rolfs, C. 1987, *Nucl. Phys.*, **A467**, 273
Kurucz, R.L. 1992, *Harvard-Smithsonian Center for Astrophysics preprint 3360*
Landré, V., Prantzos, N., Aguer, P., Bogaert, G., Lefebvre, A. and Thibaud, J.-P. 1990, *Astron. Astrophys.*, **240**, 85
Landré, V., Aguer, P., Bogaert, G., Lefebvre, A., Thibaud, J.-P., Fortier, S., Maison, J.-M. and Vernotte, J. 1989, *Phys. Rev.*, **C40**, 1972
Meynet, G. and Arnould, M. 1992, *Proceedings of the International Symposium "Origin and Evolution of the Elements"*, Paris, 22 - 25 June 1992, to appear
Nørgaard, H. 1979, in *Gross Properties of Nuclei and Nuclear Excitations VII (Hirschegg)*, ed. H. von Groote (Technische Hochschule, Darmstadt), p. 204
Prantzos, N. 1991, in *Gamma-ray Line Astrophysics*, eds. Ph. Durouchoux and N. Prantzos (A.I.P., New York), p. 129
Prantzos, N., Coc, A. and Thibaud, J.-P. 1991, *Astrophys. J.*, **379**, 729
Prantzos, N., Doom, C., Arnould, M. and de Loore, C. 1986, *Astrophys. J.*, **304**, 695
Rolfs, C.E. and Rodney, W.S. 1988, *Cauldrons in the Cosmos* (The University of Chicago Press, Chicago) (RR88)
Salpeter, E.E. 1954, *Aust. J. of Phys.*, **7**, 373
Schatzman, E. 1948, *J. Phys. Radium*, **9**, 46
Schönfelder, V. and Varendorff, M. 1991, in *Gamma-ray Line Astrophysics*, eds. Ph. Durouchoux and N. Prantzos (A.I.P., New York), p. 101
Sneden, C. 1991, *Evolution of Stars : The Photospheric Abundance Connection,* IAU Symposium 145, eds. G. Michaud and A. Tutukov (Kluwer Academic Pub., Dordrecht), p. 235
Timmermann, R., Becker, H.W., Rolfs, C., Schröder, U. and Trautvetter, H.P. 1988, *Nucl. Phys.*, **A477**, 105
Vasilevskii, V.S. and Rybkin, I.Yu. 1989, *Sov. J. Nucl. Phys.*, **50**, 411
Vogelaar, B. 1989, *Ph.D. Thesis*, California Institute of Technology (unpublished)
von Weizsäcker, C.F. 1937, *Phys. Z.*, **38**, 176
von Weizsäcker, C.F. 1938, *Phys. Z.*, **39**, 633
Ward, R.A. and Fowler, W.A. 1980, *Astrophys. J.*, **238**, 266
Wasserburg, G.J. 1985, in *Protostars and Planets II*, eds. D.C. Black and M.S. Matthews ((University of Arizona Press, Tucson), p. 703
Wauters, L. 1992, *Master Thesis*, Université Libre de Bruxelles (unpublished)
Zinner, E., Amari, S., Anders, E. and Lewis, R. 1991, *Nature*, **349**, 51

Inside the Stars, IAU Colloquium 137
ASP Conference Series, Vol. 40, 1993
Werner W. Weiss and Annie Baglin (eds.)

STELLAR DISTANCES AND HIPPARCOS

A. E. GÓMEZ
URA 335 du CNRS, Observatoire de Meudon, DASGAL,
F-92195 Meudon, France

ABSTRACT Hipparcos, the astrometric satellite of the European Space Agency, launched in August 1989, will provide absolute trigonometric parallaxes for about 100 000 stars brighter than the limiting V-magnitude 12.5. In spite of the difficulties caused by a faulty orbit, the mission is successfully going on. A lifetime of more than 4 years is expected which will allow to reach the original objectives of the mission: 2 milli-arcsec for stars brighter than V-magnitude 9 and 4 to 5 milli-arcsec for fainter stars. The observing programme is based upon a single, uniquely-defined Input Catalogue. Field and open cluster stars of almost all effective temperatures, chemical compositions and gravities are being observed. Most types of binaries and variable stars are well represented, but Hipparcos is discovering new ones. Significant parallaxes (relatives errors smaller than 20%) will be obtained for stars closer than 100 pc (about 36 000 stars). Reliable stellar structure models should fit all the possible observational constraints: the main contribution of Hipparcos parallax data will be the huge improvement in the individual luminosities. The determination of stellar masses from the study of visual binaries will also be improved. The preliminary results of the first-year data reduction are an encouraging indication of the expected quality of the final results. On the basis of a mission duration of 4 years, the final results are foreseen for 1995.

OVERVIEW OF THE HIPPARCOS MISSION

Hipparcos, the astrometric satellite of the European Space Agency (ESA), is devoted to the accurate measurement of the positions, trigonometric parallaxes and proper motions of about 120 000 selected stars down to the limiting V-magnitude 12.5. A complete description of the pre-launch status mission is given in Perryman et al. (1989a). Launched in August 1989 by Ariane, Hipparcos failed to reach its geostationary orbit because of the failure of its apogee boost motor. A revised mission was implemented and completed by the end of November 1989 (Van der Ha 1992) and the satellite became operational. Hipparcos is orbiting in a highly elliptical orbit (perigee: 500 km, apogee: 36 000 km; period: 10h 40m) and valid observations are collected during 65-70% of the time.

In the nominal mission, the target accuracy on absolute parallaxes varies from 2 milli-arcsec or better for stars brighter than V-magnitude of about 9 to 5 milli-arcsec for fainter stars. This pre-launch accuracy prediction was reviewed by Lindegren and Kovalevsky (1992) with regard to the observed performance and expected lifetime of the satellite in its present orbit. They concluded that a mission duration of 4 years, until the end of 1993, is necessary in order to reach the nominal expected accuracy. The finite duration of the mission together with the fact that the satellite is continuously and systematically scanning the sky required a preselection of the stars to be observed and a precise a priori knowledge of the stars' position and magnitude. The catalogue containing the programme stars, called the Input Catalogue (Turon et al., 1992a), was compiled within the framework of the european consortium INCA (for details see Perryman and Turon 1989). The stars were selected on the basis of 214 scientific proposals submitted to ESA in 1982 by the international community. This material was supplemented, by the INCA consortium, with a basic list of bright stars referred to as the "survey" necessary for satellite operations and data reductions. The stellar content of the Input Catalogue is given in a section below.

The data obtained by the satellite (photon counts) are processed independently by the two data reduction consortia FAST and NDAC (see Perryman et al., 1989b). The data reductions consist of re-assembling the satellite data into a system of star positions, parallaxes and proper motions. The two catalogues produced by the consortia will be merged into a single one to be released to the astronomical community.

A byproduct of the astrometric mission is the determination of the Hipparcos H_P-magnitude for each star with high accuracy. The H_P-magnitude is the magnitude corresponding to the large bandwith of the Hipparcos detector (Mignard et al., 1989). It is defined by $H_P = V + 0.408(B-V) - 0.130(B-V)^2$, the expected accuracies vary from $0^m.001$ up to H_P-magnitude 6 to $0^m.01$ at H_P-magnitude 11. Moreover, in the process of signal analysis, information on the duplicity or multiplicity of the observed stars is also obtained.

In spite of the difficulties encountered at the beginning, the mission is now successfully going on. On the basis of a mission duration of 4 years, the final Catalogue is foreseen for 1995.

TYPES OF STARS OBSERVED BY HIPPARCOS

The Input Catalogue, which contains the 118 209 programme stars, constitutes the best compromise between the scientific goals of the proposals and the satellite's technical constraints (Turon et al., 1992b). Hipparcos is observing a large variety of stars: field stars of different masses, effective temperatures, chemical compositions and gravities, most types of binary stars and stars in open clusters. Table I summarizes the principal types of stars observed by Hipparcos. As can be seen from this table, different groups of spectroscopically peculiar stars and of variable stars are represented in the observing programme.

Among the observed stars, it is to be noticed that Hipparcos is measuring all the stars brighter than V-magnitude 7, a survey of about 52 000 stars and nearby stars selected from specific catalogues such as the Gliese Catalogue (Gliese 1969, Gliese and Jahreiß 1979) and the LHS and NLTT Catalogues (Luyten 1976, 1979, 1980). As stated before, the "survey" is a basic list of bright stars distributed over the whole sky, complete to a given limiting magnitude which is a function

TABLE I Types of stars observed by Hipparcos.

A-type stars	Eruptive variable stars	R stars
A stars (H-line emission)	F-type stars	R Cr B stars
Am stars	G-type stars	RR Lyrae stars
Ap stars	H- and K- emission stars	RS CVn stars
B-type stars	He abnormal stars	S stars
Barium stars	Herbig Ae and Be stars	Semi-regular var. stars
Be stars	Horizontal branch stars	Symbiotic stars
β CMa stars	Irregular variable stars	Sub-dwarf stars (cool)
Carbon stars	K-type stars	Sub-dwarf stars (hot)
Cataclysmic variable stars	λ Bootis stars	T Tauri stars
Cepheid stars	Li rich stars	VV Cephei stars
CH like star	Long-period variable stars	Weak G-band stars
CH strong stars	M-type stars	Weak metal stars
CJ stars	Marginal barium stars	White dwarfs
CS stars	Mira Ceti stars	Wolf-Rayet stars
δ Delphini stars	O-type stars	W UMa stars
δ Scuti stars	Planetary nebulae	WW Virginis stars
Early-type H-poor stars	Pulsating variable stars	X-ray binaries

of the galactic latitude "b" and the spectral type (Crifo et al., 1985):

$$\left| \begin{array}{lll} V & \leq & 7.9 + 1.1 \sin |b| \quad \text{for spectral types earlier than or equal to G5,} \\ V & \leq & 7.3 + 1.1 \sin |b| \quad \text{for spectral types later than G5.} \end{array} \right.$$

The survey was required for satellite operations and data reductions but it constitutes a sample highly suitable for future statistical studies.

Binary stars of different spectral types are being observed: about 1 000 visual binaries, 1 000 spectroscopic binaries and 100 eclipsing binaries (Gómez et al., 1989). These figures are underestimated because they concern only the stars specifically proposed as binaries. The Input Catalogue includes about 16 000 systems of double or multiple stars, only a part of them being physical systems. Hipparcos is discovering and measuring a large number of non-single stars. After the processing of one year Hipparcos data, the FAST consortium has detected about 7 000 new double stars (Mignard et al., 1992). The data reduction task is able to recognize double stars down to a separation of $0''.12$ with Δm up to 3 mag. Most of these new detected doubles have separations smaller than about $1''.5$, corresponding probably to physical pairs. Moreover, according to Dommanget (1985), closer systems having separations between $0''.01$ and $0''.1$ will also be recognized by oscillations of their photocentre during the lifetime of the satellite (astrometric pairs).

Finally, about 1 500 stars belonging to 250 open clusters are being observed. The number of observed stars per cluster varies from 1 star to more than 200 (Mermilliod and Turon 1989).

Almost all the observed stars are closer than 1 kpc from the sun. An estimation of the expected heliocentric distance distribution for different categories of spectral type is given in Table II. About 3 000 stars are missing in the Table because they have no known spectral type and, consequently, no distance estimation. The distances (in pc) were estimated from the spectral type (and the luminosity class if available) and the apparent V-magnitude. A reddening correction was applied using the tridimensional model of the galactic interstellar extinction built by Arenou et al. (1992). Hipparcos will provide individual distances for more than 100 000 stars closer than about 500 pc, distance at which the distance relative error will be of 100%.

TABLE II Number of observed stars, as a function of heliocentric distance (in pc) and spectral type.

Spectral Type	≤ 100	100-500	500-1 000	$> 1\,000$	Total
O-B	140	6 050	2 660	1 760	10 610
A0-A9	1 800	15 240	580	180	17 800
F0-F9	13 000	12 070	110	170	25 350
G0-G9	15 610	6 180	360	210	22 360
K0-M0	4 860	24 100	3 760	480	33 200
M1-M8	490	1 260	680	270	2 700
Other	670	1 550	480	500	3 200

ACCURACY PREDICTION ON PARALLAXES

Detailed accuracy prediction was carried out before launch based on extensive computer simulations of the instrument, the observations and the data reductions (Lindegren 1989). The pre-launch accuracy prediction has been revised by Lindegren and Kovalevsky (1992) taking into account the observed performance and expected lifetime of Hipparcos in its present orbit. Table III shows the predicted standard errors (in milli-arcsec) on parallaxes as a function of the H_P-magnitude for a mission duration of 4 years. Following Lindegren and Kovalevsky (1992), the quoted expected external errors have been evaluated by increasing the formal errors by 30% which represents an upper limit to the degree of uncertainty in the accuracy prediction. Recently, the external errors on the parallaxes obtained by the Data reduction consortia after the processing of one year Hipparcos data have been estimated by various methods: comparison with spectroscopic and photometric parallaxes or using cluster stars and Magellanic Clouds stars (Arenou 1992). These preliminary results show that the external errors are close to the predicted formal errors which means that the figures quoted in Table III are probably overestimated. The quoted values are sky averages: they also depend on the ecliptic latitude. The factor to be applied to the values given in Table III varies from 1.17 on the ecliptic to 0.71 at the poles. Taking into account all these considerations, the expected accuracies on the Hipparcos parallaxes will vary between 1 milli-arcsec and 5 milli-arcsec.

TABLE III Predicted standard errors (in milli-arcsec) of parallaxes as a function of H_P-magnitude for a mission duration of 4 years.

H_P	≤ 6.5	6.5-7.5	7.5-8.5	8.5-9.5	9.5-10.5	10.5-11.5	> 11.5
σ_π	1.2	1.3	1.6	2.0	2.8	3.7	4.6

Let us compare the parallaxes expected from the Hipparcos mission with those presently available. The most complete source of ground-based trigonometric parallaxes is the New Yale Catalogue from van Altena et al. (1992). It contains 7 881 stars with accuracies ranging from 0.5 milli-arcsec to 25 milli-arcsec or higher: a typical value is 15 milli-arcsec. Hipparcos is not only measuring a larger number of parallaxes (about 100 000), but the main point is that the Hipparcos parallaxes will be absolute.

HIPPARCOS CONTRIBUTION TO THE STELLAR STUCTURE

Among the observable quantities used in modeling stellar interiors, luminosity and mass (determined from the study of visual binaries) depend on distance determination.

The main contribution of the parallaxes measured by Hipparcos will be the significant improvement on stellar luminosities. Adopting the expected average standard error on the measured parallax (2 milli-arcsec), stars closer than 100 pc will have parallax relative errors ($\frac{\sigma_\pi}{\pi}$) smaller than or equal to 20%. The corresponding expected errors on absolute magnitudes ($\varepsilon_M = 2.17\frac{\sigma_\pi}{\pi}$) due to the error in the parallax as a function of the distance are given in Table IV.

TABLE IV Expected errors on absolute magnitudes.

r(pc)	$\frac{\sigma_\pi}{\pi}$	ε_M
5	1%	$0^m.02$
10	2%	$0^m.04$
20	4%	$0^m.09$
50	10%	$0^m.2$
100	20%	$0^m.4$

Let us compare these figures with those presently available using a few examples:

- Nearby stars closer than 20 pc. Nearby stars populate the H-R diagramme essentially on the main sequence, corresponding to stars having masses between 0.3 and $2\,m_{\odot}$. The sample also contains a few stars of different types: slightly evolved stars, subgiants, giants and white dwarfs. Hipparcos is observing about 1 100 stars contained in the Gliese Catalogue of nearby stars (Gliese 1969, Gliese and Jahreiss 1979). Adopting for these stars the errors on the visual absolute magnitudes quoted in the Third version of the Gliese and Jahreiß Catalogue of nearby stars (Gliese and Jahreiss 1992), only a few of them have errors smaller than $0^m.1$. For the same stars, all the absolute magnitudes calculated with the Hipparcos parallaxes will have errors smaller than $0^m.1$. In particular, powerful tests for the input physics of the stellar interior models will become possible (Baglin 1988). Recently, Lebreton et al. (1992) have modeled a close binary system and found some inconsistencies between the adopted physical model and the observational data.

- H-R diagrammes of open clusters. H-R diagrammes of cluster stars provide well-known tests for the stellar structure theory. Hipparcos will provide a direct distance determination for about 15 galactic clusters. Table V gives the expected individual and mean parallax errors for the observed clusters closer than 250 pc. The distances were taken from the Cluster Data Base of Mermilliod (1988). The clusters are ordered by increasing distances. The individual and mean parallax relative errors are given in the fourth and fifth columns, respectively. The indicated precisions take into account the dependence on ecliptic latitude. The figures quoted for the mean parallax precision were calculated using the fact that for stars observed within a small area on the sky, like stars in clusters, a correlation of parallax results exists. In this case the expected improvement by averaging over N stars is given roughly by $\sigma_{\langle\pi\rangle} = \sigma_\pi \times N^{-0.35}$ (Lindegren 1989). Most of these clusters will have mean parallax relative errors smaller than 20%.

- Halo stars. Hipparcos does not observe globular cluster stars but it is observing different types of Population II stars, in particular about 700 sub-dwarf stars of various metallicities. At present, less than 30 sub-dwarf stars have parallax relative errors smaller than or equal to 20%. This number will be multiplied by 10 at the end of the Hipparcos mission.

On the other hand, from the 100 000 parallaxes measured by Hipparcos, mean absolute magnitudes with precisions ranging between $0^m.01$ to $0^m.2$ (Mennessier 1988) will be obtained for a large number of stellar types.

Hipparcos parallaxes will contribute to improve the determination of stellar masses from the study of visual binaries. For pairs with known orbits, the sum of masses of the components will be improved. However, for stellar structure modeling, individual masses are required. About 80 systems closer than 20 pc are observed by the satellite, but according to Popper (1980) only half of them have definitive or reliable orbits and known mass ratios. The spectral type distribution of the bright companions for these pairs shows that the concerned stars

TABLE V Cluster stars: individual and mean parallax precisions.

Name	r(pc)	N^a	$\frac{\sigma_\pi}{\pi}$	$\frac{\sigma\langle\pi\rangle}{\langle\pi\rangle}$	S.T.[b]
Hyades	48	209	11%	2%	A2
Coma Ber	79	117	17%	3%	A0
Pleiades	125	74	29%	6%	B5
IC 2602	151	34	23%	7%	B0
IC 2391	153	21	21%	7%	B3
Praesepe	158	22	37%	12%	A0
α Per	164	133	36%	6%	B2
NGC 6475	239	12	56%	23%	B8
Blanco 1	240	18	52%	19%	B5

[a] N is the number of observed stars

[b] Spectral type of the hottest star on the main sequence

have masses around the solar mass. The number of stars for each spectral type is given between parenthesis: A (2), F(8), G(12), K(6) and M(8). Moreover, only a few of these pairs will get individual masses with a relative accuracy smaller than 5%. According to Morel (1988), the Hipparcos positions available for each component during the mission will allow to derive the masses of the components for a larger number of systems. The simulations carried out by him showed that for about 110 systems individual masses will be derived with a relative accuracy better than 15% and for 25 systems better than 5%. Recently, Dommanget and Lampens (1992) discussing the expected contributions of Hipparcos to the study of double stars found less optimistic results.

CONCLUDING REMARKS

From a technical point of view, Hipparcos may continue to collect high quality measurements until roughly June 1994. This well-defined limit to the satellite lifetime is larger than the mission duration necessary to get the target accuracy on absolute parallaxes. First preliminary results obtained by both Data reduction consortia, after processing one year data, are an encouraging indication of the expected quality of the final results. Lindegren (1992) built an H-R diagramme using about 3 000 stars with relative (formal) parallax errors smaller than 10%. Nevertheless, any star in the diagramme may have a larger parallax error than given by the formal errors, since one year of observation is rather too short to separate the astrometric parameters (parallax and proper motion).

Improvements are expected for luminosities of individual and cluster stars and for luminosities and masses of stars in visual binaries. They should allow better qualitative and quantitative tests on stellar interior models.

ACKNOWLEDGMENTS

I would like to thank F. Arenou, D. Morin and D. Stefanovitch for their technical assistance in the preparation of this paper.

REFERENCES

Arenou, F., Grenon, M., and Gómez, A. 1992, *A&A*, **258**, 104

Arenou, F. 1992, *Private communication*

Baglin, A. 1988, in *Scientific Aspects of the Input Catalogue Preparation*, II, Sitges, Jan. 1988, J. Torra & C. Turon (eds), 103

Crifo, F., Turon, C., and Grenon, M. 1985, ESA SP-**234**, 67

Dommanget, J. 1985, *Ap&SS*, **110**, 47

Dommanget, J., and Lampens, P. 1992, *Ap&SS*, submitted

Gliese, W. 1969, *Veroff. Astron. Rechen-Instituts, Heidelberg*, No **22**

Gliese, W., and Jahreiß, H. 1979, *A&AS*, **38**, 423

Gliese, W., and Jahreiß, H. 1992, *Third Catalogue of Nearby Stars*, CD-ROM version

Gómez, A., Crifo, F., and Turon, C. 1989, in ESA SP-**1111**, vol. II, 89

Lebreton, Y., Auvergne, M., Morel, P., and Baglin, A. 1992, in *Inside the Stars*, I.A.U. Colloquium No **137**, Vienna (Austria), April 1992

Lindegren, L. 1989, in ESA SP-**1111**, vol. III, 311

Lindegren, L. 1992, ESA SP-**349**, (in press)

Lindegren, L., and Kovalevsky, J. 1992, *Highlights in Astronomy*, vol. 9 (in press)

Luyten, W.J. 1976, *LHS Catalogue*, Minnesota (and revised version)

Luyten, W.J. 1979, *NLTT Catalogue*, Minnesota

Luyten, W.J. 1980, *NLTT Catalogue*, Minnesota

Mennessier, M.O. 1988, in *Scientific Aspects of the Input Catalogue Preparation*, II, Sitges, Jan. 1988, J. Torra & C. Turon (eds), 113

Mermilliod, J.C. 1988, *Bull. Inf. CDS*, **31**, 175

Mermilliod, J.C., and Turon, C. 1989, in ESA SP-**1111**, vol. II, 177

Mignard, F., Foeschlé, M., Falin, J.L., Andreasen, G. K., Høg, E., Grewing, M., and Scales, D. 1989, in ESA SP-**1111**, vol. III, 205

Mignard, F., Badiale, M., Bernacca, P.L., Bernstein, H., Falin, J.L., Froeschlé, M., Hering, R., and Prezioso, G. 1992, in *Complementary Approaches to Double and Multiple Star Research*, I.A.U. Colloquium No **135**, Pine Mountain (USA), April 1992

Morel, P. 1988, in *Scientific Aspects of the Input Catalogue Preparation*, II, Sitges, Jan. 1988, J. Torra & C. Turon (eds), 235

Perryman, M.A.C., and Hassan, H. (eds) 1989a, ESA SP-**1111**, vol. I

Perryman, M.A.C., Lindegren, L., Murray, C.A., Høg, E., and Kovalevsky, J. (eds) 1989b, ESA SP-**1111**, vol. III

Perryman, M.A.C., and Turon C. (eds) 1989, ESA SP-**1111**, vol. II

Popper, D.M. 1980, *ARA&A*, **18**, 115

Turon, C., Crézé, M., Egret, D., Gómez, A., Grenon, M., Jahreiß, H., Réquième, Y., Argue, A.N., Bec-Borsenberger, A., Dommanget, J., Mennessier, M.O. et al. 1992a, ESA SP-**1136**, 7 vol.

Turon, C., Gómez, A., Crifo, F., Crézé, M., Perryman, M.A.C., Morin, D., Arenou, F., Nicolet, B., Chareton, M., and Egret, D. 1992b, *A&A*, **258**, 74

van Altena, W.F., Lee, J.T., and Hoffleit, E.D. 1992, *The General Catalogue of Trigonometric Stellar Parallaxes*, Yale University Observatory, CD-ROM preliminary version

van der Ha, J. 1992, ESA bulletin, **69**, 9

Inside the Stars, IAU Colloquium 137
ASP Conference Series, Vol. 40, 1993
Werner W. Weiss and Annie Baglin (eds.)

OBSERVABLE EFFECTS OF ROTATION ON STELLAR ABUNDANCES

SUCHITRA BALACHANDRAN
Department of Physics and Astronomy, University of North Carolina, CB #3255 Phillips Hall, Chapel Hill, NC 27599-3255.

ABSTRACT This is a review of our understanding of the effects of rotation on the stellar interior through an examination of observed surface stellar abundances. First, the evolution of angular momentum from the pre-main sequence through the giant phase is summarized. Attempts are then made to decipher the signature of abundance anomalies due to meridional circulation and rotational braking with reference to available theoretical models. Finally, special cases such as tidally-locked binaries and chromospherically-active giants are examined to determine if their rotational peculiarites have induced abundance anomalies.

INTRODUCTION

This review is a testimony to the volume of work that has been carried out in this area in the last decade, to the sophistication of our current understanding of the stellar interior, and to the belief that we are now of the threshold of being able to perform even more detailed comparison between observations and models. Until the beginning of the last decade, our knowledge of rotational velocities in stars was limited by our inablity to obtain high resolution spectra of faint objects. We knew that early-type stars exhibited a spread in rotational velocities which remained largely unchanged from the age of the Pleiades to older stars in the field (Kraft 1970). Bright, low-mass stars in the field were seen to be spinning slowly and it was assumed that these had been rotating slowly all their lives.

Very little was known about the effects of rotation on stellar composition. Observations of field giants revealed that a large fraction had smaller $^{12}C/^{13}C$ ratios (see Lambert and Ries 1981 and references therein) i.e. greater mixing has occurred, than predicted by standard dredge up theory. Meridional circulation induced mixing was examined in an effort to explain this anomaly

(Dearborn, Eggleton and Schramm 1976) and it was concluded that greater mixing was possible in the more massive stars. The large fraction of low $^{12}C/^{13}C$ ratio giants however indicated that it was the low-mass stars that were more extensively mixed (Scalo and Miller 1978). This debate could not be resolved since it is not possible to determine the masses of field redgiants with any accuracy and abundances in cluster giants could not be measured with existing spectrographs and detectors.

EVOLUTION OF ANGULAR MOMENTUM

The revolution in this field began with the discovery through photometry of rapidly rotating K dwarfs in the Pleiades cluster by van Leeuwen and Alphenaar (1982). Since those observations the database of rotational velocities in young clusters has increased by leaps and bounds. The large number of studies have resulted in an entire conference devoted to the subject (Angular Momentum Evolution of Young Stars, ed. Catalano and Stauffer, 1991) and the interested reader is referred to extensive reviews in that volume. Here I will merely touch upon key observations that reveal the status of our understanding on the subject.

On the Main Sequence

Once on the main sequence, stars of different spectral types spin-down at varying rates (see Stauffer 1991 for a review). Comparing the young clusters Alpha Persei (turn-off age ~ 0.05 Gyr) and the Pleiades (turn-off age ~ 0.07 Gyr), one sees rapidly rotating G and K dwarfs in the former (Stauffer, Hartmann and Jones 1985; 1989; Prosser 1992), but by the age of the Pleiades, the G dwarfs have spun down while rapidly-rotating K dwarfs are still observed (Stauffer and Hartmann, 1987). By the age of the Hyades (turn-off age ~ 0.6 Gyr) both G and K dwarfs have spun down but some relatively rapidly-rotating M dwarfs with velocities upto 20 km/s are still observed (Stauffer, Hartmann and Latham 1987; Radick *et al.* 1987). This can be broadly understood if the envelope is assumed to decouple from the core so that the increasing fractional mass of the envelope from G to K to M dwarfs results in the longer spin-down timescale. A rough estimate of this spin-down timescale for G, K and M dwarfs can be obtained from these three clusters: G dwarfs spin-down in a 20-30 million years, K dwarfs in 100 million years and M dwarfs in a several hundred million years.

An interesting feature is the large fraction of slowly rotating stars in both young clusters. Stauffer *et al.* (1989) pointed out that since these are seen even among G dwarfs in the very young cluster IC 2391 (turn-off age ~ 0.02 Gyr) an age spread within the cluster may be ruled out as their cause. Further, the fraction of slow rotators is too large to be accounted for as a result of near-pole-on orientations of the inclination angle. It hence appears that stars arrive on the main sequence with a genuine spread in rotational velocities. The simple spin-down picture is complicated by a number of Pleiades K dwarfs seen to be rotating at smaller velocities ($v\sin i$'s) than comparable temperature stars in the Hyades. Since the Hyades velocites are derived from rotational periods (Radick *et al.* 1987), they are true velocites and not upper limits. Unless the stellar envelope recouples with the rapidly rotating core in the intervening period, these Pleiades K dwarfs will be spinning much more slowly at the age of the Hyades. There may be other cluster to cluster variations. The younger Alpha Persei cluster contains many more rapid rotators than the Pleiades and the largest rotational velocities are around 200 km/s compared to around 100 km/s in the low-mass Pleiades stars. While it is conceivable that this is due to the difference in the ages between the two clusters, it is also possible that the angular momentum distribution in all young clusters is not similar. Additional cluster data are required to investigate such issues.

Evolution to the Main Sequence

The problem gets more complicated when we try to understand the evolution of angular momentum from the pre-main sequence phase to the main sequence. While higher mass T Tauri stars show a large range in $v\sin i$ similar to that seen in early-type main sequence stars, lower mass T Tauri stars show a surprisingly small range in $v\sin i$. The problem arises when the low mass T Tauri distribution is evolved to the main sequence (Bouvier 1991). To produce the fastest rotators seen in Alpha Persei, the most rapidly rotating T Tauri stars have to be spun up without any loss of angular momentum in the pre-main sequence phase. Such evolution will, however, not produce any of the slowly rotating stars seen in Alpha Persei and in IC 2391. Since the higher-mass T Tauris are at a later stage of evolution, their distribution may be an indication that the angular momentum loss is not uniform but depends on the velocity of the star. Pinsonneault (1992, these proceedings) suggests such a formulation. The alternative conclusion that T Tauri stars are not representative of the pre-main sequence progenitors of stars in young clusters is unsubstantiated and premature but warrants further investigation (Bouvier 1991).

From the Main Sequence to Giants

When the early-type stars evolve off the main sequence and cross the Hertzsprung gap, their rotational velocities are observed to decrease. The drop in $v\sin i$ is rapid as demonstrated by Gray (1989) and by the larger samples of de Medeiros and Mayor (1991) leading to the speculation that it is caused by the onset of the magnetic dynamo (see Gray 1991 for a review). In accordance with this scenario, Simon and Drake (1989) found that the chromospheric activity levels of giants and subgiants as measured in CIV and CII dropped across the Hertzsprung gap. Theoretical models have not yet demonstrated that such rapid braking is possible or investigated the effects on stellar composition.

EFFECTS OF ROTATION

One question raised as a result of our increased understanding of angular momentum evolution in stars is what is the effect of rapid rotation and spin-down on the structure and interior of the star? In particular, does rotation induce mixing in the stellar interior and does it thus affect stellar composition? Theoretical models have postulated rotation-induced mixing by two principle means: meridional circulation as a result of rapid rotation and mixing as a result of spin-down. In this review, I will use the abundance of lithium to test if mixing has occurred. If mixing extends to the stellar surface we expect to see changes in the lithium abundances of main sequence stars. Mixing in the interior will not be revealed till the star becomes a subgiant or giant. Using lithium abundances to look for internal mixing is fraught with problems since we do not understand the lithium depletion mechanisms fully. For instance, lithium may be depleted as a result of turbulent diffusion or gravitational settling and these effects, if relevant, must be disentangled from rotation-induced mixing.

Meridional Circulation

Rapid rotation is expected to induce meridional circulation currents. Theoretical models (Charbonneau and Michaud 1988) have suggested that meridional circulation currents may be efficient enough to mix material up to the surface in early-F and late-A stars which have been rotating rapidly all their lives. Since lithium only survives in the outer 2-3% by mass of a typical main sequence star, mixing due to meridional circulation should dilute the surface abundance.

Unfortunately the equivalent width of the Li I resonance feature, the only measurable Li feature in most stars, is extremely small in these hot stars and not measurable in rapidly rotating stars. One way around this problem is to examine stars which have just evolved off the main sequence to slightly cooler temperatures. The surface abundances of these stars are not expected to have altered as a result of their evolution but their cooler temperatures make the Li I feature stronger. Such stars are not plentiful since, once they leave the main sequence, they evolve rapidly across the Hertzsprung gap, but the sample of F stars observed by Balachandran (1990) provides a source. In that study, both lithium abundances and rotational velocities ($v\sin i$'s) were measured for about 200 stars. Thirty six of these have $M_V < 2.4$ and have evolved from main sequence temperatures greater than 7500K. Since these stars are more massive

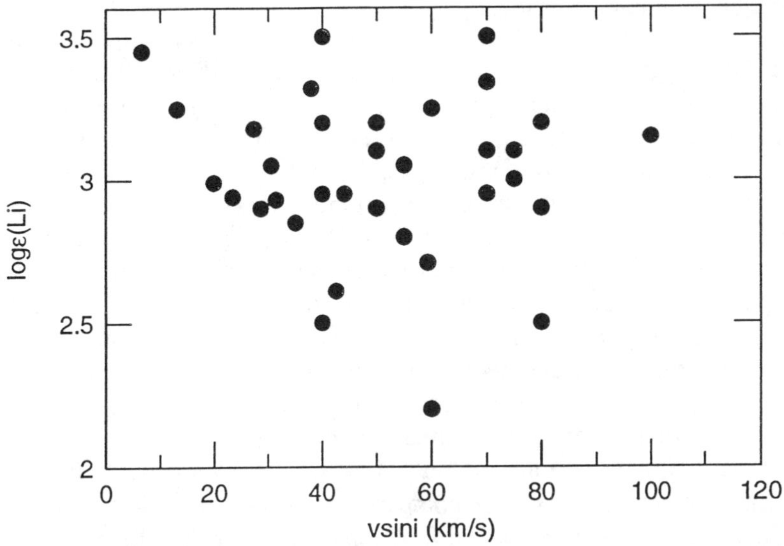

Fig. 1. Lithium abundances as a function of *vsini* in 1.6 - 1.8 $M_\odot$ stars that have evolved from early-F and late-A spectral types stars and are now seen crossing the Hertzsprung gap with temperatures between 6400K and 7000K (data from Balachandran 1990).

(1.6-1.8$M_\odot$) than lithium 'dip' stars (~ 1.35 $M_\odot$) (Boesgaard and Tripicco 1986), their abundances will not be altered by the process affecting the 'dip' stars. In Figure 1, lithium abundances in these stars are plotted as a function of *vsini* . While the sample is small it contains stars rotating as rapidly as 100 km/s and there is no evidence for a trend in lithium depletion with rotational velocity. At least in these massive stars, meridional circulation does not appear to have mixed material to the surface.

A similar conclusion was arrived at by Gilroy (1989) who examined the $^{12}C/^{13}C$ ratios in open cluster giants with ages between 50 Myr and 5 Gyr and found a large and constant $^{12}C/^{13}C$ ratio in giants more massive than 2.5 $M_\odot$ in accordance with standard dredge up theory. Clearly meridional circulation has not succeeded in mixing the ^{13}C peak outward in these stars. Smaller $^{12}C/^{13}C$ ratios were, in fact, seen in less massive stars (below 2.5 $M_\odot$) and the isotopic ratio was seen to decline with mass, reflecting perhaps the longer main sequence lives of these stars. The reason for the sharp change in $^{12}C/^{13}C$ ratio at this mass remains unknown. The observational evidence for meridional circulation related mixing is hence weak.

Rotational Braking and Mixing

The first calculation of rotational braking in the Sun performed by Endal and Sofia (1981) predicted that mixing resulting from differential rotation may extend deep enough to destroy lithium. Since then more detailed models have been calculated by Pinsonneault *et al.* (1989, 1990). According to their models, when a star is braked on the main sequence, the outer convective zone spins down first leaving the interior rotating rapidly. This differential rotation causes turbulence which leads to mixing and lithium depletion. Their models parametrize the loss of angular momentum and introduce a factor to relate the transport of angular momentum to material transport so as to match the observed lithium depletion in the Sun. With these constraints, their models make quantitative predictions for the magnitude of lithium depletion in stars depending upon the initial angular momentum of the star and its age; the depletion increases both with the magnitude of the spin-down and with the age of the star. Since these are the only detailed models available for rotational braking, all observations will be compared to their predictions.

Unlike other lithium depletion mechanisms (e.g. turbulent diffusion) in which the depletion varies with age and spectral type alone, one of the major predictions of the Pinsonneault *et al.* (1990) model is that a spread in lithium will be observed at a given temperature in an old cluster, since these present-day slow rotators are expected to have had a range in initial angular mometum similar to that seen in young clusters. Lithium abundances have been measured in two old open clusters; M67 (Hobbs and Pilachowski 1986b; Garcia Lopez *et al.* 1988) and NGC 188 (Hobbs and Pilachowski 1988), and in one intermediate-aged cluster, NGC 752 (Hobbs and Pilachowski 1986a; Pilachowski and Hobbs 1988). The results are, at best, encouraging, at worst, inconclusive. In the oldest cluster, NGC 188, the scatter is not impressive. Five of the six stars measured have an average abundance of $\log \varepsilon(\mathrm{Li}) = 2.34 \pm 0.18$, a very small dispersion indeed. Only one star has a smaller abundance of $\log \varepsilon(\mathrm{Li}) < 1.8$. The observed sample in M67 is larger. A scatter of about 0.6 dex is seen in stars around the cluster turn-off at 6200 K. Also, M67 stars, which are of the same age as the Sun, have larger than solar lithium abundances at solar temperature. In the sparsely-populated, intermediate-aged cluster NGC 752 three stars around solar effective temperature have a spread in lithium of 1.0 dex comparable to the spread predicted in the Pinsonneault *et al.* (1990) tabulations. However the lithium abundance drops sharply with temperature and some of the perceived scatter may be due to temperature uncertainties. Stronger evidence for a spread in lithium (though of smaller magnitude) is seen in the younger Hyades cluster long thought to have a tight Li-Teff relation. Recent observations by Thorburn *et al.* (1992) find a spread in lithium in stars around 6300 K. Clearly, a large sample in a well-populated old cluster would

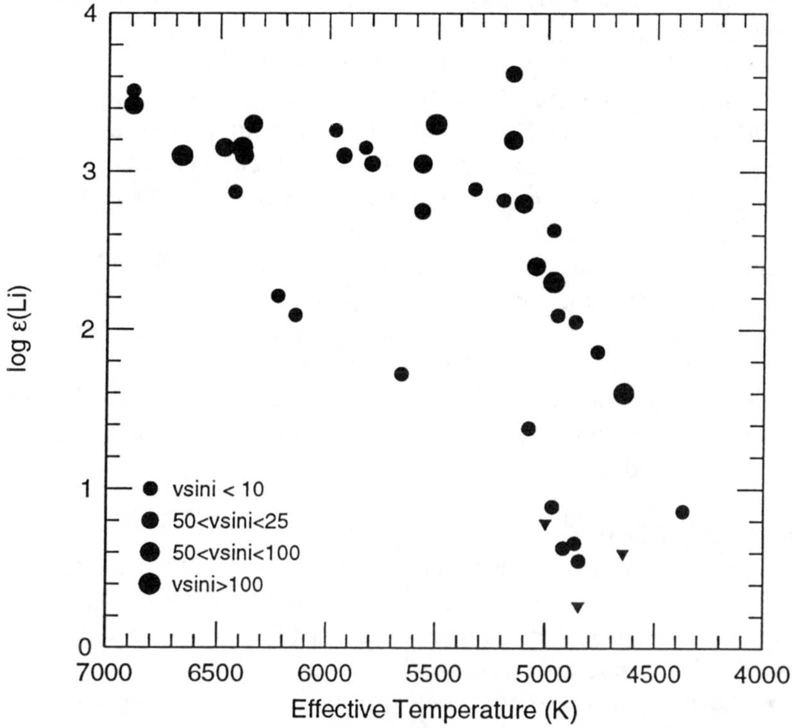

Figure 2. Lithium abundances as a function of effective temperature in the Alpha Persei cluster (from Balachandran, Lambert and Stauffer, 1988) with non-members from that sample discarded on the basis of new proper motion data from Prosser (1992). The stars are coded in *vsini* as indicated in the key.

be of use. The basic question of whether rotational spin-down is capable of generating turbulence and material mixing is yet to be answered to the skeptics' satisfaction.

While the models do not predict a spread in lithium in young clusters, the observed samples are larger and the results are intriguing. In the Pleiades cluster (Soderblom *et al.* 1992, these proceedings), the abundances appear to be fairly well-behaved. With the exception of a few stars, the lithium abundances are a fairly smooth function of temperature with the hotter stars retaining their initial lithium abundances while the cooler stars reflect depletion expected to occur on the pre-main sequence. The abundance pattern in the younger Alpha Persei cluster is different. Using the new proper motion survey of Prosser

(1992), non-member stars have been removed from our original study (Balachandran, Lambert and Stauffer 1988) and the remaining stars are plotted in Figure 2. The Li-Teff plot now exhibits two sequences of stars with the lower sequence populated exclusively by slow rotators though slow rotators are seen in the upper sequence as well. Since membership in the cluster is difficult to determine, skeptics may view any result from Alpha Persei as being clouded by the intrusion of non-members. It appears, however, that there are too many stars in the lower sequence to be accounted for by membership errors alone. The lithium abundance differences between the Pleiades and Alpha Persei, like the rotational velocity differences, may be pointing to a fundamental difference between the evolutionary history of the two clusters. If the lower lithium sequence is real (and additional data would be very useful to confirm this), Alpha Persei cannot evolve to look like the Pleiades. Whether the Alpha Persei and Pleiades lithium abundance patterns can evolve to the Hyades sequence is a very interesting and important question.

Tidally-Locked Binaries

Since the extent of lithium depletion in a star depends upon the magnitude of the spin-down in the Pinsoneeault *et al.* (1989) models, an interesting and testable consequence is that lithium depletion should cease when a binary star locks through tidal interaction. If tidal synchronization occurs early in the life of the star, i.e. the star is a short-period binary, then its lithium abundance would be anomalously high compared to other stars of the same age and effective temperature.

Soderblom *et al.* (1989) found larger than average lithium abundances in two short-period tidally-locked binaries in the Hyades, vB 22 and vB 62, and argued that the lack of spin-down in these stars may be responsible for their large abundances. However, Soderblom (1992) reported that KW 181 in Praesepe with a orbital period similar to vB 22 has a lithium abundance below the average curve; the results in Pop. I stars are mixed.

Spite, Maillard and Spite (1986) reported a large lithium abundance in BD -0°4234, a short-period binary with a metallicity of [Fe/H]=-0.7. An interesting and intriguing short-period, metal-poor binary is HD 89499. It has a circular orbit with a rotational period of 5.6 days (Ardeberg and Lindgren 1991). The secondary has not been detected. HD 89499 is particularly interesting because of both its age and its metallicity. According to Matheiu and Mazeh (1988), the orbit of such a short-period binary would have tidally-circularized around the age of the Hyades; tidal-synchronization would have occurred earlier. At that age, Li depletion would have ceased according to the Pinsonneault *et al.* (1990) models while a single star of the same temperature and metallicity would have continued to deplete Li. Since HD 89499 is old, the difference between its lithium abundance and a comparable single star will be large. Further, the

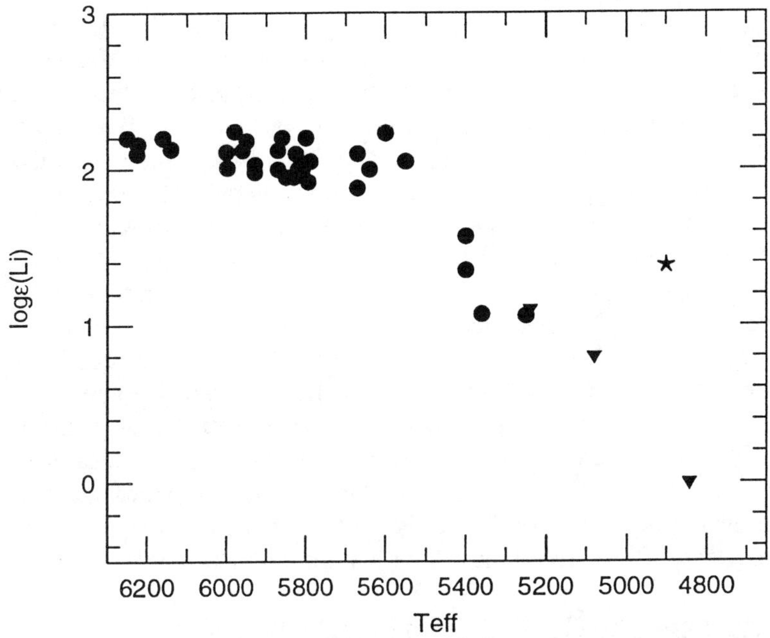

Figure 3. The lithium abundance in HD 89499 (star symbol) is shown in comparison with other stars on a lithium - effective temperature plot. The other stars are all metal-poor ([Fe/H]≤ -1.5) halo dwarfs. Effective temperatures and Li abundances for these stars are taken from Spite and Spite (1982, 1986), Spite et al. (1984), Hobbs and Ducan (1987), Rebolo et al. (1988) and Beckman et al. (1986). Filled circles are measured abundances and the inverted triangles are upper limits.

lithium abundances in Pop. II stars form a tightly defined sequence in the Li-Teff plane. At temperatures greater than about 5300 K, they form the lithium plateau first identified by Spite and Spite (1982). Observations are sparse at cooler temperatures but the overall scatter is much smaller than that seen in Pop. I stars. Recently Balachandran *et al.* (1992) observed the Li I feature in HD 89499. The abundance based on an effective temperature estimate of 4900 K is log ε(Li)=1.35 which is a factor of 10 or higher than comparable single stars of the same temeprature and metallicity (Fig. 3).

Some caution must be exercised before it can be concluded that the high Li abundance of HD 89499 indicates a connection between Li depletion and rotational braking. The largest uncertainty in the interpretation of the Li abundance in HD 89499 arises from the discrepancy between the *v*sin*i* (~ 20 km/s) measured from the spectrum and that derived from the rotational period. If the system is tidally locked, the rotational velocity of the dwarf K star must be around 6-8 km/s. The measured *vsini* is consistent with the rotational period if the star has a larger radius, i.e. it is a subgiant. In this case its Li abundance simply reflects the diluted plateau value and is consistent with abundances seen in other halo subgiants (Pilachowski, Sneden and Booth 1992) thus having no bearing on the tidally-locked character of the system. It is hence crucial to the previous interpretation that this issue be resolved. The system, however, appears to be peculiar. Strömgren photometry places the star between the main sequence and subgiants. Further, examination of M92, a globular cluster of similar metallicity, shows that the subgiant branch is at much hotter temperatures. The possibility that the star may be heavily reddened and thus warmer is inconsistent with spectroscopically determined temperatures and in conflict with the Strömgren photometry calibrations of Schuster and Nissen (1989). This star is indeed a puzzle but other short-period binaries may hold the clue to the cause of Li depletion in stars. If tidal synchronization does indeed arrest lithium depletion, then a short-period binary on the Pop. II lithium plateau may reveal the value of the primordial lithium abundance.

Chromospheric Activity, Rapid Rotation and High Li

Near-cosmic lithium abundances have been observed in a few giants (Wallerstein and Sneden, 1982, Brown *et al.* 1989, Gratton and D'Antona 1989). The high Li abundances in these stars is puzzling because their low $^{12}C/^{13}C$ ratios indicate extensive mixing. A recent study of chromospherically active giants (Fekel and Balachandran 1992) sheds new light on these stars. The authors find that some of the chromospherically active giants are single; their rapid rotation is hence not a result of angular momentum derived from their orbits. Since observational data do not show any rapid rotators later than G0 on the subgiant branch, they conclude that these giants must have dredged up angular momentum from their rapidly rotating cores. A surprising find was the large lithium abundances in many of these stars (Fig. 4) leading to the suggestion that freshly synthesized lithium is being dredged up along with the angular momentum. Previously only AGB stars had been recognized as capable of producing lithium via the ^{7}Be-transport mechanism in the hot-bottom convective envelopes of these stars (Smith and Lambert 1989; 1990). It has not been theoretically demonstrated that Li-production is possible in first ascent giants but the observational evidence is intriguing enough to warrant further study.

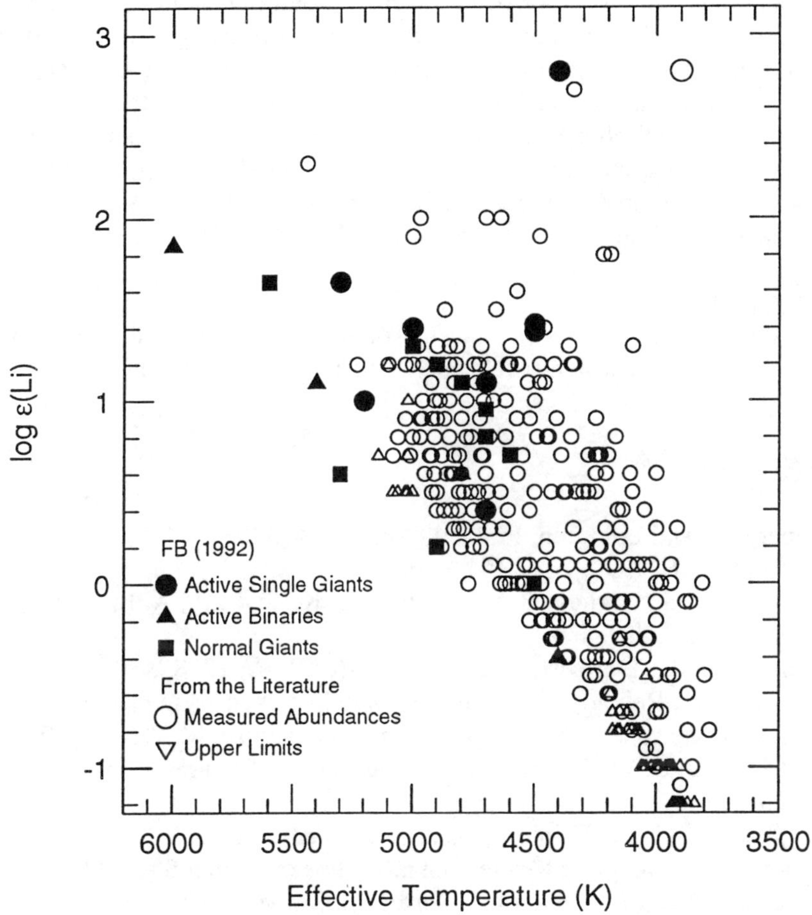

Figure 4. Lithium abundances versus effective temperature for giant stars. The filled symbols are from the sample of Fekel and Balachandran (1992). The remaining stars (open symbols) are from the sample of Brown et al. (1989) with the exception of two lithium-rich giants taken from Wallerstein and Sneden (1982) and Gratton and D'Antona (1989).

SUMMARY

We have gathered a considerable database on the rotational velocities in young stars. One major point that remains to be understood is the relationship between the rotational velocity distribution seen in young clusters and in pre-main sequence stars. We are now beginning to make inroads into understanding rotation-induced mixing. Based on theoretical models, it has been suggested that spin-down on the main sequence may be responsibe for considerable mixing and several observational tests for this model are possible. An obvious extension is to explain and understand the rotational slowing down of giants using a more rigorous theoretical framework. Conceivably such detailed understanding will lead to a more refined explanation for Li-rich giants.

Acknowledgements
I would like to thank the American Astronomical Society for the International Travel Grant Award which enabled my attendance at this colloquium.

REFERENCES

Ardeberg, A. and Lindgren, H. 1991, A&A, 244, 310.
Balachandran, S. 1990, ApJ, 354, 310.
Balachandran, S., Carney, B. W., Fullton, L. K., and Fry, A. M., 1992, BAAS, 24, 767.
Balachandran, S., Lambert, D. L. and Stauffer, J. R., 1988, ApJ, 333, 267.
Beckman, J. E., Rebolo, R., Molaro, P. 1986 in Advances in Nuclear Astrophysics, eds. E. Vangiori-Flam, J. Audouze, M. Cassé, J. P. Chièze, and Tran Thanh Van, (Gif-sur-Yvette: Editions Frontières) p. 29.
Boesgaard, A. M. and Tripicco, M. J. 1986, ApJ, 302, L49.
Bouvier 1991, in Angular Momentum Evolution of Young Stars, eds. S. Catalano and J. R. Stauffer, (Dordrecht: Kulwer).
Brown, J. A., Sneden, C., Lambert, D. L. and Dutchover, E. 1989, ApJS, 71, 293.
Charbonneau, P., and Michaud, G., 1988, ApJ, 334, 746.
de Medeiros, J. R., and Mayor, M. 1991, in Angular Momentum Evolution of Young Stars, eds. S. Catalano and J. R. Stauffer, (Dordrecht: Kulwer).
Dearborn, D. S. P., Eggleton, P. P., and Schramm, D. N. 1976, ApJ, 203, 455.
Endal, A. J. and Sofia, S. 1981, ApJ, 243, 625.
Fekel, F. C. and Balachandran, S. 1992, ApJ, in press.
Garcia Lopez, R. J., Rebolo, R. and Beckman, J. E. 1988, PASP, 100, 1489.
Gilroy, K. K. 1989, ApJ, 347, 835.
Gratton, R. G. and D'Antona, F. 1989, A&A, 215, 66.
Gray, D. F. 1989, ApJ, 347, 1021.

Gray 1991, in Angular Momentum Evolution of Young Stars, ed. S. Catalano and J.R. Stauffer, (Dordrecht: Kulwer).
Hobbs, L. M. and Duncan, D. K. 1987, ApJ, 317, 796.
Hobbs, L. M. and Pilachowski, C. 1986a, ApJ, 309, L17.
Hobbs, L. M. and Pilachowski, C. 1986b, ApJ, 311, L37.
Hobbs, L. M. and Pilachowski, C. 1988, ApJ, 334, 734.
Kraft, R. 1970, in Spectroscopic Astrophysics, ed. G. H. Herbig (Berkeley: University of California press), p. 385.
Lambert, D. L., and Ries, L. M. 1977, ApJ, 217, 508.
Mathieu, R. D. and Mazeh, T. 1988, ApJ, 326, 256.
Pilachowski, C. and Hobbs, L. M. 1988, PASP, 100, 336.
Pilachowski, C. A., Sneden, C., and Booth, J. 1992, preprint.
Pinsonneault, M. H., 1992, in Inside The Stars, eds. W.W. Weiss and A. Baglin, in press.
Pinsonneault, M. H., Kawaler, S. D., and Demarque, P. 1990, ApJS, 74, 501.
Pinsonneault, M. H., Kawaler, S. D., Sofia, S., and Demarque, P. 1989, ApJ, 338, 424.
Prosser. C. F. 1992, AJ, 103, 488.
Radick, R. R., Thompson, D. J., Lockwood, G. W., Duncan, D. K., and Baggett, W. E. 1987, ApJ, 321, 459.
Rebolo, R., Beckman, J. and Molaro, P. A&A, 172, L17.
Scalo, J. M., and Miller, G. E. 1978, ApJ, 225, 523.
Schuster, W. J. and Nissen, P. E. 1989, A&A, 221, 65.
Simon, T. and Drake, S. A. 1989, ApJ, 346, 303.
Smith, V. V. and Lambert, D. L. 1989, ApJ, 345, L75.
Smith, V. V. and Lambert, D. L. 1990, ApJ, 361, L69.
Soderblom, D. R., Oey, M. S., Johnson, D. R. H., and Stone, R. P. S. 1990, AJ, 99, 595.
Soderblom, D. R., Stauffer, J. R., Jones, B. F. and MacGregor, K. 1992, in *Inside The Stars*, eds. W.W. Weiss and A. Baglin, in press.
Spite, M., Maillard, J. P., and Spite, F. 1984, A&A, 141, 56.
Spite, F. and Spite, M. 1982, A&A, 115, 357.
Spite, F. and Spite, M. 1986, A&A, 163, 140.
Stauffer, J.R. 1991, in Angular Momentum Evolution of Young Stars, ed. S. Catalano and J.R. Stauffer, (Dordrecht: Kulwer)
Stauffer, J., and Hartmann, L. 1987, ApJ, 318, 337.
Stauffer, J., Hartmann L., and Jones, B. 1985, ApJ, 289, 247.
Stauffer, J., Hartmann L., and Jones, B. 1989, ApJ, 346, 160.
Stauffer, J., Hartmann, L. W., Jones, B. F., and McNamara, B. R. 1989, ApJ, 342, 285.
Stauffer, J., Hartmann, L., and Latham, D. 1987, ApJ, 320, L51.

Thorburn, J. A., Hobbs, L. M., Deliyanis, C. P. and Pinsonneault, M. H. 1992, Ap.J., submitted.
van Leeuwen, F. and Alphenaar, P. 1982, ESO Messenger, No. 18, p. 15.
Wallerstein, G. and Sneden, C. 1982, ApJ, 255, 577.

Inside the Stars, IAU Colloquium 137
ASP Conference Series, Vol. 40, 1993
Werner W. Weiss and Annie Baglin (eds.)

STELLAR MASSES AND RADII AS CONSTRAINTS ON STELLAR MODELS

JOHANNES ANDERSEN
Copenhagen University Observatory, Brorfeldevej 23, DK-4340 Tølløse, Denmark

ABSTRACT The current status of empirical data on stellar masses and radii of sufficient accuracy to give constraints on stellar models is reviewed. Results from the best-studied eclipsing binaries can already trace the main-sequence evolution of 1-10 $M_{\odot}$ stars in considerable detail and will be even more useful when supplemented by chemical abundance data.

Taking the deceptively simple question of the observed width of the main sequence as an example, it is shown how careful attention to the details of the data is required to reach robust conclusions about such features of modern stellar evolution models as opacity tables or convective overshooting. Only detailed modelling of specific systems with known masses, radii, and metal abundance constrain the theory strongly enough that a truly critical test is achieved. The same is true when using tidal interactions in binaries (apsidal motion, rotational synchronization, and orbital circularization) as another probe into stellar interiors.

INTRODUCTION

The most basic parameter determining the evolution of star is its *mass*, followed by its chemical composition. From these initial parameters, stellar evolution theory in principle predicts all observable properties of the star as functions of its age. Observable parameters are, most fundamentally, radius and effective temperature (or luminosity), followed by the photospheric (and chromospheric?) spectrum, rotation, surface inhomogeneities, pulsational frequencies, and perhaps neutrino flux, all subjects thoroughly reviewed at this meeting.

The best test object for stellar models is a star with independently known mass, chemical composition, and age (as input parameters to the model) and also known radius and effective temperature (to be compared with the model). The first and only star for which all these are known is the Sun: It is worth recalling that had our presence on Earth not allowed us to measure the Sun's mass, radius, and luminosity directly and with great accuracy, stellar evolution might never have figured as the subject of a scientific conference. Even today, solar seismology and neutrino experiments would be far less informative if the Sun's mass and radius were known to only $\pm 10\%$ and its metal abundance a mere guess.

For stars other than the Sun, direct determination of mass and radius for comparison with stellar models is only possible in suitable binary systems. However, in order to critically *test* modern stellar models, the data must above all be *accurate*: If uncertainties are large enough that all models can be made to fit the data by adjusting one or more free parameters, little new information is gained.

This review will demonstrate that although observable effects of such model features of current interest as opacity tables or convection prescriptions are quite subtle in main-sequence stars, the quality of the modern data, if properly utilized, is quite up to the task of testing these effects. A number of examples how the data *can* be fully utilized are given. Expressions of sympathy along the way with one or another aspect of particular models should not obscure this basic message.

BRIEF HISTORY AND CURRENT STATUS OF THE DATA

Comparison of stellar models with masses, radii, and luminosities determined from (eclipsing) binaries is an old subject: Already Schwarzschild (1958) showed a comparison of ZAMS models and B-type main-sequence binaries in the mass-luminosity diagram. For that time, the agreement (within some two magnitudes) was considered gratifying, but the errors were still large enough that neither composition nor evolutionary effects in the stars were significantly constrained.

Progress in both observations and models allowed Popper et al. (1970) to use the position of well-studied A and F-type binary components in the mass-radius plane to separate evolved and unevolved stars and use the position of the latter to constrain the metal-abundance parameter Z for the sample. For that value of Z, the locus of the unevolved stars in the mass-luminosity diagram constrained the helium abundance, Y, to a value close to that currently accepted. Still, apart from any deficiencies of stellar models at that time, uncertain factors remained in the data: No positive evidence existed to show that the stars were in fact *on* rather than just near the ZAMS; with errors of 5-10%, the mass and radius data did not allow to press the analysis any further. Moreover, it was questionable whether the stars did indeed all have the same metal abundance, and whether the adopted value of Z corresponded to what would be found by direct spectroscopic analysis.

Strömgren (1967) first proposed a detailed simulation of well-observed binaries as the most direct test of the models: For a given (known or assumed) composition, models of the two observed masses are evolved until the observed radii and luminosities of both stars are matched for a common age. As discussed already by Strömgren (1967) and later in more detail by Jørgensen (1978) and Andersen et al. (1991), errors must be no larger that 2% in mass, 1% in radius, 2% in temperature, and 25% in metal abundance for useful constraints on the models to result. Obtaining data with certifiable errors of this order *for the individual stars* is far from trivial, however. Only in the most favourable double-lined, eclipsing binaries is this currently possible, and meticulous attention to the choice of system and to the observational material and its analysis is necessary.

Data for the currently 45 binary systems (90 individual stars) for which masses and radii of both components have been determined directly and with errors below $\pm 2\%$ are listed by Andersen (1991), who also discusses the observational techniques and several applications of the data. We refer to that review and the references given there for further details on the individual binaries and the methods of analysis, and proceed in the following to discuss the relevance of this data set to the subject of this Colloquium.

EVOLUTION WITHIN THE MAIN SEQUENCE

The degree to which the basic binary parameters, mass and radius, can resolve the fine details of evolution through the main-sequence band is best displayed in the log M - log g diagram (Fig. 1). Note that g is here determined directly from M and R, independently of uncertainties in parallaxes, photometric calibrations, or spectroscopic line broadening theories; its accuracy is typically an order of magnitude better than possible by such indirect methods. As indicated by the average error bars shown in Fig. 1, evolutionary changes can be resolved to $\sim \pm 5\%$ of the width of the main-sequence band.

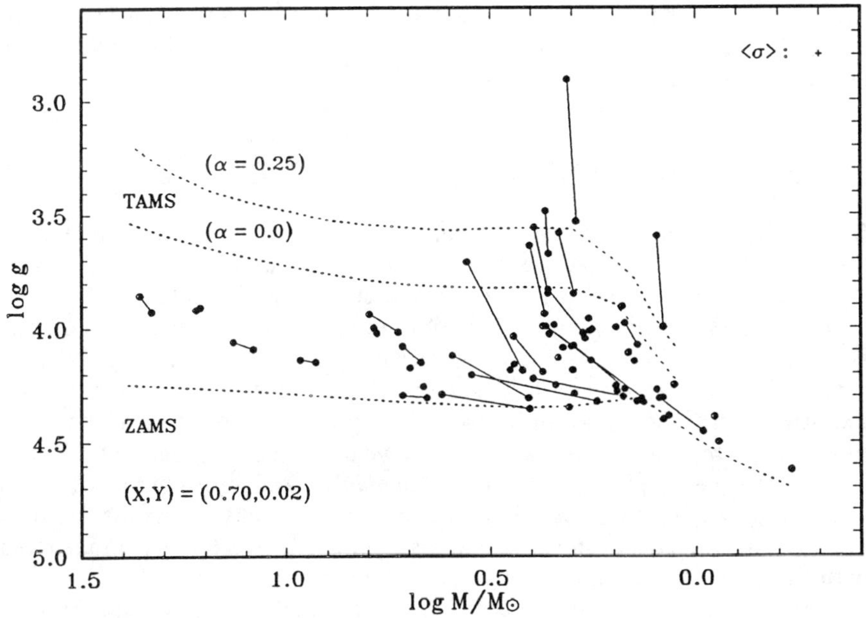

Fig. 1. Mass-gravity diagram for well-determined binary components (Andersen 1991). Lines connect members of the same system, and the average uncertainty of the points is shown. ZAMS and TAMS relations are from Claret & Giménez (1989, 1991a)

Assuming no mass loss, evolutionary tracks in the mass-gravity diagram are simple vertical lines along which the stars move (generally upward) with non-uniform speed. Appropriate stellar models must simultaneously predict the correct log g for both stars in a given system, i.e., isochrones should be parallel to the lines shown in Fig. 1. If the composition of the system is known, this is already a non-trivial test; if Z (and/or Y) may be freely adjusted, however, a much wider range of models will appear to pass the test. At present, V818 Tau (Popper & Ulrich 1988) and UX Men (Andersen et al. 1989) are the only *bona fide* main-sequence systems to which this basic test has been applied; it will be interesting to pursue it further when more metal abundance determinations have been completed (Clausen 1992).

Much of our subsequent discussion is foreshadowed already in Fig. 1, since TAMS lines (the loci for models at the point of central hydrogen exhaustion) are plotted both from models with moderate convective core overshooting ($\alpha = d/H_p = 0.25$) and from standard models ($\alpha = 0.0$; Claret & Giménez 1989, 1991a). When trying to decide whether a star in this region is still in a slow, central hydrogen-burning phase, we encounter the difficulty that it reverses its direction along the track (the "blue hook") for a brief interval before it proceeds into the Hertzsprung gap on its way to the giant branch. The log M - log g diagram (or the equivalent log M - log R diagram discussed by Stothers & Chin 1991) by itself gives no clue to the correct assignment of a given star to any of these overlapping parts of the track. More information - temperatures and/or luminosities - is needed as discussed in the next section.

THE "OBSERVED" WIDTH OF THE MAIN SEQUENCE

Both major (upwards) opacity revisions in the vicinity of a convective core and the introduction of convective overshooting increase the extent of the cores of stars with masses above $\sim 1.1\ M_\odot$. The main observable effect for near-main-sequence stars is a widening of the main-sequence band. Much current discussion is centered on this issue, and a notion seems to exist that "the width of the main sequence" is a quantity which is clearly defined and easily determined. The demonstration that this is not so is also a good illustration of some of the subtleties of the use of binary masses and radii, and we devote some attention to the subject here. While the weight of the evidence, in my opinion, at present favours models with mild overshooting over those without, the purpose of the discussion is to point out how the data can be used to discriminate between them. Models change with time; conclusions may change with them.

Open clusters are favourite test objects for stellar models in this mass range (e.g. VandenBerg 1985), and a composite colour-magnitude diagram for many clusters was used by Maeder & Meynet (1989) to demonstrate the significance of overshooting in these stars and guide the choice of overshooting parameter in their models. The seasoned cluster observer might ask to what extent field stars and binaries could contaminate the colour-magnitude diagrams of open clusters. Similarly, the binary star aficionado used to stars with well-known masses might ask whether the *real* stars do in fact have the same masses as the models used to fit them. Both possibilities are open to observational verification.

Although cluster studies are beyond the scope of this review, we note that our current radial-velocity survey (Nordström & Andersen 1991) does indeed find some cluster colour-magnitude diagrams to be so affected by binary and field stars as to severely compromise conclusions from fits to the raw diagrams. To provide evidence of a complementary nature (Andersen et al. 1990), we compiled data for well-studied eclipsing binary components in the mass range 1.5-2.5 $M_{\odot}$ (adopting somewhat softer criteria than Andersen 1991). Stars in this range have well-developed convective cores, while mass loss on the main sequence remains negligible. Compared with evolutionary tracks for the observed masses in the theoretical HR diagram, far more stars were found in rapid phases of evolution beyond the point of central hydrogen exhaustion (TAMS) than predicted by standard models. In contrast, both acceptable statistics and good fits to individual systems were found with the overshooting models by Maeder & Meynet (1988).

This interpretation was questioned by Stothers & Chin (1991) who compared standard models using both the old Cox & Stewart (1970) and the most recent OPAL opacities (Iglesias & Rogers 1991, Rogers & Iglesias 1992) for two values of the metal abundance, the latter models both with and without a small amount of overshooting. Comparing the binary data compiled by Andersen et al. (1990) with the ZAMS and TAMS loci of those models, they concluded that opacity effects alone could, in fact, account for the observed width of the main sequence. Assuming $Z = 0.03$, only a small fraction of the stars were found significantly outside the main-sequence band, consistent with the evolutionary lifetimes. They concluded that overshooting is insignificant in these stars.

Apart from the classic problem that "absence of evidence" (here, of overshooting) is not necessarily also "evidence of absence", several questions arise:

First, the metal abundance Z was a free parameter in the fit, but it is unknown whether $Z = 0.03$ does, in fact, hold for the *real* stars. Moreover, when discussing the *width* of the main-sequence band one must verify whether the *ZAMS* locus remains consistent with that of observed, unevolved stars of the appropriate metal abundance. The larger radii of metal-rich stars would further raise the ZAMS line in the $\log M$ - $\log g$ diagram (Fig. 1), already uncomfortably close to several of the observed stars. Thus, it is unlikely that $Z = 0.03$ is appropriate for the sample as a whole.

In this connection, we note that recent spectroscopic analyses of field B stars (Gies & Lambert 1992, Kilian 1992) find them to be CNO (and Si) *deficient* relative to the Sun; see also Nissen's (1992) review in this volume.

Second, the baseline models of Stothers & Chin (1991), using the outdated Cox-Stewart opacities, are not typical of recent models like those by VandenBerg (1985) or Maeder & Meynet (1988), which use the more recent Los Alamos Opacity Library (LAOL, Huebner et al. 1977). Fig. 2 compares models for a 2 $M_{\odot}$ star using both LAOL and OPAL opacities, each set computed both with and without overshooting, but with otherwise identical input physics. Clearly, in the mass range discussed above, opacity revisions now being seriously considered affect the main-sequence width significantly less than even mild overshooting.

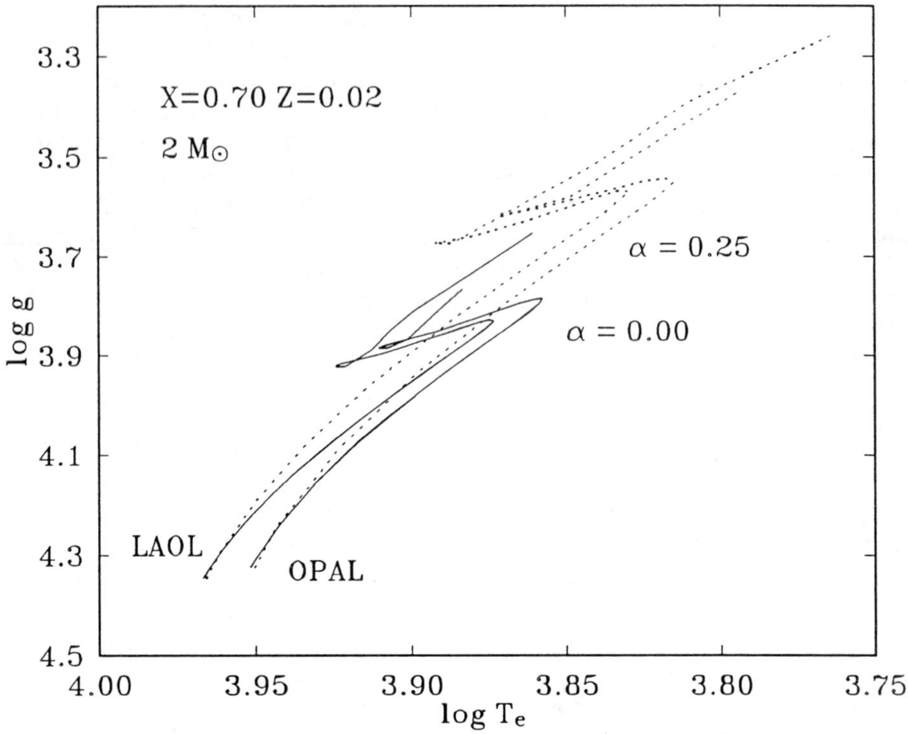

Fig. 2. Evolutionary tracks for a 2 $M_\odot$ star, with LAOL and OPAL opacities and with and without overshooting (Claret & Giménez, priv. comm.)

Third, and most important for the present discussion, is the basic ambiguity in assigning stars to slow or rapid evolutionary phases in the TAMS region. It cannot be resolved except by an actual fit of the individual binaries for the observed masses and a consistent age. This is illustrated in Fig. 3, which shows the same VandenBerg models and binary data as discussed by Andersen et al. (1990), but displayed so as to focus attention on the ZAMS and TAMS loci. A casual inspection of the figure might leave the impression that for the high (Hyades) metal abundance of the standard models shown here, only 3-4 stars are significantly outside the main-sequence band. Adding the 44 known binary components located inside this section of the main sequence to the 13 stars shown, this is an entirely plausible fraction.

Respectable agreement between the location of these stars and the tracks *for the observed masses* (stars and tracks labelled with the same symbols) is also found in most cases. However, essentially all the stars are found to be in the "blue hook" region or in the Hertzsprung gap, i.e. in very rapid phases of evolution. Note that the TAMS may be shifted towards cooler temperatures in the diagram by increasing the metal abundance even further (direction of arrow), but at the same time this makes the fit to the *individual* tracks even worse. Thus, finding a

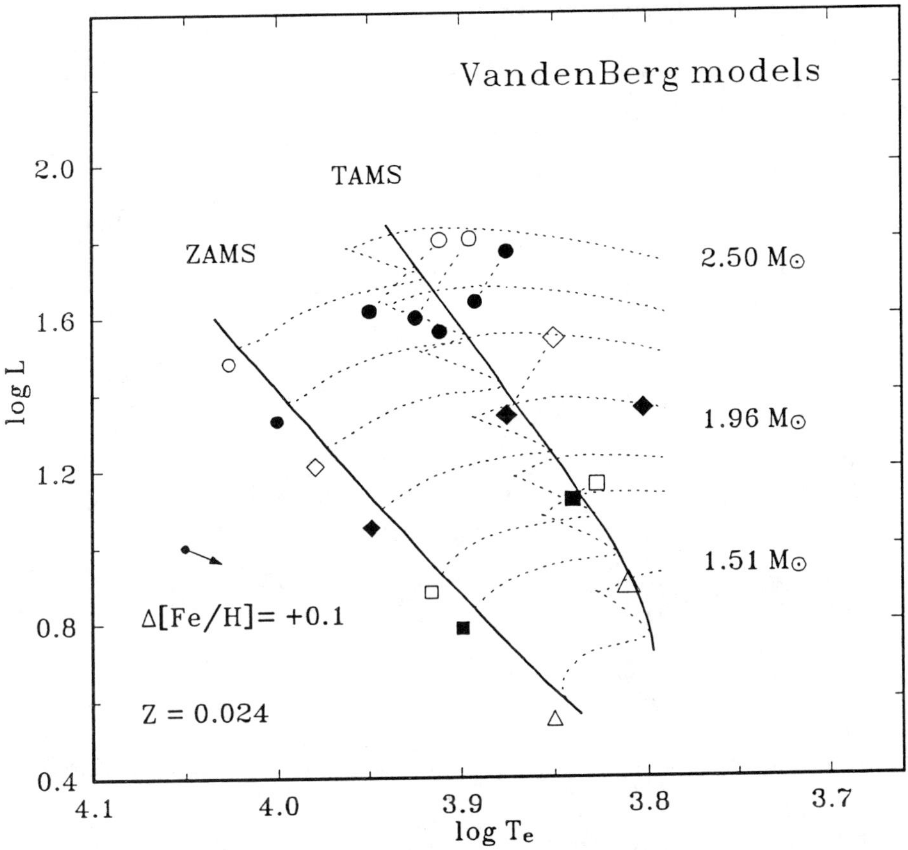

Fig. 3. Evolved binary components (log g < 3.9) of mass 1.5-2.5 $M_{\odot}$; from top to bottom: WX Cep, V1031 Ori, SZ Cen, V624 Her A, AI Hya, TZ For B, EI Cep, and RZ Cha AB. Standard evolutionary tracks for the observed masses and a Hyades composition are identified by corresponding symbols. The arrow shows how models shift for the indicated change in Z

star inside "the" main-sequence band in the HR diagram does not prove that it actually is in a slowly-evolving phase unless confirmed by a direct model fit.

In contrast, the solar-metallicity overshooting models shown in Fig. 4 not only place nearly all the stars inside the main-sequence band, but central hydrogen-burning models for the appropriate masses also fit most individual systems within the uncertainties. In this case, assuming a larger metal abundance for these young stars than that of the models (solar) actually *improves* the fit as shown by the arrow; indeed, the most evolved of these stars, TZ For, has a directly observed [Fe/H] = +0.1 (Andersen et al. 1991).

We conclude this section by re-emphasizing that its purpose is not to endorse particular models (in fact, ages from Maeder & Meynet models in this

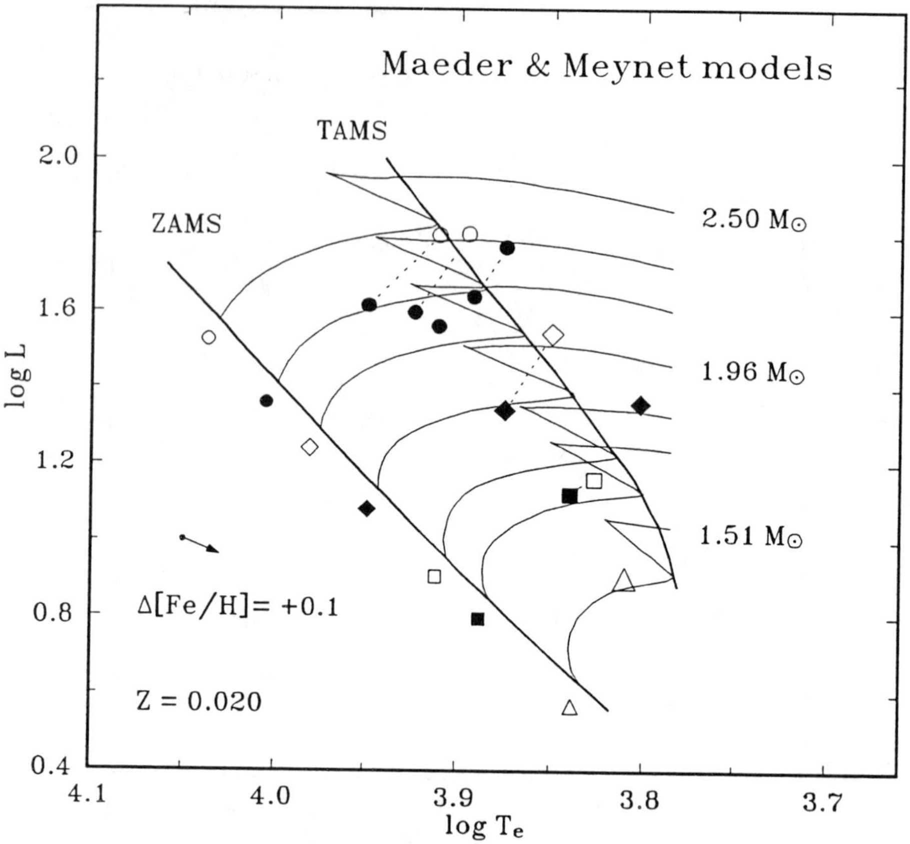

Fig. 4. Same as Fig. 3, but with solar-metallicity overshooting models by Maeder & Meynet (1988)

mass range require considerable downward revision, as shown by Schaller et al. 1992). The goal has been to show how *all* available information must be exploited to its full extent for conclusions to be solidly founded.

FITTING INDIVIDUAL SYSTEMS

The Sun is no doubt the best example of the physical insight which is possible from a refined study of a particularly well-observed star. Models computed for its precisely (and independently) known mass, chemical composition, and age must fit its precise observed radius and effective temperature. Although this is possible by adjusting the assumed helium abundance and mixing-length parameter, reproducing also the observed neutrino flux and oscillation frequency spectrum amounts to a challenge that has not yet been successfully met.

In stars other than the Sun, Sun-like oscillations have not yet been unambiguously detected (except possibly in α Cen A, Pottasch et al. 1992), and measurable neutrino fluxes are only expected at exceptional occasions (SN 1987A!). In compensation, the requirement that the two stars in a non-interacting binary system be matched by models of the *same age and chemical composition* provides a significant additional constraint on the models which becomes the most informative if the two stars have appreciably different properties and the observational data are as complete and accurate as modern techniques allow. A few examples are discussed in the following.

Assuming that the metal abundance of a binary system is known together with the component masses, the three additional parameters needed to specify a model for its present state are the mixing-length parameter $\alpha = l/H_p$, the helium abundance Y, and the age t. For the Sun, t is known, and α and Y are adjusted so that the model fits both the radius and the luminosity of the present Sun. For binary components with masses near 1 $M_\odot$, the solar mixing-length parameter is presumed valid, and Y and t are determined by fitting the radius and luminosity of one of the stars. The *test* then lies in verifying whether a model with exactly the same parameters matches the radius and luminosity of the other star for its known mass. This is trivially achieved, and no test therefore performed, if the two stars are very similar; the greater the effects of differential evolution, the more significant the test.

As lucidly explained by VandenBerg & Hrivnak (1985), the 1.2-$M_\odot$ binary AI Phe (F7 V + K0 III) is nearly ideal for the purpose, having components of nearly equal mass of which one is still near the main sequence while the other is already on the lower giant branch. In addition to good masses and radii, a photometric metallicity estimate was available, and VandenBerg & Hrivnak (1985) proceeded to model the system in detail. Inspired by that example, but puzzled by the high helium abundance that was derived, Andersen et al. (1988) provided greatly improved masses ($\pm 0.3\%$) and a (spectroscopic) [Fe/H] determination for the system. In addition, models for the precise observed masses were specially computed and carefully calibrated, using the VandenBerg (1985) code, including a solar model to determine the mixing-length parameter as outlined above.

The resulting fit is shown in Fig. 5, which merits a few comments: First, the positions of the two observed stars in the diagram are determined directly from the observed temperatures (from multicolour photometry) and radii. For a given temperature, the error in the luminosity is due only to that of the radii, so the error polygons are parallelograms as shown. Second, the range of tracks allowed by the uncertainty in the masses is indicated roughly by the width of the lines; showing only the model for the adopted mass, as is often done, can be deceptive. Third, with models for the observed metal abundance, the helium abundance found by fitting the luminosity of the hotter component is solar: $Y = Y_\odot = 0.27 \pm 0.01$. Fourth, the age of the hotter component having also been determined in the fit, the models for the subgiant primary were computed up to the same age, i.e. with *no free parameters*. Age variations of just $\pm$1-2% would suffice to make this fast-evolving model disagree with the properties of the observed star, but the VandenBerg models pass this most severe test with flying colours. The main

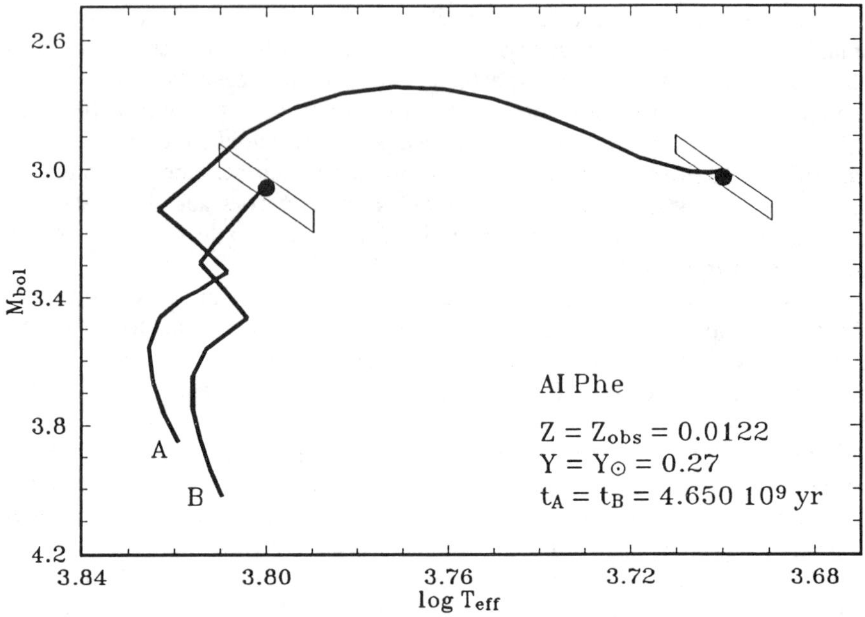

Fig. 5. Evolutionary history of AI Phe (Andersen et al. 1988). Standard VandenBerg (1985) models are shown, for the observed metal abundance, $Y = Y_{\odot} = 0.27 \pm 0.01$, and identical ages for both stars. The width of the tracks roughly corresponds to the error in the masses, the width of the error box of star A to an age uncertainty of $\sim \pm 1\%$

uncertainty in the test is now due to that in the metal abundance (only ± 0.1 dex), not the masses and radii.

By coincidence, the F-type system UX Men was found at the same time to have masses identical to those of AI Phe to within $\pm 0.1\%$(!); a metallicity estimate was also close to that for AI Phe. Nevertheless, the stars in UX Men are considerably cooler than the models which fit AI Phe, indicating a higher metal abundance. This was confirmed spectroscopically by Andersen et al. (1989), and VandenBerg (1985) models for the observed Z were again found to be in complete agreement with both observed stars. Including the Hyades binary V818 Hya (Popper & Ulrich 1988) in the sample shows standard evolution models for near solar-mass stars to perform satisfactorily over a range in Z of a factor of two.

A similar study of TZ For, a system very similar to AI Phe but of slightly higher mass ($\sim 2\ M_{\odot}$) gave a different result (Andersen et al. 1991): Standard models for the observed metal abundance fit the less-evolved F7 IV component well, but match the G8 III giant primary in neither temperature nor age. Moreover, both stars then appear to be in very rapid evolutionary phases, an unlikely if not totally impossible situation. In contrast, models with moderate overshooting (Maeder & Meynet 1988) are consistent with the position of both stars in the HR diagram for the same age. Moreover, the F star is then near the end of its main-sequence evolution and the giant in the core helium-burning ("clump") phase, i.e. both relatively long-lived; clearly a preferable interpretation.

In TZ For, the detailed results permit another test: The mass of the giant component (2.05 $M_\odot$) is intermediate between the upper mass limits for stars experiencing a helium flash, 2.3 $M_\odot$ according to standard and 1.85 $M_\odot$ for overshooting models. The maximum radii of stars undergoing He flash are, however, much larger than in stars above that limit - in TZ For in fact also larger than the Roche lobe. Had TZ For evolved according to standard models, it would have overflowed its Roche lobe in the past and probably suffered extensive mass loss.

A final example of the insight that can be derived from close attention to individual systems in crucial evolutionary phases is SZ Cen (Andersen 1991), the post-TAMS system with equal masses shown by two filled circles in Fig. 3. While increasing the metal abundance beyond the (already high) values of those models may shift the main sequence so as to include at least one of the stars, no increase in Z can get *both* stars to fit the same track *for the observed mass* (labelled by the same symbol). The overshooting models (Fig. 4) accomplish this in a natural way.

More detailed comparisons of this type will no doubt provide much useful information on these and other aspects of stellar models. Checking the presumed gradual onset of convective overshooting in (F-type) stars in the mass range 1.1-1.5 $M_\odot$ should be very informative, as the schematic treatment of overshooting in present models is probably quite unrealistic, especially in small convective cores (Zahn 1991). As discussed above, spectroscopic [Fe/H] determinations will be needed to maximize the information that can derived from such comparisons.

EFFECTS OF TIDAL INTERACTION IN BINARY SYSTEMS

Observable effects of tidal interaction in close binaries include apsidal motion, orbital circularization, and rotational synchronization. They depend very sensitively on the fractional radii of the components. Meaningful comparisons between observed and computed effects require that the radii be determined to $\pm 1\%$ or so. An extensive review of these effects is outside the scope of this paper (see e.g. Zahn 1992), but a few salient points will be mentioned.

Apsidal motion is one of the classic tests of stellar evolution models, providing a direct measure of the central density concentration in real stars (Schwarzschild 1958). Significant results of the test require well-determined apsidal-motion periods and accurate radii, mass ratios, and rotations to compute the density concentration coefficients k_2 with sufficient precision. The precise mass and position of the stars within the main-sequence band are needed to provide similarly precise theoretical predictions of k_2.

Summarizing a persistent effort on this subject, Claret & Giménez (1991b) concluded that small, but significant discrepancies remained between the computed and observed values of k_2. In subsequent work, Claret & Giménez (1992) find that when convective overshooting, mass loss, and the effects of the known axial rotation of the stars are included in the models, consistency is achieved. Thus, improvements in theory and observation are combined in this classic test to yield significant new results on stellar interiors.

For a general review of tidal synchronization and circularization mechanisms, we refer to Zahn (1992). Briefly, their timescales are roughly proportional to r^2, so rotational synchronization typically occurs 100 times faster than orbital circularization. In stars with radiative envelopes, spin-down of the surface may be decoupled from that of the interior, leading to still shorter synchronization times (Goldreich & Nicholson 1989). Both timescales depend dramatically on the stellar radii, so binaries with accurately determined properties can be used to test theories of tidal interactions in binaries.

TZ For, discussed above, turns out to be exceptionally informative also in this respect (Andersen et al. 1991): Its period (75 days) is so long that, according to standard tidal theories, several Hubble times would be needed to circularize the orbit. Still, while the orbit is in fact circular, the least-evolved component nevertheless rotates several times faster than the synchronous rate. Attempts to explain TZ For as an exceptional, initially circular case were shown to fail because *all* known giant binaries in this period range have circular orbits.

The realization that the more massive star in TZ For is in the core helium-burning ("clump") phase provides the key to understanding also its tidal history: At the tip of the giant branch, it was four times larger than now. Since the circularization time varies as r^{-8} it is reduced by a factor 6.5 10^4, to only a million years. Thus, the brief time spent by the star at the top of the giant branch sufficed for it to circularize the orbit by itself while the companion, still unevolved and with a radiative envelope, continued spinning essentially unaffected. Again, a detailed model fit to the masses and radii of TZ For was the key to understanding the common history of stellar and tidal evolution in this and similar systems.

CONCLUSIONS

Improvements in the observation and analysis of eclipsing binaries have produced a sample of nearly fifty binaries, listed by Andersen (1991), in which the masses and radii of both components are known with errors of $\pm 2\%$ or less and effective temperatures are also well-determined. A few of these stars have good spectroscopic metal abundance determinations.

Ways of comparing stellar models with the data can be devised which utilize various subsets of the available parameters mass, radius, temperature/luminosity, and metal abundance. Naturally, the fewer parameters that remain free in the fit, the more informative constraints on the models are obtained. The width of the main-sequence band is discussed as an illustration how detailed model fitting may be needed to get a reliable answer to a simple-minded question. Direct simulation of the evolution of specific systems with complete and accurate data allows a more meaningful test of stellar evolution models than comparison of mean theoretical relations with average properties for less well-observed samples of stars.

The agenda for the immediate future is to provide accurate mass and radius determinations for stars in parts of the main-sequence band that are now poorly covered (cf. Andersen et al. 1992), and to complement these with the spectroscopic data that will finally promote the metal-abundance parameter Z from fudge factor to input physics. This effort is well under way (Clausen 1992).

ACKNOWLEDGEMENTS

It is a pleasure to acknowledge the long and pleasant cooperation with J.V. Clausen, B. Nordström, D.M. Popper, A. Giménez, D. VandenBerg, and A. Claret. Observing time and financial support from ESO, the Danish Natural Science Research Council, The Danish Board for Astronomical Research, the Carlsberg Foundation, and the Smithsonian Institution are gratefully acknowledged.

REFERENCES

Andersen, J.: 1991, *A&A Rev.*, **3**, 91

Andersen, J., Clausen, J.V., Gustafsson, B., Nordström, B., VandenBerg, D.A.: 1988, *A&A*, **196**, 128

Andersen, J., Clausen, J.V., Helt, B.E., Johansen, K.T., Nordström, B., García, J.M., Giménez, A., Stefanik, R.P., Vaz, L.P.R. 1992, this volume

Andersen, J., Clausen, J.V., Magain, P.: 1989, *A&A*, **211**, 346

Andersen, J., Clausen, J.V., Nordström, B., Tomkin, J., Mayor, M.: 1991, *A&A*, **246**, 99

Andersen, J., Nordström, B., Clausen, J.V.: 1990, *ApJ*, **363**, L33

Claret, A., Giménez, A.: 1989, *A&AS*, **81**, 1

Claret, A., Giménez, A.: 1991a, *A&AS*, **87**, 507

Claret, A., Giménez, A.: 1991b, *A&A*, **244**, 319

Claret, A., Giménez, A.: 1992, this volume

Clausen, J.V. 1992, this volume

Cox, A.N., Stewart, J.N.: 1970, *ApJS*, **19**, 243

Gies, D.R., Lambert, D.L. 1992, *ApJ*, **387**, 673

Goldreich, P., Nicholson, P.: 1989, *ApJ*, **342**, 1079

Huebner, W.F., Merts, A.L., Magee, N.H., Argo, M.F.: 1977, *Los Alamos Scientific Laboratory Report*, No. LA-6760-M

Iglesias, C.A., Rogers F.J.: 1991, *ApJ*, **371**, 408

Jørgensen, H.E.: 1978, in *The HR Diagram (IAU Symp. No.* **50**), eds. A.G.D. Philip & D.S. Hayes, Reidel, Dordrecht, p. 433

Kilian, J. 1992, *A&A*, submitted

Maeder, A., Meynet, G.: 1988, *A&AS*, **76**, 411

Maeder, A., Meynet, G.: 1989, *A&A*, **210**, 155

Nissen, P.E. 1992, this volume

Nordström, B., Andersen, J.: 1991, *ESO Messenger*, **63**, 34

Popper, D.M., Jørgensen, H.E., Morton, D.C., Leckrone, D.S.: 1970, *ApJ*, **161**, L57

Popper, D.M., Ulrich, R.K.: 1986, *ApJ*, **307**, L61

Pottasch, E.M., Butcher, H.R., Van Hoesel, F.H.J. 1992, this volume

Rogers, F.J., Iglesias, C.A. 1992, *ApJS*, **79**, 507

Schaller, G., Schaerer, D., Meynet, G., Maeder, A. 1992, *A&AS*, in press

Schwarzschild, M.: 1958, *Structure and Evolution of the Stars*, Princeton Univ. Press, Princeton

Stothers, R.B., Chin, C.-w.: 1991, *ApJ*, **381**, L67

Strömgren, B.: 1967, in *Modern Astrophysics*, ed. M. Hack, Gauthier-Villars, Paris, p. 185

VandenBerg, D.A.: 1985, *ApJS*, **58**, 711

VandenBerg, D.A., Hrivnak, B.J.: 1985, *ApJ*, **291**, 270

Zahn, J.-P. 1991, *A&A*, **252**, 179

Zahn, J.-P. 1992, in *Binaries as Tracers of Stellar Formation*, eds. M. Mayor & A. Duquennoy, Cambridge Univ. Press, in press

Inside the Stars, IAU Colloquium 137
ASP Conference Series, Vol. 40, 1993
Werner W. Weiss and Annie Baglin (eds.)

ACCURATE MASSES AND RADII AS TEST DATA FOR STELLAR MODELS: STATUS AND PROGRESS ON THE COPENHAGEN PROJECT

J. ANDERSEN, J.V. CLAUSEN, B.E. HELT, K.T. JOHANSEN and B. NORDSTRÖM, Copenhagen University Observatory, Denmark

M. GARCÍA, Departamento de Física, E.I. Industriales, Madrid, Spain

A. GIMÉNEZ, Laboratorio de Física Espacial, INTA, Madrid, Spain

R.P. STEFANIK, Harvard-Smithsonian Center for Astrophysics, U.S.A.

L.P.R. VAZ, Departamento de Fisica, Univ. Fed. de Minas Gerais, Brazil

1. Basic ingredients of stellar models

The most important ingredient of a stellar model is its initial *mass*. Hence, real stars with known masses are important test objects for stellar models. Initial *chemical composition* is the other important starting parameter and should therefore also be known (Clausen, 1992). Finally, the most direct indicator of the evolution of a star is its *radius*. Together, these parameters determine the effective temperature and luminosity of the model, to be tested against a real star.

2. How accurate need the data be?

For such comparisons to give interesting results on present-day stellar models, the empirical data must be *accurate* (1-2% in M, R, and T_e; 0.1 dex in [Fe/H]): Stellar masses with errors of just $\pm 5\%$ can yield essentially no useful information on modern (near-main-sequence) stellar models.

3. What exists, and what more is needed?

Data of the required accuracy can presently only be obtained from selected, detached, *double-lined eclipsing binaries*. The current status of such data (Andersen, 1991) is summarized in Fig. 1. Here, the basic data M and R are displayed in a format designed to bring out the effects of evolution through the main-sequence band (note the size of the error bar!) as defined by (*dashed*) standard or (*dots*) overshooting models (Claret & Giménez, 1991). Data obtained within the framework of the Copenhagen (CUO) program are identified.

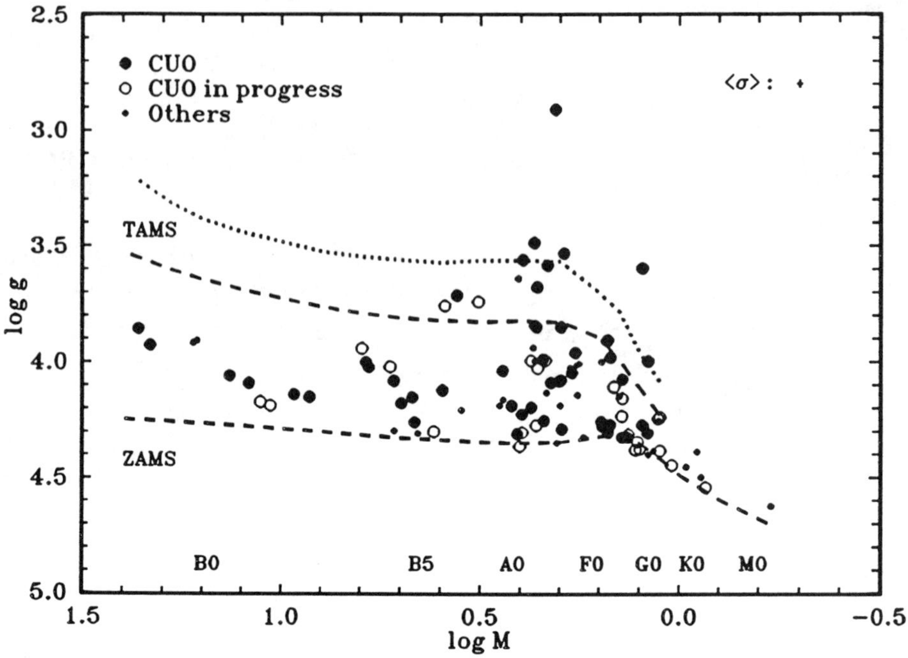

Fig. 1. log M - log g diagram for binary components with mass and radius known to $\pm 2\%$ or better.

Systems now in progress (*open circles*) are primarily:

- Early-type (OB) stars,
- Stars useful in locating the empirical ZAMS,
- Stars with [Fe/H] determination in progress, useful in testing opacity calculations.

4. The observational basis

As an example of the quality of the observational data needed to produce masses and radii which are useful for our purposes, preliminary light and radial-velocity curves for one of the new systems are shown in Figs. 2a,b.

References

Andersen, J. 1991, *Astr. Ap. Rev.*, **3**, 91.
Clausen, J.V. 1992, in *Inside the Stars*, eds. W. Weiss and A. Baglin.
Claret, A., Giménez, A. 1991, *Astr. Ap. Suppl.*, **87**, 507.

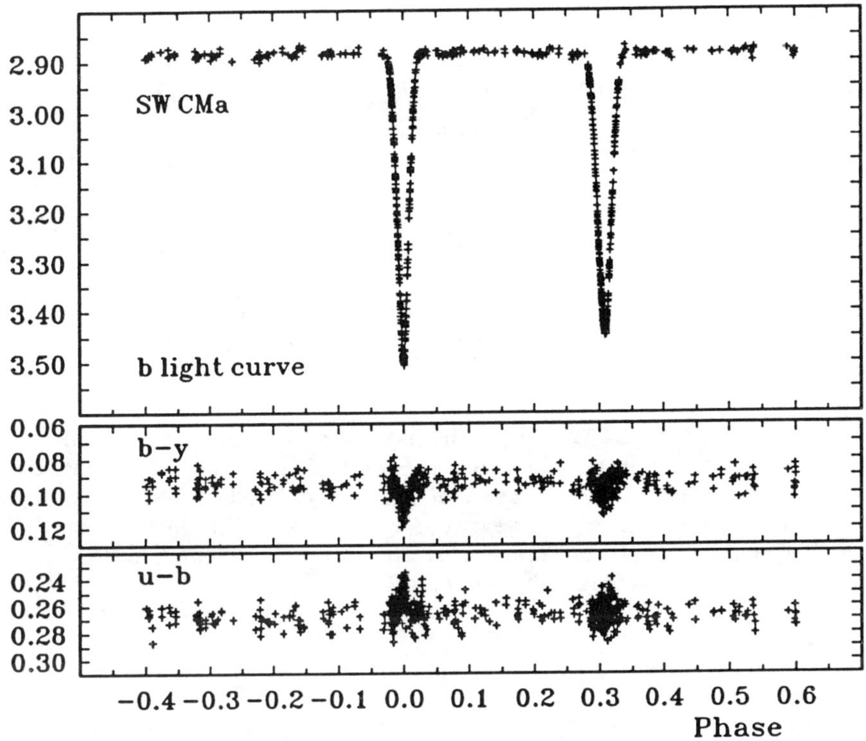

Fig. 2a. Light and colour curves of SW CMa (Danish 50-cm tel., ESO, La Silla).

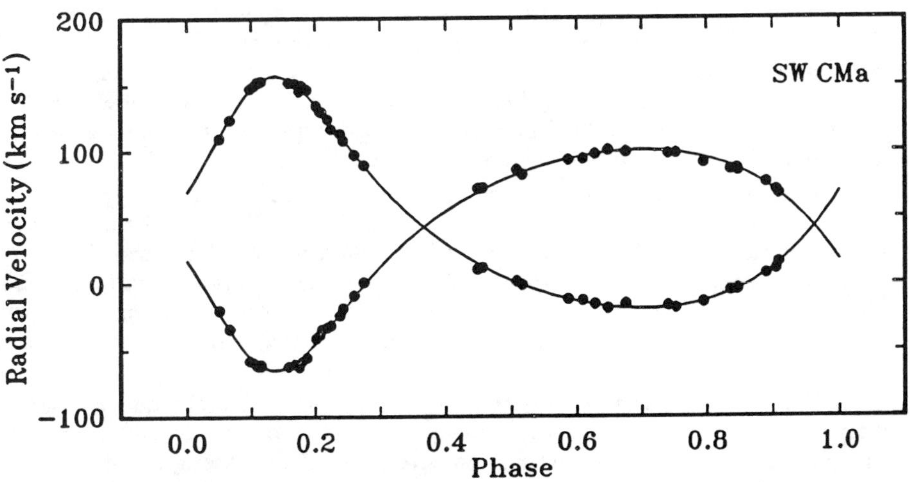

Fig. 2b. Radial-velocity curves of SW CMa (CfA echelle systems).

Inside the Stars, IAU Colloquium 137
ASP Conference Series, Vol. 40, 1993
Werner W. Weiss and Annie Baglin (eds.)

MEASUREMENTS OF MASSES AND RADII OF ECLIPSING BINARIES

FERLUGA, S. , FLOREANO, L. , MANGIACAPRA, D.

Dipartimento di Astronomia, Università di Trieste
Via G.B. Tiepolo, 11 - I 34131 Trieste, Italy

ABSTRACT We report the detection of 10 new double–line eclipsing binary systems, with first measurements of masses and radii. These results are the preliminary outcome of an observational campaign for detecting new double–line eclipsing systems, carried out at the *Observatoire de Haute Provence (OHP)* in France, with the 1.52-m telescope and its CCD high–resolution spectrograph *Aurelie*.

THE SURVEY

It is well known that only binary systems, namely double–spectrum eclipsing binaries, may provide a *direct* determination of stellar masses and radii. The number of such stars, found in one century, hardly reaches a hundred.

The purpose is to enlarge the set of known double–spectrum eclipsing binaries, by detecting the secondary lines of those systems for which only the primary spectrum was observed up to now. Our survey extends in the whole northern celestial hemisphere, up to an uneclipsed magnitude V'=10.5 , providing a total set of 20 binaries (most *sd*–type) with a faint secondary component to be searched for (often $L_2/L_1 \simeq 1\%$) .

Orbital periods of these stars are short (few days or hours); so it is not possible to take long exposures (leading to undesired line broadening because of orbital motion). On the other hand, the secondary lines are often very faint, requiring large S/N ratios. This basic observational difficulty could be overcome in most cases, thanks to the high efficiency of *Aurelie*, combined with the technique of multiple exposures (taken on different epochs at the same orbital phase).

For searching the secondary spectrum, a special technique is used by comparing couples of observations performed at two opposite quadratures. So, a given secondary line appears at opposite sides of the strong primary feature (Fig.1). This method provides immediately the secondary orbital velocity, and it is suitable for applying our special mathematical treatment, which automatically separates the spectra of the two components.

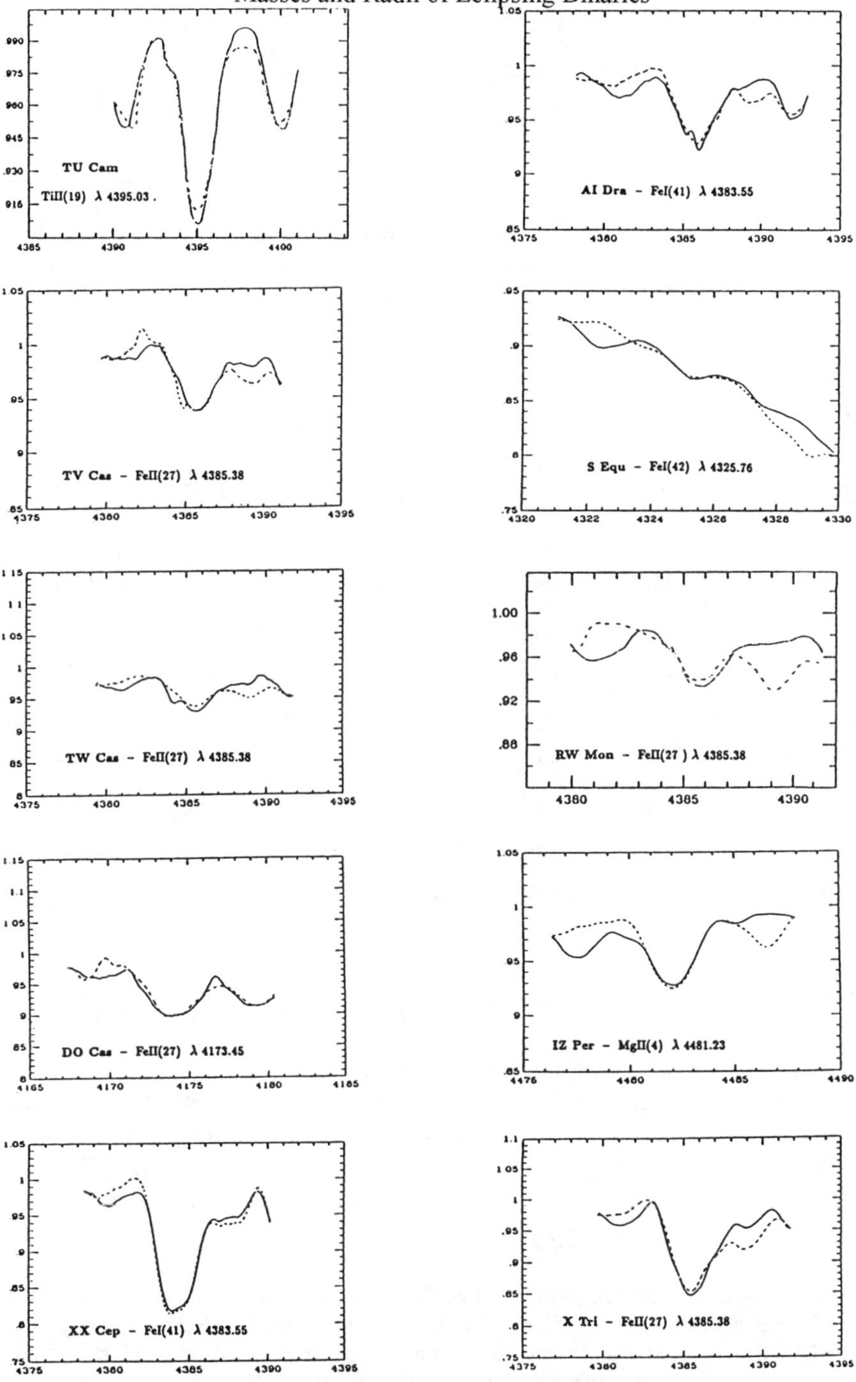

Figure 1. Secondary line detection. Continuum–normalized spectra taken at *1st* and at *2nd* quadrature (*dashed* and *solid* tracings, respectively) are shifted in wavelength, until the primary lines overlap; so the secondary contribution is revealed by difference (wavelength is in Å).

Table 1. New measured masses $m_{1,2}$ and radii $r_{1,2}$, with 1-σ errors (solar units). Orbital period P is in days; V' and V'' are the visual magnitudes (uneclipsed and eclipsed, respectively); K_1 and K_2 are the radial velocity amplitudes in km·s^{-1}.

System *type* P	V' V''	sp_1 sp_2	$K_1 \pm \Delta K_1$ $K_2 \pm \Delta K_2$	$m_1 \pm \Delta m_1$ $m_2 \pm \Delta m_2$	$r_1 \pm \Delta r_1$ $r_2 \pm \Delta r_2$
TUCam *d* 2.9	5.0 5.2	A2V F0	78.0 ± 1.4 108 ± 12	$1.21 \pm .28$ $0.88 \pm .11$	$3.52 \pm .23$ $1.13 \pm .08$
TV Cas *sd* 1.8	7.3 8.4	B9V G	87.9 ± 2.3 179 ± 7	$2.54 \pm .24$ $1.25 \pm .08$	$2.65 \pm .15$ $2.80 \pm .15$
TW Cas *sd-d* 1.4	8.3 8.9	B9V G2	87.0 ± 1.8 181 ± 3	$2.03 \pm .08$ $0.98 \pm .03$	$2.32 \pm .04$ $1.83 \pm .05$
DO Cas *c* 0.7	8.6 9.2	A5V F2	80.0 186 ± 7	$0.96 \pm .09$ $0.41 \pm .02$	$1.81 \pm .05$ $1.07 \pm .03$
XX Cep *sd* 2.3	9.1 10.3	A7V G4	28.0 ± 3.7 273 ± 20	5.70 ± 1.2 $0.58 \pm .12$	$3.41 \pm .28$ $3.03 \pm .42$
AI Dra *sd-d* 1.2	8.0 8.9	A0V G2	93.5 ± 1.1 177 ± 6	$1.76 \pm .13$ $0.93 \pm .04$	$2.02 \pm .07$ $1.82 \pm .09$
S Equ *sd* 3.4	8.0 10.8	B8V F9III	23.4 ± 0.7 189 ± 2	$2.87 \pm .06$ $0.36 \pm .01$	$2.78 \pm .05$ $3.07 \pm .12$
RWMon *sd* 1.9	9.1 11.9	B9V F9IV	74.4 ± 6.7 210 ± 2	$3.41 \pm .18$ $1.21 \pm .17$	$2.41 \pm .10$ $3.02 \pm .25$
IZ Per *sd* 3.7	7.8 9.0	B8 A4IV	52.9 ± 4.7 255 ± 7	$9.15 \pm .69$ $1.90 \pm .24$	$7.09 \pm .21$ $6.49 \pm .23$
X Tri *sd* 1.0	8.9 11.5	A5V G0V	110.0 ± 4.1 158 ± 10	$1.15 \pm .16$ $0.80 \pm .08$	$1.41 \pm .06$ $1.60 \pm .07$

20 NEW MASSES AND RADII

The 10 new detected systems provide absolute elements for 20 individual stars. Using a resolution of about 0.5Å in the range 4100–4500Å, we could generally achieve positive results for binaries up to the uneclipsed magnitude V'=9.5. The selected spectral region includes H_δ, H_γ and MgII λ4481Å, which are often our first–detected secondary lines. Moreover in this range there are many strong metallic lines (mainly Fe and Ti), which can be detected in the secondary spectrum (usually of F or G type).

Program stars were selected among those having spectroscopic elements of the primary orbit well determined in the literature (Batten *et al.*, 1989): mass function $f(m)$, orbital semiaxis $a_1 \sin i$, radial velocity amplitude K_1 . At the same time, photometric parameters should be known from numerical light–curve synthesis (Cester *et al.*, 1979): inclination i , fractional radii r_1/a and r_2/a .

The shift of the secondary lines at quadratures provides radial velocity amplitude K_2 , giving the semiaxes ratio $a_2/a_1{=}K_2/K_1$ and the mass ratio $m_2/m_1{=}K_1/K_2$. So the orbital separation $a{=}a_1{+}a_2$ is known, providing absolute radii r_1 and r_2 . Finally, the masses m_1 and m_2 are computed from the mass function $f(m)$. Results are reported in Table 1.

EXTRACTION OF THE SECONDARY

The high quality of the spectra obtained makes it possible to go beyond the simple recognition of the presence of the secondary lines. in particular, the stability of the instrumental response allows us to compare quantitatively the spectra of a given binary system at opposite quadratures, dealing with variations of about 1% of the continuum. From these variations, it has been possible to reconstruct mathematically the complete spectrum of the secondary star.

The procedure, which uses an iterative method, succeeds in putting in evidence some features that otherwise would be unrecognizable by visual inspection (Ferluga *et al.*, 1991). In principle this method of extraction could be applied to all spectroscopic binaries, provided they do not show intrinsic variability. The positive results of our survey, and the encouraging first applications of the extraction technique, suggest an extension of this research program to spectroscopic binaries in general.

ACKNOWLEDGMENTS

The authors are particularly grateful to J.P. De Greve, for invaluable suggestions.

ERRATUM – Values in Table 1 replace those reported in the *Abstract* booklet (Ferluga *et al.*, 1992: *Communic. Asteroseismol.* **43**, 4/2).

REFERENCES

Batten A. H., Fletcher J. M., Mac Carthy D. G.: 1989, *Eighth Catalogue of the Orbital Elements of spectroscopic Binary Systems*, Publ. Dominion Astroph. Observatory (Victoria, Canada) **17**, 1

Cester B., Fedel B., Giuricin G., Mardirossian F., Mezzetti M., Predolin F.: 1979, *A Catalogue of modern Lightcurve Synthesis photometric Solutions of close Binary Systems*, Mem. Soc. Astron. Italiana **50**, 553

Ferluga S., Floreano L., Mangiacapra D.: 1991, *La Lettre de l'OHP* (Publ. Obs. Haute Provence, France) **6**, 3

Inside the Stars, IAU Colloquium 137
ASP Conference Series, Vol. 40, 1993
Werner W. Weiss and Annie Baglin (eds.)

DETERMINATION OF PHYSICAL PARAMETERS OF BINARY SYSTEMS: THE PROBLEM OF SPURIOUS SOLUTIONS

F. BARONE, L. DI FIORE, S. MANCUSO, L. MILANO, G. RUSSO
Dipartimento di Scienze Fisiche dell'Università, Napoli, Italy, Istituto Nazionale Fisica Nucleare, Sez. Napoli, Italy

ABSTRACT In this paper we apply the Wilson-Price procedure to the solution of the binary system AO Cam using *simultaneously* all the available information and a statistical criterion to judge about the quality of the solutions found.

INTRODUCTION

For the solution of binary systems, we have devised a procedure (Wilson-Price Code), described in Barone *et al.* (1988, 1990). This procedure is based on the physics of the Wilson and Devinney model (1971, 1979) for light curve synthesis, but uses as fitting procedure, a global optimization algorithm, the Controlled Random Search (Price, 1976), in order to avoiding the loss of possible physical solutions and *influence* of the researcher on the final solution. With this procedure, it is possible to solve *simultaneously* light curves and radial velocity curves, also for eccentric systems, using, in this way, *all* the available information at the same time.

STATISTICAL CRITERIA OF SOLUTION

A general expression for the quality coefficient S, that expresses quantitatively the goodness of the fitting and that is the objective function to minimize, allowing the *simultaneous* solution of light and radial velocity curves is

$$S = \sum_{i=1}^{l} \sum_{j=1}^{n_i} w_{l_{i,j}} (l_{i,j_{obs}} - l_{i,j_{comp}})^2 + \sum_{k=1}^{r} \sum_{l=1}^{n_k} w_{r_{k,l}} (v_{k,l_{obs}} - v_{k,l_{comp}})^2 \qquad (1)$$

where l and r are the number of light and radial velocity curves, n_i and n_k are the number of observations of the i-th light curve and k-th radial velocity curve, $w_{l_{i,j}}$ and $w_{r_{l,k}}$ are the weights of the points of light and radial velocity curves, defined as

$$w_{l_{i,j}} = \frac{\frac{1}{\sigma_{l_{i,j}}^2}}{\frac{1}{\sigma_{tot}^2}} \quad w_{r_{k,l}} = \frac{\frac{1}{\sigma_{r_{k,l}}^2}}{\frac{1}{\sigma_{tot}^2}} \quad \frac{1}{\sigma_{tot}^2} = \sum_{p=1}^{l} \sum_{k=1}^{n_p} \frac{1}{\sigma_{l_{p,k}}^2} + \sum_{m=1}^{r} \sum_{h=1}^{n_m} \frac{1}{\sigma_{r_{m,h}}^2} \qquad (2)$$

and with $\sigma_{l_{i,j}}$ and $\sigma_{r_{k,l}}$ the corresponding errors. In this formula the fact that the luminosities and the radial velocities are not homogeneous quantitites is taken into account normalizing each curve with respect to its maximum (normalizing, of course, also the errors). It is now possible to establish a criterium to fix an upper limit S_{LIM}: every solution characterized by a quality coefficient $S < S_{LIM}$ may be considered possible, because lies within the measurement error. To evaluate S_{LIM} it is sufficient to substitute $\sigma^2_{l_{i,j}}$ to $(l_{i,j_{obs}} - l_{i,j_{comp}})^2$ and $\sigma^2_{r_{k,l}}$ to $(v_{k,l_{obs}} - v_{k,l_{comp}})^2$. Once that a minimum has been obtained, it is necessary to verify that at least 68% of the points should be in the band described by $(-S_{er_j}, S_{er_j})$ (Chauvenet criterium), defined for each curve by the relation

$$S_{er_j} = \Big(\frac{1}{n_j - nd_j}\sum_{i=1}^{n_j}\sigma^2_{i,j}\Big)^{\frac{1}{2}} \qquad (3)$$

where nd_j is the number of degrees of freedom of the j-th light curve. If this condition is satisfied we have a guarantee that the solution found has physical meaning. If this condition is not satisfied, it is possible to hypothize the presence of other physical phenomena superimposed on the light and radial velocity curves which can't be described by means the synthesis algorithm used, being it necessary to use a more complete one.

AO CAM

The system AO Cam has been studied because many different solutions exist in literature: both transit and occultation solutions were considered possible (Evans et al., 1985; Cooke and Leung, 1986). To reanalyze the system we used all the data published in literature, plus the spectral type of the primary, that should be in the range $G2 - G8$. From the analysis of the results obtained from the minimization procedure, the system AO Cam may be definitely classified as a W-type system with an overcontact configuration (10.4%), a G5 spectral type of the primary star and $\Delta T = 327\,K$. In Table 1 the AO Cam parameters are shown, while in Fig.1 the search array is reported for selected pairs of parameters. In this figure the linear correlations among the parameters are also reported, evaluated using the solutions stored in the search array. What it is very interesting is that these correlations allow a description of the solutions found by the previous investigators. With the exception of two degrees of difference in the inclination of the orbital plane of the system in the two solution of Evans *et al.*(1984), it is possible to well describe them and, at the same time, to show that what they have found are only local minima.

CONCLUSIONS

In this paper we have shown as the correct use of the statistical analysis in the definition of the quality coeffcient in the solution procedure of a system is very relevant not only to obtain a correct solution from the physical point of view, but also to understand which are the uncertainties on the solutions found in connection with the precision of the data at disposal. We have also shown that this kind of approach well fits with the Wilson-Price code that is not only able to find the global minimum, but also to show a set of possible solutions and to allow the evaluation of the correlations of the parameters around the minimum.

REFERENCES

Barone, F., Maceroni, C., Milano, L. Russo, G.: 1988, *Astron. Astroph.* **197**, 347.

Barone, F., Milano, L., Russo, G.: 1990, *An Optimization Method for solutions of Close Eclipsing Binaries* in *Active Close Binaries*, C.Ibanoglu ed., Kluwer Academic Publisher, the Netherlands, p.161-188.

Cooke, D.D. Leung, K.C.: 1985, *Astrophys. Space Sci.* **113**, 25.

Evans, E.E. III, Grossoehme, D.H., Moyer, E.J. Jr.: 1984, *Publ. Astron. Soc. Pac* **97**, 648.

Price, W.L.: 1976, *Computer J.* **20**, 367

Wilson, R.E. Devinney, E.J.: 1971, *Astrophys. J.* **166**, 605.

Wilson, R.E.: 1979, *Astrophys. J.* **234**, 1054

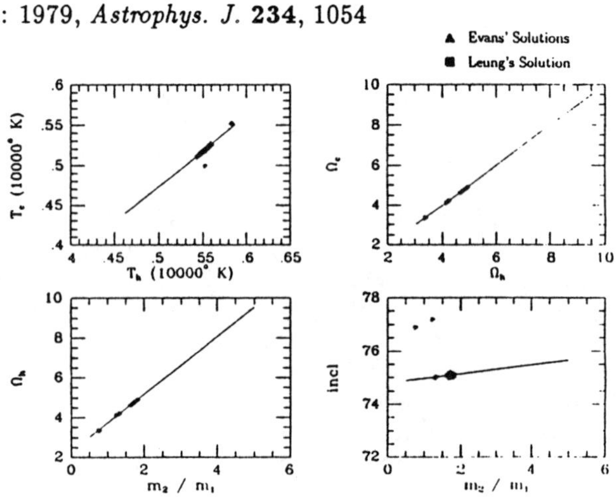

Fig.1 The Search Array for AO Cam reported for selected pairs of parameters

i (degrees)	75.09 ± 0.05	r_h(pole)	0.318
Ω_h	4.774 ± 0.057		
Ω_c	4.774 ± 0.057	r_h(side)	0.333
g_h	0.32	r_h(back)	0.369
g_c	0.32	r_c(pole)	0.407
A_h	0.5		
A_c	0.5	r_c(side)	0.431
T_h (°K)	5533 ± 39	r_c(back)	0.463
T_c (°K)	5206 ± 36	r_h(equiv)	0.342
m_c/m_h	1.71 ± 0.04	r_c(equiv)	0.436
		F	10.4 ± 0.1 %
		l_3	0.000

Table 1 AO Cam Parameters

Inside the Stars, IAU Colloquium 137
ASP Conference Series, Vol. 40, 1993
Werner W. Weiss and Annie Baglin (eds.)

V3903 SAGITARII: A MAIN-SEQUENCE (O7V+O9V) DETACHED ECLIPSING BINARY *

L.P.R. VAZ[1], N.C.S. CUNHA[1,2], E.F. VIEIRA[1,3] and M.L.M. MYRRHA[1]

[1] Depto Física–ICEx–UFMG, CP 702, 30.161 Belo Horizonte, MG, Brazil
[2] INTA–LAEFF, Villafranca, Apartado 50727, 28080 Madrid, Spain
[3] ESA, Villafranca, Apartado 50727, 28080 Madrid, Spain

ABSTRACT Photometric and spectroscopic observations of V3903 Sgr are analyzed, and absolute dimensions (masses and radii) are determined to a precision better then 3%.

1. INTRODUCTION

The eclipsing character of V3903 Sgr was discovered by Cunha *et al.* (1990). Cousins (1973) and Clariá (1976) have reported variations in the system brightness, but no light curve has been published. The system is a spectroscopic binary (Conti and Alschuler 1971, Niemela and Morrison 1988), very possibly a member of the R association Simeis 188 (Herbst *et al.* 1982). High precision photometric (*uvby*) and spectroscopic observations are used in the present work. No sign of orbital eccentricity or period variation was found in our analysis.

2. SPECTROSCOPIC ANALYSIS

V3903 Sgr was observed from 1989 to 1991 with the 1.6 m telescope and coudé spectrograph at LNA-CNPq (Brasópolis, MG, Brazil) at a mean dispersion of 18.1 Å/mm (projected slit width of 1.4 pixel). The radial velocities were measured by adjusting double lorentzian and/or gaussian curves in order to find the observed center of the lines used. More than 20 interstellar lines can be identified close to H_α. The measured radial velocities (Cunha *et al.* 1993) are shown in Fig. 1, together with data by Niemela and Morrison (1988). Our solution, given in Table I and shown in Fig. 1 as a solid line, indicates that the orbit is circular, which is confirmed by the photometric observations. The observations by Niemela and Morrison (1988) were also taken into account in this solution. Our results yield the mass ratio q $(= M_B/M_A) = 0.713 \pm 0.014$.

TABLE I Spectroscopic orbital elements for V3903 Sgr

fixed parameters	component A	component B
$P = 1^d\!.744204$	$K_A = 234.3 \pm 3.2$ km/s	$K_B = 328.6 \pm 4.7$ km/s
$T_0 = 2\,447\,754^d\!.4713$	$\gamma_A = 4.2 \pm 2.6$ km/s	$\gamma_B = 7.6 \pm 3.7$ km/s
$\omega_A = 90°$	$a_A \sin i = 11.32 \pm 0.16\ R_\odot$	$a_B \sin i = 8.07 \pm 0.11\ R_\odot$
$\omega_B = 270°$	$M_A \sin^3 i = 18.86 \pm 0.62\ M_\odot$	$M_B \sin^3 i = 13.45 \pm 0.41\ M_\odot$
$e = 0.00$	$\sigma_A = 16.4$ km/s	$\sigma_B = 23.1$ km/s

* Supported by: FAPEMIG, CNPq, FINEP, CAPES

3. PHOTOMETRIC ANALYSIS

3.1. Observations and Ephemeris

The photometric observations were obtained from 1989 to 1991, also at LNA-CNPq, with the 60 cm telescope and a mono-channel photometer equipped with a photon counting system and using a diaphragm of $39\overset{''}{.}2$ diameter (Cunha and Vaz 1993, Cunha et al. 1993). The light curves u (506 points), v (532), b (544) and y (537) have the typical rms errors of one magnitude difference between the constant comparison stars: $\sigma(\Delta u) = 0\overset{m}{.}009$, $\sigma(\Delta v) = 0\overset{m}{.}006$, $\sigma(\Delta b) = 0\overset{m}{.}005$ and $\sigma(\Delta y) = 0\overset{m}{.}005$. By using the least–squares method to minimize the O–C between the predicted and observed (Cunha *et al.* 1993) times of minima, we find the new ephemeris:

$$\text{Min I at: HJD } 2\,447\,754.4713 \underset{\pm 5}{} + 1\overset{d}{.}744204 \underset{\pm 6}{}$$

3.2. Photometric elements

The light curves were solved initially with the WINK (Wood 1971; Vaz 1986) model and the final solutions for the four colours, which yield the results of Table II, were found with the WD (Wilson 1979) model. The system configuration turned out to be detached, consistently in the two models, with both components still well inside their Roche Lobes.

TABLE II Mean elements for V3903 Sgr

i	$65\overset{\circ}{.}1 \pm 0\overset{\circ}{.}1$			
Ω_A	3.71 ± 0.07		Ω_B	3.56 ± 0.04
r_A mean	0.341 ± 0.009		r_B mean	0.300 ± 0.002
T_B/T_A	0.896 ± 0.006	($T_A = 38\,000$ K assumed)		
L_B/L_A	0.633 (y)	0.630 (b)	0.628 (v)	0.594 (u)

4. ABSOLUTE DIMENSIONS AND CONCLUSIONS

By using the observed standard indices, the luminosity ratios for the components of V3903 Sgr (Table II) and non–reddened indices for O7V and O9V stars (Crawford 1975), we find a colour excess of $E(b-y)=0\overset{m}{.}31$. This is slightly lower than $0\overset{m}{.}33$ (Herbst *et al.* 1982), but is in better agreement with colour excesses determined for other member stars of the association R Simeis 188. The absorption A_V is then $1\overset{m}{.}33$. Taking this into account, together with bolometric corrections estimated from Popper (1980), we calculate the absolute dimensions of the system and estimate its distance from us. The relevant results are shown in Table III.

Combination of the spectroscopic and the photometric solutions indicates that this is one of the rare detached systems in this spectral range, with the components still close to the ZAMS: the evolutionary models by Claret and Giménez (1989) yield an age of $3 \cdot 10^6$ yrs, with the chemical composition (X=0.70, Z=0.02). The absolute parameters for the non–interacting components of V3903 Sgr should then be representative for single stars of the same mass. More spectroscopic observations are being performed and a comprehensive analysis should appear soon (Cunha and Vaz 1993).

TABLE III Astrophysical data for V3903 Sgr

	Pri	Sec		Pri	Sec
Absolute dimensions:			Photometric data:		
M ($M_\odot$)	25.27 ± 0.84	18.02 ± 0.54	T_e (K)	$38\,000 \pm 1\,000$	$34\,000 \pm 1\,000$
R ($R_\odot$)	7.29 ± 0.21	6.42 ± 0.08	M_{bol}	$-7\overset{m}{.}80 \pm 0\overset{m}{.}13$	$-7\overset{m}{.}05 \pm 0\overset{m}{.}13$
$\log g$ (cgs)	4.115 ± 0.032	4.079 ± 0.017	$\log L/L_\odot$	4.997 ± 0.052	4.696 ± 0.052
			B.C.	$-3\overset{m}{.}40$	$-3\overset{m}{.}40$
			M_V	$-4\overset{m}{.}40 \pm 0\overset{m}{.}13$	$-3\overset{m}{.}65 \pm 0\overset{m}{.}13$
			$E(b-y)$	$0\overset{m}{.}31$	
			D(pc)	1450 ± 70	

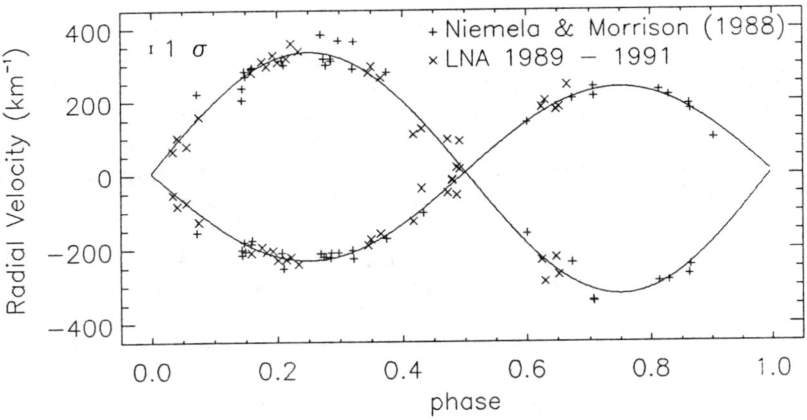

Fig. 1. Observed and theoretical radial velocity curves of V3903 Sgr. The bar at the upper left part of the figure is the mean error (σ) of 1 observation in our measurements.

REFERENCES

Claret, A. and Giménez, A.: 1989, *Astr. Ap. Suppl.* **81**, 1
Clariá, J.J.: 1976, *Inf. Bull. Variable Stars* **110**
Conti, P.S. and Alschuler, W.R.: 1971, *Ap. J.* **170**, 325
Cousins, A.W.J.: 1973, *Mon. Notes Astron. Soc. S. Afr.* **32**, 11
Crawford, D.L.: 1975, *Pub. A.S.P.* **87**, 481
Cunha, N.C.S. and Vaz, L.P.R.: 1993, to be submitted to *Astr. Ap. Suppl.*
Cunha, N.C.S., Vaz, L.P.R., Vieira, E.F. and Myrrha, M.L.M.: 1993, to be submitted to *Astr. Ap.*
Cunha, N.C.S., Vaz, L.P.R., Pôssa, C.M.M., Helt, B.E. and Clausen, J.V.: 1990, *Inf. Bull. Variable Stars* **3436**
Herbst, W., Miller, D.P., Warner, J.W., Herzog, A.: 1982, *Astr. J.* **87**, 98
Niemela, V., and Morrison, N.D.: 1988, *Pub. A.S.P.* **100**,1436
Popper, D.M.: 1980, *Ann. Rev. Astr. Ap.* **18**, 115
Vaz, L.P.R.: 1986, *Rev. Mex. Astron. Astrofis.* **12**, 199
Wilson, R.E.: 1979, *Ap. J.* **234**, 1054
Wood, D.B.: 1971, *Astr. J.* **76**, 701

Inside the Stars, IAU Colloquium 137
ASP Conference Series, Vol. 40, 1993
Werner W. Weiss and Annie Baglin (eds.)

THE BRAKING LAW OF SOLAR TYPE STARS AS DERIVED FROM CLOSE BINARY DYNAMICAL EVOLUTION

CARLA MACERONI
Osservatorio Astronomico di Roma, I-00040 Monteporzio, Italy

THE DYNAMICAL EVOLUTION OF CLOSE SOLAR TYPE BINARIES

The orbital period of a close binary system ($P \sim$ a few days), formed by two solar type components, evolves under the influence of angular momentum loss (AML) by electromagnetic braking and of spin↔orbit angular momentum transfer (AMT) by tidal coupling. Because of AMT, which tends to reestablish the spin-orbit synchronization destroyed by AML, the loss of AM takes finally place at the expenses of the orbital reservoir and produces shrinking of the orbit and *spinning-up* of the components. The variation with time of ω_K and ω , respectively the orbital and the rotational angular velocity, is expressed by a system of (stiff) ordinary differential equations and the results of the (numerical) integration strongly depend on the AML and AMT laws. A series of recent papers (Maceroni and Van 't Veer 1991, 1992, Van't Veer and Maceroni 1992, hereafter MV1, MV2 and VM) presents the resulting orbital period evolution function (PEF) with a variety of choices. A typical result, in terms of orbital period, and for a binary formed by two identical G5 V components, is shown in fig. IA. The solution shown here corresponds to the braking law a) of fig. IB , to the AMT according to Zahn (1977) and to stars rotating as rigid bodies; different assumptions have been tried in MV1and VM.

By means of PEFs similar to that of fig. IA, we computed the evolution of a given initial period distribution (IPD) into the present period distribution, PPD. This was compared with the observed distribution, after correction for detectability and selection effects. The comparison constraints the AML and AMT relations by the requirement of reproducing two remarkable observed properties of close binaries : the bimodality of the PPD of G type binaries, with a dip at short periods $0.2 < \log P < 0.4$, (Farinella et al., 1979) and their almost perfect spin-orbit synchronization (Tan et al., 1991), with a value of the synchronization parameter $f = \omega/\omega_K > 0.96$ for $P < 10^d$. In the computations we assumed a time-independent IPD, following a flat $N(\log P)$ distribution, (Abt 1983), and an age of the galactic disk of $\sim 8\ 10^9$y. The results can be summarized as follows:

- The bimodal distribution of G type binaries cannot be reproduced with a braking law following the Skumanich relation on the whole rotational range. All the distributions computed with this assumption are increasing functions of $\log P$ (Fig. II b).

- The dip at $\log P \approx 0.3$ requires a braking law for fast rotators ($\omega \geq 12\omega_\odot$) independent from ω. The amount of this ω-independent braking, and its range of validity, imply a rapid increase of $|\dot\omega|$, $\Delta \log|\dot\omega| \simeq 1.5$ for

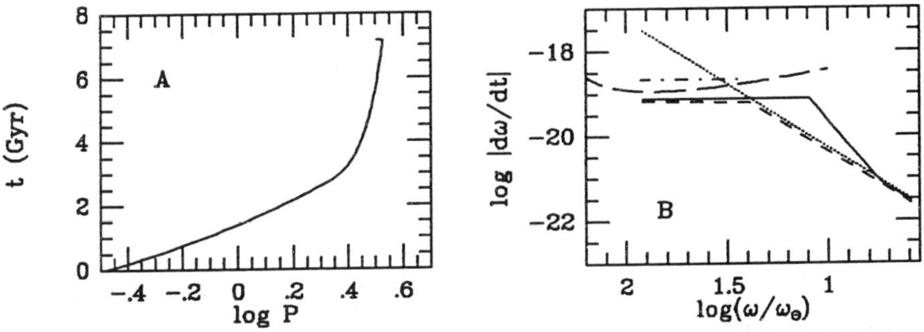

FIGURE I A) The orbital period evolution function (PEF). The time is counted backwards from contact time (t=0). B) Three braking laws corresponding to the PPD of fig. II. a) continuous line: the Skumanich ($\omega \propto t^{-1/2}$) relation for slow rotators + a $\log|\dot{\omega}|$ =const for fast rotators, + a tentative connection between the two. This relation yields the best fit of the observed distributions, b) dotted line: pure Skumanich relation, extrapolating that of slow rotators, c) dashed line: the same Skumanich relation, extrapolated to the intersection with the $\log|\dot{\omega}|$ =const line + this ω-independent law. Dashed-dotted line: braking computed from the data of G type stars in α Per and Pleiades. Long dashed line: spin-down model of Schatzman and Tutukov (1992)

$6 < \omega/\omega_\odot < 12$ (fig. IB, law a, , fig. II a).
The rate of spin-down at intermediate ω's (i.e. $12 < \omega/\omega_\odot < 25$ is much higher than that predictable by extrapolation of the Skumanich relation of single stars (Soderblom, 1983). The lower ω boundary (ω_1) of this fast rotator relation is well constrained by the dip location and the requirement $f > 0.96$. Lower values yield de-synchronized systems, because of the lower efficiency of the synchronization mechanism ($\propto \omega^4$), higher values shift the dip to too short periods. For instance, a relation made out of a Skumanich law, extrapolated towards higher ω's to the intersection with the $\dot{\omega}$ =const section and, from there on, by that constant value (short dashed line of fig. IB) is ruled out by its PPD shape (fig. II c).

DISCUSSION

If binarity and tidal coupling do not affect the AML process by changing, for instance, the magnetic field structure or the mass loss rate, as the observations of the surface activity phenomena in binaries seem to indicate (Young et al., 1989), the results of binaries can be compared with those of single stars. Therefore, In fig. IB, we compare our best braking law with a rough estimate of the spin-down of fast rotators from the data for galactic clusters (Stauffer 1991) and with a theoretical curve based on the model of spin-down of Schatzman and Tutukov (1992). In this model the rapid increase of the spin-down rate, in spite of the reduced braking efficiency with increasing rotation of the magnetic field, is explained in terms of a change in the mass loss rate, due to the transition

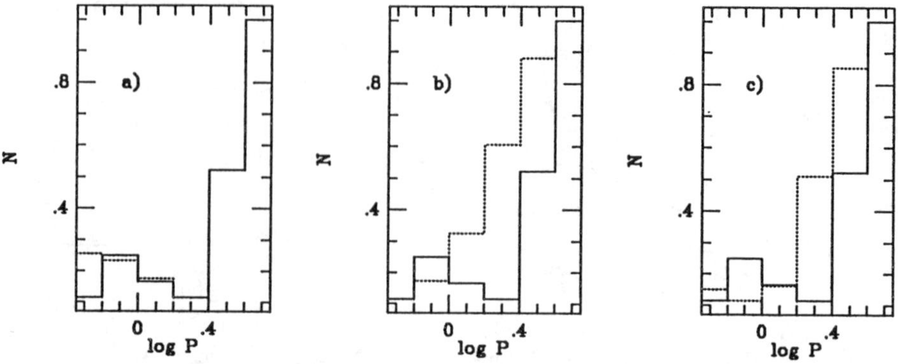

FIGURE II Comparison of the computed (dotted line) present period distribution (PPD) with the observed (corrected) one (continuous line). The three figures correspond to the three braking laws of fig. IB

from thermal or Alfvenic to centrifugal wind. This transition should take place at $\omega \sim 11\omega_{\odot}$, in good agreement with our results.

The data of young fast rotators suggest that the braking is initially acting only on the external (convective) layers, while what we know on the solar rotation suggests an almost rigid body rotation at later times (and slower rotation). We should therefore include in our models a rotation dependent core-envelope coupling. If this trend of coupling with rotation applies to binaries as well, the smaller inertial moment of the involved layers of fast rotators would produce an upward shift of the fast rotator relation. That qualitatively goes in the sense of improving the agreement between our results and those for single stars.

REFERENCES

Abt, H.A., 1983, *ARA&A*, **21**, 343

Farinella, P., Luzny, F., Mantegazza, L., Paolicchi, P., 1979, *ApJ*, **234**, 973

Maceroni, C., Van 't Veer, F., 1991, *A&A*, **246**, 91

Maceroni, C., Van 't Veer, F., 1992, in preparation

Schatzman, E., Tutukov, A., 1992, to be published on *A&A*

Soderblom, D. R., 1983, *ApJS*, **53**, 1

Stauffer, J. R., 1991 in " *Angular Momentum Evolution of Young stars*", S. Catalano and R.J. Stauffer eds., Kluwer Acad. Publ.

Tan, H., Wang, X., Pan, K., 1991, Acta Astron. Sin. **32/2**, 145

Van 't Veer F., Maceroni C., 1992 in *"Binaries as tracers of stellar formation"*, Eds. A. Duquennoy and M. Mayor, Cambridge University press, in press

Young, A., Ajir, F., Thurman, G., 1989, *PASP*, **101**, 1017

Zahn, J.P., 1977, *A&A*, **57**, 383

Inside the Stars, IAU Colloquium 137
ASP Conference Series, Vol. 40, 1993
Werner W. Weiss and Annie Baglin (eds.)

ROTATIONAL VELOCITY DISPERSION IN THE HYADES AND OTHER GALACTIC CLUSTERS FROM THE *V SIN I* OBSERVATIONS

YVES GAIGÉ
Observatoire Midi-Pyrénées

INTRODUCTION

Let us recall briefly the main points of the theory as developped by Chandrasekhar and Münch (1950) : if the axis of rotation of a star is inclined with an angle i to the line of sight, the apparent velocity is given by $y = v \sin i$ where v is the true rotational velocity. Assuming that the axes of rotation are randomly distributed in space, the frequency function $f(v)$ of the true velocities may be related to the corresponding $\phi(y)$ of the apparent ones by the integral equations :

$$\phi(y) = y \int_y^\infty \frac{f(v)}{v(v^2 - y^2)^{1/2}} dv \tag{1}$$

$$f(v) = -\frac{2}{\pi} v^2 \frac{d}{dv} v \int_v^\infty \frac{\phi(y)}{y^2(y^2 - v^2)^{1/2}} dy \tag{2}$$

The central moments of $f(v)$ can be derived directly from the observations by means of (1) without the knowledge of the function itself; for instance, the mean and the mean square deviation are given respectively by :

$$\langle v \rangle = \frac{4}{\pi} \langle y \rangle \qquad \langle (v - \langle v \rangle)^2 \rangle = 1.5 \langle y^2 \rangle - \frac{16}{\pi^2} \langle y \rangle^2 \tag{3}$$

OBSERVED DISTRIBUTIONS

Apparent distributions have been computed in different colour intervals for the Hyades cluster for which we had new rotational velocities measurements made with CORAVEL and kindly communicated by Mermilliod (1991). The velocities have been expressed in unit of their means.
Since the differentiation of the observed distribution may lead to misleading results, it would be better to assume a form for $f(v)$, work out the corresponding frequency distribution of $(v \sin i)$, and compare this with the observations. In the present study the model functions (4) have been tested to determine the probable shape of the true distribution.

$$f_{in}(v) = \begin{cases} k_{in} \left(p_i^2 - (v - v_{i0})^2\right)^n & \text{if } v_{i0} - p_i \le v \le v_{i0} + p_i \\ 0 & \text{otherwise} \end{cases} \tag{4}$$

The corresponding $\phi(y)$ are known analytically, making them suitable forms to try in practical applications. In fact it is hard to decide in favour of one shape.

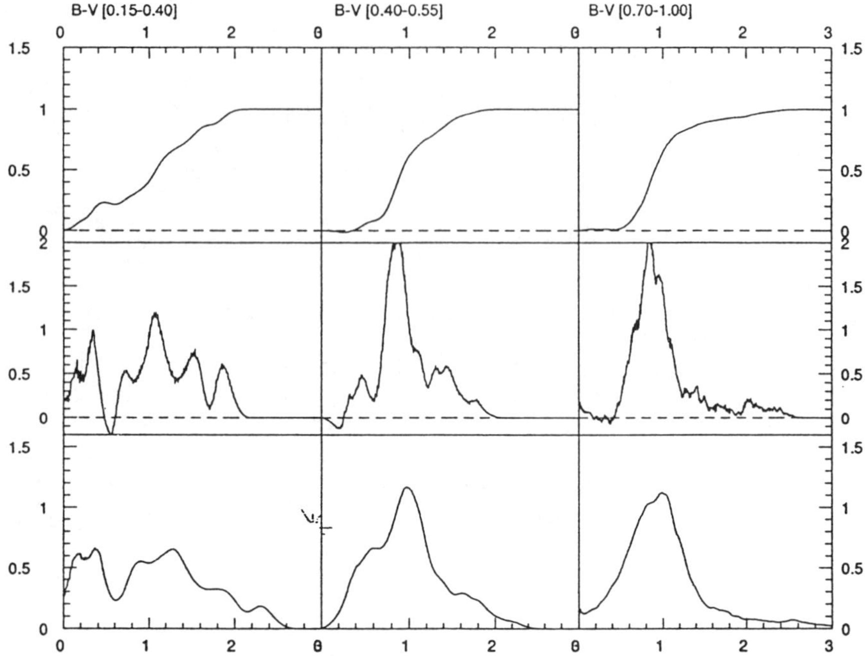

FIGURE I Observed, true and integrated true distributions for the Hyades cluster

$n = 0$ gives unrealistic apparent distribution (with an angular point) but is very easy to compute. $n = 1$ and $n = 2$ are intermediate cases between rectangular and gaussian forms, they give very similar apparent distributions. The advantage of $n = 2$ is the continuity of its derivative.

Parameters of the theorical functions may be adjusted by minimizing some distance with the observed distribution, but in the case of unimodale distribution they can be computed directly by determining the variable σ ($\sigma = \frac{\overline{v^2} - \overline{v}^2}{\overline{v}^2}$) because σ is a growing function of α (where $\alpha = \frac{p}{v_0}$). σ is also a growing function of n as α goes to ∞.

TRUE DISTRIBUTION

To get the true distribution we have to use estimators which give a smoother a curve than the histogram, and then do the derivation. This has been done for the Hyades cluster in several colour intervals.

As can be seen on figure (I) the distribution varies with effective temperature. Some peculiar features may be simply explained looking carefully the data : the negative values observed simply show the lack of $v \sin i$ between 40 and 60 $km.s^{-1}$ for B-V=[0.15,0.40] and below 20 $km.s^{-1}$ for B-V=[0.40,0.55]. This indicates that in these cases the hypothesis of random orientation for the axis of rotation is not valid : this is probably due to the small number of stars in the sample (observational bias).

B-V=[0.15, 0.40] the distribution was often assumed to be uniform, which is not exactly the case with the actual measurements.

B-V=[0.40, 0.55] this distribution is well described by the sum of two distribution of the same importance, a widely spread one and a sharper one.

B-V=[0.70, 1.00] the distribution is very similar to the precedent : at the same location, there is a slightly wider peak and there is a very faint but long wing towards high velocities.

INTEGRATED TRUE DISTRIBUTION

In order to get the integrated distribution, let us integrate the relation (2) between 0 and z. After reduction the equation becomes :

$$\int_0^z f(v)dv = \int_0^z \phi(y)dy + \frac{2}{\pi}\int_z^\infty \phi(y)\frac{d}{dy}\left(y \arcsin\left(\frac{z}{y}\right)\right) dy \tag{5}$$

It is then possible to compute the integrated distribution of the true rotational velocity with only knowing the apparent distribution $\phi(y)$, without any problems of derivation. Computations have been done for the Hyades cluster for the same colour intervals.

CONCLUDING REMARKS

In section (2) we study the rotational behaviour of a group of stars, by fitting the apparent distribution with some model functions. It has been found that in general the observed distribution can not be simply fitted by an unimodale distribution. This observed distribution varies greatly with colour.

- above $7000K$ the velocity distribution is almost uniform
- between 6400 and $7000K$ the distribution is roughly bimodale.

It has then been shown (section (3)) that the distribution of true rotational velocities may be directly computed if the observed distribution is known with an estimator giving a curve smooth enough to do the derivation. This method confirms the fact that the distribution presents one or two peaks (mainly) below $7000K$ whereas above the uniformity is questionnable, probably because of the size of the sample.

Finally, we made an improvment of the theory, by computing the integrated true distribution from the observed one without any free parameter or assumption. Contrary to the inverse problem this method can be used whatever the estimator.

REFERENCES

Chandrasekhar, S., and Münch, G. 1950 *Ap. J.*, **111**, 142.

Mermilliod, J. C. 1991 *Private Communication*

Inside the Stars, IAU Colloquium 137
ASP Conference Series, Vol. 40, 1993
Werner W. Weiss and Annie Baglin (eds.)

COMPARING CATALOGUES OF STELLAR ABSOLUTE MAGNITUDES AND ESTIMATING THEIR ACCURACY

V. MALYUTO
Tartu Astrophysical Observatory, EE2444, Toravere, Estonia

ABSTRACT The accuracy of spectral, photometric and astrometric catalogues of stellar absolute magnitudes of F-subgiants and dwarfs has been estimated from a comparison of the data for stars in common.

INTRODUCTION

If three or more independent catalogues are available, their external random r.m.s. errors may be estimated from the data residuals for objects in common to them (without resorting to any additional information) after the removal of systematic differences. Kuzmin, Malyuto and Eelsalu (1986) showed how to do this if three catalogues are available and some data may present in only two of the three catalogues. Here we analyse three catalogues of stellar absolute magnitudes based on independent spectral, photometric and astrometric measurements for F-subgiants and dwarfs of normal metal content.

OBSERVATION DATA

Designations of the catalogues are those attributed to them by the Stellar Data Centre (Strasbourg).

1. *Sp* data – Catalogue No. 3149 of MK standards. MK classes have been transformed into absolute magnitudes by us on the basis of a calibration carried out by Straizys and Kuriliene (1981).
2. *Ph* data – Catalogue No. 5014 of main astrophysical parameters from photoelectrical $uvby\beta$ photometry.
3. *Trig* data – Woolley Catalogue No. 5032 of stars within twenty-five parsecs of the Sun. The latter data are inhomogeneous, but the probable error of each value of trigonometric parallax drawn from intercomparison of different data is tabulated in the catalogue. We try to estimate the influence of these inhomogeneities on our analysis of errors of catalogues by considering two separate samples of stars: *Trig* - with $\sigma(\pi)/\pi(trig) < 0.176$ and *Trig*1 - with $\sigma(\pi)/\pi(trig) < 0.120$. The Lutz-Kelker (1973)'s corrections have been applied to all data.

ANALYSIS OF THE CATALOGUES

Only stars simultaneously occurring in two or three catalogues are considered. For six stars of low metal content ([Fe/H] < -0.3) absolute magnitudes infered from MK and/or photometric data may be systematically incorrect and we excluded those stars from comparison (estimates of metallicities for all stars in common were either taken from the *Ph* catalogue or estimated from *UBV* data given in the *Trig* catalogue). The triples and couples of stars in common number resp. $n(Sp, Ph, Trig) = 23$, $n(Sp, Ph) = 7$, $n(Ph, Trig) = 25$, $n(Sp, Ph, Trig1) = 14$, $n(Ph, Trig1) = 6$. The mean differences between M_v in the catalogues are $\delta(Sp - Ph) = 0.16 \pm 0.43$ (30 stars) and $\delta(Ph - Trig) = 0.12 \pm 0.40$ (48 stars).

After the removal of systematic differences we calculate the variances of data residuals; then we use these variances with optimum weights to deduce the external r.m.s. errors of the catalogues as Kuzmin, Malyuto, Eelsalu (1986) showed. Table explains the results of analysis of the catalogues.

TABLE External Random R.M.S. Errors of M_v Deduced From the Comparison of the Catalogues

Trig Catalogue	σ_{Sp}	σ_{Ph}	σ_{Trig}
Woolley Catalogue of stars within 25 pc of the Sun			
$\sigma/\pi < 0.176(Trig)$	0.34 ± 0.04	0.22 ± 0.03	0.32 ± 0.03
$\sigma/\pi < 0.12(Trig1)$	0.32 ± 0.04	0.24 ± 0.03	0.14 ± 0.02
Planned Hipparcos Output Catalogue, sample within 25 pc of the Sun	~ 0.33	~ 0.23	< 0.11*

* The error estimate of σ_{Trig} in the last row of Table 2 was deduced with the well-known formula $\sigma M_v = 2.17 * \sigma(\pi)/\pi(trig)$ by inserting the anticipated accuracy of Hipparcos parallaxes (0".002).

The variances of the external random r.m.s. errors are limited to 9 to 14 %. This shows that our estimates to the errors are meaningful. The most essential feature of Table is the practical coincidence of estimates to the errors of $M_v(Sp)$ and $M_v(Ph)$ calculated with the use of *Trig* and *Trig*1 catalogues respectively. It means that we may rely on these error estimates even if the *Trig*1 and especially *Trig* catalogues are somewhat heterogeneous.

CONCLUSIONS AND ANTICIPATED DEVELOPMENTS

The accuracy of spectral, photometric and astrometric catalogues of stellar absolute magnitudes of F-subgiants and dwarfs of normal metal content has been estimated from a comparison of the data for stars in common. We suppose our estimates to errors to be superior in quality to any previous ones and so to become useful for future studies of stars and our Galaxy. We conclude that the method seems to be productive and deserves further elaboration.

The most promising prospect would be the replacement of the *Trig* catalogue by data from the planned Hipparcos Output Catalogue. Trigonometric parallaxes from Hipparcos combined with the above spectral (*Sp*) and photometric (*Ph*) catalogues allow us to determine the external accuracy of astrometric estimates of M_v for stellar samples within a certain range of distances from the Sun together with obtaining more accurate estimates of errors of $M_v(Sp)$ and $M_v(Ph)$. Some anticipated results are given in Table. We may treate the *Sp* and *Ph* catalogues as probes for estimating the accuracy of various homogeneous stellar samples from the Hipparcos catalogue. The latter catalogue would provide access to more distant stars; some spectral and photometric catalogues could be involved or extended by including stars of other spectral and luminosity classes.

REFERENCES

Kuzmin, G., Malyuto, V. and Eelsalu, H. 1986, *Tartu Astrophys. Obs. Preprint,* **83**, 4.
Lutz, T.E. and Kelker, D.H. 1973, *Publ. Astron. Soc. Pac.,* **85**, 573.
Straizys, V. and Kuriliene,G. 1981, *Astrophys. Space Sci.,* **80**, 353.

Inside the Stars, IAU Colloquium 137
ASP Conference Series, Vol. 40, 1993
Werner W. Weiss and Annie Baglin (eds.)

CALIBRATION OF THE MAIN SEQUENCE STARS : ALPHA CENTAURI

NEUFORGE C.
Institut d'Astrophysique
Université de Liège,
5, avenue de Cointe,
B-4000 LIEGE-BELGIUM

ABSTRACT New low temperature opacities have been computed and applied together with the new OPAL opacities to recalibrate the α Centauri system. Our solution is in favour of a Z value of about twice $Z_{\odot}$.

INTRODUCTION

The proximity to the Sun of the binary system α Centauri allows a good estimation of the masses of the two components, which are 1.085 $M_{\odot}$ for α Cen A and 0.900 $M_{\odot}$ for α Cen B (Demarque et al., 1986), which makes it a good candidate for main sequence calibration.

Spectroscopic analyses do not agree on the metallicity, Z. French and Powell (1971), England (1980), Edvardsson, (1988) and Cayrel de Strobel (1991) favour a Z of about twice the solar value; Furenlid and Meylan (1990) find a lower value of about 1.3 times the solar metallicity.

Previous calibrations of the α Centauri system by Demarque et al. (1986) lead to a solar Z value but other authors (Flannery and Ayres, 1978; Noels et al., 1991) found an agreement with the high Z spectroscopic results.

In this work, the calibration procedure of α Centauri is a generalization of the solar one as in Noels et al. (1991). Four observables (the two luminosities and effective temperatures of α Cen A and B) allow the determination of four unknowns : helium content, Y, metallicity, age, t, and convection parameter, α, which is assumed to be the same in both stars. This hypothesis has been discussed in Noels et al. (1991). It requires, at least, accurate low temperature opacities computed for various values of Z, not available in the previous works.

CALIBRATION WITH NEW LOW TEMPERATURE OPACITIES

We calculated our own new low temperature opacities for different Z values ranging from 0.02 to 0.04. Details can be found in Neuforge (1992). As it is shown in figure 1, these atmospheric opacities are in good agreement with the

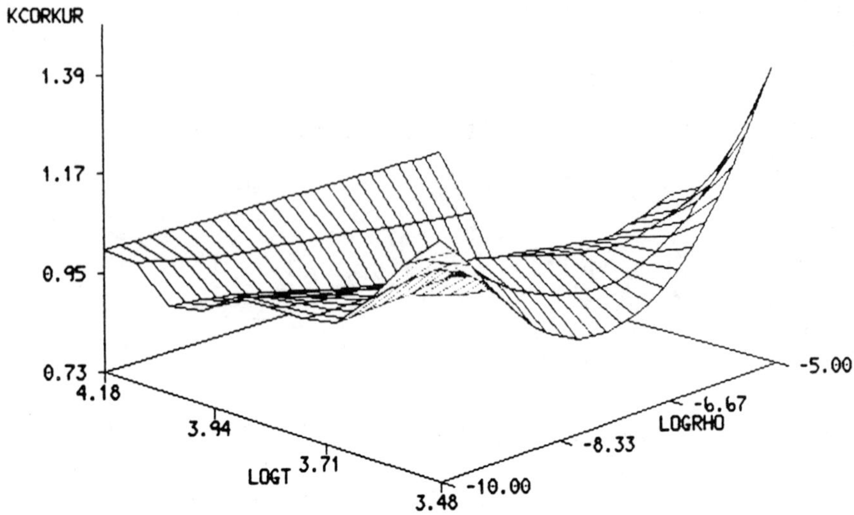

Fig. 1. Ratio between our opacities and Kurucz 's opacities as a function of lg ρ and lg T.

new atmospheric opacities of Kurucz (1992), then only available for the solar chemical composition.

With the interior opacities of Iglesias et al. (1992), OPAL, the equation of state and the nuclear reaction rates used by Noels et al. (1991), we recalibrated the Sun and found :

$$\begin{aligned} Z &= 0.02 \\ Y &= 0.281 \\ \alpha &= 2.12, \end{aligned}$$

with our own opacities while the same calibration with Kurucz 's opacities gives $Y = 0.28$ and $\alpha = 2.17$.
Our solution for the α Centauri system is

$$\begin{aligned} Z &= 0.04 \\ Y &= 0.322 \\ \alpha &= 2.25 \\ t &= 4.92 Gyr. \end{aligned}$$

DISCUSSION

The agreement found for the convection parameter in α Cen and the Sun supports our hypothesis of a unique value for the two components of α Cen. As in Noels et al. (1991), the uncertainties in luminosities don't affect significantly the solution; a displacement of the observational points in the HR diagram in the same horizontal direction (i.e. keeping constant the difference in effective temperature) only leads to a change in the convection parameter; a variation of the difference in Te affects the chemical composition but within reasonable uncertainties in Te's, our Z value remains large. The uncertainties in the masses, of the order of 0.01 $M_\odot$ (Demarque et al., 1986), lead to a limited change in Z, $0.035 \leq Z \leq 0.043$, while α ranges from 2.00 to 2.49.

CONCLUSIONS

The computation of new low temperature opacities for different Z values around the solar value allows us to assume a unique value of the convection parameter for the two components of the α Centauri system.

Using the OPAL interior opacities (Iglesias et al., 1992), it is possible to obtain a good fit between observational data and theoretical evolutionary tracks. In spite of the uncertainties in effective temperature, our solution is in favour of a high metallicity, $Z = 0.04$.

The new low temperature opacity tables are available on request (e-mail : U2126CN @ BLIULG11).

ACKNOWLEDGEMENTS

We are grateful to C.A. Iglesias, F. J. Rogers and R.L. Kurucz for sending us their data prior to publication.
We thank the belgian Fonds National de la Recherche Scientifique for financial support.

REFERENCES

Cayrel de Strobel, G. 1991, private communication.
Demarque, P., Guenther, D.B., van Altena, W.F. 1986, Ap. J., **300**, 773.
Edvardsson, B. 1988, A & A, **190**, 148.
England, M.N. 1980, Mon. Not. R. astr. Soc., **191**, 23.
Flannery, B.P., Ayres, T.R. 1978, Ap. J., **221**, 175.
French, U.L., Powell, A.T.L. 1971, Royal Obs. Bull., **173**, 63.
Furenlid, I., Meylan, T. 1990, Ap. J., **350**, 827.
Iglesias, C.A., Rogers, F.J., Wilson, B.G. 1992, preprint.
Kurucz, R.L. 1992, Rev. Mexicana Astron. Astrof., **23**, 181.
Neuforge, C. 1992, to be published.
Noels, A., Grevesse, N., Magain, P., Neuforge, C., Baglin, A., Lebreton, Y. 1991, A & A , **247**, 91.

Inside the Stars, IAU Colloquium 137
ASP Conference Series, Vol. 40, 1993
Werner W. Weiss and Annie Baglin (eds.)

ABSOLUTE PARAMETER DETERMINATION IN LOW-MASS ECLIPSING BINARIES. RADIATIVE PARAMETERS FOR BH VIR, ZZ UMA AND CR CAS.

R. CLEMENT [1], V. REGLERO [1], M. GARCIA [2], J. FABREGAT [1], A. BRAVO [2], J. SUSO [1]

1-Departamento de Matematica Aplicada y Astronomia, U. Valencia.Dr Moliner 50, E-46100 Burjasot, Valencia, Spain

2-Departamento de Fisica , E.I.Industriales, U.P.Madrid, Ronda de Valencia 3, E-28012 Madrid, Spain

ABSTRACT

A new uvby and Hβ monitoring program of low mass eclipsing binaries is currently being carried out in the framework of a 5 year observational program which also involves radial velocity determinations. The scope of this work is to provide very accurate absolute astrophysical parameters: mass, radius and effective temperatures, for main sequence late-type stars. One of the main goals is to improve the mass-luminosity relation in the low and intermediate mass range. A second objective is to perform accurate tests of the most recent grids of evolutionary models. This program is complementary to that currently being in progresss by the Copenhagen group.

In this contribution we present the photometric preliminary results obtained for three of the systems included in our long term survey: BH Vir, ZZ UMa and CR Cas for which, primary eclipses have been observed. Particular attention is paid to the determination of reddening, distances and radiative properties. A more detailed study will be carried out when the light curves and radial velocity measurements are completed

INTRODUCTION

New and more accurate observational determinations of masses, radii, temperatures and abundances are needed in order to test the new available grids of evolutionary models. Those physical quantities are poorly defined at the end of the Main Sequence. Only a few stars have astrophysical parameters determinations with the required accuracy ($< 3\%$).

Mass, radius, temperature and abundances can be derived only for double lined eclipsing binaries with accurate light and radial velocity curves. With this aim, we have started a 5 year monitoring program in order to obtain good photoelectric uvby & Hβ photometry and radial velocity measurements for six selected systems. This program is complementary to that currently carried out by Andersen et al. 1992 for early and intermediate type stars.

In this contribution we present preliminary results for three binaries: BH Vir, ZZ UMa and CR Cas.

The measurements have been carried out from March 1990 up to now at The Calar Alto Observatory (Almeria, Spain), located at 2168 m above the sea level, on seven observing runs. The telescope used was the 1.5 m reflector equipped with either a multipurpose UBVRI one-channel photoelectric photometer or, alternatively, a uvby-Hβ four-channel photometer, both using Stromgren uvby filters as well as Crawford's narrow and wide Hβ filters. More details about the instrumental configuration and reduction procedures can be found in Fabregat and Reglero 1990.

We have selected two comparisons for each binary system and obtained differential light curves within the observational accuracy of our differential photometry quoted in a few millimagnitudes (0.007 mag. in the V filter). We have measured standard stars from the lists of Perry et al. (1987) and Olsen (1991), during nights of the best photometric quality, to transform instrumental magnitudes into the standard system, standard values for the comparisons and targets are reported in Table 1.

DISCUSSION

BH Vir Photometric studies of this interesting eclipsing binary have been carried out by several authors: Kitamura et al. (1957), Koch (1967), Sadik (1978), Botsula (1978), Hoffmann (1982) and Scaltriti et al. (1985). More recently in a detailed analysis of the available light curves using the Wilson-Devinney code, Zhai et al. (1990) derived the most relevant astrophysical parameters of both components. The analysis of the photometric light curves for BH Vir is a difficult task mainly due to the large and very fast amplitude variations outside eclipses. Koch (1967) reported changes in the depth of the primary eclipse of a tenth of a magnitude, in only few days. Despite this problem, BH Vir has been included in our program due to the lack of late-type eclipsing binaries without photometric variations outside eclipse. The associated waves generated by the presence of spots in the stellar surfaces is a well know signature of late-type fast rotators.

For the present work, in order to avoid this important source of uncertainty, we have used the data obtained within a single night (JD=2448640) with measurements from phase 0.8 to 0.1 and including a well defined primary minimum. The upper limit of the ratio of luminosities Lc/Lh(V)=0.88 has been derived using the primary minimum values and an average of points outside eclipse and far away from the contact points, around phase 0.10. In this way, the observed magnitudes and colors for the combined light are nearly simultaneous (same night) with the values for the primary minimum and essentially free of variability effects (close in phase). The visual magnitudes for the primary minimum Vp=10.42 mag. and for the combined light outside eclipse Vt=9.59 mag. are in good agreement with those reported by Scaltriti et al. (1985).

Before discussing the radiative parameters of the binary, a study of the local interstellar extinction was done. An accurate determination of the reddening is needed to eliminate uncertainties in the determination of physical parameters.

TABLE I Radiative Parameters.

name	V_o	$(b-y)_o$	m_o	c_o	$H\beta$	dm_o	dc_o	E(b-y)	B-
SAO2770	9.780	0.276	0.165	0.507	2.673	0.006	0.059	0.01	0.[illegible]
SAO139643	9.518	0.367	0.185	0.344	2.609	0.018	0.026	0.01	0.[illegible]
BH Vir H	10.276	0.356	0.152	0.333	2.600	0.062	0.043	0.00	0.[illegible]
BH Vir C	10.422	0.408	0.208	0.309	2.580	0.018	0.036	0.00	0.[illegible]
SAO15242	10.120	0.368	0.163	0.333	2.595	0.059	0.053	0.00	0.[illegible]
SAO15251	9.995	0.397	0.212	0.344	2.599	0.000	0.050	0.00	0.[illegible]
ZZ UMa H	10.704	0.370	0.200	0.301	2.600	0.014	0.011	0.00	0.[illegible]
ZZ UMa C	10.501	0.424	0.180	0.362	2.583	0.067	0.104	0.00	0.[illegible]

TABLE II Radiative Parameters.

name	Mv	Fv	[Fe/H]	T_{eff}	d	Sp
SAO 2770	3.0	3.811	0.1	6604	230	F4 V
	0.3	0.034	0.2	60	30	
SAO 139643	4.3	3.748	0.0	6017	113	G0 V
	0.3	0.026	0.2	60	15	
BH Vir C	4.4	3.760	-0.5	5973	150	F9-G0 V
	0.3	0.026	0.2	60	22	
BH Vir C	4.5	3.741	-0.1	5714	150	G5 V
	0.3	0.034	0.1	100	22	
SAO 14242	4.2	3.762	-0.4	5904	151	G0 V
	0.3	0.034	0.2	60	20	
SAO 15251	4.2	3.747	0.1	5823	143	G2 V
	0.3	0.034	0.1	100	19	
ZZ UMa H	4.7	3.761	0.0	6008	158	G0 V
	0.3	0.034	0.2	60	20	
ZZ UMa C	4.5	3.733	-0.3	5630	158	G5 V-IV
	0.3	0.034	0.1	100	20	

The interstellar reddening of the system has been derived following a procedure similar to that described by Reglero et al. (1990). The two comparison stars SAO 2770, and SAO 139643 indicate a mean reddening of E(b-y)=0.01+0.004 mag. The small apparent angular distance between the comparisons and BH Vir and their location at 230 and 113 pc in front and behind BH Vir, suggests no significant interstellar extinction for BH Vir itself.

The derived physical quantities for the comparisons are listed in Table 2. The quoted errors arise from the uncertainties on the photometric calibrations and error formulae. We have used the well know standard calibration procedure for F type stars given by Crawford (1975) and extended by Olsen (1988) for early G type stars, using $H\beta$ index as a reddening free independent parameter. Effective temperatures Teff, and visual surface brightness parameters, Fv, were derived from the semiempirical calibrations of Saxner and Hammarback (1985) and Moon (1985) respectively. Both comparisons appear to be main sequence stars with solar metal abundances and located at 230 and 113 pc respectively. Their spectral types can be estimated, from the present photometric calibrations, as F4V and G0V stars respectively, with the former slightly more evolved according to its δc_o difference.

Following the decoupling procedure described by Gimenez et al (1991) and using the upper limit for the luminosity ratio Lc/Lh=0.88 discussed above, we have estimated the magnitudes, colours and indices for the hot component of BH Vir listed in Table 1.

The E(b-y)=-.003 mag. and E(b-y)=0.008 mag. computed for the hot and cool components are very close to those derived from the comparisons. Using these four independent reddening determinations of 0.007, 0.012, -0.003 and 0.008 mag. giving a mean value of E(b-y)=0.006+0.006 mag. The radiative parameters for the hot and cool components have been derived following the same procedure as described above for the comparison stars. For the cool component we have used the standard photometric calibrations given by Olsen (1984) for late type stars. Results are given in Table 2.

Reglero et al. (1990) showed that the distance (Mv) to most RS CVn systems can be estimated using the non-active star, usually the hot component. BH Vir is a more complex case. The analysis of activity carried out by several authors in the last years, suggests that the hot component is the active star but probably also the cool star is active. The indetermination induced in the absolute magnitude determinations for the presence of surface activity can be minimized with a supplementary equation DMv = Dmv. Where D is de difference in magnitudes between the hot and cool components. In other words, the distance is the same for both stars in the binary.

Effective temperatures, visual flux brightness Fv and absolute magnitudes from Table 2 led us to classify the hot component as a F9-G0 type star with a marginal metal deficiency, measured through the m_1 index (δm_o=0.062), and slightly evolved (δc_o=0.043). The cool component can be classified as a G5 main sequence star with solar composition. This photometric classification from the derived radiative parameters indicate a later spectral type for the cool star that those currently assigned in the literature. The effects of the spots on the stellar surface can not be discarded as a cause of this discrepancy.

The metal deficiency of the hot-active star is in good agreement with the relation between activity an apparent metal underabundances measured through the m_1 index, derived by Gimenez et al. (1991) for RS CVn systems. In contrast, the less active or non-active cool component shows a solar metal content. The location of both stars in the c_o-$(b-y)_o$ plane indicate their evolved nature close to the TAMS.

ZZ UMa

We have used as comparison stars for this binary system SAO 15242 and SAO 15251. Averaged standard values for both stars are reported in Table 1. The interstellar reddening computed from its calibration shows the absence of interstellar extinction in this region up to 150 pc. Both systems appear as main sequence G0V and G2V stars with solar chemical composition. The derived parameters are listed in Table 2.

Following a procedure similar to that for BH Vir, we have decoupled the hot and cold components for ZZ UMa. The estimation of the reddening for the hot component is in very good agreement with the interstellar extinction indicated by the comparisons. No indication of interstellar extinction in this region of the sky and up to 150 pc have been found in accordance with the expected values. The radiative parameters for the hot and cool components are reported in Table 2.

The hot component appears as a G0V main sequence star with solar chemical abundance. The cool star shows a later spectral type, G5, and is more evolved (δc_o=0.10) and can be located close to the TAMS.

Marginal variability outside eclipses of 0.01 mag. in the V filter has been observed in ZZ UMa as a common signature of the late-type fast rotators.

CR Cas

CR Cas has been included in the present program because of its classification as a G5 star (Leung et al.(1977)). Their uvby and Hb indices, however, are not compatible with the late type classification. A highly-reddened, E(b-y)=0.6 mag., early-B type system seems to be the most realistic approach. Our partial coverage of the light curve does not allow us to derive a photometric calibration of the radiative parameters for the components. Although, CR Cas should consequently be dropped out of the scope of the present program on low mass binaries, we will continue the photoelectric coverage of this rewarding binary.

ACKNOWLEDGMENTS

The 1.5m telescope is operated by the Observatorio Astronomico Nacional. We want to express our gratitude to the observatory staff. This work has been supported by the Spanish Comision Interministerial de Ciencia y Tecnologia (PB90-0001-C02)

REFERENCES

Andersen J., Clausen J.V., Garcia M., Gimenez A., Helt B.E., Nordstrom B.,Sefanik R.P., Vaz L.P. 1992, in *IAU Colloqium 137, "Inside the stars"*, in press.

Botsula R.A. 1978, in *Varable Star (Russian)*, **20**, 577

Crawford D.L., Mander J. 1966, *AJ*, **71**, 114

Crawford D.L., Barnes J.V. 1970, *AJ*, **75**, 978

Crawford D.L., 1975, *AJ*, **80**, 955

Fabregat J., Reglero V. 1990, *A&AS*, **82**, 531

Gimenez A., Reglero V., De Castro E., Fernandez-Figueroa M.J. 1991, *A&A*, **248**, 563

Guiricin G., Mardirossian F., Mezzetti M. 1984, *MNRAS*, **206**, 305

Hoffmann M. 1982, *A&A*, **471**, 561

Janiashvili E.B., Lavrov M.I. 1989, *Inf.Bull.onVar.Stars*, N **3289**

Leung K.C., Schneider D.P. 1977, *AJ*, **211**,844

Moon T.T. 1984, *MNRAS*, **211**, 21

Kitamura M., Nakamura T., Takahashi C. 1957, *PASP*, **9**, 191

Koch R.H. 1967, *AJ*, **72**, 411

Olsen E.H. 1984, *A&AS* **57**, 443

Olsen E.H. 1991, private communication.

Perry C.L, Olsen E.H, Crawford D.L. 1987, *PASP*, **99**, 1184

Popper D.M. 1980, *A&AAnn.Rev.* **18**, 115

Reglero V., Gimenez A., Estela A. 1990, *A&A*, **231**, 375

Sadik A.R. 1978, Phd Thesis, U. Manchester.

Saxner M., Hammarback G. 1985, *A&A*, **151**, 372

Scaltriti F., Cellino A., Busso M. 1985, *A&A*, **149**, 11

Zhai D.S., Qiao G.J., Zhang X.Y. 1990, *A&A*, **237**, 148

Inside the Stars, IAU Colloquium 137
ASP Conference Series, Vol. 40, 1993
Werner W. Weiss and Annie Baglin (eds.)

TIME-RESOLVED SPECTROSCOPY OF SHORT PERIOD RS CVn SYSTEMS

C. Lázaro and M.J. Arévalo
Instituto de Astrofísica de Canarias, E-38200, La Laguna, Spain

ABSTRACT We have initiated a programme of spectroscopic observations of RS CVn short-period group, with orbital phase resolution. The systems of this group are all eclipsing binaries with both components at the Main Sequence, and most of them have similar spectral type components. The high rotational velocities and their short orbital periods (less than 1 day) hinder the spectroscopic study of these stars. We presents the first results yielded by H_α line observations of the systems XY UMa and WY Cnc. Both systems were observed during 1991 with reasonably wide orbital phase coverage. The analysis of the spectra is made by comparison with a binary model, constructed from the observed spectra of normal stars of the same spectral type as the RS CVn system components. The model accounts for the partial contribution of each component at any orbital phase within eclipses.

OBSERVATIONS

The spectra presented here were obtained in 1991 during two observing runs using different telescopes. The first observations were made in February from 22/23 to 24/25 at the Coudé focus of the 2.2 m. Telescope at the German-Spanish Calar Alto Observatory (Almeria, Spain). We observed again on the nights November 23/24 to 26/27 at the Coudé focus of the 1.5 m Spanish telescope at the same Observatory. The 2.2 m observations cover the spectral range λ: 6531-6600 Å, with a spectral resolutions of 0.07 Å/pixel. The observations made with the 1.5 m. cover the range from λ: 6460Åto 6740Å, with a spectral resolutions of 0.54Å/pixel. The data reduction process has been performed using the software packet FIGARO for extraction and wavelength calibration.

One of the aims of this study is to follow the variations of the spectroscopic appearance of the RS CVn systems with orbital phase, covering the complete orbital cycle with good orbital resolution. This makes it convenient to maintain as low as possible the ratio beteween the exposure time of the individual spectra and the orbital period of the observed system. On the other hand, these systems are relatively faint: V= 9.5 for WY Cnc and V=9.8 for XY UMa (both at the maximum), and their orbital periods $0^d.83$ and $0^d.47$ for WY Cnc and XY UMa respectively. These periods preclude the possibility of long exposure times and limit the signal to noise ratio that may be achieved in the spectra without a severely degrading spectral and orbital resolutions. We have chosen a compromise between those opposite considerations, and thus the individual spectra spread about 0.02 in orbital phase for WY Cnc, and 0.05 in phase for XY UMa.

RESULTS

We study spectral line variations comparing the observed spectra of the RS CVn systems with a theoretical spectrum of a *normal* eclipsing binary. This theoretical spectrum is generated for any orbital phase from the observed spectra of two *normal* stars, with the same spectral type as the components of the RS CVn system. As comparison stars, we observed several bright stars with unknown active behaviour. In the model spectra presented in this work, we have used:

- HD 32147 (K5 V) with 70 Vir (G4 V), for the XY UMa spectra obtained at the 2.2 m. telescope.
- HD 32147 with HR 1656 (G4 V) for the XY UMa spectra observed in at 1.5 m. telescope.
- HD 36395 (M2 V) with 70 Vir for the WY Cnc system.

The intrinsic brightness of each component in the binary model is taken from the MARCS models atmosphere fluxes (Gustafsson et al 1975). The limb darkening and the partial contribution of each component during the eclipse phases are properly taken into account, but non-sphericity, reflection and gravity darkening effects have been neglected at this stage. The model gives the total expected spectra plus the contribution of each component. This allows to identify easily the location of the H_α emission line in the system, and to derive the ratios of line emission to photospheric fluxes for each component.

XY UMa: The spectra at fases 0.24, 0.0, show H_α excess with variable intensity (Figure 1 and Figure 2). Both component stars present emission at different phases.

WY Cnc: At the two orbital phases shown here, the system presents also a strong H_α excess (Figure 3 and Figure 4). This emission line presents two components. The strongest can be associated to the primary component, but the pointing out the origen of the other component must be based on the complete set of data.

It seems that this behaviour has been not observed previously, as Strassmeier et al. (1988) quote strong H_α absorption for the XY UMa system. Whilst giving no information on on WY Cnc. Pojmanski (1992) found H_α in XY UMa to be much weaker than in G2-G5 V stars, and strongly variable. This is indicative of the complex behaviour of these systems, adds more interest to the continuity of this kind of observations.

REFERENCES

Gustafsson,B., Bell, R.A., Eriksson,K., Nordlund, A., 1975 *Astr. Ap.*, **42**, 407.
Pojmanski 1992: Private comunication
Strassmeier, K, Hall, D.S., Zeilik, M., Nelson, E., Eker, Z., and Fekel, F., 1988 *Astr. Ap. Suppl.*, **72**, 291.

Figure 1: XY UMa , Phase=0.24

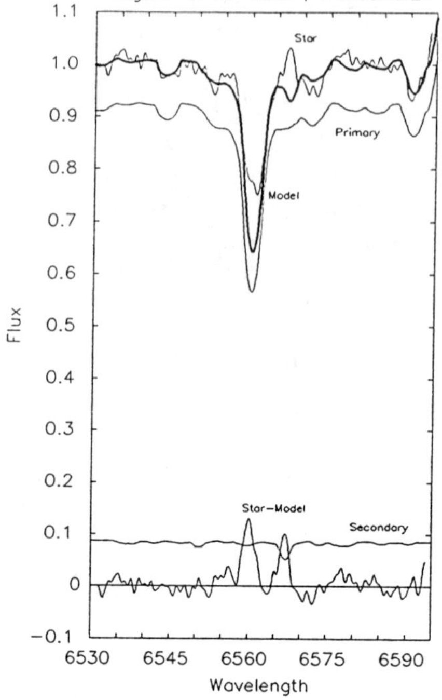

Figure 2: XY UMa , Phase=0.00

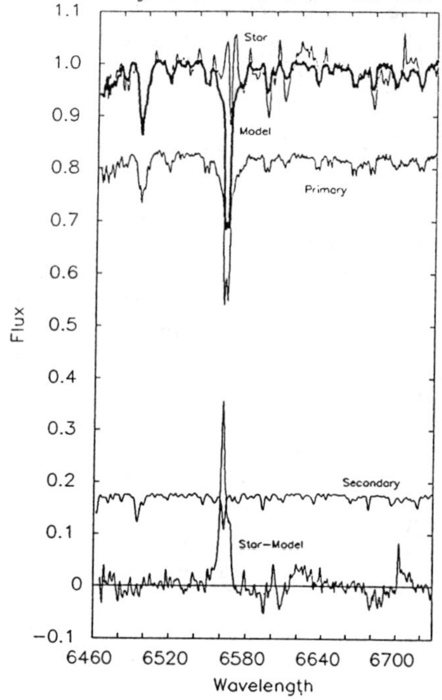

Figure 3: WY Cnc , Phase=0.56

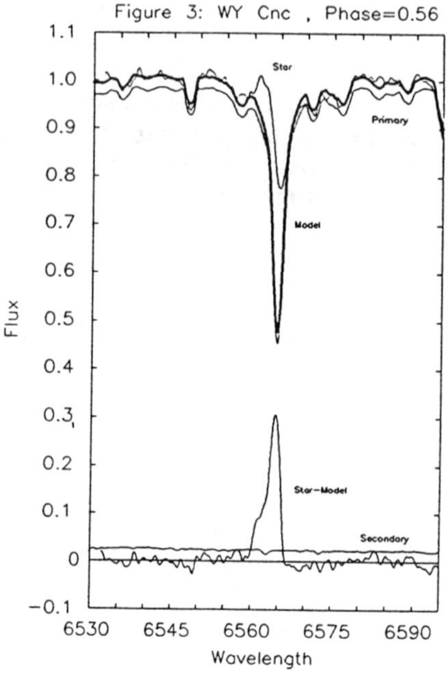

Figure 4: WY Cnc , Phase=0.62

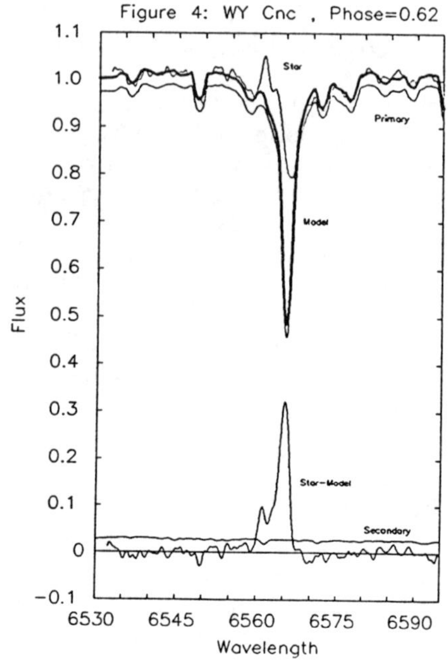

Inside the Stars, IAU Colloquium 137
ASP Conference Series, Vol. 40, 1993
Werner W. Weiss and Annie Baglin (eds.)

THE PRE – MAIN SEQUENCE

FRANCESCA D'ANTONA
Osservatorio Astronomico di Roma, Monte Porzio, Italy

ABSTRACT The Pre–Main–Sequence (pre–MS) is defined as the phase following the protostellar stage and ending with the ignition of hydrogen in the stellar core. Models in hydrostatic and thermal equilibrium such as typical "Hayashi track" models are basically adequate to describe the internal structure of pre–MS stars, although for a fraction of pre–MS objects there can be phases of accretion also at late stages, which may influence the surface abundances of light elements.

The interior evolution of pre–MS structures is focused in the two main stages of Deuterium and Lithium burning for the low mass ($M \leq 1.5 M_{\odot}$) stars. The location in the HR diagram of theoretical tracks is still subject to large uncertainty, even more apparent today that new opacities and treatment of turbulent convection are available, and the results can be internally compared. Uncertainties amount to a factor $\sim$ two for mass and age determination of individual objects, and affect the absolute location for both the D–burning and the Li–burning regions. Qualitative constraints, such as that there can not be Li–depletion at $L \geq L_{\odot}$, remain valid. Abundances of ^7Li at the surface of pre–MS stars, if measured at a stage which can definitely be considered *previous* to the possible occurrence of nuclear burning, and, consequently, previous to other depletion mechanism which may occur later on, provide constraints on galactic evolution of this element of cosmological interest.

1. DEFINITION: THE PRE–MS AND THE PROTOSTELLAR PHASES

The first phases of stellar evolution may bear important consequences for the understanding of normal stars. Great progress has been made in the latest years, both theoretically and observationally, and this review will try to focus on a few aspects, more than try to give a complete overview of the work done or in progress.

As the same definition of pre–MS stage is subject to some ambiguity, I will define as pre–MS for low masses ($M \leq 1.5 M_{\odot}$) the phase starting in the vicinity of deuterium burning in fully convective structures, and ending with the onset of hydrogen burning in the stellar core (MS).

The proto–stellar phases begin with the free–fall collapse of interstellar clouds. This highly non–homologous collapse (Larson, 1969), with the high density central regions collapsing first, soon leads to a configuration in which a central hydrostatic core gravitationally accretes the remaining diffuse matter

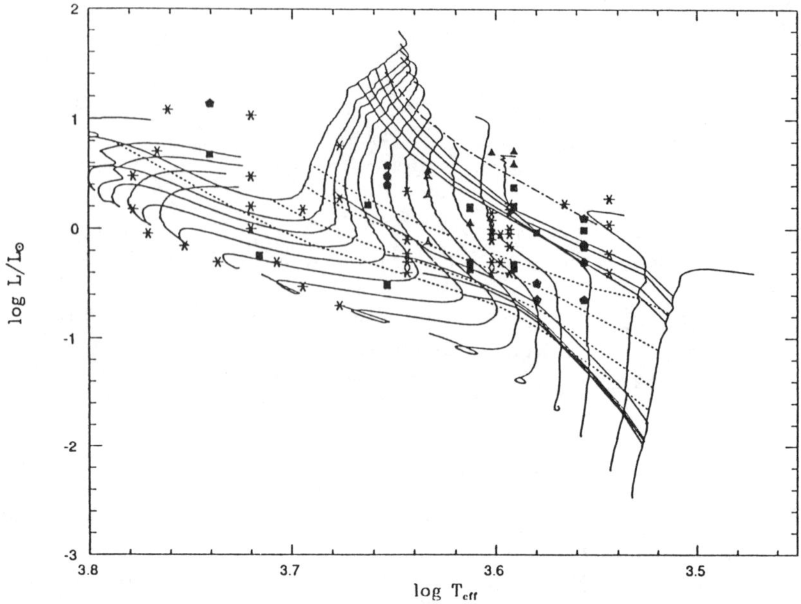

FIGURE I Representative evolutionary tracks of 1.5, 1.3, 1.2, 1.1, 1.0, 0.9, 0.8, 0.7, 0.6, 0.5, 0.4, 0.3, 0.2, 0.15 $M_\odot$, on which we show: *i*: the four upper continuous lines are the locus at which the nuclear luminosity due to D–burning reaches 20%, 60%, then goes down to 60% and 20%; the uppermost dash–dotted line shows how the region of $L_{nuc} = 20\%$ for D–burning is shifted in models having different opacities; *ii*: dotted lines: isochrones of 1, 2, 5×10^6 and 10^7yr; *iii*: the four lowest continuous lines represent, from top to bottom, beginning of Li–burning and lines at which the initial abundance is reduced by .5, 1 and 2 dex respectively; *iv*: observational data collected by Magazzù *et al.*(1992) and Basri *et al.*, 1992.

in the outer regions. The protostar is obscured by the dust surrounding the parent cloud, and it will emerge in visual light only when the main accretion phase is ended (Stahler, 1983). During the protostar phase, probably during the bipolar molecular outflow phase, the object gets rid of most of the Angular Momentum (AM) of the parent cloud. Consideration of AM evolution (see Pringle, 1989, Bodenheimer, 1991) should be a fundamental ingredient in the study of protostellar evolution. When low–mass pre–MS objects appear, at least those identified with T–Tauri's are already slowly rotating (Bouvier, 1991).

The onset of a strong stellar wind ends the accretion phase, disperses the dust, and lets the pre–MS star emerge. It is to be explained *why* the stellar wind begins, what are the physical parameters which determine the Initial Mass Function of the stars, and how the very low mass stars are formed, but this approach provides at least a phenomenological description. According to Stahler (1983), the pre–MS star appears in the HR diagram at the intersection between the Hayashi line of fully convective hydrostatic stellar models and the mass–radius relation of the hydrostatic core. The "stellar birthline" depends on the stellar mass mainly (Palla and Stahler, 1992). *D*–burning is the key process occurring in the protostar and determining its final mass–radius relation (Stahler 1988). In stars of $M \leq 2M_\odot$, D is ignited in the core and forces the star to become fully convective (Stahler, Shu and Taam, 1980). At larger masses, a more complicated behaviour sets in, due to the onset of a radiative barrier and the subsequent ignition of D in a shell (Palla and Stahler, 1990, 1991). In the absence of D–burning, the accretion timescale for $M \gtrsim 3M_\odot$ is longer than the simple Kelvin Helmoltz timescale (t_{KH}), and masses would directly appear on the MS. Introduction of D–burning then explains *why* pre–MS stars of these masses exist (see for a review on the subject Palla, 1991).

For masses $M \leq 1.5M_\odot$, the location of the birthline will depend on the details of the protostar evolution (e.g. boundary conditions and accretion rate), but, qualitatively, if the protostar phase is thermostated by D–burning, the pre–MS stars upper envelope in the HR diagram should coincide with the locus of D–burning in hydrostatic stellar models. Palla and Parigi (1992, in these proceedings) show that a $1M_\odot$ pre–MS star appears below the locus of D–burning, evolves to it by increasing L, and then follows the Hayashi line evolution. The thermal disequilibrium induced in the core during the accretion should soon be over: the star will first adjust on a hydrostatic equilibrium timescale, and then completely relax within a Kelvin-Helmoltz time, which is of the order of the former accretion timescale (see, e. g., Mercer Smith *et al.*, 1984). From this moment on, the stellar evolution occurs on the thermal timescale, apart from the momentary stop due to D–burning. When a radiative core appears and convection in the envelope progressively recedes, the star begins evolving towards the main sequence. The possible burning of other light elements (^6Li, ^7Li, ^9Be) is to be considered during this phase. Finally, hydrogen is ignited in the core and the pre–MS phase can be considered concluded.

In this review I concentrate on the following problems: *i:* the D–burning phase; *ii:* the possible influence of residual accretion on the pre–MS evolution; *iii:* the uncertainty of stellar modeling and the location of the tracks in the HR diagram; *iv:* the burning of light elements and the new data on the Lithium in pre–MS objects.

All the following discussion will be based on the results of "standard" stellar modeling (one–dimentional non rotating stars without magnetic fields, mass loss or accretion and without rotation). It is then mandatory to add a few comments regarding rotation.

Although at least those pre–MS low mass stars identified with TTauri's are slowly rotating, and the pre–MS begins when the protostellar phases have already acted to eliminate most of the Angular Momentum (AM) of the forming cloud (see, e.g. Bodenheimer, 1991), we should not, in principle, ignore the evolution of AM during the pre–MS stage. Inclusion of an initial AM compatible with the observed rotations of TTauri's does not, in itself, imply drastic changes in the structure of the star or in its observable properties, as shown, for instance, by the "1.5"–dimensional models by Pinsonneault *et al.*, 1990, which account for rotation adopting the approach by Endal and Sofia (1981). Most important, and subject to uncertainty, is the *evolution* of AM during the pre–MS phase, which, in principle, has to take into account both loss of AM at the stellar surface, due, probably, to magnetic winds, the possible advection of AM from the circumstellar disk, the establishing of gradients of rotational velocity through the structure when the radiative core appears, and its consequent transport and redistribution. The final MS stellar rotation distribution is dictated both by the initial rotation (as solid body, if the star is initially fully convective) and by the parameters which regulate loss, advection and redistribution of AM. The models by Pinsonneault *et al.* (1990) represent by now an interesting step in the exploration of the parameter space.

Further, many mechanisms which alter the surface composition of MS stars depend on rotation (e.g. meridional circulation) and also it is probable that chemical mixing is associated with the redistribution of AM: this may be an additional cause of depletion of light elements at the surface of stars (see, for a review, Demarque, 1991) and adds to the importance of exploring the evolution of rotating models. For what concerns this review, however, I notice that the timescale on which the effects of mixing associated with the AM transport takes place is of several $\times 10^7$yr, so that the AM distribution achieved within the star during pre–MS evolution might have long–term consequences during the MS lifetime, and *not*, mainly, in the pre–MS stage itself. For instance, the models by Pinsonneault *et al.*, 1991, show that the effect of rotation on Lithium depletion is not dramatic till an age of 30Myr, while it grows up to a factor 0.3 – 0.6dex (depending on $T_{\rm eff}$) already at 70Myr.

My conclusion is then that we can outline the broad lines of surface evolution of light elements in pre–MS without accounting for rotation, but rotation may be of great relevance for the interpretation of abundances and rotation evolution during the MS.

2. THE D–BURNING PHASE

As a general reference for our discussion I show in figure I the HR diagram with representative tracks of masses from 1.5 to $0.15 M_\odot$, with population I chemical inputs, and "updated" physics. The general framework in which these tracks were computed is in part described in next section and will be presented in a forthcoming paper by D'Antona and Mazzitelli (1992). The loci of D– and Li–burning are shown, as well as a few isochrones The points represent T–Tauri's in several star forming regions.

The four curves at the top of the tracks represent the main locus of D–burning. In this diagram several pre–MS appear above this region. This can be attributed in part to the influence of the disk on the apparent location of TTauri (Kenyon and Hartmann, 1990, see section 3), in part to the fact that the location of the tracks in the HR–diagram is very model–dependent (section 4), as shown by the dot–dashed line. The fact that actually TTauri's appear more or less at the "birthline" is suggestive, but it is not a definite proof that Stahler's scenario is correct. Actually, as we see in figure II, where I plot the luminosity versus time along the tracks, the D–burning stage is the first one during which the star slows down its evolution. It lasts, in fact, from $\lesssim 10^5$yr for masses $\sim 1 M_\odot$ to $\sim 10^6$yr for $M \leq 0.3 M_\odot$ or smaller: even if the pre–MS stars begin their life at larger luminosity, their luminosity function would be such that it is more probable to observe them only when they reach the D–burning line. Recently, Zinnecker *et al.* (1992), on the basis of models by Mazzitelli (1990, unpublished), have discussed how the D–burning locus could even be observed as a "peak" in the luminosity function of pre–MS stars.

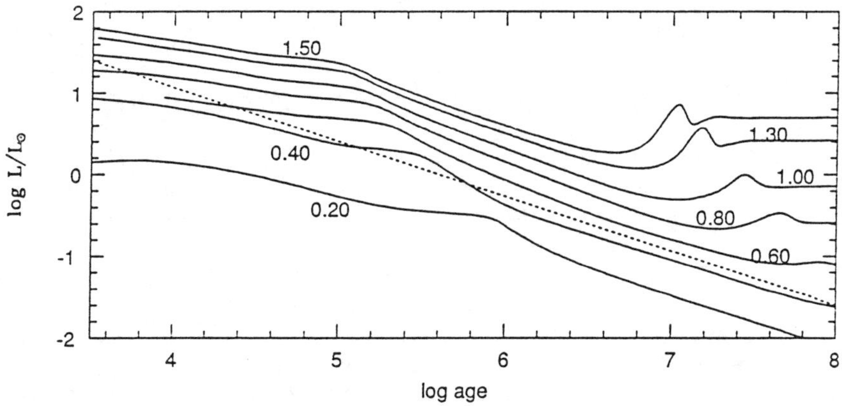

FIGURE II Luminosity evolution versus log age (in years) for models from $1.5M_\odot$ to $0.2M_\odot$. The first flattening of the curves is due to the D–burning (taken to be $X_D = 10^{-5}$ in mass fraction) and the second one is due to reaching of the main sequence. The diagonal line represents a curve of slope -2/3, which is followed by stars simply contracting at fixed $T_{\rm eff}$. The stop due to D–burning does not alter substantially this type of evolution for most stellar masses.

The exact duration depends on the abundance of D in the star forming clouds. Computations in figure II refer to $X_D = 2 \times 10^{-5}$, where X_D is the Deuterium *mass* fraction, or a number fraction $[D/H] \sim 10^{-5}$ (York and Rogerson, 1976, Boesgaard and Steigman, 1985) but values as low as [D/H]$\sim 5\times10^{-6}$ (Vidal Majar *et al.* 1983), or as large as 2.5×10^{-5} (Bruston *et al.*, 1981), as in the solar nebula, are quoted in the literature. The diagonal straight line is the typical $L \sim t^{-2/3}$ evolution expected on the basis of simple contraction along a Hayashi line at fixed $T_{\rm eff}$: D–burning does not alter substantially this evolution.

For masses $M \lesssim 0.1 M_\odot$, the D–burning lasts for a relatively long time (1–2$\times10^6$yr, see e.g. D'Antona and Mazzitelli 1985, Pozio, 1990), and its locus becomes almost vertical, at a $T_{\rm eff}$ ($\sim$ 3000K) (this value of $T_{\rm eff}$ is highly uncertain, as it depends on the opacities and other assumptions of the models). Thus, coeval pre–MS stars of low mass must appear at this $T_{\rm eff}$, with L decreasing with decreasing mass, and it is, in principle, possible to try to derive information on the Initial Mass Function (IMF) of young clouds associations from the luminosity function of these objects. This has been attempted for the ρ Oph cloud (Rieke *et al.*, 1989, Barsony *et al.*, 1989, Rieke *et al.*, 1991). In view of the difficulty of observational discovery of Brown Dwarfs (if they exist, see, e.g. Stevenson, 1991), it is very important to try to find them when they are young and more luminous, in spite of the difficulties which interpretation of these results may bear (see, e.g., Zinnecker *et al.*, 1991).

Grossman and Graboske (1973), have shown that D is not ignited in masses $M \lesssim 0.01 M_\odot$. Pozio (1990, 1991) confirmed this result and found that there exists a small mass range (identified with $M\sim 0.015 M_\odot$) at which the conditions for D–burning ($T_{center} \sim 1.1. \times 10^6$K, $\rho_{center} \sim 6$g cm^{-3}) are reached at a very low luminosity ($\sim 10^{-3.5}\ L_\odot$), so that the phase lasts much longer than usual,

$\sim 10^8$yr, mimicking a true *D-burning main sequence.* We might be able to find a signature of this event when looking for very red low luminosity BD candidates.

3. RESIDUAL MASS ACCRETION AND ITS POSSIBLE INFLUENCE

Recent observations have shown that the Classical T Tauri (CTTs), stars characterized by strong emission lines indicating both accretion and mass loss (e.g. Bertout, 1989), veiling of the lines and UV and IR excesses over a continuum of spectral type from K to M, are not the only pre-MS objects: Naked (or Weak line) T Tauri's (WTTS) have been discovered as soft X-ray sources, and afterwards identified with low mass pre-MS stars (Walter *et al.*, 1988).

The properties of CTTS are consistent with a model of a star plus a keplerian accretion disk and a boundary layer, as suggested already by Lynden Bell and Pringle (1974). Recent modeling (Adams *et al.*, 1987, Kenyon and Hartmann, 1987, Bertout *et al.*, 1988) lead to infer mass accretion rates in the range from $\sim 3 \times 10^{-9}$ to $4 \times 10^{-7} M_{\odot}$/yr (Basri and Bertout, 1989). The WTTS, among which $\sim$ 30% present small IR excesses in the K-band (2.2μ), can be modeled again with a star plus a disk, but with no boundary layer and a central cavity (possible site where planets have formed?) (Montmerle and André, 1989). Walter (1987) and Strom *et al.* (1989b) show that both CTTS and WTTS occupy the same region of the HR diagram which, interpreted with normal hydrostatic contraction models, range from ages of $\sim 3 \times 10^5$ to $\sim 3 \times 10^6$yr. This probably indicates that disk evolution is decoupled from the evolution of the central star and that the disk may disappear either early or at a very late stage.

How much the evolution can be altered by mass accretion? The D-burning luminosity which would be provided by burning of the accreted D is generally negligible, being $L_D \sim 15 L_{\odot} \times (\dot{M}/(10^{-5} M_{\odot}/yr))$ -e.g. Stahler, 1988. This could have an influence either for accretion rates much larger than observed, or at very low luminosity. Accretion itself may alter the evolution only if we have: $M/\dot{M}_{crit} \sim t_{KH}$. At large luminosity, where $t_{KH} \sim 10^5$yr, this constraint would require too large mass accretion rates. At $L \sim L_{\odot}$, we have $t_{KH} \sim 2 \times 10^6$yr, for which $\dot{M}_{crit} \sim 5 \times 10^{-7} M\odot$/yr, which is *at the upper boundary* of what observed in CTTS: in this case the evolution can be probably slightly altered, but it is clear that *such a $\dot{M}$ can not be mantained for a very long time, as low mass stars exist.* We conclude that accretion can be important for the evolution of a fraction of the CTTS (an even smaller fraction of all pre-MS stars), only for short periods of time.

The "post-T Tauri problem" (lack of TTs of age $\sim 10^7$yr in the Taurus Auriga complex, Herbig *et al.*, 1986), stressed also by the negative results of the proper motion survey by Hartmann *et al.* (1989), is probably to be interpreted in terms of existence of separated loci of star formation in Taurus, clustered at different ages (Walter, 1987).

4. THE ABSOLUTE LOCATION OF TRACKS IN THE HR DIAGRAM

I conclude that the evolution of pre-MS stars is well described by hydrostatic stellar models, which are mainly dominated by the stellar contraction. Of course the absolute location of these tracks in the HR diagram, and thus both the mass calibration of the MS, the locus of D–burning, and, as we shall see, the locus of Lithium burning, are subject to several sources of uncertainties. Mazzitelli (1989) has examined four main aspects of the input physics of stellar models which influence the HR diagram location of the tracks, namely: *i*: the convection model; *ii*: opacity; *iii*: thermodynamics; *iv*: rotation, and I refer to that paper for the general discussion. Recently we have acquired several improvements in the input physics, which eventually should lead us to a better understanding of the HR diagram location of the tracks.

4.1 The convection model

Turbulent convection in stars is generally treated according to the Mixing Length Theory (MLT, Böhm Vitense, 1958), in which the free parameter α (ratio of the mixing length l to the pressure scale height H_p) is tuned so as to fit the solar radius. Canuto and Mazzitelli (1992) have explicitly shown that this procedure, when applied to stars different from the Sun, is absolutely unsafe and likely to introduce systematic hidden errors. This is not a very new argument, but it is generally not considered at all, in the absence of a better description of turbulence.

We are presently testing the model which Canuto and Mazzitelli (1990, hereinafter CM) have proposed as an alternative to the MLT. It accounts more precisely for the nature of turbulent convection, by two main improvements: first, the entire distribution of energies of all eddies is taken into account, while the MLT is a "one–eddy" description; second, the mixing length is chosen to be the distance (z) from the point in which the flux is to be computed to the top of the convective region. Therefore, there are *no free adjustable parameters* in the models, although small adjustments can be done in the choices of a a few input values for the model, in a range allowed by experimental determinations. After the successful application of the new model to the Sun (CM), we have recently shown that it is able to reasonably fit the HR diagram of an old open cluster like M67 (D'Antona *et al.*, 1992): in particular, the model can fit the giants and clump stars location. Although it is clear that much work is still to be done before we can claim to have a better convection model, and problems exist, in particular for cool, expanded structures, for defining properly the boundary conditions and thus for using an appropriate value of z, it is necessary to continue the exploration, for example by applying it to the treatment of pre–MS structures.

It is worth to enphasize one of the characteristic of the new treatment of convection: if we compare the new tracks with tracks computed with the MLT we immediately see that the ratio $\alpha = l/H_p$ which fits the main sequence –in particular, which fits the solar $T_{\rm eff}$ – *does not provide giants having the same $T_{\rm eff}$ as the CM model.* Although this does not guarantee at all that the CM model provides a definitive description of giant structures –a compressible turbulence theory is needed for that–, nevertheless it provides another strong warning against extending the α value calibrated on the Sun to stars other than the Sun itself! Figure III shows in fact three different tracks for the solar model. The continuous line gives the evolution of a model which should in principle be the most updated possible for choice of opacities and equation of state. It also includes CM convection model. The solar $T_{\rm eff}$ at an age of 4.6×10^9yr is

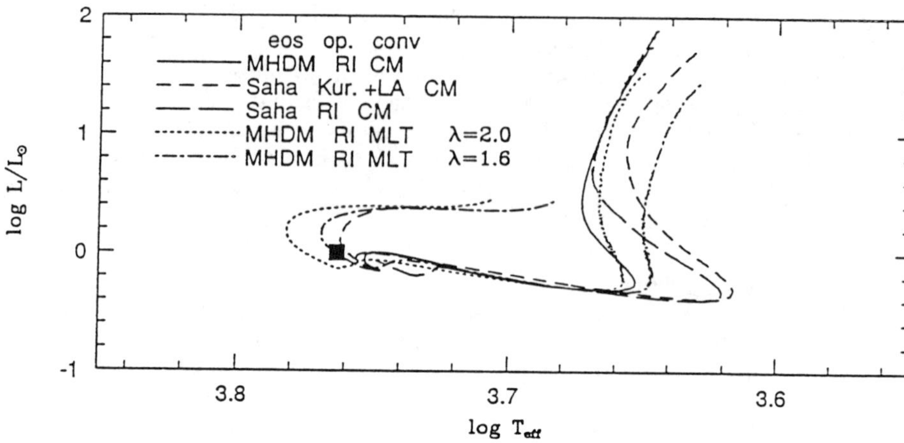

FIGURE III Five different pre–MS tracks are shown for the solar model, identified by adopted EOS, opacities (RI≡ Rogers and Iglesias, 1992; Kur.≡ Kurucz, 1991; LA≡ Los Alamos) and convection model (CM or MLT).

$\sim$ 50K smaller than the real solar $T_{\rm eff}$ (Canuto and Mazzitelli 1992). Two other evolutionary tracks are shown, having the same physical inputs and numerics, but adopting the MLT for the computation of superadiabatic convection. In order to fit the solar position (which, in this case, can be fitted better than within 50K), we need $\alpha = 1.6$. But we see that, in order to reproduce the *giant* pre–MS location of the CM model *we have to fix* $\alpha \sim 2$.

As a conclusion, a better description of turbulence seems to give *hotter* giants in pre–MS than those obtained by adopting the same α which fits the Sun. The opacity uncertainties are still too large to hope to find a test of this feature, by now.

Although convective efficiency does not alter very much the burning of light elements (mainly Lithium), it is a well known result of computations that tracks with a larger α present a somewhat larger depletion (e.g. Deliyannis *et al.*, 1990). As CM models in the giant phase correspond to MLT models with larger α, they also produce a slightly larger Li– depletion than MLT models employing the standard α calibrated on the Sun.

4.2 The equation of state

The importance of employing a good Equation of State (EOS) for the solar and stellar models has been discussed in this colloquium by Däppen (1992). In what concerns the pre–MS evolution, we simply add that use of a non–ideal gas EOS narrows the partial ionization region and then *lowers* the values of the adiabatic gradient there (e.g. Graboske *et al.*, 1975, Magni and Mazzitelli (1979), Saumon, 1990). The "standard" $1M_\odot$ track shown in figure III makes use of the Mihalas *et al.*(1988) EOS (MHDM) for the region in which it is available (densities $\rho \leq 0.01$g cm^{-3}). It is compared with a corresponding track employing a simple Saha type EOS in the same region. We see that the MHDM EOS produces *larger* $T_{\rm eff}$'s in the region where the track departs from the Hayashi track to begin approaching

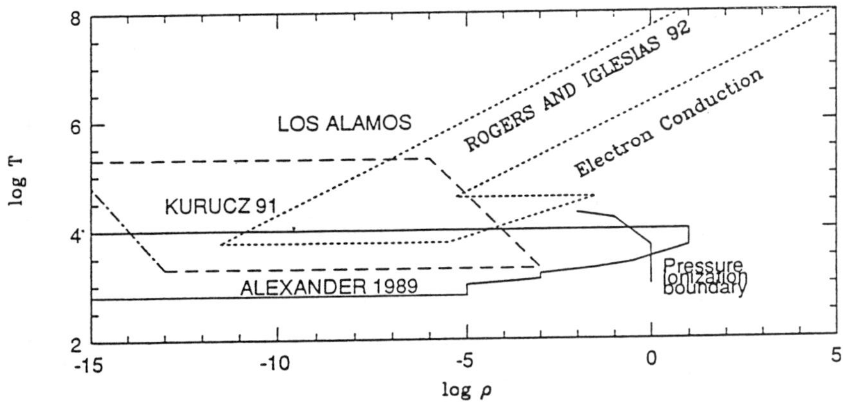

FIGURE IV Coverage of the density – temperature plane by means of available opacity tabulations

the MS. In practice, the "nose" of the track is less pronounced. Apart from the different location of masses in the HR diagram, this feature has a small but noticeable influence on the burning of Lithium: the conditions at the base of the convective envelope in the models employing the MHDM EOS, are favourable to more efficient burning.

4.3 Opacities

In spite of the great recent improvements of these latest years, this remains the softest spot of pre–MS evolution. I show in figure IV the $\rho - T$ plane where the different regions covered by recent available opacity computations are indicated. Previous to 1991, one choice was obligatory: Los Alamos (Huebner *et al.*, 1977) tables. These had to be supplemented by low temperature opacities (generally from Cox and Tabor, 1976). An important effort was done by Alexander to include molecules and grains in the computations. Actually, Alexander *et al.* (1989) complete set of low temperature opacities with updated treatment also of water vapour is, till today, the most extended available table in density, although use of these tables must take into account that the opacities at the largest densities are not to be considered, as they are past the pressure ionization boundary (figure IV). Alexander's opacities are given only for solar composition, and this prevents many applications, although it is largely sufficient for the pre–MS.

Kurucz (1991) presented opacities at low temperature for a large number of chemical compositions. These are not extended enough to completely cover the low mass external layers regions, but they are adequate for the pre–MS. Rogers and Iglesias (1992) opacities (hereinafter RI) for $T \geq 6000$K cover most of the interesting regions of the $\rho - T$ plane, although they should be extended to larger densities, if one wants to use them for low mass stars. Use of RI opacities is recommended on the basis of the improved treatment of the physics they adopt (see for a discussion Däppen, 1992, in these proceedings), and although they must be supplemented with the Los Alamos opacities in the regions where they

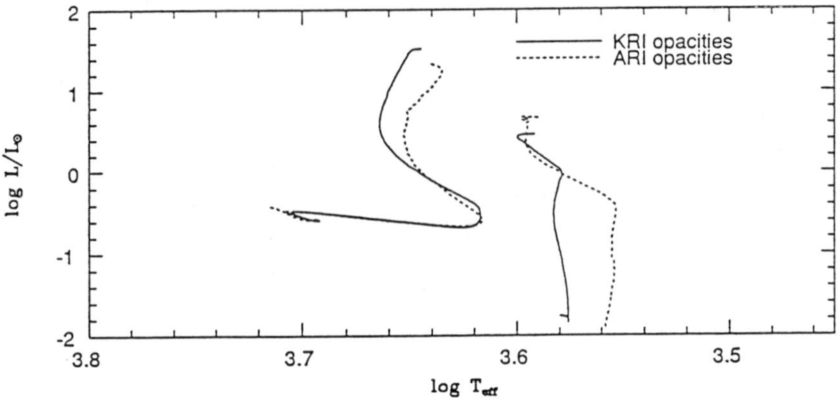

FIGURE V HR diagram for 0.8 and $0.3M_{\odot}$ tracks for different opacity inputs, as explained in the text.

are still not available. At large densities, in the regions of full ionization, any available radiative opacity must be complemented with the electron conduction, but this does not interest pre–MS evolution.

The influence of the latest available choices (ARI: Alexander at $T \leq 10000$K, plus RI; and KRI: Kurucz at $T \leq 6000$K, plus RI) for pre–MS evolution is shown in figure V, where I compare the different tracks obtained for 0.8 and $0.3M_{\odot}$ (CM convection model is adopted). The uncertainty in the $T_{\rm eff}$ determination is quite large especially for $0.3M_{\odot}$, where it becomes ~ 0.03 dex.

5. THE BURNING OF LIGHT ELEMENTS

Depletion of light elements other than D in the pre–MS phase interests mainly burning of ^6Li, ^7Li and ^9Be , which occur as soon as the internal temperatures reach respectively ~ 2, ~ 2.5 and $\sim 3.5 \times 10^6$K. These temperatures are only indicative: in very low mass stars, and in advanced phases in which the bottom of the convective layer is situated at larger densities, the temperatures of burning are smaller. A $0.08M_{\odot}$ star ignites ^7Li at $T_c \sim 1.9 \times 10^6$K and $\rho_c \sim 26$ g cm^{-3} (Pozio, 1990).

Extensive works on standard pre–MS depletion of light elements, and in particular of Lithium, are available in the literature, starting from Bodenheimer (1965), to D'Antona and Mazzitelli (1984), and up to Vandenberg and Poll (1989) and Swenson *et al.*, 1989). Depletion of ^6Li has been considered in detail by Proffitt and Michaud, 1989, and Deliyannis *et al.*, 1990 (for metal poor stars). Burning of ^9Be is considered by Deliyannis and Pinsonneault, 1990 and Ryan *et al.*, 1992. All these elements have a cosmological interest, being synthetized during the Big–Bang (see, e.g. Boesgaard and Steigman, 1985), and it is important to compare detailed models of depletion with the observations, in order, for instance, to discriminate between different models, or to test different formation scenarios.

The galactic evolution of ^7Li, in particular, is a fundamental topic which still is subject to large uncertainties (for a general discussion see, e.g., Mathews *et al.*, 1990, D'Antona, 1991, D'Antona and Matteucci, 1992). The fundamental problem in depicting galactic evolution is that it is not clear how much the observed abundances at the surface of population II stars reflect the primordial abundance. Depletion mechanisms for the light elements can be divided into two broad categories: "fast" mechanisms, which occurr on a short timescale: these are mainly burning in pre–MS ($\tau \sim 10^7$yr) and dilution as soon as convection deepens into the star evolving as a giant; "slow" mechanisms, which cause depletion on longer timescales, such as diffusion (Proffitt *et al.*, 1990, Deliyannis and Demarque, 1991), mixing associated with transport of angular momentum (Pinsonneault *et al.*, 1990) or by waves (Garcia–Lopez and Spruit 1991, Schatzman, 1991), having $\tau \sim 10^8 - 10^9$yr.

I list the few, probably firm results on ^7Li–pre–MS burning:

1. ^7Li is not touched either in stars in which the radiative core appears so early during the evolution that temperatures necessary for burning are never reached at the convective bottom (conservatively, for $M \gtrsim 1.4M_\odot$), or in objects which are so small in mass that degeneracy prevents increase of central temperature beyond burning conditions. Pozio (1991) finds that this lower limit is $M \sim 0.06M_\odot$. Rebolo *et al.*(1992) propose to use the absence/presence of the Lithium line in candidate Brown dwarfs to put an upper/lower limit to the mass of the object, and this may result a powerful tool for discriminating bona fide brown dwarfs of $M \leq 0.06M_\odot$.

2. There is *some* pre–MS depletion in the solar model, but it can not reach the factor 100 required by the obsevational value of atmospheric ^7Li in the Sun, compared with the ^7Li of the solar nebula. Most recent models (Pinsonneault *et al.*1992b, D'Antona and Mazzitelli, 1992) tend to give a somewhat larger solar pre–MS depletion than previously found, due to the complex update of interior physics and of the ^7Li+p reaction rate. Within the present uncertainties, we can conservatively say that $^7Li_{present\odot} \sim (0.1 - 0.5)^7Li_{initial}$.

3. As discussed in the introduction, theoretical modeling seems to establish that there is *one* dominant mechanism for depletion among the pre-MS stars, and it is nuclear burning only. The earlier the evolutionary stage at which we compare model results and the observations, the more probable it is that we are witnessing "standard" depletion.

4. Low metallicity stars have smaller pre–MS depletion and "standard" models consistently explain population II observations (Deliyannis *et al.*, 1990) although there are a number of other possibilities to achieve reasonable agreement with the data, which imply "long term" depletion mechanisms (Vauclair, 1988, Pinsonneault *et al.*, 1992a).

There are a few other results which are important for their consequences:

1. *No models* of pre–MS, in the present context, can predict ^7Li–burning at $L \gtrsim L_\odot$.

2. Observationally, we *do expect* a ^7Li–depletion boundary, whose precise location is subject to the above discussed uncertainties.

3. ^7Li in luminous T Tauri *must be* the initial ^7Li of the forming interstellar cloud. If we find $[Li/H]_{today} \gtrsim [Li/H]_{solarsystem}$, we get important information on the galactic evolution of ^7Li in the latest 4.6 billion years.

Typical Li–depletion boundaries from pre–MS models are shown in figure

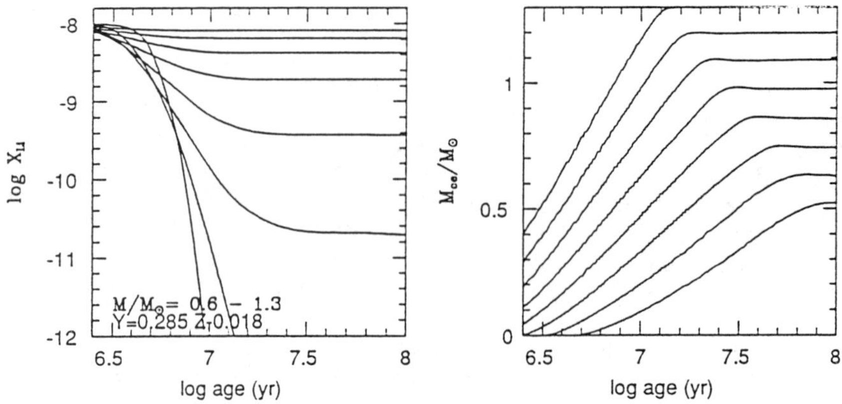

FIGURE VI Lithium abundances (left) and convective masses (right) versus age for models of standard pre–MS depletion with updated physics (D'Antona and Mazzitelli, 1992). The masses are increased by $0.1 M_\odot$ from 0.6 to 1.3.

I. Until recently, observations have seemingly implied larger abundances of ^7Li in TTauri's than in young clusters (Strom *et al.* 1989a, Basri *et al.*, 1991). Recent work by Magazzù *et al.*, 1992 –see also the updated summary given in a poster paper by Martin *et al.*(1992)– revise the picture to give a remarkably constant $\log N_{Li} \sim 3.2 \pm 0.3$, complemented by remarkable depletion only in low luminosity TTs. Although the theoretical depletion boundaries and the location of depleted stars do not match perfectly each other, this should not worry us, in view of the wide discussed uncertainties in the location of both tracks and observational data in the HR diagram. The possible conclusion from Magazzù *et al.* data is that in the last few billion years from the time of formation of the solar system there has been low or null increase in the ^7Li content. If this is confirmed, it would dismiss the models of galactic evolution in which a considerable fraction of the present ^7Li comes out from a population of slowly evolving stars, which would contribute with an efficient production of ^7Li from the time of the solar system to today (D'Antona and Matteucci, 1992).

Both Basri *et al.*(1991) and Magazzù *et al.*(1992) find no significant difference in ^7Li abundance among CTTs and WTTs, or within CTTs having very different inferred $\dot{M}$'s. *Some* TTs with very large $\dot{M}$'s have very low ^7Li. While the first point seems to indicate that probably accretion has no strong influence on the atmospheric ^7Li, the second one indicates that accretion is efficient also at stages in which ^7Li is already depleted, *at least in a few stars, and/or for a very short time* –see section 3. The effect of accretion on the final ^7Li abundance should be investigated by numerical models. Let us suppose that a $0.8 M_\odot$ can accrete for 10^6yr at a rate of $\sim 10^{-7} M_\odot$/yr and age log t=7.2. Let us take, as in figure VI, the initial abundance of ^7Li $X_{^7Li} \sim 10^{-8}$ in mass fraction. The same figure shows that, at the given age, the initial value is already reduced to $X_{^7Li} \sim 5 \times 10^{-11}$, and that the convective envelope covers $\sim 0.3 M_\odot$. The star will become a $0.9 M_\odot$ with surface abundance $\sim 2.5 \times 10^{-9}$, while its abundance in the absence of mass accretion would have been $\sim 4 \times 10^{-10}$. This case is probably extreme, but it is clear that *late accretion*, which we have seen has no considerable influence on the pre–MS structural evolution, may influence the final ^7Li abundance at the surface, at least in the sense of providing a *spread* in the resulting abundance in the young MS stars. This possibility would be very unfortunate, as the spread, in this case, can not be attributed uniquely to the advection of other depletion mechanisms, acting, during the MS phase, in different degrees from star to star.

REFERENCES

Adams, F., C., Lada, C. J., Shu, F. H. 1987, *ApJ* **312**, 788.

Alexander, D. R., Augason, G. C., Johnson, H. R. 1989, *ApJ* **345**, 1014.

Barsony, M., Burton, M. G., Russell, A. P. G., Carlstrom, J. E., Garden, R. 1989, *ApJ* **346**, L 93.

Basri, G. Bertout, C. 1989, *ApJ* **341**, 340.

Basri, G., Martin, E. L., Bertout, C. 1991, *A&A* **252**, 625.

Bertout, C. 1989, *ARA&A* **27**, 351.

Bertout, C., Basri, G., Bouvier, J. 1988, *ApJ* **330**, 350.

Bodenheimer, P. 1965, *ApJ***142**, 451.

Bodenheimer, P. 1991, in *Angular Momentum evolution of Young Stars*, Kluwer, eds. S. Catalano and J. R. Stauffer, p. 1.

Boesgaard, A. M., Steigman, G. 1985, *ARA&A* **23**, 319.

Bouvier, J. 1991, in *Angular Momentum evolution of Young Stars*, Kluwer, eds. S. Catalano and J. R. Stauffer, p. 41.

Böhm Vitense, E. 1958, *Zs. für Ap.* **46**, 108.

Bruston, P., Audouze, J., Vidal–Majar, A., Laurent, C. 1981, *ApJ* **243**, 161.

Canuto, V. M., Mazzitelli, I. 1990, *ApJ* **370**, 295 (CM).

Canuto, V. M., Mazzitelli, I. 1992, *ApJ* **389**, 724.

Cox, A. N., Tabor, J. E. 1976, *ApJS* **31**, 271.

D'Antona, F., 1991, in *Evolutionary Phenomena in the Universe*, ed. Compositori, Bologna, SIF Conference Proceedings, **32**, 43.

D'Antona, F., Matteucci, F. 1990 *A&A* **248**, 62.

D'Antona, F., Mazzitelli, I. 1984 *A&A* **138**, 431.

D'Antona, F., Mazzitelli, I. 1985 *ApJ* **296**, 502.

D'Antona, F., Mazzitelli, I. 1992, in preparation.

D'Antona, F., Mazzitelli, I., Gratton, R. 1992, *A&A*, **257**, 539.

Däppen, W. 1992, in these Proceedings.

Deliyannis, C. P., Demarque, P. 1991, *ApJ* **379**, 216.

Deliyannis, C. P., Demarque, P., Kawaler, S. D. 1990, *ApJS* **73**, 21.

Deliyannis, C. P., Pinsonneault, M. H. 1990, *ApJ* **367**, L67.

Demarque, P. 1991, in *Evolution of Stars: The Photospheric Abundance Connection*, eds. G. Michaud and A. Tutukov, Kluwer, Dordrecht, p. 71.

Endal, A. S., Sofia, S. 1981, *ApJ* **243**, 625.

Garcia Lopez, R. J., Spruit, H.C. 1991, *ApJ* **377**, 268.

Grossman, A. S., Graboske, H. C. Jr. 1973, *ApJ* **180**, 195.

Graboske, H. C. Jr., Olnes, R. J., Grossman, A. S. 1975, *ApJ* **199**, 255.

Hartmann, L. W., Jones, B. F., Stauffer, J. R., Kenyon, S. J. 1991, *AJ* **101**, 1050.

Herbig, G. H., Vrba, F. J., Rydgren, A. E. 1986, *AJ* **91**, 575.

Huebner, W. F., Merts, A. ., Magee, N. H., Argo, M. F. 1977, Los Alamos Scientific Laboratory Report, No LA-6760-M.

Kenyon, S. J., Hartmann, L. W. 1987, *ApJ* **323**, 714.

Kenyon, S. J., Hartmann, L. W. 1990, *ApJ* **349**, 197.

Kurucz, R. L. 1991, in "Stellar Atmospheres: Beyond the Classical Models", L. Crivellari, I. Hubeny, D. G. Hummer eds., NATO ASI Series, Kluwer, Dordrecht.

Larson, R. B. 1969, *MNRAS* **145**, 271.

Lynden–Bell, D., Pringle, J. E. 1974, *MNRAS* **168**,603.

Magazzú, A., Rebolo, R., Pavlenko, Ya. V. 1992, *ApJ*, in press.

Magni, G., Mazzitelli, I. 1979, *A&A* **72**, 134.

Martin, E. L., Garcia Lopez, R. J., Rebolo, R., Magazzú, A., Pavlenko, Ya. V. 1992, in these proceedings.

Mathews, G. J., Alcock, C. R., Fuller, G. M. 1990 *ApJ* **349**, 449.

Mazzitelli, I. 1989, in *Low Mass Star Formation and Pre–Main Sequence Objects*, e. B. Reipurth, ESO Conference Proceedings N. 33, 433.

Mihalas, D., Däppen, W., Hummer, D. G. 1988, *ApJ* **331**, 815.

Mercer–Smith, J. A., Cameron, A. G. W., Epstein, R.I. 1984, *ApJ* **279**, 363.

Montmerle, T., André, P. 1989, in *Low Mass Star Formation and Pre–Main Sequence Objects*, B. Reipurth ed., ESO Proceedings N. 33, 407.

Palla, F. 1991, in *IAU Symp. 147: Fragmentation of molecular Clouds and Star Formation*, ed. E. Falgarone, Kluwer.

Palla, F., Parigi, G. 1992, in these Proceedings.

Palla, F., Stahler, S. W. 1990, *ApJ* **360**, L47.

Palla, F., Stahler, S.7 W. 1991, *ApJ* **375**, 288.

Palla, F., Stahler, S. W. 1992, *ApJ* in press.

Pinsonneault, M. H., Kawaler, S., Demarque, 1990, *ApJS* **74**, 501.

Pinsonneault, M. H., Deliyannis, C. P., Hobbs, L. M., Thornburn, J. A. 1991, in *Cool Stars Workshop*, Tucson.

Pinsonneault, M. H., Deliyannis, C. P., Demarque, 1992a, *ApJS* **78**, 179.

Pinsonneault, M. H., Deliyannis, C. P., Demarque, 1992b, in these Proceedings.

Pozio, F. 1990, thesis, University of Rome, unpublished.

Pozio, F. 1991, *Mem. Soc. Astr. It.* **62**, 171.

Pringle, J. E. 1989, in *Low Mass Star Formation and Pre–Main Sequence Objects*, e. B. Reipurth, ESO Conference Proceedings N. 33 p. 89.

Proffitt, C. R., Michaud, G. 1989, *ApJ* **346**, 976.

Proffitt, C. R., Michaud, G., Richer, J. 1990, in "*Cool Stars, Stellar Systems and the Sun*, ed. G. Wallerstein, (ASP Conference Series **9**), p. 35.

Rebolo, R., Martin, E. L., Magazzú, A. *ApJ* **389**, L83.

Ryan, S. G., Norris, J. E., Bessell, M. S., Deliyannis, C.P. 1992, *ApJ* **388**, 184.

Rieke, G. H., Ashok, N. M., Boyle, R. P. 1989, *ApJ* **339**,

Rieke, G. H., Rieke, M. J., Morgan, W. T. 1991, in *Astrophysics with Infrared Arrays*, ed. R. Elston, ASP Conference Series, p. 244.

Rogers, F. J., Iglesias, C. A. 1992, *ApJS* **79** (RI).

Saumon, D. 1990, ph.D. Thesis, Rochester University.

Schatzman, E. 1991, in *Angular Momentum evolution of Young Stars*, Kluwer, eds. S. Catalano and J. R. Stauffer, p. 223.

Stahler, S. W. 1983, *ApJ* **274**, 822.

Stahler, S. W. 1988, *ApJ* **332**, 804.

Stahler, S. W., Shu, F. H., Taam, R. E. 1980, *ApJ* **241**, 637.

Stahler, S. W., Palla, F. 1990, *ApJ* **360**, L47.

Stevenson, D. J. 1991, *ARA&A* **29**, 163.

Strom, K. M., Strom, S. E., Edwards, Cabrit, Skrutskie 1989a, *AJ* **97**, 1451.

Strom, K. M.,, Wilkin, F. P., Strom, S. E., Seamon, R. L. 1989b, *AJ* **98**, 1444.

Swenson, F. J., Stringfellow, G. S., Faulkner, J. 1990, *ApJ* **348**, L33.

York, D. G., Rogerson, J. B. 1976, *ApJ* **203**, 378.

Vandenberg, D. A., Poll, H. E. 1989, *AJ* **98**, 1451.

Vauclair, S. 1988, *ApJ* **335**, 971.

Vidal–Majar, A., Laurent, C., Gry, C., Bruston, P., Ferlet, R., York, D.G. 1983, *A&A* **120**, 58.

Walter, F. M. 1987, *PASP* **99**, 31.

Walter, F. M., Brown, A., Mathieu, R. D., Myers, P. C., Vrba, F. J. 1988, *AJ* **96**, 297.

Zinnecker, H., McCaughrean, M. J., Wilking, B. A. 1992, in *Protostars and Planets III*, ed. E.H. Levy, J. I. Lunine, M.S. Mathews, (University of Arizona Press: Tucson, Arizona), in press.

Inside the Stars, IAU Colloquium 137
ASP Conference Series, Vol. 40, 1993
Werner W. Weiss and Annie Baglin (eds.)

SMALL AND INTERMEDIATE MASS STELLAR EVOLUTION - MAIN SEQUENCE AND CLOSE TO IT

NOELS A. and GREVESSE N.
Institut d'Astrophysique
Université de Liège,
5, avenue de Cointe,
B-4000 LIEGE-BELGIUM

ABSTRACT We present the standard models for small and intermediate main sequence stars and we discuss some of the problems arising with semiconvection and overshooting. The surface abundance of Li serves as a test for other physical mechanisms, including microscopic and turbulent diffusion, rotation and mass loss.

INTRODUCTION

The evolution of stars less massive than about 9 $M_\odot$ has been extensively studied during the last three decades. Recent reviews (Iben, 1991; Chiosi, 1990 and references therein) describe the physics input and the main results and problems in this mass range. Numerous tables and evolutionary tracks have been published since the pioneering works of Henyey, Iben, Kippenhahn, Weigert, Demarque... In the evolutionary tracks of Iben (1964), most of the prominent features near the main sequence, MS, still remain valid. This can be seen in the results of Becker (1981), Vandenberg (1985), Lattanzio (1986), Castellani et al. (1990)... . A complete set of evolutionary tracks has been published by Maeder and Meynet (1989) with enlightning discussions of the problems still present in the confrontation between theoretical predictions and observations.

SEMICONVECTION

The convective core reaches its maximum extent soon after the hydrostatic ajustement to nuclear reactions. The opacity coefficient can be written in a simplified way as

$$\kappa = (1 + X)[0.2 + C\rho^m T^{-n}] \tag{1}$$

where X is the mass abundance of hydrogen and C, a constant. The large sensitivity to temperature of the CNO cycle keeps the density and the temperature nearly constant with time in the nuclear burning region so, as X decreases, the opacity decreases and the convective core must shrink. In all models of mass greater than about 2 $M_\odot$ reaching the maximum extent of their convective core, there is a discontinuity of the hydrogen abundance, smaller in the convective core than in the outer layers, and of the radiative temperature gradient, ∇_{rad}, which is larger than the adiabatic value, ∇_a, on the external side of the convective core limit. Imposing the continuity of T and of ρ/μ at the inner side, i, and at the outer side, o, of the convective core limit, one can write, from equation (1)

$$\frac{\kappa_o - \kappa_i}{\kappa} = \frac{\Delta\kappa}{\kappa} = \frac{\Delta X}{1+X} + m\frac{\kappa_{tot} - \kappa_s}{\kappa_{tot}}\frac{\Delta\rho}{\rho} \tag{2}$$

where κ_s is the contribution due to scattering to the total opacity coefficient, κ_{tot}. This relation becomes

$$\frac{\Delta\kappa}{\kappa} = \Delta X\left[\frac{1}{1+X} - m\frac{\kappa_{tot} - \kappa_s}{\kappa_{tot}}1.25\mu\right] \tag{3}$$

and, taking m equal to 1, a positive value of ΔX implies a positive $\Delta\kappa$ if

$$\frac{\kappa_s}{\kappa_{tot}} > 0.24 \tag{4}$$

which is the case for models with $M \geq 2M_\odot$. This means that the outer side of the convective core limit is unstable to convection, according to the Schwarzschild criterion

$$\nabla_{rad} = \nabla_a. \tag{5}$$

When the criterion for convective instability is written from density requirements, its expression becomes the so-called Ledoux criterion

$$\nabla_{rad} = \nabla_a + \nabla_\mu \tag{6}$$

where ∇_μ is different from zero in a region of varying mean molecular weight. Ledoux (1947) has actually suggested that a μ-discontinuity is likely to be replaced by a partially mixed region. This process is known as semiconvection.

The problem is not too serious in stars of about 3 $M_\odot$ since the radiative gradient becomes smaller with time, the decrease in κ due to the smaller density and that of P/T^4 being more pronounced than the increase in L/m. As the radiation pressure is negligible, the adiabatic temperature gradient remains constant and stability towards convection is quickly restored at the outer edge of the convective core. For more massive stars, the problem becomes more serious, not because of the initial discontinuities in X and ∇_{rad}, which have similar values, but because of the increasing contribution of the radiation pressure to the total pressure. As time goes on, the radiative gradient slowly decreases as in smaller masses but the adiabatic temperature gradient decreases more rapidly. This means that a semiconvective region must take place.

It has been a long standing problem to decide which criterion (5) or (6) had to be fulfilled in the semiconvective region (Simpson, 1971; Ziolkowski, 1972; Barbaro et al., 1972; Varshawsky and Tutukov, 1973;...). Schwarzschild and Härm

(1958) already proposed a partial mixing with the convective core matter which stopped when relation (5) is satisfied while Sakashita and Hayashi (1961) discussed a processes involving relation (6). Although it seemed at first that both processes would lead to similar results for massive stars, $M > 15\ M_{\odot}$ (Stothers and Chin, 1968), Iben (1966 a,b) had drawn the attention on the drastic effects of modifying the μ-distribution during MS on the post MS phases. Chiosi and Summa (1970) found that a partial mixing cannot be maintained all the way to He burning and a fully convective intermediate zone sets in, when Schwarzschild criterion is adopted to mix the semiconvective region. More sophisticated scenarios were proposed by Stothers (1970) while Stothers and Chin (1973) showed the extreme sensititivity of the post MS phases to the μ-profiles. Other sequences of models have been computed by Stothers and Chin (1975), based on the Ledoux criterion and by Stothers and Chin (1976), with the Schwarzschild criterion for masses ranging from 5 to 60 $M_{\odot}$. They found that adopting either the Schwarzschild or the Ledoux criterion, He burning starts on the blue side of the H-R diagram for masses higher than about 17 $M_{\odot}$ while for smaller masses, results are very sensitive to the initial chemical composition, blue loops appearing more easily with the Ledoux criterion. They tried to decide between those criterions from observational tests but these appeared not to be very conclusive with a predicted excess of red supergiants with both criteria. Following a suggestion by Kato (1966), Gabriel and Noels (1976) have shown that g^{+} modes trapped in the varying μ-region are vibrationally unstable and the semiconvective mixing should be performed according to relation (5).

Chiosi and Maeder (1986) proposed a limit of 13 $M_{\odot}$ above which all stars start burning helium on the blue side of the H-R diagram when semiconvection is treated according to relation (5). For lower mass stars, the differences between relations (5) and (6) are not significant in their results.

Interesting discussions can be found in Massevich and Tutukov (1973), Iben (1974), de Loore (1980), Chiosi (1990) and references therein.

OVERSHOOTING

Another difficulty in fixing the limit of the convective region arises from a possible overshooting above or below the layer where relation (5) is satisfied. This comes from the non-vanishing kinetic energy of the convective elements at the boundary of the convective region. Relation (5) implies the annulation of the acceleration but a more correct criterion should be based on the kinetic energy of the elements. The distance to which these elements can move above the classical boundary is not known, but is probably of the order of the pressure scale height (Chiosi, 1986) although a better choice would be a fraction of the convective core radius (Zahn, 1992). Such an overshooting leads to a larger fully mixed convective core which means a longer MS duration and, in the H-R diagram, a wider MS width, especially for massive stars.

Should some overshooting be present, the discontinuities in X and ∇_{rad}

would take place in a region where ∇_{rad} is appreciably smaller than ∇_a. It is reasonable to admit that a sufficient amount of overshooting solves the semiconvection problem or, at least, makes it important for much higher masses (Langer, 1992).

The problem of core overshooting has been very recently extensively studied by Maeder and Meynet (1989) and by Stothers (1991). They take a distance of overshooting, $d = \alpha H_p$, where H_p is the pressure scale height at the convective core boundary and α is a parameter smaller than 1. Several observational tests serve as guides to derive the value of α. The conclusion of Maeder and Meynet (1989) is in favour of $\alpha = 0.25$ while Stothers (1991), regards that value as an upper limit, most of his tests being satisfied with $\alpha = 0$. It is to be expected that if α is found to be very small, semiconvection must be taken into account because of its effects on later stages of evolution. Therefore observational tests of α should be restricted to the vicinity of the MS and those involving evolved stars are likely to bias the derived value. The models used in these analyses were computed with the opacities from the Los Alamos Opacity Library (Huebner et al., 1977; LAOL) in Maeder and Meynet (1989), as in Claret and Giménez (1991a) or the Cox and Stewart (1970) opacities in Stothers (1991). The latter author argues that the best fit to the observations is, preferably to overshooting, an increase in metallicity, up to Z $\sim$ 0.04. This high value of Z is to be interpreted as a need for larger opacities in the envelopes of MS stars.

Such enhancements in the opacities have recently been found by Rogers and Iglesias (1991; OPAL). New grids of models have been computed (Stothers and Chin, 1991, Schaller et al., 1992) with an overshooting parameter chosen to match the observations. One of the best test for overshooting is the MS width (Napiwotzki et al., 1991). In Stothers and Chin (1991), the agreement is excellent with the observations, taking the OPAL opacities and Z between 0.02 and 0.03. There is no special need for overshooting here. Similar results are obtained for O- and B-type stars. From these results, Stothers and Chin favour a solar-like metal abundance and no significant overshooting. The absence of any treatment of the semiconvective region is not important as long as the models are located on the MS as was shown by Stothers and Chin (1968) and Chiosi and Summa (1970). The evolved models, being very sensitive to the μ-profile during the MS, require a semiconvective mixing starting from the zero age main sequence, ZAMS, if no overshooting is present. In Schaller et al. (1992), similar results are obtained with OPAL opacities, in the sense that the ZAMS and the terminal age main sequence, TAMS, are displaced towards the red. However the displacement is greater for the TAMS which means that a smaller distance of overshooting is needed to reproduce the observed width of the MS. Instead of $\alpha = 0.25$ with the LAOL opacities, these authors find a value of about 0.2 and conclude that differences due to opacities are not likely to suppress the overshooting. With these results, Meynet (1992) obtains very nice isochrone fittings with galactic clusters.

Another important test for overshooting is the apsidal motion test (Claret and Giménez, 1991a,b).

Li PROBLEM

We have seen that, in order to explain such an evident feature as the MS width, it was necessary to add some mixing to the convective core which is a possible solution to the problem of semiconvection. Undershooting from the convective envelope or other processes of mixing might be present in MS stars of lower masses. Recently, Alongi et al. (1991) have shown that isochrone fitting in globular clusters as well as in galactic clusters needed core overshooting and envelope undershooting. To test such processes, one of the most powerful tool is the photospheric Li abundance.

Li in Galactic Clusters

It is well known, since 1986 (Boesgaard and Tripicco, 1986a) that a gap exists in the Li abundances of the Hyades cluster in a very narrow effective temperature range, around 6400-6900 K. Figure 2.1, in Michaud and Charbonneau (1991) shows a succession of galactic clusters of increasing ages, from 0.05 Gyr for α Persei to 6.5 Gyr for NGC 188. One can see that the spectacular Li dip found in the Hyades is not unique but seems to be present in clusters older than $\sim$ 0.1 Gyr and absent in younger clusters, except for a slight tendency already showing up in the Pleiades.

Li and Be are light elements easily destroyed by (p,α) reactions at temperatures of 2.5 10^6 K and 3 10^6 K respectively.

In MS stars, a convective envelope is present at low Te, its extension increasing as Te decreases. If the lower limit of this envelope includes such high temperature layers, burning of Li at the bottom of the envelope might be at work. It has however been shown that classical convection only is not sufficient to produce the observed trend in the cool end of the Hyades (Baglin and Morel, 1984). These authors had to introduce overshooting and turbulent diffusion, each process by itself being unable to explain the shape of the abundance curve. Pinsonneault et al. (1992b) have recently shown that, with the new OPAL opacities in standard models, Li burns at the base of the convective zone and is depleted by the observed amounts in stars of masses lower than 0.9 $M_\odot$ and Te smaller than 5300 K.

To explain the Li dip and its wings in the Hyades, many physical processes have been advanced.

1. *Microscopic diffusion* (Michaud, 1986; Vauclair, 1987)

A diffusion equation is added to the problem, with a diffusion coefficient D_{12} given by Paquette et al. (1986). Other expressions of D_{12} can be found in Noerdlinger (1977) and Cox et al. (1989). This process is more efficient when the convective envelope is thin and it can be inhibited by mixing. It can lead to a depletion or to an enhancement of the surface Li abundance according to the relative values of the gravitational acceleration, g, versus the radiative acceleration, g_R, at the base of the convection zone. The diffusion timescale at

6400 K is of the order of 10^9 yr, which can explain the slight depletion in this portion of the Hyades curve (Balachandran, 1990).

2. *Meridional circulation* (Charbonneau and Michaud, 1988)

This calculation originates from a model developed by Tassoul and Tassoul (1982) who showed that meridional circulation can occur in a radiative zone surrounding the μ-gradient region. It is assumed that this circulation penetrates the convective envelope. Its effect is double. First, it may inhibit or reduce the microscopic diffusion if the rotational velocity is greater than a critical value. This can show up in reducing a possible overabundance of Li, if $g_R > g$, or underabundance, if $g > g_R$. The other aspect is an increase of the mixed region whose base can reach sufficiently hot layers where Li is destroyed. This leads to a systematic depletion of the surface abundance of Li.

3. *Turbulent diffusion* (Schatzman, 1969)

A fast rotating star with a convective envelope is likely to be spun down through winds and magnetic fields, the envelope being more affected by the braking than the core. This results in differential rotation which is known to induce turbulent mixing (Zahn, 1974). As it is still not possible to fully model the physics involved, two different approaches have been used. The first one (Schatzman, 1977; Schatzman and Maeder, 1981; Baglin et al., 1985; Vauclair, 1988; Michaud et Charbonneau, 1991) adopts a diffusion equation in which the expression of the turbulent diffusion coefficient, starting from a constant value to more sophisticated expressions introducing space and time dependencies, is parametrized to reproduce the observations. Schatzman and Baglin (1991) have recently revisited this problem and have shown that it is not possible to explain the depletion in late spectral type stars as well as in the Li dip with a process depending only on rotation.

The second approach is based on the pioneering works of Endal and Sofia (1976, 1978), introduced later in the Yale code, YREC, by Sofia et al. (1991). These authors compute rotating models in which the transport of angular momentum and the associated mixing is treated through coupled diffusion equations, with a diffusion coefficient obtained from the physics involved in the different instability mechanisms. They have however to adjust the ratio between the diffusion coefficient of matter and that of the angular momentum, to obtain coherent results for different trace elements and stars of various ages and chemical composition. Using the new OPAL opacities, Pinsonneault et al. (1992b) have found an increase of the convective envelope in standard models, which is compensated by a new lower Li (p,α) reaction rate. Rotating models are able to explain the cool part of the Li curve in the Hyades by adjusting the initial value as well as the diffusion velocity of the angular momentum.

4. *Mass loss* (Schramm et al., 1990)

The extension of the instability strip crosses the ZAMS in the mass range of 1 to 2.5 $M_\odot$. Willson et al. (1987) have suggested that pulsations could lead to mass loss for such stars during their early MS evolution. A small amount of mass loss is able to suppress the overabundance predicted by diffusion processes on the hot part of the Li dip. The work of Schramm et al. (1990) involves models computed with mass loss as the only "non standard" mechanism. It puts tight constraints on the amount of mass loss rates for MS stars around 1.3 $M_\odot$, $7\ 10^{-11} \leq \dot{M} \leq 10^{-10}\ M_\odot/yr$, in agreement with recent observations by Brown et al. (1990). The lower value is necessary to find a Li dip and the upper one, to explain the presence of Be. This light element does not seem to be depleted in stars, located in the Li dip although a slight decrease by a factor 2-4 might be present (Boesgaard and Budge, 1989).

5. *Internal gravity waves* (Garcia Lopez and Spruit, 1991)

Press(1981) suggested that an alternative way for transporting angular momentum was by internal gravity waves generated at the base of the convective zone by the convective cells acting on the radiative layers (Schatzman, 1991). Garcia Lopez and Spruit (1991) could reproduce the Li dip but only at the expense of an artificial increase in the wave flux by a factor of about 15.

Each process by itself is able to reproduce, at least parts of the Li curve in the Hyades although they probably all contribute to the depletion of Li. The reddest portion of the Li curve is mostly due to Li burning in the convective envelope with probably some additional mixing. The shoulder needs Li diffusion while the dip requires either fast diffusion or rotational mixing. The hot shoulder seems to be locked at a stable value of Li. If upward diffusion is acting, rotational mixing or mass loss should be present to inhibit the overabundance or nothing non standard happens in these stars.

Li in F-Stars

It was of course extremely tempting to try to find similar features in field F stars. Extensive work has been done by Balachandran (1990). The Li abundance in terms of rotational velocity, $v \sin i$, for about 200 F stars, shows an extremely large dispersion at low $v \sin i$ and that undepleted Li stars are found for a very wide range of $v \sin i$. In the H-R diagram, for masses between 1.0 and 1.85 $M_\odot$, the decrease in $v \sin i$ at low Te can be explained by the more important slowing down role played by the deepening of the convective envelope, although one finds slow rotators at the hot side of the sample. A clustering of Li depleted stars in the Te range of the Li dip is present, but in the same region, a number of hardly depleted stars is also visible. At the hottest side of the sample, there is no Li depletion, whatever the rotational velocity is. Similar results have been obtained by Boesgaard and Tripicco (1986b, 1987).

In all processes discussed before, only those that can lead to a spread in the Li abundance in the Li dip and to a nearly constant higher value on both wings, the largest ones being found for the hottest stars, can be retained.

Turbulent diffusion, if at work, depends on the initial value of $v \sin i$ and on the mass of the convective envelope. For early F stars, no braking should occur due to their very thin convective envelope. If one adopts a maxwellian initial velocity distribution, the present spread in $v \sin i$ reflects the initial values but whatever $v \sin i$ was and still is, no Li depletion should have occured. This can explain the observed spread in $v \sin i$ and the nearly constant value of Li, close to the meteoritic value. The spread in Li abundance inside the Li dip is explained by the wide range in initial velocities. Small initial $v \sin i$'s give small depletion whereas large inital $v \sin i$'s, because of the strong spun down, lead to high Li depletion. This means that, in the Li dip, a dispersion of the Li abundance essentially reflects a dispersion in the initial $v \sin i$'s although the present $v \sin i$'s are similar.

Li in Halo Stars

For pop II stars, constraints are different :

1. For a given Te, the lower value of Z implies a much smaller convective envelope.

2. The initial Li abundance is not necessarily that observed in undepleted pop I stars. It should be constrained by the standard Big Bang yield but uncertainties remain.

3. In the range 5500 K $<$ Te $<$ 6300 K, Spite and Spite (1982) have shown that the surface Li abundance is nearly constant, within a factor 2 at most. This value is about an order of magnitude smaller than the cosmic abundance. This is true as well for extreme halo stars (Hobbs and Thorburn, 1991). There seems to be no correlation with metallicity, in constrast with Be coming exclusively from spallation reactions and showing an increase with metallicity.

4. The ages are every large, greater than 10 Gyr. Masses involved are small, ranging from 0.6 to 0.8 $M_\odot$, which means that they are still on the MS, burning hydrogen essentially through the p-p chain. This is very important as, contrary to pop I F stars, the evolution proceeds along the MS towards higher Te's instead of leaving the MS towards smaller Te's.

The Yale and the Montreal groups have recently attempted to fit the observations with standard, diffusive and rotational models.

1. *Standard models*

The Spite plateau is quite well fitted by standard models (Demarque et al., 1991; Proffitt and Michaud, 1991a) starting with an initial, i.e. presumably primordial, abundance of Li given by $A_{Li} = 12 + \log(N_{Li}/N_H) = 2.2$ which is the value observed at the hot end of the Spite plateau. The slope at low temperatures may be adjusted by varying the Z value (Deliyannis and Demarque, 1991a).

2. *Diffusive models*

Diffusion in halo stars consists mainly of gravitational settling at the base of the convection zone which becomes very thin as Te reaches 6300 K. This shallow convection zone enhances the effect of diffusion and Li isochrones show a downward curvature at high Te. The main bulk of observations, plateau and decrease at low Te, is still quite well fitted. The initial Li abundance is larger, of the order of 2.4 (Demarque et al., 1991; Proffitt and Michaud, 1991a; Chaboyer et al., 1991). The theoretical downward curvature at high Te is a strong constraint on He diffusion. This He diffusion acts in two ways. Firstly, it reduces the hydrogen content in central regions (Stringfellow et al., 1983) and secondly, it increases the hydrogen content in the envelope thus reducing the effective temperature (Deliyannis et al., 1990). This leads to a reduction in the ages of globular clusters obtained by isochrone fittings by about 10 % (Deliyannis and Demarque, 1991b) to 15 % (Proffitt and Michaud, 1991a).

3. *Rotational models*

In the approach adopted in YREC (Pinsonneault et al., 1992a), the emphasis is put on the interaction between diffusion of angular momentum and mixing. Starting from an initial value of the Li abundance of the order of 3.1, much higher than the so-called primordial value, and equal to the value observed in early F stars, it is possible to get a curve close to the Spite plateau by adjusting some parameters, among which the initial angular momentum and the rate of spun down taken from the observations of T Tauri stars and young clusters. This could possibly account for the slight dispersion among the observations. Such a high value of the primordial Li abundance seems to be in disagreement with other elements used to test the standard Big Bang model.

The other approach of the mixing due to rotation, turbulent diffusion, has been recently applied to halo star models including gravitational settling of He and Li (Proffitt and Michaud, 1991a). The initial Li abundance is taken to be 2.5 or 2.4. The effect of turbulent diffusion is to reduce the extent of the Spite plateau without changing much the maximum at high temperature. Turbulent diffusion acts as a Li depleter at low temperature by increasing Li burning and acts against microscopic diffusion at high temperature. The very existence of the Spite plateau implies a limited amount of turbulent diffusion which is not sufficient to inhibit completely He diffusion. This should lead to a reduction in the ages of globular clusters of the order of 12 % to 15 %, down to 12 Gyr. A value near 15 Gyr would require a slightly larger value of the initial Li abundance.

Li in the Sun

Up to now, we have been dealing with different Li depletion mechanisms which could explain the observations, provided the initial abundance of Li is known. Either in halo stars, where this value should be the primordial Li abundance, or in galactic clusters, where it is taken from supposedly undepleted early F stars, this cho is questionable. The only star for which we have direct indicator of the Li abundance in the original nebula is the Sun, from the meteoritic abundance, $A_{Li,met} = 3.31 \pm 0.04$ (Anders and Grevesse, 1989). As the photospheric Li abundance is $A_{Li,\odot} = 1.16 \pm 0.1$, the depletion factor is of the order of 140. Note that in the Sun, Be is also slightly depleted by a factor of about 2 ($A_{Be,met}$= 1.42, $A_{Be,\odot} = 1.15$).

1. A small amount of Li can be burned during the pre MS phase (Bodenheimer, 1965). Even by adding overshooting, D'Antona and Mazzitelli (1984) failed to destroy much Li. New modelling of overshooting during MS by Skaley and Stix (1991) suggests that with a good choice of parameters, temperatures (2.8 10^6 K) higher than in standard models can be reached at the base of the mixed region.

2. Microscopic diffusion by itself cannot account for such a high Li depletion. Proffitt and Michaud (1991b) have recently shown that some turbulent mixing must be added. Furthermore, gravitational settling reduces the surface He abundance by about 10 %. The change in metallicity is smaller and somewhat more uncertain. Similar decreases for Y and Z at the surface were obtained by Cox et al. (1989).

3. Turbulent diffusion has been modelized by Lebreton and Maeder (1987). Although they succeeded to fit the solar Li abundance, they had problems with the ν flux and the oscillation periods. Baglin et al. (1985) showed that a similar turbulent diffusion coefficient could not fit at the same time the Hyades and the Sun. Thévenin et al. (1987) pointed out that such a fit to the solar value would lead to a Be depletion larger than the observed one. Very recently, Proffitt and Michaud (1991b) fitted the observed Li and Be abundances with models including microscopic and turbulent diffusion.

4. Using evolutionary rotating models, Pinsonneault et al. (1989) succeeded in depleting Li and Be by the proper amounts at the age of the Sun. The method is that of YREC with coupled equations describing the transport of angular momentum and isotopes and a parametrized relation giving the variation with time of the angular momentum. Five free parameters are then adjusted to fit the observations, namely the Li and Be surface abundances, starting with a reference solar model. The difference found in the diffusion coefficients could mean that mixing and transport of angular momentum are not coupled or that some other process accounts for most of the angular momentum transport. The velocity curve and the Li and Be surface abundances as functions of age can be found in Pinsonneault et al. (1989) and Demarque (1991) for different values of the initial angular momentum. The velocity curve agrees with the empiral law of Skumanich (1972). It shows the expected increase of equatorial velocity during pre MS contraction and the subsequent braking by stellar wind and magnetic field. It has to be pointed out that the velocity curve on the MS is independent of the initial conditions, which gives an upper limit to the rotational velocity on the MS. The present solar model has outer layers (r $> 0.6\ R_\odot$) rotating nearly rigidly, together with a rapidly rotating core (r $< 0.2\ R_\odot$).

5. Boothroyd et al. (1991) have recently investigated the effects of a reasonable mass loss, $\dot{M} \sim 10^{-10}\ M_\odot/yr$, in agreement with Brown et al. (1990), on the solar Li depletion, with no other mechanism taken into account. They succeed to explain the observed solar Li and Be depletions with models losing about 10 % of their initial mass during the early MS evolution. This result is nearly independant of the mass loss time scale and leads to a Y value of 0.276 only slightly smaller than their standard value. However, Swenson and Faulkner (1990) showed that mass loss was an unlikely mechanism to explain the Li depletion in the Hyades G dwarfs because these stars should have nearly similar initial masses with different mass loss rates to lead to the observed wide range of Li abundances and effective temperatures.

STANDARD SOLAR MODELS

Since the fundamental review by Bahcall and Ulrich (1988), three recent papers (Sackmann et al., 1990; Ulrich and Cox, 1991; Guenther et al., 1992) discuss the present state of the art in standard solar models and present tables with numerous solar models obtained by different groups. Uncertainties in the basic data, like the age, the Z/X ratio, the luminosity, radius and mass are discussed together with their influence on the results. All models agree to predict a Y value of $\sim 0.28 \pm 0.01$, for a Z value of $\sim 0.019 \pm 0.001$, and a chlorine neutrino flux, F_ν, of the order of 8 SNU. Only non standard solar models (Maeder, 1990) can substantially reduce this high value of F_ν, such as the mixed model recently proposed by Sienkiewicz et al. (1990) and references therein.

A correct calibration of the Sun should be based on the ratio Z/X, which is the only observed constraint on the chemical composition (Bahcall et al., 1982). Most of the standard models however are obtained with a fixed value of Z and values of Y and of the mixing length parameter α adjusted to give the luminosity and radius of the Sun at the solar age. The best value of Z/X is now 0.027 ± 0.003, obtained with the meteoritic abundance of iron.

1. It is now well known that the Y value strongly depends on the interior opacities. Values as low as 0.24 were obtained with the Cox and Stewart (1970) opacities while the present values come from the LAOL opacities (Huebner et al., 1977). The low temperature opacities, $T \leq 10\,000$ K, have a negligible effect on Y; it only changes the mixing length parameter, α, the largest values of α being obtained for the largest low temperature opacities. With the new OPAL opacities (Rogers and Iglesias, 1991), computed for the mixture recommended by Anders and Grevesse (1989), calibration of the Sun leads to a slight increase in Y, of about 0.01 (Guenther et al., 1992; Schaller et al., 1992 ; Morel et al., 1992; Neuforge, 1992).

The major change between OPAL and LAOL opacities is an increase due to metals, essentially iron, around $3\ 10^5$ K (Iglesias et al., 1992). This peak in opacity is located in the adiabatic part of the convective envelope and decreases strongly when the density increases.

Recently however, new photospheric iron abundance determinations (Holweger et al., 1990, 1991; Biémont et al., 1991; Hannaford et al., 1992) have led to an agreement with the meteoritic value. In their new opacity computations, including intermediate coupling, Iglesias et al. (1992) have adopted this new solar iron abundance, lower than the previously used value by about 40 %. The intermediate coupling results are substantially higher around $3\ 10^5$ K than the previous ones, for a given mixture. However, reducing the iron abundance somehow annihilates this increase. Nevertheless, their results clearly show the importance of iron as a substantial contributor to the opacity at high temperatures. At the solar center, $T_6 \sim 15$, $\rho \sim 150$, one finds a reduction in the opacity of about 7 %. The importance of iron had already been emphasized by Courtaud et al. (1990). We have calibrated the Sun, using the OPAL opacities with the meteoritic iron abundance and LS coupling and we find Y = 0.280 and Z = 0.02, which shows the decrease in Y, by about 0.01, for models obtained with similar physics, except for the iron abundance (Morel et al., 1992).

The ratio of the opacity computed with intermediate coupling to the opacity computed with LS coupling for the two values of the iron abundance shows that, except for the peak region, differences are extremely small and as the peak itself has little influence on the solar model, these new OPAL opacities do not affect our Y value given hereabove.

2. The equation of state (EOS) has received much attention these last years in an effort to include as correctly as possible partition functions and interactions between the components of the mixture. The chemical picture has led to the equation of state of Fontaine et al. (1977) and more recently to the so-called MHD equation of state (Hummer and Mihalas, 1988; Mihalas et al., 1988; Däppen et al., 1988) while the physical picture is used by the Livermore group (Rogers, 1986; Iglesias et al., 1987). A discussion of these approaches has been published recently by Däppen et al. (1991). Chabrier (1992) recently compared the various results obtained by Fontaine et al. (1977), Magni and Mazzitelli (1979), Mihalas et al. (1988) and his own computations (Saumon and Chabrier, 1991) for a fluid composed of pure hydrogen. Large differences occur at low temperature and medium to large densities. Under solar conditions, the physics involved in these computations lead to similar results. Differences are more striking in the adiabatic gradient at low temperature. There is a divergence of MHD for log ρ larger than about -2 and a peak in Saumon and Chabrier (1991) near log $\rho \sim 0.25$ due to a thermodynamic inconsistency. Again, in the solar regime, the differences are small.

Lebreton and Däppen (1988) and more recently Guenther et al. (1992) have investigated the effect of adopting a new EOS. With better partition functions and better interactions, they obtain a decrease of Y by about 0.01.

3. Most standard solar models predict an extension of the convective envelope, d_c, given by $d_c/R \sim 0.27$ and a temperature at the base of the order of $T_b \sim 2\ 10^6 K$.

The inversion of helioseismic data (Christensen-Dalsgaard et al., 1991) allows the determination of the sound speed inside the Sun. The transition from adiabatic layers to subadiabatic ones leads to a discontinuity in the derivative of the sound speed. The location of this feature indicates very precisely the lower limit of the mixed adiabatic region, d_a, which is $d_a/R = 0.287 \pm 0.003$. This value is somewhat larger than the predicted value given hereabove. Solar models computed with the OPAL opacities, the MHD equation of state and the high solar abundance of iron (Guenther et al., 1992; Morel et al., 1992) lead to a slightly better agreement with the helioseismic value.

4. All standard models are computed with the mixing length theory, MLT. Attempts have been made to refine the treatment of convection. Some of these have recently been proposed by Pedersen et al. (1990), Forestini (1991), Canuto and Mazzitelli (1991) and Lydon et al. (1992).

CONCLUSIONS

Observational tests seem to reveal some weakness in the physics adopted in the so-called standard models. It is our opinion that in the near future, a new generation of standard models should include many of the physical phenomena discussed here, i.e. semiconvection, overshooting, microscopic and turbulent diffusion, rotation and mass loss.

ACKNOWLEDGEMENTS

It is our pleasure to thank A. Baglin, G. Chabrier, P. Demarque and the Yale group, C. Iglesias, Y. Lebreton, A. Maeder and the Geneva group, G. Michaud, C. Neuforge, F. Rogers and E. Schatzman for sending us papers in advance of publication and for useful discussions.

REFERENCES

Alongi, A., Bertelli, G., Bressan, A., Chiosi, C., 1991, A & A **244**, 95.

Anders, E., Grevesse, N. 1989, Geochim. Cosmochim. Acta **53**, 197.

Baglin, A., Morel, P.J. 1984, in *Observational Tests of Stellar Evolution Theory*, IAU Symp. 105, ed. A. Maeder and A. Renzini, Reidel, Dordrecht, p. 529.

Baglin, A., Morel, P.J. , Schatzman, E. 1985, A & A **149**, 309.

Bahcall, J.N., Ulrich, R.K. 1988, Rev. Mod. Phys. **60**, 297.

Bahcall, J.N., Huebner, W.F., Lubow, W.H., Parker, P.D., Ulrich, R.K. 1982, Rev. Mod. Phys. **54**, 567.

Balachandran, S. 1990, Ap. J. **354**, 310.

Barbaro, G., Chiosi, C., Nobili, L. 1972, A & A **18**, 187.

Becker, S.E. 1981, Ap. J. Suppl. **45**, 475.

Biémont, E., Baudoux, M., Kurucz, R.L., Ansbacher, W., Pinnington, E.H. 1991, A & A **539**, 545.

Bodenheimer, P. 1965, Ap. J. **142**, 451.

Boesgaard, A.M., Budge, K.G. 1989, Ap. J. **336**, 798.

Boesgaard, A.M., Tripicco, M.J. 1986a, Ap. J. **302**, L49.

Boesgaard, A.M., Tripicco, M.J. 1986b, Ap. J. **303**, 724.

Boesgaard, A.M., Tripicco, M.J. 1987, Ap. J. **313**, 389.

Boothroyd, A.I., Sackmann, I.J., Fowler, W.A. 1991, Ap. J. **377**, 318.

Brown, A., Veale, A., Judge, P., Bookbinder, J., Hubeny, I. 1990, in *Cool Stars, Stellar Systems, and the Sun*, ASP Conf. Ser., Vol. 9, ed. G. Wallerstein, San Franciso, Bookerafters, p. 183.

Canuto, V.M., Mazzitelli, I. 1991, Ap. J. **370**, 295.

Castellani, V., Chieffi, A., Straniero, O. 1990, Ap. J. Suppl. **74**, 463.

Chaboyer, B., Deliyannis, C.P., Demarque, P., Pinsonneault, M.H., Sarajedini, A. 1991, Bull. AAS **22**, 1205.

Chabrier, G. 1992, private communication.
Charbonneau, P., Michaud, G. 1988, Ap. J. **334**, 746.
Chiosi, C. 1986, in *Nucleosynthesis and Stellar Evolution*, 16th Saas-Fee Course, ed. B. Hauck et al., Geneva Observatory, p. 199.
Chiosi, C. 1990, PASP **102**, 412.
Chiosi, C., Maeder, A. 1986, Ann. Rev. Astron. Astrophys. **24**, 329.
Chiosi, C., Summa, C. 1970, Astrophys. Space Sci. **8**, 478.
Christensen-Dalsgaard, J., Gough, D.O., Thompson, M.J. 1991, Ap. J. **378**, 413.
Claret, A., Giménez, A. 1991a, A. & A. Suppl. **87**, 507.
Claret, A., Giménez, A. 1991b, A. & A. **244**, 319.
Courtaud, D., Damamme, G., Genot, L, Wuillemin, M., Turck-Chieze, S. 1990, Solar Phys. **128**, 49.
Cox, A.N., Stewart, J.N. 1970, Ap. J. Suppl. **19**, 243.
Cox, A.N., Guzik, J.A., Kidman, R.B. 1989, Ap. J. **342**, 1187.
D'Antona, F., Mazzitelli, I. 1984, A & A. **138**, 431.
Däppen, W., Mihalas, D., Hummer, D.G., Mihalas, B.W. 1988, Ap. J. **332**, 261.
Däppen, W., Keady, J., Rogers, F. 1991, in *The Solar Interior and Atmosphere*,ed. A.N. Cox, W.C. Livingston, M.S. Matthews, The Univ. Arizona Press, Tucson, p. 112.
Deliyannis, C.P., Demarque, P. 1991a, Ap. J., **370**, L89.
Deliyannis, C.P., Demarque, P. 1991b, Ap. J., **379**, 216.
Deliyannis, C.P., Demarque, P., Kawaler, S.D. 1990, Ap. J. Suppl. **73**, 21.
De Loore, C. 1980, Space Sci. Rev. **26**, 113.
Demarque, P. 1991, in *Evolution of Stars : The Photospheric Abundance Convection*, ed. G. Michaud and A. Tutukov, Kluwer, p. 71.
Demarque, P., Deliyannis, C.P., Sarajedini, A. 1991, in *Observational Tests of Inflation*, ed. T. Shanks, NATO Adv. Res. Workshop, Durham, England.
Endal, A.S., Sofia, S. 1976, Ap. J. **210**, 184.
Endal, A.S., Sofia, S. 1978, Ap. J. **220**, 279.
Fontaine, G., Graboske, H.C., Van Horn, H.M. 1977, Ap. J. Suppl. **35**, 29.
Forestini, M. 1991, Ph. D. Thesis, Université Libre de Bruxelles.
Gabriel, M., Noels, A. 1976, A & A **53**, 149.
Garcia Lopez, R.J., Spruit, H.C. 1991, Ap. J. **377**, 268.
Guenther, D.B., Demarque, P., Kim, Y.C., Pinsonneault, M.H. 1992, Ap. J. **387**, 372.
Hannaford, P., Lowe, R.M., Grevesse, N., Noels, A. 1992, A & A (in press).
Hobbs, L.M., Thorburn, J.A. 1991, Ap. J., **375**, 116.
Holweger, H., Heise, C., Kock, M. 1990, A & A **232**, 510.
Holweger, H., Bard, A., Kock, A., Kock, M. 1991, A & A **249**, 545.
Huebner, W.F., Merts, A.L., Magee, N.H. Jr., Argo, M.F. 1977, Astrophysical Opacity Library, Los Alamos Sci. Lab. Rep LA 6760.
Hummer, D.G., Mihalas, D. 1988, Ap. J. **331**, 794.
Iben, I. Jr. 1964, Ap. J. **140**, 1631.

Iben, I. Jr. 1966a, Ap. J. **143**, 505.
Iben, I. Jr. 1966b, Ap. J. **143**, 516.
Iben, I. Jr. 1974, Ann. Rev. Astron. Astrophys. **12**, 215.
Iben, I. Jr. 1991, Ap. J. Suppl. **76**, 55.
Iglesias, C.A., Rogers, F.J., Wilson, B.G. 1987, Ap. J. **322**, L45.
Iglesias, C.A., Rogers, F.J., Wilson, B.G. 1992, preprint.
Kato, S. 1966, Publ. Astron. Soc. Japan **18**, 374.
Langer, N., 1992, this meeting.
Lattanzio, J.C. 1986, Ap. J. **311**, 708.
Lebreton, Y., Däppen, W. 1988, in *Seismology of the Sun and Sun-like Stars*, Tenerife, Spain, ESA-SP286, p. 661.
Lebreton, Y., Maeder, A. 1987, A & A **175**, 99.
Ledoux, P. 1947, Ap. J. **105**, 305.
Lydon, T.J., Fox, P.A., Sofia, S. 1992, Ap. J. (in press).
Maeder, A. 1990, in *Inside the Sun*, ed. G. Berthomieu and M. Cribier, Kluwer, p. 133.
Maeder, A., Meynet, G. 1989, A & A **210**, 155.
Magni, G., Mazzitelli, I. 1979, A & A, **72**, 134.
Massevich, A.G., Tutukov, A.V. 1973, in *Late Stages of Stellar Evolution*, ed. R.J. Tayler, Dordrecht, Reidel, p. 73.
Meynet, G. 1992, private communication.
Michaud, G. 1986, Ap. J. **302**, 650.
Michaud, G., Charbonneau, P. 1991, Space Science Rev. **57**, 1.
Mihalas, D., Däppen, W., Hummer, D.G. 1988, Ap. J. **331**, 815.
Morel, P., Berthomieu, G., Provost, J., Lebreton, Y. 1992, this meeting.
Napiwotzki, R., Schönberner, D., Weidemann, V. 1991, A & A, **243**, L5.
Neuforge, C. 1992, this meeting.
Noerdlinger, P.D. 1977, A & A **57**, 407.
Paquette, C., Pelletier, C., Fontaine, G., Michaud, G. 1986, Ap. J. Suppl. **61**, 177.
Pedersen, B.B., Vandenberg, D.A., Irwin, A.W. 1990, Ap. J. **352**, 279.
Pinsonneault, M.H., Deliyannis, C.P., Demarque, P. 1992a, Ap. J. Suppl. **78**, 179.
Pinsonneault, M.H., Deliyannis, C.P., Hobbs, L.M., Thorburn, J.A. 1992b, in *Cool Stars Workshop*, ed. M. Giampapa.
Pinsonneault, M.H., Kawaler, S.D., Sofia, S., Demarque, P. 1989, Ap. J. **338**, 424.
Press, W.H. 1981, Ap. J. **245**, 286.
Proffitt, C.R., Michaud, G. 1991a, Ap. J. **371**, 584.
Proffitt, C.R., Michaud, G. 1991b, Ap. J. **380**, 238.
Rogers, F.J. 1986, Ap. J. **310**, 723.
Rogers, F.J., Iglesias, C.A. 1991, preprint.
Sackmann, I.J., Boothroyd, A.I., Fowler, W.A. 1990, Ap. J. **360**, 727.
Sakashita, S., Hayashi, C. 1961, Progr. Theoret. Phys. **26**, 942.

Saumon, D., Chabrier, G. 1991, Phys. Res. A **44**, 5122.
Schaller, G., Schaerer, D., Meynet, G., Maeder, A. 1992, A & A Suppl., in press.
Schatzman, E. 1969, A & A **3**, 331.
Schatzman, E. 1977, A & A **56**, 211.
Schatzman, E. 1991, in *Solar Interior and Astmospheres*, ed. A.N. Cox, W.C. Livingston, M.S. Matthews, The University of Arizona Press, Tucson, p. 192.
Schatzman, E., Baglin, A. 1991, A & A **249**, 125.
Schatzman, E., Maeder, A. 1981, A & A **96**, 1.
Schramm, D.N., Steigman, G., Dearborn, D.S.P. 1990, Ap. J. **359**, L 55.
Schwarzschild, M., Härm, R. 1958, Ap. J. **128**, 348.
Sienkievicz, R., Bahcall, J.N., Paczynski, B. 1990, Ap. J. **349**, 641.
Simpson, E. 1971, Ap. J., **165**, 265.
Skaley, D., Stix, M. 1991, A & A **241**, 227.
Skumanich, A. 1972, Ap. J. **171**, 565.
Sofia, S., Pinsonneault, M.H., Deliyannis, C.P. 1991, in *Angular Momentum Evolution of Young Stars*, NATO Adv. Res. Workshop, Noto, Italy.
Spite, F., Spite, M. 1982, A.& A. **115**, 357.
Stothers, R.B. 1970, MNRAS **151**, 65.
Stothers, R.B. 1991, Ap. J. **383**, 820.
Stothers, R.B., Chin, C.-w. 1968, Ap. J. **152**, 225.
Stothers, R.B., Chin, C.-w. 1973, Ap. J. **179**, 555.
Stothers, R.B., Chin, C.-w. 1975, Ap. J. **198**, 407.
Stothers, R.B., Chin, C.-w. 1976, Ap. J. **204**, 472.
Stothers, R.B., Chin, C.-w. 1991, Ap. J. **381**, L67.
Stringfellow, G.S., Bodenheimer, P., Noerdlinger, P.D., Arrigo, R.J. 1983, Ap. J., **264**, 228.
Swenson, F.J., Faulkner, J. 1990, Bull. AAS **21**, 1101.
Tassoul, J.L., Tassoul, M. 1982, Ap. J. Suppl. **49**, 317.
Thévenin, F., Vauclair, S., Vauclair, G. 1986, A & A **166**, 216.
Ulrich, R.K., Cox, A.N. 1991, in *The Solar Interior and Atmosphere*,ed. A.N. Cox, W.C. Livingston, M.S. Matthews, The Univ. Arizona Press, Tucson, p. 162.
Vandenberg, D.A. 1985, Ap. J. Suppl. **58**, 711.
Varshawsky, V.I., Tutukov, A.V. 1973, Nauch. Inform. Acad. Nauk **26**, 35.
Vauclair, S. 1987, in *Atmospheric Diagnostics of Stellar Evolution : Chemical Peculiarity, Mass Loss and Explosion*, IAU Coll. 108, ed. K. Nomoto, Springer, p. 13.
Vauclair, S. 1988, Ap. J. **335**, 971.
Willson, L.A., Bowen, G.H., Struck-Marcell, C. 1987, Comments Astrophys. **12**, 17.
Zahn, J.P. 1974, in *Stellar Instability and Evolution*, ed. P. Ledoux, A. Noels, A.W. Rodgers, Reidel, Dordrecht, p. 185.
Zahn, J.P. 1992, this meeting.
Ziolkowski, J. 1972, Acta Astron. **22**, 327.

Inside the Stars, IAU Colloquium 137
ASP Conference Series, Vol. 40, 1993
Werner W. Weiss and Annie Baglin (eds.)

Evolution of massive stars: main sequence and close to it

Norbert Langer
Universitäts-Sternwarte Göttingen
Geismarlandstrasse 11, D—3400 Göttingen, F.R.G.

Abstract We discuss the status of current models for the early evolutionary stages of stars in the initial mass range 10–40 $M_{\odot}$. Effects of the pre-main sequence evolution, mass loss, internal mixing, and changes in atomic and nuclear data are outlined and confronted with several basic observational facts, which are unexplained by standard models. We conclude that especially internal mixing processes deserve much more attention in future investigations, and we show why convective mixing may be less efficient than generally assumed, but more mixing should be present in the radiative zones of at least a fraction of all massive stars.

1. Introduction: problems with massive stars

The present paper deals with the early evolution of massive stars. As we shall see, a couple of problems occur when one tries to understand and model those objects. In order to limit the discussion, we restrict ourself to stars with initilal masses less than about 40 $M_{\odot}$, or $L \lesssim 10^{5.5} L_{\odot}$. The evolution of more massive stars, which is largely dominated by mass loss, is discussed elsewhere (cf. Langer 1991, Maeder 1991, Woosley et al. 1992).

Since the developement of 'fast' computers and computer codes for calculation of the internal stellar structure and evolution in the 1950ies, very much progress has been achieved also in the understanding of massive stars. However, when inspecting an observed HR diagram for luminous stars (see e.g. Blaha and Humphreys 1989, Fitzpatrick and Garmany 1990) and comparing it with thoeretical evolutionary tracks or synthetic HR diagrams, a couple of problems are obvious (many of them are discussed also by Fitzpatrick and Garmany; cf. also Leitherer 1991):

i) Too few main sequence (MS) O and early B stars are observed; i.e. the number ratio of MS to supergiant stars is much smaller then typical computed MS to post-MS lifetime ratios.
ii) All evolutionary calculations for massive stars predict a gap between thermaly stable (i.e. quasi-stationary) MS- and post-MS positions in the HR diagram. Observations do not show any indication of a post-MS gap.

iii) Masses derived from atmosphere models (spectroscopic masses) or radiation driven wind models (wind masses) for O and early B stars are systematically smaller than corresponding masses derived from theoretical evolutionary tracks (cf. Herrero et al. 1990, 1992, Kudritzki and Hummer 1990, Kudritzki et al. 1992).

iv) Current theoretical models leave the question open, whether red supergiants (RSGs) are in general older (i.e. more evolved) or younger than blue supergiants (BSGs). Two kinds of post-MS tracks exist, one leading to He-ignition in the RSG stage and subsequent blue loops (cf. Langer 1991a), the other leading to He-ignition as BSG and subsequent redwards evolution (cf. Maeder 1990 for Z=0.005).

Thus, a survey of the HR diagram in the considered luminosity range (i.e. $10^4\,L_\odot \lesssim L \lesssim 10^{5.5}\,L_\odot$) from left to right does not leave any region without severe problems.

A second basic way of comparison between theory and observations concerns stellar surface abundances. In standard evolutionary calculations, the only way to alter the surface abundances is due to convective dredge up in the RSG stage. Consequently, the surface abundances should be the initial ones before a RSG phase is reached, and afterwards a moderate increase of the N/C ratio and a slight increase in N/O and helium is obtained (Maeder 1987). This picture is in conflict with the observed diversity of N and C line strengths in otherwise similar O and B stars (OBN/OBC-phenomenon; cf. Walborn 1988), which apparently occurs due to CNO-abundance anomalies (Schönberner et al. 1988). Furthermore, a considerable He-enrichment is found in luminous OB stars (Schönberner et al. 1988, Herrero et al. 1992). A further direct evidence for drastic nitrogen enrichment was found in the wind of the progenitor of SN 1987A (Fransson et al. 1989).

In summary, there is a large number of problems which is left by the current standard evolutionary calculations for massive stars. On the other side, the theoretical modeling of massive stars is not straightforward, since a couple of physical processes, which greatly affect the evolution of observable quantities, are quite uncertain. The most relevant of them are probably:

a) the pre-MS evolution
b) mass loss rates
c) internal mixing processes
d) atomic and nuclear data

In the following sections we will discuss the influence of these processes (in the order a-d) on the uncertainties of theoretical evolution sequences and try to single out those effects which could be relevant to the problems mentioned above.

2. Pre-main sequence evolution

We do know only very little about the pre-MS evolution of massive stars. However, what we know seems enough in order to be anxious about the validity of the initial conditions which are generally used in stellar evolution calculations. Those are: chemical homogeneity with cosmic isotope mixture, and hydrostatic and thermal equilibrium. Presumably, all these assumptions are wrong in the case of massive stars.

There is no equilibrium structure formed at the onset of nuclear burning, since the contraction time scale of a massive protostar is shorter than its accretion time scale (cf. Yorke 1986). This means that core hydrogen burning starts while the star is still accreting. Therefore no optically visible pre-MS phase of massive stars exists (see also Palla and Pargini 1993). The consequences are unclear, since realistic calculations require at least 2-D hydrodynamic, but Appenzeller (1980) has shown in a 1-D calculation, that hydrogen burning may start with a flash, and subsequently thermal and dynamical oscillations can develope.

One possible consequence of such unstable hydrogen ignition is the development of convection far beyond what will be the convective core during stable hydrogen burning. Since the time scale of conversion of ^{12}C into ^{14}N in the CN-cycle is very short (some $10^3\,yr$, cf. Clayton 1968), ^{12}C may be destroyed in a large fraction of the stellar interior. Depending on the duration of the unstable situation, also ^{16}O and He may be affected. A possible connection to the OBN/OBC-problem (cf. Sect. 1) is obvious. Note, however, that this problem still awaits a serious investigation.

Once accretion onto a massive protostar stops, it may still be situated deep inside a molecular cloud, and consequently be invisible in the optical. This idea is supported by radio observations of so called ultracompact HII regions, which are interpreted as due to luminous O stars moving through a molecular cloud (cf. Churchwell 1990). There number in the Galaxy is comparable to the number of O stars, indicating that those may spend a considerable fraction of their MS lifetime within the molecular cloud wherein they have been formed. This may contribute to the problem of missing MS stars mentioned in Sect. 1.

3. Mass loss

Mass loss rates are a major source of uncertainty in evolutionary computations for massive stars (cf. Chiosi and Maeder 1986). However, due to our restriction on stars with $M_{ZAMS} \lesssim 40\,M_\odot$ we leave out the discussion of stars whose evolution is dominated by mass loss. Especially, we do not discuss the Luminous Blue Variables (cf. Davidson et al. 1989) and the Wolf-Rayet stars (cf. Langer 1989ab) here.

For hot and luminous main sequence and supergiant stars, the radiation driven wind theory (Castor et al. 1975, Abbott 1982, Pauldrach et al. 1986, Blomme et al. 1991) yields mass loss rates as function of stellar parameters, including the metallicity Z. Leitherer and Langer (1991) have estimated the critical ZAMS mass $M_c(Z)$ below which MS mass loss has only negligible effects on the evolution (i.e. $\Delta M/M < 0.05$, ΔM being the total MS mass loss), and found $M_c(Z_\odot) \simeq 32\,M_\odot$ and $M_c(Z_\odot/10) \simeq 80\,M_\odot$. Observationally determined mass loss rates appear to be — on average — somewhat larger than the theoretical ones (cf. discussion in Langer 1989c), possibly due to the instationary character of the wind flow (Owocki et al. 1988, Ribicki et al. 1990). However, mass loss in hot MS and supergiant stages is certainly negligible at low metallicity and at most of moderate importance at solar Z for stars of the considered mass range (cf. also Maeder 1990).

Much more problematic are the mass loss rates of red supergiants, since neither precise observational determinations nor quantitative theories exist (cf. Jura 1991). Jura summarizes evidence for massive stars loosing a significant fraction of their total mass in the RSG stage. This is supported by Stencel et al. 1989, who find expanding massive shells around many supergiants. Spectral features in galactic and Magellanic Cloud supergiants seem to indicate also weaker RSG winds at lower metallicity (Jura 1991), but also this statement can not quantified yet.

Thus, RSG mass loss rates in stellar evolution calculations are more or less arbitrary. We do not know whether the most luminous RSGs do evolve into Wolf-Rayet stars (cf. Schaller et al. 1992, Woolsey et al. 1992). The most firm observational constraint comes from a supernova explosion: light curve and spectral evolution of SN 1987A in the LMC proved the existence of a H-rich envelope of $\sim 10\,M_\odot$ in the progenitor star (Hillebrandt and Höflich 1989), which, for an initial mass of $18-20\,M_\odot$, allows for a total pre-SN mass loss of $\Delta M \lesssim 3\,M_\odot$. Assuming a RSG lifetime of $1.5\,10^5\,yr$ (Langer 1991b) yields $\dot{M}_{RSG} \lesssim 2\,10^{-5}\,M_\odot\,yr^{-1}$ at $20\,M_\odot$ in the LMC.

In summary, the RSG phase is the major problem in following the time evolution of the stellar mass for the stars under consideration. Consequently uncertain are the post-RSG stages (which could be located close to the MS in the HR diagram), especially at high metallicity. Besides direct measurements of RSG mass loss rates and progress in RSG wind theories (cf. Gail 1990), the analysis of Type II supernova light curves may yield important clues for this problem in the near future.

4. Mixing!

Possibly, our ignorance about internal mixing processes are the main source of uncertainty in stellar evolution calculations. This statement is ment in the sense that the physics of the mixing processes itself is uncertain, which affects even standard evolutionary tracks.

4.1 'Convective' processes

First of all, there is the old question about which local criterion should be used to decide whether a layer is convectively stable or unstable. We do not want to repeat lengthy discussions here (cf. Langer 1991a, and references quoted there), but briefly summarize our opinion. According to the linear stability analysis the Ledoux criterion is to be used as convection criterion; layers which are stable according to the Ledoux criterion but unstable according to the Schwarzschild criterion are only vibrationally unstable ('semiconvective'), i.e. such layers are mixed only on time scales much longer than the time scale for convection (Kato 1966, Langer et al. 1983, Spruit 1992). Note that Ledoux- and Schwarzschild criterion are only different in regions of varying mean molecular weight. Furthermore, the validity of the consideration of convection as a local process is questionable, and convective motions may penetrate into layers which are stable according to local criteria ('penetrative convection'; cf. Zahn 1991, 1993).

The efficiency of all three 'convective' processes (note that semiconvection realy is a process quite different from convection), i.e. convection, semiconvection, and penetrative convection, is quite uncertain. Consequently, an efficiency parameter is ascribed to each process, designated as α_{ML}, α_{sc}, and α_{pen} in the following. α_{ML} is the usual mixing lenght parameter, and it is only important in the nonadiabatic (outer) parts of convective envelopes of cool stars, since the temperature stratification is just adiabatic otherwise. The semiconvective efficiency parameter α_{sc} describes the speed of semiconvective mixing, and $\alpha_{sc} \rightarrow 0$ is equivalent to adopting the Ledoux criterion as neutrality condition (cf. Stothers and Chin 1975), while $\alpha_{sc} \rightarrow \infty$ corresponds to the Schwarzschild criterion as neutrality condition (Stothers and Chin 1976). α_{pen} is simply a measure of the distance of penetration beyond the locally determined convective boundary in units of the pressure scale height. Note that only α_{sc} and α_{pen} have a direct effect on mixing of chemical species, while α_{ML} affects mixing only indirectly by changing the envelope structure.

It is beyond the scope of the present work to throughly discuss the possible effects of all three processes on the evolution of massive stars. The mixing lenght parameter is certainly most important for the RSGs. Smaller values of α_{ML} yield a cooler Hayashi line and somewhat deeper convective envelopes. Penetrative convection (often designated as 'overshooting') has been widely used in many evolutionary calculations in the last decade (e.g. Doom 1982ab, Bertelli et al. 1985, Stothers and Chin 1985, Maeder and Meynet 1987, Schaller et al. 1992; cf. Chin and Stothers 1991, for a more complete list), and is thus deeply discussed in the literature. Note, however, that mostly penetrative convection has been considered only for the convective core, but is at the same time ignored for other convective zones in the star. Note further that a couple of theories for penetrative convection ('non-local' convection theories) have been developed in recent years, which compute the penetration distance of convection rather than assuming it by the choice of α_{pen}. The price to be paid is usually that a couple of new parameters appear within those theories. This point, as well as laboratory experiments and 2- and 3-D hydrodynamic simulations for turbulent convection are reviewed by Zahn (1991, 1993).

Semiconvection, on the other side, has been rigorously ignored by most stellar model builders during the past, mainly due to two reasons. First, it was shown that semiconvection is much reduced in the case of mass loss (cf. e.g. Chiosi and Maeder 1986), and, second, is further suppressed by penetrative convection from convective cores (Bressan et al. 1981, Chiosi and Maeder 1986). In our opinion, several fundamental points have been missed in such reasoning. Most important, frequently only the limit of short semiconvective mixing times has been considered, i.e. $\alpha_{sc} \rightarrow \infty$, which leads to the Schwarzschild criterion as neutrality condition (see above). In this way, the effect of molecular weight gradients — which are the basic reason for the occurrence of semiconvection instead of plain convection (Kato, 1966) — are ignored, and the only remaining effect of chemical gradients is the composition dependence of the opacity coefficient and thus of the radiative temperature gradient (cf. Chiosi and Summa 1970). Due to that, the discussion of semiconvection in massive stars was limited to the H/He interface: deep inside massive stars, the opacity is dominated by electron scattering, i.e. there is no opacity effect at the He/C interface. However, semiconvection may be much more important at the He/C interface due to the tendency of growing convective cores during helium burning, in contrast to shrinking cores on the main sequence.

Evidence for the importance of semiconvection at the He/C interface in massive stars — and thus for the general tendency of molecular weight gradients to limit convective mixing — comes from the fact that the explosion of SN 1987A as a BSG, which is unexplained in standard computations, can be understood from model sequences which take this effect into account (Woosley et al. 1988, Langer et al. 1989, Langer 1991b). A second independent argument comes from Stothers and Chin (1992), who find — using the recent OPAL opacities (cf. Sect. 5) — that the supergiant distribution of the SMC cluster NGC 330 can only be understood if molecular weight gradients do choke off convection. A third argument comes from the study of Wolf-Rayet stars, which are generally regarded as uncovered He-cores (cf. Langer 1991), and thus provide a direct view 'inside the stars'. The presence of stars of spectral type WN/WC, which probably show simultaneously the products of hydrogen- and helium burning, require a slow mixing process at the He/C interface, which is provided by semiconvection (Langer 1991c). Finally, note that slow mixing in semiconvective zones is also found in laboratory experiments (Spruit 1992).

Consequently, the complete neglecting of molecular weight gradients may be a major shortcoming of most current massive star models. Note that mass loss rates — at least at low metallicity — may be much smaller than widely assumed (cf. Sect. 3), and also the evidence in favour of efficient penetrative convection appears to be largely reduced due to the new generation of opacities (cf. Sect. 5). Consequently, also semiconvection at the H/He interface is of great importance in most massive star models.

A major effect of (slow) semiconvection at the H/He interface is that He-ignition occurs in the RSG phase (cf. Langer 1991a, Stothers and Chin 1992). Consequently, nearly all BSGs should be post-RSGs, i.e. nitrogen enriched. More massive stars develop deeper convective envelopes in the RSG stage, which should lead to stronger nitrogen enhancements for more luminous BSGs (Langer 1991a); this appears to be confirmed in a recent study of galactic B supergiants by Lennon et al. (1992).

4.2 Mixing induced by differential rotation

Several problems listed in Sect. 1 still have no convincing solution due to what is mentioned in the previous sections, e.g. the missing of the post-MS gap in the observed HR diagram, the mass problem (point iii), or considerable nitrogen and helium enrichment of O stars. Rotationally induced mixing may be a way out of many of these problems. Turbulent diffusion due to differential rotation has been proposed to operate in the sun and other low mass stars (cf. e.g. Pinsonneault et al. 1989), but also in more massive stars: Maeder (1982, 1987a) and Langer (1991), using a diffusion formalism proposed by Zahn (1983), found that a considerable nitrogen enhancement in hot and luminous stars may be the result.

In a more recent calculation, also helium is found to be considerably enriched at the surface already on the main sequence, in the case of fast rotation (i.e. efficient turbulent diffusion; also this process is parametrized). Due to the incorporation of semiconvection, it was not found that the star evolved quasi-homogeneous (cf. Fig. 1). The resulting evolutionary track (computed for a $20\,M_\odot$ star with $Z = Z_\odot/4$) is displayed in Fig. 2, together with a track where

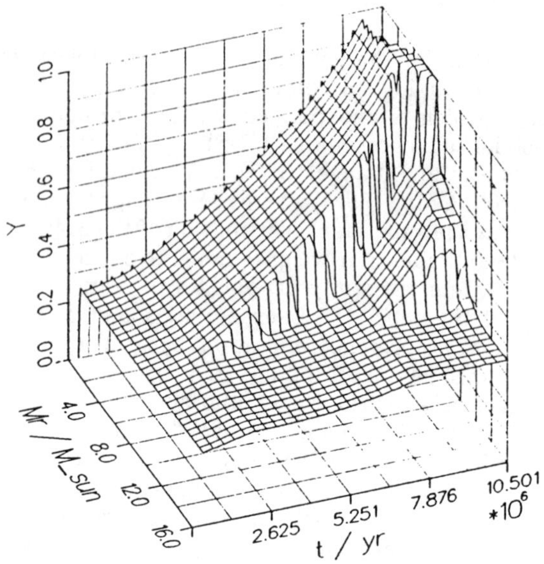

Fig. 1: *Time evolution of the internal helium profile for the* $20\,M_\odot$ *sequence with efficient turbulent diffusion of Fig. 2 during core hydrogen burning. The concective core region, an intermediate zone, and the envelope can be distinguished. Note the drastic He-enrichment in the intermediate zone and considerable He-enrichment in the envelope* ($Y_{surface}$: 0.25 → 0.35).

turbulent diffusion was switched off. It is obvious from Fig. 2, that the assumption of a variety of rotation rates (and consequently of mixing efficiencies) might contribute significantly to the solution of the problems mentioned above: the post-MS gap is closed half way, the average L/M-ratio is significantly raised, and helium and nitrogen are frequently enhanced at the stellar surface.

The physics of turbulent diffusion is far from being fully understood (despite much progress in recent years; cf. Chaboyer and Zahn 1992, Zahn 1992), and the results mentioned above have to be considered as preliminary. However, turbulent diffusion appears to be a promissing way in order to achieve a closer agreement between theory and observations.

5. Atomic and nuclear data

Clearly, uncertainties in atomic and nuclear data affect the reliablility of stellar models. There is not much room to throughly discuss this item, but as far as early evolution of massive stars is concerned, two basic points should be briefly mentioned here.

First of all, the new radiative opacities of the Livermore group ('OPAL'; Rogers and Iglesias 1992) — which are very successful in removing several long-

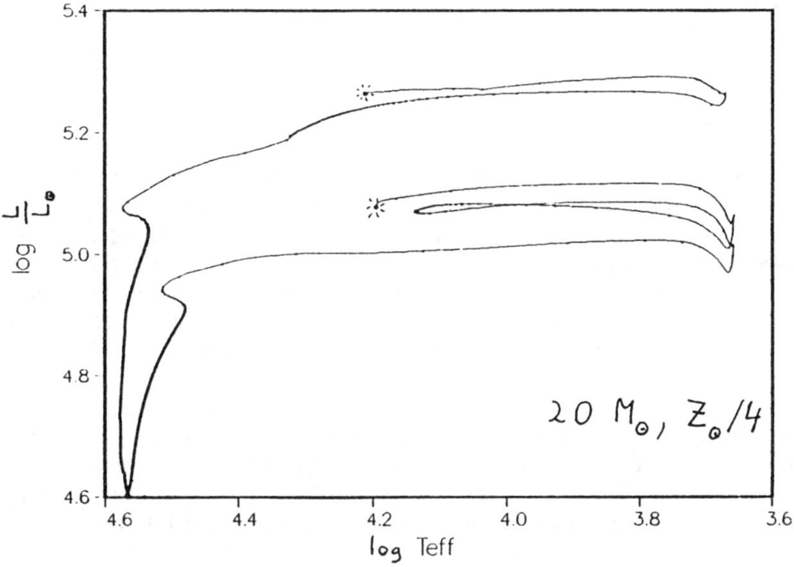

Fig. 2: *Evolutionary tracks for a $20 M_\odot$ star computed with efficient turbulent diffusion (upper track) and without turbulent diffusion (lower track) at a metallicity of $Z_\odot/4$. Semiconvection is incorporated in both sequences according as in Langer (1991b). Los Alamos opacities are used. The time between two tick marks on the tracks is 10^4 yr. Both sequences spend a considerable fraction of core helium burning as BSG and as RSG. The pre-supernova positions are indicated.*

standing discrepancies between stellar evolution theory and observations (as documented in many papers of this volume) — greatly change all massive star models (cf. Stothers and Chin 1991, Schaller et al. 1992). In particular, they weaken or even remove the requirement of penetrative convection in convective cores of massive main sequence stars (Stothers 1991, Stothers and Chin 1991). Note that consequently semiconvection at the H/He interface becomes more important (cf. Sect. 4.1).

The second item, which is very important to the post-MS evolution, but which is often not discussed at all, is the $^{12}C(\alpha,\gamma)^{16}O$ nuclear reaction rate. It determines how much carbon is transformed into oxygen during core helium burning and thus affects fundamentaly the stellar structure in this phase and beyond. A recent discussion of this point is performed by Weaver and Woosley (1992), who find that massive star nucleosynthesis is only compatible with the solar system abundance pattern for a value of the $^{12}C(\alpha,\gamma)^{16}O$ rate of 1.7 ± 0.5 times that of Caughlan and Fowler (1988). We want to add here, that a simultaneous measurement of the abundances of two of the elements helium, carbon, and oxygen in a WC or WO star could pin down the $^{12}C(\alpha,\gamma)^{16}O$ rate with even higher precision.

The recent progress gives the hope that in the near future atomic and nuclear data will no longer contribute significantly to the uncertainty of massive star models in early evolutionary phases.

6. Summary

Current theoretical models of massive stars — though successful in many respects — still contain considerable uncertainties, and still fail in explaining several fundamental observations (cf. Sect. 1). In the previous sections we have discussed our ignorance concerning the pre-MS evolution, mass loss, internal mixing, and microscopic input in massive star models. The first conclusion which immediately emerges is, that we have an enormous need for further theoretical and observational studies of massive stars; several specific items are mentioned in the text.

For the theoretical models we have shown that internal mixing processes deserve much more attention than they have obtained in the past. We have summarized indications in favour of less mixing in convection zones in the presence of molecular weight gradients (cf. Sect. 4.1) and in favour of more mixing in radiative zones, possibly induced by differential rotation (cf. Fig 2).

References

Abbott D.C., 1982, ApJ **259**, 282

Appenzeller I., 1980, in: 10th Saas-Fee Course, Swiss Society of Astronomy and Astrophysics, eds. A Maeder et al.

Bertelli G., Bressan A.G., Chiosi C., 1985 A &A **150**, 33

Blaha C., Humphreys R.M., 1989, A.J. **89**, 1598

Blomme R., Vanbeveren D., Van Rensbergen W., 1991, A &A **241**, 479

Bressan A.G., Bertelli G., Chiosi C., 1981, A &A **102**, 25

Castor J.I., Abbott D.C., Klein R.I., 1975, ApJ **195**, 157

Caughlan G.A., Fowler W.A., 1988, Atomic Data and Nuclear Data Tables **40**, 238

Chaboyer B., Zahn J.-P., 1992, A &A **253**, 173

Chin C.-w, Stothers R.B., 1991, ApJ Suppl. **77**, 299

Chiosi C., Summa C., 1970, Ap. Space Sci. **8**, 478

Chiosi C., Maeder A., 1986, ARA &A **24**, 329

Churchwell E., 1990, A &A Rev. **2**, 79

Clayton D.D.: 1968, *Principles of Stellar Evolution and Nucleosynthesis*, Univ. of Chicago Press

Davidson K., Moffat A.F.J., Lamers H.J.G.L.M. (eds.), 1989, *Physics of Luminous Blue Variables*, Proc. IAU-Colloq. 113, Kluver

Doom C., 1982a, A &A **116**, 303

Doom C., 1982b, A &A **116**, 308

Fitzpatrick E.L., Garmany C.D., 1990, ApJ , **363**, 119

Fransson C., Cassatella A., Gilmozzi R., Panagia N., Wamsteker W., Kirshner R.P., Sonneborn G., 1989, ApJ **336**, 429

Gail H.-P., 1990, Rev. Modern Astron. **3**, 156

Herrero A., Vilchez J.M., Kudritzki R.P., 1989, in: *Properties of Hot Luminous Stars*, Proc. 1st Boulder-Munich Workshop, C. Garmany, ed., A.S.P. Conf. Ser. Vol. 7, p. 50

Herrero A., Kudritzki R.P., Vilches J.M., Kunze D., Butler K., Haser S., 1992, A &A , in press

Hillebrandt W., Höflich P., 1989, Reports on Prog. Phys. **52**, 1421

Jura M, 1991, in: IAU Symp. **143** on *Wolf-Rayet stars and Interrelations with Other Massive Stars in Galaxies*, eds. K.A. van der Hucht, B. Hidayat, Kluver, p.341

Kato S., 1966, P.A.S.J. **18**, 374

Kudritzki R.P., Hummer D.G., 1990 ARA &A **28**, 303

Kudritzki R.P., Hummer D.G., Pauldrach A., Puls J., Najarro F., Imhoff J., 1992, A &A **257**, 655

Langer N., 1989a, A &A **210**, 93

Langer N., 1989b, A &A **220**, 135

Langer N., 1989c, Rev. Modern Astron. **2**, 306

Langer N., 1991, in: IAU Symp. **143** on *Wolf-Rayet stars and Interrelations with Other Massive Stars in Galaxies*, eds. K.A. van der Hucht, B. Hidayat, Kluver, p.431

Langer N., 1991a, A &A , **252**, 669

Langer N., 1991b, A &A **243**, 155

Langer N., 1991c, A &A **248**, 531

Langer N., Sugimoto D., Fricke K.J., 1983, A &A **126**, 207

Langer N., El Eid M.F., Fricke K.J., 1985, A &A **145**, 179

Langer N., El Eid M.F., Baraffe I., 1989, A &A Letter **224**, L17

Leitherer C., 1991, in: IAU Symp. **143** on *Wolf-Rayet stars and Interrelations with Other Massive Stars in Galaxies*, eds. K.A. van der Hucht, B. Hidayat, Kluver, p. 465

Leitherer C., Langer L., 1990, in: IAU-Symp. **148** on *The Magellanic Clouds*, eds. R.F. Haynes, D.K. Milne, Kluwer, p. 480

Lennon D.J., Dufton P.L., Fitzsimmous A., 1992, A &A , submitted

Maeder A., 1982, A &A **105**, 149

Maeder A., 1987, A &A **173**, 247

Maeder A., 1987a, A &A **178**, 159

Maeder A., 1990, A &A Suppl. **84**, 139

Maeder A., 1991, A &A , **242**, 93

Maeder A., Meynet G., 1987, A &A **182**, 243

Owocki S.P., Castor J.I., Ribicki G.B., 1988, ApJ **335**, 914

Palla F., Pargini G., 1993, this volume

Pauldrach A., Puls J., Kudritzki R.P., 1986, A &A **164**, 86

Pinsonneault M.H., Kawaler S.D., Sofia S., Demarque P., 1989, ApJ **338**, 424

Ribicki G.B., Owocki S.P., Castor J.I., 1990, ApJ **349**, 274

Rogers F.J., Iglesias C.A., 1992, ApJ Suppl. **79**, 507

Schaller G., Schaerer D., Meynet G., Maeder A., 1992, A &AS , in press

Schönberner D., Herrero A., Butler K., Becker S., Eber F., Kudritzki R.P., Simon K.P., 1988, A &A **197**, 209

Spruit H.C., 1992, A &A , **253**, 131

Stencel R.E., Pesce J.E., Bauer W.H., 1989, A.J. **97**, 1120

Stothers R.B., 1991, ApJ **383**, 820

Stothers R.B., Chin C.-w., 1975, ApJ **198**, 407

Stothers R.B., Chin C.-w., 1976, ApJ **204**, 472

Stothers R.B., Chin C.-w., 1985, ApJ **292**, 222

Stothers R.B., Chin C.-w., 1991, ApJ **381**, L67

Stothers R.B., Chin C.-w., 1992, ApJ **390**, L33

Walborn N., 1988, in *Atmospheric Diagnostics of Stellar Evolution*, IAU-Colloq. **108**, K. Nomoto, ed., p. 70

Weaver T.A., Woosley S.E., 1992, Physics Reports, submitted

Woosley S.E., Pinto P.A., Weaver T.A., 1988a, Proc. Astron. Soc. Austr. **7**, 355

Woosley S.E., Langer N., Weaver T.A., 1992, ApJ , submitted

Yorke H.W., 1986, ARA &A **24**, 49

Zahn J.P., 1983, in *Astrophysical Processes in Upper Main Sequence Stars*, 13th Saas-Fee Course, B. Hauck, A. Maeder, eds.

Zahn J.P., 1991, A &A **252**, 179

Zahn J.P., 1992, A &A , in press

Zahn J.P., 1993, this volume

Inside the Stars, IAU Colloquium 137
ASP Conference Series, Vol. 40, 1993
Werner W. Weiss and Annie Baglin (eds.)

NEW EVOLUTIONARY TRACKS OF PRE–MAIN-SEQUENCE STARS

Francesco Palla and Steven W. Stahler
Osservatorio di Arcetri, Largo Fermi 5, 50125 Firenze, Italy

Giacomo Parigi
Dipartimento di Astronomia, Università di Firenze, Largo Fermi 5, 50125 Firenze, Italy

ABSTRACT We present the results of new evolutionary calculations of the pre–main-sequence phase of stars in the mass range 1 to 5 $M_\odot$. The novelty of our approach is that, unlike in the standard theory of PMS evolution, the *initial conditions* of the gravitational contraction phase are not arbitrarily set, but derive from protostellar theory.

INTRODUCTION

The theory of star formation provides the appropriate conditions for stars at the beginning of the PMS phase. However, although this statement has been frequently advocated in the literature (e.g. Bisnovatyi-Kogan 1981), standard calculations have so far ignored this critical point and assumed *ad hoc* initial conditions. Since the classical works of Hayashi, Hoshi & Sugimoto (1962), Iben (1965), and Ezer & Cameron (1965), it has been a common procedure to assume that stars of all masses begin the PMS phase as fully convective objects, high up on the Hayashi track. Based on simple energy conservation arguments, Cameron (1962) estimated that $R_*^{init} \sim 50\ R_\odot\ (M_*/M_\odot)$. Thus, convection is due to, and maintained by, the large stellar radius and low effective temperatures (Bodenheimer 1966, cf. also Stahler 1988 for a general discussion).
The recent progress in the understanding of protostellar evolution justifies a critical evaluation of some of the basic assumptions of PMS theory. In fact, detailed calculations of the protostellar accretion phase have shown that not only the *initial radius*, but also the *internal structure* of stars present significant differencies from the standard models. In the following, we present new evolutionary tracks computed using the results of protostellar models. Let us start with a brief summary of the main features of protostellar evolution.

PROTOSTELLAR EVOLUTION

The key properties of protostars are determined by the action of deuterium burning. For low-mass stars ($M_* \lesssim M_\odot$), nuclear burning takes place in the center and keeps the star fully convective (Stahler 1988). For more massive stars, deuterium is burned in a subsurface shell, that can comprise a substantial fraction of the whole mass, while the interior is in a state of radiative equilibrium

(Palla & Stahler 1990). Independently of the details of the accretion flow, the protostellar evolution from low- to intermediate-mass stars is found to proceed through the following series of events: *D-burning in the center; appearance of a radiative barrier; exhaustion of central D; ignition of D-burning in a shell; rapid swelling of the star; final contraction to the main-sequence* (Palla & Stahler 1992a). These processes determine the variation of the protostellar radius with mass. The **mass–radius relation** is the fundamental input provided by star formation theory to PMS evolution. The key result is that during the accretion phase, protostars never achieve radii as large as those predicted by Cameron's estimate (*Hayashi-type* initial conditions), but typically remain a factor of 10 smaller. Also, convection is not due to surface conditions, but rather depends entirely on the nuclear activity of the star. In the assumption that accretion ends rapidly, the mass-radius relation gives a theoretical *birthline*, i.e. the locus in the H-R diagram where stars of different mass should appear when they become optically visible as PMS stars. Thus, the initial conditions are determined self-consistently.

PRE–MAIN-SEQUENCE EVOLUTION

We have followed the PMS evolution of stars of M_*=1, 2, 3.5, 4, 5 $M_\odot$. For each mass, the initial values of the radius and effective temperature are those given at the birthline. In all calculations, the chemical composition is X=0.70, Z=0.02, and the [D/H] ratio is taken to be the interstellar value 2.5×10^{-5} by number. Convection is treated using mixing length theory, with a value of the ratio l/H_p=1.5. Full details of the input physics can be found in Parigi (1992) and Palla & Stahler (1992b, in preparation). The computed set of tracks in the H-R diagram is shown in Figure 1. The main results can be summarized as follows:

M_*=1 $M_\odot$, the star begins the evolution while actively burning deuterium in the center, and *moves up* along the Hayashi track. The main D-burning phase is substantially reduced with respect to previous calculations. The effects of nuclear burning on the pulsational instability of the star have been analyzed by Parigi (1992). The amplification factors of the pulsation have been found to be 4 to 7 times smaller than that computed by Toma (1972); thus, it is unlikely that oscillations have time to develop before D-burning ceases.

$1 \lesssim M_*/M_\odot \lesssim 2$, the Hayashi phase is quite short. Convection is due to D-burning in the center, but the fraction of D available is very small and the duration of the nuclear phase is correspondingly short.

$2 \lesssim M_*/M_\odot \lesssim 4$, stars do not show a Hayashi phase at all. The initial evolution is opposite to that predicted by standard models: *stars begin the PMS phase from low luminosities and join the radiative tracks from below.*

$M_* \gtrsim 4$ $M_\odot$, stars begin the evolution already on the *radiative portion* of the evolutionary track, close to the Main Sequence. The evolutionary lifetimes can be reduced by as much as 40% with respect to previous estimates (cf. Table 1). We do not expect stars more massive than 8 $M_\odot$ to have a PMS phase at all, i.e. such stars are born directly on the ZAMS.

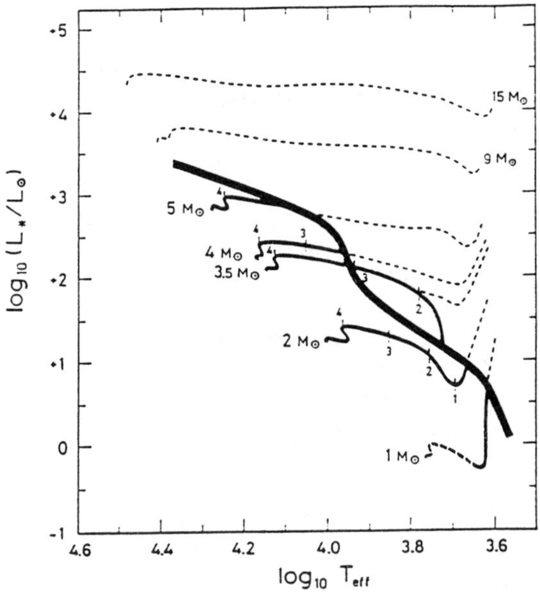

Fig. 1. PMS evolutionary tracks for stars of mass 1 $M_\odot \leq M_* \leq 5\ M_\odot$(*solid lines*). Evolutionary lifetimes are marked on each curve. The *heavy solid line* represents the "birthline". For comparison, the *dashed lines* give the standard tracks as computed by Iben (1965).

Evolutionary lifetimes (in years)

Point	1 $M_\odot$	2 $M_\odot$	3.5 $M_\odot$	4 $M_\odot$
1	1.0×10^7	1.6×10^6	—	—
2	—	3.6×10^6	2.8×10^5	—
3	—	4.6×10^6	5.3×10^5	4.3×10^5
4	—	5.4×10^6	7.1×10^5	5.6×10^5

REFERENCES

Bisnovatyi-Kogan, G.S. 1981, in *Fundamentals Problems in the Theory of Stellar Evolution*, IAU Symp. N. 93, eds. D. Sugimoto et al., D. Reidel Publ. Co., 85.
Bodenheimer, P. 1966, Ap. J., 144, 709.
Cameron, A.G.W. 1962, Icarus, 1, 13.
Ezer, D. & Cameron, A.G.W. 1965, Can. J. Phys., 43, 1497.
Hayashi, C., Hoshi, R. & Sugimoto, D. 1962, Progr. Theor. Phys. Suppl. 22.
Iben, I. 1965, Ap.J., 141, 993.
Palla, F. & Stahler, S.W. 1990, Ap. J., 360, L47.
Palla, F. & Stahler, S.W. 1992a, Ap. J., in press.
Parigi, G. 1992, Theisis, Univerisy of Florence.
Stahler, S.W. 1988, P.A.S.P., 100, 1474.
Toma, E. 1972, A&A, 19, 76.

Inside the Stars, IAU Colloquium 137
ASP Conference Series, Vol. 40, 1993
Werner W. Weiss and Annie Baglin (eds.)

THE PROPERTIES OF ACCRETION DISKS OF HERBIG AE/BE STARS AND THEIR INFLUENCE ON THE SURFACE AND THE INSIDE

P.S. THÉ
Astronomical Institute "Anton Pannekoek", University of Amsterdam, Kruislaan 403, 1098 SJ Amsterdam, The Netherlands

ABSTRACT In this paper we review the observations which lead to the believe that intermediate mass pre-main sequence Herbig Ae/Be stars are surrounded by accretion disks. A discussion is given of their properties and their contributions to surface and inside phenomena of the central stars.

INTRODUCTION

It is now well accepted that Herbig Ae/Be stars (HAEBESs) are very young intermediate mass pre-main sequence objects; age between 10^5 and 10^6 years. Their masses range from 2 to 10 $M_\odot$. Their luminosities are as high as $10^3 L_\odot$, placing them high above the main sequence. It is also well-known that HAEBESs are surrounded by an extended gaseous atmosphere, with dust at the outer regions. Recent observations have shown definitely that this circumstellar atmosphere is confined to an extended disk with a dust ring at the outside. We will give a short review of these observations and will summarize the properties of the accretion disks.

THE ACCRETION DISK

It has been shown by different authors that in the disks of HAEBESs matter is accreting towards the central star. Therefore, it makes sense to name them accretion disks.

Polarization surveys

The results of polarization surveys of matter in the surrounding of HAEBESs by several authors (e.g. Ward-Thompson et al. 1985; Aspin et al. 1985) have been theoretically explained successfully as the effect of multiple scatterting by small dust particles in flattened accretion disks surrounding the objects (Bastien and Ménard 1988).

The blueing effect

The so-called blueing effect, or colour reversal, found for several HAEBESs (Bibo and Thé 1991, and references therein) can be explained by the assumption that it is caused by the dominating blue radiation of the circumstellar disk, when the radiation of the central star is severely dimmed by a dust cloud orbiting in the outer parts of the disk. Such a phenomenon will take place only when we view the disk edge-on or nearly edge-on.

Anti-correlation between brightness variation and polarization changes
The perfect anti-correlation between a deep photometric brightness change and the large increase in linear polarization, can only be explained theoretically by assuming that it is caused by multiple scattering by small dust particles located in the outer regions of the accretion disk, when the stellar light is blocked by a cloud of dust (Grinin. et al. 1991).

Infrared excess
The observed spectral energy distributions of HAEBESs exhibit strong excess emission in the infrared of which the spectral shapes are similar to those found for strong emission-line T Tauri stars. The presene of disks around these T Tauri stars is well established from the success of detailed disk models in reproducing the excess emission (Beckwith et al. 1990), and from the observed forbidden line profiles (Strom et al. 1990). Therefore, the assumption that for HAEBESs the IR-excess is also due to a circumstellar disk is reasonable.

Millimeter continuum observations
Recent mm-continuum observations of several HAEBESs also strengthen the believe that these stars are surrounded by a disk-like extended atmosphere (Strom et al. 1990). From the observed excess mm-continuum fluxes one can derive that the circumstellar envelope should contain masses ranging from about 0.01 to 2 solar masses. If this is distributed spherically symmetric around the central star, it should not be visible at the visual wavelength region. The reason is that then the optical depth will be very large. However, by assuming that the material is confined to a disk, one can obtain the same mm-continuum fluxes, and preserve an optically thin line of sight to the star for almost all viewing angles.

THE INFLUENCE ON THE SURFACE

It has been shown by different authors that in the disks of HAEBESs matter is accreting towards the central star. Therefore, it makes sense to name them accretion disks.

Polarization surveys
The results of polarization surveys of matter in the wide surrounding of HAEBESs by several authors (e.g. Ward-Thompson et al. 1985; Aspin et al. 1985) have been theoretically explained successfully as the effect of multiple scatterting by small dust particles in flattened accretion disks surrounding the objects (Bastien and Ménard 1988).

The blueing effect
The socalled blueing effect, or colour reversal, found for several HAEBESs (Bibo and Thé 1991, and references therein) can be explained by the assumption that

it is caused by the dominating blue radiation of the circumstellar disk, when the radiation of the central star is severely dimmed by a dust cloud orbiting in the outer parts of the disk. Such a phenomenon will take place only when we view the disk edge-on or nearly edge-on.

Anti-correlation between brightness variation and polarization changes

The perfect anti-correlation between a deep photometric brightness change and the large increase in linear polarization, can only be explained theoretically by assuming that it is caused by multiple scattering by small dust particles located in the outer regions of the accretion disk, when the stellar light is blocked by a cloud of dust (Grinin. et al. 1991).

Infrared excess

The observed spectral energy distributions of HAEBESs exhibit strong excess emission in the infrared of which the spectral shapes are similar to those found for strong emission-line T Tauri stars. The presene of disks around these T Tauri stars is well established from the success of detailed disk models in reproducing the excess emission (Beckwith et al. 1990), and from the observed forbidden line profiles (Strom et al. 1990). Therefore, the assumption that for HAEBESs the IR-excess is also due to a circumstellar disk is reasonable.

Millimeter continuum observations

Recent mm-continuum observations of several HAEBESs also strengthen the believe that these stars are surrounded by a disk-like extended atmosphere (Strom et al. 1990). From the observed excess mm-continuum fluxes one can derive that the circumstellar envelope should contain masses ranging from about 0.01 to 2 solar masses. If this is distributed spherically symmetric around the central star, it should not be visible at the visual wavelength region. The reason is that then the optical depth will be very large. However, by assuming that the material is confined to a disk, one can obtain the same mm-continuum fluxes, and preserve an optically thin line of sight to the star for almost all viewing angles.

THE INFLUENCE ON THE SURFACE

From the above observational facts it is clear that HAEBESs are most probably surrounded by an accretion disk. Such a disk surrounding very young still contracting stars, which have not yet ignite thermo-nuclear reactions in their interior, must have much influence on the processes taking place on the surface and in the inside of the central star.

Some properties of accretion disks

From observations by different authors the following data of accretion disks have been collected. Dimension: $10^3 - 10^4$ AU; Mass: 0.5 - 4.0 $M_\odot$; $\dot{M}_{acc}$: 0.2 - 30 $10^{-6} M_\odot/yr$. From these data it is clear that the dimension and the mass of HAEBESs are at least an order of magnitude higher than those of T Tauri stars.

Ly-alpha emission

In a recent paper Blondel et al. (1992) describe their discovery of Ly-alpha emission in the three Herbig Ae stars: HR 5999 (A5,7 IIIe), HD 163296 (A2,3 Ve) and HD 104237 (A4,5 Ve). After making a study of the possible mechanisms which can supply the radiative energy for the emission regions, where these lines are formed, they come to the conclusion that accretion energy from a disk is large enough for this purpose. They conclude that the Ly-alpha emission in the spectra of the Herbig Ae and later type PMS stars is best explained by recombination of infalling and accreting matter.

Photometric "pulse shaped" variability

The variable Herbig Ae star HR 5999 has been observed photometrically during about 8 years under the ESO Long Term Photometry of Variables in the Strömgren photometric system. Recently the large amount of data of this star has been analysed together with those found in the litterature. This analysis reveals the fact that the large photometric variability of this star can be well-represented by a series of bursts or pulses. The mean energy (10^{41} erg) of these bursts are generally too high to be explained by passive circumstellar disturbances. Although, changes in extinction, due to dust clouds, are certainly responsible for many of the irregular variations, there must certainly be energetic phenomena closer to the star, which are really the cause of the pulse-shaped variations in brightness. Such phenomena can be caused by a sudden discrete large accretion from the disk surrounding the star. For such a phenomenon to be possible a large amount of angular momentum should be carried from the internal regions of the disk outwards. The mechanism for such a transport of angular momentum is unknown.

Reversed P Cygni profiles of Hα and MgII resonance lines

Recently a closer study have been made of the UV spectral variability of emission lines in the Herbig Ae star HR 5999, based on recemt IUE high- and low-dispersion observations taken with the long wavelength camera (LWP) and the short wavelength camera (SWP). A dramatic change in the structure of the MgII h and k lines along with some continuum flux excesses, especially at the short end of the SWP camera, have been found. Previous LWP high-dispersion observations of HR 5999 obtained since 1979 and prior to 1990 showed some variability within the same profile structure (a type III P Cygni profile for the MgII lines), however, the last four LWP images taken in September 7, 1990, September 18, 1991 and March 16-18, 1992, present reverse P Cygni profiles in the MgII lines, indicative of some active episodic accretion which can also be detected in the additional red wings of the various FeII and MnII absorption lines.

INFLUENCE ON THE INSIDE; DEUTERIUM SHELL BURNING

Recently Palla and Stahler (1992) finished their study of the influence of the accretion flow of matter in the evolution of intermediate-mass PMS objects. They have computed three evolutionary sequences with accretion rates: 10^{-5}, 3×10^{-5} and 10^{-4} $M_{\odot}/yr$. They also consider the effect of accretion onto the star through a circumstellar disk, by employing photospheric boundary conditions. In all cases, the protostar exhibits the same evolutionary phases: full convection, appearance of a radiative barrier, shell ignition of deuterium, rapid swelling , gravitational contraction, and central hydrogen ignition. It is obvious that accretion of matter through a disk is of the same importance as the accretion directly onto the star. The stellar birthline they can construct from the mass-radius relations shows good agreement with the observed distribution of PMS stars in the H-R diagram.

REFERENCES

Aspin, C., McLean, I.S., Coyne, G.V. 1985, *A&A*, **149**, 158

Bastien, P., Ménard, F. 1988, *ApJ*, **326**, 334

Beckwith, S.V.W., Sargent, A.I., Chini, R.S., Güsten, R. 1990, *AJ*, **99**, 924

Bibo, E.A., Thé, P.S. 1991, *A&AS*, **89**, 319

Blondel, P.F.C., Talavera, A., Tjin A Djie, H.R.E. 1992, *A&A*, submitted

Grinin, V.P., Kiselev, N.N., Minikulov, N.Kh., Chernova, G.P., Voshchinnikov, N.V. 1991, *Ap&SS*, **186**, 283

Herbig, G.H. 1960, *ApJS*, **57**, 587

Palla, F., Stahler, S.W. 1992, *ApJ*, in press

Pérez, M.R., Webb, J.R., Thé, P.S. 1992, *A&A*, **257**, 209

Strom, S., Keene, J., Edwards, S., Hillenbrand, L., Strom, K., Gauvin, L., Condon, G. 1990, in *Proceedings NATO Advance Research Workshop, "Angular Momentum Evolution of Young Stars"*, in press

Ward-Thompson, D., Warren-Smith, R.F., Scarrot, S.M., Wolstencroft, R.D. 1985, *MNRAS*, **215**, 537

Inside the Stars, IAU Colloquium 137
ASP Conference Series, Vol. 40, 1993
Werner W. Weiss and Annie Baglin (eds.)

THE EVOLUTIONARY CODE CESAM: NUMERICAL TECHNIQUES

P. MOREL
Cassini, URA CNRS 1362, Observatoire de la Côte d'Azur, Nice

ABSTRACT

CESAM is a consistent set of programs and routines designed for the calculations of stellar evolution. Untill now it allows the computation of the evolution from PMS or ZAMS to helium flash for stellar masses of some solar mass. It is constructed in such a way that all the physics works as external routines. The numerical techniques are based on the B-spline formalism. This formalism used both for the integration of the differential equations and for 1D and 2D interpolation schemes of various tables of physical data.

I - OVERALL STRUCTURE

CESAM has been implemented by many contributors, get together in the CNRS GDR 131. The set of partial differential equations of stellar evolution is split as usual into two ordinary differential problems: the quasi-static model, a two-point boundary values problem, and an initial value problem, the evolution of the chemical composition in time. At each time step they are solved in turn until convergence.

CESAM treats the core and the envelope as a whole system, thus avoiding the usual matching, at some level, of a core and an envelope solutions obtained with different algorithms. The outer boundary conditions of the internal structure are implemented in the $P(R,L)$, $T(R,L)$, $M(R,L)$ form, for respectively the pressure, the temperature and the mass, R and L being the external radius and luminosity of the envelope; these conditions are applied deep enough in the atmosphere so that the medium is already optically thick.

II - STATIC MODEL: B-SPLINE/COLLOCATION TECHNIQUE

The equations of the two points value problem are solved using the spline/collocation technique. In this integration method, originated by De Boor (1978), the coordinates of each unknown f_j, $1 \leq j \leq n_e$, of a sytem of n_e ordinary differential equations of order 1 are projected as $f_j(x) = \sum_{i=1}^{M} f_{i,j} N_i^m(x)$ on the B-spline basis $\{N_i^m\}_{i=1}^{M}$ of order m, on a nodal vector – to be precised – constructed on the mesh points $x_1, x_2, \ldots, x_n$. Then the derivatives become $f_j'(x) = \sum_{i=1}^{M} f_{i,j} \frac{dN_i^m(x)}{dx}$ and one has similar equations for the boundary conditions. The new unknowns, the reals $f_{i,j}$, are derived by writting that the equations are satisfied at an *ad hoc* number of so-called collocation points which are not necessarily the mesh points, as in finite difference methods. The basis beeing local, one has to deal with a block diagonal jacobian matrix for solving, with the Newton-Raphson method, the implicit set of stellar structure differential equations.

Moreover, as in Gaussian integration, with a judicious choice for the nodal vector and for the collocation points there is "superconvergency" at the mesh points where the order of precision of the solution is higher than the order of the B-splines. An another advantage of the method is the ability of retrieving exactly the numerical solution at any point simply by using the projections on the B–splines. That proves very convenient in the calculation of acoustic eigenmodes.

With the B-spline formalism it is possible to deal with discontinuities without too much difficulties, that simplify the calculations with convective mixing.

During the time evolution, according to the changes of the dependent variables, an automatic mesh refinement scheme optimizes the grid by insuring constant jumps from a mesh point to the next of a so-called "mesh distribution function" (Press *et al.*, 1986).

Following Eggleton(1971) the set of dependent variables: $\ln P$, $\ln T$, R^2, $L^{2/3}$, with respect to $M^{2/3}$ avoids the central singularity of the equations of stellar structure and it can be shown (Morel, 1992) that it also provides an optimal numerical accuracy. On the other hand, the 2/3 power of the luminosity – which avoids a singularity of the derivative of a square root on the vicinity of the origin – is not adapted to negative values of L. With CESAM the less accurate set of dependent variables: $\ln P$, $\ln T$, M, L with respect to R can be also used; there is no central singularity but an external free limit which is managed without difficulty by the automatic mesh point allocation scheme.

For the core and the envelope (calculated as a whole) a fixed number $n = 300$ of grid points is usually used with B-splines of order $m = 2$ that corresponds to $599 = 300 \times 2 - 1$ collocation points, hence the solution is of fourth-order accuracy for the quasi-static model. For the Sun, calculations have shown that higher number of mesh points or higher order for the B-spline does not lead to significant improvement of the internal accuracy (see Table I).

Many performing algorithms used in CESAM for B-spline calculations are taken from Schumaker(1981).

The three functions $P(R, L)$, $T(R, L)$, $M(R, L)$ which serve as outer boundary conditions are deduced from an atmosphere model, together with the derivatives which are needed for the iterative Newton-Raphson resolution of the implicit equations. This atmosphere model is integrated separately from the interior stratification, again with the B–spline/collocation procedure for arbitrary given $T(\tau)$ laws.

III – EVOLUTION IN TIME OF THE CHEMICAL COMPOSITION

The stiffly stable implicit BDF scheme (Hairer and Wanner 1991) is employed for the numerical integration of the set of differential equations governing the time evolution of the chemical species. The implicit equations are solved with the iterative Newton-Raphson algorithm. The choice of this scheme is guided by its properties of stiff-stability, and also because it is strictly conservative for all linear sets of the abundances. To ensure sufficient precision, one limits the changes of the chemical species during a time step. Only formula with accuracy order up to 2 are used. Due to the presence of the convective mixing the higher order schemes are almost useless.

In convective zones one assumes that the time scale of the turbulent eddies is less than the nuclear time scale, hence the time integration is performed on the mixed abundances time derivatives. A discontinuity is set on chemical composition distribution at each limit of growing convective zones. The B-spline formalism is also used for the interpolation of the chemical composition.

IV – INTERPOLATION TECHNIQUES AND ACCURACY

Without a significant loss of accuracy, tabulations and interpolations of physical data avoid important amount of calculations and allow an improvement of the numerical stability. Interpolation programs and routines with Lagrange, B-spline and Hermite schemes in 1D and 2D have been developped consistently with the numerical techniques already outlined. CESAM uses a 2D tensorial B-spline interpolation either in $(\ln T, \ln \rho)$ (LAOL tables) or in $(T_6, \ln \rho/T_6)$ (OPAL tables) and a 1D Lagrange scheme in chemical composition for the opacities, a 2D B-spline or Hermite scheme with respect to $(\ln P, \ln T)$ and a 1D five order polynomial fit at Chebychev points with respect to the chemical composition for the equation of state, a 1D B-spline piecewise polynomial scheme for the thermonuclear reactions rates. One advantage of interpolation with the B-spline formalism is the control of the order of accuracy by only one parameter but convenient dual basis are not available as it does for Hermite and Lagrange interpolations.

TABLE I

n	m	X	α	ρ_c(g cm^{-3})	T_c(10^6K)	X_c
300	2	0.70617	1.8982	152.676	15.5492	0.34233
		$\Delta X/X$	$\Delta\alpha/\alpha$	$\Delta\rho_c/\rho_c$	$\Delta T_c/T_c$	ΔX_c / X_c
150	3	1. 10^{-6}	7.7 10^{-5}	7.9 10^{-6}	-3.2 10^{-5}	1.3 10^{-4}
300	3	1. 10^{-6}	7.7 10^{-5}	6.7 10^{-6}	-1. 10^{-6}	6. 10^{-6}
600	3	2.4 10^{-5}	-3.2 10^{-6}	0.4 10^{-6}	2.3 10^{-5}	2. 10^{-4}

The accuracy of CESAM have been only investigated for calibrated Solar models. Internal accuracy is estimated from differences between models calculated with different number of shells n and with different orders of B-spline m. A comparison of these models is given in Table I where the first line gives the initial hydrogen abundance, the mixing length, the central density, temperature and hydrogen abundance for calibrated models with $n = 300$ and $m = 2$. The other lines give the relative difference of the same quantities for the other models. Solar models computed by CESAM with simplified but fixed physics have been submitted to the SMC comparison organized by GONG, with respect to the other models submitted they present relative differences smaller than 10^{-4} (Christensen Dalsgaard, 1991; Gabriel, 1991).

A more detailed discussion of the accuracy for more realistic solar models is given in Morel *et al.* (1990, 1992).

ACKNOWLEDGMENTS

I am indebted to my collegues Drs. G. Berthomieu and J. Provost for stimulating discussions and suggesting improvements in the manuscript.

REFERENCES

Christensen-Dalsgaard J. 1991, *Challenges to Theories of the Structure of Mode rate-- Mass Stars*, D.O. Gough, J. Toomre (Eds), Springer Verlag, p 11-36

De Boor C. 1978, *A Practical Guide to Splines* (Springer; third ed. 1985)

Eggleton, P. 1971 *Mon. Not. R. astr. Soc.* **151** 351-364

Gabriel M. 1991, *Challenges to Theories of the Structure of Mode rate-Mass Stars*, D.O. Gough, J. Toomre (Eds), Springer Verlag, p 51-55

Hairer E., Wanner G. 1991, *Solving ordinary differential equations II.* Springler Verlag

Morel P., Provost J., Berthomieu G. 1990, *Solar Physics* **128**, 7

Morel P.,Berthomieu G., Provost J.,Lebreton Y. 1992, these proceedings

Morel P., 1992 in preparation

Press W.H., Flannery B.P., Teukolsky S.A., Vetterling W.T. 1986 *Numerical Receipes*, Cambridge University Press, Cambridge

Schumaker L. 1981, *Splines Functions : Basic Theory*, John Whiley.

Inside the Stars, IAU Colloquium 137
ASP Conference Series, Vol. 40, 1993
Werner W. Weiss and Annie Baglin (eds.)

NEW PHYSICS, NEW EVOLUTIONARY TRACKS, NEW ISOCHRONES

P. DEMARQUE, D.B. GUENTHER and M.H. PINSONNEAULT
Center for Solar and Space Research and Department of Astonomy, Yale University, P.O. Box 6666, New Haven, CT 06511, USA

INTRODUCTION

Work has begun on the construction of a grid of theoretical isochrones based on important recent refinements in stellar physics. These refinements have already been applied to the construction of a standard solar model, which, for the first time, free of any *adhoc* assumptions, reproduces the observed p-mode oscillation spectrum within the errors of the physics (Guenther *et al.* 1992).

SCIENTIFIC GOALS

(1) To test the theory of *STELLAR STRUCTURE* and *EVOLUTION* using state-of-the art physics
(2) To provide a framework for detailed study of *INTERNAL DYNAMICS* (e.g. rotation, convection), and other *TRANSPORT MECHANISMS* (e.g. chemical diffusion, angular momentum transfer) in stars
(3) To help understand star cluster cmd's in the Galaxy and nearby systems by interpreting the *FOSSIL RECORD* left by *STAR CLUSTERS* and related systems
(3) To upgrade the building blocks for *POPULATION SYNTHESIS* for the study of *DISTANT GALAXIES* and *GALAXY EVOLUTION*.

NEW PHYSICS

(1) Solar abundance mixture from Anders and Grevesse (1989).
(2) OPAL opacities from Livermore (Iglesias and Rogers 1991).
(3) Nuclear energy generation rates from Bahcall and Pinsonneault (1992)
(4) Inclusion of Debye-Hückel corrections in the equation of state at high temperatures.
(5) In the case of stars with a convective core near the main-sequence, the effects of modest core overshooting (in the range 0.1 to $0.2H_p$) at the convective boundary are studied.
(6) In the case of stars with convective envelopes a solar calibration of the mixing length is adopted. Effects of small variations of the mixing length are explored.
(7) The Kurucz (1992) model atmospheres are used as surface boundary conditions.
(8) The color calibration grids of Green (1988) and Kurucz (1991) will be used. Improved color grids will be easy to implement.

GRID PARAMETERS

Metallicity
(1) Z= 0.0001 to 0.1
(2) For low Z (Z less than about 0.007), α-capture element abundances enhanced by 0.4 dex
Helium
(1) For Z < 0.01: Y = 0.20, 0.23 and 0.25
(2) For Z = 0.01 to 0.02: Y = 0.25, 0.30 and 0.35
(3) For Z = 0.04 to 0.10: ΔY/ΔZ = 2, 3 and 4
Initial mass
About 0.5M⊙ to 10M⊙, depending on chemical composition

EXAMPLE

The poster displayed comparisons of isochrone fits to the old open clusters M67 and NGC 188. Fig. 1 and 2 illustrate only the most important results. For the purpose of comparison, both fits use the same distance modulus (m-Mv)=9.70 and interstellar reddening E(B-V)=0.07 for M67. The solar Y and α for each opacity was also adopted, reasonable assumptions since the metallicity of M67 is close to solar. At this point, the main uncertainties (other than in the photometry) relate to the problem of the color calibration (we used here the Green et al. (1987) table), and to the treatment of convection. For recent convection improvements (which do not require the free parameter α), see D'Antona et al. (1992) and Lydon et al. (1992). Under these assumptions the derived age for M67 remains within the range $4.0^{+1}_{-0.5}$ Gyr. The OPAL age may be lower by about 0.5 Gyr.

ACKNOWLEDGEMENT

We acknowledge support from NASA grants NAG5-1486 and NAGW-2531.

REFERENCES

Anders, A. and Grevesse, N. 1989, Geoch. & Cosmoch. Acta, 53, 197.
Bahcall, J.N. and Pinsonneault, M.H. 1992 Rev. Mod. Phys. in press.
Cox, A.N. and Stewart, J.N. 1970, Ap.J.Sup., 19, 261 (CS).
D'Antona, F., Mazzitelli, I. and Gratton, R.G. 1992, A&A, in press.
Demarque, P.,Guenther, D.B. and Green, E.M. 1992, A.J., 103, 151.
Green, E.M.,Demarque, P. and King, C.R. 1987, Revised Yale Isochrones
Grevesse, N. 1984, Phys. Scripta, T8, 49.
Guenther, D.B., Demarque, P., Kim, Y.-C. and Pinsonneault, M.H. 1992, Ap.J., 387, 372.
Huebner, W.F.,Merts, A.L.,Magee, N.H. and Argo, M.F., 1977, LASL Report No. LA-6760-M (LAOL).
Iglesias, C.A. and Rogers, F.J. 1991, Ap.J., 371, 408 (OPAL).
Kurucz, R.L.1992, personal communications.
Lydon, T.J., Fox, P.A. and Sofia, S. 1992, Ap.J.,in press.

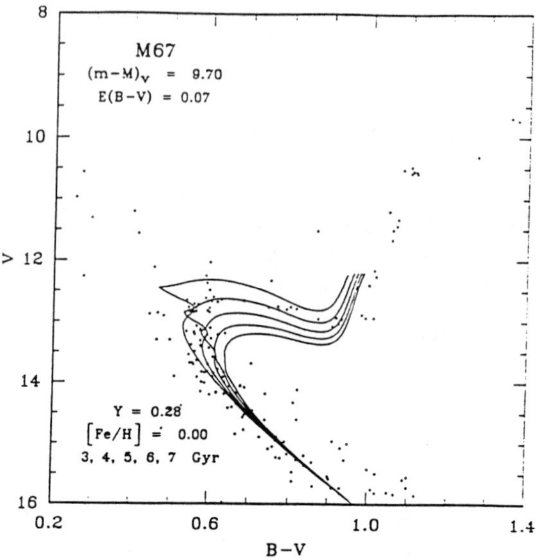

Fig.1 M67 fit. The isochrones include core overshoot (overmixing) by 0.1Hp. With the CS and LAOL opacities, convective overshoot at the core edge must be included to reproduce the observed gap in the cmd near the main sequence. The turnoff fit is good, but the giant branch does not match observations using the solar calibration for alpha.

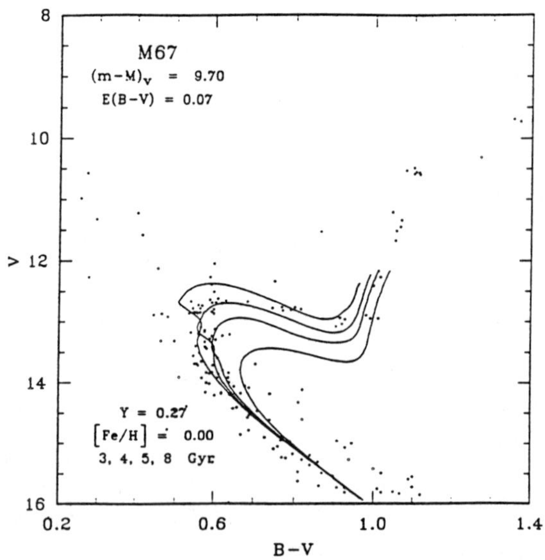

Fig. 2 Isochrones including the improvements in the physics described in the text. No convective overshoot was included. The fit around the gap suggests that the amount of convective overshoot needed to fit the gap is smaller when the improved physics is adopted. Note also that when using the OPAL opacities and the Kurucz atmospheres, the position of the theoretical giant branch is much more faithful to the observations. This illustrates the important role of low temperature opacities, in addition to the treatment of convection, in calculating reliable radii for red giant stars.

Inside the Stars, IAU Colloquium 137
ASP Conference Series, Vol. 40, 1993
Werner W. Weiss and Annie Baglin (eds.)

COULOMB CORRECTIONS AND GLOBULAR CLUSTER STELLAR EVOLUTION

CHARLES R. PROFFITT
Space Telescope Science Institute, 3700 San Martin Dr., Baltimore, MD 21218

ABSTRACT The effects of Coulomb corrections on the evolution of globular clusters stars are discussed. Coulomb corrections alter the equation of state by about 1% in most of the stellar interior, and for stars of fixed initial parameters, this results in an 8% increase in the ZAMS luminosity and an 8% decrease in the age at the main sequence turnoff. Ages for globular clusters measured by comparing to the turnoff luminosity of theoretical isochrones are lowered by $\approx 4\%$ when Coulomb effects are included.

Keywords: stellar interiors; globular clusters; equation of state

COULOMB CORRECTIONS AND THE EQUATION OF STATE

We assume the expression given by Graboske, Harwood and Rogers (1969) for the Coulomb correction term to the free energy. Corrections to other thermodynamic quantities can be derived from this expression, and explicit formulae for many of these are given by Mihalas et al. (1988), and Däppen et al. (1988). We are most interested in the pressure correction:

$$\delta P_C = -\frac{\partial F_C}{\partial V} = -\frac{2\pi^{1/2}e^3}{3(kT)^{1/2}}\left(\sum_i n_i Z_i^2 \theta_i\right)^{3/2} \tau(x) \times \left[\frac{1}{2} + \frac{3}{2}\frac{n_e\theta_e'/\theta}{\sum_i n_i Z_i^2\theta_i} + x\frac{\tau'}{\tau}\left(\frac{3}{2} - \frac{3F_{1/2}(\eta)}{2\theta_e F_{3/2}(\eta)} + \frac{n_e\theta_e'/\theta}{2\sum_i n_i Z_i^2\theta_i}\right)\right]. \quad (1)$$

The summations are over all charged particles, with $\theta_i = 1$ for ions, while for electrons $\theta_e = F_{-1/2}(\eta)/2F_{1/2}(\eta)$. F_k is the Fermi-Dirac integral of order k, and η is the degeneracy parameter. When nondegenerate, $\theta_e \rightarrow 1$, while as the electrons become degenerate $\theta_e \rightarrow 0$. The factor $\tau(x)$ is a correction for the finite size of the ions (usually near 1). Also $\theta_e' = d\theta_e/d\eta$, and $\tau' = d\tau/dx$.

For a fully ionized mixture of H and He the Coulomb correction can be written in terms of the electron density using

$$\sum_i n_i Z_i^2 \theta_i \approx n_e\left(\frac{2}{1+X} + \theta_e\right). \quad (2)$$

This formulation is especially convenient when adding Coulomb corrections to the Eggleton et al. (1973) (EFF) eos (which uses T and a function of the degeneracy parameter as the basic thermodynamic variables), as n_e can be directly written in terms of the Fermi-Dirac integrals. Thermodynamic quantities then are easily calculated. When $X = 1$, equation (2) is correct even for partially ionized plasmas. When $X = 0.765$, equation (2) overestimates the Coulomb correction by 10% when all the electrons come from singly ionized particles. Thus, when there is no He^{++} we apply 110% of the correction implied by equation (1), but this affects only a small part of the total stellar mass.

In Däppen et al. (1990), results calculated with the Mihalas et al. (MHD) and Livermore eos were compared for $Z = 0$ with X, Y, ρ and T similar to the solar center. For these parameters we match the MHD P to better than 0.1%, while the Livermore eos finds P to be about 0.15% smaller than MHD.

Our results do *not* agree with the eos results of Straniero (1988). We find consistently smaller pressures at a given ρ and T, (apparently significantly larger Coulomb corrections). Chieffi and Straniero (1989) found that Coulomb corrections had negligible effects on the time scale for main sequence evolution, but we find a substantially larger effect.

EFFECT OF COULOMB CORRECTIONS ON STELLAR MODELS

We find, for models of typical Pop. II main sequence stars, the Coulomb correction to the equation of state to be about -1% throughout most of the interior. It is significantly larger in a thin region near the surface.

The pressure at any point in a hydrostatic stellar interior balances the weight of the overlying material. When Coulomb corrections are included a higher density and/or temperature is needed to give the same pressure. To provide this extra pressure the central regions need to be slightly more compact. As a result, for models with given initial parameters, including the Coulomb correction results in slightly *increased* interior temperatures and densities, as well as increased nuclear reaction rates and luminosity. Near the ZAMS, a $0.754 M_\odot$ Pop. II model with Coulomb corrections is 8% more luminous, the central T is increased by 1%, the central ρ by 0.4%, and T_{eff} by 78 K. *The turnoff point of the track occurs 8% earlier in time at essentially the same luminosity.* The giant branch tracks are almost exactly coincident in the HR diagram.

Different results are found for solar models. To have the same luminosity at 4.5 Gyr, models with the Coulomb correction need an initial He abundance lower by about 0.01. The central density and temperature are then *lower* than in models without the Coulomb correction (see also Guenther et al. 1992).

EFFECTS ON ISOCHRONES

We will compare to the non-diffusion isochrones of Proffitt and VandenBerg (1991) (PV), which assume a scaled solar T-τ relation with $\alpha = 1.9$. If we require that the isochrones computed here be coincident with those of PV on the lower main sequence near 5100 K (effectively normalizing the T_{eff} scale using the local subdwarf calibration), then $\alpha = 1.6$ instead of 1.9 is needed.

Figure 1 compares 13 and 14 Gyr isochrones computed with Coulomb corrections and $\alpha = 1.6$ to isochrones from PV. *The age for a given turnoff luminosity is found to be ≈ 0.6 Gyr smaller when isochrones including Coulomb corrections are used.* Observationally, it is difficult to define the luminosity of the turnoff to better than 0.1 magnitudes. VandenBerg et al. (1990), suggest choosing the luminosity of a point on the main sequence below the turnoff and 0.05 mag redder in $B-V$ (≈ -0.02 in $\Delta \log T$). This approach gives ages $\approx$ 0.6–0.8 Gyr younger when Coulomb corrections are included.

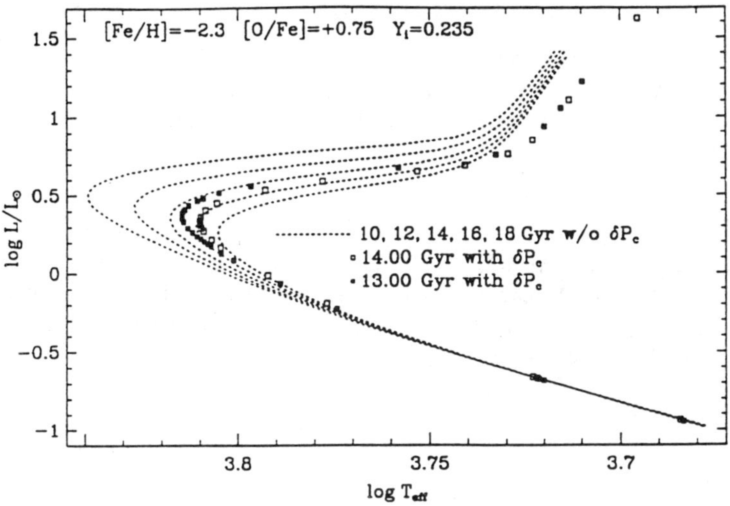

Fig. 1: Isochrones calculated with and without Coulomb corrections.

SUMMARY

- Coulomb corrections have an $\approx 1\%$ effect on the equation of state in most of the interior mass of typical main sequence Pop. II stars.
- Including Coulomb corrections decreases the evolutionary time scale for stars of fixed mass and initial composition by $\approx 8\%$.
- The ages for globular clusters estimated by comparing to the turnoff luminosity of isochrones will be decreased by $\approx 4\%$.
- The effects on isochrone shapes are large enough that Coulomb corrections should be included in all models of low mass stars.

REFERENCES

Chieffi, A. & Straniero, O. 1989, ApJS, 71, 47

Däppen, W., Lebreton, Y., & Rogers, F. 1990, *Solar Physics*, 128, 35

Däppen, W., Mihalas, D., Hummer, D. G., & Mihalas, B. W. 1988, ApJ, 332, 261

Eggleton, P. P, Faulkner , J., & Flannery, B. P. 1973, A&A, 23, 325

Graboske, H. C., Harwood, D. J., & Rogers, F. J. 1969, *Phys. Rev.*, 186, 210

Guenther, D. B., Demarque, P., Kim, Y.-C., & Pinsonneault, M. H. 1992, ApJ, 387, 372

Mihalas, D., Däppen, W., & Hummer, D. G. 1988, ApJ, 331, 815

Proffitt C. R. & VandenBerg, D. A. 1991, ApJS, 77, 473

Straniero, O. 1988, A&A Sup. 76, 157

VandenBerg, D. A., Bolte, M., & Stetson, P. B. 1990, AJ, 100, 445

Inside the Stars, IAU Colloquium 137
ASP Conference Series, Vol. 40, 1993
Werner W. Weiss and Annie Baglin (eds.)

A PECULIAR EFFECT OF CORE OVERSHOOTING ON THE INTERNAL STRUCTURE OF LOW-MASS STARS

NAMI MOWLAVI †
Institut d'Astronomie et d'Astrophysique; Université Libre de Bruxelles, C.P. 165; Av. F.-D. Roosevelt, 50, 1050 Bruxelles, BELGIUM

ABSTRACT With the aid of detailed stellar model computations, we analyse the consequences of applying a mild overshooting to $1.0\,M_\odot$ to $1.2\,M_\odot$ stars with solar metallicity. It is shown that solar models with a small convective core can be constructed, resulting in a decrease in the ^{8}B neutrino flux that can amount to 30% relative to the value obtained in models without convective core.

INTRODUCTION

It is by now accepted that convective penetration must be present at the border of convective cores (Zahn 1991). Evolutionary models taking this effect into account (Maeder and Meynet 1989) show a better agreement with observations. Such a treatment is, however, limited to intermediate-mass and massive stars, low-mass stars being supposed to have a radiative core. In fact, solar models with an "ad hoc" mixed core have been computed to try to explain the low value of the observed neutrino flux (see for example Provost *et al.* 1990, and Sienkiewicz *et al.* 1990). The origin of this assumed convective core is, however, unidentified.

The aim of this study is to show that a mixed core in a low-mass star can be the natural consequence of a mild overshooting applied at the end of the Pre-Main Sequence phase when a small convective core appears, which would otherwise disappear. Section 2 gives a brief description of the evolutionary code and the initial models, and Section 3 gives some results for the low-mass stars in the range $1.0\,M_\odot$ to $1.2\,M_\odot$. Consequences for the internal structure of the Sun are also presented. More complete results will be published elsewhere (Mowlavi and Forestini 1992).

THE MODELS

All models are computed with the hydrodynamical stellar evolution code developed by Forestini (1992). It is extended in order to include overshooting. The distance of overshooting is given by $d_{over} = \alpha_{over} \times H_p$. We limit the maximum growth of the convective core to twice its mass, so that

† Boursier I.R.S.I.A.

$$d_{over} = \min(\alpha_{over} \times H_p, d_{2M_{cc}}),$$

where H_p is the pressure scale height evaluated at the border defined by the Schwarszchild criterium ($\nabla_{RAD} = \nabla_{AD}$), and $d_{2M_{cc}}$ is the distance over which the mass of the convective core doubles. Initial models are taken from PMS models calculated until the development of a small convective core. The initial abundances are taken from Anders and Grevesse (1989), and the rates of the pp and CNO reactions are taken from Caughlan and Fowler (1988).

For each overshooting parameter, we calibrate the initial He content (Y) of the 1 $M_\odot$ star to reproduce the solar luminosity within less than 1% at the age of the Sun. These values of Y (see Table 2) are not very different for the different overshooting parameters: they vary from $Y = 0.2752$ in absence of overshooting [which is the He abundance suggested by and Grevesse (1989)] to $Y = 0.2798$ for $\alpha_{over} = 0.30$. These adopted initial values are kept for the 1.2 $M_\odot$ star.

RESULTS

We find that an overshooting of $0.05 \times H_p$ maintains a convective core in the Sun during its first 2.5 billion years, and that an as mild overshooting as $\alpha_{over} = 0.10$ is enough to keep it throughout the entire hydrogen burning phase. The Sun's lifetime is affected very little by the core mixing: with $\alpha_{over} = 0.20$, it is only 7.6% longer than the standard case without convective core.

The 1.2 $M_\odot$ star has, without overshooting, a very small convective core containing a minimum of 0.22% of the total mass of the star (at the age of 2 billion years), which grows to 5.11% toward the end of hydrogen burning (at 4.3 billion years). This growth is due to the enhancement of the energy contribution from the CNO cycle. When an overshooting of $\alpha_{over} = 0.20$ is applied, this enhancement is not present and the convective core receeds gradually from 14% in mass at the beginning of the MS phase to 5% at the age of 5.9 billion years. The time spent to burn its central hydrogen is then increased by 21.3% (see Table 1).

TABLE I Lifetimes (in 10^9 yr) of the convective core (t_{cc}) and the Main Sequence phase (t_{MS}) for 1.0 $M_\odot$ and 1.2 $M_\odot$ stars

α_{over}		0.00	0.05	0.10	0.20
1.0 $M_\odot$	t_{cc}	0.16	2.56	9.59	10.98
	t_{MS}	10.20	10.22	10.32	10.98
1.2 $M_\odot$	t_{cc}	4.83	4.89	5.25	5.97
	t_{MS}	4.92	5.05	5.30	5.97

We want to make a comment on the – arbitrary – constraint that the extended convective core has a maximum possible mass of twice the mass of the classical convective core. In the case of the solar models, this constraint limits the growth of the convective core due to overshooting. It seems reasonable to impose a maximum extension, but its amplitude is arbitrary as long as a reliable theory of overshooting has not been developed. Maeder and Meynet (1989) adopt a maximum extension in terms of the radius of the convective border, $d_{over} = \min(\alpha_{over} \times H_p, 0.25 \times R_{conv})$, which gives a slightly less extended core than ours, but it has no effect in their results as they do not predict convective cores for stars less massive than 1.1 $M_\odot$.

Finally, we present in Table 2 the mass of the convective core that is expected in the Sun when overshooting is taken into account, the hydrogen abundance at its center and the neutrino flux from ^{8}B decay relative to the case without overshooting. The flux reduction amounts up to 30% for $\alpha_{over} = 0.30$, with a convective core containing 14% of the solar mass.

TABLE II Initial He content Y, mass contained in the convective core of the Sun (M_{cc}), central hydrogen abundance in mass fraction [X_c(H)], and expected ^{8}B neutrino flux for different overshooting parameters. The flux is normalized to the value obtained in absence of overshooting ($\alpha_{over} = 0$)

α_{over}	0.00	0.05	0.10	0.15	0.20	0.25	0.30
Y	0.2752	0.2752	0.2762	0.2772	0.2784	0.2789	0.2798
M_{cc}	0.000	0.000	0.035	0.074	0.106	0.132	0.143
X_c(H)	0.369	0.387	0.461	0.493	0.512	0.530	0.530
^{8}B ν flux	1.00	1.00	1.04	0.92	0.84	0.71	0.71

ACKNOWLEDGEMENT

I would like to thank Dr. M. Forestini for his suggestion to undertake this study and for fruitful discussions. The presentation of this work at the IAU Colloquium 137 has been supported by the grant PICS 18.

REFERENCES

Anders, E. and Grevesse, N. 1989, *Geochem. Cosmochem. Acta*, **53**, 197
Forestini, M. 1991, Ph.D. Thesis (Université Libre de Bruxelles)
Maeder, A. and Meynet, G. 1989, *Astr. Ap.*, **210**,155
Mowlavi, N. and Forestini, M. 1992, in preparation
Provost, J., Berthomieu, G., Gavryuseva, E. and Gavryusev, V. 1990, *Solar Physics*, **128**, 111
Schaller, G., Schaerer, D., Meynet, G. and Maeder, A. 1992, preprint
Sienkiewicz, R., Bahcall, J.N. and Paczyński, B. 1990, *Ap. J.*, **349**, 641
Zahn, J.-P. 1991, *Astr. Ap.*, **252**, 179

Inside the Stars, IAU Colloquium 137
ASP Conference Series, Vol. 40, 1993
Werner W. Weiss and Annie Baglin (eds.)

MAIN SEQUENCE STRUCTURES INCLUDING THE OVERADIABATIC CM CONVECTION MODEL

ITALO MAZZITELLI
Istituto di Astrofisica Spaziale, Frascati, Italy

FRANCESCA D'ANTONA
Osservatorio Astronomico di Roma, Monte Porzio, Italy

INTRODUCTION

We computed the main sequence structure of masses from 2 to 0.7 $M_\odot$ for two choices of helium (Y) and metal (Z) mass fractions. Models are given for solar composition (Y=0.285, Z=0.018) and for a composition appropriate to a younger population I (Y=0.32, Z=0.03). We took particular care both in the choice of the most recent opacities presently available and in the treatment of the input physics and numerics (see also D'Antona *et al.*, 1992).

Opacities from Rogers and Iglesias (1992) were adopted where available, complemented with those by Kurucz (1991) at T$\leq$10000 K and, elsewhere, by the Los Alamos opacities (Huebner et al. 1977). The equation of state by Mihalas et al. (1988) was used at $\rho \leq 0.01$, and elsewhere the relation P=P(ρ,T) was taken from Magni and Mazzitelli (1979), while the adiabatic gradient was obtained from a Saha type e.o.s..

As for convection, traditional models adopting the Mixing Length Theory, with $\alpha = l/H_p$=1.6, calibrated on the solar models for the composition (Y=0.285, Z=0.018) (MLT models) were compared with the results obtained adopting the new model by Canuto and Mazzitelli (CM), 1991 and 1992.

RESULTS

Extensive tables of results will be presented elsewhere (Mazzitelli and D'Antona, 1993 in preparation). The outer structure of the CM model of 1.4 $M_\odot$ (figure 1) is compared with an equivalent model computed in the MLT framework (figure 2). The overadiabatic gradient profile in the CM model is peaked and narrower than in the MLT, being larger in the most outer layers, but smaller in the interior. The larger gradient gives origin to a *density inversion*, which was found also in the 1$M_\odot$ model (CM). The density in the model of 1.4$M_\odot$ increases so much when going outwards that overadiabaticity drops and then rises again, close to the convective boundary, giving rise to *a double peak of the overadiabatic gradient*, as illustrated. The $T_{\rm eff}$ of this model is ~ 200K *lower* than the corresponding MLT model. The difference between the models becomes smaller as the mass decreases, and the most external layers begin to have less influence on the $T_{\rm eff}$.

The described feature produces a peculiar shape of the Main Sequence in

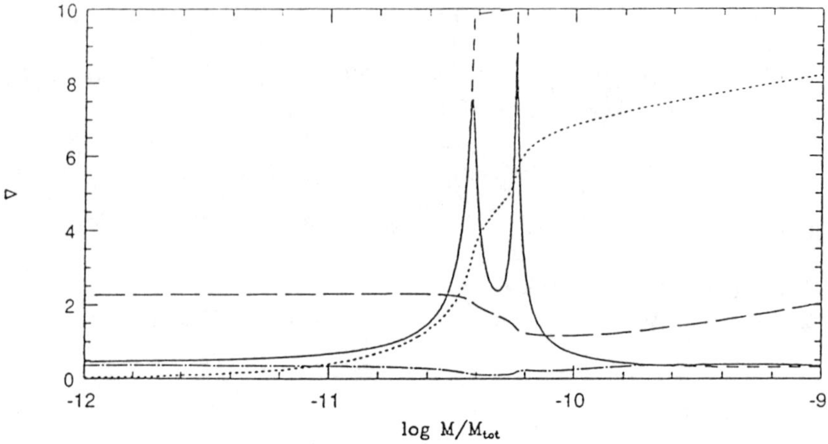

FIGURE I Structure of the external layers of a mass 1.4 $M_\odot$, having 10^8yr, in the CM framework, for $Y = 0.285$, $Z = 0.018$. The model has $\log L/L_\odot = 0.566$ and $T_{eff} = 6740$K. The continuous line is the overadiabatic gradient (∇=dlog P/dlog T), short–dashed is given the radiative gradient, and dash-dotted the adiabatic gradient. The long–dashed line is $\log\rho$, which is in a scale from -8 to -3. The dotted line is the quantity ($1-T_{eff}$ /T), in scale from 0 to 1.

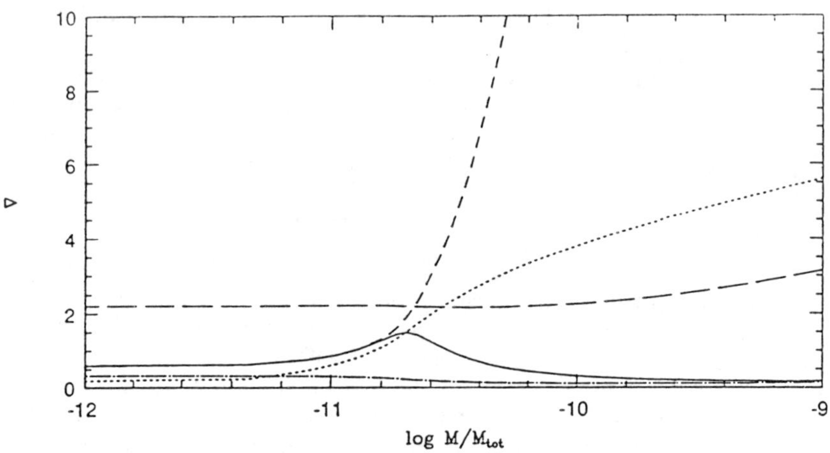

FIGURE II Structure of the external layers of a mass 1.4 $M_\odot$, having 10^8yr, in the MLT framework, with $\alpha = 1.6$. The model has $\log L/L_\odot = 0.566$ and $T_{eff} = 6910$K. Lines have the same meaning as in figure 1. Convection is here more efficient than in the CM case, and consequently the overadiabatic gradient is much smaller.

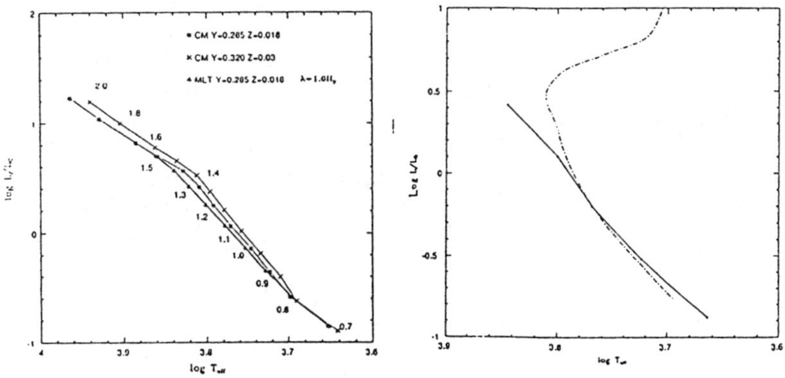

FIGURE III Left: HR diagram for the three sequences of pop. I. The models are given at an age of 10^8yr. Right: the main sequence of Z=10^{-3}, Y=0.23, age=10^8yr is shown, for masses from 1 to 0.6 $M_\odot$. A dash-dotted isochrone of age = 14×10^9yr, $Z = 2 \times 10^{-4}$ shows a small variation of slope below the turn–off, which is also due to the peculiar behaviour of CM models in this region. The opacities for the population II are from Kurucz (1991) plus Los Alamos (Huebner et al. 1977).

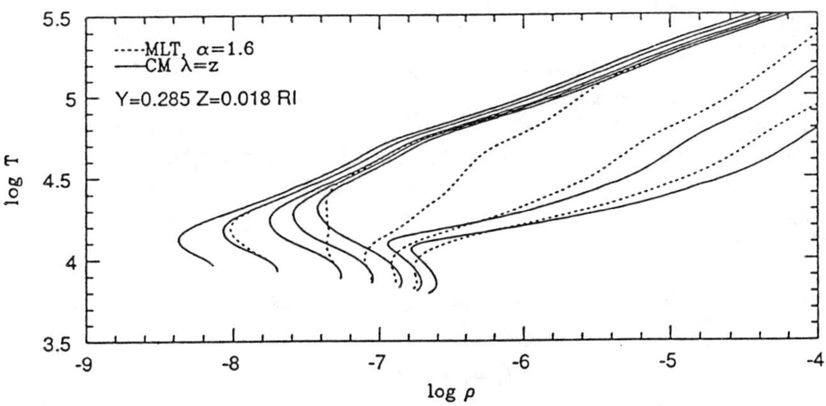

FIGURE IV Comparison of MLT tracks and CM tracks, for solar composition, in the $\log\rho, \log T$ plane. Masses are 2, 1.8, 1.6, 1.5, 1.4, 1.3 and 1.2 $M_\odot$for CM models, and 1.8, 1.6, 1.5, 1.4, 1.3 $M_\odot$for MLT models.

the HR diagram, shown in the left part of figure 3. In fact, as long as the overadiabatic gradient is double peaked, the CM models have a steeper temperature profile than MLT models; their $T_{\rm eff}$ is *smaller* and the convective envelopes are correspondingly shallower. For smaller masses, when the double peak disappears, the convective extension suddenly increases (from $\sim 10^{-8}$ to $\sim 10^{-4} M_\odot$, going from 1.4 to $1.3 M_\odot$). This transition appears as *a sudden change of slope in the HR diagram.* The location of the change of slope depends on the opacities (chemical composition), as we see comparing the Z=0.018 and Z=0.03 results, and the results for $Z = 10^{-3}$ shown in the right panel of the figure. The evolutionary tracks crossing the region in which the transition occurs also present a *non monotonic* behaviour, with a peculiar shape in the HR diagram, which is exemplified by showing, dot–dashed, an isochrone at 14×10^9yr for $Y = 0.23$, $Z = 2 \times 10^{-4}$. *Whether this small feature can be observable or not in the isochrones of some open or globular cluster is to be investigated.*

The sudden onset of convection in the framework of CM convection is also seen in the ρ – T plane (figure 4) where the "gap" between the 1.4 $M_\odot$ and 1.3 $M_\odot$ models is not noticeable in the MLT framework.

POSSIBLE TESTS OF THE CM MODEL

In the models of Y=0.285, Z=0.018 the 1.4 $M_\odot$ model is located at $T_{\rm eff} = 6740$K, in the middle of the Lithium–gap of open clusters (Boesgaard and Tripicco, 1986). The theories which are developed to explain the occurrence of the gap must be tested on CM models, and, viceversa, it is possible to use the Lithium data to constraint observationally the location, in the HR diagram, where we expect the sudden change of slope.

A careful photometry in the turn–off region of the *very metal poor* globular clusters may reveal indications of the peculiar behaviour shown in the isochrone of figure 3: this may constitute a powerful test of the atmospheric opacity computations for population II, and, eventually, even a test for the age of oldest globular clusters (D'Antona and Mazzitelli, 1993 in preparation).

REFERENCES

Boesgaard, A. M., Tripicco, M. J. 1986, *ApJ* **302**, L49.

Canuto, V. M., Mazzitelli, I. 1990, *ApJ* **370**, 295 (CM).

Canuto, V. M., Mazzitelli, I. 1992, *ApJ* **389**, 724.

D'Antona, F., Mazzitelli, I., Gratton, R. 1992, *A&A* **257**, 539.

Huebner, W. F., Merts, A. ., Magee, N. H., Argo, M. F. 1977, Los Alamos Scientific Laboratory Report, No LA-6760-M.

Kurucz, R. L. 1991b, in "Stellar Atmospheres: Beyond the Classical Models", L. Crivellari, I. Hubeny, D. G. Hummer eds., NATO ASI Series, Kluwer.

Magni, G., Mazzitelli, I. 1979, *A&A* **72**, 134.

Mihalas, D., Däppen, W., Hummer, D. G. 1988, *ApJ* **331**, 815.

Rogers, F. J., Iglesias, C. A. 1992, *ApJS* **79**.

Inside the Stars, IAU Colloquium 137
ASP Conference Series, Vol. 40, 1993
Werner W. Weiss and Annie Baglin (eds.)

CONVECTIVE OVERSHOOTING AND THE OBSERVED WIDTH OF THE MAIN-SEQUENCE BAND

R. NAPIWOTZKI, A. RIESCHICK, T. BLÖCKER,
D. SCHÖNBERNER, AND V. WENSKE
Institut für theoretische Physik und Sternwarte der Universität Kiel
Olshausenstr. 40, W-2300 Kiel, Germany

For many years we are witnessing a lively debate on the existence and extent of convective overshooting, mainly in the cores of main-sequence stars. This is an important issue, since even a small amount of overshooting increases considerably the mass of the finally hydrogen exhausted core and lenghthens the main-sequence lifetime correspondingly. The available evolutionary calculations assume either moderate overshooting, $d/H\text{p} = 0.25$, (d = overshooting distance, H_P = pressure scale height; Maeder & Meynet 1988) or strong overshooting, $d/H_\text{p} \approx 0.50$ (Bertelli et al. 1986). Presently theory is unable to quantify the exact amount of overshooting, and one has to resort to empirical determinations.

Recently, Stothers (1991) collected all available information from the literature on stellar parameters and evolutionary calculations and concluded that, within the errors, $d/H_\text{p} = 0$ is an acceptable result, with a conservative upper limit of $d/H_\text{p} < 0.2$. However, such an approach is hampered by observational errors (like distance or temperature uncertainties, rotation) that are difficult to quantify and that may mask any definitive result. Detailed investigations of detached binaries may help in this matter (Andersen et al. 1990) but the number of suitable binary systems is probably not very large.

Recently, Napiwotzki et al. (1991) proposed a new method, viz. the direct determination of surface acceleration g and effective temperature T_eff by means of model atmospheres and comparison with evolutionary calculations. This method is distance independent and allows a large number of stars to be investigated. It takes advantage of the fact that the stellar evolution is strongly accellerated after the end of the core hydrogen burning, leading to only a very small number of stars in the region above the TAMS (Terminal age main sequence). One needs, however, spectrograms of good quality. The first application of this method by Napiwotzki et al. indicated an overshoot distance intermediate between the values proposed by Maeder & Meynet (1988) and Bertelli et al. (1986). It was noted before by Lemke (1989) and by Wolff (1990) that the observed width of the main sequence in the $g - T_\text{eff}$ diagram do not agree with predictions from standard theory.

We report here on the continuation of this method: we determined g and T_eff of 16 bright B- and A-type stars of the field and computed new evolutionary tracks with the most recent opacities (Iglesias & Rogers 1991).

OBSERVATIONS AND FITTING PROCEDURE

The observed B- and A-type filed stars were randomly selected from the list of photometric β standard stars of Crawford & Mander (1966) for calibration purposes (Napiwotzki et al. 1992). The stars were observed with the Coudé spectrographs of the 2.2 m telescope at the Calar Alto Observatory, equipped with a RCA CCD, and of the McDonald Observatory, Texas, USA, equipped with a TI CCD. The (reciprocal) dispersion ranges from 7 Å/mm to 17 Å/mm, and the images were centered on H_γ or H_β.

Effective temperatures were derived by using existing photometric data ($uvby\beta$) according to the prescriptions of Moon & Dworetsky (1985). This temperature determination appears to be very accurate ($\approx \pm 150$ K) and agrees very well with that based on UV fluxes (Napiwotzki et al. 1992; Wenske & Schönberner 1992).

With $T_{\rm eff}$ fixed, surface accelerations and (projected) rotational velocities were derived by matching theoretical line profiles, computed with Kurucz's fully line blanketed model atmospheres, to the observed ones. A detailed list of results is given in Napiwotzki et al. (1992). The fit error of $\log g$ is in most cases smaller than 0.05 dex. Together with the temperature uncertainty, the total error of $\log g$ is not larger than only 0.1 dex!

MODEL CALCULATIONS

We calculated several grids of evolutionary stellar models in the mass range of 2.0 $M_\odot$ to 8.0 $M_\odot$. To determine the influence of opacities on the width of the main-sequence band three different sets of opacity tables were used: Cox & Stewart (1970; CS), Huebner et al. (1977; LA), and Iglesias & Rogers (1991; LL). The software used for all calculation was a modified version of the Kippenhahn code described e.g. in Schönberner (1979, 1983). For all sequences the chemical mixture was adopted to be (X,Y,Z) = (0.699, 0.28, 0.021). The mixing length parameter (ratio of convective mixing length to local pressure scale height) was set to 1.3. With the LL opacities model sequences for three different overshooting parameter were calculated ($d/H_{\rm P} = 0.0, 0.15, 0.25$).

DISCUSSION

In Fig. 1 we show the position of ZAMS and TAMS calculated with the LL opacities and different overshoot parameters $d/H_{\rm p}$ in the g-$T_{\rm eff}$ plane. Added are the positions of the observed field stars. It is obvious that the standard tracks calculated with $d/H_{\rm p} = 0$ are still not sufficient to fit the observations. A best fit would be possible with $d/H_{\rm p} \approx 0.15$. A comparison with our consistent calculations using the older LA and CS opacities shows that the location of the TAMS is dependent of the used opacity tables: higher values of $d/H_{\rm p}$ are necessary to fit the observations. We want to emphasize that, though we used the most recent opacities of Iglesias and Rogers (1991), the comparison with empirical data shows that the observed width of the main-sequence band is only reproducable with $d/H_{\rm p} > 0$.

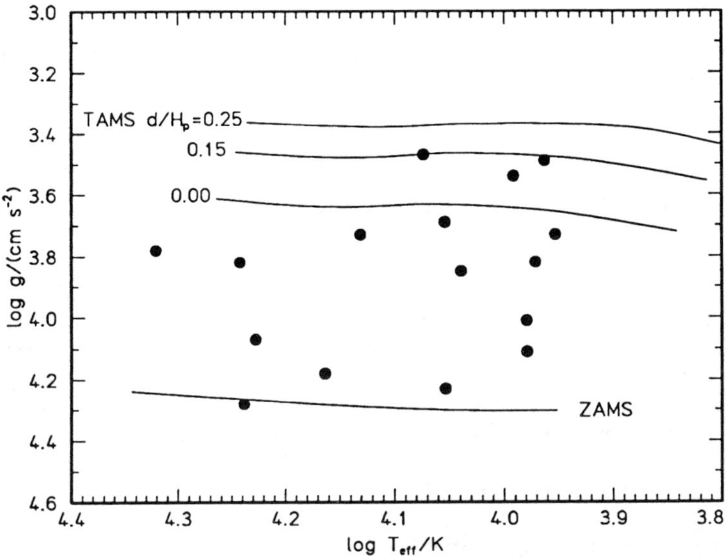

FIGURE I ZAMS and TAMS calculated with the LL opacities in the $\log T_{\rm eff}$-$\log g$ diagram together with the positions of the observed stars.

ACKNOWLEDGMENTS

We like to thank Mounib El Eid for his support in installing the new opacity tables.

REFERENCES

Andersen, J., Nordström, B. Clausen, J.V., 1990, *ApJ* **363**, L33
Bertelli, G., Bressan, A., Chiosi, C., Angerer, K., 1986, *A&AS* **66**, 191
Cox, A.N., Stewart, J.N., 1970, *ApJS* **19**, 243
Crawford, D.L., Mander, J., 1966, *AJ* **71**, 114
Huebner, W.F., Merts, A.L., Magee, N.H.Jr., Arog, M.F., 1977, Los Alamos Scientific Laboratory Rept., No. LA-670-M
Iglesias, C.A., Rogers, F.J., 1991, *ApJ* **371**, L73
Lemke, M., 1989, *A&A* **225**, 125
Maeder, A., Meynet, G., 1988, *A&AS* **76**, 411
Moon, T.T., Dworetsky, M.M., 1985, *MNRAS* **217**, 305
Napiwotzki, R., Schönberner, D., Weidemann, V., 1991, *A&A* **243**, L5
Napiwotzki, R., Schönberner, D., Wenske, V., 1992, *A&A* in press
Schönberner, D., 1979, *A&A* **79**, 108
Schönberner, D., 1983, *ApJ* **272**, 708
Stothers, R.B., 1991, *ApJ* **383**, 820
Wenske, V., Schönberner, D., 1992, these proceedings
Wolff, S.C., 1990, *AJ* **100**, 1994

Inside the Stars, IAU Colloquium 137
ASP Conference Series, Vol. 40, 1993
Werner W. Weiss and Annie Baglin (eds.)

SPIN-DOWN OF SOLAR-TYPE STARS WITH INTERNAL MAGNETIC FIELDS

Paul CHARBONNEAU and Keith B. MACGREGOR High Altitude Observatory, NCAR [1], P.O. Box 3000, Boulder, CO 80307, U.S.A.

ABSTRACT We present a selection of results from a large set of numerical simulations of the spin-down of a solar-type star containing a large scale magnetic field in its radiative interior. Our computations are *dynamical*, in that they take into account both the generation of the toroidal component by the wind-induced shear *and* its back-reaction on the azimuthal flow. Our results demonstrate the existence of classes of internal magnetic fields that can accomodate rapid spin-down near the ZAMS, *and* yield weak internal differential rotation by the solar age.

Keywords: Rotation; Magnetic Fields.

INTRODUCTION

Considerable progress has recently been made toward the goal of identifying and understanding the processes which affect the distribution and evolution of angular momentum in the interiors of the sun and solar-type stars. In particular, the response of the solar interior to the torque applied at its surface by a magnetically coupled wind has been the subject of substantial modeling efforts (see, e.g., Pinsonneault *et al.* 1989; Tassoul & Tassoul 1989; MacGregor & Brenner 1991). However, in considering the internal redistribution of angular momentum, all of these studies have been restricted to purely hydrodynamical (i.e., non-magnetic) transport mechanisms.

The potential importance of large-scale internal magnetic fields in redistributing angular momentum in the stellar radiative interiors has been stressed repeatedly by a number of authors (see, e.g., Mestel & Weiss 1987, and references therein). The mechanism itself is simple: a shear in the azimuthal flow, acting on a pre-existing large-scale poloidal magnetic field, will generate a toroidal magnetic component, with which is associated a Lorentz force in the ϕ-direction, itself acting in such a way as to oppose the driving shear. However, a consistent quantitative study of this process is still lacking. A number of questions that come to mind are: (Q1) How do internal magnetic fields affect the surface rotational evolution ? (Q2) What level of internal differential rotation can they tolerate ? (Q3) What is the strength of the toroidal field generated ? (Q4) What is the time-evolution of the internal toroidal field ? (Q5) To what degree are answers to Q1—Q4 affected by the assumed strength and geometry of the internal poloidal field ? Our purpose is to provide *quantitative* answers to the above questions, in the context of the solar spin-down problem.

---

[1] The National Center for Atmospheric Research is sponsored by the National Science Foundation.

MODEL ASSUMPTIONS AND SELECTED RESULTS

Our formulation of the spin-down problem is based on the following working assumptions: (A1) Axisymmetric system (rotation and magnetic axes parallel), (A2) Time-independent poloidal magnetic field, (A3) No fluid motion in meridional planes, (A4) Constant kinematic viscosity and magnetic diffusivity, (A5) No Coriolis forces, (A6) Static stellar envelope model, (A7) Solid-body rotation in convective envelope (hereafter "CE"), and (A8) Solid-body rotation and zero toroidal field on the ZAMS.

In view of A1, A2 and A3, we have $\mathbf{U} = \Omega(r,\theta,t)\, r \sin\theta\, \hat{\mathbf{e}}_\phi$, and $\mathbf{B} = \mathbf{B}_p(r,\theta) + B_\phi(r,\theta,t)\,\hat{\mathbf{e}}_\phi$. The spin-down problem reduces to solving the (linear but coupled) ϕ-components of the momentum and induction equations. The angular momentum loss rate is computed using the Weber & Davis (1967) axisymmetric formulation (see also Belcher & MacGregor 1976). We model angular momentum loss as a sink within the CE, rather than a torque at the surface. Solutions to the governing equations are obtained using the finite element-based numerical scheme describe in Charbonneau & MacGregor (1992*a*, appendix). We use a standard solar model computed by R. Gilliland. The model reaches the ZAMS at $t_{ZAMS} \simeq 4 \times 10^7$ yrs, at which time the core-envelope boundary is located at $R_{CE} = 0.74 R_\odot$, and the CE accounts for about 10% of the total moment of inertia.

The internal rotational and magnetic evolution can be subdivided in three more or less distinct epochs. The first is of short duration, corresponding to the buildup of the toroidal fields in the radiative interior following the shearing of the pre-existing poloidal field. The second epoch, especially for magnetic configurations having direct magnetic coupling at the core-envelope interface, is characterized by the presence of vigorous *phase mixing* (see e.g. Roxburgh, 1991). This results in the damping of the torsional Alfvén waves generated earlier. The third epoch of evolution is characterized by a state of approximate dynamical balance such that the total stresses (magnetic + viscous) at the core-envelope interface equilibrate almost exactly the wind-induced surface torque. The system essentially evolves on the Alfvén timescale (of order 10^3—10^4 yrs for a $||\mathbf{B}_p|| \simeq 1$ G) during epochs 1 and 2, but on the wind braking timescale during epoch 3. This leads to a quasi-static internal and magnetic evolution from about 10^7 yrs all the way to the solar age.

Figure 1 [Plate] shows a few snapshots of the internal evolution for three different internal poloidal fields (all of average strength 1 G). The color scale codes B_ϕ, and the contour lines the azimuthal velocity u_ϕ ($= \Omega\, r \sin\theta$; see caption for details). The portion of the envelope threaded by fieldlines that close upon themselves within the radiative zone is magnetically decoupled from the rest of the envelope, and tends to rotationally lag behind the CE, thus acting as a reservoir of angular momentum. Note the high strength of the toroidal magnetic fields generated during the early epochs of spin-down. For all configurations, toroidal field strengths reach a maximum around the Pleiades age ($t + t_{ZAMS} \simeq 70$ Myr), and decrease monotonically thereafter.

Figure 2 shows the evolution in time of the surface rotation rate for solutions of various internal field geometries and strengths. The extremely weak dependence of surface rotational evolution on field strength, for some configurations (D2 and D4), is noteworthy. In the case of D2, it is a direct consequence of the aforementioned dynamical balance at the core-envelope

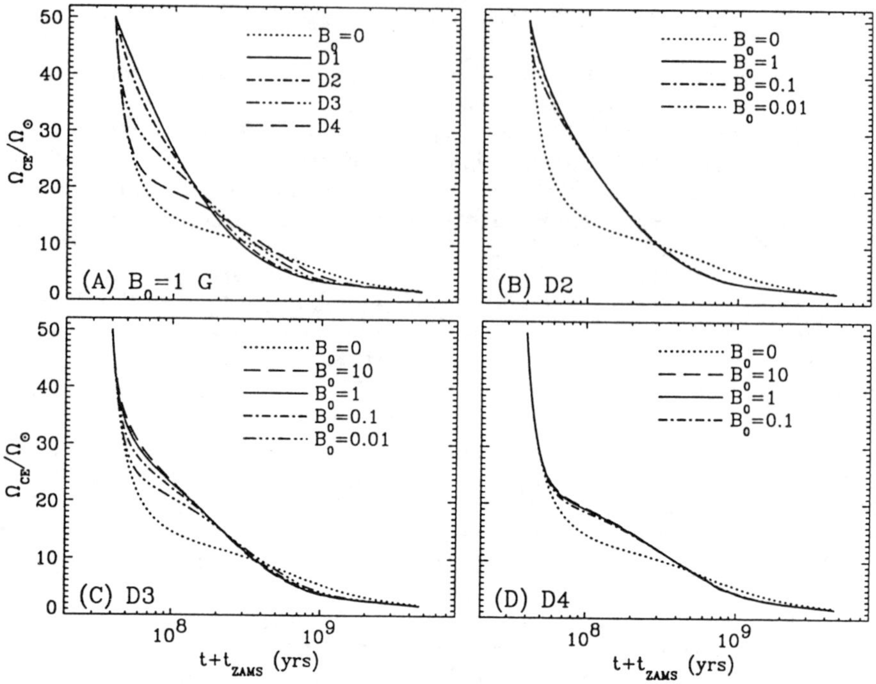

Figure 2 : Selected surface rotational evolution curves for a solar model. Part (A) shows results for a fixed poloidal fields strengths and various degrees of core-envelope magnetic coupling (see Fig. 1 [Plate]), and parts (B) (C) and (D) shows results for a fixed degree of coupling but various field strengths. Note the convergence of surface rotations rate beyond $t \sim 10^9$ yrs.

interface. The greater dependence on field strength, for the D3 configuration in the first few 10^8 yrs of evolution, reflects the influence of the poloidal field on the viscous boundary layer forming at the core-envelope interface for this field geometry. Similar results are obtained for quadrupolar internal fields.

Figure 3 shows the time evolution of internal differential rotation. The dimensionless quantity $\Delta\Omega$, is a *global* measure, obtained by integrating $\Omega(r, \theta, t) - \Omega_{CE}(t)$ over the magnetized part of the radiative core. Note on Fig. 3 how $\Delta\Omega$ reaches a maximum around $t + t_{ZAMS} \simeq 10^8$ yrs, and gradually decreases thereafter, down to $\Delta\Omega \sim 10\%$ by the solar age.

CONCLUSIONS

A detailed examination of our solutions yields the following conclusions (Charbonneau & MacGregor, 1992b; Charbonneau & MacGregor, in preparation; MacGregor, these proceedings): (C1) The presence of magnetic fields in radiative interiors is compatible with the observed rapid-down of solar analogs observed in young clusters (for a subset of poloidal field geometry). (C2) The surface rotational evolution is weakly dependent on the assumed internal poloidal field strength. (C3) The presence of magnetic fields leads to low internal angular

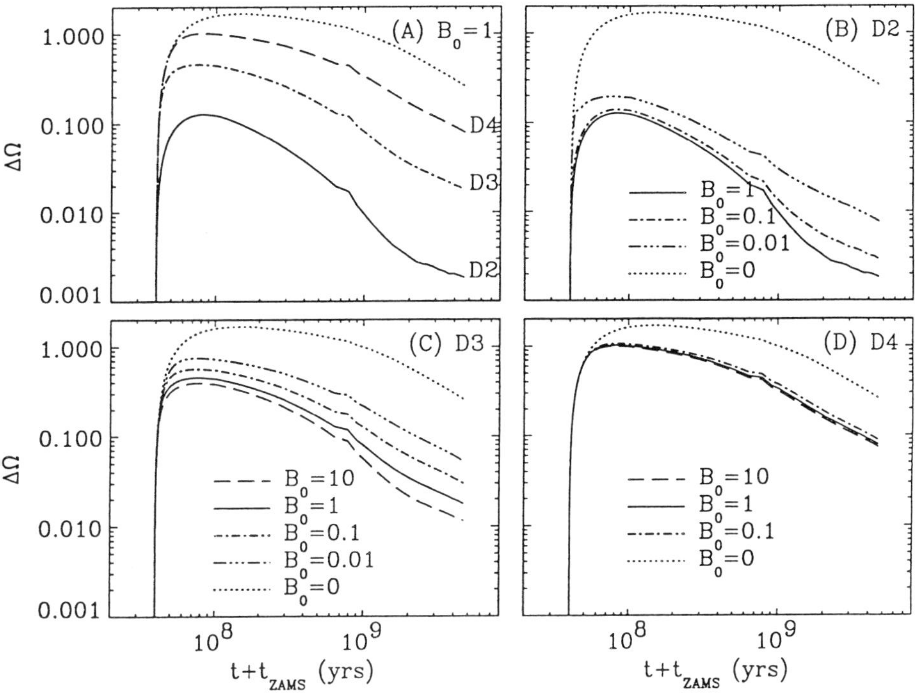

Figure 3 : Time-evolution of the internal differential rotation, for the same solutions as shown on Fig. 2. Note how even for configurations without direct magnetic coupling at the core-envelope interface (D3 and D4), $\Delta\Omega$ has fallen below 10% for most field strengths.

velocity gradients in radiative interiors by the solar age. (C4) The spin-down of solar-type stars generates strong (i.e, tens, sometimes hundreds of kG) toroidal magnetic fields in their radiative interiors. (C5) Toroidal oscillations set up at early epochs have large amplitudes in B_ϕ, very low amplitude in Ω, and are rapidly (within a few 10^5 yrs) damped via phase mixing. (C6) The present photospheric solar rotation is a very poor indicator of the strength and geometry (past and present) of the solar internal magnetic field. (C7) The present internal solar differential rotation is a much better probe of the solar internal magnetic field.

REFERENCES

Belcher, J.W., & MacGregor, K.B. 1976, *Ap. J.*, **203**, 720
Charbonneau P., & MacGregor K.B. 1992a, *Ap. J.*, **387**, 639 (paper I)
Charbonneau P., & MacGregor K.B. 1992b, submitted to ApJ.
MacGregor, K. B., & Brenner, M. 1991, *Ap. J.*, **376**, 204
Mestel, L., & Weiss, N.O. 1987, *M. N. R. A. S.*, **226**, 123
Pinsonneault, M.H., Kawaler, S.D., Sofia, S., & Demarque, P. 1989, *Ap. J.*, **338**, 424
Roxburgh, I.W. 1991, in *Angular Momentum Evolution of Young Stars*, eds. S. Catalano & J. Stauffer (Dordrecht: Kluwer Academic Publishers), p. 365.
Tassoul, J.-L., & Tassoul, M. 1989, *Astr. Ap.*, **213**, 397
Weber, E.J., & Davis, L. Jr. 1967, *Ap. J.*, **148**, 217

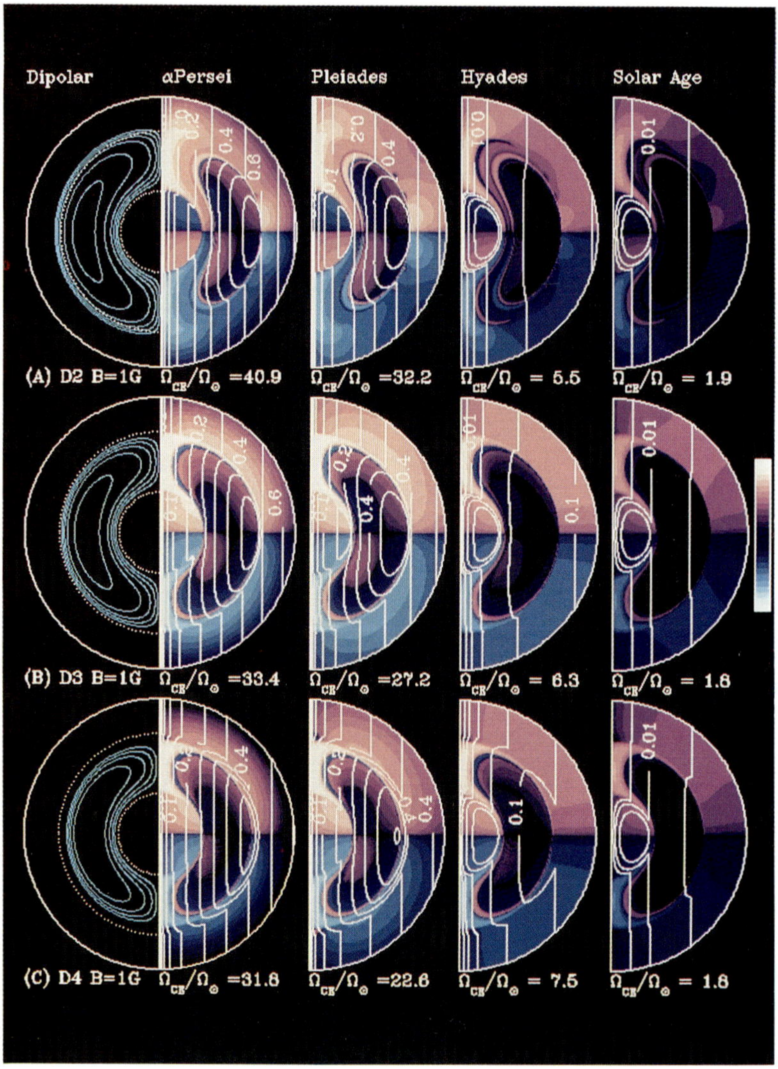

Figure 1 : Internal toroidal magnetic field and rotational profile at a few selected epochs. The blue—cyan sequence codes $log(+B_\phi)$, and the blue—pink sequence $log(-B_\phi)$ (with a lower cutoff at $|B_\phi| \leq 1$ G). The contour lines code the dimensionless azimuthal velocity u_ϕ [$= (\Omega/\Omega_0)(r\ sin\,\theta/R_*) =$ 0.005, 0.01, 0.02, 0.05, 0.1, 0.2, 0.3, 0.4, 0.5, 0.6, 0.7, 0.8, 0.9 and 1.0]. Solutions are shown for three different poloidal field configurations, the corresponding fieldlines being shown at the extreme left in each row. Note that direct magnetic coupling (i.e., $B_r \neq 0$ for some θ at $r = R_{CE}$) exists only for the D2 configuration (first row). We have adopted $t + t_{ZAMS} =$ 0.05, 0.07, 0.6 and 4.6 Gyrs for the ages of αPersei, the Pleiades, Hyades, and the Sun.

Inside the Stars, IAU Colloquium 137
ASP Conference Series, Vol. 40, 1993
Werner W. Weiss and Annie Baglin (eds.)

APSIDAL MOTION TEST: CONFRONTATION BETWEEN THEORY AND OBSERVATIONS

A. CLARET[1] AND A. GIMENEZ[1,2]
Instituto de Astrofísica de Andalucía, P.O.Box 3004, E-18080 Granada, Spain[1]
Laboratorio de Astrofísica Espacial y Física Fundamental, INTA
P.O.Box 50727, E-28080 Madrid, Spain[2]

The apsidal motion test in eclipsing binaries is one of the few observational procedures available to study the stellar interior. A careful analysis of empirical data and theoretical models has been performed since more than ten years ago by our group, and in the present communication we show the most recent results obtained from the confrontation between theory and observations.
We have carried out an extensive comparison between new stellar models (Claret & Giménez 1992) making use of the latest opacity library by Rogers & Iglesias (1992) and the best available data from double-lined eclipsing binaries (Andersen, 1991). The level at which we can theoretically reproduce the observations of classical absolute dimensions, namely, R and T_{eff} for a given mass, is shown in Figure 1, while Figure 2 shows the fulfillment of the requirements of a common age for the two components of the same system. A very similar good agreement is obtained when we take into account a moderate amount of convective core overshooting. Small differences can be easily explained in terms of metal abundance variations with respect to the adopted average value Z = 0.02. A more detailed check is shown in Figure 3 for some selected systems.

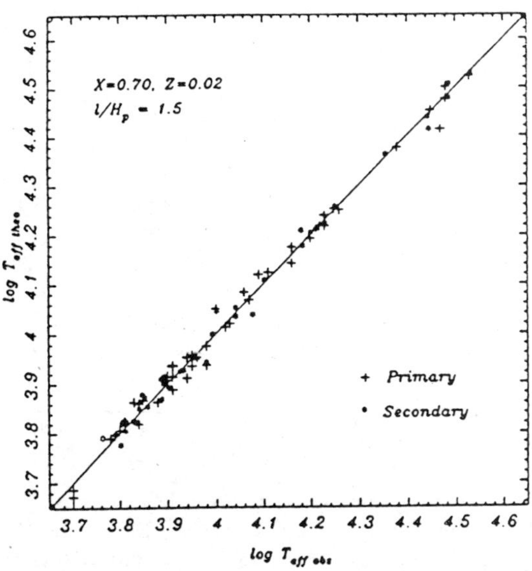

Fig. 1 Comparison between observed and theoretical effective temperatures.

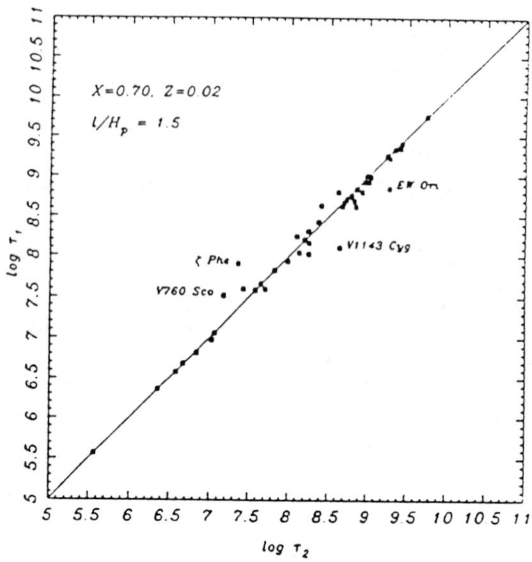

Fig. 2 Derived ages for individual components

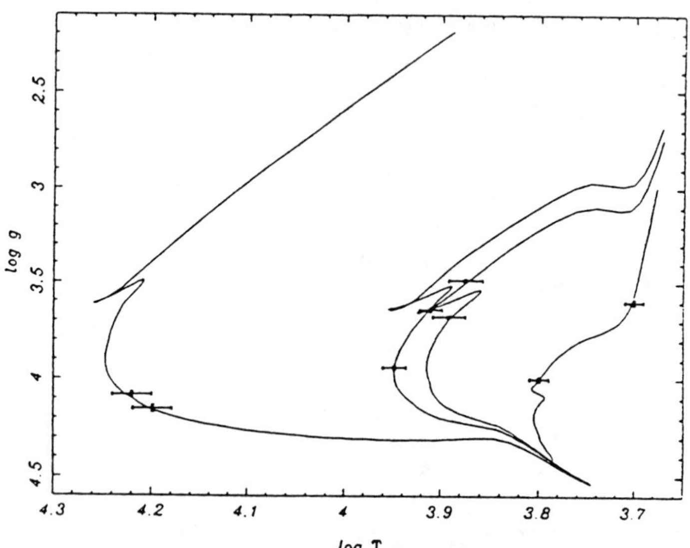

Fig. 3 Isochrones for some systems. From left to right: U Oph (5×10^7 y), WX Cep (6×10^8 y), SZ Cen (7.9×10^8 y) and AI Phe (4.3×10^9 y). For AI Phe, the corresponding isochrone was computed for Z = 0.01. Note the differences in evolutionary status.

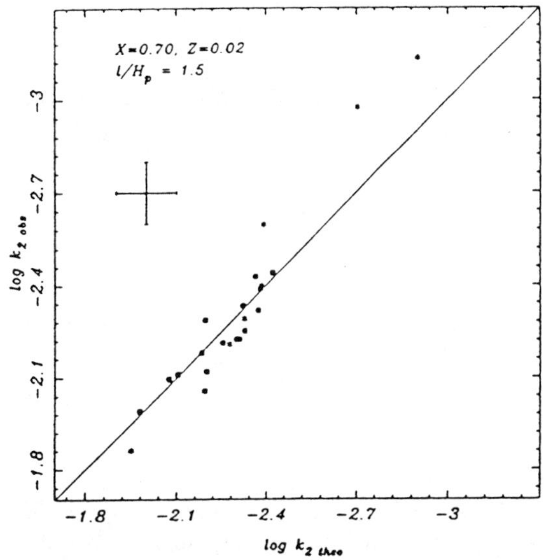

Fig. 4 Observed and predicted log k_2. Systems with low quallity determination of absolute dimensions and/or apsidal motion rates are denoted by asterisks. Typical error bar is indicated.

Concerning the derived internal structure constant, log k_2, for the eccentric systems, we show in Figure 4 the comparison with theoretically predicted values using the same evolutionary models. The previously detected systematic differences between observations and models, in the sense that real stars appear to be more centrally concentrated in mass than theoretically predicted (Claret & Giménez 1988; Giménez 1990), is for the first time not significantly measured. This result is obtained regardless of the assumption adopted of a moderate or none convective overshooting. Anyway, we have introduced stellar rotation in the evolutionary models which, concerning the apsidal motion constant, gives rise to more concentrated configurations in mass in good agreement with predictions by Stothers (1974).

We can conclude that the combination of absolute dimensions of the highest quality with accurately determined apsidal motion rates based on long term monitoring of eclipse timings, produces observational values of log k_2 which are in reasonable good agreement with theoretical predictions when the latest available opacities and the most up-to-date input physics in general have been taken into account.

REFERENCES

Andersen, J.: 1991, A&AR, vol. 3, no. 2, 91

Claret, A., Giménez, A: 1988, in *Astrophysics: Recent Progress and Future Possibilities*, Niels Bohr Institute, Copenhagen

Claret, A., Giménez, A: 1992, A&AS, in press

Giménez, A: 1990, in *Active Close Binaries*, Kluwer Academic Publishers, ed. C. Ibanoglu, pag. 137

Rogers, R. J., Iglesias, C. A.: 1992, ApJS, vol. 79, 507

Stothers, R.: 1974, ApJ, 194, 651

Inside the Stars, IAU Colloquium 137
ASP Conference Series, Vol. 40, 1993
Werner W. Weiss and Annie Baglin (eds.)

DETERMINATION OF AGE AND ORIGINAL CHEMICAL COMPOSITION OF BINARY STARS

FERLUGA, S. and ANTILLON, M.

Dipartimento di Astronomia, Università di Trieste
Via G.B. Tiepolo, 11 - I 34131 Trieste, Italy

ABSTRACT A numerical method is presented, which is able to determine quantitatively the age t and the initial abundances of helium and metals (Y_0 , Z_0) of a given binary system. Input data are simply the values of mass m , temperature $T_{\rm eff}$ and surface gravity g , of both stars. The procedure works for detached main–sequence binaries, and it does not require any preliminary hypothesis about surface abundances or age. It is assumed that the two stars evolve parallely with no interaction.

THE PROBLEM

A star with a given mass m , because of evolution, changes its physical properties (as effective temperature $T_{\rm eff}$ and surface gravity g), in function of the age t , and of the original abundance of helium Y_0 and metals Z_0. The relation between observational quantities (m ; log $T_{\rm eff}$, log g) and model parameters (Y_0, Z_0 ; t) is provided by evolutionary theory. Is it then possible to compute composition and age (Y_0, Z_0 ; t) of a given star, by knowing (m ; log $T_{\rm eff}$, log g) from observations ?

Here we present an algorithm, which finds out the solution for *individual* main–sequence binaries. In fact, the job is not possible for *single* stars, because of practical limitations (m is unknown), and also because of basic mathematical reasons (equations with infinite solutions). One may wonder why, on the contrary, it is generally possible to solve the problem for a *binary system.* This may be understood as follows.

At time t_0=0 (starting main–sequence *life*), the positions of all stellar masses in the (log $T_{\rm eff}$, log g) plane will define the *Zams* curve; while at time t_1 they will move sideways on another curve (isochrone t_1) , and so on, defining a family of isochrones t_i . For different compositions (Y_0, Z_0), there will be different overlapping families. So, a given star (single mass–point on our plane) may belong to any of the infinite isochrones t_i with different (Y_0, Z_0), intersecting at that point. Only the presence of a companion star (second mass–point) can individuate the *common* isochrone, solving the riddle.

THE *BOFER* CODE

Our computer code – called *BOFER* – performs linear interpolations among a basic grid of stellar models (Mengel *et al.*, 1979), with different initial compositions (Y_0=.20, .30 ; Z_0=.01, .04) and ages (t_i=0,.07,.1,.2,.5,1,2 Gy), using the corresponding isochrones in the (log $T_{\rm eff}$, log g) plane derived by Cester (1982).

Input values are just (m ; log $T_{\rm eff}$, log g), with their experimental errors, for the 2 component stars. The range of models, compatible with observational data and experimental errors, is determined by the *BOFER* code; so, the center of the range and its amplitude define the solution (Y_0, Z_0 ; t) and its accuracy, respectively.

Binary stars, satisfying our observational requirements, are not numerous: from the published catalogues (Popper, 1980; Andersen, 1991) and from other available material, only two dozens systems can be selected. In fact they should be *dwarf*, eclipsing systems (giving direct values of g and m), classified as *d–type* (no mass exchange). They should have *non–identical* components (2 distinct points on grid) with $\Delta m \geq 0.1\ m_\odot$, and moreover they must be *normal* binaries (no triple systems, RS CVn–types, Algols, etc.). Finally, good accuracy is required for the input data: errors on (m ; log $T_{\rm eff}$, log g) should not exceed (3%; 1%, 2%). Actually, admitted masses are from 0.9 to 3.6 $m_\odot$; extension is planned up to 25 $m_\odot$.

Due to the advanced stage of our computations, the results will become available in the immediate future.

ACKNOWLEDGMENTS

The initial idea of this method is due to B. Cester, who suggested the concept already in 1979. Two years later, the first numerical solutions were presented in a Thesis of one of the authors (Ferluga, 1981). The earliest version of the computer code (*BO-FER*) was written in collaboration with C. Boehm in 1983.

Financial support for M. Antillon is provided by a grant of the Italian Foreign Ministry.

REFERENCES

Andersen, J.: 1991, *The Astron. Astrophys. Rev.* **3**, 91

Cester, B.: 1982, private communication

Ferluga, S.: 1981, *Helium in Binary Systems* (in Italian); *Thesis*, University of Trieste (Italy)

Mengel, J.G., Sweigart, A.V., Demarque, P., Gross, P.G.: 1979, *Ap. J. Suppl.* **40**, 733

Popper, D.M.: 1980, *Ann. Rev. of Astron. and Astrophys.* **18**, 115.

Inside the Stars, IAU Colloquium 137
ASP Conference Series, Vol. 40, 1993
Werner W. Weiss and Annie Baglin (eds.)

MODELLING THE ζ HERCULIS SYSTEM

Y. LEBRETON[1], M. AUVERGNE[1], P.J. MOREL[2], A. BAGLIN[1].
[1] DASGAL, URA CNRS 335, Observatoire de Paris, 92195 Meudon Cedex, France.
[2] Département Cassini, URA CNRS 1362, Observatoire de la Côte d'Azur, 06304 Nice Cedex, France.

ABSTRACT The ζ Herculis system is one of the closest well-kown binaries. A tentative modelling shows that though the brighter star is easily described with a mass compatible with the observed one and an age of approximately 3.4 billion years, difficulties arise for the secondary, which plea for more precise observations.

OBSERVATIONAL DATA.

The ζ Herculis system (ADS 10157) is a well known visual binary detected in 1787 by Hershell. Its period being close to 34.4 years more than six revolutions have already been observed with long focus astrometric techniques. The accuracy on the sum of the masses is directed by the parallax since the orbital factor is rather well determined from the orbital elements. The most recent Sproul Observatory results (Lippincott, 1981) give a parallax $\pi = 0.087" \pm 0.003$ (plus a correction of 0.0043" to obtain an absolute value) more than 10% smaller than previously. The visual magnitude of the system is 2.81 ± 0.02 and the magnitude difference of the two components is $\Delta m_v = 2.65 \pm 0.1$, corresponding to $m_{v,A} = 2.90 \pm 0.03$ and $m_{v,B} = 5.55 \pm 0.12$.
Due to the large magnitude difference between the components the uncertainty on the mass ratio is quite large. Following Lippincott (1981) we have adopted :

$$M_A + M_B = 2.21 \pm 0.22 M_\odot$$

$$M_A = 1.43 \pm 0.14 M_\odot \text{ and } M_B = 0.78 \pm 0.08 M_\odot$$

* ζ Her A has a spectral type F9IV, a color index $(B-V)_A = 0.64 \pm 0.02$. A detailed analysis from McWilliam (1990) gives $\log T_{eff,A} = 3.76 \pm 0.01$ and $[F_e/H] = 0.0$. With a bolometric correction of -0.20 (Schmidt-Kaler, 1982) and $M_{bol,\odot} = 4.63$, one obtains $\log(L_A/L_\odot) = 0.85 \pm 0.04$.
* ζ Her B is not so well defined. Its spectral type is K0V. The color index difference between the two components has been measured by Muller (1952) and confirmed by Worley (1969) as $\delta C = 0.18 \pm 0.03$, leading to $(B-V)_B = 0.82 \pm 0.06$. Assuming the same metallicity as ζ Her A, it corresponds to $\log T_{eff,B} = 3.704 \pm 0.015$ and, with a bolometric correction of -0.37, to $\log(L_B/L_\odot) = -0.14 \pm 0.12$ leading to a very excentric position in the HR diagram (fig. 1) which cannot be attributed to a duplicity of ζ Her B, with a total mass M_B.

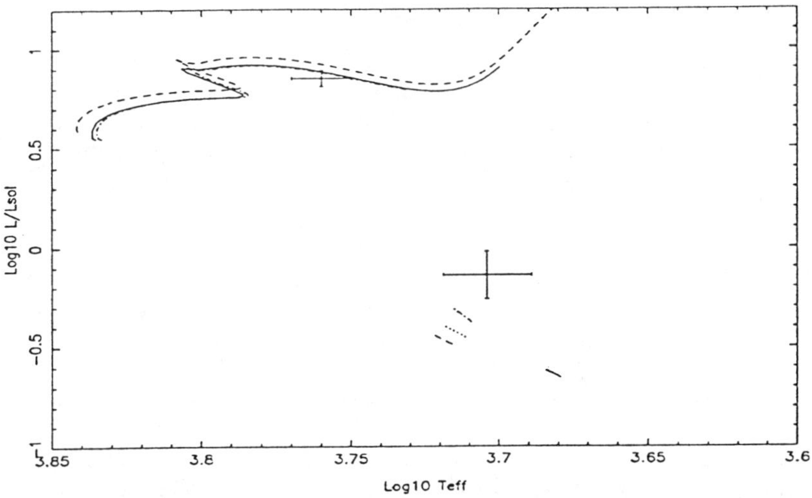

FIGURE I Evolutionary tracks and observed position of ζ Her A and B in a theoretical H-R diagram. For ζ Her A three tracks are shown: the continuous track has $M = 1.40M_\odot, Y = 0.29$, the dashed one has $M = 1.43M_\odot, Y = 0.29$ and the dotted-dashed track has $M = 1.43M_\odot, Y = 0.27$. For ζ Her B tracks followed up to 3.5 10^9 yrs. are shown. The continuous track has $M = 0.78M_\odot, Y = 0.29$, the dashed one illustrates the effect of a Y-change ($Y = 0.35$) while the dotted one gives the effect of a mass change ($M = 0.86M_\odot$). The dotted-dashed track has an extreme α value ($\alpha = 1.30$ H_p) and $M = 0.90M_\odot, Y = 0.29$.

EVOLUTIONARY MODELS.

Models were calculated with the CESAM code (Morel, 1992). We used the Rogers and Iglesias (1992) radiative opacities for temperatures greater than 10^4K and Kurucz's (1991) values for $T < 10^4K$, with a smooth transition between the two tables. Nuclear reaction rates are from Caughlan and Fowler (1988). We used the Eggleton et al.(1973) formalism for the equation of state complemented by the Coulomb correction to pressure (Christensen-Dalsgaard, 1991). The classical mixing-length theory for convection is used. The α value is considered as "universal" in the solar mass vicinity; several arguments favor this hypothesis, as the position of the red giant branch of a very wide set of clusters (Lebreton and Maeder 1986, Schaller et al. 1992) and the calibration of the α Centauri system.(Noels et al. 1991).

An overshooting process from the convective core is taken into account, with a maximum allowed distance of 0.2 H_p considered as "universal", as proposed by Schaller et al.(1992).

We used the Anders and Grevesse (1989) photospheric metal mixture with a solar total metal content $Z = 0.0194$.

The calibration of the Sun with these ingredients gives $Y_\odot = 0.291$ and $\alpha = 1.85$ H_p. This α value is used in all the models.

RESULTS.

We are left with two parameters for the modelling of ζ Her: the helium content Y and the age of the system t. As illustrated in Fig 1 an evolutionary track with a given Y-value will reach the observational point (center of the error box) for a particular value of the mass which can be expressed as a function of Y as :

$$M_A(Y)/M_\odot - 1.38 \simeq -2.0(Y - 0.291)$$

and the age is $t_A = 3.4\ 10^9$ years, almost independant of Y.
The observational errors lead to uncertainties in these determinations of $0.07M_\odot$ on the mass and $0.4\ 10^9$ years on the age. However, as the system has a solar metallicity, the most probable value for Y remains the solar one, i.e. $Y = 0.291$, leading to a mass of $1.38 \pm 0.04M_\odot$.
From the theoretical point of view, the observed position of the secondary in the HR diagram could only be explained with a very large Y and a very small mass, a larger Z and a very high mass, and/or a much smaller α. None of these possibilities is very attractive. Either observations are given with extremely optimistic error bars or some unsuspected pecularity is present.

ACKNOWLEDGMENTS

We specially thank P. Couteau for making available to us his double stars files and for discussions concerning the ζ Her orbit, J. Christensen-Dalsgaard for providing us with his CEFF equation of state routine and members of the "GDR Structure Interne, CNRS" for the implementation of the CESAM code.

REFERENCES

Anders, E., Grevesse, N. 1991, *Geochim. Cosmochim. Acta,* **53**, 197.
Caughlan, G.R., Fowler W.A., 1988, *Atomic and Nuclear Data Tables,* **40**, 284.
Christensen-Dalsgaard, J. 1991, in *Challenges to Theories of the Structure of Moderate-Mass Stars,* D.O. Gough, J. Toomre (Eds), Springer-Verlag p. 11-36.
Eggleton P. P., Faulkner J., Flannery B.P., 1973, *A&A* **23**, 325.
Kurucz, R.L., 1991, *Beyond Classical Models,* NATO ASI Series. Eds L. Crivellary, I. Hubeny, D.G. Hummer, Dordrecht, in press
Lebreton, Y., Maeder, A. 1986, *A&A* **161**, 119.
Lippincott, S.L. 1981, *PASP*, **93**, 376.
McWilliam, A. 1990, *ApJS*, **74**, 1075.
Morel, P.J., 1992, IAU Coll. 137 *Inside the Stars,* W. Weiss, A. Baglin Eds., this proceeding.
Muller, P. 1952, *Ann. Astrophys.,* **15**, 79.
Noels, A., Grevesse, N., Magain, P., Neuforge, C., Baglin, A., Lebreton, Y. 1991, *A&A* **247**, 91.
Rogers, F.J., Iglesias, C.A. 1992, *ApJS*, **79**, 507.
Schaller G., Scherer, D., Meynet, G., Maeder, A. 1992, *A&AS in press*
Schmidt-Kaler Th., 1982, in *Landolt and Bornstein,* **2**, 1
Worley, C.E. 1969, *Astron. J.*, **74**, 764.

Inside the Stars, IAU Colloquium 137
ASP Conference Series, Vol. 40, 1993
Werner W. Weiss and Annie Baglin (eds.)

PECULIAR INTRINSIC PARAMETERS OF MASSIVE STARS AS A RESULT OF AN EARLY PHASE OF HOMOGENEOUS EVOLUTION

K. P. SIMON
Institut für Astronomie und Astrophysik und Observatorium Wendelstein der Universität München, Scheinerstr. 1, D-8000 München

ABSTRACT Rotationallly induced mixing can explain several observational peculiarities of massive stars. Homogeneous evolution of differentially rotating stars including spindown due to stellar winds is studied by means of homology transformations of ZAMS models.

INTRODUCTION

Stellar parameters from careful analysis of photospheric or wind spectra are often in disagreement with masses and compositions predicted by stellar evolution theory. The immediate influence of rotation leads to age estimates of open star clusters, which are too small by almost a factor two. This was shown in a study of α Persei and other galactic clusters by Maeder (1971) and is caused by atmospheric effects rather than changes of the internal structure. However, Groenewegen et al. (1989) pointed out that O-star masses derived from spectroscopic gravity are on the average smaller than those derived from evolutionary tracks. An extended study of 25 galactic OB-stars by Herrero et al. (1992) confirmed this result and found many OB-stars to be overabundant in helium at a position in the HRD where this is not expected from theoretical evolutionary tracks. An overabundance of helium in OBN-stars was first detected by Schönberner et. al (1988) and the abundance pattern of helium and CNO was found to be consistent with the assumption that CNO cycled matter is to be seen in the photospheres of these stars. Standard evolutionary theory fails in explaining these peculiarities.

A scenario, which can explain such peculiarities, is that of homogeneous evolution due to rotationally induced mixing. This mechanism was studied by Maeder (1987) in order to explain the evolutionary status of ON blue stragglers. However, Maeder paid no attention to the fact that massive stars lose angular momentum by stellar winds and that the rotational levitation effect changes the stellar structure.

HOMOLOGY, HOMOGENEOUS EVOLUTION AND SPINDOWN

An homology transformation for massive, rotating stars can be found if we assume that a potential function Φ exists, and that all local variables depend only on this potential (barotrope stars). This assumption is consistent only with rotational laws which do not vary with distance from the equatorial plane (Poincare-Wavre theorem). The distribution of angular velocity is assumed to be known and to remain constant.

It must be noted that these assumptions are not fully self-consistent. Ignoring the existence of meridional flows and introducing a potential function makes the assumption of radiative equilibrium inconsistent with the assumption of energy transfer by radiative diffusion only. This is the well known "von Zeipel Paradox"(von Zeipel, 1924). Because the effects to be studied are much larger than the errors introduced by this inconsistency, we will ignore this problem.

We choose the only independent structural variable to be the potential normalized to the the stellar surface, φ. The independent global parameters, total mass M and mean molecular weight μ, are complemented by the rotational parameter η, which represents the equatorial angular velocity in units of the critical angular velocity.

The structural equations must be be modified in order to take the effects of rotation into account. If, for the sake of simplicity, the equipotential surfaces are assumed to be of ellipsoidal shape with the ablation of the corresponding Roche potential, we simply must replace the mass parameter M_φ by $M_\varphi/a(\eta)$ in the hydrostatic equilibrium equation. In the equation of radiative transfer the luminosity L_φ must be replaced by L_φ/I_1. I_1 is an integral over the stellar interior, which is easily evaluated for simple potentials:

$$I_1(\varphi) = -\frac{1}{4\pi G M(\varphi)} \int_{V(\varphi)} \Delta\Phi \, dV'$$

The well known "Eddington Quartic" equation

$$\frac{1-\beta_c}{\beta_c^4} = \Psi_a(T_c) M^2 \mu^4 a(\eta)^{-2} \tag{1}$$

is complemented with an equation derived from radiative transfer and thermal equilibrium

$$\frac{(1-\beta_c)^2}{\beta_c} = \Psi_b(T_c)(1+X_H)X_H X_{CN} \mu I_1(\eta)^{-1} \tag{2}$$

If we consider stars of the same central temperature to form a homology family, the functions $\Psi_a(T_c)$ and $\Psi_b(T_c)$ are homology transformation invariants, which can be found from an appropriate set of numerical ZAMS models of nonrotating stars. $I_1(\eta)$ is the integral I_1 at the surface and depends on η and the chosen rotational law. Given M, μ, and η, as well as X_H and X_{CN}, these two nonlinear equations define β_c and T_c. Knowing β_c and T_c, the total luminosity and the density follow from homology transformations.

In order to compute the loss of angular momentum we can use results from modern theory of spherically symmetric stellar winds, though rapidly rotating stars exhibit a significant ablation in conjunction with a severe variation of T_{eff} and gravity from pole to equator. It can be shown that the mass flux is almost constant on the surface and that $\dot{M}$ scales with luminosity. The angular momentum loss rate is given by:

$$\dot{J} = \int_O \frac{d\dot{M}}{do} s^2 \Omega(s) do = \frac{d}{dt}\left\{\int_V \varrho(s,z) s^2 \Omega(s) dV\right\}$$

From this we find the spin down equation ($\overset{\circ}{x}$ denotes the change rate $\frac{d}{dt}\ln(x)$):

$$\overset{\circ}{\eta}\left\{6-\frac{d\ln a}{d\ln\eta}\right\}+\overset{\circ}{\beta}_c\left\{\frac{1}{\beta_c}\right\}+\overset{\circ}{T}_c\left\{3+\frac{d\ln(\rho_c/\bar{\varrho})}{d\ln T_c}+6\frac{d\ln\bar{\omega}}{d\ln T_c}+12\frac{d\ln r_g}{d\ln T_c}\right\}=$$
$$10\,\overset{\circ}{M}\left\{\frac{2}{5r_g^2}\frac{I_2}{I_1\bar{\omega}}-1\right\}+\overset{\circ}{\mu} \qquad (3)$$

with the gyrational radius r_g, a weighted mean $\bar{\omega}$ of $\omega(s)$, and the surface integral I_2, which is unity for nonrotating stars:

$$I_2=\frac{3}{8\pi\, G\, M}\int_O s^2\omega\ \mathrm{grad}\Phi\ do$$

$\bar{\omega}$ is dependent on the density stratification and is an homology invariant like the gyration radius. For $\omega(s)\sim s^{-\alpha}$ with $\alpha\ll 1$ the parameter $\bar{\omega}$ is only slightly larger than unity. If $\overset{\circ}{M}$ and $\overset{\circ}{\mu}$ are given, the equations (1), (2) and (3) form a closed set, which determines the time development of β_c, T_c, and the rotational parameter η.

RESULTS

HRD positions of homogeneously mixed, rotating massive stars can be found by means of an homology transformation. We used scaled models of Maeder (1990), and the mass loss rates are calculated according to Kudritzki et al. (1989). If a velocity of at least 350 km/s is taken to be a precondition of the mixing process, as Maeder (1987) suggested, the mixing ends before hydrogen is exhausted. After the mixing has stopped, the stars enter an evolutionary path for helium enriched stars with an equilibrium C/N ratio, and with O essentially unchanged. Due to their increased helium content these stars are slightly overluminous. Timescales of the homogeneous phase and initial masses can be estimated and, if available from spectral analysis, the C/N ratio gives hints to the central temperature at the end of the mixed phase.

REFERENCES

Herrero, A., Kudritzki, R.P., Vilchez, J.M., Kunze, D., Butler, K. Haser, S.: 1992, *A&A*, in press

Kudritzki, R.P., Pauldrach, A., Puls, J., Abbott, D.C.: 1989, *A&A* **219**, 205

Maeder, A.: 1971, *A&A* **10**, 354

Maeder, A.: 1987, *A&A* **178**, 159

Maeder, A.: 1990, *A&AS* **84**, 139

Schönberner, D., Herrero, A., Becker, S.,Eber, F., Butler, K., Kudritzki, R.P., Simon, K.P.: 1988, *A&A* **197**, 209

Zeipel, H. von: 1924, *MNRAS* **84**, 665

Inside the Stars, IAU Colloquium 137
ASP Conference Series, Vol. 40, 1993
Werner W. Weiss and Annie Baglin (eds.)

THE BLUE STRAGGLER STARS OF GALACTIC CLUSTERS

D. SCHÖNBERNER AND R. NAPIWOTZKI
Institut für theoretische Physik und Sternwarte der Universität Kiel
Olshausenstr. 40, W-2300 Kiel, Germany

The origin and evolution of blue straggler stars (BS) in clusters is still an unsolved problem for our understanding of stellar evolution. Several rivalling theories exist, of which that of Saio & Wheeler (1980) is the most interesting one since it assumes additional mixing of fuel in the interior of BS. This theory predicts larger luminosity-to-mass ratios for a given position in the HR diagram compared to the standard evolution theory. This theory has been taylored for the old galactic cluster NGC 7789 which contains many bona fide BS, but it can also be used for other clusters of similar ages. The other theories rely on the standard theory of stellar evolution and assume (1) binary mass exchange or even merging, or (2) delayed star formation. In either case follows a "normal" luminosity-to-mass ratio. We report here on a careful spectroscopic investigation in order to discrimate between these rivalling theories.

THE OBSERVATIONS, FITTING PROCEDURE, AND INTERPRETATION

We observed B- and A-type BS in the clusters NGC 7789, M 67, NGC 752, and NGC 2632 (Praesepe) with the Coudé spectrograph of the 2.2 m telescope at the Calar Alto Observatory, equipped with a RCA CCD. The (reciprocal) dispersion was 9 Å/mm, and most images were centered on H_γ or H_β. Some images contained the He I 4471 Å line. Additional spectrograms were obtained for bright, well known stars to serve as comparisons.

Effective temperatures were derived by using existing photometric data ($uvby\beta$) according to the prescriptions of Moon & Dworetsky (1985). This temperature determination appears to be very accurate ($\approx \pm 150$ K) and agrees very well with that based on UV fluxes (Napiwotzki et al. 1992; Wenske & Schönberner 1992).

With $T_{\rm eff}$ fixed, surface accelerations, (projected) rotational velocities and Helium-to-Hydrogen ratios (if feasible) were derived by matching theoretical line profiles, computed with Kurucz's fully line blanketed model atmospheres, to the observed ones. The results are listed in Table 1. The fit error of $\log g$ is in most cases smaller than 0.05 dex. Together with the temperature uncertainity, the total error of $\log g$ is not larger than only 0.1 dex!

Earlier attempts to solve the BS riddle by a direct mass determination failed because of insufficient data quality and distance errors (note that distance enters squared into any direct mass determination). We use the luminosity to mass ratio, $L/M \sim g^{-1}T_{\rm eff}^4$ which is a distance independent quantity. Actually, we compare the relative luminosity-to-mass ratio, $(L/M)_{\rm ZAMS}/(L/M) = g/g_{\rm ZAMS}$ with $R/R_{\rm ZAMS}$ at the same temperature. $R_{\rm ZAMS}(T_{\rm eff})$ and $g_{\rm ZAMS}(T_{\rm eff})$ can be

TABLE I Temperatures, gravities, and rotation of BS

Cluster	No.	T_{eff} (K)	$\log g$ H_β	H_γ	adopted	He/H	$v \sin i$ ($\frac{\text{km}}{\text{s}}$)
NGC 7789	M 459	10800	3.67		3.67		100
	M 460/K 282	9400	3.64	3.62	3.63		80
	M 502/K 342	10600	3.95	4.01	3.98		100
	M 574/K 453	11650	4.02 4.12	4.04	4.06		120
	M 747/K 677	10650	3.55	3.59	3.57		40
M 67	F 81	12730		4.25	4.25	0.10	80
NGC 752	Hm 209	9690	4.22	4.22	4.22		50
NGC 2632	40 Cnc	9580		3.91	3.91		20
	γ Gem	9260		3.51	3.51		≤ 20
	α CMa	9970		4.30	4.30		≤ 20
	β Leo	8850		4.05	4.05		120
	α Lyr	9660		4.03	4.03		≤ 20

taken from theory (e.g. Hejlesen 1980), and the distance enters only linearly.

If we plot a diagramme g/g_{ZAMS} vs. R/R_{ZAMS}, with both ratios taken at the same T_{eff}, the ZAMS is projected into the origin, and the evolutionary tracks of different mass into virtually one single line ($T_{\text{eff}} \geq 6500$ K). The artificially mixed models of Saio & Wheeler (1980) run steeper and depend on the parameter $q_{\text{mix}} = M_{\text{mix}}/M$ which indicates the mass fraction that is kept chemically homogenous. The differences to the standard evolution is quite noticeable for $q_{\text{mix}} \geq 0.8$ (Fig. 1). Homogeneous evolution ($q_{\text{mix}} \approx 1$) proceeds nearly vertically!

For the 4 comparison stars in Table 1, stellar radii are known from parallaxes and angular diameter measurements (Code et al. 1976), and they are also shown in Fig. 1. We conclude that our method can separate stars that evolve according to the standard rules from those that behave as proposed by Saio & Wheeler, provided distances and gravities are sufficiently well known, and $q_{\text{mix}} \geq 0.7$!

According to Saio & Wheeler (1982) the brightest BS of NGC 7789 still close to the ZAMS can only be explained with $q_{\text{mix}} \approx 0.7 \ldots 0.95$. Likewise, $q_{\text{mix}} > 0.95$ is needed for F 81 in M 67, $q_{\text{mix}} \approx 0.9$ for Hm 209 in NGC 752, and $q_{\text{mix}} \approx 0.5$ for 40 Cnc in NGC 2632. To compute individual radii, we used the V fluxes of the Kurucz models for the appropiate T_{eff} and compared them to the observed fluxes from V_0, which yields angular diameters. With the cluster distances d follows then R. The cluster distances are taken from the literature. That of NGC 7789 is the most uncertain one, with $\Delta(V - M_V) = \pm 0.2$. Our mean value of $V - M_V = 11.40$ corresponds to the recent determination of Twarog & Tyson (1985). Further we used $V - M_V = 9.55$ for M 67, 8.20 for NGC 752, and 6.00 for NGC 2632. With the radii known, these BS can be also plotted into the g/g_{ZAMS}–R/R_{ZAMS} diagramme of Fig. 1. This figure shows very convincing that all investigated BS have evolved according to standard evolutionary theory! The regression line, $\Delta \log g = -0.002 - 1.663 \Delta \log R$, nearly coincides with the

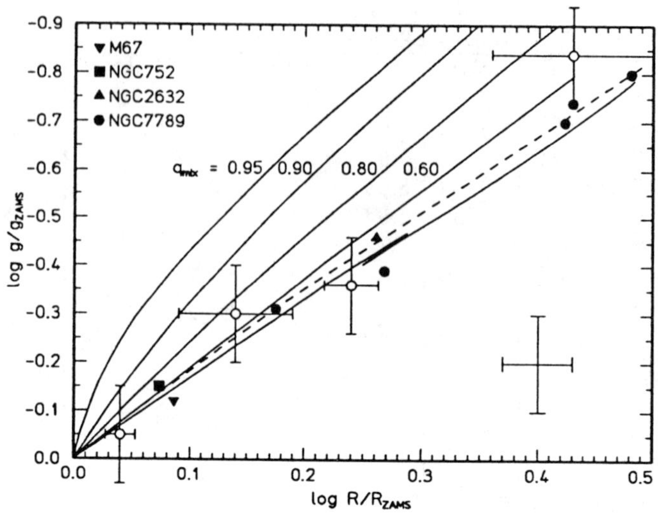

FIGURE I g/g_{ZAMS}-R/R_{ZAMS} diagramme for BS (filled symbols) and comparison stars (open symbols). A typical error bar for the BS parameters is given in the lower right corner.

predictions of this theory. Saio &Wheeler's "mixing theory" would predict a large scatter in Fig. 1 according to the varios mixing parameters assigned to these objects ($q_{\mathrm{mix}} \approx 0.7\ldots1$).

CONCLUSIONS

We disproofed by a careful spectroscopic analysis of BS stars in various open clusters the possibility of additional mixing processes as proposed by Saio & Wheeler (1980) and (Wheeler 1979). The investigated BS stars are in an evolutionary state consistent with standard evolution theory. BS stars must be the result of either (1) binary star evolution (mass exchange or merger), or (2) delayed star formation.

REFERENCES

Code A.D., Davis, J., Bless, R.C., Hanbury Brown, R., 1976, *ApJ* **203**, 417
Napiwotzki, R., Schönberner, D., Wenske, V., 1992, A&A in press
Saio, H., Wheeler, J.C., 1980, *ApJ* **242**, 1176
Twarog, B.A., Tyson, N., 1985, *AJ* **90**, 1247
Wenske, V., Schönberner, D., 1992, these proceddings
Wheeler, J.C., 1979, *ApJ* **234**, 569

Inside the Stars, IAU Colloquium 137
ASP Conference Series, Vol. 40, 1993
Werner W. Weiss and Annie Baglin (eds.)

PULSATION THEORY AND STELLAR STRUCTURE

J. CHRISTENSEN-DALSGAARD
Institut for Fysik og Astronomi, Aarhus Universitet,
DK 8000 Aarhus C, Denmark

ABSTRACT Observed periods of pulsating stars provide information about the properties of the stars. As examples I here consider double-mode pulsators, solar-like oscillations and δ Scuti stars. The ongoing expansion in the observational efforts in this area may be expected to lead to increasingly detailed tests of stellar evolution theory.

1 INTRODUCTION

The periods of the modes of oscillation of a star depend on the global properties of the star, such as mass and radius, and on its internal structure. Therefore, if one or more modes are observed to be excited one can use the observed periods to gain information about the star. It is evident that the amount of information increases with the number of modes that are observed. Extreme cases are the classical Cepheids which appear to pulsate in a single mode only, generally identified to be the fundamental radial mode; and the Sun where thousands of periods have been observed with great accuracy.

At the simplest level, the pulsation period Π can be estimated as being related to the free-fall time scale over a substantial part of the radius of the star:

$$\Pi \sim t_{\rm dyn} \equiv \left(\frac{GM}{R^3}\right)^{-1/2} , \tag{1}$$

where G is the gravitational constant, and M and R are the stellar mass and radius. A similar relation results from considering the travel time of an acoustic wave over a stellar radius; the equivalence of the two approaches is a consequence of the virial theorem, which relates the gravitational and thermal properties of the star. For many modes, particularly those of radial oscillation, equation (1) gives a reasonable approximation to the dependence of the period on the stellar mass and radius. This already provides interesting information: $t_{\rm dyn}$ varies from being of order a year for red giant pulsators, over a few days for Cepheids and a few hours for typical main-sequence stars to a few seconds for white dwarfs. Hence even the simple detection of a single oscillation period for a red star can distinguish between its being a nearby main-sequence star or a distant giant.

Equation (1) clearly involves a factor which depends on the precise type of mode, and on the details of the internal structure of the star. To quantify this it is common, particularly in studies of classical pulsating stars, to express the period in terms of the *pulsation constant*

$$Q \equiv \Pi \left(\frac{M}{M_\odot}\right)^{1/2} \left(\frac{R}{R_\odot}\right)^{-3/2} , \tag{2}$$

where $M_\odot$ and $R_\odot$ are the solar mass and radius. Thus Q provides information about the more intricate properties of stellar internal structure. The behaviour depends strongly on the type of mode. For acoustic modes, including radial oscillations, where the dominant restoring force is pressure, Q varies relatively slowly with stellar type. However, for gravity modes, where the restoring force is dominated by buoyancy, Q is quite sensitive to fine details in the evolutionary state of the star. Examples of this are discussed in Section 4 below.

In computations of stellar models, the results depend on the assumed physical properties of matter in the star; these include the equation of state, the opacity and rate of nuclear reactions. In addition, the computations require specification of the initial composition of the star; an additional parameter in most calculations specifies the efficiency of convective heat transport near the stellar surface. However, as normally carried out the computations are based on additional assumptions, of uncertain validity. Thus it is common in such "standard" stellar evolution calculations to neglect effects of microscopic diffusion, overshoot from convective envelopes or cores, and mixing resulting from rotational instabilities or from weak motion induced by convective overshoot; furthermore, significant effects on the evolution of mass loss cannot be excluded. Although calculations have been carried out that include one or more of these effects, there are substantial uncertainties in how to treat them.

Such uncertainties must be kept in mind when interpreting the observed periods. Ideally, one would want to determine the basic parameters characterizing the star (*i.e.*, its mass, age and chemical composition); to learn about the properties of stellar matter; and to investigate possible departures from standard stellar evolution theory. Even taking into account other information of a more traditional nature this ideal is evidently entirely out of reach on the basis of data for a single star; this is the case even for the Sun, despite the richness of the oscillation data and the accurate measurements of the solar mass, radius and luminosity. However, with sufficiently detailed information about stars covering a large enough area of the Hertzsprung-Russell diagram, and including the information available for the Sun, there may be hope that we can make substantial progress towards a complete test of stellar evolution theory.

A discussion of the theory of stellar pulsation is evidently entirely outside the scope of the present paper. Reference may be made to monographs on the subject (*e.g.* Unno *et al.* 1989), or to reviews on solar or stellar oscillations (see, for example, Gough & Toomre 1991; Christensen-Dalsgaard & Berthomieu 1991; Chiosi 1990). Instead, I concentrate on a few examples of the relation between the structure of stars and their pulsation periods. In Section 2 I treat the case of double-mode pulsators in the Cepheid instability strip, which provide confirmation of recent opacity calculations. Section 3 is concerned with frequency separations amongst low-degree high-order acoustic modes, which have significant diagnostic potential for the structure of stellar cores. Finally, Section 4 discusses some properties of oscillations of δ Scuti stars, which promise a great deal of information about stellar evolution, so far barely exploited.

2 DOUBLE-MODE PULSATORS: THE EFFECT OF OPACITY

While measurement of a single period in principle gives information about the mass and radius of the star in the combination M/R^3, one might expect that measurement of two periods could be used to determine the mass and radius separately. (In both cases, an assumption is of course that the nature of the modes observed has been identified). There are in fact a number of stars in the Cepheid instability strip observed to oscillate in two modes, identified as the fundamental and first overtone of radial pulsation. Petersen (1973) showed that the periods could be used to determine the masses of these stars. The result was striking: the masses so obtained were smaller by approximately a factor three than the *evolution masses* inferred by comparing evolution calculations with the location of the stars in the HR diagram. This problem has led to a great deal of effort, although until recently without much definite result (for reviews of this and other Cepheid mass problems, see Cox 1980; Simon 1987).

It is common to illustrate the problem in a *Petersen* diagram, where the ratio Π_1/Π_0 between the periods of the first radial overtone and the fundamental is plotted against $\log \Pi_0$. The observed location of a star in such a diagram is given with great precision. The instability strip is so thin that it effectively defines a mean relation between the luminosity and effective temperature. From evolution calculations, assuming a specific evolutionary state for the stars considered, one can then obtain a unique relation between mass, luminosity and effective temperature. In this way one can calculate a sequence of envelope models along the instability strip, parametrized by the stellar mass; by computing Π_0 and Π_1 for these models, one obtains a curve in the Petersen diagram.

This procedure is illustrated in Fig. 1. Here I have used a simplified relation between mass, effective temperature and luminosity (Becker, Iben & Tuggle 1977) based on evolution calculations for Population I models. In all cases envelope models extending to a fractional radius of about 0.05 were used, and the periods were found by solving the equations of adiabatic oscillations. The solid curve was obtained from models computed with the Cox & Tabor (1976) opacities. A comparison between this curve and the observed points, plotted as crosses, illustrates the extent of the discrepancy; also, it is striking that its sign depends on the period and hence the mass of the star: for low-mass models, in the δ Scuti region, the observed ratio exceeds the computed value, whereas the observations are below the computations for Cepheids.

Simon (1982) pointed out that the situation for the double-mode Cepheids could be improved if one were to increase the opacity due to heavy elements by a factor 2 – 3 in the temperature region $T = 10^5 - 2 \times 10^6$ K. On this basis he made a plea for a careful recalculation of opacities. This suggestion was taken further by Andreasen & Petersen (1988), Andreasen (1988) and Petersen (1990); they showed that by increasing the opacities by a factor 2.5 in the range $5.2 < \log T < 5.9$, the period ratios could be brought into agreement with observations for both δ Scuti stars and Cepheids; the physical basis for the modification was at the time somewhat unclear, however.

The response of Π_1/Π_0 to an opacity modification can be investigated by means of the *opacity sensitivity function* $\psi_{0,1}$ (Petersen 1992). It is assumed that the modification to the logarithm of opacity κ is solely a function $\delta \log \kappa(T)$ of T.

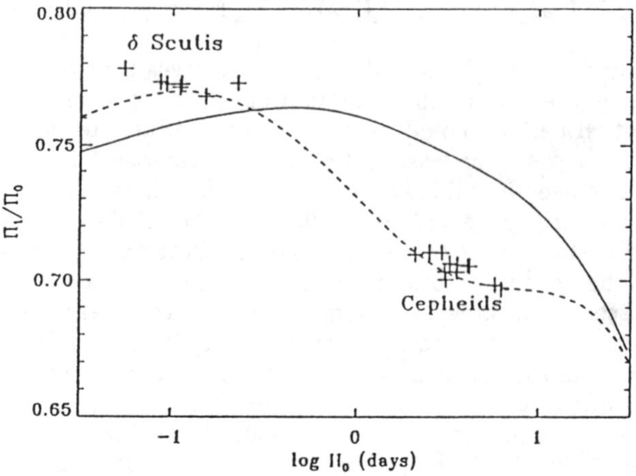

Fig. 1. Petersen diagram, plotting the ratio between the first overtone and fundamental radial period against the logarithm of the latter. The observed values are shown by crosses. The curves show the variation along the instability strip; the solid curve was based on models computed with the Cox & Tabor (1976) opacities, whereas the dashed curve used recent OPAL tables from Rogers & Iglesias (1992).

Assuming that the modification is sufficiently small, the change in the period ratio can be written as

$$\delta\left(\frac{\Pi_1}{\Pi_0}\right) = \int \psi_{0,1}(\log T)\delta \log \kappa(T) d\log T \, , \tag{3}$$

where the integral is over the model, and $\psi_{0,1}(\log T)$ gives the change to Π_1/Π_0 resulting from a delta-function change to $\log\kappa$ at T. Similar functions were used by Korzennik & Ulrich (1989) to interpret differences between observed and computed frequencies of solar oscillation in terms of errors in the opacity.

In practice $\psi_{0,1}$ is computed numerically, by making suitably localized small modifications to the opacity, at a number of points T, and recomputing the model and its periods. Opacity sensitivity functions for two models along the sequence illustrated in Fig. 1, in the δ Scuti and the Cepheid region, are shown in Fig. 2a. It is striking that $\psi_{0,1}$ has different sign for the two models around $\log T = 5.3$. This suggests that an opacity increase in this region would both increase the period ratio for the δ Scuti stars, and decrease it for the Cepheids, hence potentially bringing the models into agreement with the observations; this is evidently consistent with the results obtained by Andreasen & Petersen (1988).

As reviewed by Seaton in these proceedings, there have recently been two independent recalculations of stellar opacities, largely in response to Simon's plea; the results are in good agreement and both indicate substantial opacity increases in the temperature region relevant to the double-mode pulsators. Here I consider results obtained by the Livermore group in the OPAL project (Rogers & Iglesias 1992). Figure 2b shows the opacity change from the Cox & Tabor tables to the OPAL results, evaluated at the conditions in the two envelope

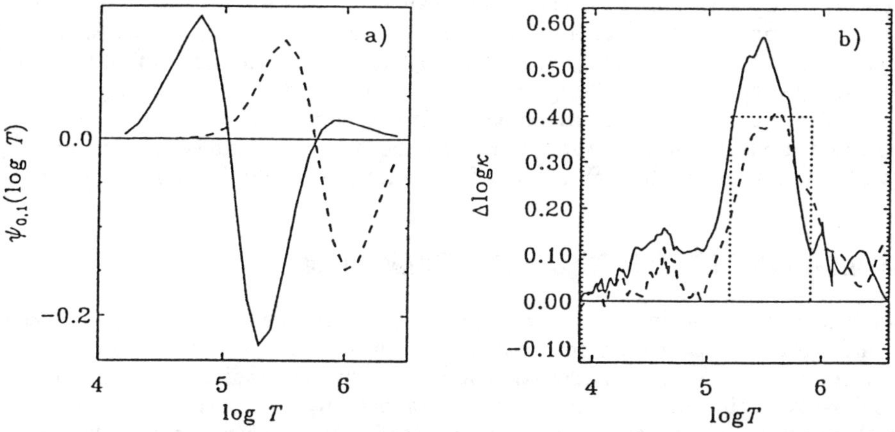

Fig. 2. (a) Opacity sensitivity functions (*cf.* eq. [3]); the solid curve is for a model with $\Pi_0 = 3.5$ days, corresponding to the Cepheid region in Fig. 1, whereas the dashed curve, for a model with $\Pi_0 = 0.09$ days, is representative for the δ Scutis. (b) The opacity increase in the OPAL tables, relative to Cox & Tabor (1976), at the conditions corresponding to the same two models. The dotted box illustrates the opacity increase proposed by Andreasen & Petersen (1988).

models considered in Fig. 2a. Evidently, there is a substantial increase in precisely the temperature range where the period ratio would be affected. The actual opacity change is compared with the change proposed by Andreasen & Petersen; although the details are different, the overall location and magnitude are evidently quite similar. This suggests that the revised opacities might in fact go a long way towards solving the double-mode problem. That this is indeed the case was found in recent calculations by Moskalik, Buchler & Marom (1992).

To illustrate the effect I have computed a second sequence of envelope models using the OPAL opacities. For consistency the relation between mass, luminosity and radius should have been based on evolution calculations with the new opacities; however, since such calculations are so far not available, I have used the same relation as for the previous calculations. The results are illustrated by the dashed line in Fig. 1; there is now generally good agreement between theory and observations.

These results provide a striking example of the detection of errors in the treatment of the basic physical properties of stellar matter through observations of pulsating stars, the errors subsequently being confirmed by a more careful analysis of the physics. It is interesting that this was possible based on observations of only two periods, but for a substantial and fairly uniform sample of stars. Evidence for the need for an opacity increase has also been obtained from analysis of the observed solar oscillation frequencies (*e.g.* Christensen-Dalsgaard *et al.* 1985; Korzennik & Ulrich 1989). In the solar case, however, the structure of the model is insensitive to the opacity at temperatures lower than 2×10^6 K where energy transport occurs through convection. Hence the solar data provide no information about the temperature region corresponding to the massive increase in opacity found in the recent calculations; this illustrates the complementary nature of stellar and solar observations, despite the richness of the latter. Dziembowski, Pamyatnykh & Sienkiewicz (1992) did indeed find that the modest opacity increases at higher temperature found in the new calculations resulted in a substantial improvement in the agreement between solar observations and models; however, some doubt has been cast on this conclusion by the recent revision of the determination of the solar photospheric iron abundance.

3 SOLAR-LIKE OSCILLATIONS IN OTHER STARS

The modes observed in the Sun are concentrated in the period range between 15 and 3 minutes; at low degree, which is the only case relevant to distant stars, such periods correspond to acoustic oscillations of high radial order. Detection of solar-like oscillations in other stars is hindered by their expected very small amplitudes; in the Sun the maximum amplitudes in velocity and relative intensity are $15\,\mathrm{cm\,sec^{-1}}$ and about 3×10^{-6}, respectively. However, there has been some evidence for oscillations resembling those of the Sun in Procyon (Brown *et al.* 1991) and α Centauri A (Pottasch, Butcher & Hoesel 1992); Hence it is of considerable interest to investigate the information content of such modes.

As the modes are of high radial order, they can be treated with reasonable accuracy by means of asymptotic theory. This shows that the cyclic frequency $\nu_{n,l}$ of a mode of degree l and radial order n is approximately given by

$$\nu_{nl} \sim \Delta\nu_0[n + {}^1\!/_2 l + {}^1\!/_4 + \alpha(\nu)] + \epsilon_{nl} \tag{4}$$

(*e.g.* Tassoul 1980). Here

$$\Delta\nu_0 = \left(2\int \frac{dr}{c}\right)^{-1}, \tag{5}$$

c being the adiabatic sound speed and r the distance to the centre of the star, is the sound travel time across a stellar diameter; α is related to the properties of the layer near the stellar surface, and ϵ_{nl} is a small correction term.

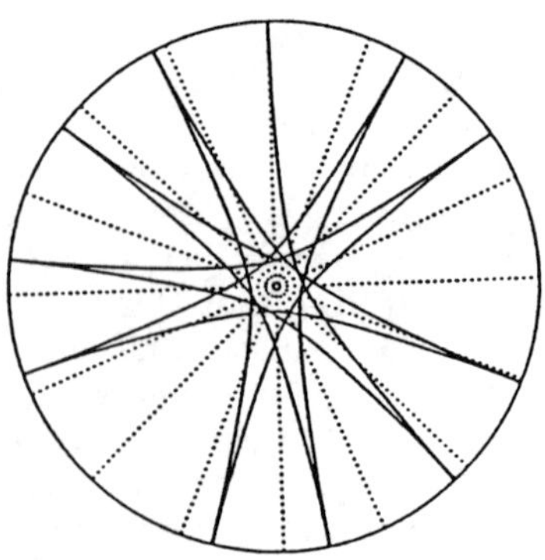

Fig. 3. The propagation of acoustic rays corresponding to a mode of degree 0 (dotted lines) and degree 2 (solid lines), in a solar model.

To leading order, and neglecting the frequency-dependence of α, equation (4) predicts a uniformly spaced frequency spectrum, consisting of superposed peaks of even and odd degree, with a basic separation of $\Delta\nu_0/2$. This separation scales roughly as $(M/R^3)^{1/2}$ and hence provides information of a similar nature to the fundamental radial pulsation period. The term in ϵ_{nl} causes a departure from this simple pattern, conveniently measured by the small separation

$$\delta\nu_{nl} = \nu_{nl} - \nu_{n-1\,l+2} \,. \tag{6}$$

According to asymptotic theory $\delta\nu_{nl}$ is largely determined by the sound-speed gradient in the stellar core. That this separation is predominantly determined by conditions in the core can be understood quite simply by considering the oscillations in terms of acoustic rays (*cf.* Fig. 3). Radial oscillations, with $l = 0$, correspond to rays that propagate radially and hence pass through the centre of the star. Consider now rays corresponding to a mode of degree 2 with almost the same frequency: near the surface they propagate almost vertically, and hence behave in a manner very similar to the rays for the radial mode; but due to their slight inclination from the radial direction they are refracted by the increase in sound speed with increasing depth, and hence do not reach the centre of the star. Thus the behaviour of the rays differ only in the core; and it is plausible that the frequency difference between the two modes provides information about the core. Although the actual behaviour of the modes is more complex, a detailed analysis largely confirms this sensitivity (see Gabriel 1989; Gough & Novotny 1990; Christensen-Dalsgaard 1992a).

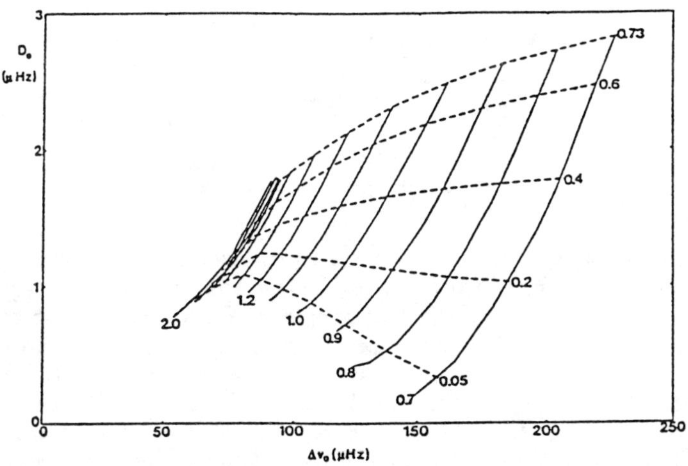

Fig. 4. Evolution tracks (solid lines) and curves of constant central hydrogen abundance X_c (dashed lines), in a $(\Delta\nu_0, D_0)$ diagram. Here $\Delta\nu_0$ is the separation between modes of the same degree, whereas D_0 is a measure of $\delta\nu_{nl}$. The stellar masses, in solar units, and the values of X_c are indicated. (From Christensen-Dalsgaard 1988).

The sound speed in stellar interiors is approximately given by the the relation valid for an ideal gas:

$$c^2 = \frac{\gamma k_B T}{\mu m_u} , \tag{7}$$

where $\gamma \simeq 5/3$, k_B is Boltzmann's constant, m_u is the atomic mass unit and μ is the mean molecular weight. Hence $\delta\nu_{nl}$ measures the combination T/μ. In the solar case, because of their dependence on temperature the separations are useful constraints on models designed to reduce the predicted neutrino flux, effectively eliminating several such models (see, for example Christensen-Dalsgaard 1992b). The dependence on mean molecular weight makes $\delta\nu_{nl}$ a measure of stellar evolution (Ulrich 1986; Christensen-Dalsgaard 1988). This is illustrated in Fig. 4, where evolutionary tracks are plotted in a $(\Delta\nu_0, D_0)$ diagram, D_0 being an average of $(4l+6)^{-1}\delta\nu_{nl}$ which according to asymptotic theory is independent of l. The dependence of $\Delta\nu_0$ on M/R^3 serves to separate the models according to mass, at least for $M \lesssim 1.3 M_\odot$. Given such a diagram, one might hope to use measured values of $\Delta\nu_0$ and D_0 to determine the mass and evolutionary state of main-sequence stars. As stressed by Gough (1987), however, such a determination is affected by uncertainties in other stellar parameters, and in the physics of stellar interiors.

TABLE I

Properties of models without overshoot, and with overshoot over a distance of 0.2 pressure scale heights. The models have the same mass and initial composition.

	No overshoot	Overshoot
Age	3.6 Gyr	3.8 Gyr
$L/L_\odot$	3.71	3.79
$T_{\rm eff}$	6340 K	6350 K

An important issue is the extent to which pulsation observations can be used to detect non-standard features in stars that are superficially similar. As an example, I consider the effect of overshoot from a convective core. This affects the hydrogen profile and hence the core structure, therefore potentially leading to detectable effects in the small frequency separations. I have selected two models from the evolution calculations of Schaller *et al.* (1992). Both have a mass of $1.25 M_\odot$ and an initial composition characterized by a hydrogen and heavy element abundance of 0.68 and 0.02. One model was computed without convective overshoot, the second included overshoot from the core over 0.2 pressure scale heights. Some properties of the models are given in Table I; clearly the stars would be indistinguishable on the basis of classical surface observations.

The profile of hydrogen abundance X, against mass fraction m/M, is shown in Fig. 5a. In the overshoot model, the region of hydrogen depletion extends substantially further than in the normal model, and as a result the central hydrogen abundance is larger. This is also reflected in the total main-sequence lifetimes of the models which are 3.95 Gyr for the normal, and 4.91 Gyr for the

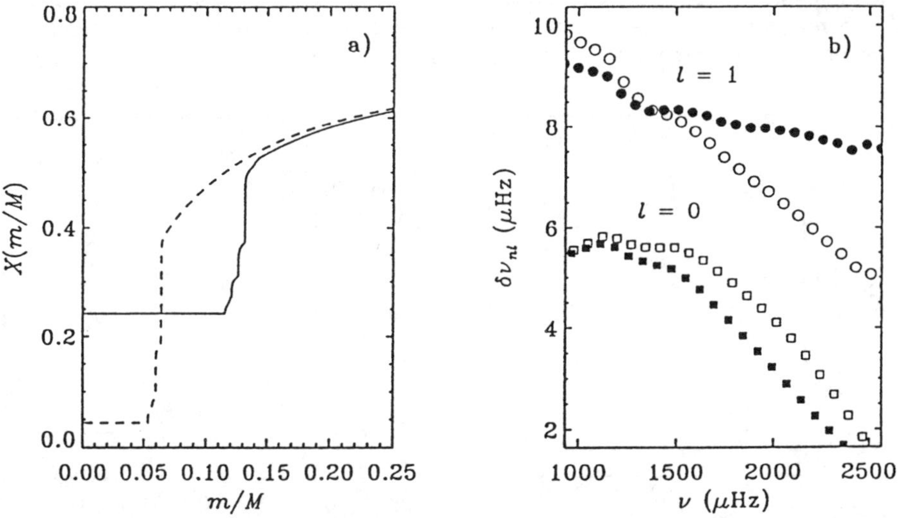

Fig. 5. (a) The hydrogen abundance in the inner portions of a model without overshoot (dashed line) and with overshoot (solid line), against mass fraction. (b) Computed frequency separations $\delta\nu_{nl}$ for $l = 0$ and 1. Open and filled symbols are for the models without and with overshoot, respectively.

overshoot, model. Hence overshoot would have a very substantial effect on age determinations based on the location of clusters in the HR diagram.

Figure 5b shows frequency separations computed for the two models. It is evident that there are quite substantial differences, both in absolute value and in the dependence on frequency. A careful analysis is required to understand how this behaviour is related to the difference in structure between the models; however, it seems plausible that one would be able to distinguish between the models on this basis.

It should perhaps be emphasized again that in reality the analysis would require that other uncertainties in the model parameters be taken into account. Hence it is perhaps unlikely that the presence of overshoot could be unambiguously detected from observations of a single star. Nevertheless, by combining data from several stars, one may hope to be able to probe the extent of overshoot and hence estimate its likely effect on the overall properties of the stars. This would be of obvious importance to our understanding of stellar evolution.

4 SOME REMARKS ABOUT δ SCUTI STARS

There is mounting evidence that many δ Scuti stars have far more complex spectra of oscillation, involving both radial and nonradial modes, than the two radial modes considered in Section 2. This realization is largely the result of extended observing campaigns, involving coordinated observations from several sites (*e.g.* Belmonte *et al.* 1991; Michel & Baglin 1991; Michel *et al.* 1992). Although the interpretation of the observed spectra is difficult (see, for example,

Mangeney *et al.* 1991) the observations in principle provide a substantial amount of information about these stars, which are in many cases in the very interesting evolutionary phase near the end of central hydrogen burning.

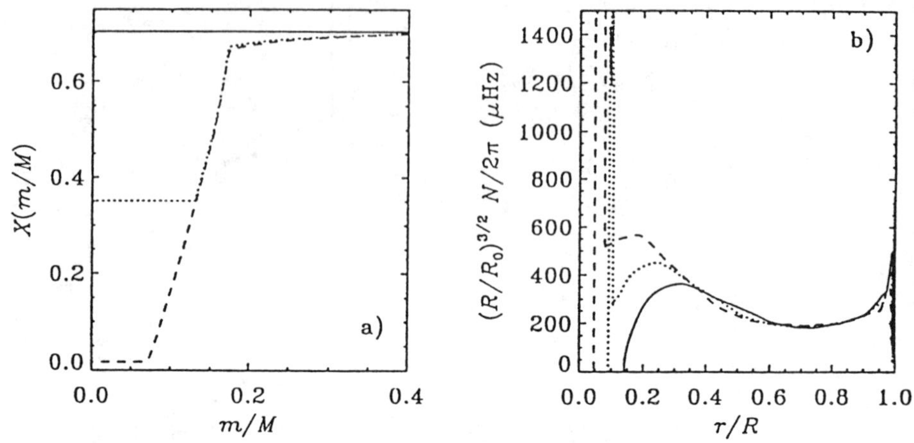

Fig. 6. (a) Hydrogen content X against mass fraction m/M for three models in a $2.2M_\odot$ evolution sequence. The solid line is for age 0, the dotted line for age 0.470 Gyr, and the dashed line for age 0.706 Gyr. (b) Scaled buoyancy frequency, expressed in cyclic frequency, for the same three models, against fractional radius r/R; for the model of age 0.706 Gyr, the maximum value of $N/2\pi$ is 2.4 mHz. In the scaling factor, R and R_0 are the radii of the current and the zero-age main sequence models, respectively.

The modes discussed in Sections 2 and 3 were either purely (in the case of the radial modes) or predominantly (for the high-order p modes) of an acoustic nature. In the case of the δ Scuti stars, however, one has to take into account also the presence of the g mode spectrum. While the p modes predominantly depend on the distribution of sound speed in the star, the behaviour of the g modes is controlled by the buoyancy frequency N, given by

$$N^2 = g\left(\frac{1}{\gamma}\frac{d\ln p}{dr} - \frac{d\ln\rho}{dr}\right) , \tag{8}$$

where g is the local gravitational acceleration, p is pressure and ρ is density. The increasing importance of the g modes, in the period range corresponding to the observations, arises from the build-up of a steep gradient in the hydrogen abundance outside the convective core in these stars. This increases the density gradient, in absolute value, in this region and hence increases N. Consequently the g mode frequencies are increased (see, for example, Dziembowski & Pamyatnykh 1991).

The changes in the hydrogen profile and the buoyancy frequency with stellar evolution are illustrated in Fig. 6, on the basis of an evolution sequence for a $2.2M_\odot$ star. The convective core is fully mixed and here, therefore, the composition is uniform. However, in stars of this mass the convective core shrinks during most of evolution, leaving behind a steep gradient in X, as shown in Fig. 6a.

This causes the sharp peak in the buoyancy frequency seen in Fig. 6b; in the convective core N is essentially zero. Note that in the outer parts of the model N follows a scaling corresponding to equation (1), *i.e.*, decreases as $R^{-3/2}$; to compensate for this I show N scaled by $(R/R_0)^{3/2}$, R_0 being the radius of the star on the zero-age main sequence.

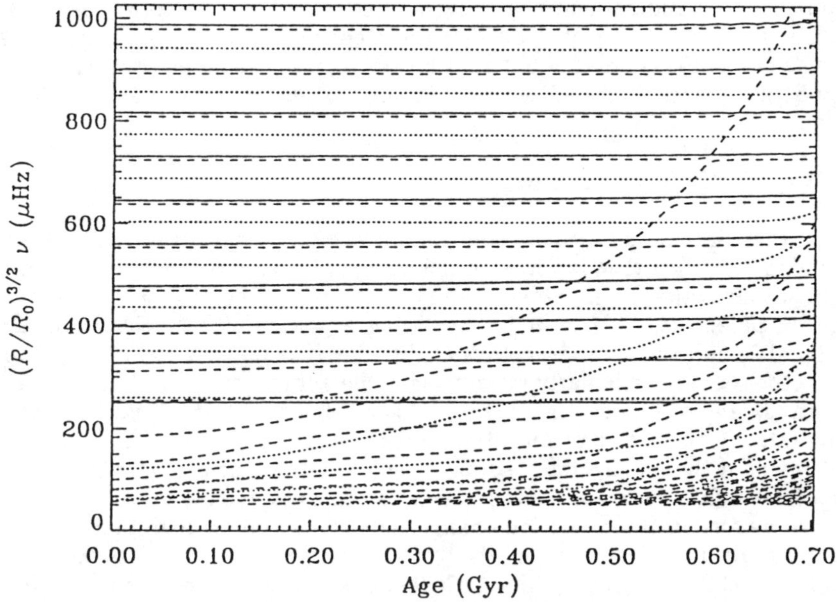

Fig. 7. Scaled oscillation frequencies, as functions of age, in a $2.2 M_\odot$ evolution sequence. Modes of the same radial order have been connected. The solid lines are for radial modes, of degree $l = 0$, the dotted lines are for $l = 1$ and the dashed lines for $l = 2$.

The behaviour of the oscillation frequencies, as functions of stellar age, are shown in Fig. 7; here again I have applied the scaling corresponding to equation (1). As a result, the frequencies of largely acoustic modes, including the radial modes, change very little with age. It should be noticed also that except at low order, these clearly exhibit the asymptotic pattern predicted by equation (4); indeed, there seems to be a pairing of the radial and $l = 2$ modes even at the lowest orders. Such a pattern of closely-spaced peaks has in fact been observed in some cases (*e.g.* Michel *et al.* 1992).

The most striking feature of the computed frequencies, however, is the interaction for $l = 1$ and 2 between the p modes and the g modes (since the g modes depend on buoyancy, they clearly cannot exist for spherically symmetric modes). As was first found by Osaki (1975), this interaction takes place through a sequence of *avoided crossings*, where the frequencies approach closely without actually crossing. At the avoided crossing the two modes exchange nature. This type of behaviour is common to a wide variety of eigenvalue problems; a very clear discussion of the phenomenon, in the context of atomic physics, was given by von Neuman & Wigner (1929). It is evident that the presence of the g-like modes in the p-mode spectrum, particularly at late evolutionary stages, complicates the analysis of observed frequencies. Dziembowski & Królikowska (1990) pointed out that mode selection might be affected by the larger energy of the modes that behave like g modes, thereby restricting the choice of modes in the identification. However, such arguments depend on the mechanisms responsible for exciting the modes and limiting their amplitudes, which are so far incompletely understood. It should also be noted that *if* g-mode like pulsations could in fact be identified, their frequencies would give strong constraints on conditions in the region just outside the stellar core. In fact, Dziembowski & Pamyatnykh (1991) pointed out that measurement of g-mode frequencies might provide a measure of the extent of convective overshoot from the core.

5 CONCLUDING REMARKS

I hope that the examples considered here have given some impression of the richness of information that is actually and potentially available in observed periods of pulsating stars. It should be noted that other types of stars offer equally exciting prospects. The rapidly oscillating Ap stars (Kurtz 1990; Shibahashi 1991), although pulsating in modes of the same nature as seen in the Sun, have distinct features; particularly interesting is the apparent association between the pulsations and the large-scale magnetic field of these stars. A very active field of research is the study of pulsating white dwarfs (*e.g.* Kawaler 1990; Winget 1991), where the modes detected have been identified as g modes; the observations have led to accurate mass determinations for some stars, as well as to the detection of evolutionary period changes at several points along the white-dwarf cooling sequence. There is also likely to be increasing activity in the study of B-star variability, including the β Cephei stars; a particularly striking feature is the fact that the instability of these stars can apparently be understood in terms of the classical opacity mechanism, when the recent OPAL opacities are used (Moskalik & Dziembowski 1992).

Detailed information about stellar interiors require the study of stars pulsating in many modes simultaneously. This immediately puts stringent constraints on the observations: the time series must be sufficiently long to ensure adequate frequency resolution; furthermore, the daily gaps in the data from a single site introduce side-band structure in the observed spectra which greatly complicates the mode identification. An additional difficulty, particularly in the case of solar-like oscillations, is the expected very small amplitudes which require that the observational techniques be pushed to the limit of sensitivity.

To avoid the daily gaps, data from several sites must be combined. A very ambitious and successful project to observe white dwarfs is the so-called Whole Earth Telescope (Winget 1991), which involves twelve sites distributed around the Earth. Coordinated observations have also yielded impressive results in the STEPHI project on δ Scuti stars (Michel & Baglin 1991). It is worth pointing out that if CCD photometry is used observations of δ Scuti stars can in fact be carried out with relatively modest means: Jones *et al.* (1992) obtained very precise measurements using a relatively small telescope in an urban setting. This points to the possibility of using the large number of existing small university telescopes for coordinated observations of selected stars; such a project is very well suited as part of student training.

The most promising prospects for the study of solar-like oscillations are probably observation from space. This eliminates the noise from the Earth's atmosphere and hence allows simple photometric measurements with the required accuracy; also, with a proper choice of orbit gaps in the data can be avoided. The EVRIS project (Baglin 1991) will observe a limited number of stars from the Russian Mars probe MARS 94. Much more extensive data will be available from the PRISMA satellite (Lemaire *et al.* 1991) which is currently undergoing Phase A studies within ESA. If approved, this project will provide detailed information about stars covering a wide range of parameters.

ACKNOWLEDGEMENTS

I am grateful to N. H. Andersen for assistance with the computations required for this paper. J. Otzen Petersen is thanked for extensive discussions about double-mode variables and for commenting on an early version of the manuscript. I thank W. Däppen and G. Houdek for help with implementing the OPAL opacities, and G. Meynet for providing me with models without and with convective overshoot. The High Altitude Observatory is thanked for hospitality during the time where the paper was written. The work reported here was supported by the Danish Natural Science Research Council.

REFERENCES

Andreasen, G. K., 1988. *Astr. Ap.*, **201**, 72.
Andreasen, G. K. & Petersen, J. O., 1988. *Astr. Ap.*, **192**, L4.
Baglin, A., 1991. *Solar Phys.*, **133**, 155.
Becker, S. A., Iben, I. & Tuggle, R. S., 1977. *Ap. J.*, **218**, 633.
Belmonte, J. A., Chevreton, M., Mangeney, A., Praderie, F., Saint-Pé, O., Puget, P., Alvarez, M. & Roca Cortés, T., 1991. *Astr. Ap.*, **246**, 71.
Brown, T. M., Gilliland, R. L., Noyes, R. W. & Ramsey, L. W., 1991. *Ap. J.*, **368**, 599.
Chiosi, C., 1990. In *Confrontation between stellar pulsation and evolution*, *A.S.P. Conf. Ser.*, **11**, p. 158, eds Cacciari, C. & Clementini, G.
Christensen-Dalsgaard, J., 1988. *Proc. IAU Symposium No 123, Advances in helio- and asteroseismology*, p. 295, eds Christensen-Dalsgaard, J. & Frandsen, S., Reidel, Dordrecht.
Christensen-Dalsgaard, J., 1992a. *Geophys. Astrophys. Fluid Dyn.*, **62**, 123.
Christensen-Dalsgaard, J., 1992b. In *The Sun: A Laboratory for Astrophysics* eds Schmelz, J. T. & Brown, J. C., Kluwer Academic Press, Dordrecht, in the press.
Christensen-Dalsgaard, J. & Berthomieu, G., 1991. In *Solar interior and atmosphere*, p. 401, eds Cox, A. N., Livingston, W. C. & Matthews, M., University of Arizona Press.
Christensen-Dalsgaard, J., Duvall, T. L., Gough, D. O., Harvey, J. W. & Rhodes, E. J., 1985. *Nature*, **315**, 378.
Cox, A. N., 1980. *Ann. Rev. Astron. Astrophys.*, **18**, 15.
Cox, A. N. & Tabor, J. E., 1976. *Ap. J. Suppl.*, **31**, 271.
Dziembowski, W. A. & Królikowska, M., 1990. *Acta Astron.*, **40**, 19.

Dziembowski, W. A. & Pamyatnykh, A. A., 1991. *Astr. Ap.*, **248**, L11.
Dziembowski, W. A., Pamyatnykh, A. A. & Sienkiewicz, R., 1992. *Acta Astron.*, **42**, 5.
Gabriel, M., 1989. *Astr. Ap.*, **226**, 278.
Gough, D. O., 1987. *Nature*, **326**, 257.
Gough, D. O. & Novotny, E., 1990. *Solar Phys.*, **128**, 143.
Gough, D. O. & Toomre, J., 1991. *Ann. Rev. Astron. Astrophys.*, **29**, 627.
Jones, A., Kjeldsen, H. Frandsen, S., Christensen-Dalsgaard, J., Hjorth, J., Sodemann, M., Thomsen, B. & Viskum, M., 1992. To be submitted to *Astr. Ap.*
Kawaler, S. D., 1990. In *Confrontation between stellar pulsation and evolution, A.S.P. Conf. Ser.*, **11**, p. 494, eds Cacciari, C. & Clementini, G.
Korzennik, S. G. & Ulrich, R. K., 1989. *Ap. J.*, **339**, 1144.
Kurtz, D. W., 1990. *Ann. Rev. Astron. Astrophys.*, **28**, 607.
Lemaire, P., Appourchaux, T., Catala, C., Catalano, S., Frandsen, S., Jones, A. & Weiss, W., 1991. *Adv. Space. Res.*, **Vol. 11, No. 4**, 141.
Mangeney, A., Däppen, W., Praderie, F. & Belmonte, J. A., 1991. *Astr. Ap.*, **244**, 351.
Michel, E. & Baglin, A., 1991. *Adv. Space Res.*, **vol. 11, No. 4**, 167.
Michel, E., Belmonte, J. A., Alvarez, M., Jiang, S. Y., Chevreton, M., Auvergne, M., Goupil, M. J., Baglin, A., Mangeney, A., Roca Cortés, T., Liu, Y. Y., Fu, J. N., Dolez, N., 1992. *Astr. Ap.*, **255**, 139.
Moskalik, P. & Dziembowski, W. A., 1992. *Astr. Ap.*, **256**, L5.
Moskalik, P., Buchler, J. R. & Marom, A., 1992. *Ap. J.*, **385**, 685.
Osaki, Y., 1975. *Publ. Astron. Soc. Japan*, **27**, 237.
Petersen, J. O., 1973. *Astr. Ap.*, **27**, 89.
Petersen, J. O., 1990. *Astr. Ap.*, **238**, 160.
Petersen, J. O., 1992. *Astr. Ap.*, in the press.
Pottasch, E. M., Butcher, H. R. & van Hoesel, F. H. J., 1992. *Astr. Ap.*, in the press.
Rogers, F. J. & Iglesias, C. A., 1992. *Ap. J. Suppl.*, **79**, 507.
Schaller, G., Schaerer, D., Meynet, G. & Maeder, A., 1992. *Astr. Ap. Suppl.*, in the press.
Shibahashi, H., 1991. In *Challenges to theories of the structure of moderate-mass stars, Lecture Notes in Physics*, vol. **388**, p. 393, eds Gough, D. O. & Toomre, J., Springer, Heidelberg.
Simon, N. R., 1982. *Ap. J.*, **260**, L87.
Simon, N. R., 1987. In *Stellar pulsation, Lecture Notes in Physics*, vol. **274**, p. 148, eds Cox, A. N., Sparks, W. M. & Starrfield, S. G., Springer, Berlin.
Tassoul, M., 1980. *Ap. J. Suppl.*, **43**, 469.
Ulrich, R. K., 1986. *Ap. J.*, **306**, L37 - L40.
Unno, W., Osaki, Y., Ando, H., Saio, H. & Shibahashi, H., 1989. *Nonradial Oscillations of Stars, 2nd Edition* (University of Tokyo Press).
von Neuman, J. & Wigner, E., 1929. *Phys. Z.*, **30**, 465.
Winget, D. E., 1991. In *White dwarfs*, p. 129, eds Vauclair, G. & Sion, E., Kluwer, Dordrecht, The Netherlands.

Inside the Stars, IAU Colloquium 137
ASP Conference Series, Vol. 40, 1993
Werner W. Weiss and Annie Baglin (eds.)

PULSATION OF ROTATING AND MAGNETIC STARS

SERGEI V. VORONTSOV
Institute of Astronomy and Department of Applied Mathematics and Theoretical Physics, University of Cambridge, Cambridge CB3 0HA, England, and
Institute of Physics of the Earth, Moscow 123810, Russia (permanent address)

ABSTRACT Rotation and a magnetic field break the spherical symmetry of a star viewed as a pulsating system, lifting the degeneracy of oscillation frequencies, and leading to (sometimes prominent) observational consequences. Theoretical studies of rotational and magnetic effects in pulsating stars are reviewed, starting with simple configurations with slow rotation and weak magnetic fields.

INTRODUCTION

The theoretical oscillation spectrum of a spherically symmetric, non-rotating and non-magnetic star is largely degenerate. The nature of this degeneracy is made transparent by the simple considerations about the completeness property of the oscillation spectrum of the spherically symmetric star.

An arbitrary displacement field in a spherical domain occupied by the star can be decomposed in terms of spheroidal vector spherical harmonics

$$\mathbf{u}_{\ell mn}(\mathbf{r}) = U_{\ell mn}(r)\hat{r}Y_{\ell m}(\theta,\phi) + V_{\ell mn}(r)\nabla_1 Y_{\ell m}(\theta,\phi), \tag{1}$$

and toroidal vector spherical harmonics

$$\mathbf{u}_{\ell mn}(\mathbf{r}) = -W_{\ell mn}(r)\hat{r} \times \nabla_1 Y_{\ell m}(\theta,\phi), \tag{2}$$

where ∇_1 is angular part of gradient operator

$$\nabla_1 = \hat{\theta}\frac{\partial}{\partial\theta} + \frac{1}{\sin\theta}\hat{\phi}\frac{\partial}{\partial\phi}, \tag{3}$$

and $\hat{r}, \hat{\theta}, \hat{\phi}$ are unit vectors in a spherical coordinate system. In linear analysis, the time dependence is separated by a common factor $\exp(i\omega t)$.

The usual p, f and g modes represent the spheroidal oscillations. Because of the spherical symmetry, their frequencies are degenerate with respect to the azimuthal order m.

Toroidal fields are horizontal and divergence-free (solenoidal). They describe torsional oscillations, with restoring forces provided by the shear stresses. With the only exception of neutron stars (McDermott, Van Horn and

Hansen 1988; Strohmayer *et al.* 1991), the shear stresses actually produce no restoring forces, and frequencies of the torsional oscillations are zero ("trivial modes"). Formally, these modes describe an arbitrary time-independent differential rotation.

In the spectrum of spheroidal oscillations, there is also one "trivial mode" with zero frequency, which is the fundamental mode of degree $\ell = 1$. This mode describes the free motion of the star as a whole. The complete oscillation spectrum thus describes all the possible motions of the star in free space: the motion of the star as a whole, its arbitrary free rotation, and oscillations around the equilibrium.

Rotation and a magnetic field make the oscillation spectrum much reacher. They destroy the spherical symmetry, lifting the degeneracy of the frequencies of p, f and g modes. The splitting of the frequency multiplets with respect to the azimuthal order m can be measured and used for the diagnostics of the internal rotation and the internal magnetic field. In addition to pressure perturbations, which provide the restoring forces for p modes, and buoyancy forces, which are the restoring forces for g and f modes, the rotation and magnetic field provide new restoring forces, which are Coriolis forces and Lorentz forces, leading to the appearance of non-trivial modes of quite different physical nature.

Pulsations of rotating and magnetic stars were an object of intensive study for decades, starting with pioneering works of Cowling and Newing (1949) and Ledoux (1951). There are a number of related topics, including stability of differentially rotating stars, angular momentum transport and interaction between rotation, pulsation and convection, which cannot be covered by this short review. For a more extensive bibliography of current literature, and for the early work, the reader is referred to the monographs of Cox (1980) and Unno *et al.* (1989), and to the papers cited below.

SLOW ROTATION

The influence of rotation on p, f and g modes can be studied using perturbation analysis, if this influence is small. "Slow" rotation means that the Coriolis forces are small compared with inertial forces, i.e. the angular velocity of rotation is small compared with oscillation frequencies, and the rotational distortion of the equilibrium configuration is small, i.e. centrifugal forces are small compared with gravity.

The rotational effects on the oscillation frequencies are usually analysed for the case of axisymmetric rotation, $\Omega = \Omega(r, \theta)$. An interesting situation when the stellar core and the envelope rotate around different axes, and the possibility of detecting such a rotation in the oscillation data, is discussed in these proceedings by Gough and Kosovichev (1992).

First-order effects of rotation

The leading-order effects on the oscillation frequencies are linear in Ω, and independent of the rotational distortion of the equilibrium stellar configuration, which is of order Ω^2. The perturbation analysis is thus straightforward, and does not require any additional assumptions.

A general expression for the frequency perturbations due to an arbitrary axisymmetric differential rotation $\Omega = \Omega(r, \theta)$ was first obtained by Hansen, Cox and Van Horn (1987):

$$\delta\omega = -\frac{(\mathbf{u}, \Omega[m - i\hat{z}\times]\mathbf{u})}{(\mathbf{u}, \mathbf{u})}, \tag{4}$$

where $\hat{z}$ is unit vector in the direction of rotation, which is taken also to be the direction of z-axis of the spherical coordinate system, $\mathbf{u}$ is the zero-order displacement field, and scalar product is defined by

$$(\mathbf{u}_1, \mathbf{u}_2) = \int \rho_0 \mathbf{u}_1^* \cdot \mathbf{u}_2 dv; \tag{5}$$

the integral is taken over the spherical volume occupied by the star, ρ_0 is the equilibrium density and an asterisk denotes complex conjugate. By separating angular dependences, the volume integrals in the expression (4) can be reduced to one-dimensional integrals containing $U(r)$ and $V(r)$ (for the θ-dependence of $\Omega(r, \theta)$ expressed in terms of harmonic functions, see Cuypers 1980).

Below in this section, we discuss the response of the oscillation frequencies to differential rotation, and thus the diagnostic capability of frequency measurements in studying internal stellar rotation. For convenience the angular velocity distribution expanded in terms of spherical harmonics

$$\Omega = \sum_i \Omega_i(r) Y_{i0}(\theta, \phi). \tag{6}$$

Such a representation of $\Omega(r, \theta)$ is a natural choice when working in spherical geometry, and appears to be the most convenient for the practical inversions of rotational splitting (Durney, Hill and Goode 1988; Gough and Thompson 1990). The frequency perturbation becomes (Vorontsov 1981)

$$\delta\omega = \sum_i \int_0^R \rho_0 r^2 \Omega_i \Big[A_{i\ell\ell}^{0mm} U (U_\ell - mU) + B_{i\ell\ell}^{0mm} V (V_\ell - mV) \\ + C_{i\ell,\ell-1}^{0mm*} V W_{\ell-1} + C_{i\ell,\ell+1}^{0mm*} V W_{\ell+1} \Big] dr \Big/ \int_0^R \rho_0 r^2 \left[U^2 + \ell(\ell+1) V^2 \right] dr. \tag{7}$$

The functions $U_\ell(r), V_\ell(r), W_{\ell-1}(r)$ and $W_{\ell+1}(r)$ represent the simple decomposition of the vector field $i\hat{z} \times \mathbf{u}$ in terms of spheroidal and toroidal vector spherical harmonics (Vorontsov and Zharkov 1981), which contains one spheroidal field of the same degree ℓ, and two toroidal fields of degree $\ell - 1$ and $\ell + 1$:

$$i\hat{z} \times \mathbf{u} = U_\ell(r) \hat{r} Y_{\ell m}(\theta, \phi) + V_\ell(r) \nabla_1 Y_{\ell m}(\theta, \phi) \\ - W_{\ell-1}(r) \hat{r} \times \nabla_1 Y_{\ell-1,m}(\theta, \phi) - W_{\ell+1}(r) \hat{r} \times \nabla_1 Y_{\ell+1,m}(\theta, \phi),$$

$$U_\ell = mV, \quad V_\ell = \frac{m}{\ell(\ell+1)} (U + V), \tag{8}$$

$$W_{\ell-1} = -i \left[\frac{(\ell+m)(\ell-m)}{\ell^2 (2\ell+1)(2\ell-1)} \right]^{1/2} [U + (\ell+1)V],$$

$$W_{\ell+1} = i\left[\frac{(\ell+m+1)(\ell-m+1)}{(\ell+1)^2(2\ell+1)(2\ell+3)}\right]^{1/2}[U-\ell V],$$

$U = U_{\ell mn}(r)$ and $V = V_{\ell mn}(r)$ describe the zero-order displacement field $\mathbf{u}(\mathbf{r})$. The angular integrals

$$\begin{aligned}
A^{qM'M}_{kJ'J} &= \int\limits_{4\pi} Y_{kq}Y_{J'M'}Y^*_{JM}\,d\Omega, \\
B^{qM'M}_{kJ'J} &= \int\limits_{4\pi} Y_{kq}\left(\nabla_1 Y_{J'M'}\right)\cdot\left(\nabla_1 Y_{JM}\right)^*\,d\Omega \\
&= \int\limits_{4\pi} Y_{kq}\left(-\hat{r}\times\nabla_1 Y_{J'M'}\right)\cdot\left(-\hat{r}\times\nabla_1 Y_{JM}\right)^*\,d\Omega, \\
C^{qM'M}_{kJ'J} &= \int\limits_{4\pi} Y_{kq}\left(\nabla_1 Y_{J'M'}\right)\cdot\left(-\hat{r}\times\nabla_1 Y_{JM}\right)^*\,d\Omega
\end{aligned} \tag{9}$$

are determined explicitly in terms of Wigner's 3-j symbols (Edmonds 1957; Luh 1973).

When Ω is a function of r alone, expression (7) for the frequency splitting reduces to (see also Gough 1981)

$$\delta\omega = m\frac{\int_0^R \rho_0 r^2\Omega(r)\left[V^2+2UV-U^2-\ell(\ell+1)V^2\right]dr}{\int_0^R \rho_0 r^2\left[U^2+\ell(\ell+1)V^2\right]dr}. \tag{10}$$

For the simplest case of uniform rotation $\Omega=\Omega_0$, it reduces further to

$$\delta\omega = m\Omega_0\frac{\int_0^R \rho_0 r^2\left(V^2+2UV\right)dr}{\int_0^R \rho_0 r^2\left[U^2+\ell(\ell+1)V^2\right]dr} - m\Omega_0, \tag{11}$$

the result first obtained by Cowling and Newing (1949) and Ledoux (1951). When the oscillations of the uniformly rotating star are viewed in the corotating frame, the last term in the expression (11) disappears: this term describes the simple kinematic effect of the transformation of the coordinate system to the inertial frame. The first term on the right-hand side describes the influence of the Coriolis forces.

When $\Omega=\Omega(r)$, $\delta\omega$ is proportional to m; the frequency separation within a particular multiplet is uniform. This frequency separation is different for different multiplets: different modes sample the stellar interior at different depths, and the observational frequency splittings can be used to infer Ω as a function of r.

The latitudinal dependence of the angular velocity of rotation does not affect the symmetry of the multiplet splitting ($\delta\omega$ remains an odd function of m), but distorts the equidistant separation of frequencies with respect to m: modes with higher $|m|$ concentrate more closely around the equatorial plane, and are thus more sensitive to the rotation near this plane. The rotational splittings thus allow one to study the internal rotation as a function of both depth and latitude. Symmetry properties of the spherical harmonics define the selection rules for the contribution of different terms of the expansion (6) to the frequency splitting. The angular integrals (9) differ from zero only if $q = M' - M$ and values of k, J', J satisfy triangle rule. Integrals A and B differ from zero only if $k + J' + J$ is even; for the integral C to differ from zero, $k + J' + J$ must be odd. Only even components in the expansion (6) thus contribute to the frequency splitting: asymmetry of the stellar rotation with respect to the equatorial plane has no influence on the oscillation frequencies. Only the components with $i \leq 2\ell$ contribute to the splitting: higher-order harmonics in the latitude dependence of the rotational velocity can not be detected with low-degree modes.

In helioseismology, reliable measurements of rotational splitting are now available for p modes in a wide range of degree ℓ. For the interpretation of the odd component of the multiplet splitting, the leading-order perturbation analysis is adequate, at least within the current accuracy of the observational data. The reconstruction of the internal differential rotation from the frequency splittings thus represents a linear inverse problem, which is studied successfully with different linear inversion techniques. For the extensive literature in this area, the reader is referred to the current reviews on helioseismology (Gough and Toomre 1991; Libbrecht and Woodard 1991; Dziembowski and Goode 1991a; Gough and Thompson 1991; Vorontsov 1992).

Second-order effects of rotation

The analysis of the effects of order Ω^2 on the oscillation frequencies is more complicated, and for realistic stellar models requires some additional simplifying assumptions. The difficulty comes from the fact that the rotational distortion of the equilibrium stellar configuration is also of order Ω^2, and contributes to the same order to the oscillation frequencies. The rotational distortion of the star thus must be specified; this problem is very difficult in general, if the self-consistent stellar model is required to be not only in dynamical, but also in the thermal equilibrium, and the stellar model itself is an evolutionary model.

After the rotational distortion of the equilibrium configuration is specified somehow, the extension of the perturbation analysis to second order is a straightforward procedure (see e.g. Saio 1981; Vorontsov and Zharkov 1981 for uniform rotation, and Vorontsov 1981; Gough and Thompson 1990; Dziembowski and Goode 1992 for arbitrary axisymmetric differential rotation). The standard stationary perturbation technique for Hermitian operators is usually used. First-order corrections to the eigenfunctions are needed to evaluate the second-order corrections to the oscillation frequencies; they include both the spheroidal and toroidal vector spherical harmonics, the spheroidal part being determined either as an expansion in terms of the unperturbed eigenfunctions of a spherically-symmetric nonrotating star, or by the direct solution of an inhomogeneous differential equation. The eigenfunctions of different modes thus acquire a mixing character, which reflects the linear coupling of different modes induced by the rotation.

The treatment of the effects of distortion of the equilibrium configuration is somewhat simplified by the fact that the contributions of the different quadratic terms to the oscillation frequencies are additive: other second-order terms are independent of the distortion. But the distortional effects can dominate in the second-order terms. The nature of the distortional effects in the frequency splittings is clearly seen for the high-degree p and f modes trapped near the stellar surface. Modes with $m = 0$ can be considered as a superposition of waves travelling around the star along meridians, and modes with $m = \pm\ell$ are formed by waves travelling along the equator. Due to the asphericity induced by the rotation, the wave travel path along the equator is larger than along the meridian, and frequencies of the tesseral modes with $m = \pm\ell$ becomes larger than those of zonal modes ($m = 0$). Together with second-order effects of the Coriolis forces, the distortion of the equilibrium configuration contributes to the multiplet splitting with an asymmetric component (even function of m), because the effect is the same for waves travelling in opposite directions.

In the context of stellar oscillations, the distortion of the equilibrium configuration by the differential rotation was considered in most detail by Dziembowski and Goode (1992). The spherically-symmetric part of the distortion was ignored, because it does not contribute to the multiplet splitting, producing only an overall shift of the multiplets. For solar p modes, it was found that the distortion dominates in the second-order rotational effects (see also Gough and Thompson 1990), and the effect itself is large enough to be taken into account in the interpretation of the even component in the frequency splittings (the even component can be produced also by other aspherical effects, including magnetic field).

The applicability of the standard perturbation technique assumes the linear interaction between different modes to be small. This condition can be violated, if the unperturbed frequencies of the modes which can interact are close enough, with the frequency difference of the order of the rotational splitting. This situation is familiar in terrestrial seismology, and can be treated by an appropriate modification of the perturbational technique for the case of quasi-degeneracy (e.g. Messiah 1961). Three-mode coupling induced by uniform rotation was considered by Vorontsov and Zharkov (1981) with an application to the computation of Jovian oscillations. Dziembowski and Goode (1992) considered accidental degeneracies in the solar p-mode spectrum; the effects of mode coupling on the rotational splitting were found to be insignificant.

WEAK MAGNETIC FIELD

A large-scale magnetic field in the star can be considered weak and its effect on the stellar oscillations can be treated within a perturbation theory, if both the dynamical effects of the magnetic field on the oscillations and the magnetic distortion of the equilibrium configuration are small. For reasonable magnetic fields, these conditions are always satisfied in stellar interiors, but can be violated near the surface of the magnetic star: we shall discuss this situation shortly in the last section of the review.

The perturbation analysis can be based directly on the oscillation equations, which include the effects of magnetic field in MHD approximation (see Gough and Taylor 1984; Dziembowski and Goode 1984; Unno *et al.* 1989; Gough and Thompson 1990; and references therein to early work). An alternative approach is the application of the variational principle (e.g. Kovetz 1966; Lynden-Bell and Ostriker 1967). When the variational principle is used, the convenient symmetric form of the Lagrangian can be written as (Vorontsov 1986)

$$\begin{aligned} L = {} & \Gamma_1 p_0 (\nabla \cdot \mathbf{u}^*)(\nabla \cdot \mathbf{u}) + \frac{1}{2}\left[\rho_0(\mathbf{u}\cdot\nabla)(\mathbf{u}^*\cdot\nabla\psi_0) + \rho_0(\mathbf{u}^*\cdot\nabla)(\mathbf{u}\cdot\nabla\psi_0)\right. \\ & \left. + (\nabla\cdot\mathbf{u}^*)(\mathbf{u}\cdot\nabla p_0) + (\nabla\cdot\mathbf{u})(\mathbf{u}^*\cdot\nabla p_0)\right] + \rho_0(\mathbf{u}\cdot\nabla\psi_1^* + \mathbf{u}^*\cdot\nabla\psi_1) \\ & + \frac{1}{4\pi G}\nabla\psi_1^*\cdot\nabla\psi_1 - \rho_0\omega^2\mathbf{u}^*\cdot\mathbf{u} \\ & + \frac{1}{4\pi}\mathbf{H}_1^*\cdot\mathbf{H}_1 - \frac{1}{8\pi}\left[(\nabla\times\mathbf{H}_0)\times\mathbf{H}_1\cdot\mathbf{u}^* + (\nabla\times\mathbf{H}_0)\times\mathbf{H}_1^*\cdot\mathbf{u}\right], \end{aligned} \tag{12}$$

where $\mathbf{H}_0$ denotes the equilibrium magnetic field, $\mathbf{H}_1$ is its Eulerian perturbation

$$\mathbf{H}_1 = \nabla \times (\mathbf{u} \times \mathbf{H}_0), \tag{13}$$

ψ_1 is the Eulerian perturbation of the gravitational potential. The first three lines in the Lagrangian correspond to the problem of adiabatic oscillations of a non-magnetic star; the last line describes the work done against the Lorentz forces. The application of the Rayleigh's principle gives the equation for the perturbations of the oscillation frequencies due to the influence of the magnetic field:

$$\delta(\omega^2)\int \frac{\partial L}{\partial(\omega^2)}dv = -\int \frac{\partial L}{\partial p_k}\delta p_k dv, \tag{14}$$

where integrals are taken over the unperturbed spherical volume occupied by the star, "parameters" p_k denote the equilibrium values of density ρ_0, pressure p_0, gravitational potential ψ_0, adiabatic exponent Γ_1, components of $\mathbf{H}_0$ and their spatial derivatives. If there is no axial symmetry of the magnetic field, or if the axis of symmetry differs from the axis of the spherical coordinate system which was chosen, zero-order eigenfunctions $\mathbf{u}$ are linear combinations of spheroidal components with different values of m; the coefficients are determined from the requirement that $\delta(\omega^2)$ is to be stationary under their variations.

Both the frequency perturbations and the magnetic distortion of the equilibrium configuration (with the only exception of force-free field, when there is no distortion) are proportional to H^2 in the leading order, and some simplifying assumptions about the equilibrium configuration must enavitably be introduced (see Gough and Thompson 1990 for a discussion). The magnetic splitting of frequency multiplets is described by an even function of m, because the influence of the magnetic field is the same for modes travelling in opposite directions around the star.

In helioseismology, the possible effects of a large-scale magnetic field concentrated in the solar interior were studied by Dziembowski and Goode (1984), Gough and Taylor (1984), Roberts and Campbell (1986), and in more detail by Gough and Thompson (1990). The effects of a toroidal field concentrated in a thin layer near the base of the convection zone were analized by Gough and Thompson (1988a, 1990) and Vorontsov (1988).

In the frequency splitting of solar p modes, the even component is really found to be significant (Jefferies *et al.* 1988; Libbrecht 1989; Libbrecht and Woodard 1990). Gough and Thompson (1988b) found evidence that the original even coefficients measured by Duvall *et al.* (1986) were due to a perturbation near the surface and attributed it to a fibril magnetic field. The analysis of this even component in recent observations indicates strongly that the dominant contribution comes from a perturbation localized near the solar surface, and this perturbation varies with the solar activity cycle (Libbrecht and Woodard 1990). By the inversion of the even component, Dziembowski and Goode (1989, 1991b) found evidence for a small additional source of perturbation, which they attribute to a steady megagauss toroidal magnetic field localized near the base of the convection zone. The dominant near-surface perturbation was attributed by Dziembowski and Goode (1991b) to the effects of a fibril magnetic field, discussed by Zweibel and Däppen (1989) (for the discussion of the effects of fibril field, see also Bogdan 1989; Bogdan and Cattaneo 1989; and references therein).

The prominent variations of the average frequencies of solar p-modes during the solar cycle have been also detected (Elsworth *et al.* 1990; Libbrecht and Woodard 1990; Woodard and Libbrecht 1991), which also indicate a perturbation localized near the surface (Libbrecht and Woodard 1990). The effects of the chromospheric magnetic field (Campbell and Roberts 1989; Evans and Roberts 1990, 1991, 1992; Jain and Roberts 1992; Wright and Thompson 1992) and of the photospheric flux tubes (Goldreich *et al.* 1991) were analyzed to explain these frequency variations.

The possible effects of a large-scale magnetic field in the oscillations of variable white dwarfs were discussed by Jones *et al.* (1989).

JOINT EFFECTS OF THE ROTATION AND MAGNETISM

The discovery of rapid oscillations of the rotating magnetic Ap stars (see Kurtz 1990 for a review) stimulated an extensive study of the joint effects of both the rotation and magnetism, when the magnetic axis is inclined to the axis of rotation (Gough and Taylor 1984; Dziembowski and Goode 1984, 1985, 1986; Kurtz and Shibahashi 1986; Gough and Thompson 1990; for a review of theoretical work, see Shibahashi 1991). The intriguing property of the oscillations of Ap stars is that the pulsation amplitudes correlate with the phase of the magnetic field in such a way that the oscillations might be interpreted as axially symmetric ($m = 0$) dipole ($\ell = 1$) p modes with symmetry axis coincident with the magnetic axis, which is oblique to the axis of stellar rotation.

In the theoretical interpretation which is now widely accepted, the uniform rotation and global dipole magnetic field are assumed; the distortion of the equilibrium spherical configuration is neglected. To study the oscillations, it is convenient to use the corotating frame, in which the magnetic field is stationary. If the variational principle is applied, the modification of the Lagrangian (12) by the effects of rotation reduces to the appearance of the simple additional term $2\rho_0\omega\Omega_0\mathbf{u}^* \cdot (i\hat{z} \times \mathbf{u})$, which describes the influence of the Coriolis forces; the angular velocity of rotation should be considered as one of the parameters p_k, with variation from 0 to Ω_0. Joint effects of the Coriolis forces and Lorentz forces lead to the splitting of a degenerate multiplet into $2\ell + 1$ frequency components, with zero-order eigenfunctions having a mixed character (linear combinations of spheroidal vector fields with different values of m). When observed from the inertial frame, each of these modes produces

$2\ell+1$ frequency components, separated by the angular velocity of rotation. The frequencies of different modes can be different in the corotating frame, and the degenerate multiplet can thus be split, in general, into $(2\ell + 1)^2$ components. Although there is only $(2\ell+1)$-fold degeneracy in the non-rotating non-magnetic star, such a multiple splitting arises because the perturbation is not stationary in the inertial frame.

But only the equidistant triplet structures are usually observed in the power spectra of rapidly oscillating Ap stars. To resolve this problem, Dziembowski and Goode (1985, 1986) argue the possibility of the selective exitation in the presence of the relatively strong magnetic field, so that only one mode is actually excited, which appears with $2\ell + 1$ frequency components in the observations from the inertial frame. The relative amplidudes of the different components depend on the relative contribution of the effects of magnetic field and rotation, and thus the mean strength of the magnetic field in the stellar interior can be measured.

Quite recently, new observations of a rapidly oscillating Ap star HR 3831 revealed that the apparent triplet structure of its oscillation frequencies is really a part of a septuplet (Kurtz *et al.* 1992). The interpretation of this complicated structure as a contribution of the lower-amplitude signal from the first-order correction to the velocity field, which contains a spheroidal vector harmonic of degree $\ell = 3$, is discussed in these proceedings by Shibahashi and Takata (1992).

"RAPID" ROTATION

When the angular velocity of stellar rotation becomes relatively high, of the order of the oscillation frequency, Coriolis forces become of the same order as inertial forces, and can no longer be considered as a small perturbation. We meet this situation when studying the low-frequency modes; the rotation itself is usually considered to be slow enough for the centrifugal distortion of the equilibrium stellar configuration to remain relatively small.

By analogy with pressure perturbations, which provide the restoring forces for acoustic waves and global p modes, and with buoyancy forces, which provide the restoring forces for gravity waves and global g and f modes, the Coriolis forces represent a new type of restoring force for waves of quite different physical nature, which can also be trapped in the star and form new global modes.

In contrast with pressure perturbations, which are usually considered to be isotropic, and with buoyancy forces, which are always vertical, the geometry of Coriolis forces makes the classification of new waves somewhat more complicated.

The first class of possible motions is represented by Rossby waves, familiar in geophysics (e.g., Pedlosky 1979). These waves have predominantly horizontal displacements, and their origin is due to the latitudinal dependence of the distribution of the vertical component of the specific angular momentum in the rotating star. The curvature of the level surfaces is thus crucial for the existence of Rossby waves; corresponding pressure and density perturbations are relatively small. When viewed in the corotating frame for a star with uniform rotation, Rossby waves propagate in the direction opposite to the rotation (retrograde waves).

Waves of another type can appear, with significant vertical displacements, which can thus be significantly affected by buoyancy. If the stratification is convectively neutral (no buoyancy forces), these modes have a mixed character: they are a combination of pure inertial waves, which arise if the Boussinesq approximation is adequate, and prograde waves (generalized Rossby waves) with restoring forces produced by additional vorticity perturbations, which are due to Lagrangian density perturbations in the rotating star (an excellent local analysis of waves in rotating stars can be found in monograph of Unno *et al.* 1989; see also Ando 1989). When the stratification is convectively stable, the inertial waves are modified by buoyancy, but their frequencies remain real in the adiabatic approximation (purely oscillatory motions). In convectively unstable layers, however, the situation is much more interesting. In the linear adiabatic approximation, the wave amplitude can grow exponentially with time. The physical meaning of this phenomenon is related with the stabilizing effect of the rotation on the convection. The interaction of the inertial waves with unstable convective modes (g^- modes) can produce overstable, or even oscillatory convective motions.

In the global analysis of the long-period oscillations of rotating stars, the main difficulty is due to the fact that a particular mode can no longer be described by a single spherical harmonic $Y_{\ell m}(\theta, \phi)$. When the effects of rotation are small, the simple perturbational analysis of p, f and g modes shows that even if we neglect the centrifugal distortion and consider only the uniform rotation, the first-order correction to the eigenfunction of a normal mode described by a spherical harmonic $Y_{\ell m}(\theta, \phi)$ contains toroidal vector fields (equation (2)) of degree $\ell - 1$ and $\ell + 1$, the second-order correction includes spheroidal fields of degree $\ell - 2$ and $\ell + 2$, and so on. When the effects of rotation are small, this linear interaction between modes of different degree, induced by the rotation, is usually small (if there are no accidental resonances between interacting modes), and the perturbational expansion converges rapidly. But that is not necessary the case for "rapid" rotation (low-frequency modes).

The breaking of the spherical symmetry thus prevents the possibility of simple variable separations (except in the case of a uniform incompressible fluid, see Bryan 1889); only the axial symmetry remains (which is always considered to be the case), so that modes can still be identified with unique values of azimuthal order m. Separation of spatial variables by using the vector spherical harmonics (equations (1,2)) results in two infinite chains of coupled differential equations, with spheroidal displacements described by $\ell = |m|, |m| + 2, \ldots$ ("even" modes) and $\ell = |m| + 1, |m| + 3, \ldots$ ("odd" modes). Such a decoupling into two series is due to the symmetry of the equilibrium configuration that remains with respect to the equatorial plane; "even" modes are symmetric, and "odd" modes are antisymmetric with respect to the equatorial plane, and modes from different series do not interact.

In this situation, different approximate techniques are used to study the low-frequency modes, which permits us at least to understand their qualitative behaviour. The most common techniques are (i) the truncation of the infinite chains of differential equations, and (ii) the application of asymptotic analysis (in sense of considering either the slow rotation or modes with high radial order). The "traditional approximation" (which neglects the horizontal component of the angular velocity of rotation) simplifies the problem significantly (e.g. Unno *et al.* 1989). The variational principle can also be generalized to study the oscillations of rotating stars (Lynden-Bell and Ostriker 1967; Ipser and Lindblom 1991).

The first prominent effect of Coriolis forces is that the degeneracy of the oscillation spectrum of a non-rotating star with respect to toroidal displacement fields (defined by equation (2): "trivial" modes with zero frequencies) is lifted. For a star with uniform rotation, these fields represent the leading term of the eigenfunctions of new modes, which are global Rossby waves with frequencies

$$\omega = \frac{2m\Omega}{\ell(\ell+1)} - m\Omega + O\left(\Omega^3\right) \tag{15}$$

in the inertial frame. In the theory of stellar oscillations, these modes were first studied by Papaloizou and Pringle (1978) and were called "r modes", and by Provost, Berthomieu and Rocca (1981) ("quasi-toroidal modes"). Papaloizou and Pringle (1978) used a technique of consequtive approximations, by putting the vertical displacements and Eulerian pressure perturbations to be zero in the leading approximation; Provost, Berthomieu and Rocca (1981) used an asymptotic technique with the rotational frequency as a small parameter. The asymptotic analysis was extended by Smeyers, Craeynest and Martens (1981) to account for gravity perturbations. All the studies are largely limited by the uniform rotation; differential rotation can lead to the Kelvin-Helmholtz instability (shear instability), and makes the problem much less tractable (Papaloizou and Pringle 1978; Dziembowski and Kosovichev 1987a,b).

An interesting property of r modes can be seen directly from expression (15): their frequencies are degenerate, at least in the leading order, with respect to the radial distribution of the eigenfunctions, i.e. with respect to the radial order n. With the only exeption of convectively neutral stratification, this degeneracy is lifted when higher-order terms of the asymptotic expansions are taken into account. It was demonstrated by Provost, Berthomieu and Rocca (1981), by using polytropic models, that the eigenfrequencies decrease with radial order when the stratification is convectively stable (polytrope of index 3), and increase with radial order when the stratification is convectively unstable (polytrope of index 1). If the absolute value of the buoyancy frequency is of the order of Ω or smaller, the asymptotic analysis becomes invalid. Even for a slowly rotating star, this situation can occure in the region with effective convection, and was analized in detail by Dziembowski and Kosovichev (1987a): a technique was developed which uses matching of the asymptotic solutions with the direct numerical solutions of the (truncated) oscillation equations in this region. For differential rotation with $\Omega = \Omega(r)$, Dziembowski and Kosovichev (1987a) argue the possibility of r-mode trapping in the vicinity of the local minimum of $\Omega(r)$.

Although the velocity field of r modes is horizontal and divergence-free in the leading order, higher-order terms in the eigenfunctions produce some pressure perturbations, and modes can be excited by the destabilizing effect of κ mechanism. Non-adiabatic analyses of Saio (1982) and Berthomieu and Provost (1983) indicate that r modes can become overstable in ZZ Ceti stars.

The asymptotic properties of high-order gravity modes (g^+ modes) in the presence of rotation were studied by Berthomieu *et al.* (1978). It was shown that the "traditional approximation" is applicable for these low-frequency modes with predominantly horizontal displacements. Results of the trancated analysis of the high-order g modes (Lee and Saio 1987a) were found to be in qualitative agreement with the asymptotic approach. Using the asymptotic analysis in "traditional approximation", Lee and Saio (1987a) demonstrated the appearance of the low-frequency cut-off of the high-order gravity modes, which is due to the stabilizing effect of the Coriolis forces, acting as additional restoring forces.

But much more interesting is the influence of rotation on the unstable gravity modes (g^- modes, or convective modes), which can be considered as an interaction between convective modes and inertial waves governed by the rotation, and describes (in general) the interaction between convection, rotation and pulsation. This problem has been an object of intensive study in recent years.

The stabilizing effect of Coriolis forces on convection makes the unstable convective modes become overstable, or even purely oscillatory. The concept of "negative-energy" modes becomes very productive in the physical understanding of the related phenomena (Lee and Saio 1989, 1990c).

If an overstable convective mode is completely stabilized by the rotation, two purely oscillatory modes appear, one with positive and another with negative energy. The kinetic energies of both modes are surely positive in the corotating frame (the studies are largely limited to uniform rotation), but one of these modes has negative kinetic energy when considered in the inertial frame: the kinetic energy of the equilibrium rotating flow is larger than that of the rotating and oscillating flow. Equipartition of kinetic and potential energies persists in the inertial frame (Lee and Saio 1990c), and the total energy of the mode is thus negative.

If energy is extracted from the negative-energy mode, its amplitude grows; the energy source is provided by the star. The radiative dissipation destabilizes the negative-energy mode, making the purely oscillatory motion overstable (Lee and Saio 1987b). Due to the resonance coupling of the convective modes in the stellar core with the stable envelope g modes, the overstable low-frequency modes can appear in rotating massive main-sequence stars (Osaki 1974; Lee and Saio 1986, 1987b, 1989, 1990c; Lee 1988).

The line-profile variations produced by low-frequency g modes excited by the oscillatory convective modes in a rapidly rotating massive main-sequence star were studied by Lee and Saio (1990a) and Lee, Jeffery and Saio (1992).

Inertial oscillations of Jupiter were studied by Lee and Saio (1990b) and Lee, Strohmayer and Van Horn (1992) as an explanation of the slowly oscillating thermal features discovered in the Jovian atmosphere by Magalhães *et al.* (1989) and Deming *et al.* (1989).

"STRONG" MAGNETIC FIELD

Even when the overall magnetic field is relatively weak, the analysis of magnetic effects in stellar oscillations by the standard perturbation technique can become invalid near the surface, if the magnetic field does not vanish there. Because of the low density, the magnetic pressure in the surface layers can become of the same order as gas pressure, and the Lorentz forces can become of the same order as inertial forces. The perturbation analysis then becomes invalid locally; this situation is exactly what we expect for the rapidly oscillating magnetic Ap stars.

The first steps in the improvement of the theoretical description were taken by Biront *et al.* (1982) using a singular perturbation technique (method of matched asymptotic expansions, which originates from boundary layer theory). The analysis of Biront *et al.* (1982) was limited to the study of the dynamical distortion of the velocity field in the outer stellar layers. This analysis was developed further by Roberts and Soward (1983) and Campbell and Papaloizou (1986), who included the energy dissipation through Alfvénic-type waves which are generated by the pulsations.

Despite of these efforts, the theoretical understanding of the rapid oscillations in Ap stars is far from being complete; there is no convincing description of the selective mode excitation, although it has been quite evident for the last ten years that the magnetic field plays a crucial role in these oscillations. Much work is needed to understand the excellent observational data which are already available, and which will undoubtedly be refined by forthcoming observations.

Author thanks Douglas Gough for carefully reading the manuscript to improve English, and for comments.

REFERENCES

Ando, H. 1989, *Publ. Astron. Soc. Japan*, **41**, 289.
Berthomieu, G., and Provost, J. 1983, *Astr. Ap.*, **122**, 199.
Berthomieu, G., Gonczi, G., Graff, Ph., Provost, J., and Rocca, A. 1978, *Astr. Ap.*, **70**, 597.
Biront, D., Goossens, M., Cousens, A., and Mestel, L. 1982, *M.N.R.A.S.*, **201**, 619.
Bogdan, T. J. 1989, *Ap. J.*, **345**, 1042.
Bogdan, T. J., and Cattaneo, F. 1989, *Ap. J.*, **342**, 545.
Bryan, G. H. 1889, *Philos. Trans. R. Soc. London,* **A 180**, 187.
Campbell, C. G., and Papaloizou, J. C. B. 1986, *M.N.R.A.S.*, **220**, 577.
Campbell, W. R., and Roberts, B. 1989, *Ap. J.*, **338**, 538.
Cowling, T. G., and Newing, R. A. 1949, *Ap. J.*, **109**, 149.
Cox, J. P. 1980, *Theory of Stellar Pulsation* (Princeton: Princeton University Press).
Cuypers, J. 1980, *Astr. Ap.*, **89**, 207.
Deming, D., Mumma, M. J., Espenak, F., Jennings, D. E., Kostiuk, T., Wiedemann, G., Loewenstein, R., and Piscitelli, J. 1989, *Ap. J.*, **343**, 456.
Durney, B. R., Hill, F., and Goode, P. R. 1988, *Ap. J.*, **326**, 486.
Duvall, T. L., Jr., Harvey, J. W., and Pomerantz, M. A. 1986, *Nature,* **321**, 500.
Dziembowski, W. A., and Goode, P. R. 1984, *Mem. Soc. Astron. Ital.*, **55**, 185.
Dziembowski, W. A., and Goode, P. R. 1985, *Ap. J. (Letters)*, **296**, L27.

Dziembowski, W. A., and Goode, P. R. 1986, in: *Seismology of the Sun and the Distant Stars,* ed. D. O. Gough (Dordrecht: Reidel), p.441.
Dziembowski, W. A., and Goode, P. R. 1989, *Ap. J.*, **347**, 540.
Dziembowski, W. A., and Goode, P. R. 1991a, in: *Solar Interior and Atmosphere,* eds A. N. Cox, W. C. Livingston and M. S. Matthews (Tucson: University of Arizona Press), p.501.
Dziembowski, W. A., and Goode, P. R. 1991b, *Ap. J.*, **376**, 782.
Dziembowski, W. A., and Goode, P. R. 1992, *Ap. J.*, in press.
Dziembowski, W., and Kosovichev, A. 1987a, *Acta Astronomica,* **37**, 313.
Dziembowski, W., and Kosovichev, A. 1987b, *Acta Astronomica,* **37**, 341.
Edmonds, A. R. 1957, *Angular Momentum in Quantum Mechanics* (Princeton: Princeton University press).
Elsworth, Y., Howe, R., Isaak, G. R., McLeod, C. P., and New, R. 1990, *Nature,* **345**, 322.
Evans, D. J., and Roberts, B. 1990, *Ap. J.*, **356**, 704.
Evans, D. J., and Roberts, B. 1991, *Ap. J.*, **371**, 387.
Evans, D. J., and Roberts, B. 1992, *Nature,* **355**, 230.
Goldreich, P., Murray, N., Willette, G., and Kumar, P. 1991, *Ap. J.*, **370**, 752.
Gough, D. O. 1981, *M.N.R.A.S.*, **196**, 731.
Gough, D. O., and Kosovichev, A. G. 1992, these proceedings.
Gough, D. O., and Taylor, P. P. 1984, *Mem. Soc. Astron. Ital.,* **55**, 215.
Gough, D. O., and Thompson, M. J. 1988a, in: *Advances in Helio- and Asteroseismology,* eds J. Christensen-Dalsgaard and S. Frandsen (Dordrecht: Reidel), p.155.
Gough, D. O., and Thompson, M. J. 1988b, in: *Advances in Helio- and Asteroseismology,* eds J. Christensen-Dalsgaard and S. Frandsen (Dordrecht: Reidel), p.175.
Gough, D. O., and Thompson, M. J. 1990, *M.N.R.A.S.*, **242**, 25.
Gough, D. O., and Thompson, M. J. 1991, in: *Solar Interior and Atmosphere,* eds A. N. Cox, W. C. Livingston and M. S. Matthews (Tucson: University of Arizona Press), p.519.
Gough, D., and Toomre, J. 1991, *Ann. Rev. Astron. Astrophys.,* **29**, 627.
Hansen, C. J., Cox, J. P., and Van Horn, H. M. 1977, *Ap. J.*, **217**, 151.
Ipser, J. R., and Lindblom, L. 1991, *Ap. J.*, **379**, 285.
Jain, R., and Roberts, B. 1992, *Ap. J.*, submitted.
Jefferies, S. M., Pomerantz, M. A., Duvall, T. L., Jr., Harvey, J. W., and Jaksha, D. B. 1988, in: *Seismology of the Sun and the Sun-Like Stars,* ed. E. J. Rolfe (ESA SP-286; Noordwijk: ESA), p.279.
Jones, P. W., Pesnell, W. D., Hansen, C. J., and Kawaler, S. D. 1989, *Ap. J.*, **336**, 403.
Kovetz, A. 1966, *Ap. J.*, **146**, 462.
Kurtz, D. W. 1990, *Ann. Rev. Astron. Astrophys.,* **28**, 607.
Kurtz, D. W., Kanaan, A., Martinez, P., and Tripe, P. 1992, *M.N.R.A.S.*, **255**, 289.
Kurtz, D. W., and Shibahashi, H. 1986, *M.N.R.A.S.*, **223**, 557.
Ledoux, P. 1951, *Ap. J.*, **114**, 373.
Lee, U. 1988, *M.N.R.A.S.*, **232**, 711.
Lee, U., and Saio, H. 1986, *M.N.R.A.S.*, **221**, 365.
Lee, U., and Saio, H. 1987a, *M.N.R.A.S.*, **224**, 513.
Lee, U., and Saio, H. 1987b, *M.N.R.A.S.*, **225**, 643.

Lee, U., and Saio, H. 1989, *M.N.R.A.S.*, **237**, 875.
Lee, U., and Saio, H. 1990a, *Ap. J.*, **349**, 570.
Lee, U., and Saio, H. 1990b, *Ap. J. (Letters)*, **359**, L29.
Lee, U., and Saio, H. 1990c, *Ap. J.*, **360**, 590.
Lee, U., Jeffery, C. S., and Saio, H. 1992, *M.N.R.A.S.*, **254**, 185.
Lee, U., Strohmayer, T. E., and Van Horn, H. M. 1992, preprint.
Libbrecht, K. G. 1989, *Ap. J.*, **336**, 1092.
Libbrecht, K. G., and Woodard, M. F. 1990, *Nature,* **345**, 779.
Libbrecht, K. G., and Woodard, M. F. 1991, *Science,* **253**, 152.
Luh, P. C. 1973, *Geophys. J.,* **32**, 187.
Lynden-Bell, D., and Ostriker, J. P. 1967, *M.N.R.A.S.*, **136**, 293.
Magalhães, J. A., Weir, A. L., Conrath, B. J., Gierasch, P. J., and Leroy, S. S. 1989, *Nature,* **337**, 444.
Messiah, A. 1961, *Quantum Mechanics* (Amsterdam: North-Holland Publ.).
McDermott, P. N., Van Horn, H. M., and Hansen, C. J. 1988, *Ap. J.*, **325**, 725.
Osaki, Y. 1974, *Astr. Ap.*, **189**, 469.
Papaloizou, J., and Pringle, J. E. 1978, *M.N.R.A.S.*, **182**, 423.
Pedlosky, J. 1979, *Geophysical Fluid Dynamics* (Berlin: Springer-Verlag).
Provost, J., Berthomieu, G., and Rocca, A. 1981, *Astr. Ap.*, **94**, 126.
Roberts, B., and Campbell, W. R. 1986, *Nature,* **323**, 603.
Roberts, P. H., and Soward, A. M. 1983, *M.N.R.A.S.*, **205**, 1171.
Saio, H. 1981, *Ap. J.*, **244**, 299.
Saio, H. 1982, *Ap. J.*, **256**, 717.
Shibahashi, H. 1991, in: *Challenges to Theories of the Structure of Moderate-Mass Stars,* eds D. Gough and J. Toomre (Berlin: Springer), p.393.
Shibahashi, H., and Takata, M. 1992, these proceedings.
Smeyers, P., Craeynest, D., and Martens, L. 1981, *Astrophys. Space Sci.,* **78**, 483.
Strohmayer, T., Ogata, S., Iyetomi, H., Ichimaru, S., and Van Horn, H. M. 1991, *Ap. J.*, **375**, 679.
Unno, W., Osaki, Y., Ando, H., Saio, H., and Shibahashi, H. 1989, *Nonradial Oscillations of Stars* (2nd ed.; Tokyo: University of Tokyo Press).
Vorontsov, S. V. 1981, *Astron. Zh.*, **58**, 1275.
Vorontsov, S. V. 1986, in: *Upper Main Sequence Stars with Anomalous Abundances,* eds C. R. Cowley *et al.* (Dordrecht: Reidel), p.37.
Vorontsov, S. V. 1988, in: *Advances in Helio- and Asteroseismology,* eds J. Christensen-Dalsgaard and S. Frandsen (Dordrecht: Reidel), p.151.
Vorontsov, S. V. 1992, *Astron. Zh.*, in press.
Vorontsov, S. V., and Zharkov, V. N. 1981, *Astron. Zh.,* **58**, 1101.
Woodard, M. F., and Libbrecht, K. G. 1991, *Ap. J. (Letters)*, **374**, L61.
Wright, A. N., and Thompson, M. J. 1992, *Astr. Ap.*, in press.
Zweibel, E. G., and Däppen, W. 1989, *Ap. J.*, **343**, 994.

Inside the Stars, IAU Colloquium 137
ASP Conference Series, Vol. 40, 1993
Werner W. Weiss and Annie Baglin (eds.)

EXCITATION MECHANISMS OF OSCILLATIONS IN STARS

YOJI OSAKI
Department of Astronomy, University of Tokyo, Tokyo 113, Japan

ABSTRACT Excitation mechanisms of oscillations in stars are discussed, in particular on the problems of the Beta Cephei pulsations and of the solar oscillations. A long-standing mystery about the excitation mechanism of the Beta Cephei pulsations seems to have finally been solved, which is due to availability of the new "OPAL" opacities. The very mechanism is the classical κ-mechanism, however, due to enhanced heavy element opacities at temperature around 2×10^5K. By using the new opacities, three independent groups have found pulsational instabilities for the fundamental radial mode and a few low-degree non-radial modes in Beta Cephei models. Stochastic excitation of solar oscillations by turbulent convection is then discussed. It is shown that the noise generation by quadrupole radiation concentrated at the top of the convection zone may well explain the frequency dependence of the observed solar acoustic power pumped into individual p-modes.

INTRODUCTION

Although most of classical pulsating variables such as Cepheid and Mira variables are giant or super-giant stars, pulsations and oscillation-related phenomena have been known in several groups of stars on the main sequence or close to it as well. They may be classified into three categories by spectral class of the relevant stars: (1) early-type pulsating variables with spectral class of O- and B-type, in which the Beta Cephei stars are the best known example of this type, (2) Delta Scuti stars and rapidly oscillating Ap stars in A- and F-type spectral range, and (3) global oscillations of the sun itself and oscillations in the sun-like stars in late spectral type. As for the excitation mechanisms of oscillations in stars are concerned, basically two different mechanisms have been known, and they are (1) the linear overstability of some particular eigenmodes of stars by the κ-mechanism, and the stochastic excitation of many modes by turbulent convection in stars with convective envelope.

In this review, I shall discuss on the excitation mechanisms of pulsations in these stars. In particular, I discuss two topics on excitation mechanisms of stellar oscillations in which great progress has been made very recently. They are the excitation mechanism of the Beta Cephei stars and the stochastic excitation of oscillations by turbulent convection in the sun.

As a matter of fact, I gave review talks twice on this subject in the past, in 1982 at pulsation conference in Boulder Colorado (Osaki 1982) and in 1985 at IAU general assembly in New Delhi (Osaki 1986). In both of these talks, I had to conclude my talks by stating that the theory of the Beta Cephei stars was unsatisfactory. This meant that something was wrong in our understanding of the stellar interior of these early type stars. Most likely place which we should blame for this insufficiency was opacity. The opacity available for a long time used to be only one opacity calculation by the Los Alamos group. Badly needed was an independent opacity calculation other than that of Los Alamos. To see this situation, let me here quote my concluding remarks of the review talk on "Beta Cephei variables" given at the IAU general assembly in New Delhi in 1985; "Something seem wrong or insufficient with our present knowledge about the internal structure of very massive stars......... As for the excitation mechanism of β Cephei variables, I suspect some kind of κ mechanism, which is still unknown but which may operate at temperature around $T \sim 2 \times 10^5$K". My suspicion has fully been confirmed by the recent works as seen above.

The newly found mechanism is the κ-mechanism due to heavy element opacities which works at the temperature around 2×10^5 K. This is the very temperature region at which the κ-mechanism was suspected to operate for excitation of the Beta Cephei pulsation. If the κ mechanism is responsible for the excitation of the Beta Cephei pulsations, one can show by a very simple argument based on the "transition zone" (see e.g., Osaki 1982) that the crucial layer for the pulsational stability is the transition layer between the quasi-adiabatic interior and the fully non-adiabatic outer layer, which is located in this case around the temperature $T \simeq (1.5 - 2) \times 10^5$ degree.

In this respect, Stellingwerf (1978) was the first to suggest a possible κ mechanism but in his case due to the helium opacity bump. He realized that there exists a small opacity bump at a temperature close to 1.5×10^5 K, which originates from the coincidence of the frequency maximum of radiation with the second ionization edge of helium at 54.4 eV. The Stellingwerf mechanism for the excitation of the Beta Cephei pulsations was examined by various people (Stellingwerf 1978; Saio and Cox 1980, Dziembowski and Kubiak 1981; Lee and Osaki 1982). These results were basically similar and they are summarized as follows; (1)although this mechanism works locally to the direction to drive the pulsation, it is still insufficient to destabilize the star as a whole; (2) the least stable stellar models by this mechanism are slightly too cool to explain the Beta Cephei pulsation. These two points suggest that what we needed was somewhat larger opacity bumps which would operate at temperature slightly higher than 1.5×10^5 of the Stellingwerf helium opacity bump.

Then came a monumental paper by Simon (1982) who argued that if the heavy element opacity was larger by a factor of 2-3 than that given by the Los Alamos opacity, it would then resolve the problem of Cepheid mass anomalies and it would also explain the excitation of the Beta Cephei pulsations. Simon (1982) thus made a plea for reexamining heavy element opacities in stars as there existed great uncertainty in the Los Alamos opacity calculations in the treatment of heavy element opacities. However, it took nearly a decade for his plea to be answered fully by two groups: the OPAL project at Lawrence Livermore National Laboratory and the "Opacity Project". As seen above, Simon's (1982) suggestion has been confirmed by the OPAL opacities, which have finally solved the long-standing problem of the excitation of the Beta Cephei pulsations.

Summary of the β Cephei excitation and impact of the new opacities

We here summarize the excitation mechanism of the Beta Cephei stars. Based on the new opacities by the OPAL code (Rogers and Iglesias 1992), the three independent groups mentioned above have examined the pulsational stability of stellar models relevant to the Beta Cephei stars. Their results are basically similar to each other and they are summarized as follows:

(1) Massive early-type stars are pulsationally unstable for the radial fundamental mode and for a few lower-degree non-radial modes having periods similar to that of the radial fundamental.

(2) The instability strip by this mechanism agrees with observed location of the Beta Cephei stars.

(3) The stability results are very sensitive to the heavy element abundance. In particular, pulsational instability is not expected in stars with low heavy element abundance such as in those stars in the Magellanic Clouds.

The new opacities have given a great impact on the stellar structure and the stellar pulsations. As seen in other talks presented in this conference, the new opacities may be able to solve various long-standing problems such as the Cepheid period ratios. It is clear that almost all calculations of the stellar structure and evolution and in particular calculations of all stellar pulsations should be reexamined based on the new opacities.

STOCHASTIC EXCITATION OF SOLAR OSCILLATIONS

Let us now discuss the excitation mechanisms of solar oscillations. Two different excitation mechanisms have been discussed for solar oscillations: (1) the linear overstability of eigenmodes due to the κ-mechanism at the hydrogen ionization zone, and (2) the stochastic excitation by turbulent convection. The latter possibility is now more favoured than the former possibility and here we mainly discuss the stochastic excitation by turbulent convection.

Observational evidence which is unfavorable for the linear overstability in the case of solar oscillations is that so many p-modes are observed with very small amplitude. On the other hand, linear overstability is the very mechanism responsible for pulsations of the Cepheid stars and the Cepheid variables show only one or two pulsation modes excited with large amplitude. If the sun were pulsationally unstable for certain modes, it would be difficult for such modes to be kept rather in such low amplitude as observed on the sun (Kumar and Goldreich 1989).

On the other hand, in the stochastic excitation mechanism, the sun is regarded as an acoustic resonator having many eigenmodes. These eigenmodes are considered to be damped oscillators and they can be excited only when they are driven externally. As it is well known, the outer 30% in radius of the sun is covered by the convective zone and the turbulent convection there generates broad-band "acoustic noise" in frequency. Oscillations at resonant frequencies are then picked up by the acoustic cavity and they give rise to many peaks in the observed power spectrum. This is the basic picture of the stochastic excitation by turbulent convection. The stochastic excitation of solar oscillations was first investigated extensively by Goldreich and Keeley (1977).

If we write the velocity of a particular mode of damped oscillator at the solar surface by v_q, we find its time dependence as

$$v_q \propto \exp(i\omega_q t - \gamma_q t), \tag{1}$$

where ω_q and γ_q are the eigenfrequency and the damping rate of the mode, respectively, and the subscript q signifies a particular eigenmode. If we inject acoustic noise having a power spectrum $P_{\text{noise}}(\omega)$ into the cavity, we obtain the correspondent response of the form (Christensen-Dalsgaard et al. 1989)

$$|A|^2 = \frac{P_{\text{noise}}(\omega)}{(\omega^2 - \omega_q^2)^2 + 4\omega^2\gamma_q^2}, \tag{2}$$

where A is the amplitude of oscillation. If the acoustic noise generated has a broad-band power in frequency, many modes are simultaneously excited at resonant frequencies of the cavity. This model can thus explain following two points very nicely: (1) millions of modes are excited, and (2) their amplitudes are expected rather low.

One can write more explicitly the energy conservation of a particular oscillation mode;

$$\frac{dE}{dt} = G - \Gamma E, \tag{3}$$

where E is the oscillation energy of a particular mode, G is the acoustic power pumped into the mode, and $\Gamma = 2\gamma_q$ is the (energy) damping rate of the mode. In steady state, we find

$$E = \frac{G}{\Gamma}. \tag{4}$$

The energy of a mode is determined by the acoustic energy generation rate at the particular mode divided by the damping rate of the mode. The central question in this model is how to determine the damping rate Γ and the acoustic noise generation rate G.

Damping rate

Several different factors contribute to the damping rate of oscillation but we may divide it into two parts

$$\Gamma = \Gamma_{\text{dyn}} + \Gamma_{\text{therm}}, \tag{5}$$

where Γ_{dyn} and Γ_{therm} stand for the damping rate due to the dynamic cause through the momentum equation and that due to non-adiabatic effects of oscillations through the thermal equation, respectively. As for the dynamical damping, the viscous dissipation by the eddy viscosity of turbulence and wave leakage at the boundary of the cavity are two of important mechanisms of damping. On the other hand, the thermal damping Γ_{therm} is due to thermal coupling of oscillation with radiation field and convective energy transport. As already discussed before, the thermal damping Γ_{therm} could be negative if modes are thermally overstable as known "negative dissipation". Most uncertain in the damping rate is the thermal coupling of oscillation with convection. Here we need time-dependent convection theory, which still remains one of the major unsolved problems in astrophysics.

Acoustic noise generation rate

Acoustic noise generation from homogeneous turbulence was first discussed by Lighthill (1952) and this mechanism is thus known as the Lighthill mechanism. According to this theory, the acoustic noise generation rate ε per unit volume from turbulence is given by

$$\varepsilon = \frac{\rho v_\lambda^3}{\lambda} M^{2n+1}, \tag{6}$$

where ρ is the mass density, v_λ is the characteristic velocity of turbulent convection, λ is the characteristic scale of the turbulence, and $M = v_\lambda/c_s$ is the Mach number of turbulence. The index n in the above equation is the multipole index. As for this index, the lowest three indices are (1) the monopole radiation $n = 0$ which occurs if there is mass source in fluid, (2) the dipole radiation $n = 1$ which occurs if external force acts on the fluid, and (3) the quadrupole radiation $n = 2$ which occurs when the turbulent Reynolds stresses act as a source term. In the original Lighthill mechanism, the quadrupole radiation is the lowest order multipole generated by turbulence.

In order to calculate the acoustic power G pumped in a particular mode in equations (3) and (4) from the acoustic noise generation rate ε, we must first resolve this into the specific power $P(k_h, \omega)$ in horizontal wavenumber and frequency and multiply the band-widths of the particular mode in horizontal wavelength and frequency and then integrate it spatially to count for the whole contribution over the convection zone. In particular, as for the contribution from the source outside the acoustic cavity, a proper care must be taken of attenuation effects of wave energy by evanescent zone between the source and the cavity.

Goldreich and Keeley (1977) was the first to try to estimate theoretically amplitudes of individual modes based on the stochastic excitation by turbulent convection. For the damping rates of oscillations, they used the turbulent viscous damping and they then estimated the forcing term based on the quadrupole radiation due to turbulent Reynolds stresses by properly taking into account eigenfunctions of modes. By combining the damping rate to the noise generation rate, they estimated theoretical amplitudes of individual modes of solar oscillations. Their results were, however, found to give too small amplitudes as compared with recent observations obtained by Libbrecht (1988). However, there existed several uncertainties in numerical factors of order unity in their formulation which can easily affect the final results by a large factor. It thus remained rather uncertain whether the stochastic excitation mechanism was quantitatively sufficient or not.

Observations and comparison with theories

Libbrecht (1988) demonstrated that the quantity G can be obtained directly from observations. In the stochastic excitation model we find from equation (3) $G = E\Gamma$ in steady state. Since the quantities E and Γ can be obtained from observations of the amplitude of the particular mode and its line width, we can obtain the acoustic power G pumped into an individual mode from observations. Libbrecht (1988) presented his results of observations for G as a function of frequency ν in his figure 3 (Libbrecht, 1988). It was found that the quantity G is insensitive to degree ℓ (or the horizontal wave number) of the mode but it is very sensitive to mode frequency. The acoustic power G pumped into individual modes has a sharp peak at frequency around 3.5 mHz which corresponds to the period of five minute and it has a power law dependence of $G \propto \nu^8$ for the low frequency side with $\nu < 3$mHz and another power law dependence $G \propto \nu^{-5.3}$ for the high frequency side with $\nu > 4$mHz.

Osaki (1990) has demonstrated that the observed frequency dependence of the acoustic power G can naturally be explained in terms of the quadrupole radiation generated by turbulent convective motion at the top of the solar convection zone. He has first argued that the dipole radiation suggested by Goldreich and Kumar (1988) may not be important because of cancellation effects between rising and sinking convective elements. He then examined the acoustic power G pumped into an individual mode based on the quadrupole radiation of the original Lighthill mechanism. As for the high frequency noise power G is concerned, it is shown that the quadrupole radiation with Kolmogoroff spectrum for turbulence can predict the power law dependence with an index of -5.5 which is in a good agreement with observations.

On the other hand, the power law dependence of $G \propto \nu^8$ for the low frequency side can be explained by the propagation effects of waves if the source of acoustic energy generation is concentrated at the top of the convection zone where waves are evanescent for modes with frequency $\nu < 3$mHz. It is shown that a simple attenuation effect of waves through the evanescent zone between the noise source and the acoustic cavity can give rise to the power law dependence of the form $G \propto \nu^{2m}$ where m is the effective polytropic index specifying the density stratification of the solar envelope. The effective polytropic index m in the relevant zone of the solar envelope is estimated to be near 4, that is in a good agreement with the observed power law dependence. A similar power-law dependence for the low frequency power was also obtained by Goldreich and Kumar (1990).

Balmforth (1992) has recently reexamined the stochastic excitation mechanism by quantitatively estimating modal amplitudes. It was shown that the theoretical results for acoustic power spectra could be made to agree with observations by adjusting various theoretical parameters. However, a following difficulty exists in Balmforth's result; if adjustable parameters in his theory are chosen such that the theoretical amplitudes at the peak frequency around 3mHz agree with observed ones, an agreement on the frequency dependence of the acoustic power between theory and observation is then broken in the sense that the theoretical acoustic power has a flatter spectrum as compared with observations. This may be related with his rather extended source distribution because much sharper power peak was obtained by Osaki (1990) who assumed a locally concentrated source at the top of the solar convection zone.

ACKNOWLEDGMENTS

It is my great pleasure to acknowledge the financial support of the Mitsubishi Foundation, which made it possible for me to attend this conference by covering the travel expenses.

REFERENCES

Balmforth, N. J. 1992, *MNRAS*, **255**, 639.

Christensen-Dalsgaard, J., Gough, D.O., and Libbrecht, K.G. 1989, *ApJ*, **341**, L103.

Cox A, N. 1987, in *Stellar Pulsation*, Lecture Notes in Physics, Vol. 274, eds. A. N. Cox, W. M. Sparks and S. G. Starrfield (Springer-Verlag, Berlin), p.36.

Cox, A. N., Morgan, S. M., Rogers, F. J. and Iglesias, C. A. 1992, *ApJ*, **393**, 272.

Dziembowski, W. and Kubiak, M. 1981, *Acta Astronomica*, **31**, 153.

Goldreich, P. and Keeley, D. A., 1977, *ApJ*, **212**, 243.

Goldreich, P., and Kumar, P. 1988, *ApJ*, **326**, 462.

Goldreich, P., and Kumar, P. 1990, *ApJ*, **363**, 694.

Iglesias, C. A. Rogers, F. J. and Wilson, B. G. 1987, *ApJ*, **322**, L45.

Kiriakidis, M., El Eid, M. F. and Glatzel, W. 1992, *MNRAS*, **255**, 1p.

Kumar, P. and Goldreich, P. 1989, *ApJ*, **342**, 558.

Lee, U. and Osaki, Y. 1982, *PASJ*, **34**, 39.

Libbrecht, K.G. 1988, in *Seismology of the Sun and Sun-like Stars*, ed. E. Rolfe, ESA SP-286 (ESA Publication Division, Noordwijk), p.3.

Lighthill, M.J. 1952, *Proc. Roy. Soc. London A*, **211**, 564.

Moskalik, P. and Dziembowski, W. 1992, *A&A*, **256**, L5.

Osaki, Y. 1982, in *Pulsations in Classical and Cataclysmic Variable Stars*, ed. J. P. Cox and C. J. Hansen (Joint Institute for Laboratory Astrophysics, Boulder), p. 303.

Osaki, Y. 1986, in *Highlights of Astronomy*, ed. J.-P. Swings (D. Reidel Publ. Co., Dordrecht), p. 247.

Osaki, Y. 1990, in *Progress of Seismology of the Sun and Stars*, eds. Y. Osaki and H. Shibahashi (Springer-Verlag, Berlin Heidelberg), p. 75.

Rogers, F. J. and Iglesias, C. A. 1992, *ApJS*, **79**, 507.

Saio, H. and Cox, J. P. 1980, *ApJ*, **236**, 549.

Seaton, M. J. 1992, in *these proceedings.*

Simon, N. R. 1982, *ApJ*, **260**, L87.

Stellingwerf, R. 1978, *AJ*, **83**, 1184.

Inside the Stars, IAU Colloquium 137
ASP Conference Series, Vol. 40, 1993
Werner W. Weiss and Annie Baglin (eds.)

MODE SELECTION AND OTHER NONLINEAR PHENOMENA IN STELLAR OSCILLATIONS

W.A.DZIEMBOWSKI
Copernicus Astronomical Center, Warszawa, Poland

ABSTRACT Nonlinear effects determining the amplitudes of unstable modes of oscillations in stellar modes, are reviewed. The two basic processes – saturation of the driving mechanism and resonant mode coupling – are discussed within the framework of the amplitude equation formalism. There are difficult problems in the theory of multiple mode interaction that must be solved to make a prediction of amplitude spectra possible. The observed spectra for δ Scu stars and other multiperiodic variables, exhibit no simple pattern that would suggest a solution of the theoretical problems.

1.INTRODUCTION

Linear stability analyses of stellar models reveal that pulsational instability, if present, occurs most often in a large number of oscillation modes simultaneously. Consequences of the instability are therefore difficult to predict. Satisfactory numerical models are available only for the radial monomode pulsation in Cepheid-type stars. We understand selection between fundamental and first overtone pulsation, but we do not understand the role of nonradial modes of high degree, which are unstable under the same condition. A sustained double-mode pulsation has not yet been reproduced in realistic models, and the cause of such behavior in stars remains controversial.

In stars in the Main Sequence (MS) and early post-MS evolutionary phases, multiperiodic variability is a common phenomenon. Up to seven modes of low degree have been tentatively identified in individual objects. The identified modes are among those which are found to be unstable in corresponding stellar models. We do not understand, however, how the mode amplitudes are determined. We do not know which of the undiscovered modes have finite amplitudes, but just too low for the detection with the present-day techniques. Thus, for instance, we cannot yet assess the reward in terms of detected modes from the planned asteroseismological observation from the space.

In this review of the status of the nonlinear theory of stellar oscillations, I focus on the mode selection problem. This is certainly the most important aspect of the theory for the main stream problems of this meeting. It would be indeed very helpful to have an *a priori* assessment of the number and kind of modes that are likely to be discovered once the amplitude resolution is greatly improved. This is not yet possible. I will, thus, present here only elements of the theory that may lead in the future to a prediction of the amplitudes. I believe that this will be an interesting and useful exercise, even though the needed theory may not be developed before the launch of first asteroseismological satellite.

We will start, however, with a different aspect of the theory. In the next section I will briefly review nonlinear effects in amplitude spectra. We will see how, with the help of the nonlinear theory, one can recover useful asteroseismological observables from such spectra. Impact of the linear nonadiabatic theory on the modal selection problem will be discussed in Section 3. Subsequently, I will outline the amplitude equation formalism which is the only practical tool for handling the nonlinear theory of *nonradial* oscillation and discuss two distinct amplitude limiting effects. Specific roles of resonances in mode selection will be reviewed in two consecutive sections. At the end we will discuss existing data on multiperiodic objects.

2. NONLINEAR OBSERVABLES

Oscillation frequencies, ω, are by far the most important observables for asteroseismology. They are not only most accurately measured but, more importantly, the easiest to interpret in terms of parameters describing stellar interiors. However, one should try to try to extract all information contained in amplitude spectra of variable stars. This is particularly important in dealing with the sparse spectra such as all (with one exception) currently available for stars.

A complete model of a pulsating star allow us to represent any observable parameter, O, in the form of the following expansion

$$\begin{aligned} O &= O_0 + \sum_k A_k \Re(Y_{(k)} \exp i(\omega_k t + \psi_k)) + \\ &+ \sum_{k,j,\pm} \lambda_{kj\pm} A_k A_j \Re(Y_{(k)} Y_{(j)} \exp i[(\omega_k \pm \omega_j)t + \psi_{kj\pm}]) + ..., \end{aligned} \quad (1)$$

where amplitudes A and frequencies ω may vary with time, but on the time scale much longer than $1/\omega$, and

$$Y_{(k)} = Y_l^m(\theta, \phi) \quad \text{for} \quad l = l_k \quad \text{and} \quad m = m_k. \quad (2)$$

The parameters occurring in this expansion may be directly compared with those occurring in the Fourier decomposition of the observed light or radial velocity curve. Linear adiabatic theory yields only ω. The amplitude ratios, A_k^a/A_k^b, and the phase differences, $\psi_k^a - \psi_k^b$, (superscripts a and b refer to various observables) may be evaluated within the framework of linear nonadiabatic theory. These quantities depend on stellar parameters, but they are also sensitive to the mode spherical harmonic degree so that they are useful for mode identification (Balona & Stobie, 1980)

Evaluation of the remaining parameters – A_k, $\lambda_{kj\pm}$, $\psi_{kj\pm}$, and corresponding higher order quantities – lies in the domain of nonlinear theory. In general, a connection between these parameters and those descrbing stellar internal structure is, rather complicated. However, in the case of the "bump" Cepheids, for instance, the phase difference, $\psi_{kj\pm} - (\psi_k \pm \psi_j)$, may be translated into information about the second overtone period (Simon & Lee, 1981; Buchler *et al.*,

1990). Furthermore, secular changes in amplitudes and phases, other than those caused by stellar evolution, may be interpret only with use of nonlinear theory.

3. GROWTH RATES

The first step toward understanding modal selection in pulsating stars, which is the determination of mode stability, belongs to the linear theory domain. Qualitatively, solution of the linear nonadiabatic problem yields growth rate, $\gamma = -\Im\omega_c$, where ω_c is the complex eigenfrequency. Since the linear theory is relatively easy and credible, it is natural to ask whether γ-values could be used as a predictors of mode amplitudes.

To get an insight to the meaning of these values it is useful to consider the following well-known expression.

$$\gamma = \frac{W}{2\omega I}, \tag{3}$$

where W denotes the net rate of mode energy gain due to nonadiabatic effects and I mode inertia. Although W is a measure of driving efficiency, the order-of-magnitude differences in γ's between various modes are caused by the differences in I (the usual normalization of eigenfunctions at the surface is assumed).

In Figure 1 we show the behavior of the integrand of W and a chosen eigenfunction for unstable radial modes in a β Cep star model. The instability results from dominance of the driving occurring in the region of the metal opacity bump at $r \approx 0.97R$ over the damping occurring below (see e.g. Dziembowski & Pamyatnykh, in this volume). The values of γ for these as well as nonradial modes of $l = 1$ and 2 degrees are shown in Figure 2 are plotted against mode frequency. In the same figure also shown are normalized growth rates defined by Stellingwerf (1978), as follows

$$\eta = \frac{W}{\int_0^1 \left|\frac{dW}{dx}\right| dx}. \tag{4}$$

This parameter is a good measure of the instability robustness and it is independent of the mode inertia

Both γ and η depend primarily on frequency. Difference in the behavior of this two quantities follows mainly from the rapid decrease of I with increasing frequency. The pattern seen in this figure are not unlike ones found in less luminous Cepheid and RR Lyrae star models, in which also first three radial modes are found unstable with the p_2 having the largest γ and p_3 being only marginally unstable.

Figure 3 shows the behavior of the growth rates in two δ Scu star models. These two models are more evolved than the β Cep model used in the previous figures. In these more evolved models all unstable nonradial modes are of dual, gravity and acoustic, character which causes that their spectra are denser than those of radial modes. The complicated behavior of γ, which for the nonradial modes exhibit local minima and maxima reflects effect of a partial mode trapping. Modes trapped in the acoustic cavity are characterized by lower I and

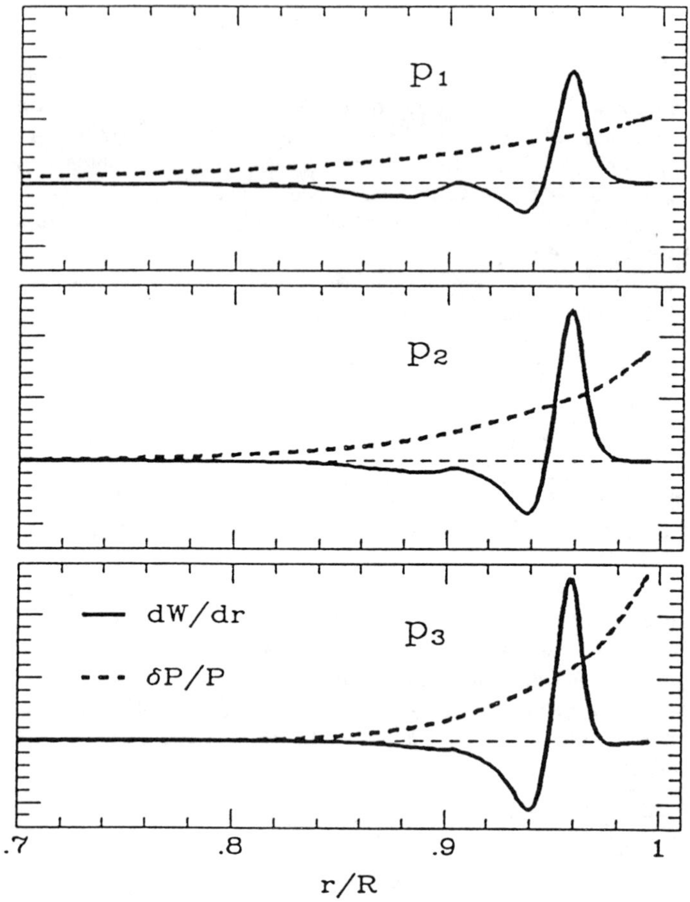

FIGURE I The differential work integral dW/dr (in arbitrary units) and the eigenfunction, $\delta P/P$, for the three first radial modes in a β Cephei star model, characterized by the following parameters $M/M_{\odot} = 12$, $\log T_{\rm eff} = 4.376$, $\log L/L_{\odot} = 4.208$, $Z = 0.03$, $X_0 = 0.7$, $X_c = 0.222$.

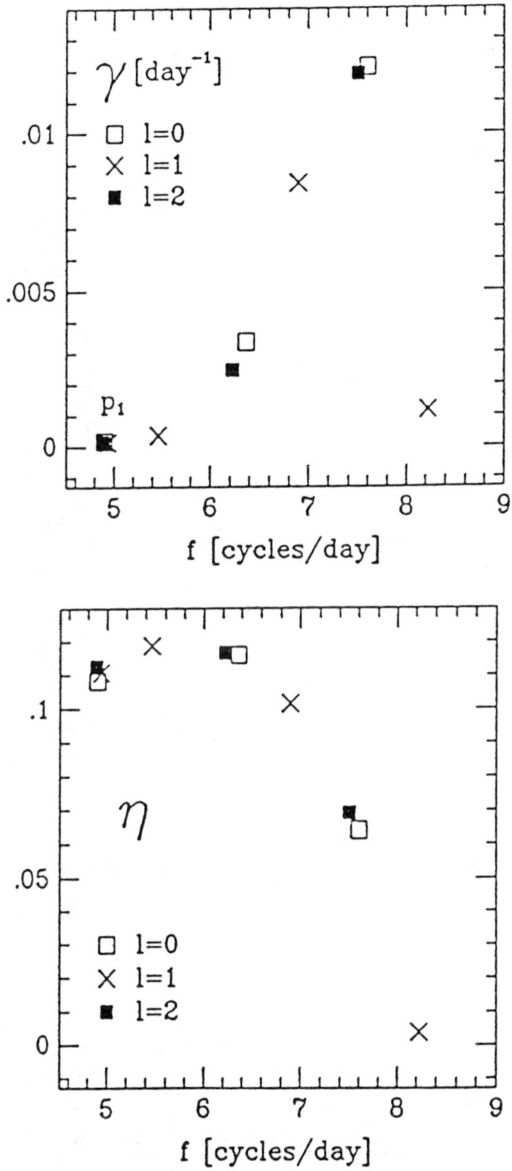

FIGURE II The growth rate, γ, and normalized growth rate, η, plotted against cyclic frequency, $f = \omega/2\pi$, for unstable modes of low spherical harmonic degree, l, in the same model as used in Figure I

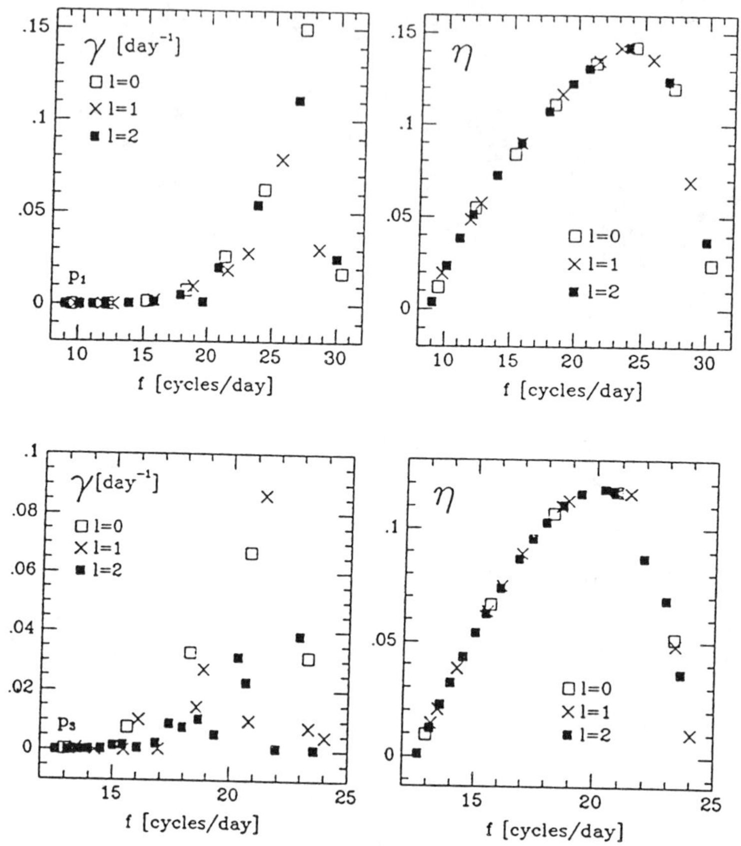

FIGURE III The growth rate, γ, and normalized growth rate, η, plotted against cyclic frequency, $f = \omega/2\pi$, for unstable modes of low spherical harmonic degree, l, in two models of δ Scu stars. The model used in the upper plots has the following parameters: $M/M_\odot = 2$, $\log T_{\rm eff} = 3.878$, $\log L/L_\odot = 1.343$, $Z = 0.02$, $X_0 = 0.7$, $X_c = 0.060$ The model used in the lower plots has the following parameters: $M/M_\odot = 2$, $\log T_{\rm eff} = 3.890$, $\log L/L_\odot = 1.490$, $Z = 0.02$, $X_0 = 0.7$, $X_c = 0$

consequently higher γ. Possibility that the trapping in the outer cavity is a crucial factor in mode selection, was first suggested by Winget *et al.*(1981) in the context of ZZ Ceti stars, a similar effect in δ Scu was discussed by Dziembowski & Królikowska (1990). The dual character modes appear also more evolved β Cep star. The frequencies of such modes in both types of variable stars depends on the details of the convective core evolution.

The instability in δ Scu and β Cep star models continues to higher l-values. There is indeed a very large number of modes which are simultaneously unstable. This gives a ground to the hope that, some day, many of them will be detected. Measurement of their frequencies would then enable us to make a detailed seismic sounding of the evolving stellar interiors. This is not certain, however. Complex nonlinear processes determine whether an unstable mode will be present in the terminal development of the instabilities, what will be its amplitude and whether it will be constant or subject, rather, to modulation. Only after considering specific nonlinear effects it may be possible to decide which, if any, of the two growth rates is a useful proxy for the amplitude.

4. AMPLITUDE EQUATION FORMALISM

Nonlinear equations for stellar oscillations are very difficult to solve. Fortunately, in many important applications the amplitude equation formalism applies, which reduces the problem from a partial differential equation system in time and three spatial variables to a system of ordinary differential equations in time. Basic assumptions underlying this simplification are, validity of the power expansion in the oscillation amplitudes, and $|\gamma| \ll \omega$ for all involved modes. It is remarkable that even for Cepheid models, the solutions of the amplitude equations agree quite well with exact nonlinear models (Buchler and Kovács 1986), in spite of relatively large amplitudes and γ's. The details of the derivation may be found for instance in papers by Vandakurov (1981) and Buchler and Goupil(1984).

4.1 Modal selection in the absence of resonances

The coupling between modes in the absence of resonances occurs only through a collective saturation of the driving mechanism. The amplitude equations have an exceptionally simple form in this case .

$$\frac{dA_k}{dt} = \tilde{\gamma}_k A_k, \tag{5}$$

where

$$\tilde{\gamma}_k = \gamma_k(1 + \sum_j \alpha_{kj} A_j^2) \tag{6}$$

We assume $\gamma > 0$ and $\alpha < 0$ for each j and k. Monomode solutions

$$A_n = \frac{1}{\sqrt{-\alpha_{nn}}} \tag{7}$$

always exists in such conditions, but they are stable only if

$$\chi_{kn} = \frac{\alpha_{kn}}{\alpha_{nn}} > 1 \tag{8}$$

for each k. There may be more than one stable monomode solution, like in the EO domain of the Cepheid instability strip where either the fundamental mode or the first overtone pulsation is possible (Stellingwerf, 1975). Choice of the mode depends in such a situation on the direction of stellar evolution.

Double mode pulsation occur represent a stable solution of Eq.(4) if

$$\chi_{1,2} < 1, \qquad \chi_{2,1} < 1 \tag{9}$$

and

$$\chi_{k,1}(1-\chi_{1,2}) + \chi_{k,2}(1-\chi_{2,1}) + \chi_{1,2}\chi_{2,1} > 1 \tag{10}$$

Ishida (1990) considered the case of three-mode coupling, which is the largest number of modes considered so far. He found a domain in the parameter space, where no stable fixed point solution exists and shown that chaotic solutions are possible there.

Investigation of the collective saturation involving much larger number of modes is a needed research project. It should be stressed that even in the case of classical pulsating stars there is a large number of high-l modes that are unstable (Dziembowski, 1977; Osaki, 1977). Presence of such modes cannot be directly detected, but they may play a role in the final amplitude development of instability. One difficulty in treating the multiple mode saturation is the fact that coefficients α involve up to third derivatives of opacity, $\kappa(\rho, T)$. Standard opacity tables are not suitable for evaluation derivatives of this high order.

4.2 Resonant mode coupling

Coupling to a stable mode is an alternative way of amplitude limitation. Here, we shall limit ourselves to considering the simplest and most important case of two-mode resonance case, assuming that for the two modes, we have

$$\omega_2 = 2\omega_1 + \Delta\omega, \tag{11}$$

where $\Delta\omega$ is a small quantity. The amplitude equations for this case are

$$\frac{dA_1}{dt} = \gamma_1 A_1 + C_1 A_2 A_1 \cos\Phi, \tag{12}$$

$$\frac{dA_2}{dt} = \gamma_2 A_2 - C_2 A_1^2 \cos\Phi, \tag{13}$$

$$\frac{d\Phi}{dt} = \Delta\omega + (\frac{C_2 A_1^2}{A_2} - 2C_1 A_2)\sin\Phi, \tag{14}$$

where $\Phi = \Delta\omega t + \Phi_2 - 2\Phi_1$. The coupling coefficients, C are integrals involving triple products of the eigenfunctions, two for mode 1 and one for mode 2. In particular, both coefficients

$$C \propto \int Y_{(1)}^2 Y_{(2)}^* \sin\theta d\phi d\theta. \tag{15}$$

There are two clearly distinct situations. One describing the case of unstable lower frequency mode, $\gamma_1 > 0, \gamma_2 < 0$, is refered to as the 2:1 resonance. The opposite one, $\gamma_2 > 0, \gamma_1 < 0$, is called the 1:$\frac{1}{2}$ or parametric resonance. The latter is much more likely to occur. Note, for instance, that in this case an unstable may be coupled to modes of various l, while in the 2:1 resonance case only to another $l = 0$ mode.

Equations (11)–(13) admit the following constant-amplitude solution

$$A_1 = \sqrt{-\frac{\gamma_1\gamma_2}{C_1C_2}(1+q^2)}, \tag{16}$$

$$A_2 = \sqrt{\frac{\gamma_1^2}{C_1^2}(1+q^2)}, \tag{17}$$

$$\tan\Phi = -q, \tag{18}$$

where $q = \Delta\omega/(\gamma_2 + 2\gamma_1)$.

Nonlinear correction to frequencies are given by

$$\frac{d\Phi_1}{dt} = C_1A_2\sin\Phi \quad \text{and} \quad \frac{d\Phi_2}{dt} = C_2\frac{A_1^2}{A_2}\sin\Phi. \tag{19}$$

They imply that frequency synchronization occurs, so that even if $\Delta\omega \neq 0$ the 2:1 resonance manifest itself only through an enhancement of λ_{11+}. Information about $\Delta\omega$ is contained in the phase difference $\phi_{11+} - 2\phi_1$ (*conf.* Equation (1)).

These fixed point solutions are not always stable. The necessary stability condition is $2\gamma_1 + \gamma_2 < 0$. In the case of the 2:1 resonance this also the sufficient condition. In the case of the 1:$\frac{1}{2}$ resonance there is an additional condition that sets a lower bound on $|\Delta\omega|$. In the unstable cases, amplitude limitation may still occur, but in a time-dependent form. The amplitudes and frequencies are then subject to periodic or chaotic modulation occurring on much shorter time scale than the evolutionary one. Regardless its form, the resonant amplitude limitation, by preventing saturation of the driving mechanism, resonances promotes multimode pulsation.

5. THE ROLE OF RESONANCE IN CLASSICAL PULSATING VARIABLES

In most cases only one mode is observed in these stars. Rapid changes in pulsation characteristics are only seldom observed. Numerical models of radially pulsating stars reproduce light and radial velocity curves for Cepheids and RR Lyr stars, in a satisfactory way. This agreement points out to the saturation of driving mechanism as the most important nonlinear effect. Effect of resonances has been invoked only to account for anomalous behaviors.

The 2:1 resonance between the fundamental mode and a higher mode has been proposed by Woltjer (1935) as the cause for the occurrence of a secondary bump in light curves of certain Cepheid. This idea was fully confirmed in modern times (Simon & Schmidt, 1976; Klapp *et al.*, 1985).

It has been suggested (Dziembowski & Kovács, 1984; Buchler & Kovács, 1988) that the same resonance may be responsible for the occurrence of double mode pulsation in certain Cepheids and RR Lyr. This could happen if the mode, that would have been the only one present in the final amplitude state, is in the 2:1 resonance with a stable mode and therefore cannot attain its full amplitude needed to saturate the driving mechanism. In such a situation, another unstable mode may grow. Moskalik (1986) proposed that a periodic modulation of light curves, observed in some RR Lyrae and known as the Blazhko effect may be caused by a similar resonance if the constant amplitude solution is unstable.

Chaotic behavior in models of luminous pulsating stars associated with occurrence of higher-order resonance was discovered by Kovács and Buchler (1988). The relevant resonances are of *half-integer* type viz., $3\omega_1 \approx 2\omega_2$. These authors propose that the deterministic chaos they found in the models is the cause of the irregular variability of the W Vir and RV Tau stars.

6. EFFECTS OF THE PARAMETRIC RESONANCE

It has been shown (Dziembowski & Królikowska, 1985) that, in Main Sequence stars, a parametric g-mode excitation occurs at very low (mmag) amplitudes of the unstable acoustic modes. The general case of three mode resonance of the type

$$\omega_a = \omega_{g1} + \omega_{g2} + \Delta\omega, \tag{20}$$

was considered in that paper. However, the two-mode formalism developed in Section 4 may be used here, because in most important cases we have

$$l_{g1} \approx l_{g2} \gg l_a, \tag{21}$$

and

$$\omega_{g1} \approx \omega_{g2} \approx \frac{\omega_a}{2}. \tag{22}$$

Excitation occurs at amplitude, A_a, which is nearly equal to the equilibrium amplitude. Thus, we get from Equation (18)

$$A_a = \sqrt{\frac{\gamma_g^2}{C_g^2}[1 + (\frac{\Delta\omega}{(\gamma_a + 2\gamma_g})^2]}. \tag{23}$$

This amplitude may be very small indeed, because the large number of potential resonant pairs allows a fine tuning of the resonance, $|\Delta\omega| \ll \omega$, and because the g-mode are very nearly adiabatic, $|\gamma_g| \ll \omega_g$. In such a situation, only evaluation of a probability distribution for A_a is feasible.

Star rotation further reduces the acoustic mode amplitude at the onset of parametric instability. Lifting the m-degeneracy of frequencies results in increasing chances of close resonances and consequently in reducing expected A_a. The effect becomes significant already at equatorial velocities of some 40 km/s (Dziembowski *et al.*, 1988).

In stars near ZAMS, for p_1, p_2 and p_3 radial modes and corresponding low-l nonradial modes, the constant amplitude solutions are most probable. In this situation, $-\gamma_g \gg \gamma_a$ and A_a is virtually independent of γ_a. There is a fair probability of periodic limit cycle and as well as chaotic solutions. The time scale of the amplitude and period variations is of the order of γ_g^{-1}.

For higher order modes and for all acoustic modes in more evolved objects, there is an increasing probability of coupling to g-modes trapped in the inner cavity. In such situations we have $-2\gamma_g < \gamma_a$, which implies not only that the constant amplitude solutions are unstable but that the interaction does not halt the amplitude growth. It may be expected that the amplitude is ultimately limited as a result of interaction with *next generation* of parametrically excited g-modes, but a theory of such multi-mode interactions remains to be developed. If this effect indeed governs nonlinear development of acoustic mode instability then, at this stage, we can only predict that modes which have relatively low amplitudes in the inner g-mode propagation zone have the best chances to attain large surface amplitudes. These are radial modes and nonradial modes that are partially trapped in the acoustic cavity.

7. NONLINEAR PHENOMENA IN MULTIPERIODIC VARIABLES

7.1 δ Scu stars

Most of these stars are low amplitude variables. Furthermore, there are many objects in the δ Scu domain of the H-R diagram which variability was searched for and was not detected. There is an evidence for period and amplitude changes occurring on the time scale of years (Breger, 1992) in several objects of this type.

Part of δ Scu stars are well studied multimode pulsators. In Table I I compiled information about amplitude spectra for few such objects. Mode identification given in this table, in most cases must, be regarded as tentative. The exception is GX Peg, in which case the identification is based on comprehensive model calculations (Goupil *et al.*,1992). Results of that work, as pointed out by its authors, give certain support for the idea of preferential excitation of modes trapped in the envelope.

Spectra of the remaining objects are much harder to understand. In particular, in 4 CnV, the only pair of modes that on the basis of the period ratio could be identified with radial modes are the two of the highest frequencies. Presence of *only* two, at most, radial modes in this broad-frequency spectrum cannot easily be explain in terms of nonlinear effects discussed in previous sections. It is also difficult to understand absence of $l = 1$ modes in the 44 Tau oscillation spectrum. I suppose, however, that an alternative interpretation of this spectrum might be possible.

Discovery of three small amplitude modes, in addition to the fundamental and first overtone pulsations, in the spectrum of AI Vel (Walraven & Balona, 1992) is a surprise. I used to believe, that the two radial modes, owing their large amplitudes, fully saturated the opacity mechanism. This discovery is a good news for asteroseismology, though, interpretation of the present spectrum is not easy.

β Cephei stars

There are many multimode objects of this type. With possible exception is 12 Lac, in all β Cep frequencies of the detected modes in individual stars occur in ranges that are significantly narrower than the distance between consecutive radial modes. Observational identification of the spherical harmonic degrees, do not reveal any obvious preference to $l = 0$ values. These properties seem to favor the saturation as the dominant nonlinear effect.

Amplitude are most often small and their rapid changes have been reported. The best known example is *Spica* (α Vir), observed as a pulsating star in 1960's

TABLE I Amplitude spectra for some δ Scu stars

star	f[c/day]	P[hours]	A[mmag]	identification
4 CnV	5.0475	4.755	7.7	$l \neq 0$
$\log T_{\rm eff} = 3.839$	5.8508	4.102	9.1	$l \neq 0$
reference:	6.9763	3.440	5.1	$l \neq 0$
Breger *et al.*	7.3778	3.253	5.2	$l = 0$, p_4
(1990)	8.5950	2.792	12.8	$l = 0$, p_5
	6.898	3.479	27.8	$l = 0$, p_1
44 Tau	7.006	3.426	21.2	$l = 2$, p_1
$\log T_{\rm eff} = 3.833$	7.304	3.286	5.9	$l = 3$, p_2
reference:	8.960	3.678	8.6	$l = 0$, p_3
Mantegazza *et al.*	9.116	2.633	14.7	$l = 2$, p_2
(1992)	9.562	2.510	6.1	$l = 3$, p_1
	11.520	2.083	6.8	$l = 0$, p_3
AI Vel	8.9627	2.678	195.0	$l = 0$, p_1
$\log T_{\rm eff} = 3.875$	9.1376	2.626	6.5	$l \neq 0$
reference:	11.5998	2.069	144.0	$l = 0$, p_2
Walraven & Balona	15.9725	1.503	8.9	$l \neq 0$
(1992)	22.5218	1.066	10.8	$l = 0$, p_6
θ^2 Tau	13.2300	1.8141	6.6	$l = 0$, p_2 or p_3
$\log T_{\rm eff} = 3.903$	13.4807	1.7803	2.6	$l = 2$
reference:	13.6956	1.7524	4.5	$l = 1$
Breger *et al.*	14.4176	1.6646	2.7	$l = 1$
(1989)	14.6165	1.6420	1.2	?
GX Peg	16.17	1.484	1.62	$l = 0$, p_2
$\log T_{\rm eff} = 3.892$	19.66	1.221	1.78	$l = 1$, $m = 1$, p_3
reference:	19.91	1.205	2.08	$l = 0$, p_3
Michel *et al.*	20.52	1.170	1.80	$l = 1$, $m = 0$, p_3
(1992)	20.78	1.155	2.18	$l = 1$, $m = -1$, p_3

and showing no detectable pulsation in 70's and early 80's (Sterken *et al.*, 1986. The occurrence of an amplitude modulation on the time scale of years cannot be used as an evidence for the role of resonances. As discussed in Section 4.1, the amplitude changes may occur if more than two modes is involved in the saturation of the driving mechanism.

Rapidly oscillating Ap stars

Kurtz (1982) demonstrated that variability in these stars is due to excitation of high-order p-modes of $l = 1$ which are symmetric about magnetic field axis. Recently, Kurtz *et al.* (1992) showed that in the case of Ap star HR3831, there is a significant contamination with harmonic of other degrees.

The driving mechanism is not yet understood. A peculiarity of the nonlinear behavior of Ap star oscillations is the occurrence of harmonics in their power spectra already at millimagnitude amplitudes. This is in stark contrast to δ Scu stars where such nonlinearities are seen only at amplitudes of tenths of a magnitude. We do not know why it is so, but it seems likely that this property of Ap star oscillations might be the clue to finding the excitation mechanism.

Variable white dwarfs

These are best studied multiperiodic variable stars. An unprecedented amplitude resolution was reached in observations made with the *Whole Earth Telescope*.

In the power spectrum of PG 1159–035 Winget *et al.* (1991) identified 101 peaks as consecutive modes in $l = 1$ and 2 multiplets. The period- spacing clearly reveals the existence of at least two trapping cavities (a consequence of the element stratification). The lack of correlation between the spacing and the amplitudes seems to disprove, in this case, the hypothesis of the preferential excitation of the modes trapped in the outer cavity. There is now a a firm evidence for amplitude changes in this star, occurring on the time scale of years.

Power spectra for some oscillating white dwarfs such as G117-B15A (Kepler *et al.*, 1992) or BPM31594 (O'Donoghue *et al.*, 1992) are dominated by nonlinear effects - harmonics, amplitude modulation. Such objects are less rewarding from the point of view of asteroseismology but they are source of valuable information about nonlinear effects in stellar oscillations that we still have to learn to use it.

ACKNOWLEDGMENTS

I am grateful to Alosha Pamyatnykh for his essential help in preparing figures for this paper. This work was supported in part by National Committee for Scientific Research grant No 2-1185-91-01.

REFERENCES

Balona, L.A., & Stobie, R.S., 1980, *MNRAS*,**190**, 231.

Breger, M., 1992, in *Nonlinear Phenomena in Stellar Variability*, ed. M. Takeuti & J.R. Buchler, special issue of *Ap&SS*, in press.

Breger, M., Garrido, R., Lin, H., Shi-yang, J., Zi-he, G., Frueh, M., & Paparo, M., 1989, *A&A*, **214**, 209.

Breger, M., McNamara, B.J., Kerschbaum, F., Lin, H., Shi-yang, J., Zi-he, G., & Poretti, E., 1990, *A&A*, **231**, 56.

Buchler, J.R. & Goupil, M.-J., 1984, *ApJ*, **279**, 394.

Buchler, J.R. and Kovács, G., 1986, *ApJ*, **318**, 232.

Buchler, J.R. and Kovács, G., 1988, in *Advanves in Helio- and Asteroseismology*, ed. J. Christensen-Dalsgaard & S. Frandsen, (Dordrecht:reidel), p.393.

Buchler, J.R., Moskalik, P., & Kovács, G., 1990, *ApJ*, **351**, 617.

Dziembowski, W., 1977, *Acta Astron.*, **27**, 95.

Dziembowski, W., & Kovács, G., 1984, *MNRAS*, **206**, 497.

Dziembowski, W., & Królikowska, M., 1985 *Acta Astron.*, **35**, 5.

Dziembowski, W., & Królikowska, M. 1990 *Acta Astron.*, **40**, 19.

Dziembowski, W., Królikowska, M., & Kosovichev, A., 1988 *Acta Astron.*, **38**, 61.

Goupil, M.J., Michel, E., Leberton, Y, & Baglin, A., 1992, *A&A*, in press.

Ishida, T., 1990, *Ap&SS***168**, 243.

Kepler S.O., *et al.*, 1992, *ApJ*, in press.

Klapp, J., Goupil, M.J., & Buchler, J.R., 1985, *ApJ*, **296**, 514.

Kovács, G., & Buchler, R.J., 1988, **334**, 971.
Kurtz, D.W., 1982, *MNRAS*, **200**, 807.
Kurtz, D.W., Kanaan, A., & Martinez, P., 1982, *MNRAS*in press.
Mantegazza, L., Poretti, E., Antonello, E., & Riboni, E., 1992, *A&A*, **256**, 113.
Michel, E., *et al.*, 1992, *A&A*, **255**, 139..
Moskalik, P., 1986, *Acta Astron.*, **36**, 333.
Osaki, Y., 1977, *PASJ*, **29**, 235.
O'Donoghue, Warner, B., & Cropper, M., 1992, *MNRAS*, in press.
Simon, N.R., & Lee, A.S., 1981, *ApJ*, **248**, 291.
Simon, N.R., & Schmidt, E.G., 1976, *ApJ*, **205**, 162.
Stellingwerf, R.F., 1975, *ApJ*, **195**, 441.
Stellingwerf, R.F., 1978, *AJ*, **83**, 1184.
Sterken, C., Jerzykiewicz, M., & Manfroid, J., 1986, *A&A*, **169**, 160.
Vandakurov, Yu.V., 1981, *Soviet.Astron. Letters*, **7**, 128.
Walraven, Th. Walraven, & Balona, L.A., 1992, preprint.
Winget, D.E., *et al.*, 1991, *ApJ*, **378**, 326.
Winget, D.E., Van Horn, H.M., & Hansen, C.J., 1981, *ApJ*, **245**, L33.
Woltjer jr., J., 1935, *MNRAS*, **95**, 260.

Inside the Stars, IAU Colloquium 137
ASP Conference Series, Vol. 40, 1993
Werner W. Weiss and Annie Baglin (eds.)

ASYMPTOTIC THEORY OF LOW-DEGREE STELLAR ACOUSTIC OSCILLATIONS

I. W. ROXBURGH
Astronomy Unit, Queen Mary and Westfield College, London E1 4NS, UK

S. V. VORONTSOV
Institute of Astronomy, University of Cambridge, Cambridge CB3 0HA, UK, and Institute of Physics of the Earth, Moscow 123810, Russia (permanent address)

ABSTRACT We extend the second-order asymptotic description developed by Tassoul (1980, 1990) to the forth order, taking into account both gravity perturbations and realistic (non-polytropic) structure of the stellar envelope. We examine the accuracy of the asymptotic description by the direct computations for a solar model.

THE ASYMPTOTIC DESCRIPTION

The govering fourth-order system of equations of linear adiabatic oscillations can be reduced to the single second-order vector differential equation

$$\frac{d^2}{dr^2}\begin{pmatrix}\eta\\P\end{pmatrix} + \mathbf{C}(r,\omega)\frac{d}{dr}\begin{pmatrix}\eta\\P\end{pmatrix} - \omega^2\mathbf{D}(r,\omega)\begin{pmatrix}\eta\\P\end{pmatrix} = 0, \tag{1}$$

where $\eta(r)$ and $P(r)$ are proportional to the Eulerian pressure and gravitational potential pertirbations, and matrices $\mathbf{C}(r,\omega)$ and $\mathbf{D}(r,\omega)$ are determined by the equilibrium model (Vorontsov 1991). We derive the uniform asymptotic approximation to the solutions of the equation (1) using Bessel functions

$$\begin{pmatrix}\eta\\P\end{pmatrix} = \left[\mathbf{Y}_0(r) + \frac{1}{\omega}\mathbf{Y}_1(r) + \frac{1}{\omega^2}\mathbf{Y}_2(r) + \frac{1}{\omega^3}\mathbf{Y}_3(r) + \ldots\right]\begin{pmatrix}\omega^{1/2}J_{\ell+1/2}(\omega\varphi)\\ \omega^{1/2}\dot{J}_{\ell+1/2}(\omega\varphi)\end{pmatrix} \tag{2}$$

and determine matrices $\mathbf{Y}_i(r)$ up to (and including) $\mathbf{Y}_3(r)$. Near the surface, where asymptotic description becomes invalid, we match these solutions with exact frequency-dependent solutions. The resulted eigenfrequency equation is

$$\int_0^R \frac{dr}{c} + \frac{1}{\omega^2}\left[A_0 + \ell(\ell+1)A_\ell\right] + \frac{1}{\omega^4}\left[B_0 + \ell(\ell+1)B_\ell + \ell^2(\ell+1)^2B_{\ell\ell}\right]$$
$$\simeq \frac{\pi}{\omega}\left[n + \frac{\ell}{2} + \frac{1}{4} + \alpha(\omega)\right] \tag{3}$$

where $\alpha(\omega)$ is frequency-dependent surface phase shift (Vorontsov 1991), and

$$A_0 = y(R) - \left[\frac{c}{2}\ln'\left(\frac{c}{hr^2}\right) + \frac{2}{3}\frac{c}{r}\right]_{r=R}, \qquad A_\ell = -\frac{1}{2}\int_0^R c'\frac{dr}{r},$$

$$B_0 = \frac{1}{2}\int_0^{r_m}\left[-cy'^2 + \frac{c}{2}\ln'\left(\frac{c}{hr^2}\right)\left(N^2 - 4\pi G\rho_0\right)'\right] dr,$$

$$B_\ell = \frac{1}{2}\int_0^{r_m}\left[c\left(cy' + N^2\right)\right]'\frac{dr}{r} - \frac{1}{2}\left[\frac{c}{r}\left(cy' + N^2\right)\right]_{r=r_m}, \tag{4}$$

$$B_{\ell\ell} = -\frac{1}{48}\int_0^{r_m}\left(c^3\right)'''\frac{dr}{r} + \frac{1}{48}\left[\frac{\left(c^3\right)''}{r} + \frac{\left(c^3\right)'}{r^2} + 2\frac{c^3}{r^3}\right]_{r=r_m},$$

$$y(r) = \frac{1}{2}\int_0^r c\left[\frac{c''}{2c} - \left(\frac{c'}{2c}\right)^2 - \frac{\ln'' h}{2} - \frac{\ln'^2 h}{4} - \frac{\ln' h}{r} - \frac{N^2}{c^2} + \frac{4\pi G\rho_0}{c^2}\right] dr,$$

$$h(r) = \exp\int_0^r\left(\frac{N^2}{g} - \frac{g}{c^2}\right) dr.$$

Constants B_0, B_ℓ, $B_{\ell\ell}$ of the fourth-order term in the eigenfrequency equation (3) are independent on the position of the matching point r_m, if it is somewhere in the envelope, where second-order asymptotic is adequate (in local sense). For the small frequency separations $\nu_{n,\ell} - \nu_{n-1,\ell+2} = (2\ell+3)D_{n,\ell}$, we get

$$2\pi\int_0^R \frac{dr}{c}\cdot D_{n,\ell} \simeq \frac{A_\ell}{\pi\nu} + \frac{1}{(2\pi\nu)^3}\left[2B_\ell + \left((2\ell+3)^2+3\right)B_{\ell\ell}\right]. \tag{5}$$

The accuracy of the asymptotic description of the absolute values of p-mode frequencies is demonstrated by Fig. 1. At higher frequencies ($\nu > 4$mHz), significant improvement is clearly seen when the fourth-order description is compared with the second-order description. But there is no improvement at lower frequencies ($\nu < 3$mHz), which indicates poor convergence of the asymptotic expansions.

The accuracy of the description of $D_{n,\ell}$-values is illustrated by Fig. 2. The improvement is seen again at higher frequencies ($\nu > 4$mHz): the fourth-order approximation tends to describe reasonably the systematic difference between the exact values of $D_{n,1}$ and $D_{n,0}$. This difference is determined by the constant $B_{\ell\ell}$, with the kernel $(c^3)'''/r$ localized sharply in the solar core. Sufficiently accurate frequency measurements would allow one to utilise the high diagnostic capability of this difference.

The failure of the asymptotic descriptions at lower frequencies is due predominantly to the effects of gravity perturbations, which are large and difficult to treat accurately by the perturbational-type analysis. This conclusion is confirmed by additional computations in Cowling approximation.

We thank J. Christensen-Dalsgaard for his solar model used in the computations. S.V.V. thanks the UK Science & Engineering Research Council, and "Zodiac" Scientific-Methodological Council for their partial support.

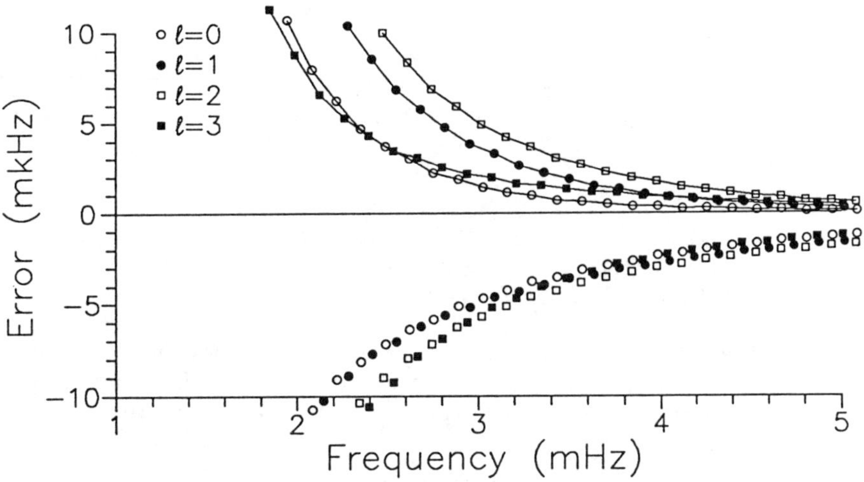

Fig. 1. Differences between asymptotic and exact eigenfrequencies, for the fourth-order (connected by solid lines) and for the second-order description. Computations were made for solar model 1 of Christensen-Dalsgaard (1982).

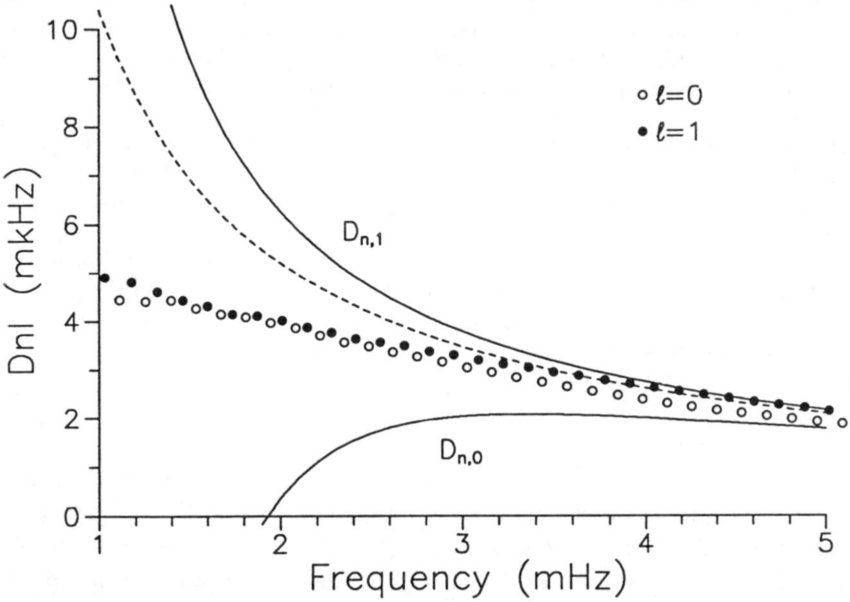

Fig. 2. Asymptotic values of $D_{n,\ell}$ resulted from the fourth-order (solid lines) and second-order (dashed line) asymptotic descriptions. The exact values of $D_{n,\ell}$ are shown by circules.

<u>REFERENCES</u>

Christensen-Dalsgaard, J. 1982, *M.N.R.A.S.*, **199**, 735.

Tassoul, M. 1980, *Ap. J. Suppl.*, **43**, 469.

Tassoul, M. 1990, *Ap. J.*, **358**, 313.

Vorontsov, S. V. 1991, *Astron. Zh.*, **68**, 808.

Inside the Stars, IAU Colloquium 137
ASP Conference Series, Vol. 40, 1993
Werner W. Weiss and Annie Baglin (eds.)

THE SUBSEISMIC APPROXIMATION FOR LOW-FREQUENCY MODES OF THE EARTH APPLIED TO LOW-DEGREE, LOW-FREQUENCY G-MODES OF NON-ROTATING STARS

I. DE BOECK, T. VAN HOOLST AND P. SMEYERS
Instituut voor Sterrenkunde, Katholieke Universiteit Leuven, Celestijnenlaan 200 B, 3001 Heverlee (Belgium)

ABSTRACT The subseismic approximation for low-frequency modes in the Earth is used in an asymptotic treatment of low-frequency *g*-modes belonging to low-degree spherical harmonics in a non-rotating star.

INTRODUCTION

In an investigation on normal modes of rotating, self-gravitating, compressible, stratified, fluid bodies, Rochester (1989) adopted the earlier suggestion of Smylie and Rochester (1981) that, for many low-frequency modes in the Earth, the contribution of the Eulerian perturbation of the pressure to the Lagrangian perturbation of that quantity is negligible. This approximation is referred to as the subseismic approximation since it is used for modes with frequencies that are low compared with the lower limit of the frequencies of acoustic modes usually studied in terrestrial seismology.

We examined the adequacy of the suggestion of Smylie and Rochester for low-frequency *g*-modes belonging to low-degree spherical harmonics in a *non-rotating* star by developing an asymptotic representation for these modes based on the subseismic approximation.

BASIC EQUATIONS

Let r be the radial distance from the center, ρ the mass density, P the pressure, Φ the gravitational potential, g the gravity, $\Gamma_1 \equiv (\partial \ln P/\partial \ln \rho)_S$ one of the generalized isentropic coefficients, and $c \equiv (\Gamma_1 P/\rho)^{1/2}$ the isentropic sound velocity. $\vec{\xi}$ is the Lagrangian displacement. We denote the Eulerian perturbation of a quantity by a prime on that quantity.

For normal modes depending on time t by $\exp(i\sigma t)$, the equations of motion, conservation of mass, and conservation of entropy can be combined to give

$$\nabla \cdot \vec{\xi} = -\frac{1}{\rho c^2}\left(P' + \frac{dP}{dr}\xi_r\right), \tag{1}$$

$$\sigma^2 \vec{\xi} = \nabla\left(\Phi' + \frac{P'}{\rho}\right) - \frac{N^2}{\rho g}\left(P' + \frac{dP}{dr}\xi_r\right)\vec{1}_r, \tag{2}$$

where $\vec{1_r}$ is the local unit vector in the radial direction, and N^2 the square of the Brunt-Väisälä frequency defined as

$$N^2 = -g\left(\frac{g}{c^2} + \frac{d\ln\rho}{dr}\right). \tag{3}$$

Adopting the subseismic approximation, we neglect P' relative to $(dP/dr)\xi_r$ in the right-hand members of Eqs. (1) and (2). For modes with a Lagrangian displacement $\vec{\xi}$ related to a single spherical harmonic $Y_\ell^m(\theta,\phi)$ as

$$\left[\frac{u(r)}{r^2}, \frac{v(r)}{r}\frac{\partial}{\partial\theta}, \frac{v(r)}{r}\frac{1}{\sin\theta}\frac{\partial}{\partial\phi}\right]Y_\ell^m(\theta,\phi), \tag{4}$$

where θ and ϕ are the angular coordinates, we derive the second-order differential equation for the function u

$$\frac{d^2u}{dr^2} - \frac{g}{c^2}\frac{du}{dr} + \left[\frac{\ell(\ell+1)}{r^2}\left(\frac{N^2}{\sigma^2} - 1\right) - \frac{d}{dr}\frac{g}{c^2}\right]u = 0. \tag{5}$$

In the usual approximation in which the Eulerian perturbation of the gravitational potential is neglected, the equation for the function u takes the form

$$\frac{d^2u}{dr^2} + \frac{1}{\rho}\frac{d\rho}{dr}\frac{du}{dr} + \left[\frac{\ell(\ell+1)}{r^2}\left(\frac{N^2}{\sigma^2} - 1\right) - \frac{d}{dr}\frac{g}{c^2}\right]u$$
$$= -\frac{\sigma^2}{S_\ell^2}\left(1 - \frac{\sigma^2}{S_\ell^2}\right)^{-1}\left(\frac{c^2}{r^2}\frac{d}{dr}\frac{r^2}{c^2}\right)\left(\frac{du}{dr} - \frac{g}{c^2}u\right) - \frac{\sigma^2}{c^2}u, \tag{6}$$

where

$$S_\ell^2 = \frac{\ell(\ell+1)c^2}{r^2}. \tag{7}$$

The left-hand members of Eqs. (5) and (6) are identical apart from the coefficient of the first derivative du/dr. The right-hand member of Eq. (6) is different from zero and consists of terms involving σ^2 in their numerator. These terms are small for small values of $|\sigma^2|$ except near $r = R$. For g^+-modes, the right-hand member of Eq. (6) also displays a singularity at the point near $r = R$ where $S_\ell^2 = \sigma^2$.

In the standard asymptotic theory, one avoids the difficulties related to the large absolute value of the right-hand member of Eq. (6) near $r = R$ by passing on to a second-order differential equation in the function v for the construction of an asymptotic approximation of the solutions from $r = R$.

Eq. (5) can be used for the construction of an asymptotic approximation of the solutions from $r = R$ as well as from $r = 0$.

ASYMPTOTIC REPRESENTATION OF LOW-FREQUENCY G-MODES

For the sake of simplification, we assume that N^2 is different from zero everywhere inside the star except at $r = 0$ and keeps the same sign. Furthermore, we consider Γ_1 to be constant.

In the frame of the subseismic approximation, the asymptotic representations of the function u constructed from $r = 0$ and from $r = R$ take the form

$$u_i = C_1 P^{-1/(2\Gamma_1)} \left(\epsilon \frac{N^2}{r^2}\right)^{-1/4} \xi_i^{1/2} J_{\ell+1/2}\left[\xi_i / \left(\epsilon\sigma^2\right)^{1/2}\right], \tag{8}$$

$$u_e = C_2 P^{-1/(2\Gamma_1)} \left(\epsilon \frac{N^2}{r^2}\right)^{-1/4} \left(2\xi_e^{1/2}\right)^{1/2} J_\nu\left[\left(2\xi_e^{1/2}\right) / \left(\epsilon\sigma^2\right)^{1/2}\right], \tag{9}$$

where

$$\epsilon = \frac{|N^2|}{N^2},\ \nu = \frac{n_e + 1}{\Gamma_1} + 1, \tag{10}$$

and

$$\xi_i(r) = [\ell(\ell+1)]^{1/2} \int_0^r \left[\epsilon \frac{N^2(r')}{r'^2}\right]^{1/2} dr', \tag{11}$$

$$2\xi_e^{1/2}(r) = [\ell(\ell+1)]^{1/2} \int_r^R \left[\epsilon \frac{N^2(r')}{r'^2}\right]^{1/2} dr'. \tag{12}$$

In these solutions, n_e is the effective polytropic index characterizing the surface layers of the star.

The degree ν of the Bessel function in the solution constructed from $r = R$ differs from the degree that is found in the standard asymptotic theory. This difference affects the equation for the eigenfrequencies. The ratio of the asymptotic eigenfrequencies determined on the basis of the subseismic approximation to the asymptotic eigenfrequencies determined by means of the standard asymptotic theory is given by

$$\frac{\left[\left(\epsilon\sigma^2\right)^{1/2}\right]_{subs.}}{\left[\left(\epsilon\sigma^2\right)^{1/2}\right]_{stan.}} = \frac{\ell + 1/2 + n_e + 2k}{\ell + 1/2 + (n_e + 1)/\Gamma_1 + 2k}. \tag{13}$$

The subseismic approximation leads to an inadequate asymptotic representation of the radial component of the Lagrangian displacement near the star's surface since this component becomes zero at $r = R$ and, for g^+-modes, displays one node less near $r = R$ than its regular number. Another consequence of the subseismic approximation is that the radial component of the Lagrangian displacement contains the factor $P^{-1/(2\Gamma_1)}$ instead of the factor $\rho^{-1/2}$.

Details and numerical results will be published in *Astronomy and Astrophysics.*

REFERENCES

Rochester, M.G. 1989, in: *Continuum Mechanics and its Applications*, eds. G.A.C. Graham and S.K. Malik, Hemisphere Publishing Corporation, New York, p. 797

Smylie, D.E. Rochester, M.G. 1981, *Physics of the Earth and Planetary Interiors* 24, 308

Inside the Stars, IAU Colloquium 137
ASP Conference Series, Vol. 40, 1993
Werner W. Weiss and Annie Baglin (eds.)

INITIAL ASTEROSEISMIC INVERSIONS

D.O.GOUGH[1] AND A.G.KOSOVICHEV[2]
Institute of Astronomy, Madingley Road, Cambridge CB3 0HA, England

INTRODUCTION

Accurate measurement of the frequencies of low-degree acoustic oscillations of sun-like stars is imminent. We report on our first calculations with proxy data, aimed at assessing the kind of physical information that is likely to be acquired from seismic analysis and the precision with which the frequencies must be measured in order to obtain that information. The results will have an important bearing on future observing strategies, for the duration of observation should be determined primarily by the precision required of the frequency measurements. Our inversions are of eigenfrequencies of modes of an evolved main-sequence star of mass $1.1M_\odot$. The modes are of degree 0, 1 and 2, with frequencies in the range 1–3 mHz. Thus, by analogy with solar oscillations, they are modes that one should expect to observe in stars similar to the sun. Figure I*a* depicts an idealized spectrum of stellar acoustic oscillations as one might expect from intensity variations such as those that could be measured from the proposed ESA spacecraft PRISMA. We report on the extent to which we have found it possible to determine the mass and radius of the stars, and on the seismic evidence for evolution having taken place in the core.

INVERSION TECHNIQUE

The frequencies ω_i can be expressed in terms of their relative small difference $\delta\omega_i^2/\omega_i^2$ from those of a standard reference model of similar mass and radius according to the linearized expression

$$\delta\omega_i^2/\omega_i^2 = \int_0^1 \left(K^i_{f,Y}\frac{\delta f}{f} + K^i_{Y,f}\delta Y\right) dx - I^i_q \delta q,$$

where $x = r/R$, $q = M/R^3$, Y is the helium abundance and f is any function of p and ρ; M and R are stellar mass and radius, in solar units, $K_{f,Y}$ and $K_{Y,f}$ are appropriate kernels and I_q is an integral over the reference model. These constraints can provide localized averages of $\delta \ln f$ and estimates of δY and δq of the kind

$$\overline{\delta \ln f} \equiv \int_0^1 \sum_i a_i(x_0) K^i_{f,Y} \delta \ln f\, dx \equiv \int_0^1 A_{f,Y}(x, x_0)\delta \ln f\, dx = \sum_i a_i(x_0)\frac{\delta\omega_i^2}{\omega_i^2}$$

[1]and Department of Applied Mathematics and Theoretical Physics, University of Cambridge

[2]and Crimean Astrophysical Observatory

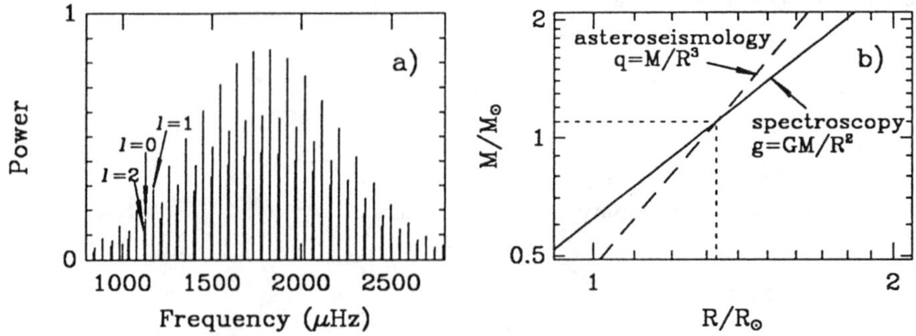

FIGURE I a) A sample of anticipated PRISMA data of observations of a star of $M = 1.1M_{\odot}, R = 1.33R_{\odot}$: p modes of l=0-2, n=6-28; b) Measuring stellar mass and radius: $M - R$ diagram

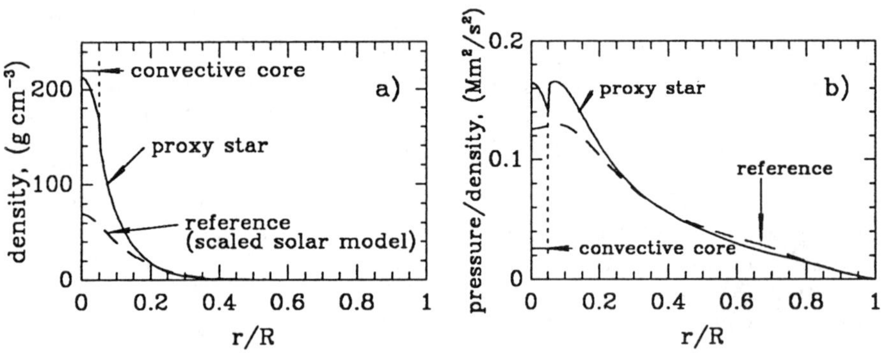

FIGURE II Parameters of the proxy star ($M = 1.1M_{\odot}, R = 1.33R_{\odot}$) and the reference model: a) ρ and b) $u \equiv p/\rho \simeq \mathcal{R}T/\mu$

by minimizing amongst coefficients $a_i(x_0)$ the functional:

$$\int_0^1 A^2_{f,Y}(x,x_0)J_f dx + \lambda_1 \int_0^1 \left(\sum_i a_i K^i_{Y,f}\right)^2 J_Y dx + \lambda_2 \left(\sum_i a_i I^i_q\right)^2 + \alpha \sum_i a_i^2 \epsilon_i^2$$

for tradeoff parameters λ_1, λ_2 and α, where ϵ_i are standard relative errors in the data. Appropriate weight functions J depend on the averages sought. For determining δq, we set $J_f = 1, J_Y = 1, \lambda_1 \neq 0$ and $\lambda_2 = 0$;
for $\overline{\delta \ln f}(x_0)$ we set $J_f = (x - x_0)^2, J_Y = 1, \lambda_1, \lambda_2 \neq 0$;
for $\overline{\delta Y}$ we set $J_f = 1, J_Y = 1, \lambda_1, \lambda_2 \neq 0$.

Some optimally localized averaging kernels $A_{f,Y}, A_{Y,f}$, with $f = u = p/\rho$, are illustrated in Figure III*a*. With only high-order low-degree modes it is not possible to localize $A_{f,Y}$ beyond $x \simeq 0.3$; to do so would require modes of lower order. The most robust quantity that can be determined is the homology frequency-scaling parameter q, which, coupled with a knowledge of surface gravity from spectroscopic analysis, determines M and R separately (see Figure I*b*).

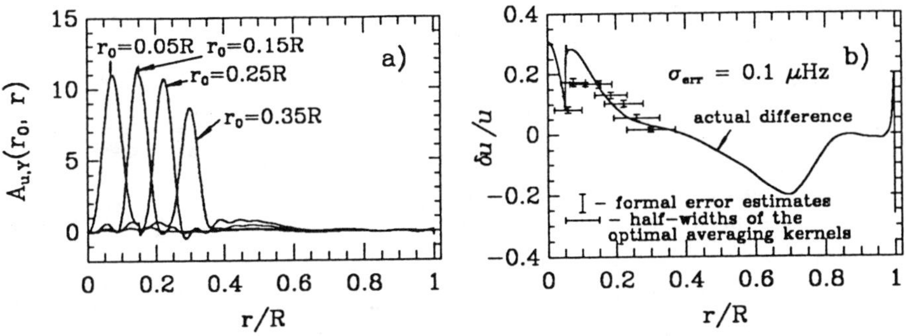

FIGURE III a) Optimal averaging kernels for $\delta u/u$; b) test inversion for relative differences in $u \equiv p/\rho$. Frequencies of the proxy star are perturbed by random errors with the standard deviation σ_{err}.

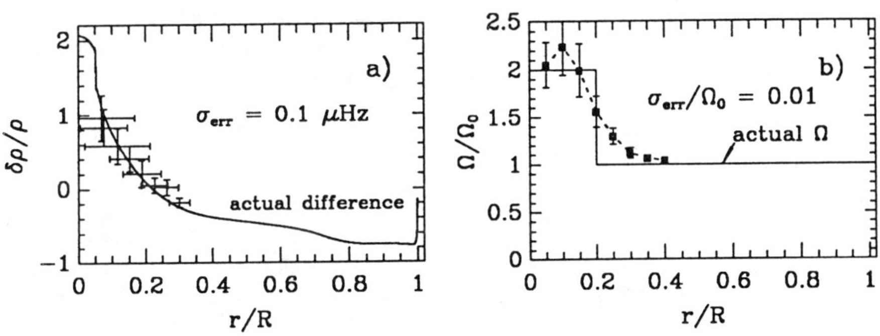

FIGURE IV Test inversions a) for relative differences in density ρ and b) for equatorial angular velocity Ω

INVERSION RESULTS

We have carried out a test inversion of artificial data obtained from a stellar model of $1.1 M_\odot$, evolved somewhat off the main sequence and with a convective core of radius $x_c = r_c/R \simeq 0.05$, and with $q = 0.468$. The reference was a solar model. The two models are illustrated in Figure II. The inversion yielded $q = 0.476$. Inversions for $\delta u/u$ and $\delta\rho/\rho$ are illustrated in Figures III*b* and IV*a* for assumed errors in the data with standard deviation 0.1 μHz. It should be appreciated that the relative difference between the models is not small, so the linearization may not have provided a good approximation. We presume that with the addition of some theoretical prejudice obtained by adopting more of the assumptions of stellar evolution theory, a closer reference model could be found and a better inversion obtained. We report that if errors with standard deviation 1 μHz are added to the data, the inversions are substantially degraded.

Finally, in Figure IV*b*, we illustrate a set of optimally localized averages of the angular velocity obtained from rotational splittings.

Inside the Stars, IAU Colloquium 137
ASP Conference Series, Vol. 40, 1993
Werner W. Weiss and Annie Baglin (eds.)

ABOUT SEISMOLOGICAL PROPERTIES OF INTERMEDIATE MASS STARS

N.AUDARD, J.PROVOST
Département Cassini, Observatoire de la Côte d'Azur, BP 229, 06304 Nice cedex 4, France

ABSTRACT Stars more massive than about $1.2M_\odot$ are characterized by a convective core, which induces at its frontier a rapid variation of the sound speed and of the Brunt-Väissälä frequency, close to a discontinuity. We present preliminary results about the search of the signature this core could have on the p-mode spectra. For a set of frequencies of three stars of 1, 1.5 and 2 $M_\odot$, we study in particular the small frequency separation $\Delta\nu_{0,2} \sim \nu_{n,l} - \nu_{n-1,l+2}$, for high order and low degree oscillation modes, which is particularly sensitive to the interior of stars. We underline characteristic behaviours of the 1.5 and 2 $M_\odot$ stars, through the comparison between computed frequencies and their approximation obtained by asymptotic and polynomial fittings, and also through second order quantities relatively to the frequencies.

I - STELLAR MODELS

The evolution of three stars of 1, 1.5 and $2M_\odot$ has been computed with the code CESAM (Morel, 1992), from age zero to an advanced stage on the main sequence, as well as the corresponding numerical frequencies of oscillations. We use the EFF equation of state, with the LAOL opacities, abundances $X_0 = 0.70$, $Z = 0.02$, and a mixing-length parameter $\lambda = 2.55$. The first five columns of table I give the characteristics of the three models closest to the solar central composition X_c; $\Omega_g = \sqrt{GM/R^3}$ is the characteristic frequency, ν_0 the equidistance between two modes of same degree and successive order.

TABLE I Characteristics of our 1, 1.5 and 2 $M_\odot$ models near $X_{c\odot}$.

mass	*age* ($10^9 yrs$)	X_c	$\Omega_g(10^4)$ (rad/s)	ν_0 (μHz)	$\Delta\nu_{0,2}asym$ (μHz)	$\Delta\nu_{0,2}(fit1)$ (μHz)	$\Delta\nu_{0,2}(fit2)$ (μHz)
$1M_\odot$	4.750	0.342	6.47	142.3	11.10	9.3	10.04
$1.5M_\odot$	1.184	0.376	3.71	78.1	16.97	4.56	5.3
$2M_\odot$	0.540	0.377	3.11	63.1	20.06	5.05	5.3

II - VALIDITY OF THE ASYMPTOTIC EXPRESSION

We analyse the frequencies of high-order low degree p-modes, by using two different approximations: the asymptotic expression derived by Tassoul (1980), and discussed by Gabriel (1989), hereafter noted $\nu_{n,l}(fit1)$, and a polynomial one (see Scherrer *et al.*(1983), and Christensen-Dalsgaard *et al.*(1988)), noted $\nu_{n,l}(fit2)$. From a set of numerically computed frequencies $\nu(num)$, and for degrees $l = 0, 1, 2, 3$, we calculate by a mean square root method the different coefficients entering in these formulations, from which we deduce the corresponding values of the small frequency separation $\Delta\nu_{0,2}(fit1)$

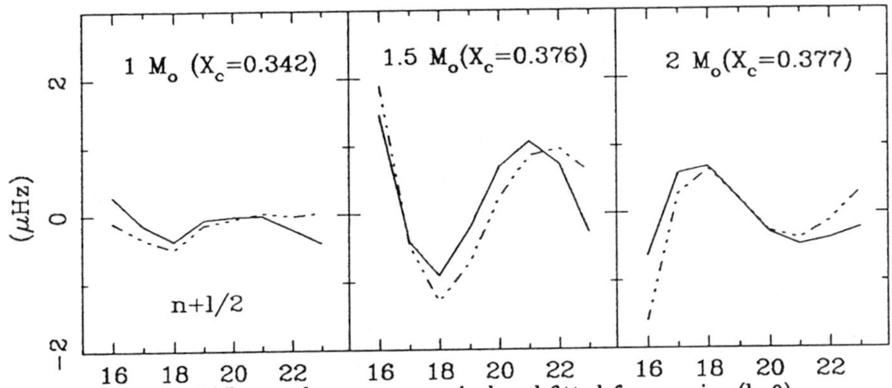

FIGURE I Difference between numerical and fitted frequencies (l=0)
−...− : $\nu(num) - \nu(fit1)$ —— : $\nu(num) - \nu(fit2)$

and $\Delta\nu_{0,2}(fit2)$ so that $\Delta\nu_{0,2}(fit1) = \nu_{n,0}(fit1) - \nu_{n-1,2}(fit1)$ and $\Delta\nu_{0,2}(fit2) = \nu_{n,0}(fit2) - \nu_{n-1,2}(fit2)$.
The general asymptotic expression provides a third formulation of $\Delta\nu_{0,2}$, that we calculate from the stellar models:

$$\Delta\nu_{0,2}(asym) = \frac{6}{4\pi^2}\left(\frac{c(R)}{R} - \int_0^R \frac{dc}{dr}\frac{dr}{r}\right)\frac{\nu_0}{\nu_{n,l}} \sim \nu_{n,l}(num) - \nu_{n-1,l+2}(num) \qquad (1)$$

Table I gives this asymptotic value of $\Delta\nu_{0,2}$ (column 6) for $n = 22$ and $l = 0$, as well as both fitted ones (columns 7 and 8). We see the three values are relatively close for the solar-aged $1M_\odot$ star, as noted by Christensen-Dalsgaard (1988), but not for other cases; this discrepancy between both fitted values and the asymptotic one questions the validity of the asymptotic expression of the frequency.
Thereafter, we make a more refined analysis by comparing directly the frequencies obtained from asymptotic and polynomial expressions to the numerical ones: $\nu_{n,l}(num) - \nu_{n,l}(fit1)$ and $\nu_{n,l}(num) - \nu_{n,l}(fit2)$. An oscillatory component appears for the 1.5 and 2 $M_\odot$ stars, not for the 1 $M_\odot$ one; Figure I represents this aspect for the three models presented in table I, with a range for the order n from 16 to 23 and the polynomial parameter $n_0 = 20$.

This peculiar behaviour of our 1.5 and $2M_\odot$ stars could be a mark of one of the effects of the convective core.

III - EFFECTS OF THE "DISCONTINUITIES"

In this final step, we underline a manifestation of the "discontinuity" of the sound speed of the 1.5 and 2 $M_\odot$ stars, which is not taken into account in the asymptotic expression used above, by examining the variation of $\nu_{n,l} - 2\nu_{n-1,l} + \nu_{n-2,l}$ for the computed frequencies. Figure II shows a great difference according to the mass: a sinusoidal behaviour appears for the three stars, but although there is two distinct oscillations for the sun-like star, only one seems to remain for the two other ones.
For the $1M_\odot$ star, as noted by Gough (1990), the smallest oscillation is related to the base of the convection zone and the largest one to the rapid variation of the adiabatic exponent γ, in the HeII ionization zone; we find that this last relation is also verified for the two other stars.

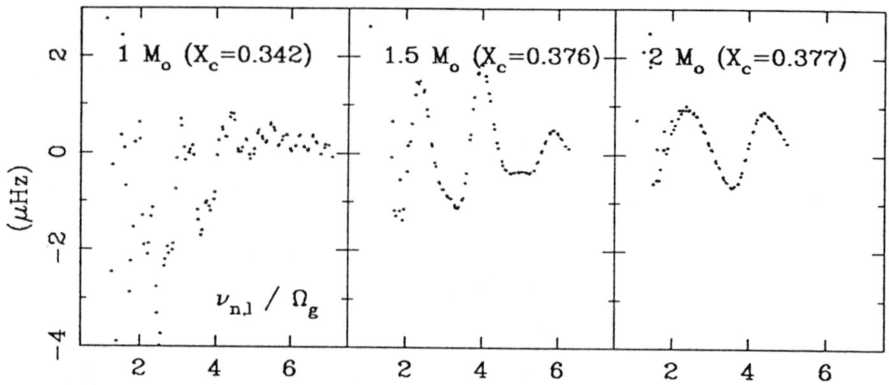

FIGURE II $\nu_{n,l} - 2\nu_{n-1,l} + \nu_{n-2,l}$ ($l = 0, 1, 2, 3$)

Further results will be published later, among others the study of the asymptotic formula of the frequency taking the rapid variation of the sound speed into account, proposed by Provost *et al.* (1992).

IV - CONCLUSION

In this study, we have shown that the classical asymptotic theory of Tassoul (1980), valid in the solar case, is inadequate to represent with accuracy the p-mode spectra of a 1.5 and a 2 $M_\odot$ stars. Nevertheless, for these two stars, the characteristic oscillation obtained from the analysis of the fittings could be a manifestation of the rapid variation of the sound speed at the frontier of their convective core; we are also analysing the oscillation exhibited by the second order quantity $\nu_{n,l} - 2\nu_{n-1,l} + \nu_{n-2,l}$.

Theoretical work is in progress to extract more accurate information of the internal structure of these intermediate mass stars from future observations.

REFERENCES

Christensen-Dalsgaard, J., 1988, in *Advances in Helio-and-asteroseismology*, Eds Christensen-Dalsgaard J. and Frandsen S., pp 295-298.

Gabriel.M., 1989, *Astron. Astrophys.* **226**, 278-283.

Gough.D.O., 1990, in *progress of seismology of the sun and stars, lecture notes in physics*, Eds Osaki.Y., Shibahashi.H., Spinger Verlag.

Morel P., 1992, these proceedings.

Provost J., Mosser B., Berthomieu G., 1992, to be submitted to *Astron. Astrophys.*

Scherrer Ph., Wilcox J.M., Christensen-Dalsgaard J., Gough D.O., 1983, *Solar phys.* **82**, 75.

Tassoul M., 1980, *Astrophys. J. Suppl.* **43**, 457-490.

Inside the Stars, IAU Colloquium 137
ASP Conference Series, Vol. 40, 1993
Werner W. Weiss and Annie Baglin (eds.)

LET US BEGIN BY MODELLING TWO OBSERVED DELTA SCUTI STARS SIMULTANEOUSLY

E. MICHEL, M.J. GOUPIL, Y. LEBRETON
Observ. de Paris-Meudon, URA 335, DASGAL, 92195 Meudon, France.

ABSTRACT We consider here two Delta Scuti stars GX Pegasi and V650 Tauri, which show very close masses and evolution stages as infered from photometrical calibration. The se two stars have been observed by the STEPHI team (Michel et al. 1990, Belmonte et al. this procedings) and they appear to be multiperiodic pulsators. For both these stars, an identification of the observed frequencies is proposed, from comparison with adiabatic eigenfrequencies computed for well suited stellar mode ls. This interpretation reveals very different pulsational behaviors despite the sta rs are expected to present very similar internal structures. We report here a first step toward constraining models by improving the fit between the observed and the calculated eigenfrequencies. The effect of overshooting and of the recent enhancement of the opacities are i nvestigated.

INTRODUCTION

With the advent of multisite campaigns of observations, an increasing number of Delta Scuti stars are found to be multiperiodic pulsators and thus good candidates for a seismological approach. In addition to more precise stellar masses and ages, seismology of these stars is expected to provide some insights into the mixing process in the vicinity of the convective core or into the mechanisms of modal selection for instance.

We start here such a seismological approach of Delta Scuti stars. Stellar models are computed which describe two observed Delta Scuti stars GX Pegasi (HD213534) and V650 Tauri (HD23643). Oscillation eigenmodes are calculated for these models and calculated eigenfrequencies are compared with observed frequencies in order to obtain a modal identification. Such identifications are presented here for both stars and the results are compared. Then, according to this identification, it is possible to use the observations to constrain the input physics of the models.

OBSERVATIONS AND MODELS

The observations which are used here are from the STEPHI network (Michel et al. 1992a for GX Pegasi; preliminary results from STEPHI team for V650 Tauri, to be published elsewhere).

Models are calculated with an updated version of the Geneva stellar evolution program (Lebreton et al. 1988). Two sets of models have been computed:

- **LA models:** In this first set of models, Los Alamos opacities were used (Weiss et al. 1990). The initial helium abundance is $Y = 0.287$ and the metallicity $Z = 0.02$. The mixing length parameter is $\alpha = 2.18$ (from solar calibration).

A sequence of models with overshooting ($d_{ov} = 0.25Hp$, with Hp the pressure scaleheight) has also been computed.

- **OPAL models:** In this second set of models, OPAL opacities were used (Rogers and Iglesias 1992). The initial helium abundance is $Y = 0.3$ and the metallicity $Z = 0.0194$. The mixing length parameter is $\alpha = 1.64$ (from solar calibration).

For both the sets of models, oscillation frequencies have been calculated using the adiabatic oscillation program kindly supplied by J. Christensen-Dalsgaard (1982).

DISCUSSION

In the case of GX Peg, the observed frequencies can be identified as two radial ($n = 2, 3$) and one nonradial ($l = 1, n = 3$) triplet ($m = 1, 0, -1$) split by rotation (see Goupil et al. 1992). In the case of V650 Tau, the comparison of computed eigenfrequencies with observed ones raises a possible simple identification: 4 radial modes ($n = 2, 4, 5, 6$) (A deeper discussion is postponed to a further paper: Michel et al 1992b, in progress).

It has already been noticed that very different pulsational behaviors can be found among Delta Scuti stars (Michel and Baglin 1991). The previous identifications given for both GX Pegasi and V650 Tauri illustrate this fact.The photometrical calibration (Kobi and North 1990) gives, for these two stars, very close global parameters (GX Peg.:$M_v = 1.48 \pm 0.1, T_{eff} = 7775 \pm 65K$, V650 Tauri.:$M_v = 1.64 \pm 0.1, T_{eff} = 8044 \pm 70K$. GX Pegasi and V650 Tau are thus expected to be very similar in mass and age (According to our models,$\Delta M \leq 0.15 M_{\odot}$, $\Delta A \leq 10^8 years$).

For both stars, differences occur between calculated frequencies and observed ones. Although they cannot be explained by numerical and observational uncertainties alone, they are not sufficient to reject the identification. They must rather be interpreted as differences between the models and the observed stars. In this point of view and in the case of V650 Tau, the comparison of the calculated and observed frequencies of 4 radial modes constitutes a very severe constraint on the modelizing of the stellar interior.

For V650 Tauri as for GX Pegasi, low order modes ($n = 2 - 4$) are found to fit the observations with a rather good accuracy while the higher order modes observed in V650 Tau show greater differences with observations.

When we change from LA to OPAL models, the fit between observed and calculated frequencies is improved in the case of GX Peg and for low order modes in the case of V650 Tauri, but it is poorer for higher order ($n = 5, 6$) modes in the case of V650 Tauri.

The effect of overshooting also has been investigated in LA models. Here again, when overshooting is taken into account, the fit is improved for low order modes but poorer for $n = 5, 6$ modes.

How can we interpret this? The $n = 5,6$ modes are more sensitive to the very outer shells than the lower order modes. Thus, if the proposed identifications are correct, the difference between the observations and the models may reflect the need to obtain a finer description of the very outer layers. Since the enhancement of the opacities is not sufficient to make the models to suit the observations, another parameter has to be taken into account. The combined effect of the chemical composition and of the opacities enhancement may help this way (work in progress).

CONCLUSION

In this paper, we compare the results obtained through comparison of observed and calculated eigenfrequencies for two Delta Scuti stars of the STEPHI's set. The modal identifications reveal very different pulsational behaviors for these stars while they are expected to have very similar internal structures.

The recent enhancement of the opacities does not change the present identification, but it seems to emphasize that , for a better fit, the description of the very outer layers must be refined.

These preliminary results have to be tested on other stars for which similar observational data are available. No doubt that the extension of such an investigation to a larger set of Delta Scuti stars will bring some new enlightment on this subject.

REFERENCES

Belmonte J.A. and STEPHI team, 1992, IAU colloquium 137 *Inside the Stars*, W. Weiss, A. Baglin Eds., this proceeding

Christensen-Dalsgaard J. 1982, MNRAS **199**, 735.

Goupil M.J., Michel E., Lebreton Y., Baglin A., 1992, submitted to A&A,

Kobi D., North P. 1990, A&A supp **85**, 999.

Lebreton Y., Berthomieu G., Provost J. and Schatzman E. 1988, A&A **200**, L5-L8

Michel, E. and STEPHI team, 1990, in *Confr. between Stellar Pulsation and Evolution*, ASP Conf. Ser. vol. 11, Eds. Cacciari C., Clementini G., 332.

Michel E., Baglin A. 1991, Adv. Space Res., vol. 11-4, 167.

Michel, E. and STEPHI team, 1992a, A&A **255**, 139.

Michel E., Goupil M.J., Lebreton Y., Baglin A., 1992b, to be submitte d to A&A,

Rogers, F.J., Iglesias, C.A., 1992, ApJ Suppl. Ser. **79**, 507

Weiss A., Keady J.J., Magee, N.H.Jr., 1990, Atomic Data and Nuclear Data Tables **45**, 209

Inside the Stars, IAU Colloquium 137
ASP Conference Series, Vol. 40, 1993
Werner W. Weiss and Annie Baglin (eds.)

THE ASTEROSEISMIC CALIBRATION OF SOLAR-TYPE STARS

D.O. GOUGH[1] AND E. NOVOTNY
Institute of Astronomy, Madingley Road, Cambridge CB3 0HA, England

INTRODUCTION

The addition of seismic parameters to stellar-model calibrations substantially increases the constraints one can place on the properties of stars. We present some preliminary calculations to assess the accuracy with which certain stellar parameters can be inferred. For simplicity we use just two of the three most basic seismic parameters characterizing the low-degree p modes that might be measured from intensity variations by instruments such as photometers planned for the ESA spacecraft PRISMA. We ascertain the accuracy of a calibration of an isolated star and of a cluster of N solar-type stars.

THE SEISMIC CONSTRAINTS

We employ two seismic parameters Δ and d_0, defined in the manner of Gough and Novotny (1990) and obtained from fits of Tassoul's (1980) asymptotic formula to the frequency set. These parameters are essentially representative values of $\Delta_{nl} = \nu_{n,l} - \nu_{n-1,l}$ and $d_{n0} = \nu_{n,0} - \nu_{n-1,2}$, where $\nu_{n,l}$ is the cyclic frequency of the mode of order n and degree l. The frequency separation Δ depends on the gross structure of the star, principally on M/R^3, which is proportional to the mean density. The separation d_0 depends strongly on the variation of sound speed in the core, and is thus sensitive to the amount of helium that has been produced by the nuclear reactions; d_0 is therefore sensitive to the age of the star. We do not use Φ, a representative value of $\nu_{n,l}/\Delta - \left(n + \frac{1}{2}l\right)$, because it measures conditions near the stellar surface and is not well modelled by theory.

As Christensen-Dalsgaard (1986) has pointed out, the mass M and age t of a late-type main-sequence star could be determined from Δ and d_0, if only the composition and mixing-length parameter α were known and if the stellar evolution theory were presumed to be correct. But if the composition and α are not known, additional astronomical data must be used. We ask how well the stellar parameters can then be determined.

Stellar models are typically defined by M, α, t and the initial hydrogen and heavy-element abundances X and Z; we denote these parameters by α_i ($i = 1, 2, ..., 5$) = ($\ln M$, $\ln \alpha$, $\ln t$, $\ln X$, $\ln Z$). They predict a_k = ($\ln R$, $\ln L$, Δ, d_0), where R and L are the radius and luminosity of the star. The problem in hand is to determine from these models how sensitively the inferred properties of stars depend on errors in the observations.

---

[1]and Department of Applied Mathematics and Theoretical Physics, University of Cambridge

ISOLATED STARS

As an example, consider a star for which $\beta_i = (\ln T_e,\ \ln L,\ Z/X,\ \Delta,\ d_0)$, where T_e is effective temperature, have been measured. In this and in the subsequent section, we consider the seismic parameters Δ and d_0 to have been determined from modes with $0 \le l \le 2$ and $14 \le n + \frac{1}{2}l \le 28$. The sensitivities of the inferred parameters $b_k = (\ln t,\ \ln M,\ \ln Y,\ \ln\alpha)$, where $Y = 1 - X - Z$, are obtained simply by transforming the partial derivatives $\partial a_k/\partial\alpha_i$ of the theory to the corresponding derivatives $\partial b_k/\partial\beta_i$ in the space of observations.

Derivatives $\partial a_k/\partial\alpha_i$ for a $1M_\odot$ stellar model of age $t = 4.6\times10^9$y in which the mixing length was taken to be α pressure scale heights are listed in Table I. They were computed by linear regression against sequences of five models with varying α_i and fixed $\alpha_j (j \ne i)$. Their transforms $\partial b_k/\partial\beta_i$, multiplied by perhaps somewhat optimistic observational errors $\delta\beta_i$, are listed in Table II. Thus the entries in Table II represent the corresponding contributions to the errors in the inferred values of b_k. Of course, each row scales linearly with the presumed error $\delta\beta_i$ in β_i. The total standard errors, for the values of $\delta\beta_i$ (assumed uncorrelated) listed in the second column, are recorded in the bottom row. It is interesting to observe in Table II that age is most susceptible to errors in d_0, as might have been expected. The accuracy of the other stellar properties depends mainly on the errors in L.

TABLE I Sensitivity of the properties a_k of a stellar model to the control parameters α_i. The unit of Δ and d_0 is μHz.

$\partial a_k/\partial\alpha_i$

a_k / α_i	$\ln R$	$\ln L$	Δ	d_0
$\ln M$	2.00	6.09	−320	−27.9
$\ln\alpha$	−0.21	0.05	39.8	0.97
$\ln t$	0.16	0.40	−31.0	−6.75
$\ln X$	−1.62	−6.25	308	31.3
$\ln Z$	−0.16	−0.79	30.8	2.79

It is a straightforward matter to transform to a five-dimensional space spanned by a different set of β_i; for example, one might replace T_e by surface gravity g. Alternatively, one might suppose that the dimension of the space of observations is greater than five, say, by adding g to the data, so that the calibration problem is formally overdetermined. One can then use this property to reduce the formal uncertainty in the inferences. Of course, one could obtain more information about the star by analysing the individual frequencies $\nu_{n,l}$ rather than only the gross parameters Δ and d_0. Such an analysis is presented by Gough and Kosovichev (these proceedings).

CALIBRATION OF A SET OF SIMILAR STARS IN A CLUSTER

The members of a star cluster are normally assumed to have the same chemical composition and the same age. These constraints can reduce the uncertainty in

TABLE II Errors δb_k for an isolated star induced by the errors $\delta\beta_i$ listed in column 2. The last row is the total error in b_k, assuming that $\delta\beta_i$ are independent.

		$(\partial b_k/\partial\beta_i)\delta\beta_i$			
β_i	$\delta\beta_i$	$\delta t/t$	$\delta M/M$	$\delta Y/Y$	$\delta\alpha/\alpha$
$\ln T_e$	0.005	−0.017	−0.031	0.082	−0.024
$\ln L$	0.1	0.005	0.153	−0.353	0.209
Z/X	0.003	−0.020	−0.001	0.040	0.012
Δ	0.2 μHz	0.003	0.003	−0.009	0.008
d_0	0.5 μHz	−0.112	−0.001	0.023	−0.033
All		0.12	0.16	0.37	0.21

the determination of Y and t dramatically. We denote the control parameters of stellar evolution theory for each star s by $\alpha_i^s = (\ln M^s, \ln\alpha^s, \ln t, \ln X, \ln Z)$ and the predicted properties by $a_k^s = (\ln R^s, \ln L^s, \Delta^s, d_0^s)$. Note that since the value of the mixing length is an assumption of the theory, the parameter α^s must be permitted to be different for each star. We consider an example in which the values of $\beta_i^s = (\ln T_e^s, \ln L^s, \Delta^s, d_0^s, Z/X)$ have been measured with standard error σ_i in a set of N solar-type stars of a cluster. We assume that the stars are sufficiently similar that their properties can be obtained by linearization about our stellar model of the previous section, using the partial derivatives in Table I. Moreover, for simplicity, we assume σ_i to be the same for all the stars. The calibration is an overdetermined problem (if $N \geq 2$), and we carry it out by minimizing amongst $b_k^s = (\ln M^s, \ln\alpha^s, \ln t, \ln Y, \ln Z)$ the deviation

$$\mathcal{X}^2 = \sum_s \sum_i \sigma_i^{-2} \left(\beta_i^s - \beta_{\mathrm{obs},i}^s\right)^2 ,$$

where $\beta_i^s\,(b_k^s)$ are the result of theory, obtained by transforming $a_k^s\,(\alpha_i^s)$; and $\beta_{\mathrm{obs},i}^s$ are measured values. We do not assume stellar evolution theory to be perfect, so the minimum value of $\mathcal{X}^2$ would not necessarily be zero in the absence of observational errors. But in the presence of random errors $\delta\beta_i^s$ in $\beta_{\mathrm{obs},i}^s$, errors δb_k^s in b_k^s arise. Let the contribution from $\delta\beta_i^s$ to the standard deviation of δb_k^s be e_{ki}^s. Then we can define the error sensitivity to be $\partial e_{ki}^s/\partial\sigma_i$. Error sensitivities for a group of 10 stars are listed in Table III. In this example, we have assumed the errors $\delta\beta_i^s$ in the intrinsic quantities β_i $(i = 1, 2, 3, 4)$ to be uncorrelated. The quantity Z/X, on the other hand, is assumed to be a cluster average, and its error is common to every star. In reality, there is also a common contribution to the error in $\ln L^s$, for example, which arises from an error in the measurement of the distance to the cluster, so that the errors $\delta\beta_2^s$ are not wholly uncorrelated. However, we have ignored that complication here, which results in the error sensitivities being independent of σ_i. The bottom row in Table III provides the total uncertainty in the calibration. Those entries were computed assuming that the errors for different i are uncorrelated.

TABLE III Error sensitivities $\partial e^s_{ki}/\partial\sigma_i$ of the calibration of a cluster of 10 stars to observational errors in $\beta^s_{\mathrm{obs},i}$ with standard deviation σ_i, the unit of σ_3 and σ_4 being μHz. The measurement errors for each star were assumed to be uncorrelated for $i = 1, 2, 3, 4$, and the error in Z/X $(i = 5)$ was taken to be common to all the stars. The bottom row is the total uncertainty in b^s_k obtained by adopting the values of σ_i listed in column 2. For clarity, we have adopted the notation: $\frac{\partial e(\ln M^s)}{\partial\sigma_i}$ for $\partial e^s_{1i}/\partial\sigma_i$, etc.

β^s_i	σ_i	$\frac{\partial e(\ln M^s)}{\partial\sigma_i}$	$\frac{\partial e(\ln \alpha^s)}{\partial\sigma_i}$	$\frac{\partial e(\ln t)}{\partial\sigma_i}$	$\frac{\partial e(\ln Y)}{\partial\sigma_i}$	$\frac{\partial e(\ln Z)}{\partial\sigma_i}$
$\ln T^s_e$	0.005	2.06	5.12	1.05	5.20	1.78
$\ln L^s$	0.1	0.49	0.66	0.02	1.12	0.38
Δ^s	0.2μHz	0.005	0.015	0.005	0.014	0.005
d^s_0	0.5μHz	0.001	0.022	0.071	0.014	0.005
Z/X	0.005	0.02	0.41	6.73	13.5	31.9
All		0.05	0.07	0.05	0.13	0.16

The redundancy amongst the measurements that has been introduced by assuming that age and composition are the same for all the stars has caused a degree of cancellation amongst the errors. Comparison of Tables II and III reveals that the uncertainties in all the calibrated quantities have been reduced, notwithstanding the increased value of the uncertainty in Z/X that has been adopted. Larger values of N lead to a further reduction, but the errors in the intrinsic variables $\ln M^s$ and $\ln\alpha^s$ soon stabilize. The errors in the global quantities $\ln t$, $\ln Y$ and $\ln Z$ continue to decline, approximately in proportion to $N^{-1/2}$. Indeed, if it is assumed that σ_i $(i = 1, 2, 3, 4)$ are independent of N, but that σ_5 is proportional to $N^{-1/2}$, as one might expect from a cluster average, then the standard erors e^s_{ki} in the global quantities b_k $(k = 3, 4, 5)$ eventually decline to zero strictly in proportion to $N^{-1/2}$.

It is straightforward to include other information in the analysis, such as g^s, as it is, also, to take correlated errors into account. In particular, in view of the high sensitivity of the inferred values of b^s_k to L^s, such a calibration might lead to a determination of the distance to the cluster that is more accurate than the methods currently in use. We shall report on our analysis in more detail elsewhere.

REFERENCES

Christensen-Dalsgaard, J. 1986. *Seismology of the sun and the distant stars* (ed. D.O. Gough, NATO ASI **C169**, Reidel, Dordrecht) p.23

Gough, D.O. & Novotny, E. 1990. *Solar Phys.*, **128**, 143

Tassoul, M. 1980. *Astrophys. J. Suppl.* **43**, 469

Inside the Stars, IAU Colloquium 137
ASP Conference Series, Vol. 40, 1993
Werner W. Weiss and Annie Baglin (eds.)

HOW MAY SEISMOLOGICAL MEASUREMENTS CONSTRAIN PARAMETERS OF STELLAR STRUCTURE?

TIMOTHY M. BROWN
High Altitude Observatory, National Center for Atmospheric Research, P.O. Box 3000, Boulder, CO 80307, USA

JØRGEN CHRISTENSEN-DALSGAARD
Institute of Physics and Astronomy, Aarhus University, Aarhus, DK

BARBARA W. MIHALAS
National Center for Supercomputing Applications, 605 E. Springfield Ave., Urbana, IL 61820, USA

ABSTRACT Pulsation frequency separations, combined with other observations, can constrain the structural parameters of Sun-like stars. As an example, we treat a hypothetical visual binary system resembling α Cen.

INTRODUCTION

Information about a star's structure is contained in its two characteristic pulsation *frequency separations* $\Delta\nu_0 \equiv \nu_{n+1,0} - \nu_{n,0}$, and $\delta_0 \equiv \nu_{n+1,0} - \nu_{n,2}$, where $\nu_{n,l}$ is the frequency of a p-mode with radial order n and angular degree l (Ulrich 1986, Gough 1987, Christensen-Dalsgaard 1988). Unfortunately, the information is somewhat confused by the lack of accurate knowledge about other parameters, especially the abundances of helium and of heavy elements. Thus we need to understand how oscillation information may be used in concert with other observable stellar properties. Here we describe single and binary systems with models which relate variable *parameters*, such as the stellar masses, initial composition, and age of the system, to larger sets of *observables*, including photometric, spectroscopic, and astrometric indices, as well as oscillation frequencies. After linearization of these relations, we employ a singular value decomposition (SVD) to calculate the errors in estimated parameters resulting from plausible errors in the observed quantities. To simplify the analysis, errors arising from inaccurate or incomplete physics in the model are ignored.

PARAMETERS, OBSERVABLES, ERRORS, AND MODELS

Here we treat two systems: a field star similar to α Cen A, and a visual binary similar to the α Cen system. The systems are considered to be observed at distances of both 1.3 and 100 pc. The model parameters and observables for these systems are shown in Table I. The standard error for the frequency separations was estimated by assuming that one-half of the modes in the range considered have sufficiently large amplitudes during a given observing run to allow their

then follow from the least-squares fit used to determine $\Delta\nu_0$ and δ_0.

TABLE I Parameters and observables for field star (top section), and for visual binary (both sections combined). M_1 and M_2 are the masses of the stars, Y and Z the abundances of helium and heavy elements, α the mixing-length parameter, τ the age of the system, D its distance, and a the separation of the binary. For the observables, m_{Vi} and $(B\text{-}V)_i$ are apparent magnitudes and color indices, π' is the parallax, a' the apparent separation of the binary, M_1/M_2 the mass ratio inferred from the orbit, and P_{orbit} the orbital period; σ is the estimated standard error in each observable.

Model Parameter	Value(s)	Observable	σ
$\log M_1$ ($M_\odot$)	0.0374	m_{V1}	0.01 mag
$\log Y$	−0.5510	$(B\text{-}V)_1$	0.01 mag
$\log Z$	−1.7000	$\log Z$	0.17
$\log \alpha$	0.40	π'	0.004 ″
$\log \tau$ (yr)	9.544	$(\delta_0)_1$	0.82 μHz
$\log D$ (pc)	0.114, 2.	$(\Delta\nu_0)_1$	0.06 μHz
$\log M_2$ ($M_\odot$)	−.0458	m_{V2}	0.01 mag
$\log a$ (AU)	1.301	a'	0.05 ″
		$(B\text{-}V)_2$	0.01 mag
		$\log(M_1/M_2)$	0.0004
		$\log P_{orbit}$	0.0001
		$(\delta_0)_2$	0.82 μHz
		$(\Delta\nu_0)_2$	0.06 μHz

The reference stellar models were computed by means of a "standard" evolution calculation with the parameters listed in Table I, chosen to obtain a pair of models roughly similar to α Cen. Measures of the average separations $\Delta\nu_0$ and $\delta\nu_0$ were found from a least-squares fit (Christensen-Dalsgaard 1988) to frequencies computed by solving the equations of adiabatic oscillation. The fit included modes of degree $l = 0 - 3$ and with $17 \leq n + l/2 \leq 29$. To compute the necessary derivative matrix, we varied each of the parameters M_1, M_2, Y, Z, α, and τ separately, by suitable small amounts. Each derivative was computed from differences centered on the reference values given in Table I.

RESULTS AND CONCLUSIONS

The results of the calculations for a few interesting cases are shown in Fig. 1. Evidently the oscillation frequency separation data can be important in constraining the model parameters, but which parameters are constrained depends strongly on the circumstances of the stellar system under study and on the available observables. Without oscillation data, the properties of isolated field stars (especially their ages) are essentially unconstrained. With such data, it becomes possible to make useful estimates of the stellar ages, and to improve significantly the estimates of the mixing length and (for nearby stars) the mass, or (for stars at large distances) the distance itself. For visual binaries similar to α Cen, the effect of adding oscillation information depends strongly on the distance to the system. For nearby systems (where astrometric data is very good), one learns mostly about Y, Z, and α, with lesser improvement in the estimate of the system's age. For distant systems, all parameter estimates improve significantly, with the most notable improvements in Y, the masses, the distance, and the semimajor axis of the orbit. Particularly for the nearby case, all parameters are well enough determined that detailed comparisons with stellar evolution theory seem feasible.

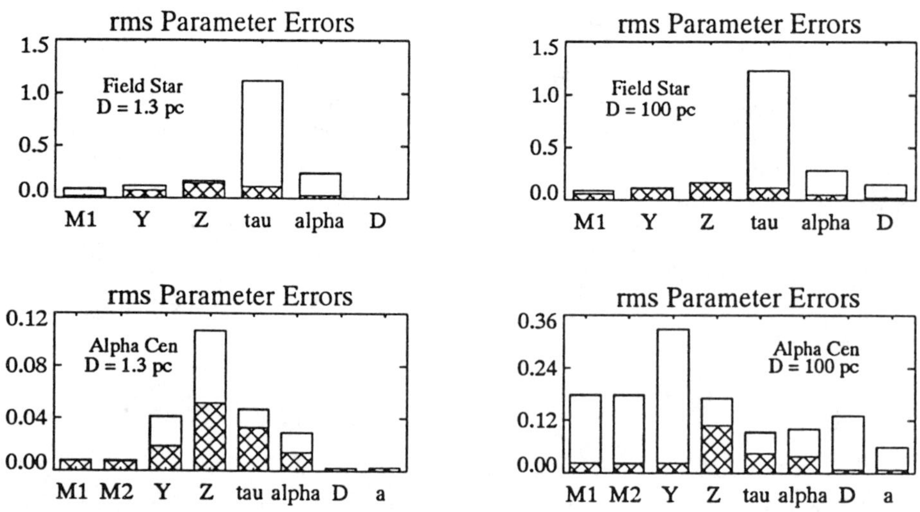

Fig. 1. RMS errors in $\log_{10}$ of the inferred structure parameters for field star (top) and visual binary (bottom), at distances of 1.3 and 100 pc. Results without oscillation data are shown as empty bars, with such data as cross-hatched bars.

REFERENCES

Christensen-Dalsgaard, J. 1988, in *Advances in Helio- and Asteroseismology,* eds J. Christensen-Dalsgaard and S. Frandsen, D. Reidel, Dordrecht, p. 295.

Gough, D.O. 1987, *Nature* **326**, 257.

Ulrich, R.K. 1986, *Ap. J.* **306**, L37.

Inside the Stars, IAU Colloquium 137
ASP Conference Series, Vol. 40, 1993
Werner W. Weiss and Annie Baglin (eds.)

DETECTING CONVECTIVE OVERSHOOT IN SOLAR-TYPE STARS

M. J. P. F. G. MONTEIRO
Astronomy Unit, Queen Mary & Westfield College, London, UK; Grupo de Matemática Aplicada da Faculdade de Ciências , and Centro de Astrofísica, Universidade do Porto, Portugal

J. CHRISTENSEN-DALSGAARD
Institut for Fysik og Astronomi, Aarhus Universitet, Denmark

M. J. THOMPSON
Astronomy Unit, Queen Mary & Westfield College, London, UK

INTRODUCTION

It is important for understanding stellar evolution to constrain observationally how overshoot occurs for stellar conditions. Simplified models of the dynamics (eg. Zahn 1991) indicate that overshoot results in a slightly subadiabatic region beyond the convectively unstable layers, followed by an almost discontinuous transition to radiative stratification. Abrupt changes such as this contribute with a characteristic periodic signal to the frequencies $\omega_{n,l}$ of modes of low degree l (Gough 1990). This signature may therefore be detectable for distant stars. Here we show that the signal is sensitive to the "severity" of the overshoot and, of practical importance for the solar case, how it may be extracted from modes of higher degree. Finally we apply our method to solar data.

To analyze the applicability of the method, we consider four stellar models, $Z_1 - Z_4$, with solar mass, radius (R) and luminosity; of these, Z_2 and Z_4 have overshoot. The bases of the nearly adiabatically stratified region in the models are at radii $r_d/R = .729$, .713, .713 and .700 respectively.

METHOD AND RESULTS

The sound speed and its gradient change rather abruptly at the base of the adiabatically stratified envelope of a solar model. Consider then the difference δc^2 between the actual sound speed squared (c^2) and a smoothed version, c_0^2. Let ω be the true frequency and ω_0 the frequency of the same mode of a star in which the sound speed is c_0. Consider the second derivative of ω with respect to n at constant l:

$$S_d = \left(\frac{d^2\omega}{dn^2}\right)_l = \left(\frac{d^2\omega_0}{dn^2}\right)_l + \left(\frac{d^2(\omega - \omega_o)}{dn^2}\right)_l \equiv S_d^0 + \delta S_d \, . \tag{1}$$

It may then be shown that

$$\delta S_d \sim -4A\tau_d^2 \left(\frac{d\omega_0}{dn}\right)_l^2 \omega \cos\left(2\tau_d\omega - \gamma\frac{L^2}{\omega} + 2\phi\right) . \tag{2}$$

Here the acoustic depth $\tau_d = \int_{r_d}^{R} c^{-1} dr$; $\gamma = \int_{r_d}^{R} r^{-2} c\, dr$; $L^2 = l(l+1)$; ϕ is the phase (treated here as a constant, though more generally it is a function of ω); and A is proportional to $\delta c^2(r_d)$ and is thus increased by overshoot (Fig. 1a). For low l this signal is periodic in frequency (the period determined by τ_d if ϕ is constant).

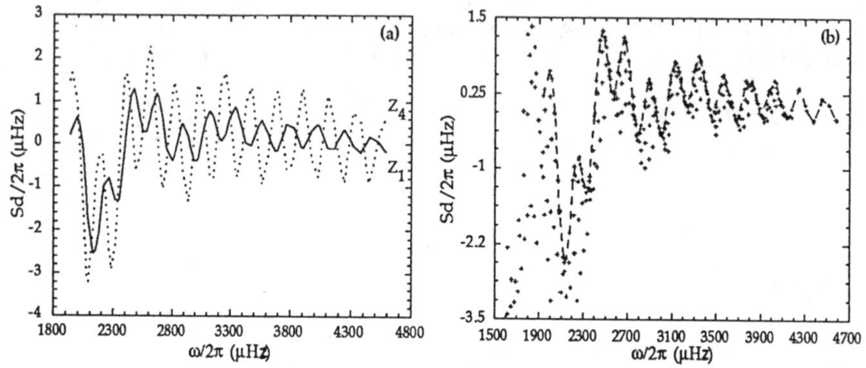

Fig. 1. Second derivatives S_d (a) for Models Z_1 (solid line) and Z_4 (dotted line), for $l = 0 - 4$; and (b) for Model Z_1 including higher degree data ($l = 5 - 20$) (the dashed line joins $l = 0 - 4$).

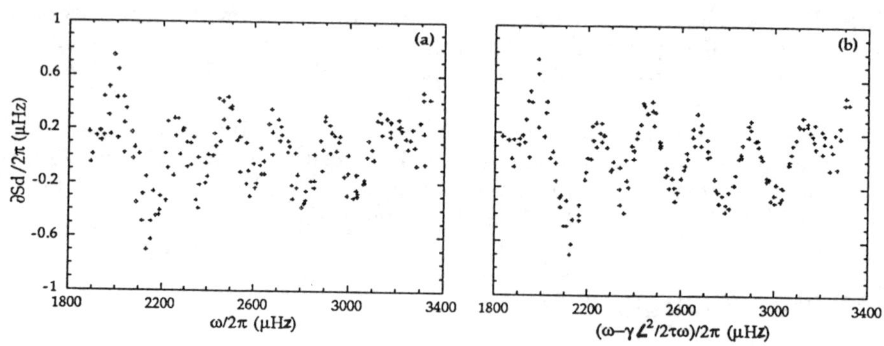

Fig. 2. (a) Second derivatives of frequency, for Model Z_1, after S_d^o has been removed. (b) The same as (a), but now the ordinate for each point has been modified (after we have found τ_d and γ by our procedure) to remove the first-order correction to the argument of 'cos' in eq. (2). The higher-degree data also now exhibit a regular oscillatory signal.

The coherent behaviour is destroyed if modes with larger l are included (Fig. 1b). Two major contributions can be identified. One affects S_d^o (eq. 1) by moving downwards the points for larger l. This component is explicitly removed by our method of isolating δS_d (see below). The second contribution is a phase shift which affects δS_d and which is accounted for by the term in γ (Fig. 2).

We determine S_d by constructing the cubic spline interpolating the points $\omega(n)$ (with the same l). S_d^0 is obtained similarly, but with a cubic spline that is forced to be smoother and hence cannot fit the frequencies exactly. Finally we make a fit, in which expression (2) is the leading contribution, to $\delta S_d = S_d - S_d^0$ to obtain A, τ_d, γ and ϕ. The results for model data and 1986 solar data (Libbrecht *et al.* 1990) are summarized in Fig. 3. Only $l = 5 - 20$ were used: the available solar data for $l = 0 - 4$ are too noisy, destroying the signal we are isolating.

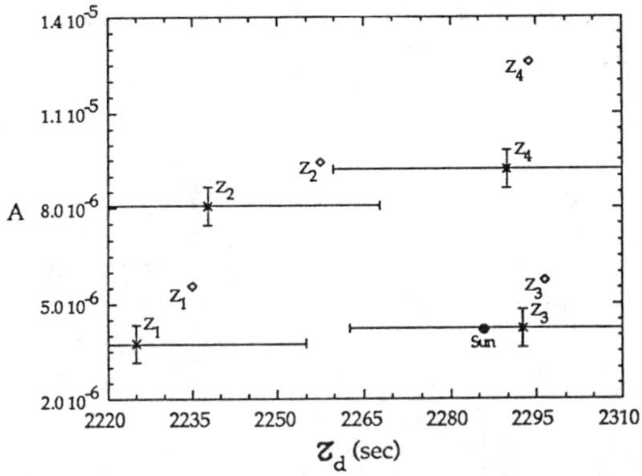

Fig. 3. The inferred amplitude A and acoustic depth τ_d for the Sun and for the models. Diamonds show results for artificial data with no errors added; crosses indicate results for data with errors, assumed gaussian with standard deviations given by Libbrecht *et al.* Error bars were derived from 15 realizations of the random errors for Model Z_3.

The solar data appear to favour models Z_1 and Z_3 which have no overshooting, rather than models Z_2 and Z_4 which have overshoot layers of depth $0.015R$ and $0.027R$ respectively. This is our principal conclusion. Amongst models with the same degree of overshoot, the relative depth determination is also fairly reliable (though note that it has rather large uncertainties in this method due to data errors). We note with interest that the Sun appears to resemble most closely Model Z_3, which has the same depth of convection zone as was inferred for the Sun by Christensen-Dalsgaard *et al.* (1991).

REFERENCES

Christensen-Dalsgaard, J., Gough, D. O. & Thompson, M. J. 1991, *Ap. J.*, **378**, 413 – 437.

Gough, D. O. 1990, in *Lecture Notes in Physics* vol **367**, 283 – 318, eds. Y. Osaki & H. Shibahashi, Springer, Berlin.

Libbrecht, K. G., Woodard, M. F. & Kaufman, J. M. 1990, *Ap. J. Suppl.*, **74**, 1129 – 1149.

Zahn, J.-P. 1991, *Astr. Ap.*, **252**, 179 – 188.

Inside the Stars, IAU Colloquium 137
ASP Conference Series, Vol. 40, 1993
Werner W. Weiss and Annie Baglin (eds.)

ASTEROSEISMOLOGICAL MODELS WITH ROTATION

Doru Marian Suran *and* Gabriela Oprescu
Astronomical Institute of the Romanian Academy
Str.Cutitul de Argint,No 5,75212 Bucharest 28, ROMANIA

Abstract The low frequency **NRP** oscillations are considered in the case of differential rotation. Partial differential equations for adiabatic oscillations are reduced to a system of ordinary equations by means of a truncated spherical-harmonic expansion. The numerical method for solving the system is described. The final model also includes line profile variations (**lpv**) and posttheoretical mass (**pth**) calculations for the complete determinations of the physical, chemical, rotational and seismological properties of the stars.

1. INTRODUCTION

We present a detailed stellar model in which low-frequency non-radial pulsations (**NRP**) are excited by differential rotation (**DR**) in turbulent convection zones. The competitivity between **NRP** and **DR** are responsible for some observational effects such as: line profile variations (**lpv**) and rotational modulation (**RM**).

To this purpose we compile a new stellar model with the following assumptions:

1. Spherical star;
2. Chemical homogeneous star (global or in shells);
3. Inert convection ($\delta L_{conv}=0$), and convective turbulence in rotation (latitude dependent heat transfer **HT**);
4. Turbulent convection in the approximation of anisotropic viscosity **AV**);
5. Stationary ($\partial/\partial=0$), differential, **slow** rotation :

$$f(r,\theta)=f_0(r)+\varepsilon[f_{10}(r)+f_{12}(r).P_2(\theta)] \tag{1}$$

6. **NRP** with differential rotation (**NRPDR**).

In this case many types of modes could be excited: *gravity*(g), *pressure ionisation* (f,p), *convective* (c), *quasi-toroidal* (r), *inertial*(i).

The characteristic interactions which could appear in the star are:

$$\text{convection+turbulence} \simeq \text{rotation} \simeq \text{pulsation} \tag{2}$$

2. STELLAR MODEL

Basic equations for the model :

$$\partial\rho/\partial t+\nabla.(\rho\mathbf{v})=0$$

$$\rho(\partial/\partial t+\mathbf{v}.\nabla)\mathbf{v}=-\nabla p-\rho\nabla\Phi+\mathbf{R}_v$$

$$\rho T(\partial/\partial t+\mathbf{v}.\nabla)S=\rho\varepsilon_N-\nabla.\mathbf{F} \tag{3}$$

$$\nabla^2\Phi=4\pi G\rho\ ,\quad \mathbf{R}_v=\rho\nu[\nabla^2\mathbf{v}+(1/3)\partial.(\partial\mathbf{v})]$$

$$\mathbf{F}=\mathbf{F}_{rad}+\mathbf{F}_{conv}=-(4\pi/3\kappa\rho).\nabla J$$

$$4\pi J=acT^4+TdS/dt$$

are solved by a two-step perturbation model:

$$\mathbf{q} = \mathbf{q}_{eq}{}^{o}+(\mathbf{q}_{conv}{}^{0}+\mathbf{q}_{rot}{}^{0}+\mathbf{q}_{nuc}{}^{0})+\mathbf{q}_{puls}(t) = \mathbf{q}_{eq}{}^{0}(r)+\mathbf{q}_{puls}(t) \tag{4}$$

where $\mathbf{q}_{eq}$- static equilibrium model, $\mathbf{q}_{conv}$-inactive convection; $(\mathbf{q}_{conv}+\mathbf{q}_{rot})$ represents static models with differential rotational in external shells; $(\mathbf{q}_{conv}+\mathbf{q}_{rot}+\mathbf{q}_{nuc})$ represents static models with differential rotation in convective cores; $\mathbf{q}_{puls}(t)$ represents the **NRP** linear adiabatic time perturbation from pulsations.

a. Equilibrium Model ($\partial/\partial t=0$, $\Omega=0$)

The equilibrium model could be defined directly from evolutionary tracks (**SES**) under the assumptions (M=*ass*, t=*ass*), or by a three-zone static model with different equations for atmosphere, envelope and interior. The mode using as input parameters :

$$(M,L,T_e,X_i) \qquad (\textit{atmosph. param.}) \tag{5}$$

calculate all needed physical and thermodinamical parameters in the *"equilibrium"* case:

$$[P_r{}^0,T_r{}^0,M_r{}^0,L_r{}^0,X_{i,r}{}^0]\ ,\ [\rho_r,\nabla_r,\kappa_r,\varepsilon_r,\ldots] \tag{6}$$

b. Static Rotational Perturbation

The basic eq.(1) is now solved under the hypothesis of static differential rotation ($\mathbf{v}=\mathbf{r}\times\Omega$). As perturbation we include **AV** and

HT hypothesis. We perturbs all the physical quantities by a second order Legendre expansion (1).

We have two different cases for the solution of the problem corresponding to envelope (**SURFF**) and/or core (**NUCLEI**) convective zones, where differential rotation acts. Using (6) as input we can derive the new set of rotational parameters (Pidatella *et al* 1986)

$$y=(\psi_{12},\psi'_{12},\psi''_{12}),(P_{10},P_{12},S_{10},S_{12},F_{r,12}),(\omega_{12},\omega_{12},\omega'_{10},\omega'_{12}) \quad (7)$$

c. **NRP** Perturbation

Our second order **NRP** works under the assumptions of linear(', $\delta \propto \exp[i(\omega t+m\varphi)]$), nonadiabatic (for $\Omega=0$) , adiabatic(for $\Omega\neq0$), and low frequency oscillations. The whole set of discrete parameters of the model is (l, m, n, j, i) where the first three are the usuals for **NRP**, the fourth is used in the rotation perturbation: $(\Omega(r,\mu)=\sum_j \Omega_j \mu_j)$; the fifth designed the number of the sheels in the star ($0\leq r_i \leq R_{star}$). In our numerical calculations we truncate the **NRP** matrices, and in order to avoid to treat too large matrices, we split the time perturbation equations of (3) into two separate systems corresponding to even and odd modes. The output parameters of the **NRP** calculations are:

$$[(y_{k,i}^{lmn})_{k=1,4},H_i^{lmn},T_i^{lmn}],[\omega^{lmn}] \quad (8)$$

where y_i are the usual **NRP** unknowns (Suran 1990).

3. RESULTS

We tested our final model in different cases , to assure the correctenss of our calculations:

a. **LNAWER+ptm** - models for radial pulsators (Suran 1986);
b. **LNAWER+ptm+lpv** -seismological models for δ Scuties (Suran 1990)
c. **LAWER+lpv**, $\Omega=\Omega(r)$-comparison with previous **NRP** rotational cases studied in the literature (Lee, Saio 1990);
d. **NRP**=0, $\Omega=\Omega(r,\theta)$- **DR** models (Pidatella *et al.* 1986).

REFERENCES

Pidatella, R.M., Stix, M., Belvedere, G, Paterno, L 1986, *Astr. Ap.*, 155, *22*
Saio, H, Lee, L. 1990 in *ESO Worshop 'Rapid Variability of OB stars'*, ed. D. Baade, *293*;
Suran M.D. 1986, *Topics in Astr. Ap. Space Sci.*, 2, *28*;
Suran M.D. 1990, in *ESO Workshop 'Rapid Variability of OB Stars'*, ed. D. Baade, *298*.

Inside the Stars, IAU Colloquium 137
ASP Conference Series, Vol. 40, 1993
Werner W. Weiss and Annie Baglin (eds.)

THEORY OF THE OBLIQUE PULSATOR MODEL FOR THE RAPIDLY OSCILLATING AP STARS

HIROMOTO SHIBAHASHI AND MASAO TAKATA
Department of Astronomy, University of Tokyo, Tokyo 113, Japan

ABSTRACT Recently, one of the rapidly oscillating Ap stars, HR 3831, has been found to have an equally split frequency septuplet, though its oscillation seems to be essentially an axisymmetric dipole mode with respect to the magnetic axis which is oblique to the rotation axis (Kurtz et al. 1992; Kurtz 1992). In order to explain this fine structure, we investigate oscillations of rotating magnetic stars by taking account of the perturbations due to the magnetic fields and the rotation. We show that the magnetic field on axisymmetric dipole modes distorts the dipole oscillation pattern to have axisymmetric octapole components so that the frequency pattern becomes a septuplet rather than a triplet, and that the additional effect of the rotation leads the frequency pattern to be asymmetric. The formula allows us to get information about the magnetic fields, the rotation, and the geometrical configuration of the star from the oscillation data.

The rapid oscillations of Ap (roAp) stars are multi-periodic and then they can provide us with much information about the inside of these stars. The Ap stars themselves are unique objects in the sense that the oscillations in these stars are influenced by strong magnetic fields. These fields are thought to play an essential role of chemical peculiarity of these stars. Seismic investigation of these stars is expected to provide us with some information about their magnetic fields, and hence asteroseismology of the roAp stars is worth of doing to understand physics of the Ap phenomena. For more details, see reviews (Kurtz 1990; Matthews 1991; Shibahashi 1991).

The amplitudes of the rapid oscillations of some roAp stars vary definitely synchronously with the magnetic field strength of the star. Hence the oscillations are interpreted as axisymmetric dipole modes with respect to the magnetic axis which is oblique to the rotation axis of the star [the oblique pulsator model of Kurtz (1982)]. According to this model, the change of the aspect angle of the oscillation induces a triplet fine structure in the frequency spectrum; each of the side components is separated exactly by the rotation frequency of the star from the central frequency component. The oblique pulsator model has been theoretically refined. It now explains the asymmetry of the amplitudes of the triplet fine structure, from which information about the magnetic field can be deduced (Dziembowski and Goode 1985, 1986; Kurtz and Shibahashi 1986).

However, Kurtz et al's (1992) recent observations of HR 3831 show that the fine structure so far thought to be a triplet is in reality a part of a septuplet. The discovery of a septuplet fine structure destabilized the fundamentals of the oblique pulsator model, and we have been faced many difficulties. In this situation, Kurtz (1992) tried to decompose the oscillation pattern in terms of spherical harmonics of degrees $l = 0, 1, 2$, and 3, and succeeded to get some reasonable fits. However, his approach is phenomenological, and needs physical justification. In this paper, we outline our formulation of the nonradial oscillations of rotating magnetic stars to derive a septuplet fine structure in the frequency spectrum (Shibahashi and Takata 1992; in preparation). Our approach is refinement of the oblique pulsator model, and we show the model explains well the observations.

We suppose that the star is rigidly rotating and that the magnetic field is a dipole field. We treat the effects of the rotation and of the magnetic field as perturbations. In doing so, we suppose that the rotation of the star is slow enough so that the effect of the rotation on oscillations is smaller than that of the magnetic field. Hereafter, we consider the oscillation mode which is principally represented by axisymmetric dipole with respect to the magnetic axis. That is, the zero-order eigenfunction is represented by the main component of $(l, m) = (1, 0)$.

After some manipulation, the eigenfunction $\boldsymbol{\xi}$ is expressed by

$$\begin{aligned} \boldsymbol{\xi} \simeq \; & (\boldsymbol{\xi}_{n,1,0} + \alpha_1 \boldsymbol{\xi}_{n,1,1} + \alpha_{-1} \boldsymbol{\xi}_{n,1,-1} \\ & + \beta_{n-1,3,0} \boldsymbol{\xi}_{n-1,3,0} + \beta_{n-1,3,1} \boldsymbol{\xi}_{n-1,3,1} + \beta_{n-1,3,-1} \boldsymbol{\xi}_{n-1,3,-1}) \\ & \times \exp\{i(\omega^{(0)} + \omega^{(1)})t\}, \end{aligned} \tag{1}$$

where the suffixes of $\boldsymbol{\xi}$, α, and β represent the radial order n, the degree l, and the azimuthal order m with respect to the magnetic axis, respectively. Here, we set the temporal dependence of eigenfunctions as $\exp(i\omega t)$ in the rotating frame, and the coefficients are

$$\alpha_{\pm 1} \propto \frac{\text{Coriolis Force } (\propto \Omega)}{\text{Lorentz Force } (\propto B^2)}, \tag{2}$$

$$\beta_{n-1,3,0} \propto \frac{\text{Lorentz Force } (\propto B^2)}{\text{Gas Press.} + \text{Grav. Force}}, \tag{3}$$

$$\beta_{n-1,3,\pm 1} \propto \frac{\text{Coriolis Force } (\propto \Omega)}{\text{Gas Press.} + \text{Grav. Force}}. \tag{4}$$

It should be noted that the previous theoretical works on the oblique pulsator model (e.g. Dziembowski and Goode 1985, 1986; Kurtz and Shibahashi 1986) discarded the first-order components represented by the coefficients β and the octapole components and took account of only the zero-order components expressed by the coefficients α and the dipole components. It is these octapole components that are responsible for the observed septuplet fine structure of the amplitude spectrum.

The spherical harmonic $Y_l^m(\theta_B, \phi_B)$ expressed by the spherical coordinates with respect to the magnetic axis (θ_B, ϕ_B) is written in terms of $(2l + 1)$ spherical harmonics of the same degree l with respect to the spherical coordinates (θ_L, ϕ_L) measured from the line-of-sight. Hence the dipole components ($\boldsymbol{\xi}_{n,1,0}$ and $\boldsymbol{\xi}_{n,1,\pm 1}$) of the eigenfunction $\boldsymbol{\xi}$ are observed as a triplet fine structure of the frequency spectrum $(\omega-\Omega, \omega, \omega+\Omega)$ and the octapole components ($\xi_{n-1,3,0}$ and $\xi_{n-1,3,\pm1}$) are observed as a septuplet $(\omega - 3\Omega, ..., \omega + 3\Omega)$ overlapping with the triplet. That is, the observable luminosity variation is formally expressed as

$$L_{\rm obs} = \sum_{m''=-1}^{1} A_{m''}^{(l=1)} \exp[i(\omega - m''\Omega)t] + \sum_{m''=-3}^{3} A_{m''}^{(l=3)} \exp[i(\omega - m''\Omega)t]. \quad (5)$$

The fine structure is asymmetric with respect to its central frequency because of contamination due to the non-axisymmetric components $\xi_{n,1,\pm1}$ and $\xi_{n-1,3,\pm1}$. These features are qualitatively consistent with Kurtz et al's (1992) recent observational results, and provide theoretical justification of Kurtz's (1992) decomposition of the amplitude spectrum of HR 3831 by a series of spherical harmonics.

After some manipulation, we derive the theoretically expected frequency pattern and relations, which are dependent of only the geometrical configuration and the ratio of the Coriolis force and the Lorentz force, among the fine structure amplitudes $A_{m''}^{(l)}$. The degree of asymmetry of the fine structure is independent on the geometrical configuration but depends on the ratio of the Coriolis force and the Lorentz force. The relative contributions of the dipole and the octapole components depend on not only the geometrical configuration and the magnetic field strength but also the limb darkening effects. These relations among $A_{m''}^{(l)}$ allow us to get information about the magnetic fields, the rotation, and the geometrical configuration of the star from the oscillation data, and hence they are promising asteroseismological tools. We are now going to apply the present formulation to Kurtz et al.'s (1992) observational data of HR 3831.

ACKNOWLEDGMENTS

One of the authors (H. S.) acknowledges the Mitsubishi Foundation, which financed the travel expenses.

REFERENCES

Dziembowski, W., and Goode, P. R. 1985, *ApJ*, **296**, L27.

Dziembowski, W., and Goode, P. R. 1986, in *Seismology of the Sun and Distant Stars*, ed. D. O. Gough (Reidel, Dordrecht), p. 441.

Kurtz, D. W. 1982, *MNRAS*, **200**, 807.

Kurtz, D. W. 1990, *ARA&A*, **28**, 607.

Kurtz, D. W. 1992, *MNRAS*, in press.

Kurtz, D. W., Kanaan, A., and Martinez, P. 1992, *MNRAS*, in press.

Kurtz, D. W., and Shibahashi, H. 1986, *MNRAS*, **223**, 557.

Matthews, J. M. 1991, *PASP*, **103**, 5.

Shibahashi, H. 1991, in *Challenges to Theories of the Structure of Moderate-Mass Stars*, ed. D. O. Gough and J. Toomre (Springer, Berlin), p. 393.

Inside the Stars, IAU Colloquium 137
ASP Conference Series, Vol. 40, 1993
Werner W. Weiss and Annie Baglin (eds.)

IS IT POSSIBLE TO DETERMINE WHETHER A STAR IS ROTATING ABOUT A UNIQUE AXIS?

D.O.GOUGH[1] AND A.G.KOSOVICHEV[2]
Institute of Astronomy, Madingley Road, Cambridge CB3 0HA, England

INTRODUCTION

Rotating stars are normally presumed to rotate about a unique axis. Would it be possible to determine whether or not that presumption is correct? This is a natural question to raise, particularly after the suggestion by T. Bai & P. Sturrock that the core of the sun rotates about an axis that is inclined to the axis of rotation of the envelope.

A variation with radius of the direction of the rotation axis would modify the form of rotational splitting of oscillation eigenfrequencies. But so too does a variation with depth and latitude in the magnitude of the angular velocity. One type of variation can mimic the other, and so frequency information alone cannot differentiate between them. What is different, however, is the structure of the eigenfunctions. Therefore, in principle, one might hope to untangle the two phenomena using information about both the frequencies and the amplitudes of the oscillations.

We consider a simple model of a star which is divided into two regions, each of which is rotating about a different fixed axis. We enquire whether there are any circumstances under which it might be possible to determine seismologically the separate orientations of the axes.

OSCILLATIONS OF A STAR WITH OBLIQUELY ROTATING CORE

We consider the stellar model illustrated in Figure Ia whose core rotates with angular velocity Ω_c about a unique axis inclined at an angle β to the angular velocity Ω_e of the envelope. We ask whether β can be determined seismologically. Assuming Ω_c and Ω_e to be independent of appropriate colatitude, the rotational splitting of high-frequency acoustic modes is a seismic average $\overline{\Omega}$ given by the triangle rule (Gough & Kosovichev, 1992) of Figure Ib.

The amplitude spectrum of intensity measurements of a quadrupole multiplet is plotted in Figure IIa, assuming the singlets have equal energy. It depends only on the magnitude of $\overline{\Omega}$ and the angle α, which is defined in Figure Ib. From the relative amplitudes of the rotationally split singlets, and with enough data to average fluctuations, α might be determined. However, that provides no information about β.

---

[1]and Department of Applied Mathematics and Theoretical Physics, University of Cambridge

[2]and Crimean Astrophysical Observatory

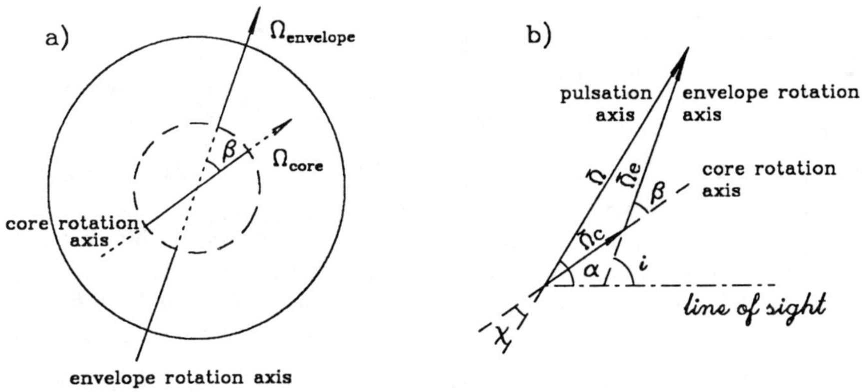

FIGURE I a) A stellar model with obliquely rotating core; b) Triangle rule for seismic averages of angular velocities

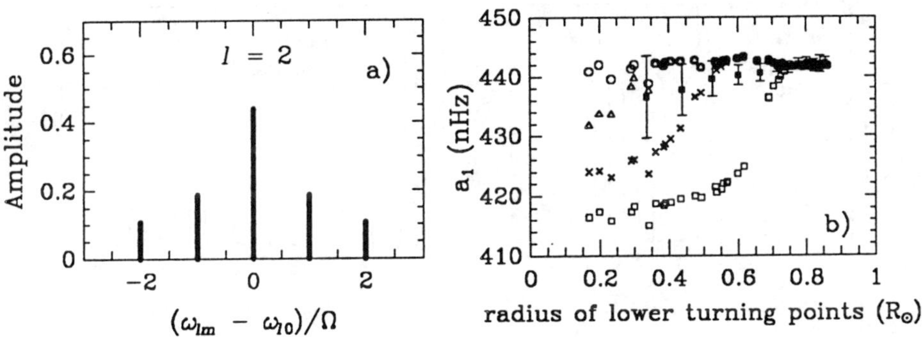

FIGURE II a) Rotational multiplets for whole-disk observations ($\alpha = 30°$); b) Averaged rotational splitting coefficients a_1 [as defined by Duvall *et al.* (1986)] with $(r_c/R_\odot, \beta)$ = 0, 0 (∘); 0.3, 40° (△); 0.5, 40° (×); 0.7, 40° (□) together with BBSO data (with error bars)

To demonstrate that β cannot be determined from frequency information, even if $v \sin i$ and the magnitude of the angular velocity of the surface were measured, we consider the case of the sun, in which an obliquely rotating core has been suggested by Bai & Sturrock (1992). In Figure II*b* we compare a variety of theoretical frequency splittings of models with cores of various radii r_c uniformly rotating with angular velocity Ω_c determined by assuming zero torque with the envelope (and assuming stress is proportional to shear). The results are at variance with the data unless either r_c or β is small. However, if the core were to be assumed to be rotating more rapidly, as in Figure III, to counter the geometrical reduction of the splitting caused by β being nonzero, the splitting data (if viewed from an inertial frame) can be reproduced. Actually, there is fine structure introduced by the earth's rotation about the sun, as illustrated in Figure IV (cf Goode & Thompson, 1992), but with a resolution of 0.1 μHz, which is better that has been achieved to date, a power spectrum is indistinguishable

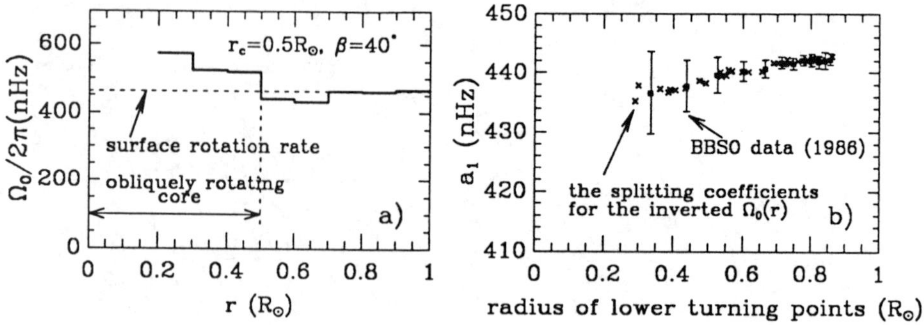

FIGURE III a) Equatorial angular velocity in the sun inferred from our inversions of BBSO data (Libbrecht & Zirin, 1986) assuming an obliquely rotating core; b) the coefficients of the rotational frequency splittings for the inverted $\Omega_0(r)$

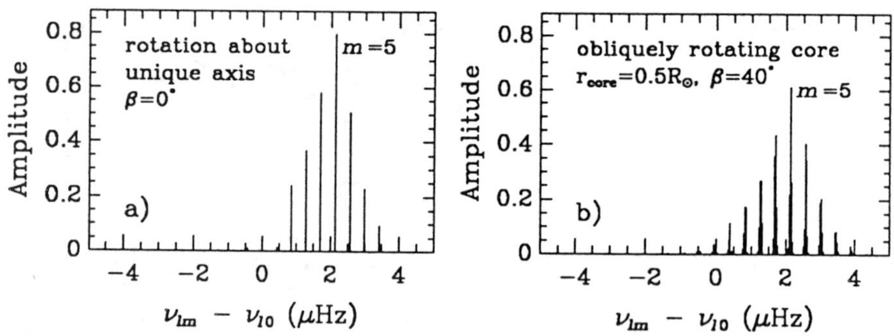

FIGURE IV Spatial filtering of a p mode multiplet of degree $l = 10$ by the Libbrecht & Zirin (1986) mapping for $l = 10, m = 5$ of artificial high-resolution Doppler observations

from that with $\beta = 0$. Therefore, we cannot yet determine whether the solar core is oblique. Since the solar system does not rotate measurably about any other star, this fine structure is not present in stellar spectra. Therefore, we cannot tell seismologically whether stellar cores rotates obliquely. However, if one is prepared to assume that a star rotates about a unique axis, from observations of both singlet frequencies and amplitudes one could in principle determine i seismologically, whatever the functional form of $\Omega(r, \theta)$.

REFERENCES

Bai, T. & Sturrock, P. 1992, *Astrophys. J.*, (in press)
Duvall Jr, T.L., Harvey, J.W. & Pomerantz, M.A. 1986, *Nature*, **321**, 500
Goode, P.R. & Thompson, M.J. 1992, *Astrophys. J.*, (submitted)
Gough, D.O. & Kosovichev, A.G. 1992, *Mon.Not.R.astr.Soc.* (in preparation)
Libbrecht, K.G. & Zirin, H. 1986, *Astrophys. J.*, **308**, 413

Inside the Stars, IAU Colloquium 137
ASP Conference Series, Vol. 40, 1993
Werner W. Weiss and Annie Baglin (eds.)

A PROCEDURE FOR TWO-DIMENSIONAL ASYMPTOTIC ROTATIONAL-SPLITTING INVERSION

T. SEKII AND D. O. GOUGH[1]
Institute of Astronomy, Madingley Road, Cambridge CB3 0HA, England

ABSTRACT Rotational splitting $\Delta\omega(n,l,m)$ of the eigenfrequencies of a star rotating with angular velocity $\Omega(r,\theta)$ about a unique axis can be represented as a weighted integral of Ω over r and θ, (r,θ,ϕ) being spherical polar coordinates about the axis of rotation. For high-frequency acoustic modes, $\Delta\omega/m$ collapses essentially to a function of $w=\omega/(l+1/2)$ and $M=m/(l+1/2)$ alone, and the weighting kernel $K(r,\theta)$ becomes asymptotically degenerate, each factor being of essentially Abel type. Therefore, formally, the splitting integral can be inverted, once a procedure has been found for extending $\Delta\omega$ over the domain of (w,M) such that the turning points (r_t,θ_t), given by $(c(r_t)/r_t,\sin\theta_t)=(w,M)$ where c is sound speed, span the star. Obtaining that representation is the most difficult stage of the inversion. We report on a procedure that treats the inverted two-dimensional Abel integral as a repeated double integral, representing the data successively along a set of parallel lines $M=$constant. The method is illustrated by an inversion of artificial data which is compared with the angular velocity from which those data were computed.

ASYMPTOTIC FORMULA FOR INVERSION

We present the results of an initial attempt to measure the angular velocity $\Omega(r,\theta)$, assumed to be about the unique axis of spherical polar coordinates (r,θ,ϕ), of a solar model from rotational splitting frequencies $f=(\omega_{nlm}-\omega_{nl0})/m$, using the asymptotic formula for high-frequency p modes (Kosovichev and Parchevskii, 1988; Gough, 1992) :

$$\Omega(r(x),\theta(y))=\frac{1}{2\pi W(x)}\frac{\partial}{\partial x}\int_{x_s}^{x}\frac{d\xi}{\sqrt{x-\xi}}\int_{x}^{\xi}\frac{W(\xi')d\xi'}{\sqrt{\xi-\xi'}}\frac{\partial}{\partial y^{1/2}}\int_0^y\frac{f(\xi,\eta)d\eta}{\sqrt{y-\eta}}\ ,$$

where

$$x=a^2,\quad y=\cos^2\theta,\quad \xi=w^2,\quad \eta=M^2\equiv 1-m^2/L^2,\quad L=l+1/2;$$

$$W=-a^{-3}d\ln r/d\ln a,\quad a=c/r\ \text{ and }\ w=\omega/L,$$

where c is sound speed and ω is the frequency of oscillation of the mode of degree l and azimuthal order m, and the subscript s denotes surface value.

[1]Department of Applied Mathematics and Theoretical Physics, University of Cambridge

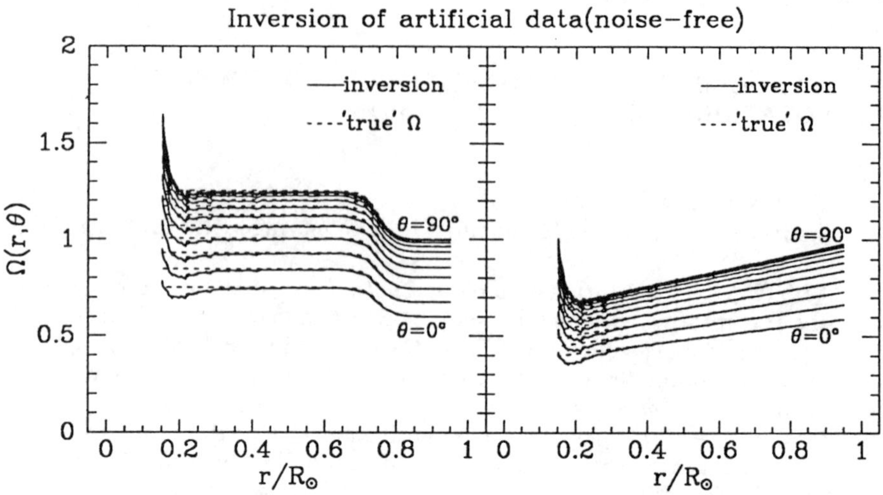

FIGURE I Inversions of $\Omega(r,\theta)$, plotted as continuous lines against $r/R_\odot$, where $R_\odot$ is the solar radius, at different values of θ distributed uniformly in $\cos\theta$. The actual values of Ω from which the data were computed are represented by the dashed lines.

NUMERICAL METHOD

Artificial data have been computed using the exact linearized expression for the rotational splitting of p modes of a solar model with cyclic frequencies $\omega_i/2\pi$ between 3.0 and 5.1 mHz. To carry out the inversions, first $f(\xi,\eta)$ was smoothed and extrapolated with respect to η in the following fashion. The smoothing was accomplished by representing discrete values f_i by the continuous function $\bar{f}$ obtained by minimizing

$$\Sigma\left(\frac{f_i-\bar{f}_i}{\delta_i}\right)^2+\alpha\int\left(\frac{d^2\bar{f}}{d\eta^2}\right)^2 d\eta$$

for a tradeoff parameter α, where f_i is the measured frequency of the i-th mode under consideration and δ_i is the assumed error in f_i. The limits of integration are the extremes of the interval within which the data lie. For modes with $l \geq 5$, extrapolation was accomplished with fourth-degree polynomials that pass through the extrême five mesh points near the ends of the interval. (Polynomials of degree $l-1$ were used for modes with $l<5$.) In all, 201 mesh points were used to span the entire interval $[0,1]$ of η. After differentiation with respect to $y^{1/2}$, smoothing with respect to ξ was carried out in a similar way. Linear extrapolation in ξ to the solar surface was carried out (by regression against the outermost five mesh points) before integration and differentiation with respect to radial variables. Radial mesh points were located at the 1380 different turning points associated with the data.

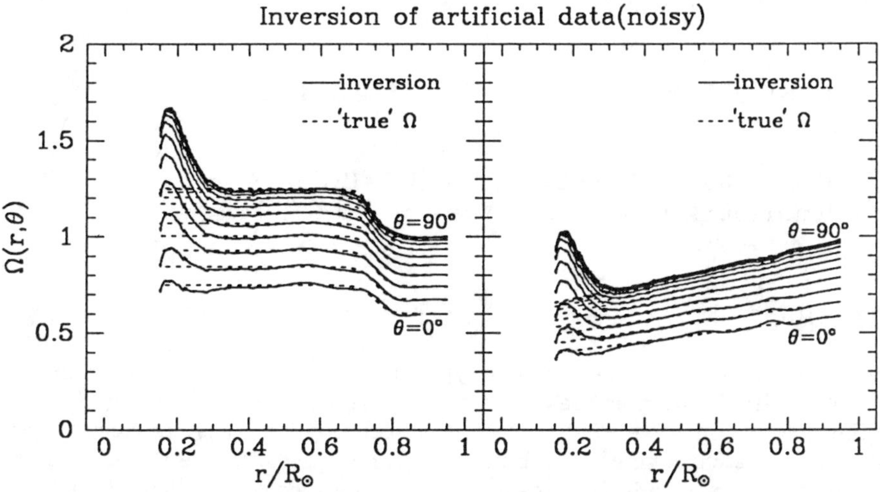

FIGURE II Inversions of the data used for figure I to which were added random errors, normally distributed with zero mean and standard deviation 2 per cent.

RESULTS

Results of inverting precise data from two simple models of Ω are illustrated in Figure I. The continuous curves are $\Omega(r, \theta)$ at eleven different values of colatitude θ from pole (lowest curve) to equator (uppermost curve), uniformly spaced in $\cos\theta$. In the smoothing, δ_i was arbitrarily set to unity for all i. The dashed curves illustrate the actual function $\Omega(r, \theta)$, measured in units of the surface equatorial value, from which the data were computed. The error near the centre arises from a breakdown in the asymptotic representation. The rotation rates obtained at the pole are the result of extrapolation.

In Figure II are shown inversions of data to which two-per-cent errors have been added. As anticipated, they are less accurate, particularly just above the depth at which the previous inversions started to deviate from the true Ω. We plan next to try improving the accuracy of the asymptotic expression at small $r/R_{\odot}$.

REFERENCES

Gough, D. O. 1992, in *Astrophysical Fluid Dynamics, Les Houches XLVII, 1987*, ed J.-P. Zahn and J. Zinn-Justin, North Holland, Amsterdam

Kosovichev A. G. and Parchevskii, K. V. 1988, *Sov. Astron. Lett.*, 14, 201

Inside the Stars, IAU Colloquium 137
ASP Conference Series, Vol. 40, 1993
Werner W. Weiss and Annie Baglin (eds.)

DUST DRIVEN WINDS FROM PULSATING STARS

E.A. DORFI, M.U. FEUCHTINGER, S. HÖFNER
Institut für Astronomie der Universität Wien, Türkenschanzstr. 17, 1180 Wien, Austria

INTRODUCTION

The cool extended atmospheres of late type giants are sites where dust formation takes place. Radiation pressure on the dust grains is an important force for driving the slow but massive winds observed in such objects. Existing calculations of dust driven stellar winds (e.g. Bowen 1988, Fleischer et al. 1991) suffer from the fact that they include approximations at various levels for different parts of the problem like the hydrodynamics or the dust formation. Furthermore they do not include time-dependent radiative transfer.

In order to overcome these insufficiencies we plan to calculate self-consistent models of dust driven winds with a full description of both the radiation hydrodynamics and the time-dependent dust formation. As a first step, however, we concentrate our investigations on the self-consistent description of the radiation hydrodynamics adopting only a simple description of the dust opacities.

BASIC PHYSICS AND NUMERICAL METHOD

Intending to calculate the dynamical behaviour of atmospheres of pulsating stars we solve the full system of radiation hydrodynamics including self gravitation and radiative transfer within the Eddington approximation. For the equation of state we assume a perfect gas law ($\gamma = 5/3$, $\mu = 1.3$), the Rosseland mean of the gas opacity is set to the constant value of 10^{-4} g/cm^3 and the dust opacity is calculated by the simple formula given in Bowen (1988).

At the inner boundary (situated a few pressure scale heights below the photosphere) the oscillation of the star is simulated by varying the radius sinusoidally. Since we start from an initially hydrostatic atmosphere the radius of the outer boundary at the beginning of the calculations is located close to the photosphere. Therefore we use a Lagrangian boundary in order to follow the expansion of the object to an outer radius of about $5 \cdot 10^{14}$ cm. At this point we fix the outer radius and employ a simple outflow boundary condition to allow mass loss from the object.

The equations of radiation hydrodynamics are discretized in conservation form on a freely moving coordinate system (adaptive grid). In this context a fixed number of grid points (typically 500) is distributed by simultaneously solving the so called grid equation (Dorfi & Drury 1987). We employ a monotonic 2nd order advection scheme and the resulting nonlinear algebraic system is solved implicitly. Shock waves are treated by the artificial tensor viscosity (Tscharnuter & Winkler, 1979) suitable for spherically symmetric problems.

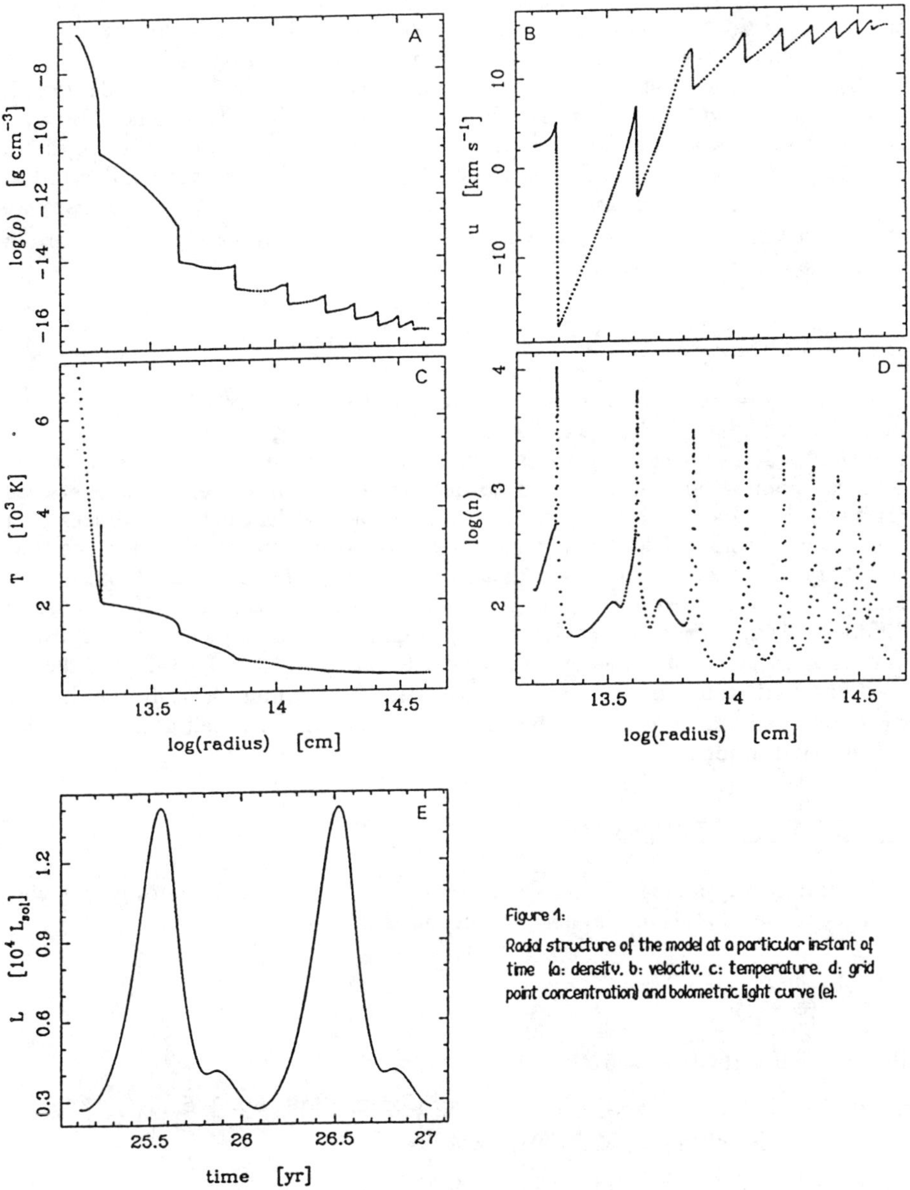

Figure 1:

Radial structure of the model at a particular instant of time (a: density, b: velocity, c: temperature, d: grid point concentration) and bolometric light curve (e).

RESULTS

In this section we present a particular model corresponding to a typical Mira variable on the AGB ($M = 1\ M_\odot$, $L = 5500\ L_\odot$, $T_{eff} = 3000$ K). The period of the piston is set to 350 days assuming a velocity amplitude of 3 km/s.

We follow the evolution until a stationary outflow is reached (after about 30 periods). The radial structure of the physical variables (cf. Fig. 1, a-d) as well as the light curve (cf. Fig. 1e) are shown in the following plots. The computational improvements can be inferred by comparing these results with the radial structures depicted e.g. in Bowen (1988, Fig. 6-8, calculated for similar initial conditions using a much simpler Lagrangian numerical scheme). In summary we state that the outflow rates ($\dot{M} = 2.4 \cdot 10^{-6}\ M_\odot/\mathrm{yr}$) and escape velocities are in the range observed for such pulsating stars.

CONCLUSIONS AND FURTHER INVESTIGATIONS

A main difference to earlier investigations is the inclusion of *time-dependent radiative transfer* which e.g. leads to non-sinusoidal light curves (Fig. 1e). In addition it is not necessary to assume adiabatic or isothermal shocks or any kind of cooling functions in order to determine the energy exchange between matter and radiation field. Furthermore Fig. 1 (a-e) clearly demonstrates that a number of unphysical spikes and sawtooth-like features typical for the flow structure of earlier calculations can be avoided by using an *accurate numerical method.* It is important to note that this method is also suitable for quite different astrophysical situations as e.g. nonlinear stellar pulsations (Dorfi & Feuchtinger, 1991) or time-dependent dust formation (Dorfi & Höfner, 1991).

The next step currently under investigation is to couple the radiation hydrodynamics and time-dependent dust formation to construct self-consistent dust driven wind-models of pulsating stars.

ACKNOWLEDGMENTS

This work is supported by the Österreichischer Fonds zur Förderung der wissenschaftlichen Forschung under project numbers P8411 and P8757.

REFERENCES

Bowen, G.H.: 1988, *ApJ* **329**, 299

Dorfi, E., Drury, L.O'C.: 1987, Journal of Comp. Phys. **69**, 175

Dorfi, E.A., Feuchtinger, M.: 1991, A&A **249**, 417

Dorfi, E.A., Höfner, S.: 1991, A&A **248**, 105

Fleischer, A.J., Gauger, A., Sedlmayr, E.: 1991, A&A **242**, L1

Tscharnuter, W.M., Winkler, K.-H.A.: 1979, Comp. Phys. Comm. **18**, 171

Inside the Stars, IAU Colloquium 137
ASP Conference Series, Vol. 40, 1993
Werner W. Weiss and Annie Baglin (eds.)

RADIAL PULSATION IN VARIABLE STARS WITH MASS LOSS

FRANK P. PIJPERS
Astronomy Unit, School of Mathematical Sciences, Queen Mary and Westfield College, Mile End Road, London E1 4NS, UK

ABSTRACT It is possible to show that cool giants with very large photospheric scale heights do not perfectly reflect pulsational waves at the photosphere. This means that for these stars the classical formulation of pulsation in which the outer boundary for the resonance cavity is assumed to be perfectly reflecting is not valid. This can have significant consequences for the eigenfrequencies of the pulsation of Long Period Variables such as Mira type variables as well as for the stability of their pulsation.

There are various problems connected with the modelling of pulsation of cool giant stars on the asymptotic giant branch. These were recently discussed by Wood (1990). Two problems are highlighted here :

1) The observed values of the pulsation constant Q do not match theoretical Q-values very well either for the fundamental or first overtone pulsation.
2) The growth rate of the fundamental mode is too large, and it has to be reduced artificially to produce stably pulsating models.

The first of these problems may be caused by inaccuracies in the observational determination of Q which is rather sensitive to the determination of the effective temperature. The spectra of these cool stars are dominated by absorbtion in molecular bands which greatly complicates determination of their effective temperature. The second problem may be due to the lack of a good time dependent theory of convection. A stable pulsation can only be obtained by introducing an artificial damping into the calculation (Wood, 1990). An order of magnitude estimate of this damping can be made from some of the results of Wood (1974). Model calculations take about 10 oscillations before a shell of some $10^{-2}\ \mathcal{M}_\odot$ is ejected. The excess energy input, E_{ex}, into the pulsations can then be calculated and is $\approx\ 5\ 10^{28}$ W. The mass loss of some $10^{-6}\mathcal{M}_\odot/yr$ carries away some energy but this energy loss term is much smaller than E_{ex}. Most of the excess energy still has to be radiated away in the wind, or carried away by the waves generated by the pulsation. At the base of the wind, the energy flux in waves is :

$$E_w \ = \ 4\pi R_*^2 \frac{1}{2} \rho_0 \langle \delta v^2 \rangle (a + v_0) \tag{1}$$

Here v_0 and ρ_0 are the outflow velocity and the density at the base of the wind and a is the sound speed. $\langle \delta v^2 \rangle$ is the mean square of the velocity amplitude of the waves. By equating E_w and E_{ex} and using a mass loss rate of $10^{-6}\mathcal{M}_\odot/yr$ an upper limit for the wave amplitude at the stellar surface can be obtained :

$$\langle \delta v^2 \rangle^{\frac{1}{2}} < 4 \text{ km/s} \tag{2}$$

The speed of sound in these cool stellar atmospheres is around 5 km/s. The excess energy in the pulsation can therefore be carried to infinity with shock waves that have a velocity amplitude which is of the order of the speed of sound.

The discrepancies 1) and 2) between observed stars and theoretical models may result from the modelling assumption that the pulsational waves are reflected perfectly at the surface of the star. This perfect reflection makes the model a closed cavity resonator whereas the real star is not a closed cavity.

For most stars the assumption that the stellar atmospheric background is hydrostatic holds quite well. The situation changes when the star has a significant stellar wind. In some cases the gas pressure may not decrease sufficiently rapidly in the stellar wind. The outer boundary is then not a pressure node for the pulsation because $p \neq 0$ and it transmits pulsational waves : the resonant cavity is open at one end. This causes damping of the pulsation and a change in frequency of the eigenmodes. The difference is very similar to the difference between open and closed organ pipes. Increasing the mass loss rate for a given stellar model is equivalent to drilling a hole of increasing size in the organ pipe.

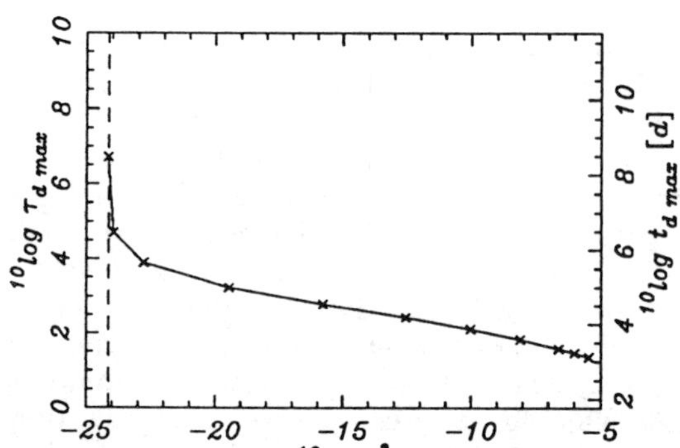

Fig. 1. The maximum damping time as function of mass loss rate. The crosses are the results for stellar polytropic models combined with a shock wave driven wind (Pijpers and Habing, 1989).

It can be shown (Pijpers, 1992) that there is a maximum value of the characteristic damping time t_d determined by the same parameters that determine the mass loss rate. Figure 1 shows this damping time for the shock wave driven wind model of Pijpers and Habing (1989). Now the sonic point of the wind is a boundary for information carried by sound waves. Outgoing waves will simply pass through the sonic point but incoming sound waves will never reach it. This provides the boundary condition for linear pulsation in a star with a stellar wind. The pulsational wave must be a purely outgoing sound wave at the sonic point. The energy of the pulsation carried away by the sound waves through the sonic point causes the pulsation to be damped.

The system of differential equations of the LAWE can be written in such a way that the background is expressed in a single complex potential function Λ. The evaluation of Λ inside the star requires the solution of the equations of stellar structure. The essentials of the core-mantle structure that is typical of AGB stars (Keeley, 1970) are represented by using a combination of polytropes. The wind is modelled using the shock wave driven wind model of Pijpers and Habing (1989).

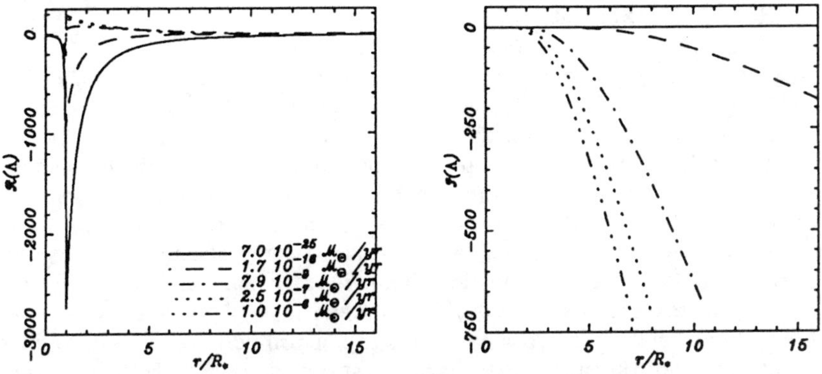

Fig. 2. The real and imaginary part of the Λ-function for various mass loss rates. The line coding is the same in both panels.

In figure 2, the real and imaginary parts of the Λ-function are shown for varous mass loss rates. The sharp dip in the real part of Λ is the stellar surface, where for a classical stellar model $\Re(\Lambda)$ would go to $-\infty$. This implies that no information could then pass through the stellar surface in either direction and it would serve as a reflective boundary for the LAWE.

The leakage of pulsation energy out of the star caused by significant mass loss may stabilize the pulsation modes of Miras and other cool giants and supergiants. The frequency of the eigenmodes in stars that have a well developed stellar wind is different from the eigenfrequencies in classical pulsation models. This may explain the discrepancy between values of the pulsation constant Q found observationally and those calculated from standard models without mass loss. The mass loss rate will act as a governing parameter for the eigenfrequencies and therefore for the pulsation constant Q.

REFERENCES

Keeley, D.A., 1970, *Ap. J.*, ,**161**, 643

Pijpers, F.P., 1992, *in preparation*

Pijpers, F.P., Habing, H.J., 1989, *Astr. Ap.*, ,**215**, 334

Wood, P.R., 1974, *Ap. J.*, ,**190**, 609

Wood, P.R., 1990, in *Proceedings of the International Colloqium From Miras to Planetary Nebulae : which path for stellar evolution ?*, ed. O. Mennessier and A. Omont, Editions Frontieres, Gif sur Yvette, p. 67

Inside the Stars, IAU Colloquium 137
ASP Conference Series, Vol. 40, 1993
Werner W. Weiss and Annie Baglin (eds.)

STELLAR DYNAMOS

F. KRAUSE
Astrophysikalisches Institut Potsdam, D-O-1591 Potsdam,
Rosa-Luxemburg-Str. 17a

ABSTRACT Energy is transported from the central regions of a star to its surface. Generally this transport is in certain layers carried on by convective motions. Because of the structure, which these motions have due to the influence of the overall rotation, the star becomes electromagnetically unstable, i.e. a large magnetic field grows from small seed fields as a result of the dynamo process. The internal structure, especially the symmetries of a star, will be, at least to some extend, reflected by the spatial structure and the time behaviour of the excited magnetic field. In this sense observations of the magnetic field on a surface of a star and the related activity phenomena can provide insight in the internal structure of the star, since characteristic parameters like thickness of the convection zone, mixing length, turnover time, profile of the differential rotation, etc. strongly influence the dynamo process.

The actual magnetic field of a star is a product of a nonlinear process. Models elaborated on the kinematical (i.e. linear) level provide insight in the excitation conditions and the linear field modes. The marginal mode, i.e. the mode which is easiest to excite, reflects properties of the nonlinear solution in case the system operates not far from the margin to the dynamo unstable region. Here the solutions show symmetries with respect to the axis of rotation and the equatorial plane, properties which are, for example, to a large extend fulfilled for the solar average magnetic field. For systems operating far from this margin irregular or even chaotic behaviour has to be expected. From observations there is a strong indication that these theoretical possibilities find their realizations within the sample of late-type stars.

INTRODUCTION

Observations of the solar surface reveal a manifoldness of phenomena closely connected with magnetic fields. At first glance these phenomena are of a small scale nature in space as well as in time. However, long time statistical studies suggest the existence of a large scale background field. This is best known from the sunspot phenomenon which has now been studied in detail over more than one century. It suggested an odd parity magnetic field - antisymmetric with respect to the equatorial plane - which consists of ring fields of opposite orientation on both sides of the equatorial plane. This toroidal magnetic field is accompanied by a dipol-type poloidal field indicated by magnetograph records of the polar caps and by structures in the corona which can be seen at solar eclipses. Both field parts form the large scale magnetic field of the Sun which oscillates with a period of about 22 years.

The oscillatory character of this magnetic field of the Sun exludes a fossil origin. An explanation has to present a mechanism which continously provides for a supply of energy. Already J. Larmor (1919), who thoroughly analyzed possible explanations in his famous paper "How could a rotating body such as the Sun become a magnet?", ruled out all alternatives and leaves the dynamo as the only one.

Since the presentation of the first conclusive spherical model of the Solar dynamo by Steenbeck and Krause in 1969 a great number of such models have been elaborated by several authors. On proper assumptions concerning the internal motions inside the Sun (i.e. convection, differential rotation, meridional motions) all relevant observational facts find an explanation.

It should, however, be noted here that the internal motions inside the Sun originate in the transport process by which the energy released in the core is brought to the solar surface. The magnetic field being produced by the induction action of these internal motions appears thus embedded in the whole complex of stellar constitution. In this general sense the dynamo problem proves to be a highly nonlinear problem which is beyond of acces even to the best to-days computers. The problem "simulating the solar dynamo" is still unsufficiently solved.

From the above adopted standpoint the global magnetic field of a star proves to be an entity which reflects by its spatial structure and its temporal behaviour the individual features of this star. Consequently, it has to be expected that stars of the same stellar type must have about the same magnetic field, if we, for the moment, refrain from the fact that the stars may differ with respect to their rotational velocities, which have a rather strong influence on the internal motions.

In this sense is the idea of the dynamo origin of the solar magnetic field strongly supported by the observational fact that late type stars also show activity cycles, i.e. they show the same magnetic activity phenomena as the Sun, in many cases also with a certain periodicity.

The situation is different for the magnetic Ap-stars: These stars have fields which significantly deviate from axial symmetry with respect to the axis of rotation, i.e. the magnetic signals recorded on the Earth are strictly periodic, thus reflecting the rotational period. Till now there are no clear indications of time variations of these fields in a co-rotating coordinate system, even if the observations of certain peculiar surface structures, which exist since the begin of this century, are included in the considerations. Thus the possibility of a frozen-in field from the birth of a star cannot be ruled out, although dynamo excitation in the star's convective core offers an alternative process (Krause 1983).

The peculiar A-stars are a small group of stars which might have a frozen-in field, the overwhelming number of stars do not show a sign of such fields. Thus it is not probable that stars get a field from the interstellar field by the condensing matter, or, if they do, the field will be destroyed by the strong turbulence during the star forming processes. However, dynamo theory revealed that turbulence and rotation make a star (or a protostar) to work as a dynamo. Consequently a star can hardly escape from being a dynamo, at least during some evolutionary phase, and will thus form its own individual magnetic field.

Magnetic fields of stars will so reflect certain internal structures, the most obvious of which is the convection zone, its thickness and its position. Further the differential rotation is of influence because of its strong amplification action. Details have to be studied with models which are as close to reality as possible.

THE BASIC PROBLEM

According to our view is the ability of a star to work as a dynamo due to internal motions which originate in the transport process which brings the energy produced in the star's core to the surface. The basic equations for describing this process are given by the conservation laws of mass, momentum and energy

$$\frac{\partial \ln \rho}{\partial t} + (\boldsymbol{u} \cdot \nabla) \ln \rho + div\ \boldsymbol{u} = 0 \tag{1}$$

$$\frac{\partial \boldsymbol{u}}{\partial t} + (\boldsymbol{u} \cdot \nabla)\boldsymbol{u} = -\frac{p}{\rho}\nabla \ln p + \boldsymbol{g} - 2\boldsymbol{\Omega} \times \boldsymbol{u} + \frac{1}{\rho}\, \boldsymbol{j} \times \boldsymbol{B} + \frac{1}{\rho}\, div\ \tau \ , \tag{2}$$

$$\frac{\partial e}{\partial t} + (\boldsymbol{u} \cdot \nabla)\ e = -\frac{p}{\rho}\ div\ \ \boldsymbol{u} + \frac{\kappa}{\rho}\Delta e + Q_{visc} + Q_{joul} \ , \tag{3}$$

and the induction equation

$$\frac{\partial \boldsymbol{B}}{\partial t} = curl\ (\boldsymbol{u} \times \boldsymbol{B}) + \eta \Delta \boldsymbol{B} \ . \tag{4}$$

Here denotes ρ the mass density, $\boldsymbol{u}$ the velocity, p the pressure, $\boldsymbol{g}$ the gravity, $\boldsymbol{\Omega}$ the angular velocity, $\boldsymbol{B}$ the magnetic field, $\boldsymbol{j}$ the current density, e the specific internal energy, κ the heat conduction, τ the viscosity tensor and η the magnetic diffusivity. Finally define Q_{visc} and Q_{joule} the heat production by viscosity and electric resistivity. These quantities are quadratic expressions in the velocity gradient and the current density.

The equations (1) to (4) have to be completed by an equation of state. Furthermore, appropriate boundary conditions have to be formulated. Physically clear conditions we have in the case where the star is considered as embedded in an empty, electrically insulating space, although this does not fully correspond to the real conditions: stars generally have a corona, i.e. the environment is electrically well conducting.

Difficulties originate in the boundary conditions since the magnetic field as a far reaching quantity cannot be described by local conditions in a correct way. Nevertheless, a certain number of models were based on the local boundary condition $\boldsymbol{B}_{tan} = 0$. It provides for some mathematical simplification, however, it has to be noted that a physical justification is not possible. Especially, no electromagnetic energy, which is produced inside by the dynamo, is transported through the surface in the environment: The pointing vector $\boldsymbol{E} \times \boldsymbol{B}$ is tangential.

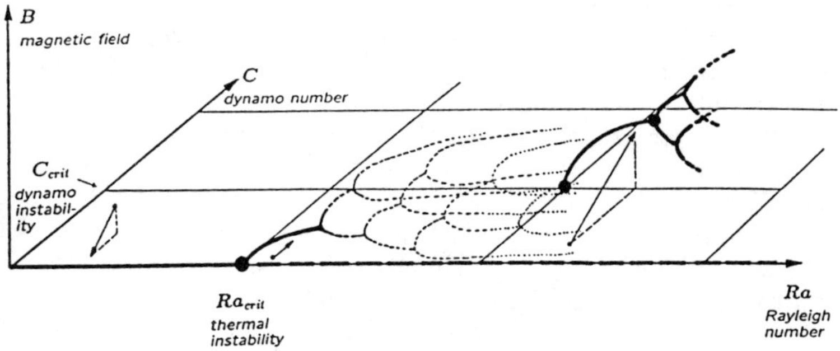

FIGURE I Stability diagram of a convection-driven dynamo. The axis to the right characterizes the temperature difference in units of the Rayleigh number Ra, the axis to the back the dynamo-active part of the convective motion by the dynamo number C. The arrows indicate the attractors, to which an arbitrary initial field developes.

In addition specify the boundary conditions that heat energy flows from the cental part of the star to the surface and leaves there by the emission of radiation.

The above presented equations have the simple solution

$$u = B = j = 0 \, , \quad \frac{\partial \rho}{\partial t} = 0 \, , \tag{5}$$

with e determined by

$$\frac{\partial e}{\partial t} = \frac{\kappa}{\rho} \Delta e \quad . \tag{6}$$

In this case the transport of energy is simply carried on by molecular conduction.

However, this solution generally proves to be unstable. First the velocity field $u = 0$ becomes unstable, and energy transport occurs by convective motions - hot material rises towards the surface, cold material sinks down. In addition, since the medium in the convection zone of the Sun or a star is electrically conducting, the magnetic field $B = 0$ also becomes unstable and dynamo excitation of a magnetic field sets in.

The 3-D-scheme in Fig. 1 shall illustrate the physical situation: The axis directed to the right characterizes the heat conduction process by the Rayleigh number R_a. At a critical value of R_a we have the first bifurcation at the thermal instability, a velocity field $u \neq 0$ is formed, the magnitude is indicated by the axis, which is directed to the background. The representation shall illustrate that first, i.e. close to the bifurcation point, the velocity shows a rather regular, periodic structure, in a greater distance it generally becomes turbulent. The dynamo number C now characterizes the electromagnetic action of this velocity field: At a critical value we have the second basic bifurcation where the magnetic field $B = 0$ becomes unstable and dynamo excitation sets in.

It is now well known that systems like the Sun or stars generally ly beyond the critical values of the parameters R_a and C. It is so clearly to be seen that the considered problem requires the solution of the full set (1) - (4) of nonlinear equations, and that we especially have to consider their irregular solutions, which describe the turbulence phenomena.

The problem appears so to be of unmanagable complexity. During the last three decennia succesfull research has been carried out on the basis of the mean-field concept (Krause and Rädler 1980). Equations for the mean fields have been developed which describe the collective phenomena of the fluctuating fields by certain averages. Parallel to this development there was also the try to calculate dynamo processes by numeric simulations (e.g. Gilman 1972, 1983, Glatzmaier 1984, 1985a,b, Chan et al. 1982, Nordlund and Stein 1989, Brandenburg et al. 1990b). Although models which describe situations close to reality are not achieved till now these investigations will be promising for the future with more and more powerfull computers (Krause 1991).

In the following we will present an overview of the results which have been achieved by the research of stellar dynamos. Most of them have been gained by using the fact that the backreaction of the magnetic field on the motion - the Lorentz force $j \times B$ - is of second order. Consequently, it makes sense first to consider the kinematical problem, especially, if the question for the origin is raised. From that platform a certain insight in the highly nonlinear dynamo problem has been elaborated in the recent years.

THE KINEMATIC PROBLEM

We will now specify the problem according to the natural conditions. As a model of a star we consider a sphere which consists of an electrically conducting fluid (ionized gas). It rotates about a certain axis. Due to gravity and the overall rotation the model shall show symmetry with respect to the equatorial plane and the axis of rotation in all non-magnetic properties. As far as it concerns turbulent quantities this requirement shall be fulfilled in the average.

The following considerations rest on the assumption of a prescribed velocity field. We study the fate of a magnetic field which exists at a certain time ($t = 0$) in the star for some reason by solving the initial value problem of the induction equation (4). We have so to determine a magnetic field $B(x,t)$ as a solution of the equations

$$\frac{\partial B}{\partial t} = curl\ (u \times B) + \eta \Delta B\ , \quad div\ B = 0 \tag{7}$$

for $r < R$ and

$$curl\ B = 0\ , \quad div\ B = 0\ , \quad B = O(1/r^3) \quad if\ r \to \infty\ , \tag{8}$$

for $r > R$. r denotes the distance from the centre and R the radius of the sphere which represents the star.

We enter with the ansatz

$$B(x,t) = e^{\gamma t}\, B(x) \quad , \tag{10}$$

into the equations (7), (8). According to general mathematical rules we find a denumerable set of eigen modes which are characterized by a set of eigen values $\gamma_1, \gamma_2, ..., \gamma_n, ...$ and corresponding eigen fields $B_1(x)$, $B_2(x), ...,$ $B_n(x), ...$. The real parts of the eigen values γ_n, $Re\{\gamma_n\}$, describe growth, resp. decay of the eigen modes. Generally we have

$$Re\{\gamma_n\} \rightarrow -\infty \;\; if \;\; n \rightarrow \infty \tag{11}$$

i.e. nearly all eigen modes decay due to Ohm's dissipation. Only a finite number of them may have non-negative growth rates: in this case the system represents a dynamo. The imaginary part of a γ_n characterizes a periodic behaviour if unequal zero.

There are in astrophysics important types of motions like differential rotation and compression (e.g. due to accretion) which provide for amplification but not for dynamo excition. In this cases the system (7), (8) degenerates and we have to consider also the ansatz

$$B(x,t) \;= t\, e^{\gamma t}\, B(x) \quad , \tag{12}$$

in order to complete the system of eigen modes. For the motions mentioned above $R_e\{\gamma\}$ would be negative: the field decays, however, first it grows proportional to time.

Research in dynamo theory over the last three decennia revealed that basically any rotating cosmical object should be a dynamo in case it shows internal motions like convection. This statement is crucial: As a consequence the large scale part of a magnetic seed field grows exponentially and the cosmic object forms its own individual magnetic field, the memory of the seed fields got lost.

The ability of being a dynamo is due to the influence of the Coriolis forces on the internal motions. They get a helical structure where generally in a certain region one kind of helicity, either lefthanded or righthanded, dominates. For example, in the convection zone of the Sun are lefthanded helical motions dominating in the northern hemisphere and righthanded in the southern.

The induction action of these motions used to be described in the frame of mean-field magnetohydrodynamics. The correlated action of the small-scale fields u', B' is described by the turbulent electromotive force $\mathcal{E}$ defined by

$$\mathcal{E} = \overline{u' \times B'} \quad , \tag{13}$$

where the overbar denotes the average. Ohm's law for the mean fields reads so

$$j = \sigma(E + u \times B + \mathcal{E}) \quad , \tag{14}$$

where from now j, E, u, B describe the mean values of these fields.

The induction action of small-scale motions with helicity leads to a turbulent emf $\mathcal{E}$ of the form

$$\mathcal{E} = \alpha B - \beta curl\; B \quad , \tag{15}$$

The first term on the righthand side represents the α-effect, i.e. an emf parallel to the magnetic field. The second one provides for an enhanced decay by turbulent diffusion. It describes the cascade of magnetic energy down to smaller and smaller scales due to the turbulent motions.

Relation (15) corresponds in this form to the idealized case of a homogeneous isotropic turbulence. α is connected with the helicity, the average of the scalar product of the fluctuating velocity field $\boldsymbol{u}'$ with its *curl*:

$$\alpha = -\frac{\tau_{cor}}{3} \overline{\boldsymbol{u}' \, curl \, \boldsymbol{u}'} , \tag{16}$$

where τ_{cor} denotes the correlation (overturn) time (Steenbeck and Krause 1969) and β is connected with the turbulence intensity by

$$\beta = \frac{\tau_{cor}}{3} \overline{\boldsymbol{u}'^2} . \tag{17}$$

In real conditions the dominance of one kind of helical motions is due to the stratification by gravity $\boldsymbol{g}$ and the influence of the rotation represented by the angular velocity $\boldsymbol{\Omega}$. According to these highly anisotropic conditions (15) has to be exchanged by the tensorial connection.

$$\mathcal{E}_i = \alpha_{ij} B_j + \beta_{ijk} \frac{\partial B_j}{\partial x_k} , \tag{18}$$

where the tensor α_{ij} has the form

$$\alpha_{ij} = \alpha_1(\boldsymbol{g} \cdot \boldsymbol{\Omega})\delta_{ij} + \alpha_2(g_i \Omega_j + g_j \Omega_i) . \tag{19}$$

Especially recent investigations revealed that the contributions by the α_2- term in (19) may be remarkable (Rädler 1983). From a discussion of an anisotropic diffusivity tensor we will refrain here.

Numerous investigations have been carried out of models which are based on the induction equation

$$\frac{\partial \boldsymbol{B}}{\partial t} = curl \, (\boldsymbol{u} \times \boldsymbol{B} + \mathcal{E}) + \eta \Delta \boldsymbol{B} , \tag{20}$$

where $\boldsymbol{u}$ describes motions like differential rotation and/or meridional motions and $\mathcal{E}$ the influence of turbulent convection according to (15) or (18), (19) or even more complicated. Often simply (15) combined with a differential rotation has been considered which is general named $\alpha\Omega$-dynamo.

The results show that generally a few growing eigen modes, i.e. those with positive $Re\{\gamma\}$, do exist. They differ especially with respect to their symmetry properties: They are either symmetric (even parity) or antisymmetric (odd parity) with respect to the equatorial plane and axisymmetric with respect to the axis of rotation or depend on the azimuth φ according to $e^{im\varphi}$. The symmetry type will accordingly be designed by Sm or Am. For the three most important types $A0$, $S0$, $S1$ the shape is illustrated in Fig. 2.

The Sun's magnetic field has, especially because of the polarity laws of the sunspots, to be classified as $A0$. For magnetic *Ap*-stars it is not clear whether or not the fields show a dominating symmetric part, however, if they do, they with high probality belong to the type $S1$ (Oetken 1977, 1979). Also for late type stars

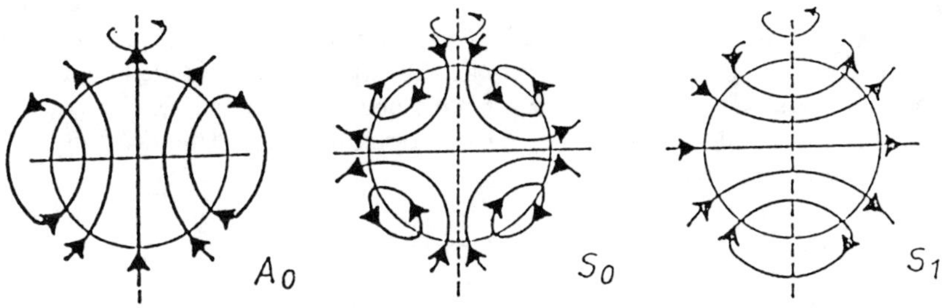

FIGURE II The three most important symmetry types of magnetic fields, which may be excited by a spherical dynamo showing symmetries with respect to the axis of rotation and the equatorial plane. In a multipole expansion the leading terms will be a dipole parallel to the axis of rotation for $A0$, a quadrupole for $S0$ and a dipole with its moment within the equatorial plane for $S1$.

the symmetry type is not clear, but significant deviations from the symmetry with respect to the axis of rotation seems to exist in some cases (Piskunov et al. 1990, Jetsu et al. 1990, Moss 1991).

Calculations of dynamo models within the kinematic theory show that fields of the three types $A0$, $S0$, $S1$ may have positive growth rates, or to say it in a different way, they take the places $B_1(x,t)$, $B_2(x,t)$, $B_3(x,t)$ within the sequence of eigen modes. So fields of these three symmetry types compete in being realized by a certain model, or a certain star.

In the frame of the kinematic theory one can find two marks for differentiating between the eigen modes: For the intensity of the induction action characterized by the dynamo number C exist a critical value, the marginal dynamo number C_{crit}, which is the border beyond of which the system is electromagnetically unstable. The related eigen mode, for which the growth rate $Re\{\gamma\} = 0$ if $C = C_{crit}$, we call the marginal mode. It is the mode which is easiest to excite. As a further mark we can use the largest growth rate. If C is in an immidiate neighbourhood of C_{crit} both these eigen modes are identical. However, for values of C which substantially exceed C_{crit} they may be different.

Our present findings allow for a solution of the initial value problem, i.e. we can describe the growth of a weak seed field. Clearly the eigen mode with the largest growth rate will dominate in a short time. That may be the marginal mode or not.

The actual magnetic fields we observe on the Sun and stars are clearly in a well developed state, in a state of saturation. This state cannot be described in the frame of a linear theory, we need information from the nonlinear investigations in order to understand the actual situation.

THE NONLINEAR DYNAMO PROBLEM

The treatment of the full dynamo problem needs the solution of the system of partial differential equations (1),.., (4), or, at least, the solution of the coupled system ,consisting of the Navier-Stokes equation and the induction equation. In order to avoid the mathematical difficulties simple modelling of the backreaction on the induction action of the motions have been introduced. The most common one is the so-called α-quenching, where a reduction of the α-effect is simply achieved by the ansatz

$$\alpha = \frac{\alpha_0}{1 + aB^k} \tag{21}$$

with a certain exponent k. In this way the induction action of the α- effect is reduced with the growing magnetic field.

Models of this kind allow the study of the extension of the eigen modes to the nonlinear regime in dependence on the dynamo number. The results allow for some general statements (Krause and Meinel 1989):

i The stability of a solution is decisive of which magnetic field will be maintained by the dynamo,

ii The extension of the marginal mode is stable in a certain neighbourhood of the marginal dynamo number,

iii The extension of a higher mode is first, i.e. in a neighbourhood of their bifurcation, unstable,

iv For values of the dynamo number that substantially exceed the marginal dynamo number bifurcations in the nonlinear regime lead to solutions of a complex structure.

According to (ii), (iii) the significance of the marginal mode is clearly to be seen.

To some extend the nonlinear extensions of the eigen modes possess the same symmetry properties, i.e. the extension is of even (odd) parity if the eigen mode is of even (odd) parity. The same is true for the symmetry with respect to the axis of rotation.

In this way the investigation of the linear problem provides insight in the symmetry properties of the nonlinear solutions, at least in some neighbourhood of the marginal dynamo number. The higher eigen modes are unstable, however, it may happen that some of them first dominate, because of their higher growth rates. Finally they decay thus demonstrating an inverse cascade of the energy (Gilbert and Sulem 1990). According to (iv) additional bifurcations leed to a break of symmetry in a certain distance from the marginal dynamo number and more irregular solutions will appear.

These findings allow for a possible scenario that predicts stable regular magnetic fields with certain symmetries (axisymmetry, equatorial symmetry) just beyond the marginal dynamo number, but for values of C that substantially exceed the marginal number, non-symmetric or even irregular fields have to be expected.

THE CONFRONTATION WITH OBSERVATIONS

The foregoing discussion revealed that it makes sense to investigate kinematic models even if the final state is investigated. Many models have so been considered (e.g. Roberts 1972, Roberts and Stix 1972, Yoshimura 1975a,b, Rädler 1980, 1986, see also the monographs Moffatt 1978, Krause and Rädler 1980, Zeldovich et al. 1983).

In connection with the explanation of the solar magnetic field many investigations concentrated on the $\alpha\Omega$-dynamo. In most cases an $A0$-type field was found as the marginal one. We thus understand the observed large-scale magnetic field of the Sun.

On the other side by changes of the internal structure also models have been found where a $S0$-type field, i.e. one with a quadrupol as leading term, are marginal (e.g. Roberts 1972). Such models may indicate that the magnetic field of the Sun not always during its history must have been of odd parity.

Furthermore, even models have been found, where fields of type $S1$ are marginal (Krause 1971, Stix 1971, Rüdiger 1980). These models provide the possibility to interprete the observational results concerning the magnetic *Ap*-stars in terms of dynamo theory (Krause 1983). Worth to note here that a field pattern as realized by meridional motions can excite a magnetic field of type $S1$ (Gailitis 1970, Moss 1990).

Very detailled investigations of spherical dynamo models have been carried out by Rädler (1980, see also Krause and Rädler 1980). Here stability maps are presented which provide insight how the magnetic fields of the different symmetry types become marginal in dependence on different influences like differential rotation, meridional motions and such turbulence parameters like α_1, α_2 in (18) and others.

A comparision with the solar magnetic field provides for a challenge to fit models best to the observed patterns. As well the modelling of the differential rotation as the different turbulence parameters provides for a diversity of possibilities.

However, things become more complicated since the differential rotation is also caused by the turbulent convection by the so-called Λ-effect (Rüdiger 1989). In this sense conclusive models require to determine all quantities like $\alpha, \alpha_1, \alpha_2, \ldots$ and those describing the Λ-effect out of the same root: the turbulent convection. In connection with reconciling the different aspects a number of difficile problems emerge (Brandenburg et al. 1990a).

Apart from theoretical determinations modern observations of solar oscillations provide for a possibility to derive the profile of differential rotation inside the Sun (Harvey 1988, Libbrecht 1988). In this way further constraints as well on the theory of the differential rotation as on the dynamo theory do now exists (Paterno, 1991). One misses the confirmation of the, already a long time existing forecasting of an increasing angular velocity with depth, which fits best to the polarity rules of the observed magnetic field. However, the oscillation data apparently allow for a radial jump of the angular velocity at the base of the convection zone. This already is sufficient for having the correct polarity relations and the migration of the toroidal field belts towards the equator (Steenbeck and Krause 1969, Roberts and Stix 1972, Belvedere et al. 1991).

BEYOND THE BREAK OF SYMMETRY

So far we discussed the dynamo from the view based on the linear theory, what means we considered the object in a state close to the dynamo instability, the realized field shows to a far extend the same properties as the marginal mode. In the following we will discuss results of nonlinear investigations and some relations to observational findings.

The break of symmetry at a certain dynamo number may appear in a way that the marginal mode of odd parity becomes unstable and the system migrates on a certain time scale to an even parity configuration, which again is unstable. This time scale proves to be much larger than the period of oscillation (i.e. the (solar) cycle period). The magnetic field has so the appearance of an odd parity oscillating field which changes after a large number of oscillations for some time the parity (Brandenburg et al., 1989a). A consideration of the energy of the excited field reveals a "grand minimum" just at the transition from one parity to the other (Brandenburg et al., 1989c). Hence one is tempted to relate this property of the model to periods of no, or minor solar activity in the past history, i.e. to the Maunder minimum.

Investigations of nonlinear models reveal a picture of high complexity (Rädler et al. 1990, Brandenburg et al. 1989b, Jennings and Weiss 1991, Jennings 1991). Jennings et al. (1990) found a model where the $A0$-type solution lose their stability to a mixed one with $A0$ and $S1$ contributions. So the possibility of modelling the observed sector structure on the Sun is feasible.

Apart from the observed field pattern there are characteristic observable parameters which may be studied in their mutual dependence. These investigations gain the more interest, since the observations of solar-type and late-type stars offer possibilities to study these dependences in a sample of objects. Those parameters are, for example, the rotational period, the cycle period and the intensity of the activity as a measure of the field strength of the excited magnetic field (Baliunas 1985, Schrijver 1991). It may so be possible in future to differentiate between different nonlinear models on a broad basis of observational data.

CONCLUSIONS

This is a try to present an insight in the development of the theory of dynamos in stars. It appears to be in a large part a well-ordered deductive theory, however, the more nonlinear processes are dominating and, consequently, nonlinear theory is needed for an appropriate description the picture becomes rather entangled. It is, indeed, the actual problem to build a passable street in the jungle of nonlinear phenomena by a cooperation of observation and theory.

REFERENCES

Baliunas, S. 1985, *Ann. Rev. Astro. Astrophys.*, **23**, 379

Belvedere, G., Proctor, M.R.E., Lanzafames, G. 1991, *Nature*, **350**

Brandenburg, A., Tuominen, I. 1991, *The Sun and Cool Stars: activity, magnetism, dynamos*, eds. I. Tuominen, D. Moss, G. Rüdiger, Springer-Verlag, 223

Brandenburg, A., Krause, F., Meinel, R., Moss, D., Tuominen, I., 1989a, *Astron. Astrophys.*, **213**, 411

Brandenburg, A., Tuominen, I., Moss, D. 1989b, *Geophys. Astrophys. Fluid Dynamics*, **49**, 129

Brandenburg, A., Krause, F., Tuominen, I. 1989c, in *Turbulence and Nonlinear Dynamics in MHD Flows*, eds. M. Meneguzzi et al., Elsevier Science Publ., 35

Brandenburg, A., Moss, D., Rüdiger, G., Tuominen, I. 1990a, *Solar Physics*, **128**, 243

Brandenburg, A., Nordlund, A., Pulkkinen, P., Stein, R.T., Tuominen, I. 1990b, *Astron. Astrophys.*, **232**, 277

Chan, K.L., Sofia, S. 1986, *Astrophys. J.*, **307**, 222

Chan, K.L., Sofia, S., Wolff, C.L. 1982,*Astrophys. J*, **263**, 935

Gilman, P.A. 1972, *Solar Physics*, **27**, 3

Gilman, P.A. 1983, *Astrophys. J. Suppl. Ser.*, **46**, 211

Glatzmaier, G.A. 1984, *J. Comput. Phys.*, **55**, 461

Glatzmaier, G.A. 1985a, *Astrophys. J.*, **291**, 300

Glatzmaier, G.A. 1985b, *Geophys. Astrophys. Fluid Dynamics*, **31**, 137

Harvey, J. 1988, Proc. *Symp. Seismology of the Sun and Sun-like Stars*, ESA SP-286, 55

Jennings, R.L. 1991, *Geophys. Astrophys. Fluid Dynamics*, **57**, 147

Jennings, R.L., Weiss, N.O. 1991, *Mon. Not. R. astr. Soc.*, **252**, 249

Jetsu, L., Huovelin, J., Tuominen, I., Vilhu, O., Bopp, B.W., Piirola, V. 1989, *Astron. Astrophys.*, **236**, 423

Krause, F. 1971, *Astron. Nachr.*, **293**, 187

Krause, F. 1983, in *Stellar and Planetary Magnetism*, ed. A. Soward, Gordon and Breach Sc. Publ.

Krause, F. 1991, in *The Sun and Cool Stars: activity, magnetism, dynamos*, eds. I.Tuominen, D. Moss, G. Rüdiger, Springer-Verlag, 1

Krause, F., Rädler, K.-H. 1980, *Mean-Field Magnetohydrodynamics and Dynamo Theory*, Akademie Verlag Berlin, Pergamon Press Oxford

Krause, F., Meinel, R. 1988, *Geophys. Astrophys. Fluid Dynamics*, **43**, 95

Larmor, J. 1919, *Rep. Brit. Assoc. Adv. Sc.*, **1919**, 159

Libbrecht, K.J. 1988, Proc. *Symp. Seismology of the Sun and Sun-like Stars*, ESA SP-286, 131

Moffatt, H.K. 1978, *Magnetic Field Generation in Electrically Conducting Fluids*, Cambridge University Press

Moss, D. 1990, *Mon. Not. R. astr. Soc.*, **243**, 537

Moss, D. 1991, in *The Sun and Cool Stars: Activity, Magnetism, Dynamos*, eds. I. Tuominen, D. Moss, G. Rüdiger, Springer-Verlag, 112

Moss, D., Tuominen, I., Brandenburg, A. 1990, *Astron. Astrophys.*, **240**, 142

Moss, D., Tuominen, I., Brandenburg, A. 1991, *Astron. Astrophys.*, **245**, 129

Nordlund, A., Stein, R.F. 1989, in *Solar and Stellar Granulation*, eds. R. Rutten and G. Severino, Kluwer Acad. Publ.

Oetken, L. 1977, *Astron. Nachr.*, **298**, 197

Oetken, L. 1979, *Astron. Nachr.*, **300**, 1

Piskunov, N.E., Tuominen, I., Vilhu, O. 1990, *Astron. Astrophys.*, **230**, 363

Rädler, K.-H. 1980, *Astron. Nachr.*, **301**, 101

Rädler, K.-H. 1983, in *Stellar and Planetary Magnetism*, ed. A. Soward, Gordon and Beach Sc. Publ., 37

Rädler, K.-H. 1986, *Astron. Nachr.*, **307**, 89

Rädler, K.-H., Wiedemann, E., Brandenburg, A., Meinel, R., Tuominen, I. 1990, *Astron. Astrophys.*, **239**, 413

Roberts, P.H. 1972, *Phil. Trans. Roy. Soc.* , **A 272**, 663

Roberts, P.H., Stix, M. 1972, *Astron. Astrophys.*, **18**, 453

Rüdiger, G. 1980, *Astron. Nachr.*, **301**, 181

Rüdiger, G. 1989, *Differential Rotation and Stellar Convection*, Akademie Verlag Berlin, Gordon and Breach

Schrijver, C.J. 1991, in *Reviews in Modern Astronomy*, **4**, ed. G. Klare, Springer-Verlag, 18

Steenbeck, M., Krause, F. 1969, *Astron. Nachr.*, **291**, 49

Stix, M. 1971, *Astron. Astrophys.*, **13**, 203

Yoshimura, H. 1975a, *Astrophys. J.*, **201**, 740

Yoshimura, H. 1975b, *Astrophys. J. Suppl. Ser.*, **294**, 467

Zeldovich, Ya. B., Ruzmaikin, A.A., Sokoloff, D.D. 1983, *Magnetic Fields in Astrophysics*, Gordon and Beach Sc. Publ.

Inside the Stars, IAU Colloquium 137
ASP Conference Series, Vol. 40, 1993
Werner W. Weiss and Annie Baglin (eds.)

OBSERVATIONAL CONSTRAINTS ON DYNAMOS IN COOL STARS

C.J. SCHRIJVER
Astronomical Institute Utrecht, P.O. Box 80000, Utrecht, The Netherlands

ABSTRACT Non–radiatively heated outer atmospheres are common among stars on the cool side of the HR diagram. These atmospheres resemble that of the Sun in which the bulk of the heating is associated with magnetic fields. Consequently it is assumed that other cool stars also generate and maintain an internal magnetic field through a dynamo action. This field gives rise to a wide spectrum of phenomena in the stellar atmosphere. The dynamo process that sustains the field thrives on the interaction of rotation and turbulent convection. In this review I formulate a set of propositions outlining our present knowledge of this interaction as distilled from stellar observations. In doing so, I outline the effects of stellar evolution on rotation rate, discuss some of the proposed "dividing lines" in the HR–diagram, and comment on the possible role of the convective turnover time in dynamo efficiency and on dynamo activity for extremely slowly and rapidly rotating stars.

I. INTRODUCTION

The highly structured solar atmosphere has fascinated astronomers from the moment we were aware of its existence. The classical, spherically–symmetric stellar models that were constructed in the first half of this century explained why, given the mass, the solar radius and luminosity have the observed values. The filamentary nature of the clearly non–radiatively heated chromosphere and corona, however, is still far from fully understood (see, for instance, proceedings edited by Ulmschneider *et al.* 1991). We do know that the solar magnetic field enables subphotospheric mechanical energy to reach the supraphotospheric plasma trapped in that field. The dissipation of that energy heats the plasma to temperatures in excess of a million Kelvin. It can be no surprise that similar processes occur in the atmospheres of stars other than the Sun. This has opened the way for testing ideas about the origin of the magnetic field, its structure, and its effects in stellar atmospheres: the dependence of activity on fundamental stellar parameters such as age, mass and radius can be analysed, while relationships with other parameters such as rotation rate and the properties of binary companions can also be studied.

The combination of solar and stellar observational material has deepened our understanding of the phenomenon that observers refer to as "atmospheric activity" while it has at the same time stimulated the development of a "dynamo theory" for the behaviour of magnetic fields in stellar interiors. In this review I summarize a selection of presently available constraints concerning stellar

Table 1. Observational constraints on dynamo theory

1 A dynamo resulting in solar-like magnetic activity appears to operate in all rotating stars with a convection zone directly beneath the photosphere.
2 Dynamo efficiency is directly dependent on the stellar rotation rate (and only indirectly on stellar age); there is no unequivocal evidence supporting a simple scaling of the dynamo efficiency with the Rossby number.
3 The atmospheric radiative losses appear to "saturate" for stars with rotation periods below about one or two days, but it remains unclear whether this reflects a saturation in the dynamo action or merely in the heating mechanism.
4 There is some evidence that the atmospheric radiative losses associated with magnetic activity observed for main–sequence stars of a given rotation period – and by inference also the dynamo efficiency – increase with increasing depth of the convective envelope from the onset of convection up to about $B - V \approx 0.45$.
5 Dynamo action is accompanied by an effective loss of stellar angular momentum, which in all stars up to and including luminosity class III is an important process in the deceleration of rotation in addition to that expected by the structural changes in radius and internal mass distribution.
6 The dividing line in the HR diagram which marks the onset of granulation – and thus of convection – in the stellar atmosphere appears to be the only such line of a fundamental physical nature where the stellar dynamo is concerned; other dividing lines delineate gradual changes (rather than fundamental ones) resulting from mass loss through the stellar magnetosphere and from changes associated with stellar evolution.
7 The presence of a companion star enhances the dynamo action in cool components of relatively close binaries not only by enforcing rapid rotation through tidal interaction, but presumably also by affecting the pattern of differential rotation resulting in an activity level well in excess of that expected from a single star of the same rotation period.
8 Activity cycles tend to increase in duration with decreasing effective temperature, while the cycle period is not a function of either rotation period or convective turnover time only.

dynamos. I do this by formulating a set of qualitative propositions (listed in Table 1) each of which is accompanied by a brief description of the evidence on which the proposition is based. The topics range from the delineation of the parameter range in which solar–like magnetic activity occurs, via the relationship between rotation and activity, to the effects of evolution and binarity.

Statements about solar–like magnetic activity in stellar interiors and atmospheres, i.e. about stellar dynamos, rely on the uniqueness of the interpretation of the diagnostics used to observe and measure stellar activity. I refer to reviews by Strassmeier (these proceedings) and Schrijver (1991) for a discussion of activity diagnostics, and concentrate on the conclusions derived from these diagnostics. First of all I define what I mean by "solar–like" in this context by formulating the quintessence of solar activity: solar activity is characterized by the existence of small (compared to the stellar surface area), long–lived (compared to photospheric convective time scales), highly structured, non–thermally heated regions above the photosphere, with plasma temperatures in the outer atmosphere reaching up to 1MK or higher. Stellar observations of spectral lines used to observe solar activity suggest solar–like activity to occur in many other stars. Evidence has even been found for the stellar equivalent of sunspots and for stellar activity cycles. Solar atmospheric activity is clearly magnetic in nature because of the close association of bright atmospheric regions with bipolar magnetic plages in the photosphere, the general appearance of bright atmospheric arches resembling the geometry of the magnetic field computed using the photospheric signal as boundary condition, and the confinement of extremely hot coronal plasma. By analogy with the Sun it is assumed that stellar activity is also magnetic in nature. Only in recent years has it become possible to measure stellar magnetic field strengths and surface filling factors more or less directly and to demonstrate the existence of a relationship between the mean value of the absolute magnetic surface flux density and atmospheric radiative losses (Saar 1991 and references therein). Schrijver (1991) combines solar and stellar data to obtain a calibration of the atmospheric radiative losses in terms of magnetic flux densities.

Knowing the relationship between atmospheric radiative losses and magnetic flux densities one can use any measure A_m of the atmospheric radiative losses (e.g. luminosity, surface flux density, or flux normalized to the bolometric flux) – specified where necessary – as an equivalent for the level of magnetic activity. Acoustic heating processes in stars with convective envelopes also give rise to substantial chromospheric and some transition–region emission (e.g. Schrijver 1987, Rutten *et al.* 1991), showing up in diagrams of observed atmospheric radiative surface fluxes versus spectral colour (flux–colour diagrams) as an empirical lower limit. This non–magnetic *basal* contribution A_b must be subtracted from observed radiative losses to obtain A_m.

II. CONVECTION AND ACTIVITY

The large body of observational material now available allows the formulation of proposition number 1: "A dynamo resulting in solar-like magnetic activity appears to operate in all rotating stars with a convection zone directly beneath the photosphere." The region in the HR–diagram in which stars exhibit solar–like activity consequently lies redward of the line at which a convection zone develops immediately beneath the stellar photosphere (cf. Fig. 1).

Although proposition 1 appears trivial in the light of current knowledge, it does put several significant constraints on dynamo theory. First of all it implies that dynamo action occurs even in stars with very shallow convective envelopes (further discussed in Section V). Secondly, it implies that dynamos also operate in the *fully* convective M–type dwarf stars (e.g. Giampapa and Liebert 1986). Hence, the existence of a discontinuity at the interface of the radiative interior and the convective envelope may be an important facet of dynamo theory, but its absence does not preclude dynamo action. Little is known presently about the dependence of activity on rotation for these very cool stars, although Giampapa and Liebert (1986) note that non–emission line late M–type dwarfs rotate slowly as compared to emission–line stars. Indications exist that the structure of the outer atmosphere of dwarf stars cooler than about M0 differs substantially from that in F–, G–, and K–type stars, particularly where the lower chromosphere is concerned (Rutten *et al.* 1989).

Activity also occurs in the young T Tauri stars (see, for instance, the review by Basri 1987a). It appears that when stars become visible through

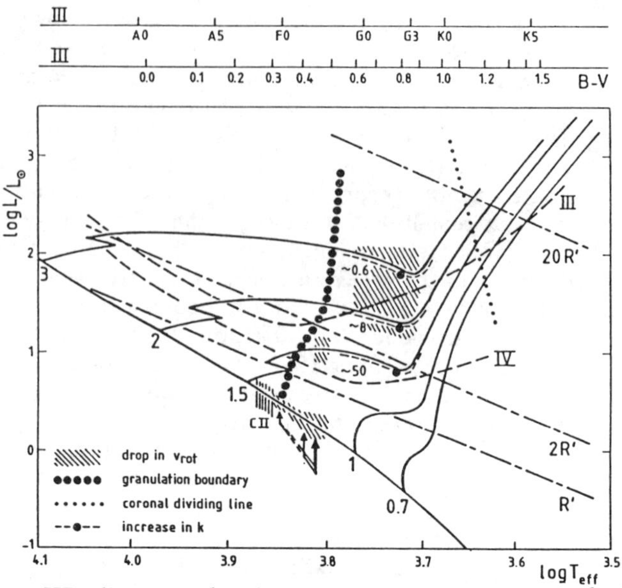

Fig. 1 HR diagram showing post main–sequence evolution (modified after Rutten and Pylyser (1988)). Stellar masses (in solar units) are given along the main sequence. Luminosity classes are delineated by long–dashed curves. The approximate position of the "granulation boundary" where the reversal in the asymmetry of photospheric lines is used as an approximate delineation of the onset of convection is marked by heavy dots (after Gray and Nagel 1989). For main–sequence stars the drop in rotational velocities (from Gray 1991) sets in close to the granulation boundary. For dwarf stars blueward of $B-V \approx 0.45$ (marked by a large arrow) the dynamo efficiency has been suggested to decrease down to $B-V \approx 0.40$ (medium–sized arrow), and possibly even down to $B-V \approx 0.3$ (small arrow; cf. Section V). Between $B-V = 0.25$ and 0.29 the chromospheric C II emission, presumably reflecting largely acoustic heating in that range, increases rapidly in strength (Simon and Landsman 1991). The rotational velocities of evolving stars drop markedly within the dashed regions (cf. Rutten and Pylyser 1988). For giants of about 2 to 3 solar masses this decrease coincides with an increase in the moment of inertia associated with an internal redistribution of mass (causing a change in the gyration radius kR without a substantial change in the stellar radius R), occurring in the intervals marked by dashed lines along the tracks (the point of steepest change in k as a function of $T_{\rm eff}$ is marked by a dot on the dashed lines). For the less massive subgiants the drop in velocities occurs near or immediately beyond the onset of surface convection. In all cases the rotation rates have dropped substantially prior to climbing the red giant branch preceding core–helium ignition. The numbers close to the evolutionary tracks give the approximate time intervals (in Myr) in which a star evolves from the granulation boundary to the upturn onto the giant branch (data from Maeder and Meynet 1988). Stars reach the coronal dividing line (dotted; from Linsky 1992) – first proposed by Linsky and Haisch (1979) – after about a fifteenfold increase in radius since leaving the ZAMS (lines of constant radius relative to an arbitrarily chosen R' are dashed–dotted).

the surrounding primordial cloud, they already exhibit very strong magnetic activity. This emission decreases rapidly as the stars age (cf. Section IV), but their chromospheric and coronal fluxes are indistinguishable from those of most active main–sequence and evolved stars (e.g. Walter *et al.* 1988). The picture is simple only for the so–called "naked T-Tauri stars" (cf. Walter 1987), however, because in other T Tau stars the emission from circumstellar material arranged in bipolar outflows and equatorial disks contributes strongly to, or even dominates, the stellar emission proper (e.g. Lemmens *et al.* 1992).

III. DYNAMO STRENGTH AND ROTATION

Rotation rate and activity are clearly related: when comparing any measure of stellar magnetic activity A_m with, for instance, rotation period P_{rot} for stars of similar colour (Fig. 2a), one sees that A_m decreases monotonically with increasing P_{rot} for stars with $P_{rot} \gtrsim 1$ day (cf. Section IV). That this is a direct physical relationship in which age plays a role only through the magnetic braking of stars (see Section VI) was already implied by Skumanich (1972) and further discussed by Middelkoop and Zwaan (1981), who pointed out that while old, single stars are inactive, old binaries that are tidally synchronized are very active (cf. Section VIII).

Proposition 1 implies that a dynamo process operates *whenever* a star rotates, thereby excluding the possibility that there is some minimally required rate of rotation to excite the dynamo. This conclusion is, admittedly, crucially dependent on the accuracy and validity of extrapolations of the relationship between rotation and activity. Schrijver *et al.* (1989), for instance, argue that extrapolation of this relationship for dwarfs suggests that stellar rotation vanishes at or very near the empirical lower–limit flux in a flux–colour diagram. In their case the main activity diagnostic was the Mt. Wilson Ca II H+K flux, in which the lower–limit flux incorporates both the basal line–core component, probably measuring purely acoustic heating, and a line–wing component. They also pointed out that the long–term Ca II H+K variability as well as the coronal soft X–ray emission disappear at or very near this lower–limit flux. Hence they conclude that it is unlikely that a solar–like cyclic dynamo operates in stars at the level of the Ca II H+K lower limit (in agreement with the explanation of basal heating as acoustic in nature, although the convection zone of a non–rotating star may in principle be capable of generating a weak background magnetic field which plays a role in the basal heating process), and conversely that activity A_m increases smoothly and monotonically with decreasing rotation period P_{rot} (at least for $P_{rot} \gtrsim 1$ day, see Section IV). Verifying whether there is a minimally required non–zero rotation velocity v_c for a dynamo to operate is more difficult for single giants, because of their generally low $v\,sin(i)$ values lying near the detection limit. Observed $v\,sin(i)$ values set an upper limit to v_c of about 1–2 km/s (Gray 1989a). Flux–colour diagrams also contain valuable information: while for dwarf stars one might argue that they may not spin down sufficiently to reach v_c, evolving single giant stars continue to slow down as long as their moment of inertia increases (cf. Section VI). Thus we can conclude that if the activity would discontinuously change from $A_m(v_c + \epsilon, T_{eff}, \ldots) > 0$ to $A_m(v_c - \epsilon, T_{eff}, \ldots) = 0$ (for small ϵ), the evolution of the moment of inertia should yield stars on the cool side of flux–colour diagrams at the basal level, separated from the "active" stars by a gap. Such a gap is not observed so that if $v_c \neq 0$, $A_m(v_c + \epsilon, T_{eff}, \ldots)$ is at most small compared to A_b.

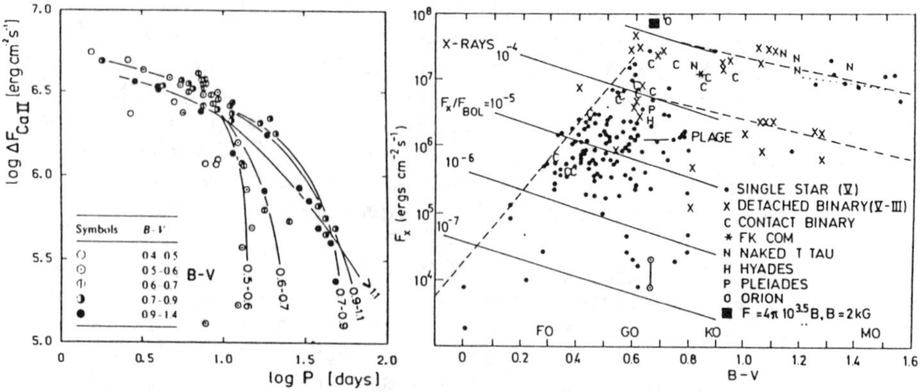

Fig. 2 a) (Left) The chromospheric Ca II H+K excess flux density, measuring magnetic activity, versus the stellar surface rotation period, for dwarfs in different colour intervals (from Rutten 1987). b) (Right) The soft X–ray surface flux density (0.2 – 4 keV) vs. colour (after Vilhu 1986).

Theoretical work of Durney and Latour (1978) suggested that there should be a relationship between activity A_m and the ratio R_o of the rotation period P_{rot} and the so–called convective–turnover time τ_c (the ratio of a typical length scale and velocity of the convection). This ratio R_o is known as the Rossby number which measures the influence of Coriolis forces on convective flows. Moreover, Noyes *et al.* (1984a) showed that R_o is also related to the dynamo number N_D (cf. Parker 1979): $N_D \simeq R_o^{-2}$. Noyes (1983) and Noyes *et al.* (1984a) argued that the observational material might support such a relationship between A_m and R_o. They find the best relationship for their sample of dwarf stars by using the ratio of chromospheric Ca II H+K flux to the bolometric flux as activity measure and the Rossby number P_{rot}/τ_c computed for the bottom of the convective envelope for a mixing length of 1.9 pressure scale heights for stars with $B - V \in [0.5, 0.9]$, while *assuming* a constant turnover time for cooler stars (an assumption that is too crude, see Gilliland 1985a).

The argumentation supporting the role of τ_c in rotation–activity relationships stands or falls with the independence of the shape of the rotation–activity relationship on spectral type. Rutten and Schrijver (1987) argue that the data for dwarfs suggest that the shape of the rotation–activity relationship is in fact colour dependent (Fig. 2a), so that no colour–dependent scaling of either rotation period or activity level, or both, can yield a simple, colour–independent relationship between rotation rate and atmospheric radiative losses. In parallel studies, Basri (1986) and Simon and Fekel (1987) pointed out that no advantage is gained by the use of the Rossby number over the comparison of Ca II H+K surface flux density and rotation period.

The defence for the relevance of a τ_c–scaling becomes even more problematic when giants are included: while Rutten (1987) argues that dwarfs and *single* giants of the same colour and rotation period are equally active, the values of τ_c for evolved stars differ from those of dwarfs (Gilliland 1985a,b, but see Basri 1987b for comments on the effective temperature scale used by Gilliland) so that a scaling with τ_c would in fact disrupt relationships. Basri (1987b), instead, noted a slight improvement of the overlap of dwarf and giant data when using

R_o, but his data included evolved binaries which appear to behave differently from single stars (cf. Section VIII). Moreover, he notes that the relationships for each of the subclasses is somewhat worse when using Rossby numbers.

These results can at best be summarized by proposition 2: "Dynamo efficiency is directly dependent on the stellar rotation rate (and only indirectly on stellar age); there is no unequivocal evidence supporting a simple scaling of the dynamo efficiency with the Rossby number." Apparently the relationship between rotation and activity in single stars involves at least the rotation rate and a basic stellar parameter such as the effective temperature (possibly indirectly related to τ_c). The functional dependence on stellar effective temperature does not allow for a simple dependence on the dimensionless Rossby number, thereby ruling out a single–parameter description $A_m(R_o)$, nor does it allow the rotation–activity relationship to be written as a product of functions of rotation rate and of basic stellar parameters only.

IV RAPIDLY ROTATING STARS

Flux–colour diagrams such as in Fig. 2b and rotation–activity diagrams including young, rapidly rotating dwarf stars support proposition 3: "The atmospheric radiative losses appear to "saturate" for stars with rotation periods below about one or two days, but it remains unclear whether this reflects a saturation in the dynamo action or merely in the heating mechanism." Radiative losses in the short–period W UMa binaries in fact appear to decrease with increasing rotation rate for rotation periods below about one day (e.g. Rutten and Schrijver 1987).

Vilhu (1987) and Vilhu *et al.* (1989) point out that the subphotospheric mechanical fluxes F_m, estimated from a mixing length model to be $\approx 1-5\%$ of the bolometric luminosity at one pressure scale height beneath the photosphere, lie between a factor of only 3 to 30 (increasing with effective temperature) above the estimated radiative losses from the outer atmosphere for stars at "saturation." They argue that F_m limits the observed radiative losses: even if the surface filling factor of the magnetic flux (close to unity for these very active stars, e.g. Saar 1991) could still increase, it could not channel more energy from (below) the photosphere into the atmosphere.

Indirect support for an actual "saturation" of the dynamo activity and not just of the atmospheric radiative losses comes from the relationship between stellar age and activity for single main–sequence stars. Magnetic activity appears to be ubuiquitously accompanied by loss of stellar angular momentum (see Section VI). This results, for instance, in a rapid decrease and convergence of rotational velocities between the ages of the α Persei and Hyades clusters. MacGregor and Brenner (1991) study the rotational evolution of solar–type stars on the main sequence by modelling stars as consisting of two rigidly rotating parts, the radiative interior and the convective envelope, between which angular momentum is transferred on a time scale τ_{ce}. Angular momentum loss is assumed to occur on a time scale $\tau_J = 3/2\, I_{conv}/(r_A^2 \dot{M})$, (with I_{conv} the moment of inertia of the convective envelope, r_A the Alfvén radius, and $\dot{M}$ the mass–loss rate) consistent with a monopolar geometry for the poloidal magnetic field component. They find good agreement with the observed rotation rates for solar–type stars in the young α Per, Pleiades, and Hyades clusters and present surface and core rotation rates of the Sun – which differ little according to helioseismological observations, with a possible exception for the innermost regions – for a time–independent core–envelope coupling time $\tau_{ce} \approx 10$ Myr – about equal to the initial spin–down time – and a mean magnetic flux density that is linearly dependent on $1/P_{rot}$ for slow rotators ($P_{rot} \gtrsim 3$ days), but independent of the surface rotation period P_{rot} for more rapid rotators, suggesting a saturation of the dynamo.

V. THE ONSET OF ACTIVITY

The fourth proposition addresses the onset of activity at the boundary between stars with and without convective envelopes in the HR–diagram (cf. Fig. 1): "There is some evidence that the atmospheric radiative losses associated with magnetic activity observed for main–sequence stars of a given rotation period – and by inference also the dynamo efficiency – increase with increasing depth of the convective envelope from the onset of convection up to about $B-V \approx 0.45$." Indirect evidence for the activation of a weak dynamo immediately blueward of the granulation boundary at about F0 on the main sequence is found in the decrease of the mean rotational velocities from that point onward (e.g. Gray 1989b), while in hotter stars the mean rotational velocity is a colour–independent fraction of the rotational breakup velocity (Kawaler 1987). Further indirect evidence is found in the increase of the range in X-ray emissions observed to occur from F0 onward (cf. Fig. 2b). Direct evidence supporting proposition 4 so far comes from a single paper by Walter and Schrijver (1987), which has been contested recently by Simon and Landsman (1991; see also Simon and Drake 1989). Simon and Landsman use the upper chromospheric C II emission lines to measure stellar activity. They argue that the C II emission of stars earlier than spectral type F5 ($B-V \approx 0.45$) is uncorrelated with rotation. A rotation–activity relationship in these stars is hard to establish accurately, because of the relatively strong contribution of the non–magnetic basal emission A_b, which Simon and Landsman do not correct for. Proper correction of the C II fluxes in their sample for the relatively strong basal flux suggests that there is indeed a relationship between rotation and activity down to at least $B-V \approx 0.40$, but that the emission level is suppressed as compared to stars of later spectral type at a given rotation period. Substantiating the claim by Walter and Schrijver (1987) that the rotation–activity relationship extends to even hotter stars requires a more extensive sample with accurate flux measurements.

Until such an extended study is completed, the hypothesis formulated by Simon and Drake (1989) and Simon and Landsman (1991) that the apparent deficiency in soft X-rays as compared to the C II and C IV emission for late–A and early–F type stars signals a strong coronal wind also remains unproven: subtraction of the relatively strong basal chromospheric flux appears to place F–type stars on or very close to the relationship between soft X–rays and C II emission defined by cooler stars, suggesting a similar atmospheric structure.

For single stars later than mid-F the dynamo efficiency appears to be independent of spectral type for stars rotating faster than about 10 days, while a colour–dependence shows up for slower rotators (Fig. 2a).

VI ANGULAR MOMENTUM LOSS BY MAGNETIC BRAKING

The next proposition addresses a problem related to the large–scale geometry of the magnetic field and the flow of plasma along field lines open to the interstellar space: a stellar wind blowing in a magnetosphere exerts a torque on the star, draining its angular momentum and thus applying a brake to the stellar rotation. Specifically: "Dynamo action is accompanied by an effective loss of stellar angular momentum, which in all stars up to and including luminosity class III is an important process in the deceleration of rotation in addition to that expected by the structural changes in radius and internal mass distribution." (Proposition 5) The rotation–activity–age relationship of main–sequence stars was touched upon in Sections IV and V, while the fundamental importance of magnetic braking for binaries is discussed in Section VIII. Here I concentrate on the angular momentum loss of stars with masses up to about $3M_\odot$ evolving off the main sequence towards the red giant branch (the strong activity of more massive stars in the giant phase is discussed by, e.g., Pasquini *et al.* 1990).

Main–sequence stars with convective envelopes (i.e. for $M \lesssim 1.3M_\odot$) lose a substantial fraction of their angular momentum during their main–sequence phase (e.g. Soderblom *et al.* 1991). More massive stars, in contrast, do not develop a solar–like dynamo untill they cross the "granulation boundary" (Fig. 1), and hence retain (most of) their angular momentum until that phase. To what extent the decrease in rotation rate after the onset of convection is affected by angular momentum loss in addition to evolutionary changes in the stellar structure depends on the time scale of the angular momentum loss relative to the time scale of stellar evolution. What is known about this?

Signatures of photospheric granulation first occur around spectral type F5 for luminosity class III stars evolving through the HR–diagram, but there is no substantial drop in rotational velocities up to about G0. Within a few subclasses in spectral type the rotational velocities then decrease by typically over an order of magnitude. Rutten and Pylyser (1988) demonstrate that, under the assumption that stars rotates rigidly, the observed change in the mean rotational velocity of giants as a function of spectral colour is in fair agreement with that expected from the change in moment of inertia, I. From this they argue that the change in I is the dominant process in slowing down these stars. Note that the increase in the radius of gyration kR which results in the drop in rotational velocities is caused by the increase in the constant k with no substantial change in radius R in that part of the evolution (cf. Fig. 1).

The drop in rotational velocities is even more pronounced for the less massive subgiants, occuring not too much blueward from the onset of convection (cf. Fig. 1). Rutten and Pylyser (1988) argue that magnetic braking is the dominant factor in slowing down these stars, because the drop occurs well before a significant change in I. Moreover the brake has a much longer time to affect these stars: subgiants of $1.5M_\odot$ take some fifty million years to pass from the granulation boundary to the red giant branch, which is much longer than the roughly one and ten million years it takes $3M_\odot$ or $2M_\odot$ stars, respectively. Note that if the core-envelope synchronization time scale of $\tau_{ce} \approx 10$Myr found for the main–sequence rotational evolution (cf. Section IV) is of the proper order of magnitude for these stars, then the evolution of subgiants may be slow enough for the entire star to be spun down and not just the convective envelope, while $2M_\odot$ stars may only lose angular momentum from the convective envelope. Whether or not this is the case, shortly after the (sub)giants begin their ascent of the red giant branch, the convective envelope reaches down to its deepest point, encompassing about three quarters of the stellar mass (e.g. Maeder and Meynet 1989; see also Böhm–Vitense (1992) on the observed increase in the N/C abundance ratio which is caused by the deepening of the convection zone, and which occurs *after* the drop in $v\,sin(i)$) so that by then the bulk of the star exhibits at most some form of differential rotation. Consequently, dredging up angular momentum from a possibly rapidly spinning stellar core is unlikely to spin up stars as they move up along the red giant branch. The assumption of rigid rotation made by Rutten and Pylyser (1988) may change the precise way in which giants spin down prior to the red giant branch, but the magnitude of the rotation they compute for stars at the upturn onto the giant branch sets an upper limit to actual rotation rates (cf. Section VII).

Gray (e.g. 1989a, 1991) argues that the drop in $v\,sin(i)$ for giants is too fast to be compatible with only structural evolutionary changes as computed by Rutten and Pylyser. I believe this criticism is largely based on the use by Gray of the intrinsically inaccurate and somewhat ambiguously plotted scale indicating spectral types rather than the colour scale in the key figure of Rutten and Pylyser (their Fig. 2a). Using Gray's (1989b) calibration of colour and spectral type, the computed drop in velocities indeed occurs within a few spectral subclasses (cf. Fig. 1) for the giants, in better agreement with Gray's data.

Gray (1989a) presents a compelling argument, however, in favour of the role of magnetic braking for LC III stars, which is based on the distribution function, $H(v_{\rm rot}, B-V)$, of observed rotational velocities. Stars of sufficiently low masses evolve away from the main sequence towards the red giant branch moving monotonically towards later spectral type. Hence the observed population of such stars at any colour stems from a similar popolation at higher effective temperatures. If no changes in the initial $H(v_{\rm rot}, B-V)$ and orientations of rotation axes occur as a function of time, and provided that selection effects associated with the intrinsically low rotational velocities of cool giants can be neglected, the behaviour of an evolving star can be studied by comparing properties of stars at different colours. In this respect Gray notes that $H(v_{\rm rot}, B-V)$ changes in shape as a function of $B-V$ from what can be approximated by a Maxwell–Boltzmann distribution for stars blueward of the granulation boundary to one with a reversed asymmetry in which the rotational velocity at which the peak in $H(v_{\rm rot}, B-V)$ occurs lies at higher values than the mean velocity for stars redward of the granulation boundary. If the changing moment of inertia I were the only effect to play a role, then this would reduce the rotational velocities in time, but $H(v_{\rm rot}, B-V)$ would not change its relative shape. Consequently, Gray argues, magnetic braking must be important. In fact he goes one step beyond this: he notes that the observed $H(v_{\rm rot}, B-V)$ for stars beyond the strong drop in velocities is compatible with all stars of a given spectral type having the same rotation velocity, with different inclination angles of the axes of rotation and some obervational scatter smearing it to the observed curve. On this he bases his rotostat mechanism (discussed below).

Gray's interpretation is, however, not unique. A similar change in the shape of $H(v_{\rm rot}, B-V)$ could be brought about by a differential braking with $|\dot{P}_{\rm rot}/P_{\rm rot}|$ decreasing with increasing $P_{\rm rot}$ in addition to evolutionary changes. This would result in a non-linear transformation of initial velocities before the drop in velocity into velocities beyond it. One can show that for the observed distribution functions the transformation can be approximated by a power law relating the velocities at time t_1, when the stars lie between F0 and G0, to those at time t_2, when the stars lie between G3 and K2 (two intervals, used by Gray 1989a, on either side of the marked decrease in rotation rate): $v(t_2) = \alpha v(t_1)^{\beta}$, with $\beta \approx 0.56$.

Can the change in shape of $H(v_{rot}, B-V)$ across the "rotational divide" mask a dependence on stellar mass, perhaps in addition to a dependence of the braking time scale on rotation rate? There is in fact a weak mass dependence in velocities blueward of the "rotation dividing line": less massive stars rotate on average less rapidly than more massive stars. Reliable distance estimates for single giants are rare, particularly for the stars redward of the divide, so that the consequences of mass–dependent effects cannot be explored further at present. On heuristic arguments, however, one may expect less massive stars to brake more strongly because of their smaller moment of inertia and their slower evolution compared to the heavier stars, while more massive stars slow down a little more than their less massive counterparts because of a somewhat larger increase in radius. Hence, the mass–dependent effects may largely cancel and thus not affect the shape of $H(v_{rot}, B-V)$ for giants too much across the "rotational divide," leaving Gray's argument favouring magnetic braking valid.

Gray argues that all giants of a given colour rotate equally fast beyond the drop in velocities (the "rotation dividing line"). He suggests the fundamental cause of this to be the so–called "rotostat" mechanism in which magnetic activity supposedly disappears once the rotation velocity drops below a colour–dependent minimum. This implies that there should be a non–zero rotation velocity $v_c(B-V)$ at which chromospheric emissions such as the Ca II H+K flux density should not differ substantially from the minimal basal value. This effect is hard to prove or disprove, because the level of the basal flux density is not known with sufficient accuracy, but with the above arguments it is clear that Gray's arguments concerning the rotational evolution of giants and concerning the rotostat certainly do not result in a unique interpretation of the data for giant stars. The alternative interpretation, combining evolutionary effects with activity–dependent magnetic braking, is equally compatible with the data. The latter moreover yields a fair prediction of the observed decrease of rotational velocities throughout the evolution of giants. Additional evidence against the idea of a minimally required velocity to start the dynamo in dwarf stars as well as in giant stars was discussed in Section III.

VII DIVIDING LINES

Several so–called "dividing lines" have been suggested as boundaries in the HR–diagram between subgroups of magnetically active stars. The "rotation dividing line" proposed by Gray (1991, and references therein) most likely delineates the combined effect of magnetic braking and evolutionary changes, as discussed in Section VI, rather than a fundamental change in the dynamo. Another well–known boundary is the "coronal dividing line" proposed by Linsky and Haisch (1979). The coronal emission from single stars redward of the dividing line is very weak (Haisch *et al.* 1991, for instance, quote an upper limit of L_X of 3 10^{25} erg/sec for Arcturus from ROSAT observations, or less than 0.7 ergs cm^{-2} s^{-1}, cf. Fig. 2b). The suggested causes for this dividing line include destabilization of coronal loops in the reduced gravity (e.g., Antiochos 1987) and a change in the field geometry allowing massive winds but no strong coronal emission. Recently, however, Haisch *et al.* (1990, 1991) argue that a decrease in rotation is responsible for the dividing line. Note that stars of masses between 0.7$M_\odot$ and 3$M_\odot$ experience about a fifteenfold increase in radius as compared to their ZAMS values upon reaching the dividing line (cf. Fig. 1).

The observed chromospheric emission from stars beyond the coronal dividing line appears to be predominantly basal (acoustic) in nature. Judge and Stencel (1991), for instance, show that ultraviolet radiation losses from giants between spectral types K1 III and M5 III (mostly first giant branch stars) decline with effective temperature, with fluxes crowding against the extrapolation of basal fluxes found for Mg II h+k by Schrijver (1987). Later spectral types (mostly second ascent or asymptotic giant branch stars) exhibit a wider range in values which Judge and Stencel suggest is related to the increased variability of these stars. Estimated wind energy fluxes are similar for the sample of first or second ascent stars, somewhat exceeding $\approx 10^{-6} F_{\rm bol}$. Because of the decreasing basal flux with decreasing effective temperature and the more or less constant wind energy flux, the energy in the wind increases in relative importance, exceeding the chromospheric radiative energy losses for stars cooler than about 3000K (around M5 III, i.e. at the edge of the range of Fig. 1).

In conlusion (proposition 6): "The dividing line in the HR diagram which marks the onset of granulation – and thus of convection – in the stellar atmosphere appears to be the only such line of a fundamental physical nature where the stellar dynamo is concerned; other dividing lines delineate gradual changes (rather than fundamental ones) resulting from mass loss through the stellar magnetosphere and from changes associated with stellar evolution."

VIII ACTIVITY IN CLOSE BINARIES

The combination of angular momentum loss in magnetically active stars and tidal interaction in binaries has the interesting result that ageing stars can actually be spun up, increasing their rotational angular momentum at the expense of the orbital angular momentum. As a consequence the orbit becomes tighter (which may eventually lead to coalescence of the components, e.g. Verbunt and Zwaan 1981), the rotation period shorter, and the cool–star component(s) more active. This effect, however, is only part of proposition 7: "The presence of a companion star enhances the dynamo action in cool components of relatively close binaries not only by enforcing rapid rotation through tidal interaction, but presumably also by affecting the pattern of differential rotation resulting in an activity level well in excess of that expected from a single star of the same rotation period." Schrijver and Zwaan (1991) have given this phenomenon, called *overactivity*, an approximate quantification demonstrating the essential role of properties of the companion star in the activity of the other star, but the details of the physical process remain unclear. Their arguments rule out processes such as magnetic interaction between the stars and interacting winds. It appears that the dynamo efficiency in the cool binary components is increased, resulting in a stronger chromospheric and coronal activity which, apart from its high level given the associated rotation period, cannot be distinguished from that of other stars. Hence it is likely that a modified differential rotation pattern is the cause of overactivity in binaries.

IX ACTIVITY CYCLES

Activity cycles are a common, although certainly not ubiquitous phenomenon among cool stars: Baliunas (1986) notes that some 60% of the 99 stars that have been monitored at Mt. Wilson during the past decades undergo (apparently) cyclic variations in the chromospheric Ca II H+K emission. The relatively inactive, slow rotators generally exhibit a smooth solar–like cycle, while more active stars vary more sharply, possible because of the excitation of different dynamo modes or of a different dynamo mechanism (Baliunas and Vaughan 1985). Noyes *et al.* (1984b) present a relationship between the cycle period $P_{\rm cycle}$ and the rotation period $P_{\rm rot}$: $P_{\rm cycle} \approx (P_{\rm rot}/\tau_c)^{1.25}$ for a sample of 13 slowly rotating dwarf stars, while Baliunas (1986) finds no such correlation for the entire sample of dwarfs. Maceroni *et al.* (1991) argue that the cycle period for main–sequence stars increases with decreasing effective temperature: F–, G–, and K–type stars have mean cycle periods of about 5.8 ± 1.2, 7.4 ± 0.9, and 9.7 ± 0.9 years, respectively. Although they argue that stars in completely different rotational regimes (relatively narrow binaries such as the RS CVn's, cataclysmic variables and main–sequence stars) have the same distribution of cycle periods (which would eliminate a dependence of $P_{\rm cycle}$ on τ_c only, because τ_c differs for these stars, cf. Section III), the data samples of other than main sequence stars are too small to test for the existence of a similar colour dependence. Maceroni *et al.* (1990) do not find a one–parameter relationship between $P_{\rm cycle}$ and $P_{\rm rot}$, but their results do not exclude a role of $P_{\rm rot}$ or τ_c in a multi–parameter description. Hence proposition 8: "Activity cycles tend to increase in duration with decreasing effective temperature, while the cycle period is not a function of either rotation period or convective turnover time only."

ACKNOWLEDGEMENTS

I thank R.G.M. Rutten and C. Zwaan for critically reading the manuscript, D.F. Gray and R.G.M. Rutten for their comments on the effects of evolution and activity on stellar rotation, and R.W. Noyes and D.R. Soderblom for comments. This work was supported by a fellowship grant from the Royal Netherlands Academy of Arts and Sciences.

REFERENCES

Antiochos, S.K., 1987, in *Cool Stars, Stellar Systems, and the Sun*, J.L. Linsky and R.E. Stencel (eds.), Springer–Verlag, p. 283.

Baliunas, S.L., 1986, in *Cool Stars, Stellar Systems, and the Sun*, M. Zeilik and D.M. Gibson (eds.), Springer–Verlag, p. 3.

Baliunas, S.L., Vaughan, A.H., 1985, Ann. Rev. Astron. Astrophys. **23**, 379.

Basri, G., 1986, in *Cool Stars, Stellar Systems, and the Sun*, M. Zeilik and D.M. Gibson (eds.), Springer–Verlag, p. 184.

Basri, G., 1987a, in *Cool Stars, Stellar Systems, and the Sun*, J.L. Linsky and R.E. Stencel (eds.), Springer–Verlag, p. 411.

Basri, G., 1987b, Ap. J. **316**, 377.

Böhm–Vitense, E., 1992, Astron. J. **1633**, 608.

Durney, B.R., Latour, J., 1978, Geophys. Ap. Fluid Dyn. **9**, 241.

Giampapa, M.S., Liebert, J., 1986, Ap. J. **305**, 784.

Gilliland, R.L., 1985a, Ap. J. **300**, 339.

Gilliland, R.L., 1985b, Ap. J. **299**, 286.

Gray, D.F., 1989a, Ap. J. **347**, 1021.

Gray, D.F., 1989b, in *FGK Stars and T Tauri Stars*, L.E. Cram and L.V. Kuhi (eds.), NASA SP-502, p.7.

Gray, D.F., 1991, in *Angular Momentum Evolution of Young Stars*, Catalano, S. and Stauffer, J.R. (eds.), Kluwer, The Netherlands, p. 183.

Gray, D.F., Nagel, T., 1989, Ap. J. **341**, 421.

Haisch, B.M., *et al.* , 1990, Ap. J. **361**, 570.

Haisch, B., Schmitt, J.H.M.M., Rosso, C., 1991, Ap. J. **383**, L15.
Hartmann, L., Baliunas, S.L., Duncan, D.K., Noyes, R.W., 1984, Ap. J. **279**, 778.
Judge, P.G., Stencel, R.E., 1991, Ap. J. **371**, 357.
Lemmens, A.F.P., Rutten, R.G.M., Zwaan, C., 1992, A. & A., in press.
Kawaler, S.D., 1987, PASP **99**, 1322.
Linsky, J.L., 1992, in proceedings of symposium *From EINSTEIN to AXAF.*
Linsky, J.L., Haisch, B.M., 1979, Ap. J. **229**, L27.
Maceroni, C., *et al.* , 1990, A. & A. **237**, 395.
MacGregor, K.B., Brenner, M., 1991, Ap. J. **376**, 204.
Maeder, A., Meynet, G., 1988, A. & A. Suppl. **76**, 411.
Maeder, A., Meynet, G., 1989, A. & A. **210**, 155.
Middelkoop, F., Zwaan, C., 1981, A. & A. **101**, 26.
Noyes, R.W., 1983, in *Solar and Stellar Magnetic Fields: Origins and Coronal Effects*, J.O. Stenflo (ed.), Reidel, Dordrecht, p. 133.
Noyes, R.W., *et al.* , 1984a, Ap. J. **279**, 763.
Noyes, R.W., Weiss, N.O., Vaughan, A.H., 1984b, Ap. J. **287**, 764.
Parker, E.N., 1979, *Cosmical Magnetic Fields; Their Origins and Their Activity*, Clarendon Press, Oxford.
Pasquini, L., Brocato, E., Pallavicini, R., 1990, A. & A. **234**, 277.
Rutten, R.G.M., 1987, A. & A. **177**, 131.
Rutten, R.G.M., Schrijver, C.J., 1987, A. & A. **177**, 155.
Rutten, R.G.M., Pylyser, E., 1988, A. & A. **191**, 227.
Rutten, R.G.M., Schrijver, C.J., Lemmens, A.F.P., Zwaan, C., 1991, A. & A. **252**, 203.
Rutten, R.G.M., *et al.* , 1989, A. & A. **219**, 234.
Saar, S.H., 1990, in *The Sun and Cool Stars: activity, magnetism, dynamos*, Tuominen, I., Moss, D., and Rüdiger, G. (eds.), Springer–Verlag, p 389.
Schrijver, C.J., 1987, A. & A. **172**, 111.
Schrijver, C.J., 1991, in *Mechanisms of Chromospheric and Coronal Heating*, P. Ulmschneider, E.R. Priest, R. Rosner (eds.), Springer–Verlag, p. 257.
Schrijver, C.J., Dobson, A.K., Radick, R.R., 1989, Ap. J. **341**, 1035.
Schrijver, C.J., Zwaan. C., 1991, A. & A. **251**, 183.
Skumanich, A., 1972, Ap. J. **171**, 565.
Simon, T., Fekel Jr., F.C., 1987, Ap. J. **316**, 434.
Simon, T., Drake, S.A., 1989, Ap. J. **346**, 303.
Simon, T., Landsman, W., 1991, Ap. J. **380**, 200.
Soderblom, D.R., Duncan, D.K., Johnson, D.R.H., 1991, Ap. J. **375**, 722.
Ulmschneider, P., Priest, E.R., Rosner, R. (eds.), 1991, *Mechanisms of Chromospheric and Coronal Heating*, Springer–Verlag.
Verbunt, F., Zwaan, C., 1981, A. & A. **100**, L7.
Vilhu, O., 1987, in *Cool Stars, Stellar Systems, and the Sun*, J.L. Linsky and R.E. Stencel (eds.), Springer–Verlag Berlin, p 110.
Vilhu, O., *et al.* , 1989, A. & A. **222**, 179.
Walter, F.M., 1987, in *Cool Stars, Stellar Systems, and the Sun*, J.L. Linsky and R.E. Stencel (eds.), Springer–Verlag Berlin, p 422.
Walter, F.M., Schrijver, C.J., 1987, in in *Cool Stars, Stellar Systems, and the Sun*, J.L. Linsky and R.E. Stencel (eds.), Springer–Verlag, p 262.
Walter, F.M., *et al.* , 1988, Astron. J. **96**, 297.

Inside the Stars, IAU Colloquium 137
ASP Conference Series, Vol. 40, 1993
Werner W. Weiss and Annie Baglin (eds.)

ACTIVITY TRACERS

KLAUS G. STRASSMEIER
Institut für Astronomie, Universität Wien,
Türkenschanzstraße 17, A-1180 Wien, Austria

ABSTRACT The vast variety of solar-like phenomena on other late-type stars, so-called activity tracers, provide an important tool for studying the structure of active stellar atmospheres and their connection to the stellar interior via strong magnetic fields. These "chromospherically active" stars include single and binary stars as well as pre- and post-main sequence objects and have rapid rotation and deep convective layers in common. They serve as astrophysical laboratories to study the vast phenomenology of activity tracers: starspots, plages, flares, prominences, which might be seen as enhanced analogs of solar activity and could be spatially resolved due to rotationally modulated indicators. In this paper we review the current observational material and discuss its impact on our knowledge of "active" atmospheres, especially in the context of stellar rotation.

ACTIVITY TRACERS THROUGHOUT THE HR DIAGRAM

Convective envelope stars: the F, G, and K stars

It is generally thought that stellar magnetic activity is determined from the interaction of stellar rotation and convection through a dynamo mechanism. The relation between various surface activity indicators and stellar (surface) rotation has been amply demonstrated (eg. Hartmann & Noyes 1987, Liebert & Probst 1987, and references therein). Similar rotation-activity relations for evolved stars also exist but are less well defined due to a much larger range of internal constitutions for a given spectral type (Basri et al. 1985, Simon & Fekel 1987, Young et al. 1989, Stepien 1989, Strassmeier et al. 1990a, Pasquini et al. 1990, Böhm-Vitense 1992). The appearance of large scale activity tracers in stellar atmospheres is only obvious for the most active stars, e.g., for components in close binaries (RS CVn's, Algols, W UMa's), for some very young stars (e.g. AB Dor, LQ Hya), and for anomalously rapidly rotating single giants (FK Comae's, and others). Figure 1 shows the position of such active stars in the H-R diagram compared to the granulation and coronal boundaries (adapted from Appourchaux et al. 1991).

Recent reviews concerning various subsets of stellar activity may be found in (chronological order): Linsky (1992) "Relations between activity indicators", Schmidt (1992) "Sunspots", Eaton (1992) "Starspot models", Strassmeier (1992) "Starspot photometry", Pallavicini (1992) "Stellar flares", Hall (1991a) "Long-term photometry", Gray (1991) "Rotation of evolved stars", Saar (1991) "Magnetic field measurements", Bruning (1991) "Mean magnetic field", Vilhu (1990) "Stellar activity", Ramsey (1990) "Optical and UV spectra of RS CVn's", Catalano (1990) "Flares in RS CVn binaries", a.o..

Radiative envelope stars: the A and B stars

The Herbig Ae/Be stars

These stars are believed to be pre-main sequence objects of intermediate mass (Finkenzeller & Mundt 1984). Extensive chromospheres and stellar winds were detected for AB Aur (A0Ve; Catala 1988), who also estimated the radiative losses from the chromosphere of AB Aur to be about 6 % of the star's bolometric luminosity, as compared to 10^{-3} for most RS CVn's. A0 stars do not have convective envelopes and thus need some other "source" for providing these stars with a chromosphere. It is usually believed that magnetic fields play the dominant role for transporting the energy to greater heights in the atmosphere, but no direct detection of a magnetic field has yet been reported. It has been proposed that a magnetic field is also the cause of the stream structure in the wind of AB Aur (Praderie et al. 1986). Indirect evidence for the existence of magnetic fields comes from the observation of periodic variations of chromospheric resonance lines such as Mg II k and Ca II K, which are thought to be formed in (corotating) stream winds (cf. Catala et al. 1991).

The chemically peculiar Ap stars

The transition between solar-type magnetic fields, i.e. isolated flux tubes embedded in non-magnetic regions, and Ap or Bp-type fields, i.e. poloidal fields covering the entire atmosphere with an usually dipolar structure, seems to occur around $T_{\rm eff} \approx 7000$ K or $\approx$ F0 (cf. Landstreet 1991), about where deep convective envelopes begin to develop. Wolff et al. (1986) finds the onset of chromospheric activity at $B - V = 0.28$ or ≈ 7300 K as deduced mainly from C IV 1549 Å line fluxes. The relationship between C IV 1549 Å flux and photospheric magnetic fields over quiet and over active solar regions was studied by Schrijver (1990) who found further evidence that a small fraction of C IV flux (the basal flux defined by Schrijver 1987) is not related to magnetic fields. Generally, no chromospheric emission is seen from Ap stars with a few exceptions: C II emission from the δ Scuti variable Θ Tuc (A7IV, $B - V = 0.23$) and possibly HR 4086 (A8V, $B - V = 0.25$), but definitively in 83 Tau (F0V, $B - V = 0.26$). For a summary see Simon & Landsmann (1991). No directly measured magnetic fields were found in non-Ap stars earlier than G0. Extensive observations of Ap stars have shown that fields are likely associated with an inhomogeneous surface distribution of certain chemical elements. Their detection through line-profile variations and consequent Doppler imaging provides an activity tracer for Ap stars (Hatzes et al. 1989, Rice & Wehlau 1990, 1991, Hatzes 1991a,b). Recently, magnetic fields were spatially resolved on the Ap star ϵ UMa by means of the Zeeman Doppler imaging technique (Donati et al. 1990) and show that the pole of the ring-like structure of excess iron and chromium abundance is consistent with the location of the pole of the magnetic field. It seems that certain surface structures are, within their uncertainties (see Rice 1991), indicative of magnetic fields on these type of stars. This seperation of elements has also been searched for on the Sun (cf. Zirin 1988). One might expect the heavier ions to sink more rapidly than the lighter. However, mixing through turbulence appears to restore a certain balance. Observations are not yet conclusive whether there is an underabundance of elements with a high first ionization potential.

STELLAR PLAGES

Integrated flux measurements

While the solar photosphere appears chaotic, the bulk of the magnetic field occurs in a limited range of structures: in active regions (plages, spots, fibrils), in unipolar regions, and ephemeral active regions. The existence of bright, chromospheric patches on stars other than the Sun was probably best demonstrated by the Ca II H & K emission survey at Mt. Wilson observatory (e.g. Wilson 1978, Baliunas et al. 1983, Baliunas & Vaughan 1985) but also by early work with IUE (e.g. Marstad et al. 1982). First evidence of correlations between enhanced chromospheric and transition region emissions and the visibility of starspots was presented by Baliunas & Dupree (1982) for λ And and by Baliunas (1988) for δ CrB. From such correlations, it seems clear now that stellar plages exist and are warmer than the underlying photosphere (cf. Mathys & Solanki 1989), but how warm remains to be determined.

But life is not that easy that we could conclude that stellar plages are generally associated with spots. They might be at some times, but not necessarily. This is nicely demonstrated by work that has been done with the IUE satellite on active stars like II Peg, AR Lac, BY Dra, V711 Tau, AU Mic, a.o.. Take for example II Peg: Rodonó et al. (1987) observed a clear enhancement of chromospheric and transition-region (TR) lines in anticorrelation with simultaneous V and $V - I$ photometry in 1981, while two years later no modulation at all was seen (Andrews et al. 1988) although the star showed a V amplitude of 0.11 mag. In 1986, clear evidence for rotational modulation in the Mg II lines was again present, however this time in the sense that higher flux was observed when the major spot concentration was out of view (Doyle et al. 1989). Strangely enough, higher temperature lines like C II and C IV (but also O I) did show a modulation in anticorrelation with the simultaneous Fine-Error-Sensor light curve. At this time II Peg showed a full peak-to-peak amplitude of 0.5 mag (Byrne 1986, Doyle et al. 1988) which was at that time the largest amplitude ever measured for a spotted star.

Similar plage-spot correlation studies were carried out also for other stars with divergent results: no clear evidence was found for V711 Tau nor AR Lac (Rodonó et al. 1987). For HK Lac, Mg II was stronger at times of spot visibility (K. Olah, priv. comm.), for IM Peg chromospheric lines (UV lines and Ca II iinfrared-triplet lines) were modulated by 25 % while transition-region lines were modulated by 50 % (Huenemoerder et al. 1990) both in phase with contemporaneous photometry.

The lesson seems clear: strict solar analogy is only valid for the less active stars while "overactive" stars like the RS CVn's cannot be viewed simply as scaled-up versions of solar-type behavior.

Spectral imaging of stellar chromospheres

From ultraviolet lines

To go beyond simple correlations between activity indicators one must spatially resolve the surface features of interest. Non-uniformities in the chromospheric brightness distribution will produce emission components that vary due to the Doppler shift produced by the star's rotation. This led to a technique called spectral imaging (cf. Neff 1992). The longitudinal position and extent of a particular stellar plage are determined by the radial velocity of the line centroid and the line width, respectively. The latitudinal position and extent are determined by the radial velocity amplitude and the total time required by the plage to cross the stellar disk. So far, this technique has been applied to only a mere handful of stars most of them not even published yet, e.g. EI Eri and HD 199178 in Neff (1991) or HD 32918 in Vilhu et al. (1991). The only star with published maps is the double-lined, eclipsing RS CVn binary AR Lac (Walter et al. 1987, Neff et al. 1989, Pagano et al. 1992).

Using a series of high-resolution Mg II k line spectra , Neff et al. (1989) derived the first "image" of the chromosphere of AR Lac (G2IV+K0IV). They identified three distinct plages on the K star and a flare on the G star (Fig. 2; adapted from Neff et al. 1989). The trailing hemisphere of the K star was globally 40 % brighter than the leading hemisphere. The G star also showed some brightness variations (by a factor of three) which the authors attributed to a large inactive region. Note that there is no direct analog for that on the Sun. A comparison with similar results from 1983 (two years earlier) suggests that the individual plages are long lived. A follow-up study for the years 1987 and 1989, also including broad-band photometry, was presented by Pagano et al. (1992). The three plages seen first in 1983 were still present but located at different longitudes (Fig. 3; adapted from Pagano et al. 1992). They had drifted at a speed of $-14° \pm 10°$ per year. Although no conclusive evidence of systematic spatial correlations with spots was found, their joint migration direction and rate suggests a common cause: magnetic fields anchored somewhere in the deep convection zone.

From optical lines

Stellar plages are also observable from optical wavelengths, mostly from the modulation of Ca II lines (H, K, 8498 Å, 8542 Å, and 8662 Å). Once again the Sun is a useful guide. A comparison of spatially resolved solar quiet-region and plage profiles for the five CaII lines was presented by Shine & Linsky (1972). Solar plages fill in the line cores, and the stronger the plage emission the more intense it is in the line wings, a fact mostly neglected in stellar work. The ratio of the core intensities of Ca II H to K increases from about 1.0 in quiet regions to about 1.1 in plages and might be a useful diagnostic for stellar plages. However, strong H and K emission is also seen from sunspots (Mattig & Kneer 1978), even when only light from the umbral regions is analyzed, and yields an H to K emission peak ratio of 1.7 (Shine & Linsky 1972). So the detection of Ca II line-core ratios greater than 1.0 does not necessarily imply that plages are the sole contributors, although they certainly contribute more light. This behavior is still to be investigated for other stars, but high-resolution line profiles ($\lambda/\Delta\lambda \approx 10^5$) are required (e.g. Pasquini et al. 1988). Probably the most promising technique to study stellar plages is to analyze rotationally modulated fluxes combined with simultaneous spectral imaging.

Magnetic field measurements

It seems clear that the bulk of magnetic fields detected on cool stars must reside in *bright* regions: in plages, facular areas, and the network. Starspots contribute so little light to the Zeeman line profile at optical wavelengths that the magnetic field would be hard to detect (Saar 1988, Donati et al. 1992; see also the review by Saar 1991).

A particularly interesting star is the K5V BY Dra binary V833 Tau. No rotational modulation of chromospheric activity indicators was detected despite the high level of chromospheric emissions and magnetic field coverage (B=2600 G, f_B=0.5, Saar 1991; $f_{spot} \approx 0.2$, Saar & Neff 1990). The star's low inclination of 10-23° (Tokovinin 1990) suggest that the activity must reside at or near the visible pole, confirming the existence of chromospherically bright regions at very high latitudes. Dark photospheric features were already claimed to exist at or near the rotation pole on V833 Tau (Olah & Pettersen 1991) and on other stars (e.g. Vogt 1988, Strassmeier 1990, Dempsey et al. 1992). Again, a phenomenon not seen on the Sun. Just recently, Donati et al. (1992) (see also these proceedings) obtained the first magnetic image of a cool star other than the Sun (V711 Tau). At one epoch they found a ring of clockwise field (≈ 300 G) surrounding a polar spot, while a region of counterclockwise field (≈ -700 G) resided near an equatorial cool spot.

For the Sun we can measure plage coverages of up to 5 % of the surface of the visible hemisphere. If we consider the most active stars one finds Ca II H and K emission higher by a factor 10 or more compared to solar-type stars. Assuming solar analogy it follows that about 65 % of a stellar surface would be covered with plages (Oranje 1983). So far, AR Lac is the only star were plages were spatially resolved and Neff et al. (1989) obtained a plage coverage of 17 % (additional to a large uniform component from the network). Still, this seems to be significantly smaller than the 65 % cited by Oranje based on an extrapolated solar analogy and the 50 % magnetic field filling factor on V833 Tau determined by Saar. Again the lesson seems clear: analogies to solar activity tracers are successful in explaining stellar observations qualitatively but simple scaling does not represent the observations quantitatively.

OBSERVATIONAL EVIDENCE FOR DIFFERENTIAL ROTATION

The solar case: from integrated sunlight

The detection of differential rotation from integrated sunlight has a long history and I am aware of only one positive detection: Keil 1990 (priv communication to Bruning 1991) from Ca II K-line measurements carried out on a daily basis for over seven years! All other searches had no positive detection (see the introduction of Bruning 1991). The main reason being the time evolution of active regions and their random appearance in longitude (LaBonte 1982). This sets the stage for stellar measurements.

Stars: from rotational modulation of disk integrated flux

A particular photometric period can be interpreted as the rotation period of the star at the latitude of a particular spot. If we observe different periods at different times one may determine the amount of differential rotation. Extensive data base on stellar rotation periods from rotational modulation exist now, e.g. from H and K observations at Mt. Wilson (Baliunas et al. 1983, Duncan et al. 1991) and from broad-band photometry with robotic telescopes (Strassmeier et al. 1989, Hooten & Hall 1990, Boyd et al. 1990). An alternative explanation for the observed period variations was proposed by Lanza et al. (1992) based upon the findings of Zappala & Zuccarello (1989) that the rotational angular velocity of young sunspot groups is on average significantly higher than for groups older than one solar rotation. Present stellar data however cannot discern between the two possibilities.

However, there is a major problem with all of these long-term studies of rotational modulation. One must assume that the lifetime of a particular activity tracer is longer, or at least comparable, to the time scale of the studied phenomenon. To detect differential rotation, for example, one must make sure to exclude evolutionary effects of individual active regions. Hall (1991a) and Hall & Busby (1990) investigated differential rotation from long-term starspot photometry. They found several stars that show evidence of differential surface rotation. Among those not listed by Lanza et al. (1992) are V478 Lyr, HR 7428, HD 37824 (Hall 1991a), but also BY Dra (Oskanyan et al. 1977), λ And (Scaltriti et al. 1984), II Peg (Bohusz & Udalski 1981), SV Cam (Zeilik et al. 1988), EI Eri (Strassmeier 1990), a.o.. But to be honest, none of these determinations is really conclusive. Moreover, there are several nondetections despite rather extended time coverages, e.g. FK Com (Jetsu et al. 1991), HD 199178 (Jetsu et al. 1990), VY Ari (Strassmeier & Bopp 1992). Nevertheless it seems well established now that, if solar-type differential rotation exists on other stars, it is smaller by a factor of ten as compared to the Sun.

Baliunas et al. (1985) likely detected differential rotation on four solar-type stars monitored by the Mt. Wilson H and K photometer: HD 206860, HD 101501, HD 190406, and HD 114710 (=β Com), while for eight other stars possible detections were reported. Recently, Donahue & Baliunas (1992) presented further data on β Com (G0V) and found strengthened evidence for differential rotation from a secular change of the seasonal rotation periods (Fig. 4; adapted from Donahue & Baliunas). They suggest that the star may have two latitudinal zones of activity. One in which the rotation periods appear to follow a Ca II H and K activity cycle, and one that does not. The equator-to-pole variation in the solar rotation is $\approx$ 30 %, compared to 5 through 21 % for the above stars (21 % for β Com), thus smaller than on the Sun. However, LaBonte (1984) and Baliunas et al. (1985)

point out that, for the Sun, the expected range of rotation-period variations would be approximately a factor ten smaller if the Sun would be observed like a star. This is because the differential rotation tracers on the Sun appear primarily at low latitudes. Bruning (1991) nicely demonstrated the visibility of differential rotation from solar mean magnetic field measurements and from a simulated stellar data set. He concluded that (1) a large number of gaps in the data coverage prevents the detection, (2) the amount of differential rotation will be underestimated because, in general, a limited range of stellar latitudes will be observed and (3) without detailed surface mapping of the active regions one cannot distinguish between pole-to-equator and equator-to-pole migration.

Stars: from surface imaging

Vogt & Hatzes (1991) found that the primary of UX Ari rotates differentially in the sense opposite to the Sun, i.e. zones near the pole of UX Ari rotate faster than zones near the equator. From three Doppler maps obtained within a five-month interval they identified two persistent features close to the (visible) pole and two near the equator. Measuring the relative rotation rates for these features enabled them to derive a differential rotation rate which is ten times smaller than the solar value (Fig. 5; adapted from Vogt & Hatzes). This is in excellent agreement with results from long-term photometry (Rodonó 1986, Hall 1991a). Although this detection is probably the most convincing of all, there is some debate whether Doppler imaging reliably recovers surface detail at or near the rotation pole (cf. Strassmeier et al. 1991, Donati 1992, these proceedings, but see also Fig. 6a,b). Figure 6 is a new Doppler map of LQ Hya obtained from 9 different lines and does *not* show a polar cap-like spot but still has a feature at very high latitude.

What about internal differential rotation?

The question probably should be "Is there any evidence for internal differential rotation from surface activity tracers?". Well, yes there is, but I'd call it very vague.

The single G8 giants HR 1362 and HD 181943 (see Strassmeier et al. 1990b; Hooten & Hall 1990 and Strassmeier 1991, respectively) are the two sole examples, at least to my knowledge, of stars with strong Ca II H and K emission and very slow rotation (P_{phtm} = 335 days, and 385 days, respectively). Their chromospheric activity levels are one to two orders of magnitude higher than expected from the empirical activity-rotation relations for giants (e.g. Strassmeier et al. 1990a; Fig. 7). Either something has spun them down while maintaining the magnetic flux, or the photometric periods are not the rotation periods and the stars are viewed at very low inclinations (if so one needs to explain the fairly large photometric amplitudes). Both stars still have substantial lithium on their surface. From an evolutionary viewpoint, they are likely in a stage of internal radial redistribution of mass caused by the rapid decrease of the thickness of the H-burning shell. Evolutionary models of cool giants normally assume rigid-body rotation (e.g. Rutten & Pylyser 1988), but see also Endal & Sofia (1979) and Tassoul & Tassoul (1989), and the new rotating stellar evolution models of the Sun (Pinsonneault et al. 1989). Durney & Robinson (1982) predict a simple dynamo theory that relates surface activity to angular rotation *and* depth-dependent differential rotation. Thus, for the case of HR 1362 and HD 181943, one may argue that the period at the layer where the dynamo is believed to function (a thin layer at the bottom of the convection zone, Parker 1975), not the surface period, determines the activity level.

Observations of stellar rotation rates in open clusters of various age (cf. Hartmann & Noyes 1987) indicate that solar-type stars reach their maximum rotation at an age of $\approx 10^{7-8}$ years and undergo rapid spin-down therafter. This rapid spin-down may cause large internal differential rotation between the radiative interior and the convective envelope, with the interior rotating more rapidly than the outside. The evolutionary models of Endal & Sofia (1981) predict approximately a factor of 5 to 10 (at most) for the difference between the rotation periods at the surface and at the radiative-convective interface.

There is some observational evidence that evolved active stars with large spots also show enhanced surface lithium abundances (Fekel 1988, Pallavicini et al. 1992). This is somewhat of a puzzle because lithium should be well mixed by convection and destroyed during their (hotter) main-sequence lifetime. Fekel & Balachandran (1992) suggest a scenario in which the surface-convection zone reaches a rapidly rotating core as the star evolves along the giant branch and dredges up to the surface high angular momentum material and freshly synthesized lithium. The rotating stellar evolution models of Pinsonneault et al. (1989) predict a rapidly rotating core and a spun-down convective layer for the Sun. It seems that the existence of Li rich giants is in agreement with theory. But what about spotted stars? Could it be that the large spotted areas of up to 20 % of the entire stellar surface (e.g. HD 12545, Strassmeier & Olah 1992) are regions of enhanced convection where most of the freshly synthesized lithium is dredged up? This could be checked by observations of rotational modulation of the lithium line.

Further evidence for the existence of internal differential rotation in binary star components comes from the connection between long-term brightness variations and orbital period changes of the eclipsing binary CG Cyg (Hall 1991b). Applegate (1992) predicts a large quadrupole moment by an alteration of the internal differential rotation gradient to predict that outer layers of a component in a binary star should rotate a little faster. According to this theory maximum luminosity should coincide with an O-C curve minimum if the star's outside spins faster than the inside but should coincide with an O-C maximum if the reverse is true. For CG Cyg, a period increase is clearly observed (Hall 1991b; Fig. 8) and occurs at maximum light, thus the sense of differential rotation is that the outside rotates faster than the inside, according to Applegate's theory.

LOCAL VELOCITY FIELDS

Photosphere

Granules are the most well-known tracers of local photospheric velocity fields but there are also larger scale velocity fields ("supergranulation"). Reviews of stellar granulation observations can be found in Gray (1988) and also in Dravins (1987). Line-bisector variations in phase with the rotation period were found on the G8 dwarf ξ Boo A (Toner & Gray 1988). This has been attributed to a surface feature carried across the stellar disk by rotation with a vertical velocity field enhanced over that of the rest of the surface. Toner & LaBonte (1990) investigated the possibility that these line bisector variations may be caused by a mass flow tangential to the surface ("moat flow"). The latter model produces an equally good fit to the observations, and clearly more observations are needed to decide between the two hypotheses.

Note that the presence of starspots with the same velocity dispersion as the surrounding photosphere will also produce a line asymmetry probably best seen in correlative analysis between a spotted star and an inactive reference star (cf. Dempsey et al. 1992). This is in agreement with the finding of Bruning & Saar (1990) that stronger asymmetry is observed for faster rotators.

Chromosphere

Neff (1992) describes a systematic redshift of 10 $\mathrm{km\,s^{-1}}$ of the observed Mg II line profiles from EI Eri, which could be either instrumental, or due to flows on a global scale in the stellar chromosphere. A similar redshift may also exist in the HD 199178 data. Crivellari et al. (1987) demonstrated the use of Ca II H-line profiles as a velocity field indicator. Strassmeier et al. (1992) monitored Hα line profile asymmetries of the young, active K star LQ Hya. Although the physical process of formation of these two lines is different, they contain similar information. In the case of LQ Hya there exists some evidence that velocity fields at the Hα-forming layers are spatially related to the surrounding of photospheric spots. This is reminiscent of the observed mass outflow around sunspots (cf. Brickhouse & LaBonte 1988). For IM Peg, Huenemoerder et al. (1990) noted that the hydrogen emission is not correlated with the starspots, but rather arises from the network. In UX Ari, however, Huenemoerder et al. (1989) detected modulation of chromospheric emission lines slightly preceeding the photometric minimum, indicating that the chromospheric emission originates close to, but outside of photospheric spots.

Corona and the stellar environment

Models of the quiescent solar corona are based on magnetic loops (Rosner et al. 1978). Solar observations confirm this picture and also detected systematic plasma flows. Vilhu (1987) and Vilhu & Walter (1987) showed that there is an upper limit to the x-ray flux, which suggests that saturation occurs when the loops fill the available coronal volume. First observational evidence for such extended geometries came from the detection of an x-ray eclipse of AR Lac. New ROSAT data will certainly expand our view of stellar coronae and their flow structure.

Several reviews on the subject of stellar flares can be found in IAU Colloquium 104 (Haisch & Rodono 1989 and Haisch et al. 1991). Flares in RS CVn binaries were recently reviewed by Catalano (1990). For this section I pick out just a few highlights relating to the general topic of activity tracers for stars on the main-sequence stars or close to it. A particularly interesting discovery was the detection of a strong stellar wind from the K-dwarf/WD pair V471 Tau (Mullan et al. 1989), probably driven by sporadic coronal mass ejection. In another context, Houdebine et al. (1990) presented evidence of a high-velocity mass ejection event ($v_r \approx 5800$ km s^{-1}) on AD Leo just starting with the onset of a particularly violent flare. There is now strengthened evidence for the existence of external plasma around the K dwarf of V471 Tau (see Young et al. 1988, 1990). Young et al. detected transient Hα emission components and their radial velocities reveal unambiguously that the emission cannot arise from the surface of the star and they suggested that it arises from plasma held in a ring or a disk orbiting the K dwarf (not the white dwarf). They also noted that these emission components are greatly enhanced immediately following the detection of a flare on the "surface" of the K dwarf. This leads to an inference of causality and suggests that this "ring" is feed by coronal mass ejection. Similar observations of transient *absorption* features were reported by Collier-Cameron & Robinson (1989) for AB Dor, a young K0 star. Modeling with a two-cloud structure at a distance between 3-4 stellar radii embedded in and corotating with a hot extended corona indicates the presence of closed magnetic loops to very great heights.

WHAT'S NEXT?

Considerable progress is being made towards concerted multiwavelength observations of activity tracers at all atmospheric layers. However, the exclusive circle of stars suitable for such techniques is indeed very small. Only objects with large-scale, I mean big, really big, spots or plages can be recognized and studied since one needs rotationally modulated indicators. At present there are about 300 spotted stars, to my knowledge. The second edition of the Chromospherically Active Binary Star (CABS) catalog (Strassmeier et al. 1988) will contain a little more than 200 close binaries with at least one active late-type stellar component. Some of these stars would be ideal targets for a space mission like PRISMA (see Catala, these proceedings) which is capable of simultaneous soft x-ray observations, UV spectroscopy, and visual and near-infrared photometry.

ACKNOWLEDGEMENTS

This work has been supported in part by FWF grant P7993, and the University of Vienna. I wish to thank Jim Neff, Bill Wehlau, and Michel Breger for a critical reading of the manuscript. I would like to thank Werner W. Weiss and the Scientific Organizing Committee for inviting me to present this review and for putting together such a stimulating meeting and the LOC for their continuous support and daily struggles. Last but not least, many thanks to numerous colleagues for sending me their work prior to publication.

REFERENCES

Andrews, A. D., et al. 1988, A&A 204, 177.

Applegate, J. H., 1992, ApJ, submitted.

Appourchaux, T., et al. 1991, PRISMA Assessment Study, ESA SCI 5.

Baliunas, S. L. 1988, in *Automatic Small Telescopes*, eds. D. S. Hayes, R. M. Genet, Fairborn Press, Mesa, p. 83.

Baliunas, S. L., Dupree, A. K., 1982, ApJ 252, 668.

Baliunas, S. L., et al. 1983, ApJ 275, 752.

Baliunas, S. L., et al. 1985, ApJ 294, 310.

Baliunas, S. L., Vaughan, A. H., 1985, ARA&A 23, 379.

Basri, G., Laurent, R., Walter, F. M., 1985, ApJ 298, 761.

Böhm-Vitense, E., 1992, AJ 103, 608.

Bohusz, E., Udalski, A., 1981, Acta Astr. 31, 185.

Boyd, L. J., et al. 1990, IAPPP 42, 44.

Brickhouse, N. S., LaBonte, B. J., 1988, SP 115, 43.

Bruning, D. H., 1991, PASP 103, 368.

Bruning, D. H., Saar, S. H., 1990, in 6th Cambridge Workshop *Cool Stars, Stellar Systems, and the Sun*, ed. G. Wallerstein, Kluwer, Dordrecht, PASPC 9, 165.

Byrne, P. B., 1986, IBVS 2951.

Catala, C., 1988, A&A 193, 222.

Catala, C., Czarny, J., Felenbok, P., Talavera, A., The, P. S., 1991, A&A 244, 166.

Catalano, S. 1990, NATO-ASI, *Active Close Binaries*, ed. C. Ibanoglu, Kluwer-Dordr., p. 411.

Collier-Cameron, A., Robinson, R. D., 1989, MNRAS 236, 57.

Crivellari, L., Beckman, J. E., Foing, B. H., Vladilo, G., 1987, A&A 174, 127.

Dempsey, R. C., et al. 1992, ApJ, in press.

Donahue, R. A., Baliunas, S. L., 1992, ApJL, submitted.

Donati, J.-F., Semel, M., del Toro Iniesta, J. C., 1990, A&A 233, L17.

Donati, J.-F., et al. 1992, A&A, in press.

Doyle, J. G., et al. 1988, A&A 192, 275.

Doyle, J. G., et al. 1989, A&A 223, 219.

Dravins, D., 1987, A&A 172, 200.

Duncan, D. K., et al. 1991, ApJS 76, 383.

Durney, B. R., Robinson, R. D., 1982, ApJ 253, 290.

Eaton, J. A. 1992, in Armagh Obs. Colloq., *Surface Inhomogeneities in Late-type Stars*, ed. P. B. Byrne, D. J. Mullan, Springer, Berlin Heidelberg New York, p. 15.

Endal, A. S., Sofia, S., 1979, ApJ 232, 531.

Endal, A. S., Sofia, S., 1981, ApJ 243, 625.

Fekel, F. C., 1988, in *A Decade of UV Astronomy with IUE*, ed. E. Rolfe, ESA SP-281, p. 331.

Fekel, F. C., Balachandran, S., 1992, ApJ, submitted.

Finkenzeller, U., Mundt, R., 1984, A&AS 55, 109.

Gray, D. F., 1988, *Lectures in Spectral Line Analysis: the F, G, K stars*, The Publisher, Ontario.

Gray, D. F., 1991, in NATO-ASI, *Angular Momentum Evolution of Young Stars*, Kluwer, p. 183.

Haisch, B., Rodonó, M., 1989, (eds), IAU Colloq. 104, *Solar and Stellar Flares*, SP 121.

Haisch, B., Strong, K. T., Rodonó, M., 1991, ARA&A 29, 275.

Hall, D. S., 1991a, in IAU Colloq. 130, *The Sun and Cool Stars: activity, magnetism, dynamos*, eds. I. Tuominen, D. Moss, G. Rüdiger, Springer, p. 353.

Hall, D. S., 1991b, ApJ 380, L85.

Hall, D. S., Busby, M. R. 1990, in NATO-ASI, *Active Close Binaries*, ed. C. Ibanoglu, Kluwer, Dordrecht, p. 377.

Hartmann, L. W., Noyes, R. W., 1987, ARA&A 25, 271.

Hatzes, A. P., 1991a, MNRAS 248, 487.

Hatzes, A. P., 1991b, MNRAS 253, 89.

Hatzes, A. P., Penrod, G. D., Vogt, S. S., 1989, ApJ 341, 456.

Hooten, J. T., Hall, D. S., 1990, ApJS 74, 225.

Houdebine, E. R., Foing, B. H., Rodonó, M., 1990, A&A 238, 249.

Huenemoerder, D. P., Buzasi, D. L., Ramsey, L. W., 1989, AJ 98, 1398.

Huenemoerder, D. P., Ramsey, L. W., Buzasi, D. L., 1990, ApJ 350, 763.

Jetsu, L., Huovelin, J., Tuominen, I., Vilhu, O., Bopp, B. W., Piirola, V. 1990, A&A, 236, 423.

Jetsu, L., Pelt, J., Tuominen, I., Nations, H. 1991, in IAU Colloq. 130, *The Sun and Cool Stars: activity, magnetism, dynamos*, eds. I. Tuominen, D. Moss, G. Rüdiger, Springer, p. 381.

LaBonte, B. J., 1982, ApJ 260, 647.

LaBonte, B. J., 1984, ApJ 276, 335.

Landstreet, J. D., 1991, in IAU Colloq. 130, *The Sun and Cool Stars: activity, magnetism, dynamos*, eds. I. Tuominen, D. Moss, G. Rüdiger, Springer, p. 342.

Lanza, A. F., Rodonó, M., Zappala, R. A., 1992, in Armagh Obs. Colloq., *Surface Inhomogeneities in Late-type Stars*, ed. P. B. Byrne, D. J. Mullan, Springer, p. 298.

Liebert, J., Probst, R. G., 1987, ARA&A 25, 473.

Linsky, J. L., 1992, in Armagh Obs. Colloq., *Surface Inhomogeneities in Late-type Stars*, ed. P. B. Byrne, D. J. Mullan, Springer, Berlin Heidelberg New York, p. 113.

Marstad, N., et al. 1982, in *Adv. in UV Astronomy: Four Years of IUE Research*, NASA Conf. Publ. 2238, p. 554.

Mathys, G., Solanki, S. K., 1989, A&A 208, 189.

Mattig, W., Kneer, F., 1978, A&A 65, 11.

Mullan, D. J., Sion, E. M., Bruhweiler, F. C., Carpenter, K. G., 1989, ApJ 339, L33.

Neff, J. E., 1991, in IAU Colloq. 130, *The Sun and Cool Stars: activity, magnetism, dynamos*, eds. I. Tuominen, D. Moss, G. Rüdiger, Springer, Berlin Heidelberg New York, p. 330.

Neff, J. E., 1992, in Armagh Obs. Colloq., *Surface Inhomogeneities in Late-type Stars*, ed. P. B. Byrne, D. J. Mullan, Springer, p. 54.

Neff, J. E., Walter, F. M., Rodonó, M., Linsky, J. L., 1989, A&A 215, 79.

Olah, K., Hall, D. S., Henry, G. W., 1991, A&A 251, 531.

Olah, K., Pettersen, B. R., 1991, A&A 242, 443.

Oranje, B. J., 1983, A&A 124, 43.

Oskanyan, V. S., Evans, D. S., Lacy, C., McMillan, R. S., 1977, ApJ 214, 430.

Pagano, I., Rodono, M., Neff, J. E., 1992, in Armagh Obs. Colloq., *Surface Inhomogeneities in Late-type Stars*, ed. P. B. Byrne, D. J. Mullan, Springer, p. 315.

Pallavicini, R., 1992, in NATO-ASI, *The Sun: a laboratory for Astrophysics*, J. Brown and J. Schmelz (eds.), in press.

Pallavicini, R., Randich, S., Giampapa, M., 1992, A&A 253, 185.

Parker, E. N., 1975, ApJ 198, 205.

Pasquini, L., Brocato, E., Pallavicini, R., 1990, A&A 234, 277.

Pasquini, L., Pallavicini, R., Pakull, M., 1988, A&A 191, 253.

Pinsonneault, M. H., Kawaler, S. D., Sofia, S., Demarque, P., 1989, ApJ 338, 424.

Praderie, F., Simon, T., Catala, C., Boesgaard, A. M., 1986, ApJ 303, 311.

Ramsey, L. 1990, in 6th Cambridge Workshop *Cool Stars, Stellar Systems, and the Sun*, ed. G. Wallerstein, Kluwer, Dordrecht, PASPC 9, 195.

Rice, J. B., 1991, A&A 245, 561.

Rice, J. B., Wehlau, W. H., 1990, A&A 233, 503.

Rice, J. B., Wehlau, W. H., 1991, A&A 246, 195.

Rodonó, M., 1986, in 4th Cambridge Workshop *Cool Stars, Stellar Systems, and the Sun*, eds. M. Zeilik, D. M. Gibson, Springer, p. 475.

Rodonó, M. *et al.* 1987, A&A 176, 267.

Rosner, R., Tucker, W. H., Vaiana, G. S., 1978, ApJ 220, 643.

Rutten, R. G. M., Pylyser, E., 1988, A&A 191, 227.

Saar, S. H., 1988, ApJ 324, 441.

Saar, S. H., 1991, in IAU Colloq. 130, *The Sun and Cool Stars: activity, magnetism, dynamos*, eds. I. Tuominen, D. Moss, G. Rüdiger, Springer, p. 389.

Saar, S. H., Neff, J. E., 1990, in 6th Cambridge Workshop *Cool Stars, Stellar Systems, and the Sun*, ed. G. Wallerstein, Kluwer, Dordrecht, PASPC 9, 171.

Scaltriti, F., et al. 1984, A&A 139, 25.

Schmidt, H. U., 1992, in Cowling Memorial Symp. *Solar System Plasma Physics*, Geophys. Astrophys. Fluid Dyn., in press.

Schrijver, C. J., 1987, A&A 172, 111.

Schrijver, C. J., 1990, A&A 234, 315.

Shine, R. A., Linsky, J. L., 1972, SP 25, 357.

Simon, T., Fekel, F. C., 1987, ApJ 316, 434.

Simon, T., Landsmann, W., 1991, ApJ 380, 200.

Stepien, K., 1989, Acta Astr. 39, 209.

Strassmeier, K. G. 1990, ApJ 348, 682.

Strassmeier, K. G., 1991, IBVS 3618.

Strassmeier, K. G., 1992, in Proc. 103rd ASP Meeting, *Robotic Telescopes for the 1990's*, ed. A. Filippenko, PASPC, in press.

Strassmeier, K. G., Bopp, B. W. 1992, A&A, in press.

Strassmeier, K. G., Hall, D. S., Boyd, L. J., Genet, R. M. 1989, ApJS 69, 141.

Strassmeier, K. G., Hall, D. S., Zeilik, M., Nelson, E., Eker, Z., Fekel, F. C. 1988, A&AS 72, 291.

Strassmeier, K. G., Olah K., 1992, A&A, in press.

Strassmeier, K. G., et al. 1990a, ApJS 72, 191.

Strassmeier, K. G., et al. 1990b, ApJ 350, 367.

Strassmeier, K. G., et al. 1991, A&A 247, 130.

Strassmeier, K. G., Rice, J. B., Wehlau, W. H., Hill, G., M., Matthews, J. M., 1992, A&A, submitted.

Tassoul, J.-L., Tassoul, M., 1989, A&A 213, 397.

Tokovinin, A. A., 1990, SvA Lett. 16, 1026.

Toner, C. G., Gray, D. F., 1988, ApJ 334, 1008.

Toner, C. G., LaBonte, B. J., 1990, in 6th Cambridge Workshop *Cool Stars, Stellar Systems, and the Sun*, ed. G. Wallerstein, Kluwer, Dordrecht, PASPC 9, 161.

Vilhu, O., 1990, in Proc. Nordic-Baltic Astronomy Meeting, eds. C.-I. Lagerkvist et al., Uppsala, Repr. HSC, p. 369.

Vilhu, O., Gustafsson, B., Walter, F. M., 1991, A&A 241, 167.

Vogt, S. S. 1988, in IAU Symp. 132, *The Impact of Very High S/N Spectroscopy on Stellar Physics*, eds. G. Cayrel de Strobel, M. Spite, Kluwer, Dordrecht, p. 253.

Vogt, S. S., Hatzes, A. P. 1991, in IAU Colloq. 130, *The Sun and Cool Stars: activity, magnetism, dynamos*, eds. I. Tuominen, D. Moss, G. Rüdiger, Springer, p. 297.

Walter, F. M., Gibson, D. M., Basri, G., 1983, ApJ 267, 665.

Walter, F. M., et al. 1987, A&A 186, 241.

Wilson, O. C., 1978, ApJ 226, 379.

Wolff, S. C., Boesgaard, A. M., Simon, T., 1986, ApJ 310, 360.

Young, A., Ajir, F., Thurman, G., 1989, PASP 101, 1017.

Young, A., Skumanich, A., Paylor, V., 1988, ApJ 334, 397.

Young, A., Rottler, L., Skumanich, A., 1991, ApJ 378, L25.

Zappala, R. A., Zuccarello, F., 1989, A&A 214, 369.

Zeilik, M., De Blasi, C., Rhodes, M., Budding, E. 1988, ApJ 332, 293.

Zirin, H., 1988, *Astrophysics of the Sun*, Cambridge Univ. Press, Cambridge.

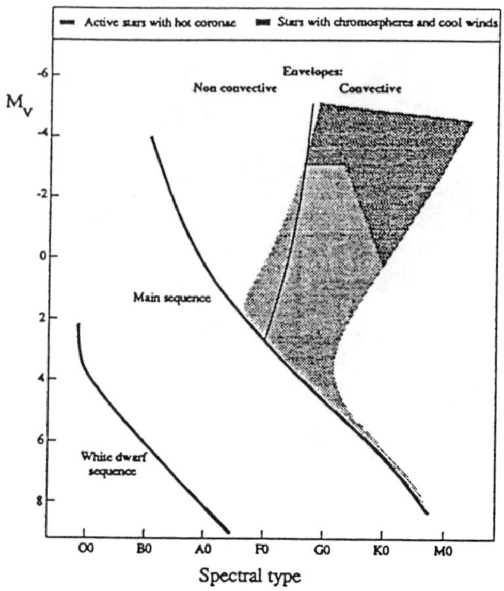

Fig. 1

Fig. 2

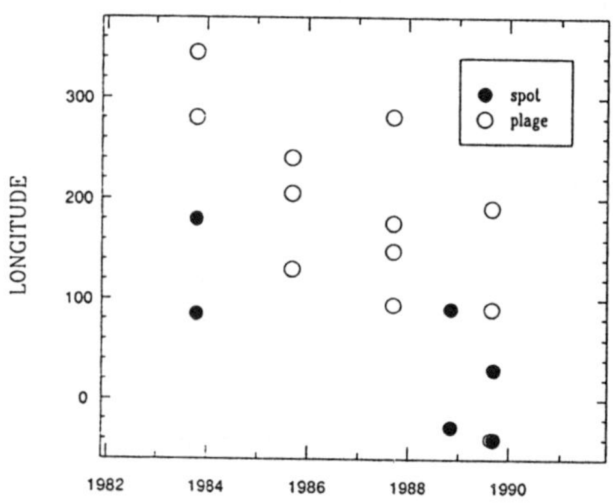

Fig. 3

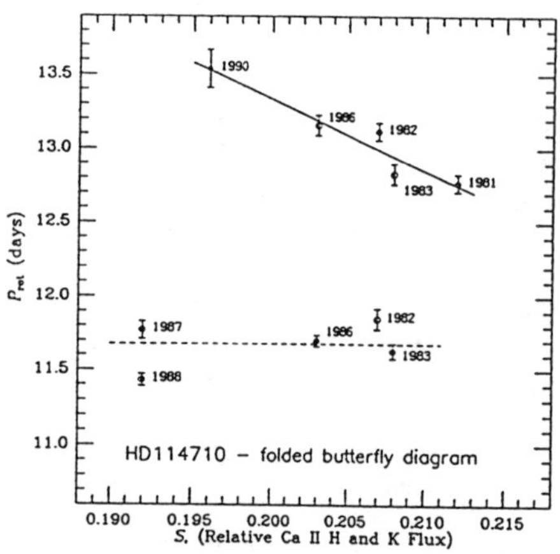

Fig. 4

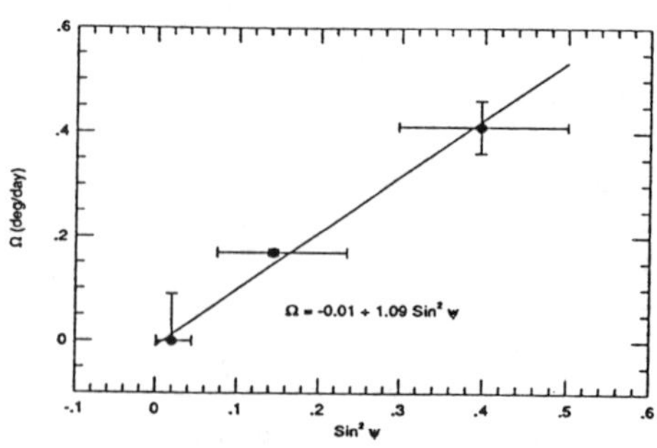

Fig. 5

Fig. 6a

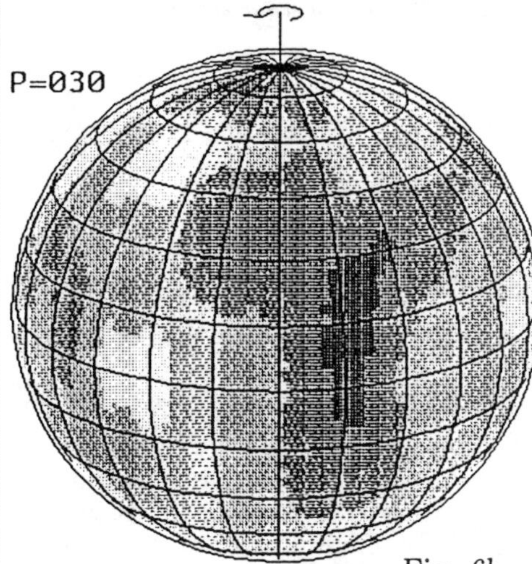

Fig. 6b

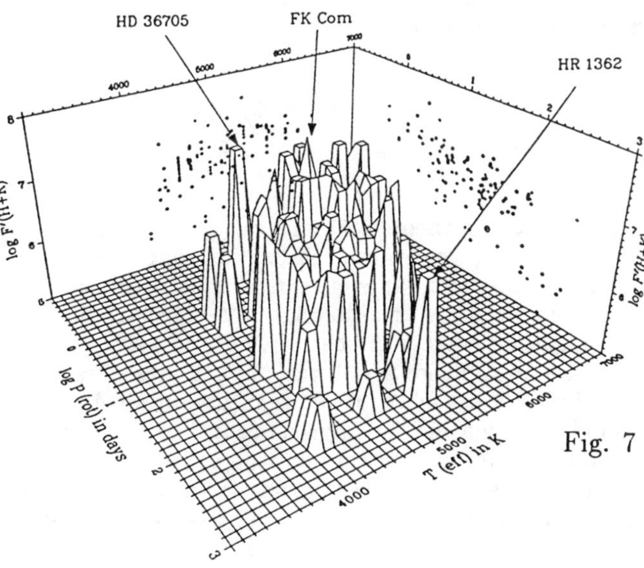

Fig. 7

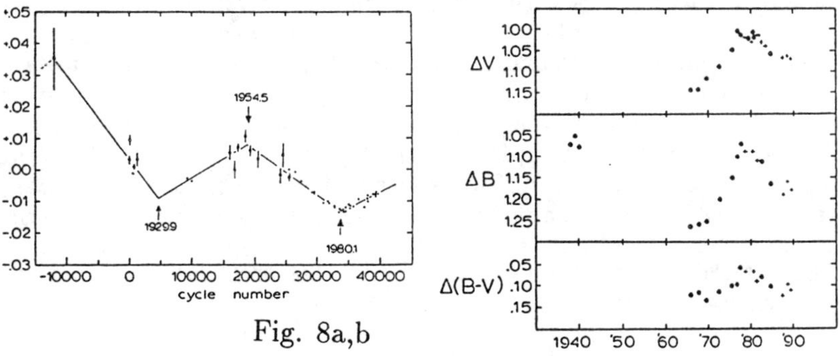

Fig. 8a,b

Inside the Stars, IAU Colloquium 137
ASP Conference Series, Vol. 40, 1993
Werner W. Weiss and Annie Baglin (eds.)

WINDS

K. B. MACGREGOR
High Altitude Observatory/NCAR,* P. O. Box 3000, Boulder, CO 80307, U.S.A.

ABSTRACT It is by now well known that most main sequence stars continuously lose mass as a consequence of the winds they emit. In addition to affecting the thermal and dynamical state of the stellar atmosphere, such mass loss can also induce changes in the interiors of stars. In the present review, we consider a few of the ways in which sustained, wind-like mass loss can alter the physical state of main sequence stellar interiors by examining the differences in internal structure, composition, and rotation between mass-losing and conservatively evolving stars.

INTRODUCTION

Numerous observational and theoretical results obtained over the past quarter century have contributed to the current realization that mass loss is a relatively common phenomenon, occurring among stars located nearly everywhere in the H-R diagram. There now exists an abundance of evidence (some of it direct, some of it circumstantial) suggesting that virtually all stars, regardless of effective temperature or evolutionary status, sustain mass loss in the form of a wind. As a research subject mass loss is frequently studied with the goal of identifying and/or elucidating the physical processes responsible for the initiation, acceleration, and maintenance of observed stellar outflows. However, in keeping with the theme of the present meeting, we herein consider how winds can affect the *interiors* of stars, specifically, stars on or near the main sequence.

In an effort to depict a variety of possible effects yet keep the analysis simple, we restrict our attention to three representative examples of the consequences of mass loss for main sequence (MS) stellar interiors. In the first, we use the evolution of massive stars ($M_* \gtrsim 15M_\odot$) during core hydrogen burning to illustrate how the removal of a significant fraction of the initial stellar mass by a wind can induce internal, structural changes. In the second, we examine the advective effect of mass loss by considering how winds can moderate surface abundance anomalies produced by the diffusive separation of

* The National Center for Atmospheric Research is sponsored by the National Science Foundation

chemical elements in stars of intermediate mass ($1.5M_\odot \lesssim M_* \lesssim 3.0M_\odot$). Finally, in the third example, we describe how the torque associated with mass loss from a rotating star can affect the physical state of the interior by considering how the internal solar rotation accommodates the shear which is continuously introduced by the wind-related braking of the sun's surface layers.

MS EVOLUTION OF MASSIVE STARS WITH MASS LOSS

Massive, hot stars having spectral types O through early B and belonging to all luminosity classes are observed to lose mass at rates in the range $10^{-8} \lesssim \dot{M}_* \lesssim 10^{-5} M_\odot$ yr^{-1} (see e.g., Cassinelli and Lamers 1987). Such mass loss is thought to be driven by the force produced when radiation from the stellar photosphere is absorbed or scattered in UV spectral lines formed in the flow. The inferred $\dot{M}_*$ values are of sufficient magnitude that the internal structure of an early-type star undergoing mass loss may change over the course of its MS lifetime τ_{MS}, as a consequence of the changing stellar mass. For example, a star with $M_* = 30M_\odot$ which loses mass at the constant rate $\dot{M}_* = 5 \times 10^{-7} M_\odot$ yr^{-1} will suffer a fractional reduction $\dot{M}_* \, \tau_{MS}/M_* = 0.1$ of its initial mass, for $\tau_{MS} = 6 \times 10^6$ years. The structural and evolutionary modifications to constant mass models arising from MS mass loss of this kind have been examined by numerous authors. Among the investigations most pertinent to the present discussion are those by Tanaka (1966), Hartwick (1967), Chiosi and Nasi (1974), Dearborn *et al.* (1978), Falk and Mitalas (1981), Brunish and Truran (1982), and Mitalas and Falk (1984). Additional references are contained in the reviews by Maeder (1984) and Chiosi and Maeder (1986).

To discern the ways in which the physical state of the interior of a mass-losing star differs from that of a conservatively evolving star, we utilize the simple, semi-analytic model described by Falk and Mitalas (1981, hereafter FM; see also Mitalas and Falk 1984; Stein 1966). From comparisons with the results of detailed model calculations, these authors found that during the phase of core hydrogen burning, the relation between stellar mass M_* and luminosity L_* could be accurately represented by

$$(L_*/L_\odot) = A \, \beta^4 \, <\mu>^4 \, (M_*/M_\odot)^3 \, . \qquad (1)$$

In equation (1), β is the ratio of gas to total pressure, $<\mu>$ is the mean molecular weight averaged over the mass of the star, and A is a constant whose value depends upon the initial stellar mass and composition.

FM further noted that the contraction of the convective core that occurs as MS evolution proceeds leads to an internal composition profile with the following characteristics. At a given time $t (< \tau_{MS})$, the hydrogen mass fraction X in the outer envelope (i.e., the region external to the initial core boundary) has the uniform value X_0. In the core, X is also uniform, with value $X_c (< X_0)$, reflecting the nuclear processing which has taken place prior to the time in question. Throughout the region comprised of material which was previously a part of the core, X has very nearly a linear dependence on mass. Following FM, we assume that the composition profile within this intermediate region is strictly linear, making possible direct evaluation of the average $<\mu(X)>$ in equation (1). The quantity β can then be obtained by solving the Eddington quartic

$$1-\beta = a\,\beta^4 \, <\mu>^4 \, (M_*/M_\odot)^2 \, , \tag{2}$$

where $a = 0.00309$ (cf. FM).

The fractional mass contained within the convective core can be estimated by assuming that the entire stellar luminosity is produced therein, and utilizing the fact that radiative and adiabatic gradients are equal at the core boundary (cf. Stein 1966). If electron scattering constitutes the primary source of opacity, it can be shown that

$$q \equiv \frac{M_{core}}{M_*} = q_0 \left(\frac{1+X_c}{1+X_0}\right) , \tag{3}$$

where the initial core mass q_0 is a known quantity, expressible in terms of specified constants characterizing the model (cf. FM). With this result, the time-evolution of X_c can be determined by integrating

$$\frac{dX_c}{dt} = -\frac{L_*}{EM_{core}} \propto \left(\frac{1-\beta}{1+X_c}\right) , \tag{4}$$

where $E(\approx 6 \times 10^{18}$ erg g$^{-1})$ is the energy released from 1 gram of material in which hydrogen has been transmuted to helium via the CNO cycle.

The final component of the model is an equation governing the time evolution of the stellar mass. In the present paper, we consider four distinct prescriptions for the mass loss rate, enumerated below:

$$dM_*/dt = 0 \tag{5a}$$

$$= 1.35 \times 10^{-7} \, (L_*/10^5 \mathrm{L}_\odot)^{1.62}, \tag{5b}$$

$$= 3.26 \times 10^{-7} \left(\frac{M_*}{M_\odot}\right) \frac{\Gamma^{1.49}}{(1-\Gamma)^{0.49}}, \tag{5c}$$

$$= 1.72 \times 10^{-6} \left(\frac{L_*}{10^5 L_\odot}\right) \left(\frac{R_*}{\mathrm{R}_\odot}\right) \left(\frac{M_\odot}{M_*}\right), \tag{5d}$$

where in 5(c), $\Gamma = 2.40\,(L_*/10^5\,L_\odot)(\mathrm{M}_\odot/M_*)$. The mass loss rates given in equations 5(b) - (d) have units $\mathrm{M}_\odot$ yr^{-1}, and are, respectively, an empirical fit to observed values obtained by Garmany and Conti (1984), the theoretical rate derived from the line-driven wind model of Castor, Abbott, and Klein (1975), and a proportionality suggested by Mc Crea (1962), scaled to yield $\dot{M}_* = 5 \times 10^{-7} \mathrm{M}_\odot$ yr^{-1} for parameters appropriate to a zero-age MS star with $M_* = 30\mathrm{M}_\odot$.

We have traced the MS evolution of a $30\mathrm{M}_\odot$ star, including mass loss, by integrating equations (4) and (5), assuming $X_0 = 0.71$. Equations (1) (with $A = 71.278$; cf. FM) and (2) were used to determine β and L_* with $<\mu>$ calculated as described above, assuming $\mu(X) = 4M_H/(5X+3)$. A summary of results is depicted in Figure 1; in each panel, the lettered curves represent models computed with the corresponding mass loss rate given in equations 5(a)-(d). For each model, the results portrayed in the figure span the time interval over which X_c is reduced from its initial value X_0 to the

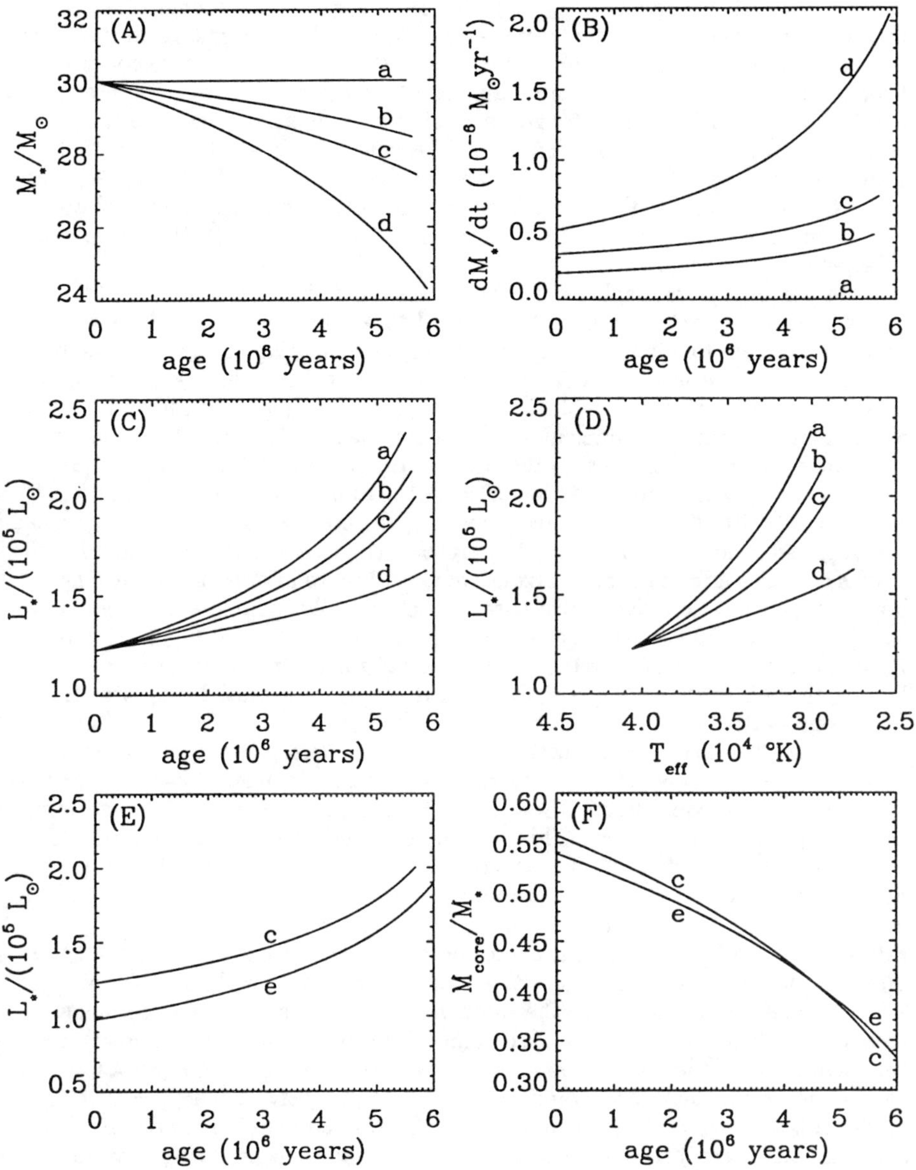

Figure 1: MS evolution of a 30 $M_\odot$ star with mass loss, as described in the text.

value $X_c = 0.05$. In panels (A) - (C), the times required for models (a) - (d) to attain this state of near hydrogen exhaustion in the core are, respectively, $5.50 \times 10^6, 5.60 \times 10^6, 5.68 \times 10^6$, and 5.87×10^6 years. We therefore conclude that the MS lifetime (i.e., the time during which hydrogen is burned in the core) of a mass-losing star is longer than that of a conservatively evolving star. Moreover, as is evident from panels (A) and (B), the rates of mass loss given by equations 5(b), (c), and (d) are (in that order) an increasing sequence at all times. Hence, for the moderate mass loss rates considered here, the enhancement of τ_{MS} (relative to a conservative star of the same initial mass) is larger for larger values of $\dot{M}_*$.

In panel (C), we display the time evolution of the stellar luminosity for each of the models. Note that at a fixed time, the luminosity of a star undergoing mass loss is smaller than that of a star having the same mass at $t = 0$ with $\dot{M}_* = 0$. Because $L_* \propto M_*^3$ (cf. eq. [1]), the model which experiences the most mass loss during MS evolution is characterized by the smallest luminosity at later times. However, since at a given time that model also has the largest value of X_c, it has a commensurably larger core mass fraction q (cf. eq. [3]). At $t = 5 \times 10^6$ years, model (a) has $X_c = 0.150$ and $q = 0.375$, while model (d) has $X_c = 0.197$ and $q = 0.390$. In panel (D), the computed evolutionary tracks are depicted in an H-R diagram. Apparent in the figure is the tendency for the tracks of stars suffering more cumulative mass loss to assume an increasingly horizontal orientation. Note also that as the tracks become less inclined, they extend to cooler effective temperatures. In model (a), $T_{eff} = 30{,}100$ K when $X_c = 0.05$, while in model (d), $T_{eff} = 27{,}500$ K for the same value of X_c. Hence, the MS width in the H-R diagram is larger for a mass-losing star than for a conservative star, other things being equal.

For model (c), $M_* = 27.43 M_\odot$ when $X_c = 0.05$ near the termination of core hydrogen burning. It is of interest to compare the properties of this model with those of a model having $M_* = 27.43 M_\odot$ and $\dot{M}_* = 0$, evolved to the same central hydrogen abundance. This is done in panels (E) and (F), wherein the curves pertaining to the latter, conservatively evolving star are labelled (e). As noted above, model (c) attains $X_c = 0.05$ at $t = 5.68 \times 10^6$ years; in model (e) hydrogen is depleted to this level at $t = 6.00 \times 10^6$ years. Inspection of panel (E) reveals that the luminosity of the mass-losing model exceeds that of the $\dot{M}_* = 0$ model at all times. In particular, L_* for model (c) is larger than L_* for model (e) even when both stars have the same values of M_* and X_c. The primary reason for this behavior can be ascertained from panel (F), which shows the time evolution of the convective core mass fraction for both models. As can be determined from the figure, when $M_* = 27.43 M_\odot$ and $X_c = 0.05, q = 0.342$ in model (c), while $q = 0.331$ for model (e). Thus, in comparison to a star of the same mass and central hydrogen abundance, a star with $\dot{M}_* \neq 0$ has a more massive convective core. Because $L_* \approx \bar{\epsilon} M_{core}$, where $\bar{\epsilon}$ is the mass-averaged specific energy generation rate, the luminosity of model (c) is greater than that of model (e).

We close this section by noting that an additional, well-studied consequence of mass loss on the internal structure of massive stars is its effect on the occurrence, structure, and lifetime of semi-convective regions. The reader is referred to the previously cited papers for discussion of this topic.

ABUNDANCE ANOMALIES IN MS STARS OF INTERMEDIATE MASS

Definitive observational evidence for the existence of winds from MS stars with spectral types A-F is presently unavailable. However, theoretical attempts to account for the chemical peculiarities exhibited by some such stars may provide an indication of the occurrence of mass loss from them.

The stars that constitute the class of metallic-line objects designated "Fm-Am stars" have the following attributes in common (Cayrel *et al.*, 1991; Boyarchuk and Savanov 1986; Preston 1974). They are MS stars with spectral types between about F0 and A0, effective temperatures between 7000 K and 10,000 K, and masses in the range $1.5 \lesssim (M_*/M_\odot) \lesssim 3.0$. Spectroscopic analyses have revealed that approximately 20% of all stars with these properties exhibit anomalous photospheric abundances. The classical pattern of abundance anomalies observed to be characteristic of the Fm-Am stars as a group consists of underabundances (relative to solar photospheric abundances) of the elements Ca and Sc, and overabundances of virtually all elements heavier than Fe. Underabundant species are typically deficient by a factor $\sim$ 5-10, while overabundant species can be enhanced by as much as a factor $\sim$ 10-100, although substantial dispersion exists among both elements and stars. The Fm-Am stars do not have detectable surface magnetic fields, and, when compared with chemically normal stars in the same spectral type range, they have significantly lower rotational velocities. This latter characteristic may be related to the fact that the vast preponderance of metallic-line stars are members of binary systems (Abt and Levy 1985).

The chemical peculiarities observed in Fm-Am stars are thought to be the result of selective, radiatively-driven diffusion of some species in the outer layers of the stellar interior (Michaud 1970; Watson 1970, 1971; see also the reviews by Michaud 1977, 1987, 1991; Vauclair and Vauclair 1982). Stars having these spectral types arrive on the MS in a uniformly mixed state, with subphotospheric convection zones due to both H and He. These adjacent, superficial convective regions are quite thin; for $M_* \geq 1.5M_\odot$, the mass fraction contained in the outer, H convection zone is $M_{conv}/M_* \lesssim 10^{-9}$ (see, e.g., Michaud *et al.*, 1976). Over time He settles gravitationally; for stars with rotational velocities low enough that meridional circulation is unimportant, after $\sim 10^6 - 10^7$ years of MS evolution, the initial He abundance in the outer envelope is diminished by a factor of about 3, and the associated inner convection zone disappears. Upward diffusion of heavy elements can subsequently take place in the stable layers just beneath the remaining H convection zone.

At the base of the H convection zone, the prevailing physical conditions are such that the dominant ionization stages of many chemical elements have bound-bound transitions with frequencies situated near the peak of the stellar flux distribution. For abundant species (roughly, those lighter than Fe), the upward acceleration arising from the absorption of radiation in such lines is generally smaller than gravity, being limited in magnitude by large line optical depths. Ions of these elements undergo gravitational settling. However, for species whose intrinsic abundances are sufficiently small that they can be accurately described as trace constituents of the gas (i.e., elements heavier than Fe), the radiative acceleration associated with optically thin lines can exceed gravity by a large amount.

Consider an ion of element A with mass $m_A = Am_H$ and concentration $c_A = N_A/N_H \ll 1$. In the absence of large concentration gradients and neglecting thermal diffusion, the velocity of diffusion of A relative to the background electron-proton gas is approximately (Chapman and Cowling 1970; Aller and Chapman 1960)

$$u_D \approx -D(g - g_{rad})\,\frac{m_A}{kT} \approx -(g - g_{rad})\,\tau_{coll}\ . \tag{6}$$

In equation (6), $D(\sim v_{th,A}^2\ \tau_{coll})$ is the coefficient of diffusion, g and g_{rad} are, respectively, the gravitational and radiative accelerations, τ_{coll} is the A-proton collision time, and $u_D \gtrless 0$ implies motion in the $\pm\hat{e}_r$ direction. According to (6), species for which $g_{rad} > g$ in the region beneath the H convection zone are pushed upward, while those with $g_{rad} < g$ sink under the influence of gravity. Upwardly diffusing species, upon entering the convection zone, are quickly mixed throughout the volume by the turbulent motions therein. The time evolution of the relative abundance of such an element A in the surface layers of the star is governed by the equation of continuity

$$\frac{\partial}{\partial t}(c_A N_H) + \nabla \cdot (c_A N_H \mathbf{u}_D) = 0. \tag{7}$$

By integrating equation (7) over the volume of the convection zone and applying Gauss theorem, it follows that

$$M_{conv}\ \frac{\partial c_A}{\partial t} = 4\pi\ R_{conv}^2\ (\rho c_A u_D)\Big|_{r=R_{conv}}\ , \tag{8}$$

where ρ is the total mass density, and M_{conv} and R_{conv} are the mass and base radius of the convection zone. Hence, the concentration of element A within the convection zone grows at a rate which depends upon the magnitude of the current diffusive flux into it from below.

The theory outlined above is remarkably successful in reproducing the pattern of abundance anomalies observed in Fm-Am stars. That is, species for which the *computed* g_{rad} is greater than g below the H convection zone are generally *observed* to be overabundant, and vice versa. However, the magnitudes of the abundance enhancements obtained from equations (6) and (8) together with calculated g_{rad} values are often significantly larger than those observed; overabundances by factors in excess of 10^4 can be produced in times $\lesssim \tau_{MS}$ (Michaud *et al.*, 1976). According to (8), the H convection zone functions as a reservoir which, in the absence of additional effects, upward diffusion from deeper layers continuously fills. The rate of filling is decreased only if g_{rad} becomes $< g$ (e.g., by saturation), or if the region in which $g_{rad} > g$ for a particular element becomes depleted of that element.

An alternative possibility for limiting the magnitudes of diffusively produced overabundances is to drain the reservoir from above by mass loss. Assume that a typical star of this type undergoes steady, spherical mass loss at a rate $\dot{M}_*$ which is insufficient to alter the internal stellar structure. Conservation of mass then requires that a fluid parcel at radius r within the star have the small, outward velocity $u_W = \dot{M}_*/(4\pi\ r^2\rho)$. Hence, the continuity equation (7) for species A must be amended by making the substitution $\mathbf{u}_D \rightarrow \mathbf{u}_D + \mathbf{u}_W$, so that integration over the convection zone volume leads to the result

$$M_{conv}\ \frac{\partial c_A}{\partial t} = 4\pi R_{conv}^2\ \ [\rho c_A(u_D + u_W)]\Big|_{r=R_{conv}} - \dot{M}_* c_A\ , \tag{9}$$

where it has been assumed that the composition of the wind is the same as that of the convection zone. Note that equation (9) contains the sink term $-\dot{M}_* c_A$, describing the rate at which ions of species A are removed from the convection zone as a result of advection by the wind flow. When the evolution of surface abundances for Fm-Am stars is computed using equation (9) instead of (8), it is found that the concentrations of elements having $g_{rad} > g$ beneath the H convection zone grow only for a time $\sim M_{conv}/\dot{M}_*$ before declining at later times. Consider the results of Michaud *et al.* (1983) for a star with $M_* = 1.8\mathrm{M}_\odot, T_{eff} = 7800$ K, $\log g = 4.15$, and $M_{conv}/M_* \approx 10^{-9}$. These authors find that for mass loss rates in the range $\dot{M}_* = 10^{-14}–10^{-15}\mathrm{M}_\odot\ \mathrm{yr}^{-1}$, the surface abundances of species such as Ba, Eu, and Hg (for which $g_{rad} > g$) cease to increase some $10^5 - 10^6$ years after the disappearance of the He convection zone. At this stage, the diffusive overabundances of these elements are ~10-50, in accord with observational determinations. Moreover, for such modest rates of mass loss, the wind-induced, upward advection of *all* species (arising from the first term on the RHS of equation (9)) is sufficiently slow that elements having $g_{rad} < g$ below the H convection zone (e.g., Ca and Sc) remain underabundant in the stellar surface layers for times $\sim \tau_{MS}$. Hence, models based on selective, radiative diffusion in the presence of steady mass loss appear to be capable of accounting for both the morphology and magnitude of the observed chemical peculiarities in Fm-Am stars.

As noted at the outset of this section, winds from MS stars in the spectral type range A-F have yet to be directly observed. If the rates of mass loss from these stars are typically as low as the values inferred from the application of diffusion theory to Fm-Am stars, then the current dearth of wind detections is not surprising. In this regard, recent observations at optical and radio wavelengths (Lanz and Catala 1992; Brown *et al.*, 1990) have established upper limits of between 10^{-9} and $10^{-10}\mathrm{M}_\odot\ \mathrm{yr}^{-1}$ for the rates of mass loss from a number of A and F dwarf stars.

INTERNAL ROTATION OF SOLAR-TYPE STARS

Direct observations of the solar wind from satellites in Earth orbit have established that the Sun loses mass at a rate $\dot{M}_\odot \sim 10^{12}\ \mathrm{g\ s^{-1}} \sim 10^{-14} M_\odot\ \mathrm{yr}^{-1}$. If the current value of $\dot{M}_\odot$ is typical of its magnitude over the entire course of the Sun's MS lifetime, then an amount of mass $\sim 10^{-4} M_\odot$ would be lost during that period, assuming $\tau_{MS} \sim 10^{10}$ years. Such a reduction of the solar mass is too small to engender structural changes of the type discussed previously. Moreover, although a hydrogen convection zone exists in the outer layers of the solar interior, the physical conditions which prevail at its base ($M_{conv}/M_\odot \approx 1.5 \times 10^{-2}, T \approx 2 \times 10^6$ K) are not conducive to the segregation of chemical elements by diffusion, as in the Fm-Am stars (e.g., $M_{conv}/M_* \sim 10^{-9}, T \approx 3 \times 10^4$ K for $M_* = 1.8 M_\odot$). There remains, however, one physical property of the Sun's interior which is profoundly affected by the solar wind, namely, its rotational state.

Because the Sun rotates, the wind that emanates from the outer, coronal layers of the solar atmosphere removes angular momentum as well as mass. Near the base of the flow, the solar magnetic field enforces a state of approximate corotation of the plasma with the Sun. The resulting torque exerted on the Sun by the magnetized wind opposes the solar rotation, and is of sufficient magnitude to significantly reduce the solar angular velocity in a time $\lesssim \tau_{MS}$ (Weber and Davis 1967). Spacecraft observations of the wind plasma and interplanetary magnetic field yield estimates in the range $0.2 - 0.3 \times 10^{30}$ dyne cm sr^{-1} for the total angular momentum flux associated with outflow in the solar equatorial plane (Pizzo *et al.*, 1983). Integrating this result over all solid angles, it follows that the rate at which the Sun's rotation is braked by the solar wind is about $\dot{J}_\odot \approx 3 \times 10^{30}$ dyne cm, implying an e-folding time for angular momentum loss (assuming rigid rotation) $J_\odot/\dot{J}_\odot \sim 10^{10}$ years. Note that the application of this torque to the surface layers of the Sun gives rise to stresses within the sheared fluid which act to redistribute the angular momentum contained therein. In the convection zone, the viscosity associated with turbulent fluid motions is sufficiently great (e.g., $\nu \sim 10^{12}$ cm^2 s^{-1}) that the deceleration of the topmost layers is communicated to the bottom in a time $(R_\odot - R_{conv})^2/\nu \sim$ a few years. Hence, angular momentum redistribution in response to rotational braking takes place virtually instantaneously (i.e., in comparison to the angular momentum loss time scale) within the convection zone.

In contrast, the time required for the radiative interior of the Sun to respond to a shear imposed at its outer boundary is much less certain. This is due in part to the fact that the mechanisms by means of which the solar wind torque is transmitted throughout this region are at present not well understood. Evidence derived from the analysis of helioseismological data suggests that within that portion of the radiative interior bounded by $r \approx 0.4R_\odot$ and the base of the convection zone ($r \approx 0.7R_\odot$), the solar angular velocity is only weakly dependent on depth and latitude, with a value about equal to that of the photosphere (see, e.g., Gough 1991, and references therein). Recent measurements of the rotational splittings of p-modes having low angular degree (i.e., $\ell = 1, 2$) further suggest that at great depth the angular velocity increases, attaining a mean value over the region $0.0 \lesssim (r/R_\odot) \lesssim 0.2$ some 4.6 times that of the surface (Toutain and Frölich 1992). Note that if near-uniform rotation proves to be an accurate characterization of the rotational state of the bulk of the solar radiative interior, then angular momentum redistribution therein must be efficient, occurring over a time interval shorter than the evolutionary or spin-down time scales.

Such a state of affairs may not have prevailed early in the Sun's MS lifetime. Observations of rotation among solar-type stars in young, open clusters indicate that shortly after their arrival on the MS, many such stars sustain appreciable angular momentum loss (see, e.g., Stauffer 1991, and references therein). Indeed, from analyses of rotational velocity data for the most rapidly rotating G dwarf stars in the α Persei and Pleiades clusters, deceleration from velocities in excess of 50 km s^{-1} to values < 20 km s^{-1} in a time $\sim$ few $\times 10^7$ years is implied. As in the case of the Sun, this spin-down is presumably a consequence of the torque exerted on an individual star by the magnetically coupled wind it emits. However, the comparatively short rotational braking time inferred from observations is indicative of an enhanced rate of angular momentum loss; $\dot{J} \sim 10^4 \dot{J}_\odot$ for rigid rotation, while $\dot{J} \sim 10^3 \dot{J}_\odot$ if only the outer, convective envelope is braked. The latter, smaller value for $\dot{J}$ is more easily accounted for within the framework of standard, hydromagnetic stellar wind theories (Charbonneau 1992). This suggests that for stars of solar mass, the processes responsible for angular momentum transport from the core to the convection zone may operate with a time scale which is longer than the braking time, during the earliest phases of MS evolution.

Most extant computational models for the evolution of the Sun's internal and surface rotation assume rigid rotation of the convection zone, and treat angular momentum redistribution in the radiative interior by hydrodynamic means only (Endal and Sofia 1981; Pinsonneault *et al.* 1989; Tassoul and Tassoul 1989). Among the processes which have been utilized in studies of angular momentum transport in the non-convective portion of the solar interior are meridional circulation, Ekman flow, waves, and turbulent diffusion associated with the development of shear flow instabilities (Tassoul 1978; Endal and Sofia 1978; Zahn 1983). An alternative possibility is that angular momentum is redistributed inside the Sun by hydromagnetic means (see, e.g., Mestel and Weiss 1987; Roxburgh 1991). In particular, given the high electrical conductivity of the fluid of which the solar radiative interior is comprised, it seems reasonable to consider the consequences of the presence of a large-scale, primordial magnetic field $\mathbf{B}$ within that volume. From the induction equation, it is apparent that the local interaction between the velocity shear arising from wind-induced, differential rotation and the poloidal component $\mathbf{B}_p(= [B_r, B_\theta]$ in spherical polar coordinates) of such a field leads to the production of a toroidal magnetic field component B_ϕ. Associated with the growth of B_ϕ is the development of an azimuthally-directed Lorentz force which acts in such a way as to restore the sheared fluid to a state of uniform rotation. Note that the impulsive application of a transverse shear to a particular fluid volume element threaded by $\mathbf{B}_p$ gives rise to a field B_ϕ which propagates along $\mathbf{B}_p$ at the local Alfvén speed $u_A(= B_p/\sqrt{4\pi\rho})$. As a result of the Lorentz force exerted on each fluid parcel traversed by the disturbance, angular momentum is effectively redistributed along magnetic lines of force. Numerical simulations of the impulsive spin-up of a viscous, electrically conducting fluid indicate that for most poloidal field configurations, a state of nearly uniform rotation prevails within a time which is short compared to evolutionary time scales for the Sun or solar-type stars (Charbonneau and MacGregor 1992a; see also Moss, Mestel, and Tayler 1990).

Elsewhere in this volume, Charbonneau and MacGregor (1992c) present selected results from calculations performed as part of an effort to trace the rotational evolution of an internally magnetized 1 $M_\odot$ star, from the time of its arrival on the zero-age MS to the age of the present-day Sun. These computations pertain to a spherical star containing an axisymmetric poloidal magnetic field whose axis is aligned with the stellar axis of rotation. The convective envelope is taken to rotate rigidly at all times, and within the radiative interior the fluid velocity $\mathbf{u}$ and magnetic field $\mathbf{B}$ are assumed to have the forms $\mathbf{u} = u_\phi(r,\theta,t)\hat{e}_\phi$ and $\mathbf{B} = \mathbf{B}_p(r,\theta) + B_\phi(r,\theta,t)\hat{e}_\phi$. Starting from an initial state of uniform rotation, angular momentum is continuously removed from the convection zone at a rate in accord with the MHD wind model of Weber and Davis (1967). The evolution of u_ϕ and B_ϕ in response to the shear imposed at the periphery of the radiative interior is then calculated by numerical solution of the azimuthal components of the time-dependent induction and momentum equations. The reader is referred to the aforementioned paper by Charbonneau and MacGregor for additional details concerning the computational method and results.

In Figure 2, we show a sample of results relevant to the coupling between the convective envelope and radiative interior for a model with initial angular velocity Ω fifty times that of the present-day Sun, and in which the poloidal field $\mathbf{B}_p$ has dipolar symmetry with average strength in the range 0.1 to 10 G. For the column of panels labelled D2, $\mathbf{B}_p$ has a non-vanishing component B_r normal to the interface $r = R_{conv}$ over a restricted range of θ, while for the panels labelled D3, $B_r = 0$ at $r = R_{conv}$ for all θ. Depicted in the figure are: the induced toroidal field B_ϕ, at $r = R_{conv}$, $\theta = 45°$ (panels [A] and [B]); the torque (dJ/dt) applied to the convection zone by the wind, and the total magnetic (τ_B) and viscous (τ_ν) stresses exerted on the surface at $r = R_{conv}$ (panels [C] and [D]); a global measure of the differential rotation $\Delta\Omega$ (cf. Charbonneau and MacGregor 1992c) within the magnetized portion of the radiative interior (panels [E] and [F]). All quantities are shown as functions of time after the time of arrival on the zero-age MS. Note that for both the D2 and D3 configurations, B_ϕ exhibits an initial phase of approximately linear growth as time increases. During this period, the torque applied to the outer stellar layers by the wind is, for the most part, unbalanced by either the magnetic (in the D2 case) or viscous (in the D3 case) stresses which couple the envelope to the core. As is apparent from panels (E) and (F), the resulting spin-down of the convection zone is accompanied by an increase in $\Delta\Omega$, a quantity where instantaneous magnitude is proportional to the difference between the volume-averaged angular velocity of the radiative interior and the angular velocity of the envelope (cf. Charbonneau and MacGregor 1992c).

At later times, the rotational evolution of the envelope for both the D2 and D3 configurations is controlled by a near balance between the rate of angular momentum removal by the wind, and the rate of angular momentum resupply from the underlying radiative interior through the action of magnetic and/or viscous stresses. Of particular interest in this regard is the case in which the core and convection zone are directly coupled magnetically (e.g., D2). Inspection of panels (A) and (C) reveals that at the time when such an approximate balance of forces commences, the magnitude of B_ϕ is larger for smaller assumed B_p. This behavior can be understood by noting that when viscous and resistive dissipation are both negligible, the azimuthal component of the momentum equation is simply,

$$\rho \frac{\partial}{\partial t} \left(\Omega r^2 \sin^2\theta\right) = \frac{1}{4\pi} \mathbf{B}_p \cdot \nabla \left(r \sin\theta \, B_\phi\right) , \tag{10}$$

where we have written $u_\phi(r,\theta,t) = \Omega(r,\theta,t)\, r\sin\theta$. By integrating equation (10) over the core volume and applying Gauss' theorem, it follows that,

$$\frac{\partial}{\partial t} J_{core} = \frac{\pi}{4} R_{conv}^3 \ < B_r B_\phi > , \tag{11}$$

where J_{core} is the total angular momentum of the radiative interior and $< B_r B_\phi >$ is a weighted average with respect to θ at $r = R_{conv}$. The right-hand side of equation (11) represents the rate at which angular momentum is transported from the core to the envelope by the magnetic stress exerted at the interface. It is this quantity which, at later times, balances the torque applied to the envelope by the wind. Because for the solutions depicted the magnitude of dJ/dt does not depend strongly on the strength or geometry of the internal poloidal field (cf. panels [C] and [D]), the dynamical balance

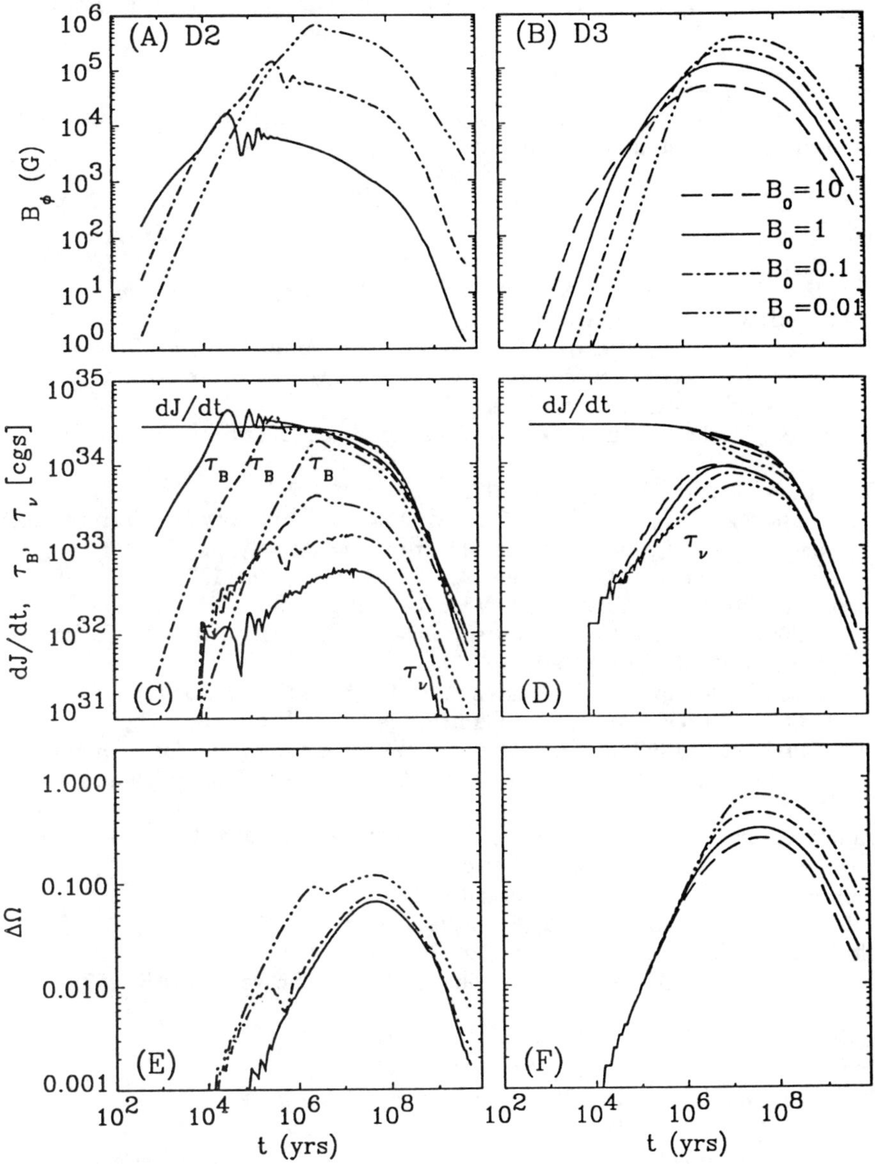

Figure 2: MS rotational evolution of an internally magnetized, 1 $M_\odot$ star as described in the text.

$R^3_{conv} < B_r B_\phi > \approx \dot{J}_{conv}$ requires a larger value of B_ϕ for a smaller assumed value of B_p.

Acknowledgement

I am grateful to Paul Charbonneau for numerous, substantive discussions concerning the content of this paper.

REFERENCES

Abt, H.A., & Levy, S.G. 1985, *Ap. J. Suppl.*, **59**, 229

Aller, L.H. & Chapman, S. 1960, *Ap. J.*, **132**, 461

Boyarchuk, A.A., & Savanov, I.S. 1986, in *Upper Main Sequence Stars with Anomalous Abundances*, eds. C. Cowley, M.M. Dworetsky, and C. Megessier (Dordrecht: Reidel), p. 433

Brown, A., Vealé, A., Judge, P., Bookbinder, J.A., & Hubeny, I. 1990, *Ap. J.*, **361**, 220

Brunish, W.M., & Truran, J.W. 1982, *Ap. J.*, **256**, 247

Cassinelli, J.P., & Lamers, H., J.G.⊥, M. 1987, in *Exploring the Universe with the IUE Satellite*, ed. Y. Kondo (Dordrecht: Reidel), p. 139

Castor, J.I., Abbott, D.C., & Klein, R. 1975, *Ap. J.*, **195**, 157

Cayrel, R., Burkhart, C., & Van't Veer, C. 1991, in *Evolution of Stars: The Photospheric Abundance Connection*, eds. G. Michaud and A. Tutkov (Dordrecht: Reidel), p. 99

Chapman, S., & Cowling, T.G. 1970, *The Mathematical Theory of Non-Uniform Gases* (Cambridge: Cambridge Univ. Press), p. 257

Charbonneau, P. 1992, in *Proceedings of the 7th Cambridge Workshop on Cool Stars, Stellar Systems, and the Sun*, eds. M.S. Giampapa and J.A. Bookbinder, in press

Charbonneau, P., and MacGregor, K.B. 1992a, *Ap. J.*, **387**, 639

————————-1992b, *Ap. J. (Letters)*, in press

————————-1992c, these proceedings

Chiosi, C., & Maeder, A. 1986, *Ann. Rev. Astr. Ap.*, **24**, 329

Chiosi, C., & Nasi, E. 1974, *Astr. Ap.*, **34**, 355

Dearborn, D.S.P., Blake, J.B., Hainebach, K.L., & Schramm, D.N. 1978, *Ap. J.*, **223**, 552

Endal, A.S., & Sofia, S. 1978, *Ap. J.*, **220**, 279

Endal, A.S., & Sofia, S. 1981, *Ap. J.*, **243**, 625

Falk, H.J.,, & Mitalas, R. 1981, *M.N.R.A.S.*, **196**, 225

Garmany, C.D., & Conti, P.S. 1984, *Ap. J.*, **284**, 705

Gough, D.O. 1991, in *Angular Momentum Evolution of Young Stars*, eds. S. Catalano and J. Stauffer (Dordrecht: Kluwer), p. 271

Hartwick, F.D.A. 1967, *Ap. J.*, **150**, 953

Lanz, T., & Catala, C. 1992, *Astr. Ap.*, **257**, 663

Maeder, A. 1984, in *Observational Tests of the Stellar Evolution Theory*, eds. A. Maeder and A. Renzini (Dordrecht: Reidel), p. 299

Mc Crea, W.H. 1962, *Q.J.R.A.S.*, **3**, 63

Mestel, L., & Weiss, N.O. 1987, *M.N.R.A.S.*, **226**, 123

Michaud, G. 1970, *Ap. J.*, **160**, 641

Michaud, G. 1977, in *Highlights of Astronomy, Vol. 4, Part II*, ed. E.A. Miller (Dordrecht: Reidel), p. 177

—————-1987, *Physica Scripta*, **36**, 112

—————-1991, in *Evolution of Stars: The Photospheric Abundance Connection*, eds. G. Michaud and A. Tutkov (Dordrecht: Reidel), p.111

Michaud, G., Charland, Y., Vauclair, S., & Vauclair, G. 1976, *Ap. J.*, **210**, 447

Michaud, G., Tarasick, D., Charland, Y., & Pelletier, C. 1983, *Ap. J.*, **269**, 239

Mitalas, R., & Falk, H.J. 1984, *M.N.R.A.S.*, **210**, 641

Moss, D.L., Mestel, L., and Tayler, R.J. 1990, *M.N.R.A.S.*, **245**, 550

Pinsonneault, M.H., Kawaler, S.D., Sofia, S., & Demarque, P. 1989, *Ap. J.*, **338**, 424

Pizzo, V., Schwenn, R., Marsch, E., Rosenbauer, H., Mühlhäuse, K.-H., Neubauer, F.M., 1983, *Ap. J.*, **271**, 335

Preston, G.W. 1974, *Ann. Rev. Astr. Ap.*, **12**, 257

Roxburgh, I.W. 1991, in *Angular Momentum Evolution of Young Stars*, eds. S. Catalano and J. Stauffer (Dordrecht: Kluwer), p. 365

Stauffer, J. 1991, in *Angular Momentum Evolution of Young Stars*, eds. S. catalano and J. Stauffer (Dordrecht: Kluwer) p. 117

Stein, R.F. 1966, in *Stellar Evolution*, eds. R.F. Stein and A.G.W. Cameron (New York: Plenum), p. 3

Tanaka, Y. 1966, *Pub. Astron. Soc. Japan*, **18**, 47

Tassoul, J.-L., & Tassoul, M. 1989, *Astr. Ap.*, **213**, 397

Tassoul, J.-L. 1978, *Theory of Rotating Stars* (Princeton: Princeton University Press)

Toutain, T., & Fröhlich, C. 1992, *Astr. Ap.*, **257**, 287

Vauclair, S., & Vauclair, G. 1982, *Ann. Rev. Astr. Ap.*, **20**, 37

Watson, W.D. 1970, *Ap. J. (Letters)*, **162**, L45

—————-1971, *Astr. Ap.*, **13**, 263

Weber, E.J., & Davis, L. 1967, *Ap. J.*, **148**, 217

Zahn, J.-P. 1983 in *Asrtophysical Processes in Upper Main Sequence Stars* (Geneva: Publ. Observatoire Genève), p. 253

Inside the Stars, IAU Colloquium 137
ASP Conference Series, Vol. 40, 1993
Werner W. Weiss and Annie Baglin (eds.)

STELLAR ACTIVITY: CONSTRAINTS EXPECTED FROM SPACE EXPERIMENTS

CLAUDE CATALA
Laboratoire de Recherche Spatiale, Observatoire de Paris, Section de Meudon, 92195 Meudon Principal Cedex, France

ABSTRACT

Dynamo theory needs to be guided and tested by observations. However, solar observations alone cannot provide sufficient tests of the dynamo models, and constraints obtained from stellar observations are still few, indirect, and poorly related to the properties of dynamos.

This review discusses which parameters must be observed on stars other than the Sun to usefully guide the theoretical efforts and test the resulting models. We show that most of these observables, which characterize the stellar activity phenomena, are accessible only from space, and that space-based experiments are one of the best tools to trigger progress in the field of solar and stellar activity in the next few decades.

INTRODUCTION

One of the major problems of stellar physics concerns the generation of magnetic fields in the interior of the sun and stars by dynamo action, and the effects of such magnetic fields in stellar atmospheres, known by the name of stellar activity.

The most widely accepted idea is that the sun and solar-type stars produce their magnetic fields by an $\alpha\omega$-type dynamo, for which the large scale magnetic field is maintained by the composition of two types of motions of the conducting plasma: turbulent motions, commonly assumed to be due to convection or overshooting, and differential rotation (Cowling 1980).

Two complementary theoretical approaches have been used to attack this problem:

- *the mean-field theory:* the collective action of small-scale motions is represented by a mean current parallel to the large scale magnetic field, the so-called α-effect (e.g. Steenbeck et al. 1966; Moffat 1978; Krause and Rädler 1980). Although the mean-field theory has been successful in reproducing some aspects of the solar cycle, like e.g. the solar cycle period and the butterfly diagram (Steenbeck and Krause 1969), it is only a linear theory, and does not take into account the feedback of the magnetic field B on the plasma motions through the Lorentz force, that limits the growth of B. Thus, the mean-field theory has two major limitations: first, it is not capable of predicting the intensity of the resulting magnetic field; second, we cannot be sure that the modes excited in the linear regime are still the dominant ones in the nonlinear situation. In other words, even the spatial large scale structure of the magnetic field derived from a linear analysis may be difficult to compare with the real situation in the sun and stars. However, several possibilities exist to take into account the non-linear effects in a parameterized way (e.g. α–quenching). If the resulting parameterization can be guided and tested by appropriate observational constraints, there is hope that some meaningful models may be produced in the future.

- *the nonlinear numerical simulations:* these simulations aim at solving the full set of nonlinear MHD equations in 2D or 3D (Gilman and Miller 1981; Gilman 1983; Glatzmaier 1984; Brandenburg 1990). The problems encountered by the mean–field theory disappear, but a serious limitation remains: due to the limited (although quite large) capacity and power of modern computers, the spatial resolution reachable by these simulations is rather low, and consequently the small scales that play a decisive role in the stellar dynamo process are not resolved. However, we can hope that due to the rapid improvements in computer technology, we are not so far from being able to produce meaningful numerical experiments that can be used for understanding stellar dynamo.

The improvements in the theory that will certainly be achieved in the coming years will be of little impact if they are not accompanied by a solid observational basis, that is needed both as a guide and as a source of constraints for the models. Typically, one needs to determine *scaling laws* describing the dependence of the dynamo process upon several basic characteristic parameters of the stars, chosen to probe various regimes of the magneto–hydrodynamics in the stellar interiors.

Unfortunately, the observational constraints that can be placed on dynamo theories are still very poor. The only star for which we have a significant knowledge concerning the surface magnetic field is the Sun. The large scale spatial structure of the solar magnetic field, and in particular the latitude of appearance of sunspots, as well as the characteristics of the solar cycle, have been often used as constraints for the dynamo models. However, the Sun alone, in spite of the luxury of details that it can offer to us on all spatial scales, represents only one point in parameter space, and will never provide us with the scaling laws that are needed to guide and constrain the dynamo models.

In this paper, we shall extend the review of Mangeney and Praderie (1983), by discussing what kind of observables are needed for providing these constraints, and how these observables can be obtained. We show that most of the needed observables can be obtained only by space–based instruments, and that a mission like PRISMA (Lemaire et al. 1991), which will observe a large sample of stars in visible photometry, in UV and XUV spectro–photometry, continuously for large intervals of time, will certainly bring decisive elements for our understanding of solar and stellar dynamos.

WHAT OBSERVABLES DO WE NEED?

The observables

In order to guide and test the dynamo theories, we need to find out if there is a relation between the output of dynamo, i.e. the characteristics of the emergent magnetic field (magnetic flux, horizontal structure of the magnetic field, cyclic behavior) on one hand, and what we think are the ingredients of the dynamo, i.e. convection and rotation, on the other hand. If such a relation exists, we must derive it as accurately as possible. In other words, we need to determine scaling laws between these two classes of observables, that must be verified by the theoretical models.

1. The output of the dynamo

First of all, scaling laws concerning the *magnetic flux* are of utmost necessity for guiding the modelling efforts. The magnetic flux indeed provides a measurement of the efficiency of the dynamo process, and its dependence upon the basic stellar parameters would yield tests of the dynamo models.

The output of all types of dynamo theories also include some prediction about the large scale magnetic field structure at the stellar surface. A success of the mean–field theory, for instance, has been to reproduce the latitude of emergence of sunspots during the solar cycle (Steenbeck and Krause 1969). Therefore, the observation of the *latitudinal distribution of active regions* in other stars, and more precisely any trend that could be observed between this distribution and the basic stellar parameters would provide a guide and a decisive test for dynamo theories.

The *north/south asymmetry* in the number of sunspots is a well–known property of solar activity, that depends on the degree of nonlinearity of the dynamo, and that can be compared to the predictions of the models (Vizoso and Ballester 1990). A useful guide would be available if we could have this type of information for a large number of stars spanning the HR diagram.

Nonlinear dynamo theories can also predict the degree of non–axisymmetry of the resulting magnetic field. The existence of *active longitudes* on the Sun is already a useful constraint, but here again, we need a systematic exploration of the stellar parameter space to derive the trends of the departures from axisymmetry with the basic characteristics of the stars (rotation, mass, differential rotation, etc...).

Finally, the *period of the activity cycle* can also be predicted by dynamo theories, as well as the evolution of the magnetic field structure and intensity along the cycle. Very useful scaling laws involving these two parameters could be obtained if they could be estimated for a large number of stars.

2. The ingredients of the dynamo

In addition to the magnetic field structure and its intensity, another class of observables are needed for allowing us to guide and test the theoretical developments: we need to determine accurately enough the basic stellar parameters that presumably control the dynamo process. The first one of these observables is *rotation.* Differential rotation indeed generates a toroidal magnetic field from a poloidal one, by stretching the magnetic field lines. Rotation and differential rotation also enter the α–effect because of their infuence on the mean helicity of the small scale motions. It is usually assumed that differential rotation $\delta\Omega/\ell$ (where ℓ is a characteristic length for the variation of rotation rate) increases with Ω (that is ℓ does not increase much with Ω). In that case, we can look for scaling laws involving Ω only. However, it would be valuable to introduce the *differential rotation* in the analysis, *both in depth and in latitude*, whenever possible. Differential rotation indeed is directly involved in the $\alpha\omega$ dynamo, and therefore could be used in the derivation of scaling laws, were we able to observe it in a large number of stars.

In recent years, the evidence that the site of solar dynamo is at the bottom of the convection zone has accumulated (e.g. Stenflo 1990). It is not clear whether the convective motions themselves are responsible for the α–effect at that location, or whether the turbulence induced by overshooting or simply by the rotational shear plays the most important role. In all cases, the depth at which the dynamo action occurs, i.e. the depth of the convection zone, has a great importance in the process. This is why introducing estimates of the *depth of the convection zone* in the scaling laws can be very fruitful as a guide for stellar dynamo models.

The traditional description of convection by the mixing length theory may not be sufficient to properly take into account the small scale motions acting to produce the poloidal magnetic field through the α–effect. Therefore, it is also very important to improve the *modelling of convective velocity fields.* Such a progress in the theory of convection also needs accompanying observational tests.

Finally, since most of the observables that are needed depend on stellar *mass and age*, it is also necessary to determine these two quantities with the highest accuracy.

The scaling laws

1. Which magnetic tracers must we choose?

Since the direct measurement of the magnetic flux is possible only in some very few particular cases (very slow rotators for the Robinson method, see e.g. Saar 1990; bright fast rotators with large scale magnetic field for the Zeeman–Doppler technique, see e.g. Donati and Semel 1990), we have to rely on well–chosen magnetic tracers. The first question is thus: which tracers are best suited for a measure of the magnetic flux? Table I presents the most commonly used magnetic activity tracers, and summarizes the various problems encountered when using them.

The presence of an important photospheric contribution for visible lines makes them inconvenient to use for determining the chromospheric flux. This well–known problem disappears at shorter wavelengths.

For optically thick lines, it is not straightforward to translate the measured flux into an emission measure, which is a useful quantity for building empirical models of the atmosphere (Judge 1990). Even though this problem may be overcome by a careful treatment of the line transfer, it is always preferable to use optically thin lines. From Table I, it can be seen that all the relevant optically thin lines are in the UV and XUV ranges.

TABLE I the magnetic tracers

layer	lines	photospheric contribution	optical thickness	basal flux	stars
chrom.	Hα	yes	thick	strong?	very active cool stars
chrom.	Ca II H & K	yes	thick	strong	F to M stars
chrom.	Mg II h & k	no	thick	strong	F to M stars
chrom.	Si II 1810 A	no	thin	weak	F to M stars
chrom.	C II 1335 A	no	thin	weak	F to M stars
t.r. + chrom.	Lyα	no	thick	??	A7 to M stars
t.r.	C IV 1550 A	no	thin	weak	F to M stars
t.r.	He II 1640 A	no	thick	??	F to M stars
t.r.	N V 1240 A	no	thin	??	F to M stars
corona	XUV lines	no	thin	weak	all spectral types

Schrijver (1987) suggested the wide existence for cool stars of a non–magnetic energy flux (e.g. an acoustic flux), superimposed on the magnetic heating, witnessed by the presence of a lower limit for the flux in the magnetic activity tracers of a large sample of cool stars. This basal flux, if present, must be subtracted if one wants to study the scaling laws concerning magnetic activity. Since the determination of this basal flux is a complicated matter, lines with weak basal fluxes will be considered as more convenient magnetic tracers. Rutten et al. (1991) have studied the basal fluxes as a function of $B - V$ color for the usual magnetic tracers listed in Table I. Their work demonstrates that UV lines (except Mg II h & k) have very low basal fluxes, and therefore must be chosen rather than visible lines.

Lines formed at coronal temperatures (in the XUV range) have a major advantage: stars all over the HR diagram (and not only cool stars) show signs of coronae (Vaiana et al. 1981), and their X–luminosity seems to follow the same type of dependence on an effective Rossby number as the cool stars (Mangeney and Praderie 1984). If we can have access to the XUV range, stars of all spectral types can be included in the analysis, and the dependence of this type of activity upon the basic stellar parameters can be studied over a very wide range of stellar properties.

In conclusion, it is clear that the access to the UV and XUV ranges provided by space experiments brings enormous advantages, and allows us to choose the best magnetic tracers.

2. How does the magnetic flux scale with stellar parameters?

A significant test of the dynamo theories would come from the answer to the question: is dynamo efficiency related to the dynamo number N_D, which is the square of the ratio between the magnetic diffusion time and the magnetic field amplification time, $N_D = (\tau_{diff}/\tau_A)^2$?

Under several assumptions on the length scales involved in the α–effect, it can be shown that N_D scales as Ro^{-2}, where Ro is the Rossby number, measuring the ratio of the inertial forces to the Coriolis forces. The initial question has thus been translated by several authors into the following one: are magnetic tracers related to Ro, and if yes, how?

Many attempts have been made to answer this question, by looking for correlations between the emission flux in some magnetic activity tracers (see Table I) and either the rotation period, or a Rossby number, calculated differently from one author to another, or even simply the age (Noyes et al. 1984; Hartmann et al. 1984; Mangeney and Praderie 1984; Simon et al. 1985; Marilli et al. 1986; Basri 1987; Rutten and Schrijver 1987). The goal of this review is not to give a detailed account of all these (sometimes discrepant) results. A fair summary of these analysis would be that they are not entirely conclusive, in that sense that they do not demonstrate clearly which parameter, if any, controls the magnetic tracer fluxes.

There are three observational requirements to fulfill in order to improve the results of this type of analysis. First, the star sample must be as large and unbiased as possible, and must cover the HRD as completely as possible. This requirement implies in particular the access to the XUV range, because it is the only wavelength domain where hot stars are "active", as mentioned above. Second, we need to choose the best magnetic tracers available, which are in the UV and XUV, as discussed in the previous section. Finally, the calibration of these tracers in terms of magnetic flux (which relies only on solar measurements) must be improved, which implies the simultaneous observation of the tracers and of the magnetic flux in a sub–sample of these stars at different masses. This effort has been initiated (e.g. Marcy 1984; Saar and Schrijver 1987), but definitely needs to be continued. Thus methods for direct measurements of the magnetic flux must be strongly developed, even if they are limited to a small number of targets.

3. How does the structure of the magnetic field scale with stellar parameters?

Both the mean–field theory, and the nonlinear dynamo theories predict that the horizontal structure of the magnetic field depends on the properties of the dynamo, and in particular on the dynamo number. Therefore, we must try to correlate our estimates of the magnetic horizontal scale with N_D. This point is very important, and basically untouched so far. This is certainly where further efforts should concentrate in the near future.

What we need to study are the relationships between the various characteristics of the horizontal large scale structure of the magnetic field and the basic stellar parameters, including mass, age, rotation period, Rossby number, etc... The horizontal large scale structure of the magnetic field can be characterized by the existence of preferential latitudes for the emergence of the toroidal magnetic field that will give an estimate of the horizontal scale of the dynamo wave, or by north/south asymmetries and active longitudes that will describe some aspects of the nonlinearity of the dynamo.

We note that the calibration of magnetic tracers is no longer a problem here, since we are not interested in the magnetic flux itself, but simply in its spatial distribution.

HOW CAN WE OBTAIN THE NEEDED OBSERVABLES?

Activity cycles

The measurement of cycle periods requires systematic observations of the same stars over very long periods of time ($\geq$ 10 years), much longer than the typical life–time of a space mission (a couple of years). These observations must rather rely on ground–based techniques, e.g. long–term monitoring of the Ca II H & K flux, than on space–based observations. A database of cool stars with measured cycle periods already exists (Wilson 1978; Vaughan et al. 1981), and a good selection of targets for a space mission like PRISMA should include objects from this database.

Magnetic flux at stellar surface

1. Magnetic tracers

The magnetic flux in the stellar atmosphere can be derived from the emission flux of convenient magnetic tracers. We have shown that these best tracers are in the UV and XUV domains, and therefore can be obtained only from space. These are the Mg II, C II, Si II, C IV, N V, He II, Lyα lines in the UV, and the coronal lines in the XUV. The instruments on board the PRISMA mission will provide accurate measurements of fluxes in these UV and XUV lines. As mentioned above, for a good calibration of these tracers in terms of magnetic flux, some of the studied stars will have to be observed simultaneously from the ground with the Robinson or the Zeeman–Doppler techniques, depending on their rotation rate. The lines mentioned above probe different temperature regimes, and thus different layers of the stellar atmosphere. Therefore, the simultaneous measurements of these line fluxes with PRISMA will provide a tool for studying the height dependence of the magnetic energy deposition, from the base of the chromosphere up to the corona. For those stars with a simultaneous direct measurement of the photospheric magnetic flux by the Robinson or Zeeman–Doppler techniques, the situation will be ideal, since we will have a simultaneous measure of a) the magnetic flux at the stellar photosphere and b) the radiative losses in various layers above the photosphere.

2. Mean value of the activity level

For a meaningful analysis of the dependence of the tracers of magnetic energy flux upon any kind of stellar parameter, it is of course necessary to determine first a mean value of the activity level. It can be rather dangerous to restrain on observations obtained at a particular rotational phase, since we may face a very active longitude at that moment, or on the contrary a very quiet region, not representative of the real activity level of the star.

Figure I illustrates this point by showing the amplitude of rotational modulation for some typical stars, superimposed on the original plot R'_{HK} versus Ro

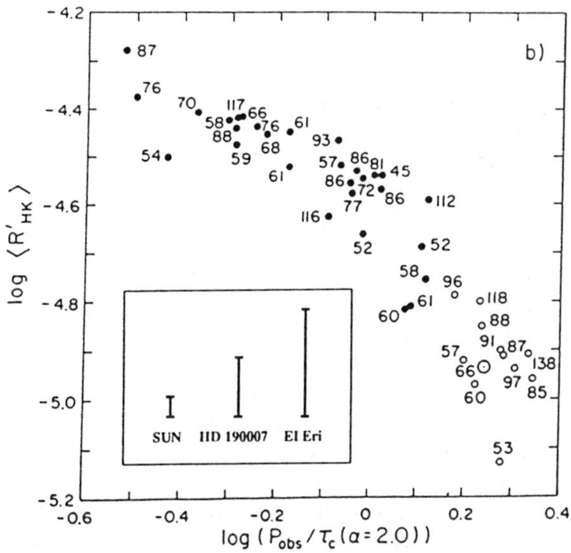

FIGURE I Amplitude of the rotational modulation for three types of stars (insert): the Sun near activity maximum, HD 190007 = active K dwarf, EI Eri = RS CVn system. The rotational modulation amplitudes are superimposed on the original plot showing the dependence of the Ca II H & K flux upon the Rossby parameter (adapted from Noyes et al. 1984)

by Noyes et al. (1984). This example shows that, without a careful averaging of the emission flux variations during the rotation (this averaging procedure was actually done in Noyes et al.'s paper), an artificial scattering of the data will appear and eventually hide possible scaling laws.

Thus, we conclude that an important requirement for the observations is a good rotational phase coverage, which implies *a monitoring of the stars on a time scale comparable to or greater than their rotation period.* Such a requirement does not necessarily imply the use of a space–based instrument, since ground–based multi–site campaigns will lead to similar results. However, the probability of success of such campaigns is rather low, and a space mission is by far the most efficient solution for obtaining a continuous coverage. The design of PRISMA makes it a perfect tool for the kind of observations described here.

Distribution of active regions, surface rotation, surface differential rotation

In the absence of spatial resolution at the stellar surface, the missing spatial information can be obtained by Doppler Imaging techniques. These methods rely on the stellar rotation that continuously modifies the aspect of the visible stellar disk, as well as the projected velocity of any given area at the stellar surface. A precise description of these methods is beyond the scope of this paper, but an excellent review on the subject can be found in Donati (these proceedings). The important point here is that they require a *good phase coverage of the stellar rotation.* Besides, since the magnetic structures can in principle grow and evolve on time scales shorter than the rotation period (this is indeed the case for the Sun), it is important that the data are obtained in a *monitoring as continuous as possible* to distinguish between rotation–induced variations and intrinsic variations of the lines.

Doppler Imaging techniques have been successfully applied for obtaining photospheric images (Vogt 1988; Piskunov et al. 1990; Strassmeier et al. 1991, Donati et al. 1992, Jankov and Foing 1992). These photospheric images have been obtained for extremely active stars only, for which the photospheric magnetic field appears to be on very large scales. On the other hand, in order to derive scaling laws concerning the horizontal structure of the magnetic field, we need to observe a sample of stars including moderately active stars in addition to very active stars. If we assume that the structure of the magnetic field in single moderately active stars looks like that of the Sun, then we expect that the magnetic field will be structured on small spatial scales at the photosphere (spots), whereas the scales at the level of the chromosphere, the transition region and the corona will be much larger. Figure II illustrates this point by showing two simultaneous images of the Sun, one at the photosphere (K1) and the other at the chromosphere (K3). Clearly, the small scale photospheric structures would be impossible to detect by Doppler Imaging, which provides only a coarse resolution of the stellar surface, while the large scale structures of the chromospheric image would be easily mapped by this technique, if the Sun was a faster rotator. The situation is even better in the transition region, where the active regions appear on larger scales. Therefore, the study of the magnetic field structure in single, moderately active stars will be possible only through chromospheric or transition region imaging. Since most of the convenient lines are in the UV, it is necessary to use a UV space mission providing a sufficient spectral resolution, and having the capability to monitor continuously its targets for long intervals of time, like PRISMA.

Chromospheric or transition region imaging has been limited so far to very simple analysis: recognizing and tracking emission components in line profiles (Gondoin 1986), or multi–Gaussian fitting (Neff et al. 1989). However, provided the S/N and resolving power are sufficiently high, more sophisticated Doppler Imaging methods, similar to those used for photospheric Doppler Imaging, can be applied to chromospheric and transition region data. Time–resolved, high spectral resolution, moderate– to high–S/N ratio data strings of chromospheric and transition region lines could be translated into ab–initio images of these layers. Simulations of such image reconstructions are under way, and preliminary results suggest that spectra with $R \approx 30000$ and S/N ratios of the order of 30–50 can yield useful chromospheric images of moderately active stars rotating with a period of 3–10 days (Donati and Catala, in preparation).

In particular, a very useful information that can be derived from such images is the presence or absence of preferential latitudes for the activity, easier to obtain for the active regions at chromospheric and transition region levels than for photospheric spots. This information is directly related to the large scale structure of the toroidal magnetic field (as are the latitudes of appearance of sunspots), and can provide strong constraints on the dynamo models, as discussed earlier.

The detection of north/south asymmetries may be more problematic: when a star is seen from a direction in its equatorial plane, Doppler Imaging cannot solve the north/south ambiguity; when it is seen from a direction near one pole,

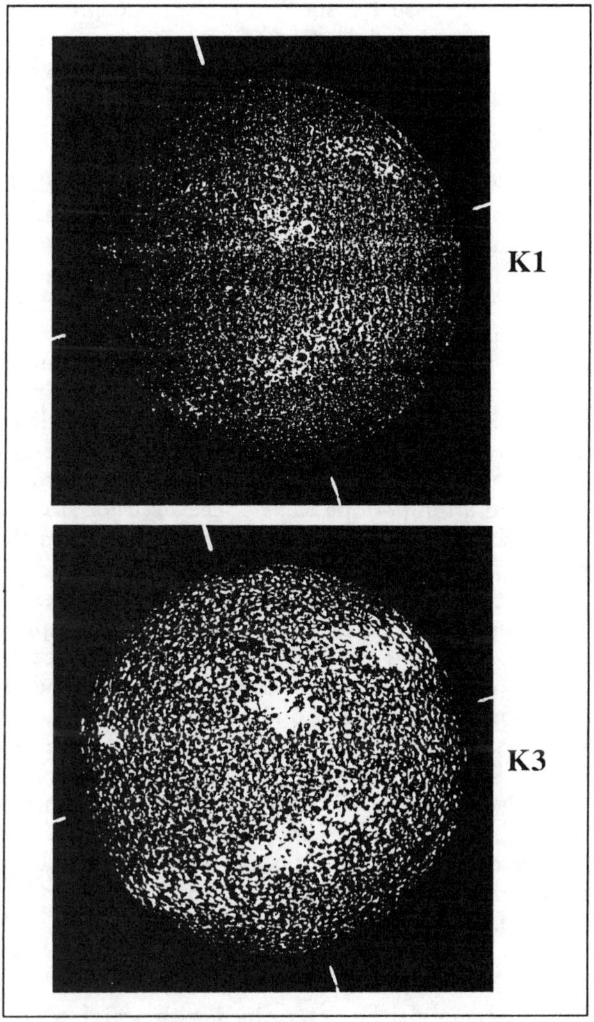

FIGURE II Quasi-simultaneous images of the Sun at photospheric (K1) and chromospheric (K3) levels, obtained with the Meudon Observatory spectroheliograph. Note the difference in the spatial scales of the magnetic structures between the photosphere and the chromosphere.

the other hemisphere is invisible. However, in some favorable cases, e.g. if the viewing angle is intermediate, and if the active regions are restricted to low latitudes (which is the case for the Sun), then the presence of a strong north/south asymmetry in the distribution of active regions can be detectable. Such observations could be used in replacement of the asymmetry of the number of starspots to constrain the nonlinearity of the dynamo.

Doppler images can also reveal the presence of active longitudes, if they are very prominent. In fact, the departures from axisymmetry are more easily detectable than latitudinal distribution of active regions. It is well–known indeed that the uncertainties in the Doppler image reconstruction are always more important in latitude than in longitude.

A by–product of the monitoring of UV and XUV lines is a confirmation of the surface rotation period, that can be obtained from the ground in principle (by Ca II H & K or photometric modulation). If the departures from axisymmetry are strong enough, the UV and XUV lines show a rotational modulation, and the rotation period can be obtained simply by a Fourier analysis of line–flux time series. Also, since starspots modulate the light curve with the rotation period, high precision photometric monitoring from space can also yield a precise measurement of the rotation period, if the star can be monitored for several rotation cycles. Thus, rotation periods can be measured for stars for which the rotational modulation amplitude is too low to be observed from the ground.

Finally, if chromospheric Doppler images corresponding to several successive rotation cycles can be obtained, surface differential rotation can be measured. Such a method has already been successfully applied to photospheric images of the RS CVn system UX Ari (Vogt and Hatzes 1990). The surface differential rotation can be useful in two different ways as a test of dynamo theories. First, we can assume that the surface latitudinal differential rotation prevades throughout the convection zone, as in the Sun, and therefore constrain its value at the base of the convection zone, presumably the site of the dynamo. Second, the mechanism by which the latitudinal differential rotation is maintained probably implies some feedback of the dynamo–generated magnetic field, and is not understood. Obtaining a significant number of measurements of this phenomenon for different types of stars spanning the HR diagram would provide a very useful guide to understand this problem.

Mass and age

These two basic parameters must be determined with a high accuracy, in order to analyze the evolution of active phenomena in time for different masses. The traditional methods for deriving them rely on stellar evolution models. For single main sequence stars, the mass is determined from the location in the HR diagram. The estimate of the age poses a more serious problem. Reliable estimates of stellar ages can be obtained only for members of open clusters. For field stars before and after the main sequence, theoretical evolutionary tracks can be used, but are not necessarily reliable enough to provide meaningful ages. On the main sequence, the situation is even worse, and age estimates for field stars must rely on measurements of surface Lithium abundances and basically untested theory of turbulent transport in radiative interiors.

As widely discussed in these proceedings, a major breakthrough is to be expected from space–based asteroseismological techniques. In particular, the measurements of low–degree p–modes frequencies and separations will provide very accurate estimates of mass and age of all types of stars, by examining their location in the asteroseismological HR diagram. The resulting gain in accuracy and reliability of mass and age determinations will have a tremendous impact on the analysis of stellar activity.

Modelling of convection

We have already insisted on the major role played by convective velocity fields. The description by the mixing length theory seriously limits the progress of $\alpha\omega$ dynamo theories. Asteroseismological observations will probably allow us to improve our modelling of convection well beyond the mixing length theory, and thus to introduce better descriptions of the convective velocity fieds into the dynamo theories.

Internal differential rotation

The rotational splittings of low–degree p–modes, which in principle can be measured for stars that rotate fast enough, can provide us with a crude estimate of the internal differential rotation. The rotational splittings of a given mode can indeed be written as

$$2\pi\Delta\nu_{n\ell m} = -m\int_0^R \Omega(r)K_{n\ell}(r)r^2dr$$

where $K_{n\ell}(r)$ is the rotational kernel, a function of the mode amplitudes depending only on n and ℓ. Thus, the quantity $2\pi\Delta\nu_{n\ell m}/m$ corresponds to the average of the internal rotation rate $\Omega(r)$, weighted with the normalized function $r^2K_{n\ell}(r)$. Although this function tends to favor the regions near the stellar surface, the contribution of the central regions to the rotational splittings can be significant if the rotation rate near the center is several times higher than the surface rotation rate. In these cases, the comparison between the surface rotation rate $\Omega_{\rm surf}$ and the rotational splittings can tell us how different from $\Omega_{\rm surf}$ the internal rotation rate can be, unfortunately without giving us more details about $\Omega(r)$. Although rather crude, this procedure can allow a comparison of the amount of differential rotation in different types of stars (e.g. young versus old stars; convective versus radiative stars, etc...).

Depth of the convection zone

The frequencies of low–degree p–modes can also yield an estimate of the depth of the convection zone. This estimate can perhaps be obtained directly through inversion techniques, but this requires extremely accurate measurements of the frequencies which may or may not be possible.

On the other hand, asteroseismology will necessarily lead to decisive improvements in stellar structure modelling, and the depth of the convection zone will thus be determined rather reliably through modelling in all cases.

SUMMARY AND CONCLUSION

In this review, we have shown that dynamo theories need meaningful observational guides and tests, in particular scaling laws followed by the characteristics of the magnetic field, which include the magnetic flux and the horizontal large scale structure of the magnetic field.

The major requirements for obtaining these scaling laws are *the use of UV and XUV spectroscopy and spectrophotometry, the continuous monitoring of targets over one or several rotation periods*, and the possibility to obtain *asteroseismological measurements*. This demonstrates the need for space–based instruments, dedicated to this type of observations. A mission like PRISMA, whose goal is precisely to provide a continuous monitoring of a large sample of stars, in high precision visible photometry, and in UV spectroscopy and spectrophotometry, is extremely well suited for these purposes, and is therefore expected to bring a major breakthrough in our study of stellar magnetic activity.

ACKNOWLEDGMENTS

I am very grateful to Françoise Praderie, André Mangeney, Elisabeth Ribes, Evry Schatzman, and Jean–François Donati for fruitful discussions. I am indebted to Elisabeth Ribes, Irina Soru–Escaut, and Gérard Servajean for providing me with the material necessary for producing Fig. II.

REFERENCES

Basri, G. 1987, *ApJ* 316,377

Brandenburg, A., Nordlund, A., Pulkkinen, P., Stein, R.T., Tuominen, I. 1990, *A&A* 232,277

Cowling, T.G. 1981, *ARA&A* 19,115

Donati, J.–F., Semel, M. 1990, in *The Sun and Cool Stars: Activity, Magnetism, Dynamos*, IAU Coll. 130, Helsinki, eds. I. Tuominen, D. Moss, G. Rüdiger, *Lecture Notes in Physics* No 380, p. 326

Donati, J.–F., Brown, S.F., Semel, M., Rees, D.E., Dempsey, R.C., Matthews, J.M., Henry, G.W., Hall, D.S. 1992, *A&A*, in press

Gilman, P.A., Miller, J. 1981, *ApJS* 46,211

Gilman, P.A. 1983, *ApJS* 53,243

Glatzmaier, G.A. 1984, *J. Comput. Phys.* 55,461

Gondoin, P. 1986, *A&A* 160,73

Hartmann, L., Baliunas, S.L., Duncan, D.K., Noyes, R.W. 1984, *ApJ* 279,778

Jankov, S., Foing, B.H. 1992, *A&A*, in press

Judge, P.G. 1990, *ApJ* 348,279

Krause, F., Rädler, K.–H. 1980, *Mean–Field Magnetohydrodynamics and Dynamo Theory*, Akademie–verlag Berlin, Pergamond Press, Oxford

Lemaire, P., Appourchaux, T., Catala, C., Catalano, S., Frandsen, S., Jones, A., Weiss, W. 1991, *Adv. Space Res.* Vol. 11, No. 4, p. 141

Mangeney, A., Praderie, F. 1983, in *Active Phenomena in the Outer Atmosphere of the Sun and Stars*, France–Japan Seminar, Paris, eds. J.C. Pecker, Y. Uchida, Observatoire de Paris Press, p. 96

Mangeney, A., Praderie, F. 1984, *A&A* 130,143

Marilli, E., Catalano, S., Trigilio, C. 1986, *A&A* 167,297

Moffat, H.K. 1978, *Magnetic Field Generation in Electrically Conducting Fluids*, Cambridge University Press

Neff, J.E., Walter, F.M., Rodono, M., Linsky, J.L. 1989, *A&A* 215,79

Noyes, R.L., Hartmann, L.W., Baliunas, S.L., Duncan, D.K., Vaughan, A.H. 1984, *ApJ* 279,763

Piskunov, N.E., Tuominen, I., Vilhu, O. 1990, *A&A* 230,363

Rutten, R.G.M., Schrijver, C.J. 1987, *A&A* 177,155

Rutten, R.G.M., Schrijver, C.J., Lemmens, A.F.P., Zwaan, C. 1991, *A&A* 252,203

Saar, S.H. 1990, in *The Sun and Cool Stars: Activity, Magnetism, Dynamos*, IAU Coll. 130, Helsinki, eds. I. Tuominen, D. Moss, G. Rüdiger, *Lecture Notes in Physics* No 380, p. 389

Saar, S.H., Schrijver, C.J. 1987, in *Cool Stars, Stellar Systems, and the Sun*, Linsky J.L., Stencel R.E. (eds.), Springer, New–York, p. 38

Schrijver, C.J. 1987, *A&A* 172,111

Simon, T., Herbig, G., Boesgaard, A.M. 1985, *ApJ* 293,551

Steenbeck, M., Krause, F., Rädler, K.-H. 1966, *Z. Naturforsch.* 21a,1285

Steenbeck, M., Krause, F. 1969, *Astron. Nachr.* 291,49

Stix, M. 1991, *Geophys. Astrophys. Fluid Dynamics* 62, 211

Strassmeier, K.G., Rice, J.B., Wehlau, W.H., Vogt, S.S, Hatzes, A.P., Tuominen, I., Piskunov, N.E., Hackman, T., Poutanen, M. 1991, *A&A* 247,130

Vaiana, G.S., Cassinelli, J.P., Fabbiano, G., Giacconi, R., Golub, L., Gorenstein, P., Haisch, B., Harnden, F.R., Jr., Johnson, H.M., Linsky, J., Maxson, C.W., Mewe, R., Rosner, R., Seward, F., Topka, K., Zwaan, C. 1981, *ApJ* 245,163

Vaughan, A.H., Baliunas, S.L., Middelkoop, F., Hartmann, L.W., Mihalas, D., Noyes, R.W., Preston, G.W. 1981, *ApJ* 250,276

Vizoso, G., Ballester, J.L. 1990, *A&A* 229,540

Vogt, S.S 1988, in *The Impact of very high S/N Spectroscopy on Stellar Physics*, IAU Symp. 132,253, Cayrel de Strobel G., Spite M. (eds.), Kluwer, Dordrecht

Vogt, S.S, Hatzes, A.P. 1990 in *The Sun and Cool Stars: Activity, Magnetism, Dynamos*, IAU Coll. 130, Helsinki, eds. I. Tuominen, D. Moss, G. Rüdiger, *Lecture Notes in Physics* No 380, p. 297

Wilson, O.C. 1978, *ApJ* 226,379

Inside the Stars, IAU Colloquium 137
ASP Conference Series, Vol. 40, 1993
Werner W. Weiss and Annie Baglin (eds.)

EVOLUTION OF STELLAR ACTIVITY IN EARLY POST-MAIN-SEQUENCE PHASES

A. MAGGIO, S. SCIORTINO
Istituto e Osservatorio Astronomico, Palermo, Italy

L. BIANCHI
Osservatorio Astronomico di Torino, Italy

F.R. HARNDEN JR.
Harvard-Smithsonian Center for Astrophysics, USA

R. ROSNER
Department of Astronomy and Astrophysics and Enrico Fermi Institute, University of Chicago, USA

RESULTS AND DISCUSSION

We present preliminary observational evidences on the variation of the activity level in late type stars, during the evolutionary phases on the main sequence and beyond. We have selected a sample of 51 stars (Fig. 1), lying mostly along evolutionary tracks between 1.3 and 1.7 solar masses, which have been observed in soft X-rays with the *Einstein* Observatory, and in UV with IUE (Maggio *et al.* 1990; Haisch *et al.* 1990). Two ROSAT targets, and four new IUE observations are also included.

We find that the radiative emission from the outer atmospheres of stars with $M > 1.6 M_{\odot}$ seems to behave differently than for stars with lower mass.

On the main sequence, the X-ray luminosity of most stars with B-V < 0.42 (spectral type F3) is relatively low, at $L_x \sim 3 \times 10^{28}$ $erg\ s^{-1}$ (Fig. 2). In the early evolutionary phases beyond the main sequence, the X-ray luminosity of the higher mass stars tend to increase sistematically up to $\sim 10^{30}$ $erg\ s^{-1}$, while the lower mass stars show an initial moderate increase followed by a drop, at B-V $\sim$ 0.6, below our sensitivity threshold.

A mass-dependent behavior is shown also by the C IV (1550 Å) emission line strength, a transition region diagnostic. For $M > 1.6\ M_{\odot}$ stars near the main sequence, the C IV luminosity is $L_{CIV} \sim 10^{29}$ $erg\ s^{-1}$. While evolution proceeds, it appears to decrease steadly, reaching our detection limit near the base of the red giant branch. On the other hand, for the $M \leq 1.6\ M_{\odot}$ stars, the C IV emission shows a behavior similar to what found in X-rays, although with a fairly large scatter in the data.

To understand the above observational picture, at least qualitatively, our working hypothesis is the following: On the main sequence, the earlier, more massive stars in our sample have too shallow convection zones for a magnetic dynamo mechanism to be efficient. The chromospheres and even the transition regions of these stars are dominated by mechanical heating, but not the coronae (see also Simon and Drake 1989; Simon and Landsman 1991). This explains the relatively high UV emission with respect to the X-ray emission. These stars retain an high angular momentum during the main sequence lifetime, since they are not affected by magnetic braking. While evolving, these stars develop a convection zone which triggers the onset of an efficient magnetic dynamo, which in turn provides the required coronal heating. At the same time, the mechanical heating is theoretically expected to decline (Narain and Ulmschneider 1990). On the other hand, for the lower mass stars the magnetic dynamo is already at work on the main sequence, and their rotational velocity is known to decrease with age, together with their activity level (Pallavicini *et al.* 1981; Maggio *et al.* 1987). Upon leaving the main sequence, the depth of the convection zone start to increase, but the rotation rate keeps declining, and eventually drops below some threshold value where the dynamo ceases to be effective.

To test this hypothesis we have searched for correlations between the activity level, and dynamo-related parameters, such as the rotational velocity and the Rossby number (Durney, Mihalas and Robinson 1981). The latter has been computed by properly estimating the convective turnover time from theoretical evolutionary models (Rucinsky and Vandenberg 1986).

We find that the X-ray luminosity is not correlated with the rotation rate for the whole sample. A significant correlation is found excluding the $M > 1.6\ M_\odot$ stars on the main sequence, the $M \leq 1.6\ M_\odot$ more evolved stars with $L_x < 10^{28.5}\ erg\ s^{-1}$, and also the most intense X-ray source (the reddest high-mass star in Fig.2). According to our hypothesis, the first two classes do not correlate because the dynamo mechanism is not effective, while for the latter object the influence of the increased convection zone depth on the efficency of the $\alpha - \omega$ dynamo cannot be neglected. In fact, a significant correlation is found between the X-ray luminosity and the Rossby number, by excluding only the sources with the lowest X-ray emission level (Fig. 3).

REFERENCES

Durney, B.R., Mihalas, D., and Robinson, R.D. 1982, *PASP*, 93, 537.

Haisch, B.M., Bookbinder, J., Maggio, A., Vaiana, G.S., and Bennett, J.O. 1990, *ApJ*, 361, 570.

Maggio, A., *et al.* 1987, *ApJ*, 315, 687.

Maggio, A., Vaiana, G.S., Haisch, B.M., Stern, R.A., Bookbinder, J., Harnden, F.R., Jr., and Rosner, R. 1990, *ApJ*, 348, 253.

Narain, U., and Ulmschneider, P. 1990, *Space Sci.Rev.*, 54, 377.

Pallavicini, R., Golub, L., Rosner, R., Vaiana, G.S., Ayres, T., Linsky, J.L. 1981, *ApJ*, 248, 279.

Rucinsky, S.M., and Vandenberg, D.A. 1986, *PASP*, 98, 669.

Simon, T., and Drake, S.A. 1989, *ApJ*, 346, 303.

Simon, T., and Landsman, W. 1991, *ApJ*, 380, 200.

Vandenberg, D.A. 1985, *ApJS*, 58, 711.

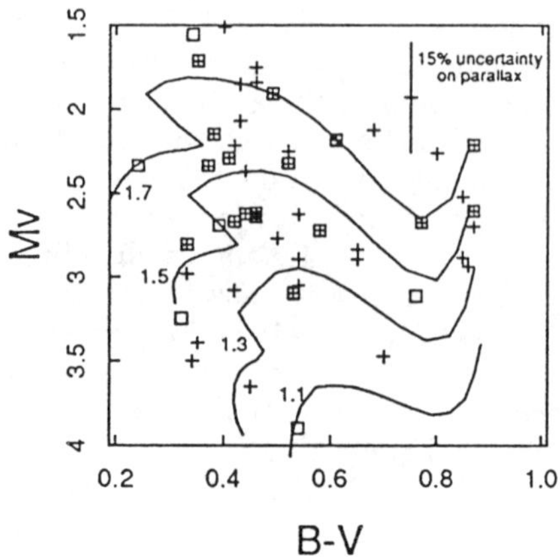

Fig. 1. H-R diagram for the stars in the selected sample. Only stars within a distance of 50 pc, and B-V color less than 0.9 have been included. Typical uncertainty on the absolute magnitude, due to an error of 15% on the parallax is also shown. Squares indicate stars observed in X-rays, while crosses are used for IUE observations. The theoretical evolutionary tracks are from Vandenberg (1985), and refer to stars with masses in the range 1.1 - 1.7 $M_{\odot}$. These tracks have been used to estimate the star mass by interpolation. Given the uncertainty on the parallaxes, this estimate may be in error by about $\pm\ 0.2 M_{\odot}$.

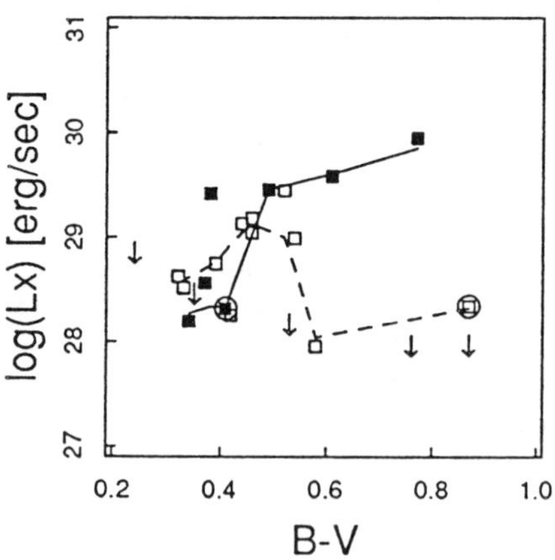

Fig. 2. Scatter plots of X-ray luminosity vs. B-V color. Squares are used for detections, and arrows for upper limits; filled symbols identify $M\ > 1.6\ M_{\odot}$ stars; circles indicate the new ROSAT observations. The lines are the result of a robust locally weighted regression, which takes into account both detections and upper limits, applied separately to the high-mass (solid line) and low-mass (dashed line) subsamples.

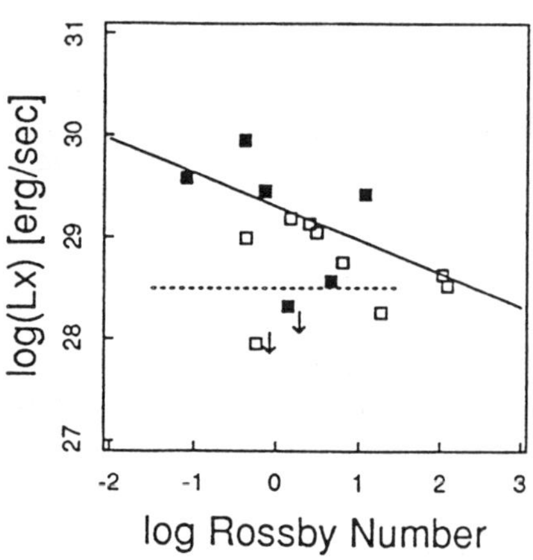

Fig. 3. X-ray emission vs. Rossby number. The symbols are as in Fig. 2. The best fit power law for all stars with $\log(L_x) >\ 28.5\ erg\ s^{-1}$ (those below the dotted segment have been excluded), irrespective of their mass, yields a slope of $-0.34^{+0.06}_{-0.06}$ with a correlation coefficient of $-0.74^{+0.10}_{-0.16}$. We judge the fit quality reasonably good, given the uncertainties in the evaluation of the Rossby number.

Inside the Stars, IAU Colloquium 137
ASP Conference Series, Vol. 40, 1993
Werner W. Weiss and Annie Baglin (eds.)

PHOTOSPHERIC ACTIVITY AMONG EARLY F–TYPE STARS

L. MANTEGAZZA[1], E. PORETTI[2], E.ANTONELLO[2], F. ZERBI[1]
[1]Universitá di Pavia, Via Bassi 6 I–27100 Pavia
[2]Osservatorio Astronomico di Brera, Via Bianchi 46 I–22055 Merate

INTRODUCTION

In the last decade stars showing variability of unclear origin have been detected among early F–type stars mainly as secondary results in the context of observational programmes devoted to the study of δ–Scuti and CP stars.

For this reason only a small number of these objects has been thoroughly studied: their positions in HR diagram is shown in fig.1.

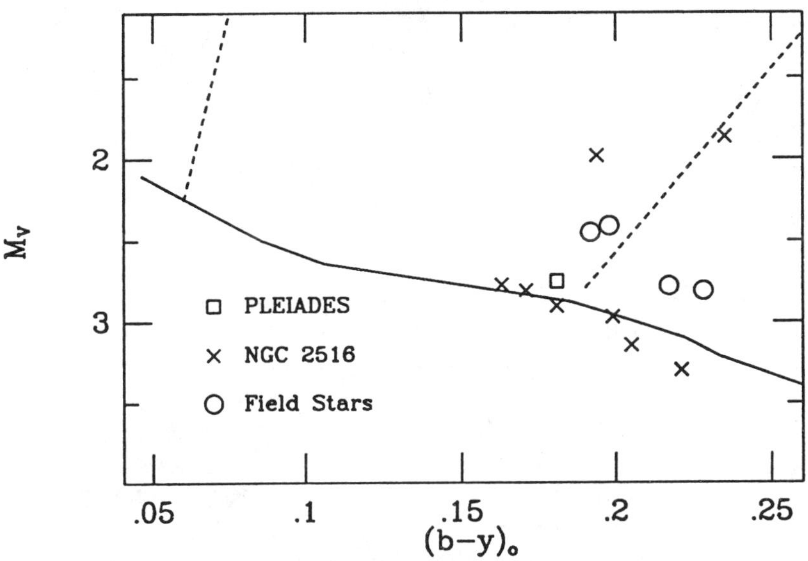

FIGURE 1: HR diagram of the observed stars.

The stars in NGC 2516 have been singled out in a survey work by Antonello and Mantegazza (1986), HD 164615 has been studied by Abt *et al.*(1983), the variability in 9 Aur has been discovered by Krisciunas *et al.*(1990,1991) and HD 23375 in Pleiades has been pointed out by Breger (1972). The solid line in Fig.1 is the ZAMS and the dashed lines are the borders of the δ–Scuti stars instability strip.

We recently discovered two objects of this type, HD 224638 and HD 224945, in our δ–Scuti stars observing programme, and we exploited the opportunity to investigate their behaviour (Mantegazza and Poretti 1991).

PHOTOMETRIC BEHAVIOUR

In September–October 1991 we made differential photometry of the two stars with respect to HD 225086 in the Johnson's *B* filter at La Silla Observatory (ESO). We collected 1000 and 985 measurements for HD 224638 and HD 224945 respectively during 20 non–consecutive nights.

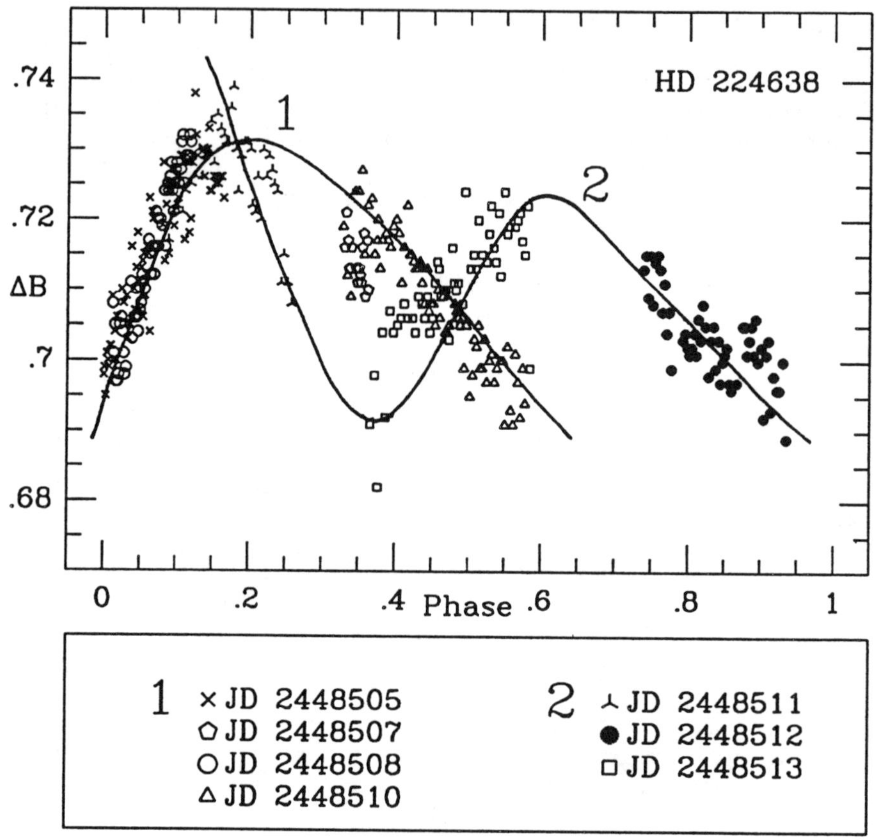

FIGURE 2: HD 224638. Phased light curve. P=1.47 d

The light curves show variations of 0.10 and 0.08 mag respectively with apparent time–scales of a few days. These values are comparable with those of the other stars in fig.1. For both stars the mean magnitude remains constant during the whole observing run.

Least squares spectra computed for both stars exhibit a complex structure confined at frequencies below 4 c/d.

Unfortunately the analysis by means of different techniques (Least squares,

PDM and autocorrelation functions) doesn't supply reliable periods. This happens because the light curves do not repeat exactly after few cycles.

As shown in fig.2 for a few nights the light curve of HD 224638 is well phased with a 1.47 d period but suddendly a phase shift of about 0.2 p occurs at JD 2448511.

HD 164615 (Abt *et al.*1983) shows a similar, even if perhaps more regular behaviour, but this may be due to the poor sampling (159 measurements scattered over 362 nights) rather than to physical reasons.

The simplest way to explain this behaviour is to ascribe it to spots carried across the visible disk by rotation. Moreover all the available observational data regarding all the objects of our sample do not seem to support alternative models.

MODELS AND CONCLUSIONS

Rotational spotted structure is related to convection through dynamo theory. Therefore spots are expected to be present since the onset of convection at F0 (Simon 1987). Observational evidence of spots is anyway available for stars cooler than F7 only (Radick 1982;1983a;1983b)

In spite of an intense chromospheric activity revealed by their spectroscopy, the absence of spots in stars between F0 and F7 has stimulated the developement of other non–radiative models (Wolff 1987; Giampapa and Rosner 1984).

A complicated spotted structure with short lifetimes, such as that suggested by the observations discussed here, is expected in a dynamo behaviour for early F–type stars, since they possess thin convection layers (Giampapa and Rosner 1984).

Further investigations of the variability among early F–type stars could therefore resume the dynamo theory even for stars with spectral types between F0 and F7.

REFERENCES

Abt, H.A., Bollinger, G. and Burke, E.W.,jr., 1983, *Ap.J*, **272**, 196

Antonello, E. and Mantegazza, L., 1986,*A & A*, **164**, 40

Breger, M., 1972, *Ap.J*, **176**, 367

Giampapa, M.S. and Rosner R., 1984, *Ap.J*, **286**, L19

Krisciunas, K. and Guinan, E., 1990, Inf. Bull. Var. Stars n.3511

Krisciunas, K. *et al.*, 1991, Inf. Bull. Var. Stars, **3672**

Mantegazza, L. and Poretti, E., 1991, Inf. Bull. Var. Stars n.3690

Radick, R.R., *et al.*, 1982, *P.A.S.P.*, **94**, 934

Radick, R.R., *et al.*, 1983a, *P.A.S.P.*, **95**, 300

Radick, R.R., *et al.*, 1983b, *P.A.S.P.*, **95**, 631

Simon, T.,1987, in *Lectures Notes in Physics*, **291**, 265

Wolff, S.C., 1987, in *Lectures Notes in Physics*, **291**, 223

Inside the Stars, IAU Colloquium 137
ASP Conference Series, Vol. 40, 1993
Werner W. Weiss and Annie Baglin (eds.)

ENERGY LOSSES BY STELLAR ACTIVITY PHENOMENA

MARIA KATSOVA
Sternberg State Astronomical Institute,
Moscow University,
119899 Moscow V-234 Russia

ABSTRACT For active late-type dwarfs, the total energy losses including both chromospheric and coronal radiation as well as the energy losses caused by stellar flares reach $3\cdot10^{-3}$ L_{bol}. To maintain such high energetic losses requires a high efficiency of the dynamo action in the convection zone. Mass loss by active late-type stars can probably approach the rate of 10^{-11} $M_\odot$/yr if we take into account the large-scale structure of their coronae and transient events. The phase of activity can continue for up to 1% of the lifetime of these stars on the main sequence due to braking by such a stellar wind.

ENERGY LOSSES IN OUTER ATMOSPHERES

Solar cycle is a result of the coupling between the middle-scale and large-scale magnetic fields. The energy losses of active regions relating to the middle-scale magnetic fields are determined by thei X-ray radiation. The behavior of the L_x/L_{bol} ratio versus colour index B-V shown in Fig.1 (where values of the L_x/L_{bol} for the red dwarfs are taken from Pallavicini et.al, 1990; Katsova & Tsikoudi 1992) exhibits the saturation effect similar to that found earlier by Vilhu & Walter 1987. It means that practically the whole volume of the inner corona ($r<1.5R_*$) is occupied by hot coronal loops.

The strong spottedness and the winds of active late-type stars manifest theirselfs also at the far-IR (Katsova & Tsikoudi 1992; Mullan et.al,1992).

For active red dwarfs, the power expending during flares is close to the X-ray luminosity and can reach it, so that $L_x/L_{bol}\approx L_{fl}/L_{bol}\approx10^{-3}$ (Gershberg & Shakhovskaya 1983).

Energy losses in the chromospheric lines of the dG and early dKe stars exceed as a rule the coronal energy losses due to emission features on the middle and small scale like flocculae and boundaries of chromospheric networks. For the later active stars, coronal energy losses begin to dominate.

THE LARGE-SCALE CORONAL STRUCTURES AND STELLAR WIND

In the presence of the large-scale magnetic fields structures like coronal holes and streamers form, and the rate of mass loss by late-type dwarfs can reach $1\text{-}2\cdot10^{-11}$ $M_\odot$ /yr; that is a factor 10-100 greater than in the solar case (Badalyan & Livshits 1992). We are estimated wind energy losses at the level of its formation (Fig.2).

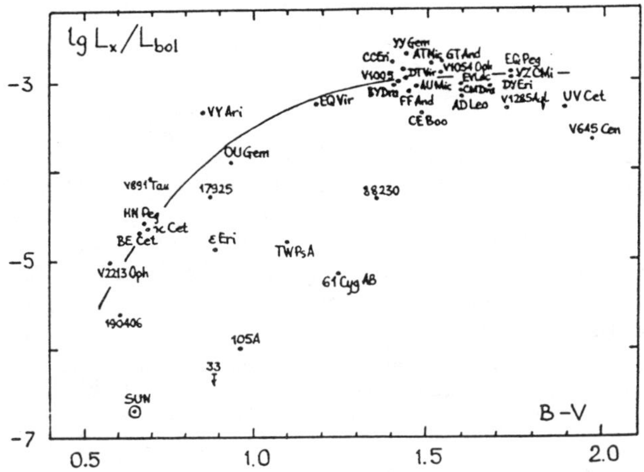

Figure 1

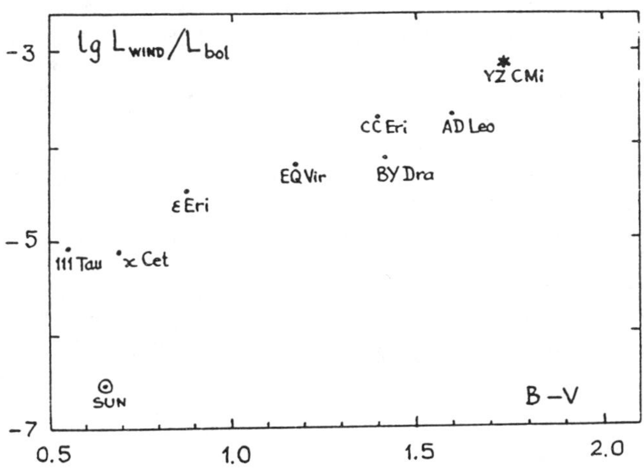

Figure 2

Thus, the stellar activity cycle is arranged in such a manner that an active late-type dwarfs lose on the different space scales up to $3\cdot10^{-3}$ L_{bol} due to both stationary and nonstationary processes (see Table).

TABLE The energy losses as regards to the bolometric luminosity for stars of different spectral types) L_{act}/L_{bol}:

spectral type	dG2	dK5	dM4e
chromophere	$2 \cdot 10^{-5}$	$6 \cdot 10^{-5}$	$2 \cdot 10^{-4}$; 10^{-3} (flares)
corona	$2 \cdot 10^{-7}$	$3 \cdot 10^{-4}$	10^{-3}
wind	$5 \cdot 10^{-8}$	$3 \cdot 10^{-5}$	$3 \cdot 10^{-4}$; $8 \cdot 10^{-4}$(transient)

SOME PROBLEMES WITH THEORY

i - How is it possible to provide such a high ratio of $L_{act}/L_{bol} \approx 3 \cdot 10^{-3}$? It needs to create not only middle-scale magnetic field, but also the large-scale structures, i.e. connection between toroidal and poloidal components of the field. The dynamo mechanism has to be extremely intensive. The energy of the mean magnetic field of the convective zone can be estimated after Vainstein et al(1980)

$$E_{MF} = 4\pi R^3 \omega_0 (B_\varphi B_r \;/4\pi)$$

where $\omega_0 = v_{rot}/R_*$ is angular velocity of the rotating star. For $B_\varphi = 1000$ G and $B_r = 10^{-3}\, B_\varphi$ we obtain $E_{MF} = 10^{30}$ erg/s and lg $E_{MF}/L_{bol} = -3.6$ for the Sun. This value lies close to the boundaries of the dynamo possibility.

ii - Mass loss of about $10^{-11}\, M_\odot$/yr exists apparently in active late-type dwarfs in the form of a quasi-stationary wind as well as non-stationary events like transients. This restricts the time scale of the existence of activity as well as the redistribution of the angular momentum and braking due to a magnetic stellar wind can be estimated as

$$\tau = 0.6(R_*/r_A)^2 \cdot M \cdot dM/dt^{\;-1}.$$

For the radius of the Alfven sphere $r_A = 10R_*$ and $M_* = 0.5M_\odot$ this becomes $3 \cdot 10^8$ years. This is 2 order of magnitude less than the lifetime on the main sequence for the dK5 stars which is $40 \cdot 10^9$ years.

Therefore, the energetic losses during the phase of stellar activity of the late-type dwarfs have to be taken into account when dynamo theory and structure of the convective zone are considered.

REFERENCES

Badalyan O.G., Livshits M.A. 1992, Soviet Astron. v.36(1)
Gershberg R.E., Shakhovskaya N.I. 1983, Ap.Sp.Sc. v.95, p.235
Katsova M.M., Tsikoudi V. 1992, in preparation
Mullan D.J. et al, 1992 Armagh Observ.Prepr.Ser. No 137
Pallavicini R., Tagliaferri G.; Stella L. 1990, As.Ap.v.228,p.403
Vainstein S., Zel'dovich Ya., Ruzmaikin A. The turbulent dynamo in astrophysics. Moscow. Nauka. 1980
Vilhu O., Walter F.M. 1987, Ap.J. v.321, p.958

Inside the Stars, IAU Colloquium 137
ASP Conference Series, Vol. 40, 1993
Werner W. Weiss and Annie Baglin (eds.)

LITHIUM ABUNDANCE AND ACTIVITY FOR 57 RS CVn SYSTEMS

Liu Xuefu[1], Zhao Gang[2], Tan Huisong[3] and Lu Fangjun[1]

1. Dept. of Astronomy, Beijing Normal Univarsity
2. Beijing Observatory, Chinese Academy of Science
3. Yunnan Observatory, Chinese Academy of Science

ABSTRACT High-resolution (0.145A/diode, high signal-to-noise (>100) Reticon spectra of 57 RS CVn systems, which were observed with the coudé spectrographs of the McDonald 2.1m telescope and Yunnan 1m telescope, were used to analyse a correlation between Li abundance and chromospere activity. Li abundances of 57 RS CVn stars, which include 76 detected components have been determined through the determination of equivalent widths of Li doublet (6707.761A + 6707.912A) and Kuruz's model atmosphere. The model metal abundance and line analysis of 10 RS CVn stars were determined from the present spectral lines data. The main results are the following: 1, Lithium abunances of 76 RS CVn components for 57 RS CVn systems are obviously reduced with the the effective temperatures gradually decreasing. 2, Li abundance has staistically decreasing trend in RS CVn stars with the rotational velocities vsini increasing. 2, The position of 168 RS CVn stars in the (U-B,B-V) Figure show that they can be devided into two groups. The most of them seem to be not poor-metal stars. Li abundances of these stars seem to have no obious correlations with their stellar chromosphery activity levels, but 7 RS CVn systems, in which cool component is more active show that the more active component has a lower Li abundance.

Inside the Stars, IAU Colloquium 137
ASP Conference Series, Vol. 40, 1993
Werner W. Weiss and Annie Baglin (eds.)

ROSAT WFC EUV OBSERVATIONS OF AB DOR (AND OTHER LOCAL GROUP STARS)

BARRY J. KELLETT & GORDON E. BROMAGE
Rutherford Appleton Laboratory, United Kingdom

INTRODUCTION

The Rosat Observatory was launched on June 1st, 1990, and after a brief Performance Verification phase, performed an almost complete sky survey. One of the two main telescopes on the Observatory is the UK Wide Field Camera (WFC). The WFC is the first extreme ultraviolet (EUV) instrument to survey the whole sky, and during the survey detected several hundred active late-type stars.

The WFC (Wells *et al.*, 1990) was built by a consortium of five UK institutes, four Universities and the Rutherford Appleton Laboratory. During the all-sky survey, two filters were used. These allow broad-band photometry to be performed. The S1A filter (lexan/carbon/boron) and S2A filter (lexan/beryllium) were alternated on a one day basis. The S1A filter covers the energy range 0.085–0.180 KeV, centred on 0.124 KeV, while the softer S2A filter covers the range 0.070–0.110 KeV (centred on 0.090 KeV) (10% efficiency) [see Kent *et al.* (1990) for a full description of the WFC filters].

Because of the way that the survey was performed any given EUV source would be observed for a minimum of 5 days. However, the data would be composed of brief "snapshots" of the source lasting 30–80 seconds, separated by one Rosat orbit (about 96 minutes). Therefore, the typical exposure time is about 1000 seconds in each of the two survey filters.

AB Doradus (HD36705)

AB Dor is a rapidly rotating young star which reveals its magnetic-related activity in a variety of ways. In particular, it is a strong X-ray source, with flares observed by the *Einstein* HRI, EXOSAT ME, and *Ginga* LAC [Vilhu & Linsky (1987), Collier-Cameron *et al.* (1988), Vilhu *et al.* (1991)]. *Ginga* was also able to show some marginal evidence for X-ray modulation with the 0.514 day rotation period of AB Dor.

In the Rosat All-Sky survey, AB Dor is nearly "circumpolar" and was visible for about 2 months (in the WFC) of the 6 month survey. During much of the survey, AB Dor would have been viewed from about the same point in the Rosat orbit. Unfortunately, this point is quite close to the southern auroral particle zones of the Earth, and consequently, the WFC detector was switched off for much of the potential survey time of AB Dor, and with quite a high background rate during much of the remaining time. This leads to a good deal of scatter in the data (although *some* of the scatter must also be due to the activity of AB Dor itself). However, AB Dor was detected as a very significant source in both the survey filters. The question of whether the survey data shows any modulation on the 0.514 day rotation period of AB Dor is more difficult to establish because of the problems mentioned above. But it is possible that some modulation was present for at least part of the time. This conclusion was also reached by Kürster *et al.* 1992, from the Rosat XRT survey data.

PZ Tel (HD174429)

Another well known fast rotator is PZ Tel. Unfortunately, the WFC survey of PZ Tel was badly affected by data loss (similar to AB Dor), and only received a relatively short exposure (see Table 2). It is also much further away than AB Dor (68 pc versus 20 pc). Consequently, no EUV flux was detected from PZ Tel

"Speedy Mic" – the new fast rotator

During the WFC survey, many previously unknown late-type stars were found to be EUV emitters. Optical follow-up spectroscopy has confirmed many of them to be very active indeed. One of the most interesting of the early discoveries was a new source in the southern constellation *Microscopium.* The new source was quickly christened "Speedy" for reasons that will become apparent. The source was clearly variable during the survey. In fact, detailed analysis has shown that the S1 and S2 detections of this source are due *entirely* to the presence of a long-lasting flare on the star, with *NO* quiescent emission being detected from Speedy (Bromage *et al.*, 1992).

The WFC position lies very close to a catalogued 9th mag K0 star (HD197890). Optical follow-up spectra of Speedy taken with the 1.9m SAAO telescope with the RPCS detector clearly indicated broad emission lines of Ca II H + K in the spectrum of Speedy. We estimate a (deconvolved) FWHM of ~230 km/s, which is faster than AB Dor.

Further optical monitoring of Speedy has failed to show any signs of binarity, so we have concluded that this is a very rapidly rotating single K dwarf. It also shares the same common proper motion shown by AB Dor and PZ Tel. In fact all 3 K dwarfs are members of the Pleiades or Local group (see Innis *et al.*, 1986, Bromage *et al.*, 1992).

The Local Group K dwarfs

In recent years, 4 other K dwarfs have been associated with the Local Group (see Anders *et al.*, 1991). These stars are listed in Table 1, along with the stars already discussed. Although these 4 K dwarfs all show much smaller vsin i's than the stars already discussed above (based on their relative photometric periods, given in Table 1, for example), all 4 were clearly detected in the survey. Table 2 summaries the WFC EUV observations of these ~K0V stars. In all 4 cases, detailed analysis of the data has shown that these detections are "quiescent" emission from these stars; ie. there are no significant variations present in the data which might indicate the presence of flares.

TABLE 1: Local Association K Stars

HD Number (Name)	V	Sp.Typ.	P_{phot} (d)	P_{orb} (d)	d_{pc}
17433 (VY Ari)	6.87	G9Ve	17.4	13.20	21
36705 (AB Dor)	6.95	K1–2III–IVp	0.514	–	20
82558	7.50	dK0	1.66	–	13
102077 (V838 Cen)	8.97	K0–1Vp	1.84	–	42
139084 (V343 Nor)	8.14	K0V	4.2	–	55
174429 (PZ Tel)	8.45	K0III–Vp	0.945	–	68
197890 ("Speedy")	9.3	K0V	0.338	–	~40

TABLE 2: WFC Survey Results

Star	S1A Filter		S2A Filter	
	Survey (s)	Cts/Sec	Survey (s)	Cts/Sec
VY Ari	1261.2	0.094	849.0	0.096
AB Dor	13442.	0.075	12539.	0.090
HD82558	1955.2	0.033	2106.7	0.028
V838 Cen	954.2	0.025	861.8	0.030
V343 Nor	1261.5	0.021	1780.6	0.040
PZ Tel	589.5	<0.009	710.3	<0.034
Speedy	1126.0	0.030	1302.7	0.072

CONCLUSION

The WFC EUV survey results presented here show that the Local Association K dwarfs are strong EUV emitters. The stars in the association are thought to be 50–70 million years old, and it would seem it is these stars relatively young age that make them so active; since the "slow" rotators, like V343 Nor and VY Ari, appear just as strong as the rapid stars, like AB Dor and Speedy. In fact, as already mentioned, Speedy Mic was *NOT* seen in quiescence, whereas the 4 K dwarfs added to this study were *ALL* clearly detected in quiescence.

ACKNOWLEDGMENTS

The Wide Field Camera is supported by the UK Science and Engineering Research Council. The WFC instrument and associated Ground Systems were developed by a consortium of five UK institutes: the University of Leicester, SERC's Rutherford Appleton Laboratory, Imperial College STM London, the University of Birmingham, and the Mullard Space Science Laboratory of University College London. The authors would like to thank all WFC personnel at the Consortium Institutes involved in the design and development of the Instrument. This work utilised STARLINK software and facilities at the Rutherford Appleton STARLINK node.

REFERENCES

Anders, G.J., Innis, J.L., Coates, D.W., & Thompson, K., 1991, *MNRAS*, **252**, 408.

Bromage, G.E., *et al.*, 1992, *PASP Conf. Ser. (Proc. Tucson Workshop, Oct. 1991)*, in press.

Collier-Cameron, A., *et al.*, 1988, *MNRAS*, **231**, 131.

Innis, J.L., Thompson, K., & Coates, D.W., 1986, *MNRAS*, **223**, 183.

Kent, B.J., *et al.* 1990, *Proc. S.P.I.E.*, **1344**, 255.

Kürster, M., *et al.*, 1992, *PASP Conf. Ser. (Proc. Tucson Workshop, Oct. 1991)*, in press.

Vilhu, O., & Linsky, J.L., 1987, *PASP*, **99**, 1071.

Vilhu, O., Tsuru, T., & Collier-Cameron, A., 1991, *"Iron Line Diagnostics in X-ray Sources"*, Eds. Treves, A., Perola, G.C., & Stella, L., 30.

Wells, A., *et al.* 1990, *Proc. S.P.I.E.*, **1344**, 230.

Inside the Stars, IAU Colloquium 137
ASP Conference Series, Vol. 40, 1993
Werner W. Weiss and Annie Baglin (eds.)

THE MUSICOS PROJECT: MULTI–SITE CONTINUOUS SPECTROSCOPY

CLAUDE CATALA, JACQUES BAUDRAND, TORSTEN BÖHM
Observatoire de Paris, Section de Meudon, 92195 Meudon Principal Cedex, France

BERNARD H. FOING
Institut d'Astrophysique Spatiale, BP 10, 91371 Verrières-le-Buisson Cedex, France

INTRODUCTION

Many scientific programs, most of them linked to stellar physics (such as asteroseismology, stellar rotational modulation, surface structures, Doppler imaging, Zeeman-Doppler imaging, variable stellar winds) require a continuous spectroscopic coverage during several days.

MUSICOS (for **MU**lti–**SI**te **CO**ntinuous **S**pectroscopy) is an international project for setting up a network of high resolution spectrometers coupled to telescopes of the 2m class, well distributed around the world, and partly dedicated to continuous spectroscopy.

The strategy to reach this objective was defined during two workshops organized at Paris-Meudon Observatory in 1988 and 1990, and consists of three steps: 1) organize multi-site spectroscopic campaigns using resident instruments on various telescopes around the world and transportable fiber–fed spectrographs where adequate spectroscopic equipment is not available; 2) design and develop a cross-dispersed echelle spectrograph, well suited for the scientific programs that require multi–site observations; 3) propose this MUSICOS spectrograph for duplication at several collaborating sites.

THE MUSICOS SPECTROGRAPH

We have now completed phase 2 of this strategy. We have constructed the MUSICOS spectrograph, which is now ready, and installed at the "Bernard Lyot" 2m telescope at Pic du Midi. This spectrograph has been designed to meet the requirements of scientific programs that need multi-site observations. Figure I shows the optical layout of the spectrograph. It is a cross–dispersed spectrograph, covering most of the visible domain in 2 exposures (380–540 nm and 540–870 nm), with a resolving power of 38000. The spectrograph is fed by 2 fibers, giving the possibility to record simultaneously 2 spectra with interleaved orders. Other characteristics of this spectrograph include high stability and easy transportability. The cost of the spectrograph was kept low (< 100,000 USD, including detector). Finally, we developed a dedicated automatic reduction software for the MUSICOS spectrograph. Table I indicates the performances of this instrument, as obtained at Pic du Midi 2m "Bernard Lyot" telescope, for an A0

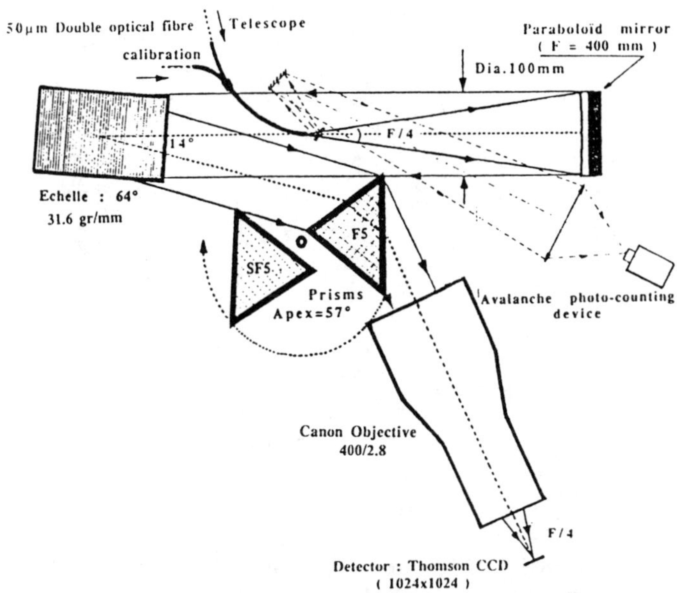

FIGURE I Optical layout of the MUSICOS spectrograph

star with $m_V = 7$, with a thick coated CCD, in one hour.

TABLE I

λ (nm)	S/N	CCD Q.E.
400	125	0.12
500	320	0.30
600	405	0.45
700	380	0.40
800	300	0.30

The instrument, its dedicated reduction software, as well as its performances on the sky are fully described in Baudrand and Böhm (1992).

THE MUSICOS 1989 CAMPAIGN

In December 1989, we organized a vast multi–site spectroscopic campaign, with the philosophy presented above. Three scientific programs were followed: 1) search for nonradial pulsations in the Be star 48 Per; 2) search for co–rotating structures in the wind of the Herbig Ae star AB Aur; 3) Doppler Imaging and flare patrol for the RS CVn system HR 1099.

The campaign was organized around a core program, involving telescopes in Hawaii, Kitt Peak, ESO La Silla, France OHP, Crimea, and China. Two spectrographs had to be transported to remote sites for this campaign: the fiber–fed ISIS spectrograph was transported from Meudon to the 2.16m telescope in Xinglong, while the ISIS–bis spectrograph travelled to Hawaii, to be mounted on the UH 2.2m telescope.

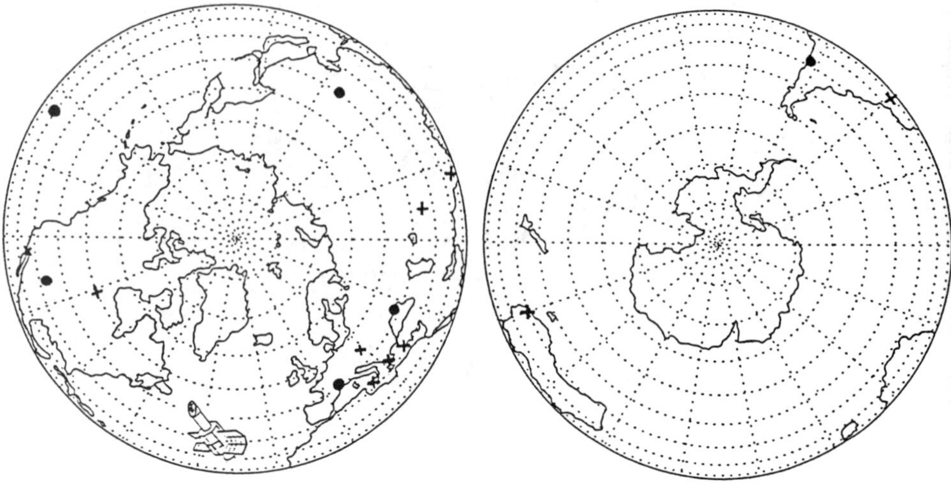

FIGURE II Distribution of the sites involved in the MUSICOS 1989 campaign. The filled symbols correspond to sites of the core program, while "pluses" indicate additional sites.

In addition to this core program, a large number of astronomers all over the world decided to contribute to this campaign, after it was announced in an IAU circular. Altogether, up to 17 telescopes were pointed toward the targets, including IUE. Figure II shows the distribution of the sites on the Earth.

The MUSICOS 1989 campaign was a great success, with some major scientific impact. The full organization of the campaign, as well as some preliminary results can be found in Catala and Foing (1990). The final scientific results will be published shortly (Foing et al., 1992).

We are currently planning a new campaign for December 1992, involving the newly developed MUSICOS spectrograph, that we intend to transport to the 2.2m telescope in Hawaii. Our two other fiber–fed spectrographs, ISIS and ISIS–bis will be transported to India and China.

REFERENCES

Baudrand, J., Böhm, T. 1992, *A&A*, in press

Catala, C., Foing, B.H. 1990, in *Proceedings of the second MUSICOS workshop*, Meudon, C. Catala and B.H. Foing (eds.), Observatoire de Paris press

Foing, B.H., Char, S., Jankov, S., Catala, C., Ayres, T., Neff, J.E., Simon, T., Zhai, D.S., Cutispoto, G., Catalano, S., Rodono, M., Akan, C., Baudrand, J., Cao, H., Cuby, J.G., Czarny, J., Dreux, M., Felenbok, P., Ghosh, K., Guérin, J., Huang, L., Hubert, A.M., Jiang, S., Jinxin, H., Keskin, V., Mavridis, L., Nolthenius, R., Zihe, G. 1992, *A&A*, submitted

Inside the Stars, IAU Colloquium 137
ASP Conference Series, Vol. 40, 1993
Werner W. Weiss and Annie Baglin (eds.)

SHORT-TERM SPECTRAL VARIABILITY IN THE HERBIG Ae STAR AB AUR: PRELIMINARY ANALYSIS OF CHROMOSPHERIC LINES [1].

TORSTEN BÖHM
Département d'Astrophysique Extragalactique et de Cosmologie, Observatoire de Paris, Section de Meudon, 92195 Meudon Principal Cedex, France

CLAUDE CATALA
Laboratoire de Recherche Spatiale, Observatoire de Paris, Section de Meudon, 92195 Meudon Principal Cedex, France

INTRODUCTION

The Herbig Ae stars are PMS objects of intermediate mass. Their location in the H-R diagram indicates that they are in the radiative phase of their quasi-static contraction toward the main sequence, i.e. that they do not possess outer convective zones, according to the standard stellar evolution theory (Iben, 1965; Gilliland, 1986). In spite of the expected absence of subphotospheric convective envelopes, these stars show remarkable signs of activity: emission in the Mg II h and k lines, presence of the C IV resonance lines at 1550 A and He I 5875.7 A line, Ca II IR triplet in emission, etc... Considering that stellar activity, witnessed by the same type of indicators in other parts of the H-R diagram, is generally attributed to dynamo magnetic fields and/or acoustic waves generated in the convection zone, these active phenomena are quite paradoxical in the Herbig Ae stars.

The main question concerns the origin of their activity: is this activity linked to phenomena occurring within the stars, like e.g. dynamo-generated magnetic fields, or to an external agent, like e.g. a boundary layer between an accretion disk and the stellar surface? We already have some indirect clues that the activity of the Herbig Ae stars might be of magnetic origin (Praderie et al., 1986; Catala et al. 1986) observed a rotational modulation of lines formed in the wind of AB Aur, prototype of the Herbig Ae stars. By analogy with the solar wind, they proposed that the modulation might be due to the corotation of azimuthal structures in the wind, controlled by a surface magnetic field.

However, the link between the stellar wind modulation and a structured magnetic field is far from being firmly established. In particular, we have no clear indication that the source of the modulation is near the stellar surface, since the lines for which such a modulation has been observed are formed far in the wind. Are we seeing a corotating structure linked to the stellar surface by a magnetic field, or some other type of variable phenomenon occurring in the

---

[1] Based on observations obtained with the "Bernard Lyot" telescope, at Pic du Midi Observatory, France.

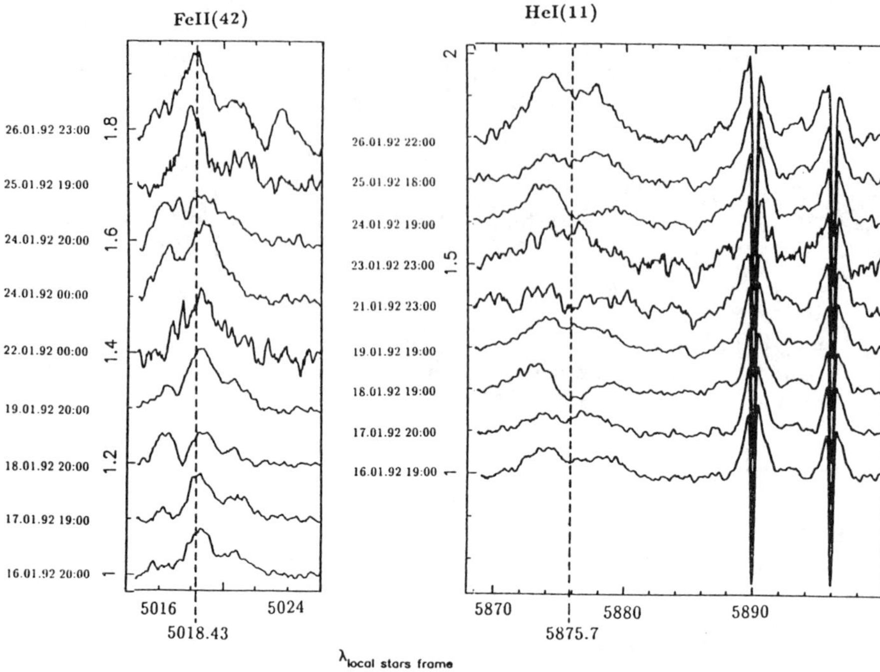

FIGURE I Profiles of the He I and Fe II lines

wind itself, like e.g. inhomogeneities passing periodically on the line of sight?

To answer this question, we have searched for the same type of modulation in a line which is formed closer to the stellar surface, at the base of the wind. From previous line profile analysis, we know that a chromosphere is present at the base of AB Aur's wind (Catala et Kunasz, 1987; Catala, 1988). The He I 5875.7 A line, quasi-absent in main sequence A-type stars and photospheric in earlier MS stars, is obviously of chromospheric origin in AB Aur.

OBSERVATIONS

The MUSICOS echelle spectrograph (Baudrand and Böhm, 1992) covers the whole visible domain, in two exposures (390–540, and 540–890 nm). In January 1992 we monitored AB Aur with this instrument at the 2m Bernard Lyot Telescope (Pic du Midi, France) at high resolution (R=38000). Nine spectra were obtained in both wavelength domains during 11 nights. We observed remarkable variations of several spectral lines, like the He I 5875.7 A (multiplet 11) and the Fe II 4923.9 A, 5018.4 A and 5169.0 A lines (multiplet 42). Both groups of lines show intense emission profiles which are strongly variable with a timescale shorter than one day (Fig. I)!

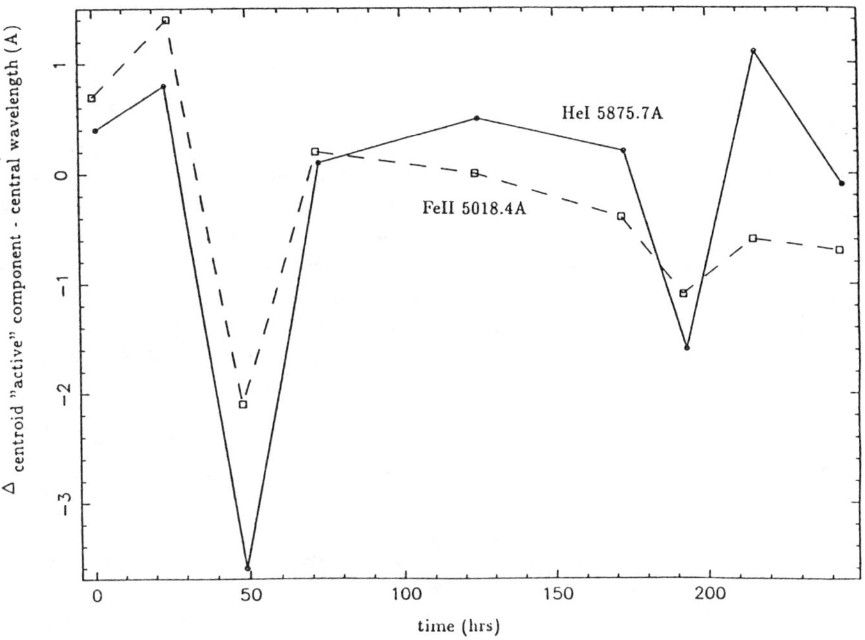

FIGURE II Covariation of He I and Fe II lines. Plotted is the wavelength shift of the active component centroid of each line against time.

ROTATIONAL MODULATION?

As a possible interpretation of these variations, we propose that the line profiles consist of a "variable" emission component superimposed on a "quiet" profile, also including emission; the variable component, which varies both in intensity and position, may be due to an active region which is corotating with the star.

For the HeI 5875.7A line, we determined an "asymmetry parameter", defined as $A = F_a \times \Delta\lambda_a$, where F_a is the integrated flux in the active component, and $\Delta\lambda_a$ is the wavelength shift of the active component centroid with respect to the line center in the stellar rest frame. The variations of A are compatible with a sinusoidal modulation with period P=37hrs, which may correspond to the star's rotation period (vsini about 90 kms^{-1}), although we do not have enough data to decide whether the observed variations are truely periodic. More observations are needed to verify this hypothesis.

COVARIATION OF HE I AND FE II LINES

The different shapes of the He I and Fe II lines suggest that they are formed in different regions. However, the variations of fluxes and centroids of the "variable" component of both groups of lines (He I and Fe II triplets) are strongly correlated (Fig. II)!

Therefore, if the observed variability is due to rotational modulation, the co-variation of these lines formed in different regions indicates that the same azimuthal structuration exists in these regions, and that the chromosphere is globally structured.

CONCLUSION

The correlation of the variations of the chromospheric He I and Fe II lines and their modulation by the star's rotation, if confirmed by further analysis, would indicate the presence of a globally structured chromosphere.

The regions of formation of these lines are probably very close to the photosphere, and thus the source of the structure responsible for the observed variations seems to be at the stellar surface, rather than further out in the wind.

These results are only preliminary. If confirmed, they will provide a further clue for a magnetically structured atmosphere in AB Aur.

REFERENCES

Baudrand, J., Böhm, T. 1992, *A&A* in press

Catala, C., Felenbok, P., Czarny, J., Talavera, A., Boesgaard, A.M. 1986, *ApJ* 308,7

Catala, C., Kunasz, P.B. 1987, *A&A* 174,158

Catala, C. 1988, *A&A* 193,222

Gilliland, R.L. 1986, *ApJ* 300,339

Iben, I., Jr. 1965, *ApJ* 141,993

Praderie, F., Simon, T., Catala, C., Boesgaard, A.M. 1986, *ApJ* 303,311

Inside the Stars, IAU Colloquium 137
ASP Conference Series, Vol. 40, 1993
Werner W. Weiss and Annie Baglin (eds.)

SEARCH FOR MAGNETIC STARS AT EARLY STAGES OF EVOLUTION

YU.V.GLAGOLEVSKIJ
Special Astrophysical Observatory, Nizhnij Arkhyz,357147, Russia

Young stars, as a rule, are too faint for measurements of magnetic field either by photographic method with the use of Zeeman analizer, or photoelectrically from hydrogen lines. That is why it is necessary to look for indirect ways of magnetic field detection, for example, by measurement of polarization. Ae/Be Herbig stars without a magnetic field are surrounded by a gaseous envelope in the form of a globe or a spheroid, flattened along the rotational axes (as dependent on stellar rotation velocity), and also by a gaseous-dust accretion disc in the plane of equator. There are powerful flows in gaseous envelopes of stars, connected with mass loss and accretion. If a star is a magnetic oblique rotator (as a magnetic star of the Main Sequence), then the gaseous envelope may acquire the shape of alongated ellipsoid with the major axes coincident with that of dipole (Dolginov et al., 1979). From the poles there arises a jet flow controlled by a magnetic field, as in He-r and He-w stars, having already reached the Main Sequence (Barker et al., 1982). Calculations show (Dolginov et al., 1979), that maximum polarization in the extended envelope $p \approx 4\%$ arises when the ratio of ellipsoid axes is $a \approx 2.5b$. The electric vector of the dominating oscillation of the light wave is perpendicular to the plane through the axis of symmetry of the ellipsoid and the line of sight. Naturally, the magnetosphere rotates together with the star, involving the gaseous envelope, resulting in the variation of the degree and direction of polarization. Additional polarization is created by the polar jets, where the direction of the dominating oscillations of the electric vector is perpendicular to the axis of the polar stream, and value of maximal polarization may reach 5% along the beam.

The gaseous envelope is formed mainly around the star and dust is chiefly contained in the disc. The scatter of light in the gaseous component occurs in the optical wavelength range, but in the dust one - in the infrared range. As a consequence of the fact that the electric vector in the former case is oscillating perpendicular to the axis of symmetry of the envelope, and in the latter its oscillations develop along the axis, the linear polarization changes by $\alpha = (90^{\circ} - \beta)$ from the optical to infrared spectral range. Here β - is the angle between the magnetic field dipole axis and the axis of rotation. The data presented show that in a case of a dipole field in an Ae/Be Herbig star, one may expect periodical variability of the linear polarization in the optical spectral region and constant polarization in the infrared region. It is possible to try to measure the inclination of the dipole axis.

Let us see what results of previous research can be used for selection of samples in search for magnetic fields. Young stars of Ae/Be Herbig type become visible near the "birth line" found by Palla and Stahler (1990), when a star loses a large part of its surrounding envelope. It turns out that for rapidly evolving massive stars of He-r type this occurs when they have already got to the Main Sequence. That is why the sample of stars for investigation should not include early B-stars. We found that the intersection boundary of the "birth line" with the Zero age Main Sequence is likely to fall on Sp = B5 - B7, but not on B2, found by Palla and Stahler (1990). This assumption can be made on the basis of Fig. 1, constructed from the data of Table 6.1 from (Shevchenko, 1989). Spectral classes are on the horizontal axis, and on the vertical axis are the amplitudes of the photometric variability Δm. The points are the average values Δm in the narrow intervals of spectral types. It is well seen that at Sp = B0 - B7 values $\Delta m = 0.4$, then Δm grow (first it was noted by Finkenceller and Mundt (1984)) and reach maximum at Sp ≈ A3, then they drop to $\Delta m \approx 1^m$. The average values of amplitudes Δm for stars TTau equal 1^m too, and it shows that the late Be Herbig stars and TTau stars are active approximately at the same degree. It is possible to make the following assumptions:

1) The nature of activity of TTau and Ae/Be Herbig stars is the same and there is no boundary near $M = 1.5M_\odot$ (Sp ≈ A5 - F0), where some investigators (on the basis of theoretical arguments) assume transition from convective to nonconvective atmospheres at early stages of evolution. Finkenceller and Mundt (1984) reported their opinion that, unlike the opinion of some theoreticians, young Ae/Be Herbig stars pass the convective stage of evolution. The same opinion was expressed by Palla and Stahler (1990), and also by Kuan and Kuhi (1975) for explanation of activity and thestellar wind nature.

2) The majority of Be Herbig stars are in the limits of the Main Sequence band. Their activity has stopped and the photometric variability becomes small. The later stars (Sp ≈B5 - B7) are visible at large distances (Δt years) from the Zero age line (they get rid of the surrounding gaseous-dust envelope relatively earlier), and nonstationary processes in stars are in full swing.

One may suppose that in the first approximation, the stronger nonstationarity, the farther the star by time Δt from the Zero age line; at the same time the amplitude Δm is larger the larger the luminosity L. In this case $\Delta m \sim (L \times \Delta t)$. In Fig. 1 is shown the dependence $(L \times \Delta t)$+const upon Sp. The values of L and Δt are estimated from the average dependence between the

Zero age line and "birth line" (Palla and Stahler, 1990). The calculated dependence are shown in Fig. 1 for three predictions: when the "birth line" intersects the Zero age line at points Sp = B2, B5 and B7. The best fit of calculated boundary and the experimental one is at Sp = B5 - B7. As to the shape of the curve, it can be easily fit with the experimental one by selection of the "birth line" shape. But it has no sense because, first, there are very little data for confident determination of the maximum and the right part of the dependence in Fig. 1 and, second, the dependence $\Delta m \sim (L \times \Delta t)$ maybe not so simple as we supposed. Magnetic stars at early stages of evolution are possible to find rather among Ae Herbig stars than among Be Herbig stars. It is doubtful that the separation boundary of convective and nonconvective atmospheres falls at Sp = B5-B7.

REFERENCES

Barker,.P.K., Brown,D.N., Bolton,C.T., Landstreet,J.D.: 1982, in *Advances inUltraviolet Astronomy*, NACA, p. 589.

Dolginov,A.Z., Gnedin,Yu.N., Silantiev,N.A.: 1979, in *The Diffusion and Polarisationof the Radiation in Cosmic Medium.* Moskwa, Nauka.

Finkenceller,U., Mundt,R.: 1984, *Astr. Ap. Suppl.*, **55**, 109.

Kuan,P., Kuhi,L.V.: 1975, *Ap. J.*, **199**, 148.

Palla,F., Stahler,S.V.: 1990, *Ap. J. Lett.*, **360**, L47.

Shevchenko,V.S., 1989, in *Ae/Be Herbig Stars.* Tashkent.

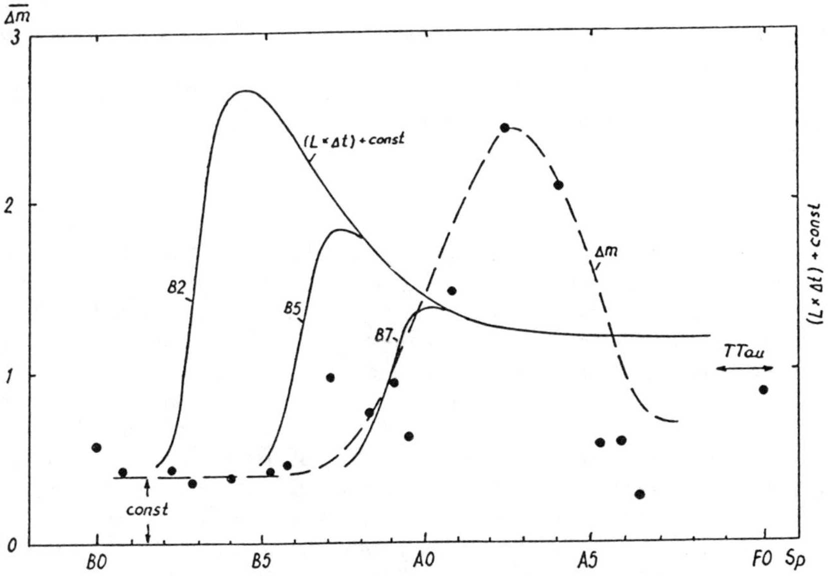

Fig. 1

Inside the Stars, IAU Colloquium 137
ASP Conference Series, Vol. 40, 1993
Werner W. Weiss and Annie Baglin (eds.)

PRISMA THE INSTRUMENTS

ANDREW JONES
Institute of Physics and Astronomy, Århus University, DK-8000 Århus C, Denmark
span solar::ajones, internet ajones@solar.stanford.edu

INTRODUCTION

Our understanding of Solar structure has increased dramatically in the last couple of decades thanks mainly to the opening of new windows of observation providing high quality data to theoreticians with access to powerful computing facilities. Two of the new windows were UV and X-ray images of the Sun, allowing a detailed view of the upper solar atmosphere, and the development of very high resolution spectrometers allowing us to exploit the solar oscillations to probe the internal structure of the Sun. It is the goal of PRISMA to extend these techniques to other stars, which using the Sun as a calibration point will allow us to explore stellar structure and evolution in ways not possible now.

In this poster I will present a possible selection of instruments able to achieve this goal, and explain some of the rationale in their design. A more general overview is presented by T. Appouchaux also in these proceedings. It must be stressed that these are not the definitive instruments to be flown on PRISMA, but rather result from a study to show the feasibility of such a mission. Should PRISMA be chosen as the next ESA medium sized mission, an 'Announcement of Opportunity' will be issued by ESA and the responses of all people interested in constructing the instrument will be considered.

For more information about PRISMA please send an electronic mail request to:**prisma@obs.aau.dk [129.142.17.96]**

ASTEROSEISMOLOGY

There are two main techniques for measuring the solar oscillations, Doppler velocity measurements with very high stability and resolution spectrometers, and intensity measurements with photometers. For ground based observations the spectroscopic techniques are preferred, as the fluctuations in the transparency Earth's atmosphere are some 100 times larger than the solar intensity fluctuations. From space this problem disappears, and the relative simplicity of the photometric instruments lends itself to satellite operation. In the case of stellar measurements there are further factors in favour of using photometric techniques. The optical layout of the two photometers on PRISMA is very similar, as they both have to fulfil the stringent conditions of providing a wide field of view, and a flat image. At the same time it must be possible to achieve very good photometric precision for all stars in the field. In order to satisfy these requirements, a three mirror off axis design was chosen. Both photometers use simple active optics to stabilise the stellar image on the detector. This reduces the noise due to spacecraft jitter though recent calculations by British Aerospace show that the satellite platform may be very stable.

The rôle of the large photometer is to provide seismic data for many stars in a $1.5° \times 1.5°$ field. To monitor many stars at once a Charge Coupled Device (CCD) detector has been selected. As there is only a limited data transmission bandwidth, only the position and intensity of the stars is transmitted. We have proposed a parallel architecture for the onboard data reduction system, this has the twin advantages of high throughput, and increased redundancy.

The small photometer is dedicated to obtaining very high quality time series for bright stars. A coelostat allows the SP to point independently of the other instruments, so it can view the same star for many months. A photo diode detector is used as this provides a large dynamic range, and relatively simple electronics.

ACTIVITY MONITORING

Monitoring various lines in the stellar atmosphere gives a picture of the star above the photosphere, and allows us to trace the rotation, and magnetic flux. Simultaneous multi wavelength measurements allow us to track surface inhomogeneities such as star spots and plages, whilst the ability to perform Doppler imaging provides information on the position and area of individual active regions.

The UVS is a Cassegrain telescope feeding a cross dispersed echelle spectrometer. The telescope selects a star by rotation of the primary mirror about its focus. A slit jaw camera and actuated secondary stabilises the image. The spectrometer is a conventional design covering 120–285 *nm* with a resolving power of about 29000. An off-axis paraboloid collimates the beam onto a small echelle grating, whilst a concave grating is used both as a cross-disperser and camera objective. The echelle working in 41st.–94th. orders gives a dispersion range from 5.2 mm/nm at long wavelengths to 12 mm/nm at the short wavelength limit. A spiral anode detector with 1000×1000 pixels, and a centring accuracy of better than $20\mu m$ record the spectrum.

The XUVT images the same $1.5° \times 1.5°$ field as the LP and UVS. This instrument has undergone several revisions since the publication of the PRISMA Assessment study report(Appourchaux et. al. 1991). The major change is to increase the collecting area by using a multi mirror design. The reason to use multiple mirrors rather than a single larger is driven by the current state of the art in x-ray multi layer mirrors. The reflectivity of the mirror is very sensitive to the surface smoothness, and the ability to control the thickness of the multi layers. Currently this limits the maximum diameter of such mirrors to about 25 – 30 *cm*, whilst still maintaining a reflectivity of about 30% at 17 *nm*. The current design uses four 20 cm diameter mirrors, though we have also thought about substituting objective gratings to disperse the image, if this technology proves to be sufficiently mature. A 'Solar Blind' 256×256 anode mesh MCP detector with 50 μm square pixels is planned. This arrangement will allow the detection of solar type activity on stars to 7th. magnitude.

DISCUSSION

The question of whether we need to go to space to make these measurements needs to be asked. For the UV and X-ray measurements the answer is obvious in that the Earth's atmosphere is not transparent at these wavelengths. However, we have just said above that in the case of the Sun we can make measurements from the ground. This is also to some extent true for stellar measurements. There is currently intense work on both Doppler velocity and photometric searches for solar type oscillations on other stars, but in the foreseeable future these techniques show little sign of producing more than an indication that such oscillations exist, rather than providing a tool to probe stellar structure. In the last 6 months the first decisive evidence for solar type oscillations has emerged from velocity observations of α Cen A (Pottasch et. al. 1992). For the first time the results seem to show a resolved spectrum of oscillations above the noise. Unfortunately the Doppler techniques are limited to slowly rotating stars with sharp, unblended absorption lines.

In January 1992 a multi telescope CCD photometric campaign was conducted to study a field in the M67 open cluster. The results of these observations are still being analysed, but it appears that an ultimate noise level of less than 10 μmag. can be achieved for a 5 night 7 telescope campaign. This is indeed exciting, however, it exposes many of the problems of ground based, multi site precision photometry. The measurements of brighter stars are not photon noise limited, but rather scintillation sets the noise limit. The problems of obtaining a good duty cycle are also apparent, as is demonstrated in a separate presentation by Søren Frandsen at this meeting.

On PRISMA the photometers are designed to achieve a precision of 1 ppm in 1 month on 8th and 6th magnitude stars for the LP and SP respectively .For brighter objects this level is reached faster. Thus, very high quality oscillations spectra will be obtained on many objects sampling the HR diagram, and allowing an extensive exploration of the physics of stellar interiors.

ACKNOWLEDGMENTS

I am grateful to the PRISMA science team of: T. Apporchaux, C. Catala, S. Catelano, M. Fridlund, S. Frandsen, C. Frölich, D. Gough, P. Lemaire, G.Tondello, and W.Weiss.

REFERENCES

Appourchaux T., Catala C., Catalano S., Frandsen S., Jones A., Lemaire P., Pace O., Volonté S., Weiss W. 1991 'PRISMA Probing Rotation and Interior of Stars: Micro variability and Activity – Report on the Assessment Study' ESA SCI(91)5

Pottasch E. M., Butcher H. R., van Hoesel F. H. J. 1992 'Solar like oscillations on a Centauri A' To appear in Astronomy and Astrophysics.

Inside the Stars, IAU Colloquium 137
ASP Conference Series, Vol. 40, 1993
Werner W. Weiss and Annie Baglin (eds.)

COMPARISON OF STELLAR OSCILLATIONS AND ACTIVITY DATA TO INFER INTERNAL ROTATION

GAETANO BELVEDERE
Istituto di Astronomia, Università di Catania, Italy

ABSTRACT Recently it has beeen suggested that the latitude distribution of the main surface features of solar activity is intimately related to the angular velocity profile inside the Sun through the working of a MHD dynamo in the boundary layer between the convective and the radiative zones (Belvedere et al. 1991).

Although the present observational capabilities are not very encouraging, here we want to point out, in the framework of the analogy to the Sun (solar-stellar connection), that space observations of surface distribution and latitudinal migration of active regions on stellar surfaces, which could be carried out in this decade with more sophisticated techniques, may conversely allow us to infer the rotation profile, and consequently the angular momentum distribution, in stellar interiors. This methodology may in principle be considered alternate or complementary to the classical one based on observation of acoustic oscillations.

PROLOGUE: WHAT WE HAVE LEARNED FROM THE SUN

Assuming the internal rotation profile inferred by the most recent helioseismological data (Harvey 1988; Morrow 1988; Brown et al. 1989; Dziembowski et al. 1989; Libbrecht 1989), Belvedere et al. (1991) carried out a non linear $\alpha - \omega$ dynamo model in the boundary layer betweeen the convective and radiative zones (about 0.65 $R_\odot$ to 0.70 $R_\odot$), reproducing the essential features of the solar activity cycle: the equatorward migration of sunspots and most faculae, which belong to the equatorial activity belt (latitude $\lambda < 35°$) and the poleward migration of polar faculae, filaments and large scale magnetic flux, which are characteristic of the polar activity belt ($\lambda > 40° - 50°$). This means that the two latitudinal belts of solar activity, which are distinguished by opposite latitudinal migration of tracers, would appear as a natural and direct consequence of the internal rotation profile, via the working of dynamo in the boundary layer. Specifically, it is a consequence of the rigid rotation rate $\omega_o = 2.7 \times 10^{-6}$ rad/s below the convection zone ($r = 0.65 R_\odot$) being precisely the value at the surface at the critical latitude $\lambda_o = 35°$, that defines the boundary between the two latitudinal belts. Since no radial variation of the surface rotation (a function of latitude) has been evidenced in the convection zone, we should have in the boundary layer $\partial\omega/\partial r > 0$ for $\lambda < \lambda_o$ and $\partial\omega/\partial r < 0$ for $\lambda > \lambda_o$. Accordingly, the $\alpha - \omega$ dynamo criterion (Parker 1955; Yoshimura 1975a), based on the sign of the product $\Gamma = \alpha \partial\omega/\partial r$ (with α negative in the boundary layer (Yoshimura 1975b; Glatzmaier 1985a,b) in the northern emisphere), allows equatorward propagation of dynamo waves for $\lambda < \lambda_\circ$ and poleward propagation for $\lambda > \lambda_\circ$.

Indeed, this picture has received a substantial support by the non linear $\alpha - \omega$ dynamo calculations performed by Belvedere et al.(1991) in a thin spherical shell ($d = 0.05R_\odot$) representing the boundary layer, with full latitude-time resolution. In fact, they have shown the existence of periodic (dynamo wave-like) stable solutions with both equatorward and poleward migrating branches, for dynamo numbers $|D| = \alpha\omega_\circ d^3/\eta^2$, where η is the turbulent viscosity, of the order of the unity. This is a basic result: the simulation shows that horizontally propagating dynamo waves can exist even in a very thin layer.

SURFACE ACTIVITY - A PROBE OF INTERNAL STELLAR ROTATION

Although many uncertainties do still exist even in the case of the Sun, nevertheless we are tempted to suggest that, conversely, observation of latitudinal distribution and migration of active regions on late main sequence (G,K,M) slowly rotating stars, by photometric and spectro scopic methods, may in principle allow us to infer the internal rotation profile and angular momentum distribution in a conceptually simple and direct way, offering a tool to future investigation of their dependence on stellar parameters such as spectral type, average rate of rotation and age.

What a theorist needs from observers

For this purpose we need accurate measurements of: (i) angular velocity at the equator or at some fixed latitude; (ii) latitudinal differential rotation profile at the surface, over a suitable latitude interval; (iii) latitude drift of activity tracers over a suitable time span. All necessary data may in principle be given by photometric and spectroscopic observations (Baliunas & Vaughan 1985; Hartmann & Noyes 1987; Rodonò 1987).

This way it should be possible to determine the surface latitude (if any) at which the direction of migration changes its sign and the corresponding value of the surface angular velocity, that are the essential data in order to deduce the internal rotation profile. The angular momentum profile would consequently be computed taking into account the density profile given by theory of stellar structure. Of course, all this is based on the analogy to the Sun - that seems to be fairly supported in the context of the solar-stellar connection (Belvedere 1985; Noyes 1985; Belvedere 1990) - namely on the fact that late spectral type main sequence stars have outer convection zones, whose extent is theoretically known, and suitable boundary layers. Similar patterns of convection are also plausible, at least for slow rotators. Unfortunately, the present available photometric and spectroscopic data of surface distribution of stellar active regions (Rodonò 1986; Vogt et al. 1987; Vogt & Hatzes 1991) only refer to binary, fast rotating, hyperactive stars of RS CVn and BY Dra type - whereas analogy to the Sun requires single, slowly rotating, mild activity main sequence stars - and show a definite tendency of active regions to be concentrated at high latitudes. Moreover, the sensitivity and resolution power of present observational capabilities are far from making it possible to perform high precision measurements, like those we suggest here, not only for solar type single stars with relatively weak activity, but even for strong activity RS CVn and BY Dra type stars.

PERSPECTIVES

However, future improvement of observational techniques based on the analysis of both light curves and line profiles (to this regard we mention the Zeeman-Doppler Imaging Method (Semel 1989; Donati et al. 1989) that is an extension of the Doppler Imaging Method (Vogt et al. 1987) and allows surface magnetic cartography), as well as observation from space with large instruments, may make it possible in next future. We refer in particular to the 170 cm diameter Spectrum Ultraviolet telescope orbital station, SUV, a collaborative project proposed by USSR and Italy to perform, besides other things, observations of stellar chromospheres (Rodonò et al. 1991). As matter of fact, one of the reasons to organize the Catania "Space Projects" meeting, came out from the theorist's curiosity to explore the future observational capabilities in connection to some theoretical requirements.

Further, comparison between surface activity data and acoustic oscillations data for a suitable sample of stars, may offer the opportunity to test the validity of our internal rotation probing method, whose basic principles are outlined here, and eventually the reliability of the boundary layer dynamo. This may be performed within the present decade in the framework of a single space observation project, Probing Rotation and Interior of Stars: Microvariability and Activity , PRISMA , which is under consideration of ESA (Lemaire et al. 1991).

If reliable on both theoretical and observational sides, this methodology might allow us to get information on the dependence of internal rotation and angular momentum distribution on mass, average rotational rate and age. Of course, application to other stars of results of dynamo theory obtained for the Sun is certainly not to be considered definitive, but only tentative. We are well aware that there are several uncertainties even in the solar case. Anyway, the present attempt may provide a way which future observations may confirm or refute.

REFERENCES

Baliunas S.L., Vaughan A.H.: 1985, *Ann. Rev. Astron. Astrophys.* **23**, 379

Belvedere G.: 1985, *Solar Phys.* **100**, 363

Belvedere G.: 1990, in: Berthomieu G., Cribier M. (eds.) *Proc. IAU Coll.* **121**, Inside the Sun, Kluwer, Dordrecht, p. 371

Belvedere G., Proctor M.R.E., Lanzafame, G.: 1991, *Nature* **350**, 481

Brown T.M., Christensen-Dalsgaard J., Dziembowski W., Goode P.R., Gough D.O., Morrow C.A.: 1989, *Astrophys. J.* **343**, 526

Donati J.F., Semel M., Praderie F.: 1989, *Astron. Astrophys.* **225**, 467

Dziembowski W.A., Goode P.R., Libbrecht, K.G.: 1989, *Astrophys. J.* **337**, L53

Glatzmaier G.A.: 1985a, *Geophys. Astrophys. Fluid Dyn.* **31**, 137

Glatzmaier G.A.: 1985b, *Astrophys. J.* **291**, 300

Hartmann L.W., Noyes R.W.: 1987, *Ann. Rev. Astron. Astrophys.* **25**, 271

Harvey J.W.: 1988, in Rolfe E.J. (ed.), Seismology of the Sun and Sun-like Stars, *ESA SP-286*, p. 55

Lemaire P., Appourchaux T., Catala C., Catalano S., Frandsen S., Jones A., Weiss W.: 1991, in Catalano S. & Stauffer J. (eds.), *Proc. Nato ARW*, Angular Momentum Evolution of Young Stars, Kluwer, Dordrecht, p.397

Libbrecht K.G.: 1989, *Astrophys. J.* **336**, 1092

Morrow C.A.: 1988, in Rolfe E.J. (ed.), Seismology of the Sun and Sun-like Stars, *ESA SP-286*, p. 91

Noyes R.W.: 1985, *Solar Phys.* **100**, 385

Parker E.N.: 1955, *Astrophys. J.* **122**, 293

Rodonò M.: 1987, in Schroter E.H. & Schussler M. (eds.) Solar and Stellar Physics, Lecture Notes in Physics **292**, 39

Rodonò M. et al.: 1986, *Astron. Astrophys.* **165**, 135

Rodonò M., Tanzi E.G., Boyarchuk A.A., Steshenko N.V.: 1991, in Catalano S., Stauffer J. (eds.), *Proc. Nato ARW*, Angular Momentum Evolution of Young Stars, Kluwer, Dordrecht, p. 405

Semel M., 1989: *Astron. Astrophys.* **225**, 456

Vogt S.S., Hatzes A.P.: 1991, in Tuominen I., Moss D., Rudiger G. (eds), *Proc. IAU Coll.* **130**, The Sun and Cool Stars: Activity, Magnetism, Dynamos, Lecture Notes in Physics **380**, 297

Vogt S.S., Penrod G.D., Hatzes A.P.: 1987, *Astrophys. J.* **321**, 496

Yoshimura H.: 1975a *Astrophys. J.* **201**, 740

Yoshimura H.: 1975b *Astrophys. J.* **29**, 467

Inside the Stars, IAU Colloquium 137
ASP Conference Series, Vol. 40, 1993
Werner W. Weiss and Annie Baglin (eds.)

ASTEROSEISMOLOGY FROM THE GROUND

S. FRANDSEN
Institute of Physics and Astronomy, Århus University, Denmark

ABSTRACT In stars which pulsate in more than one mode, be it acoustic or gravity modes, radial or non-radial, we have a chance to study the interior structure by use of seismic techniques. Generally the greater the number of modes observable the greater the chances will be to obtain valuable direct information about the stellar interior not accessible by any other technique.

The observational problem of acquiring high quality data divides the stars where seismic techniques can be applied into two groups: A group of stars with intensity amplitudes in the range of millimagnitudes and the stars with very small amplitudes around 10 μmagnitudes or less. The first group contains the white dwarfs, the rapidly oscillating Ap stars, the δ-Scuti stars and the β-Cephei stars. The modes are excited directly or indirectly by some form of κ-mechanism due to an opacity source which vary from class to class. The second group is made up of stars with outer convective envelopes which generate acoustic noise that may excite resonant modes, often a very large number.

This paper is a discussion of the possibilities for observing these stars with ground based techniques, and also a statement about what can not be done from the ground.

INTRODUCTION

For a general idea about the subject I recommend the papers by Brown (1990a) and Baade (1991). Also the proceedings from meetings in Aarhus, Tenerife and Cambridge contain articles that are relevant.

The paper is organized in four sections. The first defines the observational arena for asteroseismology. The techniques to be applied depends on the characteristics of the type of stars to be studied. The important parameters are the period P, the semi-amplitude A, the magnitudes of the objects under study m_V and the number of modes in the pulsation spectra. The second section addresses the Doppler techniques and the following the photometric techniques. The final section is a small discussion of the shortcomings of ground based observations and the necessity to observe from space.

THE MAIN AREAS OF GROUND-BASED ASTEROSEISMOLOGY

Seismic techniques can be applied to a number of different types of stars. For each category the set of modes define a specific observational and theoretical scenario, which is special and different among the types of objects. Ground-based observations contribute mainly in the following areas, which in some cases represent the full data needed, in others only supplementary data to space observations.

- 1 Spectroscopy and Doppler velocity studies due to the weight and size of the equipment to perform such studies.
- 2 Long term monitoring (stellar cycles/evolutionary changes)
- 3 Faint objects where photon noise dominates scintillation noise, i.e. white dwarfs. This depends obviously on telescope size.
- 4 Future interferometric measurements (differentiel speckle photometry).
- 5 Doppler imaging which demand very high resolution and high S/N ratios.
- 6 Simultaneous support observations with space experiments.

What's needed? This question is related to the physics of the objects, but generally the data required are more time consuming and complicated to obtain than for classical variables or for single stars in general. A list of properties of the data could look like

- 1 Very low noise measurements (no $1/f$ noise for βCephei or δ-Scuti stars)
- 2 Long time series with few interruptions to produce high quality spectra.
- 3 Doppler imaging to help mode identifation.
- 4 Devoted studies of a small set of key objects (compare with the work on the Sun!).
- 5 Development of new, more sensitive instrumentation for extreme precision in Doppler studies. The lowest noise obtained until date is 30cms^{-1} (Pottasch et al. 1992). The goal is to reach 1cms^{-1}.

In order to meet the above demands new techniques and devoted observations are required. Somebody must invest an effort in organizing networks of modern state of the art instrumentation. Small or medium size telescopes allocated for extended periods to these studies will be needed and to make life acceptable for the observers, they should be highly automized.

The stars that are the main candidates for seismic studies can be put into five classes, which have characteristics as given in Table I.

Solar type stars is yet an area to be explored due to the minute amplitudes of the oscillations. But the white dwarfs observed by the WET consortium (Nather et al. 1990) show oscillation spectra of similar richness representing extremely valuable data for analyzing the interior structure of the white dwarfs. A spectrum is shown as Figure I. The other classes present spectra with fewer

TABLE I Classes of variables

Class	Period	Amplitude	Magnitudes	Observ. modes	Occur.
δ-Scuti	0.5-7h	1-50 mmag	$2 < m < 15$	1-5	single/clust.
β-Cephei	3-6h	10-100 mmag	$2 < m < 15$	1-2	single/clust.
roAp's	4-15m	10-1000 μmag	$4 < m < 8$	1-10	single
Wh. dw.	3-40m	1-100 mmag	$13 < m < 16$	1-100	single
Solar *'s	3-40m	1-100 μmag	$0 < m < 6$	30-50[a]	single[b]

[a]Full disk modes only
[b]Assuming velocity measurements only

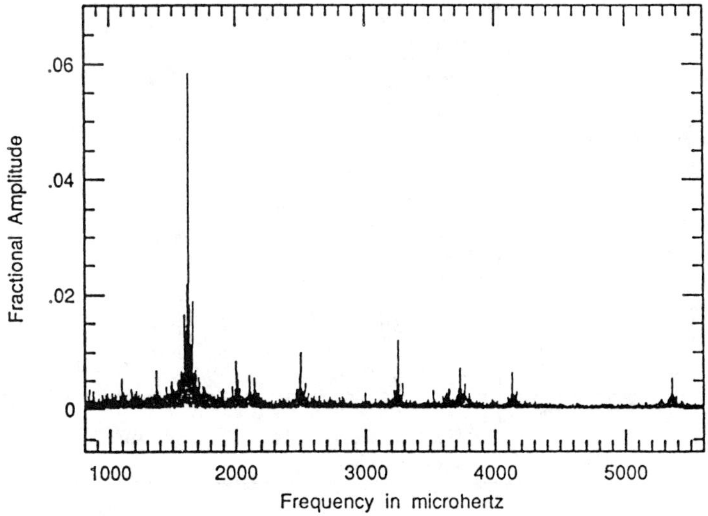

FIGURE I Part of amplitude spectrum of the white dwarf G29-38 from Winget et al. 1990

modes. Given stars enough this still opens up a very promising field of interpretation. Examples of δ-Scuti type spectra are found in Michel and Baglin (1991), the STEPHI program, and Mantegazza et al. (1992).

VELOCITY STUDIES

Velocity studies have been carried out by the extension of solar techniques or by modifications of classical spectrographs to obtain the stability needed.

In Table II a list of groups is given that have attempted to observe Doppler velocities due to solar type observations or suggested techniques yet to be tested to do so.

TABLE II Instrumentation based on spectroscopy

Group/Reference	Technique	Usage
1 Channel/1 line		
a. Nice group	Na resonance cell	αCen, Procyon
Schmider 1988, 1990		
b. Birmingham group	K resonance cell	
Innis et al. 1990		Procyon
c. Imperial College	F-P filter	Ap stars
Reay et al. 1986	Michelson Interferometer	
d. Harvey Butcher group	F-P filter	αCen
Pottasch et al. 1992	fiber + scrambler	
Multichannel/Multiline		
a. Brown et al.	HRS + fiber + scrambler	αCen, Procyon
1991	+ calibration lamp	
b. Libbrecht	HRS + fiber	γEqu
1988	+ calibration lamp	
c. Frandsen	HRS + fiber	βHyi
1987	core/continuum ratio	
d. Connes	HRS + fiber + scrambler	none
1985	+ F-P + 2 lasers	
Intermediate solution		
a. Frandsen et al.	HHS interferometer	none
1992	+ cross-disperser	

Non multiplexing techniques

Although one can easily show that a proper S/N-ratio can not be obtained with any acceptable investment of telescope time by a filter type instrument, it still seems reasonable to try to detect oscillations in a few bright solar-type stars by this technique. References to such experiments can be found in Table II. The most convincing power spectrum is presented by Pottasch et al. (1992) and is

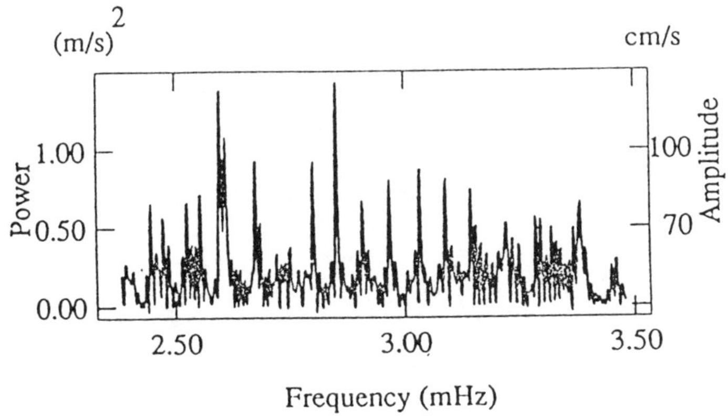

FIGURE II Power spectrum of αCen A observed by Pottasch et al. 1992. Equally spaced peaks seems to be present as seen on the Sun

given here as Figure II. Still, one does not have a completely solid proof of the existence of *p*-modes in a star, but if the peaks identified by Pottasch et al. was detected in an independent experiment, this would probably be enough to convince most people about the reality of the modes.

Based on the results obtained so far, it is evident, that future work in this direction for solar type stars hold no promise. One must use more efficient instrumentation. Filter type instruments, although, represent an interesting solution to study slower, larger fluctuations in giants or radial velocity changes due to planets.

Multiplexing techniques

It is a straight forward idea that a high resolution spectrograph might be employed, provided it could be calibrated well enough. Simple as it sounds, in practice it has proven very difficult. Similar difficulties have been encountered in the search for planets around stars.

Nevertheless this is in the end the only efficient way to observe Doppler velocities and several groups are working on the perfection of the technique to have measurements of a quality close to the theoretical noise limit. Still much need to be done. The ultimate solution has been formulated by Connes (1985), but the design contain a number of technical innovations, that has to be made to work. Other less ambitious projects use fiber coupled spectrographs built for precise radial velocity measurements and include the MUSICOS project (Baudrand and Böhm, 1992), the AFOE (Noyes et al. 1992) and the VELUX spectrograph (Andersen et al. 1992).

Figure III, a modified version of a figure from Connes (1985), demonstrates that even with modest telescopes solar type *p*-modes can be detected in slow rotators with a modest effort given the right instrumentation. One can even go to rather faint magnitudes. More discussions about detection limits can be found in Appourchaux (1991) and Brown (1990b).

A drawback of spectrographs is that they have to be matched to the tele-

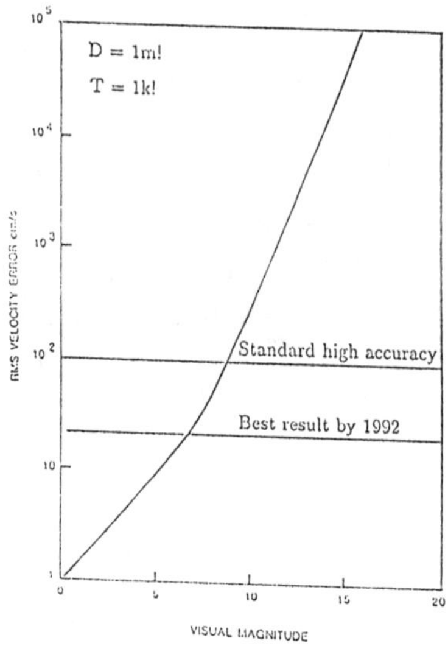

FIGURE III Observational limits

scopes. Also they tend to become large instruments, which are hard to make stable, even when used in a highly differential mode. It can be very difficult to locate the source of instrumental noise, which come from subtle mechanical and optical imperfections of the instrument.

An intermediate solution in terms of efficiency is a multiplexed interferometer as presented by Frandsen et al. (1992). This combines the stability of the filter techniques with some multiplex gain. It also provides a flexible instrument that can be used at many different telescopes without modifications. It involves some new technology, which yet have to be shown to work.

It can be stated that progress in this field of Doppler techniques is taking place, but slowly, and it will take some years before the technique will provide the very high quality data needed. In addition to the technical problems, multisite campaigns need to be set up to get nice, clean spectra, which poses an organizational challenge.

INTENSITY STUDIES

The intensity oscillation amplitudes for solar type stars are very small, only ppm's, whereas the other classes have amplitudes in the mmag range. To enable real seismology long observing periods are needed for all variables and the less interruptions the better.

As holes in the time series are difficult to avoid, even for large groups of telescopes like WET, very good S/N data is required to enable a deconvolution of the window function.

The smallest detected amplitudes so far are of the order a few tenth of a mmag. Recently attempts to detect solar type oscillations in stars in M67 are

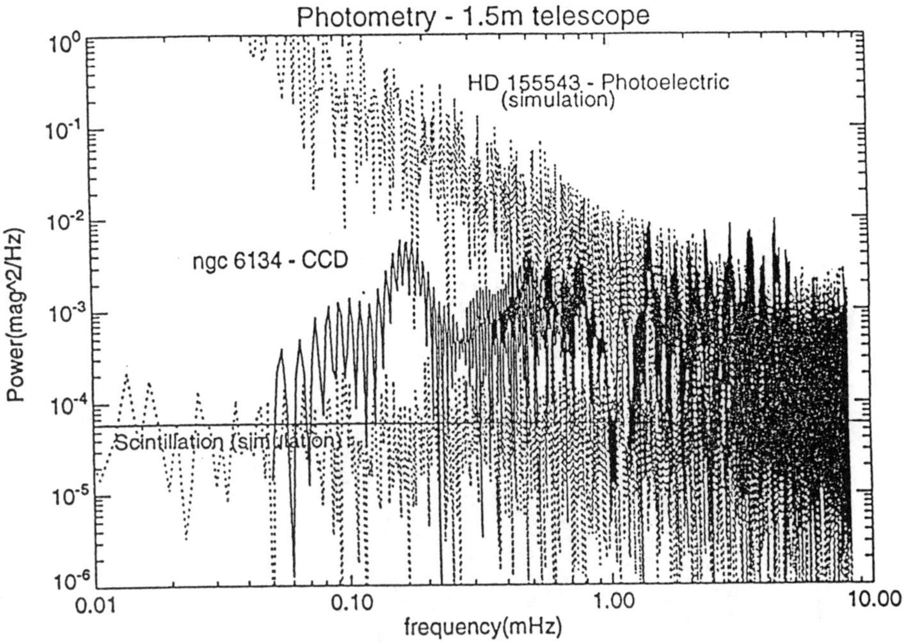

FIGURE IV A figure from Kjeldsen and Frandsen (1992). The curves have been scaled to correct for the difference in observing time. The photoelectric simulation matches the observed power spectrum as given by Belmonte et al. (1990) There is no big difference in the 3 mHz range, but in the δ-Scuti range (0.1mHz) the differences become very large between CCD and photoelectric photometry

expected to lead to an order of magnitude improvement in detection sensitivity (Gilliland et al. 1990).

The use of CCD detectors in high precision photometry has expanded the possibilities due to the stability of this detector type, and the option to make simultaneous, differential photometry for many objects. The limiting factor now is the available photons or the scintillation. For stars above the scintillation limit there is not much that can be done to improve S/N, because the noise decreases only slowly with the diameter of the telescope.

An investigation of photometry with CCD's by Kjeldsen and Frandsen (1992) demonstrate the superiority of CCD measurements especially at very small amplitudes and at low frequencies. Figure IV show actual data and simulated data where it is evident that especially the $1/f$ noise is much smaller for CCD data than photoelectric data. At some observatories conditions are so stable that photoelectric work does not add any atmospheric noise (except scintillation) to the photon noise, but this is normally not the case.

TABLE III Classes of variables

Observatory	D	Period allocated	% contr.
Kitt Peak	4.0m	11-16 Jan	23.9
CFHT	3.6m	12-16 Jan	18.7
AAT	3.9m	14-17 Jan	8.1
Calar Alto	3.6m	14-16 Jan	4.9
Nordic Opt.	2.5m	12-17 Jan	7.7
Palomar	5.0m	11-17 Jan	36.7
Kavalur	2.3m	xx-xx Jan	0.0

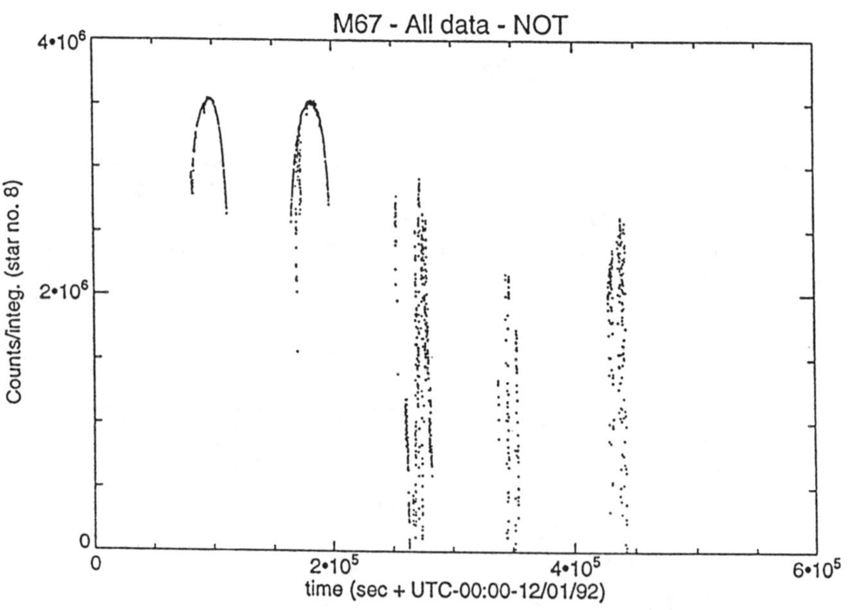

FIGURE V Record of the data observed by the NOT 2.5m during the January 92 campaign. Fluxes are given for a particular star.

Photometry from the ground

With large format CCD's the situation has improved for making high precision differential photometry: Using the formulas from Kjeldsen and Frandsen (1992), and choosing an effective filter width of $\Delta\lambda = 100\text{nm}$ and an integration time of $t = 60\text{s}$, we get for the magnitude m for which scintillation and photon noise is equal the result

$$m = 5.1 + 5/3 \log(D) \quad (1)$$

where the diameter of the telescope D is in cm. For stars brighter than this scintillation dominates. Scintillation has the fairly slow dependence on diameter ($D^{-2/3}$), and it is advantageous to use smaller telescopes and longer observing periods instead of applying for costly time on large telescopes.

Another advantage of using a small telescope is that the field of view increases. If we look at the number of stars in a CCD frame of $30 \times 30\text{mm}^2$, assuming the density of stars to be uniform in space and 10000 stars brighter than $m = 6$, we get for the number of objects in the field of view bright enough to be scintillation dominated

$$n_{obj} = 9.18 \times 10^2 (d/f)^2 D^{-1} \quad (2)$$

where d is the detector size in cm and the f the f-ratio of the telescope, typically $6 < f < 15$. For a small telescope like the 50cm telescopes at La Silla with f=13.6 and a detector with d=3cm we find in the field an average of 0.9 stars per frame. The example is for a fairly slow telescope. A modern telescope with f=6 would give several stars per frame. The increase of detectors from around 1cm^2 to of the order 10cm^2 is seen to be quite important. For a given f-ratio the chances of finding suitable comparison stars decrease linearly with telescope diameter.

For a given star the largest telescope might not be the optimal instrument if comparison stars are not available in the field. Generally, for a small telescope, we should be able to find bright comparison stars, in the sense that they contribute no more to the noise than the program star. Of course, near the galactic plane, the situation is better than for the average sky, and near the poles worse.

Large telescope campaigns

It is interesting to explore how far one can come with photometric techniques. The most ambitious project so far is the latest campaign (Jan. 92) on M67 organized by Gilliland and Brown (1991), where a number of large telescopes was pointed towards a small group of stars for 5-6 nights. The data accumulated is given in the final column of Table III as the weighted percentage of the data from each site. The total number of useful nights was around 20.

Among the participating telescopes was the Nordic Telescope at La Palma, and Figure V shows the raw magnitudes for one of the stars from 5 nights of data.

Based on the reduction of data from Kitt Peak the extrapolated noise in the interesting frequency range is 7 μmag, very close to the level that was estimated as the value for a successful campaign. The result still depends on whether merging the data can be done without losses. It is also assuming that the data

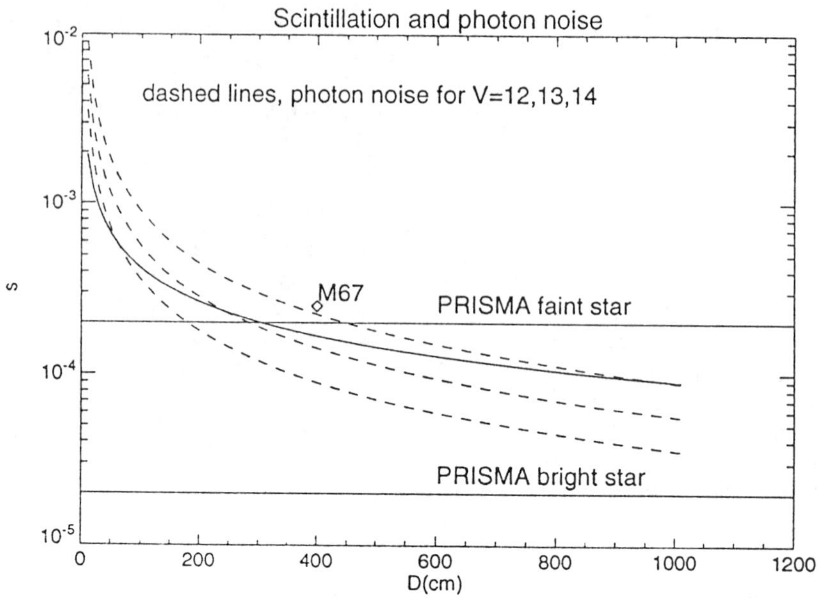

FIGURE VI The noise per frame ($t = 60s$) as function of the diameter D is plotted for scintillation noise (full drawn) and photon noise (dashed). The preliminary result for the Kitt Peak 4m is indicated by the symbol. The horizontal lines present the noise for the brightest and the faintest star observed with the Large Photometer on PRISMA, where ppm accuracy will be reached.

from other telescopes can be used as efficiently as the Kitt Peak data. Data from several sites have been reduced and looked at and no serious problems discovered so far.

The results obtained can be compared with the curves in Figure VI, which at the same time illustrate how difficult it will be to do much better. In this figure the noise level due to scintillation and counting statistics is plotted against aperture of a telescope. Magnitudes typical for the selected M67 stars are plotted as dashed lines.

Results from long time series

As modes often occur in δ-Scuti stars or white dwarfs with very small frequency differences, like a few μHz, the modes can only be resolved when long time series are obtained, where long is of the order a week for white dwarfs and several weeks for δ-Scutis.

The gain in quality of the power spectra measured is evident when one looks at the WET results for white dwarfs (Winget et al. 1990), where hundred of modes have been identified. Piece of a spectrum was shown as Figure I.

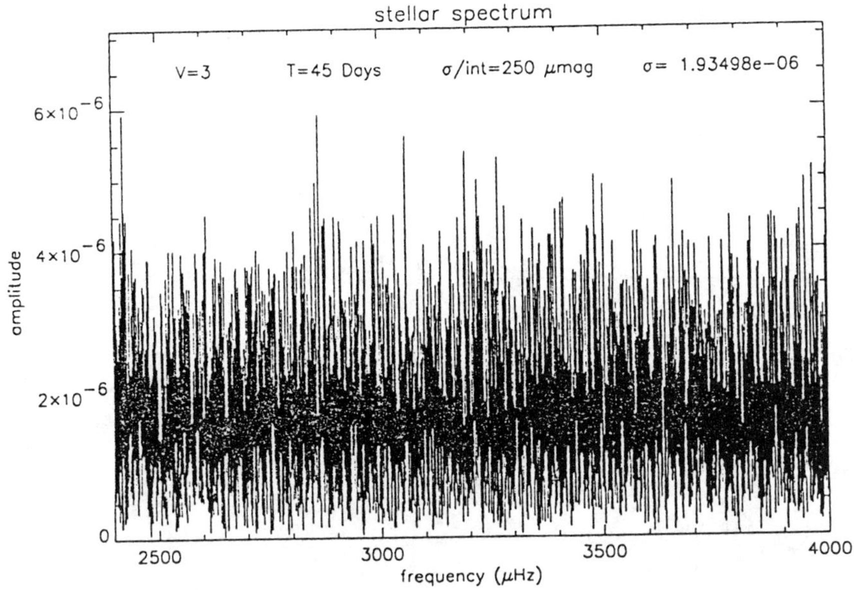

FIGURE VII A simulated amplitude spectrum obtained by a network of 4m telescopes. Integration time 60s. The noise level per integration is indicated as well as the r.m.s. noise in the spectrum. A solar spectrum is in there, but the noise dominates to the extent, that no signal is seen at all.

BEYOND THE ATMOSPHERE

Even with a network of large telescopes intensity fluctuations due to *p*-modes in stellar analogs to the Sun can not be detected due to the scintillation in the atmosphere. The best way to illustrate this is by showing a simulation (Figure VII) of 45 days of data as observed by a network of 4m class telescopes using a window function similar to what was obtained during the aforementioned M67 campaign. The noise per measurement is the typical value for a 60s exposure obtained at the Kitt Peak 4m during the M67 campaign. Hidden in the noise is a simulated solar power spectrum, which clearly can not be detected.

In contrast to this an example of a simulation of a spectrum obtained with the large photometer on the PRISMA (report 1991) payload is shown as Figure VIII. It is evident, that the advantage of going to space is enormous. The same improvement in quality will be the case for all stars bright enough to be within the range of the PRISMA instruments. This selection of stars is very large as the faint limit for ppm noise level is 8th magnitude. Even at 5 magnitudes below this level, the noise level is still very small and will allow extremely interesting data to be collected for the higher amplitude variables.

Seismology as used to explore the Sun can only be applied to stars in a

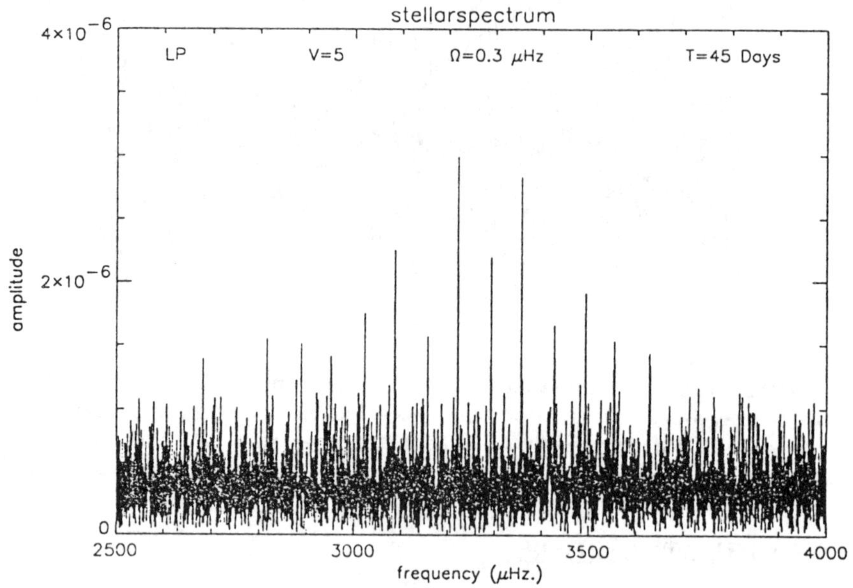

FIGURE VIII A simulated amplitude spectrum of the Sun at $V = 5$ as observed by the Large Photometer on PRISMA. Indicated is the rotational rate Ω and the length of the observing period T. The r.m.s. noise level is $\sigma = 3.1 \times 10^{-7}$.

wider context by observing from space with a reduced noise level and continuous viewing exceeding one month. Only then can we measure frequencies precisely and study the separation between close frequencies, the shape of the peaks and the splitting due to rotation in slowly rotating stars.

But among all the interesting objects only a fraction can be among the targets of one space mission. The observations from space must be limited to a sample of objects, which are of particular interest in astrophysics. Everything should be done to improve ground based techniques to supplement and complement space observations, and to avoid more observations from space than absolutely necessary due to the high cost and long planning involved.

REFERENCES

Andersen, T., Dybdal, S., Frandsen, S. and Reiz, A. 1992, *The NOT Newsletter* **6** (VELUX)

Appourchaux, T. 1991, *A&A* **222**, 361

Baade, D. 1991, in *Rev. in Mod. Astr.* **4**, Springer

Baudrand, J. and Böhm, T. 1992, *A&A* accepted (MUSICOS)

Belmonte, J.A., Peres Hernández, F. and Roca Cortes, T. 1990, *A&A* **331**, 383

Brown, T.M. 1990a, *The Sun and Asteroseismology*, preprint

Brown, T.M. 1990b, in *CCDs in Astronomy, Tuscon Arizona, ASP Conf. Ser.* **8**, 335

Brown, T.M., Gilliland, R.L., Noyes, R.W. & Ramsey, L.W. 1991, *ApJ* **368**, 599

Connes, P. 1985, *Astrophys. Sp. Sci.* **110**, 211

Frandsen, S. 1987, *A&A* **181**, 289

Frandsen, S., Douglas, N., Butcher, H. 1992, *A&A*, submitted

Gilliland, R., Brown, T.M., Duncan, D.K., Suntzeff, N.B., Lockwood, G.W., Thompson, D.T., Schild, R.E., Jeffrey, W.A. and Penprase, B.E. 1991, *AJ* **101**, 541

Gilliland, R. and Brown, T.M. 1991, *Observing guide: M67 Stellar Oscillations Network Campagin*, Private communication

Innis, J.L., Isaak, G.R., Speake, C.C., Brazier, R.I. & Williams, H.K. 1990 preprint

Kjeldsen, H. and Frandsen, S. 1992 *PASP*, accepted

Libbrecht, K.G. 1988, "The Impact of very high S/N Spectroscopy in Stellar Physics", IAU Symp. 132, eds. Cayrel de Strobel, G. & Spite, M., 83

Mantegazza, L., Poretti,E, Antonello, E., Riboni, E. and Zerbi, F. 1992, These Proceedings

Michel, E. and Baglin, A. 1991, *Adv. Space Res.* **11**, 70

Nather, R.E, Winget, D.E, Clemens, J.C., Hansen, C.J. and Hine, B.P. 1990, *ApJ* **361**, 309 (WET)

Noyes, R.W, Brown, T.M., Horner, S., Korzennik, S. and Nisenson, P. 1992, These Proceedings (AFOE)

Pottasch, E.M., Butcher H.R. & van Hoesel, F.H.J. 1992, *A&A*, accepted

PRISMA, *Report on the Assessment Study*, January 1991, ESA SCI(91) 5

Reay, N.K., Ring, J., Pietraszewski, K.A.R.B., Bell, C.R. & Leeper, M. 1986, *SPIE Vol 627 Instrumentation in Astronomy VI*, 27

Schmider, F.-X., Fossat, E., Grec, G. & Gelly, B. 1988, *Proc. Symp. Seismology of the Sun and the Sun-like Stars*, Tenerife, Spain, ESA SP 286, 605

Schmider, F.-X. 1989, Thesis, Universite de Paris VII

Winget, D.E. and the WET consortium (27 authors) 1990, *ApJ* **357**, 630

Inside the Stars, IAU Colloquium 137
ASP Conference Series, Vol. 40, 1993
Werner W. Weiss and Annie Baglin (eds.)

LOW-LEVEL STELLAR VARIABILITY

M. GRENON
Geneva Observatory, CH–1290 Sauverny, Switzerland

ABSTRACT The capability of ground-based photometry to detect and monitor low-level stellar variations is discussed in the particular case of amplitudes in the range of few magnitudes to few percent. The technical and atmospherical constraints are briefly recalled. The variability level of the so-called constant stars is discussed as function of the spectral type, the luminosity class and the effective temperature. The photometric aspects of the Hipparcos mission are presented here since this satellite allows a systematic detection of microvariables and the obtention of light curves across the whole HR-diagram with an unprecedented efficiency.

INTRODUCTION

A global description of the variability in the HR-diagram at the milimagnitude level is still missing. In the following, we intend to describe what is now known from ground based photometry and what the expectations from space photometry with the Hipparcos satellite are. The emphasis will be put on variability detection and light curve obtention as a by-products of ground and space programs.

For specific objects, differential photometry allows precise determinations of the amplitudes at the sub-milimagnitude level provided that the periods are short with respect of the night duration. Multi-site observations partly overcome this difficulty. For longer period stars, the variability characterisation results from absolute photometry, the detection threshold is then no better than a few milimagnitudes but there is no more limitation regarding the period which may range from few minutes to several years.

A nomenclature problem arises for the classification of variables according to their amplitude range. Classical variables show amplitudes (A) which are of the order of a magnitude or a substantial fraction of it, the term of microvariable is currently and somewhat improperly applied to variables with amplitudes in the range of few percents. What should be the designation for variables with A in the range of few 10^{-3} or 10^{-4} mag ? Whatever will be the accepted designation, we will investigate here the variability of so-called microvariable and constant stars.

THE DETECTION THRESHOLD

For many years, ground-based observers considered it to be fashionable to produce colours and magnitudes with an accuracy not better than a few percent. The adopted integration times, the number of comparison stars, the thermal control of the instruments and adopted reduction procedure, lead ultimately to errors, in the case of the classical UBV photometry, of 0.02 mag on (B–V) and 0.03 mag on V.

As a matter of fact, these errors may be reduced by one order of magnitude, provided a number of precautions are taken for the construction of the photometers and for the data collection and reduction. When observing from the ground, the most limiting factor, magnitude independent, is the scintillation. To reach to 0.005 mag level requires integration times not shorter than 10 seconds for a 70 cm telescope. Too often integrations are stopped when the photon noise is below a specified level. The atmospherical extinction changes seriously limit the photometric accuracy. During typical photometric nights, the visual extinction may vary by 0.02–0.03 at the Zenith. After events like the *El Chichon* or more recently the *Pinatubo* eruption, the atmospherical extinction becomes anisotropic and variable. As an example, four months after the *Pinatubo* eruption, up to 8 comparison stars per hour were necessary to monitor the extinction at the 0.002 mag level. The thermal control of the detector is critical too. Thermal drifts induces sensitivity changes of the detectors and pass-bands shifts of the combination filter plus detector. A temperature stabilisation of ± 0.1° C ensures a gain stability of ± 0.4 milimagnitude. Photon-noise stars to be significant only for rather faint stars or extreme colour stars. The ultimate accuracy is also dependent on the stability of comparison stars themselves. Too often lists of standards are contaminated by microvariables (up to 40% in a recent VRI standard) and observers are tempted to use stars with a large number of observations as comparison, whereas it is precisely because they were suspected variables that they receive more attention.

Provided that the above conditions are fulfilled, it is possible to achieve absolute photometry at the 0.002–0.003 mag level with a one meter telescope. For more details on accuracy photometry see e.g. Harvey (1988) and Young (1984).

THE GENEVA PHOTOMETRY ACCURACY

Since the installation of the Swiss telescope at La Silla in 1976 and the use of a new generation photometer, the accuracy of photometric data in the Geneva system was seriously improved, because the technical improvements and the quality of the Chilean skies. The observational strategy is to observe a star twice at a one month's interval approximately, and then for a third time one year later. Colours and magnitudes are means over 3 to 12 minutes observations. This procedure allows a detection, if not a description, of the short and long-term variables.

The distributions of residuals on V magnitude is given Fig. I. The mode of the distribution is in the range 2 to 4 milimagnitudes. Stars with larger σV are either faint stars or microvariables. Bright stars are suspected as microvariables

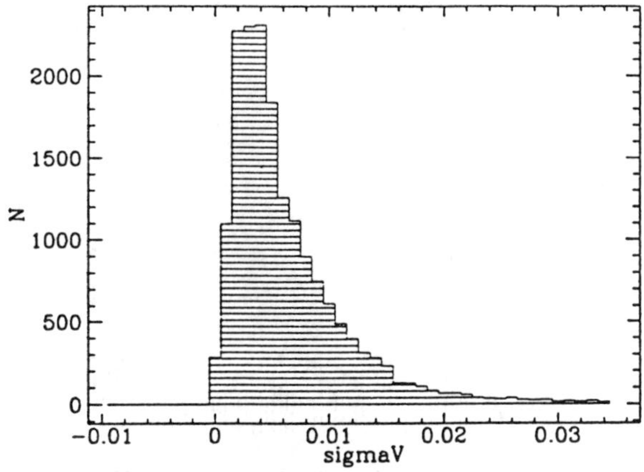

FIGURE I The distribution of residuals on V mag for 29 000 stars with three or more measurements in Geneva photometry.

if $\sigma V > 0.006$ mag. Possible variables are almost systematically re-observed to confirm their status and a number of 6–10 observations are generally performed.

The photometric catalogue of Geneva contains now 40 000 stars and among them 29 000 have been observed at least three times. Since all spectral types and luminosities are represented, a global investigation of the stellar stability may be attempted across the whole HR-diagram.

THE RELATION SPECTRAL-TYPE-AMPLITUDE

Since early-type and luminous stars are often heavily reddened, a straightforward way to reveal the instability domains as a function of the temperature, is to plot the residuals in V versus the spectral type.

Figure II shows σV as a function of the digitised type irrespective to the luminosity class. Non variable stars show $\sigma V = 0.0048$ mag and a mean standard deviation of the colours of 0.00371 mag. The classical instability strip is centred on F3 for main sequence stars and extends towards lower temperature with increasing luminosity and amplitude. Stars earlier than B1 appear quite stable, with the exception of WR stars which are all microvariables. The most striking feature is the B1–B2 range with many stars showing standard deviation between .04 and .17 mag. This κ-driven instability concerns a very narrow range of T_{eff} if one takes into account the uncertainty on the spectral type itself. The domain B8–A0 contains the magnetic and peculiar Ap, Bp stars. Their amplitudes in V are half those of B1–B2 stars. The noise increases at K0 is conspicuous as well as that of M stars with σV increasing with advancing M subtype.

When the standard deviation on colours (σcol) is considered , see Fig. II, the behaviour as a function of the spectral type is fairly different. At given σV, the corresponding σcol decreases with increasing T_{eff}; this effect is a consequence of the loss of sensitivity to T_{eff} of the colour indices towards high T_{eff}.

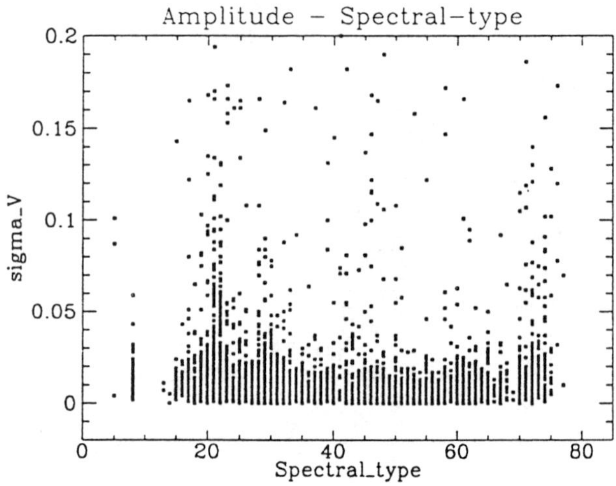

FIGURE II The distribution of the standard deviation in V as function of the spectral type. Coding of types: 08=WR; 15=O5; 20=B0; 30=A0; 40=F0; 50=G0; 60=G0; 70=M0.

THE RELATION LUMINOSITY-TEMPERATURE-AMPLITUDE

If the stellar physical properties are homogeneous in a given range of spectral type and luminosity class, it is equivalent to derive the mean variability level from large sets of observations on few representative stars or from a large collection of stars with few measurement per star. The number of stars in the Geneva catalogue with $n > 2$ observations per star permits the second alternative.

For the discretisation in terms of luminosity, the following classes are retained: Ia0–Ia, Ib,II, II–III, III, III–IV, IV, IV–V, V, V–VI and VI. In terms of spectral type, bins of one to three sub-types are considered. The number of stars per box is very variable but high for the main sequence with a maximum of 880 for B8–B9V stars. The minimum standard deviation is observed for late F and early G main sequence stars and is equal to 0.0048 mag. This value is adopted as the measurement error. For a given box, the intrinsic stellar standard deviation on V is computed as follows:

$$\sigma V_i = (\sigma V_{obs}^2 - 0.0048^2\ )^{1/2} \tag{1}$$

where σV_{obs} is the mean observed standard deviation in that box. In order to minimise the contamination by eclipsing binaries, all stars with $\sigma V > 0.2$ mag or $\sigma col > 0.1$ are rejected from the samples. The distribution of variability domains in Fig. IV, is then that of low amplitude and "constant" stars; the mean amplitude is largely underestimated for stars inside the instability strip and for late M giants.

The high stability areas in the HR-diagram are: the F2–G5 main sequence, the F1–G9 class IV–V, the G0–G9 subgiants and the III–IV first giant branch. The bright G2–K2 giants of class II are remarkably stable too with a mean rms < 5 mmag. On the blue side of the instability strip, the evolved B5–A6 stars are very stable provided they are less luminous than the Ia, Ib supergiants.

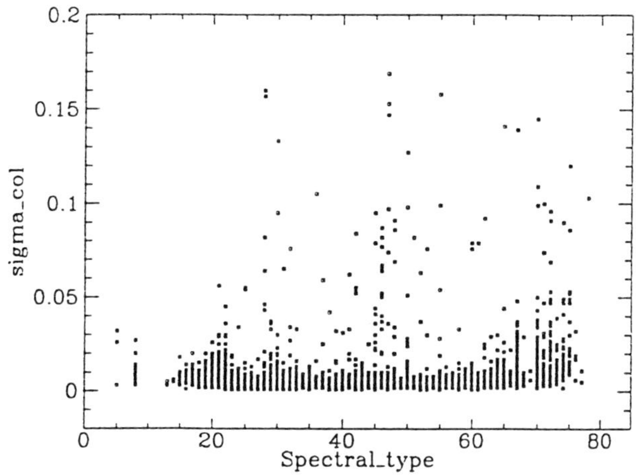

FIGURE III The distribution of the standard deviation on Geneva colours as function of the spectral type. Codes for the type as in Fig. III.

The κ-mechanism is quite efficient for B1–B2 dwarfs and giants with typical rms of 15 to 40 mmag. Its efficiency tends to disappear at higher luminosity or to be displaced to lower temperatures. All luminous supergiants are microvariables or variables, especially in the vicinity of the instability strip.

The microvariability level of A5–G5 class III giants is of the order of 7 mmag outside the instability strip. The variability is partly due to the presence of a higher than normal proportion of double stars with composite spectra.

On the red side of the instability strip, there is a gentle increase of the variability with the temperature decrease down to the early M stars. At M1–M2, there is a sudden increase corresponding to the development of TiO absorption bands. The old mild metal poor stars and the population II subdwarfs seem less stable than main sequence population I of the same mass.

THE LATE TYPE STARS VARIABILITY

For most red dwarfs and part of supergiants, an MK classification is not available. For stars with a negligible interstellar reddening, the calibrations of the Geneva photometry provide estimates of Mv, T_{eff}, [M/H] or I_{TiO} with a precision sufficient for an unambiguous attribution to the various branches in the HR-diagram.

The HR-diagram of Fig. V results from an automatic photometric classification of field stars and contains 7 915 stars found to lie within 150 pc where the reddening may be neglected. Stars below the main sequence and the red giant branch have composite spectra or colours and spurious Mv. M giants are not plotted here since their absolute magnitudes are still undefined.

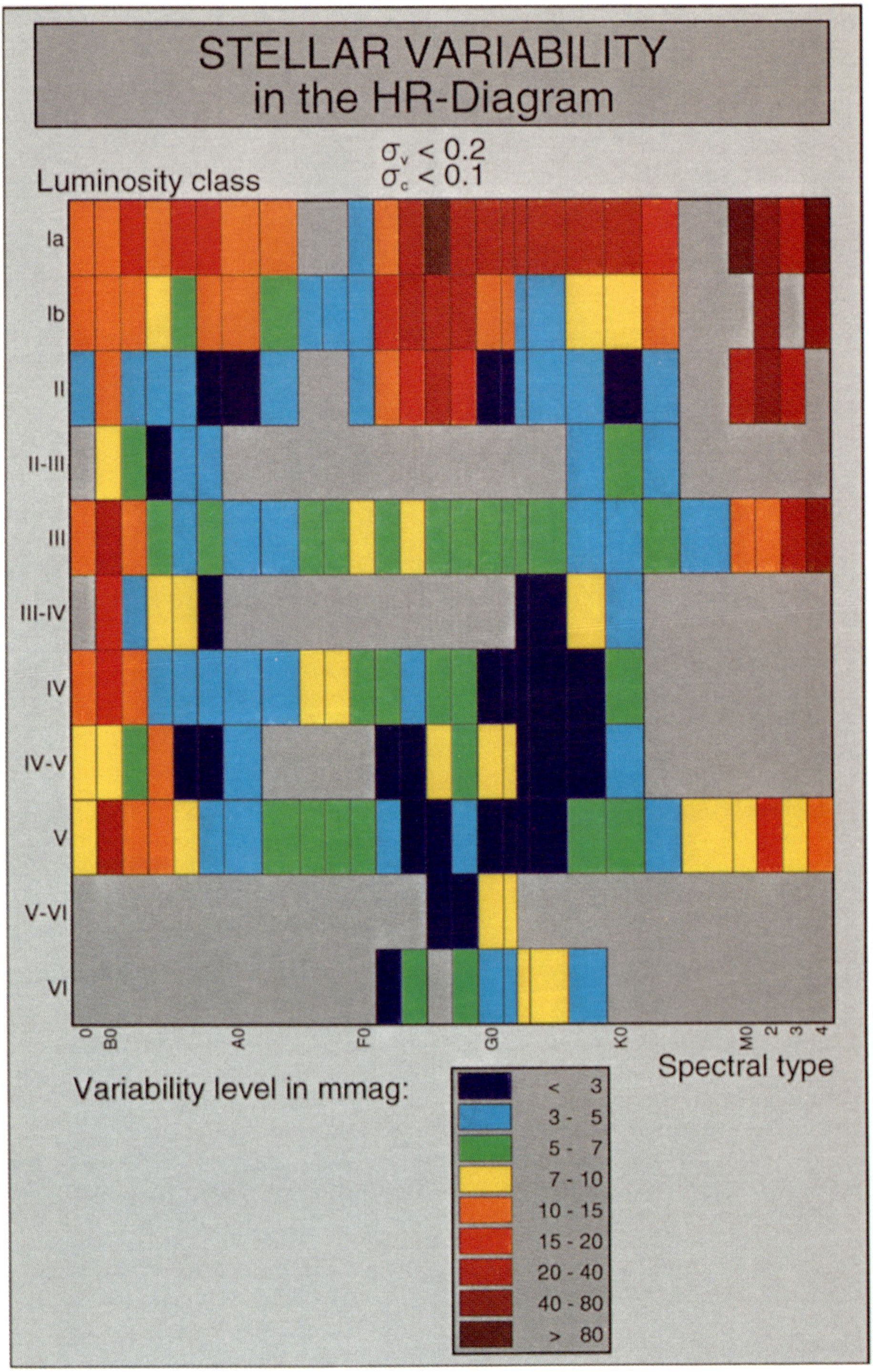

FIGURE IV The intrinsic low level variability in the HR-diagram. Stars with $\sigma V > 0.2$ or $\sigma col > 0.1$ are excluded. The diagram is based on the properties of 29 000 stars with 3 or more measurements in Geneva photometry.

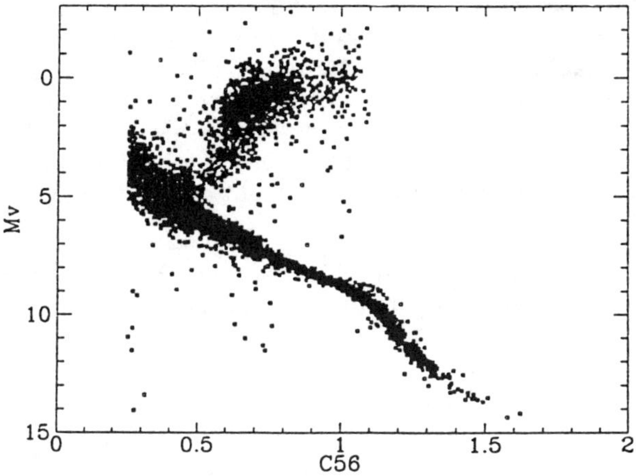

FIGURE V The HR-diagram of 7 915 field stars within 150 pc from Geneva calibrations. The blue limit of C56 index (≡ (B2–V1)) corresponds to the F5 type.

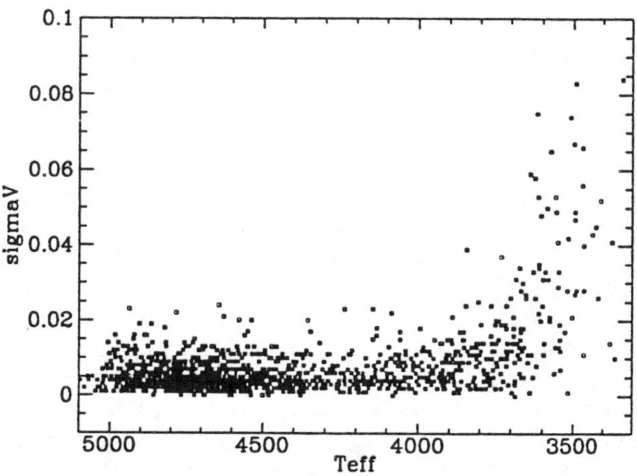

FIGURE VI The distribution of σV for 1 536 red giants with Mv in the range −2 to +2 and d < 150 pc.

The distribution of σV as a function of T_{eff} is given in Fig. VI for the red giants. The minimum amplitude occurs in the temperature interval 4500–4700 °K. The slight variability increase in the range 4800- 5100°K is probably due to the higher fraction of binaries at the left of the first giant branch and of the clump, not distinguished here. The increase of σV below 4 000°K is fairly proportional to the amount of TiO absorption. The intrinsic dispersion σV_i is given in Table I for red giants with $\sigma_V < 0.2$ and $\sigma_{col} < 0.1$ mag.

TABLE I The intrinsic standard deviation in V for red giants, in units of 10^{-4} magnitude with the number N of stars in each interval of T_{eff}.

T_{eff}	$\sigma V_i \pm$ pe	N
4900 – 5000	35 ± 4	130
4800 – 4900	23 ± 3	170
4700 – 4800	0 ± 2	227
4600 – 4700	6 ± 2	210
4500 – 4600	10 ± 3	147
4400 – 4500	5 ± 3	100
4300 – 4400	24 ± 5	61
4200 – 4300	27 ± 5	49
4100 – 4200	33 ± 5	63
4000 – 4100	41 ± 5	63
3900 – 4000	53 ± 5	63
3800 – 3900	82 ± 9	61
3700 – 3800	99 ± 10	53
3600 – 3700	193 ± 23	50
3500 – 3600	272 ± 34	28

A large body of main-sequence stars was for several purposes observed in Geneva photometry. They were selected according to a minimum parallax, a limiting magnitude and, most frequently, to a minimum proper-motion. The sample of 2 870 stars with Mv > 5.0 is representative of the old disc and halo. Nearby stars may be partially younger. The intrinsic variability is fairly constant in the temperature range 4 200°K–5 200°K, with rms of the order of 4 to 6 mmag. A discontinuity seems to appear around 4 200°K at the location of the Mazzitelli-gap (1972) sometimes referred as the Wielen gap (1974). Between 3 500 and 4 200°K the typical rms is 8 mmag. The increase at lower temperatures is partly due to a photon-noise effect on faint late M stars, not corrected here.

The relation between the intrinsic σV_i and the effective temperature is given in Fig. VII for the low main sequence, parallel with the number of stars per interval of temperature.

The change in the temperature-luminosity relation in the main sequence with its associated change of variability level occurs at a temperature corresponding to the spectral type K7V.

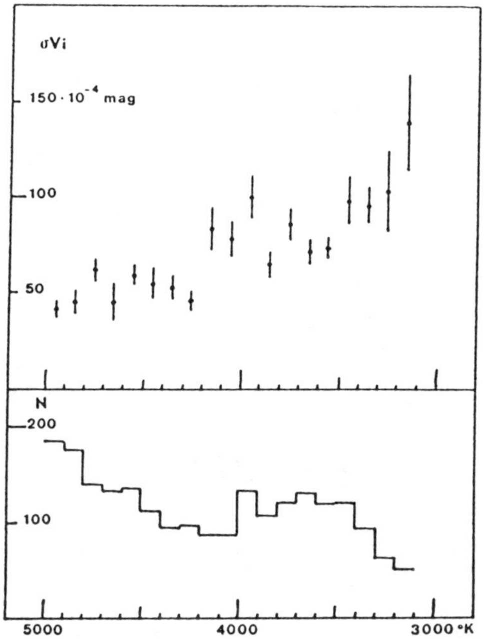

FIGURE VII The relation σV_i versus T_{eff} for 2200 K and M dwarfs, within 150 pc and with N > 2 measurements (top). Below, the number of stars per T_{eff} interval shows the presence of the Mazzitelli-gap at 4 000 – 4 200°K.

A physical study of the variability along the main sequence would imply to consider other indicators of the stellar activity as the variability in the UV, the effects of star spots through the luminosity changes and the TiO intensity variations. Also, it is necessary to distinguish single stars from binaries and to evaluate the age and overall metallicity effects. These aspects are beyond the scope of this presentation.

For a discussion of the physical processes responsible for the microvariability in late -type stars, see Gray, 1991.

LIGHT CURVES OF LOW AMPLITUDE VARIABLES

The luminosity changes of low level variables are often far from being simple. They may be erratic , wind driven, in the case of early type luminous stars, semi-periodic with phase shifts when produced by transit of groups of star spots, or multiperiodic in the case of low amplitude pulsating B stars.

The characterisation of the variability requires frequently several hundred measurements collected over long intervals of time, when the searched effects are close to the detection limit. As an example, the eclipsing binary RZ Eri was monitored for 11 years before Burki et al. (1992) were in the position to disentangle the reflection from the starspot effect in the secondary component of this system.

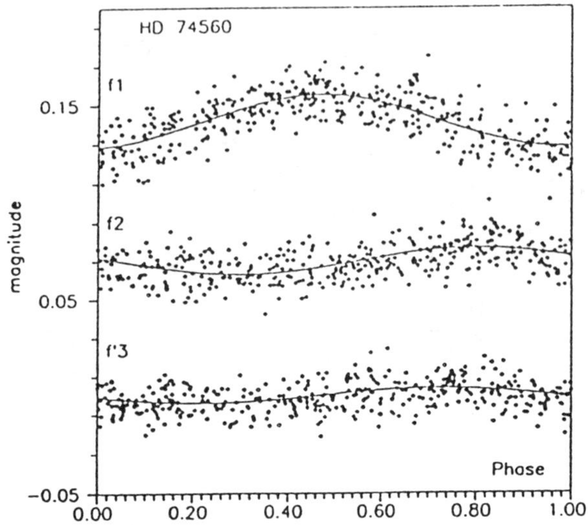

FIGURE VIII Phase diagrams for the visual magnitude data of HD 74 560 by Waelkens (1991).

The analysis of slowly pulsating B type stars was undertaken by the Leuven group for a small number of candidates. The phase diagram for the visual magnitude data of HD 74 560, by Walkens (1991) is a good illustration, cf. Fig. VIII, of the ground-based photometry capability and limitations. No less than 408 observations, collected between 1976 and 1989, were necessary to discover the three non-radial modes with P=1.551, 1.645 and 1.737 days and A=0.027, 0.013 and 0.008 mag respectively.

A second example is the light-curve of a multiperiodic B8V star shown in Fig. IX. This star studied by North (1992) has four determined periods up to now, i.e. P1 = 0.8044, P2 = 0.847, P3 = 0.9032 and P4 = 0.9591 d. Even with a set of 435 measurements, there remain mismatches between the observed light-curve and that computed from the provisional periods, phases and amplitudes.

Similar problems arise for the description of the stellar cycles among late K and M dwarfs. The size of the amplitudes, the number of necessary measurements, clearly indicate that the limits of absolute ground based photometry are met.

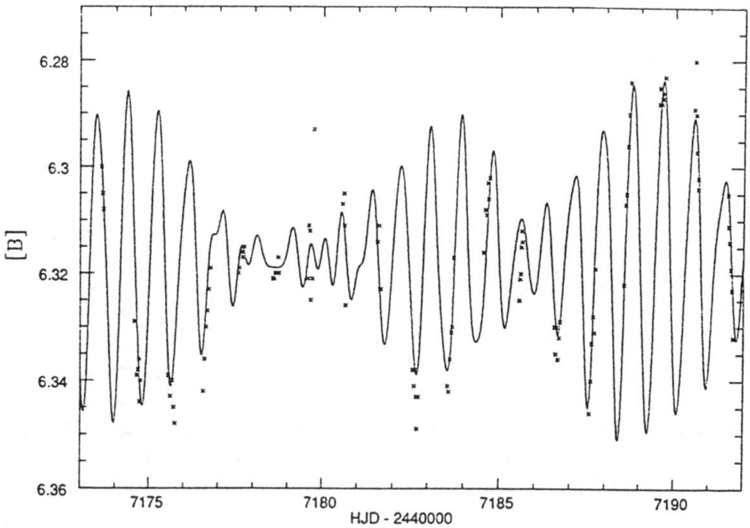

FIGURE IX The light curve in B magnitude of the multiperiodic B8V star, HD 37151 by North (1992).

The number of cases which may be studied in such detail is necessarily very limited. This is why mass production of accurate magnitudes from the space is so attractive.

THE HIPPARCOS PHOTOMETRIC SURVEYS

During the development of the satellite, the detection chains were modified in order to optimise the photometric outputs. An accurate calibration procedure was set up to allow high accuracy photometry with the main mission detector. The broad band, H_p, extends from 380 to 900 nm and is defined in order to optimise the signal-to-noise ratio and hence the detection of luminosity changes (see Grenon, 1992).

The star mapper chains were equipped with blue and visual filters denoted B_T and V_T. Their effective wavelengths and widths optimise both the satellite attitude determination and the accuracy on star colours.

With the main-mission detector, the 118 000 program stars are with regularity observed throughout the mission. The list of monitored stars is published in the *Hipparcos Input Catalogue* by Turon et al. (1992).

All stars brighter than B_T or $V_T = 10.5$ are systematically observed with the Tycho chains. They belong to an *Input Catalogue* of $2 \cdot 10^6$ stars computed from the *Hubble Space Telescope Guide Star Catalogue* (GSC).

About 500 000 objects will have B_T and V_T determined by the Tycho experiment. For about 500 000 additional stars only one magnitude, and then no colour, will be determined.

THE HIPPARCOS TIME-SAMPLING

The sampling by the satellite is particular. Stars are observed 9 times during a Field of View (FOV) crossing of 20 sec duration. Mean magnitudes, precision and epochs are stored after a detailed calibration process.

Stars are observed in two FOVs, separated by an angle of 58 degrees. Depending on its location on the sky, a given star is observed in groups of 4 to 6 transits. The time intervals between transits in the preceding and following FOVs is 21 min, and 107 min between transits in the following and preceding FOVs.

According to the galactic latitude β, groups of transits are obtained at a rate of 9 to 25 groups per year. The maximum rate is for β=45°. As the satellite precesses, about half of the program stars may occupy the node of the scanned great circle. Up to 50–60 transits may then be observed consecutively. A mean number of 200 observations per star is expected from the Hipparcos mission.

THE ACCURACY ON H_P, B_T, V_T MAGNITUDES

Due to the absence of scintillation and in spite of the small telescope size, the photometric accuracy of the H_p magnitude is very high. The accuracy is photon-noise limited and magnitude dependent. Tycho B_T, V_T data are four times less accurate because of a shorter integration time and a higher background signal.

TABLE II The expected accuracy on H_p as a function of the instrumental H_p-magnitude, and on B_T and (B_T-V_T) as function of the Tycho B_T magnitude.

H_p, B_T		:	6	7	8	9	10	11
σ_{H_p}	1 transit	:	.006	.009	.015	.024	.043	.08
	5 hours	:	.002	.004	.006	.010	.017	.03
σ_{B_T}	1 transit	:	.03	.04	.06	.12	.24	.48
	5 hours	:	.009	.012	.017	.035	.07	.14
$\sigma_{B_T-V_T}$	5 hours	:	.013	.017	.024	.049	.10	.20

The typical accuracy for a given transit and for a group of transits over 5 hours is given in Table II.

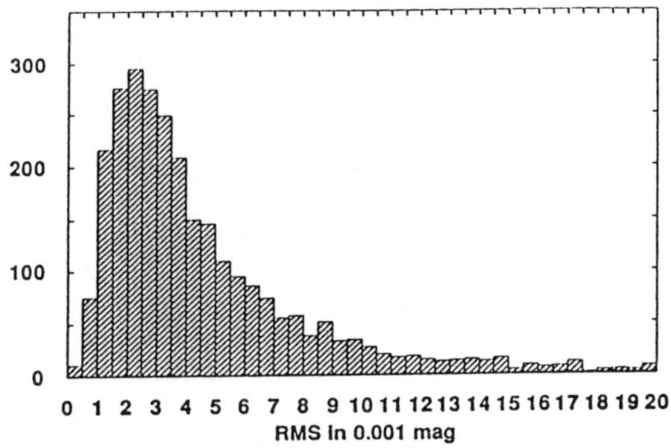

FIGURE X The residuals on H_p magnitude, with one point per group of transits, from F. Mignard et al. (1992).

The irradiation by solar and cosmic particles reduces the transmission of the main mission optics at a rate of +0.013 mag between two consecutive groups of transits. This transmission change is chromatic and, after a 4.5 y mission, magnitudes of (B–V)=0.0 stars will be increased by +0.85 mag and those of the reddest stars by less than +0.20 mag. A continuous calibration, in-orbit, is performed by the data reduction consortia, using large sets of standard stars. The accuracy of the procedure is such that the zero point of the magnitude scale is daily defined with an error rate inferior to 0.001 mag. The distribution of residuals on H_p magnitude (one point per group), see Fig. X, has a modal rms of 2 mmag. It demonstrates the stability of the main mission photometric system, in spite of the large transmission variations. The tail is due to the presence of faint stars and of microvariable stars. The precision of Hipparcos H_p magnitudes, averaged over 5 hours, is higher than that of Geneva photometry for stars brighter than 6.5, similar to that of Strømgreen photometry by Olsen for 7.0 mag stars and comparable to that of UBV photometry for 9.5 mag stars.

No significant aging affect the Tycho B_T and V_T channels. The performances and properties of the Tycho system remain unaltered all along the mission.

THE DETECTION OF VARIABLE STARS WITH HIPPARCOS

Up to now, the minimum detection amplitude Amin was 0.2–0.3 mag for surveys on photographic plates. From Tycho, the lowering of Amin to few percent will enormously increase the number of newly discovered or confirmed variables. The major gain from the Hipparcos mission will, because of its accuracy, come from the main-mission photometry.

TABLE III The minimum detectable amplitude as function of the instrumental magnitude H_p at a confidence level of 99%.

H_p	:	6	8	10	12
Amin	:	.007	.017	.048	.20

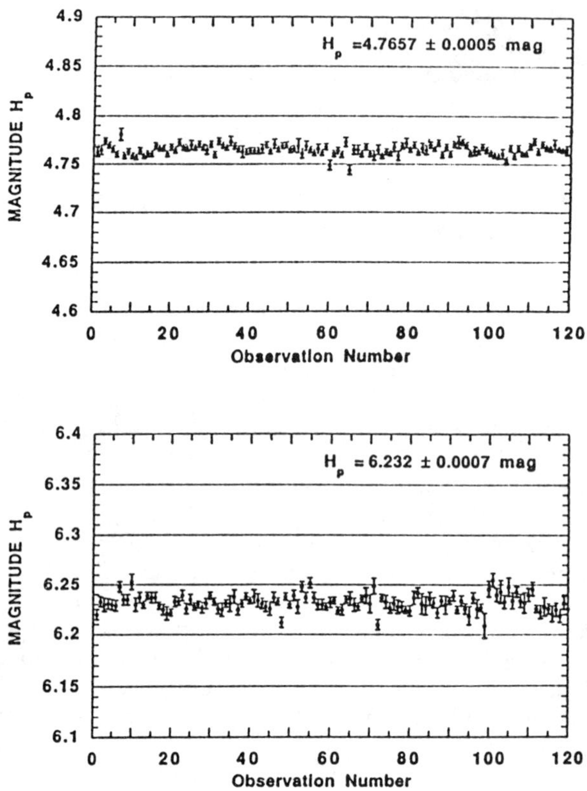

FIGURE XI Evolution of the instrumental magnitude in the case of a 4th and 6th magnitude star. From F. Mignard, FAST Consortium.

In Tab. III, Amin is given as a function of H_p in the case of a sinusoidal variation.

For bright stars, Amin is similar to that detectable with accurate ground-based photometry. The decisive advantage of Hipparcos photometry is the number of measurements per star, about 200, and the duration of the monitoring, i.e. 3.5–4.5 years. The time distribution of the observations is adequate for an unbiased detection of periods in the range of few minutes to several years. The Hipparcos sampling law is possibly not optimum for multiperiodic stars. It appears better suited for the detection of rotationally induced variability alike the Bp phenomenon or effects of stellar activity. Transits of spot groups of size similar to those of the Sun, are barely detectable for bright nearby stars.

In Fig. XI, examples of the evolution of the observed magnitude with the time are shown. Systematic departures from the mean value are possibly real. The stability of the photometric system is such that secular luminosity changes of the order of 0.001 to 0.002 magnitude per year are detectable. A few rapidly evolving stars may be discovered. More frequently, slow luminosity changes will result from opacity changes in circumstellar envelopes.

The detectable amplitude with Tycho was computed by Mauder & Høg, 1987. For B_T = 8, 9, 10 the detectable amplitude at the 99% confidence level is equal to .06, .12 and .24 mag respectively. All classical variables are indeed detectable by Tycho. A spectacular increase of the number of confirmed variables is expected, in particular among luminous stars and semi-regulars.

A major revision of our knowledge of the stellar instabilities is a foreseen by-product of the Hipparcos mission.

ACKNOWLEDGMENTS

We are indebted to L. Weber for the preparation of the colour plate.

REFERENCES

Burki, G., Kviz, Z., North, P. 1992, *A&A*, **256**, 463

Gray, D.F. 1991, in *Angular Momentum Evolution of Young Stars*, vol. 183,

Grenon, M. 1992, *A&A*, **258**, 88

Harvey, J.W. 1988, in *Advances in Helio– and Asteroseismology* vol. 497

Mauder & Høg. E. 1987, *A&A*, **185**, 349

Mazzitelli, I. 1972, *Mem. Soc. Astr. Ital.*, **45**, 541

Mignard, F., Froschlé, M., Falin, J.L. 1992, *A&A*, **258**, 142

North, P. 1992, private communication

Turon, C. et al. 1992, *The Hipparcos Input Catalogue*, Eds. ESA SP–1136

Waelkens, C. 1991, *A&A*, **246**, 453

Wielen, R. 1974, *Highlights of Astronomy*, **3**, 395

Young, A.T. 1984, NASA Conf. Proc. **2350**, p.8, Eds. W. Borucki and A. Young

Inside the Stars, IAU Colloquium 137
ASP Conference Series, Vol. 40, 1993
Werner W. Weiss and Annie Baglin (eds.)

THE SPACE BONUS

WERNER W. WEISS
Institute for Astronomy, Türkenschanzstrasse 17, A-1180 Wien

ABSTRACT We will discuss the opportunities opened by space experiments to make measurements that cannot - or only marginally - be made from ground.

THE ADVANTAGES OF SPACE EXPERIMENTS

- Time base of uninterrupted data sets
- Uninterrupted data sets
- No atmospheric noise
- Thermal and noise stability
- Unlimited Access to electromagnetic spectrum
- Experimental science (comet probes, planetary explorer missions, etc.)
- Microgravity

From the list above the first four are particularly important for high quality asteroseismic data. The time base of a data set, expressed for example as number of observed cycles, determines the frequency resolution. For ground based observations, this time base is frequently limited by day/night cycles, and changing seasons. Properly chosen orbits allow a space borne experiment to overcame these limits almost entirely.

Figure 1 illustrates the influence of the time base on the frequency resolution for the pulsating CP2 star HR 1217. A frequency analysis of photometric data (Kurtz et al., 1989), indicates that HR 1217 probably pulsates with at least 6 frequencies. There is strong evidence that rotational side lobes have also been observed. These 21 frequencies (presented in Figure 10 of their paper) were used to compute synthetic data which cover different number of cycles. The corresponding frequency spectra are presented in the insert of our Figure 1. The main figure presents an exanded view of the spectrum, centered on the group of frequencies around 229.3 c/d (2.65 mHz). At least 20000 cycles have to be observed for a full frequency resolution which is determined by $\Delta\nu \approx 1/(P \cdot n)$, where P is the pulsation period and n the number of continuously observed cycles. As we have learned in this conference, frequency resolutions of the order of a μHz are needed to answer certain questions concerning the structure and evolution of stars. Very long and homogeneous data sets, spanning up to several years, are required to achieve this extreme high resolution in frequency.

But it is not enough to obtain a long time strings. Gaps in the data cause a problem for the interpolation of harmonic functions, because the number of cycles which have not been observed can only be estimated. Hence, harmonic functions with different frequencies can be fitted equally well through the same data set.

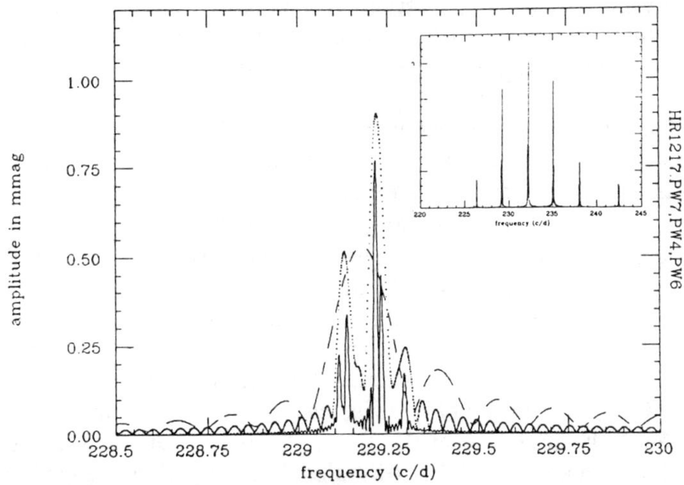

Figure 1: Increasing frequency reolution for data sets with 1500 (dashed line), 6000 (dotted line) and 25000 (full line) cycles.

The spectral window, also called window function, is usually presented to illustrate this ambiguity of frequency solutions, called aliasing. Figure 2 gives an example for spectral windows as a function of the duty cycle. The simulations are based on noise free data of one sinusoidal cycle. Obviously, a duty cycle close to 100% is desirable in order to eliminate ambiguities in the identification of intrinsic frequencies.

The aliasing problem due to the day/night cycle is very disturbing for ground based observations. The only way out of this problem is to observe the target star from

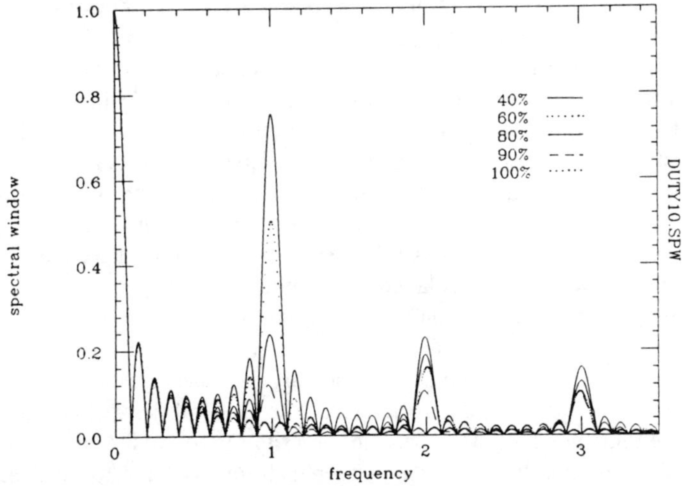

Figure 2: Spectral window for one period as a function of the duty cycle.

several sites well distributed at various longitudes. There are many difficulties merging data obtained from such international observatory campaigns, including differences in instrumentation and climate. Space borne experiments can be freed from such limitations.

Last, but not least, space experiments do not suffer scintillation noise from the terrestrial atmosphere which usually is the dominant source of noise in recent network data. This noise is well described by a formula given by Young (1967),

$$\sigma_{atm} = 0.08 \cdot D^{-2/3} \cdot X^{1.8} \cdot e^{-h/2000} \cdot t^{\cdot}$$

with the telescope aperture D in cm, sealevel of the observatory h in m, the air mass X and the integration time t (in seconds).

It is a simple exercise to estimate the stellar magnitude for a give telescope where scintillation and photon noise for a ground based observatory network equals the photon noise of a particular space borne photometric instrument. Such a diagram is given in Figure 3. The inclined bold line indicates the magnitude range where a 40 cm space borne telescope (the case of the Large Photometer of PRISMA) gives better photometric data than a network of at least six telescopes, distributed in longitude and which have an aperture of 10 meters, whereas the horizontal bold line indicates the magnitude range for which such a telescope network would give the needed accuracy in less time than the space borne photometer. The basis for figure 3 is a noise level of $\sigma = 10^{-6}$ (right hand side ordinate) and $5 \cdot 10^{-7}$ (left hand side ordinate) in the power spectrum for a star with magnitude *mag(V)* and observed with 100% duty cycle.

For the fainter stars observed by the Large Photometer of PRISMA, the noise level of 10^{-6} (due to photon noise in this case) is achieved in about 30 days. It would need a continuously working 10 m telescope network in order to reduce the scintillation noise to the same level, which would be the dominating noise source for ground based observations. If frequencies have to be determined with comparable accuracy, but in a shorter time (for example to investigate the damping mechanism of modes or to prove chaotic excitation) a dedicated ground based network of 50 m diameter telescopes, or even larger, would be needed to perform the job which PRISMA can do with a 40 cm aperture.

We can conclude that asteroseismic data of phenomena which have to be studied in time scales of weeks and with a precision of 10^{-6} or better cannot be obtained from ground and are therefor a domain of space experiments.

The last properties of space experiments listed at the beginning do not need detailed discussion. Stellar astrophysisists are presently not in the position of having the possibility to perform *hands-on* experiments with stars, contrary to planetary system scientists who have successfully sent a probe to planets, comets and who are doing experiments with the interplanetary medium.

Microgravity experiments may bring important insights in the near future to fundamental physics and hence also to astrophysics. I refer in this context to various space experiments under consideration which are designed to prove far more precisely the validity of the equvivalence principle of the General Relativity Theory than has been possible until now. Another benefit from the microgravity environment in space can come for astrophysics when very larger instruments have to be deployed which would be impossible, or extremely expensive to build on ground (large radio telescopes, interferometers, gravitational wave detectors, e.g.).

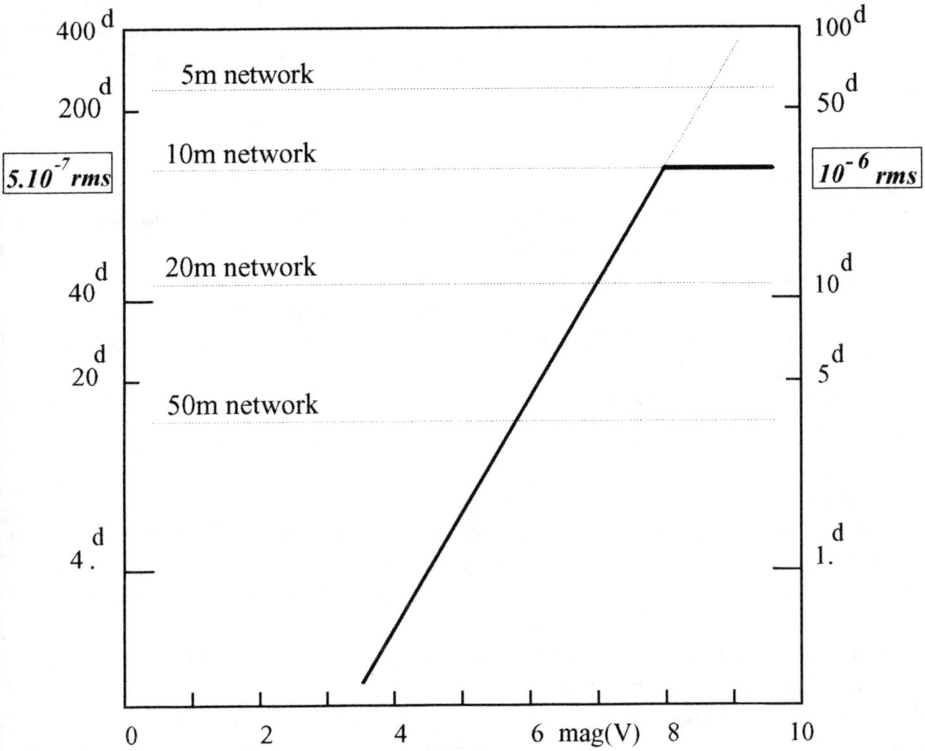

Figure 3: Comparison of a space borne 40 cm photometric telescope with ground based telescope networks in the 10 m size class.

The last aspect of space experiments listed here has been already widely used in the past by the astrophyiscs community. There is no need to dwell more on the enormous amount of new scientific insights which has been gained from this side. The investigation of stellar activities located in various optical depths cannot be done without having access to the UV and X-ray region of the electromagnetic spectrum (see also C. Catala, this volume).

ASTEROSEISMOLOGY FROM SPACE

EVRIS will be the first true space experiment devoted to asteroseismology and is described in this volume (Baglin et al., 1992). During the cruise phase of nearly 300 days on the MARS-94 space probe, EVRIS will continuously monitor about 15 bright target stars with a single-channel photometer. A photon statistics limited noise level of few 10^{-6} will allow astroseismic investigations of solar type stars for the first time. But, as described by Baglin et al. (1992, op. cit.), stars other than of solar type will also be observed.

A considerably larger and more sophisticated instrument, PRISMA, will benefit from the experience gained from EVRIS and will push the limits of observational astrophysics even further. In addition to the unique possibilities for asteroseismology, PRISMA will simultaneously monitor stellar activities by spectrometric and photometric means, from the optical to the near XUV spectral region. PRISMA is descibed also in this volume (Appourchaux et al., 1992).

REFERENCES

Appourchaux, T., Gough, D., Hoyng, P., Catala, C., Frandsen, S., Fröhlich, C., Jones, A., Lemaire, P., Tondello, P., Weiss, W. W., 1992, this volume

Baglin, A., Weiss, W. W., Bisnovatyi-Kogan, G., 1992, this volume

Kurtz, D. W., Matthews, J. M., Martinez, P., Seeman, J., Cropper, M., Clemens, J. C., Kreidl, T. J., Sterken, C., Schneider, H., Weiss, W. W., Kawaler, St. D., Kepler, S. O., 1989, M.N.R.A.S. **240**, 881

Young, A. , 1967, Ap. J. **72**, 747

Appendix:

Cost estimate for building a dedicated network of six identical 10 m telescopes, of which four telescopes can be set up at already existing sites, where for two other telescopes such sites will have to be developed. The estimate is based on discussions with M. Tarenghi and on published data.

The cost for building one 8m telescope for KPNO was estimated to about 70 M$, one additonal copy for CTIO would cost 60 M$ (NOAO newsletter). The cost for the first Keck 10m telescope was about 93 M$, the estimate for the copy is the same (Physics Today). The Japanese 7.5m telescope will cost about 300 M$, but this instrument is fully developed and built by industries.

No.	item	M $
1	6x10 telescopes (including duplication bonus)	500
2	Headquater building, hardware	5
3	HQ salaries per year	16
4	HQ expenses per year	16
5	development of 2 new sites	40
6	6 sites, installation	180
7	sites, operation per year (10-15 people per site for installation during 2 years, 6-8 peoples to run each site)	60
8	Detector development (2 per telescopes)	20

To set up a working observatory network at six excellent sites will therefor cost about 745 M$ (budget items 1, 2, 5, 6, 8). The costs to run this network for two years can be estimated to 184 M$ (items 3, 4, 7).

Inside the Stars, IAU Colloquium 137
ASP Conference Series, Vol. 40, 1993
Werner W. Weiss and Annie Baglin (eds.)

RESULTS FROM IPHIR AS GUIDELINE FOR ASTEROSEISMOLOGY

THIERRY TOUTAIN and CLAUS FRÖHLICH
Physikalisch-Meteorologisches Observatorium Davos, World Radiation Center
CH-7260 Davos Dorf, Switzerland

ABSTRACT The long uninterrupted observation of the Sun as a star with the IPHIR experiment can be used to set limits to the significance of the determination of p-mode parameters as a function of time and the length of the time series of measurements.

AMPLITUDE VARIATION

The detailed analysis of the long uninterrupted time series of IPHIR on the USSR PHOBOS Mission (Toutain & Fröhlich, 1992) can be used to determine the potentiality for asteroseismology.

Results of a bi-variate spectral analysis of individual IPHIR modes with a sine wave of constant frequency and amplitude show that the mode frequency and amplitude are highly variable on time scales typical for the excitation mechanism. Examples for the amount and kind of variation of the phase and amplitude of a running fourier transform of 8 days length is illustrated in Figure 1 for l=0 which show the coherence^2, the phase and the gain. The coherence^2 times 100 corresponds to the part explained in percent in the IPHIR series with the constant sine wave. The heavy lines indicate regions with coherence^2>0.5. The gain corresponds directly to the amplitude and its variation is large: it may change between two subsequent observations 5 days apart easily by a factor of 5. There is no obvious difference in the amplitude variation between the l=0 and 1 mode (not shown here) indicating that interference between the unresolved modes does not cause the modulation, but the excitation-damping mechanism. This is in contrast to the conclusion of Ehgamberdiev et al. (1991) from Earth-bound observations that the amplitude modulation is a combination of interference and excitation-damping with the influence of the latter being much lower than the former.

FREQUENCY DETERMINATION

In order to determine the line parameters of the modes (centroid frequency, amplitude, splitting) a model has to be used to fit the observed power spectra. The stochastic excitation suggests a Lorentzian line shape for the power defined by a centroid frequency, a linewidth, an amplitude and a noise level. Moreover, the power spectra have a χ^2 statistics with 2 degrees of freedom yielding a very complicated power-spectrum

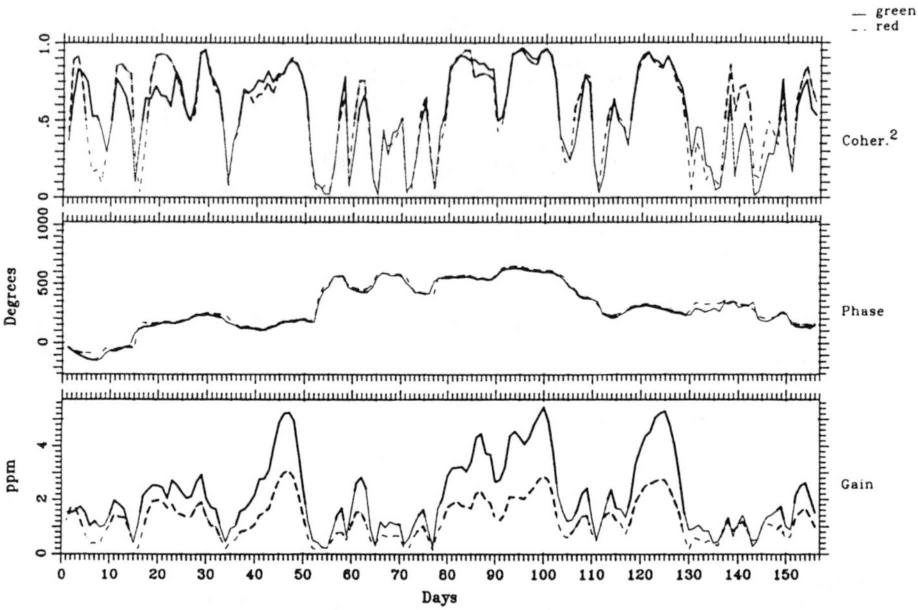

Figure 1 Bi-variate analysis of the l=0, n=21 mode from 8 days stretches of IPHIR data separated by one day against a sine-wave with amplitude 1ppm and the frequency found from the whole series of 160 days.

as shown in Fig. 2. In order to determine the line parameters a Lorentz line has to be fitted to the spectrum by e.g. a maximum likelihood method taking into account the proper statistics. In the case of real p-modes the situation is further complicated by the fact that lines are overlapping. Thus more than one Lorentzian has to be fitted at a time. For a doublet l=0 and 2 two central frequencies, a rotational splitting, two amplitudes and a noise level are fitted; for a l=1 mode a central frequency, a splitting, an amplitude and a noise level. For details see Toutain & Fröhlich (1992). Because of the statistical behavior rather long series of observation are needed to determine centroid frequencies with a decent significance.

In order to study the dependence of the uncertainty on the length of time series, the IPHIR time series is analyzed using 10, 20, 40, and 80 days of observations by dividing the original one (160 days) in 16, 8, 4 and 2 series. The results of the determination of the centroid frequencies of the strongest l=0...2 lines are shown in Fig.3 with n=21 and 20 respectively. The shaded area is centered at the frequency determined from 160 days and the width matches the corresponding statistical uncertainty. Triangles are showing the results from Lorentzian fits, asterisks the ones of center of gravity determinations which are used if Lorentzian fits are not converging. In the case of 10 and 20 days the fitting does indeed not always converge. It works well if the initial value for the determination is close enough to the line which is in general not known a priori. In the solar case, however, the initial values are very close and the good accuracy achieved cannot be directly applied to asteroseismology. One might

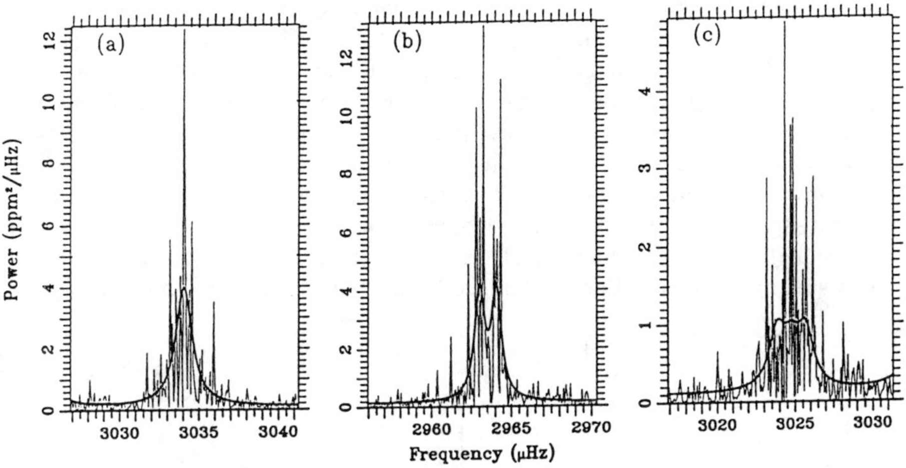

Figure 2 Examples of three p-mode lines for l=0,1 and 2 and n=21,20 and 20 respectively (from left to right) with the Lorentz lines fitted.

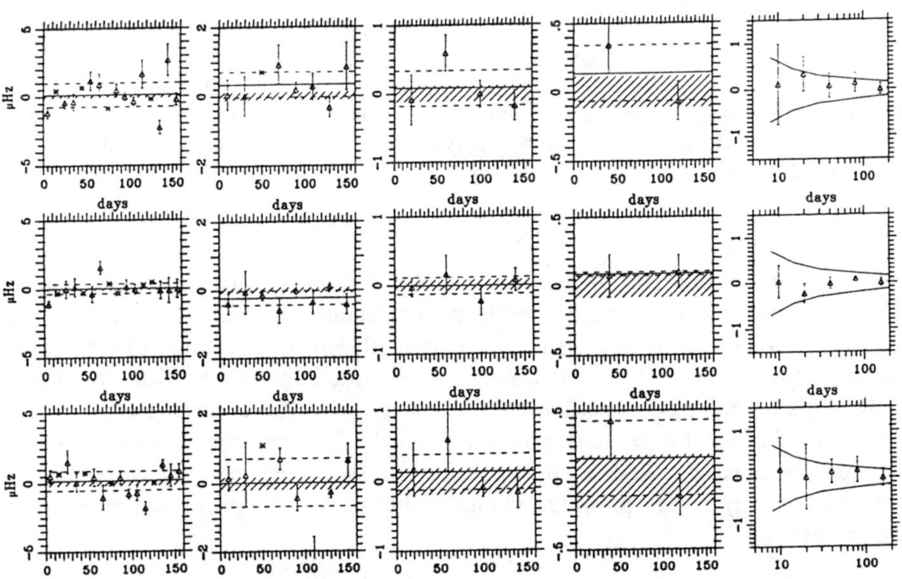

Figure 3 Determination of the centroid frequency (around 3mHz) for l=0...2 (from top to bottom) for 10, 20, 40 and 80 days (from left to right). The right most column shows the uncertainty as a function of observing time.

improve the situation somehow by fixing some of the parameters such as the line-width, but the determination would then need even more a priori knowledge. The right most column in Fig.3 summarizes the achievable uncertainties as function of observing time with the error bars indicating the range of the different determinations.

Simulations have been performed for a l=0 line and 200 different χ^2 distributions. The result of these simulations are shown as 1σ limits (full lines) in the right most column of Fig.3. These indicate that the IPHIR data represent such a statistical behavior (the triangles, lie within the limits) and the uncertainty varies about as the inverse of the square root of the observation time. The influence of the method on the frequency determination is shown in Fig.4. For all three methods the starting value has been chosen as 0.5 μHz below the correct value The method of searching the center of gravity seems to depend most on the initial value used - its final value is still closer to the initial one than the true one. Thus, this method is not recommended; the two others are similar with a slight preference for maximum likelihood method.

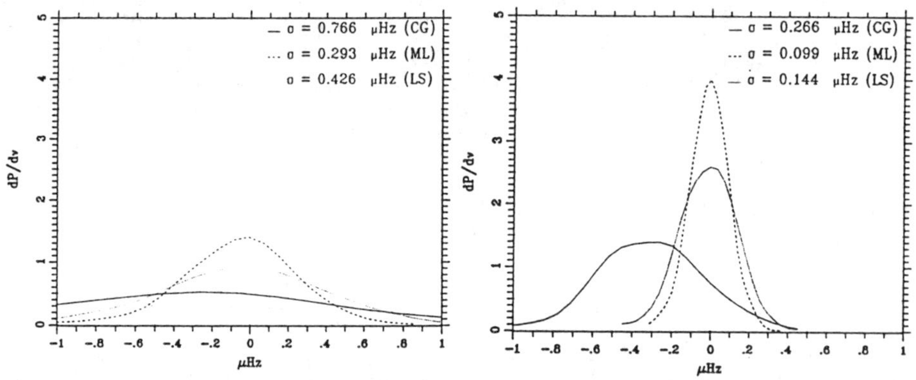

Figure 4 Distribution of the results from 1000 simulated frequency determinations from observations of 1 month (left) and 1 year (right) with three different methods: center of gravity, maximum likelihood and non-linear least squares.

CONCLUSIONS

The results of the IPHIR analysis and the simulations indicate that an accuracy of ±1 μHz at the 3σ level can only be achieved with about 3 months of observation. This is only true if the signal to noise is similar to the IPHIR lines analyzed here, namely about 30 for 6 months of observation. If it is less one might have to use even longer series. It is interesting to note, however, that the accuracy is not much dependent on l although the signal to noise is quite lower for the l=2 line. This is probably due to the simultaneous fit with l=0. Again, this combined fitting may not be possible in asteroseismology as the a priori knowledge of the behavior is much lower than the one for the Sun.

REFERENCES

Ehgamberdiev, S., S.Khalikov, M.Lazrek, E.Fossat, 1991, *A&A*, **253**, 252.
Toutain, Th., C.Fröhlich, 1992, *A&A*, **257**, 287.

Inside the Stars, IAU Colloquium 137
ASP Conference Series, Vol. 40, 1993
Werner W. Weiss and Annie Baglin (eds.)

SOLAR-LIKE OSCILLATIONS ON α CENTAURI A

EDWARD M. POTTASCH, HARVEY R. BUTCHER and FRANS H.J. van HOESEL
Kapteyn Observatory, Mensingheweg 20, 9301 KA Roden
The Netherlands.

ABSTRACT Detection of individual modes of oscillation near 5 minutes period is proposed, based on the comb-like signature which is similar to that found in the sun. The mean peak amplitudes (a_p) of the strongest individual modes are $75 \leq a_p \leq 120$ cm/s, and the mean separation between alternate modes is 55.3 ± 3.6 μHz. Apparent modal lifetimes are on the order of 2 days, but we suggest this may be caused by interference of adjacent modes.

1. INTRODUCTION

For the past decade serious attempts (e.g. Brown and Gilliland 1990, Brown et al. 1991, Gelly et al. 1986, Noyes et al. 1984) have been made to detect solar-like p-modes on bright solar-like stars. It is expected that p-modes measured on solar-like stars will result in the determination of two parameters, $\Delta\nu_0$ and D_0, which are closely related to the sound speed integral across the solar radius and the sound speed gradient inside the nuclear burning core respectively (cf. Christensen-Dalsgaard 1984, 1988).

2. OBSERVATIONS

Our observations of αCenA were made during 6 consecutive nights, 2 - 7 April 1990, using a single channel, Fabry-Perot (F-P) based, velocity monitor with the 3.6-m telescope of the European Southern Observatory at La Silla, Chili. The instrument makes use of a specially stabilized version of the ET-50 tunable etalon marketed by Queensgate Instruments, Ltd. (Butcher and Hicks 1986).

For the present observations, the F-P interferometer was combined with an order isolation interference filter to yield a narrow band filter, of 0.12 Å FWHM tunable over ± 50 km/s about the FeI absorption line at λ=5576 Å. When the filter is tuned to the line flank to provide maximum sensitivity to radial velocity, the system response was determined to be $\Delta I/I = 4.44 \times 10^{-5}$ per m/s, where I is the intensity signal measured by the Si photodiode detector, and was found to be linear with velocity over the range required.

The design goal for stability of the bandpass of the interferometer of less than 1 m/s for time scales between a few seconds and many minutes has been achieved on the one hand by incorporation of low noise etalon servo electronics, and on the other by characterizing the system response to temperature, pressure, and optical illumination variations, then stabilizing these environmental factors to the required degree. No separate, stabilized reference has been employed (except during test measurements to verify performance).

The velocity data time series discussed here consist of 0.2 seconds integrations taken in repeated I_1 - I_2 - I_3 - I_4 sequences, where I_1 and I_4 are intensity signals with the etalon band tuned to the blue flank of the absorption line, and I_2 and I_3 are the signals on the red flank. These raw measures have been combined ('dechop') to generate a velocity sensitivity parameter $R(t)=((I_1+I_4)-(I_2+I_3))/(I_1+I_2+I_3+I_4)$.

3. ANALYSIS

The 'dechop' procedure transforms the intensity measurements into relative velocities. The resulting velocity curve shows the diurnal motion of the earth's rotation, which during a typical observation period of nine hours on αCenA from La Silla, shows a velocity difference of about 500 m/s. For comparison we note that typical solar modes of oscillation have amplitudes in the order of 20-40 cm/s, and periods in the order of 5 minutes. The earth's rotational trend has been removed from the data by subtracting a 6th order polynomial fit. Individual bad points (clouds, shutter closure) are replaced with linearly predicted points. Finally the data are smoothed by subtracting the convolution product with a Gaussian function from the data (full width at half height ≈ 45 minutes). This filtering process should therefore leave any solar like oscillation signal from αCenA intact.

The power spectrum of the complete data set does not show a statistically significant excess power in the 3 mHz region. Since it is expected that the power spectrum is almost periodic in frequency we calculated the power spectrum of a certain range within the power spectrum, as described by Gelly et al. (1986).

4. RESULTS

4.1. Power spectra

N random fluctuations, each with an rms amplitude of A, produce a signal of $A/\sqrt{N}$. Our stellar data show a rms (σ_m) of 8.4 m/s per 51.2 seconds integration time. The amplitude (a_s) for a continuous sine function to reach this power level in a 'white' spectrum therefore is: $\sigma_s = \sigma_m/\sqrt{N}) \approx 15$ cm/s. Power spectra of two-nights sequences show slight (but not significant) enhanced power in the 3 mHz region. The relative error of a peak P(ν) in the power spectrum at low frequencies (3 mHz region) for peaks with a power less than 0.375 $(m/s)^2/\Delta\nu$ (complete data series) or 1.0 $(m/s)^2/\Delta\nu$ (two-nights) is ≥ 49%. Assignment of possible p-modes is discussed by Pottasch et al. (1992).

4.2. Power spectra of power spectra

The p-mode structure in the power spectrum consists of first order spacings [ν(n,*l*) - ν(n-1,*l*)], and second order spacings [ν(n,*l*) - ν(n-1,*l*+2)]. Each peak in the Fourier domain is furthermore convolved with the Fourier transform of the window function (1/24 hours spacing). The resulting sidelobe structure can to some extent be removed by the CLEAN procedure. If the power spectrum shows a solar-like periodicity in frequency then the power spectrum of the power spectrum will reveal: 1) the distance $0.5\Delta\nu_0$ and its higher harmonics (where $0.5\Delta\nu_0 \approx \langle\nu(n,l) - \nu(n-1,l)\rangle/2$) and 2) the frequency separation $\Delta\nu_0$ will be visible as a minor peak accompanied by its higher harmonics. We find that $0.5\Delta\nu_0 = 55.3 \pm 3.6$ μHz.

5. DISCUSSION AND CONCLUSION

The power spectrum of the complete data set does not improve the signal to noise ratio as would be expected for modes with lifetimes in the order of 3 to 6 days, as on the sun (Libbrecht 1988). The best signal-to-noise ratio appears in two-nights data segments. When we have a closer look at the power spectra of these segments it becomes apparent that only a few major peaks occur in more than one two-nights power spectra. On the Sun (Ehgamberdiev et al. 1991) and possibly on Procyon (Brown et al. 1991) it has become evident that the occurrence of p-modes in the power spectrum, shows a large scatter from day to day. This daily scatter in amplitude is believed mainly to be caused by interference between unresolved modes contributing to each observed peak in the power spectra. The intermittent occurrence of peaks in the successive two-nights power spectra is therefore probably caused by the combination of a short measurement interval and the particular separations of adjacent modes. Indeed the power spectrum of the complete data set shows linewidths that correspond to longer lifetimes. If interference is the cause of amplitude variations on timescales of two nights, then on days of largest amplitude the interference is likely to be constructive. Therefore we estimate the amplitudes of the individual modes consequently to be in the range: 75 cm/s ≤ amplitude ≤ 120 cm/s.

When adopting a value for $0.5\Delta\nu_0 = 55.3 \pm 3.6$ μHz (αCenA) and a solar value for $\Delta\nu_0 = 135$ μHz, the corresponding radius $R_{\alpha CenA} = 1.17\ R_\odot \pm 0.05\ R_\odot$, since $\Delta\nu_0$ is roughly proportional to $(M/R^3)^{0.5}$ (where $M_{\alpha CenA} = 1.085\ M_\odot$ [7]). Furthermore $L_{\alpha CenA}/L_\odot = 1.33 \pm 0.11$, for T_{eff}(αCenA) = 5710 K and T_{eff}(Sun) = 5770 K.

Christensen-Dalsgaard (1984) showed that first and second order frequency spacings can be used to infer the age or mass of a star ($\Delta\nu_0$ - D_0 diagram). We find that $[\nu(n,0) - \nu(n-1,2)]/6 \approx D_0 \approx 1.18 \pm 0.27$. The $\Delta\nu_0$ - D_0 diagram yields an age for αCenA of 7.2 ± 2.6 Gy.

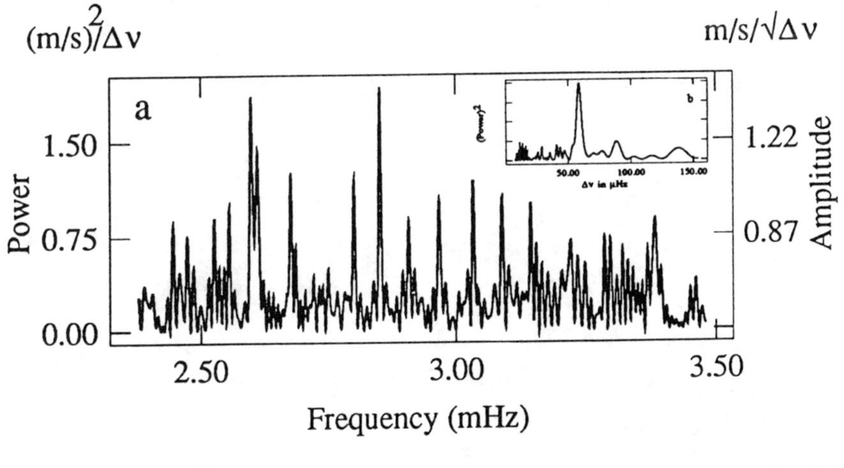

Fig.1. (a) CLEANed power spectra of two consecutive nights of observation. It is suggested that individual peaks are likely constructed by interference with neighbouring peaks. Therefore it is likely that the mode amplitudes (a_p) are in the order of $75 \le a_p \le 120$ cm/s. (b) The power spectrum of the power spectrum indicates that the p-modes are periodic in frequency space, where the spacings between the individual modes $0.5\Delta\nu_0 = 55.3 \pm 3.6$ μHz.

REFERENCES

Belmonte J.A., Pérez Hernández F., Roca Cortés T. 1990, *A&A* 231, 383

Brown T.M., Gilliland R.L. 1990, *ApJ* 350, 839

Brown T.M., Gilliland R.L., Noyes R.L., Ramsay L.W. 1991, *ApJ* 368, 599

Butcher H.R., Hicks T.R. 1986, in *Proceedings of the NATO Advanced Research Workshop on Seismology of the Sun and distant Stars*, ed. D.O. Gough, p. 347

Christensen-Dalsgaard J. 1984, in *Proceedings of the workshop on space research prospects in stellar activity and variability*, ed. Mangeney, F. Praderie, p. 11

Christensen-Dalsgaard J. 1988, in *Proceedings IAU Sym. 123: "Advances in Helio- and Asteroseismology"*, ed. Christensen-Dalsgaard, Frandsen, p. 295

Demarque P., Guenter D.B., van Altena W.F. 1986 *ApJ* 300, 773

Ehgamberdiev S., Khalikov S., Lazrek M., Fossat E. 1991, *A&A* 253, 252

Gelly B., Grec G., Fossat E. 1986, *A&A* 164, 383

Libbrecht G. 1988, in *Proceedings of the ESA-IAC Sym. Seismology of the sun & sun-like stars*, ed. Rolfe E.J., SP-286, p. 3

Noyes R.W., Baliunas S.L., Belserene E., Duncan D.K., Horne J., Widrow L. 1984, *ApJ* 285, L23

Pottasch E.M., Butcher H.R., van Hoesel F.H.J. 1992, *A&A* (in press).

Inside the Stars, IAU Colloquium 137
ASP Conference Series, Vol. 40, 1993
Werner W. Weiss and Annie Baglin (eds.)

RADIAL AND NONRADIAL MODE INSTABILITY IN B-TYPE STARS

W.A.DZIEMBOWSKI AND A.A.PAMYATNYKH[1]
Copernicus Astronomical Center, Warszawa, Poland

Recently, three independent groups(Cox *et al.* 1992; Kiriakidis *et al.* 1992; Moskalik and Dziembowski, 1992), using opacity tables published by Iglesias and Rogers (1991), demonstrated that β Cep star models are pulsationally unstable. The instability is driven by the classical κ- mechanism acting in the layer with temperatures near 2×10^5K where there is a bump in metal opacity. The groups reported results of calculations made for rather narrow ranges of stellar parameters and oscillation modes. We conducted an extensive search for unstable modes in complete evolutionary models of B-type stars of luminosity classes III – V. Our aim was to determine the domain of instability and examine its role not only in β Cep stars but also in variable stars located in the nearby areas of the H-R diagram.

An unexpected new aspect of our calculations is the use of the improved opacity data. In a very recent work Iglesias, Rogers and Wilson (1992) showed that effects of spin-orbit interactions significantly enhance opacity in the critical region for driving the pulsations in β Cep stars. These effects and improved information about the solar metal mixture have been included in the updated opacity tables kindly provided to us via electronic mail by Dr. Rogers. The consequences of this change in opacity for stability of B-type star models are indeed quite important. Contrary to previously announced results, that only the fundamental mode is unstable, we now find the first two radial overtones to be unstable, as well. Thus, the discrepancy between the theoretical prediction and the mode identification suggested by observers has been removed. Furthermore, no longer is a high metal abundance ($Z > 0.03$) required to explain the occurrence of pulsation in most of the objects. In fact, the theoretical instability domain in the H-R diagram, based on the models calculated with $Z = 0.02$, agrees better with the observational β Cep domain than that based on the models $Z = 0.03$.

The pattern of appearance of unstable modes in a selected sequence of stellar models is shown in Figure I. Frequencies plotted there are expressed in the $\sqrt{4\pi G\bar{\rho}}$ units. Only the expansion phase of the Main Sequence evolution is covered. Models in the contracting phase have similar pulsation properties except that the nonradial mode spectrum is somewhat denser. This effect becomes much more pronounced in models with hydrogen exhausted cores. However, since the crossing of the instability strip in this evolutionary phase is very fast, these models cannot be regarded as realistic for most β Cep stars.

---

[1]On leave from: Institute of Astronomy, Moscow, Russia

The behavior of radial modes ($l = 0$) is reminiscent of that in the Cepheid instability strip. With decrease of effective temperature, we first encounter an overtone instability. However, the effect of luminosity changes is exactly the opposite. In the β Cep instability strip, the overtones are unstable only in the more luminous objects. One may notice in Figure I that the onset of the instability in nonradial modes of low degree does not depend on l in a visible way. In models calculated with $Z = 0.02$, the instability domain is naturally smaller. In particular, the second overtone (p_3) is always stable.

In Figure I one can see the familiar behavior of the low-order g-modes whose frequencies grow during Main Sequence evolution to enter the low-order p-mode frequency range at some time. Such modes become unstable remaining largely trapped in deep interior. Their frequencies are sensitive to the evolution of the convective core boundary, which is still not well understood. Thus, detection of such modes in β Cep stars would be great importance for asteroseismology.

In the plots for $l = 6$ and 8 modes, a separate instability domain appears at low frequencies. The unstable modes are high-order g-modes which are propagating rather than evanescent waves in the driving zone. Nevertheless the physical mechanism of the instability remains the same. Occurrence of two separate instability domains is consequence of the fact that the modes in the intermediate period range have relatively low amplitudes in the driving zone. Instability of these long-period modes of rather high l may be relevant to the line-profile variability discovered in several Be stars (see e.g. Baade 1984). However, they are not likely to be detectable by means of photometry.

For stars with lower effective temperatures and luminosities the instability shifts toward lower l's. We found lots of unstable modes with $l = 1$ and 2 and periods ranging up to 3 days, in sequences of $5M_\odot$ star models calculated with $Z = 0.02$ and 0.03. It seems virtually certain that this high-order g-mode instability explains the origin of long-period variability discovered by Waelkens and Rufener(1985) in certain mid-B stars.

The possibility that the same mechanism may drive oscillations in Luminous Blue Variables has been already suggested by Moskalik and Dziembowski (1992) who noted that the instability strip widens in the supergiant domain of the H-R diagram. Thus, it is not unlikely that the κ-mechanism acting in the layer of the metal opacity bump is the ultimate cause of all of the intrinsic variabilities discovered in B-type stars.

REFERENCES

Baade, D., 1984, *A&A*, 134, 105.

Cox, A.N., Morgan, S.M., Rogers, F.J, and Iglesias, C.A. 1992, *ApJ*, in press.

Iglesias, C.A., and Rogers, F.J.,and Wilson, B.G. 1991, *ApJ*, in press.

Iglesias, C.A., and Rogers, F.J., 1991, *ApJ*, 371, 408.

Kiriakidis, M., El Eid, M.F., and Glatzel, W., 1992, *MNRAS*, 255, 1p.

Moskalik, P., and Dziembowski, W.A., 1992, *A&A*, 256, L5.

Waelkens, C., & Rufener, F, 1985, *A&A*, 152, 6.

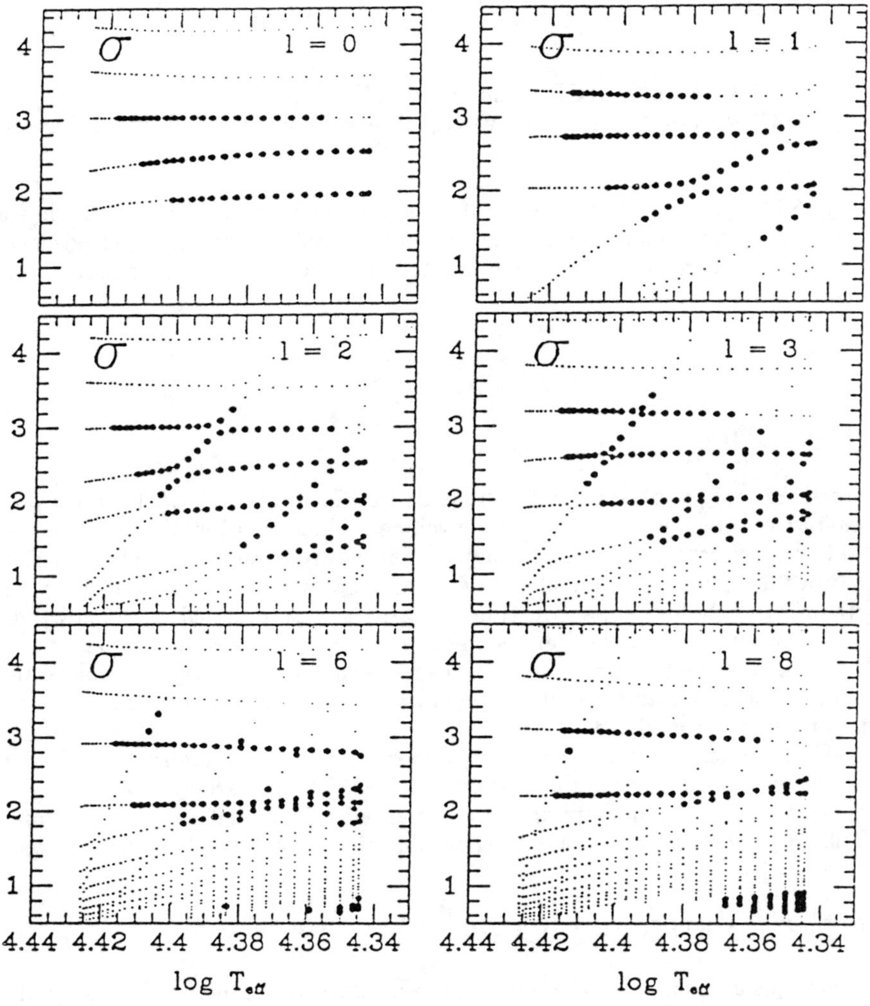

FIGURE I Dimensionless frequencies, σ, of low order and low degree, l, modes in models of a $12M_{\odot}, Z = 0.03$ star in their Main Sequence evolutionary phase. The small and the big dots correspond to stable and unstable modes, respectively.

Inside the Stars, IAU Colloquium 137
ASP Conference Series, Vol. 40, 1993
Werner W. Weiss and Annie Baglin (eds.)

PHOTOMETRIC DETECTION OF LOW–ORDER NRP IN Be STARS

K. PAVLOVSKI, Ž. RUŽIĆ and V. VUJNOVIĆ
Hvar Observatory, Zagreb University, Kačićeva 26, 41000 Zagreb, Croatia

ABSTRACT Period analysis of Be stars 28 Cyg and EW Lac has shown that their photometric variations are multiperiodic. Dominant periods have counterpart in low–order NRP modes identified in the independent line profile studies.

INTRODUCTION

Be stars are early–type stars with the emission lines in spectra which are produced in an extended circumstellar envelope. The emission line profiles are variable on different time scales, ranging from hours to decades. The emission sometimes disappears completely, producing a normal B-type spectrum. Shell spectrum could be developed on some occasions characterized by narrow and deep absorption cores in the Balmer lines and singly ionized metal lines.

Systematic photometric observations have shown that large fraction of the Be stars are light variables. Most of them are found to be short–term variables, with periodicity around 1 day.

Many models were hypothesized to explain origin of emission lines, i.e. extended envelopes around these stars, and also their variable nature. The most promising at the moment are non–radial pulsations (NRP) which can force variable mass outflow and change considerably the optical depth of the envelope.

NRP: PHOTOMETRIC DETECTION

Spectroscopic diagnostics of the NRP in the Be stars are based on the line profile studies. Both low–order and high–order NRP modes are revealed (Walker 1991).

In photometry detection of NRP is not unambiguous. Namely, periods of light variations found are close to what is expected for rotational modulation (Cuypers 1991). Hence, our detection of NRP is based on the following considerations:

1. Identification of the spectroscopic periods in photometric data
2. Multiperiodicity which is expected in the case of NRP
3. Wavelength dependence of light amplitudes

We concentrated on 28 Cyg and EW Lac, two Be stars for which large data bases, both spectroscopic and photometric, exist at the Hvar Obs. and elsewhere. Particularly intensively observed was 28 Cyg over last few years in satellite UV and ground–based optical region. In a periodogram analysis we isolated the dominant frequency (period). In extensive photometry of 28 Cyg carried out in the summer 1985 at the Hvar Obs. we found a dominant frequency $f_{ph} = 1.54$ c/d (Fig. 1). This confirmes our previous finding on shorter data set (Pavlovski & Ružić 1990). These results agrees perfectly with the period of low-order NRP mode $l = 2$ found in line-profile analysis by Peters & Penrod (1988) and Hahula & Gies (1991), $f_{sp} = 1.55$ c/d.

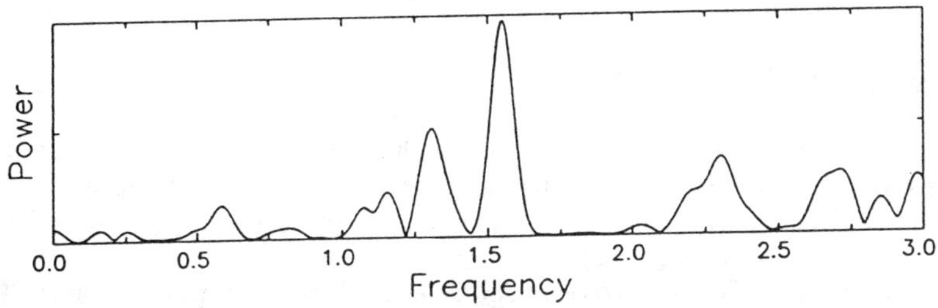

Fig. 1. CLEANed Fourier power spectrum of the B-band light variations of 28 Cyg in 1985 Hvar observations. Two prominent peaks remained on the frequencies $f_1 = 1.55$ c/d and $f_2 = 1.30$ c/d. The former peak corresponds to period of the low-order $l = 2$ NRP mode identified in line-profile studies.

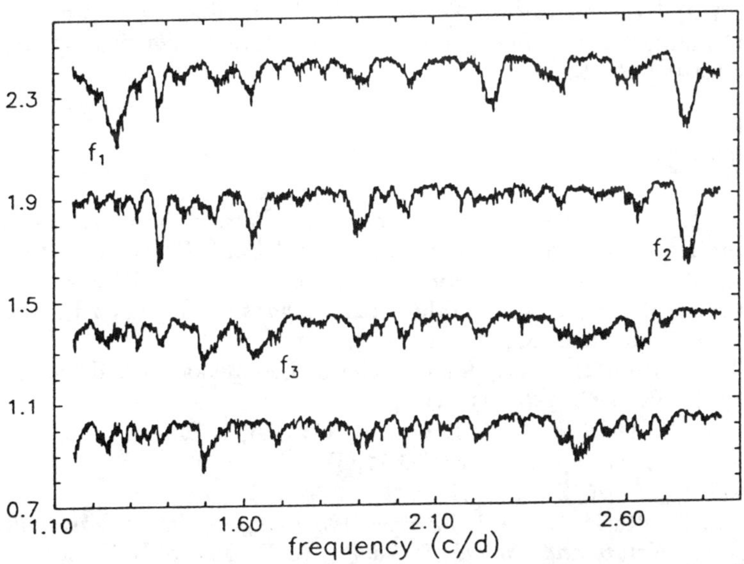

Fig. 2. PDM θ-statistics for the multisite world-wide V passband observations of the Be star EW Lac in 1983, for subsequent steps of prewhitening. Following frequencies are identified: $f_1 = 1.28$ c/d, $f_2 = 2.79$ c/d, and $f_3 = 1.62$ c/d (Pavlovski et al. 1992).

However, successive prewhitening and/or CLEANing show that additional significant frequencies are present and we performed non-linear least-square fitting which support multiperiodicity. Obviously, light amplitudes in V and B passbands are similar, with a larger amplitude in the U passband in agreement with theoretical calculations (Watson 1988).

Our recent reanalysis of the multisite observations of EW Lac in 1983 (Pavlovski et al. 1992) confirmed previously claimed multiperiodicity (Pavlovski 1987). Further analysis of photometric observations in separate seasons has revealed a correlation between pulsational amplitudes and variable shell activity (Pavlovski & Ružić 1991). Photometric periods are recently confirmed from detail investigation of changes in He I λ6678 line by Floquet et al. (1991, 1992).

A PULSATION MECHANISM

While discussing a possible pulsation mechanism, it would be interesting to notice close similarities of the multiperiodic characteristics for the Be stars 28 Cyg and EW Lac, and *Slowly Pulsating B Stars*, group of B–type variables found and coined by Waelkens (1991).

In an analysis of extensive observational data accumulated at ESO over years, Waelkens found that most of the stars in his sample are multiperiodic. He further argued that these stars pulsate in high–radial low–order g–modes. His conclusion is based on not yet published calculations by Degryse. Having applied her calculations as communicated by Waelkens (1991) to the periods found by us for the Be stars, the models point toward medium radial wavenumbers ($k \sim 10$ for $l = 1$ and $k \sim 20$ for $l = 2$). It should be interesting to work out which physical parameter(s) differenciate(s) the non–emission *slowly pulsating* B stars from pulsating *Be* stars.

REFERENCES

Cuypers, J. 1991, in *Proc. Workshop on Rapid Variability of OB Stars: Nature and Diagnostic Value* , ed D. Baade, ESO Conf. Proc. 36, p. 83.

Floquet, M., Hubert, A. M., Hubert, H. 1991, in *Proc. Workshop on Rapid Variability of OB Stars: Nature and Diagnostic Value* , ed D. Baade, ESO Conf. Proc. 36, p. 39.

Floquet, M., Hubert, A. M., Janot-Pacheco, E., Mekkas, A., Hubert, H., Leister, N. V. 1992, *Astr. Ap.*, (in press).

Hahula, M. E., Gies, D. R. 1991, *Bull. A.A.S.*, **23**, 1379.

Pavlovski, K. 1987, *Ap. Space Sci.*, **134**, 317.

Pavlovski, K., Ružić, Ž. 1990, *Astr. Ap.*, **236**, 393.

Pavlovski, K., Ružić, Ž. 1991, in *Proc. Workshop on Rapid Variability of OB Stars: Nature and Diagnostic Value* , ed D. Baade, ESO Conf. Proc. 36, p. 191.

Pavlovski, K., Ružić, Ž., Pavlović, M., Jeong, J. J., Nha, I.-S. 1992, *Ap. Space Sci.*, (in press).

Peters, G. J., Penrod, G. D. 1988, in *Decade of UV Astronomy with IUE*, ESA SP–281, Vol. 2, p. 117.

Waelkens, C. 1991, *Astr. Ap.*, **246**, 453.

Walker, G. A. H. 1991, in *Proc. Workshop on Rapid Variability of OB Stars: Nature and Diagnostic Value* , ed D. Baade, ESO Conf. Proc. 36, p. 27.

Watson, R. D. 1988, *Ap. Space Sci.*, **140**, 255.

Inside the Stars, IAU Colloquium 137
ASP Conference Series, Vol. 40, 1993
Werner W. Weiss and Annie Baglin (eds.)

NEW LIMITS ON THE INSTABILITY STRIP OF THE HRD: OBSERVATIONS FROM A NORTHERN SKY SURVEY OF Ap STARS FOR RAPID VARIABILITY

Matthew J. Nelson
Dept. of Astronomy, University of Wisconsin, Madison, Wisconsin.

Tobias J. Kreidl
Lowell Observatory, Flagstaff, Arizona.

ABSTRACT The search for pulsational variables that occupy the hotter realms of the instability strip in the Hertzsprung-Russel diagram (HRD) have been traditionally restricted to the regions where models indicated that pulsation is tractable. Recent surveys of stars have revealed, however, that stars that were thought not to lie in the instability strip do indeed pulsate. This raises questions about the extent of the instability strip.

While this has implications for all hot pulsating stars, we use here primarily our survey results of Ap stars to compare with the models. We report results here from a high-speed photometric survey of 120 Ap stars that was conducted between 1985 and 1991 to search for rapid variability. The absence of pulsations in the hotter Ap stars (roughly B8-A3) is noted and deemed significant on the basis of the number of stars observed in this temperature range, as well as the overall quality of the data. This, however, does not preclude their existence, especially since HD 218495 was recently discovered to be a rapidly oscillating Ap (roAp) star (Martinez, Kurtz and Kaufmann 1991), and has a spectral type of about A3.

Color-magnitude diagrams of the survey stars are presented, with the known roAp stars included for reference. The diagrams are presented in the Strömgren and Geneva systems. The color-magnitude diagrams demonstrate the completeness of the survey in covering Ap stars at a wide range of temperatures. We find no obvious means of using color indices to differentiate roAp stars from non-pulsating Ap stars.

SURVEY RESULTS

The Geneva Color System

The Geneva [g] index seemed the most promising for selecting roAp stars from stable Ap stars. Figure 1 is the Geneva [g],Δ diagram. The [g] index is a temperature and luminosity indicator, while the Δ parameter provides a measure of the Balmer discontinuity. This diagram separates quite well the hotter SiCr stars from the cooler SrCrEu stars, which in turn overlap to some extent the Am group. Although many roAp stars have unusually high [g] indices (around 0.2) one must point out that at least one survey star (β CrB) with a high [g] index was observed very intensively, and is apparently non-variable. In addition, for [g]≤0.16, the roAp stars no longer stand out from the non-variable stars of our survey. Hence, we feel the Geneva [g],Δ diagram, while more encouraging than the others, still has its limitations.

The Strömgren Color System

Figure 2 shows the Strömgren [c1],[m1] diagram, and is included to provide a comparison of our results in the Geneva color system with another photometric system. In the Strömgren system the [m1] parameter is a metallicity/peculiarity indicator, and the [c1] parameter is a luminosity indicator. The roAp stars populate the lower right portion of the diagram. In this survey the [c1],[m1] color system does not show any distinction between the roAp stars and stable Ap stars. Many survey stars which show no variability overlap the region occupied by the roAp stars, tending to disagree with the color distinctions found by Matthews (1990) for this color diagram.

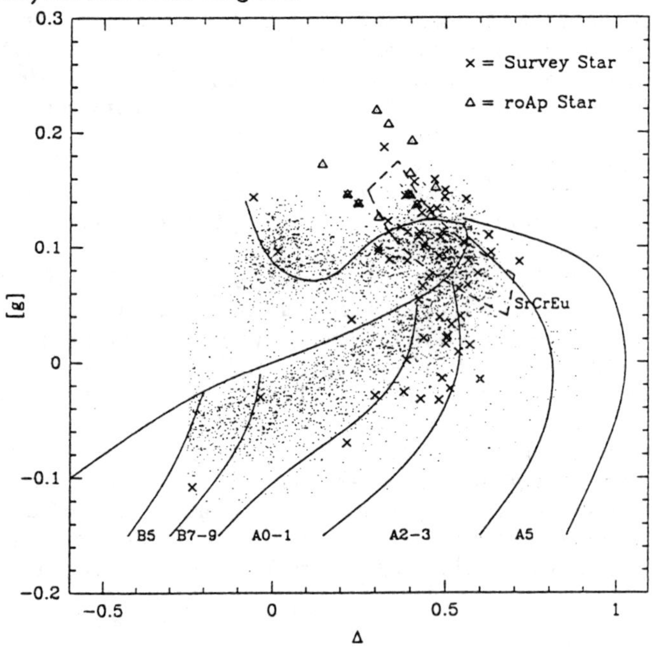

Figure 1. The Geneva [g],Δ diagram. Temperature ranges B5 through A5 are indicated on the diagram as adjacent bands. The part of the diagram populated by SrCrEu peculiarity stars is indicated as a dashed box. The Open symbols indicate stars from this survey, the closed symbols indicate known roAp stars, and the small dots are the stars in the Geneva system with colors in the same range as the survey stars.

CONCLUSIONS

The primary result of this study is that various combinations of photometric indices, while pointing towards the roAp stars having the characteristic signatures of cool, SrCrEu stars, still fail to isolate the roAp stars from nonvariable Ap stars. Differentiating roAp stars from non-variable stars may be possible with spectroscopic studies, and a study of roAp and stable-to-oscillation stars via high-resolution spectroscopy should be undertaken to look for subtle signature differences. Matthews (1990) came to the same conclusion. About one week of observing time on a 2 meter class telescope should be sufficient for significant headway. Future photometric surveys should concentrate on cooler Ap stars, in particular the SrCrEu types.

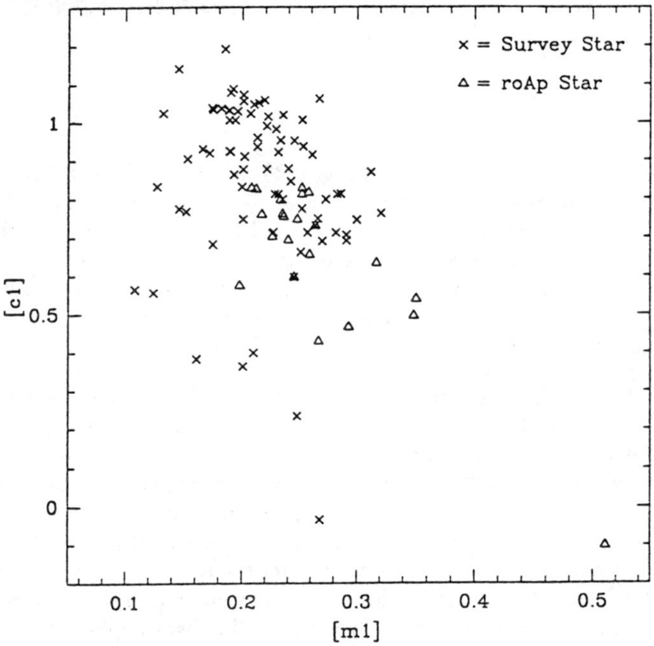

Figure 2. The Strömgren [c1],[m1] diagram. The symbols have the same meaning as in Figure 1.

ACKNOWLEDGEMENTS

MJN acknowledges the support of the High Speed Photometer project (NASA grant NAS 5-24487). TJK acknowledges partial support through the NSF grant AST-9716971 and the Lowell Observatory endowment.

REFERENCES

Martinez, P., Kurtz, D. W., and Kaufmann, G. M. 1991, *M.N.R.A.S*, **250**, 666.
Matthews, J. M. 1990 "Progress of Seismology of the Sun and Stars," *Lecture Notes in Physics 367*,

Inside the Stars, IAU Colloquium 137
ASP Conference Series, Vol. 40, 1993
Werner W. Weiss and Annie Baglin (eds.)

β Cas: SIMULTANEOUS STROMGREN PHOTOMETRY

E. RODRIGUEZ, A. ROLLAND, P. LOPEZ DE COCA, R. GARRIDO
Instituto de Astrofísica de Andalucía. Apartado 3004. 18080 Granada. Spain.

S.F. GONZALEZ-BEDOLLA
Instituto de Astronomía. UNAM. Apartado Postal 70-264. CP-4510. México D.F. Mexico

The observations were carried out in August, 1983 using the 75 cm telescope at Sierra Nevada Observatory in Spain. This telescope is equipped with a six-channel *uvbyβ* photometer for simultaneous measurements in *uvby* and the narrow and wide H_β channels, respectively. The observations were collected during four nights in the four *uvby* filters and one more night was devoted to measuring β Cas in the *n* and *w* bands of the H_β Crawford system. In all cases, a calibrated neutral filter was used for β Cas. The comparison star was C1=HR 9085 (V=5.^{m}55, F0III).

The analysis of frequencies was carried out using the Discrete Fourier Transform method, as described in López de Coca et al. (1984), to the observational *uvby* points collected. The periodograms showed a principal peak at ν=9.91 cd^{-1}. After prewhitening for this frequency, the resulting periodograms did not show any trace of another peak, suggesting the monoperiodic nature of this star, in agreement with Antonello et al. (1986). Under the assumption of monoperiodicity of β Cas, the classical O-C method can easily be applied. Nine times of maxima were obtained for the four nights by using the method described in Rodríguez et al. (1990), where each light maximum has been derived as an average over the four *uvby* bands. When the O-C method was applied, the least squares fit converged to a linear ephemeris with the following elements: T_o=2445568.^{d}5097 (±0.^{d}0007) and P_o=0.^{d}10091 (±0.^{d}00002). This linear ephemeris satisfactorily reproduces our data of β Cas.

To derive the physical parameters such as effective temperature, T_e, gravity, log g, absolute magnitude, M_v, and pulsation constant, Q, the method described in López de Coca et al. (1990) has been used. Considering the indices of b-y=0.^{m}216, m_1=0.^{m}177, c_1=0.^{m}785 (Hauck and Mermilliod 1990) and β=2.^{m}721 (obtained from our data), the following values of T_e=7170 K, log g=3.62 and M_v=1.^{m}6 can be obtained for β Cas. The Q value corresponding to the period of P=0.^{d}10091 is 0.^{d}024. Values of −0.^{m}004 and 0.2 can also be obtained for δm_1 and [Me/H], respectively, using the reference lines given by Philip and Egret (1980) and Nissen's (1981) calibration for metal abundances. From a comparison with the (Δm_1^*,β) grids from Rodríguez et al. (1991), a m_1 index variation of about 0.^{m}002 reversed, as compared with the light curve variation, must be expected over the cycle of pulsation of β Cas.

The lack of other pulsation frequencies excludes methods based on period ratios or frequency differences in order to determine the nature of radial or nonradial pulsation in this star and to identify the pulsation mode. However, Garrido et al. (1990) have shown, using *uvby* photometry, that the "amplitude ratio versus phase shift" diagrams for the (*v*,*y*) and (*b*,*y*) pairs of bands are discriminant between radial and low order nonradial pulsations for δ Sct stars. It is also true from the amplitude ratio, Δ(b-y)/Δy, versus phase shift, $\phi_{b-y}-\phi_y$, diagrams.

In Table I are listed the phase shifts, in degrees, and amplitude ratios between the different bands and colour indices, derived from the results of the Fourier analysis. In agreement with other δ Sct pulsators (Garrido and Rodríguez 1990; Rolland et al. 1991; Rodríguez et al. 1992), it is shown that the maximum light in the *v*, *b* and V bands occurs after that corresponding to the *u* band in about 0.03 cycles. From Table I it can also be seen that the curves for the b-y index is very close to the light curve whereas the c_1 index curve is shifted with respect to V curve in the sense that the maximum in c_1 occurs around 0.09 cycles after the maximum in V. These effects can be explained in terms of temperature and gravity variations along a pulsational cycle (Garrido and Rodríguez 1990). These effects can also be seen in Figure 1, where synthetic light and colour indices curves are shown. These synthetic curves were constructed using the results from the Fourier analysis.

It is also possible to see in both Figure 1 and Table I that the maximum light in the *y* filter occurs before both maxima in the *v* and *b* filters. In addition, the maximum in b-y also occurs after that corresponding to the *y* band. These effects suggest nonradial pulsation in β Cas. Furthermore, a close inspection of the phase shifts and amplitude ratios in Table I and the diagrams given by Garrido et al. (1990) leads to nonradial pulsation with l=1 as the most reliable mode identification, although the observed Δ(b-y)/Δy=0.25(±0.02) ratio seems too high. This disagreement disappears if we consider a specific model with the appropiate T_e, log g and Q values (7200 K, 3.6, $0.^d024$) of this star, by using the Kurucz's (1979) models, and an increasing of the α_T(b) parameter by 0.3 (Watson 1988) is applied. The radial number, k, can be estimated from the constant of pulsation Q. A comparison with theoretical Q values given by Fitch (1981) for l=1 indicates the second overtone, k=2, corresponding to the obtained Q value of $0.^d024$.

Nonradial pulsation is not unusual for low amplitude δ Sct stars, especially when more than one frequency is excited. Observations have shown that a number of these variables pulsate with a large number of simultaneously excited nonradial p modes. However, nonradial pulsation has been also found for low amplitude δ Sct stars where only one pulsation frequency is present. An example is τ Peg. At present, τ Peg is the only δ Sct star known that shows both only one frequency and nonradial pulsation. Breger (1991) finds that this star pulsates in a nonradial p_3 or p_4 mode with l=2. Thus, the detection of only one frequency of pulsation is not an indication of radial pulsation for stars with low amplitude. Similar to τ Peg, β Cas may be a second example of low amplitude δ Sct star with a single pulsation frequency that shows nonradial pulsation.

REFERENCES

Antonello, E., Pastori, L., Poretti, E. 1986, Inf. Bull. Var. Stars, No. **2958**
Breger, M. 1991, Astr. Ap., **250**, 107
Garrido, R., García-Lobo, E., Rodríguez, E. 1990, Astr. Ap., **234**, 262
Garrido, R., Rodríguez, E. 1990, Ap. Space Sci., **169**, 205
Hauck, B., Mermilliod, M. 1990, Astr. Ap. Suppl., **86**, 107

Kurucz, R.L. 1979, Ap. J. Suppl., **40**, 1
López de Coca, P., Garrido, R., Rolland, A. 1984, Astr. Ap. Suppl., **58**, 441
López de Coca, P., Rolland, A., Rodríguez, E., Garrido, R. 1990, Astr. Ap. Suppl., **83**, 51
Nissen, P.E. 1981, Astr. Ap., **97**, 145
Philip, A.G.D., Egret, D. 1980, Astr. Ap. Suppl., **40**, 199
Rodríguez, E., López de Coca, P., Rolland, A., Garrido, R. 1990, Rev. Mex. Astron. Astrofís., **20**, 37
Rodríguez, E., Rolland, A., López de Coca, P., García-Lobo, E., Sedano, J.L. 1992, Astr. Ap. Suppl., **93**, 189
Rodríguez, E., Rolland, A., López de Coca, P., Garrido, R. 1991, Astr. Ap., **247**, 77
Rolland, A., Rodríguez, E., López de Coca, P., García-Pelayo, J.M. 1991, Astr. Ap. Suppl., **91**, 347
Watson, R.D. 1988, Ap. Space Sci., **140**, 255

TABLE I Phase shifts and amplitude ratios

u-v (degrees)	u-b (degrees)	u-y (degrees)	v-b (degrees)	v-y (degrees)	b-y (degrees)	(b-y)-y (degrees)	c_1-y (degrees)
13.1	12.0	10.9	−1.1	−2.2	−1.1	−5.3	−34.0
2.2	2.1	2.3	1.7	1.9	1.8	3.4	5.2
u/v	u/b	u/y	v/b	v/y	b/y	(b-y)/y	c_1/y
0.84	1.00	1.26	1.19	1.49	1.26	0.25	0.58
3	4	5	4	5	4	2	5

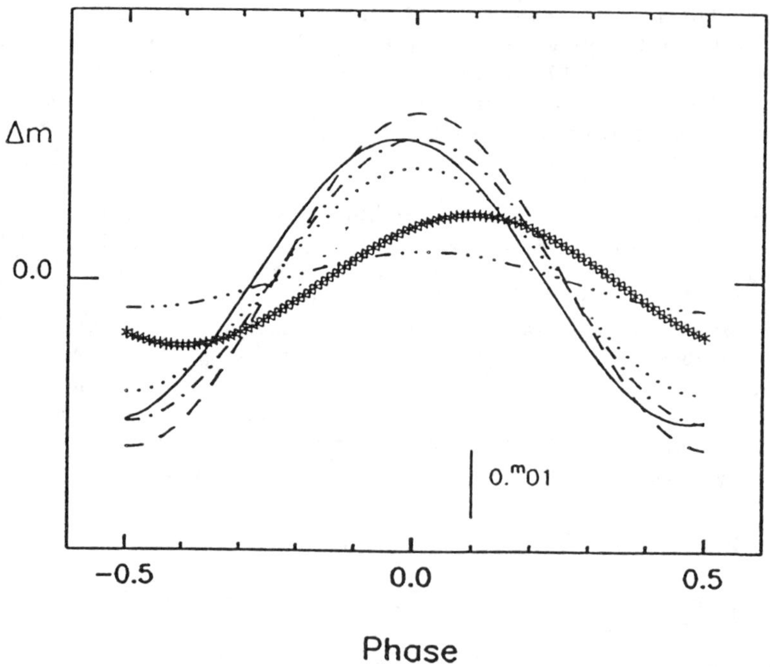

Fig. 1. Synthetic light curve and colour index variations of β Cas versus phase. u:—, v:- - -, b:-.-.-, V:....., (b-y):-...-...-, c_1:*****.

Inside the Stars, IAU Colloquium 137
ASP Conference Series, Vol. 40, 1993
Werner W. Weiss and Annie Baglin (eds.)

METHODOLOGICAL ASPECTS IN THE FREQUENCY ANALYSIS OF MULTIPERIODIC δ SCUTI STARS

L. MANTEGAZZA
Dipartimento di Fisica Nucleare e Teorica, Università di Pavia – 27100 Pavia – Italy

E. PORETTI, E. ANTONELLO, E. RIBONI
Osservatorio Astronomico di Brera, Via E. Bianchi, 46 – 22055 Merate – Italy

DATA COLLECTION

At Merate Observatory the study of δ Sct stars began in the sixties and spectroscopic and photometric campaigns were continuously undertaken in order to clarify the controversial points. As an obvious extension of the research, the observation of δ Sct stars was proposed for telescope–time allocation at European Southern Observatory, in order to take advantage of the ESO facilities and of the considerably better sky of La Silla. We put in our observing programme δ Sct stars showing cycle–to–cycle variations and, possibly, an amplitude larger than 0.05 mag in order to have a better signal–to–noise ratio; they are listed in the table. It is important to notice that we always tried to collect a number of measurements as large as possible (from 1988 we collected more than 1000 measurements per star), because it is a well–established fact that periodicities with

Star	Site	Observing period	Survey [hours]	Reference
HD 37819	Merate	Jan. 1986	32	A&A 1987, **181**, 273
HR 1225	ESO	Nov. 1987	38	A&A 1989, **220**, 144
o^1 Eri	ESO	Nov. 1987	38	A&A 1989, **220**, 144
HR 547	ESO	Nov. 1987	22	A&A 1989, **220**, 144
HD 16439	Merate	Dec. 1988–Jan. 1989	54	A&A 1990, **230**, 91
HD 101158	ESO	Apr. 1989	62	A&A 1991, **245**, 136
V974 Oph	ESO	Apr. 1989	64	work in progress
X Cae	ESO	Nov. 1989	54	A&A 1992, **255**, 153
44 Tau	Merate	Dec. 1989–Feb. 1990	117	A&A 1992, **256**, 113
BD+2°1867	ESO	Jan.–Feb. 1991	100	work in progress
BD+2°1867	Merate	Jan.–Feb. 1991	43	work in progress
HD 224639	ESO	Sep.–Oct. 1991	120	work in progress
HD 18878	Merate	Nov. 1991–Jan. 1992	150	work in progress

an amplitude of a few thousandths of magnitude can be present in δ Sct stars. Moreover, when several modes are excited, an adequate frequency resolution becomes necessary and, therefore, a sufficient time baseline is requested (Poretti and Mantegazza 1992). A poor data sampling also gives an apparent good fitting for several solutions, all leaving a small residual rms, and there is no possibility to choose among them: different authors can pick up different solutions, generating the conflicting interpretations often found in δ Sct star literature.

THE FREQUENCY ANALYSIS METHOD

In the frequency analysis we used the simultaneous least–squares fitting procedure proposed by Vaníček (1971), which does not require any prewhitening since the amplitudes and phases of the known constituents (i.e. the frequencies already identified) are recalculated for each new trial frequency. Moreover, before searching for new frequency the values of the known frequencies can be refined by means of a non–linear sine wave least–squares fitting routine; for example, the program PERDET (Breger 1989) or the program MTRAP (Carpino et al. 1987). In spite of all these precautions, we generally performed some parallel analysis, considering not only the highest peak but also its aliases. Indeed, when dealing with a rich power spectrum a particular care must be taken in the iden-

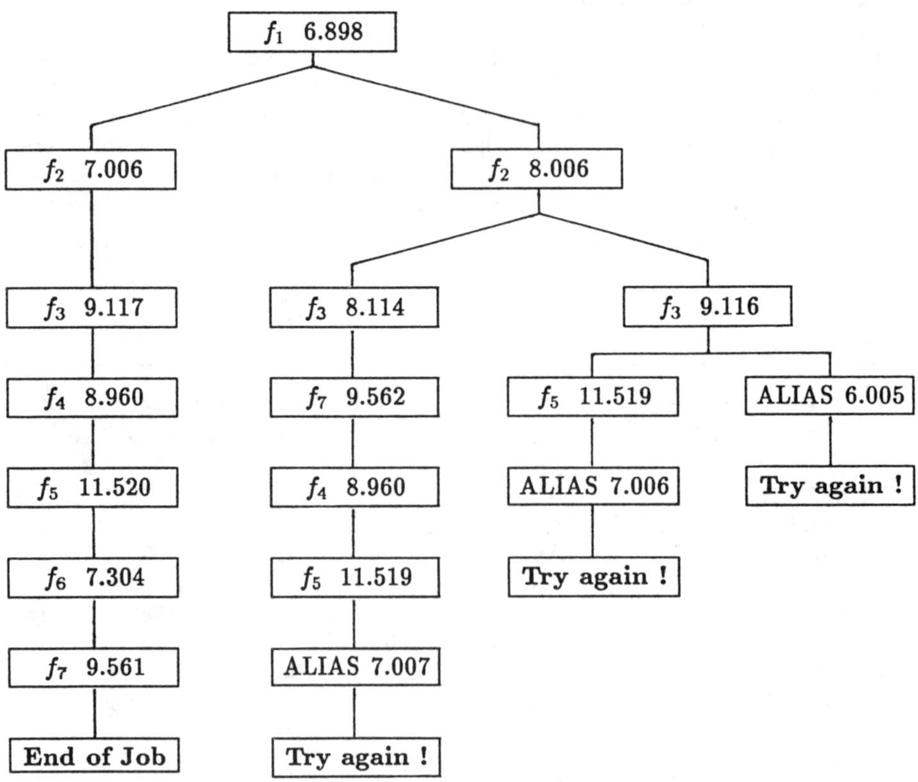

tification of the true frequency because the aliases at ±1 c/d can be as strong as the true peak, especially if the datasest is collected in a single–site campaign.

These parallel analyses can supply additional confirmations on the goodness of the choices. Let us discuss the application of this procedure to the 2434 V measurements collected on 44 Tau. This star was considered a double–mode radial pulsator (Wizinowich and Percy 1979) until Poretti et al. (1992) showed that seven frequencies were necessary to fit its light curve in a satisfactory way (f_1=6.898 c/d, f_2=7.006 c/d, f_3=9.117 c/d, f_4=8.960 c/d, f_5=11.520 c/d, f_6=7.304 c/d, f_7=9.561 c/d). The second power spectrum (i.e. that obtained by considering f_1 as a known constituent) suggested as possible values both 7.006 c/d and 8.006 c/d. When choosing the former, the analysis revealed the above reported seven frequencies, without any doubt on the identifications of each of these terms (but in any case we performed additional tests to verify them). On the other hand, if we consider for f_2 the value of 8.006 c/d we obtain a more complicated third power spectrum, where the two frequencies 8.114 c/d and 9.116 c/d are both possible. Again, when continuing with the value f_3=9.116 c/d we quickly meet the frequency 6.005 c/d, i.e. an alias of the adopted value for f_2. If we continue with the value f_3=8.114 c/d we find the f_7, f_4, f_5 terms and finally we meet 7.007 c/d, i.e. another alias of the adopted value for f_2. All this process is better evidenced in the figure. Therefore, we suggest that the absence of aliases of previously identified frequencies in the successive power spectra indicates that the right choices were done.

FINAL CONSIDERATIONS

The variability of the mode amplitudes is one of the most recent results in the study of δ Sct stars. This phenomenon gives a strict constraint to the observer: indeed, it is necessary to obtain a good solution for each observing season in order to compare their results and establish the amount of the mode amplitude variability. For these reasons we consider necessary: a) the collection of large datasets, well sampled in time and covering a time baseline as long as possible; b) a careful frequency analysis of these measurements, as described in the previous section; c) a careful comparison between all the competitive solutions and between the behaviours of their residuals.

REFERENCES

Breger M., 1989, *Comm. Astrosesismology* **6**

Carpino M., Milani A., Nobili A.M., 1987, *A&A* **181**, 182

Poretti E., Mantegazza L., 1992, *The Messenger*, in press

Poretti E., Mantegazza L., Riboni E., 1992, *A&A* **256**, 113

Vaníček P., 1971, *Ap&SS* **12**, 10

Wizinowich P., Percy J.R., 1979, *PASP* **91**, 53

Inside the Stars, IAU Colloquium 137
ASP Conference Series, Vol. 40, 1993
Werner W. Weiss and Annie Baglin (eds.)

RADIAL PULSATIONS IN 63 HERCULIS

BÜLENT UYANIKER
Physics Department, Middle East Technical University
06531 Ankara, Turkey.

ABSTRACT Constructing a theoretical model of 1.88 $M_\odot$ pulsation properties of 63 Her have been analyzed. For the model, in question, pulsation periods in the fundamental mode and first two overtones have been derived by solving the linearized pulsation equations. The model has been found to be vibrationally stable.

INTRODUCTION

The variability of 63 Her (HR 6391) was first dicovered by Breger (1969). He announced it to pulsate with a period of $0^d.077$. Since then, it has been observed by other astronomers (Reed and Welch, 1988; Akan, 1991). Breger and Bregman (1975) reported that 63 Her oscillates in the radial fundamental mode with a period of $0^d.077$. Reed and Welch found that the star pulsates possibly in three distinct frequencies corresponding to $0^d.088$, $0^d.052$ and $0^d,130$. Akan obtained six frequencies which are 0.686, 1.383, 10.312, 18.588 and 19.870 c/d. In their analysis, Reed and Welch have concluded that$0^d.08$ corresponds to radial pulsation in the fundamental mode. They also analyzed the Breger's data and obtained $0^d.105$ and $0^d.051$ for the radial fundamental mode and for the second overtone, respectively. Belmonte and Michel (1991) gave, $0^d.250$, $0^d.157$, $0^d.130$, $0^d.080$, $0^d.077$, $0^d.050$ and $0^d.038$ of periods for 63 Her, while Manganey at. al. (1991) obtained six frequency solution due to their pulsation analysis.

In the present study, our intention is to investigate the pulsational properties of 63 Her in connection with stellar evolution so that we have constructed a theoretical model for 63 Her and evolved it until to satisfy the observed position in the H-R diagram. Then we solved the linerized pulsation equations for radial adiabatic oscillations.

THE MODEL AND PULSATION ANALYSIS

We have constructed a theoretical model using Ezer (1969)'s evolutionary code for a mass of 1.88 $M_\odot$. We have also chosen solar-type chemical composition of X=0.739, Z=0.021, and the value of $\ell/H=1.3$ for mixing-length parameter. Basic properties of the model are given in Table I.

To obtain the linear set of equations governing the radial adibatic oscillations classical linear theory is applied (Ledoux and Walraven, 1958). Linearized equations are then solved using a pulsation code (Uyanıker and Kırbıyık, 1992). Basic results of pulsation analysis are displayed in Table II.

In Table II. obtained periods as well as the corresponding frequencies are displayed. Last column of the table shows the ratio of surface to the center value of the eigenfunction σ'. Eigenfunctions for the fundamental mode and first two overtones are plotted in Figure 1. The vibrational stability coef-

Table I: Physical properties of the model

$M/M_\odot$	$R/R_\odot$	$L/L_\odot$	$\log T_e$	Age (yr)
1.88	2.45	19.2	3.89	1.112×10^9
T_c	ρ_c (gm/ cm^3)	ρ_{ph} (gm/ cm^3)	X_c	Y_c
2.45×10^7	101	7.91×10^{-9}	0.071	0.908

Table II: Pulsation properties of the model in question

Mode	P(day)	$\sigma^2 \times 10^{-7}$	ν (μhz)	$(\delta r/r)_s$ / $(\delta r/r)_c$
F	0.110	4.34	104.8	2.179×10^2
1 H	0.077	8.90	150.1	-9.590×10^2
2 H	0.061	14.40	191.0	1.065×10^3

ficient, σ', is computed as Boury et. al. (1975). In Figure 2, behaviour of σ' is given with respect to the radius fraction. Note that the dip taking place almost at the surface corresponds approximately 0.98 of the total radis. This irregularity of the σ' may be due to the effects of the medium belonging to that portion of the star. In that region the temperature is about 3.5×10^4 °K, so we conclude that at this portion of the star a hydrogen ionization zone exists.

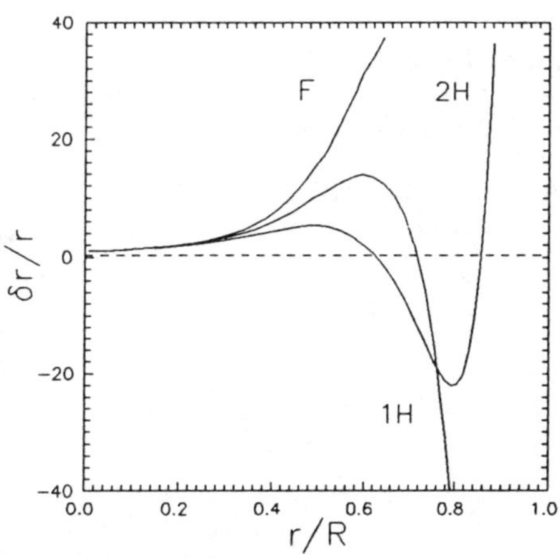

Figure 1: Change of $\delta r/r$ as a function of the fractional stellar radius.

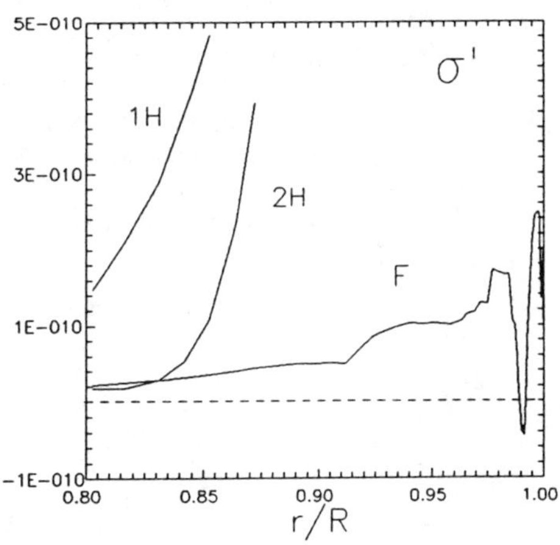

Figure 2: Change of vibrational stability coefficient with respect to r/R.

DICSUSSION

We have found a period of $0^d.11$ for the fundamental mode. Furthermore, we obtained $0^d.077$ and $0^d.061$ for the first overtone and second overtone, respectively. Considering the Breger's data (analyzed by Reed and Welch) the period we found for the fundamental mode is compatible with the observations. It is interesting to note that $0^d.077$ we obtained for the first overtone is exactly the same with the reported observational fundamental periods (Breger, 1969; Baglin et.al., 1973). One possible explanation may be that, what they observe as the fundamental radial mode appears as the first overtone of the radial mode.

Periods obtained in the present study agree well with observations although there are differences in identifying with same modes. Oscillations checked for vibrational instability showed that all modes are vibrationally stable against radial pulsations. We have also found that the hydrogen ionization zone taking place almost at the surface is responsible from the generation of pulsations in 63 Her.

REFERENCES

Akan, M. C. : *Private Communication*

Baglin, A.; Breger, M.; Chevalier, C.; Hauck, B.; Le Contel, J. M.; Serayan, J. P.; Valtier, J. C. : 1973 *Astron. Astophys.*, **23**, 221.

Belmonte. J. A. and Michel, E. : 1881, *Delta Scuti Star Newsletter*, **3**, 16.

Boury, A.; Gabriel, M.; Noels, A.; Scuflaire, R.; Ledoux, P. : 1975, *Astron. Astrophys.*, **41**, 279.

Breger, M. : 1969, *Astrophys. J.*, **74**, 166.

Breger, M. and Bregman, J. N. : 1975, *Astrophys. J.*, **14**, 399.

Ezer, D. : 1969, *Write-up for Stellar Evolution Program*, (Private Copy).

Ledoux, P. and Walraven, T. H.: 1958, *Hdb. der Physik, ed.* Flügge, S.,**51**, 353.

Manganey, A.; Dappen, W.; Praderie, F. and Belmonte, J. A. : 1991, *Astron. Astrophys.*, **244**, 351.

Reed, L. G. and Welch, G. A. : 1988, *Astron. J.*, **95**, 1510.

Uyanıker, B. and Kırbıyık, H. : 1992, *Astophys. and Space Sci.* (in press).

Inside the Stars, IAU Colloquium 137
ASP Conference Series, Vol. 40, 1993
Werner W. Weiss and Annie Baglin (eds.)

STEPHI: A NEW APPROACH TO δ SCUTI ASTEROSEISMOLOGY

J.A. BELMONTE, T. ROCA CORTÉS, I. VIDAL, F.X. SCHMIDER
Instituto de Astrofísica de Canarias, 38200, La Laguna, Tenerife, Spain.

E. MICHEL, A. BAGLIN, M. CHEVRETON, M. AUVERGNE, A. MANGENEY, M.J. GOUPIL, G. MASSACRIER, T. SERRE
Observatoire de Paris-Meudon, 92195 Meudon, France.

M. ÁLVAREZ
Instituto de Astronomía de la U.N.A.M., Ap. Postal 877, Ensenada, BC, Mexico.

S.Y. JIANG, Y.Y. LIU, J.N. FU
Beijing Observatory, Chinese academy of Sciences, Beijing, China.

N. DOLEZ
Observatoire Midi-Pyrénées, 31400 Toulouse, France.

J.P. SAREYAN, J.C. VALTIER
Departement d'Astrophysique, Université de Nice, 06034 Nice, France.

THE STEPHI NETWORK

The STEPHI network (STEllar PHotometry International) is a multinational cooperation stablished in order to get as perfect data as possible on several δ Scuti stars, within a reasonable period of time. STEPHI is currently integrated by nearly 15 people, belonging to the first four institutions cited above. However, astronomers of other institutions have eventually contributed to the network. The idea was born in 1986 and the network has been working to full operation since August 1989. A campaign, observing one or two δ Scuties is normally undertaken every 15 months. They are usually order of three weeks long, avoiding full-moon periods. Up to now, four campaigns have been performed, the last ending February 1992.

Very good outcomes on five δScuties [63 Herculis in June 1987 (Belmonte et al, 1991), GX Pegasi in September 1989 (Michel et al, 1992a), v650 Tauri in November 1990 (Belmonte and Michel, 1991; Michel et al, 1992b) and, finally, BN Cancri and BU Cancri in February 1992] have been obtained. Indeed, some very useful aditional information on comparison stars (see e.g. Belmonte et al, 1990) and sky conditions (transparency, brightness, scintillation, etc...) has been obtained as well. Aditionally, some key information about mostly all multiperiodic (four or more periods) δ Scuti stars studied up to March 1992, with good quality data, is presented in this report.

A preliminary analysis of last campaign data has been performed. Both stars show a very probable multiperiodic behaviour [they were already known δ Scuti stars (Kovàcs, 1981; Bossi et al, 1982)]. Dominant periods found for BNCnc are 52 and 59 minutes, whilst the highest peak found for BUCnc corresponds to a period of 76 minutes. A much deeper analysis is currently undertaken at the different intitutions.

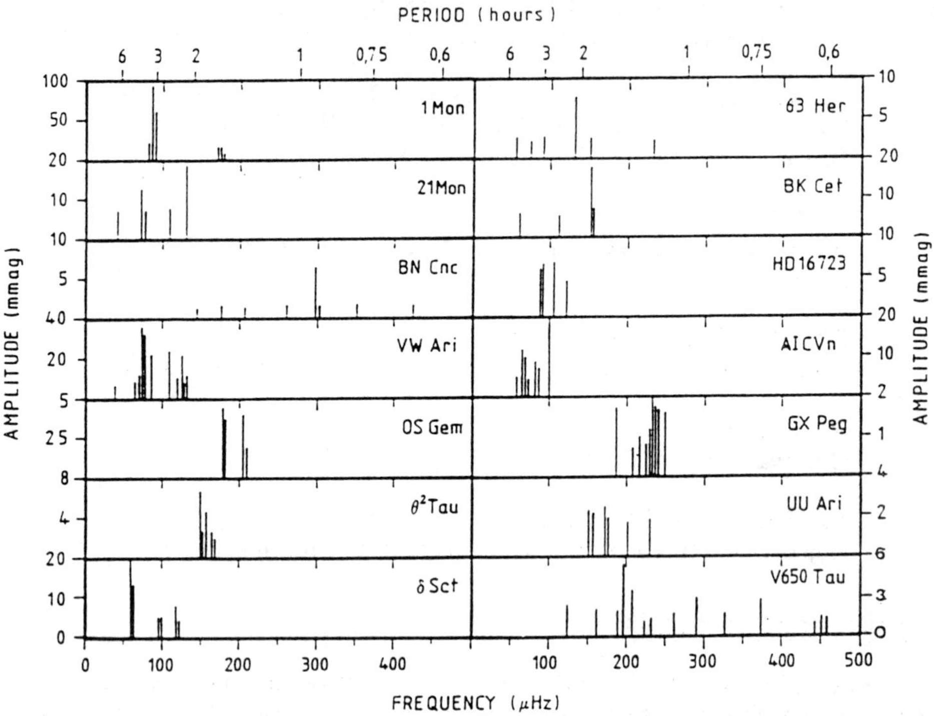

Fig. 1 Up-to-now, δ Scuti star models have had the lack of great simplicity, mainly due to the few observational information avaliable (most stars known were mono or biperiodic). However, this information has improved enormeously during the last few years with the organization, not always easy, of longer campaigns and, specially, of world observing networks. This has permited to obtain large coverages of high quality data of several δ Scuties. The figure shows the amplitude "clean" spectra of most δ Scuties where more than four periods have been discovered so far (see also: Mantegazza et al, 1992). These are the kind of data, theoreticians should start to take into account to improve modelling. This is specially important since stars with apparently similar properties (i.e. same simple modelling) can show extremely different oscillatory behaviour. For more information on each star, you may consult: *Balona et al, 1980* (1 Mon); *Stobie et al, 1977* (21 Mon); *Kovàcs, 1981* (BN Cnc); *Rucinski, 1978* or *Mc Namara and Horan, 1984* (VW Ari); *Boglia and Conconi, 1985* (OS Gem); *Breger et al, 1987 and 1989* (θ^2 Tau); *Mc Namara et al, 1989* (δ Scu); *Belmonte et al, 1991* (63 Her); *Lampens and Rufener, 1990* (BK Cet and HD16723); *Breger, 1990* and *Breger et al, 1990* (AI CVn); *Michel et al, 1992a* (GX Peg); *Ostermann et al, 1991* (UU Ari) and, finally, *Michel et al, 1992b* (v650 Tau).

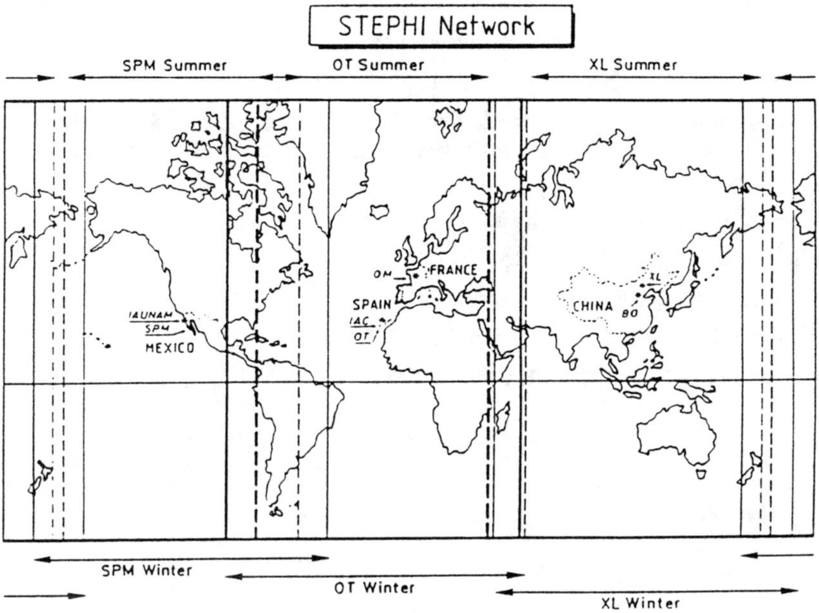

Fig. 2 The STEPHI network. The location of the three observatories: San Pedro Mártir (SPM, Baja california, Mexico), Teide (OT, Izaña, Tenerife, Spain) and Xing-Long (XL, Hopeh, China), allows a theoretical 100% coverage in winter and some 85% in summer. However, due to bad weather conditions and minor technical problems, coverages from 40% to 60% are commonly reached. Meudon provides the network with three identical four-channel Chevreton's photometers (Belmonte et al, 1991).

REFERENCES

Balona L.A. & Stobie R.S., 1980. *M.N.R.A.S.*, **190**, 931.

Belmonte et al, 1990. *A&A*, **231**, 303.

Belmonte et al, 1991. *A&A*, **246**, 71.

Belmonte J.A. & Michel E., 1991. *Delta Scuti News.* **3**, 16.

Boglía P. & Conconi P., 1985. *A&A*, **149**, 15.

Bossi et al, 1982. *Astrophys. Spa Sci.* **85**, 25.

Breger M., 1990. *A&A*, **240**, 308.

Breger et al, 1987. *A&A*, **175**, 117.

Breger et al, 1989. *A&A*, **214**, 209.

Breger et al, 1990. *A&A*, **231**, 56.

Kovàcs G., 1981. *Act. Astron.* **31**, 75.

Lampens S.R. & Rufener F., 1990. A&A Sup. **83**, 145.

Mantegazza et al, 1992. This proceedings.

Mc Namara B.J. & Horan S.J., 1984. *Ap. J.* **282**, 741.

Mc Namara et al, 1989. *In IAU Col. 111: the Use of Pulsating Stars in Fundamental Problems of Astronomy. E.G. Schmidt (ed).* Cambridge Univ. Press, 276.

Michel et al, 1992a. *A&A*, **255**, 139.

Michel et al, 1992b. This proceedings.

Ostermann et al, 1991. *A&A*, **245**, 543.

Rucinski S.M., 1978. *Act. Astron.* **28**, 545.

Stobie et al, 1977. *M.N.R.A.S.*, **179**, 389.

Inside the Stars, IAU Colloquium 137
ASP Conference Series, Vol. 40, 1993
Werner W. Weiss and Annie Baglin (eds.)

HIGH- AND LOW-DEGREE MODES IN DELTA SCUTI STARS: 1- AND 2-D FOURIER ANALYSES OF τ PEG

EDWARD J. KENNELLY[1]
Department of Geophysics & Astronomy, University of British Columbia, Vancouver, British Columbia V6T 1W5 Canada

JAYMIE M. MATTHEWS[1,2]
Département de physique, Université de Montréal, C.P. 6128, succ. A, Montréal, Québec H3C 3J7 Canada

GORDON A. H. WALKER[1]
Department of Geophysics & Astronomy, University of British Columbia, Vancouver, British Columbia V6T 1W5 Canada

PROBING THE INTERIORS OF δ SCUTI STARS

Stellar seismology of δ Scuti stars offers the best prospect for determining the interior properties of normal non-magnetic A-F stars on or near the main sequence. The discovery of rich eigenfrequency spectra in the light variations of several δ Scuti stars (e.g., θ^2 Tau: Breger *et al.* 1989; GX Peg: Michel *et al.* 1990) should make this possible for low-degree modes ($\ell \leq 4$). However, due to cancellation effects, modes of higher ℓ cannot be studied through photometry or radial velocity measurements.

The detection of rapid line-profile variations among rapidly rotating δ Scuti stars (e.g., Walker *et al.* 1987; Kennelly *et al.* 1992) has extended our sensitivity to much higher modes. These variations are consistent with moderate- to high-degree modes, where $|m| \sim \ell$. (In the simplest case, they could be pure sectorial modes with $|m| = \ell$ and pulsation amplitudes strongly confined to the stellar equator.) To use these data effectively, we need an objective way to identify the modes and their time evolution.

Therefore, we introduce here a new technique to analyse the profile variations of δ Scuti stars: **a two-dimensional Fourier transform** in both time and "Doppler space". This technique has three main advantages over conventional photometric and RV studies:

---

[1] Visiting Astronomer, Canada-France-Hawaii Telescope, operated by the National Research Council of Canada, the Centre National de la Recherche Scientifique de France and the University of Hawaii

[2] Visiting Observer, 0.6-m Air Force Telescope at Mauna Kea Observatory, Institute for Astronomy, University of Hawaii

- **Direct determination of both frequency and mode of oscillation** is possible from the 2-D transform.
- **Modes of both high- and low-degree** ($0 \leq \ell \leq 16^+$) **can be resolved**, limited only by the instrumental resolution and the intrinsic width and rotational broadening of the line profile.
- **Complex patterns** in the profile variations are decomposed into Fourier terms, to study multi-mode beating or mode growth/decay.

THE MULTI-MODE BEHAVIOUR OF τ PEG

High-resolution high-S/N coudé spectra from CFHT and simultaneous UBV photometry from the U. Hawaii 0.6-m telescope were obtained for this rapidly rotating δ Scuti variable [$v\sin i \simeq 145$ km/s; sp. A5 V] . The spectra cover a wavelength range of 4460 - 4525 Å and span about 7 hr. RV variations were derived from these spectra by a cross-correlation technique.

The RV and light curves are shown below, along with their respective Fourier amplitude spectra. We find a frequency near 18.5 c/d (agreeing with the published value of 18.4052 c/d (Breger 1991), given the limited frequency resolution of this data set) and a velocity-to-light amplitude ratio $K/\Delta V \simeq 53$ km s^{-1}mag^{-1} (consistent with a *non*-radial mode of low ℓ).

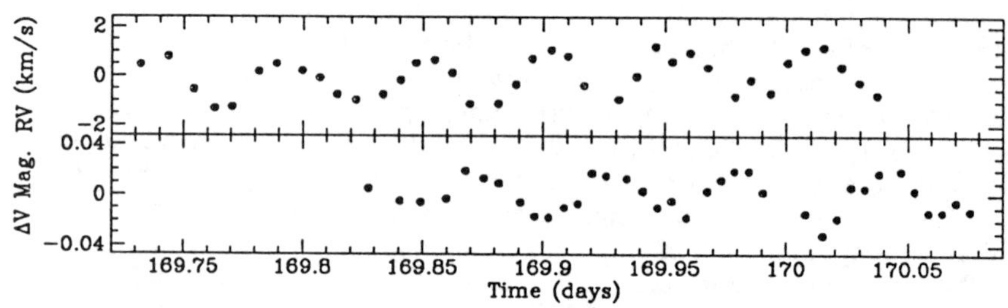

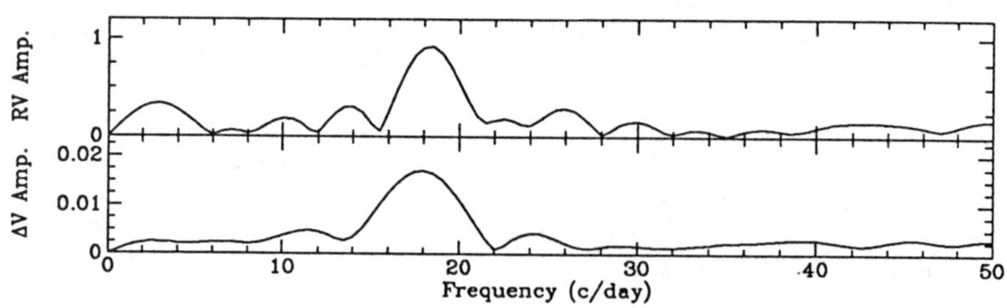

A contour plot of the profile variations of the unblended Fe II λ4508Å line (in which wavelength has been translated into angular position on the stellar equator) is presented in the next figure (left panel). Several "bumps" progress from blue to red through the profile, and their amplitudes appear to be modulated with time. 2-D Fourier analysis of these variations reveals a pattern of at least four peaks at frequencies between 18 and 30 c/d and apparent $|m|$ values near 3, 6-7, 11 and 15 (right panel). (The power below 10 c/d is likely due to slow instrumental changes, etc.)

The small peak with low $|m|$ at 18.5 c/d undoubtedly corresponds to the mode causing the light and RV oscillations. Low-degree modes are more coherent across the entire profile and lead to large integrated effects despite small local amplitudes. High-degree modes of larger amplitude do not contribute significantly to RV variations.

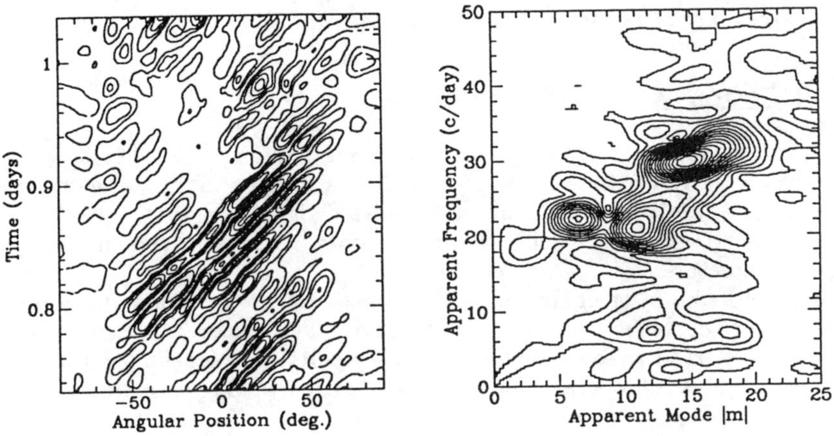

One cannot simply translate this 2-D transform directly into mode identifications because the spectral window, temporal and spatial harmonics and possible mode interactions must first be taken into account. Note also that the "apparent frequency" includes effects of the star's rotation, while the "apparent mode" is merely a transformation assuming that $|m| = \ell$. We are currently investigating the detailed properties of the 2-D transform to obtain more definitive mode values for τ Peg.

Similar data are being analysed for two other intriguing δ Scuti variables: θ^2 Tau and 97 Tau.

REFERENCES

Breger, M. 1991, *Communications in Asteroseismology* **29**, 1

Breger, M., Garrido, R., Huang Lin, *et al.* 1989, *Astron. Astrophys.* **214**, 209

Kennelly, E.J., Yang, S., Walker, G.A.H., and Hubeny, I. 1992, *PASP* **104**, 15

Michel, E., Belmonte, J.A., Alvarez, M., *et al.* 1990, in *Confrontation Between Stellar Pulsation and Evolution*, eds. C. Cacciari and G. Clementini (ASP Conf. Series **11**), p. 332.

Walker, G.A.H., Yang, S. and Fahlman, G.G. 1987, *Ap.J.* **320**, L139

Inside the Stars, IAU Colloquium 137
ASP Conference Series, Vol. 40, 1993
Werner W. Weiss and Annie Baglin (eds.)

OBSERVATIONAL CONSTRAINTS ON MODE EXCITATION IN δ-SCUTI STARS IN OPEN CLUSTERS

S. FRANDSEN AND H. KJELDSEN
Institute of Physics and Astronomy, Århus University, Denmark

SEISMOLOGY WITH δ-SCUTI STARS

The special opportunities offered by δ-Scuti stars are the following:
- They regularly oscillate in several modes, and with high precision observations one has measured of the order 5 modes (Michel and Baglin 1991).
- The excitation is a very delicate and therefore a very sensitive measure of the internal structure.
- They are found in open clusters which constitute a very rich scenario for tests of the theory of stellar structure and evolution. This makes it feasible to observe several stars simultaneously and gain more results than for single stars.
- The brightness is moderately high and still they live long enough to be present in a variety of clusters and in the field as well.

EXCITATION PROBLEMS

Some of the questions we would like to see answers to are the following:
- When do modes become excited? What is the relation between excitation and rotation and how does the chemical composition influence excitation?
- If modes are excited, why exactly the few modes we see among the large set of possible modes?
- Why do we see amplitude changes in some stars?

THE OLD OPEN CLUSTERS

In this paper we display results based on an analysis of 5 old open clusters. We have used high precision time series CCD photometry to observe a large number of stars of which many of the brighter are situated in the instability strip. The clusters included are: NGC 1513, NGC 2423, NGC 2660, NGC 6134 and NGC 6802. The technique has been gradually refined during the project and we have a fairly inhomogenous material for the various clusters. The main achievement can be described in two paragraphs:
a. The number (and probably type) of variables found seem to vary from cluster to cluster. This may serve as a probe of the internal constitution of the stars (needs to be confirmed).
b. In one case (NGC 6134) we do have data enough to make an individual analysis of the variables. As shown elsewhere this leads to a precise knowledge of the modes excited, except that proper models of the stars do not exist yet.

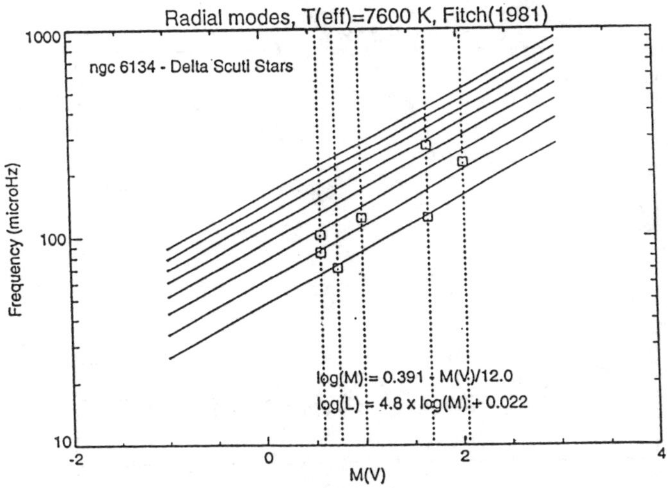

FIGURE I Figure demonstrating how modes may be identified when observing stars in a cluster. There is a freedom to move the points in the horizontal direction to fit the bottom line which is the fundamental frequency at a given luminosity.

TECHNIQUES FOR MODE IDENTIFICATION

The mode spectrum for an evolved δ-Scuti star is very dense and very complicated (Dziembowski and Krolikowska 1990). During evolution the p-modes switch character and change into g-modes, which complicates the analysis and the identification very much. The tools suggested to be used for mode identification are:
- First of all long time series with few missing windows to resolve the frequency spectrum and to allow precise frequency determination.
- Time series of the colour of a variable to study phase shifts, which depend on the l-value of the mode.
- Doppler imaging of the surface in fast rotators to identify the surface pattern characteristics of modes.
- Differential analysis of variables in cluster (see next)

A NEW SCHEME FOR MODE IDENTIFICATION

1. In a cluster we can fix accurately the positions of the stars in a colour-luminosity diagram except for a zero point on each axis.
2. For simplicity we assume for NGC 6134 that the 5 stars have the same effective temperature, which is nearly the case. We also assume the simple models computed by Fitch (1981) are adequate. Some of the stars oscillate in the fundamental mode. Fitting these frequencies to the theoretical we can determine the zero point for the luminosity precisely (see Figure I). Now all overtones can be identified by reading them off the figure, f.ex. as being a p_5 mode. With a more numerous set of variables the chances are that also a metal content and an age can be determined due to the constraints imposed by a large set of fre-

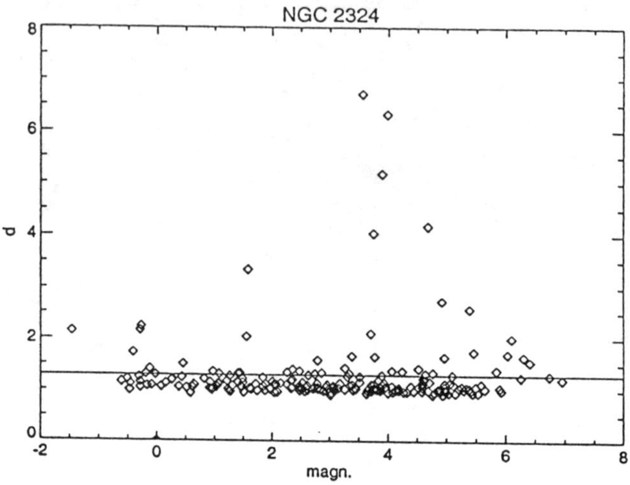

FIGURE II NGC 2324. There are a small number of variables, but the time string is too short to allow an analysis of the variables detected. Note once more the large number of constant stars.

quencies. Also approximate values for these parameters are known from classical techniques used in clusters, which provide a first good estimate.

THE SEARCH FOR VARIABILITY

We search for variables by calculating a quantity *d*. This is the ratio between the r.m.s. scatter around the mean of the time series of magnitudes for a given star divided by another r.m.s. scatter obtained by taking differences between successive measurements.

To search for variables we plot *d* against magnitude and then take a note of all stars above a certain value of *d* depending on the number of frames available for a given cluster. In each diagram the brighter stars have been through a selection where stars affected by CCD defects or other disturbing sources have been rejected. An example is given as Figure II. The fraction of variables detected is very small compared to expected (30% for field stars) in most clusters.

REFERENCES

Breger, M. 1979, Publ. Ast. Soc. Pac. **91**, 5

Dziembowski, W. and Królikowska, M. 1990, Acta Astron. **40**, 19

Fitch, W.S. 1981, Astroph. J. **249**, 218

Michel, E. and Baglin, A. 1991, Adv. Space Res. **11**, 70

Inside the Stars, IAU Colloquium 137
ASP Conference Series, Vol. 40, 1993
Werner W. Weiss and Annie Baglin (eds.)

CCD ASTEROSEISMOLOGY OF OPEN CLUSTERS

INMACULADA VIDAL, JUAN A. BELMONTE
Instituto de Astrofísica de Canarias, c/Via Láctea s/n 38200, La Laguna, Tenerife, Spain.

In the last few years Asteroseismology has contributed to the knowledge of stellar interior, in some extend. This contribution has been done with - observations, providing the characteristics of the acuostic modes of stellar oscillations; and with theory relating them to inner parameters, locating stars in the JCD diagram (Christesen-Dalsgaard J., 1989). At the beginning only classical pulsating stars had high enough amplitudes able to be detected. But techniques have been improved and now the goal is the detection of oscillations in all kind of stars, specially solar-like ones. Up to now, few cases of possible detections have been reported, the best evidences are: Belmonte *et al*, (1990) in an FV star, Brown *et al* (1991) on the subgigant Procyon, and Pottasch *et al* (1992), in α Cen.

In order to reach precision high enough, CCD photometry has been used in the last few years, as an alternative and more powerful way. This new technique has already been applied by other groups such as R. Gilliland and T. Brown in the USA (Gilliland *et al*,1991) and S. Frandsen and H. Kjeldsen in Denmark (Kjeldsen and Frandsen, 1992) with promising results. Our group has two data series one of NGC6802 (observed with the 2.5m INT at ORM, Canary Islands, Spain), and the other of a δ Scuti, BQCnc, in Praesepe cluster (observed with the 1.m JKT at the same observatory). With the last one, we expect to compare limits and characteristics between CCDs and photometers. At present we are finishing up with the reductions of NGC6802. In order to obtain the light curves of stars in the field, we have used aperture photometry (only when the crowdness of the field is not very high) and a more general PSF fitting. Some initial results on NGC6802 analysis are presented in this report.

Besides, with a 2.5 meter telescope, observing a 15th magnitude star, the noise level is dominated by counting noise. The selected stars range of magnitudes is 15 to 17, and one 13th magnitude star, so photon noise will be the most important component of noise in spectra.

REFERENCES

Belmonte J.A., Pérez Hernández F. and Roca Cortés T., 1990. *Astron. Astr.* **231**, 383.

Belmonte J.A., Chevreton M., Mangeney A., Praderie F., Saint-Pé O., Puget P., Alvarez M., and Roca-Cortés T., 1991. *Astron. Astr.* **246**, 71-83.

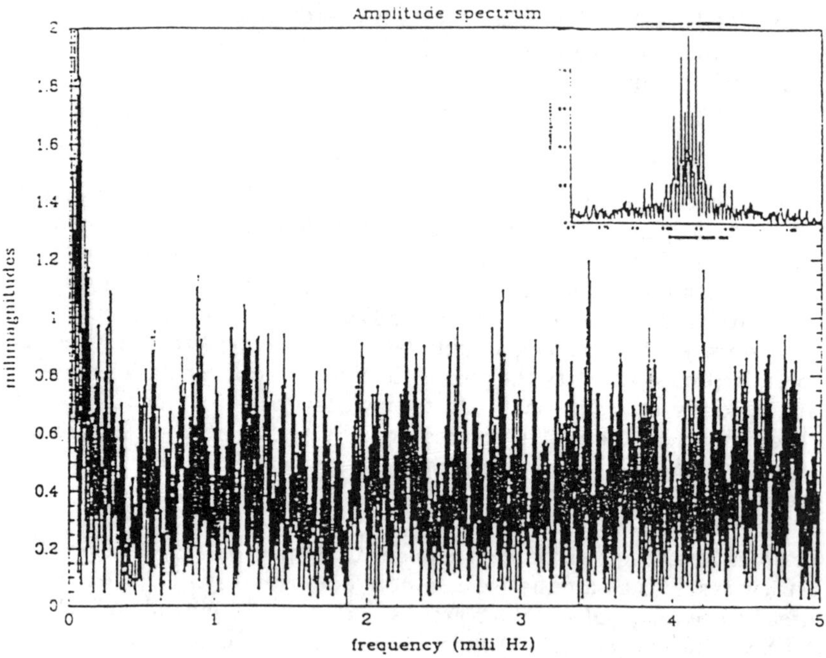

FIGURE I Amplitude spectrum of a 15th magnitude star in NGC6802. Residual series are tipycally 6 hours long, with an exposure time of 50 sec. It has been obtained via an iterative sinewave fitting procedure (Belmonte *et al*, 1991) applied to five series (9,10,11,12,14 June 1990). Noise level is about 530 micromagnitudes, althouhg expected peaks for solar-like stars are well below this level, (around 50 μmag for the best expectations) still some information could be obtained from this kind of spectra. Spectral window is plotted as well. The gaps in data make it wide, so possible peaks are dificult to detect.

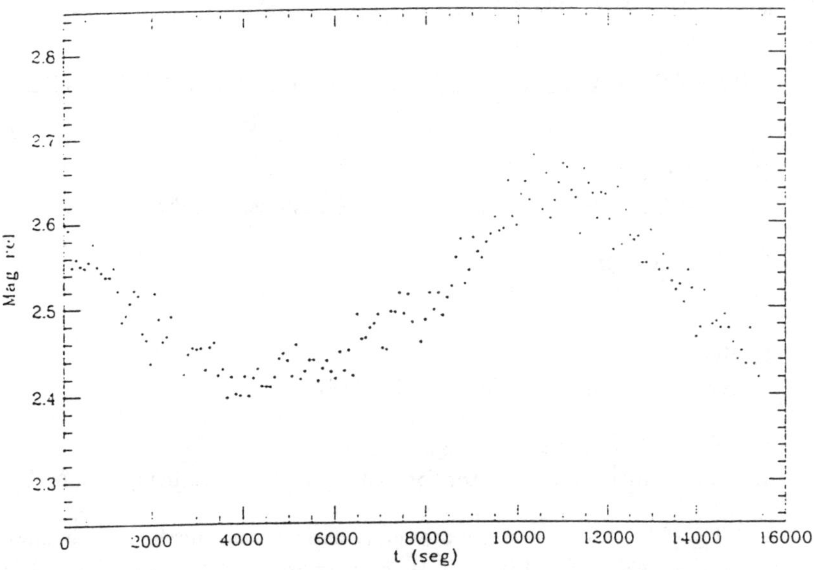

FIGURE II Light curve of a variable star found in NGC6802. From its amplitude and aproximate period we suppose it is a W UMa variable. This curve has been obtained with the computer of Nordic Optical Telescope, at the ORM, using Hans Kjeldsen's reduction programs, called MOMF, which are based on a combine PSF fit and aperture photometry algorithms.

Brown T.W., Gilliland R.L., Noyes R.L. and Ramsey R.W., 1991. *Ap. J.*, **368**, 599.

Christensen-Dalsgaard. J., 1989. *'Lecture Notes on Stellar Oscillations'* Univ. de Aarhus.

Gilliland R.L., Brown T.W., Duncan D., Suntzeff N.B., Schild R., Jeffrey W.A. and Penprase B. , 1991. *Astron. J.*, **101**, 541.

Kjeldsen H. and Frandsen S., 1992. *P.A.S.P., in press.*

Pottasch E.M., Butcher H.R. and van Hoesel F.H.J., 1992, *submitted.*

Stetson P.B. 1987. *P.A.S.P.* **99**, 191.

Inside the Stars, IAU Colloquium 137
ASP Conference Series, Vol. 40, 1993
Werner W. Weiss and Annie Baglin (eds.)

THE AFOE—A NEW INSTRUMENT FOR ASTEROSEISMOLOGY

R. W. NOYES
Harvard-Smithsonian Center for Astrophysics, Cambridge, MA, USA

T. M. BROWN
High Altitude Observatory, Boulder, CO, USA

S. HORNER
University of Chicago, Chicago, IL, USA

S. KORZENNIK, P. NISENSON
Harvard-Smithsonian Center for Astrophysics, Cambridge, MA, USA

ABSTRACT We describe technical aspects and performance characteristics of an Advanced Fiber Optic Spectrograph (AFOE), planned for the Mt. Hopkins 1.5 m telescope for asteroseismology of sun-like stars.

Determining the oscillation spectra of sun-like stars by Doppler techniques requires measuring velocity fluctuations to better than 1 m s^{-1} within about one night's observation. Straightforward considerations (*e.g.* Brown 1990) imply that the photon noise-limited precision attainable by measuring a single spectrum line is approximately $\delta v_{rms} \simeq cw/[\lambda d(N_{pix}I_c)^{1/2}]$, where δv_{rms} is the rms Doppler noise, c, w, d, and λ are the speed of light, the apparent linewidth (including instrumental broadening), the relative line depth, and the line wavelength, N_{pix} is the number of detector pixels spanning the line, and I_c is the continuum intensity measured in photoelectrons per pixel. For parameter values that are typically found in bright-star spectroscopy ($\lambda = 500$ nm, $w = 0.01$ nm, $d = 0.6$, $N_{pix} = 3$, $I_c = 10^5$), one obtains $\delta v_{rms} = 11$ m s^{-1}. For bright stars and moderate-sized telescopes, an exposure I_c can be reached in a few minutes.

This level of precision is still inadequate for seismology of sun-like stars. The Advanced Fiber Optic Echelle (AFOE) spectrograph achieves a substantial gain in precision through several approaches: (1) We increase N_{pix} by increasing the number of lines measured. This is achieved with a cross-dispersed echelle, coupled with a large-format (2048^2 15-μ pixel) CCD. (2) We minimize the apparent linewidth w by maximizing the spectral resolution. Spectral resolution must be traded off against spectral coverage; this is optimized for maximum S/N when observing slowly-rotating Sun-like stars. This results in coverage in one exposure of about 55% of the wavelength range between 470 and 610 nm, with

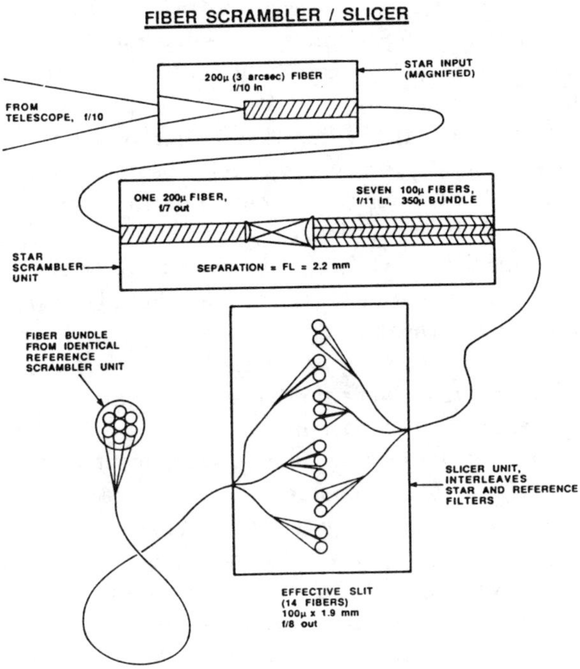

Figure 1: The Fiber Scrambler/Slicer receives the f/10 star image in a 200-μ (3 arcsec) fiber [*top*] and relays it to the star-scrambler unit [*center*], where complete scrambling is effected by interchanging positional and angular coordinates of the star image (see text), and where the light is re-imaged at f/11 into a bundle of seven 100-μ fibers. An identical arrangement is provided for the reference image. The total of 14 100-μ fibers are then brought to the AFOE and interleaved [*bottom*] along the entrance slit, where they produce an f/8 effective slit image. The two input fibers may be switched between star and reference to difference out any residual dissimilarities between the two optical paths.

a resolution of 58000. (3) We maximize I_c for a given exposure time by use of a fiber-optic image slicer to avoid slit losses, even in indifferent seeing; the image slicer(see Figure 1) maps the 200 μ (3 arcsec) telescope entrance aperture onto a 100 μ wide spectrograph entrance slit. We also use high-reflectance optical coatings wherever possible. The peak efficiency of the system (including telescope transmission, fiber coupling, spectrograph efficiency, and detector QE) is about 1.5%. This system on a 1.5m telescope should allow Doppler measurements with photon-limited precision of 80 cm s^{-1} in a 60s exposure on a V=0 G-type star (*i.e.* 50 cm s^{-1} in 5 hours total integration on a similar V=5 star). We have already obtained near photon-limited precision with the Penn State Fiber Optic Echelle (FOE) on the KPNO 2.1m telescope (Brown *et al.* 1991, Brown 1990) so we are confident that the expected level of precision is realistic.

A 1 m s^{-1} Doppler shift represents a motion of the spectrum on the detector of only 3×10^{-4} of a 15-micron pixel. Not only must the precision of individual Doppler measurements permit detecting such shifts, but measurement stability must be maintained to this level over times significantly longer than the period of the oscillation modes being measured. Our approach is to provide as great mechanical stability as practical in the AFOE, and to use continuous wavelength calibration to correct for residual instrumental drifts.

To maximize mechanical stability, the spectrograph will be mounted on a vibration-isolated optical table, connected to the telescope by an optical fiber, and enclosed in a thermally-controlled housing within a dedicated room. The CCD detector will be continuously cooled thermoelectrically, to avoid transient behavior accompanying filling with cryogens.

To calibrate out residual instrumental drifts, the AFOE incorporates two distinct continuous calibration methods: (1) An iodine vapor absorption cell may be inserted into the beam of starlight, imposing a spectrum of thousands of molecular lines which act as a well-defined wavelength reference (*e.g.* Marcy and Butler 1992); (2) A separate optical fiber will bring light from a reference emission-line source into the spectrograph adjacent to the stellar input fiber. To minimize effects of different optical paths through the spectrograph, the fiber image slicer (Figure 1) will interleave the stellar and reference light within each echelle order. Any remaining differences may be calibrated out by periodically interchanging the fibers carrying stellar and reference light.

One important potential source of noise in Doppler measurements with the AFOE is variable illumination of the collimator due to imperfect seeing or guiding. This problem is not completely eliminated by the scrambling action of a single optical fiber, since high-quality fibers do not alter the off-axis angle of light rays that pass through them. To solve this problem, the AFOE incorporates a double fiber scrambler (Brown, 1990), in which a pair of fibers is coupled together using a pair of microlenses separated by their common focal length; the off-axis angle in the first fiber is then mapped into radial position on the face of the second, and subsequently scrambled within the second fiber. The double scrambler will be incorporated into the fiber image slicer (see Fig. 1), to minimize light losses.

The Sun is a useful object for calibrating the AFOE precision and stability, since it is very bright and has a very well-known radial velocity relative to the Earth. To allow regular observations of the Sun, a heliostat and optical fiber feed will bring unimaged sunlight into the spectrograph in the same configuration and brightness as a $V \approx -5$ star. Solar observations will: provide an independent stellar wavelength standard; allow characterization of pure instrumental drifts (since the high signal level will virtually eliminate photon noise); and through detection of solar p-modes provide a known calibration source for stellar oscillation mode amplitudes.

REFERENCES

Brown, T.M. 1990, in *CCD's in Astronomy*, A.S.P. Conference Series **8**, p. 335.

Brown,T.M., Gilliland, R.L., Noyes, R.W., and Ramsey, L.W. 1991, *Ap. J.* **368**, 599.

Marcy, G.W., and Butler, R.P. 1992, *P.A.S.P.* **104**, 270.

Inside the Stars, IAU Colloquium 137
ASP Conference Series, Vol. 40, 1993
Werner W. Weiss and Annie Baglin (eds.)

SOHO: A PRISMA PRECURSOR

P. LEMAIRE
I.A.S., Bat.121, Université de Paris XI, 91405 Orsay Cedex

ABSTRACT

Probing the interior of the Sun, a slow rotator G2 V main sequence star, and connecting the atmospheric structure to inside phenomena is one of the objectives of the SOHO mission. An overview of the helioseismology and of the coronal instruments is given. It furnishes a guide line for missions that are dedicated to the probing of stellar interiors in using microvariability and activity as tracers, such as the PRISMA study.

INTRODUCTION

SOHO (SOlar and Heliospheric Observatory) is an ESA/NASA solar mission to be launched in 1995 at the L_1 Sun-Earth Lagrangian position. Continuous observations will be performed during more than 2 years.

The mission objectives are:

- to infer the internal solar structure through helioseismology sounding technique. Continuous observations of the global doppler velocity oscillation and the global irradiance variation will provide informations on the deep core from oscillation modes l = 0, 1 and 2. Coarse resolution on the solar disc in irradiance gives access to intermediate levels. High resolution on the solar disc in doppler velocity is a probe of the upper part of the internal structure where the convection shapes and permanently disturbs the structuration.
- to observe the outer part of the solar atmosphere and provide criteria in order to infer coronal heating mechanisms and processes that impulse and drive the solar wind. Combining the solar global view from the chromosphere to the corona with detailed analysis of resolved structures will constrain any theoretical consideration. Observations on and outside the solar disc will draw the complete (three dimensions) outline of the coronal structuration.
- to make in-situ measurement of the interplanetary plasma.

HELIOSEISMOLOGY SEGMENT

Global doppler velocity measurement (GOLF- Gabriel et al, 1989)
o use of a sodium resonance cell to provide accurate absolute wavelength reference

o full Sun doppler velocity and global magnetic field measurements
o expected velocity accuracy: 1 mm/s in the 10^{-7} - 6. 10^{-3} Hz frequency range
o oscillation modes l = 0 - 4
o continuous observations during 2 years

Solar Oscillation Imager (SOI - Scherrer et al, 1989)
o Michelson Doppler Imager instrument
o full solar image analysis: line of sight velocity, transverse velocity, line and continuum intensity, and detailed magnetic field
o 2 and 0.7 arcsec pixels (1920 arcsec solar diameter)
o line of sight velocity sensitivity: < 1 m/s
o oscillation modes up to l = 1000
o continuous observation 2 months/year (partial otherwise)

Solar monitor experiment (VIRGO - Fröhlich et al, 1989))
o active cavity radiometers, photometers and low resolution imager
o frequencies, amplitude and phases of intensity oscillations mode in the frequency range 10^{-7}- 10 mHz
o sensitivity better than 10^{-7} magnitude
o oscillation modes l = 0 - 7
o continuous observations during 2 years

Combining the helioseismology segment provides:
o high accuracy determination of modes in the range l = 0 to 1000 measured in doppler velocity
o high accuracy determination of modes in the range l = 0 to 7 measured as brightness variations. Comparison between velocity and brightness oscillations gives further constraints on the generation and the propagation of oscillations.
o a full set of measurements to derive internal structure and dynamics of the solar interior.

SOLAR ATMOSPHERIC SEGMENT

Full disc Extreme ultraviolet Imager (EIT - Delaboudiniere et al, 1989)
o Multi-wavelength imager at 17.1, 19.5, 28.4 and 304 nm in upper chromosphere and coronal lines
o 3 arcsec pixels
o full solar disc field-of-view

Coronal Diagnostic Spectrometer (CDS - Patchett et al, 1989)
o grazing incidence telescope with a grazing incidence spectrometer (15 - 80 nm spectral range) and a normal incidence spectrometer (2 bands in 24 - 80 nm range)
o resolving power 2000 - 9000 ($\lambda/\Delta\lambda$)
o angular resolution ~ 2 arcsec
o capability of small field scanning (240 x 240 arcsec) and wide field (up to the full solar disc and .5 solar radius outside)
o diagnostic of spectral lines in the 10^4- few 10^6 K range

Ultra violet spectrometer (SUMER - Wilhelm et al, 1989)
o normal incidence telescope and spectrometer in the 50 - 160 nm spectral range
o resolving power: 20000 - 40000 ($\lambda/\Delta\lambda$)
o angular resolution ~1 arcsec
o capability of small field scanning (300 x 300 arcsec) and large field scanning (full solar disc and .5 solar radius outside)
o detailed profile of spectral lines in the 10^4-10^6 K temperature range

UltraViolet Coronograph (UVCS - Kohl et al, 1989)
o scanning normal incidence coronograph from 0.2 to 9 solar radii above the limb with spectrometric observations in the 49.5- 61. nm and 102.- 125. nm wavelength bands.
o resolving power 500 - 9000 ($\lambda/\Delta\lambda$)
o angular resolution from 10 arcsec to arcmin
o diagnostic of density, velocity and temperature

White light coronograph (LASCO - Michels,Brueckner et al, 1989)
o 3 complementary coronographs extending the measurements within fields of 0.1- 2.0 (C1), 0.5 - 5.0 (C2), 2.0 - 29.0 (C3) solar radii above the solar limb
o angular resolution (pixel size) 5.6" (C1), 11.4" (C2) and 56" (C3)
o measurements of coronal structure and evolution.

Combining coronal segment provides:
o a global and detailed view of the corona
o long term and short term evolutions of coronal structures
o constraints on heating mechanisms and acceleration of the solar wind
o rotational rate at several heights above the limb
o access to detailed processes acting in the small scale structures in quiet and active solar area

CONCLUSION

Combining the heliosismology and atmospheric segments, SOHO will provide:
o a complete sounding of the Sun from the deep core to the most external layers
o a unique tool to test stellar theories at a given age, chemical composition, mass and luminosity
o a set of fine diagnostics on small area of the Sun and on the full Sun

REFERENCES

THE SOHO MISSION , Scientific and technical aspects of the instruments, ed.V. Domingo, ESA SP-1104, 1989

Inside the Stars, IAU Colloquium 137
ASP Conference Series, Vol. 40, 1993
Werner W. Weiss and Annie Baglin (eds.)

E.V.R.I.S. ON BOARD MARS 94: THE FIRST SPACE EXPERIMENT DEVOTED TO STELLAR SEISMOLOGY.

A. BAGLIN[1], W. WEISS[2], G. BISNOVATYI-KOGAN[3].
[1] DASGAL, URA CNRS 335, Observatoire de Paris, 92195 Meudon Cedex, France.
[2] Institut für Astronomie, Universität Wien, Wien, Österreich.
[3] IKI, Moscou, Russia.

ABSTRACT The EVRIS experiment is an exploratory mission devoted to stellar seismology. It will observe approximately ten bright stars, for 20 days each, during the cruise of the Russian "MARS 94" mission. The photometer will be able to detect amplitudes of modes as small as a few 10^{-6} magnitude.

SCIENTIFIC OBJECTIVES.

The generalisation of helioseismology to stars is a challenge of primary importance for basic stellar physics. The rationale for such an aim has already been developed several times (i.e. Hudson et al. 1986, Praderie et al. 1989, Baglin, 1990). Looking at nearby stars will provide information on different stellar structures and evolutionary phases, for instance, on the structure of convective cores which governs the age of stars more massive than the Sun, on the laws of rotation in faster rotators and on the transfer of angular momentum processes. But the major difficulties when going from the Sun to stars is the lack of photons and the lack of spatial resolution.

WHY A SPACE MISSION?

The need to go to space has already been extensively documented (see i.e. Mangeney and Praderie 1984, Hudson et al. 1986, Baglin 1990, Weiss 1992). Two methods are available to observe faint stellar oscillations (Harvey, 1989). The spectroscopic one can be used from the ground, but it needs extremely large telescopes for long periods, and in any case, it is limited to slowly rotating objects. Indeed, the aforementioned scientific objectives cannot be achieved with this technique. The photometric approach suffers from atmospheric absorption and fluctuations, and even with recent CCD developments (see i.e. Frandsen, this conference), the threshold of 10^{-6} cannot yet be reached from the ground. From space, if the trajectory is sufficiently clean, the stability of the environment permits to reach the accuracy of the photon noise.
Since the first proposals in 1981, our philosophy has been to design an experiment as simple as possible, which could be launched as soon as possible, as an

exploratory mission. The tremendous success of the IPHIR experiment on the PHOBOS mission has been a strong encouragement and an excellent guide.

INSTRUMENT SPECIFICATIONS.

A compromise has to be made between the necessity of collecting as many photons as possible and the weight and cost of the mission. The condition to detect a coherent signal in a pure flux of photons is:

$$N_p > A^{-2} \tag{1}$$

where A is the relative amplitude, N_p the number of photons counted per second by the detector and $q = \min(T, t)$ with T the total duration of the record, and t the life-time of the signal. Using the solar values as references (A being a few 10^{-6} and t a few days), condition (1) implies $N_p > 10^6$.

The International Scientific Committee for the MARS 94 mission has accepted a small asteroseismology experiment of a total weight of less than 7 kg., installed on a pointing platform, operated during cruise. These constraints have lead to the following choices:

* a diameter of the pupil of 7 cm, a large bandpass of the order of 2000 Å, centered in the visible region, and a total efficiency of the chain of detection of 0.1.; values which limit the investigation to objects brighter than $m_v \approx 3.5$;

* a large diaphragm (radius 8') to include the diffraction pattern at an accuracy of 10^{-5};

* a photometer with a small PM (Hamamatsu) working in counting mode.

TARGETS.

The constraints are numerous. They apply essentially to:

* the visual magnitude, as already discussed; if we accept to lower the detection threshold then some fainter objects could be observed;

* the luminosity class (main sequence or subgiants), to keep simple objects and time scales of pulsations around a few hours and less;

* binaries, only with a much fainter component, to avoid confusion;

* the visibility from the spacecraft; as the instrument must point towards the antisolar direction, it is installed below the solar panels, and only one half a hemisphere will be visible;

* the surrounding sky, which has to be as clean and as well known as possible, to reduce the external noise generated by the parasite sources entering and leaving the diaphragm due to pointing fluctuations.

Taking into account all these limitations, an interesting set of targets is available (Fig.1), with a wide variety of masses, ages, chemical compositions and states of rotation.

OBSERVING STRATEGY.

We will observe first a known variable to check the instrument, then stars of different types. If after 5 days, no signal is obtained on a target, the observation

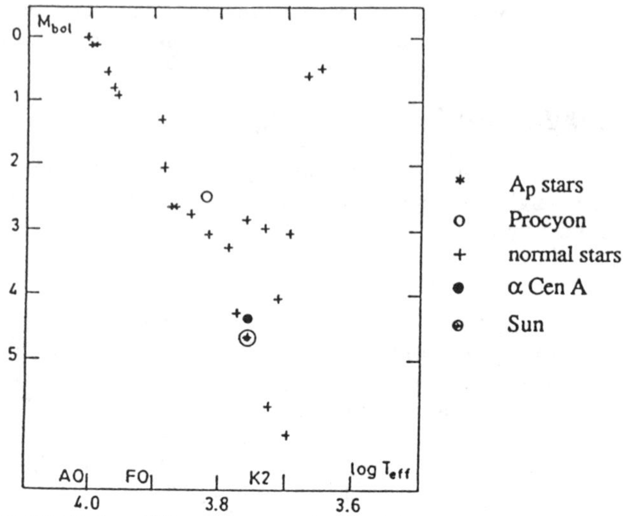

FIGURE I HR diagram of possible targets.

will be stopped and another object will be pointed; on the contrary, if the spectrum is very rich, the decision can be made to continue longer on the same target. Results will be obtained almost in real time, every 5 days during the telemetry sequences. Reductions will be made independently by the French, the Austrian and the Russian teams, and then the results will be made available to the scientific community.

ACKNOWLEDGMENTS

We want to thank all Co-investigators of EVRIS and the whole technical team that is now manufacturing the experiment and its pointing platform.

REFERENCES

Baglin, A. : 1990, *Solar Phys.*, **133**, 155.
Baglin, A.: 1991, *Adv. Space Res.*, **11**, 133.
Frandsen, S. : 1992, this conference
Harvey, C. : 1988, in *Advances in Helio-and Asteroseismology* (J. Christensen-Dalsgaard and S. Frandsen eds.), p. 497.
Hudson, H.S., Brown, T.M., Christensen-Dalsgaard, J., Cox, A.N., Demarque, P., Harvey, J.W., Mc Graw, J.T., Noyes, R.W.: 1986, A concept for an Asteroseimology Explorer. Proposal submitted to NASA.
Praderie, F., Mangeney, A., Lemaire, P., Puget, P., Bisnovatyi-Kogan, G. : 1988, in *Advances in Helio-and Asteroseismology* (J. Christensen-Dalsgaard and S. Frandsen eds.), p. 549.
Mangeney, A., Praderie, F. : 1984, *Proceedings of the workshop on Space Prospects in Stellar Activity and Variability*, Paris Observatory Press.
Weiss W. : 1992, this conference.

Inside the Stars, IAU Colloquium 137
ASP Conference Series, Vol. 40, 1993
Werner W. Weiss and Annie Baglin (eds.)

WAVELET ANALYSIS OF SOME PULSATING STARS

KÁROLY SZATMÁRY and JÁNOS GÁL
Dept. of Exp. Physics, JATE University
H-6720 Szeged, Dóm tér 9, Hungary

The detection of a variable period is very important as it gives information on the evolutionary state or on the binary nature of a pulsating star. If a pulsator is moving in a binary system, its light curve is frequency-modulated by the orbital period (light-time effect). Mode switching or chaos also may be the cause of the changes in the amplitude and period of the light variations.
The determination of period variability is difficult. Usually the conventional Fourier-spectrum cannot give any information about a possible period variation. The O-C diagram is useful to detect a variable period, but it does not give the amplitude and phase values.
The wavelet analysis (Grossmann et al. 1989), which is local both in time and frequency space seems well suited for pulsating stars, especially if one or more periods vary in time. For the application of this method is necessary a long and almost continuous observation. The shortest period variables (white dwarfs, see Goupil et al. 1990; and roAp stars) and the long period pulsators (SR and Mira stars) are the most suitable objects to the wavelet analysis.
The wavelet analysis is a new procedure in astrophysics therefore we have investigated how to use it for detecting of changes of the stellar pulsation. We have generated some test data series, which are connected with interesting phenomena (eg. frequency modulation, amplitude modulation, phase jump, mode switching). In these cases the conventional Fourier analysis of the whole data set cannot give required information. The wavelet maps of these tests suggest that the wavelet analysis is a useful method to determine of variable phenomena of stellar pulsation.
We demonstrate the difference between the Fourier-spectrum and the wavelet map in the case of the semiregular stars Y Lyn (Szatmáry and Vinkó 1992) and Z UMa.

DISCRETE FOURIER TRANSFORMATION

The power spectrum for the frequency f_k:

$$F(f_k) = \left(2/N\ C_{jk}\right)^2 + \left(2/N\ S_{jk}\right)^2,$$

where N is the number of data; $m(t_j)$ is the magnitude at the time t_j, and

$$C_{jk} = \sum_{j=1}^{N} m(t_j)\ cos(2\pi f_k t_j), \qquad S_{jk} = \sum_{j=1}^{N} m(t_j)\ sin(2\pi f_k t_j).$$

DISCRETE WAVELET TRANSFORMATION

The power spectrum for the frequency f_k:

$$W(f_k, \tau) = f_k \left(C^{\tau}_{jk}\right)^2 + f_k \left(S^{\tau}_{jk}\right)^2,$$

where τ is the time shift and

$$C^{\tau}_{jk} = \sum_{j=1}^{N} m(t_j)\ cos(2\pi f_k t_j)\ exp\left[-[(t_j - t_0 - \tau) f_k]^2/2\right],$$

$$S^{\tau}_{jk} = \sum_{j=1}^{N} m(t_j)\ sin(2\pi f_k t_j)\ exp\left[-[(t_j - t_0 - \tau) f_k]^2/2\right].$$

The window is a gaussian with a halfwidth of f_k and it is shifted with τ. t_0 is the first time of the observation.

REFERENCES

Goupil, M.J., Auvergne, M. and Baglin, A. 1990, in *Confrontation between Stellar Pulsation and Evolution*, ed. C. Cacciari and G. Clementini, Astron. Soc. of the Pacific, Conf. Series Vol. 11, p. 578.

Grossmann, A., Kronland-Martinet, R. and Morlet, J. 1989, in *Wavelets: Time-Frequency Methods and Phase Space*, ed. J.M. Combes, A. Grossmann and Tchamitchian Ph., Springer, p.2.

Szatmáry, K. and Vinkó, J. 1992, *MNRAS* (in press)

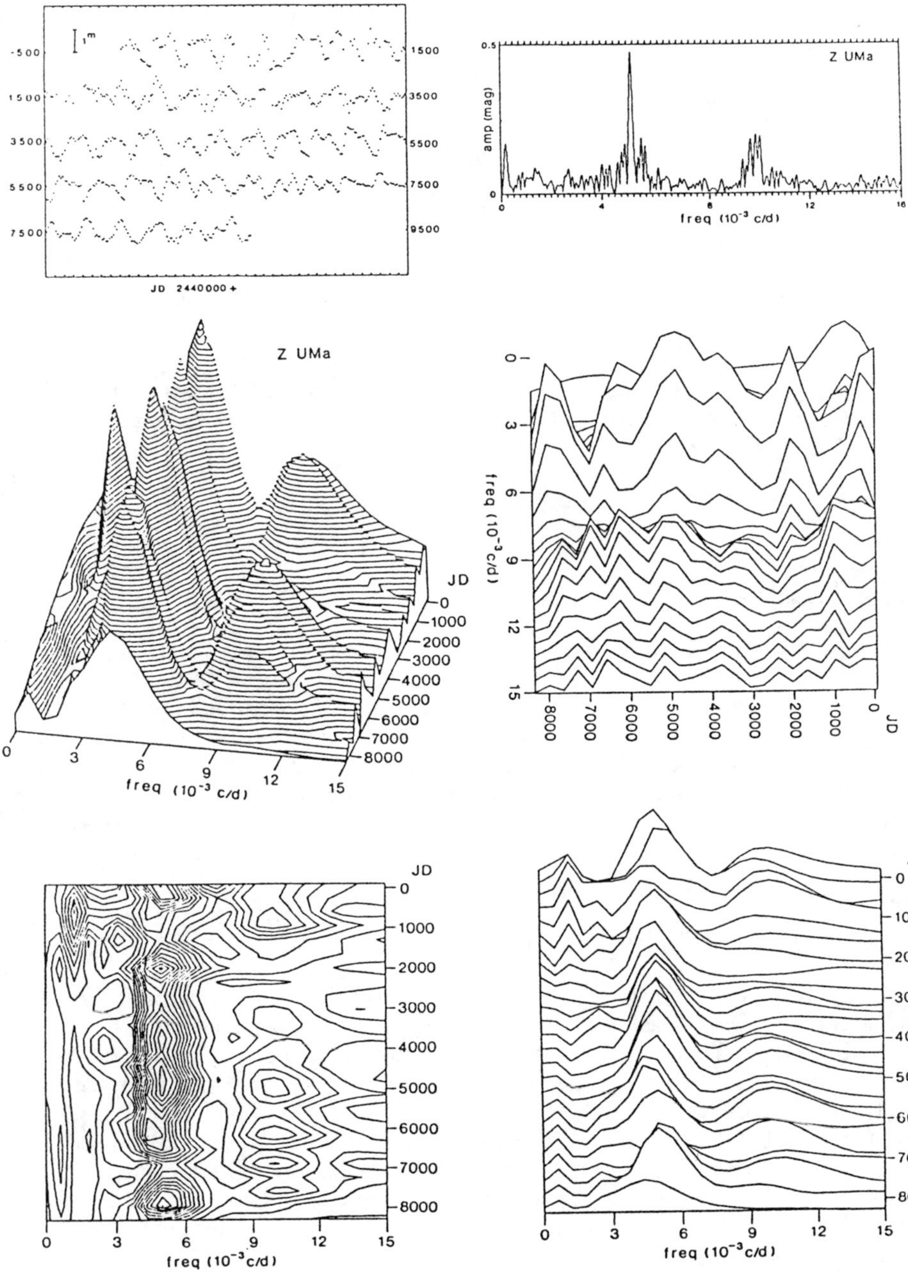
1m
-500
1500
3500
5500
7500
1500
3500
5500
7500
9500
JD 2440000+
Z UMa
amp (mag)
freq (10^-3 c/d)
Z UMa
JD
freq (10^-3 c/d)
freq (10^-3 c/d)
JD
JD
freq (10^-3 c/d)
JD
freq (10^-3 c/d)

Inside the Stars, IAU Colloquium 137
ASP Conference Series, Vol. 40, 1993
Werner W. Weiss and Annie Baglin (eds.)

DATA ANALYSIS TECHNIQUES FOR DETERMINING HIGH PRECISION DOPPLER SHIFTS USING IODINE ABSORPTION CELLS

S.D. HORNER
Department of Astronomy and Astrophysics, The University of Chicago and High Altitude Observatory, National Center for Atmospheric Research [1]

T.M. BROWN
High Altitude Observatory, National Center for Atmospheric Research [1]

ABSTRACT Some details of the data analysis used to determine high precision Doppler shifts from data obtained using an absorption cell are briefly discussed.

INTRODUCTION

High precision Doppler measurements require stable wavelength reference due to instabilities inherent in spectrographs. One solution is to use an absorption cell placed in the optical path of the starlight (cf. Campbell and Walker 1979, and Marcy and Butler 1992). The reference spectrum is then subject to the same temporal changes in the spectrograph that cause shifts of the stellar spectrum on the detector. If an iodine absorption cell is used as the wavelength reference source for observations, a precision of $\sim 10\ m\ s^{-1}$ can be obtained (Marcy 1991).

DATA ANALYSIS PROCEDURE

If a stable wavelength reference can be established, the principal difficulty with measuring Doppler shifts at the meter–per–second level is that the spectral lines are slightly undersampled, and the shifts must be measured to the 0.01 or 0.001 pixel level. The conventional wisdom is that the optimized spectrograph PSF should result in spectral lines with a FWHM of 2 pixels. Line profiles of this width contain small but significant components at spatial frequencies higher than the detector's sampling frequency. These components lead to problems in interpolating the line, which in turn result in shift errors that are roughly sinusoidal with shift. This problem is more serious for the

[1] The National Center for Atmospheric Research is Sponsored by the National Science Foundation

iodine than for stellar spectrum, because of the iodine's smaller intrinsic line width.

Initially, we attempted to analyze the data by cross–correlating the observations of the program stars with the I_2 spectrum superimposed, against the I_2 spectrum *alone*, then against the stellar spectrum *alone*. The blending of the two superimposed spectra resulted in large errors in determining the shift, as Figure 1 shows using artificial data.

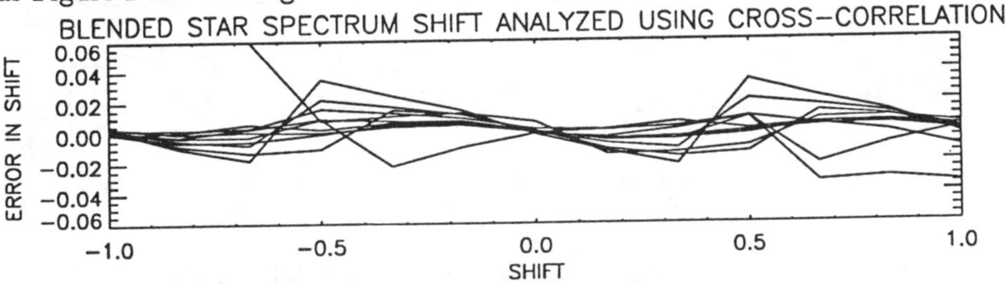

Fig. 1. The error in determining the Doppler shift (in pixels), versus the shift for artificial data consisting of stellar and I_2 spectra using the cross–correlation method. The errors seen result mostly from shift–dependent blending of stellar lines with iodine lines.

Another technique for analyzing spectra taken with an iodine absorption cell is the "least squares" technique. We can represent the star times the iodine spectrum as

$$D = S(\lambda + \delta_1)\, I(\lambda + \delta_2)\,. \tag{1}$$

Where D is the data with I_2 spectrum superimposed over the star spectrum, S is the star spectrum, I is the I_2 spectrum, and δ_1 and δ_2 are the shifts of the stellar and I_2 spectra respectively. By performing a Taylor series expansion on Equation (1), and dropping all terms above first order, we have, for small shifts δ_1 and δ_2,

$$D = [S(\lambda) + \delta_1 \frac{\partial S}{\partial \lambda}]\, [I(\lambda) + \delta_2 \frac{\partial I}{\partial \lambda}]\,. \tag{2}$$

Then we determine δ_1 and δ_2 using a least squares procedure, minimizing the difference between the observed spectrum of star with I_2 and the result of Equation (2) calculated using reference spectra of star alone (S) and I_2 alone (I). We then shift shift S and I by δ_1 and δ_2 respectively using an interpolation scheme (we are currently using Fourier interpolation), and iterate.

These techniques are complicated by the existence of cosmetic defects on the detector, which demand the application of a window function to remove the bad pixels. The bad columns of some CCDs can be aligned parallel to and between orders, so they are entirely on unused portions of the detector. For other detectors, especially CCDs with large numbers of bad columns, it is more practical to arrange the detector so the orders run perpendicular to the bad columns. The bad columns then compromise a few pixels in each order rather than rendering an entire order unusable. These problems will be reduced when the current generation CCDs are more widely available on spectrographs. For those of us using early CCDs, the window function must be dealt with carefully. With the cross-correlation method, it is unclear how to properly account for the window function. This, however, adds additional error to the Doppler shifts. An advantage of the least squares technique is that the bad pixels can simply be left out of the summation, so only the data from the pixel is lost, and no additional error source is introduced. Figure 2 shows the result of analyzing the artificial data using a realistic window function and the least squares method.

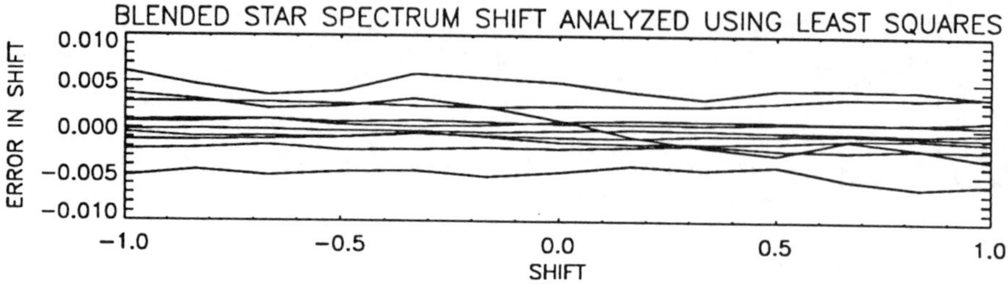

Fig. 2. The error in determining the Doppler shift (in pixels), using the least–squares method versus the shift with a realistic window function applied to the artificial data consisting of stellar and I_2 spectra.

A disadvantage of using I_2 in the absorption cell is the density and narrowness of the spectral lines. The I_2 spectrum is essentially saturated with lines and molecular band structures from 5000Å to 6000Å. The location of the continuum is thus lost. The spectra must be flattened to remove the spectrograph blaze function and flat fielding errors due to the different temperatures of the star and flat fielding source. Incorrectly flattening the spectra leads to unacceptably large errors (see Figure 2). By iterating the least squares procedure, and using the low–order trends in the residuals to correct the continuum level, the errors can be reduced to an acceptable level, as Figure 3 shows.

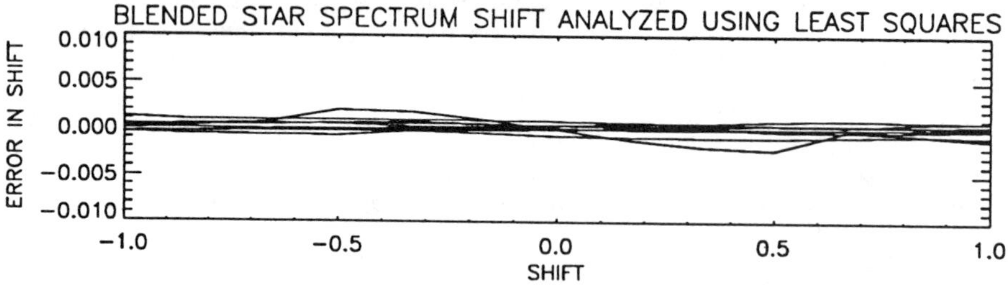

Fig. 3. The error in determining the Doppler shift (in pixels), using the least–squares method versus the shift with a realistic window function applied to the artificial data and using the residuals to correct the normalization of the continuum.

REFERENCES

Campbell, B., and Walker, G.A.H. 1979, *Pub. A.S.P.*, **91**, 540.

Marcy, G.W. 1991, *private communication.*

Marcy, G.W., and Butler, R.P. 1992, *Pub. A.S.P.*, **104**, 270.

Inside the Stars, IAU Colloquium 137
ASP Conference Series, Vol. 40, 1993
Werner W. Weiss and Annie Baglin (eds.)

IMPACT OF OBSERVATIONS ON PREJUDICE AND INPUT PHYSICS

D.O.GOUGH
Institute of Astronomy, Madingley Road, Cambridge CB3 0HA, England
and Department of Applied Mathematics and Theoretical Physics,
University of Cambridge

ABSTRACT My brief from the principal organizers of this colloquium is to draw your attention to some of the prejudices upon which the observations that have been discussed here have had or should have had some impact, and to remark on how that might revise our views about physics: it should not be a balanced summary of the meeting, but a prejudiced review.

It appears that the most outstanding prejudice that has come across throughout this meeting is that we, the astrophysicists (I use the term quite literally), believe that we know where we are heading and how to get there. It is very encouraging that this is more than a prejudice; it is also the judgement of our principal guest, M S Longair, who is here representing the rest of astronomy. The reason for our confidence, I believe, is that our understanding of the insides of stars is based on the solid foundations laid down by great men, Eddington and Chandrasekhar, upon whose work a more elaborate theory was firmly built by Hoyle, Schwarzschild, Kippenhahn and his collaborators, and by Iben, once high-speed electronic computers became available. However, the subject then appeared to become moribund: so successful were the pioneers that it appeared to the outsider that there was nothing more to do, except merely to clear up a few minor details. But the astute realized that those details were possibly important clues to more profound understanding, and laboured to unravel the evidence that has led to new discoveries, many of which have been under discussion this week.

As is always the case in man's endeavour to understand the physical world, the way was led by the observers, whose great advances particularly in photometry and spectroscopy have pinned down the surface properties of stars much more precisely; moreover, A Gomez has given us an inkling of what further information will soon become available, both photometric and astronometric, when the Hipparcos data are released. But more exciting yet is the possibility of seismic information, which truly probes the stellar interiors. Professor R A Lyttleton once remarked: 'If a modern astronomer were to meet a nineteenth-century chimney sweep, he would deduce that the sweep were made entirely of carbon.' Helioseismology has, and asteroseismology soon will render that remark outdated.

Seismology is the newest of the powerful observational tools to become available to astrophysicists. A great deal has already been learnt about the inside of the sun from seismic observation, and corresponding advances in our perscrutation of other stars will, in the future, take place. Seismic calibration of stellar models will be the first technique to bear fruit, provided it is carried out intelligently with insight such as that afforded by J Christensen-Dalsgaard's penetrative review. Inversions that answer specific questions about internal structure will subsequently become possible, once suitable data are available. Initial inversions of artificial stellar data indicate that a high degree of precision is required, and I am sure that such precision is attainable by campaigns such as PRISMA, notwithstanding any dubiety expressed by the sceptics. I recall, after a lecture such as this about the sun delivered at a Joint Discussion of the IAU General Assembly in Grenoble only fifteen years ago, that there was expressed not only scepticism but even outright disbelief that helioseismology would ever bear fruit. The principal objection, which stemmed from one of this week's lecturers (who now appears to have been at least partially converted), was that only high-overtone oscillations have been and were likely to be observed, and that since they form an harmonic sequence, only one new piece of information can ever be obtained from seismology (I paraphrase). My reply, of course, was that with adequate accuracy the one-per-cent deviations of the low-degree overtone frequencies from the harmonic sequence, and the similar small deviations of higher-degree modes from the predictions of some theoretical solar model, would reveal the true solar interior. The prejudice of the sceptics was that such accuracy could never be achieved. The reality, however, demonstrated by the outstanding progress made by the observers, is now that K G Libbrecht and M F Woodard illustrate data with error bars to ± 1000 standard deviations in order that the uncertainty can be seen, and that G R Isaak offers us the promise of low-degree frequencies accurate to a part in 10^5 in the not-too-distant future. Even after stripping the almost useless first two digits from each datum, one still has plenty in hand to make inferences that teach us some real physics.

What physics have we learnt so far? Perhaps the most broadly relevant to the immediate subject-matter of this meeting is the importance of the contribution to stellar opacity of *LS* coupling in radiatively induced atomic transitions. It was the discovery in 1985 that the sound speed throughout much of the radiative interior of the sun is up to one per cent greater than it was in the solar models of the day that led to the realization that the opacity immediately beneath the convection zone had probably been underestimated by up to some 20 per cent. Encouraged by several helioseismologists, principally W Däppen, the opacity in this region was reassessed at Livermore by C A Iglesias and F J Rogers. And indeed, at temperatures and densities representative of the outer layers of the radiative region of the solar interior, the seismological prediction was confirmed. But in addition, as we heard in M J Seaton's review, the more careful new computations have also yielded values of the opacity at temperatures of several hundred thousand Kelvin that are up to three times those obtained previously at Los Alamos. Although not relevant to the sun, this result has far-reaching implications for a variety of other stars.

There has been an almost universal prejudice amongst stellar modellers that the tables of opacity with which they have been supplied are not wholly reliable. That is not to belittle the work of the Los Alamos group; their achievement was truly a major advance of the time. However, in trying to understand certain quite specific properties of stellar pulsation, some workers, such as R F Stellingwerf, have been moved to suggest artificial modifications to the tables.

The ramifications of the new computations by Iglesias and Rogers, however, are more diverse: we have heard already from Christensen-Dalsgaard, W A Dziembowski, G Michaud and their colleagues that with the new opacities there is no longer a problem in explaining simultaneously the period ratios of double-mode Cepheids and δ Scuti stars, that the twenty-five-year-old problem of finding a viable mechanism to excite pulsations in β Cephei stars has been resolved, and that the observed lithium gap is more closely reproduced by calculations of element segregation against diffusion. New tables from the ambitious Opacity Project will soon be available; as Seaton has described, they are somewhat similar to the Livermore values where the latter are available. Of course, no good scientist will believe them unquestioningly to the last digit. However, much of the prejudice developed against the earlier tables has evidently already been dissolved.

Another arena in which seismic observation is teaching us physics is the study of the thermodynamic properties of plasmas. That has come about from an investigation of the adiabatic compressibility of the solar plasma under conditions inducing the second ionization of helium. The initial motivation was to determine the primordial solar helium abundance, a quantity of considerable cosmological importance. But, of course, the transition from adiabatic compressibility to helium abundance can be made only via the equation of state, which, as Däppen has explained, is inadequately known for modelling the sun with helioseismic accuracy. Gone now is the prejudice, held previously by many a stellar modeller, that fine details contributing only a few tenths per cent to the free energy of a gas are astrophysically irrelevant. Indeed, a recent yet unpublished collaboration involving V A Baturin and S V Vorontsov has produced a calibration of a grid of solar envelope models against the seismic data of Libbrecht and Woodard. The equation of state includes a free parameter representing the finite size of bound charged species, and with plausible choices of heavy-element abundance, the calibration yields an effective radius of several Bohr radii. I am not claiming that that radius has been definitively measured, for there may be other aspects of the microscopic properties of the plasma that have been inadequately modelled. What is exciting about the investigation, however, is that the effect has been detected. And it is often the case that once a phenomenon such as this has been detected, a clear measurement of it follows soon afterwards. That demonstrates the existence of yet another field in which astronomy can provide a delicate testbed for more fundamental physics.

Although the full implications of the uncertainties in the equation of state have yet to be investigated, it is appearing to be the case that the helium abundance in the solar convection zone is substantially lower than the value that properly calibrated standard evolutionary models require. Michaud has explained why that should be. But the settling he has calculated with C R Proffitt is rather less than is required to explain the difference. Therefore there still remains some discrepancy to be removed. We must not forget, however, that there are also other inconsistencies to resolve. So long as the solar neutrino problem, in particular, remains unexplained, one should regard the theoretical models with a due measure of suspicion.

Despite that by virtue of its relation to the rest of physics the solar neutrino problem is potentially the most important issue we face at present, I shall not discuss it extensively, because R K Ulrich has already explained part of the controversy in more detail than I have space here. (One might recall in passing, however, D R O Morrison's controversial claim that the controversy does not exist.) Instead I shall address a recent protest against a prejudice which Ulrich did not discuss because, as he pointed out in response to a question, it is not controversial and was therefore inappropriate material for his lecture. It concerns an issue raised recently in a preprint by R L Kurucz questioning the rates of reactions such as p(p,$e^+\nu$)D. Together with the other deuterium-producing reaction, p(pe^-, ν)D, it is the bottleneck of the p-p chain, and thus controls the rate of evolution of all late-type main-sequence stars (though the control is not as strong as one might naively think). Thus, it can hardly be irrelevant to our measuring of the ages of the globular clusters, which is accomplished by comparing the position of the main-sequence turnoff on the H-R diagram with theory, and which, as Longair stressed in his introduction, plays a major role in the story of the value of Hubble's constant H_0. Moreover, according to N Langer, these two are the only important reactions of the chain whose cross-sections have not been measured experimentally at any energy: the rates are purely the product of theory. That a serious misgiving should have been voiced is therefore reason enough to discuss the issue here. In principle, so far as I can understand it, the computation of the mean reaction rate should contain first the evaluation $\mathcal{B}$ of the probability of a single reaction, which involves a two-body or a three-body barrier penetration calculation (in the presence of the rest of the plasma), followed by a statistical-mechanical average $\mathcal{A}$ over the ensemble of all possible states. If the plasma were dilute, its presence would produce but a small perturbation to what $\mathcal{B}$ would have been were there to have been only the two reacting particles, p+p, in the Universe (or p+p+e in the case of the pep reaction), and the operations $\mathcal{B}$ and, in particular, the component $\mathcal{A}_e$ of $\mathcal{A}$ over electron states would commute, enabling one to perform the simpler calculation $\mathcal{B}\mathcal{A}_e$ in place of $\mathcal{A}_e\mathcal{B}$. The same would be true also of all the other nuclear reactions in the pp chain and the CNO cycle, the assumed commutation of $\mathcal{A}_e$ and $\mathcal{B}$ providing the basis for the Salpeter electron-screening corrections. What Kurucz has questioned is the validity of the commutation assumption. He points out that the admittedly unlikely event of the electron, considered in a semi-classical sense, being close to the two protons near the instant of maximum repulsion provides so great a perturbation in a dense plasma that it must be computed prior to the averaging, and that when that is done the outcome for the deuterium-producing reactions differs from the usual calculation $\mathcal{B}\mathcal{A}_e$ by so much that the neutrino flux predicted for Kamiokande and the ^{37}Cl detector is reduced to the observed value. Although Kurucz is undoubtedly formally correct about the order in which the calculations should be carried out, I find it difficult to believe that the process he discusses could have so great an effect, because only a relatively minor contribution to the barrier-penetration probability integral comes from small proton-proton separations. Kurucz does not quote explicitly the factors by which the reactions are augmented, though one can deduce from the information he provides that the rate of the combined deuterium-producing reactions is increased by a factor in excess of 5. As Kurucz points out, the main outcome is to decrease the temperature in the core of the sun, leading to a reduction in the ^{7}Be and ^{8}B neutrino fluxes (assuming, implicitly, that the process does not also substantially augment the ^{7}Be- and ^{8}B-producing reactions). This is partially offset by a decrease in the initial hydrogen abundance required to produce the observed photon luminosity at the appropriate age, which contributes a slight enhancement of the neutrino flux, but this is rather small compared with the influence of the modification to the temperature. [J Christensen-Dalsgaard commented immediately after this lecture that he disbelieved that Kurucz's argument could reduce the neutrino flux. He reported that he had carried out a computation in which he had augmented all the nuclear reaction rates by multiplying the screening correction in Salpeter's formula (which is computed as $\mathcal{B}\mathcal{A}_e$) by the same constant factor, and that in that case the resulting neutrino flux had *increased*. The formal result is quite easy to understand: by balancing the reactions in the chain it is straightforward to show that, because of the charge dependence of the screening, the ppII and ppIII branches are enhanced relative to ppI by more than the reduction due to the decline in temperature. However, enhancement of the $\mathcal{B}\mathcal{A}_e$ screening correction by identical factors to mimic $\mathcal{A}_e\mathcal{B}$ is hardly what Kurucz had in mind. One might add, in passing, that Salpeter's screening of the target does not take account of the incident proton, whose effect is to decrease, not increase, the screening factor, and therefore to *reduce* the neutrino flux.]

In an extensively circulated response to Kurucz's argument, J N Bahcall and E E Salpeter have pointed out that because electrons are not well localized they cannot influence the barrier penetration probability by as much as Kurucz claims. From the semiclassical viewpoint the electron is too rarely in the vicinity of the protons near the instant of maximum repulsion. Bahcall and Salpeter did not address explicitly the extent to which $\mathcal{A}_e$ and $\mathcal{B}$ fail to commute. However, if one takes into account the spread of the electron wave function as Bahcall and Salpeter require, it appears that even if the barrier penetration probability were to be increased to unity when the electron, considered as a particle, were near the colliding protons, after averaging over phase space the influence on the mean reaction rate should not be very great. However, it would nonetheless certainly be worthwhile investigating the extent to which fluctuations in the electron screening (other than those induced by the relatively slow movement of the protons, which has already been considered) do modify the Salpeter formula and its recent extensions.

There have been many other prejudices expressed at this meeting, none of which I have space to address in any detail. One that I cannot refrain from just mentioning, however, is the presumed universality of the so-called mixing-length parameter α. I find that prejudice difficult to understand, particularly because different workers use different formulae to relate heat flux to temperature gradient, often without even stating which one they have used, yet they are not unhappy then to compare the values of α they prefer. The desire for a parameterless formula has been expressed several times at this colloquium. That is quite understandable. Nonetheless, I regard it to be a more pressing goal to find a more faithful representation of the physics, even if that does require calibration against observation. That is, of course, only my prejudice. I even go so far as to consider it essential that for a theory of convection (and not only convection, for that matter) to be trustworthy, it should work also in the laboratory. That opinion is almost heretical, most stellar modellers claiming that conditions in stars are so different from those on Earth that no meaningful comparison can be made. Yet, aside from the final assumption that the mixing length is proportional to the pressure (or density) scale height, the astrophysical formulation of the mixing-length theory does not explicitly demand that the convection be in a star, and in any case the approximations to the equations of fluid motion that are used are actually more suited to terrestrial than to stellar conditions. My plea is that because we do not have an adequate theory of convection we calibrate α across the H-R diagram by comparing stellar models with observation, finding how it depends on luminosity, effective temperature and chemical composition. Then at least we shall have some useful foundation with which to compare future theories. We might then be using astronomy as a useful 'experimental' tool to test macroscopic physics. Needless to say, the results of the calibration should be accompanied by an explicit statement of precisely which formulae for the heat flux and the Reynolds stress have been used.

The importance of convective overshoot and semiconvection to stellar evolution has been stressed several times, notably by J P Zahn and A Noels, but also by others. After many years of disagreement, Zahn now appears to agree with Roxburgh's formulation of overshoot. However, as Roxburgh emphasized earlier (admittedly in a different context) we must not be misled by that: 'When people agree it does not necessarily mean that they are right.' With overshooting from cores seeming to take on a lesser role than was previously believed, the problems of semiconvection are again coming under scrutiny. The prejudices concerning the final state that is attained when gradients of temperature and molecular weight are in opposition are, as is astrophysically popular, based on neutrality with respect to some stability condition. As I understand it, preferences range from the so-called 'Ledoux criterion', which is a restatement of the *linear* convective stability criterion introduced into the astronomical literature by K Schwarzschild, and the 'Schwarzschild criterion', which is what that convective stability criterion would have been were gradients in chemical composition to have been ignored, and which, before (and even after) Ledoux's paper to point it out was published, had been mistaken for the criterion introduced by K Schwarzschild, and furthermore which M Schwarzschild and R Härm considered to be the approximate outcome of the *nonlinear* development of the instability. It might be of interest at this point just to throw in the prejudice of the fluid dynamicists: that the mean stratification adjusts itself into thin horizontal layers of essentially homogeneous isentropic convection separated by diffusive interfaces across which the jumps in temperature and chemical inhomogeneity are such as to render the ratio of their fluxes equal to the square root of the ratio of the corresponding microscopic diffusion coefficients, irrespective of any other conditions controlling the flow. That prejudice is well confirmed by laboratory experiment. And it would provide a powerful tool for predicting the stratification of chemical abundances in stars if indeed it really could be trusted under all conditions. Unfortunately, however, the only dynamical analysis of the phenomenon of which I am aware, and which admittedly is far from being satisfactory, predicts that under the very different conditions encountered in stars the behaviour of this phenomenon is actually quite different from laboratory experience.

Can any of the prejudices be tested? Christensen-Dalsgaard indicated how one might differentiate seismologically between two ideal stellar models, one with and one without overshoot, using the small separation $d_{n,l} = \nu_{n,l} - \nu_{n-1,l+2}$ between the frequencies $\nu_{n,l}$ of acoustic modes of high order n and low degree l. However, some care must be taken to take into account asphericity, which also contributes to $d_{n,l}$ and which is great enough in some stars to be quite substantial. It is also necessary to seek more subtle signatures which one might hope to use for distinguishing between regions of overshoot and of semiconvection. It is worth remarking also that there are two posters addressing the possibility of detecting overshoot beneath convective envelopes. The conclusions appear to differ. It is interesting that those authors to whose hearts overshooting is the more dear appear to have invested enough effort to find evidence in favour of it, whereas the others did not. This is yet another example of how prejudice has an impact on physics.

Any session on stellar dynamos is always so steeped in prejudice that one runs the risk of forgetting what a dynamo actually is. If one is to try selling one to a national electricity generating organization, one obvious prerequisite is that, after subtracting internal ohmic decay, the electrical output from the mechanical work invested exceeds zero. Most of the dynamo theorists I know who are not also astrophysicists accept that prerequisite too. But stellar physicists, in general, do not. Even the existence of a dynamo so defined can hardly be demonstrated in stars, since the decay time of a large-scale magnetic field in the radiative zone of a star is either greater than or comparable with the lifetime of that star. F Krause argued that the reversal of the exterior field on any timescale much shorter is proof of the presence of a dynamo, and C Schrijver seemed implicitly to have taken that for granted. Implicit also in the discussions seemed to be that the dynamo exists in the body of the convection zone. Yet dynamical modellers of the solar convection zone have failed to reproduce simultaneously the field reversals, the butterfly diagram and the so-called torsional oscillations of the photospheric layers. Moreover, they finally seem to have capitulated in the face of the suggestion that the angular velocity throughout the convection zone inferred from seismology appears to be a function only of latitude, whereas the models require it to be approximately a function of only distance from the rotation axis, at least at distances greater than the radius of the base of the convection zone. Consequently, they have retreated just beneath the base of the convection zone, where it was suggested by E A Spiegel and N O Weiss that the dynamo must reside. If that were the case, then one would expect such dynamos to be absent from fully convective stars, for in those stars convection zones have no base. That is why P Podsiadlowski asked Schrijver whether what he had called dynamo activity was absent in fully convective stars. What he was really asking, of course, was whether there is any evidence for the dynamo being located immediately beneath the convection zone. Schrijver's answer was: 'No.' So where does that leave us?[1] Something one might reexamine is the interpretation of the seismological evidence. At present the solar rotational splitting data are expressed in terms of a severely truncated expansion which certainly does not preclude the possibility that in the equatorial regions the angular velocity in the convection zone is a function of only distance from the rotation axis. If one were to adopt that functional form as a constraint, one would find that the rotation in

[1] After the meeting I learned from Podsiadlowski that F. d'Antona had told him later that ROSAT data showed much lesser X-ray emission from late-type main-sequence stars that are expected to be fully convective, suggesting that such stars do have much weaker magnetic fields.

the polar regions would be inferred to be slower than was previously expected, and not so smoothly varying with position. Whether one finds that to be less or more credible than the alternative I mentioned previously depends solely on one's prejudice. So perhaps Krause is right. Indeed, in the poster by Libbrecht *et al* the rather closely spaced plot of the contours of constant angular velocity deduced seismically is actually not very unlike that wanted by the dynamical modellers of the convection zone.

It is perhaps worth pointing out that having prejudices is not unique to stellar physics. Indeed, Longair, in his wide-ranging introductory lecture, presented us with observational evidence which, when naively interpreted, leads to the following values of the deceleration parameter, Hubble's constant and the density parameter: $q_0 = 3.5, H_0 = 50 - 100 \text{ km s}^{-1}\text{Mpc}^{-1}, \Omega \simeq \alpha$. (The last value is my paraphrase, α now being the fine structure constant, not to be confused with the mixing-length parameter which is not a constant.) Yet Longair clearly gave us the impression that he believed that $q_0 = 0.5 \pm 0.5$ and $\Omega = 1$, even if he didn't say so explicitly. The latter, so far as I can see, is based purely on theology, or perhaps on aesthetics. Yet perhaps that is more reliable a criterion than some scientists might readily accept. Knowledge comes from a variety of different sources, and one should not reject any evidence without first weighing it carefully. That $\Omega > \alpha$ provides the case for the existence of dark matter. Given that the matter has not been detected directly, it must surely be the case too that whether one believes it to be hot or cold is largely a matter of prejudice. Longair clearly described the case against it being hot, and left one with the impression that he thought it must be cold. This, indeed, is the popular belief. From my outsider's naive understanding of Longair's talk, however, it seems that if it is to account for the properties of large-scale structure, it might be simpler for us if dark matter were to be luke warm (which might actually be a mixture of hot and cold).

Before concluding, at the request of W Weiss I include a provocative remark. I have chosen to draw your attention to the poster by R Vera, which explains how there might be neutron stars in the cores of main-sequence stars, and how, in particular, the existence of one at the centre of the sun would explain, amongst other things, the low neutrino flux. This is reminiscent of an idea by S W Hawking, who many years ago suggested that the neutrino problem is solved by a black hole in the solar core. I had often wondered whether, if there were such a condensed body in the sun, it would remain at the centre. For might not it lead to an overstability, though of a kind rather different from that discussed earlier this week by Y Osaki? Consider the condensed body to receive a small perturbative impulse, leading to motion and a consequent asymmetry in the accretion flow around it. The compression of the reacting gas in the wake would induce a localized increase in the nuclear energy generation rate, leading to an expansion of the heated gases, as in a jet engine. The thrust would be insufficient to counter gravity, so the body would still accelerate towards the centre of the sun, and oscillate much as it would in the absence of the engine. But the direction of the thrust is evidently such as to augment the amplitude of oscillation. The increase cannot continue forever, however. Once the body rises far from the core, energy generation in the wake ceases, and the motion is retarded by accretion of matter having relatively low momentum. Thus a balance between driving in the core and damping in the envelope is reached (as is also the case for the oscillation of many other kinds of star). The period of oscillation depends on the amplitude at which that balance occurs, and if the trajectory were to extend almost to the surface of the sun it would be about 160 minutes. Therefore one might hope to test the theory by looking for an oscillation with this period. Early investigations suggested that indeed it might be present, but today there are only very few who believe the evidence to be convincing. So perhaps this story must be abandoned.

Finally, permit me to state my prejudice about prejudice. I believe that one should never approach a new scientific problem with an unbiassed mind. Without prior knowledge of the answer, how is one to know whether one has obtained the right result? But with prior knowledge, on the other hand, one can usually correct one's observations or one's theory until the outcome is correct. At that point one usually stops looking for more mistakes. One can then be content that one's understanding is confirmed, and receive a boost to one's ego (which is well understood in Vienna). However, there are rare occasions on which, no matter how hard one tries, one cannot arrive at the correct result. Once one has exhausted all possibilities for error, one is finally forced to abandon a prejudice, and redefine what one means by 'correct'. So painful is the experience that one does not forget it. That subsequent replacing of the old prejudice with a new one is what constitutes a gain in real knowledge. And that is what we, as scientists, continually pursue.

Inside the Stars, IAU Colloquium 137
ASP Conference Series, Vol. 40, 1993
Werner W. Weiss and Annie Baglin (eds.)

KEY ISSUES: A ROUND TABLE DISCUSSION

T. M. BROWN[1], P. DEMARQUE[2], R. NOYES[3], F. PRADERIE[4], I. W. ROXBURGH[5], E. SCHATZMAN[6] (CHAIR)
[1] High Altitude Observatory, National Center of Atmospheric Research, P. O. Box 3000, Boulder, CO 80307, USA
[2] Yale University, Department of Astronomy, P.O. Box 6606, New Haven, CT 06511, USA
[3] Smithonian Astrophysical Observatory, 60 Garden Street, Cambridge, MA 02138, USA
[4] DESPA, URA CNRS 264, Observatoire de Paris, 92195 Meudon Cedex, France
[5] School of Mathematical Science, Queen Mary College, Mile End Road, London, E14NS, UK [6] DASGAL, URA CNRS 335, Observatoire de Paris, 92195 Meudon Cedex, France

INTRODUCTION by Evry Schatzman

We have taken part to an exceptionally rich colloquium, characterized by a large amount of information in all fields of physics, and a remarkable collection of observational data. Our understanding of what is going on inside the stars has changed in a radical manner during the last years, on one side because the observations are bringing new kinds of information, and on the other side because theory in its development is taking into account a number of processes, some of them having been completely ignored only ten years ago.

It is impossible to draw here a complete list of problems. There are many cases where some inconsistency could be found in the theory itself or some contradiction between observational data and theory. I recommended that such a list should be drawn carefully. The participants to the round table discussion will give their contribution, helping to open the way to new fields of research and to new discoveries.

In the following, you will find the conclusions of Pierre Demarque, Ian Roxburgh, Bob Noyes, Timothy Brown, Françoise Praderie and myself.

INTERNAL DYNAMICS OF STARS by Pierre Demarque

I want to say a few words about the internal dynamics of stars, which has been a topic of growing interest in the last few years. This interest is likely to continue to grow in the next decade and we can expect dramatic advances in the areas of convection, rotation and non-radial pulsation.

Numerical models of stellar convection will become increasingly realistic. For example, the study of spectral line bisectors will lead to a better understanding of the outer convective layers of cool stars. For these stars, models of surface convection zones free of the arbitrary mixing length parameter and based on sounder fluid dynamics will become available, yielding an improved envelope structure and (at last!) permitting the calculation of stellar radii from first principles. A better understanding of penetrative convection and overmixing (convective overshoot) will follow. The role of magnetic fields in convective patterns and transport will also be increasingly studied.

Thanks to the recent observations of rotational surface velocities, of spectroscopic abundances of the light elements Li, Be and B, and of the products of CNO processing mixed to stellar surfaces, progress has already been made in exploring the processes of angular momentum transfer and redistribution in stars. The light elements Li and Be probe the rotational history of stellar envelopes, while the CNO isotopes, which are processed by nuclear reactions in the deep interior probe the degree of mixing from the core to the surface. More such abundance observations, particularly in star clusters, young and old, metal-rich and metal-poor, will constrain the theory. Together with the rotational velocities, which reveal the characteristic timescales of the surface braking due to the stellar winds, they provide in the interior stringent tests of angular momentum transfer mechanisms and of their mixing efficiency. Abundance spreads would also be particularly significant.

Finally, there is the exciting prospect of using seismic techniques to probe stellar internal dynamics. In this context, the PRISMA space mission, proposed to ESA and featured at this colloquium, is very promising. My main comment here is that, like all techniques in astrophysics, seismology will yield best scientific results only if it is done in parallel with detailed observations of chemical abundances and other basic parameters (such as parallaxes) of the target stars, and with the help of theoretical modeling.

SOME PROBLEMS IN STELLAR STRUCTURE AND EVOLUTION
by Ian W. Roxburgh

Convection

The first point I wish to make is that convection is a global phenomenon. Since the convective "mixing time" is very much smaller than both the radiative and evolutionary time scales, a convective zone should be considered as a global system; what happens at one part of the zone effects what happens at another and the whole zone achieves some sort of global (statistical) equilibrium. Local theories, such as the mixing length model, cannot capture this essential feature of convection and some different approach is needed.

Numerical simulations (cf Stein and Nordlund 1989) indicate that "convective plumes" extend over most of the convective region and penetrate into the surrounding stable layers. "Theories" incorporating such plumes need to be developed and applied to models of convective zones, and perhaps more importantly to convective cores. Such high velocity plumes may be of considerable importance in the mixing of chemical elements beyond the boundary of convectively unstable regions, since the cumulative effect of deep penetration by plumes occupying only a small fraction of the volume could produce effective mixing of chemical elements in the evolutionary time scale of a star.

A model of convection incorporating downward moving plumes was developed in the early 1960's by Hoyle (cf Faulkner et al 1965) who observed that downward moving material becomes increasingly thermally isolated from its surroundings and may penetrate all the way from the top to the bottom of the layer; this model was modified by Roxburgh and Vickers (see Vickers 1970) who incorporated the change in the area occupied by such plumes as a function of depth, but this "theory" has been little used since that time. A "plume" model was also developed by Schmidt et al (1984), [see also Zahn 1991] but again this has not been developed to the stage where it can be used in modelling stellar convective regions. Upward moving plumes may also be important; in stellar interiors the decrease of the "thermal conduction" ($4acT^3/3\kappa\rho$) with decreasing temperature implies that it is the upward moving elements that become increasingly thermally isolated from the surrounding regions rather than the downward moving elements.

Convective overshooting — the integral constraint

Next I wish to comment on the so called "Roxburgh criterion" for convective overshooting which follows from integrating the entropy equation over the volume of the convective zone (Roxburgh 1976, 1978, 1989)

$$\int_0^{r_c} (L_{rad} - L_{nuc}) \frac{1}{T^2} \frac{dT}{dr} dr = \int_0^{r_c} \frac{\Phi}{T} 4\pi r^2 dr > 0 \tag{1}$$

To obtain this result turbulent averages have been replaced by functions of the mean field; recent numerical calculations, admittedly in a parameter regime removed from that applicable to stellar interiors, suggests that this approximation is reasonable (Roxburgh and Simmons 1992). I should however emphasise that the integral condition was derived to give an estimate of the uncertainty in standard stellar models (where the boundary of convective cores is determined by the condition $L_{rad} = L_{nuc}$), not to provide a "correct" determination of the extent of convective penetration. Since the viscous dissipation Φ is positive definite the condition with $\Phi = 0$ provides a useful estimate of the upper limit on the extent of convective overshooting. To illustrate the point consider stars with very small convective cores and expand L_{rad} and L_{nuc} in powers of r as

$$L_{rad} = A_0 r^3 - A_2 r^5 + \dots, \quad L_{nuc} = B_0 r^3 - B_2 r^5 + \dots \tag{2}$$

The "classical" core boundary where $L_{rad} = L_{nuc}$ is then at $r_0 = [(A_0 - B_0)/(A_2 + B_2)]^{1/2}$. Expanding $T = T_0 - T_2 r^2 + T_4 r^4 + \dots$ ($T_2 < 0$), and evaluating the integral condition (with $\Phi = 0$) gives the core boundary at $r = r_c$ where

$$r_c^2 < \frac{7}{5}\left(\frac{A_0 - B_0}{A_2 - B_2}\right) = \frac{7}{5} r_0^2 \tag{3}$$

Hence the maximum extension of the core due to convective overshooting is $0.18\, r_0$.

In modelling the extension of convective cores due to overshooting several authors extend the unstable core by some fraction of the local scale height; since the scale height tends to infinity as r tends to zero this can give an overshoot region much larger than the limit from the integral constraint, and give solar models in which a convective core remains up to the present time. In the solar models I have computed with overshooting limited by the integral constraint, a ^{3}He and ^{12}C driven convective core lasts considerable longer than in models without overshooting (due to the transport of ^{3}He and ^{12}C into the centre), but nevertheless the core dies away at an age of about 10^8 years.

Mixing — the effect of chemical composition gradients

Next I wish to stress a point I have made several times before: the existence of instabilities is not sufficient to ensure mixing of chemical elements; mixing requires sufficient energy input to overcome the stabilising effect of a chemical composition gradient (cf Roxburgh 1984, 1985, 1991). This can be demonstrated as follows. Consider a "blob" of fluid raised sufficiently slowly through a distance $\mathbf{z}$ in the vertical direction so that the pressure temperature inside the blob are the same as the surroundings; the density excess δr at $\mathbf{z}$, and the total work per unit mass W done in interchanging two blobs a distance $\mathbf{k}$ apart, are

$$\frac{\delta\rho}{\rho} = \frac{\delta\mu}{\mu} = -\frac{\mathbf{z}\cdot\nabla\mu}{\mu}, \quad \mathrm{W} = 2\int_0^{\mathrm{k}} \mathrm{g_z}\frac{\delta\rho}{\rho} = -(\mathbf{k}\cdot\mathbf{g})\frac{\mathbf{k}\cdot\nabla\mu}{\mu} \tag{4}$$

where $\mathbf{g}$ is the acceleration due to gravity. Any process that mixes regions of a star has to provide this energy. If the major part of the star is to be mixed then the distance k is of order of magnitude the radius of the star the total energy required is of the order $\mathrm{W} \approx \Delta\mu/\mu(\mathrm{GM}^2/\mathrm{R})$. With $\Delta\mu \approx (d\mu/dt)\mathrm{t_0}$, where $\mathrm{t_0}$ is the age of the star, mixing would require a continuous rate of input of energy per unit mass per unit time of approximately $\Phi(1/\mu)d\mu/dt$, where Φ is the absolute value of the gravitational potential. Most energy sources, eg. instabilities driven by differential rotation and meridional circulation, do not satisfy this condition in stellar interiors, although they can provide enough energy in the outer layers. One effective mixing mechanism is convection, the energy source being ultimately the nuclear reactions and the rate of production of energy per unit mass from nuclear sources is considerably in excess of that required to overcome the growing composition gradient.

^{3}He instability in solar type stars

Another outstanding problem is the ^{3}He instability in solar type stars (Dilke and Gough, 1972, Christensen-Dalsgaard et al 1974), in which a non radial g-mode is excited by nuclear reactions, the energy input over an oscillation period from the burning of ^{3}He exceeding the energy losses by radiation (^{12}C can also drive such an instability). An estimate of the energy available is obtained from calculating the rate of production of excess ^{3}He at the time instability sets in and assuming this to be converted into kinetic energy. This is more than sufficient to overcome the stabilising effect of the composition gradient and at least on energetic grounds this instability is capable of causing mixing.

The consequences of this instability have not been fully explored and it is usual to ignore it when calculating evolutionary models of the sun and stars. This is not reasonable. One possibility is that the instability leads to finite amplitude oscillations where the excess rate of burning of ^{3}He during an oscillation maintains the distribution near the marginally stable state; this requires quite large amplitudes. Resulting models of the present Sun have a higher central hydrogen abundance, lower central temperature and a reduced Solar Neutrino Flux (Roxburgh 1985, 1987). Another possibility is that the oscillation breaks down into mild turbulence which diffuses the excess ^{3}He into the centre where it is burnt (Roxburgh 1984, 1991), the diffusion being such as to maintain the star near marginal instability. Evolutionary models of the Sun including this diffusion do not show any major changes over standard solar models. Another possibility (Dziembowski, private communication) is that the unstable mode couples to stable modes and the oscillation has quite small amplitudes with negligible effect on the internal structure and evolution.

The non linear evolution of this instability and its possible effect on solar and stellar evolution needs further investigation.

Magnetic fields in stellar interiors

Magnetic fields may or may not be important in stellar evolution but we cannot simply ignore their prescence we need to quantify their effects. The matter inside stars is essentially a fully ionised plasma and even if by some "fluke" a star had no initial magnetic field, a field would be generated from the turbulent motions and/or differential rotation by the "battery" effect of the electron partial pressure. (cf Biermann 1950, Roxburgh 1966). In the surface layers of the Sun and many other stars, magnetic fields play a dominant role, and much effort has been expended in trying to understand the generation of the solar magnetic field by a dynamo action induced by differential rotation and the helicity of the turbulent motions in the outer convective zone.

But what about stellar interiors? Many stars are thought to have a central convective core during most of their main sequence phase, and even smaller mass stars like the Sun probably had a small convective core in their early main sequence phase. Why should not a dynamo operate in such convective cores? Indeed we might expect a turbulent dynamo to operate in this phase and build up a strong field even in the absence of rotation. Results from studies of homogeneous MHD turbulence (cf Durney, de Young and Passot 1990, Durney, de Young and Roxburgh 1992) indicate that even in the absence of helicity, kinetic energy of the turbulence is converted into magnetic energy, and that this magnetic energy cascades to large scale modes. With significant helicity the magnetic energy can exceed the kinetic energy of the turbulence. What happens to such a field generated in a stellar convective core? The perceived wisdom in the case of the solar convective envelope is that magnetic fields "float" to the top of the outer convective zone due to buoyancy (the density inside a flux tube being less than that of the surroundings), but a stellar convective core is surrounded by a stable region so where does the flux go to? Does it accumulate at the top of the convective core? If so what effect does it have on the structure and dynamics of the boundary region? Does it inhibit convective penetration or enhance mixing by producing a time varying magnetic force. What limits the input of kinetic energy into magnetic energy? Could the field become large enough to suppress the convection? These problems require further study.

Magnetic fields and differential rotation

Another problem that has concerned me for some time is the effect of "phase mixing" in the redistribution of angular momentum in differentially rotating stars and in spin down of stars due to angular momentum loss in a stellar wind (cf Roxburgh 1985,1991). In the absence of magnetic instabilities differential rotation generates a large toroidal field from a small poloidal field, this toroidal field reacts back on the differential rotation and the system undergoes torsional oscillations with a period given by the local Alfvén travel time along the poloidal field. If we neglect the effect of any poloidal motions induced by the oscillating field then the problem is governed by the two evolution equations

$$\frac{\partial(\rho\Omega r^2 \sin^2\theta)}{\partial t} = \frac{1}{4\pi}\mathbf{B}_p \cdot \nabla(B_\phi r \sin\theta) + \mathrm{div}(\rho\nu r^2 \sin^2\theta \cdot \nabla\Omega) \qquad (5)$$

$$\frac{\partial \mathbf{B}_\phi}{\partial t} = r\sin\theta\,(\mathbf{B}_p \cdot \nabla\Omega) + r\sin\theta\,\mathrm{div}\left(\frac{\eta}{r^2\sin^2\theta}\nabla(r\sin\mathbf{B}_\phi)\right) \qquad (6)$$

where $\mathbf{B}_p(r,\theta)$ is the poloidal field, $\mathbf{B}_\phi(r,\theta)$ the toroidal field, $\Omega(r,\theta)$ the angular velocity, (r,θ,ϕ) are spherical polar coordinates ν the kinematic viscosity and η the magnetic diffusivity. With η and ν zero these equations only contain derivatives of Ω and $\mathbf{B}\phi$ along the poloidal field lines $(\mathbf{B}\phi\cdot\nabla)$ and not perpendicular to them; poloidal field lines are therefore uncoupled and have different oscillation periods; neighbouring field lines therefore rapidly get out of phase. This "phase mixing" creates large gradients in the toroidal field and rotation, thereby causing enhanced dissipation of energy. Since dissipation converts magnetic and kinetic energy into heat, and since the source of energy is the differential rotation, we might expect the system to evolve towards the lowest energy state. But for a given angular momentum the state of lowest energy is that of uniform rotation, so we expect such phase mixing to lead to uniform rotation. Detailed numerical calculations are in progress by Gerado Ocana and myself (cf Roxburgh 1991). Charboneau and MacGregor have also undertaken similar calculations (see these proceedings) although they appear to have replaced $(\mathbf{B}\cdot\nabla\mathbf{B})_\phi$ by $(\mathbf{B}_p \cdot \nabla)\mathbf{B}_\phi$ rather than $[\mathbf{B}_p \cdot \nabla(r \sin\theta\ \mathbf{B}_\phi)]$ / $(r \sin\theta)$ and to have dropped a term involving the derivative of the density gradient from equation (5). Numerical calculations by Ocana and Roxburgh will be published in the near future.

STUDY OF STELLAR ACTIVITY MECHANISMS USING PRISMA
by Robert Noyes

A major focus of PRISMA is study of the nature of stellar magnetic activity. This can be done through detailed comparison of stellar behavior observed by PRISMA with analogous behavior observed on the sun. Although PRISMA is not expected to remain operational over a typical stellar magnetic activity cycle, nevertheless it can tell us much about magnetic activity through repeated observations of total luminosity, magnetic activity indicators, and p-mode frequencies in particular stars.

It is now well known that the solar luminosity varies with the envelope of the activity cycle, being about 0.1% larger near activity maximum than near activity minimum. Younger stars, such as those of the age of the Hyades (approximately 600 million years), also to show a variation, but with opposite sign relative to their overall magnetic activity level (e.g. Radick et al 1990). This may be because their much larger dark starspots overwhelm the increased brightness due to the associated plages, in contrast to the case for the sun. Another activity-related phenomenon seen on the sun is a slight shift of p-mode frequencies toward higher frequencies when magnetic activity increases; the frequency-dependence of the shift implies an activity-related perturbation of structure just below the surface (Libbrecht and Woodard 1990). The frequency shifts are closely correlated with individual episodes of surface magnetic activity averaged over the sun, rather than the 11-year cycle timescale itself (Woodard et al 1991). We do not know, however, how this interesting behavior manifests itself on other stars. It seems likely that as younger stars undergo their much larger episodes of magnetic activity growth and decay, the p-mode frequencies will be found to follow in step, but with a much larger fractional change than seen in the sun. However the fact that on younger stars the associated luminosity changes are seen to be oppositely-directed from those on the sun suggests that there could be some surprises. In any case, following the p-mode frequency changes to the expected much larger amplitudes in more active stars may provide more information on the nature of stellar activity and its subphotospheric roots than we can learn from the sun alone.

Also, although from groundbased monitoring data we know a lot about magnetic activity in younger solar-like stars (e.g. for one solar-mass stars of Hyades age the rotation period is about a week, the emission from chromospheric plages is some three times greater, the spot area is some 30 times greater, etc.; see Noyes et al 1991), we do not know with confidence for younger solar-like stars how their differential rotation varies with latitude, or anything at all about the variation of their rotation with depth. PRISMA should be able to address such issues, through long term monitoring of selected stars or clusters at the millimagnitude level, and (for fast enough rotators) resolving rotational splitting of l=1 p-modes.

We may hope that PRISMA will be operational long enough to explore many aspects of stellar magnetic activity, manifested through changes of observed parameters as activity changes on individual stars. The finding that solar p-mode frequencies respond to individual episodes of activity rather than simply to the 11-year envelope of activity is encouraging from an observational point of view, for it suggests that an operational life of even a few years may allow PRISMA to obtain important data on analogous frequency changes in other stars. A focussed campaign of PRISMA observations of a few particularly interesting stars should monitor not only changing p-mode frequencies but of course spot modulation (through broadband photometry) and chromospheric and coronal activity (through the ultraviolet spectrometer and XUV telescope). Naturally it should be accompanied by intensive and coordinated groundbased study of time variations in the same stars.

ON SEISMOLOGY OF SUN-LIKE STARS by Timothy M. Brown

Surely it is reasonable to be optimistic about asteroseismology's favorable impact on stellar evolution theory. The availability of new and precise information about the interior structure of distant Sun-like stars cannot fail to improve our understanding of these objects. Having said this, however, one must add that seismological data, just like other sorts of data, will have their limitations; these should be recognized from the outset.

First, our current understanding suggests that the observable mode frequencies of a star will provide 2 or 3 independent and well-defined numbers about that star. By themselves, these are too few even to define the evolutionary state of the star. Evidently more information will be required just to characterize these stars, let alone to test the physical content of our theories describing them. Second, for the foreseeable future, we will not have oscillations data concerning a very large number of stars. Whether observed from space or from the ground, the resources demanded by such measurements are simply too great to permit detailed study of more than (say) a few hundred objects.

In view of these facts, it seems clear that for seismological measurements to be worthwhile, they must be able to answer pointed astrophysical questions. This in turn implies that the target list for PRISMA is extraordinarily important. It will be vital to include on the list stars whose special circumstances (as members of star clusters or of binary systems, for example) give us information beyond that available from seismology. Moreover, we must whenever possible use more easily-observed stars (δ Scutis, roAp stars) to answer our questions, resorting to Sun-like stars only when necessary. Indeed, my opinion is that PRISMA (and seismology of Sun-like stars generically) can truly succeed only in combination with a vigorous and wide-ranging program of ground-based observations.

WHAT IS NEEDED NEXT? by Françoise Praderie

This meeting was one with an unusually large audience. The theme "Inside the stars" proved to have a very large federative capacity. It is that same theme which supports the space mission PRISMA. It does it by proposing to complement the theoretical approaches to stellar interiors by the collection of a systematic set of yet unavailable data. PRISMA is a mission dedicated for the stellar community, with a focus directed on the interior of stars. So, to reply to the question: what is needed next?, I will try to show that, from the experimental side, PRISMA is one of the answers.

The mission proposed is a discovery type mission for stellar physics. Not in the sense of unveiling an unknown part of the γ ray sky, but in the sense of strengthening the theoretical corpus describing stellar interiors by observational data.

In my view, PRISMA is as fundamental for stellar physics as HIPPARCOS. As some may remember, Hipparcos was for a long time the object of only moderate interest on the part of the stellar community, because the arguments given for the need to go to space were subtle and therefore did not look compelling. PRISMA sometimes suffered from similar reserves, but it seems now that they have been overcome.

Today, and in particular thanks to this meeting, the novelty of going to space to study the interior of stars has become clear. Firstly the Sun opened the paths, with experiments like IPHIR, and soon the instruments on SOHO. Second, the asteroseismology instruments on PRISMA take advantage of using the temporal stability of space, a poorly exploited property, emphasized by G. Isaak for quite a long time. Finally X and UV activity tracers will be observed simultaneously on a time scale of order of, or longer than, the rotation period of stars.

Of course the mission will need complementary ground-based observations, such as magnetic field strength in the photosphere, survey of the activity cycle, and photometry of large spots. It will also benefit from basic stellar data as provided by Hipparcos.

But going to space will allow unique performances: for example obtain a high S/N ratio in asteroseismology in about one day, or bypass the sensitivity of IUE by a fairly considerable factor.

One specific aspect of PRISMA is that three types of instruments will point in the same direction of the sky. Why? To give experimental support to the knowledge of interiors, two approaches will be used: a)the direct sounding of interiors up to the core, b)the indirect record of the magnetic field through effects in the outer layers at various altitudes. Both approaches require long time series with a high duty cycle, and both are strongly related, since the oscillation modes depend on the activity state of the star. The connection between theory and observables is at a different level of achievement in both cases. A lot remains to be done so that activity observables give access to the generation process of the magnetic field. Dynamo theorists will have to guide the PRISMA team, until the reverse situation announced by C. Catala (data to guide the theory) becomes effective.

In conclusion, PRISMA is an opportunity for the stellar physics community. The impact of the stellar physics community on this mission is needed, since it is time to reinforce the scientific objectives, and also to let other communities (planetary, mm, high energies, etc...) know that these objectives are considered of prime value by stellar physicists.

Let us be confident that the general rule will apply: More will be discovered by PRISMA than has been foreseen by its proposers...

CONCLUSIONS by Evry Schatzman

Two lessons can be drawn from the colloquium. The first one, which is not the least important is a lesson of methodology. The second lesson concerns the future of internal constitution studies. The number of unsolved problems is incredible; from the theoretical point of view, we can show the directions of future work; let's bet that large observational programs will bring unexpected data which will simultaneously help us to understand the inside of the stars and certainly raise new problems.

As far as methodology is concerned, I would like to draw attention to the following points:

- Models, taking into account just one physical process generally do not agree with the observations. This provides evidence for the presence of other physical processes.
- In each phenomenon which we consider are involved a variety of physical processes. It appears necessary now to take into account all physical processes simultaneously. Stars have to be considered globally.
- We should resist to the temptation of using too much phenomenology. Laws of physics are universal, and a law written in a phenomenological form, applying to a particular case, should be checked by applying it to all possible other cases.

I know quite well that such remarks are completely useless, and that any one of us has its own working method. However, I think that future theoretical work should be more rigourous.

It is in such a spirit that we should look at the unsolved problems. Let us consider some examples:

- The helium content of the Sun. Helioseismology gives Y = 0.23 in the convective zone. Models give for the inside Y = 0.27 or 0.28. Does gravitational separation explain the difference? And then, what are the constraints on the transport processes? What are the constraints on the hydrodynamics in the radiative zone?
- Abundance of lithium in fast rotating stars. We have the exemple of α Per fast rotators, with a high lithium abundance. Are the present models of lithium burning compatible with these abundances?
- The data on magnetic fields of solar like stars is becoming more and more important. This raises the problem of a better understanding of the origin of stellar magnetic fields and of their evolution.

Finally, let me recall the need of more and more precise observational data. Despite the fact that it does not necessarily involve new spectacular instrumentation, such high quality information is of unesteemable value. Each step of the accuracy performances raises new theoretical problems: this is the way to a better understanding of the inside of stars.

But I would like to end these remarks by a come back to the introduction of Malcolm Longair. The theory of stellar structure gives the key to chronometry of the Universe. At the present epoch, there is a tendency to develop more and more extragalactic research. It is true that this is a field with a large potential number of discoveries. But, as long as galaxies are made of stars, our understanding, our system of datation, our survey of the depths of space depend on our knowledge of stars.

Astrophysics is not made of stellar studies only, but there would not be any astrophysics without it!

REFERENCES

Biermann, L. 1950, Zeitschrift für Naturforschung A, **56**, 65

Charboneau P. and MacGregor K. B., 1992, in: Inside the Stars, eds. W. W. Weiss and A. Baglin, these proceedings

Christensen-Dalsgaard, J., Dilke, F. W. W. and Gough, D. O., 1974, *MNRAS* **169**, 429

Dilke, F. W. W. and Gough, D. O., 1972, *Nature* **240**, 262

Durney, B. R., de Young, D. S. and Passot, T. P., 1990, *ApJ* **362**, 702

Durney, B. R., de Young, D. S. and Roxburgh, I. W., 1992, *On the Generation of Large Scale and Turbulent Magnetic Fields in Solar Type Stars, Solar Physics*, in press

Faulkner, J., Griffiths, K., and Hoyle, F., 1965. *MNRAS* **129**, 363

Libbrecht, K. G. and Woodard, M. F. 1990, *Observations of solar cycle variations in solar p-mode frequencies and splittings.* in: Oji Int. Seminar on Progress of Seismology of the Sun and Stars, eds. Y. Osaki and H. Shibahashi, Springer, Berlin

Noyes, R. W.,Baliunas, S. L, and Guinan, E. F. 1991, *What can other stars tell us about the sun?*, in: The Solar Interior and Atmosphere, ed A. Cox, W. Livingston, and M. Matthews, U. Ariz. Press, p. 1161

Radick, R. R., Lockwood, G. W., and Baliunas, S. L. 1990, *Science* **247**, 39

Roxburgh, I. W., 1966, *MNRAS* **132**, 201

Roxburgh, I. W., 1976, *The Internal Structure of the Sun and Solar Type Stars*, in: Basic Mechanisms of Solar Activity, eds. V. Bumba and J. Kleczeck, D. Reidel, Dordrecht, p. 451

Roxburgh, I. W., 1978, *A&A* **65**, 281

Roxburgh, I. W., 1984, *On Turbulent Mixing*, in: Observational Tests of the Stellar Evolution Theory, eds. A. Maeder and A. Renzini, D. Reidel, Dordrecht, p. 519

Roxburgh, I. W., 1985, *Solar Physics* **100**, 21

Roxburgh, I. W., 1987, in: The Internal Solar Angular Velocity, eds. B. R. Durney and S. Sofia, Kluwer, Dordrecht, p. 1

Roxburgh, I. W., 1989, *A&A* **211**, 361

Roxburgh, I. W., 1991, *Angular Momentum Transport, Magnetic Fields and Mixing*, in: Angular Momentum Evolution of Young Stars, eds. S. Catalano and J. R. Stauffer, Kluwer, Dordrecht, p. 365

Roxburgh, I. W., and Simmons, 1992, Inside the Stars, eds. W. W. Weiss and A. Baglin, these proceedings

Schmidt, J. H. M. M., Rosner, R. and Bohn, H. U., 1984, *ApJ* **283**, 31

Stein, R. and Nordlund, Å, 1989, *ApJ* **342**, L95

Vickers, G., 1970, *Problems of Convection in Stars*, PhD Thesis, University of London

Woodard, M.F., Kuhn, J.R, Murray, N. and Libbrecht, K.G. 1991, *ApJ* **373**, L81

Zahn, J. P., 1991, *A&A* **252**, 179

Inside the Stars, IAU Colloquium 137
ASP Conference Series, Vol. 40, 1993
Werner W. Weiss and Annie Baglin (eds.)

ABSOLUTE MAGNITUDE DETERMINATION OF Be STARS: THE INADEQUACY OF THE 2200 Å BUMP

V. DOAZAN
Observatoire de Paris, 61, Av. de l'Observatoire, 75014 Paris, France

A. DE LA FUENTE
INTA-VILSPA, Apartado 50727, 28080 Madrid, Spain

M. BARYLAK
ESA IUE Observatory, Apartado 50727, 28080 Madrid, Spain

N. CRAMER
Observatoire de Geneve, CH-1290 Sauverny, Suisse

ABSTRACT We present far UV observations of Pleione made with the IUE satellite at two epochs: in 1979, when the star exhibited a strong shell spectrum and, in 1991, when it showed a Be-type spectrum. Between these two epochs, the 2200 Å bump changed dramatically, thus, rendering it inadequate for interstellar extinction/distance determination.

INTRODUCTION

Pleione (BU Tau, 28 Tau, HD 23862, B8Ve, $vsini = 320 kms^{-1}$) is the only Be star wich has been observed in both the Be-shell and the Be phases in the far UV. These unique observations gave us the opportunity to test the validity of the methods which use the 2200 Å bump to determine the interstellar extinction/distance of Be stars.

OBSERVATIONS

The study of low (6 Å) and high resolution (0.2 Å) spectra of Pleione obtained with the International Ultraviolet Explorer (IUE) during the strong Be-shell phase of 1979 and the well-developed Be phase of 1991 shows that :

1. the 2200 Å bump varies dramatically
2. the far UV radiative energy flux increases considerably between the Be-shell and the Be phases in the 1250-3000 Å wavelength range, (see Doazan et al., 1991)
3. the slope of the far UV continuum changes strongly.

Fig. 1

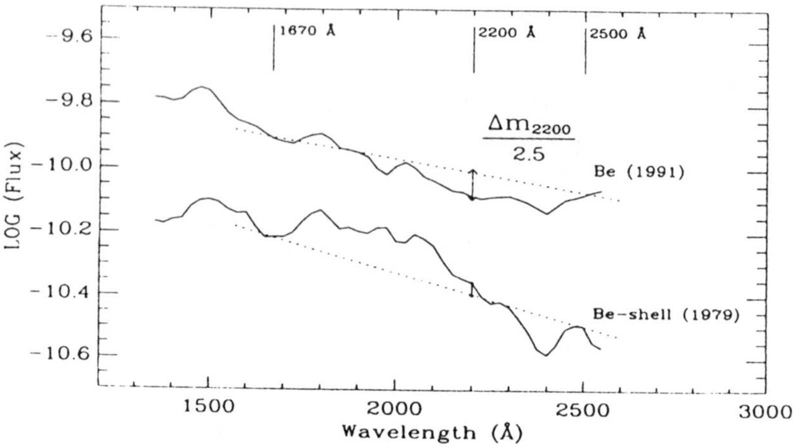

Fig. 1 presents the spectral region of Pleione used for determining the interstellar extinction of Be stars from the observed 2200 Å bump. It is plotted at the two epochs: strong Be-shell (Jan. 2, 1979: SWP3780 & LWR3341) and well-developed Be phase (Aug. 21, 1991: SWP42302 & LWP21051). Note the large increase in absolute flux between these two epochs (no offset is applied).

We see immediately that the *pattern of the 2200 Å spectral region has completely changed between these two phases.* Moreover, the bump is in absorption in the Be phase, but its appears in weak emission in the Be-shell phase.
The study of the high resolution far UV spectra of Pleione obtained at the same epochs shows that the entire spectral region (1550-2550 Å) used in the determination of the 2200 Å bump is crowded by shell absorption lines during the Be-shell phase. We identified more that 1300 lines in this bump. By contrast, during the Be phase very few spectral lines are observed in that region.

CONCLUSION

The above far UV observations of Pleione lead to the two unavoidable conclusions:

1. the 2200 Å bump cannot be used reliably for the determination of the interstellar extinction/distance of Be stars - as was done in several far UV surveys

2. any absolute magnitude determination of Be stars must be based on observations made in the broadest possible spectral range and must necessarily include the far UV spectral region, where most of the variable radiative energy flux is emitted.

REFERENCES

Doazan, V., Sedmak, G., Barylak, M., Rusconi, L.: 1991, *A Be star Atlas of far UV and optical high resolution spectra*, ESA SP-1147

Inside the Stars, IAU Colloquium 137
ASP Conference Series, Vol. 40, 1993
Werner W. Weiss and Annie Baglin (eds.)

ASTEROSEISMOLOGY OF THE DBV WHITE DWARF GD 358 WITH THE WHOLE EARTH TELESCOPE

D. E. Winget[1 2], R. E. Nather[1], J. C. Clemens[1], J. Provencal[1], S. J. Kleinman[1], P. A. Bradley[1], C. F. Claver[1], J. S. Dixson[1], M. A. Wood[3], A. D. Grauer[4 5], B. P. Hine[6], C. J. Hansen[7], P. Bergeron[8], P. Birch[9], M. Candy[9], E. Leibowitz[10], T. M. K. Marar[11], S. Seetha[11], B. N. Ashoka[11], D. O'Donoghue[12], B. Warner[12], D. A. Buckley[12], P. Tripe[12], G. Vauclair[13], N. Dolez[13], M. Chevreton[14], A. Kanaan[15], S. O. Kepler[15 16], and T. Augusteijn[17]

ABSTRACT We report the analysis of 154 hours of nearly continuous high-speed photometric data on the pulsating DB white dwarf (DBV) GD 358 obtained during the Whole Earth Telescope (WET) run of May 1990. The Fourier transform (FT) of the light curve is dominated by power in the range from 1200 – 1700μHz, with more than 180 significant peaks in the total transform. We also see significant power at the sums and differences of the dominant frequencies, indicating the importance of nonlinear behavior. We can use this data to obtain an accurate total stellar mass, and surface He layer mass. The implied surface He layer mass, if correct, provides a significant and surprising challenge to stellar evolution theory, as well as the theory of chemical mixing.

---

1 Department of Astronomy McDonald Observatory, University of Texas
2 Alfred P. Sloan Research Fellow
3 Department of Physics and Space Sciences, Florida Institute of Technology
4 Department of Physics and Astronomy, University of Arkansas
5 Visiting Astronomer, Kitt Peak National Observatory
6 NASA Ames Research Center
7 J.I.L.A., University of Colorado
8 Departement de Physique, Université de Montréal, Canada
9 Perth Observatory, Bickley, Western Australia
10 Department of Physics and Astronomy, Ramat Aviv, Israel
11 Indian Space Research Organization, Bangalore, India
12 Department of Astronomy, University of Cape Town, South Africa
13 Observatoire Midi-Pyrenees, France
14 Observatoire de Meudon, France
15 Instituto de Fisica, Universidade Federal do Rio Grande do Sul, Brazil
16 Visiting Astronomer, Cerro Tololo Interamerican Observatory
17 European Southern Observatory Chile

We observed the complex pulsator GD 358 with the Whole Earth Telescope (hereafter WET, see Nather *et al.* 1990 for a description of the WET) during May 1990. The data were reduced and analyzed using the techniques described in Nather *et al.* (1990), and Winget *et al.* (1991).

We display a portion of the FT of the light curve in Figure 1 in a multi-panel form. The top panel has been rescaled to accommodate, at least partly, the large dynamic range of the power spectrum. The data were nearly continuous during the central portion of the run; this is evidenced by the extremely low power levels in the side lobes of a single frequency sampled as the data (see Figure 1 insert). The FT is dominated by power in the range from $1000 - 2000\mu Hz$. This power consists of triplets with a typical internal uniform frequency spacing of $6.2 \pm 0.1\mu Hz$.

The triplets are consistent with $l = 1$ non-radial g-modes with radial overtones from $k = 13 - 18$, approximately equally spaced in period. We have indicated tentative radial overtone numbers (based on preliminary numerical modeling by P. A. Bradley 1992, private communication) with numerical labels in the figure. We have placed arrows over the expected locations of overtones 10,11,12,19 and 21. With the exception of three other small groups of power, essentially all of the significant power in the light curve can be accounted for in the context of this set of $l = 1$ triplets, and sums and differences of these frequencies.

The mean period spacing of the triplets is consistent with the theoretical expectations based on extensive numerical calculations by one of us (P. A. Bradley) for consecutive radial overtones of $l = 1$ nonradial g-modes, and a mean stellar mass near 0.60 solar masses. The deviations from uniform period spacing indicate compositional stratification with a surface He-layer fractional mass much less than $1.0x10^{-6}$.

The fine structure shows some variation from triplet to triplet within the large amplitude modes; preliminary theoretical calculations by C.J. Hansen (private communication) indicate that this structure is consistent with differential rotation in the surface He layer, as suggested by G. Vauclair, and W. Dziembowski. These asteroseismological results have important consequences for prior shell-burning and mass-loss evolutionary stages, as well as white dwarf spectral evolution theory and theories of convective mixing. If we use the mean frequency splitting, and assume that the observed modes are indeed $l = 1$, we have an inferred rotation period of 0.93 days.

Finally, this star presents us with a striking example of nonlinear behavior in the form of sums and differences of the dominant power. A portion of the frequencies that represent sums of the dominant frequencies are indicated in the figure. It is unclear, at present, if the sums and differences represent parametric resonances and mode coupling, or are pulse-shape artifacts, or some combination of both. Even at this preliminary stage in the analysis it is interesting to note that the sums are dominated by the mode labelled 17, even though this is *not* the mode with the highest observed amplitude in this data set. The multiplet fine-structure of the sums will provide some clues in this regard.

ACKNOWLEDGEMENTS

We are grateful for support from CNRS France, SERC UK, the National Science Foundation grants AST85–52456, AST86–00507, AST87–12249, AST88–13572, NASA grant NGT–50210 and from the National Geographic Society grant 3547–87.

REFERENCES

Nather, R. E., Winget, D.E., Clemens, J.C., Hansen, C.J., and Hine, B.P. 1990, Ap. J., **361**, 309.

Winget, D.E., *et al.* 1991, *Ap. J.*, **378**, 326.

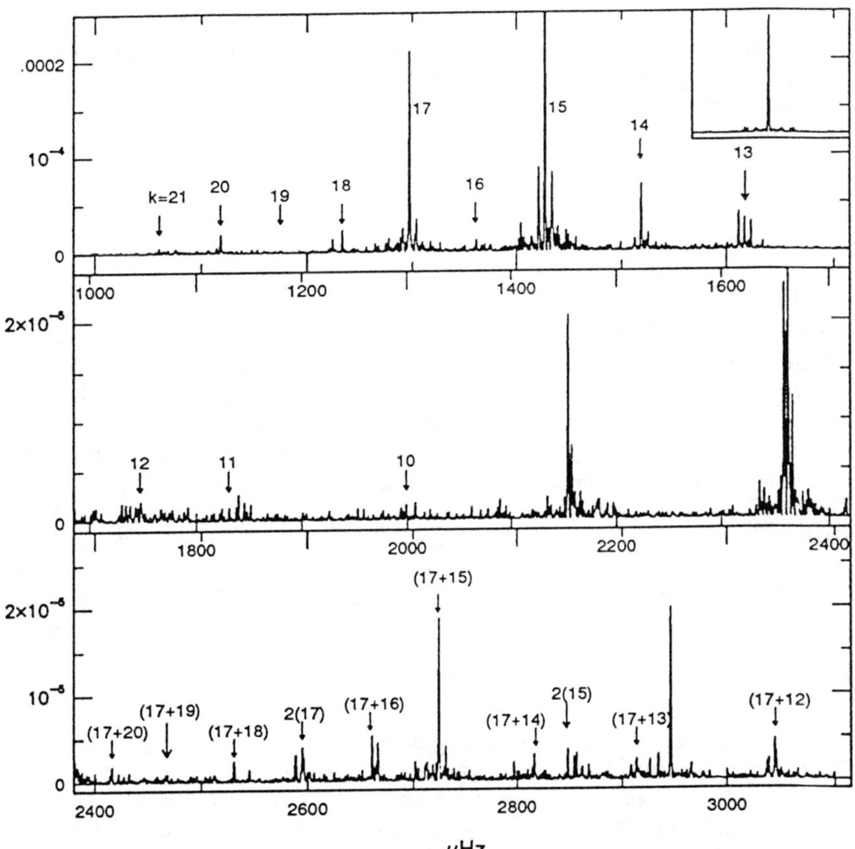

FIGURE 1: A portion of the power spectrum of GD358. Note the different vertical scales in the panels to accomodate the extreme dynamic range. The spectral window is given in the upper corner.

Inside the Stars, IAU Colloquium 137
ASP Conference Series, Vol. 40, 1993
Werner W. Weiss and Annie Baglin (eds.)

CHEMICAL EVOLUTION OF ALGOL-TYPE STARS

Marek J.SARNA
Astronomy Centre, University of Sussex, Falmer, Brighton BN1 9QH, UK. and N.Copernicus Astronomical Center, Polish Academy of Sciences, Warsaw, Poland.

ABSTRACT The evolution of the chemical abundances at the surface of both components of Algol–type binaries is examined. First, we have examined some mixing processes which can affect element distribution in components of Algol–type stars. Second, the C abundance determinations from theoretical models are compared with values known from observational analysis. We conclude that there is no significant difference in chemical composition in components between case AB and early case B mass transfer. We have found a real difference between systems which evolve from initial mass ratios of 10/4 and 10/9.

INTRODUCTION

It is well known that mass transfer between the components of a close binary affects their surface chemical composition (Paczyński 1971). Evolutionary calculations show that nuclear processed layers are exposed at the surface of the loser (primary – the initially more massive star – donor) and transferred to the gainer (secondary – mass accreting star). Chemical abundance analyses (C, N and O) are available for a few Algol–type systems (Cugier and Hardorp 1988, Cugier 1989, Parthasarathy, Lambert and Tomkin 1983, Tomkin and Lambert 1989). In this paper we shall be interested in the Algol–type binaries and discuss the importance of the different mixing processes in their components. Because available observational data concerns short period Algols ($1^d - 3^d$) we have decided to consider mass transfer in case AB and early case B.

MIXING PROCESSES

We have considered the following mixing mechanisms: ordinary convection, Eddington–Sweet circulation in the whole star and/or near the surface, circulation caused by the accretion effect and mixing due to inversion of the molecular weight (μ) in the outer layers (instability due to Ledoux criterion, thermohaline mixing). The main conclusion may be summarized as follows. Only convection, thermohaline mixing and circulation caused by the accretion effect are important for binary systems in the semi–detached phase. Other processes have either a very long timescale in comparison to the mass transfer timescale or affect only the thin outer layers of the envelopes.

ASSUMPTION AND INITIAL CONDITIONS

Our calculations were carried out under the following assumptions:
1) The initial total mass ($M_T = M_{1,0} + M_{2,0}$) and mass ratios ($q_0 = M_{1,0}/M_{2,0}$) are: $M_T = 6\ M_\odot$ and $q_0 = 10/4, 10/9$.
2) All evolutionary calculations were carried out using a standard Henyey-type code (Paczyński 1970). Both components of the binary systems were calculated simultaneously. Only conservative evolution is considered.
3) We included two of the different mechanisms of mixing in our numerical code: the accretion effect and the thermohaline mixing.

RESULTS

Constructing conservative evolutionary models means that we assume the total mass and the total angular momentum of a binary system to be conserved during the process of the mass transfer between the components of the system. We will depart from these principles only in the case when a thick convective envelope has developed in the primary component. In this case loss of angular momentum by a magnetic stellar wind is turned on automatically. The models presented in this part describe the evolution of a few systems with different periods and mass ratios. The chemical evolution for the carbon (C) on the stellar surface for both components in a system exchanging mass in early case B ($M_{1,0} = 4.28\ M_\odot, q_0 = 10/4, P_0 = 1^d.7$) is presented in the figure below.

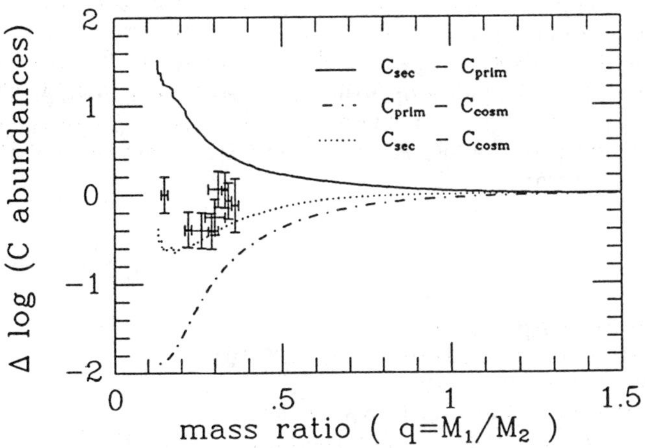

Fig. 1. The carbon abundance of the donor and gainer of the system $M_{1,0} = 4.28\ M_\odot$, $q_0=10/4$, $P_0 = 1^d.7$, as a function of mass ratio. The nine errorbars presented the carbon abundance for secondaries in Algol-type stars. Note that evolution is from right to left.

We can see that during the first fast phase of mass transfer to about $t=6.3\times10^4$ yrs and $q=1.4$ there are no differences in the element abundances between components and cosmic values. Next carbon gradually decreases. During the slower phase of mass transfer the primary presents a very high depletion (-2 dex) of carbon abundance while the secondary shows an intermediate value from −0.4 to −0.6 dex (in the mass ratio range 0.4–0.1). In the secondary envelope accretion effects at first play an important role (to $t=7.\times10^5$yrs); next the thermohaline mixing determines the element distribution in the envelope. We have not found (Sarna 1992) significant differences in the value of the carbon depletion between systems evolved from case AB and B (for this same q_0). However, we have found real differences between systems evolved from initial mass ratios of 10/4 and 10/9. For a system that has undergone mass exchange in case B with orbital parameters: $M_{1,0} = 3.16\ M_\odot, q_0 = 10/9, P_0 = 1^d.617$, when we consider the carbon profile then we notice that depletion of this element is relatively small and equals −0.1 to −0.2 dex for mass ratios in the range 0.6 – 0.1. The depletion is two times less than in case B with $q_0 = 10/4$ (see Fig.1). For the model evolving by case AB mass transfer, with $q_0 = 10/9$ and $M_T = 6\ M_\odot$ the convective layers have developed when hydrogen rich matter was transferred from the mass–losing star. In the convective envelope of the primary $d\mu/dr \sim 0$, and therefore in the secondary envelope the thermohaline convection turns on at the end of mass transfer (because $\tau_{diff} \sim \overline{\mu}/\Delta\mu$).

CONCLUSIONS

The results of our evolutionary investigations can be summarized as follows.
1) The conservative models have explained well the observed carbon abundances in secondary components of Algols.
2) There is no significant difference in the chemical composition between case AB and early case B mass transfer. We have found real differences between systems evolving from initial mass ratios of 10/4 and 10/9. The systems with $q_0=10/4$ show twice the depletion of carbon of those with $q_0=10/9$. Therefore, we conclude that the "correlation" between carbon depletion and the mass ratio probably shows only the bimodality of the initial mass ratio distribution for proto–Algol–type stars.

REFERENCES

Cugier, H. 1989, Astr. Ap., 214,168.
Cugier, H. and Hardorp, J. 1988, Astr. Ap., 202,101.
Paczyński, B. 1970, Acta Astr., 20,47.
Paczyński, B. 1971, Ann. Rev. Astr. Ap., 9,183.
Parthasarathy, M., Lambert, D. L. and Tomkin, J. 1983, M.N.R.A.S., 203,1063.
Sarna, M.J. 1992, M.N.R.A.S.,(in press).
Tomkin, J. and Lambert, D. L. 1989, M.N.R.A.S., 241,777.

Inside the Stars, IAU Colloquium 137
ASP Conference Series, Vol. 40, 1993
Werner W. Weiss and Annie Baglin (eds.)

STAR FORMATION AND THE ORIGIN OF STELLAR MASSES

PH. PODSIADLOWSKI AND N. M. PRICE
Institute of Astronomy, Madingley Road, Cambridge, England

ABSTRACT We present a new model to explain stellar mass distributions in different stellar environments. In our model, the protostar phase is terminated, when the protostellar core embedded in a molecular clump experiences a collision with another star or protostellar clump, which ejects the protostellar core from its parent clump. Such dynamical interactions are necessarily important, if stars preferentially form in dense clusters. We show that, in a simple model, the initial mass function approaches a simple, asymptotic form, which strongly resembles observed mass functions. The model has important consequences for star formation in different environments. We also discuss the implications of the model for our understanding of pre-main-sequence stellar evolution.

1. COLLISIONS AND THE INITIAL MASS FUNCTION (IMF)

Modern observations of star-forming regions show that stars preferentially form in very dense, but ultimately unbound clusters (for references see Podsiadlowski and Price 1992). This suggests that collisions between young stellar objects (YSOs) are important. We have studied two types of collisions: collisions between two protostellar cores (i.e., protostars embedded in dense envelopes from which they accrete) and collisions between a protostellar core and a star (i.e., a YSO which is no longer embedded in a massive envelope). The cross section for collisions between protostellar cores is mainly determined by the geometric size of their envelopes. The most likely outcome, if the collision is supersonic, is that the protostellar cores are separated from their envelopes, thus terminating the protostar accretion phase. A collision between a protostellar core and a star will also separate a protostar from its envelope, if the kick velocity imparted to the protostar (relative to its envelope) by a star passing through the core exceeds the central escape velocity. We estimate that, for characteristic core properties, the cross section for the latter collisions is of order 10 % of the cross section for direct core-core collisions. In the simulations we present below, direct core-core collisions dominate initially, until the number of stars significantly exceeds the number of protostellar cores.

To illustrate how such collisions lead to a mass spectrum, we adopt a very simple model for the star-forming process. We assume that stars form at a constant rate (per unit gas mass), α, and that protostars accrete mass at a constant rate, $\dot{M}$, until they are ejected (the individual rates for core-core collisions and core-star collisions are β and βZ $[Z \sim 0.1]$, respectively). In Figure 1 we

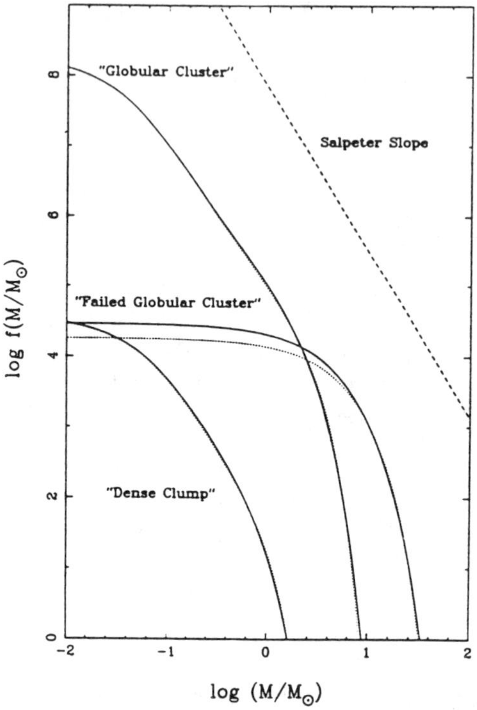

Figure 1 – IMFs resulting from a collisional model for three star-formation environments, a globular cluster (GC), a "failed" GC and a dense molecular clump ("Clump"), respectively (solid curves as indicated). The "GC" and the "Clump" IMFs both have power-law segments, resembling a Salpeter IMF (shown as a dashed curve). The initial gas masses M_g, in the three cases were 10^6, 10^6 and $5 \times 10^2 M_\odot$, the star–formation rates, α, were 10^{-7}, 10^{-8} and $10^{-8}/M_\odot$/yr, the collisions rates, β, were 10^{-9}, 5×10^{-10} and 2×10^{-7}/yr (Z=0.1 in all cases), the protostar accretion rates, $\dot{M}$, were $10^{-5}, 10^{-5}$ and $10^{-6}\, M_\odot$/yr. At the end of the simulation (after 10^8, 10^7 and 5×10^7 yrs, respectively), the gas mass fractions were $\sim$ 50, $\sim$ 70 and $\sim$ 70%, respectively.

present the IMFs that result from this simple model for different star-forming environments, where we used parameters which may be representative for a dense molecular clump, a bound (surviving) globular cluster and a "failed" globular cluster, respectively.

One of the major results of our calculations is that the IMF asymptotically approaches a simple and universal distribution provided that star formation is relatively inefficient (which here means that the timescale on which gas is locked up in stars is longer than the core-star collision time). This asymptotic solution is of the form:

$$f(m)\, dm \propto \frac{1-(1+m)e^{-m}}{m^2} g\, dm \tag{1}$$

where $m \equiv M/M_{\rm sc}$ is a dimensionless mass and $g \equiv \exp[-(M/M_{\rm co})^2]$ is the exponential cut-off factor (M is the mass of the star, and $M_{\rm co}$ and $M_{\rm sc}$ are the cut-off mass and the scaling mass respectively). For $M_{\rm sc} < M < M_{\rm co}$, the distribution approaches a Salpeter-like power law (Salpeter 1955), $f(M)dm \propto m^{-2.35}dm$, and the whole IMF exhibits all the gross features of Scalo's mass function (Scalo 1986). While the cut-off mass depends only on intrinsic parameters ($M_{\rm co} \sim \dot{M}/\sqrt{\beta Z \alpha M_{\rm g}}$, where $M_{\rm g}$ is the total mass of the cloud or clump), the scaling mass decreases with time, t, ($M_{\rm sc} \simeq \dot{M}/[\beta Z \alpha M_{\rm g} t]$). Thus the distribution more closely approaches a simple power law to lower masses the longer star formation proceeds. Although this simple model is not fully realistic, the very existence of a universal asymptotic solution may be sufficient to guarantee that the overall distribution, averaged over various populations, may still reflect the basic characteristics of the solution.

2. APPLICATIONS

Our model has important implications for star formation in different stellar environments. For example, the cut-off of the distribution at high masses depends on the density of the star-forming region. In high-collision environments like globular clusters, massive stars may be unable to form (see Fig. 1). This may avoid the problem of self-pollution with heavy elements produced in the first supernovae without requiring very fast and efficient star formation in globular clusters. On the other hand, if failed globular clusters are the environment in which halo stars and population III stars are formed, their IMF should be relatively flat, i.e., biased towards high masses (see Fig. 1).

The model may also be relevant to the formation of binary and multiple systems, since core-star collisions in which the protostar is dislodged from the center of its core, but remains bound to it should be much more common than collisions which eject the protostar. After a protostar has been displaced from the center of a core, a new protostar can form in the center, which may ultimately lead to the formation of a binary.

3. PRE-MAIN-SEQUENCE EVOLUTION

Our model can be tested most directly — at least in principle — by comparing its predictions with the mass distributions observed in individual star-forming clusters. This requires knowledge of the mass – luminosity relation of accreting protostars to construct the protostar luminosity function (PLF; alternatively one can use this method to test protostar theory). Protostar theory is rather uncertain, since the structure of protostars depends on the accretion rate, $\dot{M}$, and is sensitive to the surface boundary condition. We have constructed a simple protostar luminosity function based on the following assumptions: (1) Protostars lie on Hayashi tracks. (2) Before deuterium burning, the mass – radius relation is determined by the condition that the accretion timescale is of the order of the Kelvin-Helmholtz time. (3) During deuterium burning, the mass – radius relation is approximately given by $R \simeq 5\, R_\odot\,(M/M_\odot)$. With these assumptions, we can derive a theoretical birthline in the Hertzsprung-Russell diagram and a PLF. While our theoretical birthline is in good agreement with the observed birthline (also see Stahler 1988), our PLF differs substantially from the one derived by Stahler and Fletcher (1992, preprint); most importantly, it has no peak at low masses ($M \sim 0.1\, M_\odot$). This difference can be entirely attributed to different initial conditions for the accretion process. We emphasize that this discrepancy has to be resolved, before a meaningful comparison between our star-formation scenario and observations of star-forming clusters can be made.

REFERENCES

Podsiadlowski, Ph. and Price, N. M. 1992, Nature, in press.

Salpeter, E. E. 1955, ApJ, 121, 161.

Scalo, J. M. 1986, Fund. Cosmic Phys., 11, 1.

Stahler, S. 1988, ApJ, 332, 804.

Inside the Stars, IAU Colloquium 137
ASP Conference Series, Vol. 40, 1993
Werner W. Weiss and Annie Baglin (eds.)

A NEW MODEL FOR THE STRUCTURE AND EVOLUTION OF MAIN SEQUENCE STARS.

RAFAEL A. VERA

Departamento de Fisica. Facultad de Ciencias. Universidad de Concepcion. Casilla 3C. Concepcion. Chile.

ABSTRACT From both astronomical facts and theory it is inferred that the primary energy of main sequence stars most probably comes from gravitational work around a neutron star. The new mechanisms involved can in principle account for the observed facts, including those unconsistent with conventional models.

INTRODUCTION

It is well recognized that the densest states of matter in the universe are those of neutron stars (NSs) and black holes (BHs). The total gravitational (G) energy released from collapse of matter around a NS or a BH is about two orders of magnitude larger than that of nuclear fusion (NF) of the same mass of H. Then it is reasonable that the energy observed in the universe comes, primarily, from G collapse around NSs. Since most of this energy comes from main sequence stars (MSSs), this most probably occurs just inside of them.
From the works of Vera (1981a,b, 1986) on gravity, it is also inferred that stars would normally be formed by condensation of gas around black remains from previous stellar evolution. Condensation around dead stars and NSs would rapidly regenerate new stars of high density, i.e., MSSs or blue stragglers. Condensation around planets and other bodies of similar densities would form voluminous red stars. The identification of these stars in HR diagrams of star clusters is simple.

THE NEW MAIN SEQUENCE STAR MODEL

In it, matter would collapse around a central NS. The high energy photons and plasma released in this region, similarly to the case of a drop of water over hot iron, would generate an interphase that would temporarily stop the collapse until the energy in excess is dispersed. Neutrons of elements heavier than H falling into the NS, would be preferentially captured by it, according to "nuclear stripping". The high energy protons

rejected by the NS can dissociate other heavy nuclei. This lower density material, with regenerated nuclear latent energy, would support convection currents transporting heat and nuclear latent energy to upper levels, where they can be liberated. this would keep high star density and superficial temperatures. Nuclear fusion of H would would generate neutrons that would be captured, later on, by the NS.
The high differential rotation around the NS would prevent non axial unsymmetries. The strong magnetic field generated by the rotation of the positively charged NS, would delay, preferentially, the non axial fall of electrons thus increasing, temporarily, the positive charges of the rotating core. This, turn, would increase the magnetic field and so on. These fields and the high energy photons and particles generated during collapse would react against it. Then the collapse pulses amplitudes would be somewhat larger about the polar regions. Thus this mechanism would be primary source of all of them: energy, magnetic fields, high energy particles and photons, convection currents and coherent pressure wavefronts. Through the time, the NS mass would increase at the cost of a decrease of that of its external envelope thus producing a slight temperature increase, i.e., becoming bluer. The amplitude of the pressure pulses travelling away from the core would normally be absorbed within the envelope thickness unless that some natural oscillation modes can be excited thus producing variable stars. Eventually, these pulses could trigger coherent thermonuclear explosions within the envelope. Depending on the masses, they could either eject rather spherical shells of gas (planetary nebulas) or feed back more powerful collapses, with more complete envelopes ejections thus leaving rather naked NSs that may also continue capturing matter in pulsed ways (pulsars).

Pre main sequence stages.
Old planets or similar bodies covered by new gas would become low density (rather giant) red stars with a central core of heavier elements. The gradient of atomic masses in these cores would prevent convection in them. Thus, stars would contract themselves similarly to conventional models, but with somewhat higher central densities. During this contraction, the endoenergetic nuclear reactions of heavy elements in the core below would absorb energy from gravitational contraction and NF of H. Sooner or later a macronuclei or NS is likely to start growing there. The increasing G energy released around the NS would initiate new convective currents that would speed up the transition from low density red stars up to bluer MSSs. This is consistent with the low number of stars observed in the HR diagram between red giants and MSSs. During these changes, pulsed collapses can also excite natural oscillation modes thus producing periodical variable stars.

Main tests:

For reason of space, only a few ones can be mentioned here:

1.- The low neutrino luminosity of the Sun would. A NS of about 0.01Mo would account for it.

2.- The theoretical mass-luminosity, equal to the product of the average rate of neutron generation and the G energy yield per neutron, is proportional to $M^{3.67}$ in agreement with current literature.

3.- The main internal magnetic field would be produced through the above mechanism so that magnetic unsymmetries during sunspot cycles would come from axial oscillation of the NS relative to the envelope.

4.- The similitude of pulse shapes of supernovas and those observed in many stellar phenomena, like variable stars, pulsars, supernovas and X-ray burstars would come from the similitude of collapses under different conditions.

5.- The coherence of stellar surface oscillations and or pulsations are clearly consistent with central NSs. The symmetry of planetary nebulas also shows that some coherent thermonuclear explosion was triggered by a pressure wave coming from nearly point central source.

6.- The smaller Sun quadrupole moment, derived from difference between the theoretical and experimental values of the perihelium shift of mercury. This shows up the dominant effect of the rather spherical G field of the central NS.

7.- Narrow cosmic jets of high energy particles. They would be produced by the above mechanism occurring around BHs. The small cosmic ray peak with magnetic rigidity of 1.6 GV reported by Mc Donald (1959) would correspond to He fall (Vera 1981b).

8.- The blue progenitor of SN1987A would correspond, to the contrary of conventional models, to the latest stages of MSS evolution. Observe that this would come out from evolution of some red giant, after formation a NS which may grow up through more that one stellar cycle untill becoming a BH.

REFERENCES

McDonald and B. Webber. 1959, Phys. Rev. 115, 194.

Vera, R. A. 1981a, in Proceedings of the Einstein Centennial Simposium on Fundamental Physics, ed. S.M. Moore, A. Rodriguez-Vargas, A. Rueda, G. Viollini (Universidad de los Andes, Bogota, Universita di Roma), p. 597.

Vera, R. A. 1981b, Int. J. of Th. Phys. 20, 19.

Vera, R. A. 1986, in Proceedings of the IV Marcel Grossmann Meeting on General Relativity, ed. R. Ruffini, Elsevier Science Publications B V, p. 1743.

Inside the Stars, IAU Colloquium 137
ASP Conference Series, Vol. 40, 1993
Werner W. Weiss and Annie Baglin (eds.)

THE EFFECT OF ELECTRON BEAMS ON THE HYDROGEN EMISSION IN IMPULSIVE FLARES ON THE SUN AND STARS

V.V. ZHARKOVA AND V.A.KOBYLINSKY
Physics Department, Kiev University, Glushkova 6, Kiev 252022, Ukraine

INTRODUCTION

Flares are very frequent phenomena on the sun and on other stars and are actively investigated with precise observational techniques. These studies are particularly stimulated by recent X-ray and UV observations on spacecraft (Skylab, SMM). From all these observations the flares on the sun and stars seem to be of the same physical nature but of different energetic power. A number of investigations devoted to comparison between the solar and stellar flares led to the conclusion that they have the similar mechanisms for the energy release and transport and for their manifestations in different ranges of electromagnetic radiation.

The hydrogen emission is easy to observe in these events, and thus it is convenient for diagnostic analysis of physical conditions in the flares. However, the impulsive events were considered to be observed in X-rays so only. Zharkova & Kobylinsky (1992, 1993) have recently shown that the hydrogen continuous radiation of the solar flares is sensitive to the electron beam parameters. We also develop an approach to detection of the flares of non-thermal or thermal origin by observing the hydrogen Balmer and Pashen continuum emissions with high temporal resolution.

FORMATION OF THE HYDROGEN EMISSION IN IMPULSIVE FLARES

The hydrogen emission in elementary flare bursts is originated during a very short period of the electron beam injection. We consider the first few seconds of the injection, during which the hydrogen emissions in UV and in the optical spectrum are appear to be correlate with the hard X-rays. The state of the flaring plasma is described simultaneously by gas-dynamic equations and radiative transfer, statistical equilibrium and particle conservation equations. As a first approximation we use a solution of the gas-dynamic problem obtained by Somov *et al.* (1981) for an appropriate set of the beam parameters F_0 and γ. Then we calculate the hydrogen line and continuum intensities by solving the full non-LTE problem. The technique is described in details by Zharkova & Kobylinsky (1991).

We consider the electron beam effects on the flaring atmosphere for the initial energy flux of electrons with energy above 20 keV on a top boundary $F_0 = 10^9$ and 10^{11} erg $cm^{-2}s^{-1}$ and for the X-ray power index γ=3 and 5. These values are in agreement with the solar observations and the return current stability limit of an electron beam (Aboudarham & Henoux, 1986; Zharkova

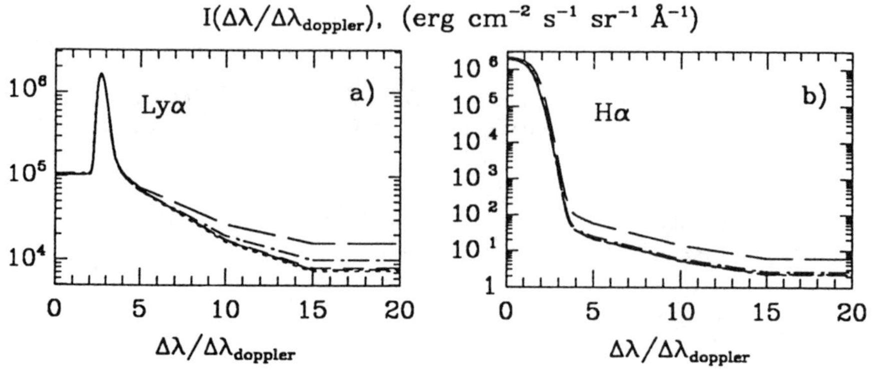

FIGURE I The hydrogen Lyα and Hα line intensities computed for the gas-dynamic flare models at time 2.5 s: a pure thermal model (dotted curves); models with energetic electrons: γ=3, F_0 =10^9 erg cm^{-2}s^{-1} (short dashes); γ=3, F_0 =10^{11} (long dashes); γ=5, F_0 =10^9 (dot-short-dashed lines); γ=5, F_0 =10^{11} (dot-long-dashed lines)

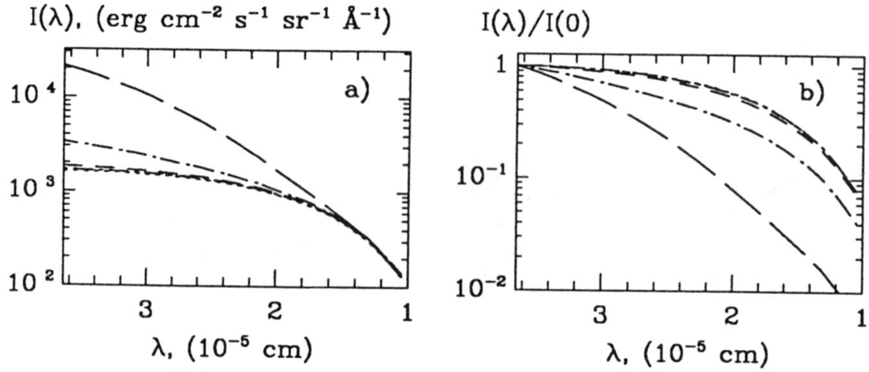

FIGURE II The intensity distributions (a) and slopes (b) of the Ba-continuum at 2.5 s for the same flare models as in Figure I.

& Kobylinsky, 1992). For other stars the higher values of F_0 =10^{12}-10^{13} erg cm^{-2}s^{-1} should be used. In Figure I we illustrate the hydrogen Lyα and Hα lines intensities and in Figure II the intensity distribution and slope in the head of the Ba continuum computed for the gas-dynamic models at 2.5 s after the beginning of electron beam injection. Immediately after the beam injection the non-thermal effects are confined in the high chromosphere. After the gas-dynamic response of the chromosphere ($t > 4$ s) the influences of the energetic electrons on the line profiles and continuum intensity distributions and their slopes depend strongly on physical conditions in the flares and on beam parameters.

Hydrogen Lyα and Lyβ line profiles vary rather similarly. During the first second the line core intensities decrease for low F_0 (10^9) and increase for higher F_0 (10^{11}) as compared with the pure thermal model. In contrast the wing intensities depend strongly on the spectral index γ and are less sensitive to the initial flux F_0, increasing for smaller γ and higher F_0. At 2.5 s after the injection the Ly-line profiles show a significant self-absorption in higher intensities at smaller γ. After 4 s the Ly-line profiles begin to return to their form without the self-absorption. The Hα-line profiles (see Figure I) become broader and their central intensities become lower with F_0 increasing and γ decreasing. The most significant effects occur in the line wings, where for the large fluxes (10^{11} erg cm^{-2}s^{-1}) the intensity increases by factor 2 as compared with the thermal model. We conclude that the beam electrons affect the hydrogen lines mainly in their wings, and that rather precise observations are needed to detect the effects.

The hydrogen continuum intensities are much more sensitive to non-thermal impacts. The Ly-head intensity is grows slightly with increasing F_0 and decreasing γ for all the time after the injection. The maximum variations occur in the first second, and then they gradually decay. The Ba (Figure II) and Pa continuum intensity distributions are greatly affected by the non-thermal impacts. In the first second they engender strong rising of the Ba and Pa-head intensities, particularly, for high F_0 and small γ. The head intensity slopes are also increased by the non-thermal effects. It is interesting that for the beams with initial fluxes $F_0=10^{11}$ erg cm^{-2}s^{-1} and $\gamma=3$ we see almost a straight line in Figure II in spite of the varying slope over wavelength for other parameters. Later ($t > 4$s) the effect in the Ba and Pa continuum intensities is smaller because of the increasing role of the thermal processes due to the gas-dynamic processes.

CONCLUSIONS

The non-thermal effects in the hydrogen continua give an opportunity to study behaviour of electron beams in deep layers to which they penetrate into the chromosphere. The non-thermal processes are probablely responsible for a temporal correlation between bursts observed in the hard X-rays and in UV-spectra during the beam injections in the impulsive solar flares. The events of smaller scales on the sun (micro- and nanoflares) and of greater scales on flare stars are candidates for the coronal heating mechanism and solar and stellar wind origin. If the solar flares and these flare-like phenomena have the same non-thermal origin we can achieve a better understanding of the fundamental problems on both the sun and the stars.

The authors are very grateful to the LOC of the IAU Colloquium 137 and to Prof. W.Weiss for financial support and the opportunity to take a part in the meeting.

REFERENCES

Aboudarham,J. & Henoux,J.-C.: 1986, *Astron.Astrophys.* **156**, 73
Somov,B.V., Spektor,A.R. & Syrovatskii, S.I.: 1981, *Solar Phys.* **73**, 145
Zharkova,V.V. & Kobylinsky,V.A.: 1989, *Soviet Astron.Lett.* **15**, 366
Zharkova,V.V. & Kobylinsky,V.A.: 1991, *Soviet Astron.Lett.* **17**, 34
Zharkova,V.V. & Kobylinsky,V.A.: 1992, *Kinemat. and phys. of space bodies* **8** No 3 (in Russian)
Zharkova,V.V. & Kobylinsky,V.A.: 1993, *Solar Phys.* (submitted)

Inside the Stars, IAU Colloquium 137
ASP Conference Series, Vol. 40, 1993
Werner W. Weiss and Annie Baglin (eds.)

SPECTROSCOPIC OBSERVATIONS OF THE Be STAR PHI-PER

MAHENDRA SINGH
Uttar Pradesh State Observatory Manora Peak
Naini Tal- 263 129, India

MASAHIRO MON and TOMOKAZU KOGURE
Department of Astronomy, Faculty of Science
Kyoto University Sakyo-Ku Kyoto 606, Japan

MASAKAZU SUZUKI
Kanazawa Institute of Technology
P.O.Kanazawa-South,Ishikawa 921 Kanazawa,Japan

ABSTRACT: *Based on a series of high dispersion Coude spectra taken with 188-cm telescope of Okayama Astrophysical Observatory over an extensive period of time (nearly 20 years),spectroscopic behaviours were measured and discussed. Synchronous variations of V/R (intensity ratio of violet to red emission peaks)variability with the radial velocity curves (RV) for Balmer and HeI 3888 shell absorption lines were confirmed. Using the peak separations of the Balmer emission lines , envelope extension was also estimated. Main difficulty was found in explaining the envelope extension of the primary star of the system which overflows the Roche-lobe limit when we consider the binary nature of the system.*

INTRODUCTION

Phi-Per is a well known spectroscopic binary Be Shell star with the spectral type B2 IV ep (Lesh, 1968).Radial velocity variations of this star were first noted by Campbell(1902) and were attributed to binary motion by Lundendorff (1910) and Cannon (1910).Lockyer (1925,1926) was the first to notice its V/R variability synchronous with phase.Poeckert (1981) and Suzuki (1980) have derived the orbital parameters of the system and found that the masses of the systemic components are $20M_\odot$ *and* $4M_\odot$ *with the separation 1.42 AU.*

In this paper we present our new spectroscopic observations (Coude spectra) taken with 188-cm telescope of Okayama Astrophysical Observatory . The dispersion was 10A/mm. All the observational data were folded for the orbital period of 126.696 days with the epoch $T_0 = JD(Hel)$ *2424473.500 (Hickok,1969).*

The radial velocity (RV) and V/R (intensity ratio of violet to red emission components) variability for the Balmer and HeI 3888 shell absorption lines were determined. Using the peak separations of emission components of Balmer lines the extension of the envelope was estimated.

OBSERVATIONS and MEASUREMENTS

Nearly 100 Coude spectrograms taken with 188-cm reflector of Okayama Astrophysical Observatory (1968-1986) were analysed. PDS(PE) microdensitometer measurements were done at Kwasan Observatory, Kyoto University. The data were processed at the Computer Centre of the Kanazawa Institute of Technology and Kyoto with the help of a software developed by one of our authors M.Suzuki. University. Computer plots of intensity curve in units of both the radial velocity and wavelengths were obtained for individual spectra. From these plots RV values at different points of the line profiles were measured. Correspondingly,intensity values for the emission components were also measured to study the V/R variability. The RV curve for the Balmer (H_β, H_γ, H_δ) and HeI 3888 shell lines are shown in Figures 1 and 2. A plot of log V/R against phase is shown in Fig 3.The peak separations was varying between 150 to 250 km/s. The envelope extension R_e was determined using $R_e = 4GM_1/V_p^2$ (Poeckert,1981),where M_1 is the mass of the primary and V_p is the peak separation of the emission line.

RESULTS and DISCUSSIONS

From Figs. 1,2 and 3 ,it is clear that the radial velocity and V/R both vary with phase synchronously.There is no phase lag.Occasionally this star exhibits quadrupole peaks with two peak separations (Kogure et. al. 1991) suggesting some complicated structure of the envelope which can be interpreted in terms of a double ring structure. If we calculate the envelope extension using Poeckert(1981) expression, we find that the envelope extends upto $265R_\odot$ for a peak separation of 200 km/sec and goes beyond the Roche-lobe of the primary star of the system.

REFERENCES

Campbell, W.W. 1902,Astrophys.J.**16**,195.
Cannon, J.B. 1910,J.R.A.S.,Canada **4**,195.
Hickok, F.R. 1969,M.Sc.Thesis,university of Toronto,Canada.
Kogure, T.,Mon, M.,Suzuki, M.,Singh, M.1991,
Pac.Rim Colloq.on New Frontiers in Binary Stars Research
held on Nov 5-13,1990 at Seoul and Taejon, Korea(in press).
Lesh, J.R. 1968, Astrophys. J.Suppl. **17**,371 .
Lockyer, W.J.S. 1925, M.N.R.A.S. **85**,580.
Lockyer, W.J.S., 1926, M.N.R.A.S. **86**,474 .
Lundendorff, H.1910, Astr.Nachr**186**,17.
Poeckert, R., 1981,P.A.S.Pacific, **93**,297.
Suzuki, M., 1980, Publ.Astron.Soc.Japan, **32**,331.

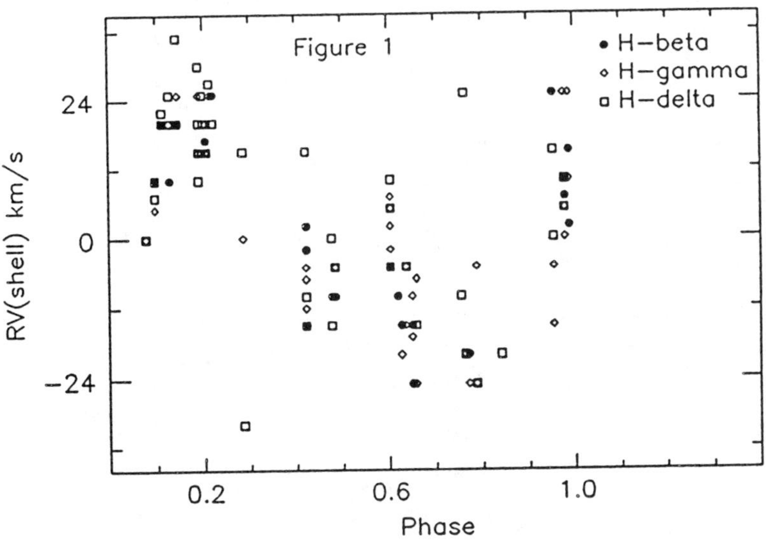

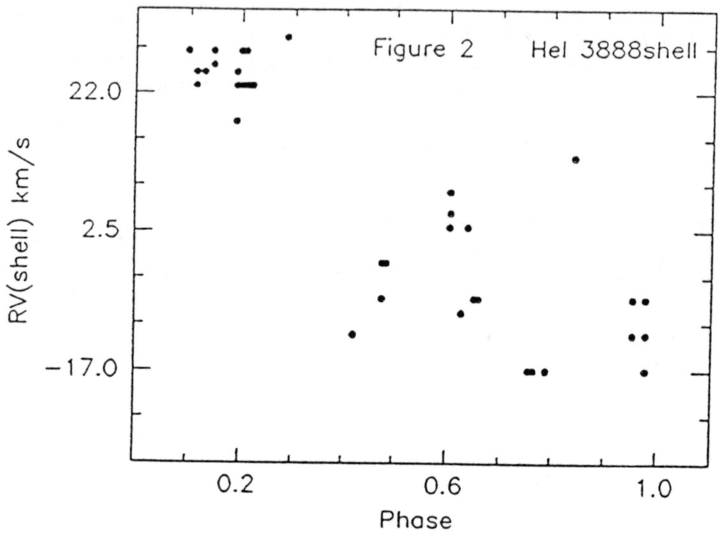
Figure 2
HeI 3888shell
22.0
2.5
−17.0
RV(shell) km/s
0.2
0.6
1.0
Phase

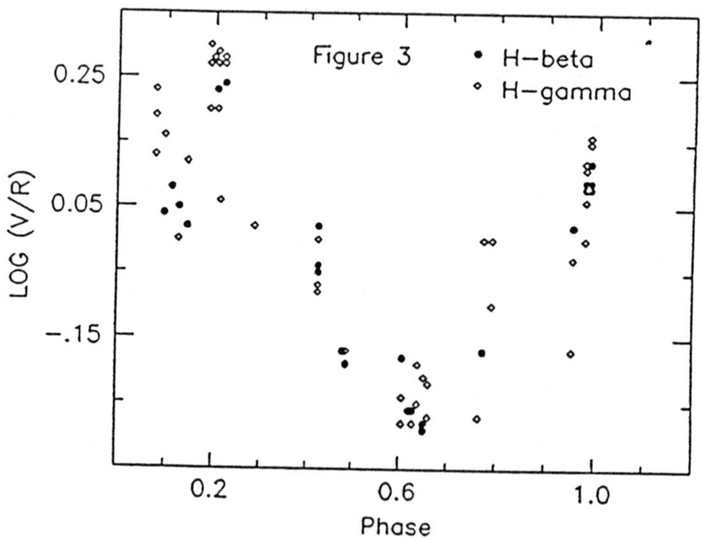
Figure 3
H−beta
H−gamma
0.25
0.05
−.15
LOG (V/R)
0.2
0.6
1.0
Phase

Inside the Stars, IAU Colloquium 137
ASP Conference Series, Vol. 40, 1993
Werner W. Weiss and Annie Baglin (eds.)

STELLAR EVOLUTION AND MASS TRANSFER IN BINARIES

A. A. BOJARCHUK
Institute of Astronomy, Moscow, Russia.

V. M. CHECHETKIN, O. A. KUZNETZOV, YU. P. POPOV
Keldysh Institute of Applied Mathematics, Moscow, Russia.

ABSTRACT The problem of angular momentum loss in binaries and its influence on stellar evolution is considered. The results of 2D hydrodynamics computer simulation of mass transfer is presented.

More then 30% stars exist in binary systems. Therefore the evolution of star in interacting binary systems is of great importance for theory of stellar evolution. For example, the Supernova of type IA is appear to occur as a result of evolution in interacting binary system (Iben and Tutukov 1984; Chechetkin *et al* 1980). The main problem for theory of stellar evolution in binaries is mass transfer and loss of angular momentum. When the stellar envelope reaches its Roche lobe, the star begins to loss the matter and if loss of mass is significant, stellar evolution changes as compared with evolution of single star (Masevich and Tutukov 1988). The size of Roche lobe depends on components mass ratio and distance between the components. Due to loss of angular momentum the distance between components is decreased. This effect can lead to acceleration of stellar evolution.

We present a full hydrodynamical model for simulating of mass transfer in binaries consisting of the giant or subgiant (hereafter the secondary) and a dwarf (hereafter the primary). For instance, we can apply our models for investigation of a star wind in Symbiotic stars.

We have chosen the system with parameters:

$\mathfrak{M}_1 = \mathfrak{M}_\odot$ $\mathfrak{M}_2 = 4 \cdot \mathfrak{M}_\odot$

A (distance between components) = $575 \cdot R_\odot$

orbital velocity of secondary = 8.1 km/sec

$R_2 = 191 \cdot R_\odot$ (moderate filling of Roche lobe)

We took the follows boundary conditions on the surface of secondary: a_b = 1 km/sec, c_b = 43 km/sec. We obtain the stationary shock wave and non-stationary hot tongues behind the primary, moving down-stream. Flow lines shows that it takes particles three revolution to reach the primary and the flow lines that form the disk initiate on a side opposite to the primary (see Fig.1).

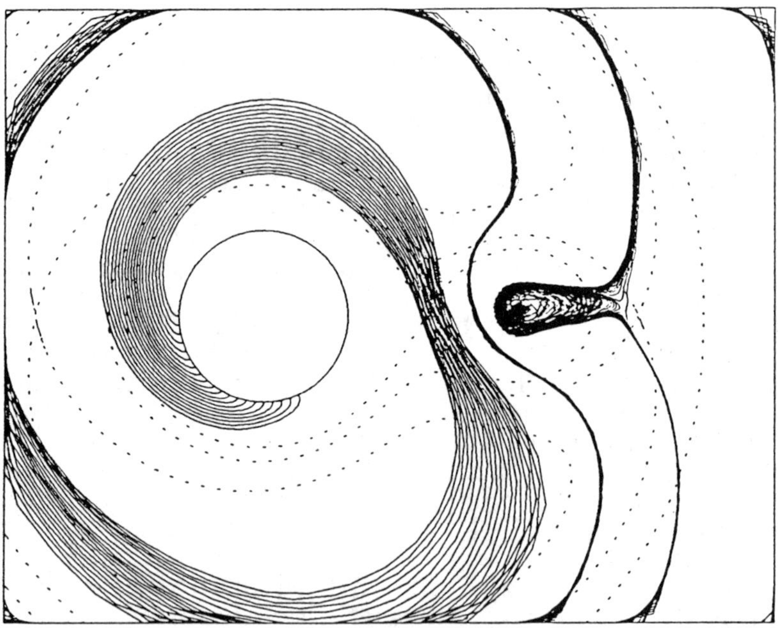

Fig.1

We define the angular momentum as $J = \Omega \cdot A^2 \cdot \mathfrak{M}_1 \cdot \mathfrak{M}_2 / \mathfrak{M}$ and momentum loss rate as $\dot{J} = \dot{\mathfrak{M}} \cdot r \cdot V_\varphi$, where r – distance from the center of mass of the system, V_φ –

tangential velocity in laboratory system. And we defined the time scale for momentum loss as $\tau = J/\dot{J}$. Fig.2 represents dependency $\tau(t)$ for input flow from secondary and output flow (far from center mass of the system) for the value of mass loss $\dot{\mathfrak{M}} = 10^{-5}\ \mathfrak{M}_\odot$/year.

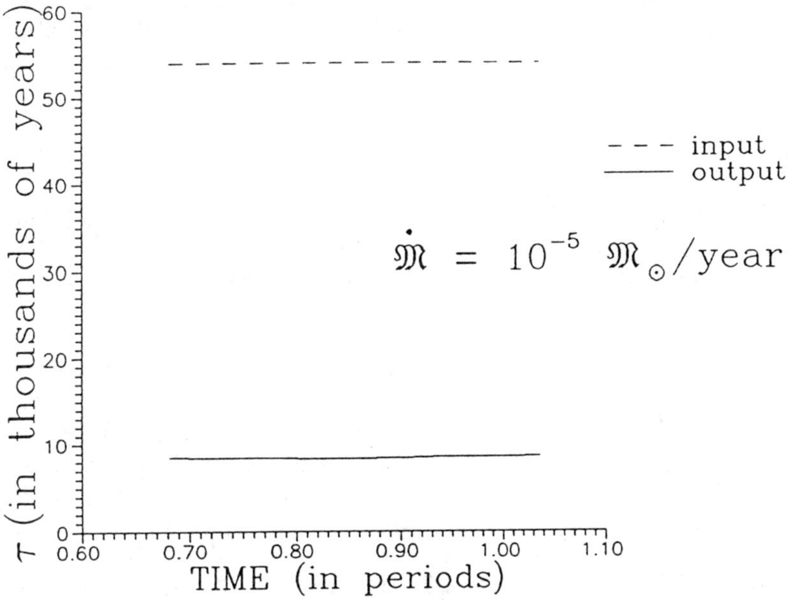

Fig.2

REFERENCES

Chechetkin, V.M., Gerstein, S.S., Imshennik, V.S., Ivanova, L.N., Khlopov, M.Yu. 1980, *Astrophys. and Space Sci.*, **67**, p.61.

Iben, I., Jr., and Tutukov, A.V. 1984, *Ap.J.Suppl.*, **54**, p.335.

Masevich, A.G. and Tutukov, A.V., 1988, Evolution of Stars – Theory and observation, Moscow, Nauka.

Inside the Stars, IAU Colloquium 137
ASP Conference Series, Vol. 40, 1993
Werner W. Weiss and Annie Baglin (eds.)

THE NAMING OF STARS AND THE STUDY OF PROTOSTARS

D. R. O. MORRISON
CERN, Geneva 23, Switzerland

ABSTRACT The logic of naming stars could be based more on physical principles as happened in the naming of resonances in Particle Physics. A possible system might be based on extensive Monte Carlo studies of protostars followed by standard Stellar Evolutionary Model calculations.

INTRODUCTION

This IAU Colloquium on "Inside the Stars" has been excellent and I have learnt a great deal. However as a relative outsider but with experience of a closely allied branch of Science, I hope you will allow me to make two comments that may be constructive.

THE NAMING OF STARS

To an outsider, it is hard to find a consistent logic in the names given to stars. In Particle Physics a similar situation arose in the 1960's and early 1970's, when a large number of new particles and resonances were discovered. Names were given to them in an almost random way, using both latin and greek letters and being based on factors such as the name of the discoverer. Finally the Particle Data Group, based in LBL, Berkeley, proposed a complete and logical naming system based on the physical properties of the particles such as their baryon number, strangeness, mass, spin, parity, etc. Despite some early grumbling, this has been universally adopted and surprisingly quickly. It is now greatly appreciated and is now essential in the physical understanding of the subject. It is proposed that the stars be similarly renamed in terms of their physical properties. Ultimately this will help in the understanding of the evolution of stars.

HOW TO CLASSIFY STARS?

Although this conferences fulfilled its aim of discussing "Inside the Stars", I was slightly surprised that there was no review talk about how stars are formed - how are the initial conditions of the Evolutionary Models reached, are they all the same or are there subsequently important differences?

Understanding stars in terms of their observational properties appears to have been moderately successful, but there seems to be a feeling that it is not sufficiently perfect to allow stars to be named in a scientific manner. In High Energy Physics, one is accustomed to groups of physicists doing vast series of

Monte Carlo calculations. This suggests another approach. While Stellar Evolutionary Models are quite successful as we have heard at this Colloquium, the description of the formation of stars is rather less successful. An understanding of this may be the real solution to the problem of how to understand and to classify stars.

It is considered that stars are formed from protostars which are large clouds of material which mainly under gravity, coalesce to give 1, or 2 or 3 or more centres. Some of these centres are sufficiently large that they give stars. The result is that about 65% of stars are in binary systems and less than 15% are single stars as our Sun. This shows that multiple star systems are normal. There clearly must be interactions between the various centres, for example our Sun probably interacted with its surroundings, absorbing material and rotating strongly, ie it was a T Tauri star in its early history. Other effects such as nearby supernova explosions, colliding galaxies etc. would also significantly influence the protostar.

It is hoped that a major project can be started to do Monte Carlo calculations of all possible types of protostars, to follow their development to stars, to study the interactions of these stars with any other star(s) in the protostar system, and then to follow their evolution up to the present. This is a major project since there are many possibilities, but it would be a very rewarding project

Later comparison with observations would yield important information. Finally it would be possible to name stars in a relatively simple way according to their basic physical properties and life-history. Then progress and understanding would be much faster.

Inside the Stars, IAU Colloquium 137
ASP Conference Series, Vol. 40, 1993
Werner W. Weiss and Annie Baglin (eds.)

PRISMA: A NEW SPACE MISSION FOR STELLAR PHYSICS

T. APPOURCHAUX
ESA/ESTEC, P.O.Box 299, 2200AG Noordwijk, The Netherlands

D. GOUGH
Institute of Astronomy, Madingley Road, Cambridge CB3 OHA, England

P. HYOYNG
SRON, Sorbonnelaan 2, 3584 CA Utrecht, The Netherlands

C. CATALA
DESPA - Observatoire de Paris-Meudon, 92100 Meudon, France

S. FRANDSEN
Aarhus Universiteit, Langelandsgade, 8000 Aarhus C, Denmark

C. FRÖHLICH
PMOD/WRC, POB 173, 7260 Davos Dorf, Switzerland

A. JONES
Aarhus Universiteit, Langelandsgade, 8000 Aarhus C, Denmark

P. LEMAIRE
IAS, BP 10, 91371 Verrieres-le-Buisson Cedex, France

G. TONDELLO
Universita di Padova, Via Gradenigo 6A, 35100 Padova, Italy

W. W. WEISS
Institute für Astronomie, Türkenschanzstraße 17, 1180 Wien, Austria

INTRODUCTION

PRISMA (Probing Rotation and Interior of Stars: Microvariability and Activity) is a new space mission of the European Space Agency. PRISMA is currently in a Phase A study with 3 other competitors. PRISMA is the only ESA-only mission amongst those four and only one mission will be selected in Spring 1993 to become a real space mission.

The goal of the Phase A study is to determine whether the payload of PRISMA can be accommodated on a second unit of the X-ray Multi-Mirror (XMM) bus; and whether the budget of the PRISMA mission can be kept below 265 MAU ('88 Economic conditions). The XMM mission is an approved cornerstone and is in a Phase A together with PRISMA.

This paper will mainly focus on the scientific objectives that can be achieved with PRISMA. Figure I gives an overview of the PRISMA scientific objectives. The scientific objectives can be classified into three areas of interest:

- Stellar structure and evolution
- Stellar atmospheres and magnetic fields
- Stellar dynamics and stellar dynamos

These three subjects are developed in section 1. A short overview of the observing programme is also given in section 2. The payload is outlined in section 3. And we conclude on the future of the PRISMA mission in the epilogue.

SCIENTIFIC OBJECTIVES

Stellar structure and evolution

The impressive success of the theory of the structure and evolution of stars in reproducing classical astronomical observations of stellar surfaces has convincingly demonstrated that the theory is more-or-less right. We can therefore safely presume that at least the gross properties of stars have been correctly explained. However, there remain apparently minor discrepancies between theory and observation which might be symptomatic of serious flaws in some quite important aspects of our modelling of stellar interiors, yet which hardly influence the surface properties at all. Only by probing beneath the stellar surface can those superficially subtle aspects of the internal structure be measured.

It is the objective of PRISMA to measure such properties. Although stars are optically opaque, they are acoustically transparent. Therefore, by measuring and analysing the frequencies of acoustic oscillations, a wealth of qualitatively new information about the internal structure of stars can be acquired. This information provides us with not only measures of gross properties of stars, such as mean density, but also with data from which one can ascertain important detailed features of the internal structure that control evolution.

Acoustic modes are confined within a cavity of propagation whose upper boundary is very near the surface of the star. Consequently the modes can be detected by observing modulations of the photosphere. PRISMA aims at measuring variations in the radiative intensity induced by the oscillations. Those variations can be analysed to determine the characteristic frequencies of modes of low degree, all of which penetrate deeply into the star, some reaching even to the very centre.

Different aspects of the oscillation spectrum manifest different properties of the stellar interior. For example, the mean frequency separation between modes of differing order but the same degree, the easiest of all seismic signatures to measure, essentially determines the mean density of the star, which is proportional to M/R^3. When coupled with a spectroscopic determination of surface gravity, GM/R^2, the mass M and radius R of the star can then be separately determined. The mean frequency displacement between spectra of modes of different degree measures sound-speed variation in the energy-generating core, and can thus be used as a direct signature of chemical inhomogeneity produced by

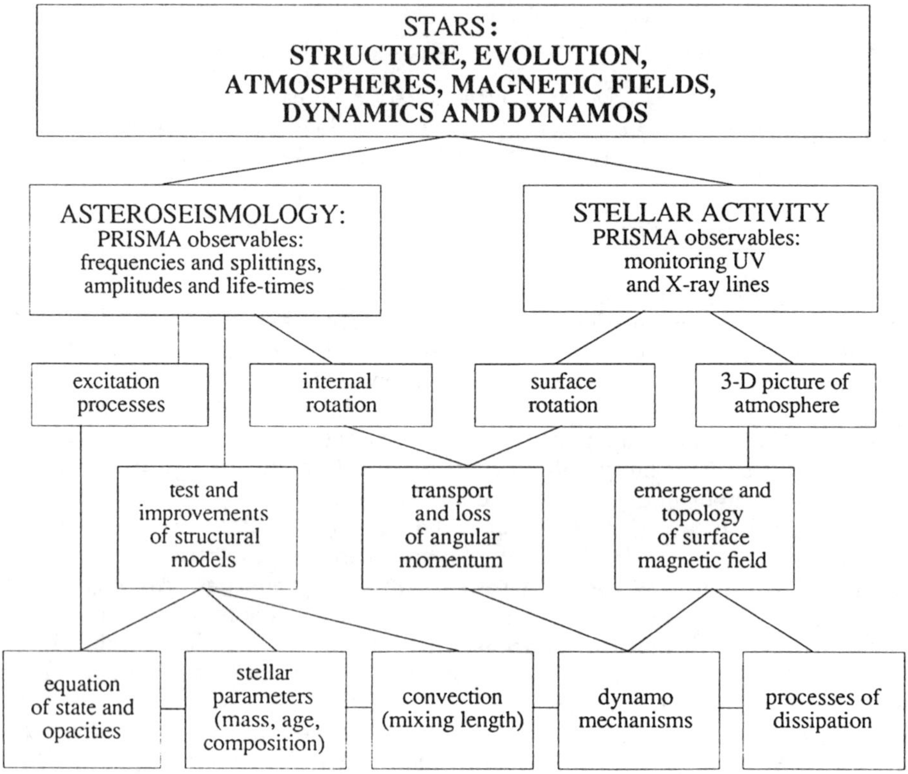

FIGURE I Scientific objectives of PRISMA

the nuclear reactions. The degree of inhomogeneity is a measure of age. But more important still are the sophisticated analyses that provide measurements of the variation of density throughout the core, or the degree of overshoot at the base of the envelope of a late-type main-sequence star. Both those diagnostics are important to our understanding of the redistribution of chemical elements, a knowledge of which is essential for a proper appreciation of the implications of chemical abundance analyses of stellar atmospheres based on classical spectroscopic data. Material redistribution influences the chemical and structural evolution of the star, and is therefore a crucial element in the comparison of theory with observation in the H-R diagram.

An important component of the PRISMA investigation is the calibration of theoretical models of open clusters, aided by data from HIPPARCOS. Not only will this permit a determination of the cosmologically important internal stellar helium abundance, but it will also yield direct measures of the ages of the clusters, and thereby improve our ability to diagnose evolutionary properties of stars. The information so acquired will make our determinations of the ages of globular clusters, inferred through theoretical modelling, more secure, even though stars in those clusters will not be directly accessible to PRISMA. That will strengthen the limits that can be placed on the age of the Galaxy.

Stellar atmospheres and magnetic fields

It is now accepted that stellar activity is the manifestation of the interaction of magnetic fields with the plasma in the atmospheres of stars. The magnetic field is believed to be generated by a magnetic dynamo seated in the convecting envelope. The basis for this picture is the Sun, where buoyant magnetic fields produced by the dynamo are seen to rise through the photosphere to form bipolar active regions. The polarity and latitude of emergence of active regions exhibit a cyclic behaviour with a period of 22 years, the solar cycle. The entire solar surface is covered with the decay products of old active regions. Consequently, the surface magnetic flux is organised in flux tubes and highly fragmented. Flux tubes guide energy from subphotospheric layers, which heats the outer atmosphere to temperatures of 10^6 K. As the photospheric flow field carries the magnetic field along, there ensues a hierarchy of magnetic instabilities, the most impressive of which is a large solar flare. The theoretical basis for these phenomena is much less secure than the observational evidence. Despite vigorous efforts, we do not understand how the solar dynamo works, although differential rotation and helical convection are known to be important ingredients. Starting from a poloidal field the former generates a strong toroidal field, which is converted in turn into a poloidal field by the helical convection (the α-effect), and the result is a periodic dynamo. But the details of this process, and in particular the many subtle nonlinear interactions remain to be filled in. There is an urgent need to test theories of magnetic dissipation and of the magnetic dynamo with a larger sample of stars than just our Sun. This has been the drive behind the 'Solar Stellar Connection', which field is now about ten years old, and which has yielded ample evidence that solar-type magnetic activity exists in all rotating stars sufficiently cool to have a convective envelope. Such a test is therefore well possible but requires a concerted effort.

A major aim of PRISMA is to provide the necessary observations of stellar activity in a systematic fashion, such that the H-R diagram is well covered. As stars rotate while being observed by the XUV, UV and optical monitors of PRISMA, the modulation in the signals caused by the irregular distribution of magnetic fields over the stellar surface can be used to determine the surface rotation and to construct a crude magnetic image of the surface of the star. Either by Doppler imaging using spectral lines, or modulation imaging using broad-band signals (such as provided by white- light and XUV monitors), the combined signals will allow plage mapping, revealing the spot contents of plages on stars, and also the emission scale height of the stellar coronae. With this information we can study the mechanisms heating the outer atmospheres of cool stars in much more detail than in present-day theories in which results depend on too many independent variables.

Stellar dynamics and stellar dynamos

The cross-fertilization of stellar activity and asteroseismology is particularly evident for the problem of the internal dynamics and magnetic dynamos of stars. An important goal of the PRISMA mission is the determination of the mean internal angular velocities of stars. Together with measures of the surface rotation, that will provide information about the spatial variation of the rotation in the interior, which is crucial to our understanding of the distribution of angular momentum. Temporal variations can be inferred from observations of cluster stars, for which we shall have improved seismic age determinations. Small acoustic frequency variations associated with activity have been observed on the Sun, and appear to be due to magnetically induced modifications to the subphotospheric outer reflecting surface of the acoustic cavity. We therefore anticipate much greater variations in more active stars. Rotation, differential rotation, and latitude distribution of stellar activity, combined with information on the depth of the convection zone (from refined stellar models) and on the cycle period (from ground-based measurements) are the essential ingredients for modelling stellar dynamos. The object is to derive scaling laws and to test these with the help of a sufficiently large sample of stars. An example of an interesting class of objects to test dynamo theories are early F stars close to the granulation boundary. These stars have shallow convection zones, and are expected to have substantial radial differential rotation due to efficient magnetic breaking. At the same time, these stars should feature relatively strong global oscillations, allowing a good determination of their internal structure and rotation. For stars with short magnetic cycles there is even the perspective that we may observe different phases of the cycle.

OBSERVING PROGRAMME

To provide powerful tests for the models of stellar structure and evolution, and to constrain theories of angular momentum transport and of magnetic field generation by dynamo action, several requirements must be imposed on the choice and number of targets to be observed by PRISMA.

In the first place, the targets must correctly sample the stellar parameter space (mass, temperature, age). Secondly, it is necessary to observe stars with different rotation rates. It can easily be seen that a significant sampling of the (spectral type - luminosity class - rotation rate) space cannot be reached with a number of targets smaller than about 100. To be able to observe more than 100 targets in a two-year mission, with exposure times of 1 month per target, the strategy is adopted in which several stars are observed simultaneously.

A better test of stellar structure models can be achieved if one or several basic stellar parameters are known independently. This is why a very high priority must be given to possible targets for which independent information is available, or for which additional constraints can be placed on the models:

- Members of open clusters, for which a good estimate of the age is available.
- Members of some binary systems, for which the masses, and sometimes the radii, are well known.
- Stars observed by HIPPARCOS, for which we have a good determination of the absolute luminosity.
- Stars observed by IUE and ROSAT
- Stars observed in the Mount Wilson CaII program

Besides these prime targets which will drive the instrument and strategy definition, the scientific return of the mission will be maximized if a sample of very important targets are included: these are the classical pulsating variables, such as δ Scuti, RR Lyrae, Cepheids, λ Boötis, and also rapidly oscillating Ap stars, Am stars and Maia variables. These targets should not drive the major requirements of the mission, but the philosophy would be to observe them with PRISMA whenever possible.

MODEL PAYLOAD AND OBSERVABLES

The model payload for PRISMA comprises 4 instruments. There are 2 photometric instruments:

- A Large Photometer (LP), with an aperture equivalent to 41 cm diameter, has a 1.5° x 1.5° field of view. The LP uses a 3-mirror off-axis design to reduce the effects of straylight. It collects simultaneously the broad band visible flux of stars brighter than magnitude 8 to detect micromagnitude flux variations.
- A Small Photometer (SP), with an aperture equivalent to 14 cm diameter, has a 3° x 3° field of view. It has the same purpose and design as the LP except that micromagnitude flux variations will be detected on stars brighter than 6th magnitude. An internal mirror allows a wide selection of fields offset from the line of sight of the other instruments.

The observables of the Large and Small Photometers are summarized in Table I.

And there are 2 spectroscopic instruments:

- An Ultra Violet Spectrometer (UVS), optically aligned with the Large Photometer, has a 60 x 45 cm aperture and covers the 120-285 nm wavelength range that includes chromospheric and transition region lines. A

TABLE I PRISMA observables: Photometric segment

Observables	Range	Accuracy	Scientific keywords
Frequencies	0.02 mHz-0.1 Hz	0.1 μHz	Stellar structure, g modes, He content, neutrino physics
Amplitudes	10^{-6}-10^{-1}	10^{-7}-10^{-5}	Excitation, stellar spots and plages, non-linear pulsators?, convection
Linewidths	1-10 μHz	0.05 μHz	Excitation, convection, surface structure
Frequency separations	>0.3 μHz	0.1 μHz	Stellar parameters, stellar structure, rotation, magnetic fields

high-resolution cross-dispersed echelle system allows observation of one star at a time in a 1.5° x 1.5° field of view (Resolution 0.004-0.01 nm).

- An eXtreme Ultra Violet Telescope (XUVT) images the same 1.5° x 1.5° field as the Large Photometer, in 2 bands sensitive to coronal temperatures (possibly centered at 14 nm and 17 nm).

The observables of the UV Spectrometer and XUV Telescope are summarized in Table II.

TABLE II PRISMA observables: Spectroscopic segment

Observables	Range	Accuracy	Scientific keywords
UV flux	10^{-13}-10^{-17} W m^{-2}	5%	3D pictures
UV line-profiles	120-285 nm	0.004-0.01 nm (resolution)	Doppler imaging
X-ray flux	10^{-14}-10^{-17} W m^{-2}	20%	Corona, flares

A more detailed description of the PRISMA payload can be found elsewhere in these proceedings (Jones, 1992).

EPILOGUE

The combination of asteroseismology and stellar activity is a wholly new concept in stellar investigation. Helioseismology has already reaped enormous rewards, and has led to a much more secure and precise knowledge of the internal structure and rotation of the Sun. This has led in turn to important consequences for solar dynamo models. The PRISMA mission aims to extend that understanding to other stars. It has been demonstrated already by helioseismology and its theoretical asteroseismic extensions that from the low-degree modes that will be available we can increase in our knowledge of other stars really significantly, particularly in their cores, in which thermonuclear energy is being generated. Indeed, the PRISMA mission can do that for a range of stars. Therefore, not only will that provide us simply with a greater knowledge of the structures of those stars, but, taken together, the sum of the information on stellar activity and seismology will substantially improve our knowledge of the mechanisms of magnetic activity, magnetic braking and especially stellar evolution, and with that, significantly increase our understanding of the nature of the Universe.

The future for the PRISMA mission is bright. This mission has no precursor and it will return new science. Not only will we be able to understand the insides of the stars, but in the long term it will bring us a better comprehension of the outsides of the stars. We may quote Malcolm Longair who said in his opening talk:

In my view, in the case of projects such as PRISMA, it is not a question of 'If it will fly' but rather 'When it will fly'. The scientific significance seems to me unassailable.

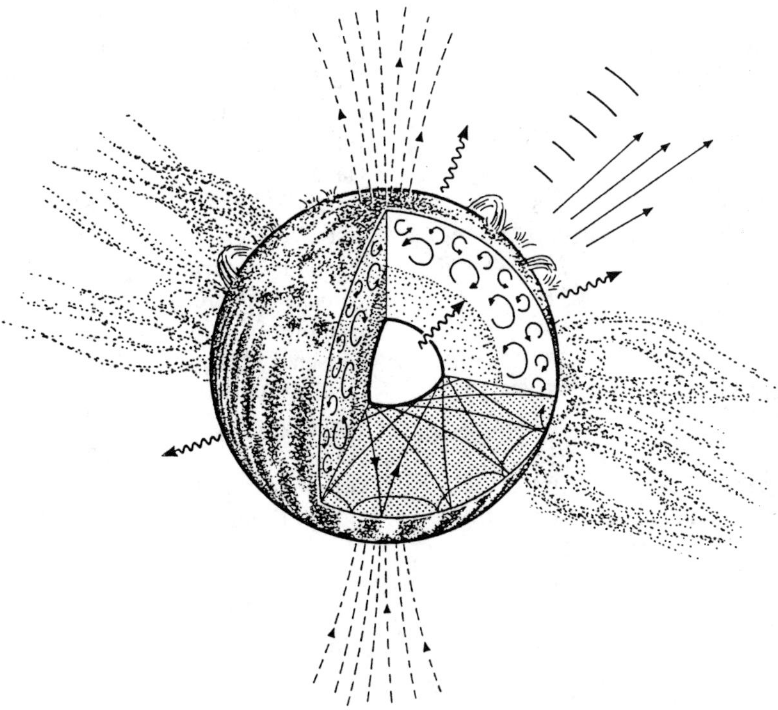

CITATION INDEX
(bold face: Review)